saur

Peter Wennrich (Founder/Begründer)

International Encyclopedia of Abbreviations and Acronyms in Science and Technology

Compiled by Michael Peschke

Internationale Enzyklopädie der Abkürzungen und Akronyme in Wissenschaft und Technik

Bearbeitet von Michael Peschke

Vol 1/Band 1

A–Da

K · G · Saur

München · New Providence · London · Paris 1996

Die Deutsche Bibliothek – CIP-Einheitsaufnahme

International encyclopedia of abbreviations and acronyms in science and technology = Internationale Enzyklopädie der Abkürzungen und Akronyme in Wissenschaft und Technik / Peter Wennrich (Begr.). Bearb. von Michael Peschke. – München ; New Providence ; London ; Paris : Saur.
ISBN 3-598-22970-4
NE: Wennrich, Peter [Begr.]; Peschke, Michael [Bearb.]; Internationale Enzyklopädie der Abkürzungen und Akronyme in Wissenschaft und Technik
Vol. 1
Bd. 1. A – Da. – 1996
ISBN 3-598-22971-2

∞

Gedruckt auf säurefreiem Papier/
Printed on acid-free paper

A Reed Reference Publishing Company
Printed in the Federal Republic of Germany

Datenerfassung und Satz / Computer-controlled keyboarding, data preperation and automatic data processing by
Michael Peschke, Berlin

Druck/Printed by Strauss Offsetdruck, Mörlenbach
Binden/Bound by Buchbinderei Schaumann, Darmstadt

ISBN 3-598-22970-4 (8 Volume Set)
ISBN 3-598-22971-2 (Vol 1)

Contents / Inhalt

Vol 1 / Band 1

A - Da

Preface

Abbreviations, Acronyms, and Initials in all their various forms confront us today more than ever. We encounter them in all areas of daily life, both at home and at work: they are often the basis of documentation and communication and frequently themselves contain the central information in a given context. Decoding them is, however, often difficult, sometimes even impossible because adequate reference sources are either not available or not readily accessible.

As the use and the invention of abbreviations is increasing world-wide, no one reference work can hope to answer all queries concerning them correctly. The present *International Encyclopedia of Abbreviations in Science and Technology* with appr. 680,000 entries in 8 volumes attempts to encompass the most important acronyms from the most diverse general and scientific fields. It does, however, limit itself to languages which use the Roman alphabet.

The International Encyclopedia also lists common vernacular abbreviations in general use. Thus it goes beyond official terminology. Historical abbreviations have also been included. Where the use of punctuation within the abbreviation is erratic, the most commonly-used form has been given preference. Further information of this nature can be found in the "Notes for the User".

Notes for the User

1. The abbreviations are presented in alphabetical order from a to z. Ä, æ, ç, ö, ö, , ü are arranged as ae, c, oe, ue. Symbols and (diacritic) signs such as . - () [] ^ ^° & + are of no importance in regard to the alphabetical order.

1. 1. Alphanumeric sequences are listed in the alphabetical place of the respective letter(s).

1.2. Roman numerals are arranged in their value as literal character. Arabs numerals in alphanumeric sequences are arranged in their value as number. Abbreviations beginning with a number are arranged after the character z.

2. Styles of type are used as follows:

2.1. boldface for the abbreviation

2.2. plain for the remaining text

3. Within each entry (beginning with the abbreviation protruding to the left), the meanings are likewise listed a) in alphabetical order of the label indicating the language in which the abbreviation is used, b) in alphabetical order.

3.1. Each abbreviation is followed by the label indicating the area or sphere in which the abbreviation is used and/or the pertinent language.

3.2. The sign of equality (=) normally separates explanatory remarks (usually abbreviated) from each other as well as from the remaining text. In connection with the grammalogue of a language the colon (:) refers to a different expression in another language of the respective meaning.

3.3. Any additional explanations are enclosed in parentheses ().

3.4. Brackets [] are components of abbreviations as well as of meanings. However, parentheses are also used for marking a word that can be substituted for the preceding word, and for marking that part of a word, which, if omitted, will result in a different meaning or spelling.

Language Codes

A	Afrikaans	l	Latin
a	Danish	N	Norwegian
c	Czech	n	Dutch
d	German	P	Polish
e	English	p	Portuguese
F	Finnish	R	Russian
f	French	r	Swedish
H	Hungarian	S	Slovenian
h	Hebrew	s	Spanish
i	Italian	z	Turkish

Berlin, August 1995

Michael Peschke

Vorwort

Abkürzungen, Akronyme und Initialen in ihren mannigfachen Erscheinungsformen begegnen uns heute überall und mehr denn je. Sie treten in allen Bereichen des täglichen und beruflichen Lebens auf; sie sind oft Grundlage von Dokumentation und Kommunikation und vielfach die entscheidenden Informationsträger. Ihre Bedeutung zu entschlüsseln, ist gleichwohl oft schwierig, manchmal unmöglich, weil geeignete Hilfsmittel fehlen oder diese nur schwer greifbar sind.

Weil der Gebrauch und die Erfindung von Abkürzungen und Abbreviaturen in der ganzen Welt ständig zunehmen, vermag niemand allein auf diesem Gebiet alle Fragen zutreffend zu beantworten. Die vorliegende *Internationale Enzyklopädie der Abkürzungen und Akronyme in Wissenschaft und Technik*, die ca. 680.000 Einträge in 8 Bänden verzeichnet, versucht, möglichst umfassend die wichtigsten Akronyme aus den unterschiedlichsten Fach- und Lebensbereichen zusammenzustellen. Dabei beschränkt sie sich freilich auf die "lateinisch" schreibende Welt.

Die Internationale Enzyklopädie berücksichtigt auch alltagssprachlich übliche und gebräuchliche Formen. Sie enthält also nicht nur die offiziellen Versionen. Auch Abkürzungen, die der Vergangenheit angehören, sind aufgeführt. Abkürzungen, die teils mit, teils ohne Punkt geschrieben werden, sind nach der jeweils überwiegenden Verwendung verzeichnet. Im übrigen geben die "Hinweise für den Benutzer" alle Erläuterungen über die Anlage der Einträge wie des gesamten Werkes.

Hinweise für den Benutzer

1. Die Abkürzungen werden in der alphabetischen Reihenfolge a...z gebracht. Ä, æ, ç, ö, ö, , ü werden wie ae, c, oe, ue eingereiht. Zeichen wie . - ( ) [ ] ˆ ^ ° & + sind für die alphabetische Reihe ohne Bedeutung.

1. 1. Alphanumerische Folgen stehen an der von dem (den) Buchstaben bestimmten alphabetischen Stelle.

1.2. Römische Zahlzeichen werden nach ihrem Buchstabenwert eingereiht. Arabische Zahlzeichen innerhalb alphanumerischer Folgen werden nach ihrem Zahlenwert geordnet. Abkürzungen, die mit einer arabischen Ziffer beginnen, stehen nach dem Buchstaben Z.

2. Die Schriftarten werden wie folgt verwendet:

2.1. halbfett für die Abkürzung

2.2. Grundschrift für den übrigen Text

3. Innerhalb einer (mit einer links herausgerückten Abkürzung beginnenden) Wortstelle sind die Bedeutungen a) nach ihrem Bedeutungsbereich (Sortierung Sprache), b) in ihrer alphabetischen Reihenfolge aufgeführt.

3.1. Der Abkürzung folgt die Angabe der Sprache.

3.2. Das Gleichheitszeichen (=) trennt im allgemeinen abgekürzte erläuternde Hinweise voneinander und vom übrigen Text. Der Doppelpunkt (:) in Verbindung mit einem Sprachkürzel verweist auf anderssprachige Bezeichnungen der entsprechenden Bedeutung.

3.3. Sonstige Erläuterungen stehen in runden Klammem ().

3.4. Eckige Klammern [] schließen Bestandteile von Abkürzungen und von Bedeutungen ein. Bei den Bedeutungen werden diese Klammern mitunter auch dazu benutzt, ein Wort zu kennzeichnen, das an die Stelle des vorangehenden gesetzt werden kann, bzw. Wortteile zu markieren, die, läßt man sie weg, eine andere Bedeutung oder eine andere Schreibung ergeben.

Sprachenkürzel

A	afrikaans	l	lateinisch
a	dänisch	N	norwegisch
c	tschechisch	n	holländisch
d	deutsch	P	polnisch
e	englisch	p	portugiesich
F	finnisch	R	russisch
f	französich	r	schwedisch
H	ungarisch	S	slowenisch
h	hebräisch	s	spanisch
i	italienisch	z	türkisch

Berlin, August 1995

Michael Peschke

A - Da

A

A d: Abfluß[menge]
- d: Abforderung
- d: Abfrage
- d: Abfrageseite
- d: Abschnitt
- d: Absetzer
- d: Abstammungsachse
- d: Abstandsachse
- d: Abteilung
- d: Abzweigstück
- d: Additionsreaktion
- d: Adenin
- d: Aggregat
- d: Akkumulator
- d: Akten
- d: Aktivierungsenergie
- d: Alarm
- d: Alkohol
- d: Alp[en]
- d: Alphabet[faktor]
- d: alterungsbeständig
- d: Amateur
- d: Ammeter
- d: Ampere[meter]
- d: Amplitude[nmodulation]
- d: Amt
- d: Androezeum
- d: Androzeum
- d: Anfang
- d: Anfangsaktivität
- d: angelassen
- d: Angora
- d: Angriffspunkt
- d: Angström[einheit]
- d: Anhänger
- d: Anordnung
- d: Anpassung[sschaltung]
- d: Anschluß
- d: Anspannungszeit
- d: Antenne
- d: Anthrazit
- d: Anweisung
- d: Anzeige
- d: Anzeigefunktion
- d: Apfel
- d: apothekenpflichtig
- d: April
- d: Arbeit
- d: Archaeophyt
- d: Archiv für das Studium der Neueren Sprachen (journ.)
- d: Argent
- d: Argon
- d: Arithmetik
- d: Arm
- d: Armature
- d: Armee
- d: Art
- d: Arterie[n]
- d: Artillerie
- d: Äthanol
- d: Atom[gewicht]
- d: Ätzalkalität[szahl]
- d: Aufbau (journ.)
- d: Aufbruch
- d: Aufgabe
- d: Aufklärung
- d: Auftastsignal
- d: Auftrieb[skraft]
- d: Aufzeichnung
- d: Augmentation
- d: Augsburg (Kfz)
- d: August
- d: Ausbeute
- d: Ausbildung
- d: Ausgabe
- d: Ausgabestelle
- d: Ausgang
- d: Auslands-
- d: Auslaß
- d: Auslaufstrecke
- d: Auslösung
- d: Auspuff
- d: Aussaat
- d: Aussaat[jahr]
- d: Ausschuß
- d: Außendruck
- d: Außenkabel
- d: Aussteuerung[sfaktor]
- d: Austastsignal
- d: Austenit
- d: Austritt
- d: Auswahlprüfung
- d: Auswertung
- d: Autobahn
- d: Axiallader
- d: [Bundes-]Autobahn
- d: Österreich (Kfz)
- e: Abbey
- e: Ablative
- e: Able
- e: Abnormal
- e: About
- e: Absent
- e: absolute temperature
- e: Absorbance
- e: Absorbancy
- e: Absorptivity
- e: Abstracts
- e: Abundant
- e: Academical
- e: Academician
- e: Academy
- e: Acceleration
- e: Acceptance
- e: Acceptor
- e: Access
- e: Accessed
- e: Accessory
- e: Accumulator
- e: Accusative
- e: Ace
- e: Acetum
- e: Acid
- e: Acoustic
- e: acoustic source
- e: Acquiescence
- e: Acre
- e: Act
- e: Act[ing]
- e: Actin
- e: Actinomycin
- e: Action
- e: Activity
- e: Actual
- e: Actual Weight of an Aircraft
- e: Acute
- e: Adaptable
- e: Adder
- e: Address
- e: Adenine
- e: Adenosine
- e: Adequate
- e: Adjective
- e: Adjunct
- e: adjusted
- e: Adjustment
- e: Adjutant
- e: Administration
- e: Admiral
- e: Admiralty
- e: Adrenalin[e]
- e: Adriamycin
- e: Adulterer
- e: adulteress
- e: Advance
- e: Aerial
- e: Aerology
- e: Aeronautical
- e: Aeronautics
- e: Aerospace
- e: Affect
- e: Affinity
- e: Affirmation
- e: Affirmed
- e: Affix
- e: After
- e: Aftercooled
- e: Age
- e: Agency
- e: Agglomerate
- e: Air
- e: Air-launched
- e: Airborne

- e: Aircraft
- e: Airman
- e: Airplane
- e: Akinetic
- e: Alanine
- e: Alaska Steamship Company
- e: Albedo
- e: Albumin
- e: Alcoa Steamship Company
- e: Alcohol
- e: Alfa
- e: Alive
- e: Allergy
- e: Allocator
- e: Alpha
- e: Alphabetic
- e: Alternate
- e: Altimeter
- e: Alto
- e: Alveolar
- e: Alveolar Gas
- e: Ambassador
- e: Amber
- e: Ambient
- e: Ambiguity
- e: Ambitendency
- e: Ambulance
- e: Ambulatory
- e: Amended
- e: America
- e: America[n]
- e: Americanization
- e: Americanize
- e: Amethopterin
- e: Ampere[s]
- e: Amphetamine
- e: Amphibian
- e: Amphibious
- e: Amphophil
- e: Ampicillin
- e: Amplification
- e: Amplifier
- e: Amount
- e: Anaesthetics
- e: Analog[ue]
- e: Analysis
- e: Analyzer
- e: Anaphylaxis
- e: Anchorite
- e: Anchor Line
- e: Androecium
- e: Anesthetic
- e: Angel
- e: Anglican
- e: Angling
- e: Angstrom(unit)
- e: angstrom unit
- e: Animal
- e: Anion
- e: Anisean
- e: Annual
- e: Annular
- e: Annus (= year)
- e: Anonymous
- e: Answer
- e: Antarctic
- e: Antenna
- e: Anterior
- e: Anthracite
- e: Anti-Aircraft
- e: Anticlockwise
- e: Antigen
- e: Anus
- e: Apostle
- e: Apples
- e: Applique
- e: Approved
- e: Approximate
- e: April
- e: Aqua
- e: Aqueous
- e: Arab
- e: Archaeophyt A
- e: Architect
- e: Arctic
- e: Are
- e: Argon
- e: Arithmetic
- e: Armament
- e: Armoured
- e: Army
- e: Arrest
- e: Arrival
- e: Arrive
- e: Article
- e: Articulated
- e: Artillery
- e: Asbestos[wire]
- e: Asian
- e: Asked
- e: Asparagine
- e: Aspect Ratio
- e: Assault
- e: Assembly
- e: Assented
- e: Assessment
- e: Assist
- e: Assistant
- e: Associate
- e: Association
- e: Asymmetric[al]
- e: Asynchronous
- e: Athletic
- e: Atlantic
- e: Atlas
- e: Atmosphere
- e: Atom
- e: Atomic
- e: Atomic Weight
- e: Attack
- e: Attendance
- e: Attenuation
- e: Attribute
- e: Audit
- e: Audited
- e: Auditor
- e: August
- e: Auricular
- e: Australian
- e: Austria
- e: Authentic
- e: Author
- e: Automatic
- e: Automation
- e: Automobile
- e: Autosome
- e: Auxiliary
- e: Availability
- e: Blood Type
- e: Bond Rating
- e: Brassière Cup Size
- e: Chemical Activity
- e: Fraunhofer Line Due to Oxygen
- e: Hail
- e: Includes Extra[s]
- e: Linear Acceleration
- e: Mean Sound Absorption Coefficient
- e: School Grade Meaning Excellent
- e: Shoe Size Narrower Than b
- e: Signals of Both Stereo Channels in Phase
- e: Total Acidity
- F: Alas
- f: Alpe[s]
- f: Arrivé
- i: Abasso
- i: Alpi
- i: arrivare
- l: absolvn
- l: Adhibendus
- l: albus
- l: Ante
- l: Auris
- l: Austria (Kfz)
- r: avbeta
- s: abajo
- s: Alteza
- s: aprobado

A. d: Akademie
- d: Antiphon
- d: Antrag
- d: Anwalt
- e: Amere[s]
- l: Arteria

a d: Abstand
- d: Achsenabstand
- d: achtern
- d: am, an, an der
- d: analog
- d: Anker
- d: Arbeitskontakt
- d: Atto (10^{-18})
- d: aufrecht
- d: Ausladung
- d: Axis
- e: abbreviation
- e: absolute
- e: absorption
- e: acceleration in feet per second
- e: accept[ed]
- e: accidental
- e: accomodation
- e: account
- e: acre[s]
- e: acronym
- e: active
- e: addendum
- e: adult
- e: after[noon]
- e: allele
- e: altitude
- e: altitude intercept
- e: amateur
- e: ampere
- e: amplitude
- e: anisotrop
- e: anneal[ed]
- e: annealing
- e: anode
- e: anthrac ite
- e: arc
- e: area
- e: argent
- e: arterial
- e: artery
- e: astragal
- e: at
- e: atomic[weight]
- e: attenuation constant
- e: atto (10^{-18})
- e: autonomous
- e: available
- e: aviation
- e: aviator
- e: axis
- e: azimuth[al]
- e: azure
- e: specific absorption coefficient
- e: specific rotation
- l: annus
- l: arterta
- s: antes

a. d: alt
- d: am
- d: an
- d: anonym
- d: Arch
- d: asymmetrisch
- d: auch
- d: auf
- d: aus
- d: außen
- d: axial
- e: asymmertric[al]
- l: anno

A- d: A Negative
- d: Aussen-
- e: A-minus
- e: A-negative
- e: atomic
- e: Negative Terminal of A Battery

A+ d: A Plus
- d: A Positive
- e: A-plus
- e: A-positive
- e: Positive Terminal of A Battery

A° e: Absolute Temperature
- l: anno

a' e: all
- e: a prime
- e: minute angle

a'' e: double prime
- e: second [angle]

Ä d: Ämter
- d: Äther

ä. d: äußerlich

Å a: aas

A0A2 d: Modulation, getastet

A1 d: Amplitudenmodulation, Telegrafie (unmodulierter Träger)
- d: erster Aortenton

A-1 e: air personnel officer
- e: angstrom unit
- e: excellent
- e: first rate
- e: first class
- e: Lloyds Register symbol indicating first rate equipment
- e: personnel section of an air force staff
- e: skyraider single-engine general- purpose attack aircraft flown from aircraft carn
- e: top quality
- e: tops
- e: very best

A-106 e: Agusta Anti-Submarine Warfare single-engine singleseat helicopter

A-109 e: Agusta high-performance eight-seat twin-engine helicopter

A1A d: Amplitudenmodulation, Morsetelegrafie

A1B d: Amplitudenmodulation (Fernsehschreibtelegrafie)

A/1C e: Airman First Class

A-1 Skyraider e: Douglas single-engine attack aircraft

A2 d: Amplitudenmodulation, Telegrafie (unmodulierter Träger)

A-2 e: air intelligence officer
- e: almost A-1 in quality

A² e: Ann Arbor
- e: aortic second sound
- e: Asian influenza virus

A-2 e: intelligence section of an air force staff
- e: just short of the best

A/2C e: Airman; Second Class
- e: Airman Second Class

A²C² e: see AACC

A2-Register d: zweites Akkumulator-register

A3 d: Amplitudenmodulation; Telefonie (ZSB; voller Träger)

A-3 e: air operations and training officer
- e: operations and training section of an air force staff
- e: Skywarrior twin-engine turbojet tactical all-weather attack aircraft operating f
- e: training and operations

A-32 e: single-seat single-engine jet fighter-interceptor
- r: Lansen [A-32 Lance]

A-37 e: radar-homing or television-guided air-to-surface missile

A3A d: Amplitudenmodulation; Telefonie (ESB (Einseitenband); verminderter Träger)

A3B d: Amplitudenmodulation; Telefonie (USB (unteres Seitenband); verminderter Träger)

A/3C e: Airman; Third Class

A3E d: Amplitudenmodulation; Telefonie (ZSB (Zweiseitenband); voller Träger)

A3h e: Amplitudenmodulation; Telefonie (ESB - Einseitenband); voller Träger)

A3J d: Amplitudenmodulation; Telefonie (ESB (Einseitenband); unterdrückter Träger)

A³RV e: Advanced Aerial Armed Reconnaissance Vehicle

A-3 Skywarrior e: Douglas carrier- based twin-engine jet reconnaissance and light bombing plane

A-4 e: air material and supply officer
- e: material and supply section of an air force staff
- e: Skyhawk single-engine turbojet attack aircraft operating from aircraft carriers
- e: supply and material

A4 e: Amplitudenmodulation; Faksimile

A-5 e: planning
- e: supersonic twinengine turbojet all-weather attack aircraft operating from aircra

A5D5W e: Alcohol 5%, Dextrose 5% in Water

A 5 MP e: Adenosine-5-Monophosphoric Acid

A5P e: Adenosine-5-Phosphat

A-6 e: communications

A-60 e: Saab twin-engine twoplace jet trainer-utility aircraft also called the Saab 105

A-68 e: protein found in the brains of Alzheimers victims

A-6A e: Intruder twin-engine turbojet long-range carrierbased low-altitude attack aircraft

A-6 Intruder e: Grumman carrierbased twin-engine jet lowlevel attack bomber

A7 e: Amplitudenmodulation (MFT)

A-7 Corsair II e: Ling-Temco-Vought carrier-based single-engine jet light-attack bomber

AA d: Abnehmeranlage
- d: Abrechnungsart
- d: Abrechnungsautomat
- d: Absolutadresse
- d: absolute Adresse
- d: Alkali-Alkalität
- d: Allylalkohol
- d: alphanumerische Adressendatei
- d: Analogausgabe
- d: Anfangsadresse
- d: Anionenaustauscher
- d: Arbeiter und Angestellte
- d: Arbeitsamt
- d: Arbeitsanteil
- d: Arbeitsanweisung
- d: Arbeitsausschuß
- d: Archäologischer Anzeiger (journ.)
- d: Atlas Autocode
- d: Aufgabeart
- d: Auftragsart
- d: Ausbildungsabteilung
- d: Ausführungsanweisung[en]
- d: Ausgangsanzeige[r]
- d: Auslandsabteilung
- d: Auswerteart
- d: Automatikbefehl Aus
- d: Automatisierungsanlage
- d: Gemeinschaft der Anonymen Alkoholiker
- e: Abampere
- e: Abbreviated Analysis
- e: Able and Available
- e: Abort Accept
- e: Absolute Address
- e: Absolute Addressing
- e: absolute alcohol
- e: Absolute Altitude
- e: Abstracting Automatic
- e: Academic Alliances
- e: Academy Award
- e: Academy of Aphasia
- e: Accelerated Assemblies
- e: Accelerometer Assembly
- e: Access Address
- e: Access Authorization
- e: Accompanied by Adult
- e: Accountable Activity
- e: Accounting Automaton
- e: Accuracy in Academia
- e: Acetylacrolein
- e: Achieved Availability
- e: Achievement Age
- e: Acoustics Associates
- e: Acquisition Aiding
- e: Acrylic Acid
- e: Acting Appointment
- e: Acting Army
- e: Action Aid
- e: Activation Analysis
- e: Active Accounts
- e: Active Alkali
- e: Active Army
- e: Active Assets
- e: Activity Account
- e: Acts of the Apostles
- e: Actual Address
- e: Actual Argument
- e: Actual Availability
- e: Acute Angle
- e: Acute Appendicitis
- e: Add Address
- e: Addicts Anonymous
- e: Additional Addressing
- e: Address Accumulator
- e: Address Adder
- e: Address Algorithm
- e: Address Allocation
- e: Address Appendix
- e: Address Arithmetic
- e: Address Array
- e: Address Assignment
- e: Adenylic Acid
- e: Adjustment Assistance
- e: Adjuvant Arthritis
- e: Administration on Aging
- e: Administrative Assistant
- e: Adoption Act
- e: Adoption Agency
- e: Adrenal Autoantibody
- e: Adult Accompaniment

- e: Adult Authority
- e: Advance Airlines
- e: Advanced Algebra
- e: Advanced Analytical
- e: Advertising Association
- e: Aerosol Analyzer
- e: Aerotronic Associates
- e: Affected Areas
- e: Affiliate Assembly
- e: Affirmative Action
- e: After All
- e: After Arrival
- e: Aerial Array
- e: Air Almanac
- e: Air America, Inc.
- e: Air Armament
- e: Air Attache
- e: Aircraft Artificer
- e: Airlift Association
- e: Airman Apprentice
- e: Airplane Avionics
- e: Airship Association
- e: Air to Air
- e: Alarm Apparatus
- e: Alcoholics Anonymous
- e: Alert Availability
- e: Algebraic Adder
- e: Algebraic Addition
- e: Alkalyzing Agent
- e: All Abnormal
- e: All after
- e: All Along
- e: Allyl Alcohol
- e: Alopecia Areata
- e: Alphanumeric Address
- e: Alternate Area
- e: Aluminum [Company of] America
- e: Aluminum Association
- e: Always Aflot
- e: American Airlines
- e: American Airlines, Inc.
- e: American Airways
- e: American Anthropologist
- e: American Army
- e: American Assembly
- e: American Association
- e: Amino Acid
- e: Aminoacyl
- e: Amplitude of Accommodation
- e: Analog[ue] Adder
- e: Analythical Abstracts (journ.)
- e: Anglo-American
- e: Angular Accelerometer
- e: Angular Aperture
- e: Aniline Association
- e: Anisyl Acetone
- e: Ann Arbor
- e: Annual Audit
- e: Ansett Airways
- e: Antenna Array
- e: Anterior Aorta
- e: Anterograde Amnesia
- e: Anthranilic Acid
- e: Anti+Aircraft
- e: Anti-Aero
- e: Antibody Activity
- e: Anticipated Answer
- e: Anticipatory Avoidance
- e: Antioxidant Activity
- e: Antiproton Accumulator
- e: Apollo Applications
- e: Approach Aid
- e: Appropriate Authority
- e: Appropriations and Allocations
- e: Approving Authority
- e: Approximate Absolute
- e: Aptitude Area
- e: Arachidonic Acid
- e: Arbitrary Access
- e: Arboricultural Association
- e: Archaeology Abroad
- e: Architectural Association
- e: Argument Association
- e: Arithmetic[al] Algorithm
- e: Arithmetic[al] Average
- e: arithmetic average
- e: Arlington Annex
- e: Armament Artificer
- e: Armature Accelerator
- e: Armored Ambulance
- e: Army Act
- e: Arrival Angle
- e: Arrival Approved
- e: Art and Architecture (journ.)
- e: Artificial Aerial
- e: Arts Anonymous
- e: Ascending Aorta
- e: Aspergillus Asthma
- e: Assembly Area
- e: Assets Accounting
- e: Assistant Adjutant
- e: Assistant Administrator
- e: Associate Administrator
- e: Associate in Accounting
- e: Associate in Arts
- e: Astrological Association
- e: Astronautica Acta (journ.)
- e: Atlantic Area
- e: Atlas Agena
- e: Atmospheric Applications
- e: Atomic Absorption
- e: Atomic Age
- e: Attack Assessment
- e: Audible Alarm
- e: Audit Agency
- e: Australia Antigen
- e: Authors Alteration
- e: Authors Alternation
- e: Auto Acquisition
- e: Autoanalyzer
- e: Automatic Abstracting
- e: Automatic Account[ing]
- e: Automatic Acquisition
- e: Automatic Answer
- e: Automatic Approval
- e: Autonomous Area
- e: Auxiliary Accumulator
- e: Auxiliary Address
- e: Auxiliary Algorithm
- e: Auxiliary Assembly
- e: Avenue of Approach
- e: Average Acceleration
- e: Average Adjuster
- e: Average Audience
- e: Aviation Annex
- e: Bond Rating
- e: Brassiere Cup Size
- l: Argenteum Astrum
- R: Aviatsionnayja Armiya
- s: Aerolineas Argentinas

A. A. d: Armeeabteilung
- d: Aufklärungsabteilung
- e: Anti-Aircraft
- e: Automobile Association

Aa. d: Arterien
- l: arteriae

a. A. d: alte Ausgabe
- d: alter Art
- d: anderer Ansicht
- d: auf Abruf
- d: auf Anfrage
- d: auf Antrag

aa d: andere
- d: Arbeit - Arbeitskontakt
- e: acetic acid
- e: adjectives
- e: Alveolar-Arterial
- e: always afloat
- e: aminoacetone
- e: ana
- e: arctic-alpine
- e: arteries
- e: ascorbic acid
- e: equal parts
- l: arterias

a.A. d: auf Anhänger
- d: auf Anordnung

A.A. e: Air Almanach
- e: Aproximate Absolute (Temperature)
- e: Armament Artifier
- e: Army Act
- e: Artificial Aerial

aA e: attoampere

a. a. l: ad acta

A & A d: Antike und Abendland (journ.)
- d: Arbeitsschutz und Arbeitsmedizin
- e: Advertise and Award
- e: Amendments and Additions
- e: America and Australia
- e: Art and Archeology (journ.)
- e: Arts & Architecture (journ.)

a & a e: abbreviations and acronyms
- e: additions and amendments
- e: aid and attendance

A/A e: Acting Assistant
- e: Advice Allotment
- e: Advice of Allotment
- e: Aerodrome to Aerodrome
- e: Agnostics/Atheists
- e: Air-to-Air
- e: Air Abort
- e: Airborne Alert
- e: Airport and Airways
- e: Analysis of Accounts
- e: Angle of Attack
- e: Any Acceptable
- e: Astronautics and Aeronautics

a/A e: arterial to alveolar oxygen ratio

a/a e: airdrome to airdrome
- e: antiaircraft

A-A e: Analog[ue]-Analog[ue]
- e: Asian-African

a-a e: air-to-air

AÄ d: Adressenänderung

ä. A. d: ältere Ausgabe

ÄA d: Änderungsauftrag

AAA d: atmosphärische Abbremsanalyse
- e: Acetoacetanilide
- e: Action Against Allergy
- e: Action on Alcohol Abuse
- e: Active Acquisition Aid
- e: Active Antenna Array
- e: Activity Assembly Area
- e: Acute Anxiety Attack
- e: Advance Australia Award
- e: Advanced Attack Aircraft
- e: Aerosol Association of Australia
- e: Agency Activity Analysis
- e: Agricultural Adjustment Act
- e: Agricultural Adjustment Administration
- e: Agricultural Adjustment Agency
- e: Agricultural Aircraft Association
- e: All African Award
- e: All American Aviation
- e: Allegheny Airlines
- e: Allergy Association Australia
- e: Allied Airborne Association
- e: Allocation Assessment and Analysis
- e: Alternative Antenna Array

- e: Amateur Astronomers Association
- e: American Academy of Achievement
- e: American Academy of Actuaries
- e: American Academy of Advertising
- e: American Academy of Allergy
- e: American Accounting Association
- e: American Acupuncture Association
- e: American Airship Association
- e: American Allergy Association
- e: American Ambulance Association
- e: American Antarctic Association
- e: American Anthropological Association
- e: American Association of Anatomists
- e: American Astronomers Association
- e: American Automobile Association
- e: Americans Against Abortion
- e: Annals of Archaeology and Anthropology
- e: Anti-Aircraft Armament
- e: Anti-Aircraft Artillery
- e: Anti-Aircraft Assistant
- e: Antique Airplane Association
- e: Apollo Access Arm
- e: Approved As Amended
- e: Archives of American Art
- e: Army Appropriation Account
- e: Army Audit Agency
- e: Aromatic Amino Acids
- e: Assistant Air Attachè
- e: Associate in Applied Arts
- e: Association of Accounting Administrators
- e: Association of Attenders and Alumni (Academy of International Law-Den Haag)
- e: Association of Authors Agents
- e: Astronaut-Actuated Abort
- e: Australia Asia Airlines
- e: Australian American Association
- e: Australian Antarctic Adventures
- e: Authorized Accounting Activity
- e: Automated Agency Accounting
- e: Automatic Accounting Apparatus
- e: Awaiting Aircraft Availability

AaA e: Art and Archaeology (journ.)
Aaa e: Alaska
aaa e: abdominal aortic aneurism
- e: acquired aplastic anemia
- e: amalgam
- e: Amalgam[ation]
- e: androgenic anabolic agent

AA & A e: Armo[u]r and Ammunition
aa & a e: armor, armament and ammunition
AAAA e: Aerial Agricultural Association of Australia
- e: Always Accessible / Always Afloat
- e: American-African Affairs Association
- e: American Association Against Addiction
- e: American Association for the Advancement of Agnosticism
- e: American Association for the Advancement of Atheism
- e: American Association of Advertising Agencies
- e: American Association of Audio Analgesia
- e: Antique Aeroplane Association of Australia
- e: Army Aviation Association of America
- e: Australian Aerial Agricultural Association
- e: Australian Association for Applied Anthropology

AAAAD e: Airborne Antitank Armor Air Defense
AAAAPSF e: American Association for Accreditation of Ambulatory Plastic Surgery Facilities
AAA AQ Btry e: Anti-Aircraft Artillery Automatic Weapons Battery
AAA AW e: Anti-Aircraft Artillery Automatic Weapons
AAA AW Btry (SP) e: Anti-Aircraft Artillery Automatic Weapons Battery (Self-Propelled)
AAAB e: American Association of Architectural Bibliographers
AAABG e: Australian Association of Animal Breeding and Genetics
AA Abn e: Anti-Aircraft Artillery Airborne
AAA Bty e: Anti-Aircraft Artillery Battery
AAAC e: All-Aluminium Alloy-Conductor
- e: American-Arab Affairs Council
- e: American Association of Accompanists and Coaches
- e: Anti-Aircraft Artillery Command
- e: Antimicrobial Agent Associated Colitis
- e: Association of Australian Acoustical Consultants
- e: Australian Army Aviation Corps
- e: Australian Association of Agricultural Consultants

AAACE e: American Association for Adult and Continuing Education
- e: American Association of Agricultural College Editors

AAACOM e: Army Anti-Aircraft Command
AAA Committee e: Associated Audio Archives Committee
AAAD e: Aromatic Amino Acid Decarboxylase
AAAE e: American Association for Agricultural Education
- e: American Association of Academic Editors
- e: American Association of Airport Executives
- e: Asian Automotive and Accessories Exhibition
- e: Association of Arts Administration Educators

AAAEE e: American Afro-Asian Educational Exchange
AAAF e: Anglo-American Air Force
AAAFDC e: Anti-Aircraft Artillery Fire Direction Center
AAAG e: Acting Assistant Adjutant General
- e: Annals of the Association of American Geographers

AA Ag e: Associate in Arts in Agriculture
A.A. Ag e: Associate of Arts in Agriculture
AAA & GM e: Anti-Aircraft Artillery and Guided Missile Equipment
AAA & GM Ctr e: Anti-Aircraft Artillery and Guided Missile Center
AAA & GM S e: Anti-Aircraft Artillery and Guided Missile School
AAA Gn Bty e: Anti-Aircraft Artillery Gun Battery
AAA gun crm e: Anti-Aircraft Artillery Gun Crewman
AAAI e: American Academy of Allergy and Immunology
- e: American Association for Artificial Intelligence
- e: Association of Australian Aerospace Industries

AAAID e: Arab Authority for Agricultural Investment and Development
AAA IFC e: Anti-Aircraft Artillery Integrated Fire Control
AAA IFC PL e: Anti-Aircraft Artillery Integrated Fire Control Platoon
AAA IFC rdr eqp e: Anti-Aircraft Artillery Integrated Fire Control Radar Equipment
AAAIMH e: American Association for the Abolition of Involuntary Mental Hospitalization
AAAIP e: Advanced Army Aircraft Instrumentation Program[me]
AAAIS e: Advanced Army Airborne Indicating System
- e: Advanced Army Aircraft Instrumentation System
- e: Anti-Aircraft Artillery Intelligence Service

AAAIWA e: Automobile, Aerospace and Agricultural Implement Workers of America
AAAL e: American Academy of Arts and Letters
- e: American Association for Applied Linguistics

AAALAC e: American Association for Accreditation of Laboratory Animal Care
AAALMG e: Anti-Aircraft Artillery Light Machine Gun
AAALO e: Anti-Aircraft Artillery Liaison Officer
AAAM e: Advanced Air-to-Air Missile
- e: American Anthropological Association Memoirs
- e: American Association for Automotive Medicine
- e: American Association of Aircraft Manufacturers
- e: Association for the Advancement of Automotive Medicine

AAAMA e: Air-to-Air Armament Mission Analyses
AAAMS e: Advanced Anti-Armor Missile Systems
AAA Msl e: Anti-Aircraft Artillery Missile
AAAN e: Affiliated Advertising Agencies Network
- e: American Academy of Applied Nutrition

AA An e: Annals of Archaeology and Anthropology
AAANA e: American Academy of Ambulatory Nursing Administration
AA and A e: Armament and Ammunition
- e: Armor

AA and AS e: Anti-Aircraft and Anti-Submarine
AAANY e: Amateur Astronomers Association of New York
AAANYC e: Amateur Astronomers Association of New York City
AAAOB e: Anti-Air Artillery Order of Battle
AAAOC e: Anti-Aircraft Artillery Operation Center
AAAOCS e: Anti-Aircraft Artillery Officer Candidate School
AAAOM e: American Association for Acupuncture and Oriental Medicine
AAA opns & intel sp e: Anti-Aircraft Artillery Operations and Intelligence Specialist
AAAOR e: Anti-Aircraft Artillery Operations Room
AAAP e: Airborne Associate Array Processor
- e: American Association of Applied Psychology

AAAR e: American Association for Aerosol Research
- e: Association for the Advancement of Aeronautical Research

AAARC e: Anti-Aircraft Artillery Reception Center
- e: Anti-Aircraft Artillery Remote Control

AAARCAT e: Anti-Aircraft Artillery Remote-Controlled Aerial Target

AAARC Rep Fman e: Anti-Aircraft Artillery Remote Control Repair Foreman

AAARC Rep Hlpr e: Anti-Aircraft Artillery Remote Control Repair Helper

AAARC Rep Insp e: Anti-Aircraft Artillery Remote Control Repair Inspector

AAARC Repm e: Anti-Aircraft Artillery Remote Control Repairman

AAA rdr e: Anti-Aircraft Artillery Radar

AAA rdr crm e: Anti-Aircraft Artillery Radar Crewman

AAA Rep Fman e: Anti-Aircraft Artillery Repair Foreman

AAA Rep Hlpr e: Anti-Aircraft Artillery Repair Helper

AAA Repm e: Anti-Aircraft Artillery Repairman

AAARG e: American Atheist Addiction Recovery Groups

AAARRD e: Anti-Aircraft Artillery Radar

AAARRDCRM e: Anti-Aircraft Artillery Radar Crewman

AAARTC e: Anti-Aircraft Artillery Replacement Training Center

AAAs e: Acute Anxiety Attacks

AAAS e: American Academy of Arts and Sciences
- e: American Academy of Asian Studies
- e: American Association for the Advancement of Science
- e: Armored Anti-Aircraft System
- e: Associate of the American Antiquarian Society
- e: Australian Association for the Advancement of Science
- e: Automated Attendance Accounting System
- e: Automatic Activation Analysis System

A.A.A.S. e: Associate in Arts and Science

AAA & S e: Associate in Arts and Science

AAASA e: Association for the Advancement of Agricultural Sciences in Africa

AAAS Bull e: American Association for the Advancement of Science Bulletins

AAA Slt e: Anti-Aircraft Artillery Searchlight

AAASS e: American Association for the Advancement of Slavic Studies

AAAST e: African Association for the Advancement of Science

AAASTS e: Anti-Aircraft Artillery Service Test Station

AAATC e: American Association for the Advancement of Tension Control
- e: Anti-Aircraft Artillery Training Center

AAATDC e: Association for the Advancement of Appropriate Technologies for Developing Count

AAAUS e: Association of Average Adjusters of the United States

AAAV e: Advanced Amphibious Assault Vehicle
- e: Avian Adeno-Associated Virus

AAAWC e: Alternate Anti-Air Warfare Commander

AAA wire e: Alumin[i]um Alloy Anodizing Wire

AAB d: Allergiker- und Asthmatiker- Bund
- d: allgemeine Ausführungsbestimmungen
- d: automatische Angebots- und Auftragsbearbeitung
- d: Automatisierungsanlagenbau
- e: Actualizing Assessment Battery
- e: Adaptive Angle Bias
- e: Air-to-Air Banner
- e: Air Assault Brigade
- e: Aircraft Accident Board
- e: Aircraft Armament Bulletin
- e: Alliance Agricole Belge
- e: American Association of Bioanalysts
- e: Analytical Abstract
- e: Anti-Aircraft Battery
- e: Army Air Base
- e: Army Artillery Board
- e: Army Aviation Board
- e: Associate in Arts in Business
- e: Association of Applied Biologists
- e: Atlas-Agena B
- e: Aviation Armament Bulletin

AA/B e: Anti-Aircraft Balloon

AABB e: American Association of Blood Banks
- e: Associated Air Balance Council

AABC e: American Association of Biofeedback Clinicians
- e: Associated Air Balance Council

AABCP e: Advanced Airborne Command Post

AABEVM e: Association of American Boards of Examiners in Veterinary Medicine

AABFS e: Amphibious Assault Bulk Fuel System

AABGA e: American Association of Botanical Gardens and Arboretums

AABI e: Antilles Air Boats Incorporated

AABL e: Advanced Atmospheric Burst Location

AA bln e: Anti-Aircraft Balloon

AABM e: Airborne Anti-Ballistic Missile[s]
- e: American Academy of Behavioral Medicine
- e: Association of American Battery Manufacturers
- e: Australian Association of British Manufacturers

aaBm e: analytical anatomy by the Braille method

AA Bn e: Anti-Aircraft Battalion

AABNCP e: Advanced Airborne Command Post

AABP e: Acetylaminobiphenyl
- e: Aptitude Assessment Battery Program[me]
- e: Aptitude Assessment Battery Programming

AABS e: All Attitude Bombing System

aabshil e: aircraft anti-collision-beacon-system high-intensity light[ing]

AABSHIL e: Aircraft Anticollision Beacon System High-Intensity Light

AABT e: American Association of Behavioral Therapists
- e: Association for the Advancement of Behavior Therapy

AA Bty e: Anti-Aircraft Battery

AA Bus e: Associate in Arts in Business

AABW e: American Association of Book Wholesalers
- e: Antarctic Bottom Water

AABY e: As Ammended By

AAC d: Audio Aktiv Comparativ
- e: Abbreviated Address Calling
- e: Abort Advisory Channel
- e: Acoustical Absorption Coefficient
- e: Acoustical Attenuation Constant
- e: Acquisition Advice Code
- e: Activity Address Code
- e: Actual Acquisition Cost
- e: Adaptive Antenna Control
- e: Advanced Airborne Computer
- e: Aerated Autoclaved Concrete
- e: Aeronautical Advisory Council
- e: Aeronautical Approach Chart
- e: Afro-Asian Center
- e: Agreement and Account of Crew
- e: Agricultural Advisory Council
- e: Air-Air-Cooling
- e: Air Approach Control
- e: Airborne Armament Control
- e: Aircraft Armament Change
- e: Airworthiness Advisory Circular
- e: Alameda Administration Center
- e: Alaskan Air Command
- e: All-Aluminium Conductor
- e: All-American Canal
- e: Allanto-Amnio-Chorion
- e: Allied Air Co-Operation
- e: Alphanumeric Accounting Computer
- e: Aluminum Anodizers Council
- e: Alumnae Advisory Center
- e: American Academy of Criminalistics
- e: American Alumni Council
- e: American Archery Council
- e: American Association of Chiropractors
- e: American Atheist Center
- e: American Cement Corporation
- e: Aminoacetylcatechol
- e: Ampere[s] Alternating Current
- e: Amplitude Absorption Coefficient
- e: Anglo-American Code
- e: Antarctic Circle
- e: Anti-Aircraft Cannon
- e: Anti-Aircraft Command
- e: Anti-Aircraft Corps
- e: Anti-Aircraft Command
- e: Anti-Aircraft Common
- e: Antibiotic-Associated Colitis
- e: Antimicrobial Agents and Chemotherapy (journ.)
- e: Arabinosylazacytidine
- e: Arithmetic and Controls
- e: Army Air Corps
- e: Army Ammunition Command
- e: Army Audiovisual Center
- e: Army Aviation Center
- e: Arsenic Atmosphere Czochralski
- e: Assembly and Checkout
- e: Assembly Area Command
- e: Assets Availability Code
- e: Assistant Area Commander
- e: Association of American Colleges
- e: Association of Analytical Chemists
- e: Atomic Absorption Coefficient
- e: Attack Aircraft Carrier
- e: Australian Agricultural Company
- e: Australian Air Corps
- e: Australian Association of Chiropractors
- e: Automatic Amplitude Control
- e: Automatic Aperture Control
- e: Automatic Approach Control
- e: Automatic Auto-Collimator
- e: Automotive Advertisers Council
- e: Auxiliary Air Control
- e: Auxiliary Artillery Corps
- e: Aviation Armament Change
- i: Auto Avia Contruzione
- p: Associação Academica de Coimbra

A.A.C. e: Acting Armo[u]rer Corporal
- l: anno ante Christum
aac e: affirmative action clause
- e: average annual cost
a. a. C. l: anno ante Christum
AACA e: American Association of Certified Allergists
- e: Amphibious Auto Club of America
- e: Antique Automobile Club of America
- e: Associate of the Association of Cost Accountants
- e: Automotive Air Conditioning Association
AACAHPO e: American Association of Certified Allied Health Personnel in Ophthalmology
AA-C and Ref Tech e: Associate in Air-Conditioning and Refrigeration Technology
AACAP e: American Academy of Child and Adolescent Psychiatry
- e: Association of American Colleges Arts Program[me]
AACB e: Aeronautics and Astronautics Coordination Board
- e: Australian Association of Clinical Biochemists
AACBM e: American Association of Creamery Butter Manufacturers
AACBP e: American Academy of Crown and Bridge Prosthodontics
AACC e: Administrative Area Control Centre
- e: Affirmative Action Coordinating Center
- e: Airport Associations Coordinating Council
- e: All-Altitude Control Capability
-e: American Association for Contamination Control
- e: American Association of Cereal Chemists
- e: American Association of Clinical Chemistry
- e: American Association of Clinical Chemists
- e: American Automatic Control Council
- e: Army Air Corps Centre
- e: Army Aviation Control Center
- e: Association of Agricultural Computer Companies
- e: Automatic Approach Control Complex
- e: Automatic Approach Control Coupler
aacc e: all-attitude control capability
AACCA e: Associate of the Association of Certified and Corporate Accountants
AACCN e: American Association of Critical-Care Nurses
AACCP e: American Association of Colleges for Chiropody Podiatry
AACD e: Antenna Adjustable Current Distribution
AACDP e: American Association of Chairmen of Departments of Psychiatry
AACE e: Airborne Alternate Command Echelon
- e: American Association for Cancer Education
- e: American Association of Cost Engineers
- e: Association for Adult Continuing Education
AACEP e: Aircraft Alerting Communications Electromagnetic Pulse
AACF e: Anti-Aircraft Co-Operation Flight
- e: Army Area Calibration Facilities
AACFD e: Automatic Account Card Feed[ing] Device
AACFT e: Army Aircraft
AACG e: Acute Angle Closure Glaucoma
- e: American Association for Crystal Growth
Aach d: Aachen
Aach. Bl. d: Aachener Blätter (journ.)
AACHIR e: Augusta Area Committee for Health Information Resources
AACHP e: American Association for Comprehensive Health Planning
a. a. Chr. l: anno ante Christum
a. a. Chr. n. l: anno ante Christum natum
AACHS e: Afro-American Cultural and Historical Society
- e: German geographical placename equivalent of Aix-la-Chapelle on the Belgian-Dutch
AACHSA e: Associate of the Australian College of Health Service Administrators
AACHT e: American Association for Clinical Histocompatibility Testing
AACI e: Accredited Appraiser; Canadian Institute
- e: American Association for Conservation [of] Information
- e: American Association of Ceramic Industries
A.A.C.I. e: Associate of the Australian Chemistry Institute
AACI e: Association of American Cancer Institutes
AACIA e: American Association for Clinical Immunology and Allergy
AACIL e: All-Articles Configuration Inspection Log
AACJC e: American Association of Community and Junior Colleges
AACL e: Affect Adjective Check List
- e: Anxiety Adjective Check List
- e: Associate of the Amateur Cinema League (of America)
AACLAME e: Australian Advisory Council on Language and Multicultural Education
AACM e: Advanced Air Cycle Machine
- e: American Academy of Compensation Medicine
- e: Associate of the Australian College of Music
- e: Automatic Armor Cluster Munition
- e: Average Absolute Control Movement
AACN e: American Association of Colleges of Nursing
a. a. C. n. l: anno ante Christum natum
AACO e: American Association of Certified Orthoptists
- e: Assault Airlift Control Office[r]
AACOBE e: Army Automation Command Operating Budget Estimate
AACOBS e: Australian Advisory Council on Bibliographical Services
AACOM e: American Association of Colleges of Osteopathic Medicine
- e: Army Area Communications
AACOMS e: Army Area Communication[s] System
AACON e: Agency Accounting Control System
AACP e: Advanced Airborne Command Post
- e: American Academy for Cerebral Palsy
- e: American Academy for Child Psychiatry
- e: American Academy of Child Psychiatry
- e: American Academy of Clinical Psychiatrists
- e: American Association for Correctional Psychology
- e: American Association of Colleges of Pharmacy
- e: American Association of Colleges of Podiatry
- e: American Association of Community Psychiatrists
- e: American Association of Computer Professionals
- e: American Association of Correctional Psychologists
- e: Anglo-American Council on Productivity
- e: Anti-Aircraft Command Post
- e: Army Acquisition Corps Program[me]
- e: Associate of the Association of Computer Professionals
A.A. & C.P. e: Anchors and Chains Proved
AACPDM e: American Academy for Cerebral Palsy and Developmental Medicine
AACPM e: American Association of Colleges of Podiatric Medicine
AACPP e: Association of Asbestos Cement Pipe Producers
AACPS e: American Association of Clinic Physicians and Surgeons
AACR e: All America Cables and Radio
- e: American Association for Cancer Research
- e: Anglo-American Cataloguing Rules
AA-C & Ref Tech e: Associate in Air-Conditioning and Refrigeration Technology
AACRF e: Anti-Aircraft Close Range Fire
AACS e: Aberdeen-Angus Cattle Society
- e: Advanced Automatic Compilation System
- e: Airborne Astrographic Camera System
- e: Airspeed and Altitude Computer Set
- e: Airways and Air Communication[s] Service
- e: American Academy of Cosmetic Surgery
- e: Anti-Aircraft Control Station
- e: Anti-Armor Capabilities Study
- e: Army Airways Communication[s] Service
- e: Army Airways Communication[s] System
- e: Army Alaska[n] Communication[s] System
- e: Asynchronous Address Communication[s] System
- e: Attitude and Antenna Control System
- e: Attitude and Articulation Control Subsystem
- e: Auxiliary Attitude Control System
AACSB e: American Assembly of Collegiate Schools of Business
AACSC e: Army Automation and Communication Steering Committee
AACSCEDR e: Associate and Advisory Committee to the Special Committee on Electronic Data Ret
AACSE e: American Association of Classified School Employees
AACSL e: American Association for the Comparative Study of Law
AACSM e: Airways and Air Communications Service Manual
AACSR e: Airways and Air Communication[s] Service Regulation[s]
- e: Aluminium Alloy Conductor; Steel Re-Inforced
AACSRON e: Airways and Air Communication[s] Service Squadron
AACSWG e: Airways and Air Communication[s] Service Wing
AACT e: American Academy of Clinical Toxicology
- e: Armenian Assembly Charitable Trust
AACTE e: American Association of Colleges for Teacher Education
AACTF e: Army Aircraft

AACTS e: Automatic Anechoic Chamber Test System
AACTVO e: American Association of Cable TV Owners
AACU e: American Association of Clinical Urologists
AACUBO e: American Association of College and University Business Officers
AACUO e: Association for Affiliated College and University Offices
AACVPR e: American Association of Cardiovascular and Pulmonary Rehabilitation
AAD d: Adressenaddierer
- d: Allgemeiner Anlagedienst
- d: Auftastimpuls-Auslösediode
- e: Abbreviated Address Dialling
- e: Acetoacetate Decarboxylase
- e: Active Acoustic Device
- e: Address Adder
- e: Admission and Disposition
- e: Adrenalectomized Alloxan Diabetic
- e: Advanced Academic Degree
- e: Advanced Airborne Demonstrator
- e: Advanced Ammunition Depot
- e: Air Ammunition Depot
- e: Aircraft and Armament Development
- e: Aircraft Assignment Directive
- e: Alaskan Air Depot
- e: Alloxazine Adenine Dinucleotide
- e: American Academy of Dentists
- e: American Academy of Dermatology
- e: Anti-Aircraft Defences
- e: Anti-Aircraft Director
- e: Army Air Defense
- e: Army Automation Directorate
- e: Army Aviation Digest
- e: Assembly and Disassembly
- e: Association of American Dentists
- e: Atlas-Agena D
- e: Australian-Antarctic Discordance
- e: Average Absolute Deviation
Aad d: Alpha-Aminoadipinsäure
aad e: active acoustic[al] device
AADA e: Advanced Air Depot Area
- e: Airport and Airway Development Act
- e: Anti-Aircraft Defended Area
- e: Army Air Defense Area
- e: Auxiliary to the American Dental Association
AADB e: Army Air Defense Board
AADC e: Advanced Airborne Digital Computer
- e: Advanced Avionics Digital Computer
- e: Afro-Asian Documentation Centre
- e: Air-to-Air Doppler Clutter
- e: Air Aide-De-Camp
- e: Airborne Antiarmor Defense Concept
- e: All-Application Digital Computer
- e: American Air Defense Command
- e: American Association of Dental Consultants
- e: Amino Acid Decarboxylase
- e: Anti-Aircraft Defense Command[er]
- e: Anti-Aircraft Director Center
- e: Approach and Departure Control
- e: Army Air Defense Command[er]
- e: Arnold Air Development Center
- e: Asthma and Allergic Disease Center
- i: American Association of Dental Consultants
AADCC e: Army Air Defense Command and Control
AADCCS e: Army Air Defense Control and Coordination System
AADCM e: Awaiting Action Deck Court Material
AADCOM e: Army Air Defense Command[er]
AADCP e: Army Air Defense Command Post
AADCS e: Automatic Air Data Calibration System
AADD e: Auxiliary Active Digital Display
- e: Aviation and Air Defense Division
AADE e: American Academy of Dental Electrosurgery
- e: American Association of Dental Editors
- e: American Association of Dental Examiners
- e: American Association of Diabetes Educators
AA Def e: Anti-Aircraft Defense
AA Def HQ e: Anti-Aircraft Defense Headquarters
AA de L e: Academia Argentina de Letras
AADEOS e: Advanced Air Defense Electro-Optic Sensor
AADGE e: Allied Air Defense Ground Environment
AADGP e: American Academy of Dental Group Practice
AADHS e: Advanced Avionics Data Handling System
AA Dip e: Architectural Association Diploma
AADIS e: Army Air Defense Information Service
- e: Army Air Defense Information System
- e: Automatic Air Defense Information System
A/Adj e: Assistant Adjutant
AADL e: Artillery Ammunition Development Laboratory
AADM e: American Academy of Dental Medicine
A Adm e: Associate in Administration
AADMC e: Army Aviation Depot Maintenance Center
AADMS e: Advanced Academic Degree Management System
AADN e: American Association of Doctors Nurses
AADOO e: Army Air Defense Operations Office[r]
AADP e: Advanced Avionic Display Processor
- e: American Academy of Denture Prosthetics
- e: Anti-Aircraft Defended Point
- e: Army Aviation Development Plan
AADPA e: American Academy of Dental Practice Administration
AA & DPD e: Appropriation Accounts and Data Processing Division
AADPRT e: American Association of Directors of Psychiatric Residency Training
AADPSP e: Activity Automatic Data Process Security Plan
AADR e: Alternate Address
- e: American Academy of Dental Radiology
- e: American Association for Dental Research
AADS e: Advanced Air Defense System
- e: Advanced Army Defense System
- e: Aero/Acoustic Detection System
- e: American Association of Dental Schools
- e: American Association of Dermatology and Syphilology
- e: Anti-Aircraft Defense System
- e: Area Air Defense System
- e: Army Air Defense School
- e: Army Air Defense Site
- e: Army Air Defense Staff
- e: Army Air Defense System
- e: Ascent Air Data System
- e: Automated Acoustic Detection System
- e: Automatic Aircraft Diagnostic System
- e: Automatic Applications Development System
AADSF e: Advanced Automated Directional Solidification Furnace
AADS System e: Access Area Digital Switching System
AADT e: Annual Average Daily Traffic
- e: Association of Advanced Dental Technicians
AADTS e: Association of Advisors in Design and Technical Studies
A-ADU d: Amplituden-Analog/Digital-Umsetzer
AADV e: Acquisition Aid Vehicle
AADW e: Advanced Air Defense Weapon
AAE d: Amtsanschlußeinrichtung
- d: automatische Anrufbeantwort[ungs]-einheit
- d: automatische Anrufbeantwortungsein-richtung
- d: automatische Anrufeinrichtung
- d: automatische Anruferzeugung
- d: automatische Anschalteinrichtung
- d: automatische Antworteinheit
- e: Abort Advisory Equipment
- e: Accredited Airport Executive
- e: Active Assistive Exercise
- e: Acute Allergic Encephalitis
- e: Aeronautical and Astronautical Engineering
- e: Aerospace Ancillary Equipment
- e: Aerospace Auxiliary Equipment
- e: American Association of Endodontists
- e: American Association of Engineers
- e: Ancillary Armament Equipment
- e: Apparent Activation Energy
- e: Appropriation and Expense
- e: Architectural and Engineering
- e: Armament and Electronics
- e: Army Acquisition Executive
- e: Army Aviation Element
- e: Army Aviation Engineers
- e: Asia Australia Express
- e: Assembly and Equipment
- e: Assembly and Erection
- e: Australasian Antarctic Expedition
- e: Auto-Answer Equipment
- e: Automated Analysis Equipment
- e: Automatic Adaptive Equalization
- e: Automatic Answering Equipment
- e: Average Absolute Error
- e: Azimuth and Elevation
aae e: above airport elevation
- e: average annual earnings
AAEA e: African Adult Education Association
- e: American Agricultural Editors Association
AAEC e: Attitude Axis Emergency Control
- e: Australian-American Engineering Corporation
- e: Australian Atomic Energy Commission
AAECo e: Australian Aircraft and Engineering Co.
AAECS e: Auxiliary Area Environmental Control System
AAED e: Academic American Encyclopedia Database
- e: Active Airborne Expendable Decoy
- e: American Academy of Esthetic Dentistry
- e: American Association of Entrepreneurial Dentists

AAEDC e: American Agricultural Economics Documentation Center
AAEE e: Aeronautical and Aircraft Experimental Establishment
- e: Aeroplane and Armament Experimental Establishment
- e: Airplane & Armament Experimental Establishment
- e: American-Asian Educational Exchange
- e: American Academy of Environmental Engineers
- e: American Association of Economic Entomologists
- e: American Association of Electrical Engineers
- e: American Association of Electro-myography and Electrodiagnosis
- e: American Association of Electro-myography and Electrodiagnostics
A & AEE e: Aircraft and Armament Experimental Establishment
A Ae E e: Associate in Aeronautical Engineering
A.Ae.E. e: Associate in Aeronautical Engineering
AAEEH e: American Association of Eye and Ear Hospitals
AAEF e: Australian Air Expeditionary Force
- e: Australian American Education Foundation
- e: Automated Analytical Electrophoresis Facility
AAEFA e: Army Aviation Engineering Flight Activity
AAEGTS e: Auxiliary Area Emergency Gas Treatment System
AAEH e: Association to Advance Ethical Hypnosis
AAEI e: American Association of Exporters and Importers
AAELSS e: Active-Arm External-Load Stabilization System
AAEM e: Acetoacetoxyethyl Methacrylate
- e: American Academy of Environmental Medicine
- e: American Association of Electro-diagnostic Medicine
AAERR e: Aerophysics and Aerospace Engineering Research Report
AAES e: Advanced Aircraft Electrical System
- e: American Association of Engineering Societies
- e: American Association of Evangelical Students
- e: Anti-Aircraft Experimental Section
- e: Australian Agricultural Economics Society
- e: Automated Aircrew Escape System
AAESA e: Alabama Association of Elementary School Administrators
- e: Army Acquisition Executive Support Agency
AAESDA e: Association of Architects, Engineers, Surveyors and Draughtsmen of Australia
AAESWB e: Army Airborne Electronics & Special Warfare Board
AA-EVP e: American Association - Electronic Voice Phenomena
AAEW e: Atlantic Airborne Early Warning
AAEWR e: Advanced Aircraft Early Warning Radar
AAF d: Antiatelektase-Faktor
- e: Accounting and Finance
- e: Acetic-Alcohol-Formalin
-e: Acetylamino Fluoren[c]e
- e: Agglutination Activating Factor
- e: Agricultural Aids Foundation
- e: Allted Air Forces
- e: American Advertising Federation
- e: American Air Filter
- e: American Airforce
- e: American Architectural Foundation
- e: American Astronautical Federation
- e: Aminioallantoic Fluid
- e: Amino Acid Formula
- e: Amnioallantoic Fluid
- e: Anterior Auditory Field
- e: Anti-Aircraft Fire
- e: Anti-Armo[u]r Fuze
- e: Army Air Field
- e: Army Air Force
- e: Army and Air Force
- e: Ascorbic Acid Factor
- e: Atlantic Amphibious Force
- e: Australian Air Force
- e: Auxiliary Air Force
- e: Average Adjustment Factor
- f: Academie d'Agriculture de France
a-a-f e: acetic-alcohol-formalin
aaf e: acetylaminofluorine
A.A.F. e: Armament Artificer
AAFA e: Air-Air Forced Cooling
- e: Architectural Aluminium Fabricators Association
- e: Associate in Arts in Fine Arts
AA/FA e: Ambient Air / Forced-Air-Cooled
A & AFA e: Army and Air Force Act
- e: Asthma and Allergy Foundation of America
AAFAIS e: Army and Air Force Air Intelligence School
AAFB e: Andrews Air Force Base
- e: Army Air Force Board
- e: Army Air Force Bulletin
- e: Army and Air Force Base
- e: Atypical Acid-Fast Bacilli
- e: Auxiliary Air Force Base
AAFBF e: Anderson Air Force Base Flightline
AAFBS e: Army Air Forces Bombardier School
AAFBTC e: Army Air Forces Basic Training Center
AAFBU e: Army Air Forces Base Unit
AAFC e: Airborne Audio Frequency Coder
- e: Anti-Aircraft Fire Control
- e: Army Air Forces Center
- e: Arnmy Air Force Classification Center
- e: Association of Advertising Film Companies
aafc e: antiaircraft fire control
AAFCC e: Army Air Forces Classification Center
AAFC Center e: Army Air Forces Classification Center
AAFCE e: Allied Air Forces; Central Europe
AAFCFTC e: Army Air Forces Central Flying Training Command
AAFCO e: Association of American Feed Control Officials
- e: Association of American Fertilizer Control Officials
AAFCOP e: Alternate Air Force Command Post
AAFCS e: Advanced Automatic Flight Control System
- e: Anti-Aircraft Fire Control System
AAFCTTC e: Army Air Forces Central Technical Training Command
AAFCWF e: Army and Air Force Central Welfare Fund
- e: Army and Air Force Civilian Welfare Fund
AAFd e: Army Airfield
AAFDBI e: American Association of Fitness Directors in Business and Industry
AAFE e: Advanced Application[s] Flight Experiment
- e: American Association of Feed Exporters
AAFEA e: Australian Airline Flight Engineers Association
AAFEB e: Anaerobic Attached-Film Expanded-Bed
AAFEC e: Army Air Forces Engineering Command
AAFEESS e: Automated Armed Forces Examining and Entrance Station System
AAFEFTC e: Army Air Forces Eastern Flying Training Command
AAFEMPS e: Army and Air Force Exchange and Motion Picture Service
AAFE Program[me] e: Advanced Applications Flight Experiments Program[me]
AAFEs e: Advanced Applications Flight Experiments
A & AFES e: Army and Air Force Exchange Service
AAFETTC e: Army Air Forces Eastern Technical Training Command
AAFG e: Amino Acid Formula with Glutamate
AAFGL e: Auxiliary Air Force General List
AAFGS e: Army Air Forces Gunnery School
AAFH e: Academy of American Franciscan History
AAFI e: Air-Assisted Fuel Injection
- e: Allied Air Forces in Italy
AAFIF e: Automated Air Facilities Intelligence File
- e: Automated Air Facility Information File
AAFIR e: Army Air Force Intelligence Report
AAFIS e: Advanced Avionics Fault Isolation System
- e: Army Air Forces Intelligence School
- e: Army and Air Force Intelligence Staff
AAFM e: American Association of Feed Microscopists
- e: Army Air Force Manual
AAFMAA e: Army and Air Force Mutual Aid Association
AAFMC e: Army Air Forces Material Center
AAFMG e: American Association of Foreign Medical Graduates
AAFMPS e: Army and Air Force Motion Picture Service
AAFMTO e: Army Air Force; Mediterranean Theater of Operations
AAFNE e: Allied Air Forces; Northern Europe
AAFNS e: Army Air Forces Navigation School
AAFNTIR e: Army Air Force Nontechnical Intelligence Report
AAFO e: American Association for Functional Orthodontics
AAFOIC e: Army Air Forces Officer in Charge
AAFP e: American Academy of Family Physicians
- e: American Association of Financial Professionals

AAFPF e: American Academy of Family Physicians Foundation
AAFPFS e: Army Air Forces Pre-Flight School
AAFPFSP e: Army Air Forces Pre-Flight School Pilot
AAFPOA e: Army Air Forces; Pacific Ocean Areas
AAFPRS e: American Academy of Facial Plastic and Reconstructive Surgery
AAFPS e: Army and Air Force Pilot School
- e: Army and Air Force Postal Service
AAFRS e: American Academy of Facial, Plastic and Reconstructive Surgery
AAFS e: Academy of Ambulatory Foot Surgery
- e: American Academy of Forensic Sciences
- e: American Association of Foot Specialists
- e: Amphibious Assault Fire System
- e: Amphibious Assault Fuel System
- e: Atomic Absorption Flame Spectrometer
AAFSAT e: Army Air Forces School of Applied Tactics
AAFSC e: Additional Air Force Speciality Code
- e: Army Air Forces Service Command
AAFSE e: Allied Air Forces; Southern Europe
AAFSETC e: Army Air Forces Southeast Training Command
AAFSS e: Advanced Aerial Fire Support System
- e: Armys Airborne Fire-Support System
AAFSSO e: Advanced Aerial Fire Support System Office
AAF STD e: Army Air Forces Standard
AAFSWPA e: Allied Air Forces; South West Pacific Area
AAFT e: Army Air Force Translation
AAFTAC e: Army Air Forces Tactical Center
AAFTC e: Army Air Forces Training Center
- e: Army Air Forces Training Command
AAFTI & E e: Army Air Forces Troop Information and Education
AAFTIR e: Army Air Force Technical Intelligence Report
AAFTO e: Army Air Force Technical Order
AAFTS e: Advanced Automatic Film Titles System
- e: Army Air Forces Technical School
- e: Automated Adaptive Flight Training System
AAFTTC e: Army Air Forces Technical Training Command
- e: Army and Air Forces Technical Training Center
AAFTTS e: Army Air Forces Technical Training School
AAFU e: All-African Farmers Union
- e: Augmented Assault Fire Units
AAFV e: Australian Armed Forces; Vietnam
AAFWB e: Army and Air Force Wage Board
AAFWFTC e: Army Air Forces Western Flying Training Command
AAFWSB e: Army Air Force Weather Service Bulletin
AAFWSM e: Army Air Force Weather Service Manual
AAFWTTC e: Army Air Forces Western Technical Training Command
AAG d: Antennenanpassungsgerät
- e: Acquisition Advisory Group
- e: Aeromedical Airlift Group
- e: Aeronautical Assignment Group
- e: Air Adjutant-General
- e: Annals of American Geographers (journ.)
- e: Anti-Aircraft Gun
- e: Area Advisory Group
- e: Army Artillery Group
- e: Assistant Adjutant General
- e: Association of American Geographers
- e: Australian Association of Gerontology
A.A.G e: Air Adjutant General
- e: Assistant Adjutant General
- e: Assistant Attorney General
ÄAG d: Änderungs- und Anpassungsgesetz
AAGBA e: American Angora Goat Breeders Association
AAGC e: American Association for Gifted Children
AAGCM e: Awaiting Action General Court Material
AAGE e: All Aspect Gunsight Evaluation
- e: Army Advisory Group on Energy
AAGF e: Anti-Aircraft Gun Fire
AAGFE e: American Association of Gravity Field Energy
AAGFO e: American Academy of Gold Foil Operators
AAGIWA e: American Association of Grain Inspection and Weighing Agencies
AAGL e: American Association of Gynecological Laparoscopists
- e: Anti-Aircraft Gun-Laying
AA & GM e: Anti-Aircraft and Guided Missile
AAGM e: Anti-Aircraft Guided Missile
AA & GM Branch e: Anti-Aircraft and Guided Missile Branch
AAGMC e: Anti-Aircraft Artillery and Guided Missile Center
AA & GM Ctr e: Anti-Aircraft Artillery and Guided Missile Center
AAGMS e: Anti-Aircraft Guided Missile School
- e: Anti-Aircraft Guided Missile Site
- e: Anti-Aircraft Guided Missile Station
- e: Anti-Aircraft Guided Missile System
AAGNC e: Australian Association for Geriatric Nursing Care
AAGO e: Associate of the American Guild of Organists
AAGp e: Aeromedical Airlift Group
AAGP e: American Academy of General Practice
- e: American Association for Geriatric Psychiatry
AAGR e: Air-to-Air Gunnery Range
- e: Anti-Aircraft Gunnery Range
A Agr e: Associate in Agriculture
A. Agr. e: Associate in Agriculture
AAGR e: Average Annual Growth Rate
A Agri e: Associate in Agriculture
A.Agri. e: Associate in Agriculture
AAGS e: All-America Gladiolus Selections
- e: American Association for Geodetic Surveying
- e: Association of African Geological Surveys
AAG Stn e: Anti-Aircraft Gun Station
AAGT e: American Army General Test
AAGUS e: American Association of Genito-Urinary Surgeons
aagw e: air-to-air guided weapon[s]
AAH d: Analogausgabe mit Halteverstärker
- e: Academy of Accounting Historians
- e: Advanced Attack Helicopter
- e: Air Arc Heater
- e: American Academy of Homiletics
- e: Amphipathic Alpha Helix
- e: Association for Applied Hypnosis
- e: Australian Auxiliary Hospital
- e: Automated Attitude Hold
aah e: anti-armor helicopter
AAHA e: American Academy of Health Administration
- e: American Animal Hospital Association
- e: American Association of Handwriting Analysts
- e: American Association of Hospital Accountants
- e: Awaiting Action of Higher Authority
AAHB e: Anti-Aircraft Hostile Battery
AAHC e: American Albino Horse Club
- e: American Association of Healthcare Consultants
- e: American Association of Hospital Consultants
- e: Association of Academic Health Centers
AAHCM e: Association for the Advancement of Health Care Managers
AAHD e: American Academy of the History of Dentistry
- e: American Association of Hospital Dentists
AAHDC e: American Association of Hospital Dental Chiefs
AAHDS e: American Association of Health Data Systems
AAHE e: American Association for Higher Education
- e: American Association of Housing Educators
- e: Associate in Arts in Home Economics
- e: Association for the Advancement of Health Education
AAHED e: Association of Appliance and Home Entertainment Distributors
AAHH e: Air Arc Hester Housing
AAHL e: Australian Animal Health Laboratory
AAHM e: American Academy of Homeopathic Medicine
- e: American Association for the History of Medicine
- e: Association of Architectural Hardware Manufacturers
AAHN e: American Association for the History of Nursing
AAHP e: American Association for Humanistic Psychology
- e: American Association of Homeopathic Pharmacists
- e: American Association of Hospital Podiatrists
Aahp e: Army artificial heart pump
AAHPA e: American Association of Hospital Purchasing Agents
AAHPER e: American Alliance for Health, Physical Education and Recreation
- e: American Association for Health, Physical Education and Recreation
AAHPERD e: American Alliance for Health, Physical Education, Recreation and Dance
AAHPhA e: American Animal Health Pharmaceutical Association
AAHPS e: Australian Association for the History and Philosophy of Science
AAHQ e: Advanced Allied Headquarters
- e: Allied Air Headquarters
- e: Army Artillery Headquarters
AA(HS) e: Airman Apprentice (High School)
AAHS e: American Association for Hand Surgery
- e: American Aviation Historical Society
AAHSLD e: Association of Academic Health Sciences Library Directors

AAHT e: Anti-Armor Helicopter Troop
AAI e: Adolescent Alienation Index
- e: African-American Institute
- e: Afro-American Institute
- e: Agricultural Ammonia Institute
- e: Air-to-Air Identification
- e: Air-to-Air Intercept
- e: Air-to-Air Interrogation
- e: Airborne Alert Indoctrination
- e: Aircraft Accident Investigation
- e: Akron Art Institute
- e: Alfred Adler Institute
- e: All Attitude Indicator
- e: Amateur Astronomers, Incorporated
- e: American Association of Immunologists
- e: American Association of Inventors
- e: American Audio Institute
- e: Angle of Approach Indicator
- e: Angle of Attack Indicator
- e: Anti-Aircraft Instructor
- e: Arab American Institute
- e: Army Adaptation Inventory
- e: Army Analysis of Intelligence
- e: Asphalt Ageing Index
- e: Associate of the Auctioneers Institute
- e: Association Adjustment Inventory
- e: Association of Art Institutions
- e: Atlantic Art Institute
- e: Authorized Active Inventory
- e: AUTODIN / AUTOVON Interface
- e: Azimuth Angle Increment
aai e: air-to-air identifilcation
- e: angle-of-approach indicator
AAIA e: American Association for International Aging
- e: Associate of the Association of International Accountants
AAIAL e: American Academy and Institute of Arts and Letters
AAIAN e: Association for the Advancement of Instruction about Alcohol and Narcotics
AAIB e: Aircraft Accident Investigation Board
- e: American Association of Instructors of the Blind
AAIC e: Accounting Authority Identification Code
- e: Air Accidents Investigation Committee
- e: Allied Air Intelligence Center
- e: American Amateur Inventors Club
- e: Australian Advertising Industry Council
- e: Australian Artificial Intelligence Conference
AAICJ e: American Association for the International Commission of Jurists
AAICP e: Air-to-Air Identification Control Panel
AAICS e: Automatic Aircraft Intercept Control System
AAICU e: Alabama Association of Independent Colleges and Universities
AAICV e: Amphibious Armored Infantry Combat Vehicle
AAID e: American Academy of Implant Dentistry
- e: American Academy of Implant Dentures
- e: American Association of Industrial Dentists
- e: Angular Accelerometer Input Device
- e: Arithmetic Array Identification
- e: Arithmetic Array Identifier
AAIE e: American Association of Industrial Editors
- e: American Association of Industrial Engineers
- e: Association for the Advancement of International Education
- e: Association of Applied Insect Ecologists
AAIEE e: Associate of the American Institute of Electrical Engineers
AAIEx e: Associate of the Australian Institute of Export
AAIFF e: Air-to-Air Identification Friend or Foe
AAIG e: Anti-Aircraft Instructor Gunnery
AAIH e: American Academy of Industrial Hygiene
AAII e: Associate of the Australian Insurance Institute
- e: Association for the Advancement of Invention and Innovation
AAIIP e: Alfred Adler Institute for Individual Psychology
AAIL e: Airborne Ion Laser
AAIM e: American Association of Industrial Management
AAIMME e: Associate of the American Institute of Mining and Metallurgical Engineers
AAIMS e: An Analytical Information Management System
AAIN e: American Association of Industrial Nurses
- e: Anglo-American Industrial Newsletter
AAIP e: Academic Administration Internship Program[me]
- e: Advanced Avionics Integration Program[me]
- e: Advanced Avionics Interface Program[me]
- e: American Academy of Implant Prosthodontics
- e: Associate of the American Institute of Physics
- e: Association of American Indian Physicians
AAIPS e: American Association of Industrial Physicians and Surgeons
- e: Automated Air Information Production System
A AIR SC e: Army Air Support Control
AAIS e: Administrative Analysis; Information and Statistics
- e: American Association for Information Sciences
- e: Anti-Aircraft Information System
- e: Anti-Aircraft Artillery Information Service
- e: Associate Administrator for Information Systems
AAISW e: American Association of Industrial Social Workers
AAIT e: American Association of Inhalation Therapists
AAITC e: Australian Automobile Industry Training Committee
AAITO e: Association of African Industrial Technology Organizations
AAIV e: American Association of Industrial Veterinarians
AAJ e: American Alpine Journal
- e: American Association for Justice
- e: Arab Airways, Jerusalem
- e: Augmented Air Jet
- e: Australian Anthropological Journal (journ.)
AAJAC e: Automatic Anti-Jam Circuit
AAJC e: American Association of Junior Colleges
- e: Automatic Anti-Jam Circuit
AAJCS e: Anglo-American Joint Chiefs of Staff
AAJE e: American Association for Jewish Education
AAJFA e: dry-type self cooled/ forced-air-cooled
AAJR e: American Academy for Jewish Research
AAJS e: American Association for Jesuit Scientists
- e: Associate in Arts in Judaic Studies
AAK d: algorithmisches Auswahlverfahren zur Konstruktion
- d: Amplituden-Adreßkodesystem
- d: Anti-Antikörper
A. A. K. d: Arbeitsgemeinschaft Allergiekrankes Kind
AAKF e: American Amateur Karate Federation
AAKP e: American Association of Kidney Patients
AAL e: Absolute Assembly Language
- e: Academy of Art and Literature
- e: Account Access Layer
- e: Acoustical Absorption Loss
- e: Additional Authorization List
- e: Adelaide Airways Limited
- e: Aircraft Approach Light
- e: Aircraft Approach Limitation
- e: Aircraft Armament Laboratory
- e: Aircraft Assignment Letter
- e: Ames Aeronautical Laboratory
- e: Angle of Attack Limiter
- e: Anterior Axillary Line
- e: Anti-Aircraft Light
- e: Arctic Aeromedical Laboratory
- e: Arctic Aerospace Laboratory
- e: Arctic Approach Limitation
- e: Asia America Line
- e: Association of Architectural Librarians
- e: Association of Assistant Librarians
- e: Australian Acoustical Laboratory
- e: Australian Air League
- e: Authorized Allowance List
- e: Aviation Armament Laboratory
aal e: above aerodrome level
AALA e: Anti-Aircraft Light Automatic
- e: Asian-American Librarians Association
- e: Associate in Arts in Liberal Arts
AALAS e: American Association of Laboratory Animal Science
AALASO e: Afro-Asian Latin-American Students Organization
AALC e: Active Army Locator System
- e: Advanced Airborne Launch Center
- e: African American Labor Center
- e: Amphibious Assault Landing Craft
- e: Amplified Automatic Level Control
AALE e: Associate in Arts in Law Enforcement
AALIM e: American Academy of Legal and Industrial Medicine
AALIPS e: Arthur Adaptation of the Leiter International Performance Scale
AALIS System e: Army Automated Library and Information Support System
AAlI e: American Association of Individual Investors
- e: Association for the Advancement of Invention and Innovation
AALL e: American Association of Law Libraries
AALM e: Advanced Air-Launched Missile
- e: Advanced Air-Launched Motor
AAlMCo e: American Association of Insurance Management Consultants
AAL Mg e: Anti-Aircraft Light Machine Gun
AALO e: Anti-Aircraft Liaison Officer
- e: Army Air Liaison Officer

AALPP e: American Association for Legal and Political Philosophy
AALPS e: Automated Air-Load Planning System
AALP System e: Automated Air-Load Planning System
AALRF e: Anti-Aircraft Long-Range Fire
AALS e: Acoustic Artillery Location System
- e: Active Army Locator System
- e: American Association of Language Specialists
- e: Association for Arid Lands Studies
- e: Association of American Law Schools
- e: Association of American Library Schools
AALT e: Automatic Azimuth Laying Theodolite
AALUE e: Asymptotically Admissible Linear Unbiased Estimator
AALW e: Assembled Air-Launched Weapon
AAM d: angeborener Auslösemechanismus
- e: Academy of Ancient Music
- e: Acoustic[s] Analysis Memo
- e: Acting Air Marshal
- e: Additional Addressing Module
- e: Address Arithmetic Microprocessor
- e: Advisory Agricultural Meteorologist
- e: Air-to-Air Missile
- e: Airborne Activity Monitor
- e: Aircraft Availability Model
- e: Alphabetical Accounting Machine
- e: American Academy of Mechanics
- e: American Academy of Microbiology
- e: American Association of Microbiology
- e: American Association of Museums
- e: Amplitude and Angle Modulation
- e: Anti-Antimissile Missile
- e: Army Aircraft Maintenance
- e: Assembly and Maintenance
- e: Audio Alarm Module
- e: Australian Air Mission
- e: Automatic Accounting Machine
- e: Automatic Address Modification
- e: Auxiliary Aiming Mark
A-AM e: Afro-American Museum
AA/M e: Associate Administrator for Management
AAMA e: African American Museums Association
- e: American Academy of Medical Administrators
- e: American Agricultural Marketing Association
- e: American Amusement Machine Association
- e: American Architectural Manufacturers Association
- e: American Association of Medical Assistants
- e: American Automobile Manufacturers Association
- e: Architectural Aluminum Manufacturers Association
- e: Asian American Manufacturers Association
AAMAA e: Army Aviation Mission Area Analysis
AA mach gn[r] e: Anti-Aircraft Machine Gun
A-A MAN e: Afro-American Men Against Narcotics
AAMAP e: Army Automation Master Plan
AAMB e: Army Automation Memorandum Budget
AAMBER e: Australian Antarctic Marine Biological Ecosystem Research
AAMBP e: Association of American Medical Book Publishers
AAMC e: American Association of Medical Clinics
- e: Army Air Materiel Command
- e: Army Artillery and Missile Center
- e: American Association of Medico-Legal Consultants
- e: Army Aviation Materiel Command
- e: Association of American Medical Colleges
- e: Australian Army Medical Corps
AAMCH e: American Association for Maternal and Child Health
AAMD e: American Association on Mental Deficiency
AAME e: Acetylarginine Methyl Ester
- e: American Antarctic Mountaineering Expedition
- e: American Association of Microprocessor Engineers
AAMES e: American Association for Middle East Studies
AAMFT e: American Association for Marriage and Family Therapy
AAMG e: Anti-Aircraft Machine Gun
aamg e: antiaircraft machine gun
AA-MGaVO d: Ausführungsanweisung zur Muster-Garagenverordnung
AAMGE e: Air-to-Air Missile Guidance Element
AA Mg Pl e: Anti-Aircraft Machine Gun Platoon
AAMHPC e: American Association of Mental Health Professionals in Corrections
AAMI e: Age-Associated Memory Impairment
- e: All-Aspect Maneuvering Index
- e: American Association of Machinery Importers
- e: American Association of Microcomputer Investors
- e: Associate of the Australian Marketing Institute
- e: Association for the Advancement of Medical Instrumentation
- e: Association of Allergists for Mycological Investigation
AAMID e: Accoplishment of Assigned Mission Impeded by Deadline
AAMIH e: American Association for Maternal and Infant Health
AAML e: Arctic Aeromedical Laboratory
- e: Army Aviation Materials Laboratory
AAMM e: Anti-Anti-Missile Missile
AAMMC e: American Association of Medical Milk Commissioners
AAMMP e: Active Army Military Manpower Program[me]
AAMP e: Advanced Architecture Microprocessor
- e: American Academy of Maxillofacial Prosthetics
- e: American Association of Meat Processors
- e: Army Aviation Modernization Plan
AAMR e: American Academy on Mental Retardation
AAMRDL e: Army Air Mobility Research and Development Laboratory
AAMRL e: American Association of Medical Record Librarians
AAMS e: Advanced Antitank Missile System
- e: Airborne Auxiliary Memory System
- e: Army Aircraft Maintenance Shop
- e: Army Armor School
- e: Army Artillery and Missile School
- e: Automated Azimuth Measuring System
AAMSA e: Army Aviation Maintenance Support Activity
AA/MSF e: Associate Administrator for Manned Space Flight
AAMSI e: American Association for Medical Systems and Informatics
AAMSU e: Army Air Movement Support Unit
AAMSW e: American Association of Medical Social Workers
AAMT e: American Association for Music Therapy
AAMTAP e: Army Aircraft Mobile Technical Assistance Program[me]
AAMU e: Army Advanced Marksmanship Unit
AAM Unit e: Azimuth Angle Measuring Unit
AAMus e: Associate in Arts in Music
AAMVA e: American Association of Motor Vehicle Administrators
AAMWS e: Australian Army Medical Womens Service
AAMX e: Aceto-Acet-Meta-Xylidid
- e: Acetoacetat-Meta-Xylidide
AAn d: Aachen
AAN d: Ausrüstungsanordnung[en]
- d: Ausstattungsanweisung
- e: Advance Acquisition Notification
- e: Advance Alteration Notice
- e: Aliphatic Ammonium Nitrate
- e: American Academy of Neurology
- e: American Academy of Nutrition
- e: American Association of Neuropathologists
- e: American Association of Nurserymen
- e: Aminoacetonitrile
- e: Amino Acid Nitrogen
- e: Analgesic-Associated Nephropathy
- e: Assignment Action Number
- e: Associate in Arts in Nursing
AAn e: American Anthropologist (journ.)
aan e: assignment action number[s]
A.A.N. e: Associate in Arts in Nursing
AANA e: Abridged and American Nautical Almanac
- e: American Association of Nurse Anesthetists
AANB e: Alpha-Amino-N-Butyric acid
AANCP e: Advanced Airborne National Command Post
A and A e: Additions and Amendments
- e: America and Australia
A and AEE e: Aeroplane and Armament Experimental Establishment
- e: Aircraft and Armament Experimental Establishment
A and C e: Addenda and Corrigenda
A. and C.P. e: Anchors and Chains Proved
A and D e: Accounting and Disbursing
A and E e: Aerospace and Electronic
- e: Aerospace and Equipment
- e: Agriculture and Engineering
- e: Airframe and Engine
- e: Appropriation and Expense
- e: Architecture and Engineering
- e: Armamants and Electronics
AANDE e: Armament and Electronics
A and E e: Armaments and Electronics
- e: Azimuth and Elevation
AANDEMAINTSQ e: Armament and Electronics Maintenance Squadron
A and F e: Accounting and Finance
- e: Agriculture and Forestry
A and H e: Accident and Health
- e: Alive and Healthy

A and I e: Abstracting and Indexing
- e: Accident and Indemnity
- e: Arts and Industry

A and L Program[me] e: Allocation and Loading Program[me]
- e: Assets and Liabilities Program[me]

A and M e: Agricultural and Mechanical
- e: Ancient and Modern
- e: Architectural and Mechanical

A and Mech e: Agricultural and Mechanical

A and P e: Apogee and Perigee
- e: Atlantic and Pacific
- e: Auscultation and Percussion

A and R e: Antioptimal and Right
- e: Assembly and Repair

A and T e: Assemble and Test

A and TT e: Alcohol and Tobacco Tax

A and V e: Audible and Visual Indicator

A and W e: Alive and Well
- e: Antioptimal and Wright

A and WI e: America and West Indies

AANI e: Associate of the Australian Naval Institute

AANNT e: American Association of Nephrology Nurses and Technicians

AANP e: Aircraft Accident Notification Procedures
- e: American Association of Naturopathic Physicians
- e: American Association of Neuropathologists

AAnS d: Auslandsanschaltesatz
- d: Auslandsanschaltsatz

AANS e: American Academy of Neurological Surgery
- e: American Association of Neurological Surgeons
- e: American Association of Neurological Surgery
- e: Australian Army Nursing Service

AANSW e: Archives Authority of New South Wales

AAnthr e: American Anthropologist (journ.)

AAnw. d: Amtsanwalt[schaft]

AAnw d: Ausführungsanweisung

AAnz d: Amtlicher Anzeiger

AAO e: Academy of Applied Osteopathy
- e: Administrative Arrangements Order
- e: Advanced Arithmetic[al] Operations
- e: Advanced Assembly Outline
- e: Aircraft Approach Overlay
- e: American Academy of Ophthalmology
- e: American Academy of Optometry
- e: American Academy of Osteopathy
- e: American Association of Ophthalmology
- e: American Association of Orthodontists
- e: Amino Acid Oxidase
- e: Anglo-Australian Observatory
- e: Anti-Aircraft Officer
- e: Anti-Air Output
- e: Army Acquisition Objective
- e: Artillery Air Observer
- e: Ascorbic Acid Oxidase
- e: Assembly and Checkout
- e: Association for the Advancement of Ophthalmology
- e: Astronaut Activities Office
- e: Australian Academy of Osteopathy
- e: Australian Army Order
- e: Authorized Acquisition Objective
- e: AUTOVON (Automatic Voice Network) Assistance Operator
- R: Abastumanskaya Astrofzicheskaya Observatoriya

aao e: amino-acid oxidase

aaO d: am angeführten Ort

a.a.O. d: am angeführten Ort
- d: an anderen Orten

a. a. O. d: am angegebenen Ort

AAOA e: Ambulance Association of America
- e: American Academy of Otolaryngologic Allergy
- e: Auxiliary to the American Osteopathic Association

AAOC e: American Association of Osteopathic Colleges
- e: Anti-Aircraft Operations Center

AAOCS e: Anti-Aircraft Artillery Officer Candidate School

AAOD e: Army Aviation Operating Detachment

AAODC e: American Association of Oil Well Drilling Contractors

AAOE e: Airborne Antarctic Ozone Experiment
- e: American Association of Osteopathic Examiners

AA of A e: Ambulance Association of America

AAOG e: American Association of Obstetricians and Gynecologists

AAOGAS e: American Association of Obstetricians, Gynecologists and Abdominal Surgeons

AAOGP e: American Academy of Orthodontics for the General Practitioner

AAOH e: Asian Association of Occupational Health

AAOHN e: American Association of Occupational Health Nurses

AAOI e: American Association of Inventors

AAOM e: American Academy of Occupational Medicine
- e: American Academy of Oral Medicine
- e: American Association of Orthomolecular Medicine
- e: American Association of Orthopaedic Medicine

AAOME e: American Association of Osteopathic Medical Examiners

AAOMS e: American Association of Oral and Maxillofacial Surgeons

AAONMS e: Ancient Arabic Order of Nobles of the Mystic Shrine

AAO & O e: American Academy of Ophthalmology and Otolaryngology

AAOO e: American Academy of Ophthalmology and Otolaryngology

AAOP e: American Academy of Oral Pathology
- e: American Academy of Orthotists and Prosthetists
- e: Anti-Aircraft Observation Post
- e: Anti-Aircraft Observation Post

AAOPB e: American Association of Pathologists and Bacteriologists

AAOPS e: American Association of Oral and Plastic Surgeons

AAOR e: American Academy of Oral Roentgenology
- e: Anti-Aircraft Artillery Operations Room

AAOrthMed e: American Association of Orthopedic Medicine

AAOS e: American Academy of Orthopaedic Surgeons
- e: American Academy of Orthopaedic Surgery
- e: American Association of Osteopathic Specialists

AAOT e: American Association of Orthoptic Technicians

Aap d: Antennenaufstellungsplan

AAP d: Apothekenabgabepreis (DDR)
- d: automatischer Arbeitsplatz
- e: Academy of American Poets
- e: Adaptive Array Processor
- e: Advance Acquisition Plan[ning]
- e: Advanced Acquisition Plan
- e: Advanced Air Park
- e: Advise if Able to Proceed
- e: Aerial Aspect Processor
- e: Aerodynamics Advisory Panel
- e: Affirmative Action Plan
- e: Affirmative Action Program[me]
- e: Aggregation-Attachment Pheromone
- e: Agribusiness Accountability Project
- e: Aircraft Actually Possessed
- e: Aircraft Assembly Plant
- e: Airlock Adapter Plate
- e: Algorithm for Arithmetic[al] Operation
- e: Allied Administrative Publication
- e: Allied Army Procedures
- e: Allied Army Publications
- e: Allied Authorized Publication
- e: Allowance Appendix Page
- e: Alpha Antiprotease
- e: American Academy of Pediatrics
- e: American Academy of Pedodontics
- e: American Academy of Periodontology
- e: American Academy of Psychoanalysis
- e: American Academy of Psychotherapists
- e: American Association for Parapsychology
- e: American Association of Pathologists
- e: American Association of Psychiatrists
- e: American Association of the Professions
- e: American Aviation Publications
- e: Aminoantipyrine
- e: Analog[ue] Aerial Positioner
- e: Analog[ue] Antenna Positioner
- e: Analog[ue] Auto Pilot
- e: Analysis and Production
- e: Angina Awareness Program[me]
- e: Antenna Aspect Processor
- e: Apollo Applications Program[me]
- e: Approach Astrophysics Payload
- e: Army Air Personnel
- e: Army Ammunition Plant
- e: Army Apprenticeship Program[me]
- e: Army Automation Program[me]
- e: Army Avionics Program[me]
- e: Associate Administrator for Airports
- e: Association for Advancement of Psychoanalysis
- e: Association for Advancement of Psychology
- e: Association for Applied Psychoanalysis
- e: Association for Astrological Psychology
- e: Association for the Advancement of Psychoanalysis
- e: Association for the Advancement of Psychology
- e: Association for the Advancement of Psychotherapy
- e: Association of Academic Physiatrists
- e: Association of American Physicians
- e: Association of Applied Psychoanalysis
- e: Association of Aviation Psychologists
- e: Associative Array Processor
- e: Atmospheric Analysis and Prediction
- e: Attached Applications Processor
- e: Auxiliary Acceleration Pump

aap. e: advise if able to proceed
- e: air at atmosphere pressure

AAp e: Angle of Approach

AAPA e: American Academy of Physician Assistants
- e: American Academy of Podiatry Administration
- e: American Alfalfa Processors Association
- e: American Association of Pathologists Assistants
- e: American Association of Physical Anthropologists
- e: American Association of Physicians Assistants
- e: American Association of Port Authorities
- e: American Association of Psychiatric Administrators
- e: Asian American Psychological Association
- e: Australian Air Pilots Association

AAPB e: American Association of Pathologists and Bacteriologists
- e: Association for Applied Psychophysiology and Biofeedback

AAPBA e: Amino(aminophenyl)-benzamide
AAPBG e: Army Automation Program[me] Budget Guidance
AAPBS e: Australian Association of Permanent Building Societies
AAPC e: Advertising Agency Production Club
- e: All-African Peoples Conference
- e: American Association of Professional Consultants
- e: Application of Autonomous Passive Classification
- e: Australian Aluminium Production Commission

AAPCC e: Adjusted Average Per Capita Cost
- e: American Association of Poison Control Centers
- e: American Association of Psychiatric Clinics for Children

AAPCEM e: Association of Air Pollution Control Equipment Manufacturers
AAPCK e: Acetyl(alanyl)phenylalanyl-chloromethyl Ketone
AAPCM e: Association of American Playing Card Manufacturers
AAPCO e: Association of American Pesticide Control Officials
AAPD e: Absolute Average Percent Deviation
- e: American Academy of Pediatric Dentistry
- e: American Academy of Physiologic Dentistry
- e: Automated Astronomic Positioning Device

AAPE e: American Academy of Physical Education
- e: Average Absolute Percentage Error

AAPEP e: Adolescent and Adult Psychoeducational Profile
AAPERS e: Active Army Personnel Reporting System
AAPF e: Academy of American Poets Fellowship
AAPFCO e: Association of American Plant Food Control Officials
AAPG e: Affirmative Action Planning Guide
- e: Allowance Appendix Package
- e: American Association of Petroleum Geologists
- e: Armament and Avionics Planning Guidance

AAPH e: American Association for Partial Hospitalization
- e: American Association of Professional Hypnologists
- e: American Association of Professional Hypnotherapists

AAPHD e: American Association of Public Health Dentistry
- e: American Association of Public Health Dentists

AAPHI e: Associate of the Association of Public Health Inspectors
AAPHP e: American Association of Public Health Physicians
AAPHR e: American Association of Physicians for Human Rights
AAPI e: Australian Architectural Periodicals Index
AAPICU e: American Association of Presidents of Independent Colleges and Universities
AAPIER e: American Association for Public Information, Education and Research
AAPIU e: Allied (or Army) Air Photograph Interpretation Unit
- e: Allied Aerial Photographic Interpretation Unit

AAPL e: American Academy of Psychiatry and the Law
- e: American Association of Petroleum Landmen
- e: American Association of Professional Landmen
- e: An Additional Programming Language
- e: An Array Processing Language
- e: Apple Computer Incorporated

AAPLE e: American Academy for Professional Law Enforcement
- e: American Association for Professional Law Enforcement

aapm e: amphiapomict
AAPM e: American Association of Physicists in Medicine
- e: Army Aviation Planning Manual

A/APM e: Assistant Air Provost Marshal
AAPMC e: Antibiotic-Associated Pseudomembranous Colitis
AAPMR e: American Academy of Physical Medicine and Rehabilitation
AAPO e: Advanced Aircraft Programs Office
- e: All-African Peoples Organization
- e: Apollo Applications Program[me] Office

AAPOR e: American Association for Public Opinion Research
AAPP e: Affirmative Action Program[me] Plans
- e: Airborne Auxiliary Power Plant
- e: American Association of Police Polygraphists
- e: Association of Amusement Park Proprietors
- e: Australian Association of Psychology and Philosophy
- e: Auxiliary Airborne Power Plant

AAPPES e: Army Automation Planning, Programming and Evaluation System
- e: Army Auto Plan and Progress Evaluation System

AAPPP e: American Association of Planned Parenthood Physicians
AAPPS e: American Association of Podiatric Physicians and Surgeons
AAP/PSP e: Association of American Publishers / Professional and Scholarly Publishing Divis
AAPR e: Army Aviation Program[me] Review
AAPRCO e: American Association of Private Railroad Car Owners
AAPRD e: American Academy for Plastics Research in Dentistry
AAPRM e: American Association of Passenger Rate Men
AAPRSO e: Army Aviation Personnel Requirements of Sustained Operations Study
AAPS e: Active Aircraft Plume Suppression
- e: Advanced Automative Power System
- e: Airborne Angular Position Sensor
- e: Alternative Automotive Power System
- e: American Association for the Promotion of Science
- e: American Association of Pharmaceutical Scientists
- e: American Association of Phonetic Sciences
- e: American Association of Plastic Surgeons
- e: Annals of the American Academy of Political and Social Science (journ.)
- e: Association for Ambulatory Pediatric Services
- e: Association of Alternate Postal Systems
- e: Association of American Physicians and Surgeons
- e: Automated Astronomic Positioning System

AAPs e: Army Ammunition Plants
- e: Associative Array Processors

AAPSC e: American Association of Psychiatric Services for Children
AAPSD e: Alternative Automotive Power System Division
AAPSE e: American Association of Professors in Sanitary Engineering
AAP/SHORAD e: Air-Augmented Propulsion for Short-Range Air Defense
AA psn e: Anti-Aircraft Artillery Position
AAPSP e: Advanced Automative Power Systems Program[me]
AAPSS e: American Academy of Political and Social Science
- e: American Academy of Political and Social Sciences

AAPSW e: Associate of the Association of Psychiatric Social Workers
AAPT e: American Association of Philosophy Teachers
- e: American Association of Physics Teachers
- e: Association of Asphalt Paving Technologists

AAPTC e: Assistant Airport Traffic Controller
AAPTO e: American Association of Passenger Traffic Officers
AAPTS e: Association of Americas Public Television Stations
AAPTSR e: Australian Association for Predetennined Titme Standards and Research
AAPU e: Airborne Auxiliary Power Unit
AAPWISM e: American Association of Public Welfare Information Systems Management
AAPY e: American Association of Professors of Yiddish
AAQ e: Active Asynchronous Quenching
- e: Architectural Association Quarterly
- e: Armed Aircraft Qualification

AAQM e: Acting Assistant Quartermaster
A.A.QM.G. e: Acting Assistant Quartermaster General

AA & QMG e: Assistant Adjutant and Quartennaster General
AAQS e: Ambient Air Quality Standard
AAR d: Anfangsadreßregister
- d: Antigen-Antikörperreaktion[en]
- e: A Address Register
- e: Abby Aldrich Rockefeller
- e: Accumulation Area Ratio
- e: Address Addend Register
- e: Administrative Adjustment Report
- e: Administrative Appeals Reports
- e: Affirmative Action Register
- e: After Action Report
- e: After Action Review
- e: Against All Risks
- e: Air-to-Air Refueling
- e: Air Attack Radar
- e: Air Augmented Rocket
- e: Airborne Attack Recorder
- e: Aircraft Accident Record
- e: Aircraft Accident Report
- e: Alliance for Aging Research
- e: Alternate Acquisition Radar
- e: American Academy in Rome
- e: American Academy of Religion
- e: Amino Acid Racemization
- e: Applied Agricultural Research, Inc.
- e: Approved Auto Repair
- e: Army Area Representative
- e: Association of American Railroads
- e: Association of Authors Representatives
- e: Australian Associated Resources
- e: Automatic Alternative Routing
- e: Automotive Affiliated Rcpresentatives
- e: Auxiliary Address Register
- e: Average Annual Rainfall

Aar e: Aarhus
- e: Australia antigen radioimmunoassay

Aar. d: Aarhus
A/AR e: Aero/Acoustic Rotor
a.a.r. e: all average recoverable
aar e: antigen-antiglobulin reaction
AARA e: American Ambulance and Rescue Association
AARAD e: Aspect Angle Radiation Code
aa rating e: average-audience rating
AARB e: Advanced Air Refueling Boom
- e: Australian Road Research Board

AARBPLA d: Änderungen der Arbeitsplandatei
AARC e: Alliance for Acid Rain Control
- e: American Association for Respiratory Care
- e: Ann Arbor Railroad Company
- e: Anti-Aircraft Regional Command
- e: Army Aeronautical Research Center
- e: Association for the Advancement of Released Convicts
- e: Atlantic Aerospace Rescue and Recovery Center
- e: Atlantic Air Rescue Center
- e: Australian Aeronautical Research Committee
- e: Australian Applied Research Centre
- e: Australian Automobile Racing Club

AARCAT Det e: Anti-Aircraft Artillery Remote-Controlled Aerial Target Detachment
AARch e: Acta Archaeologica (journ.)
A Arch e: Associate in Architecture
AARCO e: Afro-Asian Rural Construction Organiisation
AARD e: American Academy of Restorative Dentistry
AARDC e: Army Air Reconnaissance for Damage Assessment in the Continental
AARDCO e: Association of American Railroad Dining Car Officers
AARDL e: Artillery Ammunition Rocket Development Laboratory
AARDS e: Australian Advertising Rate and Data Service
aaRdT d: allgemein anerkannte Regeln der Technik
AARE e: Australian Association for Research in Education
AAREC e: Adaptive Agile Radar ECCM (Electronic Counter-Countermeasures)
aarg a: aargang
AARG e: Annual Average Rate of Growth
Aarh. d: Aarhus
Aarh e: Aarhus
AARL e: Advanced Applications and Research Laboratory
- e: Advanced Automation Research Laboratory
- e: Army Aeromedical Research Laboratory
- e: Army Aeronautical Research Center
- e: Army Aircraft Radio Laboratory
- e: Australian Academic and Research Libraries

AARM e: Adaptive Assembly Robotic Machine
- e: Advanced Anti-Radiation Missile

AARMA e: Assistant Army Military Attaché
AARMTO e: All Australian Register of Massage Therapists Organizations
AAROM e: Active Assistive Range of Motion
AARP e: American Association of Retired Persons
- e: Ancient American Rocket Pioneers
- e: Annual Advance Retainer Pay
- e: Atomic Air Raid Precautions

AARPLS e: Advanced Airborne Radio Position Location System
AARPS e: Air-Augmented Rocket-Propulsion System
AARPUT e: Average Aptitude Requirement per Unit Time
AARR e: Ann Arbor Railroad
- e: Argonne Advanced Research Reactor

AARRC e: American Air Raid Relief Committee
- e: Army Aircraft Requirements Review Committee
- e: Atlantic Aerospace Rescue and Recovery Center

AARRO e: Afro-Asian Rural Reconstruction Organization
AARS e: Accelerated Accounting and Reporting System
- e: Accrual Accounting and Reporting System
- e: Advanced Airborne Radar System
- e: All-America Rose Selection[s]
- e: American Association of Railroad Superintendents
- e: American Association of Railway Surgeons
- e: Army Aerial Reconnaissance System
- e: Army Aircraft Repair Ship
- e: Army Amateur Radio System
- e: Automated Attendance Reporting System
- e: Automatic Address Recognition Subsystem
- e: Automatic Address Recognition System

AARST e: American Association of Radon Scientists and Technologists
AART d: allgemein anerkannte Regeln der Technik
- e: American Association for Rehabilitation Therapy
- e: American Association for Respiratory Therapy
- e: American Association of Religious Therapists
- e: American Association of Retired Teachers

AARTA e: American Association of Railroad Ticket Agents
AARTC e: Anti-Aircraft Replacement Training Center
AARTS e: American Council on Education Registry Transcript System
AARU e: Agricultural Aviation Research Unit
- e: Association of Arab Universities

AARV e: Aerial Armo[u]red Reconnaissance Vehicle
- e: Armo[u]red Artillery Resupply Vehicle

aarv e: aerial armored reconnaissance vehicle
AARWBA e: American Auto Racing Writers and Broadcasters Association
AAS d: Allgemeiner Ausschuß für Segelsport
- d: Allgemeines Adaptions-Syndrom
- d: Allmählichkeits- und Abwasserschäden
- d: Atomabsorptions-Spektralphotometer
- d: Atomabsorptions-Spektrophotometrie
- d: Atomabsorptionsspektroskopie
- d: atomare Absorptions-Spektralphotometrie
- d: Ausnutzung des Arbeitsspeichers
- e: Abort Advisory System
- e: Academy of Applied Science
- e: Achievement Anxiety Scale
- e: Activity Accreditation Schedule
- e: Adjusted Air Speed
- e: Advanced Accounting System
- e: Advanced Active Sonobuoy
- e: Advanced Administration System
- e: Advanced Administrative System
- e: Advanced Aerial System
- e: Advanced Air Station
- e: Advanced Antenna System
- e: Advanced Automated System
- e: Advanced Automation System
- e: Advanced Avionic System
- e: Aeromedical Airlift Squadron
- e: Air Armament School
- e: Airborne Aerial System
- e: Airborne Antenna System
- e: Aircraft Airworthiness Section
- e: Airport Advisory Service
- e: Alert Area Supervisor
- e: All-American Selections
- e: American Academy of Sanitarians
- e: American Amaryllis Society
- e: American Antiquarian Society
- e: American Association of School Administrators
- e: American Astronautical Society
- e: American Astronomical Society
- e: American Autobahn Society
- e: Analog Alarm Section
- e: Ancient Astronaut Society
- e: Angular Acceleration Susceptibility
- e: Annual Authorizations Service
- e: Annual Average Score
- e: Anti-Aircraft Searchlight
- e: Aortic Arch Syndrome
- e: Apollo Abort System
- e: Architectural Acoustics Society
- e: Army Air Service

- e: Army Apprentices School
- e: Army Attachè System
- e: Army Aviation School
- e: Arnold Air Society
- e: Ascending Activating System
- e: Asian and African Studies
- e: Aspirator Air System
- e: Associate in Applied Science
- e: Association for Academic Surgery
- e: Association for Asian Studies
- e: Association for the Advancement of Science
- e: Association of Academies of Science
- e: Atomic Absorption Spectrometer
- e: Atomic Absorption Spectrophotometry
- e: Atomic Absorption Spectroscopy
- e: Attack Assessment System
- e: Attendance Accounting Subsystem
- e: Australian Academy of Science
- e: Australian Acoustic Society
- e: Australian Air Services
- e: Australian Air Society
- e: Australian Association of Surgeons
- e: Automated Accounting System
- e: Automatic Addressing System
- e: Automatic Address Selector
- e: Automatic Address Substitution
- e: Automatic Adjusted Suspension
- e: Automatically-Adjustable Shock-absorber
- e: Automatic Announcement Subsystem
- e: Automatic Array Scaling
- e: Auxiliary Ambulance Service
- e: Azimuth Alignment System
- l: Academiae Americanae Socius
- l: Acta Apostolicae

a.a.S. d: auf angeführter Seite

AAs e: Alcoholics Anonymous members
- e: American Atheists
- e: Asian Americans
- e: Authors Slterations

aas e: arithmetic assignment statement

AA'S e: Authors Alternations

AASA e: Academy of Arts and Sciences of the Americas
- e: Administrative Assistant to the Secretary of the Army
- e: Advances in Alcohol and Substance Abuse
- e: Afro-American Student Association
- e: American Association of Surgeons Assistants
- e: Associate of the Australian Society of Accountants
- e: Astronomical and Astrophysical Society of America

aa-sat e: advanced anti-satellite

AASB e: Alabama Association of School Boards
- e: American Association of Small Business

AASBEVM e: Association of American State Boards of Examiners in Veterinary Medicine

AASC e: Acupuncture Association of Southern California
- e: Aerospace Applications Studies Committee
- e: African-American Scholars Conference
- e: Allied Air Support Command
- e: American Association of Small Cities
- e: American Association of Specialized Colleges
- e: American Association of State Climatologists
- e: Army Area Signal Center
- e: Army Area Supply Center
- e: Army Automation Steering Committee
- e: Association for the Advancement of Science in Canada

AASCM e: Awaiting Action Summary Court Martial

AASCO e: Association of American Seed Control Officials

AASCU e: American Association of State Colleges and Universities

AASD e: American Academy of Stress Disorders
- e: American Association of Social Directories
- e: Anti-Aircraft Self-Destroying

AASDJ e: American Association of Schools and Departments of Journalism

AASDMCC e: American Association for Small Dredging and Marine Construction Companies

AASE e: Airborne Arctic Stratospheric Expedition
- e: American Academy of Safety Education
- e: American Academy of Sanitary Engineers
- e: American Association of Special Educators
- e: Army Aviation Support Element
- e: Associated Australian Stock Exchanges
- e: Association for Applied Solar Energy

AASEC e: American Association of Sex Educators and Counselors

AA Sec e: Anti-Aircraft Sector

AASECT e: American Association of Sex Educators, Counselors and Therapists

AASF e: Advanced Air Striking Force
- e: Army Aviation Support Facility
- e: Associate Administrator for Space Flight

AASFE e: American Association of Sunday and Feature Editors

AASG e: American Association of Students of German
- e: Association of American State Geologists

AAS & GP e: American Association of Soap and Glycerin Producers

AASH e: Adrenal Androgen Stimulating Hormone
- e: American Association for the Study of Headache
- e: Australian Association of Speech and Hearing

AASHO e: American Association of State Highway Officials

AASHTO e: American Association of State Highway and Transportation Officials

AASI e: Advertising Agency Service Interchange
- e: American Academy for Scientific Interrogation
- e: American Association for Scientific Interrogation
- e: Anti-Aircraft Searchlight
- e: Associate of the Ambulance Service Institute

A'asia[n] e: Australasia[n]

AASIP e: Appalachian Airport Safety Improvement Program[me]

AASIR e: Advanced Atmosphere Sounder and Imaging Radiometer

AASK e: Aid to Adoption of Special Kids

AASL e: American Antiquarian Society Library
- e: American Association of School Librarians
- e: American Association of State Librarians
- e: American Association of State Libraries

A & ASL e: American & Australian Steamship Line

aasl e: anti-aircraft searchlight

AASLD e: African Association for the Study of Liver Diseases
- e: American Association for the Study of Liver Diseases

AASLH e: American Association for State and Local History

AASLS e: Anti-Aircraft Searchlight Station

AASLT e: Air Assault

AASM d: Assoziation Afrikanischer Staaten und Madagaskars
- e: Abigail Adams Smith Museum, New York City
- e: Advanced Air-to-Surface Missile
- e: Association of American Steel Manufacturers
- e: Association of Aviation and Space Museums

AASMB e: Australian Association of Stud Merino Breeders

AASMD e: Airborne Anti-Ship Missile Defense

AASME e: Associate of the American Society of Mechanical Engineers

AASMMA e: Advanced Avionic System for Multi-Mission Application

AASMS e: Advanced Air-to-Surface Missile Seeker

AASND e: American Association for the Study of Neoplastic Diseases

AASNS e: Asian-Australasian Society of Neurological Surgeons

AASO e: Administrative Aircraft Standardization Office
- e: Assigned Activity Standardization Office
- e: Association of American Ship Owners

AASOR e: Annual of the American Schools of Oriental Research (journ.)

AASP e: Advanced Automated Sample Processor
- e: American Academy of Sports Physicians
- e: American Association for Social Psychiatry
- e: American Association of Senior Physicians
- e: American Association of Stratigraphic Palynologists
- e: Army Automation Security Program[me]
- e: Association of African Studies Programs

AASPA e: American Association of School Personnel Administrators

AASPB e: American Association of State Psychology Boards

AA Spectrometry e: Atomic Absorption Spectrometry

AASPRC e: American Association of Sheriffs Posses and Riding Clubs

AASPS e: Annals of the American Academy of Social and Political Science (journ.)

AASq e: Aeromedical Airlift Squadron

AASR e: Abhazian Autonomous Soviet Socialist Republic
- e: Adjarian Autonomous Soviet Socialist Republic
- e: Advanced Army System Requirements
- e: Airport and Airways Surveillance Radar
- e: Ancient [and] Accepted Scottish Rite

AASRC e: American Association of Small Research Companies

AASRE e: American Association of Schools of Religious Education

AASRI e: Arctic and Antarctic Scientific Research Institute

AASS e: Advanced Acoustic Search Sensor
- e: Advanced Airborne Surveillance Sensor
- e: Afro-American Students Society
- e: American Academy of Spinal Surgeons
- e: American Association for Social Security

- e: Army Area Signal System
- e: Automatic Abort-Sensing System
- l: Americanae Antiquarianae Societatis Socius

AASSA e: Association of American Schools in South America

AASSC e: Association for the Advancement of Scandinavian Studies in Canada

AAS Sh e: Anti-Aircraft Smoke Shell

AASSP e: Arkansas Association of Secondary School Principals

AASSREC e: Asian Association of Social Science Research Councils
- e: Association of Asian Social Science Research Councils

AAST e: American Association for the Surgery of Trauma

AASTA e: Anti-Aircraft Station
- e: Army Aviation Systems Test Activity

AASTC e: Associate in Architecture-Sydney Technical College

AASTD e: Association for the Advancement of the Science and Technology of Documentation

AASU e: Afro-American Student Union
- e: All Africa Students Union

AASW e: Airborne Anti-Submarine Warfare
- e: American Association of Scientific Workers
- e: American Association of Social Workers
- e: Australian Association of Scientific Workers
- e: Australian Association of Social Workers

AASWA e: American Association for the Study of World Affairs

AASWI e: American Aid Society for the West Indies

AASWS e: Antimassed Armor Strike Weapon System

AAT d: Aktionsanzeigeteil
- d: Alpha-1 Antitrypsin
- d: Aminoazotoluol
- d: Anionenaustauscher
- d: Aspartat-Amino-Transferase
- e: Academic Aptitude Test
- e: Accelerated Apprenticeship Training
- e: Achievement Anxiety Test
- e: Activation Acceptance Team
- e: Advanced Avionics Testbed
- e: Aerodynamic Accounting Technique
- e: After Acid Treatment
- e: Air Abrasive Trimming
- e: Air Traffic Service
- e: Alanine Aminotransferase
- e: Alpha-Antitrypsin
- e: American Academy of Thermology
- e: American Academy of Transportation
- e: Analytic Approximation Theory
- e: Anglo-Australian Telescope
- e: Anti-Aircraft Tank
- e: Anti-Aircraft Tehnician
- e: Assembly and Test
- e: Association of Accounting Technicians
- e: Attitude Acquisition Technique
- e: Attitude Angle Transducer
- e: Auditory Apperception Test
- e: Australian Antarctic Territory
- e: Austrian Airtransport
- e: Automated Assessment Tool
- e: Automatic Aerial Timer
- e: Automatic Answer Trunk
- e: Automatic Antenna Timer
- e: Average Access Time

aat e: acute abdominal tympany
- e: alpha-I antitrypsin
- e: auditory attending task

A-AT e: Anglo-Australian Telescope

AATA e: American Association of Teachers of Arabic
- e: American Automobile Touring Alliance
- e: Animal Air Transportation Association

AATB e: Advanced Amphibious Training Base
- e: Army Aviation Test Board

AATC e: Advanced Air Training Command
- e: Alaskan Arctic Training Center
- e: All-Arms Training Center
- e: Anti-Aircraft [Artillery] Training Center
- e: Army Air Transport Command
- e: Army Aviation Test Command
- e: Automatic Air Traffic [Controlsystem]

AATCAN System e: Army Air Traffic Control and Navigation System

AATCC e: Airhead Air Traffic Coordination Center
- e: American Association of Textile Chemists and Colorists

AATCLC e: American Association of Teachers of Chinese Language and Culture

AATCM e: Academy of Air Traffic Control Medicine

AATCO e: Army Air Traffic Coordinating Office

AATCS e: An Automatic Test Control System

AATD e: Australian Association of Teachers of the Deaf
- e: Aviation Applied Technology Directorate

AATDC e: Army Air Transport and Development Centre

AATE e: American Association of Teachers of Esperanto
- e: Aminoadenosine Triacid Ester
- e: Association of Australian Teachers of English
- e: Atlantic Association of Teacher Educators
- e: Avionics Automatic Transmission Line

AATEA e: American Association of Teacher Educators in Agriculture

AA Tech e: Associate in Automotive Technology

AATEFL e: Australian Association for the Teaching of English as a Foreign Language

AA Ter Ed e: Associate in Arts in Terminal Education

AATF e: Active Air Target Fuse
- e: American Association of Teachers of French
- e: Anechoic Acoustic Test Facility

AATG e: American Association of Teachers of German

AATH e: American Association of Teaching Hospitals

AATI e: American Anti-Terrorism Institute
- e: American Association of Teachers of Italian

AA Tk e: Anti-Aircraft Tank

AATL e: Avionics Automatic Transmission Line

AATM e: American Academy of Tropical Medicine

AATMS e: Advanced Air Traffic Management Systems

AATNU f: Administration de l'assistance technique des Nations Unies

AATO e: All Africa Teachers Organization
- e: Army Air Transport Organization

AATOC e: Airhead Air Traffic Coordination Center

AATOE e: American Association of Theatre Organ Enthusiasts

AATOF e: Anti-Aircraft Throw-Off Firing

AATP e: American Academy of Tuberculosis Physicians

AATPG e: Analog[ue] Automatic Test Program[me] Generator

AATPO e: Association of African Trade Promotion Organizations

AA tps e: Anti-Aircraft Troops

AATPS e: Australian Association of Temporary Personnel Services

AATR e: Apollo Applications Test Requirements
- e: Association of Auto and Truck Recyclers

AATRACEN e: Anti-Aircraft [Artillery] Training Center

AATRI e: Army Air Traffic Regulatio[s] and Identification

AATRIS e: Army Air Traffic Regulation and Identiftcation System

AA-TRNS d: Aminoacyl-Transferribo-nucleinsäure

AATS e: Alternate Aircraft Takeoff System
- e: American Academy of Teachers of Singing
- e: American Association for Thoracic Surgery
- e: American Association of Theological Schools
- e: American Association of Trauma Specialists
- e: Armament Auxiliaries Test Set
- e: Automatic Altitude Trim System
- e: Automatic Anti-Theft System
- e: Azores Air Transport System

AATSEEL e: American Association of Teachers of Slavic and Eastern European Languages

Aatseel Bull e: American Association of Teachers of Slavic and East European Languages Bulletin (journ.)

Aatseel Jour e: American Association of Teachers of Slavic and East European Languages Journal (journ.)

AATSP e: American Association of Teachers of Spanish and Portuguese

AATT e: Advances in Automatic Testing Technology
- e: American Association for Textile Technology
- e: American Association of Teachers of Turkish
- e: Australian Army Training Team
- e: Australian Association for Theatre Technology

AATTA e: Arab Association of Tourism and Travel Agents

AATTC e: American Airlines Technical Training Corporation

AAT & TC e: Anti-Aircraft [Artillery] Training and Test Center

AATU e: Association of Air Transport Unions

AATUF e: All-African Trade Union Federation

AATVA e: American All-Terrain Vehicle Association

AATW e: Advanced Anti-Tank Weapon

AAU e: Acoustic Add-on Unit
- e: Add-on Audio Unit
- e: Address Arithmetic Unit
- e: Administrative Area Unit
- e: Al-Azhar University
- e: Altitude Alerting Unit
- e: Angular Accelerometer Unit
- e: Anti-Aircraft Unit
- e: Army Administrative Unit
- e: Associated Aviation Underwriters
- e: Association of African Universities

- e: Association of American Universities
- e: Association of Atlantic Universities
- e: Automatic Answering Unit
- e: Auxiliary Air Unit

AAUA e: American Association of University Administrators
AAUAPDD e: American Association of University Affiliated Programs for Persons with Developm
AAU/BNA e: Association of Atlantic Universities / Blackwell North America
AAUBO e: Association of Atlantic University Business Officers
a. a. u. c. l: anno ab urbe condita
AAUCS e: Australian-Asian Universities Cooperation Scheme
AAUN e: American Association for the United Nations
- e: Australian Association for the United Nations

AAUP e: American Association of University Professors
- e: Association of American University Presses

AAUPB e: American Association of University Professors; Bulletin (journ.)
AAUPF e: American Association of University Professors Foundation
AAUP-UAES e: American Association of University Professors of Urban Affairs and Environmental Sciences
AAUQ e: Associate in Accountancy-University of Queensland
AAUS e: American Association of University Students
AAUTA e: Australian Association of University Teachers of Accounting
AAUTC e: Army Aviation Unit Training Command
AAUTI e: American Association of University Teachers of Insurance
AAUW e: American Association of University Women
AAUWEF e: American Association of University Women Educational Foundation
AAV d: automatisierte Ausleihverbuchung
- e: Adenovirus-Associated Virus
- e: Advanced Aerospace Vehicle
- e: Aggregate Abrasion Value
- e: Airborne Assault Vehicle
- e: Alternative Access Vendor
- e: Amphibious Assault Vehicle
- e: Anti-Afterburn Valve
- e: Anti-Aircraft Volunteer
- e: Assault Amphibian Vehicle
- e: Assessed Annual Value
- e: Association of Avian Veterinarians

aav e: adeno-associated virus
AAVA e: American Association of Veterinary Anatomists
- e: Army Audio Visual Agency
- e: Australian Automatic Vending Association

AA Values e: Arithmetic Average Values
AAVB e: American Association of Veterinary Bacteriologists
AAVC e: Anomalous Atrioventricular Conduction
- e: Australian Army Veterinary Corps

AAVCA e: Australian Affiliation of Voluntary Care Agencies
AAVCS e: Airborne Automatic Voice Communications System
- e: Automatic Aircraft Vectoring Control System

AAVCT e: American Academy of Veterinary and Comparative Toxicology
AAVD e: American Academy of Veterinary Dermatology
- e: Automatic Alternate Voice Data

AAVIM e: American Association[s] for Vocational Instructional Materials
AAVLD e: American Association of Veterinary Laboratory Diagnosticians
AAVM e: Acting Air Vice Marshal
AAVMC e: Association of American Veterinary Medical Colleges
AAVN e: American Academy of Veterinary Nutrition
- e: American Association of Veterinary Nutritionists
- e: Army Aviation
- e: Australian Audio-Visual News

AAVNC e: Army Aviation Centre
AAVP e: American Association of Veterinary Parasitologists
- e: American Association of Veterinary Pathologists
- e: Association of American Volunteer Physicians

AAVPT e: American Academy of Veterinary Pharmacology and Therapeutics
AAVRB e: Australian Audio-Visual Reference Book
AAVRO e: American Association of Vital Records and Organizations
AAVRPHS e: American Association for Vital Records and Public Health Statistics
AAVRS e: All-Attitude Vertical Reference System
AAVS e: Aerospace Audio-Visual Service
- e: American Anti-Vivisection Society
- e: Aspirator-Assisted Vacuum System
- e: Automatic Aircraft Vectoring System

AAVSB e: American Association of Veterinary State Boards
AAVSED e: Automatic Air-Valving Surface Effects Device
AAVSO e: American Association of Variable Star Observers
AAVT e: Association of Audio-Visual Technicians
A Av Tech e: Associate in Aviation Technology
AAW d: Anzeiger der Akademie der Wissenschaften (journ.)
- e: Advertising Association of the West
- e: Aeromedical Airlift Wing
- e: Air Acetylene Welding
- e: American Academy of Wine
- e: American Association of Woodturners
- e: Anti-Air Warfare

A & AW e: Art and Archaeology; Washington (journ.)
AAWA e: American Automatic Weapons Association
AAWB e: American Association of Workers for the Blind
AAWBOT e: Aviation Anti-Submarine Warfare Basic Operational Trainer
AAWC e: American Association of Workers for Children
- e: Anti-Air Warfare Center
- e: Anti-Air Warfare Commander
- e: Anti-Air Warfare Coordinator
- e: Australian Advisory War Council

AAWCJC e: American Association of Women in Community and Junior Colleges
AAWD e: American Association of Women Dentists
- e: Association of American Women Dentists

AAWEX e: Anti-Air Warfare Exercises
AAWEXINPT e: Anti-Air Warfare Exercise in Port
- e: Anti-Air Warfare Exercises in Port

AAWF e: Auxiliary Aviation Weather Facility
AAWg e: Aeromedical Airlift Wing
AAWH e: American Association for World Health
ÄAWI d: Ärzte- und Apotheker-Wirtschafts-Institut
AAWL d: Abhandlungen der Akademie der Wissenschaften und der Literatur (journ.)
AAWM e: American Association of Waterbed Manufacturers
AAWO e: Afro-Asian Workers Organization
- e: Association of American Weather Observers

AAWORD e: Association of African Women for Research and Development
AAWP e: American Association for Women Podiatrists
AAWPI e: Association of American Wood Pulp Importers
AAWR e: American Association of Women Radiologists
AAWRATS e: Anti-Air Warfare Readiness Assessment Training System
AAWS e: Airborne Alert Weapon System
- e: American Association of Wardens and Superintendents
- e: Anti-Air Warfare Systems
- e: Automatic Aviation Weather Service
- e: Automatic Attack Warning System
- e: Auxiliary Aircraft Warning Service

AAWs e: Anti-Armor Weapons
AAWS-H e: Advanced Antitank Weapon System - Heavy
AAWS-M e: Advanced Antiarmor Weapon System - Medium
- e: Advanced Antitank Weapon System-Medium

AAWSSC e: Army Atomic Weapon Systems Safety Committee
AAWSUP e: Anti-Air Warfare Support
AAWTC e: Assistant Airways Traffic Controller
AAWTIP e: Anti-Air Warfare Training in Port
AAWV e: American Association of Wildlife Veterinarians
AAWW d: Anzeiger der Akademie der Wissenschaften in Wien (journ.)
AAWWS e: Airborne Adverse Weather Weapons System
AAXICO e: American Air Export & Import Company
AAYM e: American Association of Youth Museums
AAYPP e: American Association of Yellow Pages Publishers
AAZ d: Allgemeine Automobil-Zeitung
AAZK e: American Association of Zoo Keepers
AAZM s: Aqueducto y Alcantarillado de la Zona Metropolitana
AAZN e: American Association for Zoological Nomenclature
AAZPA e: American Association of Zoological Parks and Aquariums
AAZV e: American Association of Zoo Veterinarians
AB d: Adreßbus
- d: Allgemeine Ausführungsbestimmungen
- d: Allgemeine Bedingungen
- d: Aluminium Bronze
- d: Amplitudenbegrenzer

- d: Amtsbürgermeister[ei]
- d: anaerobes Becken
- d: Anfangsbedingung
- d: Anfangsbestand
- d: Anodenbatterie
- d: Anwendungsberater
- d: Aperturblende
- d: Arbeitsbereich
- d: Arbeitsbewertung
- d: Arbeitsbreite
- d: Arbeitsgemeinschaft Bäuerliche Landwirtschaft
- d: armierter Beton
- d: Artilleriebeobachtung
- d: Aschaffenburg
- d: Aufsichtsbehörde
- d: Ausfallspeicherbank
- d: Ausführungsbestimmung[en]
- d: Ausgang begrenzt
- d: Ausgangsblock
- d: Auslauf Beizbecken
- d: Ausnahmebedingung
- d: Außenbeamter
- d: Außenhandelsbank
- d: Aussetzbetrieb
- d: aussetzender Betrieb
- d: Autobahn
- e: Abbreviation
- e: Abdomen
- e: Able-Bodied
- e: Able-Bodied Seaman
- e: Aboriginal
- e: Abort
- e: Absorbed Burden
- e: Abstract
- e: Abstracting Board
- e: Accessories Bulletin
- e: Accidental Breakdown
- e: Accident Benefits
- e: Accumulator and Buffer
- e: Accumulator Bit
- e: Acquisition Beacon
- e: Acrylonitrile-Butadiene
- e: Acute Bisectrix
- e: Adapter, Bulkhead
- e: Adapter Booster
- e: Adaptive Behaviour
- e: Add Binary
- e: Additional Benefits
- e: Additional Bit
- e: Address[ed] Bit
- e: Address Base
- e: Address Blank
- e: Address Buffer
- e: Address Bus
- e: Address Byte
- e: Adjustment Bond
- e: Administrative Battalion
- e: Administrative Bulletin
- e: Admiralty Board
- e: Advisory Board
- e: Aerial Bomb
- e: Aeronautical Board
- e: After Body
- e: Aid to the Blind
- e: Air-Blast
- e: Air Bag
- e: Air Base
- e: Air Bearing
- e: Air Berlin
- e: Air Blast
- e: Air Board
- e: Air Bomber
- e: Air Brake
- e: Air Break
- e: Air Brick
- e: Air Bug
- e: Airman; Basic
- e: Alberta
- e: Alert Building
- e: Algorithmic Block
- e: All Before
- e: Alphabetical Block
- e: Alternative Broadcasting
- e: Aminobenzamide
- e: Aminobenzophenone
- e: Ammunition Bearer
- e: Analecta Biblica
- e: Anheuser-Busch Incorporated
- e: Anode Battery
- e: Antenna Supports
- e: Anterior Burster
- e: Anti Body
- e: Antigen-Binding
- e: Apex Beat
- e: Application Block
- e: Applied Biosystems, Inc.
- e: Arc Brazing
- e: Architectural Barriers
- e: Arctic Bibliography (journ.)
- e: Area Bombing
- e: Argument Byte
- e: Arithmetic Bit
- e: Arithmetic Bus
- e: Armor Board
- e: Army Book
- e: Art Bulletin (journ.)
- e: Aryan Brotherhood
- e: Assault Breaker
- e: Assembly Bill
- e: Assistant Barrister
- e: Associate in Business
- e: Association of Brewers
- e: Atlantic Beach
- e: Audio Bandwidth
- e: Auto Beacon
- e: Automated Banking
- e: Automated Bibliography
- e: Automatic Blowdown
- e: Aviation Battalion
- e: Aviation Boatswains Mate
- e: Avionics Bay
- e: Bachelor of Arts
- l: Artium Baccalaureus
- s: Aktiebolag

Ab d: Abbauland
- d: Abfall
- d: Abflacher
- d: Abflachschaltung
- d: Abnormal
- d: Abschreibung
- d: Abtaster
- d: Albumin
- e: Abbot
- e: Abner
- e: Abraham
- e: Abridgment
- e: Alabamine
- e: Alabamium
- p: Abade

ab. d: abends
- d: abgefertigt
- d: abgegangen
- d: abgesandt
- d: abonnieren
- d: abonniert
- e: Aberation

Ab. d: Aberration
- d: Abitur
- d: Abiturient[in]
- e: Abscess
- e: Air-break

ab d: albite
- e: abnormal
- e: abortion
- e: about
- e: above
- e: abundant
- e: afterburner
- e: airbrake
- e: alcian blue
- e: ambient brine
- e: Anchor Bolt
- e: antibody
- e: asbestos body
- e: asthmatic bronchitis
- e: at bats
- e: Axiobuccal
- s: abril

A. B. d: Arbeitsbataillon
- d: Armierungsbataillon
- d: Artilleriebeobachter
- d: Augsburger Bekenntnis

A.B. d: Armeebefehl
- l: Artium Baccalureus

a.B. d: auf Befehl
- d: auf Bestellung
- d: außer Betrieb

a.b. e: above bridges
- e: as before

A/B e: Acid / Base
- e: Aid to the Blind
- e: Air-Freight Bill
- e: Air Breather
- e: Aircraft Bulletin
- e: Airnman Basic
- e: Answerback
- r: Akttebolag

a/b e: acid-base
- e: airborne

A & B e: Antigua & Barbuda

a & b e: applejack and benedictine
- e: assault and battery

A + B e: Signals of Both Stereo Channels in Phase

A-B e: Signals of Both Stereo Channels out of Phase

AB1 e: Aviation Boatswains Mate; First Class

AB2 e: Aviation Boatswains Mate; Second Class

AB-204 e: Agusta-Bell gunship twin-engíne helicopter

AB-205 e: Agusta-Bell ten-place troop-transport helicopter

AB-206 e: Agusta-Bell five-seat turbine-powered helicopter

AB3 e: Aviation Boatswains Mate; Third Class

AB-47 e: Agusta-Bell three-place utility helicopter

ABA d: Abwasserbehandlungsanlage
- d: Allgemeine Bedingungen für Anschlußbahnen
- d: Azobenzenearsonate
- e: Abscisic Acid
- e: Acrylonitrile-Butadiene-Acrylate
- e: Active Buffer Address
- e: Adaptive Bounding Algorithm
- e: Airborne Assault
- e: Air Brake Association
- e: Allergic Bronchopulmonary Aspergillosis
- e: American, British, Australien
- e: American Board of Anesthesiology
- e: American Booksellers Association
- e: American Bus Association
- e: American Business Association
- e: Aminobutyric Acid
- e: Amplifier Buffer Attenuator
- e: Annual Budget Authorization
- e: Antiquarian Booksellers Association

- e: Applied Behaviour Analysis
- e: Appropriation and Budget Activity
- e: Armadillo Breeders Association
- e: Associate in Business Administration
- e: Association for Behaviour Analysis
- e: Association of Black Anthropologists
- e: Australian Biotechnology Association
- e: Australian Blue Asbestos
- e: Australian Booksellers Association
- e: Australian Braille Authority
- e: Authorized Bond Allotment
- l: Antoniani Benedictini Armeni

ab-a e: abampere
aba e: antibacterial activity
ABAA e: Airborne Anti-Aircraft
- e: American Beverage Alcohol Association
- e: Antiquarian Booksellers Association of America

ab ab. l: ab absurdo
ABABOL e: Algorithmic and Business Oriented Language
ABAC e: A Basic Coursewriter
- e: Abraham Baldwin Agricultural College
- e: Antiquarian Booksellers Association of Canada
- e: Appropriation and Budget Account Code
- e: Association of Balloon and Airship Constructors

ABACC e: Alcoholic Beverages Advertising Code Council
Abaco e: Great and Little Abaco islands in the Baharnas north of New Providence Island
ABACPD e: American Bar Association Center for Professional Discipline
ABACPR e: American Bar Association Center for Professional Responsibility
abact e: abacterial
ABACUS e: A Better Administrative Computer Utility System
- e: Agents and Brokers Automated Computer Users System
- e: Air Battle Analysis Center Utility System

Abacus e: Architecture and Building Aids Computer Unit; Strathclyde
ABACUS e: Architecture and Building Aids Computer Unit of Stratclyde (University)
- e: Automatic, Block-schematic Advanced Control User-oriented System
- e: Autonetics Business and Control United Systems

ABAD e: Air Base Air Defense
- e: Air Battle Analysis Division

ABADRL e: Arthropod-Borne Animal Diseases Research Laboratory
ab aet. l: ab aeterno
AB-AF e: Alcian Blue-Aldehyde Fuchsin
ABAFA e: Association of British Adoption and Fostering Agencies
ABAG e: Association of Bay Area Governments
A-bahn d: Autobahn
ABAI e: American Bell Association International
- e: American Board of Allergy and Immunology
- e: American Boiler and Affiliated Industries

ABAJ e: American Bar Association Journal
- e: Antiquarian Booksellers Association of Japan

ABAK i: Association di Biblioteka i Archivo di Korsow
ABAkü d: Amtsbezeichnungs- Abkürzung
ABAL d: Armee-Bekleidungs- und Ausrüstungslager
- e: Aminobutyraldehyde

abamp e: Abampere
ABAMP e: Absolute Ampere
aband e: Abandoned
AbÄndG d: Abänderungsgesetz
ABAND LT HO e: Abandoned Lighthouse
ABANDT e: Abandonment
ABAO d: Arbeitsschutz- und Brandschutz-Anordnung[en]
AbAP e: Antibody-Against-Panel
ABAR d: Ablaufanforderungsregister
- e: Advanced Battery Acquisition Radar
- e: Alternate Battery Acquisition Radar
- e: Auxiliary Battery Acquisition Radar

ABA-Raumregulierungssystem
- d: allein-bei-allen-Raumgliederungssystem

ABARE e: Australian Bureau of Agricultural and Resource Economics
- e: Australian Bureau of Agriculture and Resource Economics

ABAR Unit e: Advanced Battery Acquisition Radar Unit
ABAS d: Ablaufanalysesystem
- d: Ablaufanforderung[sauswahl]steuerung

AbAS d: Abwurfanrufsucher
ABAS e: American Board of Abdominal Surgeons
- e: American Board of Abdominal Surgery
- e: Australian Buying Advisory Service

ABASS e: Air Base Augmentation Support Set
- e: Assembly for Behavioural and Social Sciences

ABAT e: Abatement
abat e: abattoir
ABAT e: Air Base Advisory Team
ABATU e: Advanced Base Air Task Unit
- e: Advanced Base Aviation Training Unit

ABAV e: Air Bleed Actuator Valve
ABAW d: Abhandlungen der Bayerischen Akademie der Wissenschaften (journ.)
Abb. d: Abbildung[en]
- d: Abbindung

ABB d: Allgemeine Blitzschutzbestimmungen
- d: Arbeitsbefreiungsbescheinigung
- d: Arbeitsgemeinschaft für Blitzschutz und Blitzableiterbau
- d: Arbeitsstelle für betriebliche Berufsausbildung
- d: Ausschuß für Blitzableiterbau eV.
- e: Abbreviation
- e: Akron Barberton Belt
- e: American Board of Bioanalysis
- e: Anterior Basal Body
- e: Applied Biochemistry and Biotechnology
- e: Array of Building Blocks
- e: Artificial Breeding Box
- e: Axisymmetric Blunt Body

Abb e: Abbess
- e: Abbey
- e: Abbot

abb e: abbreviate[d]
- e: ablating blunt body
- e: automatic back bias
- i: abbassamento
- i: abbonamento
- i: abbuono

Abb' f: Abbaye
Abb. l: Abbas
ABBA e: American Bee Breeders Association
- e: American Board of Bio-Analysis
- e: American Board of Bio-Analysts

Abbaustr. d: Abbaustrecke
ABBB e: Association of Better Business Bureaus
ABBC e: Association of Bottled Beer Collectors
ABBE e: Advisory Board on the Built Environment
ABbF d: Anweisung für den Bau von bituminösen Fahrbahndecken
ABBF e: Advanced Beef Breeds Federation
- e: Association of Bronze and Brass Founders

Abbild. d: Abbildung[en]
ABBIM e: Association of Brass and Bronze Ingot Manufacturers
ABB/INFORM e: Abstracted Business/Information
Abbl. d: Abblasung
abbl. d: abblenden
ABBM e: Automatic Baseband Monitor
ABBMM e: Association of British Brush Machinery Manufacturers
ABBMS e: American Board of Bloodless Medicine and Surgery
A.B.B.O. e: Associate of the British Ballet Organisation
ABBO e: Associate of the British Ballet Organization
ABBOS d: Ausführungsbestimmungen zur Eisenbahnbau- und -betriebsordnung
ABBOStrab d: Ausführungsbestimmungen zur Bau- und Betriebsordnung für Straßenbahnen
Abbot e: Vickers self propelled fortress
Abbotsford e: British Columbias Matsqui Institution
Abbot[t] e: Abbotson
Abbr. d: Abbreviation
- d: Abbreviatur[en]

ABBR e: Abbreviate
abbr e: abbreviated
- e: abbreviation

ABBRA e: American Boat Builders and Repairers Association
abbrev e: Abbreviate[d]
- i: abbreviatura

ABBREV e: Abbreviation
abbrevia e: abbreviations
abbreviaz i: abbreviazione
abbrev[s] e: abbreviation[s]
abbrevto e: abbreviomania[c][al][ly]
ABBRON e: Abbreviation
ABBRP e: American Board of Bionic Rehabilitative Psychology
ABBR PO e: Abbreviated Purchase Order
ABBS e: Apple Bulletin Board System
- e: Australian Bibliography and Bibliographical Services
- e: Australian Bird Banding Scheme

ABBSchG d: Ausführungsbestimmungen zum Berufsschulgesetz
ABBV d: Allgemeine Bedingungen für Beschaffungsverträge
ABBVO d: Autobahnbetriebs- und -verkehrsordnung
ABBWA e: American Black Book Writers Association
ABC d: Abrechnungscomputer
- d: Aconite, Belladonna und Chloroform
- d: Ammoniumbicarbonat
- d: anwendungs- und bedienungsorientiertes Computerkonzept
- d: anwendungs- und bedienungsorientiertes Computersystem
- d: Arbeitnehmer-Beratungs-Center
- d: atomar, bakteriologisch, chemisch
- e: Aberrant Behavior Center
- e: A Better Chance [Program]
- e: Abridged Building Classification [for Architects, Builders and Civil Engineers]

- e: Abstracts in Bio-Commerce
- e: Acceptable Biological Catch
- e: Accounting Budget and Cost System
- e: Accounting and Budgetary Control
- e: Aconite, Belladonna and Chloroform
- e: Action for Brain-injured Children
- e: Active Bioprosthetic Composition
- e: Activity-Based Cost
- e: Adaptable Board Computer
- e: Adaptive Bus Controller
- e: Administration by Competency
- e: Adriatic Base Command
- e: Advance Base Components
- e: Advance Baseline Configuration
- e: Advanced-Booking Charter
- e: Advanced Ballistics Computer
- e: Advanced Ballistics Concepts
- e: Advanced Biomedical Capsule
- e: Advanced Blade Concept Advancing
- e: Advanced Booking Charters
- e: Aerated Bread Company
- e: African Bibliographic Center
- e: After Bottom Center
- e: Agricultural Biotechnology Center
- e: Agricultural Business and Commerce
- e: Air-Blast Cooled
- e: Air Balance Consultants
- e: Air Bath Chamber
- e: Air Blast Cooled
- e: Airborne Command
- e: Airborne Computer
- e: Airborne Control
- e: Air Bridge to Canada
- e: Air Bubble Craft
- e: Aircraft Builders Council
- e: Aircraft of Bomber Command
- e: Alcoholic Beverage Control
- e: Ale, Bread and Cheese
- e: Alkyl Benzene Sulfonate
- e: Allied Banking Commission
- e: Allocations for Budgetary Control
- e: Allowable Biological Catch
- e: Already Been Converted
- e: Alum, Blood and Charcoal
- e: Alum, Blood and Clay
- e: Amateur Book Collector (journ.)
- e: America, Britain, Canada
- e: American-British-Canadian
- e: American-British Conversation
- e: American Bibliographical Center
- e: American Blimp Corporation
- e: American Blood Commission
- e: American Book[prices]] Current
- e: American Book Collector (journ.)
- e: American Book Council
- e: American Botanical Council
- e: American Broadcasting Company
- e: American Broadcasting Corporation
- e: American Business Conference
- e: Angular Baseline Corrector
- e: Animal Birth Control
- e: Answer Back Code
- e: Anterior Bulbar Cell
- e: Antigen-Binding Capacity
- e: Any Boy Can [Program]
- e: Apparel Business Control [System]
- e: Approach By Concept
- e: Architectural Barriers Committee
- e: Argentinia, Brazil, Chile
- e: Asphaltenic Bottom Cracking
- e: Aspiration Biopsy Cytology
- e: Assessment Biological and Chemical
- e: Assessment of Basic Competencies
- e: Assisi Bird Campaign
- e: Assistant Beach Commander
- e: Associated Bottlers Company
- e: Associated Builders and Contractors
- e: Association of Biotechnology Companies
- e: Association of Bituminous Contractors
- e: Association of Black Cardiologists
- e: Association of British Climatologists
- e: Atanasoff-Berry-Computer[s]
- e: Atomic, Biological, Chemical
- e: Audit Bureau of Circulation[s]
- e: Augmented Bibliographic Citation
- e: Australia Bulk Carrier
- e: Australian Band Council
- e: Australian Bankruptcy Cases
- e: Australian Broadcasting Commission
- e: Australian Broadcasting Corporation
- e: Australian Business Computer
- e: Auto Bill Calling
- e: Auto Body Computer
- e: Automatic Background Control
- e: Automatic Bandwidth Control
- e: Automatic Bar Checker
- e: Automatic Bass Compensation
- e: Automatic Beam Control
- e: Automatic Bending Center
- e: Automatic Bias Compensation
- e: Automatic Bias Control
- e: Automatic Bill Calling
- e: Automatic Binary Computer
- e: Automatic Blade Control
- e: Automatic Blank Character
- e: Automatic Blip Counter
- e: Automatic Block Controller
- e: Automatic Boiling-Column Reactor
- e: Automatic Branch Control
- e: Automatic Bridge Control
- e: Automatic Brightness Control
- e: Avidin-Biotin Complex
- e: Axio-Buco-Cervical
- e: Brazil, Chile
- e: Line Antwerp Bulk Carriers
- i: Association of Bridal Consultants
- p: Academia Brasileira de Ciencias

abc e: abecedarium
- e: acid-balance control
- e: aconite, belladonna, chloroform
- e: advance-booking charter
- e: advanced base camp
- e: alarms by carner
- e: alphabet
- e: already been chewed
- e: alum, blood, charcoal
- e: atomic, biological, chemical
- e: automatic bandwith control
- e: automatic bass control
- e: automation of bibliography through computerization
- e: automotive boosters clubs
- e: axiobuccocervical

A.B.C. e: Alphabetical [Railway Guide]

AbC e: American-born Chinese

AB & C e: Atlanta, Birmingham and Coast

ABC³ e: Airborne Battlefield Command and Control Center
- e: Airborne Battlefield Command and Control Center

ABCA e: American-British-Canadian-Australian
- e: American Black Chiropractors Association
- e: American Building Contractors Association
- e: American Business Communications Association
- e: Antique Bottle Collectors Association
- e: Army Bureau of Current Affairs

ABC Abw Btl d: Atomares, Bakteriologisches und Chemisches Abwehrbataillon

ABC Abw Kp d: Atomare, Bakteriologische und Chemische Abwehrkompanie

ABC Abw Lehr Btl d: Atomares, Bakteriologisches und Chemisches Abwehrlehrbataillon

ABC Abw-S d: Atomare, Bakteriologische und Chemische Abwehr-Schule

ABC -AIR -STD e: American British-Canadian Airforces Standard

ABCAL e: Associated British Cables

ABC-anomaly e: Abashian-Booth-Crowe anomaly

ABC-ASP. e: American-British-Canadian-Army Standardization Program[me]

ABCB e: Air-Blast Circuit Breaker
- e: American Board of Clinical Biofeedback
- e: American Bottlers of Carbonated Beverages
- e: Australian Broadcasting Control Board

ABCBs e: Air-Blast Circuit Breakers

ABCC e: Airborne Battlefield Command and Control
- e: Airborne Command Center
- e: Airborne Communications Center
- e: Alternative Birth Crisis Coalition
- e: American Board of Clinical Chemistry
- e: American Business Computers Corp.
- e: Association of British Correspondence Colleges
- e: Atomic Bomb Casualty Commission
- e: Automatic Brightness and Contrast Control

A-B CC e: Arab-British Chamber of Commerce

ABCCC e: Airborne Battlefield Command and Control Center

ABCCH d: Abc der Chemie

ABC-Clio e: American Bibliographical Center Clio Press

ABCCTC - e: Advanced Base Combat Communication Training Center

ABCD e: Abbot Biological and Chemical Data
- e: Able Seaman Clearance Diver
- e: Accelerated Business Collection and Delivery
- e: Action for Boston Community Development
- e: Advanced Base Construction Depot
- e: Agency for Business and Career Development
- e: Agrophysics Breeding Control Device
- e: America, Britain, China, Dutch East Indies
- e: American Society of Bookplate Collectors and Designers
- e: Association for Bridge Construction and Design
- e: Association of Better Computer Dealers
- e: Association of Biomedical Communication Directors
- e: Asymmetry, Border, Colour and Diameter
- e: Atomic, Biological, Chemical and Damage
- e: Atomic, Biological, Chemical and Damage Control
- e: Atomic, Biological and Chemical Defense
- e: Automated Biological and Chemical Data

abcd e: above and beyond the call of duty
- e: airway [opened], breathing [restored], circulation , definitive
- e: atomic, biological, chemical and damage
- e: awaiting bad conduct discharge

ABCDCAL e: Alcoholic Beverage Control Department California

ABCD Control e: Atomic, Biological, Chemical and Damage Control

ABCD System e: Automated Biological and Chemical Data System
ABCE e: Adult Basic and Continuing Education
- e: Ann Baker College of English
ABCF e: Adaptive Break-in Control Function
- e: As-Built Configuration File
ABC-f.m e: Australian Broadcasting Commission-frequency modulation [system]
ABCFRC e: Auke Bay Coastal Fisheries Research Center
ABCH e: American Board on Clinical Hypnosis
ABCI e: Advanced Business Communications Incorporated
- e: Airport Business Center Incorporated
- e: Automotive Booster Clubs-International
ABCIA e: American Board of Clinical Immunology and Alergy
ABCIL e: Antibody-mediated Cell-dependent Immune Lympholysis
ABC Islands e: Aruba, Bonaire, Curaçao Islands
ABC-Krieg d: atomarer-bakteriologischer chemischer Krieg
ABCL e: American Birth Control League
- e: As-Built Configuration Lists
ABCM e: Adriamycin, Bleomycin, Cyclophosphamide, Mitomycin C
- e: Air Burst Contact Maker
- e: Antilock Brake Control Module
- e: Associate of the Bandsmans College of Music
- e: Association of British Chemical Manufacturers
- e: Association of Building Component Manufacturers
ABCMC e: Automotive Battery Charger Manufacturers Council
ABCN e: Airodrome Beacon
ABC-NAVY-STD e: American-British-Canadian Navies Standard
ABC Number e: Alpha Block Control Number
ABCO e: Advanced Base Components
A/B Comd e: Airborne Command
ABCOP d: Abc der Optik
ABCOULOMB e: Absolute Coulomb
ABCP e: Airborne Command Post
- e: American Board of Cardiovascular Perfusion
- e: Association of Blind Chartered Physiotherapists
ABCR e: As-Built Configuration Record
- e: Association for Beautiful Colorado Roads
- e: Association of Beverage Container Recyclers
- e: Atomic, Biological, Chemical and Radiological
ABCRETT e: American Board of Certified and Registered Encephalographic Technicians and Technology
ABCRF e: American Business Cancer Research Foundation
ABCRS e: American Board of Colon and Rectal Surgery
ABCS e: Abel Bibliographic Control System
- e: Advanced Brake Control System
- e: Advisory Board for Cooperative Systems
- e: American Board of Cosmetic Surgery
ABCs e: Antigen-Binding Cells
ABCS e: Automatic Base Communication System
- e: Automatic Base Communication Systems
- e: Automatic Blip Counter System
- e: Automatic Broadcast Control System
- e: Avionics Bay Cooling System
ABCSC e: American-British-Canadian Stores Catalogue
ABCSI e: American Bowling Computer Service Incorporated
ABCSP e: American-British-Canadian Standardization Program[me]
ABCST e: Automatic Broadcast
ABC-Staaten d: Argentinien, Brasilien, Chile
ABCStd e: American-British-Canadian Standard[s]
ABC-tv e: Australian Broadcasting Commission-television
ABCU e: Association of Burroughs Computer Users
ABCW e: Atomic, Biological, Chemical Warfare
ABC-Waffen d: atomare, biologische und chemische Waffen
ABC Washdown e: Atomic, Biological and Chemical Washdown
ABCWIU e: Aluminum, Brick and Clay Workers International Union
ABD d: Abadan (Flughafen)
Abd. d: Abdankung
Abd d: Abdomen
abd d: abdominal
Abd. d: Abend
ABD d: Anschlußsteuerung Bedieneinheit
- e: Abbreviated Dial
- e: Abdicated
- e: Abduction
- e: Abductor
- e: Aboard
- e: Access Block Diagram
- e: Adhesive Bonding
- e: Adriamycin, Bleomycin, Dacarbazine
- e: Advanced Base Depot
- e: Advanced Base Dock
- e: Airborne Ballistics Division
- e: All But Dissertation
- e: Alloy Bulk Diffusion
- e: American Board of Dermatology
- e: Aminobenzamidine
- e: Annular Base Drag
- e: Apparent Bulk Density
- e: Area Business Databank
- e: Army Budget Directive
- e: Association of British Detectives
- e: Australian Business Directory
- e: Automated Block Diagramming
- e: Average Body Dose
- e: Azobenzene Derivative
abd e: abandoned
- e: alum, blood and charcoal
ABDA d: Arbeitsgemeinschaft der Berufsvertretungen Deutscher Apotheker
- e: American-British-Dutch-Australian
- e: Australian-British-Dutch-American
ABDACOM e: Advanced Base Depot Area Command
- e: American-British-Dutch-Australian Command
- e: American-British-Dutch-Australian Supreme Command
ABDAIR e: American-British-Dutch-Australian Air Operational Command
ABDARM e: American-British-Dutch-Australian Army Operational Command
A/B-Datei d: Auftrags-Bedarfsdatei
ABDB e: Active Duty Base Date
ABDC e: After Bottom Dead Center
ABDE e: Airport Bird Detection Equipment
ABDER e: Abduction; External Rotation
abdest. d: abdestillieren
ABDIR e: Abduction; Internal Rotation
ABD-Kreis d: Archiv-, Bibliotheks- und Dokumentations-Kreis
ABDL e: Automatic Binary Data Link
ABDMS e: American Board of Dental Medicine and Surgery
ABDNSHP e: Abandon Ship
ABDNSP e: Abandon Ship
Abdnt e: Abandonment
abdom d: Abdomen
- d: abdominal
- e: abdomen
- e: abdominal
Abdom Surg e: Abdominal Surgery (journ.)
ABDP e: Association of British Directory Publishers
ABDPH e: American Board of Dental Public Health
Abdr. d: Abdruck[e]
ABDR e: Army Battle Damage Repair
- e: Association of Blood Donor Recruiters
Abdr. gen. d: Abdruck genehmigt
Abd.Rk. d: Abderhaldensche Reaktion
Abdr. vorg. d: Abdruck vorgesehen
abds. d: abends
ABDS e: Accounting and Budget Distribution System
- e: Air Base Defense Simulator
ABDs e: All But Their Dissertations
ABDS e: Associate of the British Display Society
ABDSA e: Association of British Dental Surgery Assistants
ABD/SCADS e: Air Base Defense / Sensor Communications and Display System
ABDSP e: Anza-Borrego Desert State Park
ABD Technique e: Alloy Bulk Diffusion Technique
ABD Transistor e: Alloy Bulk Diffusion Transistor
ABDUC e: Abduction
ABDV e: Adriamycin, Bleomycin, Dacarbazine, Vinblastine
- e: Arrhenatherum Blue Dwarf Virus
ABE d: Allgemeine Bedingungen für Bahnamtliche Expreßgutbestätter
- d: Allgemeine Betriebserlaubnis
- d: Austausch von Betriebserfahrungen
- e: Abnormal End
- e: Acetone, Butanol and Ethanol
- e: Acute Bacterial Endocarditis
- e: Adult Basic Education
- e: Air-Based Electronics
- e: Air-Breathing Engine
- e: Airborne Bombing Evaluation
- e: Air Breathing Engine
- e: Air Burst Effect
- e: Alarm Bell
- e: American Board of Endodontics
- e: Annunciator Bell
- e: Arithmetic Building Element
- e: Army Background Experiment
- e: Association of British Editors
- e: Augmented Ballast Expulsion
- e: Automatic Block End
- f: Association Belge d'Électrodéposition
AbE d: Abzweig-Einsatz
A/BE e: Airborne Equipment
ABEA e: American Broncho- Esophagological Association
ABEC e: Amphenol-Borg Electronics Corporation
- e: Annular Bearing Engineers Committee
AB Ed e: Bachelor of Arts in Education
ABEDA e: Arab Bank for Economic Development in Africa

ABEI d: Allgemeine Bedingungen für Entwicklungsverträge mit Industriefirmen
- e: Aminobutylethylisoluminol
ABEL d: Auftragsbelegerstellung
Abel e: Abelard
ABEL e: Air-Breathing Electric Laser
ABEM e: American Board of Emergency Medicine
- e: American Board of Environmental Medicine
A/Ben e: Airdrome Beacon
ABEND e: Abnormal End
ABEP e: Adult Basic Education Program[me]
- e: American Board of Examiners in Psychotherapy
ABEPH e: American Board of Examiners in Psychological Hypnosis
ABEPP e: American Board of Examiners in Professional Psychology
ABEPSGP e: American Board of Examiners of Psychodrama, Sociometry and Group Psychotherapy
aber d: Aberration
Aber. e: Aberdeen
ABergV d: Allgemeine Bergbauverordnung
ABERT e: Automatic BIT (Binary Digit) Error Rate Test
- p: Associação Brasileira de Emissors de Radio e Televisão
ABERU e: Airborne Emergency Reaction Unit
Abes e: Abyssinia[n]
abes e: aerospace business environment simulator
ABES e: Air Breathing Engine System
- e: American Broncho-Esophagological Association
- e: Association for Broadcast Engineering Standards
ABESPA e: American Boards of Examiners in Speech Pathology and Audiology
abess d: abessinisch
ABEST d: Auftragsbestand
ABest d: Ausführungsbestimmung[en]
ABET e: Accreditation Board for Engineering and Technology
ABETS e: Airborne Beacon Electronic Test Set
- e: Airborne Electrical Test Set
ABEXed e: Absorbance Expanded
ABez. d: Amtsbezirk
ABF d: Arbeiter- und Bauern-Fakultät (DDR)
- d: Arbeitszeiterfassung, Betriebsdatenerfassung und Fertigungssteuerung
- d: automatisches Betriebsführungssystem
- e: Adaptive Beam Forming
- e: Advance Booking Fare
- e: Advanced Beam-Former
- e: Aerial Beam-shape Factor
- e: Air Base Flight
- e: Air Burst Fuze
- e: Aircraft Battle Force
- e: American Beekeeping Federation
- e: Ammonium Biflouride
- e: Anaerobic Bacterial Flora
- e: Antenna Beam-shape Factor
- e: Army Benevolent Fund
- e: Associated British Foods
- e: Association of British Factors
- e: Audio Bandpass Filter
- e: Australian Bridge Federation
- e: Availability Balance File
- e: Average Branching Factor
Abf. d: Abfahrt
- d: Abfassung
- d: Abfertigung
- d: Abfindung
- d: Abflug
- d: Abforderung
- d: Abfuhr
- d: Abfüllung
Abf d: Abferkelergebnis
abf e: absolute binody final
ABFA e: American Board of Forensic Anthropology
- e: Azobisformamide
ABFAAB e: Airborne Force, Army Air Base
abfab e: absolutely fabulous
abfarad d: absolute farad
ABFC e: Advanced Base Functional Component
ABFCR e: American Business Foundation for Cancer Research
ABFCS e: Advanced Base Functional Component System
ABFD e: Affordable Basic Floppy Disk
A/BF & D e: Airborne Fill-and-Drain
ABFDS e: Aerial Bulk Fuel Delivery System
AbfG d: Abfallbeseitigungsgesetz
AbfH d: Abfertigungshilfsstelle
ABFI d: Allgemeine Bedingungen für Forschungsverträge mit Industriefirmen
Abfk d: Abfertigungskasse
Abf.-Kl. d: Abfuhrklausel
Abfl. d: Abflug
A.B.Fla d: Ausbildungsvorschrift für die Flugabwehr
ABFOR e: American-British Forces
ABFP e: American Board of Family Practice
- e: American Board of Forensic Psychiatry
- e: American Board of Forensic Psychology
Abf.-Progr. d: Abfuhrprogramm
Abfr. d: Abfrage[einrichtung]
- d: Abfragung
AbfrSt d: Abfragestelle
AbfrVS d: Abfrageverbindungssatz
Abf.-S. d: Abfindungssumme
ABFS e: Auxiliary Building Filter System
ABFSWS e: Associated Blacksmiths, Forge and Smithy Workers Society
abft e: Abaft
Abführ. d: Abführung
ABFV e: Anti-Backfire Valve
Abf.-Verg. d: Abfertigungsvergütung
Abf.-Vorschr. d: Abfertigungsvorschrift
ABG d: allgemeine Betriebsgenehmigung
- d: Allgemeines [Preußisches] Berggesetz
- d: Allgemeines Baugesetz
- d: Arbeitsgemeinschaft Berliner Galerien
- d: Archiv für Begriffsgeschichte (journ.)
- d: Aufbaugesetz
- d: Ausgangsbaugruppe
- e: Abnormal Blood Gas
- e: Air Base Group
- e: Antibacklash Gear
- e: Arterial Blood Gas
- e: Association of British Geodesists
- e: Aural Bearing Generator
- e: Axiobuccogingival
Abg. d: Abgabe[n]
- d: Abgang
- d: Abgas[e]
- d: Abgeordneter
- d: Abguß
- d: Abholung
abg. d: abgeändert
- d: abgefaßt
- d: abgekürzt
- d: abgeordnet
- d: abgerissen
- d: abgeschlossen
- d: abgestempelt
- d: abgestürzt
ABGA e: Australian Blueberry Growers Association
ABGB d: Allgemeines Bürgerliches Gesetzbuch
ABGC e: Australian Banana Growers Council
abgedr. d: abgedrängt
- d: abgedruckt
- d: abgedrückt
abgef. d: abgefahren
- d: abgefaßt
- d: abgefertigt
abgeg. d: abgegangen
abgeh. d: abgeheftet
- d: abgehoben
abgek. d: abgekürzt
abgel. d: abgelichtet
abger. d: abgerechnet
- d: abgereist
- d: abgerissen
AbGes. d: Abänderungsgesetz
abges. d: abgesagt
- d: abgesandt
abgeschl. d: abgeschlagen
- d: abgeschlosse
abgest. d: abgestempelt
abgezw. d: abgezweigt
AbgG d: Abgabengesetz
ABGG e: Amanitinylazobenzoylglycylglycine
Abg.-H. d: Abgangshafen
- d: Abgeordnetenhaus
abgk. d: abgekürzt
ABgm. d: Amtsbürgermeister[ei]
ABGMT e: Arizona Bureau of Geology and Mineral Technology
Abgn. d: Abgabenachricht
AbgO d: Abgabenordnung
ABGP e: Air Base Group
ABGR e: Australian Biographical and Genealogical Record
Abg.-St. d: Abgangsstation
ABGTS e: Auxiliary Building Gas Treatment System
AbgVO d: Abgabeverordnung
ABGWIU e: Aluminium, Brick and Glass Workers International Union
ABH d: Allgemeine Bedingungen für die Haftpflichtversicherung
- e: Above Burst Height
- e: Advanced Base Hospital
- e: Alpha-Benzene-Hexachloride
- e: Association for the Bibliography of History
- e: Association of British Hispanists
- e: Automatic Bobbin Handling
- e: Average Busy Hour
Abh. d: Abhandlung[en]
- d: Abhang
- d: Abholer
- d: Abholung
abh. d: abhängen
- d: abhängig
- d: abheben
- d: abholen
- d: abhören
ABHA e: Associate of the British Hypnotherapy Association

Abh.Akad.Wiss. d: Abhandlungen der Akademie der Wissenschaften (journ.)
Abhd. d: Abholdienst
Abh. Deut. Akad. Wiss. d: Abhandlungen der Deutschen Akademie der Wissenschaften
ABHENRY e: Absolute Henry
ABHES e: Accrediting Bureau of Health Education Schools
ABHH e: Association of Baptist Homes and Hospitals
ABHHA e: American Baptist Homes and Hospitals Association
ABHJ e: Angle Bulkhead Jack
ABHM e: American Board of Homeopathic Medicine
- e: Association of Builders Hardware Manufacturers
Abhn. d: Abhandlungen
ABHP e: American Board of Health Physics
ABHS e: American Board of Hand Surgery
Abh.-Sch. d: Abholschein
ABHX e: Air Blast Heat Exchanger
ABI d: Aktion Bildungsinformation
- d: Arbeiter- und Bauern-Inspektion (DDR)
- d: Arbeitsgemeinschaft für betriebliches Informationswesen
- d: Arbeitskreis Belegschaftsinformation
- e: Abstracted Business Information
- e: Add Binary Immediate
- e: Advance Book Information
- e: Advanced Biotechnologies Inc.
- e: Agile-Beam Illuminator
- e: American-British Intelligence
- e: American Bankruptcy Institute
- e: American Bell, Incorporated
- e: American Butter Institute
- e: Application Binary Interface
- e: Assignment of Beneficial Interest
- e: Associate of the Bookkeepers Institute
- e: Association of British Investigators
- e: Audit Bureau of Circulations
- e: Authorized Break of Inspection
- e: Automated Behavioral Intelligence
- e: Automated Broker Interface
- e: Auxiliary Building Isolation
- e: Aviation Billet Indicator
- i: Associazione Bibliotecari Italiani
Abi d: Abitur
ABIA e: American Boardsailing Industries Association
- e: Association of British Introduction Agencies
- e: Australian Boating Industry Association
ABIAS e: Air Bag Impact Attentuation System
ABIC e: Adaptive Behaviour Inventory for Children
- e: Army Battlefield Interface Concept
ABICC e: Associate of the British Institute of Certified Carpenters
ABICOMP e: Association of the Brazilian Computer and Peripherals Industries
ABIH e: American Board of Industrial Hygiene
ABII e: American Bell International Incorporated
abi/inform e: abstracted business information/information needs
ABIIS e: Automated Blood Inventory Information System
ABIL e: Airborne Beacon Interference Locator
ABILS e: Automated Bulk Items List System
ABIM e: Abridged Index Medicus
- e: American Board of Internal Medicine
- e: Association of British Insecticide Manufacturers
ABIMS e: American Board of Industrial Medicine and Surgery
ABIN d: Arbeitsgemeinschaft Betriebsfunk der Industrie- und Nahverkehrsbetriebe
ab in. l: ab initio
AB in CE e: Bachelor of Arts in Civil Engineering
AB in Ch E e: Bachelor of Arts in Chemical Engineering
AB in EE e: Bachelor of Arts in Electrical Engineering
AB in H Ec e: Bachelor of Arts in Home Economics
AB INIT l: Ab Initio
AB in J e: Bachelor of Arts in Journalism
AB in ME e: Bachelor of Arts in Mechanical Engineering
AB in Sec Ed e: Bachelor of Arts in Secondary Education
AB in TH e: Bachelor of Arts in Theology
ABIOL e: Advanced Base Initial Outfitting List
ABIP e: Australian Books in Print
ABIPC e: Abstract Bulletin of the Institute of Paper Chemistry
ABIPP e: Associate of the British Institute of Professional Photography
ABIR e: All-Band Intercept Receiver
ABIRA e: American Biographical Institute Research Association
ABIRD e: Aircraft-Based Infrared Detector
ABIS d: Automatisierung in der bibliothekarischen Sacherschließung
- e: Anglo-Brazilian Information Service
- e: Apollo Bioenvironmental Information System
- e: Associate of the British Interplanetary Society
- e: Audit Base Inventory System
ABISA e: Association of Burglary Insurance Surveyors-Australia
ABISL e: Advanced Base Initial Support Lists
Abit. d: Abitur
- d: Abiturient[in] [nen]
ABIT e: Aircraft Blast Interaction Tests
ABJ d: Abidjan (Flughafen)
- e: Adhesively Bonded Joint
- e: Adhesively Bonded Junction
- e: Angle Bulkhead Jack
- e: Association of Broadcasting Journalists
ABJS e: Association of Bone and Joint Surgeons
ABK d: Alliierte Bankkommission
- d: Arbeitsbegleitkarte
- d: Auftragsbegleitkarte
- e: Airborne Identification Kit
Abk. d: Abkommen
- d: Abkunft
- d: Abkürzung
abk. d: abkürzen
Abk. d: Abkürzung[en]
ÄBK d: Ärzteberatungskommission
ABKA e: American Boarding Kennels Association
abkdrt. d: abkommandiert
AbKfz d: Arbeitskraftfahrzeug
ABKI e: Airborne Identification Kit
Abk.-L. d: Abkürzungsliste
Abk.-Liste d: Abkürzungsliste
Abkühl. d: Abkühlung
Abkürz[.] d: Abkürzung[en]
Abk.-Verz. d: Abkürzungsverzeichnis
ABL d: Allgemeine Bedingungen für die Ausführung von Leistungen
- d: Amtsblatt
- d: Anwaltsblatt
- d: Automatisches Biologisches Laboratorium
- e: Above Baseline
- e: Accepted Batch Listing
- e: Adaption Binary Load
- e: Air Blast Landing
- e: Airborne Laser
- e: Alameda Belt Line
- e: All Busy Low
- e: Allegheny Ballistic[s] Laboratory
- e: Allocated Baseline
- e: Alternative-Based Language
- e: American-British Laboratory
- e: American Biotechnology Laboratory
- e: Ammunition Base Load
- e: Animated Biological Laboratories
- e: Antigen-Binding Lymphocyte
- e: Aquatic Biological Laboratory
- e: Architectural Block-diagram Language
- e: Archives of Business and Labour
- e: Armament Boresight Line
- e: Armoured Box Launcher
- e: Army Biological Laboratory
- e: Asian Business League
- e: Assembly Breakdown List
- e: ATLAS (Abbreviated Test Language for Avionics Systems) Basic Language
- e: Atmospheric Burst Locator
- e: Automated Biological Laboratory
- e: Automated Biology Library
- e: Automatic Black Level
- e: Automatic Bootstrap Loader
- e: Axiobuccolingual
abl. d: abladen
- d: ablagern
- d: ablegen
- d: ablehnen
- d: ableiten
- d: ablichten
- d: abliefern
- d: ablösen
Abl. d: Ablage[erung]
- d: Ablagerung
- d: Ablativ
- d: Ablauf
- d: Ablaut
- d: Ableger
- d: Ablehnung
- d: Ableitung
- d: Ablesung
- d: Ablösung
Abl d: Ablenkung
- d: Ablieferung
- d: Abschaltlampe
- e: Ablative
- e: Atlas Basic Language
A.B.L. d: Armeebetriebsstofflager
abl s: abril
ABLA e: American Blind Lawyers Association
Ablager d: Ablagerung
ABlAHK d: Amtsblatt der Alliierten Hohen Kommission in Deutschland
ABlAllKdtr. d: Amtsblatt der Alliierten Kommandatura Berlin
ABLAT e: Ablative
Ablauf- Planungsforsch. d: Ablauf und Planungsforschung (journ.)
ABLB e: Alternate Binaural Loudness Balance
- e: Alternative Binaural Loudness Balancing

AblB d: Ablieferungsbuch
ABLBMP d: Amtsblatt des Bundesministers für das Post- und Fernmeldewesen
ABLC e: Amphotericin B Lipid Complex
- e: Automatic Backlight Compensation
ABLCHG e: Airborne Launching
ABLCWS e: Altogether Builders, Labourers and Constructional Workers Society
ABl. d. EKD d: Amtsblatt der Evangelischen Kirche in Deutschland
ABLE e: A Better Language Experiment
- e: Ability Based on Long Experience
- e: Activity Balance[d] Line Evaluation
- e: Adult Basic Learning Examination
- e: Advanced Bio-mechanical Linkage Enablement
- e: Advanced Blown Lift Enhancement
- e: Agricultural Biological Literature Evaluation
- e: Agricultural Biological Literature Exploitation
- e: Amazon Boundary Layer Experiment
- e: Association for Biology Laboratory Education
- e: Asymptotically Best Linear Estimate
- e: Atmospheric Boundary Layer Experiment
- e: Audit Basic Learning Examination
- e: Autonetics Base-Line Equipment
- e: Azimuth, Bias, Level Equalization
able. e: activity balance line evaluation
ABlEGKS d: Amtsblatt der Europäischen Gemeinschaft für Kohle und Stahl
Ableit. d: Ableiter
- d: Ableitung
ABlELF d: Amtsblatt für Ernährung, Landwirtschaft und Forsten
ABLES e: Airborne Battlefield Light Equipment System
ABLEurGem d: Amtsblatt der Europäischen Gemeinschaften
Abl.-F. d: Ablieferungsfrist
ABLF e: Atmosphere Boundary Layer Facility
ABLfELF d: Amtsblatt für Ernährung, Landwirtschaft und Forsten
Abl.-Fr. d: Ablieferungsfrist
Abl.-Frist d: Ablieferungsfrist
Abl-G d: Ablenk[ungs]generator
ABLG e: Anti-Blacklash Gear
ABLISS e: Association of British Library and Information Science Schools
ABLJ e: Adjustable Buoyancy Life Jacket
ABlKR d: Amtsblatt des Kontrollrates in Deutschland
ABLO e: Australian Botanical Liaison Officer
ABLP e: Air Bearing Lift Pad
Abl.-Pfl. d: Ablieferungspflicht
Abl.-Pl. d: Abladeplatz
- d: Ablagerungsplatz
ABL Post d: Amtsblatt des Bundesministers für das Post- und Fernmeldewesen
ABLS e: Abelson Lymphosarcoma
- e: American Board of Laser Surgery
- e: Association of British Library Schools
- e: Atlas Biomedical Literature System
- e: Bachelor of Arts in Library Science
A.B.L.S. e: Arts Bachelor in Library Science
Abl.-S. d: Ablösesumme
- d: Ablösungssumme
Abl.-Sch. d: Ablieferungsschein
- d: Ablösungsschein
ABLSS e: Advanced Ballistic-type Lodistic Spacecraft System
Abl.-Term. d: Ablieferungstermin
ABLUE e: Asymptotically Best Linear Unbiased Estimator
ABLV e: Air-Breathing Launch Vehicle
ABM e: Abducens Motoneuron
- e: Abeam
- e: Acquisition Bus Monitor
- e: Acute Bacterial Meningitis
- e: Adjusted Balance Method
- e: Advance[d] Bill of Material
- e: American Building Maintenance
- e: Amino-form Bind Medium
- e: Anti-Ballistic Missile
- e: Apogee Boost Motor
- e: Assistant Beach Master
- e: Association of Breastfeeding Mothers
- e: Asynchronous Balanced Mode
- e: Atomic Beam Method
- e: Automated Batch Mixing
- e: Automatic Batch Mixing
- e: Automatic Billing Machine
- e: Avian Basal Medium
- e: Aviation Boatswains Mate
- e: Treaty Anti-Ballistic Missile Treaty
A.B.M. e: Associate in Business Management
ABm d: Amtsbaumeister
ABMA d: automatische Brandmeldeanlage
- e: American Boiler Manufacturers Association
- e: American Brush Manufacturers Association
- e: Army Ballistic Missile Agency
ABMath e: Bachelor of Arts in Mathematics
ABMC e: American Battle Monuments Commission
- e: American Business Media Council
ABMCP e: Aviation Boatswains Mate; Catapult
ABMD e: Advanced Ballistic Missile Defense
- e: Air Ballistics Missile Division
ABMDA e: Advanced Ballistic Missile Defense Agency
- e: Associated Building Material Distributors of America
ABME e: American Board of Master Educators
- e: Asynchronous Balanced Mode Extended
- e: Bachelor of Arts in Mechanical Engineering
ABMEC e: Association of British Mining Equipment Companies
ABMEWS e: Anti-Ballistic-Missile Early Warning System
ABMG e: American Board of Medical Genetics
ABMH e: Aviation Boatswains Mate; Handler
abmho d: Absolute Mho
AbMI e: Antibody Mediated Immunity
ABMIS e: Airborne Ballistic Missile Intercept System
ABMLAMS e: American Board of Medical-Legal Analysis in Medicine and Surgery
ABMLS e: Accrediting Bureau of Medical Laboratory Schools
ABMM e: American Board of Medical Microbiology
- e: Antiballistic-Missile Missile
ABMP e: American Board of Medical Psychotherapists
- e: Associated Bodywork and Massage Professionals
ABMPH e: Aviation Boatswains Mate; Plane Handler
ABMPM e: Association of British Manufacturers of Printers Machinery
ABMPTP e: Association of Black Motion Picture and Television Producers
ABMR e: Academy of Behavioral Medicine Research
- e: Atlantic Ballistic Missile Range
- e: Australian Board of Mineral Resources
ABMRF e: American Business Mens Research Foundation
ABMS e: Advanced Ballistic Missile Systems
- e: American Board of Medical Specialities
- e: American Bureau of Metal Statistics
- e: Antiballistic Missile System
- e: Artillery Ballistic Meteorological System
ABMT e: American Board of Medical Toxicology
ABN e: Aerodrome Beacon
- e: American Board of Nutrition
- e: American Broadcasting Network
- e: Aseptic Bone Necrosis
- e: Associated Broadcast News
- e: Association of British Neurologists
- e: Australian Bibliographic Network
- e: Azobisiso Butyronitril
Abn. d: Abnahme[stelle]
- d: Abnehmer
- d: Abnormität
- d: Abnutzung
- e: Aberdeen
abn. d: abnehmen
- d: abnorm[al]
ab'n e: abbreviation
abn e: airborne
AbnA d: Abnahmeamt
ABNA e: Achievable Benefit Not Achieved
Abn AAA e: Airborne Anti-Aircraft Artillery
ABNCF e: Airborne National Command Force
ABNCP e: Airborne Command Post
ABND e: Abandon
ABNDASC e: Airborne Direct Air Support Center
abndn[d] e: abandon[ed]
ABNDT e: Abundant
ABNI e: Available but Not Installed
ABNINF e: Airborne Infantry
ABNL d: Abnormal
ABNLSC e: Additive Bernoulli Noise Linear Sequential Circuit
ABNM e: American Board of Neurological Microsurgery
- e: American Board of Nuclear Medicine
abnml d: abnormal
ABNN e: American Board of Neuroscience Nursing
ABNO e: All But Not Only
ABNOC e: Airborne Operations Center
ABNOMS e: American Board of Neurological and Orthopaedic Medicine and Surgery
abnor d: abnormal
ABNORM e: Abnormal
AbnPrPl d: Abnahmeprüfplatz
ABNRADARARADIOINSTLTM e: Airborne Radar and Radio Installation Team
ABNRADIOMAINTTM e: Airborne Radio Maintenance Team
ABNS e: American Board of Neurological Surgery
- e: American British Numismatic Society
ABNSIGBN e: Airborne Signal Battalion

ABNT p: Associaçao Brasileira de Normas Tecnicas
Abnutz. d: Abnutzung
AbnVPl d: Abnahmeverteilplatz
ABO d: Absatz- und Bezugsorganisation
- d: allgemeine Bauordnung
- e: Absent Bed Occupancy
- e: Absent Bed Occupant
- e: Accessory Boring Organ
- e: Administration by Objectives
- e: Admiralty Berthing Officer
- e: Advanced Byte-Oriented
- e: Agents of Biological Origin
- e: American Board of Ophthalmology
- e: American Board of Opticianry
- e: American Board of Orthodontics
- e: American Board of Otolaryngology
- e: Anti-Boat-Obstacle
- e: Apparent Body Orientation
- e: Army Budget Office
- e: Association of British Orientalists
- e: Astable Blocking Oscillator
- e: Aviators Breathing Oxygen
Abo. d: Abonnement
- d: Abonnent
abo d: aboriginal
- e: accumulated benefit obligation
A.Bo. d: Ankerboje
A.B.O. d: Apotheken-Betriebsordnung
Abo e: Aborigine
Åbo F: Turku
ABOA e: Aminobenzoic Acid
- e: Australian Bibliography of Agriculture
ABOC e: American Board of Optometry Certificate
ABOD d: Arbeiten zur Bayerisch-Österreichischen Dialektgeografie (journ.)
- e: Advanced Base Ordnance Depot
Abo Eng e: Aboriginal English
Aboessen d: Abonnementessen
ABOG e: American Board of Obstetrics and Gynecology
abohm e: Absolute Ohm
ABOHN e: American Board for Occupational Health Nurses
ABOI e: Association of British Oceanic Industries
ABOIP e: Amended Basis of Issue Plan
ABOIPFD e: Amended Basis of Issue Plan, Feeder Data
abol e: abolish[able]
ABOL e: Abolished
abol e: abolisher
- e: abolishment, abolition[ary]
- e: abolitionism
- e: abolitionist
ABOL e: Adviser Business-Oriented Language
ABOM e: American Board of Orthopaedic Microneurosurgery
- e: Assistant Base Operations Manager
A-bomb e: Atomic Bomb
ABOMS e: American Board of Oral and Maxillofacial Surgery
ABOOW e: Assistant Battalion Officer-of-the-Watch
ABOP e: American Board of Oral Pathology
Abopreis d: Abonnementpreis
ABOPS e: Association of Business Officers of Preparatory Schools
ABOR e: Aborigine
- e: Abortion
Abordn.-Best. d: Abordnungsbestimmungen
abortus e: aborted fetus
abortus pill e: abortion-inducing pill, developed in France as Mifepristone, generic name: RU-48
ABOS e: Advanced Banking On-line System
- e: American Board of Oral Surgery
- e: American Board of Orthopedic Surgery
abos e: Aborigines
ABOSS e: Advanced Bombardment Sub-System
ABOTA e: American Board of Trial Advocates
ABP d: Anrufbearbeitungsprogramm
- d: Arbeitsbeschaffungsprogramm
- e: Absolute Boiling Point
- e: Acetyl Benzoyl Peroxide
- e: Actin-Binding Protein
- e: Active Band-Pass
- e: Actual Block Processor
- e: Adapter-Binding Post
- e: Adriamycin, Bleomycin, Prednisone
- e: Advanced Business Processor
- e: Air Bearing Platform
- e: Airborne Beacon Processor
- e: Aldosterone-Binding Protein
- e: Alphanumeric Block Printer
- e: American Board of Pathology
- e: American Board of Pediatrics
- e: American Board of Pedodontics
- e: American Board of Periodontology
- e: American Board of Prosthodontics
- e: American Broadcasting Paramount
- e: Aminobiphenyl
- e: Androgen Binding Protein
- e: Arabinose Binding Protein
- e: Associated Book Publishers
- e: Associated British Ports
- e: Associated Business Publications
- e: Association for Birth Psychology
- e: Association of Black Psychologists
- e: Assumed Binary Point
- e: Asteroid Belt Probe
ABp d: Allesbahnpost
Abp e: Archbishop
abp e: arterial blood pressure
ABPA e: Acoustical and Board Products Association
- e: Advanced Base Personnel Administration
- e: Allergic Bronchopulmonary Aspergillosis
- e: American Board Products Association
- e: American Book Producers Association
ABPANC e: American Board of Post Anesthesia Nursing Certification
ABPC e: Abelson Plasmacytoma
- e: American Book Publishers Council
- e: Associated British Picture Corporation
- e: Association of British Pewter Craftsmen
ABPD d: Alpha-Beta-Pseudokoinzidenz-Differenzverfahren
- e: American Board of Pediatric Dentistry
- e: American Board of Podiatric Dermatology
ABPDC e: American Board of Professional Disability Consultants
ABPD-Method e: Alpha-Beta Pseudo-Coincidence Difference Method
ABPDV d: Alpha-Beta-Pseudokoinzidenz-differenz-Verfahren
ABPD Verfahren d: Alpha-Beta-Pseudokoinzidenzdifferenz-Verfahren
ABPE e: Acute Bovine Pulmonyra Emphysema
ABPF e: Audio Bandpass Filter
ABPG e: Advanced Base Proving Ground
ABPH e: American Board of Psychological Hypnosis
ABPI e: Ankle / Brachial Pressure Index
- e: Association of the British Pharmaceutical Industry
ABPL e: Abelson Plasmacytoid Lymphosarcoma
ABPLM e: Asynchronous Bipolar Pulse Length Modulation
- e: Asynchronous Bipolar Pulse Length Modulator
ABPM e: American Board of Preventive Medicine
- e: Association of Business Product Manufacturers
ABPMR e: American Board of Physical Medicine and Rehabilitation
ABPN e: American Board of Psychiatry and Neurology
- e: Association of British Paediatric Nurses
ABPO e: Advanced Base Personnel Officer
- e: American Board of Podiatric Orthopedics
ABPP e: American Board of Professional Psychology
- e: Aminobromophenylpyrimidinone
- e: Amyloid Beta Protein Precursor
ABPPAC e: American Book Publishers Political Action Committee
ABPR e: African Book Publishing Record
- e: American Book Publishing Record
- e: Association of British Picture Restorers
Abpr. d: Abpraller
ABPS e: Air-Breathing Propulsion Subsystem
- e: Airborne Beacon Processing System
- e: Air Breathing Propulsion System
- e: American Board of Plastic Surgery
- e: American Board of Podiatric Surgery
- e: Automated Bill Payment System
A/BPS e: Airborne Propellant System
ABPsi e: Association of Black Psychologists
ABPSS e: Associate of the British Psychological Society
ABPSTS e: Association of Blind and Partially Sighted Teachers and Students
ABPU e: Advanced Base Personnel Unit
ABPV d: Allgemeine Bergpolizeiverordnung
ABPW e: Agency Broadcast Producers Workshop
ABPWC e: Association of Business and Professional Women in Construction
ABQ e: Albuquerque (airport)
ABQAUR e: American Board of Quality Assurance and Utilization Review
ABQAURP e: American Board of Quality Assurance and Utilization Review Physicians
ABQI e: Antipyrylbenzoquinoneimine
ABR d: Abortus-Bang-Ringprobe
- d: Abrufnummer
- d: Ausgabebaugruppe Register
- e: Abbreviation
- e: Aberrant Banding Region
- e: Abnormal Banding Region
- e: Abreast
- e: Abridged
- e: Absolute Bed Rest
- e: Acceptable Biological Removal
- e: Accounting and Business Research
- e: Acrylate-Butadiene Rubber
- e: Active Business Records
- e: Additional Billet Requirements
- e: Adhesive Bonding Repair
- e: Airborne Resupply
- e: American Bankruptcy Reports
- e: American Board of Radiology
- e: American Book Review

- e: Amphibian Boat Reconnaissance Aircraft
- e: Answer-Bid-Ratio
- e: Army Ballistic Research
- e: Asahi Brand Research
- e: Association for Biomedical Research
- e: Auditory Brainstem Response
- e: Australian Book Review
- e: Automatic Backup and Recovery
- e: Automatic Band Rate
- e: Automatic Bit Rate

abr. d: abrechnen
- d: abreisen
- d: abrichten
- d: abrunden
- d: abrüsten

Abr. d: Abrechnung
- d: Abreise
- d: Abriß
- d: Abruf
- d: Abrundung
- d: Abrüstung

abr e: abridgement
- e: abridgment

Abra e: Abraham

ABRA e: American Blood Resources Association

ABRAC e: Agriculture Biotechnology Research Advisory Committee

ABRACADABRA e: Abbreviations and Related Acronyms Associated with Defense, Astronautics, Business and Radioelectronics (journ.)

ABRACE s: Associacao Brasilieira de Computadores Electronicos

ABRALOC e: Acoustic Beacon Ranging and Location

Abram e: Abraham

abras e: abrasions

Abrasive Eng Soc Mag e: Abrasive Engineering Society Magazine

Abrasiv Eng e: Abrasive Engineering

Abr.-B. d: Abrechnungsbuch

ABRB e: Advanced Base Receiving Barracks

Abr.-Buch d: Abrechnungsbuch

ABRC e: Advisory Board for the Research Councils

Abrd. d: Abrundung

ABRD e: Advanced Base Receiving Depot
- e: Advanced Base Repair Depot
- e: Advanced Base Reshipment Depot

AbrdVO d: Abrundungsverordnung

ABRE e: Air Battalion Royal Engineers

Abreib. d: Abreibung

Abres e: Advanced Ballistic Reentry System

ABRES e: Advanced Baltistic Re-Entry System

ABRET e: American Board of Registration of Electroencephalographic Technologists

abrev s: abreviarura

abrév f: abrévtation

AB/RGW d: allgemeine Bedingungen des RGW

AB & RI e: Animal Breeding and Research Institute

AbrIMed e: Abridged Index Medicus

Abr Index Med e: Abridged Index Medicus

ABRL e: Aviation Base Responsibility List

ABRM e: Anterior Byssus Retractor Muscle

ABRMS e: American Board of Ringside Medicine and Surgery

Abr.-Nr. d: Abrechnungsnummer
- d: Abrufnummer

ABRO e: Animal Breeding Research Organization

AbrPA d: Abrechnungspostamt

AbrPl d: Abrechnungsplatz

ABR-Probe d: Abortus-Bang-Ringprobe

ABRS e: Adolescent Behaviour Rating Scale
- e: Aquatic Based Recreation Survey
- e: Australian Biological Resources Study

Abr Sci Pubs e: Abridged Scientific Publications

ABRST d: Abrechnungsstelle

ABRSV e: Abrasive

ABRSV RES e: Abrasive-Resistant

Abr SW e: Airbreak Switch

ABRT e: Abort

Abr.-Term. d: Abrechnungstermin

ABRTREQ e: Abort Request

ABRV e: Advanced Ballistic Reentry Vehicle

AbrVO d: Abrundungsverordnung

ABS d: Abrechnungscomputer
- d: Absatzwirtschaft
- d: Abscisinsäure
- d: Acrylnitril-Butadien-Styrol
- d: Allgemeine Berufsschule
- d: Anodenbasisschaltung
- d: Anschlußsteuerung Bildschirmeinheit
- d: Antiblockiersystem
- d: Aortenbogen-Syndrom
- d: Aufbauschneiden
- d: Ausgabe-Blattschreiber
- d: Ausgabebaugruppe Steuerung
- d: Ausgabeblattschreiber
- d: Ausschuß für biologische Sicherheit
- e: Abdominal Surgery
- e: Able-Bodied Seaman
- e: Absend Subscriber
- e: Absent
- e: Absolute
- e: Absorp
- e: Abstract
- e: Acrylonitrile-Butadiene-Styrene
- e: Active Boom Suspension
- e: Acute Brain Syndrome
- e: Adaptive Behaviour Scale
- e: Adaptive Braking System
- e: Additional Budget Submissions
- e: Admitting Blood Sugar
- e: Aerial Base Spring
- e: Affects Balance Scale
- e: Air Base Simulator
- e: Air Base Squadron
- e: Air Base Survivability
- e: Airbome Backing Store
- e: Air Breathing System
- e: Albumin-Buffered Saline
- e: Alkyl Benzene Sulphonate
- e: Alkylbenzyl Sulfonates
- e: Allied Business System
- e: Altitude Barometric Switch
- e: Amalgamated Book Services
- e: American Beethoven Society
- e: American Begonia Society
- e: American Behavioural Scientist (journ.)
- e: American Bible Society
- e: American Biological Society
- e: American Board of Surgery
- e: American Break Shoe
- e: American Breeder Service
- e: American Broadcasting System
- e: American Bryological Society
- e: American Bureau of Shipping
- e: Animal Behaviour Society
- e: Antenna Base Spring
- e: Anti-Bursa Serum
- e: Anti Blocking System
- e: Antilock Braking System
- e: Antique Boat Society
- e: Antiskid Braking System
- e: Approved to British Standard
- e: Armoured Boarding Steamer
- e: Army Broadcasting Service
- e: Asset-Backed Security
- e: Associated Biomedic Systems Inc.
- e: Associated Blacksmiths of Scotland
- e: Associated Broadcasting Services
- e: Associate in Business Science
- e: Associate of the Building Societies Institute
- e: Association of Broadcasting Staff
- e: Association on Broadcasting Standards
- e: At Bed Side
- e: Atlantic Base Section
- e: Auckland Botanical Society
- e: Australian Building Specification
- e: Australian Bureau of Statistics
- e: Automated Bioassay System
- e: Automatic Bar and Chucking
- e: Automatic Beam-current Stabilization
- e: Automatic Bibliographic Services
- e: Automatic Bomb Sight
- e: Automatic Braking System
- e: Auxiliary Building Sump
- e: Average Busy Season

Abs. d: Absatz
- d: Abschaltung
- d: Abschnitt
- d: Absender
- d: Absendestelle
- d: Absetzung
- d: Absicherung
- d: Absicht
- d: Absorption
- e: Absolvent
- e: Abstracts

abs. d: abseits
- d: absetzen
- d: absolut
- d: absolvieren
- d: absondern
- d: abstrakt

a. b. S. d: an berufsbildenden Schulen

abs e: absolute value
- e: absorb[ent]
- e: Air Break Switch

Abs e: Absorbent

AbS e: Analysis by Synthesis

ABs e: Aryan Brotherhood members

ABSA e: Annual of the British School at Athens (journ.)
- e: Association of British Secretaries in America
- e: Australian Bird Study Association

abs alt e: Absolute Altitude

ABSAME e: Association for Behavioral Sciences and Medical Education

ABSAP e: Airborne Search and Attack Plotter

abs art e: abstract art

ABSB e: Air Burst / Surface Burst

ABSBH e: Average Busy Season Busy Hour

ABSC e: Applied Behavioural Studies Centre
- e: Associate of the British Society of Commerce
- e: Automatic Bass Compensation
- e: Automatic Blip-Scan Counter

absc e: abscissa

abs.-c e: Asbestos-Covered

Abscam e: Arab scam

Abs cat e: Abyssinian cat

Absch. d: Abschluß

Abscheid. d: Abscheider
- d: Abscheidung

Abschl. d: Abschlag
- d: Abschluß

abschl. d: abschlagen
- d: abschleppen
- d: abschließen[d]
Abschl.-Ausz. d: Abschlagsauszahlung
Abschl.-D. d: Abschleppdienst
- d: Abschlußdatum
Abschn. d: Abschnitt
Abschr. d: Abschreibung
- d: Abschrift
ABS CLG e: Absolute Ceiling
ABSCM e: Association of Boys and Students Clothing Manufacturers
abs.conv. e: absolutely convergent
abscoulomb e: Absolute Coulomb
ABSCS e: Automatic Blip-Scan Counter System
ABSD e: Advance Base Sea Dock
- e: Advanced Base Section[al] Dock
- e: Advanced Base Supply Depot
- e: Army Base Supply Depot
abs e e: Absolute Error
ABSEL e: Association for Business Simulation and Experiential Learning
Absen d: Absender
abse. re. l: absente reo
abs exp e: abstract expressionism
AbsG d: Absicherungsgesetz
Abs.-Ges. d: Absicherungsgesetz
ABSI e: Adaptive Behaviour Scale for Infants and Early Childhood
- e: Associate of the Boot and Shoe Institution
AbsichG d: Absicherungsgesetz
ABSiE e: American Broadcasting Station in Europe
ABSIE e: American Broadcasting System in Europe
ABSIOL e: Advanced Base Initial Outfitting List
ABSIRS e: American Bureau of Shipping Information Retrieval System
ABS-Kunststoff d: Akrylonitril- Butadien-Styrol-Kunststoff
ABSL e: Ambient Background Sound Level
ABSLA e: Approved Basic Stock Level of Ammunition
ABSLDR e: Absolute Loader
ABSLY e: Absolutely
ABSM e: Associate of the Birmingham School of Music
- e: Association of British Steriliser Manufacturers
ABSMGP e: Associate of the British Society of Master Glass Painters
ABSO e: Auxiliary Business Service Organization
ABSOL e: Absolute
absol. d: absolut
abso[l] e: absolute[ly]
absol f: absolument
ABSORP e: Absorption
Absorpt. e: Absorption
Absp. d: Absperrung
Abspalt. d: Abspaltung
Abspaltbark. d: Abspaltbarkeit
Abs Pap ACS e: Abstracts of Papers of the American Chemical Society
Abs.-Pl. d: Absatzplan[ung]
Abspr. d: Absprache
- d: Absprung
ABS-PS e: Adaptive Behavior Scale-Public School
ABSq e: Air Base Squadron
ABSS e: Advanced Beach Signal Station
AbsSt d: Absendestelle
ABS System e: Automatic Beam-current Stabilizing System
abst. d: abstammen
- d: abstellen
- d: abstimmen
- d: abstufen
Abst. d: Abstammung
- d: Abstand
- d: Abstieg
- d: Abstimmung
- d: Abstufung
- d: Abwehrstelle
A.B.St. d: Artilleriebeobachtungsstelle
abs t e: absolute temperature
abstat unit e: absolute electrostatic unit
Abstd. d: Abstand
abs. Temp. d: Absolute Temperature
AbStg d: Abschaltesteuerung
ABSTI e: Advisory Board on Scientific and Technical Information
abstr. d: abstrakt
Abstr. d: Abstraktion
- d: Abstrich
abstr e: abstract
ABSTR e: Abstract[ing]
Abstr Comput Lit e: Abstracts of Computer Literature (journ.)
Abst.Zd. d: Abstandszünder
ABSV e: Absolute Value
ABSv e: Air Base Service
ABS VM e: Absolute Voltmeter
ABSVOLT e: Absolute Volt
abs vot e: absentee voting
ABSVS e: Auxiliary Building Special Ventilation System
absw e: air-break switch
ABSW e: Air Break Switch
- e: American Baptist Seminary of the West
- e: Association of British Science Writers
Absz. d: Abzisse
abs z e: absolute zero
ABT d: Anschlußsteuerung für Bildschirm und Tastatur
- d: Arbeitsstelle für Bibliothekstechnik
- e: About
- e: Absorbing Trunk
- e: Acetate Base Tape
- e: Advanced Booster Technology
- e: Air-Blast Transformer
- e: Airborne Tracking
- e: All Body Type
- e: Allied Board of Trade
- e: American Board of Toxicology
- e: American Board of Trade
- e: Applied Business Technology Corp.
- e: Associate in Business Technology
- e: Association of Book Travelers
- e: Association of Building Technicians
- e: Atlantic Booster Test
- e: Australian Broadcasting Tribunal
- e: Automatic Battery Test
- e: Automatic Block Transfer
- e: Automatic Block Transport
- e: Automatic Bus Terminal
- e: Automatic Bus Transfer
- e: Auxiliary Ballast Tank
Abt. d: Abteil[ung]
- d: Abtrieb
- d: Abtrift
abt e: Abort
- e: Abundant
- e: after beginning of time
ABTA e: American Bridge Teachers Association
- e: Australia-British Trade Association
ABTAPL e: Association of British Theological and Philosophical Libraries
AbtBf d: Abteilungsbauführer
ABTD e: American Book Trade Directory
- e: Australian Book Trade Directory
- e: Automatic Bulk Tape Degausser
Abt.-Dir. d: Abteilungsdirektor
Abteuflstg. d: Abteufleistung
Abteuflstg. j. M. u. Sch. d: Abteufleistung je Mann und Schicht
ABTF e: Airborne Task Force
- e: Assault Battalion Task Force
ABTICS e: Abstract and Book Title Index Card Service
Abt. Kdr. d: Abteilungskommandeur
Abtl. d: Abteilung
Abt.-L. d: Abteilungsleiter
Abtlg[n]. d: Abteilung[en]
ABTM e: American Board of Tropical Medicine
Abt'n d: Abteilungen
ABTNOMS e: American Board of Thoracic Neurological Orthopaedic Medicine and Surgery
ABTOAD e: Air Blast Time-of-Arrival Detector
AbtPr d: Abteilungspräsident
Abtr. d: Abtragung
- d: Abtreibung
- d: Abtrennung
- d: Abtretung
- d: Abtrieb
- d: Abtrift
ABTR e: Association for Brain Tumor Research
ABTrans e: Air-Blast Transformer
Abtrenn. d: Abtrennung
ABTS e: American Board of Thoracic Surgery
ABTSA e: Association of British Tree Surgeons and Arborists
ABTSS e: Airborne Transponder Subsystem
ABTT e: Associate of British Theatre Technicians
ABTU e: Advanced Base Torpedo Unit
- e: Advanced Base Training Unit
- e: Air Bombers Training Unit
- e: Arms Basic Training Unit
- e: Army Basic Training Unit
ABU d: Allgemeine Bedingungen für Unfallversicherung
- d: Arbeitsgemeinschaft Betriebsuntersuchungen
- e: Administrative Base Unit
- e: Advanced Business Unit
- e: Alarm Buzzer
- e: American Board of Urology
- e: Annunciator Buzzer
- e: Answer-Back Unit
- e: Asia-Pacific Broadcasting Union
- e: Asian Broadcasting Union
A.B.U. d: Aktionsgemeinschaft Bessere Umwelt; Köln
ABuA d: Arbeit, Beruf und Arbeitslosenhilfe (journ.)
ABUAHP e: American Board of Urologic Allied Health Professionals
abul e: abulia
- e: abulic
ABull e: Art Bulletin (journ.)
ABus e: Associate in Business
abut e: Abutment
ABV d: Abschnittbevollmächtigter (DDR)
- d: Allgemeine Bedingungen der Vertrauensschadenversicherung
- d: Arbeitsgemeinschaft Baufachverlage
- d: Arbeitsgemeinschaft Berufsständische Versorgungseinrichtungen
- d: automatische Bildverarbeitung

- d: automatischer Blockierverhinderer
- e: Above
- e: Absolute Value
- e: Adriamycin, Bleomycin, Vinblastine
- e: Air Blast Value
- e: Air Blast Valve
- e: Air Bubble Vehicle
- e: Armed Boarding Vessel
- e: Auxiliary Building Ventilation

ABvBO d: Ausführungsbestimmungen zur vereinfachten [Eisenbahn-] Bau- und Betriebsordnung

ABvBOS d: Ausführungsbestimmungen zur vereinfachten [Eisenbahn-] Bau- und Betriebsordnung

ABVD e: Adriamycin, Bleomycin, Vinblastine, Dacarbazine

ABVERS d: Abbau der Verschmutzung

abvolt d: absolute Volt

ABVS e: Advisory Board on Veterinary Specialties

ABV/T d: Abschnittsbevollmächtigter der Transportpolizei (DDR)

ABVT d: Allgemeine Bedingungen für die Versicherung von Gütertransporten
- e: American Board of Veterinary Toxicology

ABW d: Anschlußsteuerung Blockdruckwerk
- d: Außenbogenweiche
- e: Acidified Boiling Water
- e: Advanced Base Workshop
- e: Air Base Wing
- e: Anterior Bite Wing
- e: Armo[u]red Brigade Workshop
- e: Autoband Welder
- e: Automated Batch Weighting

Abw. d: Abwärme
- d: Abwehr
- d: Abwerbung
- d: Abwertung
- d: Abwesenheit
- d: Abwicklung
- d: Abwurf[waffe]

abw. d: abwärts
- d: abwechseln[d]
- d: abwehren
- d: abweichen[d]
- d: abweisen
- d: abwerben
- d: abwerfen
- d: abwerten
- d: abwertend
- d: abwesend
- d: abwickeln

Abw d: Abwasser
- d: Abweichung

a. b. W. d: auf besonderem Wege

ABWA e: American Bottled Water Association
- e: Associated Business Writers of America

AbwAG d: Abwasserabgaben-Gesetz

abwatt d: Absolute Watt

ABWC e: Automatic Bandwidth Control

ABWG e: Air Base Wing

ABWHE e: Association of Black Women in Higher Education

ABWIK e: Assault and Battery with Intent to Kill

ABWL e: Army Bacteriological Warfare Laboratory

AB WORKS e: Airborne Workshop

ABWR e: Advanced Boiling Water Reactor

ABWRC e: Army Biological Warfare Research Center

Abw.-St. d: Abwehrstelle
- d: Abwicklungsstelle

ABXL e: Abaxial

ABY e: Acid Bismuth Yeast
- e: Albany (airport)

Aby e: Abyssinia

ABZ d: Abrechnung, Buchung und Zahlbarmachung
- e: Albendazole

abz. d: abzahlbar
- d: abzahlen
- d: abziehbar
- d: abziehen
- d: abzüglich
- d: abzugsfähig

Abz. d: Abzeichen
- d: Abzug

ABz d: Amtsbezirk

Abz.-G. d: Abzahlungsgeschäft

AbzG d: Abzahlungsgesetz

Abz.-Gesch. d: Abzahlungsgeschäft

Abzk d: Abzugsknopf

Abz.Vor. d: Abzugsvorrichtung

Abzw. d: Abzweig
- d: Abzweigstelle

Abzw d: Abzweigung

Abzwn d: Abzweigungen

ABZWVT d: Abzweigverteiler

AC d: Aachen (Kfz)
- d: Abgeordnetenkonvent
- d: Aeroklub
- d: Affinitätschromatographie
- d: Antrittskonvent
- d: Anwendungscode
- d: Arbeitsplatzcomputer
- d: Automobilclub
- e: Abcoulomb
- e: Abdominal Circumference
- e: Aboce Clouds
- e: Above Clouds
- e: Absence Card
- e: Absolute Code
- e: Absolute Coding
- e: Absolute Convergence
- e: Absolute Counter
- e: Absorber Circuit
- e: Absorption Capacity
- e: Absorption Circuit
- e: Absorption Coil
- e: Abstract Code
- e: Abstract Computer
- e: Acceleration Command
- e: Accelerator
- e: Acceptance
- e: Accept[ed]
- e: Acceptor Circuit
- e: Access
- e: Access Coding
- e: Access Control
- e: Access Cycle
- e: Accessory Cells
- e: Accommodation Convergence
- e: Account
- e: Accountants and Controllers
- e: Account Card
- e: Account Code
- e: Account Current
- e: Accounting Calculation
- e: Accounting Card
- e: Accounting Carriage
- e: Accounting Computer
- e: Accumulating Counter
- e: Accumulator
- e: Accumulator Contents
- e: Accuracy Check
- e: Accuracy Control
- e: Acetic Acid
- e: Acetylcholin[e]
- e: Acetylcysteine
- e: Acid
- e: Acid Concentrator
- e: Acoustical
- e: Acoustic Coupler
- e: Acoustics
- e: Acquisition Costs
- e: Acromioclavicular
- e: Acting
- e: Actinium
- e: Action Chart
- e: Action Code
- e: Activation Coefficient
- e: Activator
- e: Active Capital
- e: Active Cirrhosis
- e: Active Component
- e: Active Computer
- e: Active Concentration
- e: Activity Captain
- e: Activity Code
- e: Actual Cost
- e: Acupuncture Clinic
- e: Adapter Cable
- e: Adapter Cluster
- e: Adaptive Control
- e: Add[er] Carry
- e: Adder Circuit
- e: Adding Circuit
- e: Adding Counter
- e: Additional Claim
- e: Additional Code
- e: Addressable Clock
- e: Address Calculation
- e: Address Card
- e: Address Carry
- e: Address Chain
- e: Address Character
- e: Address Check[ing]
- e: Address Code
- e: Address Coding
- e: Address Comparing
- e: Address Complete
- e: Address Computation
- e: Address Constant
- e: Address Conversion
- e: Address Counter
- e: Addressed Core
- e: Adelbert College
- e: Adelphi College
- e: Aden Colony
- e: Adenylate Cyclase
- e: Adherent Cell
- e: Adjacent Channel
- e: Adjusted Charge
- e: Adjustment-Calibration
- e: Administrative Central
- e: Admiral Commanding
- e: Adopted Child
- e: Adrenal Cortex
- e: Adrenocorticoid
- e: Adriamycin
- e: Adriamycin, Cyclophosphamide
- e: Adrian College
- e: Advance Calling
- e: Advanced Certification
- e: Advertising Council
- e: Advice of Charge
- e: Advisory Circular (journ.)
- e: Advisory Committee
- e: Advisory Council
- e: Aerated Concrete
- e: Aerial Combat
- e: Aerial Current
- e: Aerodynamic Center

- e: Aeronautical Center
- e: Aerospace Center
- e: Aerospace Corporation
- e: Affinity Chromatography
- e: After Christ
- e: Agribusiness Council
- e: Aiming Circle
- e: Air-Cooled
- e: Air Canada
- e: Air Command
- e: Air Commodore
- e: Air Component
- e: Air Compressor
- e: Air Conduction
- e: Air Control
- e: Air Controller
- e: Air Controlman
- e: Air Corps
- e: Air Council
- e: Aircraft Cannon
- e: Aircraft Carrier
- e: Aircraft Commander
- e: Aircraft Control
- e: Aircraftsman
- e: Aircrewman
- e: Airfield Construction
- e: Airframe Change
- e: Airfreight Container
- e: Airways Control
- e: Airworthiness Certificate
- e: Airworthiness Committee
- e: Alabama College
- e: Alaskan Command
- e: Albion College
- e: Albright College
- e: Aldershot Command
- e: Alert Condition
- e: Algebraic Calculation
- e: Algebraic Curve
- e: Algorithm Control
- e: Alien Cell
- e: Alkali Cellulose
- e: Allegheny College
- e: Allens Creek
- e: Allergic Conjunctivitis
- e: Alliance College
- e: Allied Chemicals
- e: Allied Commission
- e: Allocation Carry
- e: Allocation Counter
- e: Allowable Cost
- e: Allyl Chloride
- e: Alma College
- e: Alphabetic Character
- e: Alphabetic Code
- e: Alphabetic Coding
- e: Alphabetic Collator
- e: Alphanumeric Character
- e: Alphanumeric Code
- e: Alphanumeric Coding
- e: Alphanumeric Command
- e: Altered Course
- e: Alternate Call
- e: Alternate Cancellation
- e: Alternate Code
- e: Alternate Current
- e: Alternating Current
- e: Alteration Cancellation
- e: Altitude Compensator
- e: Altitude Control
- e: Alverno College
- e: Amat.illo College
- e: Ambulance Corps
- e: American Conditions
- e: Amherst College
- e: Ammonium Citrate
- e: Ammunition Column
- e: Ammunition Corporal
- e: Amortized Cost
- e: Amphibian Command
- e: Amphibious Car
- e: Amphibious Corps
- e: Analog[ue] Channel
- e: Analog[ue] Circuit[ry]
- e: Analog[ue] Computation
- e: Analog[ue] Computer
- e: Analog[ue] Computing
- e: Analysis Console
- e: Analytical Chemist
- e: Analytic Chemist
- e: Anastomosing Cell
- e: AND Circuit
- e: Anderson College
- e: Andrew College
- e: Anisotropic Constant
- e: Annhurst College
- e: Annual Conference
- e: Annulment of Certification
- e: Anodal Contraction
- e: Anode Circuit
- e: Another Copy
- e: Answer Complete
- e: Answer Construct
- e: Antarctic Convergence
- e: Antecubital
- e: Antenna Current
- e: Anterior Chamber
- e: Anterior Commissure
- e: Anterior Connective
- e: Anterior Cortical
- e: Anti-Centre
- e: Anti-Char
- e: Anti-Collision
- e: Anti-Concrete
- e: Anti-Corrosive
- e: Anti-inflammatory Corticoid
- e: Anticoagulant
- e: Anticoincidence Counter
- e: Anticomplementary
- e: Antioch College
- e: Antiphologistic Corticoid
- e: Aortic Closure
- e: Aperture Card
- e: Apical Cell
- e: Appeal Case
- e: Appeal Court
- e: Appeals Council
- e: Appellate Court
- e: Appliance Computer
- e: Application Control
- e: Application Controller
- e: Apprenticeship Committee
- e: Approach Chart
- e: Approach Control
- e: Approach Coupler
- e: Approximate Calculation
- e: Approximate Computation
- e: Approximation Calculus
- e: Aquinas College
- e: Arcadia College
- e: Arc Cutting
- e: Architect of the Capitol
- e: Arctic Circle
- e: Area Code
- e: Area Commander
- e: Area Coverage
- e: Arithmetic Check[ing]
- e: Arithmetic Computation
- e: Arithmetic Control
- e: Arkansas College
- e: Armament Command
- e: Armament Control
- e: Armaments Command
- e: Arm Circumference
- e: Armo[u]red Cable
- e: Armo[u]red Car
- e: Armo[u]red Carrier
- e: Armo[u]red Cruiser
- e: Armstrong College
- e: Army Circular
- e: Army Contract
- e: Army Corps
- e: Army Council
- e: Array Calculation
- e: Asbury College
- e: Ashland College
- e: Aspect Card
- e: Asphalt Cement
- e: Asphalt Composition
- e: Asphaltic Concrete
- e: Assembler Code
- e: Assistant Cashier
- e: Assistant Censor
- e: Assistant Chief
- e: Assistant Clerk
- e: Assistant Commandant
- e: Assistant Commissioner
- e: Assistant Controller
- e: Associate Contractor
- e: Associate in Commerce
- e: Assumption College
- e: Astable Circuit
- e: Astronomical Constant[s]
- e: Asynchronous Computer
- e: Athens College
- e: Atlanta Constitution
- e: Atlantic Charter
- e: Atlantic Council
- e: Atlas-Centaur
- e: Atomicity Controller
- e: Atriocarotid
- e: Attack Characterization
- e: Attack Console
- e: Attention Command
- e: Attenuation Characteristics
- e: Attitude Control
- e: Audio Center
- e: Audit Compliance
- e: Auditory Cortex
- e: Augmentation Concentration
- e: Augusta College
- e: Augustana College
- e: Aurora College
- e: Austin College
- e: Australian Corps
- e: Australian Cruiser
- e: Authors Correction
- e: Authorization and Consent
- e: Authorization under Consideration
- e: Author Catalogue
- e: Auto-Canon
- e: Autocarrier
- e: Autocoder
- e: Autocollimator
- e: Autocorrelation
- e: Automatic Calculator
- e: Automatic Carriage
- e: Automatic Check[ing]
- e: Automatic Checkout
- e: Automatic Clearing
- e: Automatic Code
- e: Automatic Coding
- e: Automatic Computer
- e: Automatic Configuration
- e: Automatic Control[ler]
- e: Automation Center
- e: Automobile Club
- e: Auxiliary Command
- e: Auxiliary Console
- e: Availability Check
- e: Availability Code
- e: Average Consumer
- e: Averett College
- e: Aviation Cadet

- e: Awaiting Connection
- e: Axial Centrifugal
- e: Axiocervical
- e: Axiom of Choice
- e: Azacytidine
- e: Azusa College
- l: Acidus
- l: Anno Christi
Ac d: Aktinium
- e: Acre
- e: Acute
- e: Acyl
- e: Alto-Cumulus
ac e: absolute ceiling
- e: acetate
- e: acetyl
- e: acetyl-choline
- e: acoustical[ly]
- e: active
- e: aerial current
- e: aerodynamic center
- e: air condition[ed]
- e: air conditioning
- e: air cool
- e: air cooled
- e: aircraft
- e: also completed
- e: analog computer
- e: anodal closure
- e: antenna current
- e: anti-clutter
- e: anticorrosive
- e: antiphlogistic corticoid
- e: area controller
- e: asbestos cement
- e: asphalt concrete
- e: auriculocarotid
- e: average cost
- e: azimuth comparator
A. C. d: Allgemeiner Convent
- e: Admiralty Constabulary
- e: Artillery Clerk
- e: Artillery Controller
- f: année courante
- l: Ante Christum
- s: Año Cristo
a.c. e: all colors
- l: anni currentis
- l: anno currente
- l: ante cibos
- l: ante cibum
a. C. l: ante Christum
a C i: avanti Cristo
Ac. l: Acidum
A & C e: Abatement and Control
- e: Addenda and Corrigenda
- e: Arithmetic and Controls
- e: Automation and Control
A/C e: Account [Control]
- e: Air Commodore
- e: Assemble and Checkout
- e: Aviation Cadet
a/c e: account
- e: account current
- e: air conditioning
- e: aircraft
- e: alter course
- p: ao cuidado de
A-C e: Allis-Chalmers
- e: Anchored Catheter
a-c e: alternating-current
ac- d: allcyclisch
AC1 e: Air Controlman; First Class
AC 1 e: Aircraftsman First Class
AC-119 e: Fairchild-Hiller armed gunship
AC-130 e: Lockheed armed gunship
AC2 e: Air Controlman; Second Class
AC 2 e: Aircraftsman Second Class
AC2MP e: Army Command and Control Master Plan
AC2S e: Army Command and Control System
AC3 e: Air Controlman; Third Class
AC 3 e: Aircraftsman Third Class
AC-47 e: DC-3 Douglas
ACAA e: Agricultural Conservation and Adjustment Administration
- e: American Coal Ash Association
- e: Army Concepts Analysis Agency
- e: Asthma Care Association of America
- e: Automatic Chemical Agent Alarm
ACAAC e: American College Admissions Advisory Center
ACAACCC e: American College Admissions Advisory and Career Counseling Center
ACAAD e: Authorized Commanders Atomic Air Defense
ACAAE e: Australian Council on Awards in Advanced Education
ACAAI e: Air Cargo Agents Association of India
ACAAS e: Automatic Chemical Agent Alarm System
ACAB e: Air-Conditioning Advisory Bureau
- e: Allied Central Air Bureau
- e: Army Contract Adjustment Board
ACABQ e: Advisory Committee on Administrative and Budgetary Questions
Acac e: Acetylacetonate
ACAC e: Acetylacetone
- e: Admiral Commanding; Aircraft Carriers
- e: Air Contact Area Commander
- e: Air Control Area Commander
- e: Allied Container Advisory Committee
- e: American College Admissions Center
- e: Association of College Admission Counsellors
- e: Australian College of Aesthetics and Cosmetology
- e: Automated Direct Analog[ue] Computer
ACACA e: Army Command and Administration Communication Agency
ACACE e: Advisory Council for Adult and Continuing Education
AC Acid e: Anthracenecarboxylic Acid
ACACN e: American Council of Applied Clinical Nutrition
ACACS e: Air Cycle Air-Conditioning System
ACAD e: Advanced Computer Aided Design
- e: Air Containment Atmosphere Dilution
- e: Alarm Communications and Display System
- e: American Conference of Academic Deans
acad e: academic
- e: academic[ian]
- e: academical
Acad e: Academy
- e: Acadia
- l: Academia
ACADA e: Advanced Chemical Agent Detector Alarm
Acad aper e: Academy of Motion Picture Arts and Sciences aperture
ACAD CL YR e: Academic Class Year
ACADE e: Association for Computer Art and Design Education
Academic e: Academic Press
Acad Fran f: Académie Française
ACADI f: Association des Cadres Dirigeants de l'Industrie
Acadia e: old name for French-speaking Canada and Nova Scotia
Acadia[n] e: Novia Scotia[n]
Acad Ins B-L f: Académie des Inscriptions et Belles-Lettres
Acad mask e: Academy of Motion Picture Arts and Sciences mask
Acad Med e: Academy of Medicine
Acad Mgmt e: Academy of Management
Acad Mus e: Academy of Music
Acad Natur Sci Phila Proc e: Academy of Natural Sciences of Philadelphia Proceedings (journ.)
Acad of Nat Sci Jour e: Academy of Natural Sciences; Journal
Acad Pr e: Academic Press
Acad Pr Ark e: Academic Press of Arkansas
ACADS e: Alarm Communications and Display Segment
Acad Sci f: Académie des Sciences
Acad Sci St. Louis Trans e: Academy of Science of St. Louis Transactions (journ.)
Acad Sin e: Academia Sinica
Acad St Cec e: Academia di Santa Cecilia, Rome
Acad Therapy e: Academic Therapy Publications
Acad U e: Acadia University
AC & AE e: Association of Chemical and Allied Employees
ACAE e: Association of Chinese and American Engineers
ACAF e: Amphibious Corps, Atlantic Fleet
- e: Australian Citizen Air Force
- e: Automatic Circuit Assurance Feature
AcAF e: Acetylaminofluorene
ACAG e: Allied Control Authority for Germany
ACAH e: Accreted Crystaline Anthropoid Homologue
- e: Acylcholine Acyl-Hydrolase
- e: Autoimmune Chronic Active Hepatitis
ACAHA e: Association of Chief Administrators of Health Authorities
ACAI e: Alkali and Clean Air Inspectorate
ACAL e: Aircraft Change Application List
- e: Aircraft Configuration Allowance List
ACALM e: American College of Animal Laboratory Medicine
ACAM e: American College of Advancement in Medicine
- e: Apollo Computer Address Matrix
- e: Asian Centre of Agricultural Machinery
- e: Augmented Content-Addressed Memory
AC-AMDF e: Auto-Correlation - Average Magnitude Difference Function
ACA Memory e: Associative Content Addressable Memory
ACAMPS e: Automated Communications and Message Processing System
ACAMR e: Associate Committee on Aviation Medical Research
ACAMS e: Automatic Continuous Air Monitoring System
ACAN e: Action Committee Against Narcotics
- e: Army Command and Administrative Network
- e: Army Communications Administrative Network
AC and WS e: Aircraft Control and Warning Station[s]

ACAnes e: American College of Anesthetists
a. cant. e: after cant frames
Acanth e: Acanthocephala
acanthite e: silver sulfide
ACAO e: Allied Civil Affairs Office
- e: Assistant County Advisory Officer
Acap e: Acapulco
ACAP e: Advanced Combat Air Patrol
- e: Advanced Composite Airframe Program[me]
- e: Advanced Computer for Array Processing
- e: Analyst Capability
- e: Army Child Advocacy Program[me]
- e: Army Cost Analysis Paper
ACAP e: Automatic Circuit Analysis Program[me]
- e: Aviation Consumer Action Process
- e: Aviation Consumer Action Project
ACAPA e: American Concrete Agricultural Pipe Association
ACAPE e: Aerospace Computer Automatic Program[me] Evaluator
Acap gold e: Acapulco gold
a capp i: a cappella
ACAPS e: Analog[ue] Circuit Analysis and Partitioning System
- e: Automated Chemical Analysis for Process Solutions
- e: Automated Costing and Planning System
Acapulco s: Acapulco de Juárez
ACAR e: Alumin[i]um Cable; Alumin[i]um Reinforced
- e: American Cable and Radio
ACARD e: Advisory Council for Applied Research and Development
ACARS e: Aircraft Communications Addressing and Reporting System
- e: Arinc [Corporation] Communications Addressing and Reporting System
ACARTSD e: African Centre for Applied Research and Training in Social Development
Ac Arty e: Accompanying Artillery
ACAS e: Advisory, Conciliation and Arbitration Service
- e: African Commission on Agricultural Statistics
- e: Airborne Collision-Avoidance System
- e: Aircraft Collision Avoidance System
- e: American Computer Appraisal Service
- e: Analytical Chemistry and Applied Spectroscopy
- e: Anchored Cell Analysis and Sorting
- e: Army Commissary Automation System
- e: Army Crisis Action System
- e: Associate of the Casualty Actuarial Society
- e: Association for Central Asian Studies
- e: Association of College Auxiliary Services
- e: Atlantic Coast Air Service
- e: Automatic Collision Avoidance System
- e: Automatic Component Assembly System
- e: AUTOVON Centralized Alarm System
ACAs e: Arms Control Associations
AC/AS e: Assistant Chief of Air Staff
A.C.A.S. e: Assistant Chief of the Air Staff
ACASD e: Air Clutch Antislack Device
A.C.A.S.(I.) e: Assistant Chief of the Air Staff (Intelligence)
A.C.A.S.(Ops.) e: Assistant Chief of the Air Staff (Operations)
A.C.A.S.(O.R.) e: Assistant Chief of the Air Staff (Operational Requirements)
ACASP e: Air Conditioning Analytical Simulation Package
A.C.A.S.[S.] e: Assistant Chief of the Air Staff [Signals]
ACASS e: Automated Credit Authorization Subsystem
ACAST e: Advisory Committee for the Application of Science
- e: Advisory Committee on Applications of Science and Technology (UNESCO)
- e: Advisory Committee on the Application of Science and Technology to Development
A.C.A.S.(T.) e: Assistant Chief of the Air Staff (Training)
ACASTD e: Advisory Committee on the Application of Science and Technology to Development
ACAT e: Acquisition Category
- e: Advanced Computer Audit Techniques
- e: Advanced Conformal Antenna Technique
- e: Air Conditioner Air Transportable
acata e: acatalectic[al]
ACATA e: Association for Computer-Assisted Text Analysis
ACATT e: Aviation Combined Arms Team Trainer
ACAU e: Armo[u]red Cavalry Assault Vehicle
- e: Automatic Calling and Answering Unit
ACAV e: Alternating Current Anode Voltage
- e: Armored Cavalry Assault Vehicle
ACAVS e: Advanced Cab and Visual System
ACAW e: Air Communications and Weather
- e: Aircraft Control and Warning
ACB e: Acceptance and Certification Branch
- e: Access[-method] Control Block
- e: Accounting and Business Research
- e: Adapter Control Block
- e: Aerial Control Bay
- e: Air-Cushion Barge
- e: Air Circuit Breaker
- e: Air Crew Bulletin
- e: Airfield Construction Branch
- e: Airman Classification Battery
- e: Align Control Box
- e: Allocate Control Block
- e: Alternating Current Bridge
- e: American City Bureau
- e: American Crisis Biographies
- e: Aminochlorobenzophenone
- e: Amphibious Construction Battalion
- e: Annoyance Call Bureau
- e: Antenna Control Bay
- e: Antibody-Coated Bacteria
- e: Antrum-Corpus Boundary
- e: Aortocoronary Bypass
- e: Application Control Block
- e: Army Classification Battery
- e: Army Communications Board
- e: Assam Carbon Black
- e: Association of Clinical Biochemists
- e: Asynchronous Communications Base
- e: Australian Computer Bulletin (journ.)
- e: Auxiliary Controller Bus
- f: Association Canadienne des Bibliothèques
Ac B d: Aktinium B
- e: Accessory Bundle
- e: Actinium B
acb e: aortocoronary saphenous vein bypass
- e: arterialized capillary blood
- e: asbestos cement board
ACBA System e: Aircrew Body Armor System
ACBB e: American Council for Better Broadcasts
ACBC e: Alternating Current Bridge Circuit
ACBCT e: Automatic Circuit Board Card Tester
AC/BD e: Acrylonitrile / Butadiene
ACBD e: Active Commission Base Date
ACBE e: Air Contrast Barium Enema
ACBEI e: Association for Community Based Educational Institutions
ACBFC e: Academy of Comic-Book Fans and Collectors
ACBG e: Aortacoronary Bypass Graft
ACBGEN e: Application Control Block Generation
ACBIS e: Academic Collective Bargaining Information Service
ACBM e: Advisory Committee for Biology and Medicine
- e: Atomic Cesium Beam Maser
- e: Aviation Chief Boatswains Mate
ACBMAG e: Aviation Chief Boatswains Mate; Arresting Gear and Barriers
ACBMCP e: Aviation Chief Boatswains Mate; Catapult
ACBMGA e: Aviation Chief Boatswains Mate; Gasoline System
ACBMPH e: Aviation Chief Boatswains Mate; Plane Handler
ACBNM e: Acute Care Bed Need Methodology
ACBNZ e: Associated Credit Bureau of New Zealand
ACBO e: Assistant Counterbattery Officer
- e: Association of Chief Business Officials
ACB of A e: Associated Credit Bureaus of America
ACBP e: Alternating Current; Beam [Voltage] Pushing
ACBR e: Accumulator Buffer Register
ACBRA e: American CB Radio Association
ACBS e: Accrediting Commission for Business Schools
ACBs e: Adapter Control Blocks
ACBS e: Alert Crew Billet Security
ACBSI e: Associate of the Chartered Building Societies Institute
ACB Signal e: Access Barred Signal
ACBT e: Automatic Circuit Board Tester
ACBWS e: Automatic Chemical Biological Warning System
Ac C d: Aktinium C
ACC e: Academic Computation Center
- e: Academic Computer Center
- e: Academic Computing Center
- e: Acceleration
- e: Accept[ance]
- e: Acceptable Container Condition
- e: Access
- e: Accessories
- e: Accessory
- e: Accident
- e: Accommodation
- e: Accompanied
- e: Account
- e: Accounting Classification Code
- e: Accra, Ghana (airport)
- e: Acetylcysteine
- e: Acid Copper Chromate
- e: Acinic Cell Carcinoma
- e: Acknowledge Control
- e: Action Change Card
- e: Adapter Common Card
- e: Adaptive Control [with] Constraint[s]
- e: Adenoid Cystic Carcinoma
- e: Adjustable Chain Clutch

- e: Administrative Committee on Coordination
- e: Admiralty Corrosion Committee
- e: Adrenocortical Carcinoma
- e: Advanced Carbon-Carbon
- e: Advanced Communications Course
- e: Advanced Computer Communications
- e: Advanced Concepts Center
- e: Aerial Control Console
- e: Aeroplane Construction Committee
- e: Agricultural Credit Corporation
- e: Airborne Control Computer
- e: Air Center Commander
- e: Air Chief Commandant
- e: Air Control Center
- e: Air Control Commission
- e: Air Coordinating Committee
- e: Aircraft Carrier
- e: Air Crew Change
- e: Airport Consultants Council
- e: Airspace Control Center
- e: Alarm Control Center
- e: Allied Chemical Corporation
- e: Allied Control Center
- e: Allied Control Commission
- e: Allied Control Council
- e: Alpena Community College
- e: Alternating Current Circuit
- e: Alternating Current Coil
- e: Alternating Current Conductivity
- e: Altocumulus Castellatus Clouds
- e: Alveolar Cell Carcinoma
- e: Amateur Computer Club
- e: Ambulatory Care Center
- e: Ambulatory Care Clinic
- e: American Chamber of Commerce
- e: American College of Cardiology
- e: American College of Chemosurgery
- e: American College of Cryosurgery
- e: American Copper Council
- e: American Craft Council
- e: American Crystallographic Community
- e: American Cyanamid Company
- e: Aminocyclopropane-Carboxylic Acid
- e: Amphibious Command Car
- e: Analytical Calibration Curve
- e: Annual Capital Charge
- e: Annual Contributions Contract
- e: Anodal Closing Contraction
- e: Anodal Closure Clonus
- e: Anodal Closure Contraction
- e: Antarctic Circumpolar Current
- e: Antenna Control Console
- e: Antibody-Containing Cell
- e: Aortic Cross Clamping
- e: Application Control Code
- e: Approach Control Center
- e: Approved Capital Costs
- e: Area Control Center
- e: Area Coordination Center
- e: Area Coverage Change
- e: Argonne Code Center
- e: Armament Control Computer
- e: Armored Column Cover
- e: Armored Combat Command
- e: Army Catering Corps
- e: Army Chemical Center
- e: Army Chemical Corps
- e: Army Commanders Conference
- e: Army Communications Command
- e: Army Component Command
- e: Army Cooperation Command
- e: Army Corps Commander
- e: Artillery Control Console
- e: Asbestos Claims Council
- e: Asbestos Compensation Coalition
- e: Ashland Community College
- e: Asian Cultural Council
- e: Assault Crisis Center
- e: Associated Communications Corporation
- e: Associated Computer Consultants
- e: Association of Chiropractic Colleges
- e: Association of Computer Clubs
- e: Association of Computer Consultants
- e: Astronaut Control Console
- e: Asynchronous Communications Control[ler]
- e: Atlantic Christian College
- e: Atlantic Coast Conference
- e: Attack Control Concept
- e: Attack Control Console
- e: Auburn Community College
- e: Audio Control Center
- e: Aural Comprehension Course
- e: Australasian Corrosion Centre
- e: Australian Chiropody Council
- e: Australian Computer Conference
- e: Automatic Camera Control
- e: Automatic Carrier Control
- e: Automatic Chroma Circuit
- e: Automatic Chroma Control
- e: Automatic Chrominance Control
- e: Automatic Claiming and Cancelling
- e: Automatic Climate Control
- e: Automatic Colo[u]r Control
- e: Automatic Combustion Control
- e: Automatic Computer Control
- e: Automatic Console Control
- e: Automatic Contrast Control
- e: Automatic Control Center
- e: Automatic Control Certified
- e: Automatic Control Console
- e: Automatic Course Control
- e: Automation Classification Code
- e: Automotive Composities Consortium
- e: Auxiliary Crew Compartment
- e: Average Correlation Coefficient[s]
- e: Aviation Control Center
- i: Airport Consultants Council

acc. e: accelerated
- e: accelerator
- e: accepted
- e: according
- l: acceptus

acc e: accept
- e: accident[al] [ly]
- e: accommodate[d]
- e: accompaniment
- e: accompanist
- e: account[ed]
- e: accounting
- e: accumulator
- e: astronomical great circle course
- l: altocumulus castellatus

Acc e: Account[ant]
- e: Accumulate
- e: Accusative
- e: Additive Card Code

Ac C e: Actinium C

A-C-C e: Appleton-Century-Crofts

ACCA e: Accelerated Capital Cost Allowance
- e: Aeronautical Chamber of Commerce of America
- e: Agricultural Central Cooperation Association
- e: Air Conditioning Contractors of America
- e: American Clinical and Climatological Association
- e: American College of Cardiovascular Administrators
- e: American College of Clinic Administrators
- e: American Cotton Cooperative Association
- e: Antenna Counterbalance Cylinder Assembly
- e: Associated Colleges of the Chicago Area
- e: Asynchronous Communications Control Attachment

ACC-ACO e: Army Communications Command Advanced Concepts Office

Accad i: Accadèmia

Accad di Mil i: Accadèmici di Milano

Accad Nav i: Accadèmia Navale

Accad Sta Cec i: Accadèmia di Sunta Cecilia

ACCAP e: Autocoder to COBOL (Common Business Oriented Language) Conversion Aid Program[me]

Ac cast e: Altocummulus Castellatus

ACCAT e: Advanced Command and Control Architectural Testbed

Acc & Aud e: Accountant and Auditor

ACCB e: Aircraft Change Control Board
- e: Aircraft Configuration Change Board
- e: Aircraft Configuration Control Board
- e: Airframe Change Control Board
- e: Alternating Current Circuit Breaker
- e: Atlas Configuration Control Board

ACCBT e: Aircrew Classification Battery Test

Acc Bty e: Accompanying Battery

ACCC e: Ad-hoc Committee for Competitive Communications
- e: Advisory Council on College Chemistry
- e: Alternate Central Computer Complex
- e: Alternate Command and Control Center
- e: Area Chemist Contractors Committee
- e: Association of Canadian Community Colleges
- e: Association of Child Care Centres
- e: Association of Community Cancer Centers

ACCCE e: Association of Consulting Chemists and Chemical Engineers

Acc Chem Res e: Accounts of Chemical Research

AC & CCI e: American Coke and Coal Chemicals Institute

AC&CCI e: American Coke and Coal Chemicals Institute

ACC/COM e: Air Coordinating Committee Communications Subcommittee

ACC/COP e: According to the Custom of the Port

ACCD e: Accelerated Construction Completion Date
- e: Accident
- e: Advance Corporate Contract Directive
- e: Aerospace Communication and Controls Division
- e: Aircraft Compatibility Control Drawing

ACCDCE e: Accordance

acc dec e: acceptable deception[s]-half truth[s]

acc. def. e: accessory defenses

Acc Dis e: Accidental Discharge

ACCE e: Acceptance
- e: American College for Continuing Education
- e: American Council for Construction Education

ACCEC e: Australian Chamber of Commerce Export Council

ACCEd e: Associate of the College of Craft Education
AC-CEF e: Allis-Chalmers Critical Experimental Facility
ACCEL e: Accelerated
- e: Accelerometer
- e: Automated Circuit Card Etching Layout
accel e: accelerate[d]
- e: accelerating
- e: acceleration
- e: accelerator
- i: accelerando
ACCELS e: Automated-Circuit Card-Etching Layout System
- e: Automated Circuit Etching Layout System
ACCEM e: Advanced Composite Cost Estimating Model
ACCENT System e: Autogenetically-Controlled Cesium Electro-Nuclear Thrust System
accepon f: acceptation
ACCEPT e: Access Electronic Payment Terminals
- e: Automated Cargo Clearance and Enforcement Processing Technique
ACCES e: Associated Computer Consultants Exchange System
- e: Army Civilian Career Evaluation System
ACCESS e: Access Characteristics Estimation System
- e: A Complete Computerized Examination System
- e: Action Center for Educational Service and Scholarships
- e: Afloat Consumption Cost and Effectiveness Surveillance System
- e: Air Canada Cargo Enquiry and Service System
- e: Aircraft Communication Control and Electronic Signalling System
- e: Aircraft Communication Electronic Signalling System
- e: American College of Cardiology Extended Study Services
- e: Applications of Cost-Saving Concepts through Equipment and System Salvage
- e: Architects Central Constructional Engineering Surveying Service
- e: Argone Code Center Exchange and Storage System
- e: Army Commissary Computer Entry Store System
- e: Assembly Concept for Construction of Erectable Space Structures
- e: Australian Contribution to the Cost of Education for Students Scheme
- e: Automated Catalog[ue] of Computer Equipment and Software Systems
- e: Automated Communications Electronic Service System
- e: Automated Computer Communications Electronic Service System
- e: Automated Computer Controlled Editing Sound System
- e: Automated Control and Checking of Electrical Systems Support
- e: Automatic Card Control Entrance Security System
- e: Automatic Central Communications Electronic Switching System
- e: Automatic Checking and Control for Electrical Systems Support
- e: Automatic Computer-Controlled Electronic Scanning Set
- e: Automatic Computer-Controlled Electronic Scanning System
- e: Automatic Crane Control Storage System
access e: accessories
Access e: Accessory
A.C.C.E.S.S. e: A Cooperative Community Educational School System
ACCF e: Area Communications Control Function
- e: Association of Community College Facilities
ACC/FWC e: Automatic Combustion Control and Feedwater Control
ACCG e: Accumulating
- e: American Committee for Crystal Growth
ACCGAT e: Accumulator Gating
ACCGS e: Air Cadet Central Gliding School
AcCh e: Acetylcholine
ACCH e: American College of Clinical Hypnosis
- e: Association for the Care of Childrens Health
- e: Association for the Care of Children in Hospitals
ACCHAN e: Allied Command Channel
AcChR e: Acetylcholine Receptor
Acc. Hrs. e: Accumulative Hours
ACCI e: Accident[al] Injury
- e: Adult Career Concerns Inventory
- e: American Corporate Counsel Institute
- e: American Cottage Cheese Institute
- e: Association of Crafts and Creative Industries
Accid e: Accident[al]
accis e: accismus
ACCIS e: Account Current Company Information System
- e: Advisory Committee for the Coordination of Information
- e: Agricultural Chemical Computerized Information System
- e: Air Command and Control Improvement System
ACCL e: American College of Computer Lawyers
- e: American Council of Commercial Laboratories
- e: Anodal Closure Clonus
- e: Army Coating and Chemical Laboratory
ACCLAIM e: Automated Circuit Card Layout and Implementation
ACCLAIMS e: Army Communications Security Commodity, Logistical and Accounting Management System
ACCLRM e: Accelerometer
ACCM e: Acoustic Counter-Countermeasures
- e: Advanced Concept Cost Model
- e: Advisory Council on Calibration and Measurement
- e: American College of Clinic Managers
- e: Asynchronous Communications Control Module
aCcm e: Abcoulomb Centimeter
aCcm² e: Abcoulomb per Square Centimeter
aCcm³ e: Abcoulomb per Cubic Centimeter
ACC-MD e: Accumulator/Multiplicand Divisor
ACCME e: Accreditation Council for Continuing Medical Education
ACC/MET e: Air Coordinating Committee Meteorological Subcommittee
ACCMIS e: Army Command and Control Management Information System
ACCML e: Army Chemical Corps Medical Laboratories
ACCM P-H e: Appleton-Century-Crofts Medical
ACCMS e: American Center for Chinese Medical Sciences
ACCN e: Accommodation
- e: Arms Control Computer Network
- e: Audit Central Control Network
Accn e: Accumulation
ACCNET e: Army Command and Control Network
ACCNT e: Accounting
ACCO e: American College of Chiropractic Orthopedists
- e: American Cyanamid Company
- e: Associated Christian Colleges of Oregon
- e: Association of Child Care Officers
acco i: accompagnamento
AcCoA e: Acetyl Coenzyme A
Acc of F e: Accuracy of Fire
ACCOM e: Accommodate
- e: Accompany
- e: Aircraft Communicator
accom e: accommodation
- e: accompaniment
accom ad lib e: accompaniment ad libitum
ACCOMMODON e: Accommodation
accom oblto e: accompaniment obligato
ACCOMP e: Academic Computer Group
accomp e: accompaniment
- e: accomplish[ed]
ACCOMPG e: Academic Computer Group
ACCOMP Group e: Academic Computer Group
ACCOMPL e: Accomplish
ACCON e: Acoustic Control
ACCOR e: Australian Coal Corporation
ACCORD e: According
- e: Analog[ue] Computer Checkout Routine Digitally
- e: Army Computer Capabilities Online Repository and Disseminator
- e: Association for Co-Ordinated Rural Development
ACCORDCE e: Accordance
ACCORDG e: According
ACCORDS e: Acoustic Correlation and Detection System
ACCOS e: Automatic Computer Calculation of Optical Systems
ACCOY e: Accompany
ACCP e: Alternating-Current Current Pushing
- e: American College of Chest Physicians
- e: American College of Clinical Pharmacology
- e: American College of Clinical Pharmacy
- e: American Council on Chiropractic Physiotherapy
- e: Army Correspondence Course Program[me]
ACCPA e: Army Chemical Center Procurement Agency
ACCPAC e: Accounting Package
Accp App e: Accepted Applicants
Accpce e: Acceptance
accpt e: accept[ance]
accpy e: accompany
AccR e: Accelerating Relay
ACCR e: Accumulator
- e: Accurue[d]
- e: American Council on Chiropractic Roentgenography
- e: American Council on Chiropractic Roentgenology
- e: Annual Cost of Capital Recovery
ACCRA e: Australian Chart and Code for Rural Accounting
accrd int e: accrued interest

Acc Rec e: Accepted Record
ACCRED e: Accredited
accres i: accrescendo or accrescimento
ACCRY e: Accessory
- e: Accuracy
ACCS e: Access
- e: Advanced Communications Control System
- e: Aerospace Command and Control System
- e: Afloat Command and Control System
- e: Airborne Command and Control System
- e: Airborne Command Control Squadron
- e: Air Command and Control System
- e: Aircraft Communications System
- e: Altitude Coordinate Converter System
- e: Amphibious Command and Control System
- e: Armored Crashworthy Crew Set
- e: Army Command and Control Study
- e: Army Command and Control System
- e: Associate of the Cooperation of Certified Secretaries
- e: Attitude Coordinate Converter System
- e: Automated Calibrated Control System
- e: Automated Calibration Control System
- e: Automated Circulation Control System
- e: Automated Command and Control System
- e: Automated Communications and Control System
- e: Automatic Calling Card Service
- e: Automatic Checkout and Control System
ACCSA e: Allied Communications and Computer Security Agency
ACC/SCN e: Administrative Committee on Coordination - Subcommittee on Nutrition
ACCSIC e: Atomic Collision Cross Sections Information Center
ACCSL e: Accessorial
ACCSq e: Airborne Command Control Squadron
ACCT e: Access Control
- e: Access Time
- e: Accident
- e: Account
- e: Accountancy
- e: Ad-hoc Committee for Competitive Telecommunications
- e: Alliance for Coal and Competitive Transportation
- e: American Council for Competitive Telecommunications
- e: Association of Community College Trustees
acct e: accountant
- e: accounting
ACCTC e: Army Catering Corps Training Centre
ACCTCY e: Accountancy
ACCTD e: Accented
acctg e: accounting
acctg sp e: accounting specialist
ACCTID e: Account Identifier
Acc Tks e: Accompanying Tanks
acctncy e: accountancy
ACCTNG e: Accounting
acctnt e: accountant
ACCTS e: Accounts
accts pay e: accounts payable
accts rec e: accounts receivable
ACCT STR e: Accountable Strength
ACCTVS e: Association of Closed Circuit Television Surveyors
ACCU e: Accumulator
- e: Association of Catholic Colleges and Universities
- e: Audio Central Control Unit
- e: Award Central Control Unit
Accu d: Akkumulator
a-c cu e: alternating-current control unit
accu e: automatic combustion-control unmanned
ACCUGRAF e: Accurate Graphics
accum e: accumulate
- e: accumulator
ACCUMULON e: Accumulation
ACCUR e: Accurate
accur l: accuratissime
ACCUS e: Accusative
ACCV e: Alternating Current Control Voltage
ACCV e: Armored Cavalry Cannon Vehicle
ACCW e: Alternating-Current Continuous Wave[s]
ACCWM e: Air Cleaner Cold Weather Modulator
ACCWS e: Assistant Chief, Chemical Warfare Service
- e: Auxiliary Component Cooling Water System
accy e: Accessory
ACCY e: Accountancy
- e: Accuracy
ACD d: allgemeine Computer-Datenbank
- d: Antikoagulierende Citrat-Dextrolösung
- e: Absolute Cardiac Dullness
- e: Access Control Document
- e: Accessory Control Document
- e: Accordion
- e: Acid-Citrate Dextrose
- e: Action-Chart Diagramer
- e: Adapter Control Detector
- e: Addressed Cable Delivery
- e: Adjustable Center Distance
- e: Administrative Commitment Document
- e: Administrative Contract Document
- e: Admiralty Chart Datum
- e: Advanced Circuit Design
- e: Advanced Copies Delivered
- e: Aerial Control Display
- e: Aerodynamic Configuration Drivers
- e: Air Condensate Drain
- e: Aircraft Damage
- e: Aircraft Depot
- e: Alarm Control & Display
- e: Allergic Contact Dermatitis
- e: Allied Chemical Corporation
- e: Allied Civil Defense
- e: Alphabetic Coding Device
- e: Alphabetic Collating Device
- e: Alternating Current Dialling
- e: Alternating Current Dump
- e: American Center for Design
- e: American College Dictionary
- e: American College of Dentists
- e: Analysis and Computation Division
- e: Antenna Control Display
- e: Anterior Chest Diameter
- e: Anticoagulant Citrate Dextrose
- e: Arms Control and Disarmament
- e: Army Communications Division
- e: Army Corps Depot
- e: Assembler Card Deck
- e: Assistant Command Director
- e: Associative Computer Device
- e: Asymptotic Conical Dipole
- e: Attack Center Display
- e: Australian College of Dentistry
- e: Australian College of Dermatologists
- e: Authorized Computer Distributors
- e: Automatic Call Distribution
- e: Automatic Call Distributor
- e: Automatic Chart Display
- e: Automatic Circuit Design
- e: Automatic Contour Digitizer
- e: Automatic Control Distribution
- e: Aviation Commission Date
- l: Archaeologiae Christianae Doctor
Ac D d: Aktinium D
AcD e: Actinium D
acd e: accord
- e: acid-citrate-dextorse
- e: active duty commitment
- e: arms control through defense
- e: axiodistocervical
ACDA e: American Component Dealers Association
- e: Aviation Combat Development Agency
ACDAR e: Acoustic Detection and Ranging
ACDB d: Autoclub des Deutschen Beamtenbundes
- e: Airport Characteristics Data Bank
- e: Army Corporate Database
ACDC e: Administrative Communications Distribution Center
- e: Advanced Coherent Deception Countermeasure
- e: Army Combat Development[s] Committee
- e: Australian Counter Disaster College
A[.]C[.] / D[.]C[.] e: Alternating Current / Direct Current
ACDCM e: Archbishop of Canterburys Diploma in Church Music
ACDCR e: Alternating Current / Direct Current Receiver
ACDCS e: Automated Classified Document Control System
ACDCT e: Alternating Current / Direct Current Tube
ACDCV e: Alternating Current / Direct Current Valve
ACDDS e: Advanced Cartographic Data Digitizing System
ACDEC e: Army Combat Development Experimentation Center
ACD-ESS e: Automatic Call Distributor - Electronic Switching System
ACDI e: Agricultural Cooperative Development International
- e: Asynchronous Communications Device Interface
ACDIAL e: Alternating Current Dialling
ac dis e: academic dismissal
ACDIS e: Association for Development of Computer-based Instructional Systems
ACDL e: Asynchronous Circuit-Design Language
ACDM e: Assessment of Career Decision Making
- e: Association of Chairmen of Departments of Mechanics
- e: Axial Vector Dominance Model
ACDMS e: Automated Control of [a] Document Management System
Acdnt e: Accident
ACDP e: Antenna Control and Display Panel
- e: Association of Compact Disk Publishers
- e: Automatic Centralized Data Processing
ACDPI e: American Cultured Dairy Products Institute
- e: Association for a Competitive Data Processing Industry
ACDPS e: Automated Cartographic, Drafting and Photogrammetric System
ACdr. e: Air Commodore
ACDR e: Alpha-Cedrene

ACDRB e: Active Contract Data Review Board
ACDRU e: Arms Control and Disarmament Research Unit
ACDS e: Accept Data State
- e: Advanced Combat Direction System
- e: Advanced Command Data System
- e: Advisory Committee on Dangerous Substances
- e: Anglo-Continental Dental Society
- e: Attitude Control and Determination Subsystem
- e: Australian College of Dental Surgeons
- e: Automated Cargo Document System
- e: Automatic Call Distribution System
- e: Automatic Comprehensive Display System
- e: Automatic Computer Display System
ACD System e: Automatic Call Distribution System
Acdt e: Accident
ACDT e: Accumulated Data Tape
- e: Alternating Current Discharge Tube
ACDTR e: Airborne Central Data Tape Recorder
acdu e: active duty
ACDUBLI e: Active Duty Obligation
ACDUINS e: Active Duty Under Instruction
AC Dump e: Alternating Current Dump
ACDUOBLI e: Active Duty Obligation
ACDUTRA e: Active Duty for Training
ACE d: adrenocorticaler Extrakt
- d: Auto-Club Europa
- e: Absorption of Conversion Electrons
- e: Academic Courseware Exchange
- e: Accelerated Cathode Excitation
- e: Accelerated Co-Pilot Enrichment
- e: Acceptance Checkout Equipment
- e: Accounting, Cost, Estimating
- e: Acetic
- e: Acknowledge Enable
- e: Activity Civil Engineer
- e: Adaptive Computer Experiment
- e: Add Command Execution
- e: Adrenal Cortical Extract
- e: Advanced Clean Emission
- e: Advanced Compilation Equipment
- e: Advanced Compound Engine
- e: Advanced Computational Element
- e: Advanced Computing Environment
- e: Advanced Control Experiment
- e: Advanced Custom Emitter Coupled Logic
- e: Adverse Channel Enhancement
- e: Advisory Centre for Education
- e: Aerial Combat Evaluator
- e: Aerosol Climatic Effects
- e: Aerospace Contract Engineers
- e: After the Christian Era
- e: Agricultural Communicators in Education
- e: Airborne Communications and Electronics
- e: Airborne Cooperational Equipment
- e: Air Cargo Exhibition
- e: Air Collection and Enrichment
- e: Air Combat Element
- e: Air Combat Emulator
- e: Air Combat Engagement
- e: Air Conditioning Equipment
- e: Aircraft Catering Equipment
- e: Aircraft Condition Evaluation
- e: Air Crash Equipment
- e: Air Crew Error
- e: Air Cushion Equipment
- e: Airspace Control Element
- e: Alcohol-Chloroform-Ether
- e: Allied Command; Europe
- e: Alternate Command Elements
- e: Alternating Current Electrocoagulation
- e: Alternating Current Energy
- e: Alternating Current Engineering
- e: Altimeter Control Equipment
- e: Altitude Control Electronics
- e: Altitude Control Equipment
- e: Ambush Communication Equipment
- e: American Chemical Exchange
- e: American College of Ecology
- e: American College of Epidemiology
- e: American Council on Education
- e: American Council on Exercise
- e: American Council on the Environment
- e: Analytic Comnputer Equipment
- e: Analytic Computer Equipment
- e: Angiotensin Converting Enzyme
- e: Animated Computer Education
- e: Area Control Error
- e: Armored Combat Earthmover
- e: Army Certificate of Education
- e: Army Combat Engineers
- e: Army Corps of Engineers
- e: Assessment of Combat Effectiveness
- e: Assistant Chief Engineer
- e: Assistant Chief of Engineers
- e: Associate Credit Executive
- e: Associated Corpuscular Emission
- e: Associate of the College of Engineering
- e: Association for the Conservation of Energy
- e: Association of Collegiate Entrepreneurs
- e: Association of Communication Engineers
- e: Association of Computer Educators
- e: Association of Conservation Engineers
- e: Association of Consulting Engineers
- e: Asynchronous Element
- e: Atlantic City Electric [Company]
- e: Atmospheric Control Experimentation
- e: Attitude Control Electronics
- e: Audio-Connecting Equipment
- e: Audit, Control and Evaluation
- e: Australian College of Education
- e: Australian College of English
- e: Automated Computing Engine
- e: Automated Cost Estimates
- e: Automated Cost Estimating
- e: Automated Credit Enquiry
- e: Automatic Cable Equalization
- e: Automatic Calling Encoder
- e: Automatic Calling Equipment
- e: Automatic Card Ejection
- e: Automatic Circuit Exchange
- e: Automatic Clutter Eliminator
ace e: adjusted current earnings
- e: aerospace control environment
- e: alcohol-chloroformether
- e: automatic checkout equipment
- e: automatic computer evaluation
- e: automatic computing engine
- e: automatic computing equipment
- e: automatic continuous evaporation
- e: automatic control equipment
- e: automatic controlled exposure
- e: automatic control of execution
- e: autumn circulation experiment
- e: auxiliary control element
- e: auxiliary conversion equipment
- e: avenger control electronics
- e: average cumulative error
- e: aviation and computer enthusiasts
- e: aviation construction engineer[s]
- e: awards for cablecasting excellence
ace. e: angiotensin converting enzyme
A.C.E. e: Automatic Continuity Equipment
ACEA e: Air Line Communication Employees Association
- e: Assistant Civil Engineer Adviser
- e: Association of Consulting Engineers Australia
- e: Australian Coal Exporters Association
ACEAA e: Advisory Committee on Electrical Appliances and Accessories
ACE-ACCIS e: Allied Command Europe Automated Command Control and Information System
ACEARTS e: Airborne Countermeasures Environment and Radar Target Simulation
- e: Airborne Countermeasures Environment and Radar Target Simulator
ACEB e: Army Clothing and Equipment Board
- f: Association Canadienne des Écoles Bibliothecaires
ACEBD e: Airborne and Communications Electronics Board
Ace Bks e: Ace Books
ACEC e: Ada Compiler Evaluation Criteria
- e: Advisory Council for Energy Conservation
- e: American Consulting Engineers Council
- e: Area of Critical Environmental Concern
- e: Army Communication[s] and Electronic[s] Command
- e: Association of Consulting Engineers of Canada
- e: Australian Computer Education Conference
- f: Ateliers de Constructions Électriques de Charleroi
AC & EC e: Army Communications and Equipment Coordination
ACED e: Advanced Communications Equipment Depot
- e: Aerospace Crew Equipment Development
AC Ed e: Associate in Commercial Development
- e: Associate in Commercial Education
ACEE e: Adult Cost per Entered Employment
- e: Air Combat Engagement Experiment
- e: Aircraft Emission Estimator
- e: Aircraft Energy Efficiency
ACEEE e: American Council for an Energy Efficient Economy
ACEH e: Acid Cholesteryl Ester Hydrolase
ACEHSA e: Accrediting Commission on Education for Health Services Administration
acei e: angiotensin converting enzyme inhibitor
ACEI e: Association of Consulting Engineers of Ireland
ACEID e: Asian Centre of Educational Innovation for Development
ACEJ e: American Council on Education for Journalism
ACEL e: Aerospace Crew Equipment Laboratory
- e: Air Crew Equipment Laboratory
ACELF f: Association Canadienne des Educateurs de Langue Française
ACEM e: Actinium Emanation
- e: Aviation Chief Electricians Mate
a cemb i: a cembalo
ACEMIS e: Army Communications and Electronics Management Information System
- e: Automated Communications and Electronics Management Information System

ACEN e: Access Enable
ACENET e: Allied Command Europe Communications Network
ACENZ i: Association of Consultant Engineers of New Zealand
ACE/OCS e: A Cost Estimating/Order Control System
ACEORP e: Automotive and Construction Equipment Overhaul and Repair Plant
ACEP e: Advisory Committee on Export Policy
- e: American College of Emergency Physicians
- e: Arms Control Education Project
ACEPD e: Automotive and Construction Equipment Parts Depot
ACEQ e: Association of Consulting Engineers of Québec
ACER e: Australian Council for Educational Research
ACERA e: Air Carrier Economic Regulation Act
ACERC e: Army Coastal Engineering Research Center
ACEREP e: Allied Command Europe Report
ACERP e: Advanced Communications-Electronics Requirements Plan
ACERS e: Allied Command Europe Reporting System
ACerS e: American Ceramic Society
ACES e: Acceptance Checkout and Evaluation System
- e: Acceptance Control Equipment System
- e: Access
- e: Acoustic Containerless Experiment System
- e: Advanced Concept Ejection Seat
- e: Advanced Concept Escape System
- e: Advisory Council on Education Statistics
- e: Air Carrier Engineering Service
- e: Air Collection & Enrichment System
- e: Air Collection Engine System
- e: Air Combat Expert Simulation
- e: Annual Cycle Energy System
- e: Anti-Submarine Composite Engineering Squadron
- e: Army Communications Electronics School
- e: Army Communications Equipment Support
- e: Army Continuing Education System
- e: Assurance Control Economics System
- e: Australian Council for Educational Standards
- e: Automated Camera Effects System
- e: Automated Circulation and Enquiry System
- e: Automated Code Evaluation System
- e: Automated Reliability and Maintenance Management System Control Executiv
- e: Automatically Controlled Electrical System
- e: Automatic Checkout and Evaluation System
- e: Automatic Checkout Equipment Sequencer
- e: Automatic Control Evaluation Simulator
ACESA e: Arizona Council of Engineering and Scientific Associations
- e: Australian Commonwealth Engineering Standards Association
- e: Australian Computer Equipment Suppliers Association
ACESC e: Acceptance Checkout Equipment Spacecraft
ACESIA e: American Council for Elementary School Industrial Arts
ACESIS e: Army Corps of Engineers Socioeconomic Information System
ACESS e: Accessory
- e: Advanced Cabin Entertainment and Services System
- e: Advanced Computer Environmental Systems Support, Inc.
ACET e: Acetate
- e: Acetone
- e: Advisory Committee on Electronics and Telecommunications
- e: Australian Council for Education through Technology
- e: Automatic Cancellation of Exended Targets
- f: Association Canadienne des Entreprises de Telecommunications
Acet e: Acetylene
acet a e: acetic acid
ACETA e: Australian Commercial and Economics Teachers Association
ACE Technique e: Advanced Custom Emitter Coupled Logic Technique
ACE Test e: American Council on Education Test
acetl e: acetylene
ACETS e: Air-Cushion Equipment Transportation System
AcetV d: Acetylenverordnung
acetyl-co e: acetyl-coenzyme A
ACEU e: Aerocontrol Electronics Unit
ACEVAL e: Air Combat Evaluation
ACF d: Azetatfaser
- e: Abbreviated Cost Form
- e: Absolute Cumulative Frequency
- e: Absorption Colo[u]r Filter
- e: Academic Computer Facility
- e: Accelerated Card Feed[ing]
- e: Access Control Facility
- e: Access Control Field
- e: Access Cost Factor
- e: Accessory Clinical Findings
- e: Account Card Feed[ing]
- e: Acid Concentrator Feed
- e: Active Contract File
- e: Acute Care Facility
- e: Address Census File
- e: Address Code Format
- e: Advanced Communication[s] Function
- e: Agricultural Cooperative Federation
- e: Air Combat Fighter
- e: Air Cushion Ferry
- e: Alternate Command Facility
- e: Alternate Communication[s] Facility
- e: Alternating Current Field
- e: American Car and Foundry
- e: American Coal Foundation
- e: Analog[ue] Computer Facility
- e: Area Computing Facility
- e: Area Coverage File
- e: Army Council Form
- e: Association of Consulting Foresters
- e: Australian Conservation Foundation
- e: Authorization Control Facility
- e: Auto-Correlation Function
- e: Automatic Card Feeding
- e: Automatic Control Features
- e: Automatic Processing Equipment Control Facility
- e: Availability Control File
- e: Axisymmetrical Conical Flow
- f: Académie Canadienne Française
- f: Automobile-Club de France
A.C.F. e: Automatic Cop Feeder
ACFA e: Advanced Complementary [-symmetry] Metal Oxide Semiconductor Frame Aligne
- e: Australian Cystic Fibrosis Association
ACFAS f: Association Canadienne-Française pour t'Avancement des Sciences
ACFAT e: Aircraft Carrier Firefighting Assistance Team
ACFB e: Australian Canned Fruits Board
ACFC e: Aviation Chief Fire Controlman
ACFD e: Account Card Feed[ing] Device
- e: Association of Canadian Faculties of Dentistry
ACFE e: American Car & Foundry Electronics
ACFEA e: Air Carrier Flight Engineers Association
ACFEL d: Alternating Current-Film-schicht-Elektroluminiszenz
- e: Arctic Construction and Frost Effects Laboratory
ACFF e: Affinity Cross-Flow Filtration
- e: Alternating Current Flip-Flip
ACFG e: Automatic Continuous Function Generation
- e: Automatic Continuous Function Generator
ACFHE e: Association of Colleges for Further and Higher Education
ACFI e: Advisory Committee on Flight Information
- e: American Car and Foundry Industries
ACFL e: Access Floor
Ac flocc e: Altocummulus floccus
ACFM e: Actual Cubic Feet per Minute
- e: Association of Cereal Food Manufacturers
ACFMR e: Australian Cancer Fund for Medical Research
ACF/NCP e: Advanced Communication Function for Network Control Program[me]
ACFNY e: Asthmatic Childrens Foundation of New York
ACFO e: Alternating Current Fan-Out
- e: American College of Foot Orthopedics
- e: American College of Foot Orthopedists
- e: Assistant Chief Fire Officer
AC foam e: Ambient Cure Foam
ACFP e: Advanced Computer Flight Plan
ACFR e: American College of Foot Roentgenologists
ACFS e: Alternating Current Field Strength
- e: American College of Foot Specialists
- e: American College of Foot Surgeons
- e: Australian Centre for Foreign Studies
ACFT e: Aircraft Flying Training
ac ft e: acre feet
- e: acre foot
acft e: aircraft
Acft Asgmt O e: Aircraft Assignment Order
Acft C e: Aircraft Carrier
Acft CrmnBad e: Aircraft Crewman Badge
acft insp e: Aircraft Inspector
acft instr mech e: Aircraft Instrument Mechanic
Acft Obsr Tng Wg e: Aircraft Observer Training Wing
acft str e: Aircraft Strength
ACF/VTAM e: Advanced Control Function/ Virtual Telecommunication Access Method
ACG e: Activated Charcoal Granules
- e: Acycloguanosine
- e: Adaptive Control Geometrical
- e: Address Coding Guide
- e: Adjacent Charging Group
- e: Advanced Concepts Group
- e: Air-Core Gauge

- e: Airborne Coordinating Group
- e: Air Cargo Glider
- e: Airlines Carriers of Goods
- e: Alpha Control Guidance
- e: Alternating Current Generator
- e: American College of Gastroenterology
- e: Angio Cardiographie
- e: Angle Closure Glaucoma
- e: Apexcardiogram
- e: Apexcardiograph[y]
- e: Area Coordination Group
- e: Assistant Chaplain General
- e: Assistant Commissary General
- e: Auto Car Guard
- e: Automatic Caution Guard
- e: Automatic Character Generation
- e: Automatic Code Generation
- e: Automatic Code Generator
- e: Automatic Control Gear
- e: Automatic Correlation Guidance
- e: Automotive Component Group

AcG e: Accelerator Globolin
Ac-G e: Accelerator Globulin
acg e: apex cardiogram
ACGA e: American Cranberry Growers Association
- e: American Cut Glass Association
- e: Australasian Computer Graphics Association

ACGB e: Aircraft Corporation of Great Britain
ACGBI e: Automobile Club of Great Britain and Ireland
ACGC e: American Consolidated Gold Corporation
- e: Annual Computer Graphics Conference

ACGF e: Australian Citrus Growers Federation
- e: Autocovariance Generating Function

ACGG e: Associate Committee of Geodesy and Geophysics
ACGIH e: American Conference of Gonvernmental Industrial Hygienists
ACGL e: Associate of City & Guilds of London
ACGM e: Advisory Committee on Genetic Manipulation
- e: Aircraft Carrier General Memorandum

ACGME e: Accreditation Council for Graduate Medical Education
ACGP e: American College of General Practice
ACGp e: Area Coordination Group
ACGPOMS e: American College of General Practitioners in Osteopathic Medicine and Surgery
ACGS e: Aerial Caltographic and Geodetic Squadron
- e: Aerospace Cartographic and Geodetic Service
- e: American-Canadian Genealogical Society
- e: American Council on German Studies
- e: Assistant Chief of the General Staff

ACGSq e: Aerial Cartographic and Geodetic Squadron
ACGSS e: Army Command and General Staff School
ACH d: Abteilungschef
- e: Acknowledge Hold
- e: Adrenal Cortical Hormone
- e: Advanced Chain Home
- e: Aftercoming Head
- e: Air Change per Hour
- e: Air Cleaner Housing
- e: Aircrafthand
- e: Aircraft Hangar
- e: Alternating Current Hum
- e: Alumin[i]um Chlorohydrate
- e: American Center for Homeopathy
- e: Area Combined Headquarters
- e: Arm-Chest-Height
- e: Association for Computers and the Humanities
- e: Attempts per Circuit per Hour
- e: Australian Camp Hospital
- e: Automated Clearing House
- e: Automatic Cartridge Handler

ACh d: Acetylcholin
- e: Automatic Check[ing]

Ach e: Acetylcholine
- e: Analog[ue] channel

ach e: actual obtained achievement
- e: arm, chest, height

ACHA e: American Catholic Historical Association
- e: American College Health Association
- e: American College of Hospital Administrators
- e: American Council of Highway Advertisers

A Channel e: Left Stereo Channel
AC & HBR e: Algoma Central and Hudson Bay Railway
ACHCA e: American College of Health Care Administrators
ACHCR e: American Council for Health Care Reform
ACHE e: Action Committee for Higher Education
- e: Air-Cooled Heat Exchanger
- e: Alabama Commission on Higher Education
- e: American College of Healthcare Executives
- e: American Council of Hypnotist Examiners

AchE e: Acetylcholinesterase
ache. e: acetylcholinesterase
A Chem e: Associate in Chemistry
ACHEMA d: Ausstellung chemischer Apparate
- e: Air Cooled Heat Exchanger Manufacturers Association

A Chem S e: American Chemical Society
AChemS e: Association for Chemoreception Sciences
ACHEX e: Aerosol Characterization Experiment
AC-HF e: Alternating Current-High Frequency
ACHI e: Application Channel Interface
- e: Association for Childbirth at Home; International

achiev e: achievement
ACH-Index e: Arm, Chest, Hip Index
- e: Arm Girth, Chest Depth, Hip Width Index

A Ch J e: American Chemical Journal
ACHL e: American Council for Healthful Living
ACHLIS e: Australian Clearing House for Library and Information Science
AChM e: Acetylcholine Mustard
ACHNHP e: Appomattox Court House National Historical Park
ACHO e: American College of Home Obstetrics
ACHOBS e: Assistant Chief Observer
ACHOD e: Acoustic Helicopter Overflight Detector
ACHOO e: Autosomal-dominant Compelling Helio-Ophthalmic Outburst
ACHP e: Advisory Council for Historic Preservation
- e: Advisory Council on Historic Preservation

ACHPER e: Australian Council for Health, Physical Education and Recreation
ACHPIRST e: Australian Clearinghouse for Publications in Recreation Sport and Tourism
ACHQ e: Area Combined Headquarters
ACHR e: American Catholic Historical Researches (journ.)
AChR e: Acetylcholine Receptor
a. Chr. l: ante Christum
AChRAb e: Acetylcholine Receptor Antibody
a. Chr. n. l: ante Christum natum
Achrom e: Achromatism
ACHS e: Associate of the Society of the Chiropodists
- e: Association of Community Home Schools
- e: Automatic Checkout System

AChS e: American Chemical Society
A Ch S e: Associate of the Society of Chiropodists
ACHSA e: American Correctional Health Services Association
Achskl. d: Achselklappe
Achskm d: Achskilometer
Achsst. d: Achselstück
ACHTR e: Advisory Committee for Humid Tropics Research (UNESCO)
ACHVA e: Air Conditioning, Heating and Ventilating
achvit e: achievement
ACI e: Acid
- e: Acoustic Comfort Index
- e: Actual Cost Incurred
- e: Acute Coronary Insufficiency
- e: Additive Controlling Ignition
- e: Address Compilation Instruction
- e: Address Computation Instruction
- e: Adenylate Clyclase Inhibitor
- e: Adjacent Channel Interference
- e: Adjusted Calving Interval
- e: Adult-Child Interaction [Test]
- e: Advanced Chip Interconnect
- e: Advanced Circuits International
- e: Age Controlled Item
- e: Air-Controlled Interception
- e: Airborne Controlled Intercept[ion]
- e: Air Cargo Incorporated
- e: Air Combat Information
- e: Air Combat Intelligence
- e: Air Commuter, Incorporated
- e: Air Conditioning Institute
- e: Air Control Intercept
- e: Air Council Instruction
- e: Aircraft Condition Inspection
- e: Air Curtain Incinerator
- e: Allocated Configuration Identification
- e: Allocated Configuration Item
- e: Alloy Casting[s] Institute
- e: Alteration Cancellation Installation
- e: Altered Commercial Item
- e: Alternating Current Inductance
- e: Alternating Current Induction
- e: Alternating Current Inductivity
- e: Alternating Current Input[s]
- e: Altitude Command Indicator
- e: American Canvas Institute
- e: American Carpet Institute
- e: American Casting Institute
- e: American Concrete Institute
- e: American Cryogenics Incorporated
- e: Amplifier-Control Intercommunications

- e: Analytical Condition Inspection
- e: Annealed Cast Iron
- e: Anticlonus Index
- e: Army Council Instruction
- e: Assist Card International
- e: Associated Colleges of Indiana
- e: Associate of the Clothing Institute
- e: Associate of the Commerce Institute
- e: Association for Conservation of Information
- e: Association of Commerce and Industry
- e: Asynchronous Communications Interface
- e: Atlas Chemical Industries
- e: Attitude Control[s] Indicator
- e: Audio Cassette Interface
- e: Australian Chemical Institute
- e: Australian Chemistry Institute
- e: Australian Consolidated Industries
- e: Austrian Cultural Institute
- e: Author Comfort Index
- e: Automated Car Identification
- e: Automatic Card Identification
- e: Automatic Car Identification
- e: Automatic Closure and Interlock
- e: Automatic Control Instrumentation
- e: Automation Center International
- f: Association Canadienne de l'Informatique
- f: Association Cartographique Internationale

AC & I e: Acquisition, Construction and Improvement
aci e: adult correctional institution
- e: airborne-controlled interception
- f: assure contre l'incendie

ACIA e: Associated Cooperage Industries of America
- e: Associate of the Catering Institute of Australia
- e: Associate of the Corporation of Insurance Agents
- e: Asynchronous Communication[s] Interface Adapter
- e: Aviation Career Incentive Act

ACIArb e: Associate of the Chartered Institute of Arbitrators
ACIAS e: American Council of Industrial Arts Supervisors
ACIAs e: Asynchronous Communications Interface Adapters
ACIASAO e: American Council of Industrial Arts State Association Officers
ACIATE e: American Council of Industrial Arts Teacher Education
ACIB e: Arithmetic Checker in Bus
- e: Associate of the Corporation of Insurance Brokers

acic e: acicular
ACIC e: Aeronautical Chart[ing] and Information Center
- e: Air Corps Information Circular
- e: Allied Captured Intelligence Center
- e: American Committee for International Conservation
- e: Apollo Contractor Information Center
- e: Associate of Canadian Institute of Chemistry
- e: Australian Chemical Industry Council
- e: Automatic Combat Intelligence Center
- e: Auxiliary Combat Information Center

ACICO e: Assistant Combat Information Center Officer
ACICS e: Australian Consolidated Industries Computer Service
ACICU e: Arkansas Council of Independent Colleges and Universities
ACID e: Acceleration, Cruising, Idling, Deceleration
- e: Aircraft Identification
- e: Association of Canadian Industrial Designers
- e: Attempted Corporate Integration of Dividends
- e: Automatic Circuit Design
- e: Automatic Classification and Interpretation of Data
- l: Acidus

acid e: acidosis
- e: acidulated drop
- e: automatic classification and interpretation ofdata
- e: hallucinogenic drug such as LSD-25

ACID Generation e: Automatic Cross-referencing and Indexing Document Generation
ACID Generator e: Automatic Cross-referencing and Indexing Document Generator
acid phos e: acid phosphatase
acid p'tase e: acid phosphatase
ACIDS e: Aircraft Integrated Data System
ACID Symbols e: Arithmetic, Coding, Information and Digit Symbols
ACIF Test e: Anticomplement Immunofluorescence Test
ACIG e: Advanced Computer Image Generator
ACIGS e: Assistant Chief of the Imperial General Staff
ACIGY e: Advisory Council for the International Geophysical Year
ACII e: Associate of the Chartered Insurance Institute
ACIIW e: American Council of the International Institute of Welding
ACIL e: Aberration-Compensated Input Lens
- e: American Council of Independent Laboratories
- e: Automatic Controlled Instrument Landing

ACILA e: Associate of the Chartered Institute of Loss Adjustors
ACIM e: Accident Cost Indicator Model
- e: Advanced Common Intercept Missile
- e: American Committee on Italian Migration
- e: Auxiliary Computer Input Multiplexer
- e: Availability Centred Inventory Model
- e: Axis-Crossing Interval Meter

ACIMD e: Advanced Common Intercept Missile Demonstration
ACIMS e: Aerial Colour Infrared Management System
- e: Aircraft Component Intensive Management System
- e: Aircraft Component Management System

ACIN e: Australian Curriculum Information Network
ACINF e: Advisory Committee on Irradiated and Novel Foods
ACINT e: Acoustic Intelligence
ACIO e: Air Combat Intelligence Office[r]
ACIOA e: Advisory Committee on International Oceanographic Affairs
ACIOP e: Atlantic Command Intelligence Operating Procedures
ACIP e: Active Certificate Information Program[me]
- e: Advisory Committee on Immunization Practices
- e: Aerodynamic Coefficient Identification Package
- e: Aerodynamic Coefficient Instrumentation Package
- e: American College of International Physicians
- e: Analytical Condition Inspection Program[me]
- e: Attack Center Indicator Panel
- e: Aviation Career Incentive Pay
- e: Avionics Communications and Information Processing

ACIPCO e: American Cast Iron Pipe Company
ACIR e: Adder Center Input Register
- e: Automotive Crash Injury Research
- e: Aviation Crash Injury Research

ACIRC e: Air Circulating
- e: Asian Community Information Resource Centre

ACIRD e: Advisory Committee on Isotopes and Radiation Development
ACIRL e: Australian Coal Industries Research Laboratories
ACIS e: Advanced Credit Information System
- e: Aeronautical Chart and Information Squadron
- e: Africa Church Information Service
- e: Air Cargo Integrated System
- e: Aircraft Crew Interphone System
- e: American Council for International Studies
- e: Arms Control Impact Statement
- e: Army Central Intelligence Service
- e: Army Combat Identification System
- e: Associate of the Chartered Institute of Secretaries
- e: Association for Computing and Information Sciences
- e: Automated Claims Information System
- e: Avionics, Control and Information Systems

ACISQ e: Aeronautical Chart and Information Squadron
ACI System e: Automatic Car Identification System
ACIT e: Air Cannon Impact Tester
ACITS e: Advisory Committee on Information Technology Standardization
ACIU e: Advanced Central Interface Unit
- e: Allied Central Interpretation Unit

ACIV e: Associate of the Commonwealth Institute of Valuers
ACIWLP e: American Committee for International Wild Life Protection
ACJ e: Air Correction Jet
- e: American Council for Judaism
- e: Associate in Criminal Justice
- e: Attitude Control Jet

ACJHSIS e: Arkansas Criminal Justice Highway Safety Information System
ACJP e: Air Correction Jet-Primary
- e: Airways Corporations Joint Pensions

ACJS e: Academy of Criminal Justice Sciences
- e: Air Correction Jet-Secondary

ACK e: Acknowledge[d]
- e: Alternate Coding Key
- e: Altitude Conversion Kit
- e: Artificial Cable Kit
- e: Assistant Cook
- e: Automatic Colo[u]r Killer
- e: Automatic Course Keeping

ack e: accidentally killed
- e: acknowledge[ment]

AcK e: Actinium K
Ack-Ack e: Aluminum Company of America
ack-ack e: antiaircraft

ACK-ADK e: Automatic Course Keeping-Automatic Depth Keeping
ACKGT e: Acknowledgment
ACKI e: Acknowledge Input
ACKNE e: Acknowledge
ACKNOWL e: Acknowledgement
ACKNT e: Acknowledgment
ACKO e: Acknowledge Output
- e: Acknowledgment
ACKT e: Acknowledgement
ackt e: acknowledgment
ACKTX e: Automatic Circuit Exchange
ACL d: Abacus Computer Leasing GmbH
- e: Access Control List
- e: Action-Centred Leadership
- e: Adjective Check List
- e: Administration Communication Language
- e: Advanced CMOS Logic
- e: Aeronautical Computer[s] Laboratory
- e: Aircraft Circular Letter
- e: Aircraft Control Link
- e: Aircraft Load
- e: Air Cushion Landing
- e: All Comment Listing
- e: Allowable Cabin Load
- e: Allowable Cargo Load
- e: Allowable Cleanliness Level
- e: Allowable Container Load
- e: Alternate Concentration Limit
- e: Alternating Current Line
- e: Alternating Current Loss
- e: Altimeter Check Location
- e: Aminocaprolactam
- e: Analytical and Computer Laboratory
- e: Analytical Chemistry Laboratory
- e: Anterior Cruciate Ligament
- e: Anti-Collision Light
- e: Antigen-Carrier Lipid[e]
- e: Application Control Language
- e: Armo[u]red Cable; Lead-Sheath
- e: Ascent Closed Loop
- e: Association for Computational Linguistics
- e: Association of Cinema Laboratories
- e: Astro Communication Laboratory
- e: Atlantic Container Line
- e: Atlas Commercial Language
- e: Atlas Computer Laboratory
- e: Attained Competency Level
- e: Audit Command Language
- e: Australian Centre for Languages
- e: Australian College Libraries
- e: Authorized Consumption List
- e: Automated Coagulation Laboratory
- e: Automatic Carrier Landing
- e: Automatic Cartridge Loader
- e: Automatic Coding Language
- e: Automatic Current Limiting
- e: Aviation Circular Letter
- e: Avionics Cooling Loop
acl e: air-cushion landing
aCl e: aspiryl chlonde
ACLA e: American Clinical Laboratory Association
ACLAIS e: Association for Commonwealth Literature and Language Studies
ACLAM e: American College of Laboratory Animal Medicine
ACLAN e: American Comparative Literature Association Newsletter
ACLANT e: Allied Command Atlantic
ACLB e: Aircraft Launching Bulletin
- e: Association of Christians in Local Broadcasting
ACLC e: Adaptive Communication Live Controller
- e: Automatic Current Limiting Curve
ACLCS e: Airborne Command-Launch Control Subsystem
ACL Curve e: Automatic Current Limiting Curve
ACLD e: Above Clouds
- e: Air-Cooled
ACLDB e: Army Central Logistics Data-Bank
- e: Australian Corporate Law Data-Base
AcLDL e: Acetylated Low-Density Lipoprotein
ACLDV e: Air Cleaner Duct and Valve
ACLE e: Automatic Clutter Eliminator
Ac Lent l: Altocummulus Lenticularis
AC-LF e: Alternating Current - Low Frequency
ACLG e: Air Cushion Landing Car
ACLICS e: Airborne Communications Location, Identification and Collection System
ACLIS e: Alaska Conference on Library and Information Service
- e: Australian Council of Libraries and Information Services
ACLM e: American College of Legal Medicine
ACLMRS e: Advisory Committee for Land Mobile Radio Services
A/CLNR e: Air Cleaner
ACLO e: Agena Class Lunar Orbiter
- e: Air Cadet Liaison Officer
- e: Association of Cooperative Library Organizations
ACLOS e: Automatic Command to Line of Sight
ACLP e: Above Core Load Pad
- e: Above Core Load Plane
- e: Air Cushion Launch Platform
ACLPS e: Academic Clinical Laboratory Physicians and Scientists
ACLR e: Access Control Logging and Reporting
ACLS e: African Communications Liaison Service
- e: Aircraft Carrier Landing System
- e: Air Cushion Landing System
- e: All-Weather Carrier Landing System
- e: American Council of Learned Societies
- e: Automated Carrier Landing System
- e: Automated Control and Landing System
- e: Automatic Carrier Landing System
- e: Auxiliary Contractor Logistic Support
ACLSN e: American Council of Learned Societies Newsletter
ACLS News. e: American Council of Learned Societies Newsletter (journ.)
ACLSP e: Automated Carrier Landing Systems Project
ACLS System e: Advanced Cardiac Life Support System
ACLSV e: Armoured Combat Logistics Support Vehicle
ACLSVF e: Armoured Combat Logistics Support Vehicle Family
ACLT e: Accelerate
ACLTR e: Accelerator
ACLU e: American College of Life Underwriters
- e: Atlantic Container Line Unit
ACLV e: Air-Cushion Logistic Vehicle
- e: Apple Chlorotic Leafspot Virus
aclv e: accrued leave
ACM d: Arbeitsgemeinschaft Christlicher Motorradfahrer
- e: Access Control Machine
- e: Accumulator
- e: Acoustic Countermeasure[s]
- e: Acquisition and Control Module
- e: Active Countermeasures
- e: Additional Crew Member
- e: Addressable Core Memory
- e: Address Calculation Machine
- e: Address Code Multiplex
- e: Address Complete Message
- e: Adriamycin, Cyclophosphamide, Methotrexate
- e: Advanced Circuit Module
- e: Advanced Concepts Missile
- e: Advanced Consumer Marketing
- e: Advanced Conventional Munition
- e: Advanced Cruise Missile
- e: Aerial Combat Maneuvres
- e: Aerodynamic Configured Missile
- e: Aerospun Cluster Munition
- e: Air Carrier Mechanics Association
- e: Air Chief Marshal
- e: Air Combat Man[o]euvering
- e: Air Commerce Manual
- e: Air Court Martial
- e: Aircraft Colouring and Marking
- e: Air Cycle Machine
- e: Alarm Control Module
- e: Alkaline Contaminant Material
- e: Allocated Configuration Management
- e: Alterable Control Memory
- e: Alternating Current Mains
- e: Alternating Current Meter
- e: Alternating Current Motor
- e: Alumina Ceramic Manufacturers Association
- e: American College of Medicine
- e: Amplitude Comparison Monopulse
- e: Amsterdam Centre for Mathematics and Computer Sciences
- e: Analog[ue] Command Module
- e: Anatomy Covering Material
- e: Anatomy Covering Memo
- e: Annual Corrective Maintenance
- e: Anti-Cruise Missile
- e: Antiarmour Cluster Munition
- e: Arab Common Market
- e: Artificial Compression Method
- e: Asbestos-Containing Material
- e: Asbestos-Covered Metal
- e: Associated Colleges of the Midwest
- e: Association for Computing Machinery
- e: Association of Computer Manufacturers
- e: Associative Communications Multiplexer
- e: Astrocyte-Conditioned Medium
- e: Audio Center Module
- e: Australian Chamber of Manufactures
- e: Authorized Controller Material
- e: Automatic Clutter Mapper
- e: Automatic Clutter Mapping
- e: Automatic Coating Machine
- e: Automatic Colo[u]r Monitor
- e: Automatic Control Module
- e: Automating Calculating Machine
- e: Auxiliary Core Memory
- e: Aviation Chief Metalsmith
- e: Axon Cylinder Membrane
acm e: active countermeasure[s]
- e: anatomy-covering material
- e: anatomy-covering memo
- e: auxiliary mine layer
A/cm d: Ampere / Zentimeter
AC/M e: Aircraft Meteorological
a-c-m e: albumin-calcium-magnesium
ACMA e: Acidproof Cement Manufacturers Association
- e: A Contractor Managed Account
- e: Advanced Civil / Military Aircraft

- e: Agricultural Cooperative Managers Association
- e: Air Carrier Mechanic Association
- e: Alliance Cabinet Makers Association
- e: Alumina Ceramic Manufacturers Association
- e: American Cast Metals Association
- e: American Cutlery Manufacturers Association
- e: Army Class Manager Activity

ACMAC e: Association for Computing Machinery Accreditation Committee

ACMAS e: Audiovisual Conference of Medical and Allied Sciences
- e: Automated Classified Material Accountability System

ACMB e: Applications Configuration Management Board

ACMC e: Advanced Cruise Missile Combustor
- e: Area Combined Movements Center
- e: Automotive Chemical Manufacturers Council

ACMD e: Advanced Concepts and Missions Division
- e: Assistant Chief Medical Director

ACMDOSq e: Air Commando Squadron

ACME e: Acme Screw Thread
- e: Advanced Computer for Medical Research
- e: Advisory Council on Medical Education
- e: Aircraft Component Mating Evaluation
- e: Antenna Contour Measuring Equipment
- e: Application of Computers to Manufacturing Engineering
- e: Association of Consulting Management Engineers
- e: Attitude Control and Maneuvering Electronics
- e: Automated Classification of Medical Entities
- e: Automatic Control and Maneuvering Electronics
- e: Automatic Correction of Multiple Errors

acme e: advanced computer for medical

ACMEE e: Australian Conference on Management Education for Engineers

ACMEL e: Associated Continental Middle East Lines

ACMER e: Advanced Computer for Medical Research

ACMES e: Army Countermine Mobility Equipment System
- e: Attitude Control and Maneuver[ing] Electronics System

ACMES System e: Automatic Chinese Message Switching System

ACMF e: Air Corps Medical Forces
- e: Allied Central Mediterranean Force
- e: American Corn Millers Federation

ACMFS e: Automated Combat Mission Folder System

ACMGA e: American College of Medical Group Administrators

ACM-GAMM e: Association for Computing Machinery-German Association for Applied Mathematics

ACMH e: Advisory Committee on Major Hazards

ACMHA e: American College of Mental Health Administration

ACMHN e: Australian Congress of Mental Health Nurses

ACMI e: Air Combat Maneuvering Instrumentation [System]
- e: American Cotton Manufacturers Institute
- e: American Cystoscope Makers, Inc.
- e: Art and Craft Materials Institute

ACMIS e: Automated Career Management and Information System

ACML e: Association of Canadian Map Libraries

ACMLA e: Association of Canadian Map Libraries and Archives

ACMLC e: Army Chemical Center

ACMM e: Associate of the Conservatorium of Music; Melbourne
- e: Aviation Chief Machinists Mate

ACMMC e: Aviation Chief Machinists Mate; Carburetor Mechanic

ACMMF e: Aviation Chief Machinists Mate; Flight Engineer

ACMMH e: Aviation Chief Machinists Mate; Hydraulic Mechanic

ACMMI e: Aviation Chief Machinists Mate; Instrument Mechanic

ACMMP e: Aviation Chief Machinists Mate; Propeller Mechanic

ACMMT e: Aviation Chief Machinists Mate; Gas Turbine Mechanic

ACMN e: Aircrewman

ACMNPV e: Autographa Californica Multiply-embedded Nuclear Polyhedrosis Virus

ACMO e: Afloat Communications Management Office
- e: Alaska Coastal Management Office
- e: Assistant Chief of Mission Operations
- e: Assistant Countermortar Officer
- e: Authorized Controlled Material Order

ACMOS e: Advanced Complementary Metal Oxide Semiconductor

ACMP e: Accompany
- e: Advanced Cruise Missile Program[me]
- e: Anthropology Case Materials Project
- e: Assistant Commissioner of the Metropolitan Police

acmp e: Accompany[ing]

ACMPM e: Air Combat Maneuvering Performance Measurement

ACMPS e: Automated Communications and Message Processing System

ACMR e: Advanced Computer for Medical Research
- e: Air Combat Maneuvering Range
- e: Attitude Control and Maneuver Rate

ACMRR e: Advisory Committee on Marine Resources Research

ACMRU e: Audio Commercial Message Repeater Unit

ACMS e: Advanced Configuration Management System
- e: Air Call Medical Services
- e: Air Combat Maneuvering Simulator
- e: Air Conditioned Microclimate System
- e: American Chinese Medical Society
- e: American Coordinated Medical Society
- e: Application Control Management System
- e: Army Command Management System
- e: Australian Clay Minerals Society
- e: Automated Career Management System
- e: Automated Configuration Management System

ACMSC e: Association for Computing Machinery Standards Committee

ACM-SC e: Association for Computing Machinery-Standards Committee

ACM-SIBCAI e: Association for Computing Machinery-Special Interest Group on Computer-Assisted Instruction

ACM-SIGACT e: Association for Computing Machinery-Special Interest Group on Automata and Computerability Theory

ACM-SIGARCH e: Association for Computing Machinery-Special Interest Group on Architecture of Computer Systems

ACM-SIGART e: Association for Computing Machinery-Special Interest Group on Artificial Intelligence

ACM-SIGBDP e: Association for Computing Machinery-Special Interest Group on Business Data Processing

ACM-SIGBIO e: Association for Computing Machinery-Special Interest Group on Biomedical Computing

ACM-SIGCAI e: Association for Computing Machinery-Special Interest Group on Computer-Assisted Instruction

ACM-SIGCAS e: Association for Computing Machinery-Special Interest Group on Computers and Society

ACM-SIGCOMM e: Association for Computing Machinery-Special Interest Group on [Data] Communication[s]

ACM-SIGCOSIM e: Association for Computing Machinery-Special Interest Group on Computer Systems Installation Management

ACM-SIGCPR e: Association for Computing Machinery-Special Interest Group on Computer Personal Research

ACM-SIGCRS e: Association for Computing Machinery-Special Interest Group on Computerized Retrieval Services

ACM-SIGCSE e: Association for Computing Machinery-Special Interest Group on Computer Science Education

ACM-SIGDA e: Association for Computing Machinery-Special Interest Group on Design Automation

ACM-SIGDOC e: Association for Computing Machinery-Special Interest Group on Documentation

ACM-SIGGRAPH e: Association for Computing Machinery-Special Interest Group on Computer Graphics

ACM-SIGIP e: Association for Computing Machinery-Special Interest Group on Information Publishing

ACM-SIGIR e: Association for Computing Machinery-Special Interest Group on Information Retrieval

ACM-SIGLASH e: Association for Computing Machinery-Special Interest Group on Language Analysis and Studies in the Humanities

ACM-SIGMAP e: Association for Computing Machinery-Special Interest Group on Mathematical Programming

ACM-SIGMICRO e: Association for Computing Machinery-Special Interest Group on Microprogramming

ACM-SIGMINI e: Association for Computing Machinery-Special Interest Group on Minicomputers

ACM-SIGMOD e: Association for Computing Machinery-Special Interest Group on Management of Data

ACM-SIGNUM e: Association for Computing Machinery-Special Interest Group on Numerical Mathematic[s]

ACM-SIGOPS e: Association for Computing Machinery-Special Interest Group on Operating System
ACM-SIGPLAN e: Association for Computing Machinery-Special Interest Group on Programming Languages
ACM-SIGSAM e: Association for Computing Machinery-Special Interest Group on Symbolic and Algbraic Manipulation
ACM-SIGSIM e: Association for Computing Machinery-Special Interest Group on Simulation
ACM-SIGSOC e: Association for Computing Machinery-Special Interest Group on Social and Behavioural Science Computing
ACM-SIGUCC e: Association for Computing Machinery-Special Interest Group on University Computer Centers
ACM-SIGUOL e: Association for Computing Machinery-Special Interest Group on User Online Interaction
ACMT e: Advanced Cruise Missile Technology
- e: Aerial Combat Maneuvering Training
- e: American College of Medical Technologists
- e: Automatic Configuration Management Tool
ACM Trans Comp e: Association for Computing Machinery Transactions on Computer Systems
ACM Trans Comput Syst e: Association for Computing Machinery Transactions on Computer Systems
ACM Trans Database Syst e: Association for Computing Machinery Transactions on Database Systems
ACM Trans Graphics e: Association for Computing Machinery Transactions on Graphics
ACM Trans Math Softw e: Association for Computing Machinery Transactions on Mathematical Software
ACM Trans Off Inf Syst e: Association for Computing Machinery Transactions on Office Information System
ACM Trans OIS e: Association for Computing Machinery Transactions on Office Information System
ACM Trans Program[me] Lang Syst e: Association for Computing Machinery Transactions on Programming Languages
ACMV e: Assist-Control Mechanical Ventilation
ACMVS e: Air Combat Maneuvering Visual System
ACN e: Accession Number
- e: Acetone Cyanohydrin
- e: Acetonitrile
- e: Acrylonitrile
- e: Action Control Number
- e: Activity Classification Number
- e: Activity Control Number
- e: Acute Conditioned Necrosis
- e: Advance Change Notice
- e: Agricultural Communications Networks
- e: Air Consignment Note
- e: All Concerned Notified
- e: Alternating Current Network
- e: American Chain & Cable
- e: American College of Neuropsychiatrists
- e: American College of Nutrition
- e: Anthocyanin
- e: Artificial Cloud Nucleation
- e: Asbestos Cloth Neck
- e: Assignment Control Number
- e: Authorized Code Number
- l: Ante Christum Natum
acn e: activities control number
- e: Automatic Celestial Navigation
A.C N. l: Ante Christum Natum
ACNA e: Advisory Council of the National Arboretum
- e: Advisory Council on Naval Affairs
- e: Analog[ue] Computer for Net Adjustment
- e: Arctic Institute of North America
ACNAM e: Associations Council of the National Association of Manufacturers
ACNAS e: Admiral Commanding North Atlantic Station
- e: Advanced Cableship Navigation Aid System
ACNB e: Australian Commonwealth Naval Board
ACNBC e: Associate Committee on the National Building Code
ACNE e: Action Committee for Narcotics Education
- e: Alaskans Concerned for Neglected Environments
ACNEE e: Action Committee for Narcotics Education and Enforcement
ACNET e: Alternating Current Network
AC Network e: Alternating Current Network
ACNFP e: Advisory Committee on Novel Foods and Processes
ACNGS e: Allens Creek Nuclear Generating Station
ACNHA e: American College of Nursing Home Administrators
ACNI e: Automatic Called Number Identification
ACNIL i: Azienda Comunale Navigazione Interna
ACNM e: American College of Nuclear Medicine
- e: American Cordage and Netting Manufacturers
ACNMR e: Alternating Current Normal Mode Rejection
ACNMS e: Advisory Committee on Nuclear Materials Safeguards
ACNN e: Air Commander, North Norway
ACNO e: Assistant Chief of Naval Operations
ACNO (Air) e: Assistant Chief of Naval Operations
ACNO (Alt Op) e: Assistant Chief of Naval Operations (Fleet Operations and Readiness)
ACNOCOMM e: Assistant Chief of Naval Operations for Communications and Cryptology
ACNO(COMM)/DNC e: Assistant Chief of Naval Operations (Communications) / Director, Naval Communication
ACNO (Int) e: Assistant Chief of Naval Operations (Intelligence)
ACNO (Log) e: Assistant Chief of Naval Operations (Logistics)
ACNO (MA) e: Assistant Chief of Naval Operations (Marine)
ACNO (NR) e: Assistant Chief of Naval Operations (Naval Reserve)
ACNO (OP) e: Assistant Chief of Naval Operations (Operations and Readiness)
ACNO (P) e: Assistant Chief of Naval Operations (Personnel)
ACNOT e: Accident Notice
- e: Assistant Chief of Naval OperationsTransportation
ACNP e: American College of Neuropsychopharmacology
- e: American College of Nuclear Physicians
ACNP (A) e: Assistant Chief of Naval Personnel (Administration)
ACNPV e: Autographa Californica Nuclear Polyhedrosis Virus
ACNS e: Academic Computing and Network Services
- e: Advanced Communications Network Service
- e: Advisory Committee on Nuclear Safety
- e: Assistant Chief of Naval Staff
- e: Associated Correspondents News Service
ACNS(A) e: Assistant Chief of Naval Staff (Air)
ACNS(A & W) e: Assistant Chief of Naval Staff (Air and Warfare)
ACNS(P) e: Assistant Chief of Naval Staff (Plans)
ACNS(W) e: Assistant Chief of Naval Staff (Warfare)
ACNTS(Air) e: Assistant Chief of Naval Technical Services (Air)
ACNTS(S) e: Assistant Chief of Naval Technical Services (Ships)
ACNUR s: Alto Comisionado de las Naciones Unidas para las Refugiadns
ACNY e: Advertising Club of New York
ACO e: Abort Once around Cutoff
- e: Acceptance Checkout
- e: Accounting Control Office
- e: Action Cutout
- e: Adaptive Control [with] Optimization
- e: Adaptive Control Optimized
- e: Administrative Contracting Office[r]
- e: Admiralty Compass Observatory
- e: Advance Contracting Officer
- e: Adviser on Combined Operations
- e: Agricultural Climatological Office
- e: Air Cargo
- e: Air Communication Officer
- e: Air Control Officer
- e: Aircraft Control Operator
- e: Alpha Cutoff
- e: Alternating Current Output[s]
- e: American Academy of Optometry
- e: American College of Orgonomy
- e: American College of Otorhinolaryngologists
- e: American Council of Otolaryngology
- e: Annual Cost of Ownership
- e: Anodal Closing Odour
- e: Area Clearance Officer
- e: Armament Concepts Office
- e: Assembly and Checkout
- e: Association of Conservation Officers
- e: Atomic Coordination Office
- e: Attack Cut-Out
- e: Austin Community College
- e: Australian College of Ophthalmologists
- e: Australian College of Organists
- e: Auto Call Originator
- e: Automatic Call Originate
- e: Automatic Call Origination
- e: Automatic Call Originator
- e: Automatic Cutout
- e: Average Calculating Operation
aco e: anodal closing odor
a co s: a cargo
ACOA e: American Construction Owners Association
- e: Associate Committee on Aerodynamics

ACoA e: Anterior Communicating Artery
ACOB e: Arithmetic Checker Out Bus
ACOC e: Air Command Operations Center
- e: Allied Command Operations Center
- e: Area Communication Operations Center
- e: Associated Contractor Orginated Change
- e: Automatic Control Operations Center
AC/OC e: Air Co-Operation Command
ACOCA e: Army Communication Operations Center Agency
ACOCC e: Atlantic Commander Operational Control Center
Aco-Con e: Acoustic Control
ACOCS e: Army Customer Order Control System
ACODAC e: Acoustic Data Capsule
ACODS e: Army Container-Oriented Distribution System
ACOE e: Automatic Checkout Equipment
ACOEM e: American College of Occupational and Environmental Medicine
ACOEP e: American College of Osteopathic Emergency Physicians
ACOF e: Attendant Control of Facilities
AC of AS e: Assistant Chief of Air Staff
AC of N e: Assistant Controller of the Navy
ACOFO e: American College of Foot Orthopedists
AC of Ord e: Assistant Chief of Ordnance
Ac of S e: Academy of Sciences
ACOFS e: Assistant Chief of Staff
AC of S e: Assistant of Staff
ACOFS (Maint) e: Assistant Chief of Staff (Maintenance)
ACOFS (Mov) e: Assistant Chief of Staff (Movements)
ACOFS (RQMT & ACQN) e: Assistant Chief of Staff (Requirements and Acquisitions)
ACOFS (STOR & DISTR) e: Assistant Chief of Staff (Storage and Distribution)
ACOFS WPD e: Assistant Chief of Staff War Plans Division
AC of T e: Assistant Chief of Transportation
acog e: Aircraft on Ground
- e: aircraft on ground
ACOG e: American College of Obstetricians and Gynecologists
ACOHA e: American College of Osteopathic Hospital Administrators
ACOI e: American College of Osteopathic Internists
ACOL e: Antiproton Collector
- e: Application Control Language
ACOLI e: Advance Circuit Order and Layout Information
A Coll H e: Associate of the College of Handicraft
acol[s] e: acolyte[s]
ACOM e: Advisory Committee on Objective Measurement
- e: Association for Computer Operations Management
- e: Australian College of Occupational Medicine
- e: Automatic Code Machine
- e: Automatic Coding Machine
- e: Aviation Chief Ordnanceman
- e: Awards Committee
A Com e: Air Commodore
- e: Associate in Commerce
ACOMA d: automatische Codiermaschine
ACO/MGE e: Acceptance and Checkout/Maintenance Ground Equipment
A-Com-in-C e: Air-Commodore-in-Chief
A Comm e: Air Commodore
A. Comm. e: Associate in Commerce
ACOMMW e: Aerospace Communications Wing
ACOMPLINE e: A Computerized London Online Information Network
ACOMPLIS e: A Computerized London Information Service
A Compt e: Army Comptroller
ACOMR e: Advisory Committee on Oceanic Meteorological Research
ACOMS e: American College of Oral and Maxillofacial Surgeons
- e: Army Communications Objectives Measurement Survey
- e: Automated Collection Management System
ACOMT e: Aviation Chief Ordnanceman, Turret Mechanic
ACON d: akademisches Computernetz
Acon e: Aconcagua
A/COND e: Air Conditioning
ACONS e: Aerospace Control Squadron
a-contact e: Make Contact [of the contactor]
ACOOG e: American College of Osteopathic Obstetricians and Gynecologists
ACOP e: Airborne Corps Operation Plan
- e: American College of Optometric Physicians
- e: American College of Osteopathic Pediatricians
- e: Approved Code of Practice
- e: Army Customer Order Program[me]
- e: Association of Chief Officers of Police
Ac op l: Altocumulus Opacus
A-COPE e: Adolescent-Coping Orientation for Problem Experiences
ACOPP e: Abbreviated COBOL Pre-Processor
- e: Adriamycin, Cyclophosphamide, Oncovin, Procarbazine, Prednisone
ACOPS e: Advisory Committee on [Oil] Pollution of the Sea
ACORD e: Action through Creative Organization, Research and Discussion
- e: Advanced Concepts for Ordnance
- e: Advisory Committee on Energy Research and Development
- e: Advisory Council on Research and Development
- e: Advisory Council on Research and Development for Fuel and Power
- e: Automatic Component Ordering
ACORDE e: A Consortium on Restorative Dentistry Education
ACORE e: Advisory Council for Orthopaedic Resident Education
ACORN e: Acronym-Oriented Nut
- e: Associate Content Retrieval Network
- e: Australian Confederation of Operating Room Nurses
- e: Automated Coder of Report Narrative
- e: Automated Conversion Routine
- e: Automatic Checkout and Recording Equipments
- e: Automatic Checkout and Recording Network
- e: Automatic Coder of Report Narrative
ACOS e: A Common Operational Software
- e: A Computer Series
- e: Advanced Computer-Oriented System
- e: Advisory Committee on Safety
- e: American College of Osteopathic Surgeons
- e: Application Control Operating System
- e: Arms Control Observation Satellite
- e: Assistant Chief of Staff
- e: Australian Chamber of Shipping
- e: Automated Cloud Observation System
- e: Automatic Checkout Set
ACOSH e: Appalachian Center for Occupational Safety and Health
ACOSS e: Active Control of Space Structures
ACOST e: Advisory Committee on Science and Technology
ACOT e: Assistant Chief of Transportation
ACOTP e: Assistant Captain of Transportation
ACOUS e: Acoustic[s]
acous e: acoustical
acous coup e: acoustic coupler
acousid e: acoustic seismic intrusion detector
acoust e: acoustic[al]
- e: acoustical[ly], acoustic[s]
ACOUST e: Acoustics
Acoustic Modem e: Acoustic Modulator-Demodulator
ACOUSTINT e: Acoustical Intelligence
ACP e: Abnormal Control Plasma
- e: Acceptance Checkout Procedure
- e: Accomplishment / Cost Procedure
- e: Acetophenone
- e: Acetyl Carrier Protein
- e: Acid Calcium Phosphate
- e: Acid Phosphatase
- e: Acoustic Communication Program[me]
- e: Action Control Point
- e: Action Current Potential
- e: Acyl Carrier Protein
- e: Adaptive Control Process
- e: Additive Colo[u]r Process
- e: Advance Command Post
- e: Advanced Circuit Program[me]
- e: Advanced Computational Processor
- e: Advanced Cooperative Project
- e: Advisory Committee on Pesticides
- e: Aerospace Computer Program[me]
- e: African, Caribbean and Pacific countries
- e: Agricultural Conservation Program[me]
- e: Air-Conditioning Pack
- e: Airborne Command Post
- e: Air Carcinogen Policy
- e: Air Control Point
- e: Aircraft Communication Procedure[s]
- e: Airlift Command Post
- e: Airline Carriers Passengers
- e: Airline Control Program[me]
- e: Airman Commissioning Program[me]
- e: Alarm Control Panel
- e: Align Control Panel
- e: All Cards Printed
- e: Allied Communications Procedures
- e: Allied Communications Publication[s]
- e: Allocation Control Point
- e: Alphabetical Code Punching
- e: Alternate Care Plan
- e: Alternate Command Post
- e: Alternating Current Potential
- e: Alternating Current Power
- e: Alternating Current Probe
- e: Alternative Coated Paper
- e: Alternative Complement Pathway
- e: Altimeter Check Point
- e: American College of Pharmacists

- e: American College of Physicians
- e: American College of Podopediatrics
- e: American College of Prosthodontists
- e: American College of Psychiatrists
- e: Ammunition Control Point
- e: Analog[ue] Computer Program[me]
- e: Analytical Computer Program[me]
- e: Anchors and Chains Proved
- e: Ancillary Control Processor
- e: Animal Care Panel
- e: Annotated Card Program[me]
- e: Anodal Closing Picture
- e: Apparent Candle Power
- e: Area Characteristic Plan[ning]
- e: Area Command Post
- e: Area Community Physician
- e: Area Coordinating Paper
- e: Arithmetic and Control Processor
- e: Armament Control Panel
- e: Armo[u]red Car; Personnel
- e: Army Capabilities Plan
- e: Army Controllership Program[me]
- e: Asphaltic Concrete Pavement
- e: Aspirin, Caffeine, Phenacetin
- e: Associate Client Program[me]
- e: Associated Construction Publications
- e: Associate of the College of Preceptors
- e: Associates of Clinical Pharmacology
- e: Association-Storing Processor
- e: Association for Child Psychoanalysis
- e: Association of Child Psychotherapists
- e: Association of Clinical Pathologists
- e: Association of Computer Professionals
- e: Association of Correctional Psychologists
- e: Association of Coupon Processors
- e: Astronaut Control Panel
- e: Atmospheric Contamination Potential
- e: Attack Center Panel
- e: Atypical Chest Pain
- e: Audio Centre Equipment
- e: Audio Control Panel
- e: Audit Control Point
- e: Automated Calibration Procedure
- e: Automated Chemistry Program[me]
- e: Automated Communications Publications
- e: Automated Computer Program[me]
- e: Automatic Colt Pistol
- e: Automatic Command Processing
- e: Auxiliary Checkpoint
- e: Auxiliary Control Panel

Acp e: Accepted
acp e: acoustical paper
- e: azimuth change pulse

A.C.P. e: Airdrome Control Point
A & CP e: Access and Control Point
a & cp e: anchors and chains proved
A-CP e: Anti-Concorde Project
a-c-p e: aspirin-caffeine-phenacetin
aCp e: automatic Colt pistol
ACPA e: Activated Carbons Producers Association
- e: Adaptive Controlled Phased Array
- e: Adjusted Compensation Payment Act
- e: Affiliated Chiropodists-Podiatrists of America
- e: Affitliated Chiropodists-Podiatrists of America
- e: American Chronic Pain Association
- e: American Cleft Palate Association
- e: American College of Physicians Assistants
- e: American Concrete Pavement Association
- e: American Concrete Pipe Association
- e: American Concrete Pumping Association
- e: Associate of the Institution of Certified Public Accountants
- e: Association of Computer Professionals Australia
- e: Association of Computer Programmers and Analysts
- e: Audio Capture and Playback Adapter

A-CPA e: Asbestos-Cement Products Association
Ac-PAF d: Wechselstrom-Plasma-anzeigefeld
ACPAI e: Affiliated Conference of Practicing Accountants International
ACPATT e: All Commands Process as Attached
ACPC e: Aircraft Camera Parameter Control
- e: Aminocyclopentanecorboxylic [acid]

ACPCA e: American Cleft Palate-Craniofacial Association
ACPC Acid e: Aminocyclopentanecorboxylic Acid
ACPCP e: Australian College of Private Consulting Psychologists
ACPD e: Army Control Program[me] Directive
ACPDP e: Alternating Current Plasma Display Panel
ACPDS e: Advisory Committee on Personal Dosimetry Services
ACPE e: American College of Physician Executives
- e: American Council on Pharmaceutical Education
- e: Analytical Computer Program[me] Evaluation
- e: Assocation for Continuing Professional Education
- e: Association for Clinical Pastoral Education
- e: Australian Chemical Processing and Engineering
- e: Australian College of Physical Education

ACPEL d: Alternating Current-Pulverschicht-Elektrolumineszenz
ACPERS e: Army Civilian Personnel System
ACPF e: Acoustic Containerless Processing Facility
- e: Amphibious Corps, Pacific Fleet
- e: Australian Cancer Patients Federation

ACPH e: Acid Phosphatase
- e: Air Change per Hour

ACPI e: American City Planning Institute
- e: Assistant Chief Patrol Inspector
- e: Automatic Cable Pair Identification
- e: Aviation Crime Prevention Institute

ACPIC e: American Council for Private International Communications
ACPL e: Acoustical Plaster
- e: Atlas Crew Procedures Laboratory
- e: Atmospheric Cloud Physics Laboratory

ACPM e: Acoustic Containerless Processing Module
- e: Activity Career Program[me] Manager
- e: Advisory Committee for Program[me] Management
- e: Aerospace Computer Program[me] Model
- e: American College of Preventive Medicine
- e: American Congress for Preventive Medicine
- e: Associate Contractor Program[me] Manager
- e: Associate of the Confederation of Professional Management
- e: Attitude Control Propulsion Motor[s]

ACPMME e: Association of Concentrated and Powdered Milk Manufacturers
ACPMR e: American Congress of Physical Medicine and Rehabilitation
ACPO e: Acting Chief Petty Officer
- e: Associate Contractor Projects Office

ACPOC e: Association of Childrens Prosthetic-Orthotic Clinics
ACPP e: Adrenocorticopolypeptide
- e: Advisory Committee on Polar Programs
- e: Aircraft Crashworthiness Program[me] Plan
- e: Association for Child Psychology and Psychiatry

ACPPA e: American Concrete Pressure Pipe Association
ACPPD e: Average Cost Per Patient Day
ACPR e: Advanced Core Performance Reactor
- e: Advanced Core Pulsed Reactor
- e: Advanced Critical Pulse Reactor
- e: American College of Podiatric Radiologists
- e: Annular Core Pulse Reactor

ACPRA e: American College Public Relations Association
AC Process e: Arbed Cockerill Process
ACPRS e: Ames Cubic Precision Ranging System
ACPS e: Air-Conditioning and Pneumatic System
- e: Alkaline Calcium Petroleum Sulfonate
- e: Arab Company for Petroleum Services
- e: Armament Control Processor Set
- e: Attitude Control Propulsion System
- e: Automatic Crash-Program Systems
- e: Auxiliary Control Propulsion System

ACP's e: Allocation Control Points
ACPSEM e: Australasian College of Physical Scientists and Engineers in Medicine
Ac. Psn. e: Accepted Position
ACPT e: Accept
- e: Acid Phosphatase with Tartrate
- e: Asian Confederation of Physical Therapy

Acpt e: Acceptance
ACPTA e: Australian Commonwealth Post and Telegraph Association
ACPTC e: Association of College Professors of Textiles and Clothing
ACPTD e: Accepted
ACPTR e: Acceptor
ACPTT e: Air Combat Part Task Trainer
ACPU e: Auxiliary Computer Power Unit
ACP-Virus e: Adeno-Conjunctival-Pharyngeal Virus
ACPX e: Advanced Circuit Packaging Extended
- e: Advanced Computer Program[me] Experimental

ACPY e: Accompany
ACQ e: Acquire
- e: Acquisition
- e: Acquittal
- e: All Courses and Quadrants
- e: Annual Contracted Quantity
- e: Aquire

acq adj e: Acquisition Adjustment
ACQE e: Acquisition and Control Query Executive
Acq Libr e: Acquisition[s] Librarian
ACQM e: Assistant Chief Quartermaster
- e: Automatic Circuit Quality Monitor[ing]

ACQN e: Acquisition
ACQOT e: Aviation Cadet Qualifying Test
ACQR e: American Catholic Quarterly Review (journ.)

ACQ RDR e: Acquisition Radar
acq rdr rep hlpr e: acquisition radar repair helper
ACQ RDR REPM e: Acquisition Radar Repairman
ACQSEL Switch e: Acquisition Select Switch
ACQSN e: Acquisition
AC & QT e: Acceptance, Conforming and Qualification Test
ACQT e: Aviation Cadet Qualifying Test
ACQU e: Acquisition
acquis e: Acquisition[s]
ACR e: Abandon Call, Retry
- e: Abandon Call and Retry
- e: Absolute Category Rating
- e: Abstracts of Classified Reports
- e: Accelerated Carrier Return
- e: Accelerated Cost Recovery
- e: Acceptable Contrast Ratio
- e: Access Control Register
- e: Accumulator
- e: Accumulator Register
- e: Across
- e: Active Cavity Radiometer
- e: Actual Cost Report
- e: Adaptive Character Reader
- e: Additional Clock Resolution
- e: Address Control Register
- e: Address Conversion Register
- e: Address Correction Requested
- e: Adenomatosis of the Colon and Rectum
- e: Adjacent Channel Rejection
- e: Administrative Communications Requirements
- e: Advanced Capabilities Radar
- e: Advanced Cargo Rotorcraft
- e: Advanced Combat Rifle
- e: Advanced Confidential Report
- e: Advanced Converter Reactor
- e: Advanced Cracking Reactor
- e: Aerial Combat Reconnaissance
- e: Aerial Coupler Receiver
- e: Aerial Coupling Regulator
- e: Aerodrome Control Radar
- e: Aeroelastically Conformable Rotor
- e: Air Carrier
- e: Air Control and Reporting
- e: Air Control[led] Radar
- e: Air Control Room
- e: Aircraft Checkers Report
- e: Aircraft Control Room
- e: Air Crew Rescue
- e: Airfield Control Radar
- e: Alaskan Communications Region
- e: Algoma Central Railway
- e: Allied Commission on Reparations
- e: Allowance Change Request
- e: Alternate Central Processing Unit Recovery
- e: Alternating Current Rectifier
- e: Alternating Current Relay
- e: Alternating Current Resistance
- e: Alternating Current Ringing
- e: American Academy in Rome
- e: American College of Radiology
- e: American College of Rheumatology
- e: American Computer Referal
- e: American Criminal Reports (journ.)
- e: Ammunition Condition Report
- e: Annual Confidential Report
- e: Antenna Coupler Receiver
- e: Antenna Coupling Regulator
- e: Anti-Circular Run
- e: Anti-Constipating Regimen
- e: Anti-Constipation Regimen
- e: Anticonstipation Regimen
- e: Applied Communications Research, Incorporated
- e: Applied Computer Research
- e: Approach Control Radar
- e: Area Coordination Review
- e: Armo[u]red Cruiser
- e: Assistant Chief for Research
- e: Association for Consumer Research
- e: Association of Computer Retailers
- e: Atmospheric Corrosion Resistant
- e: Audio Cassette Recorder
- e: Automatic Call Recorder
- e: Automatic Call Recording
- e: Automatic Card Reader
- e: Automatic Carriage Return
- e: Automatic Cassette Recorder
- e: Automatic Character Reader
- e: Automatic Character Recognition
- e: Automatic Compression-Release
- e: Automatic Compression Regulator
- e: Auxiliary Computer Room
acr e: acre
Acr e: Acrylic
AC/R e: Alternating Current/Ripple
AC & R e: American Cable and Radio
ACRA e: American Car Rental Association
- e: American Collegiate Retailing Association
- e: American Craft Retailers Association
- e: Associate of the Corporation of Registered Accountants
- e: Association of College Registrars and Administrators
- e: Auto Collision Repair Association
A/C Ratio e: Albumin/Coagulin Ratio
ACRB e: Army Council of Review Boards
- f: Aero-Club Royal de Belgique
AC & RC e: Active Components and Reserve Components
ACRC e: Air Compressor Research Council
- e: Arkansas Cancer Research Center
- e: Audio Centre Receiver
acrd e: accrued
ACRD e: Anticipated Card Release Date
- e: Automatic Compression-Release Device
ACRDC e: Australian Code of Residential Design and Construction
ACRDL e: Army Chemical Research and Development Laboratory
ACRE e: Acting Commander; Royal Engineers
- e: Action Committee for Rural Electrification
- e: Adjutant to the Corps of Royal Engineers
- e: Advanced Chemical Rocket Engineering
- e: Advisory Committee on Releases to the Environment
- e: Atlantic City Remodelers Exposition
- e: Automatic Call Recording Equipment
- e: Automatic Checkout and Readiness Equipment
A/C rec e: Accounts Receivable
ACREC e: American College of Real Estate Consultants
acre ft e: acre foot
ACRE-FT/D e: Acre-Feet per Day
A.C.rel. e: Alternating-Current Relay
ACRES e: Airborne Communication Relay Station
- e: Australian Centre for Remote Sensing
AC-RF e: Alternating Current-Radio Frequency
ACRFAET e: Aircraft Crash Rescue Field Assistance and Evaluation Team
ACRFLT e: Across Flats
ACRFT e: Aircraft
acrg e: acreage
ACRH e: Argonne Cancer Research Hospital
ACRI e: Air Conditioning and Refrigeration Institute
- e: American Cocoa Research Institute
ACRiLIS e: Australian Centre for Research in Library and Information Science
ACRIM e: Active Cavity Radiometer Irradiance Monitor
- e: Association for Correctional Research and Information Management
ACRIP e: Adrenocortical Renin Inhibitory Peptide
- e: Australian Computer Research in Progress
ACRIS e: Agency for International Development Consultant Registry Information Syste
- e: Aperture Card Raster Input Scanner
- e: Aperture Card Remote Imaging System
ACRIT e: Advisory Committee for Research on Information Transfer
- e: Automatic Crystal Resistivity Indexing Tester
ACRL e: Association of College and Research Libraries
ACRL AAS e: Association of College and Research Libraries Asian and African Section
ACRL ANSS e: Association of College and Research Libraries Anthropology and Sociology
ACRL ARTS e: Association of College and Research Libraries Art Section
ACRL AS e: Association of College and Research Libraries Art Section
ACRL ASS e: Association of College and Research Libraries Anthropology and Sociology
ACRL BIS e: Association of College and Research Libraries Bibliographic Instruction Science
ACRL CLS e: Association of College and Research Libraries College Libraries Section
ACRL C & RL e: Association of College and Research Libraries College and Research Libraries
ACRL EBSS e: Association of College and Research Libraries Education and Behavioral Sciences
ACRL LPSS e: Association of College and Research Libraries Law and Political Science Sciences
ACRL RBMS e: Association of College and Research Libraries Rare Books and Manuscripts
ACRL SEES e: Association of College and Research Libraries Slavic and East European Specialist
ACRL STS e: Association of College and Research Libraries Science and Technology Section
ACRL ULS e: Association of College and Research Libraries University Libraries Section
ACRL WESS e: Association of College and Research Libraries Western European Specialist
ACRL WSS e: Association of College and Research Libraries Womens Studies Section

ACRM e: Air Cushion Rig Mover
- e: American College of Radio Marketing
- e: American Congress of Rehabilitation Medicine
- e: Aviation Chief Radioman
AC/RM e: Air-Conditioning Room
ACRMA e: Air-Conditioning and Refrigerating Machinery Association
ACRMP e: Automation Communication Resource Management Plan
ACRN e: Accounting Classification Reference Number
- e: Accounting Code Reference Number
ACRO e: Aircraft Control Room Officer
acro e: acrophobe
- e: Acrophobia
ACRODABA e: Acronym Data Base
acrol e: acrolect[ic][al][ly]
ACROM e: Associative-Capacitive Read-Only Memory
acromeg e: acromegally
acron e: acronym
acronymiz e: acronymization
- e: acronymizing
- e: acronymizor[s]
ACRONYMS e: Acceptable Contractions of Randomly Organized Names Yielding Meritorious Spontan
- e: A Convenient Reduction of Nomenclature Yielding Mnemonic Syllables
Acropolis e: Acropolis Hill, Athens
across e: acrostic
ACRP e: Advisory Committee on Radiological Protection
- e: Airborne Communications Reconnaissance Platform
- e: Airborne Communications Reconnaissance Program[me]
- e: Armament Control Relay Panel
- e: Army Cost Reduction Program[me]
ACRQ e: Access Request
ACRR e: Annular Core Research Reactor
- e: Atlantic City Radio Regulations
ACRRT e: American Chiropractic Registry of Radiologic Technologists
ACRS e: Accelerated Capital Recovery System
- e: Accelerated Cost Recovery Schedule
- e: Accelerated Cost Recovery System
- e: Acceptor Ready State
- e: Across
- e: Active Contrast Reduction System
- e: Advisory Committee on Reactor Safeguards
- e: Air Cushion Recovery System
- e: Air Cushion Restraint System
- e: Association for Correctional Research and Statistics
- e: Automatic Chemical Reaction System
ACRs e: Access Control Registers
acrs e: acreage
ACRSE e: American Council on Rural Special Education
ACRT e: Applied Circuit Technology
- e: Aviation Chief Radio Technician
ACRTC e: Advanced Cathode Ray Tube Controller
ACRU e: Advisory Committee on Resource Use
- e: Applied Climate Research Unit
AC-rubber e: Anticrystallizing Rubber
ACRV e: Armo[u]red Command and Reconnaissance Vehicle
- e: Artillery Command Reconnaissance Vehicle
- e: Assured Crew Return Vehicle
- e: Audio Center Receiver
- e: Automated Command Response Verification
ACRW e: Aircrew
- e: American Council of Railroad Women
ACRY[L] e: Acrylic
ACS d: Antireticulär-Cytotoxisches Serum
- d: Automobilclub der Schweiz
- e: Abridged Call Sign
- e: Absorption Cross Section
- e: Academic Computer Service[s]
- e: Academic Computing Services
- e: Accelerated Climatic Simulator
- e: Access
- e: Accounting Control System
- e: Accumulator Switch
- e: Accuracy Control System
- e: Acetylstrophanthidin
- e: Acoustic Computer Signal
- e: Acquisition and Command Support
- e: Activity Characteristics Sheet
- e: Adaptive Code System
- e: Adaptive Control[ling] System
- e: Additional Core Storage
- e: Addressable Core Store
- e: Adhesive Component System
- e: Administrative Computing Service
- e: Administrative Control System
- e: Admiralty Computing Service
- e: Adrenocorticosteroid
- e: Advance Count Switch
- e: Advanced Cardiovascular Systems
- e: Advanced Ceramic System
- e: Advanced Clothing Subsystem
- e: Advanced Communication[s] Service[s]
- e: Advanced Communication[s] System
- e: Advanced Computer Services
- e: Advanced Computer System[s]
- e: Advanced Control System
- e: Advanced Count Switch
- e: Advanced Cryptographic System
- e: Advice Called Subscriber
- e: Aerodrome Control Service
- e: Affiliated Computer Systems
- e: Affinely Connected Space
- e: Afloat Correlation System
- e: Agena Control System
- e: Agricultural Cooperative Service
- e: Agriculture Cooperative Service
- e: Air Capable Ship
- e: Air Conditioning Sensor
- e: Air Conditioning System
- e: Air Contact System
- e: Aircraft Carrier Squadron
- e: Aircraft Communication[s] System
- e: Aircraft Control and Surveillance
- e: Aircraft Control System
- e: Airfield Construction Squadron
- e: Airplane Configuration System
- e: Airways Communication Station
- e: Alarm and Control System
- e: Alaska[n] Communications System
- e: Allied Chiefs of Staff
- e: Alternate Core Spray
- e: Alternating Current, Synchronous
- e: Alternating Current Supply
- e: Alternating Current Synthesizer
- e: Altitude Control System
- e: American Cable Supply Corp.
- e: American Cancer Society
- e: American Carbon Society
- e: American Ceramic Society
- e: American Chemical Society
- e: American College of Surgeons
- e: American Colonization Society
- e: American Communication Services
- e: American Community Schools
- e: American Cryonics Society
- e: American Crystal Sugar
- e: Analog[ue] Computer Simulation
- e: Analog[ue] Computer System
- e: Analog[ue] Control System
- e: Analysis Computer System
- e: Analysis of Coping Style
- e: Anisotropically Conductive Silicone
- e: Annament Control System
- e: Annealed Copper-covered Steel
- e: Anodal Closing Sound
- e: Anterior Convex Side
- e: Antireticular Cytotoxic Serum
- e: Aperture Current Setting
- e: Apollo Command Systems
- e: Apollo Communications System
- e: Applied Colo[u]r System[s]
- e: Applied Computer Sciences
- e: Applied Computer Solution
- e: Approximate Cubic Search
- e: Armament Control System
- e: Armo[u]red Crew Seat
- e: Army Calibration System
- e: Army Camouflage School
- e: Army Communications Service
- e: Army Communicative Systems
- e: Army Community Services
- e: Art Center School
- e: Artillery Computer System
- e: Asbestos Cement Sheet
- e: Assembler Control System
- e: Assembly Control System
- e: Assembly Cost System
- e: Assessment of Cognitive Skills
- e: Asset Control System
- e: Assistant Chief of Staff
- e: Associated Computer System
- e: Associate in Commercial Science
- e: Association of Clinical Scientists
- e: Association of Council Secretaries
- e: Astro Communications System
- e: Asynchronous Communications Server
- e: Atmosphere Climate Study
- e: Atmosphere Control System
- e: Attack Center Switchboard
- e: Attitude Command System
- e: Attitude Control and Stabilization
- e: Attitude Control System
- e: Audio Communications System
- e: Audio Conducted Susceptibility
- e: Australia Container Service
- e: Australian Cancer Society
- e: Australian Card Service
- e: Australian Ceramic Society
- e: Australian Chamber of Shipping
- e: Australian Computer Society
- e: Australian Construction Services
- e: Australian Customs Service
- e: Autocorrelation Sequence
- e: Autograph Card Signed
- e: Automated Circulation System
- e: Automated Commercial System
- e: Automated Communications Set
- e: Automated Communications System
- e: Automatic Call Selector
- e: Automatic Cartographic System[s]
- e: Automatic Checkout System[s]
- e: Automatic Cheque Sorter
- e: Automatic Coding System[s]
- e: Automatic Control Subsystem[s]
- e: Automatic Control System[s]
- e: Automatic Counter System[s]
- e: Automation Composition System
- e: Autonomous Control Subsystem

- e: Auxiliary Control Station
- e: Auxiliary Cooling System
- e: Auxiliary Core Storage
- e: Auxiliary Core Store
- e: Average Calculating Speed
- e: Azimuth Control System

A.C.S. e: Admiral Commanding, Submarines
- e: Alaska[n] Communications System
- e: Associate in Commercial Science

A/C S e: Aircraft Security
A-CS e: Alignment Countdown Set
a-c s e: alternating-current synthesizer
ACs e: appeal cases
ÄCS d: Äthylchlorsilan
ACSA e: Acoustical Society of America
- e: Allied Communications Security Agency
- e: Allied Communications Support Area
- e: Army Communications Systems Agency
- e: Association of Collegiate Schools of Architecture
- e: Australian Corps of Signals Association

ACSAC e: Assistant Chief of Staff for Automation and Communications
ACS/AF e: Assistant Chief of Staff/Air Force
ACSA(P) e: Assistant Chief Scientific Adviser (Projects)
ACSAP e: Automated Cross-Section Analysis Program[me]
ACSAS e: Advanced Conformal Submarine Acoustic Sensor
- e: Automated Configuration Status Accounting System

ACSB e: Amplitude Companded Single Sideband
- e: Apollo Crew Systems Branch
- e: Aviation Command Screening Board

ACSC e: Accrediting Commission for Specialized Colleges
- e: Air Carrier Service Corporation
- e: Air Command and Staff College
- e: American Council on Schools and Colleges
- e: AMTRAK Commuter Services Corporation
- e: Applied Communications Systems Center
- e: Armament Control System Checkout
- e: Armaments Cooperation Steering Committee
- e: Army Computer Systems Command
- e: Atlanta Cancer Surveillance Center
- e: Australian Computer Science Conference
- e: Automated Contingency Support Capability

A-C Scale e: Anti-Caucasian Scale
ACSCCIM e: Assistant Chief of Staff for Command and Control Information Management
ACSCE e: Army Chief of Staff for Communications Electronics
ACSC-E e: Assistant Chief of Staff for Communications-Electronics
ACSCI e: Association for Computer- Based Systems for Career Information
ACS Conference e: Australian Computer Science Conference
ACSCOT e: American College of Surgeons Committee on Trauma
ACSCP e: Association of California State College Professors
ACSD e: Academic Computer Services Division
- e: Army Communication Service Division
- e: Assistant Controller of Stores Department
- e: Automatic Color-Scanned Device

ACS/DCI e: American Chemical Society/Division of Chemical Information
ACSDO e: Air Carrier Safety District Office[r]
ACSE e: Application Control Service Element[s]
- e: Association of Consulting Structural Engineers

ACSEA e: Allied (or Air) Command South East Asia
ACSEB e: Aviation Clothing and Survival Equipment Bulletin
ACSED e: Automated Computer Science Education
ACSES e: Automated Computer Science Education System
ACSET e: Advisory Committee on the Supply and Education of Teachers
ACSF e: Artificial Cerebrospinal Fluid
- e: Attack Carrier Striking Force

ACSFOR e: Assistant Chief of Staff for Force Development
ACSG e: Area Coordination Subgroup
AC-SG e: Alternating Current Signal Generator
ACSGp e: Area Coordination Subgroup
ACSGRP e: Area Coordination Subgroup
ACSH e: American Council on Science and Health
ACSI e: Affiliated Computer System Incorporated
- e: American Communication Services Incorporated
- e: Assistant Chief of Staff, Installations
- e: Assistant Chief of Staff for Intelligence
- e: Automotive Cooling Systems Institute

ACSIF e: Alaska[n] Communication System Industrial Fund
ACSIG e: Alignment Countdown Set Inertial Guidance
ACSig O e: Assistant Chief Signal Officer
ACSIL e: Admiralty Centre for Scientific Information and Liaison
ACSIM e: Assistant Chief of Staff for Information Management
ACSI-MATIC e: Assistant Chief of Staff, Intelligence [Automatic]
ACSIM-C4 e: Assistant Chief of Staff for Information Management-Command, Control, Communication
ACSIM Technique e: Accelerated Constrained Simplex Technique
ACSINT e: Assistant Chief of Staff for Intelligence
ACSIR e: Advisory Council for Scientific and Industrial Research
ACSIRO e: Australian Commonwealth Scientific and Industrial Research Organization
ACSL e: Advanced Continuous Simulation Language
- e: Advanced Continuous System Language
- e: American Computer Science League
- e: Australian Company Secretarys Letter

AcSM e: Academy Sergeant-Major
ACSM e: Advanced Conventional Stand-off Missile
- e: American College of Sports Medicine
- e: American Congress on Surveying and Mapping
- e: Apollo Command and Service Module
- e: Assemblies, Components, Spare-parts and Materials
- e: Associate of the Camborne School of Mines

acsm e: acoustic support measure[s]
ACSMA e: American Cloak and Suit Manufacturers Association
ACSMH e: Association of Clerks and Stewards in Mental Hospitals
ACSN e: Advanced Change Study Notice
- e: Appalachian Community Service Network
- e: Association of Collegiate Schools of Nursing

ACSO e: Access Opening
- e: Aircraft Capable of Satellite Operations
- e: Australian Council of School Organisations

ACSOC e: Acoustical Society of America
ACSP e: Advanced Control Signal Processor
- e: Advisory Council on Scientific Policy
- e: Aircraft Cross-Servicing Program[me]
- e: Alternating-Current, Spark Plug
- e: Army Central Service Point
- e: Army Chemical Supply Point
- e: Association of Collegiate Schools of Planning

ACSPA e: Australian Council of Solaried and Professional Associations
A/Cs Pay e: Accounts Payable
ACSPIC e: Applied Computing and Software Production and Inventory Control
ACS-PNL e: Access Panel
ACS-PRF e: American Chemical Society-Petroleum Research Fund
ACSQ e: Airborne Communications Squadron
ACSR e: Advanced Combat Surveillance Radar
- e: Alumin[i]um Cable; Steel-Reinforced
- e: Alumin[i]um Conductor; Steel-Reinforced
- e: Arizona Cactus and Succulent Research

ACSRC e: Assistant Chief of Staff for Reserve Components
A/cs Rec e: Accounts Receivable
ACSS e: Air Combat and Surveillance System
- e: Air Command and Staff School
- e: Alternating Current Signalling System
- e: Analog[ue] Computer Sub-System
- e: Army Chief of Support Services
- e: Association of Colleges and Secondary Schools
- e: Augmented Contact Support Set
- e: Automated Colo[u]r Separation System
- e: Automated Contingency Support System
- e: Automated Contract Specification System

ACS/S & A e: Assistant Chief of Staff; Studies and Analysis
ACSSAVO e: Association of Chief State School Audio-Visual Officers
ACSSB e: Amplitude Companded Single Sideband
ACS-SIA e: Australian Computer Society-Software Industries Association
ACSSM e: Associate Committee on Soil and Snow Mechanics
ACSSN e: Association of Colleges and Secondary Schools for Negroes
ACSSO e: Australian Council of State School Organisations
ACSSR e: Ammunition Consolidated Stock Status Report
ACSSRB e: Administrative Centre of Social Security for Rhine Boatmen

ACST e: Access Time
- e: Acoustic[s]
- e: Advisory Council on Science and Technology
- e: Army Clerical Speed Test
- e: Australian College of Speech Therapists

acst e: acoustic[al] [s]
ACSTFA e: Advisory Committee on Science and Technology and Foreign Affairs
ACSTI e: Advisory Committee for Scientific and Technical Information
ACSTIS e: Advanced Circular Scan Thermal Imaging System
acst plas e: acoustical plaster
acst t e: acoustical tile
ACSTT e: Advisory Committee on the Supply and Training of Teachers
ACSUS e: Association for Canadian Studies in the United States
ACSV d: automatisch computergesteuerte Schneideeinrichtung für Videobänder
ACSW e: Academy of Certified Social Workers
- e: Advanced Conventional Standoff Weapon
- e: Advanced Crew-Served Weapon

ACSWS e: Admiralty Civilian Shore Wireless Service
acsyn e: aircraft synthesis
ACSYNT e: Aircraft Synthesis
ACSYS e: Accounting Computer System
AC-system e: Aircab-system
ACSZ d: Aryl Cyclic Sulfonium Zwitterion
ACT e: Acceleration Time
- e: Access to Careers in Technology
- e: Accountant
- e: Account Control Table
- e: Accumulation Time
- e: Accumulator; Temporary
- e: Acetate Cloth Tape
- e: Achievement through Counselling and Treatment
- e: Acoustical Tile
- e: Acoustic Charge Transport
- e: Acquisition; Control of Test
- e: Acting
- e: Actinidin
- e: Action for Childrens Television
- e: Activated Clotting (or Coagulation) Time
- e: Active Cleaning Technique
- e: Active Control Technique (or Technology)
- e: Activities
- e: Activity
- e: Activity Completion Technique
- e: Actuary
- e: Actuating
- e: Actuation
- e: Actuator
- e: Adaptive Computer Technologies
- e: Adaptive Control of Thought
- e: Address Conversion Table
- e: Adjuvant Chemotherapy
- e: Adrenocorticotrophin
- e: Advance Corporation Tax
- e: Advanced Capability Tanker
- e: Advanced Career Training
- e: Advanced Chassis Technology
- e: Advanced Color Technology, Inc.
- e: Advanced Communications Technology
- e: Advanced Composite Technology
- e: Advanced Computer Technique[s]
- e: Advanced Computer Technology
- e: Advanced Concepts Team
- e: Advanced Concepts Test
- e: Advanced Concept Tire
- e: Advanced Concept Train
- e: Advanced Conversion Technology
- e: Advanced Core Test
- e: Advertised Computer Technologies
- e: Advisory Council on Technology
- e: Aerial Crosstalk
- e: Agricultural Central Trading
- e: Air-Charged Temperature
- e: Air-Cooled Triode
- e: Air Cargo Transport
- e: Air Combat Tactics
- e: Air Combat Training
- e: Air Control Team
- e: Air Council for Training
- e: Aircraft Commander Time
- e: Aircrew Classification Test
- e: Air Cushion Trailer
- e: Air Cushion Transporter
- e: Airman Classification Test
- e: Airport Control Tower
- e: Algebraic Compiler and Translator
- e: Allied Chemical Technology
- e: Alpha Counter Tube
- e: Alternating Current Transmission
- e: Alternating Current Trigger[ing]
- e: Alternative Community Telephone
- e: Alumina Ceramic Test
- e: American College of Theriogenologists
- e: American College of Toxicology
- e: American College Testing[program]
- e: American Council on Transplantation
- e: Analog[ue] Circuit Technique
- e: Analog[ue] Circuit Technology
- e: Analog[ue] Control Technology
- e: Analogical Circuit Technique
- e: Anglo-Canadian Telephone Co.
- e: Annual Change Traffic
- e: Antenna Cross Talk
- e: Anti-Comet-Tail
- e: Antichymotrypsin
- e: Anticoagulant Therapy
- e: Anticompromise Technique
- e: Apparatus Carrier Telephone
- e: Applied Computer Techniques Limited
- e: Area Correlation Tracker
- e: Armo[u]red Cavalry Trainer
- e: Army Chemical Typewriter
- e: Army Communicative Technology
- e: Asset Control Techniques
- e: Assignment Control Trainee
- e: Associated Container Transportation
- e: Associate of the College of Technology
- e: Association for Communications and Technology
- e: Association for Commuter Transportation
- e: Association of Christian Therapists
- e: Association of Civilian Technicians
- e: Association of Classroom Teachers
- e: Association of Communications Technicians
- e: Association of Cytogenetic Technologists
- e: Assure Competitive Transportation
- e: Attention Control Training
- e: Augmented Catalytic Thruster
- e: Australian Capital Territory
- e: Auto-Call-Telex
- e: Auto-lock Channel Tuning
- e: Autocoder to Common Business Oriented Language Translator
- e: Automated Contingency Translator
- e: Automated Continuity Tester
- e: Automatically Controlled Transportation
- e: Automatic Cable Tester
- e: Automatic Cannon Technology
- e: Automatic Capacitance Testing
- e: Automatic Capacitor Tester
- e: Automatic Channel and Time
- e: Automatic Checkout Technician
- e: Automatic Checkout Techniques
- e: Automatic Circuit Test[er]
- e: Automatic Circuit Testing
- e: Automatic Code Translation
- e: Automatic Code Translator
- e: Automatic Component Test[er]
- e: Automatic Component Testing
- e: Automatic Credit Transfer
- e: Average Calculating Time
- e: Aviation Classification Test
- e: Azimuth Control Torquer

act. e: acting
Act e: Actinium
- e: Activator
- e: Active

act e: action
- e: activated
- e: activated coagulation time
- e: activation
- e: actor
- e: actual
- e: actuarial
- e: actuate
- e: advanced coronary treatment
- e: aspartate carbamoyl transferase
- e: atropine coma therapy

Act. e: Actress
ACTA e: Activate Test Article
- e: Advanced Cargo / Tanker Aircraft
- e: Advanced Combat Training Academy
- e: Air Coach Transport Association
- e: Alternative Carrier Telecommunications Association
- e: American Cardiology Technologists Association
- e: American Cement Trade Alliance
- e: American Corrective Therapy Association
- e: Associated Container Transportation Australia
- e: Association of Chart and Technical Analysts
- e: Automated Calibration Temperature Activated
- e: Automatic Centrifugal Training Apparatus
- e: Automatic Computerized Transverse Axial

a cta s: a cuenta
ACTAC e: Association of Community Technical Aid Centres
Acta Chem Scand l: Acta Chemica Scandinavia (journ.)
Acta Crystallog l: Acta Crystallographica (journ.)
Acta Math Acod Sci Hung l: Acta Mathematica Academiae Scientiarum Hungaricae (journ.)
Acta Metall l: Acta Metallurgica (journ.)
Acta Neurochir. l: Acta Neurochirurgica (journ.)
Acta Neurochir.; Suppl. l: Acta Neurochirurgica; Supplementum (journ.)
Acta Neuropathol. l: Acta Neuropathologica (journ.)
Acta Neuroveg. l: Acta Neurovegetativa (journ.)
Acta Oto-Laryngol l: Acta Oto-Laryngologica (journ.)
Acta Phgys. Austr. Suppl. e: Acta Physica Austriaca Supplementum (journ.)
Acta Phys l: Acta Physica (journ.)
Acta Phys. Austr. l: Acta Physica Austriaca (journ.)

Acta Phys Pol l: Acta Physica Polonica (journ.)
Acta Phys Sin l: Acta Physica Sinica (journ.)
ACTAR e: Acoustics of the Target
ACTAS e: Army Consideration of Tactical Air Support
ACTA Scanner e: Automatic Computerized Transverse Axial Scanner
ACTB e: Aircrew Classification Test Battery
ACTC e: Advanced Computer Technique Corporation
- e: Air Class Teachers Certificate
- e: Air Commerce Type Certificate
- e: Art Class Teachers Certificate
ACT/CONV e: Activation / Conversion
ACT Corporation e: Advanced Computer Techniques Corporation
ACTCS e: Active Thermal Control System
- e: Air Conditioning and Temperature Control System
Act. ct. e: Actual Count
act.ct e: actual count
ACTD e: Attitude Control Torquing Device
- e: Automatic Telephone Call Distribution
Act. Def. e: Active Defense
Act Dir e: Acting Director
ACTDS e: Automatically Cued Target Detecting System
ACTDU e: Active Duty
ACTE e: Actuate
ACTe e: Anodal Closure Tetanus
ACTE e: Automatic Checkout Test Equipment
ACTEA e: Accrediting Council for Theological Education in Africa
ACTEDS e: Army Civilian Training, Education and Development System
ACTEL e: Alternating Current Thin-film Electroluminescence
ACTER e: Active Filter
- e: Anti-Countermeasures Trainer
ACTEW e: Australian Capital Territory Electricity and Water
ACTF e: Activities File
- e: Altitude Control Test Facility
ACTFA e: Atlantic City Table of Frequency Allocations
ACTFL e: American Council on the Teaching of Foreign Languages
ACTG e: Acting
- e: Actuate
- e: Actuating
- e: Advance Carrier Training Group
- e: AIDS (Acquired Immune Deficiency Syndrome) Clinical Trials Group
actg chf e: acting chief
Actg-Lt e: Acting-Lieutenant
actg sec e: acting secretary
Actg Sub-Lt e: Acting Sub-Lieutenant
ACT gun e: Anti-Comet Tail Gun
ACTH d: Adrenokortikotropes Hormon
- e: Adrenocorticotrophic Hormone
ACTH e: Arbitrary Correction To Hit
ACTheol e: Australian College of Theology
ACTH-LI e: Adrenocorticotrophin-Like Immunoreactivity
ACTI e: Advanced Computer Training Institute
- e: Advisory Committee on Technology Innovation
- e: Association of Chemists of the Textile Industry
ACTIAC e: Arms Control Technical Information and Analysis Center
ACT/IC e: Active in Commission
ACTIFS e: Active Control for Total In-Flight Simulator
ACTIL e: Australian Cotton Textile Industries Ltd
ACTIME e: Agency for the Coordination of Transport in the Mediterranean
ACTIMED e: Agency for the Coordination of Transport in the Mediterranean
ACTION e: Accepting Challenges of Today in Our New-world
ACTIONS e: Allocation and Control Through Identification of Ongoing Situations
ACTIS d: Angewandte Computertechnik für Informationssysteme GmbH
act/is e: active-in service
ACTIS e: Advanced Circular-scan Thermal Imaging System
- e: Acquired Immune Deficiency Syndrome Clinical Trials Information Service
- e: Auckland Commercial and Technical Information Service
ACTIV e: Activate[d]
- e: Activation
activ e: activity
ACTIVE e: Advanced Computer Training In a Versatile Environment
ACTL e: Actual
- e: American College of Trial Lawyers
ACTM e: Ashridge Centre for Transport Management
- e: Audio Center Transmitter
ACTMC e: Army Clothing, Textile and Material Center
ACTN e: Action
- e: Adrenocortico-Trophin
actn e: adrenocorticotrophin
actnt e: accountant
ACTNZ e: Associated Container Transportation New Zealand
ACTO e: Action Officer
- e: Advanced Control Test Operation
- e: Army Communicative Technology Office
- e: Association of Chief Technical Officers
acto e: automatic computing transfer oscillator
ACT-OC e: Active-Out of Commission
act/oc e: active-out of commission
ACTOL e: Air Cushion Take-Off and Landing
Act Opns e: Active Operations
ACTOR e: Askania Cline Theodolite Optical-Tracking Range
ACT-OS e: Active-Out of Service
ACTP d: Adrenokortikotropes Polypeptid
- e: Adrenocorticotrophic Polypeptide
- e: Advanced-Composite Thermoplastic
- e: Advanced Computer Techniques Project
- e: Advanced Computer Technology Project
- e: Advanced Control Technology Program[me]
- e: Alternating Current Testing Plant
- e: Australian Capital Territory Police
ACTPC e: Acting Pay Clerk
ACTPO e: Accountable Property Officer
ACT Program[me] e: American College Testing Program[me]
ACTR e: Actuator
- e: Air Corps Technical Report
- e: American Council of Teachers of Russian
Ac tra l: Altocumulus Translucidus
ACTRA e: Association of Canadian Television and Radio Artists
ACTRAC e: Accurate Tracking
ACTRAM e: Advisory Committee on the Transport of Radioactive Materials
ACTRAN e: Analog[ue] Computer Translator
- e: Autocoder-to-Common Business Oriented Language Translation (or Translating) Service
ACTRAN Service e: Autocoder to COBOL Translating Service
ACTREG e: Australian Capital Territory Regulations
ACTREP e: Activities Report
actrl e: acoustic trial[s]
ACTRL e: Acoustic Trials
ACTRS e: Association of Catholic TV and Radio Syndicators
ACTRSWD e: Asian Centre for Training and Research in Social Welfare and Development
ACTRUS e: Automatically Controlled Turbine Run-Up System
Ac-Try d: Acetyltryptophan
ACTS e: Acoustic Control and Telemetry System
- e: Acquisition, Cataloguing, Technical System
- e: Activated Carbon Treatment System
- e: Active Control Torque System
- e: Activity Code and Text Systems
- e: Advanced Channel Testing System
- e: Advanced Communications Technology Satellite
- e: Advanced Communications Tracking Satellite
- e: Advisory Commission on Textbook Specifications
- e: Air-Cushion Takeoff System
- e: Air Corps Tactical School
- e: Air Crew Training System
- e: Airline Computer Tracing System
- e: All-Channel Television Society
- e: American Christian Television Service
- e: American Christian Television System
- e: American Coalition for Traffic Safety
- e: Analog[ue] Conditioning and Test System
- e: Application Control and Teleprocessing System
- e: Arc Current Time Simulator
- e: Area Communications Terminal Subsystem
- e: Army Criteria Tracking System
- e: Association of Cable Television Suppliers
- e: Association of Career Training Schools
- e: Attitude Control and Translation System
- e: Automated Coin Toll Service
- e: Automated Commitment Tracking System
- e: Automated Component Trading System
- e: Automated Computer Time Service
- e: Automated Configuration Tracking System
- e: Automated Custom Terminal System
- e: Automatic Coin Telephone Service
- e: Automatic Common Business Oriented Language Translation System
- e: Automatic Computer Telex Services
- e: Automatic Computer Telex System
ACTSA e: Advanced Computer Tape Standards Association
ACTSECNAN e: Acting Secretary of the Navy
ACT SO e: Afro-Academic Cultural Technological and Scientific Olympics

ACTS/PROP e: Attitude Control and Translation System / Propulsion
ACTS/SCE e: Attitude Control and Translation System/Stabilization and Control Electronics
ACTSU e: Association of Computer Time-Sharing Users
ACTT e: Acetate
- e: Approved Course of Teacher Training
- e: Association of Cinematograph and Television Technicians
ACTU e: Automatic Control of Training Unit
ACTV e: Advanced Compatible Television
- e: American Coalition for Traditional Values
actv e: activate
- e: armored cavalry towing vehicle
Actv e: Activation
- e: Activity
act. val. e: actual value
ACTVNANAL e: Activation Analysis
actvt e: activate
Actvtr e: Activator
ACTVTY e: Activity
act. wt e: actual weight
ACTWU e: Amalgamated Clothing and Textile Workers Union
AC & TWU e: Atlantic Communication and Technical Workers Union
ACTY e: Activity
- e: Actuary
ACTYS e: Adaptable Computerized Type Setting
ACU d: Arbeitsgemeinschaft Computergestützter Unterricht
- e: Abilene Christian University
- e: Acceleration Compensation Unit
- e: Acceleration Control Unit
- e: Acknowledgement-signal Unit
- e: Acknowledgement Unit
- e: Activity Credit Unit
- e: Acute Care Unit
- e: Address Control Unit
- e: Advanced Connector Unit
- e: Aerial Control Unit
- e: Air-Conditioning Unit
- e: Airborne Control Unit
- e: Air Cleanup Unit
- e: Aircraft Control Unit
- e: Alarm Control Unit
- e: Align Control Unit
- e: Alternating Current Control Unit
- e: Analog[ue] Calibration Unit
- e: Analog[ue] Control Unit
- e: Analysis Control Unit
- e: Annunciator Control Unit
- e: Antenna Control Unit
- e: Armament Control Unit
- e: Assault Craft Unit
- e: Association of College Unions
- e: Association of Commonwealth Universities
- e: Association of Computer Users
- e: Autocycle Union
- e: Automatic Calling Unit
- e: Automatic Call Unit
- e: Automatic Control Unit
- e: Auxiliary Conditioning Unit
- e: Availability Control Unit
- e: Avionic Control Unit
- e: Avionics Cooling Unit
- l: Altocumulus
A-CU d: arsenhaltiges Kupfer
AcU e: Actinium Uranium
acu e: acuity
- e: aculeat
- e: acumen
- e: acuminate
- e: acupressure
- e: acupuncture
- e: acute
- e: arithmetic and control unit
- e: arithmetic computer
ACUA e: Association of Cambridge University Assistants
- e: Association of College and University Auditors
- e: Automatic Calling Unit Adapter
ACUB d: Agrochemischer Untersuchungs- und Beratungsdienst
ACUCA e: Association of Christian Universities and Colleges in Asia
ACUCAA e: Association of College, University and Community Arts Administrators
ACUCHE e: Association of Computer Units in Colleges and Institutions of Higher Education
ACU-ERI e: American Conservative Union Education and Research Institute
ACUG e: Association of Computer User Groups
ACUHO e: Association of College and University Housing Officers
ACUHO-I e: Association of College and University Housing Officers-International
ACUI e: Association of College Unions International
- e: Automatic Calling Unit Interface
ACUIB e: Association of Canadian University Information Bureaus
- e: Association of Canadian University Information Bureaux
ACUIIS e: Association of Colleges and Universities for International Intercultural Studies
ACUMG e: Association of College and University Museums and Galleries
Acuña e: Ciudad Acuña
ACUNS e: Association of Canadian Universities for Northern Studies
ACUNY e: Associated Colleges of Upper New York
ACUO e: Association of College and University Offices
- e: Avionics Cooling Unit Operator
ACUP e: Association of Canadian University Presses
- e: Association of College and University Printers
ACUPAE e: American Council for University Planning and Academic Excellence
acupunc e: acupuncture
- e: acupuncturist
ACURAD e: Acoustic Underwater Range Determination
ACURIL e: Association of Caribbean University and Research Institute Libraries
ACURL e: Association of Caribbean University and Research Libraries
ACURP e: American College of Utilization Review Physicians
ACUS e: Atlantic Council of the United States
ACUSA e: Assistant Chief of United States Airstaff
ACUSNY e: Association of Colleges and Universities of the State of New York
ACUSYST e: Automated Culture System
ACUT e: Automatic Calling Unit Terminator
ACUTA e: Association of College and University Telecommunications Administrators
ACUTE e: Accountant[s] Computer Users Technical Exchange
- e: Association of Canadian University Teachers of English
ACV d: Arbeitsgemeinschaft Chemische Verfahrenstechnik
- e: Access Control Verification
- e: Actual Cash Value
- e: Address Control Vector
- e: Air Control Valve
- e: Air-Cushion Vehicle
- e: Air Cushion Vessel
- e: Airway Closing Volume
- e: Alarm Check Value
- e: Alarm Check Valve
- e: Alfalfa Cryptic Virus
- e: Alternating Current Volt[s]
- e: Alternating Current Voltage
- e: Amount of Critical View
- e: Armo[u]red Combat Vehicle
- e: Armo[u]red Command Vehicle
- e: Armoured Cannon Vehicle
- e: Armoured Cavalry Vehicle
- e: Associate; College of Violinists
- e: Associate of the College of Violinists
- e: Auxiliary Aircraft Carrier (or Tender)
ACVB d: aorto-koronarer Venen- Bypass
- e: Aorto-Coronary Venous Bypass
ACVC e: Ada Compiler Validation Capability
- e: Ada Compiler Validation Criteria
- e: Armament and Combat Vehicle Center
- e: Army Commercial Vehicle Code
ACVD e: Acute Cardiovascular Disease
- e: American College of Veterinary Dermatology
ACVE e: Accelerometer Calibration Vibration Exciter
ACVEN e: Advisory Committee on Vehicle Emission and Noise
ACVF e: Advanced Composite Vertical Fin
- e: Autocovariance Function
ACVIM e: American College of Veterinary Internal Medicine
Ac vir e: Altocumulus Virga
ACVL e: Association of Cinema and Video Laboratories
ACVM e: Adaptive Control Voltage Module
- e: American College of Veterinary Microbiologists
ACV-MP e: Acyclovir Monophosphate
ACVO e: American College of Veterinary Ophthalmologists
ACVP e: Additive Colo[u]r Viewer Printer
- e: American College of Veterinary Pathologists
ACVR e: American College of Veterinary Radiology
ACVs e: Air Cushion Vessels
ACVS e: American College of Veterinary Surgeons
- e: Australian College of Veterinary Scientists
- e: Automatic Computer Voltage Stabilizer
- e: Automatic Continuity Verifiers
ACVSF e: Air-Cooled Vault Storage Facility
ACVT e: Activate
- e: American College of Veterinary Toxicologists
- e: Armoured Combat Vehicle Technology
ACV Test e: Airway Closing Volume Test

ACVTP e: Acyclovir Triphosphate
- e: Armored Combat Vehicle Technology Program[me]
AC VTVM e: Alternating-Current Vacuum Tube Voltmeter
ACV[T]Y e: Activity
ACW d: Ausgabecodewandler
- e: Access Control Word
- e: Airborne Collision Warning
- e: Air Control [and] Warning[system]
- e: Aircraftswoman
- e: Alternating Continuous Wave[s]
- e: Alternating Current Winding
- e: Amateur Cine World (journ.)
- e: American Chain of Warehouses
- e: Anti-Clockwise
- e: Anticarrier Warfare
AC/W e: Acetone/Water
AC & W e: Air Communications and Weather
- e: Aircraft and Warning
- e: Aircraft Control and Warning
acw e: automatic car wash
ACWA e: Amalgamated Clothing Workers of America
- e: American Clean Water Association
- e: Assistant Casework Supervisor
- e: Associate of the Institute of Cost and Works Accountants
- e: Automatic Car Wash Association
ACWAI e: Automatic Car Wash Association International
ACWC e: Advisory Committee on Weather Control
- e: American Council of Women Chiropractors
ACWE e: American Cotton Waste Exchange
ACWF e: Army Central Welfare Fund
ACWG e: Aircraft Control and Warning Group
ACWIS e: American Committee for the Weizmann Institute of Science
ACWL e: Army Chemical Warfare Laboratory
ACWO e: Aircraft Control and Warning Officer
ACWP e: Actual Cost for Work Performed
ACWRON e: Aircraft Control and Warning Squadron
ACWS e: Air Control and Warning Station
- e: Aircraft Control and Warning Squadron
- e: Aircraft Control and Warning System
- e: All-Canada Weekly Summaries
- e: Assistant Casework Supervisor
ACWs e: Access Control Words
AC & WS e: Aircraft Control and Warning Station[s]
ACWSS e: Aircraft Control and Warning System Station
ACW System e: Air[craft] Control and Warning System
ACWT e: Average Customer Wait Time
AcX d: Aktinium X
Ac X e: Actinium X
AC/XRT e: Adriamycin, Cyclophosphamide / X-Ray Therapy
ACY e: Akron, Canton & Youngstown
- e: American Cyanamid Company
- e: Anthocyanin
- e: Atlantic City (airport)
ACy e: All Cycles
AC & Y e: Akron, Canton & Youngstown
ACYC e: Anticyclonic
acyl-co A e: coenzyme A ester
acyro e: acyrologia
- e: acyrologic[al]
- e: acyrology
AcZ d: Acetylzahl
ACZ d: Agrochemisches Zentrum
ACZA e: Ammoniacal Copper Zinc Arsenate
ACZCS e: Advanced Coastal Zone Colour Scanner
AD d: Ablaufdiagramm
- d: Abwehrdienst
- d: Adressendefinition
- d: Ahorn
- d: Alarmdrucker
- d: Amateurfunk-Diplom
- d: Amplitudendiskriminator
- d: Analog-Digital
- d: Anode
- d: antarktische Divergenz
- d: Arbeitsgemeinschaft Druckbehälter
- d: Arbeitsplatzdatei
- d: Archivdatei
- d: Archiv für Diplomatik (journ.)
- d: assoziierter Deskriptor
- d: Außendienst
- d: Außendurchmesser
- d: Austauschdienst
- e: Abbreviated Dialling
- e: Abdominal Diameter
- e: Above Deck
- e: Absolute Delay
- e: Absolute Determination
- e: Absolute Deviation
- e: Accelerated Depreciation
- e: Access Door
- e: Accession Document
- e: Accessions Document
- e: Accidental Discharge
- e: Accident Dispensary
- e: Account Directory
- e: Account Distribution
- e: Accounting and Disbursing
- e: Accounting Department
- e: Accounts Department
- e: Accrued Dividend
- e: Achievement Drive
- e: Acknowledgment Due
- e: Acoustic Decoupler
- e: Acquisition Data
- e: Acquisition Director
- e: Actinomycin D
- e: Action Driver
- e: Active Dosimeter
- e: Activity Discard
- e: Adamantane
- e: Adapt
- e: Adaptation
- e: Adaptor
- e: Add Decimal
- e: Addendum
- e: Adder
- e: Addict
- e: Additional Data
- e: Address[able] Data
- e: Address Decoder
- e: Address Decoding
- e: Address Defined
- e: Addressed Device
- e: Adduct[s]
- e: Aden Airways
- e: Adenoidal Degeneration
- e: Adenoid Degenerative
- e: Adenovirus
- e: Adipose Fin
- e: Adjuvant Disease
- e: Administrative Department
- e: Administrative Directive
- e: Administrative District
- e: Admiralty Dockmaster
- e: Admission Directive
- e: Adoption Directory
- e: Adult
- e: Advanced Development
- e: Advantage[s]
- e: Advertising
- e: Advisory Direction
- e: Aerial Delivery
- e: Aerodrome
- e: Aerodynamic Decelerator
- e: Aeronautical Data
- e: After Date
- e: After Digital
- e: Afterdischarge
- e: Aggregate Demand
- e: Agricultural Decisions (journ.)
- e: Agricultural Department
- e: Airborne Designator
- e: Aircraft Depot
- e: Aircraft Depth
- e: Aircraft Division
- e: Air Data
- e: Air Defense
- e: Air Density
- e: Air Depot
- e: Air Despatch
- e: Air Director
- e: Air Division
- e: Airdrome
- e: Air Duct
- e: Airframe Design
- e: Airworthiness Directive
- e: Alarm Device
- e: Alcohol Dehydrogenase
- e: Aleutian Disease
- e: Algebraic Dependency
- e: Allergic Disease
- e: Allowable Deficiency
- e: Alloy-Diffused
- e: Alloy-Diffusion-Technique
- e: Alloy Diffusion
- e: Alphabetic[al] Data
- e: Alphanumeric Data
- e: Alphanumeric Device
- e: Alternate Days
- e: Alternate Drop
- e: Altitude Deviation
- e: Altitude Differential
- e: Alzheimers Disease
- e: Ambulance Driver
- e: American Documentation (journ.)
- e: Ammunition Depot
- e: Ampere Demand [Meter]
- e: Amplifier Detector
- e: Amplitude Discriminator
- e: Analgesic Dose
- e: Analog[ue]/Digital
- e: Analog[ue] Data
- e: Analog[ue] Device[s]
- e: Andorra
- e: Androstenedione
- e: Androstenoide
- e: Anodal Deviation
- e: Anodal Duration
- e: Anterior Deltoid
- e: Anti-Disturbance
- e: Antidiarrhea
- e: Antigenic Determinant
- e: Apollo Development
- e: Application Documentation
- e: Applied Dynamics Limited
- e: Approximate Digestibility
- e: Architectural Design
- e: Architectural Digest

- e: Area Director
- e: Area Discriminator
- e: Area Drain
- e: Arithmetic Device
- e: Arm/Destruct
- e: Armament Depot
- e: Armament Division
- e: Armo[u]red
- e: Armoured Division
- e: Army Department
- e: Army Depot
- e: Army Digest
- e: Army Driver
- e: Array Declaration
- e: Array Declarator
- e: Array Descriptor
- e: Art Digest (journ.)
- e: Art Director
- e: Articular Disc
- e: Artillery Division
- e: Artificer Driver
- e: As Drawn
- e: Assembler Deck
- e: Assembly District
- e: Assembly Drawing
- e: Assignment Date
- e: Assistant Director
- e: Associate Degree
- e: Associate Director
- e: Assured Destruction
- e: Atlantic & Danville
- e: Atomic Defence
- e: Atomic Drive
- e: Attack Director
- e: Attendant
- e: Attention Display
- e: Autograph Document
- e: Automatic Depositor
- e: Automatic Detection
- e: Automatic Dictionary
- e: Automatic Display
- e: Automatic Divide
- e: Autosomal Dominant
- e: Availability Date
- e: Available Data
- e: Avalanche Diode
- e: Average Depth
- e: Average Deviation
- e: Average Diameter
- e: Average Dose
- e: Aviation Delay
- e: Avoidable Delay
- e: Awaiting Disconnection
- e: Axiodistal
- e: Azide Dextrose-broth
- e: Azimuth Drive
- e: Doctor of Arts
- l: Anno Domini
- R: Aviatsionnaya Diviziya

Ad d: Adresse
- e: Adapter
- e: Address
- e: Adenosine
- e: Adeodato
- e: Adlai
- e: Administration
- e: Administrative
- e: Administrator
- e: Adrenal
- e: Adverb
- e: Aedh

ad d: andalusite
- e: active duty
- e: adapted
- e: adjoint
- e: advance[d]
- e: advertisement
- e: advice
- e: after drain
- e: air-dried
- e: ampere demand[meter]
- e: armo[u]red
- l: ante diem

A.D. e: Admiralty Design
- e: Awning Deck
- l: Agnus Dei
- l: Anno Domini

a. D. d: an der Donau
- d: auf Dienstleistung
- d: auf Dienstreise
- e: air-Dry
- e: aldershot District
- d: außer Dienst

a. d. d: an dem
- d: an der
- d: auf dem
- d: auf der
- d: aus dem
- d: aus der
- l: auris dexter

a d f: a droit

A & D e: Admission and Discharge
- e: Alcohol and Drug
- e: Architects and Designers
- e: Assembly and Disassembly
- e: Atlantic & Danville

a & d e: accounting and disbursing
- e: ascending and descending
- e: control alcohol and drug control

A-D e: Advance-Decline

a-d e: advance-decline line

A/D e: Advance/Decline
- e: Advanced Design
- e: Air Depot
- e: Altitude/Depth
- e: Analog[ue]-to-Digital
- e: Arm / Destruct
- e: Assembly / Disassembly
- e: Awaiting Delivery

a/d e: avoirdupois

ÄD d: äquatoriale Divergenz

ä.D. d: äußerer Durchmesser

ad 2 vic. l: ad duas vices

ADA d: Absolutdruckaufnehmer
- d: Adenosindesaminase
- d: Adresse für Ausgabegerät
- d: Allgemeine Dienstanweisung [für das Post- und Fernmeldewesen]
- d: Allgemeines Dienstalter
- d: Amtsdienstalter
- d: Anzeiger für Deutsches Altertum und Deutsche Literatur (journ.)
- d: Außendienstangestellter
- e: Academy of Dispensing Audiologists
- e: Acetonedicarboxylic Acid
- e: Action Data Automation
- e: Active Duty Agreement
- e: Activity Discard Acknowledgement
- e: Actuarial Data Assembly
- e: ADA (Programmiersprache benannt nach Ada Byron)
- e: Address Adder
- e: Address Adding
- e: Adenosine Deaminase
- e: Adjusted Daily Average
- e: Advanced Development Analysis
- e: Advanced Digital Avionics
- e: Advanced Disk Array
- e: Advance Deposit Agreement
- e: Advance Design Aids
- e: Advisory Area
- e: Aerojet Differential Analyzer
- e: After Date of Award
- e: Airborne Data Automation
- e: Aircraft Defense Analysis
- e: Air Data Assembly
- e: Air Defense Area
- e: Air Defense Artillery
- e: Airlines Deregulation Act
- e: Aldrindicarboxylic Acid
- e: Alphanumeric Data Acquisition
- e: Aluminium Dihydroxyaminoacetate
- e: American Dairy Association
- e: American Dehydrators Association
- e: American Dermatological Association
- e: American Diabetes Association
- e: American Dietetic Association
- e: Ammonium Dihydrogen Arsenate
- e: Amplifier Detector Assembly
- e: Analog[ue]-Digital-Analog[ue]
- e: Analog[ue] Drive Assembly
- e: Analog[ue] Driver Amplifier
- e: Angle Data Assembly
- e: Angular Differentiating-Integrating Accelerometer
- e: Anthraquinone Disulfonic Acid
- e: Applied Decision Analysis
- e: Arming Device Assemblies
- e: Assistant Defence Advisor
- e: Assistant Director of Artillery
- e: Assistant District Attorney
- e: Association of Drainage Authorities
- e: Atomic Development Administration
- e: Atomic Development Authority
- e: Audio Distribution Amplifier
- e: Australian Dental Association
- e: Australian Department of Agriculture
- e: Auto Directional Antenna
- e: Automated Differential Agglutination
- e: Automated Dispensing Analyzer
- e: Automated Document Analysis
- e: Automatic Damper Arm
- e: Automatic Data Acquisition
- e: Automatic Data Aids
- e: Automatic Document Analysis
- e: Automatic Dynamic Analyzer
- e: Automobile Dealers Association
- e: Avalanche Diode Amplifier
- e: Average Daily Allowance
- e: Average Daily Attendance
- e: Average Deviation Adjustment
- e: Azide Dextrose Agar
- e: Azimuth Drive Assembly
- e: Azodicarbonamide

AdA d: Akademie der Arbeit

AD-A e: Air Density-A (Satellite)

ada e: airtight drywall approach
- e: american dental association

Ada e: computer programming language promoted by the Department of Defense

ADA# e: American Diabetes Association diet number

ADAA e: Air-Driven Air Amplifier
- e: American Dental Assistants Association
- e: Australian Dictionary of Acronyms and Abbreviations

ADAAC e: Automatic Data Acquisition and Computer

ADAB d: Allgemeine Dienstanweisung für die Bundesbahnbeamten

ADABD e: Air Defense Artillery Board

ADAC d: Allgemeiner Deutscher Automobilclub
- e: Acoustic Data Analysis Center
- e: Air Defence Artillery Commander
- e: Air Defense Artillery Complex

- e: Airfoil Design and Analysis Center
- e: All-Digital Attack Center
- e: Analog[ue]-Digital-Analog[ue] Converter
- e: Applications in Design Automation Committee
- e: Assistant Director of Army Contracts
- e: Astrodata Data Acquisition and Control
- e: Automated Direct Analog[ue] Computer
- e: Automatic Data Automatic Correction
- e: Automatic Diagnosis Automatic Correction
- e: Automatic Direct Analog[ue] Computer
- e: Aviation Development Advisory Committee

AD/AC e: Air Defense Aircraft
- e: Automatic Detection / Automatic Classification

adac e: automatic direct analog computer
adacc e: automatic data acquisition and computer complex
ADACIOM e: Associated Drug and Chemical Industries of Missouri
ADACS e: Attitude Determination and Control System
- e: Automatic Data Acquisition and Control System

A-DACT e: Adriamycin; Dactinomycin
ADACT e: Advise Action Taken
ADACX e: Automatic Data Acquisition and Computer Complex
ADAD e: Air Defense Alerting Device
ADAD e: Air Defense Artillery Director
ADADA e: Advise Approximate Date
ADADS e: Army Depot Automatic Diagnostic System
- e: Assistant Director Army Dental Service

ADADs e: Automatic Dialing and Announcing Devices
ADAE e: Advanced Diploma in Art Education
ADAEA e: Alcohol and Drug Abuse Education Act
ADAGE e: Air Defense Air-to-Ground Engagement
ADAGP f: Association pour la Diffusion des Arts Graphiques et Plastiques
ADAH e: Assistant Director of Army Health
ADAHF e: American Dental Association Health Foundation
ADAI e: Apollo Documentation Administration Instruction
ADAIS e: Aerodynamic Data Analysis and Integration System
aDAl d: allgemeines Dienstalter
ADA-L e: Action Data Automation- Large
ADAL e: Action Data Automation Language
- e: Authorized Dental Allowance List

ADALCON e: Advise All Concerned
ADALINE e: Adaptive Linear
ADALINE Element e: Adaptive Linear Element
ADALINE Neuron e: Adaptive Linear Neuron
AD/ALT e: Advanced Destroyer/Aircraft Lightweight Torpedo
ADAM d: Allgemeine Dialog-Anwendungs-methode
- d: automatische Datenmeßeinrichtung
- d: automatischer Datenmeßsender
- e: Adaptive Arithmetical Method
- e: Adaptive Digital Avionics Module
- e: A Data Management [System]
- e: Advanced Data Access Method
- e: Advanced Data Management
- e: Advanced Direct-Landing Apollo Mission
- e: Agriculture Department Automated Manpower
- e: Air-Base Damage Assessment Model
- e: Air-Delivered Attack Marker
- e: Air-quality Data Monitoring
- e: Air Defense Antimissile
- e: Air Deflection and Modification
- e: Air Deflection and Modulation
- e: All-Digital Answering Machine
- e: Analog[ue] Data Acquisition Module
- e: Angular Distribution Auger Microscopy
- e: Aperture Distribution and Maintenance
- e: Area Denial Artillery Munition
- e: Artillery-Delivered Antipersonnel Mine
- e: Artillery Delivered Antipersonnel Munition
- e: Associometrics Data Management
- e: Automated Data Management
- e: Automated Deposition of Advanced Materials
- e: Automated Digital Antenna Measurement
- e: Automatic Digital Aerial Measurement
- e: Automatic Direct Access Management
- e: Automatic Distance and Angle Measurement
- e: Automatic Dynamic Analysis Measurement
- e: Axisymmetric Duct Aeroacoustic Modelling

A'dam d: Amsterdam
adam e: adamantine
- e: Alzheimers Disease Association of Maryland
- e: associometrics data management[system]
- e: Automatic Document Abstracting Method

ADA/MH e: Alcohol, Drug Abuse and Mental Health
ADAMHA e: Alcohol, Drug Abuse and Mental Health Administration
adamite e: basic zinc arsenate
adaml e: advise by admiral
- e: advise by airmail

adamm e: area defense anti-missile missile
Adam's e: Adam's Bridge
- e: Adam's Peak

ADAMS e: Advanced Action Manipulator System
- e: Advanced Design Aluminium Metal Shelter
- e: Airborne Data Acquisition Multifunction System
- e: Airborne Data Analysis and Monitoring System
- e: Application Development and Management System
- e: Automated Dynamic Analysis of Mechanical System

ADAM System e: A Data Management System
- e: Advanced Data Management System
- e: Associometrics Data Management System
- e: Automated Data Management System
- e: Automated Digital Aerial Measurement System

AD AN l: Ad Annum
ADANDAC e: Administrative and Accounting
adandac e: administrative and accounting purposes
a.d. and c. e: advice duration and charge
AD and C e: Advise Duration and Charge
AD and D e: Accidental Death and Dismemberment
ADAnw. d: allgemeine Dienstanweisung
ADAOD e: Air Defense Artillery Operations Detachment
ADAOO e: Air Defense Artillery Operations Office[r]
ADAP e: Active Duty Assistance Program[me]
- e: Adaptor
- e: Airport Development Aid Program[me]
- e: Alzheimers Disease-Associated Protein
- e: American Dental Assistants Program[me]
- e: Analog[ue]-Digital Automatic Program[me]
- e: Assistant Director of Administrative Planning

adap e: adapted
Adap e: Adapter
A.D.A.P. e: Assistant Director of Army Psychatry
ADAPCP e: Alcohol and Drug Abuse Prevention and Control Program[me]
ADAPS e: Armament Delivery Analysis Programming System
- e: Assistant Director of Army Postal Services
- e: Automatic Display and Plotting System

ADAPSO e: Association of Data Processing Service Organizations
ADAPSO/SIA e: Association of Data Processing Service Organizations/ Software Industry Association
ADAPSP e: Association of Data Processing Service Organizations Panels
ADAPT e: Accent on Developing Abstract Processes of Thought
- e: Active Duty Assistance Program[me] Team
- e: Adaptable
- e: Adaptation
- e: Adapter
- e: Adaption
- e: Adaption of Automatically Programmed Tools
- e: A Diagnostic and Prescriptive Technique
- e: Advanced Data Adapter/Processor Tester
- e: Advanced Development Aims Processor Transponder
- e: Aerospace Data Adapter Programmable Tester
- e: Alcohol and Drug Abuse Prevention Team
- e: Alcohol and Drug Abuse Prevention Treatment
- e: Analog[ue] Digital Automatic Program[me] Tester
- e: Arizona Data Accessing Programming and Training
- e: Advanced Research Project Agency Data-base Access and Presentation Termin
- e: Automated Data Analysis and Planning Technique
- e: Automated Data Analysis and Presentation Techniques
- e: Automated Data Analysis using Pattern-recognition Techniques
- e: Automatic Data Acquisition and Processing Techniques
- e: Avco Data Analysis and Prediction Techniques

adapticom e: adaptive communication
ADAPTS e: Air-Deliverable Anti- Pollution Transfer System
- e: Air-Deployable Anti-Pollution Transfer System

- e: Analog[ue]-to-Digital-to-Analog[ue] Process and Testing System
- e: Automatic Data Acquisition and Processing for Transportation System

ADAR e: Advanced Data Acquisition Routine
- e: Advanced Design Array Radar
- e: Advanced Development Array Radar
- e: Air Defense Area
- e: Analog[ue]-to-Digital-to-Analog[ue] Recording

ADARA e: American Deafness and Rehabilitation Association

ADARCO e: Advise Data of Reporting in Compliance with Orders

ADARD e: Acid Deposition and Atmospheric Research Division

ADARE e: Advise Date of Receipt

ADARS e: Adaptive Antenna Receiver System
- e: Airborne Data Acquisition and Recording System
- e: Army Defense Acquisition Regulation Supplement

ADAR System e: Analog[ue] Data Reduction System

ADAS d: Abkommen über deutsche Auslandsschulden
- e: Acid Deposition Assessment Staff
- e: Action Data Automation Small
- e: Action Data Automation System
- e: Adaptive Digital Acquisition Sampling
- e: Advanced Digital Avionics System
- e: Agricultural Development Advisory Service
- e: Airborne Data Acquisition System
- e: Airborne Dynamic Alignment System
- e: American Dental Association Specifications
- e: Analog[ue] Data Acquisition System
- e: Architecture Design and Assessment System
- e: Architecture Development and Assessment System
- e: Army Digital Avionics System
- e: Automated Data Acquisition System
- e: Automatic Data Acquisition System
- e: Automatic Data Analysis System
- e: Automatic Disk Allocation System
- e: Auxiliary Data Annotation Set
- e: Auxiliary Data Annotation Set (or System)
- e: Auxiliary Data Annotation System

ADA-S e: Action Data Automation- Small

ADASC e: Auto Dismantlers Association of Southern California

adash e: advise date of shipment

ADASP e: Air Defense Annual Service Practice
- e: Automatic Data and Select Program[me]

ADASSA e: Aerodynamic Design and Analysis System for Supersonic Aircraft

ad ast. l: ad astra

ADAT d: automatisierte, dateigesteuerte Testabwicklung
- e: Allied Directorate of Air Transport
- e: Army Dependents Assurance Trust
- e: Assistant Director of Army Telegraph
- e: Automatic Data Accumulation and Transfer
- e: Automatic Data Accumulator and Transfer

ADATE e: Automatic Digital Assembly Test Equipment

ADATM e: Artillery-Delivered Anti-Tank Mine

Adatom d: adsorbiertes Atom

ADATS e: Air Defense Anti-Tank System
- e: Air Defense Artillery Threat Simulator
- e: Army Development and Acquisition of Threat Simulators
- e: Assistant Director of Auxiliary Territorial Service
- e: Automated Data and Telecommunications Service

ADAU e: Air Defense Alert Unit
- e: Auxiliary Data Acquisition Unit

A-D-Auflösung d: Analog-Digital-Auflösung

adaval e: Advise Availability

ADAV & RS e: Assistant Director Army Veterinary and Remount Services

ADAW d: Abhandlungen der Deutschen Akademie der Wissenschaften zu Berlin (journ.)

ADAWS e: Action Data Automation and Weapons System
- e: Assistant Director of Army Welfare Services

A-day e: Assault Day

ADB d: Adreßbuch des Deutschen Buchhandels
- d: Adreßbuch des deutschsprachigen Buchhandels
- d: Adressen- und Datenbus
- d: Allgemeine Deutsche Binnentransport-Versicherungsbedingungen
- d: Allgemeine Deutsche Biographie
- d: Anlagendatenbank
- d: Arbeitsgemeinschaft Deutscher Betriebsingenieure
- e: Accidental Death Benefit
- e: Acoustic Distribution Box
- e: Address Bus
- e: Adjusted Debit Balance
- e: Aerodynamic Data Book
- e: African Development Bank
- e: Air Defense Board
- e: Apollo Data Bank
- e: Apple Desktop Bus
- e: Arctic Drift Barge
- e: Asian Development Bank
- e: Australian Dictionary of Biography
- e: Average Daily Balance

adb a: automatisk databehandling

adb. d: adiabatisch

A. D. B. d: Allgemeine Deutsche Burschenschaft
- d: Allgemeiner Deutscher Beamtenbund
- d: Allgemeiner Deutscher Burschenbund

AdB d: Arbeitsgemeinschaft der Buchhandlungen

adB e: acceleration decibel[s]

ADBD e: Active Duty Base Date

ADBE e: Adobe Systems Incorporated

ADBF e: Australian Deer Breeders Federation
- e: Azurophil-Derived Bactericidal Factor

ADBM e: Apple Desktop Bus Microcontroller
- e: Association of Dry Battery Manufacturers

ADBMS e: A Data Base Management System
- e: Available Database Management System

ADBNL d: Arbeitsgemeinschaft Deutscher Beauftragter für Naturschutz und Landschaftspflege

ADBPA f: Association pour le Développement des Bibliothèques Publiques en Afrique

a. d. Bs. d: aus dem Besitz

ADBS e: Advanced Data-Base System

ADBT e: Access Division Binding Time

ADC e: Accounting Detail Card
- e: Accretion Disk Corona
- e: Acoustic Device; Countermeasure
- e: Acoustic Dimension Compiler
- e: Acquired Immune Deficiency Syndrome Dementia Complex
- e: Acquisition, Development and Construction
- e: Action Delay Character
- e: Active Diffusion Control
- e: Active Duty Commitment
- e: Actual Device Code
- e: Actuation Data Communication
- e: Adaptive Data Compression
- e: Address-Complete Charge
- e: Add with Carry
- e: Advanced Data Code[r]
- e: Advanced Delivery of Correspondence
- e: Advanced Design Change
- e: Advanced Development Concept
- e: Advanced Display Computer
- e: Advanced Drawing Change
- e: Advance Delivery of Correspondence
- e: Advancing Developing Countries
- e: Advise Duration and Charge
- e: Aerial Dish Control
- e: Aerodrome Control
- e: Aerodrome Defence Corps
- e: Aerodynamic Data Correlation
- e: Aerophysics Development Corporation
- e: Aerospace Defense Command
- e: Agricultural Development Center
- e: Agricultural Development Council
- e: Aid to Dependent Children
- e: Aiming Date Chart
- e: Airborne Digital Computer
- e: Aircraft Directives Configuration
- e: Air Data Computer[s]
- e: Air Data Computing
- e: Air Data Converter
- e: Air Defense Center
- e: Air Defense Command
- e: Air Defense Computer
- e: Air Defense Controller
- e: Air Development Center
- e: Air Diffusion Council
- e: Air Direction Center
- e: Airdrome Defense Corps
- e: Alaska Defense Command
- e: Albumin, Dextrose, Catalase
- e: Alloy Data Center
- e: Allyl Diglycol Carbonate
- e: Almost-Developed Country
- e: Ambulance Design Criteria
- e: Ampere[s] Direct Current
- e: Ampex Disk Controller
- e: Analog[ue]-Digital Computer
- e: Analog[ue]-Digital Control
- e: Analog[ue]-to-Digital Conversion
- e: Analog[ue]-[to-]Digital Converter
- e: Analog[ue] Derived Clock
- e: Analytic Decisions Corporation
- e: Analytic Drag Control
- e: Anchorage Dependent Cell
- e: Anodal Duration Contraction
- e: Antenna Dish Control
- e: Anthracenedicarboxaldehyde
- e: Apollo Display Console
- e: Apparent Depth of Compensation
- e: Applied Data Communications
- e: Area Damage Control
- e: Area Data Center

- e: Area Defense Counsel
- e: Areas of Deeper Convection
- e: Arginine Decarboxylase
- e: Armament Development Center
- e: Army Dental Corps
- e: Asian Development Center
- e: Assembler Card Deck
- e: Assistant Defense Counsel
- e: Assistant Director of Ceremonies
- e: Assistant Director of Contracts
- e: Assistant Division Commander
- e: Assistive Device Center
- e: Asynchronous Data Channel
- e: Asynchronous Digital Combiner[s]
- e: Audio/Data Communications
- e: Australian Diabetic Council
- e: Authorized Data Chain
- e: Automated Document Composer
- e: Automatic Data Channel
- e: Automatic Data Coding
- e: Automatic Data Collecting
- e: Automatic Data Collection
- e: Automatic Data Collector
- e: Automatic Data Computer
- e: Automatic Data Computing
- e: Automatic Depth Control
- e: Automatic Deviation Control
- e: Automatic Digital Calculator
- e: Automatic Digital Computer
- e: Automatic Drift Control
- e: Automatic Drip Coffee
- e: Automatic Drive Control
- e: Average Daily Census
- e: Aviation Development Council
- e: Axiodistocervical
- e: Azodicarbonamid

adc e: active-duty commitment
- e: Adopted Child
- e: animal damage control

A. D. C. d: Allgemeiner Deputierten-Convent

AdC e: Adrenal Cortex
- e: Armo[u]red Car

AD & C e: Ammunition Distribution and Control

A/DC e: Analog[ue]-to-Digital Converter

ADCA d: Allgemeine Deutsche Creditanstalt
- e: Advanced Design Composite Aircraft
- e: Advisory Dental Council of Australia
- e: American Dexter Cattle Association
- e: Aminodecephalosporanic Acid
- e: Aminodeoxyclavulanic Acid
- e: Australian Department of Civil Aviation

ADCAD e: Airways Data Collection and Distribution

ADCANUS e: Air Defense, Canada-United States

ADCAP e: Advanced Capability

ad cap. l: ad captandum

ad caps. l: ad capsulam

ADCAPT e: Alternating Current-Direct Current Chip Automatic Production Tester

ADCAR e: Automated Document Control and Retrieval System

ADCAS e: Automatic Data Collection and Analysis System

ADCAT e: Automatic Data Correction and Transfer

ADCC e: Actual Development Cost Certification
- e: Air Defence Cadet Corps
- e: Air Defense Command and Control
- e: Air Defense Command Center
- e: Air Defense Command Computer
- e: Air Defense Control Center
- e: Antibody-Dependent Cell-mediated Cytotoxicity
- e: Applied Data Communications
- e: Area Damage Control Center
- e: Associated Day Care Centers
- e: Asynchronous Data Communications Channel
- e: Asynchronous Data Communications Controller
- e: Atlantic Development Council Canada

ADCCC e: Air Defense Command Commendation Certificate

ADCCCS e: Air Defense Command, Control and Coordination System

ADCCM e: Advanced Cooperative Countermeasure

ADC COMNET e: Air Defense Command Communications Network

ADCCP e: Advanced Data Communication[s] Control Protocol
- e: Advanced Data Communication[s] Control Procedure

ADCCS e: Air Defense Command and Control System
- e: Air Defense Control and Coordination System

Adc Cse e: Advanced Course

ADCEO e: Assistant Division Communications Electronics Officer

ADCEP e: Advanced Structural Concept and Evaluation Program[me]

ADCI e: American Die Casting Institute

ADCII e: American Die Casting Institute, Incorporated

ADCIR e: Administrative Circular

ADCIS e: Association for the Development of Computer-based Instruction[al] Systems

ADCL e: Accredited Dosimetry Calibration Laboratories

ADCLAMPLIFIER e: Automatic Direct Current Level Amplifier

ADCLS e: Advanced Data Collection and Location System

ADCMC e: Antibody-Dependent Cell-Mediated Cytotoxicity

ADCN e: Advanced Design Change Notice
- e: Advanced Drawing Change Notice

ADCNM e: Assistant Deputy Chief of Naval Material

ADCNO e: Assistant Deputy Chief of Naval Operations

ADC/NORAD e: Air Defense Command/North American Air Defense

ADCO e: Air Defense Communications Officer
- e: Alcohol and Drug Control Office[r]
- e: As Design Changes Occur

ADCOC e: Air Defense Command Operation Control
- e: Area Damage Control Center

ADCOINS e: Air Defense Command Interoperability System

ADCOM e: Administrative Command
- e: Administrative Committee
- e: Advance[d] Command
- e: Advanced Communications
- e: Aerospace Defense Command
- e: Analog[ue]-to-Digital Computer
- e: Association of Data Center Owners and Managers

Ad Comd e: Administrative Command

AD Comd e: Air Defense Command

ADCOMINPAC e: Administrative Command, Minecraft; Pacific Fleet

ADCOMPHIBSLANT e: Administrative Command, Amphibious Forces; Atlantic Fleet

ADCOMPHIBSPAC e: Administrative Command, Amphibious Forces; Pacific Fleet

ADCOMR e: Aerospace Defense Command Region

ADCON e: Address Constant
- e: Administrative Control
- e: Advance Concepts for Terrain Avoidance
- e: Analog[ue]-to-Digital Conversion
- e: Analog[ue]-to-Digital Converter
- e: Area Damage Control

adcon e: advance concepts
- e: advise or issue instructions to all concerned
- e: analog-to-digital computer

ADCONSEN e: Advice and Consent of the Senate

A-D Conversion e: Analog[ue]-Digital Conversion

A-D Converter e: Analog[ue]-Digital Converter

ADCOP e: Acquisition and Distribution of Commercial Products
- e: Air Defense Command Post
- e: Associate Degree Completion Program[me]

ADCP e: Acoustic Doppler Current Profiler
- e: Acquisition and Distribution of Commercial Products
- e: Advanced Control Programmer
- e: Advanced Data Communication Protocol
- e: Advanced Data Communications Control Procedure
- e: Air Defense Command Post

ADC/PL e: Advanced Data Collection - Position Location

ADC Program[me] e: Analog[ue] Digital Control Program[me]

ADCR e: Air Defense Command Regulation
- e: All Digital Control Run
- e: Applicable Document Contractual Record

ADCS e: Advanced Data Communication for Stores
- e: Advanced Defense Communications Satellite
- e: Air Data Computer Set
- e: Air Data Computing System
- e: Air Defence Control Ship
- e: AUTODIN Coordination Station
- e: Automated Document Control System
- e: Automatic Data Collecting System
- e: Automatic Data Collection System
- e: Automatic Data Correlation System
- e: Automatic Digital Circuit Switch

ADCs e: Analog[ue]-to-Digital Converters

ADCSCD e: Assistant Deputy Chief of Staff for Combat Developments

ADCSLOG e: Assistant Deputy Chief of Staff for Logistics

ADCSLOG-SA e: Assistant Deputy Chief of Staff for Logistics for Security Assistance

ADCSOPS e: Assistant Deputy Chief of Staff for Operations and Plans

ADCSOPS (JA) e: Assistant Deputy Chief of Staff for Operations and Plans (Joint Affairs)

ADCSP e: Advanced Defense Communication[s] Satellite Program[me]
- e: Advanced Defense Communication[s] Satellite Project

ADCSPC e: Air Data Computer Static Pressure Compensator

ADCSRDA e: Assistant Deputy Chief of Staff for Research, Development and Acquisition
ADCU e: Alarm Display and Control Unit
- e: Alternate Detection and Control Unit
- e: Analog[ue]-to-Digital Conversion Unit
- e: Association of Data Communications Users
ADC Unit e: Analog[ue]-to-Digital Conversion Unit
ADCUPD e: Air Defense Command Unit Proficiency Directive
Ad Curtain e: Advertisement Curtain
ADCUS e: Advise Customs
ADCVR e: Analog[ue]-to-Digital Converter
ADD d: addiere[n]
- d: Addierer
- d: Addition
- d: Adressendecoder
- d: Adressendefinition
- d: Arbeitskreis Dezentrale Datentechnik
- d: Arbeitskreis Dezentrale Datenverarbeitung
- d: Arbeitskreis Digitale Datensysteme
- e: Abstracts of Declassified Documents
- e: Accessibility and Dissemination of Data
- e: Acoustic Deception Device
- e: Acoustic Detection Device
- e: Acoustic Discrimination of Decoys
- e: Activated Dough Development
- e: Addenda
- e: Addendum
- e: Adder
- e: Addis Ababa (airport)
- e: Additive
- e: Address Decoding
- e: Address Defined
- e: Address Definition
- e: Advanced Development Design
- e: Aerospace Defense Division
- e: Aerospace Digital Development
- e: After Diversity Demand
- e: Airborne Deception Device
- e: Airborne Digital Decoder
- e: Air Defense Development
- e: Air Defense District
- e: Airstream Direction Detector
- e: Alphanumeric Digital Display
- e: American Dialect Dictionary
- e: American Doctorial Dissertation (journ.)
- e: Analog[ue]-Digital Display
- e: Analog[ue] Data Digitizer
- e: Analog[ue] Digital Designer
- e: Androsta-1,4 Diene-3,17-Dione
- e: Armaments Design Department
- e: Arming Decision Device
- e: Artificer Deep Driver
- e: Ascent Descent Director
- e: Attention Deficit Disorder
- e: Auditory Discrimination in Depth
- e: Automatic Data Descriptor
- e: Automatic Data Distribution
- e: Automatic Digital Depth
- e: Automatic Digital Display
- e: Automatic Distribution of Documents
- e: Automatic Document Distribution
- e: Automatic Drawing Device
- R: Aviastiia Dalnego Deistvija
Add. d: Additionswerk
add. e: addenda
- e: addendum
- e: address
- e: airborne digital decoder
- e: automatic drawing device
- l: addendum
Add e: Addition[al]
- e: Address
add e: adduction
- e: adductor
a.d.d. e: after-diversity demand
a. d. D. d: an der Donau
- d: an diesem Datum
- d: auf dem Dienstweg
- d: auf der Durchreise
AD & D e: Accidental Death and Dismemberment
add`ed e: addressed
ADDA e: Air Defense Defended Aerea
- e: American Design Drafting Association
- e: Attention Deficit Disorder Association
- e: Australian Database Development Association
- e: Automatic Decentral Data Acquisition
AD/DA e: Analog[ue] Digital / Digital Analog[ue]
ADDAC e: Analog[ue] Data Distributor and Computer
- e: Analog[ue] Data Distributor and Control
ADDACE e: Analog[ue]-to-Digital, Digital-to-Analog[ue] Conversion Equipment
ADDAM e: Adaptive Dynamic Decision-Aiding Method
ADDAR e: Automatic Digital Data Acquisition and Recording
ADDAS e: Airborne Digital Data Acquisition System
- e: Automatic Digital Data Assembly System
ADDC e: Air Defense Direction Center
- e: Alignment and Diagnostic Display Console
- e: Analog[ue]-Discrete Data Converter
- e: Analog[ue]-to-Digital Data Converter
ADDCS e: Aircraft Decontaminating, Deicing, Cleaning System
ADDDS e: Automatic Direct Distance Dialling System
ADDD System e: Automatic Direct Distance Dialling System
adde e: addresse
ADDED e: Additional Education
add'ed e: addressed
ADDEE e: addressee
ad. def an. l: ad defectionem animi
ad. deliq. l: addeliquium
ADDELREP e: Additional Delay in Reporting
ADDER e: Automatic Digital-Data- Error Recorder
- e: Automatic Digital-Data-Error Recording
ADDEV e: Advanced Development
ADDF e: Abu Dhabi Defense Force
- e: Address Field
- e: Amplitude Density Distribution Function
- e: Automated Deferred Discrepancy File
Addg e: Adding
add. Gew.-% d: additives Gewicht- Prozent
ADD-H e: Attention Deficit Disorder with Hyperactivity
ADDH e: Attention Defitcit Disorder with Hyperactivity
ADD-HA e: Attention Deficit Disorder with Hyperactivity
ADDI e: Airborne Dual Detector Indicator
ad diag e: admitting diagnosis
ADDIC e: Alcohol and Dependency Intervention Council
addict. e: addiction
Addis e: Addis Ababa, Ethiopia
addit e: additional
ADDIV e: Air Defense of the Division
Addl e: Additional
AD/DL e: Aircraft Division / Department List
ADDL e: Aircraft Dummy Deck Landing
- e: Anti-Digit Dialling League
ADDLASS e: Air Deployable Drifting Linear Array SONAR System
ADDM e: Addendum
- e: Automated Drafting and Digitizing Machine
ADDMS e: Automatic Depth / Deployed Moored Sweep
ADDN e: Addition
- e: Add Numeric
addnl e: additional
ADDNR e: Assistant and Deputy Director of Naval Recruiting
ADDNS e: Additions
Addo e: Addo Elephant Park near Port Elizabeth, South Africa
ADDON e: Add On
ADDP e: Air Defense Defended Point
ADDPEP - e: Aerodynamic Deployable Decelerator Performance Evaluation Program[me]
ADDPLA e: Additional Places
ADDR e: Adder
- e: Address[ing]
- e: Address Register
Addr Alg e: Adder; Algebraic
- e: Algebraic Adder
Addrb e: Addressable
- e: Adressable
ADDRESOR e: Analog[ue]-to-Digital Data-Reduction System for Oceanographic Research
ADDRESS e: Automated Design of Damage Resistant Structures
ADDRI e: Apollo Document Distribution Requirements Index
ADDROUT e: Address Out
ADDRS e: Analog[ue]-to-Digital Data Recording System
Adds e: Address
ADDS e: Advanced Data Display System
- e: Advanced Deep Diving Submersible
- e: Airborne Designation and Discrimination Study
- e: Airborne Detection Discrimination Sensor
- e: Air Defense Demonstration System
- e: Air Development Delivery System
- e: American Digestive Disease Society
- e: American Diopter and Decibel Society
- e: Apollo Data Descriptions Standards
- e: Apollo Documentation Description Standards
- e: Apollo Document Descriptions Standards
- e: Applied Digital Data System[s]
- e: Army Data Distribution System
- e: Assistant Director of Dental Services
- e: Astro Digital Doppler Speedometer
- e: Automated Digital Data System
- e: Automated Digital Design System
- e: Automatic Data Digitizing System
- e: Automatic Data Distribution System
- e: Automatic Digital Design System
- e: Automatic Direct Distance Dialling System
- e: Automatic Document Distribution Service
ADDSB e: Address Disable
add's'd e: addressed
ADDSRTS e: Automated Digitized Document Storage, Retrieval and Transmission System

Add-Sub e: Add-Substract
ADDT e: Additive
- e: All Digital Data Tape
- e: Angular Distribution Data Tape
- e: Angular Distribution Data Type
ADDTL e: Additional
ADDU e: Additional Duty
- e: Alcohol and Drug Dependence Unit
ADDUNIFALW e: Additional Uniform Allowance
Add./Vorw.-Kopplg. d: Addition/ Vorwärtskopplung
ADE d: Adresse für Eingabegerät
- d: allgemeine Datenerfassung
- d: allgemeines Datenerfassungsprogramm
- d: Anschlußeinheit für Datenendplatz
- d: Anschlußeinheit für Datenendplätze
- d: Anschlußeinheit für Datenerfassung und -ausgabe
- e: Action at a Distance Electrodynamics
- e: Acute Disseminated Encephalitis
- e: Acute Disseminated Encephalomyelitis
- e: Address Enable
- e: Address Error
- e: Advanced Data Entry
- e: Aerial Delivery Equipment
- e: After Delivery Economies
- e: Aircraft Data Entry
- e: Air Defense Element
- e: Air Defense Emergency
- e: Air Defense Evaluation
- e: Air Density
- e: Air Density Explorer
- e: Alpha Disintegration Energy
- e: Alphanumeric Display Equipment
- e: Analog[ue]-to-Digital Encoder
- e: Animal Disease Eradication
- e: Antibody-Dependent Enhancement
- e: Apparent Digestible Energy
- e: Approved Data Element
- e: Armaments Design Establishment
- e: Armored Division Equivalent
- e: Army Department Establishments
- e: Arrhythmogenic Dose of Epinephrine
- e: Assistant Division Engineer
- e: Association for Documentary Editing
- e: Atomic Defense Engineering
- e: Attenuation Distortion Equivalent
- e: Audible Dippler Enhancer
- e: Audible Doppler Enhancer
- e: Authorized Data Element
- e: Automated Data Exchange
- e: Automated Debugging Environment
- e: Automated Design Engineering
- e: Automated Drafting Equipment
- e: Automated Draughting Equipment
- e: Automatic Data Encoding
- e: Automatic Data Entry
- e: Automatic Data Evaluation
- e: Automatic Design Engineering
- e: Automatic Drafting Equipment
- e: Automatic Drawing Equipment
a. d. E. d: an der Elbe
Ade e: Adenine
ADEA e: Age Discrimination and Employment Act
- e: American Driver Education Association
- e: Army Development and Employment Agency
- e: Assistant Director of Expense Accounts
- e: Australian Diabetes Educators Association
ADEC d: Adreßkodierung
- e: Address Decoder
- e: Address Decoding
- e: Aiken Dahlgren Electronic Calculator
- e: Association for Death Education and Counseling
- e: Australian Drug Evaluation Centre
ADED e: Air Defense Effectiveness Demonstration
- e: Air Delivery Equipment Division
ADEDA e: Advise Effective Date
ADEDS e: Advanced Electronic Display System
ADEDU e: Assistant Director for Education
ADEE e: Air Defense Electronic Environment
- e: Air Defense Experimental Establishment
- e: Association for Dental Education in Europe
ad effect l: ad effectum
ADEFSq e: Air Defense Squadron
ADEG e: Auxiliary Display Equipment Group
ADEH e: Alumin[i]um Diethyl Hydride
A de JC s: Antes de Jesu Cristo
ADEK d: allgemeiner Datenerhebungskatalog
ADEKRA d: Arbeitsgemeinschaft Deutscher Kraftwagenspediteure
ADEKS e: Advanced Design Electronic Key System
ADELE d: automatische Datenerfassung durch Lochkarteneingabe
ADELEI d: automatische Datenerfassung durch Lochkarteneingabe
Adélie f: Adélie Land
ADELL e: Adult Education and Lifelong Learning
ADEM d: Anwendungs-Datenerfassungs-Matrix
- e: Acute Disseminated Encephalomyelitis
- e: Adaptively Data Equalized Modulator-Demodulator
- e: Automatically Data-Equalized Modulator-Demodulator
ADEMS e: Advanced Diagnostic Engine Monitoring System
- e: Airborne Display Electrical Management System
- e: Automated Data Entry Measurement System
aden e: augmented deflector exhaust nozzle
ADEO e: Advance Development Engineering Order
ADEOS e: Advanced Earth Observation Satellite
- e: Advanced Earth Observing Satellite
- e: Air Droppable, Expendable Ocean Sensor
ADEP e: Air[craft] Depot
A Dep e: Air[craft] Depot
ADEP e: Air Depot
- e: Asset Data Evaluation Program[me]
ADEPE d: Arbeitsgemeinschaft der Pelzwirtschaft
ADEPO e: Automatic Dynamic Evaluation by Programmed Organizations
ADEPREP System e: Army Deployment Reporting System
- e: Army Development Reporting System
ADEPT e: A Display Expedited Processing and Tutorial
- e: A Distinctly Empirical Prover of Theorems
- e: Advanced Development Prototype
- e: Aerospace Draftsmans Education and Proficiency Training
- e: Agricultural and Dairy Educational Political Trust
- e: Air Force Depot Equipment Performance Tester
- e: Algebraic and Differential Equations Processor and Translator
- e: Analog[ue] Data Extractor and Plotting Table
- e: Automated Direct-Entry Packaging Technique
- e: Automatic Data Extractor and Plotting Table
- e: Automatic Dynamic Evaluation by Programmed Testing
- e: Automatic Dynamic Evaluation by Programming Testing
ADEPT System e: A Display-Expedited Processing and Tutorial System
ADER e: Automatic Data Extraction Routine
ADERI e: American Disability Evaluation Research Institute
ADES d: akustisches Datenerfassungssystem
- e: Advanced Diagnostic Executive System
- e: Air Defense Engagement System
- e: Air Defense Engineering Service
- e: Analysis, Design and Evaluation System
- e: Angle-Dispersed Electron Spectroscopy
- e: Association of Directors of Education; Scotland
- e: Automated Data Entry System
- e: Automatic Data Entry System
- e: Automatic Digital Encoding System
A de S f: Académie des Sciences
a des i: a destra
A de T S e: Alex de Tocqueville Society
ADEU e: Automatic Data Entry Unit
ad eund. grad. l: ad eundem gradum
AdEW d: Arbeitsgemeinschaft der Elektrizitätswerke
ADEW e: Air Defense Early Warning
ADEWS e: Air Defense Electronic Warfare System
ADEX e: Advanced Anti-Submarine Warfare Exercise
- e: Advanced Data Entry Executive
- e: Air Defense Exercise
ADF d: adaptives Digitalfilter
- d: Aktion Demokratischer Fortschritt
- d: Arbeitsgemeinschaft Dermatologische Forschung
- d: automatische Datenverarbeitung in der Fertigungsindustrie
- d: automatisierte Datenerfassung in Fertigungsbetrieben
- e: Acid Detergent Fibre
- e: Acoustic Depth Finder
- e: Acquisition Data Facility
- e: Adapter Definition File
- e: Advanced Disk File
- e: Aerial Direction Finding
- e: Airborne Direction Finder
- e: Air Defense Force
- e: Air Development Force
- e: Air Direction Finder
- e: Alaska Defense Frontier
- e: Alcohol and Drug Foundation
- e: Algorithm Development Facility
- e: All Dielectric Filter
- e: Alternating Double Filtration
- e: American Dairy Federation
- e: American Duty Free
- e: Anterior Dendritic Field
- e: Application Development Facility
- e: Approved Deposit Fund
- e: Approximate Degrees of Freedom

- e: Asian Development Fund
- e: Assured Destruction Force
- e: Australian Defence Force
- e: Automated Design Facility
- e: Automatic Direction Finder
- e: Automatic Display Finder
- e: Automatic Document Feed[er]
- e: Auxiliary Detonating Fuze

adf e: after deducting freight
a.d.f. e: automatic direction finding
ADFA e: Australian Defence Force Academy
- e: Australian Dried Fruits Association
- e: Automatic Direction Finder Approach
- e: Automatic Direction Finder Aproach

ADFAP e: Automatic Direction Finder Approach
- e: Automatic Direction Finding Approach

adfc e: adiabatic film cooling
ADFC e: Air Defense Filter Center
ADFG e: Alaska Department of Fish and Game
- e: Arbitrary Diode Function Generator

ADFI e: American Dog Feed Institute
ad fin. l: ad finem
ADFL e: Association of Departments of Foreign Languages
ADFO e: Assistant Director; Flight Operations
ADFOR e: Adriatic Force
ADFP e: Atroxin-Defibrinated Plasma
ADFR e: Automatic Direction Finder; Remote-Control[ed]
ADFS d: Arbeitskreis für die Dokumentation Sozialwissenschaftlicher Forschung
- e: American Dentists for Foreign Service
- e: Australian Documentary Facsimile Society
- e: Automatic Direction Finding System

ADFSC e: Automatic Data Field System[s] Command
ADFT e: Artillery Direct Fire Trainer
ADFU e: Air Defense Firing Unit
ADFW e: Assistant Director of Fortifications and Works
ADG e: Advance Development Group
- e: Aerial Directive Gain
- e: Aeronautical Development Group
- e: Air-Driven Generator
- e: Aircraft Delivery Group
- e: Air Defense Group
- e: Air Density Gauge
- e: Antenna Directive Gain
- e: Assistant Director General
- e: Attack Display Group
- e: Attitude Display Group
- e: Automatic De-Gaussing
- e: Automotive Development Group
- e: Auxiliary Deception Generator
- e: Average Daily Gain
- e: Aviation Depot Group
- e: Axidistogingival

a. d. G. d: aus dem Gebiet
- d: aus der Gegend

adg e: Advancing
ADGA e: American Diary Goat Association
ADGAVS e: Assistant Director-General; Army Veterinary Services
ADGB d: Allgemeines Deutsches Gestütbuch
- e: Accessory Drive Gear Box
- e: Air Defence of Great Britain

ADGBC e: Air Defence of Great Britain Command
A & D G C e: Alliance and Dublin Consumers Gas Company
adge e: air-defense ground environment
ADGE e: Air Defence Ground Environment
ad gen. l: ad generalia
AD-GerÄt d: Antennen-Diversity-Gerät
ADGL e: Adrenal Gland
ADGM e: Assistant Director of Guided Missiles
ADGMS e: Assistant Director-General of Medical Services
Adgo d: Allgemeine Deutsche Gebührenordnung
adgo e: adagio
- i: adàgio

ADGp e: Air Defense Group
a. d. Gr. d: aus diesem Grund
- d: aus diesen Gründen

AD GR ACID l: Ad Gratum Aciditatem
ad grat. acid l: ad gratam acidatem
ad grat. gust. l: ad gratum gustum
ADGRU e: Advisory Group
ADGT e: Assistant Director-General of Transportation
ADH d: Alkoholdehydrogenase
- d: Allgemeiner Deutscher Hochschulverband
- d: Antidiuretisches Hormon
- d: Arbeitsgemeinschaft Deutscher Hochschulsport
- d: D-Arabitol Dehydrogenase
- e: A-Digit Hunter
- e: Academy of Dentistry for the Handicapped
- e: Adhere
- e: Adherent
- e: Alcohol Dehydrogenase
- e: Analog[ue] Data Handling
- e: Antidiuretic Hormone
- e: Arbitrary Device Handler
- e: Assistant Director of Hygiene
- e: Association of Dental Hospitals
- e: Automatic Data Handling

Adh. d: Adhäsion
A. d. H. d: Anmerkung des Herausgebers
a. d. H. d: aus dem Hause
adh e: Adhesive
ÄDH d: Äpfelsäure-Dehydrogenase
ADHA e: American Dental Hygienists Association
- e: Analog[ue] Data Handling Assembly

AdhÄs. d: Adhäsion
ADHase e: Alcohol Dehydrogenase
ADHC e: Adult Day Health Care
- e: Air Defense Hardware Committee

adhca e: advise this headquarter of complete action
ADHD e: Attention Deficit Hyperactivity Disorder
ADHEL e: Air Defense High Energy Laser
ADHGB d: Allgemeines Deutsches Handelsgesetzbuch
ADHI e: Area Defense Homing Interceptor
adhib l: adhibeatur
Adhibend l: Adhibendus
ad h.l. l: ad hunc locum
ad hoc l: ad hoc
ad hom. l: ad hominem
ADHOS e: Alternative to Dedicated Hospital Ship
AD Hq e: Advanced Divisional Headquarters
A. d. Hrsg. d: Anmerkung des Herausgebers
ADHS e: Analog[ue] Data Handling System
ADHs e: Antidiuretic Hormones
ADI d: Anwenderverband Deutscher Informationsverarbeiter e.V.
- d: Auslands- und Dolmetscherinstitut
- e: Academy of Dentistry International
- e: Acceptable Daily Intake
- e: Accounting Department Instructions
- e: Acoustical Door Institute
- e: Address Incomplete
- e: Automatic Ground Control System Computer Discrete In
- e: Air Defense Initiative
- e: Air Defense Institute
- e: Air Defense Interceptor
- e: Air Distribution Institute
- e: Alarm Data Interface
- e: Alien Declared Intention
- e: Allied Distribution
- e: Allowable Daily Intake
- e: Alternate Digit Inversion
- e: Alternating Direction Implicit
- e: Alternating Direction Iterative
- e: Altitude Direction Indicator
- e: Alzheimers Disease International
- e: American Defense Institute
- e: American Designers Institute
- e: American Directors Institute
- e: American Documentation Institute
- e: Analog[ue] Display Indicator
- e: Analog Devices, Incorporated
- e: Antidetonant Injection
- e: Antidetonation Injection
- e: Apollo Document Index
- e: Applied Dynamics International
- e: Approved Driving Instructor
- e: Area of Dominant Influence
- e: Assembly Decay Indicator
- e: Assistant Director of Intelligence
- e: Association for Direct Instruction
- e: Attitude Direction Indicator
- e: Attitude Director Indicator
- e: Attitude Display Indicator
- e: Automatic Data Input
- e: Automatic Derivation of Invariants
- e: Automatic Direction Indication
- e: Axiodistoinclusal
- s: Agencia para el Desarrollo Internacional

adi e: adiabat[ic]
Adi e: Adiuretin
AdIA d: Aerodynamisches Institut Aachen
ADIA e: African Development Information Association
- e: American Diamond Industry Association
- e: Automatic Documentation in Action

ADIC e: American Dental Interfraternity Council
- e: Analog[ue]-to-Digital Conversion
- e: Analog[ue]-to-Digital Converter
- e: Automated Digital Interior Communications

ADIC System e: Analog[ue]-to-Digital Conversion System
ad id l: ad idem
ADIE e: Acquisition Data Input Equipment
- e: Acquistion Data Input Equipmnent
- e: Automatic Dialling and Indicating Equipment
- e: Automatic Dialling and Indication Equipment

A Dies Tech e: Associate in Diesel Technology
ADIF e: Autocrine Differentiation-Inhibiting Factor
- e: Avionics Development and Integration Facility

ADIG d: Allgemeine Deutsche Investmentgesellschaft
ad ig. l: ad ignorantiam
ADIK d: Aussprache, Daten, Instruktionen, Konstanten

adil e: air defense identification line
ADIL e: Air Defense Identification Line
adimd e: advise immediately by dispatch
ADINA e: Automatic Dynamic Incremental Nonlinear Analysis
ADINET e: Administrative Network
ad inf l: ad infinitum
ad init. l: ad initium
ADINSP e: Administrative Inspection
ad int. l: ad interim
Ad Intel Cen e: Advanced Intelligence Center
ADINTELCEN e: Advanced Intelligence Center
ADIO e: Analog[ue]-Digital Input-Output
Adion d: adsorbiertes Ion
ADIOS e: Advanced Digital Inertial Optical Sensor
- e: Analog[ue] Digital Input-Output System
- e: Asian Dust Input to the Oceanic System
- e: Automatic Diagnosis Input-Output System
- e: Automatic Digital Input-Output System
ADIP e: Advanced Developing Institutions Program[me]
- e: Air Defense, Interdiction and Photographic
- e: Alloy Development for Irradiation Performance
- e: Alternating Direction Implicit Procedure
- e: Automated Data Interchange-systems Panel
ADipA e: Associate Diploma in Arts
ADIPE e: Aircraft Design-Induced Pilot Error
adipu e: advise whether individual may be properly used in your installation
ADir. d: Amtsdirektor
A.-Dir. d: Archivdirektor
Adirondacks e: Adirondack Mountains, New York
ADIRS e: Australasian Drug Information Services Drug Information Retrieval System
- e: Australian Drug Information Services Drug Information Retrieval System
ADIRU e: Air Data Inertial Reference Unit
ADIS d: administratives Informationssystem
- e: A Data Interchange System
- e: Advanced Driver Information System
- e: Airborne Digital Instrumentation System
- e: Air Defense Integrated System
- e: Association for the Development of Instructional Systems
- e: Attitude Director Indicator System
- e: Australasian Drug Information Service
- e: Automatic Data Interchange System
ADI Signal e: Address Incomplete Signal
ADISP e: Aeronautical Data Interchange System Panel
- e: Automated Data Interchange Systems Panel
ADISS e: Advanced Defense Intelligence Support System
ADIT e: Alien Documentation, Identification and Telecommunications
- e: Alliance Defense Industry and Technology
- e: Analog[ue]-to-Digital Integrating Translator
- e: Analog[ue]-to-Digital Integration Translator
- e: Automated Data on Instruction Technology
- e: Automatic Detection and Integrated Tracking
adit e: analog digital integrating translator
ADITAS d: Außenwerbungs-Dispositions- und Tafelabrechnungssystem
ADITC e: Australian Drilling Industry Training Committee
ADITS e: Aircraft Diagnostics and Integrated Test System
ADIU e: Address and Data Interface Unit
- e: Airborne Data Insertion Unit
- e: Armament and Disarmament Information Unit
A DIV e: Air Division
ADIZ e: Air Defense Identification Zone
- e: Atlantic Defense Identification Zone
ADJ d: Arbeitsgemeinschaft deutscher Junggärtner
- e: Adjective
- e: Adjoining
- e: Adjourned
- e: Adjunct
- e: Adjust[able]
- e: Adjusted
- e: Adjuster
- e: Adjusting
- e: Adjustment
- e: Angle Deception Jamming
Adj. d: Adjunkt
a.d.J. d: aus dem Jahre
adj e: adjacent
- e: adjectival
- e: adjoint
- e: adjourn[ed]
adj. e: adjustieren
Adj e: Adjutant
ADJAG e: Assistant Deputy Judge Advocate General
ADJBLE e: Adjustable
ADJD e: Adjourned
- e: Adjusted
ADJF e: Adjustment File
AdjG d: Adjektivgruppe
ADJG e: Adjutant-General
ADJ L e: Adjoining Landowner
ADJMOM e: Adjoint Gamma-Ray Moments
ADJN e: Adjourn
ADJOING e: Adjoining
ADJ/PPR e: Ajusted Permanent Pay Record
ADJS e: Angle Deception Jamming System
Adj Sec e: Adjutants Section
ADJ SP e: Adjustable Speed
ADJSPD e: Adjustable Speed
ADJT e: Adjustable
- e: Adjutant
adjt e: adjoint
ADJV d: Allgemeiner Deutscher Jagdschutzverein
ADK d: Adreßdekodierschaltung
- d: Arbeitsgemeinschaft der Deutschen Kautschukindustrie
- d: Arbeitsgemeinschaft Deutscher Kraft- und Wärmeingenieure
- d: Autodrehkran
- e: Adenylate Kinase
- e: Attach-Detach Kit
- e: Automatic Depth Keeping
A dk e: Awning Deck
ADKA d: Arbeitsgemeinschaft Deutscher Krankenhausapotheker
ADKe d: Autodrehkräne
ADKI d: Arbeitsgemeinschaft Deutscher Konstruktionsingenieure
ADKPZ d: Arbeitskreis Deutscher Klein- und Pelztierzüchter
AdKV d: Arbeitsgemeinschaft der Kleinverlage
ADL d: Allgemeine Deutsche Lehrer-Zeitung
- d: Arbeitsgemeinschaft Deutscher Landwirte
- d: Arbeitsgemeinschaft Deutscher Lochkartenfachleute
- d: Arbeitsgemeinschaft für Datenverarbeitung und Lochkartentechnik
- d: Arbeitsgemeinschaft für Elektronische Datenverarbeitung und Lochkartentechnik
- d: Automatische Datenverarbeitung und Logistik GmbH
- e: Acceptable Defect Level
- e: Acoustic Delay Line
- e: Active Duty List
- e: Adelaide (airport)
- e: Aerial Dummy Load
- e: Airborne Data Link
- e: Airborne Data Loader
- e: Aircraft Data Line
- e: Antenna Dummy Load
- e: Apollo Documentation List
- e: Architecture Description Language
- e: Area Dental Laboratory
- e: Armament Data Line
- e: Armament Datum Line
- e: Armament Development Laboratory
- e: Artificial Delay Line
- e: Assistant Director of Labour
- e: Atmospheric Devices Laboratory
- e: Authorized Data List
- e: Automated Display Language
- e: Automated Drafting Language
- e: Automatic Data Link
- e: Automatic Data Logger
- e: Automatic Drafting Language
- e: Average Decreasing Line
- e: Avionics Development Laboratory
adl e: activities of daily living
- e: automatic data link[ing]
ad l. l: ad libitum
ADLAR e: Adaptive Lens Array
ADLAT e: Advanced Low Altitude Technique
ADLATAD e: Advise Latest Address
ADLB e: Armo[u]red-Vehicle Launched Bridge
ADLC e: Advanced Data Link Control[ler]
- e: Advanced Data Link Controlling
- e: Antibody-Dependent Lymphocyte-mediated Cytotoxicity
Adler e: Adler Planetarium
ADLI e: Advanced Deck-Launched Interceptor
ADLIB e: A Design Language for Indicating Behavior
ad lib. l: ad libitum
ad libit. l: ad libitum
ADLIB Management System e: Adaptive Library Management System
ADLIPS e: Automatic Data-Link Plotting System
ADLIP System e: Automatic Data Link Plotting System
ad lit. l: ad litteram
ADLLV d: Allgemeiner Deutscher Lehrer- und Lehrerinnenverein
ADLM e: Account Data List Management
- e: Acoustic Delay Line Memory
ADLM e: Aerial Delivered Land Mine
ADLMS e: Air-Delivered Land Mine System
Adln l: Adductor Longus
ADL-Nachr. d: ADL-Nachrichten

ADLN Muscle e: Adductor Longus Muscle
ADLO e: Air Defense Liaison Office[r]
- e: Association of Direct Labour Organizations
ad loc. l: ad locum
ADLOG e: Advanced Logistical [Command]
ADLR e: Assistant Director of Light Railways
- e: Automated Direct Labor Reporting
ADLRI e: Arthur D. Little Research Institute
ADLS e: Airborne Data Link System
- e: Air Dispatch Letter Service
- e: Anterior Dorsolateral Scale
- e: Automatic Drag-Limiting System
ADLT e: Airborne Data Link Terminal
ADLTY e: Adultery
ADM d: adaptive Deltamodulation
- d: Amplitudendemodulation
- d: Arbeitsgemeinschaft der Deutschen Milchkontrollangestellten
- d: Arbeitsgemeinschaft des Deutschen Motorsports
- d: Arbeitskreis Deutscher Marktforschungsinstitute
- d: Außendienstmitarbeiter
- e: Academy of Dental Materials
- e: Acoustic Digital Memory
- e: Acquisition Decision Memorandum
- e: Activity Data Method
- e: Adaptable Data Manager
- e: Adaptive Delta Modulation
- e: Add-on Module
- e: Add Magnitude
- e: Add Memory
- e: Administrative Medicine
- e: Admiral[ty]
- e: Admissions
- e: Admit[ted]
- e: Adriamycin
- e: Advanced Data Management
- e: Advanced Deployment Model
- e: Advanced Development Memory
- e: Advanced Development Model
- e: Advance Decoy Missile
- e: Advanced Microdevices
- e: Aerodrome
- e: Affiliated Dress Manufacturers
- e: Agarose Diffusion Method
- e: Aggregate Data Manager
- e: Aid in Decision Making
- e: Air Decoy Missile
- e: Air Defense Missile
- e: Air Defense Mission
- e: American Demographics Magazine (journ.)
- e: Ammonium Dimolybdate
- e: Amplitude Demodulation
- e: Annals of Discrete Mathematics
- e: Annual Delegate Meeting
- e: Apollo Data Manager
- e: Application Data Management
- e: Application Description Manual
- e: Arc Data Monitor
- e: Area Defense Missile
- e: Area Denial Munition
- e: Arrow Diagramming Method
- e: Assistant Deputy Minister
- e: Assistant District Manager
- e: Asynchronous Disconnected Mode
- e: Atomic Demolition Munition
- e: Augmented Data Manipulation
- e: Augmented Data Manipulator
- e: Automated Data Management
- e: Automated Depot Maintenance
- e: Automated Drafting Machine
- e: Automatic Degreasing Machine
- e: Automatic Detection Mark
- e: Automatic Digital Microphotometer
- e: Automatic Display Mode
- e: Automatic Distribution of Microfiche
- e: Average Daily Membership
Adm. d: Administration
- d: Administrator
- d: Admiral
- d: Admiralität
- d: Admission
- d: Admonition
adm. d: administrativ
- d: admittiert
adm e: administration
- e: administrator
- e: admission
- e: air-launched decoy missile
- e: airdrome
Adm e: Administrative
ADMA e: Advanced Direct Memory Access
- e: Agricultural Development and Marketing Authority
- e: Aircraft Distributors and Manufacturers Association
- e: Alkyldimethylamine
- e: American Drug Manufacturers Association
- e: Amplitude Domain Multiple Access
- e: Area-Dominant Military Aircraft
- e: Australian Display Manufacturers Association
- e: Automatic Damper Manufacturers Association
- e: Aviation Distributors and Manufacturers Association
- e: Aviation Distributors and Manufacturers Association
adma e: automatic drafting machine
ADMACR e: Austrian Documentation Centre for Media and Comunication Research
admad e: advise method and date of shipment
Ad. Man. e: Advertisement Manager
AD MAN MED l: Ad Manus Medici
AdmArzt d: Admiralarzt
admass e: advertising & mass media effect
ADMC e: Actuator Drive and Monitor Computer
- e: Air Defense Missile Command
- e: Antibody-Dependent Macrophage-mediated Cytotoxicity
AdmCen e: Administration (or Administrative) Center
ADMCEN e: Administration Center
AdmCen e: Admiralty Center
ADM (C & P) e: Assistant Deputy Minister (Construction-Engineering and Properties)
ADMD e: Adjust Mode
- e: Administration Management Domain
- e: Administrative Division
- e: Administrative Management Domain
admd e: After Diversity Maximum Demand
ADME e: Absorption, Distribution, Metabolism, Excretion
- e: Assistant Director of Mechanical Engineering
ADM(F) e: Assistant Deputy Minister (Finance)
ADMG e: Admitting
- e: Air Defense Machine Gun
- e: Assistant Deputy Military Governor
- e: Assistant to the Deputy Military Governor
ADMI e: American Dry Milk Institute
ADMIG e: Australian Drug and Medical Information Group
ADMIN e: Administrate
- e: Administration
- e: Administratrix
- e: Automated Document Management Information Network
admin e: administrative
- e: administrator
Admin. Engr. e: Administrative Engineer
ADMINI e: Administrative Instruction[s]
ADMINID e: Administrative Identification
ADMININSP e: Administrative Inspection
ADMININST e: Administrative Instructions
AdminInstr e: Administrative Instructions
ADMIN MOD e: Administration Module
ADMINO e: Administrative Office[r]
- e: Administrative Order
Admin O e: Administrative Order[s]
ADMINORD e: Administrative Order
ADMINPLAN e: Administrative Plan
Admin & Plng. e: Administration and Planning
ADMINR e: Administrator
ADMINREP e: Administrative Report
Adm Inst e: Administrative Instructions
Adm Instns e: Administrative Instructions
Adminstr e: Administrator
Adminstrv e: Administrative
ADMINSUP e: Administrative Support
ADMIR e: Automatic Diagnostic Maintenance Information Retrieval
ADMIRAL e: Automatic and Dynamic Monitor with Immediate Relocation, Allocation and Loading
ADMIRE e: Adaptive Decision-Maker in an Information Retrieval Environment
Admire e: Automatic Diagnostic Maintenance Information Retrieval
admire. e: automatic diagnostic maintenance information retrieval
ADMIREL e: Automatic and Dynamic Monitor with Immediate Relocation
ADMIRES e: Advanced Management Information and Retrieval System
- e: Automatic Diagnostic Maintenance Information Retrieval System
ADMIS e: Admissions
- e: Aircrafts Departing at Minimum Intervals
- e: Automated Data Management Information System
ADMIT e: Aeronautical Depot Maintenance Industrial Technology
Admix e: Administratix
ADMIX e: Administratrix
ADML e: Adenovirus Major Late
- e: Admiral
- e: Advanced Design Methods Laboratory
- e: Automatic Documentation and Mathematical Linguistics
- e: Average Daily Member Load
Adml e: Admiral[ty]
ADMM e: Administrative Memo
Admn e: Administration
ADMO e: Air Defense Management Office
- e: Assistant Director of the Meteorological Office
- e: Australian Directory of Music Organizations
Admo e: Administrative Officer
Adm O e: Administrative Order
admon s: administración
Admor s: Administrador

admos e: automatic device for mechanical order selection
AdmOStArzt d: Admiraloberstabsarzt
ad mov. l: ad movearur
ADMP e: Aerodynamic Damping Moment in Pitch
ADMR e: Absorbance-Detected Magnetic Resonance
- e: Australian Directory of Music Research
admr e: Administrator
Adm Radiol e: Administrative Radiology
ADMRL e: Application Data Material Readiness List
- e: Automatic Data Material Requirements List
ADMRX e: Administratrix
ADMS e: Advanced Data Management System
- e: Air Defense Missile Squadron
- e: Analog[ue] and Digital Monitoring System
- e: Application Data Management Services
- e: Assistant Director of Medical Services
- e: Asynchronous Data Multiplexer Synchronizer
- e: Atmospheric Diffusion Measuring System
- e: Automated Data Management System
- e: Automated Depot Maintenance Study
- e: Automated Document Management System
- e: Automatic Digital Message Switch[er]
- e: Automatic Digital Message Switching
adms e: administrator
ADMSC - e: Automatic Digital Message Switching Centers
Adm Sci Q e: Administrative Science Quarterly
ADMSE e: Automatic Digital Message Switch Equipment
Adm Serv e: Administrative Services
ADMSG e: Advise by Message
ADMSLBN e: Air Defense Missile Battalion
admsn e: admission
adm sp e: administrative specialist
ADMSPT e: Administrative Support
Admst. d: Admiralstab
AdmStArzt d: Admiralstabsarzt
ADMSTR e: Administrator
ADMT e: Assistant Director of Military Transport
- e: Association of Dental Manufacturers and Traders
ADMTR e: Administrator
ADMTRX e: Administratrix
ADMTY e: Admiralty
ADMU e: Air Data Measuring Unit
ADMUWI d: Additionsmultiplikations-widerstand
ADMV d: Allgemeiner Deutscher Motorsport-Verband
ADMWg e: Air Defense Missile Wing
ADMX e: Administratrix
Admy e: Admiralty
ADN d: Allgemeiner Deutscher Nachrichtendienst (DDR)
- e: Accession Designation Number
- e: Actual Device Number
- e: Address complete; No-charge
- e: Adenine
- e: Adiponitrile
- e: Alcohol and Drug News
- e: Array Declarator Name
- e: Associate Degree in Nursing
- e: Avionics Decision Notice
A.D.N. e: Associate Degree in Nursing
adn s: ácido desoxirribonucleico
a-DNA e: alternate deoxyribonucleic acid
ADNA e: Assistant Director of Naval Accounts
ADNAC e: Air Defense of North American Continent
ad naus. l: ad nauseam
ADNC e: Air Defense National Center
- e: Air Defense Notification Center
- e: Assistant Director of Naval Construction
ADNE e: Assistant Director of Naval Equipment
ADNET e: Action Data Network
- e: Administrative Distributed Network
ad neut. l: ad neutralizandum
ADNI e: Assistant Director of Naval Intelligence
ADNO e: Assistant Director of Naval Ordnance
ADNOC e: Abu Dhabi National Oil Company
adnok e: advise if not correct
- e: axiodistoclusal
ADNV d: Allgemeiner Deutscher Neuphilologen-Verband
ADO d: Adenosin
- d: allgemeine Dienstordnung
- e: Additional Day Off
- e: Address Out
- e: Administration Duty Officer
- e: Advanced Development Objective
- e: Air Defence Officer
- e: Air Defense Operations
- e: Air Drop Operator
- e: Airport District Office
- e: Alleged Discrimination Official
- e: Animal Disease Occurrence
- e: Area Dental Officer
- e: Army Distribution Objective
- e: Assistant District Officer
- e: Associated Disbursing Officer
- e: Association of Dispensing Opticians
- e: Audio Decade Oscillator
- e: Automatic Ground Control System Computer Discrete Out
- e: Automotive Diesel Oil
- e: Automotive Distillate Oil
- e: Avalanche Diode Oscillator
- e: Axiodisto-Occlusal
a.d.O. d: an der Oberfläche
a. d. O. d: an der Oder
ADo d: Anschlußdose
Ado e: Adenosine
- i: Adagio
A & DO e: Analog[ue] and Discrete Output
ADOBE e: Atmospheric Diffusion of Beryllium [Program[me]]
ADOBE Program[me] e: Atmospheric Dispersion of Beryllium Program[me]
ADOC - e: Air Defense Operations Center
- e: Automatic Defense Operation Center
ad oc. l: ad oculos
ADOCBL e: Adenosylcobalamin
ADOCS e: Advanced Digital / Optical Control System
ADOD e: Assistant Director of Operations Division
ADOF e: Arbitrary Degree of Freedom
- e: Assistant Director of Ordnance Factories
AD of A e: Assistant Director of Artillery
AD of D e: Assistant Director of Dockyards
ADOFP e: Armo[u]red Division Ordnance Field Park
AD of R e: Assistant Director of Railways
AD of S e: Army Depot of Supply
AD of ST e: Assistant Director of Sea Transport
AD of S&T e: Assistant Director of Supplies and Transport
AD of T e: Assistant Director of Transport
AD of W e: Assistant Director of Works
A-Dog e: Alternate Diesel or Gas-Turbine
ADOGA e: American Dehydrated Onion and Garlic Association
adoit e: automatically-directed out-going intertoll trunk
ADOIT e: Automatically Directed Outgoing Intertoll Trunk
ADOIT Test Circuit e: Automatically Directed Outgoing Intertoll Trunk Test Circuit
ADOK d: Aufbau-Dokumentation
ADOL e: Adolescence
- e: Adolescent
ADOM e: Acid Deposition and Oxidant Model
- e: Army Depot Operations Management
ADONIS e: Article Delivery Over Network Information Services
- e: Automatic Digital On-Line Instrumentation System
- e: Automatic Digital On-Line Instruments System
- e: Automatic Digital Online Information System
ADOP e: Additive Operational Project
adop e: adopt[ion]
ADOP e: Adriamycin, Oncovin, Prednisone
- e: Advanced Distributed Onboard Processor
ADOPE e: Automatic Decisions Optimizing Predicted Estimates
ADOPT e: Adoption
- e: Advanced Optics Technology
- e: Automatic Design Optimization Techniques
AdöR d: Anstalt des öffentlichen Rechtes
ADOS e: Advanced Diskette Operating System
- e: Area Distribution Officers
- e: Army Depot of Supply
- e: Assistant Director of Ordnance Service
- e: Astronautical Defensive-Offensive System
- e: Authorization for Disposal of Overhead Supplies
- e: Automated Document Ordering System
ADoS d: Anschlußdosenstecker
ADoS8 d: Anschlußdosenstecker; achtpolig
ADOSt. d: Allgemeine Dienstordnung für die Staatsbehörden
ADOT e: Automatically Directed Outbound Trunk
- e: Automatically Directed Outgoing Trunk
- e: Automatic Digital Optical Tracker
ADOW e: Automatic Dial Order Wire
adoxograph e: adoxographer
- e: adoxographic[al][ly]
- e: adoxography
ADP d: Adenosindiphosphat
- d: Adenosindiphosphorsäure
- d: alphanumerischer Datenerfassungsplatz
- d: antiduretisches Prinzip
- d: Arbeitsdienstpflicht
- d: Arbeitsgemeinschaft für Zucht und Prüfung deutscher Pferde
- e: Academy of Denture Prosthetics
- e: Acceptance Data Package
- e: Accountability Data Package

- e: Acid Dew Point
- e: Acoustic Data Processor
- e: Add Decimal Packed
- e: Adenosine Diphosphate
- e: Administrative Data Processing
- e: Administrative Detail Program[me]
- e: Advanced Data Processing
- e: Advanced Development Plan[s]
- e: Agent Development Program[me]
- e: Agricultural Development Project
- e: Airborne Data Processor
- e: Air Data Package
- e: Air Data Probe
- e: Air Defence Position
- e: Air Delivery Platoon
- e: Air Driven Pump
- e: Airport Development Program[me]
- e: Alclometasone Dipropionate
- e: Alphabetic Duplicating Punch
- e: Alphanumeric Data Printer
- e: Alternative Defense Posture
- e: American Data Processing, Incorporated
- e: Aminodiphenyl
- e: Ammonium Dehydrogen[e] Phosphate
- e: Angle Detector Parallel
- e: Angular Distribution Pattern
- e: Animal Disease and Parasite
- e: Apollo Dynamic Programs
- e: Apparatus Drawings Project
- e: Approach Deterioration Parameter
- e: Area Distribution Panel
- e: Armament Delivery Point
- e: Army Depot Police
- e: Assistant Director of Pathology
- e: Assistant Director of Port
- e: Assistant District Postmaster
- e: Association for Denture Prosthesis
- e: Association of Database Producers
- e: Association of Directory Publishers
- e: Assumed Decimal Point
- e: Atmospheric Dynamics Program[me]
- e: Automated Data Processing
- e: Automatic Data Plotter
- e: Automatic Data Plotting
- e: Automatic Data Processing
- e: Automatic Data Processor
- e: Automatic Delection Procedure
- e: Automatic Destruct Program[me]
- e: Automatic Digital Processor
- e: Automatic Duplicator Printer

a/dp e: acid dip
ADPA - e: Accounting and Data Processing Abstracts (journ.)
- e: Air Data Probe Assemblies
- e: American Defense Preparedness Association
- e: Assistant Director of Prevention of Accident
- e: Automobile Dealers Parts Association

ADPACS e: Automated Data Processing and Communications Service
ADPase e: Adenosine Diphosphatase
ADPASS e: Advanced Patrol Sensor System
ADPB e: Air Defense Planning Board
ADPBCT e: Automatic Data Processing Budget Control Totals
ADPBUD e: Automatic Data Processing Budget
ADPC e: Abu Dhabi Petroleum Company
- e: Agricultural Development Planning Center
- e: Automatic Data Processing Center

ADP/CIS e: Automation of Data Processing / Computerization of Information Systems
ADPCM d: adaptive Differenz-Pulscode-modulation
- d: adaptives Differenz-Pulscode-Modulationsverfahren
- e: Adaptive Differential Pulse-Code Modulation
- e: Adaptive Differential Pulse-Code Modulator
- e: Adaptive Differential Pulse Code Modulated
- e: Association for Data Processing and Computer Management

ADPCMs e: Adaptive Differential Pulse-Code Modulators
ADPCMT e: Adaptive Differential Pulse Code Modulated Transcoder
ADPE e: Association of Data Processing Employees
- e: Automatic Data Processing Engineering
- e: Automatic Data Processing Equipment
- e: Auxiliary Data Processing Equipment

ADPEA e: Australian Data Processing Employees Association
ADP-EFS e: Automatic Data Processing Electronic Financial Services
ADPEP e: Automated Data Preparation by Electronic Photocomposition
- e: Automated Data Preparation Evaluation Program[me]

ADPES e: Angle-Dispersed Photoelectron Spectroscopy
- e: Automatic Data Processing by Equipment Systems

ADPE/S e: Automatic Data Processing Equipment/System
- e: Automatic Data Processing Equipment and Software

ADPES e: Automatic Data Processing Equipment Services
ADPESO e: Automatic Data Processing Equipment Selection Office
ADPFB e: Automatic Data Processing Field Branch
ADPG d: Adenosin-Diphosphat- Glucose
- e: Adenosine Disphosphoglucose
- e: Air Defense Planning Group

ADPI e: American Dairy Products Institute
ADPIP e: Automatic Data Processing International Program[me]
ADPKD e: Autosomal Dominant Polycystic Kidney Disease
ADPL e: Algorithmic Processor Description Language
- e: Assistant Director for Policy and Liaison
- e: Association of Data Processing Librarians
- e: Automated Drawing Parts List
- e: Average Daily Patient Load

ADPLAN e: Advancement Planning
- e: Advertising Planning

ADPLL e: All Digital Phase Locked Loop
ADPLO e: Automatic Data Processing Liaison Officer
ADPLS e: Automated Drawing Parts List System
ADPM d: Adaptive Delta-Puls- Modulation
ADPM e: Automatic Data Processing Machine
ADPMG e: Assistant Deputy Postmaster-General
ADPMIS e: Automatic Data Processing Management Information System
ADPMO e: Automatic Data Processing Modification Order
ADPO e: Advanced Development Program[me] Office
- e: Automatic Data Processing Operations

adpo e: Aircraft Depot
ad pond. om. l: ad pondus omnium
ADPP e: Assembly Detail Purchased Parts
- e: Automatic Data Processing Program[me]

ADPPB e: Automatic Data Processing Production Branch
ADPP & DB e: Automatic Data Processing Planning and Development Branch
ADPP & PB e: Automatic Data Processing Programming and Processing Branch
ADPPRS e: Automatic Data Processing Program[me] Reporting System
ADPR e: Adenosine Diphosphate Ribose
- e: Assistant Director for Plans and Resources
- e: Assistant Director of Public Relations

ADPREP e: Automatic Data Processing Resource Estimating Procedures
ADP Research e: Animal Disease and Parasite Research
ADP-Rib e: Adenosine Diphosphate Ribose
ADPRIN e: Automatic Data-Processing Intelligence Network
ADPRO e: Automatic Data Processing Requirements Office
ADPRT e: Adenosine Diphosphate Ribosyltransferase
ADPS e: Acid Deposition Planning Staff
- e: Active Delay Phase Shift
- e: Advanced Digital Processing System
- e: Advanced Personnel Data System
- e: Advanced Planning Data Sheet
- e: Apollo Document Preparation Standards [handbook]
- e: Army Data Processing System
- e: Assistant Director of Personnel Services
- e: Assistant Director of Postal Services
- e: Automated Data Processing System[s]
- e: Automatic Data Processing Security
- e: Automatic Data Processing Services
- e: Automatic Data Processing Subsystem
- e: Automatic Data Processing System[s]
- e: Automatic Display Plotting System

ADPs e: Artillery Destruction Programs
ADPSC e: Automatic Data Processing Service Center
ADPSD e: Automatic Data Processing Systems Development
ADPSE e: Automatic Data Processing Systems and Equipment
ADPSO e: Association of Data Processing Service Organizations
- e: Automatic Data Processing Selection Office

ADPSR e: Architects/Designers/Planners for Social Responsibility
ADPSS e: Assistant Director of Printing and Stationary
ADPSSEP e: Automated Data Processing System Security Enhancement Program[me]
ADPSSO e: Automatic Data Processing System Security Officer
ADPSYM e: Adaptive Process Symposium
ADPT e: Adenosine Phosphoribosyl-transferase
Adpt d: Adapter
- e: Adaptor

adpt e: adapted
- e: adapter

ADP-T e: Automated Data Processing Telecommunications
ADPTC e: Automatic Data Processing Training Center

ADP/TCS e: Automatic Data Processing Telephone Computing Services
ADP/TM e: Automatic Data Processing and Telecommunications Management
ADPTOS e: Automatic Data Processing Tactical Operation System
Adptr d: Adapter
ADPTR e: Adaptor
ADPTS e: Assistant Director of Physical Training and Sports
ADPU e: Airborne Data Processing Unit
- e: Airborne Digital Processing Unit
- e: Association of Public Data Users
- e: Automatic Data Processing Unit
ADQC e: Almost Difference Quasi-ternary Code
- e: Almost Differential Quasi-ternary Code
ADQ Code e: Almost Differential Quasiternary Code
ADQT e: Adequate
ADR d: Additionsregister
- d: Adressenregister
- d: Adressenteil
- d: Adreßkonstante
- d: Arbeitsgemeinschaft Deutscher Rinderzüchter
- e: Accelerated Depreciation Range
- e: Accepted Dental Remedies (journ.)
- e: Accident Data Recorder
- e: Accident Data Recording
- e: Achievable Data Rate
- e: Acid-Detergent Residue
- e: Adder
- e: Add Register
- e: Address
- e: Address Register
- e: Adriamycin
- e: Advanced Dated Remittances
- e: Advanced Development Report
- e: Advance Deviations Report
- e: Adverse Drug Reaction
- e: Advisory Route
- e: Aeronautical Data Report
- e: Airborne Data Recorder
- e: Airborne Digital Recorder
- e: Aircraft Destination Record
- e: Aircraft Direction Room
- e: Aircraft Discrepancy Report
- e: Air Defense Region
- e: Air Design Review
- e: Airfield Damage Repair
- e: Alphanumeric Data Recorder
- e: Alphanumeric Data Recording
- e: Alternate Data Retry
- e: Alternative Dispute Resolution
- e: American Depositary Receipt
- e: Ammunition Disposition Request (or Report)
- e: Analog[ue]-[to]-Digital Recorder
- e: Analog[ue] Data Recorder
- e: Analog[ue] Data Recording
- e: Analytical Data Reflection
- e: Angle Data Recorder
- e: Angle Data Recording
- e: Applied Data Research
- e: Applied Data Research, Incorporated
- e: Army Density Report
- e: Asset Depreciation Range
- e: Audit Discrepancy Report
- e: Austin Data Recorder
- e: Australian Design Rule
- e: Automatical Digital Relay
- e: Automatic Data Recorder
- e: Automatic Data Recording
- e: Automatic Data Relay
- e: Automatic Data Retrieval
- e: Automatic Diagnostic and Recovery
- e: Automatic Dialog Replacement
- e: Automatic Dialogue Replacement
- e: Automatic Digital Relay
- e: Automatic Digit Recognizer
- e: Automatic Digit Recognizing
- e: Automatic Direct Reader
- e: Automatic Direct Reading
- e: Automatic Distortion Reduction
- e: Automatic Dividend Reinvestment
- e: Average Daily Rate
- e: Aviation Design Research
- f: Association pour le Dévelopement de la Recherche
adr a: addresse
- r: address
Adr d: Adenalin[e]
Adr. d: Adressat
- d: Adresse
- e: Adriatic
a. d. R. d: an der Ruhr
a-d r e: analog-to-digital recorder
ADR. e: Assistant Director of Remounts
ADRA e: Animal Diseases Research Association
- e: Army Dollar Resource Allocation
- e: Automatic Dynamic Response Analyzer
- e: Automotive Dismantlers and Recyclers Association
adrac e: automatic digital recording and control
ADRAN [System] e: Advanced Digital Ranging [System]
ADRAT e: Advanced Deep-Running Acoustic Torpedo
ADRB e: Addressable
ADRC e: Air Defence Radar Control
- e: Alzheimers Disease and Related Conditions
- e: Alzheimers Disease Research Center
- e: Animal Drug Research Center
- e: Automatic Data Range Changer
- e: Automatic Data Rate Changer
- e: Automatic Digital Relay Center
Adr Chk e: Address Check
ADRCP e: Alzheimers Disease Research Centers Program[me]
ADRD e: Automatic Data Rate Detection
- e: Automatic Data Rate Detector
ADRDA e: Alzheimers Disease and Related Disorders Association
ADRDE e: Advise Reason for Delay
- e: Air Defense Research and Development Establishment
A Dr & Dgn e: Associate in Drafting and Design
ADRE e: Automatic Dynamic Range Expansion
ADRECS e: Advanced Recovery System
ad ref. l: ad referendum
Adrema d: Adressiermaschine
ADREN e: Address Enable
adren d: Adrenal[in]
- e: adrenal
- e: adrenalin
adrenals e: adrenal glands
ADREP e: Accident-incident Data Reporting
- e: Aircraft-accident Data Reporting Panel
- e: Automatic Data-processing Resource Estimating Procedures
ADREPP e: Aircraft-accident Data Reporting Panel
ADRES e: Aircraft Data Recording Evaluation System
- e: Army Data Retrieval [Engineering] System
- e: Automatic Dynamic Range Expansion System
A-D Resolution e: Analog[ue]-Digital Resolution
ADRG e: Applied Demographic Research Group
- e: Assistant Deputy Registrar General
- e: Automatic Data Routing Group
ADRI e: Add Delete Rework Instruction
- e: Automatic Dead-Reckoning Instrument
- in: Angkatan Darat Republik Indonesia
Adria d: Adria
- e: Adriatic Sea
Adriatic e: Adriatic Sea
ADRIES e: Advanced Digital Radar Imagery Exploitation System
ADRIS d: automatisierte Dokumenten-Recherche und Information über Standards (DDR)
- e: Advanced Degree Requirements Information System
- e: Association for the Development of Religious Information Systems
- e: Automatic Dead Reckoning Instrument Systems
ADRK e: Auxiliary Display Request Keyboard
ADRM e: Aerodrome
ADRM e: Analog-to-Digital Remastering of musical recordings
ADRMP e: Automated Dialler and Recorded Message Player
ADRMPS e: Auto-Dialled Remote Message Players
ADRN e: Advanced Document Revision Notice
- e: Advance[d] Drawing Release Notice
ADROBN e: Airdrome Battalion
ADROIT e: Automated Data Retrieval and Operations Involving Timeseries
ADRP e: Acoustic Data Reduction Program[me]
- e: Airdrop
- e: Autosomal Dominant Retinitis Pigmentosa
ADRPM e: Acoustic Detection Range Prediction Model
ADR-Prüfung d: All-Deutsche Rosen-Prüfung
Adr Reg d: Adressenregister
- d: Adreßregister
ADRS e: Adaptive Data Reporting System
- e: Address
- e: A Departmental Reporting System
- e: Airborne Digital Recording System
- e: Analog[ue]-to-Digital Data Recording System
- e: Asset Depreciation Range System
- e: Automated Data Retrieval System
- e: Automatic Data Reporting System
- e: Automatic Document Request Service
- e: Automatic Document Request System
Adr S e: Adriatic Sea
ADRs e: Asset Depreciation Ranges
ADRSA e: Assistant Data Recording System Analyst
ADRSS e: Automated Data Reports Submission System
ADR System e: Analog[ue]-to-Digital Data Recording System
ADRT e: Analog[ue] Data Recorder Transcriber
- e: Analog[ue] Data Recording Transcriber
- e: Assistant Director of Railway Transport
adr tel f: adresse telegraphique
ADRU d: Ausgabe Druckliste

ADRUSAR e: Army Density Report; United States Army Reserve
ADRV d: Arbeitsgemeinschaft Deutscher Rechenzentrumsverbände
A-DRV e: Atomic Drive
AD Ry e: Assistant Director of Railways
ADS d: Administrationssupervisor
- d: Adressenspalte
- d: Adressensteuerung
- d: Adresse Spalte
- d: Adsorption
- d: Aktion Demokratischer Studenten
- d: Allgemeine Deutsche Seeversicherungsbedingungen
- d: Allgemeiner Deutscher Sprachverein
- d: Amplitudendichtespektrum
- d: antidiuretische Substanz[en]
- d: Arbeitsgemeinschaft Deutscher Schafzüchter
- d: Arbeitsgemeinschaft Deutscher Schweinezüchter
- d: Arbeitsgemeinschaft Deutscher Studentenbühnen
- d: Arbeitsgemeinschaft Deutscher Studentenschaften
- d: Arbeitsgemeinschaft Deutscher Studentensport
- d: Auftragsdatenspeicher
- d: Ausgangsdatenstrom
- d: automatisiertes Datenerfassungssystem
- d: automatisiertes Datenverarbeitungssystem
- e: Accelerated Declassification System
- e: Accessory Drive System
- e: Accounting Data System
- e: Accumulated Data Set
- e: Accurately Defined System
- e: Accurately Defined System[s]
- e: Acoustic Doppler Sounder
- e: Activated Data Sheet
- e: Activity Data Sheet
- e: Address
- e: Address Data Strobe
- e: Address Display Subsystem
- e: Address Display System
- e: Address Scan[ning]
- e: Administration and Data Server
- e: Administration of Designed Services
- e: Administrative Data Systems
- e: Adsorbed
- e: Advanced Data Scaler
- e: Advanced Data System[s]
- e: Advanced Debugging System
- e: Advanced Declassification Schedule
- e: Advanced Deep-dive System
- e: Advanced Digital System
- e: Advanced Display System
- e: Advanced Diving System
- e: Advanced Dosimetry System
- e: Advanced Dressing Station
- e: Advertising Dimensions Standards
- e: Aerial Delivery System
- e: Aerial Demonstration Squadron
- e: Aeronautical Design Standard
- e: Aerospace Data Systems
- e: Agent Distributor Service
- e: Agricultural Development Service
- e: Aircraft Development Service
- e: Air Data Sensor
- e: Air Data Subsystem
- e: Air Data System
- e: Air Defense Sector
- e: Air Defense Ship
- e: Air Defense Squadron
- e: Air Defense System
- e: Air Development Service
- e: Air Development Station
- e: Air Development System
- e: Air Dried Sheet[s]
- e: Air Drop System
- e: Airport Data System
- e: Algorithm Description Section
- e: All-Digital Simulator
- e: Allied Data Sciences
- e: Alternate Delivery System
- e: Alternate Device Support
- e: Alternative Delivery System
- e: Alzheimers Disease Society
- e: American Daffodil Society
- e: American Dahlia Society
- e: American Data Systems
- e: American Dental Service
- e: American Dental Society
- e: American Denture Society
- e: Ammunition Distribution System
- e: Analog & Digital Systems
- e: Analog[ue]-[to]-Digital Subsystem
- e: Analog[ue]-[to]-Digital System
- e: Angle Data Subsystem
- e: Angle Detector Sequential
- e: Animal Defence Society
- e: Anker Data Systems
- e: Annual Demographic Survey
- e: Anomaly Dynamics Study
- e: Anticoincidence Detection System
- e: Anti-ice / De-ice System
- e: Antidiuretic Substance[s]
- e: Applicant Data System
- e: Application Design Service
- e: Application Development Systems
- e: Applied Decision System[s]
- e: Arctic Drift Station
- e: Arctic Drilling System
- e: Area, Date, Subject
- e: Arithmetic Design System
- e: Army Dental Service
- e: Array Declarator Subscript
- e: Assault Data System
- e: Assistant Director of Survey
- e: Association of Diesel Specialists
- e: Aston Dark Space
- e: Atmospheric Diving Suit
- e: Atmospheric Diving System
- e: Attitude Display System
- e: Audio Distribution System
- e: Audio Subsystem
- e: Autodesk Development System
- e: Automated Data System[s]
- e: Automated Declassification System
- e: Automated Design System
- e: Automated Diagnostic System
- e: Automated Dispatch System
- e: Automated Documentation System[s]
- e: Automated Drafting Services
- e: Automated Drafting System
- e: Automated Duplicating System
- e: Automatic Data Set
- e: Automatic Data Surveillance
- e: Automatic Data System
- e: Automatic Defense System
- e: Automatic Depressurization System
- e: Automatic Development System
- e: Automatic Digital Switch
- e: Automatic Dispatch System
- e: Automatic Door Seal
- e: Autopilot Disengage Switch
- e: Aviation Data Service, Incorporated
- e: Aviation Depot Squadron
- e: Azimuth Determining System
- f: Academie des Sciences
- l: Adversus
a. d. S. d: an der Saale
ads e: antibody deficiency syndrome
- e: autograph document signed
- e: automated data sheet
A/DS e: Audio/Digital System[s]
ADSA e: Air-Derived Separation Assurance
- e: American Dairy Science Association
- e: American Dental Society of Anesthesiology
- e: Atomic Defense Support Agency
ad. saec. l: ad saeculum
ADSAF e: Automatic Data System for the Army in the Field
Adsaf e: Automatic Data System within the Army in the Field
ADSAI e: American Dermatologic Society of Allergy and Immunology
AD Salv e: Assistant Director of Salvage
AD/SAM e: Air Defense / Surface-to-Air Missile
ADSAP e: Advise as Soon as Possible
ADSARM e: Advanced Defense Suppression Antiradiation Missile
ADSAS e: Air-Derived Separation Assurances System
ADSAT e: Anomalous Dispersion Spherical Array Target
- e: Assistant Director of Supplies and Transport
ad sat. l: ad saturandum
- l: ad saturationem
ADSATIS e: Australian Defence Science and Technology Information System
- e: Australian Defense Science and Technology Information Service
ADSC e: Active Duty Service Commitment
- e: Advanced Section Communication Zone
- e: Air Defense Software Committee
- e: Air Defense Systems Command
- e: Army Distinguished Service Cross
- e: Association of Drilled Shaft Contractors
- e: Automatic Data Service Center
- e: Automatic Digital Switching Center
- e: Average Daily Service Charge
A/D-Schalter d: Analog/Digital- Schalter
Ad Sci e: Advancement of Science
ADSCOM e: Advanced Shipboard Communications
ADSCP e: Advanced Defense Satellite Communications Project
ADSCS e: Aided Display Submarine Control System
ADSC Zone e: Advanced Section Communication Zone
ADSD e: Active Duty Service Date
- e: Air Defense Systems Directorate
adsd e: addressed
adsda e: advise earliest date
ADSDP e: Automated Data System Development Plan
ADSE e: Addresse
- e: Alternative Delivery Schedule Evaluator
- e: American Dental Society of Europe
ad sec. e: ad sectam
ADSEC e: Advance[d] Section
- e: Air Defense Systems Engineering Committee
ADSEL e: Address Selective
ADSEL radar e: Addressable Secondary Surveillance Radar
ADSEP - e: Automatic Data Set Editing Program[me]
ad sep. l: ad separatum
ADSF e: Automated Directional Solidification Furnace
ADSFC e: Automatic Data Field System Command

ADSG e: Alternative Delivery Schedule Generator
- e: Atomic Defense and Space Group
ADSHIPDA e: Advise Shipping Data
- e: Advise Shipping Date
ADSHPDAT e: Advise Shipping Date
ADSI e: Adapted Delivered Source Instruction
- e: Administrative Design Service Information
- e: Agricultural Development Service, Inc.
ADSIA e: Allied Data Systems Interoperability Agency
ADSID e: Air-Delivered Seismic Instruction Detector
- e: Air Defense Systems Integration Division
adsid e: alr-delivered acoustic-implant seismic-intrusion detector
AD Sigs e: Advanced Signals
- e: Assistant Director of Signals
ADSIGS e: Advanced Signals Assistant Director of Signals
ADSIM e: Advanced Simulation
A & DSL e: Administrative and Direct Support Logistics
ADSL e: Assembly Department Shortage List
- e: Authorized Depot Stockage List
ADSM e: Air Defense Suppression Missile
- e: American Defense Service Medal
- e: Army Distinguished Service Medal
- e: Automated Data Systems Manual
ADSMG e: Advice by Message
ADSMO e: Air Defense System Management Office
ADSN e: Accounting and Disbursing Station Number
ADSO e: Aerospace Defense Systems Officer
- e: Assistant Division Supply Officer
- e: Automatic Display Switching Oscilloscope
ADS/O e: Application Development System/Online
ADSOC e: Administrative Support Operations Center
ADS-Ofen d: Albert De Sy-Ofen
ADSOL e: Analysis of Dynamical Systems On-Line
ADSORB d: Adsorbant
Adsorpt e: Adsorption
ADSOS e: Advance Services of Supply
ADSOT e: Automatic Daily System Operability Test
ADSp d: Allgemeine Deutsche Spediteurbedingungen
ADSP e: Adaptive Digital Signal Processor
- e: Advanced Digital Signal Processor
- e: Automatic Dispatching Stick Repeater
ADSPEC e: Additional Specialty
ad spec. l: ad specialia
ADSPN e: Advise Disposition
ADSq e: Air Defense Squadron
ADSR e: Attack / Decay / Sustain / Release
ADSS e: Advanced Digital Simulation System
- e: Advanced Software System
- e: Aircraft Damage [Sensing] System
- e: Air Data Screening System
- e: Air Defense Suppression System
- e: Analysis of Digitized Seismic Signals
- e: Army Decision Support System
- e: Atomatic Data Switching System
- e: Australian Defense Scientific Service
- e: Automated Data Subsystem
- e: Automatic Data Switching System
ADSS Radar e: Addressable Secondary Surveillance Radar
a. ds. T. d: an demselben Tag
ADST d: Anforderung an die Datensteuerung
- e: Assistant Director of Sea Transport
- e: Atlantic Daylight Saving Time
- e: Automatic Digital Information Network Digital Subscriber Terminal
ADS & T e: Assistant Director of Supplies and Transport
adstadis e: Advise Status and/or Disposition
ADSTAP System e: Advancement, Strength and Training Plan System
ADSTAR e: Advance Document Storage and Retrieval
- e: Automatic Document Storage And Retrieval
ADSTKOH e: Advise Stock On Hand
ADSTP e: Administrative Data System[s] Teleprocessing
ADSU e: Advanced Direct Support Unit
- e: Air Data Sensor Unit
- e: Airstream Direction Sensing Unit
ADSUM d: Adressensuchmethode
ADSUP e: Automatic Data System[s] Uniform Practice[s]
ADSVB d: Allgemeine Deutsche Seeversicherungsbedingungen
AD Svy e: Assistant Director of Survey
ADSW e: Advanced Defense Suppression Weapon
A-D Switch e: Analog[ue]-to-Digital Switch
ADSWSO e: Air Defense Special Weapons Support Organization
adsym e: automobile defog-defrost system model
ADT d: Adenosintriphosphat
- d: Arbeitsgemeinschaft Deutscher Textilingenieure
- d: Arbeitsgemeinschaft Deutscher Tierzüchter
- e: Abstract Data Type
- e: Accelerated Development Test[s]
- e: Accepted Dental Therapeutics
- e: Active Disk Table
- e: Active Duty [for] Training
- e: Actual Departure Time
- e: Adenosine Triphosphate
- e: Admission / Discharge / Transfer
- e: Advanced Design Team
- e: Advanced Development Technology
- e: Advanced Dispenser Technology
- e: Advanced Driver Training
- e: Aerated Drain Tank
- e: Aerial Demonstration Team
- e: Agar-gel Diffusion Test
- e: Aided Tracking
- e: Airborne Data Terminal
- e: Airborne Digital Timer
- e: Air Data Transducer
- e: Alaskan Daylight Time
- e: Alphanumeric Display Tube
- e: Alternate-Day Treatment
- e: American District Telegraph
- e: Amphibious Training Demonstrator
- e: Any Desired Thing
- e: Application-Dedicated Terminal
- e: Applied Drilling Technology
- e: Approved Departure Time
- e: Articulated Dump Truck
- e: Assistant Director for Training
- e: Assistant Director of Torpedoes
- e: Assistant Director of Transport
- e: Assured Depot Task
- e: Asynchronous Data Transceiver
- e: Asynchronous Data Transfer
- e: Atlantic Daylight Time
- e: Atomic Damage Template
- e: Attribute Distributed Tree
- e: Auditory Discrimination Test
- e: Automated Dithionate Test
- e: Automatic Damage Template
- e: Automatic Data Translator
- e: Automatic Data Transmission
- e: Automatic Data Transmitter
- e: Automatic Debit Transfer
- e: Automatic Detection and Tracking
- e: Autonomous Data Transfer
- e: Average Data Technique
a.d.T. d: an demselben Tage
adt e: advertisement
- e: average daily traffic
AD/T e: air detector/tracker
ADTA e: Aircraft Development Test Activity
- e: Air Data Transducer Assembly
- e: American Dental Trade Association
- e: Augment Docking Target Adapter
- e: Aviation Development Test Activity
ÄDTA d: Äthylendiamintetraazetat
ADTAC e: Air Defense Tactical Air Commander
- e: Automatic Digital Tracking Analyzer Computer
ADTAK e: Advanced Action Taken
ADTAM e: Air-Delivered Target-Activated Munitions
AdTb d: Altdeutsche Textbibliothek (journ.)
ADTB e: Apollo Docking Test Device
ADTC e: Air Defense Technical Center
- e: Armament Development and Test Center
ADTD e: Anticipated Data Transmission Date
- e: Apollo Docking Test Device
- e: Association of Data Terminal Distributors
ADTDS e: Air Defense Tactical Data Systems
ADTe e: Anodal Duration Tetanus
ÄDTE d: Äthylendiamintetraessigsäure
- e: Äthylendimintetraessigsäure
ADTEC e: Army Ballistic Missile Defense Agency Data Processing Testbed and Evalua
ADTECH e: Advanced Decoy Technology
AD Technique e: Alloy-Diffusion Technique
ADTELP e: Advice by Teletype
ad tert. vic. l: ad tertium vicem
adtevac process e: adsorption/ temperature/vacuum process
ADTF e: At-Depth Test Facility
- e: Aviation Development Test Facility
ADTG e: Application Development Task Group
ADTIC e: Arctic, Desert, Tropic Information Center
ADTLP e: Army-wide Doctrinal and Training Literature Program[me]
AD/TMD e: Air Defense/Theater Missile Defense
ADTN e: Aminodihydroxytetrahydro-naphthalene
ADTn e: Assistant Director of Transportation
ADTng e: Assistant Director of Training
ADTP e: Accelerated Development Test Program[me]
AD-TR e: Armament Development Technical Report
ADTr e: Assistant Director of Transportation

ADTRANSISTOR e: Alloyed Diffused Transistor
ADTS e: Air-Dried and Tub-Sized
- e: Airborne Data Transfer System
- e: Air Data Test System
- e: Amplitude Degradation Test System
- e: Automated Data and Telecommunications Service
- e: Automatic Data and Telecommunications Service
- e: Automatic Data Test System
- e: Automatic Data Transfer System
- e: Avionics Depot Test Station
ADTSC e: Auto Dealers Traffic Safety Council
ADTSEA e: American Driver and Traffic Safety Education Association
adtu e: automatic digital test unit
ADTU e: Auxiliary Data Translator Unit
ADTV e: Advertise[ment]
ADU d: Analog-Digital-Umschalter
- d: Analog-Digital-Umsetzer
- d: Analog-Digital-Umsetzung
- d: Analog-Digital-Umwandler
- d: Analog-Digital-Umwandlung
- d: automatischer Datenumsetzer
- e: Acceleration-Deceleration Unit
- e: Accumulation and Distributing Unit
- e: Accumulation and Distribution Unit
- e: Accumulation Distribution Unit
- e: Adapter Unit
- e: Address Decoding Unit
- e: Aircraft Delivery Unit
- e: Air Distribution Unit
- e: Alignment Display Unit
- e: American Depot Unit
- e: Ammonium Diuranate
- e: Analog[ue] Delay Unit
- e: Analog[ue] Display Unit
- e: Angular Dialling Unit
- e: Angular Display Unit
- e: Annunciator Display Unit
- e: Arc Detector Unit
- e: Automatic Data Unit
- e: Automatic Dialling Unit
- e: Auxiliary Display Unit
a. d. U. d: an der Unstrut
A.d.Ü. d: Anmerkung des Übersetzers
ADUKAM d: Auswert[e]programm für Durchlaufzeit- und Kapazitätsauslastungs-Analysen mit Hilfe
adult e: adulterant
- e: adulterate
- e: adulteration
ADULT e: Adultery
ADUM e: Automated Data Unit Movement
A-D-Umsetzer d: Analog-Digital- Umsetzer
A-D-Umsetzung d: Analog-Digital-Umsetzung
A/D-Umsetzung d: Analog/Digital-Umsetzung
ADUS e: Avionics Data Utilization System
ad us. l: ad usum
ADUSA d: Analog/Digital-Umwandlung mit statistischer Auswertung
ad us. ext. l: ad usum externum
ADUT e: Advanced Development Unit Test
ADV d: Abteilung für Datenverarbeitung
- d: allgemeine Datenverarbeitung
- d: Allgemeine Durchführungsverordnung
- d: Arbeitsgemeinschaft der Volksgesundheitsbewegung
- d: Arbeitsgemeinschaft Deutscher Verkehrsflughäfen
- d: Arbeitsgemeinschaft für Datenverarbeitung
- d: Arbeitskreis Datenverarbeitung
- d: Atlas der deutschen Volkskunde
- d: Ausbildungsgesellschaft für Datenverarbeitung
- d: automatische Datenverarbeitung
- d: automatische Datenverarbeitung[sanlage]
- d: automatisches Daten-Verarbeitungs-system
- d: automatisierte Datenverarbei-tung[sanlage]
- e: Acid Degree Value
- e: Acreage Diversion
- e: Actinomycin-D, Dacarbazine, Vincristine
- e: Adenovirus
- e: Advanced Development Vehicle
- e: Advantage
- e: Adverbial
- e: Advice
- e: Advise
- e: Advisory
- e: Advocate
- e: Airborne Digital Voltmeter
- e: Air Defense Variant
- e: Air Defense Version
- e: Air Diverter Valve
- e: Aleutian Disease Virus
- e: Anti-Drainback Valve
- e: Arc-Drop-Voltage
- e: Atmospheric Dump Valves
- l: Adversus
Adv d: Advent
- d: Adventivarten
- d: Adverb
- e: Advection
- e: Advertisement
Adv. d: Advokat
- e: Adviser
- e: Advisor
AdV d: Arbeitsgemeinschaft der Vermessungsverwaltungen der Länder der Bundesrepublik Deutschland
adv e: advance[d] [s]
- e: advance[d] Guard
- e: advancement
- e: adverb[ial]
- e: advertising
- e: advice
- e: advise
A/DV e: Arterio / Deep Venous
a/dv e: arterio/deep venous [injection]
ad v. l: ad valorem
adv. l: adversum
ADVA d: automatische Datenverarbeitungsanlage
advac e: advise acceptance
adv. Activ. Anal. e: Advances in Activation Analysis (GB)(journ.)
Adv Aerosol Phys e: Advances in Aerosol Physics
advail e: advise availability
ADVAL e: Advise Availability
ad val. l: ad valorem
ADVALT e: Advice of Allotment
Advan Agron e: Advances in Agronomy
ADVANCE e: Airborne Doppler Velocity Altitude Navigation Compass Equipment
ADVANCEM e: Advancement
Advance Sci e: Advancement of Science
Advan Phys e: Advances in Physics
Adv Appl Mech e: Advances in Applied Mechanics
ADVAST e: Advanced Station
Adv At Mol Phys e: Advances in Atomic and Molecular Physics
Advb d: Adverb
ADVB d: Arbeitsgemeinschaft der Datenverarbeitungs-Bildungszentren
advb e: adverb
- e: adverbial
Adv BOD e: Advance Base Ordnance Depot
Adv Bse e: Advanced Base
Adv Cert in Ed e: Advanced Certificate in Education
Adv Cert in Mus Ed e: Advanced Certificate in Music Education
Adv Chem Phys e: Advances in Chemical Physics
adv chgs e: advance charges
Adv Clr Sta e: Advaced Clearing Station
Adv Dep e: Advanced Depot
ADVDISC e: Advance Discontinuance of Allotment
ADVDLA-DEP e: Advance Payment of Dislocation Allowance to Dependents
ADV DStO d: Allgemeine Durchführungs-verordnung zur Dienststrafordnung
ADVE d: Automatische Datenverarbeitungseinheit
advec e: advection
advect e: advection[al][ly]
- e: advective
- e: adventurer
Adv Electron Electron Phys e: Advances in Electronics and Electron Physics (journ.)
adven e: adventure
- e: adventure[r]
Adv. Engr. e: Advisory Engineer
Adv Eng Software e: Advances in Engineering Software (journ.)
ADV-EN-SEQ e: Advance-Enable-Sequence
adversat e: adversative
ADVERT e: Advertisement
advert e: advertising
advertique e: advertising antique
adverts e: advertisements
Adv Flt Trng e: Advanced Fleet Training
Adv Food Res. e: Advances in Food Research (journ.)
adv frt e: advance freight
AdvG d: Adverbialgruppe
ADVG d: automatisierte Datenverarbeitung
advg e: Advancing
- e: Advertising
Adv. Genet. e: Advances in Genetics (journ.)
Adv. Geophys. e: Advances in Geophysics (journ.)
ADVG NW d: automatisierte Datenver-arbeitung in Nordrhein-Westfalen
Adv. Heat Transfer e: Advances in Heat Transfer (journ.)
Adv HQ e: Advanced Headquarters
ADVICE e: Analytical Determination of the Values of Information to Combat Effectiveness
A/DV Injection e: Arterio/Deep Venous Injection
Adv. Inorg. Chem. Radiochem. e: Ad-vances in Inorganic Chemistry and Radiochemistry (journ.)
Adv Intel Cen e: Advanced Intelligence Center
AdvIntelCen e: Advanced Intelligence Center
AD-Virus e: Adenoid Degeneration Virus
ad virus e: adenovirus
ADVISE e: Area Dental Visual Identification Security Equipment

ADVISER e: Advanced Interactive Symbolic Evaluator of Reliability
- e: Airborne Dual-channel Variable Input Severe Environmental Recorder/ Reproducer
- e: Automatic Document Version Improving System for Easy Revising
ADVISOR e: Advanced Integrated Safety and Optimizing Computer
ad vitr. l: ad vitrum
advl d: Adverbial
- e: adverbial
Adv LG e: Advanced Landing Ground
ADVM e: Adaptive Delta Modulation Voice Modulator-Demodulator
- e: Adaptive Delta Voice Modulation
Adv Magn Reson e: Advances in Magnetic Resonance
ADVMOS e: Advanced Military Occupational Specialty
AdvMsgCen e: Advanced Message Center
advmt e: Advertisement
adv mtr e: advertising matter
ADVN e: Advance (or Advancement)
advncd cse e: advanced course
Adv. Nucl. Phys. e: Advances in Nuclear Physics (journ.)
Adv. Nucl. Sci. Technol. e: Advances in Nuclear Science and Techgnology (journ.)
Adv. Obstet. Gynecol. e: Advances in Obstetrics and Gynecology (journ.)
ADVON e: Advanced Echolon
- e: Advanced Operations Unit
Adv Ord Test Sta e: Advanced Ordnance Test Station
Adv. Org. Chem. e: Advances in Organic Chemistry (journ.)
Adv. Phys. e: Advances in Physics (journ.)
Adv. Plasma Phys. e: Advances in Plasma Physics (journ.)
Adv. Pmt. e: Advance Payment
adv poss e: adverse possession
Adv Quantum Chem e: Advances in Quantum Chemistry (journ.)
advr e: advisor
Adv. Radiat. Biol. e: Advances in Radiation Biology (journ.)
ADVRS e: Assistant Director of Veterinary and Remount Services
ADVRY e: Advisory
ADVS d: automatisches Datenverarbei-tungssystem
- e: Advise
- e: Assistant Director of Veterinary Services
advs. l: adversus
Adv. Sci. e: Advancement of Science (journ.)
Adv. Space Sci. Technol. e: Advances in Space Science and Technology (journ.)
ADVSR e: Advisor
ADVST e: Advance Stoppage
Adv. Struct. Res. Diffr. Metho e: Advances in Structure Research by Diffraction Methods (journ.)
ADVSY e: Advisory
Adv. Syst. Engr. e: Advanced Systems Engineer
ADVT e: Advanced Development Verification Test
- e: Advertisement
- e: Advertising
advt e: advertise[r]
ADVT-C e: Advanced Development Verification Test-Coordinator
ADVT-G e: Advanced Development Verification Test-Government
ADVTG e: Advantage
- e: Advertising
ADVTGE e: Advantage
Adv. Theor. Phys. e: Advances in Theoretical Physics (journ.)
ADVTM e: Advisory Team
ADVTNG e: Advanced Training
Adv. Tracer Methodol. e: Advances in Tracer Methodology (journ.)
ADVUL e: Air Defense Vulnerability [Simulation]
Adv. X-Ray Anal. e: Advances in X-Ray Analysis (journ.)
ADVY e: Advisory
ADW d: Allgemeine Deutsche Warennummerung
- d: Analog-Digital-Wandler
- d: Analog-Digital-Wandlung
- d: Arbeitsgemeinschaft Deutscher Werbungsmittler
- e: Aerial Distribution Wire
- e: Air Defence Warning
- e: Air Defence Weapon
- e: Air Defense Wing
- e: Air Denial Weapon
- e: Air Depot Wing
- e: Arithmetic Data Word
- e: Assault with Deadly Weapon
- e: Assistant Director of Works
- e: Automated Data Wiring
AdW d: Akademie der Wissenschaften
A. D. W. d: Allgemeiner Deutscher Waffenring
a. d. W. d: an der Weinstraße
- d: an der Weser
- d: an der Wupper
ADWA e: Atlantic Deeper Waterways Association
A-D-Wandler d: Analog-Digital- Wandler
A-D-Wandlung d: Analog-Digital-Wandlung
ADWAR e: Advanced Directional Warhead
ADWC e: Air Defense Weapons Center
ADWCP - e: Automated Digital Weather Communications Program[me]
ADWCR e: Air Defense Weapons Center Regulation
ADW(E&M) e: Assistant Director of Works (Equipment and Maintenance)
ADWE & M e: Assistant Director of Works; Electrical and Mechanical
ADWEPS e: Air Defense Weapons Cost Effectiveness Study
A/D Wg e: Air Depot Wing
ADWKP e: Air Defense Warning Key Point
ADWS e: Air Defense Warning System
- e: Air Defense Weapon System
- e: Assistant Director of Warlike Stores
- e: Automated Digital Weather Switch
- e: Automatic Digital Weather Switch
ADWSS e: Air Defense Weapon Simulation System
ADX e: Add Index
- e: Add Index Register
- e: Advanced Development Experimental
- e: Air Defence Exercise
- e: Asymmetric Data Exchange
- e: Automatic Data Exchange[r]
- e: Automatic Digital Exchange
adx e: automatic data exchange
ADXC e: Alaskan DX Certificate
ADX rats e: Adrenalectomized rats
ADX Register e: Add Index Register
ADY e: Additional Duty
Ady e: Admiralty
Adyg e: Adygey
ADYN e: Amplidyne
ADZ d: Adresse Zeile
- e: Air Defense Zone
adz e: advise
Adzh e: Adzhar
AE d: Abfindungserklärung
- d: Abfrageeinheit
- d: Ablenkeinheit
- d: Abrechnungseinheit
- d: Abschirmung; extern
- d: Abstimmeinheit
- d: Abtasteinschub
- d: Adresseneingang
- d: algebraische Eingabe
- d: Analogeingabe
- d: Analysatoreinrichtung
- d: Anfangsergebnis
- d: Angström-Einheit
- d: Angstromeinheit
- d: Annoncenexpedition
- d: Anrufeinheit
- d: Anschlußeinheit
- d: Antitoxineinheit
- d: Anwendungsentwicklung
- d: Anzeigeeinheit
- d: Arbeitsebene
- d: Arbeitseinheit[en]
- d: Asche-Elementaranalyse
- d: astronomische Einheit
- d: Auftragseingang
- d: Auftragselement
- d: Ausfuhrerklärung
- d: außergewöhnliche Ereignisse
- d: Auswerteeinheit[en]
- d: Automatikbefehl Ein
- d: Automatik Extern
- d: Automatisierungseinrichtung
- d: aviäre Enzephalomyelitis
- d: Axonendanschwellung
- e: Abort Electronics
- e: Above the Elbow
- e: Absolute Electrometer
- e: Absolute Error
- e: Absolute Expression
- e: Absorbed Energy
- e: Accidental Earth
- e: Accidental Error
- e: Accommodation Endorsement
- e: Account Executive
- e: Accrued Expenditure
- e: Accumulated Error
- e: Acoustic Emission
- e: Acousto-Electric
- e: Acrodermatitis Enteropathica
- e: Active Enhancement
- e: Activity End
- e: Added Entry
- e: Additional Expenses
- e: Address Error
- e: Addressing Error
- e: Administrative Entity
- e: Adult Education
- e: Aeon
- e: Aerial
- e: Aeroelectronic
- e: Aeromedical Evacuation
- e: Aeronautical Engineer[ing]
- e: Aerospace Education
- e: Aerospace Environment
- e: Affect Elaboration
- e: After End
- e: Age Equivalent
- e: Aggregate Expenditure
- e: Agricultural Engineer[ing]
- e: Airborne Electronics
- e: Airborne Equipment
- e: Air Ejector

- e: Air Electrical
- e: Air Engineer
- e: Air Engineering
- e: Air Explorer
- e: Algebraic Equation
- e: Algoma Eastern Railway
- e: Algorithmic Element
- e: All England
- e: Almost Everywhere
- e: Alphanumeric Expression
- e: Alternative Energy
- e: American English
- e: American Ephemeris
- e: Aminoethyl
- e: Ammunition Exminer
- e: ammunition ship
- e: Ancillary Equipment
- e: Angle of Elevation
- e: Anoxic Encephalopathy
- e: Antarctic Expedition
- e: Apollo Engineering
- e: Appearance Energy
- e: Application Engineer
- e: Application Engineering
- e: Application Entity
- e: Applications Explorer
- e: Applied Entomology
- e: Apportioned Effort
- e: Arithmetic Expression
- e: Armed Experimental
- e: Army Education
- e: Array Element
- e: Array Expression
- e: Artificial Erythrocyte
- e: Assault Echelon
- e: Assault Engineer
- e: Assimilation Efficiency
- e: Assistant Editor
- e: Assistant Engineer
- e: Associate Editor
- e: Associate Engraver
- e: Associate in Education
- e: Associate in Engineering
- e: Asymmetric Epoxidation
- e: Atlas Explorer
- e: Atmospheric Entry
- e: Atmospheric Explorer
- e: Atomic Emission
- e: Atomic Energy
- e: Attenuation Equalizer
- e: Audit Entry
- e: Auroral Electrojet
- e: Australian Encyclopaedia
- e: Australian Engineers
- e: autoimmune Encephalomyelitis
- e: Automatic Electric
- e: Automatic Equipment
- e: Automation Device
- e: Automotive Engineer
- e: Autumnal Equinox
- e: Auxiliary Engineering
- e: Auxiliary Equation
- e: Auxiliary Equipment
- e: Average Error
- e: Avian Encephalomyelitis
- e: Aviation Electricians Mate
- e: Aviation Engineer
- e: Aviation Engineering
- e: United Arab Emirates
- r: Aktiebolaget Atomenergi
- R: Atomnaya Energiya

A. E. d: allerhöchste Entscheidung
- d: allerhöchste Entschließung
- d: allerhöchster Erlaß

A.E. e: Aeronautical Engineer
- e: Agricultural Engineer
- e: Architectural Engineer
- e: Associate in Education
- e: Associate in Electronics
- e: Associate in Engineering

a.E. d: als Ersatz

a. E. d: am Ende

ae d: Aegirine
- e: aircraft equipment
- e: air escape
- e: arithmetic element
- l: aetatis

aE d: aktuelle Evaporation

A/E d: Ausgabe/Eingabe
- e: Absorptivity Emissivity
- e: Architect/Engineer
- e: Architectural Engineering

A & E d: Azimuth und Elevation
- e: Aerospace and Electronic
- e: Aerospace and Equipmnent
- e: Agricultural and Engineering
- e: Agriculture and Engineering
- e: Aircraft and Engineering
- e: Aircraft and Engines
- e: Analysis and Evaluation
- e: Appeal and Error
- e: Appropriation and Expense
- e: Architect and Engineer
- e: Architectural and Engineering
- e: Architecture and Engineering
- e: Armament[s] and Electronics
- e: Assembly and Equipment
- e: Assembly and Erection
- e: Azimuth and Elevation

a & e e: accident and emergency

A-E e: Architect-Engineer
- e: Architectural and Engineering
- e: Astro-Eugenics

AE1 e: Aviation Electricians Mate, First Class

AE2 e: Aviation Electricians Mate, Second Class

AE3 e: Aviation Electricians Mate, Third Class

AEA d: Automatisierungs- und Elektroenergie-Anlagen
- d: Vereinigung Volkseigener Betriebe Automatisierungs- und Elektroenergieanlagen
- e: Abort Electronics Assembly
- e: Active Element Array
- e: Activity End Acknowledgement
- e: Actors Equity Association
- e: Actual Expenses Allowable
- e: Adult Education Association
- e: Advanced Engine; Aerospike
- e: Advanced Engine Aerospace
- e: Aerial Elevation Angle
- e: Aerospace Education Association
- e: Aft End Assembly
- e: Aggregate Expense Analysis
- e: Agricultural Education Association
- e: Agricultural Engineers Association
- e: Aircraft Electronics Association
- e: Aircraft Engineers Association
- e: Air Efficieny Award
- e: Air Entraining Agent
- e: Alcohol, Ether, Acetone
- e: American Economic Association
- e: American Education Association
- e: American Electrical Association
- e: American Electrology Association
- e: American Electrolysis Association
- e: American Electronics Association
- e: American Engineering Association
- e: American Evaluation Association
- e: Antenna Elevation Angle
- e: Area Edcuation Agency
- e: Assignment Eligibility and Availability
- e: Associate in Engineering Administration
- e: Association for European Airlines
- e: Atlantic Education Association
- e: Atomic Energy Act
- e: Atomic Energy Authority
- e: Auger Electron Analysis
- e: Autoerotic Asphyxiation
- e: Automatic Error Analysis
- e: Automation Economic Analysis
- e: Automotive Electric Association

AEAA e: Ascent Engine Arming Assembly

AEAC e: Alternate Emergency Action Center
- e: Atomic Energy Advisory Committee
- e: N-(2-Aminoethyl)Aminodeoxy- Cellulose

AEACA e: Acetyl Epsilon Aminocapronic Acid

AEACII e: Atomic Energy Advisory Committee on Industrial Information

AEACP e: Airborne Emergency Alternate Command Post

AEADST e: Advanced Electronic and Digital Sensor Technology

AEAF e: Allied Expeditionary Air Force

AE-Amp e: Above Elbow Amputation

AEAO e: Airborne Emergency Actions Officer

AEAPS e: Auger Electron Appearance Potential Spectroscopy

AEAR e: American Ear Association for Research

AEAS e: Air Equipment and Support
- e: Automatic Equalization/Analyzation System

aea sol e: alcohol-ether-acetone solution

AEA Solution e: Alcohol, Ether, Acetone Solution

Ae Atk e: Aerial Attack

AEB d: Abwassereinleitungsbedingungen
- d: Asiatische Entwicklungsbank
- d: außerunterrichtliche Erziehung und Bildung (DDR)
- e: Acquired Epidermolysis Bullosa
- e: Advanced Engine Bell
- e: Aera Electricity Board
- e: Aerial Exploitation Battalion
- e: Aft Equipment Bay
- e: Airborne and Electronics Board
- e: American Egg Board
- e: Apollo Engineering Bulletin
- e: Area Electricity Board
- e: Associated Examining Board
- e: Atomic Energy Board
- e: Atomic Energy Bureau
- e: Automatic End of Block
- e: Auxiliary Equipment Building
- e: Average Extent of Burning

AE-B e: Atmosphere Explorer B

AEBA e: Agricultural Economics Bulletin for Africa
- e: Automatic Emergency Broadcast Alerting

AEBE e: Approach End Barrier Engagement

AEBG d: Allgemeines Eisenbahngesetz

AEBIG e: Association of Special Libraries and Information Bureaux Economic and Business Information Group

AEBR e: Airborne Electron Beam Recorder

AEBRD e: Electrically Activated Bank Release Device

AEBSTA e: Atomic Energy Bureau of Science and Technics Agency

AEC e: Academy of Electrical Contracting
- e: Action in Error Condition
- e: Adaptive Echo Cancellation
- e: Additional Extended Coverage

- e: Adult Education Centre
- e: Adult Education Council
- e: Advanced Extension Course
- e: Advance Exit Common
- e: Aerial Embarkation Center
- e: Aeronautical Research Council
- e: Aft End Cone
- e: Aft Events Controller
- e: Agricultural Economics [Division]
- e: Agricultural Engineering Centre
- e: Agricultural Executive Committee
- e: Agricultural Executive Council
- e: Aircraft Radio Corporation
- e: Air Eligibility Code
- e: Air Emplaced Classifier
- e: Airship Experimental Center
- e: Alaska Engineering Commission
- e: Alcohol Education Centre
- e: Altitude Engine Control
- e: Aluminum Extruders Council
- e: American Economic Council
- e: American Education Coalition
- e: American Electrical Cases
- e: American Engineering Council
- e: American Express Card
- e: Aminoethyl Cellulose
- e: Aminoethyl Cysteine
- e: Analog[ue] Electronic Computer
- e: Arab Electronic Company
- e: Architects Emergency Committee
- e: Architectural and Engineering Construction
- e: Arithmetic Element Controll[er]
- e: Army Education[al] Center
- e: Army Educational Corps
- e: Army Electronics Command
- e: Army Extension Courses
- e: Asia Electronic Conference
- e: Assembled Electronic Component
- e: Associate Enforcement Counsel
- e: Association of Education Committees
- e: Association of Electronic Cottagers
- e: Association of Episcopal Colleges
- e: At Earliest Convenience
- e: Atlas Educational Center
- e: Atomic Energy Commission
- e: Australian Education Centre
- e: Australian Energy Commission
- e: Automatic Editing Control
- e: Automatic Electrical Control[ler]
- e: Automatic Electrode Changer
- e: Automatic Electrode Changing
- e: Automatic Energy Control
- e: Automatic Envelope Compression
- e: Automatic Error Correction
- e: Automatic Exciter Control
- e: Automatic Exposure Control
- e: Average Electrode Current
- e: Aviation Engineer Command

A & EC e: Atlantic & East Carolina
aec e: attitude engine control
- e: Micro Curie

AECA e: Arms Export Control Act
AEC-A e: Atomic Energy Commission; Albuquerque Operations Office
AEC-AFSWP-TP e: Atomic Energy Commission-Armed Forces Special Weapons Project-Technical Publications
AEC-AI e: Atomic Energy Commission, Argonne, Illinois
AEC-ANM e: Atomic Energy Commission; Albuquerque, New Mexico
AEC-ASC e: Atomic Energy Commission; Aiken, South Carolina
AECB e: Arms Export Control Board
- e: Atomic Energy Control Board

AEC-BC e: Atomic Energy Commission; Berkeley, California
AECC e: Aeromedical Evacuation Control Center
- e: Aerospace Energy Conversion Committee
- e: Atomic Energy Commission Classified
- e: Automobile Emissions Control by Catalyst

AEC-CC e: Atomic Energy Commission; Canoga Park, California
AECCG e: African Elephant Conservation Coordinating Group
AECD e: Allergy Eczematous Contact Dermatitis
- e: Atomic Energy Commission Declassified
- e: Atomic Energy Commission Document
- e: Auxiliary Emission Control Device

AEC-DASA-TP e: Atomic Energy Commission-Defense Atomic Support Agency Technical Publication
AEC-DD e: Atomic Energy Commission-Department of Defense
AEC-DNA-TP e: Atomic Energy Commission-Defense Nuclear Agency Technical Publication
AEC-DoD e: Atomic Energy Commission-Department of Defense
AECE e: Airborne Engineer Contraction Equipment
AEC-FOA e: Atomic Energy Commission; Fernal Office Area
AECG e: Ambulatory Electrocardiogram
AECGIS e: Automated Electrocardiograph Interpretive System
AEC-HW e: Atomic Energy Commission; Hanford, Washington
AECI e: African Explosives and Chemical Industries
AECIA e: Australian Electronics Consumer Indusry Association
AEC-II e: Atomic Energy Commission; Idaho Falls, Idaho
AECL e: Aircraft and Equipment Configuration List
- e: Anglo-European Container Line
- e: Atomic Energy Centre-Lahore
- e: Atomic Energy of Canada, Limited

AEC-LN e: Atomic Energy Commission; Las Vegas, Nevada
AEC-LOC e: Atomic Energy Commission; Lockland Aircraft Reactors Operations
AECM e: Albert Einstein College of Medicine
- e: Association of European Candle Manufacturers
- e: Atomic Energy Commission Manual

AEC-NASA e: Atomic Energy Commission-National Aeronautics and Space Administration
AEC-NY e: Atomic Energy Commission; New York Operations Office
AECO e: Aeromedical Evacuation Control Officer
- e: American Electromedics Corporation

AECOM e: Army Electronic[s] Command
AEC-OR e: Atomic Energy Commission; Oak Ridge Operations Office
AEcoS e: Aquatic Ecosystem Simulator
AEC-OT e: Atomic Energy Commission; Oak Ridge, Tennessee
AECP e: Advance Engineering Change Proposal
- e: Airman Education and Commissioning Program[me]
- e: Altitude Engine Control Panel
- e: Army Extension Course Program[me]

AEC-PP e: Atomic Energy Commission; Pittsburgh, Pennsylvania
AEC-PR e: Atomic Energy Commission; Pittsburgh Naval Reactors Operations Office
AECPR e: Atomic Energy Commission Procurement Regulations
AEC-RW e: Atomic Energy Commission; Richland, Washington
AeCS d: Aeroclub der Schweiz
AECS e: Advanced Engine Control System
- e: Advanced Environmental Control System
- e: American Electro Chemical Society
- e: Apollo Environmental Control System
- e: Association of European Correspondence Schools
- e: Australia Europe Container Service
- e: Automated Environmental Control System

A.E.C.S. e: Automatic Evaporation Control System
aec sec e: Inverse Secant (Arc Secant)
AECT e: Association for Educational Communications and Technology
- e: Automatic Exposure Control Technique

AECTC e: Archives of Environmental Contamination and Toxicology
AEC/TIC e: Atomic Energy Commission / Technical Information Center
AECTRC e: Advanced Environmental Control Technology Research Center
AEC-UN e: Atomic Energy Commission; Upton; New York
AeCvD d: Aeroclub von Deutschland
AED d: Abfrageeinrichtung für Datenverkehr
- d: Arbeitsgemeinschaft für Entwicklungsplanung und Datenverarbeitung
- d: Atomkernenergie-Dokumentation
- e: Academy for Educational Development
- e: Active Electronic Decoy
- e: Advanced Electronics Design
- e: Advanced Engineering Data
- e: Aerodynamic Equivalent Diameter
- e: Aeromedical Education Division
- e: Aeronautical Engineering Department
- e: Aeronautical Engineering Duty
- e: Aerospace Electrical Division
- e: Aircraft Explosive Device
- e: Air Equipment Department
- e: Algorithmic Language Extended for Design
- e: Alphanumeric Entry Device
- e: Ammunition Engineering Directorate
- e: Analelectrodiabetic
- e: Analog[ue] Event Distributor
- e: Analysis and Evaluation Division
- e: Antiepileptic Drug
- e: Armament Engineering Directorate
- e: Associated Equipment Distributors
- e: Association for Educational Development
- e: Association of Electronic[s] Distributors
- e: Astro-Electronics Division
- e: Atomic Emission Detector
- e: Atomic Energy Division
- e: Australian Education Directory
- e: Automated [Systems] Engineering Design
- e: Average and Excess Demand
- e: Avionics Electrical Distribution
- l: Artium Elegantium Doctor

AEd e: Associate in Education

A.E.D. l: Artium Elegantium Doctor
AEDA e: Ammunition, Explosives and other Dangerous Articles
AEDB e: Apollo Engineering Development Board
- e: Apollo Engineering Documentation Board
AEDC e: Air Engineering Development Center
- e: American Educational Computer, Inc.
- e: Arnold Engineering Development Center
AEDCAP e: Automated Engineering Design Circuit Analysis Program[me]
AEDCM e: Advanced Electrochemical Depolarized Concentrator Module
AEDD e: Air Engineering Development Division
AEDE e: Airplane Economic Design Evaluator
AEDF e: Advanced Engineering Data Form
AEDNET e: Automated Engineering Design of Networks
AEDO e: Aerological Duty Officer
- e: Aircraft Engineering District Office
AEDP e: Advanced Electrical Development Package
- e: Association for Educational Data Processing
- e: Automated External Defibrillator-Pacemaker
- e: Automatic Engineering Design Program[me]
AEDPS e: Automated Engineering Documentation Preparation System
AEDR e: Avionics Equipment Design Review
AED-RCA e: Astro-Electronics Division-RCA
AED/R & S e: Associated Equipment Distributors Research and Services
AEDS e: Advanced Electric Distribution System
- e: Advanced Engineering Data Set
- e: Airport Engineering Data Sheet
- e: Analog[ue] Event Distribution System
- e: Association for Educational Data Systems
- e: Association for Electronic Data Systems
- e: Association of Educational Data-Systems
- e: Association of Electronic Data Systems
- e: Atmospheric Electric Detection System
- e: Atomic Energy Detection System
AEDST e: Australian Eastern Daylight Saving Time
AEDU e: Admiralty Experimental Diving Unit
AEE d: Analogeingabe; elektronische
- e: Absolute Essential Equipment
- e: Additional Expediting Expense
- e: Airborne Evaluation Equipment
- e: All Electronic Equipment
- e: Alliance for Environmental Education
- e: Anomalously Enriched Element
- e: Army Entrance Examination
- e: Assistant Executive Engineer
- e: Association for Experiential Education
- e: Association of Energy Engineers
- e: Atomic Energy Establishment[s]
- e: Average Excitation Energy
- s: Asociación Electrotécnica Española
AeE e: Aeronautical Engineer
AE.E. e: Associate in Engineering
Ae.E. e: Association for Experimental Education
- e: Atomic Energy Establishment
AE-E e: Atmosphere Explorer E
AEEA d: Aminoethylethanolamine
AEEB e: Atomic Energy Establishment Bombay
AEEC e: Airlines Electronic Engineering Committee
AEEE e: Army Equipment Engineering Establishment
AEEGS e: American Electroencephalographic Society
AEEL e: Aeronautical Electronic[s] and Electrical Laboratory
- e: Aeronautical Engineering and Electronic Laboratory
- e: Arctic Environmental Engineering Laboratory
AEELS e: Airborne ELINT Emitter Location System
AEEM e: Airborne Electronic Equipment Modification
AEEMS e: Automatic Electric Energy Management System
AEEN f: Agence Européenne pour l'Energie Nucléaire
AeEng e: Aeronautical Engineer
AEEP e: Association of Environmental Engineering Professors
- e: Association of European Engineering Periodicals
- e: Automotive Energy Efficiency Program[me]
AEER e: Adult Entered Employment Rate
AEERB e: Army Enlisted Education Review Board
AEERL e: Air and Energy Engineering Research Laboratory
AEET e: Atomic Energy Establishment; Trombay
AEETF e: Armys Electronic Environmental Test Facility
AEEW e: Atomic Energy Establishment; Winfrith Heath
AEF d: Ausschuß für Einheiten und Formelzeichen
- e: Advanced Electronics Field
- e: Advertising Educational Foundation
- e: Aeromedical Evacuation Flight
- e: Aerospace Education Foundation
- e: Airborne Equipment Failure
- e: Aircraft Engineering Foundation
- e: Air Experience Flight
- e: Albert Einstein Foundation
- e: Allied Expeditionary Force
- e: Allogeneic Effect Factor
- e: American Economic Foundation
- e: American Expeditionary Force
- e: Aviation Engineer[ing] Force
- e: Aviation Engineer Force
AEFA e: American Education Finance Association
- e: Army Experimental Flight Activity
- e: Aviation Engineering Flight Activity
AEFC e: Alkaline Electrolyte Fuel Cell
- e: Atlantic Estuarine Fisheries Center
AEFDV e: Acute Encephalography and Fatty Degeneration of the Viscera
AEFEO e: Automotive Emissions and Fuel Economy Office
AEFF e: Assurance Engineering Field Facility
A Effect e: Alienation Effect
AEFLIC e: Allied Expeditionary Force Long-lines Control
AEFLLC e: Allied Expeditionary Force Long Lines Control
AEFR e: Aurora, Elgin & Fox River
AEFS e: Antiexposure Flight Suit
- e: Arctic Environmental Field Station
AEG d: Allgemeine Elektrizitäts-Gesellschaft
- d: Allgemeines Eisenbahngesetz
- d: Arbeitnehmererfindungsgesetz
- e: Active Element Group[ing]
- e: Acute Erosion Gastritis
- e: Aeromedical Evacuation Group
- e: All Edges Gilt
- e: Analytic Ephemeris Generator
- e: Association of Engineering Geologists
- e: Association of Exploration Geochemists
- e: Association of Special Libraries and Information Bureaux Electronics Group
- e: Atlantic Environmental Group
- e: Atrialelectrogram
- e: Aviation Evaluation Group
- e: Air Encephalogram[s]
- l: Ad Eundem Gradum
Aeg e: Aegean
aeg. l: aeger
Aegean e: Aegean Sea
Aegeans e: Aegean Islands
AEGIMRDA e: Army Engineer Geodesy, Intelligence and Mapping Research and Development Agency
AEGIS e: Active Electronic Gimballess Inertial System
- e: Agricultural, Ecological and Geographical Information System
- e: Aid for the Elderly in Government Institutions
- e: American European Government Information Services
- e: An Existing Generalized Information System
- e: Assessment of Effectiveness of Geologic Isolation Systems
- e: Astro Electronic Gimballess Inertial System
Aegis e: U.S. Navy cruiser class
AEG Kernreakt. d: AEG Kernreaktoren (journ.)
AEGM e: Association of Electronic Guard Manufacturers
AEGp e: Aeromedical Evacuation Group
Aeg S e: Aegean Sea
AEG-Tfk d: Allgemeine Elektrizitäts-Gesellschaft-Telefunken
AEGTS e: Annulus Exhaust Gas Treatment System
AEH d: Anhydroenneaheptitol
- e: Aerial Effective Height
- e: Antenna Effective Height
- e: Archives of Environmental Health
- e: Atmospheric Electricity Hazard
AEHA e: Army Environmental Health Agency
- e: Army Environmental Hygiene Agency
- s: Anuario Español e Hispano- Americano
AEHL e: Aeromedical Environmental Health Laboratory
- e: Army Environmental Health Laboratory
AEHP e: Atmospheric Electricity Hazards Protection
AEI e: Acclimatization Experiences Institute
- e: Acrylic Eye Illustrator
- e: Aerial Exposure Index
- e: Aerospace Education[al] Instructor
- e: Albert Einstein Institution
- e: Allow Enable Intercept
- e: Alternate Energy Institute
- e: American Enterprise Institute

- e: Annual Efficiency Index
- e: Armament Enhancement Initiative
- e: Armor Enhancement Initiative
- e: Associated Electrical Industries
- e: Association of Electrical Industries
- e: Audio End Instrument
- e: Australian Education Index
- e: Automatic Electrical Inspection
- e: Automatic Electrical Instruction
- e: Automatic Error Interrogation
- e: Azimuth Error Indicator
- f: Association des Ecoles Internationales
- i: Associazione Elettrotecnica Italiana

aei e: average efficiency index

AEIAF e: Albert Einstein International Academy Foundation

AEIB e: Activation Engineering Information Bulletin
- e: Association for Education in International Business

AEIC e: Advanced Earned Income Credit
- e: Association of Edison Illuminating Companies

AEICP e: Association of Entertainment Industry Computer Professionals

AEIDC e: Arctic Environment Information and Data Center

AEIMS e: Administration Engineering Information Management System
- e: Administrative Engineering Information Management System

AEIP e: Allied Electrical [Industry] Publications

AEIPPR e: American Enterprise Institute for Public Policy Research

AEIROF e: Anodically Electrodeposited Iridium Oxide Film

AEIS e: Associate of the Educational Institute of Scotland
- e: Association of Electronic Industries in Singapore

AEIT e: Academy for Electronic Information Technology

AEJ e: Adult Education Journal
- e: Alumin[i]um Extension Jacket

AEJI e: Association of European Jute Industries

AEJMC e: Association for Education in Journalism and Mass Communciation

AEK d: alkalisierendes Eiweiß-Kohlenhydrat
- d: Anschalteeinheit für Kartentelefon
- d: Atomenergiekommission
- e: Aircraft Ejection Kit
- e: All-Electric Kitchen

AEKG d: Archiv für Elsässische Kirchengeschichte (journ.)

AEL d: algebraische Eingabelogik
- d: Arbeitsgemeinschaft für Elektrizitätsanwendung in der Landwirtschaft
- e: Acceptor Energy Level
- e: Actuarial Engine Life
- e: Acute Erythroleukemia
- e: Admiralty Engineering Laboratory
- e: Advanced Engineering Laboratory
- e: Aerobiology and Evaluation Laboratory
- e: Aeronautical Engine[ering] Laboratory
- e: Aerospace Electronics Laboratory
- e: Aircraft Engine Laboratory
- e: Aircraft Equipment List
- e: Airplane and Engine License
- e: Alberta Environment Library
- e: Allowable Expense Level
- e: Allowance Equipage List
- e: Alumin[i]um Electrical Lead
- e: American Electronic Laboratories
- e: Americanism Educational League
- e: Animal Educational League
- e: Appalachian Environmental Laboratory
- e: Armament and Electronics Laboratory
- e: Army Electronics Laboratories
- e: Association of Environmental Laboratories
- e: Association of Equipment Lessors
- e: Atomic Energy Level
- e: Audit Entry Language
- e: Audit Error List
- e: Authorized Equipment Listing
- e: Automation Engineering Laboratory
- e: Average Effectiveness Level
- e: Aviation Electric Limited

AELA e: Australian Equipment Lessors Association

AELC e: Accident-Experience Learning Curve
- e: Aerospace Engine Life Committee
- e: Architect-Engineers Liaison Commission

AELD e: Ascent Engine Latching Device

AELDC e: Atomic Energy Levels Data Center

AELECTECH e: Associate in Electrical Technology

AELECTRTECHN e: Associate in Electronics Technology

A El Ed e: Associate in Elementary Education

AELI e: American Export Lines, Inc.

AELJ e: Atomic Energy Law Journal

AELMRP e: Atomic Energy Labor Management Relations Panel

AELO e: Aeromedical Evacuation Liaison Officer

AELP e: Allied Electrical Publications
- e: Automatic Error Localization Program[me]

AELRO e: Army Electronics Logistics Research Office

AELS e: Airborne Electronics Laser System

AELW e: Airborne Electronics Warfare

AELW School e: Airborne Electronic Warfare School

AEM d: Ausfuhrerklärung und -anmeldung
- e: Absorption Edge Modulator
- e: Accelerated Evaluation Method
- e: Acoustic Emmission Monitor[ing]
- e: Advance Engineering Memorandum
- e: Aeronautical Mobile
- e: Aircraft and Engine Mechanic
- e: American Energy Month
- e: Analytical Electron Microscopy
- e: Application Explorer Mission
- e: Architect-Engineer-Manager
- e: Arsenal Exchange Model
- e: Association of Electronic Manufacturers
- e: Atomic Emission Monitoring
- e: Attack Evaluation Model
- e: Augmented Energy Management
- e: Automatic Environment Monitoring
- e: Automobile Engineering and Manufacturing
- e: Aviation Electricians Mate

A.E.M. e: Air Efficiency Medal

AEMA e: American Environmental Manufacturers Association
- e: Asphalt Emulsion Manufacturers Association
- e: Australian Electrical Manufacturers Association

AEMB e: Airborne Electromechanical Bombing
- e: Alliance for Engineering in Medicine and Biology

AEMC e: Acryloyloxyethyl N-Methyl-carbamate
- e: Albert Einstein Medical Center
- e: American Electro Metal Corporation

AEMCC e: Air and Expedited Motor Carriers Conference

AEMCO e: Aircraft Engineering Maintenance Company

AEME e: Australian Electrical and Mechanical Engineers

AEM-ED e: Association of Electronic Manufacturers; Eastern Division

AEMIE f: Association Européene de Médecine Interne d'Ensenmble

AEMIS e: Aerospace and Environmental Medicine Information System

AEMO e: Advance Engineering Material Order

AEMP e: Associated European Management Publishers
- e: Atmospheric Electromagnetic Pulse

AEMS e: Agro-Environmental Monitoring System
- e: Aircraft Engine Management System
- e: American Engineering Model Society
- e: Armament Electronic Maintenance Squadron
- e: Automated Edge Match System
- e: Automated Electrophoresis Microscope System

AEMSA e: Army Electronics Material Support Agency

AEMSM e: Association of European Metal Sink Manufacturers

A & EMSq e: Armament and Electronic Maintenance Squadron

AEMT e: Association of Electrical Machinery Trades
- e: Automated Electronic Maintenance Training
- e: Automatically Erectable Modular Torus

AEMTM e: Association of European Machine Tool Merchants

AEMU e: Absolute Electromagnetic Unit

AEN e: Adaption Error Note
- e: Address Enable
- e: Advance Evaluation Note
- e: Attenuation Equivalent Nettiness
- f: Affaiblissement Equivalent pour la Netteté

Aen d: Abfrageeinheiten

AEn d: Abstimmeinheiten

A En e: Associate in English

aen. l: aeneus

AENC e: Avian Embryo Nutrient Cartridge

A & E Network e: Arts & Entertainment Network

AENG e: Airways Engineer

A[.]Eng[.] e: Associate in Engineering

A Eng Co-ord e: Assistant for Engineering Co-ordination

A Eng Elect e: Associate in Engineering Electronics

aengl. d: altenglisch

A Engr e: Associate in Engineering

AENORS e: Anticipated Engine Not Operationally Ready Supply

A Ent O e: Area Entertainments Officer

AEO e: Acousto-Electric Oscillator
- e: Advance Engineering Order
- e: Airborne Electronics Operator
- e: Air Electronics Officer
- e: Air Engineer[ing] Office[r]
- e: All Engines Operating
- e: Ammunition Executive Office
- e: Appeal Examining Officer
- e: Area Education Officer
- e: Area Engineering Officer
- e: Army Energy Office
- e: Assistant Experimental Officer
- e: Association of Education Officers

AEOB e: Advanced Engine Overhaul Base
AEOC e: Aminoethylhomocysteine
- e: Aquatic Ecosystem Objectives Committee

AEODP e: Allied Explosive Ordnance Disposal Publication
AEOG e: Air Ejection Off Gas
AEOI e: Atomic Energy Organisation of Iran
Aeol. e: Aeolian (or Aeolic)
AEOO e: Aeromedical Evacuation Operations Office[r]
AEOR e: Arithmetic Element Output Register
AEOS e: After Engineering Operating Station
- e: Astronomical, Earth and Ocean Sciences

AEOSS e: Advanced Electro-Optical Sensor Simulation
AEOTR e: Advanced Electro-Optical Tracker / Ranger
AEP d: Absolute Electrode Potential
- d: akustisch evoziertes Potential
- d: Arbeitsgemeinschaft Extraterrestrische Physik
- e: Accrued Expenditure Paid
- e: Acoustic Evoked Potential
- e: Adaptive Escalator Predictor
- e: Addo Elephant Park
- e: Adult Education Program[me]
- e: Advanced Energy Project
- e: Aggregate Exercise Price
- e: Aircraft Equipment Procedures
- e: Air Evacuation Patients
- e: Airport Electronic Project
- e: Allied Engineering Publication
- e: Allied Equipment Publication
- e: Alternative Education Project
- e: American Electric Power
- e: Aminoethylpiperazine
- e: Annual Engineering Plan
- e: Annual Execution Plan
- e: Anterior Extreme Position
- e: Apollo Experiment Pallet
- e: Apollo Extension Program[me]
- e: Aqueous Extraction Process
- e: Army Equipment Policy
- e: Artificial Earth Pulse
- e: Artificial Endocrine Pancreas
- e: Association of Educational Psychologists
- e: Atomic Electric Project
- e: Atomic Energy Project
- e: Auditory-Evoked Potential
- e: AUTODIN Enhancement Program[me]
- e: Automated Environmental Prediction
- e: Automatic Electronic Production
- e: Automatic End Point
- e: Automatic Extracting Program[me]
- e: Average[d] Evoked Potential
- e: Avionics Evaluation Program[me]
- s: Asociación Electrotécnica Peruana

AE & P e: Ambassador Extraordinary and Plenipotentiary
aep e: average evoked potential
AEP Acid e: Aminoethylphosphonic Acid
AEPC e: Appalachian Electric Power Company
- e: Army Equipment Policy Committee

AEPD e: 2-Amino-2-Ethyl-1.3 Propanediol
AEPEM e: Association of Electronic Parts and Equipment Manufacturers
AEPF d: Arbeitsgemeinschaft für empirisch-pädagogische Forschung
AEPG e: Army Electronic[s] Proving Ground
AEPI e: Aerospace Engineering Process Institute
- e: American Educational Publishers Institute
- e: Atmospheric Emissions Photometric Imaging

AEPIC e: Architecture and Engineering Performance Information Center
AEPL e: An Extensible Programming Language
- e: Approved Equivalent Parts List

AEPP e: Association of Existential Psychology and Psychiatry
AEPR e: Acoustic Electron Paramagnetic Resonance
AEPRT e: All Equipment Production Reliability Test
AEPS e: Advanced Electronic Publishing System
- e: Advanced Extravehicular Protective System
- e: Aircraft Electrical Power System
- e: Aircrew Escape Propulsion System
- e: American Electric Power Service
- e: American Electroplaters Society
- e: Apollo Telescope Mount Electrical Power System
- e: Automated Environmental Prediction System

AEPs e: Allied Engineering Publications
aep[s] e: auditory-evoked potential[s]
AEP'S e: Average Evoked Potentials
AEPSC e: American Electric Power Service Corporation
- e: Atomic Energy Plant Safety Committee

AEPW e: Aircraft Emergency Procedures over Water
AEq e: Age Equivalent
AEQI e: Agricultural Environmental Quality Institute
AEqP e: Allied Equipment Publications
AER d: Abnorme Erlebnisreaktion
- d: Adressenerweiterungsregister
- d: Analogeingabe mit Relaisdurchschaltung
- d: analoger Einzelregler
- d: Analogwerterfassung
- e: Abbreviated Effectiveness Report
- e: Academic Evaluation Report
- e: Acoustic Evoked Response
- e: Address Extension Register
- e: Aerial
- e: Aerial Effective Resistance
- e: Aerodynamic[s]
- e: Aeronautic[s]
- e: Aeronautical
- e: Aeronautical Engineering Report
- e: Aeroplane
- e: After Engine Room
- e: Agranular Endoplasmic Reticulum
- e: Agricultural Economic Report
- e: Airborne Extended Range
- e: Air Equivalence Ratio
- e: Airman Effectiveness Report
- e: Alarm Equipment Rack
- e: Albumin Excretion Rate
- e: Alteration Equivalent to a Repair
- e: Alumin[i]um Efficient Radiator
- e: Alumin[i]um Express Representative
- e: American Economic Review
- e: Antenna Effective Resistance
- e: Apical Ectodermal Ridge
- e: Approach End Runway
- e: Army Emergency Relief
- e: Army Emergency Reserve
- e: Association for Education by Radio
- e: Asynchronous Exit Routine
- e: Atomic Energy Review
- e: Auditory Evoked Response
- e: Australian Economic Review
- e: Average Evoked Response
- e: Azimuth Elevation Range

aer. e: Aerodrome
aer e: aldosterone excretion rate
AERA e: Airborne Electronics Research Activity
- e: American Educational Research Association
- e: American Electric Railway Association
- e: American Electronic Research Association
- e: Associate Engraver of the Royal Academy
- e: Automotive Engine Rebuilders Association

AERATIO e: Absorptivity Emissivity Ratio
A/E Ratio e: Asorptivity / Emissivity Ratio
AERB e: Army Educational Requirements Board
- e: Army Education Review Board

AERC e: Aircraft Engine Record Card
- e: Association of Ecosystem Research Centers
- e: Atlantic Educational Research Council

aercab e: advanced aircrew escape/ rescue capability
AERCAB e: Advanced Escape / Rescue Capability
AERCOM e: Aeronautical Communications
AERCW e: Auxiliary Essential Raw Cooling Water
AERD e: Agricultural Engineering Research and Development
- e: Atomic Energy Research Department

AERDA e: Army Electronics Research and Development Activity
AERDC e: Agricultural Extension and Rural Development Centre
AERDDP e: Australian Energy Research, Development and Demonstration Projects
AERDL e: Army Electronics Research and Development Laboratory
- e: Army Engineer Research and Development Laboratories

Aer. E. e: Aeronautical Engineer
AERE e: Alden Electronic and Recording Equipment
- e: Association of Environmental and Resource Economists
- e: Atomic Energy Research Establishment

aerea e: aeration
AEREA e: American Electric Railway Engineering Association
AERE Rep. e: Atomic Energy Research Establishment Report
AERF e: Atlas Economic Research Foundation
AERG e: Advanced Environmental Research Group
- e: Alternative Education Resource Group
- e: Army Engineer Reactors Group

AERI e: Agricultural Economics Research Institute
- e: Agricultural Economics Research Institution
- e: Automotive Exhaust Research Institute

AERIC e: Applied Economic Research and Information Centre

AERIS e: Airborne Electronic Ranging Instrumentation System
- e: Airways Environmental Radar Information System
- e: Automatic Electronic Range Instrumentation System
- e: Automatic Electronic Ranging Information System

aerl e: aerial

AERL e: Aero-Elastic Research Laboratory
- e: Arctic Environmental Research Laboratory
- e: Avco-Everett Research Laboratory

aerlgl e: Aerological

AERM e: Aerographers Mate
- e: Alignment Error Rate Monitor

AERM 1c e: Aerographers Mate First Class

AERM 2c e: Aerographers Mate Second Class

AERM 3c e: Aerographers Mate Third Class

Aer Méx s: Aero México

AERNL e: Aeronautical

AERNO e: Aeronautical Equipment Reference Number

Aero d: Aeronautik
- e: Aeronautic[s]

aero. d: aeronautisch

AERO e: Aerodynamic
- e: Aerographer
- e: Aerospace
- e: Alternative Energy Resources Organization
- e: Association of Electronic Reserve Officers
- e: Automatic[s] Earnings Recomputation Operation
- e: Azimuth, Elevation and Range Overtake

aero e: aeronautical

Aerobee e: Aerojet/Bumblebee

AEROCOM e: Aeronautical Communications

Aero Commander e: U-4 transport aircraft

AERODF e: Aerospace Defense Flight

AERODS e: Aerospace Defense Squadron

AERODW e: Aerospace Defense Wing

aerodyn e: Aerodynamic[s]

Aero E e: Aeronautical Engineer

Aero.E. e: Aeronautical Engineering

Aero Eng e: Aerospace Engineering

Aero Eng R e: Aeronautical Engineering Review

Aer Of e: Aerological Officer

Aero F & F e: Aerospace Facts and Figures

AEROFLOT R: Aero Flotilla

AEROG e: Aerologist

AEROHEAT e: Aerodynamic Heating

AEROIS e: Aerospace Intelligence Squadron

Aero J e: Aeronautical Journal

AEROL e: Aerological

Aerol e: Aerologist

aeromed e: aeromedical

aeromod e: aerodynamic modelling

aeromus e: aeronautical museum

Aeron. d: Aeronautik

aeron. d: aeronautisch

Aeron e: Aeronaut
- e: Aeronautic[s]
- e: Aeronautical

aéron f: aéronautique

Aeronaut Eng Rev e: Aeronautical Engineering Review

AERONAVES s: Aeronaves de México

AERONL e: Aeronautical

Aeron Res Rep e: Aeronautical Research Report

Aero O/Y F: Finnair

Aerop s: Aeropuerto

aeropost e: aerodynamic postprocessing

AERO R Bn e: Aeronautical Radiobeacon

AERO R Rge e: Aeronautical Radio Range

AEROS e: Advanced Earth Resources Observation System
- e: Aerometric and Emissions Reporting System
- e: Aeronomy Satellite
- e: Artificial Earth Research and Orbiting Satellite

AEROSAT e: Aeronautical Communications Satellite System
- e: Aeronautical Satellite

AEROSG e: Aerospace Support Group

AEROS-NATE e: Aeronomy Satellite - Neutral Atmosphere Temperature Experiment

Aerosol Sci e: Aerosol Science

Aerosp e: Aerospace

AEROSPACE e: Aeronautics and Space

AEROSPACECOM e: Aerospace Communication[s]

Aerosp Eng e: Aerospace Engineering

Aerosp Med e: Aerospace Medicine

Aerosp Technol e: Aerospace Technology

AEROSS e: Aerospace Support Squadron

AEROSSq e: Aerospace Support Squadron

AEROTAL s: Aerolineas Territoriales de Colombia

aerotel e: airplane hotel

Aerovias "Q" s: Aerovias Cubana

AERP e: Advanced Equipment Repair Program[me]

aerp e: Aeroplane

A & ERRC e: American and English Railroad Cases

AERS e: Airborne Environmental Reporting System
- e: Airborne Equipment Repair Squadron
- e: Airborne Expendable Rocket System
- e: Aircraft Equipment Requirement Schedule
- e: Atlantic Estuarine Research Society

AERT e: Acceptable Environmental Range Test
- e: Advanced Environmental Research and Technology
- e: Association for Education by Radio-Television

AERTJ e: Association of Education by Radio-Television Journal

AERU e: Agricultural Economics Research Unit

AES d: Auger-Elektronenspektrometrie (oder -spektrometer)
- d: Auger-Elektronenspektroskopie
- e: Abrasive Engineering Society
- e: Acrylonitrile Ethylene Styrene
- e: Active Electromagnetic System
- e: Advanced Engineering Services
- e: Advanced Extravehicular Suit
- e: Aeromedical Evacuation Squadron
- e: Aerospace and Electronic Systems
- e: Aerospace Electrical Society
- e: Aerospace Electronics Systems
- e: Agricultural Economics Society
- e: Agricultural Estimates
- e: Agricultural Experiment Station
- e: Agricultural Extension Service
- e: Air and Earth Shock
- e: Air and Energy Staff
- e: Aircraft Ejection Seat
- e: Aircraft Electrical Society
- e: Airways Engineering Society
- e: Alkylethoxylated Sulfate
- e: All-Electronics Show
- e: Alternative Economic Strategy
- e: American Electrochemical Society
- e: American Electroencephalographic Society
- e: American Electromechanical Society
- e: American Electronical Society
- e: American Electroplaters Society
- e: American Eletrotechnical Society
- e: American Entomological Society
- e: American Epidemiological Society
- e: American Epilepsy Society
- e: American Equilibration Society
- e: American Ethnological Society
- e: American Eugenics Society
- e: Analog[ue] Event System
- e: Analysis and Evaluation Staff
- e: Aortic Ejection Sound
- e: Apollo Earth-Orbiting Station
- e: Apollo Experiment Support
- e: Apollo Extension System
- e: Applications Environment System
- e: Architectural Engineering System
- e: Area Electronic Supervisor
- e: Army Education Scheme
- e: Army Engineer School
- e: Army Excess Property
- e: Army Exchange Service
- e: Array Element Study
- e: Artificial Earth Satellite
- e: Artillery Equipment School
- e: Astronomical Explorer Satellite
- e: Atlantic Estuarine Society
- e: Atmospheric Environment Service
- e: Atomic Emission Spectroscopy
- e: Audio Engineering Society
- e: Auger Electron Spectrometry
- e: Auger Electron Spectroscope (or Spectroscopy)
- e: Australian Economic Statistics
- e: Australian Educational Secretariat
- e: Australian Entomological Society
- e: Automatic Emission Spectroscopy
- e: Automatic External Standard
- e: Automatic External Standardization
- e: Auxiliary Encoder System
- e: Auxiliary Encoding System
- e: Avionics Expert System

AE's e: Arithmetic Elements

AE & S e: Air Equipment and Support
- e: Authorized Equipment and Stores

A & ES e: Arson and Explosion Squad

AESA e: American Educational Studies Association
- e: Association of Environmental Scientists and Administrators

AESAP e: Army Entertainment Scholarships and Awards Program[me]

AESBNW e: Association of Engineers and Scientists of the Bureau of Naval Weapons

AESBOW e: Association of Engineers and Scientists of the Bureau of Weapons

AESC e: American Engineering Standards Committee
- e: Automatic Electronic Switching Center
Aescul g: Aesculapius
AESD e: Acoustic Environment[al] Support Detachment
AESE e: Association of Earth Science Editors
- e: Association of Electrical Station Engineers
AESF e: Array Element Successor Function
AESFS e: American Electroplaters and Surface Finishers Society
AESHS e: Alfred E. Smith High School
AES(I) e: Association of Engineers and Scientists (Independent)
AESIR e: Aerospace Instrumentation Range Station
AESIS e: Australian Earth Sciences Information System
AESJ e: Atomic Energy Society of Japan
AESKULAP d: Auftragserfassungssystem mit Kommissionierung und Lagerbestandsführung
AESLA e: Associate of the Experimental Station Laboratory Association
AESM e: Albert Einstein School of Medicine
AESMC e: Automotive Exhaust Systems Manufacturers Council
AESMD e: Aircraft Escape System Maintenance Data
AESO e: Airborne Electronic Sensor Operator
- e: Aircraft Environmental Support Office
AESOP e: Accounts Enquiry Sales and Order Processing
- e: An Evolutionary System for On-line Processing
- e: An Experimental Structure for On-Line Planning
- e: Artificial Earth Satellite Observation Program[me]
- e: Automated Educational Services Online Processing
- e: Automated Engineering and Scientific Optimization Program[me]
AESOPS e: Army Materiel Systems Analysis Agency Evade Sustained Operations Performing
AESP e: Auxiliary Engineering Signal Processor
AESq e: Aeromedical Evacuation Squadron
AESQ e: Air Explorer Squadron
AESR e: Aeronautical Equipment Service Record
- e: Army Equipment Status Report
AESRS e: Army Equipment Status Reporting System
AESS e: Aerospace and Electronics Systems Society
- e: Aircraft Ejection Seat System
- e: American Ethnic Science Society
- e: Automatic Electronic Switching System
AESSO e: Assistant Embarkation; Supply and Stores Officer
AESt d: Allgemeiner Europäischer Stückguttarif
AEST e: Aeromedical Evacuation Support Team
AESTE e: Association for the Exchange of Students for Technical Experience
aesth e: aesthete
- e: aesthetic[ian] [s]
aesu e: Absolute Electrostatic Unit
AESU e: Aerospace Environmental Support Unit
AET d: Aminoethylisothiuronium
- d: Arbeitsgemeinschaft Entstörungstechnik
- d: Arbeitskreis Energietechnik
- d: Archiv für Eisenbahntechnik (journ.)
- d: Ausschuß elektrotechnische Terminologie
- d: S-(2-Aminoäthyl)-isothiouronium-brombid-Hydrobrombid
- e: Absorption Equivalent Thickness
- e: Acoustic Emission Testing
- e: Actual Elapsed Time
- e: Actual Equipment Trainer
- e: Actual Evapotranspiration
- e: Actual Exposure Time
- e: Advanced Energy Technology
- e: Aeromedical Evacuation Technician
- e: Aircraft Equipment Trainer
- e: All Events Trace
- e: Apparent Elastic Thickness
- e: Approximate Exposure Time
- e: Army Extension Training
- e: Associate in Electrical Technology
- e: Associate in Engineering Technology
- e: Atrial Ectopic Tachycardia
- e: Australian Eastern Time
- e: Auto Exhaust Testing
- e: Automatic Exchange Tester
aET d: aktuelle Evapotranspiration
aet e: absorption-equivalent thickness
A.E.T. e: Associate in Electrical Technology
- e: Associate in Electronic Technology
aet. l: aetatis
AETA e: American Embryo Transfer Association
AETAC e: Aviation Electronic Technicians Mate; Combat Aircrewman
AETC e: Accessory and Equipment Technical Committee
AETCA e: Aviation Electronic Technicians Mate; Combat Aircrewman
AETD e: Aero-Electronic Technology Department
AETDA e: Aminoethyltricosadiynamide
AETE e: Aerospace Engineering Test Establishment
AETF e: Azimuth Error Test Feature
- e: Azimuth Error Test Fixture
AETFAT f: Association pour l'Etude Taxonomique de la Flore d'Afrique Tropicale
A. et GR d: Ascherson und Graebner
AETI e: Apollo Engineering and Technology Index
AETIS e: Army Extension Training Information System
A. et J. KERN d: Anton und Josef Kerner
AETL e: Approved Engineering Test Laboratory
- e: Armament and Electronics Test Laboratory
- e: Army Engineer Topographic Laboratories
AETM e: Aviation Electronic Technicians Mate
AETMS e: Airborne Electronic Terrain Map System
AETN e: American Educational Television Network
AEtP e: Allied Electronics Publication[s]
AEtPs e: Allied Electronic Publications
AETR e: Advanced Electronically Tuned Radio
- e: Advanced Engineering Test Reactor
- e: Advanced Epithermal Thorium Reactor
AETS d: 2-Acetamidoethane Thiolsulfonate
- e: Army Extension Training System
- e: Association for the Education of Teachers in Science
AETT e: Acetyl Ethyl Tetramethyl Tetralin
- e: Association for Educational and Training Technology
AETU e: Atomic Energy Technical Unit
AEU d: Arbeitskreis Evangelischer Unternehmer
- e: Absolute Electrical Unit
- e: Accrued Expenditure Unpaid
- e: Altitude Encoder Unit
- e: Amalgamated Engineering Union
- e: American Ethical Union
- e: Annual Estimated Usage
- e: Army Exhibit Unit
- e: Asia Electronics Union
- e: Asian Electronics Union
AEÜ d: Archiv für Elektrische Übertragung (journ.)
- d: Archiv für Elektronik und Übertragungstechnik (journ.)
AEUS e: Absolute Electrical Unit Scale
AEV d: Arbeitsgemeinschaft der Elektroverbände eV
- d: Arbeitsgemeinschaft Erdölgewinnung und -Verarbeitung
- d: Aufnahmeentzerrverstärker
- d: Ausbilder-Eignungsverordnung
- d: autonome Energieversorgung
- e: Aerothermodynamic Elastic Vehicle
- e: Armored Engineer Vehicle
- e: Avian Encephalomyelitis Virus
- e: Avian Erythroblastosis Virus
AEVA e: Australian Equine Veterinary Association
aevac e: air evacuation
AEVMOSFET e: Angle-Evaporated Vertical-channel Metal-Oxide Semiconductor Field Effect Transistor
AEVO d: Arbeitserlaubnis-Verordnung
- d: Ausbilder-Eignungsverordnung
AEVöD d: Ausbilder-Eignungsverordnung öffentlicher Dienst
AEVS e: Automatic Electronic Voice Switch
AEW d: Aargauisches Elektrizitätswerk
- d: Alb-Elektrizitätswerk
- d: Amt für Energiewirtschaft
- d: Arbeitsgemeinschaft der Elektrizitätswerke
- d: Arbeitsgemeinschaft Ernährungswirtschaftlicher Werbestellen
- e: Admiralty Experiment[al] Works
- e: Air[craft] Early Warning
- e: Airborne Early Warning
- e: Airborne Electronic Warfare
- e: Air Electronic Warfare
- e: Air Evacuation of Wounded
- e: American Education Week
- e: American Energy Week
- e: Association of Electrical Wiremen
AEWA e: Airborne Early Warning Aircraft
AEW acft e: Air Evacuation Wounded Aircraft
AEWB e: Army Electronic Warfare Board
AEW & C e: Airborne Early Warning and Control
AEWCAP e: Airborne Early Warning Combat Air Patrol
AEWCON e: Airborne Early Warning and Control
AEW & Con Wg e: Airborne Early Warning and Control Wing
AEW & CSq e: Airborne Early Warning and Control Squadron

AEWES e: Army Engineers Waterways Experiment Station
AEWF e: Airborne Early Warning Fighter
AEWIB e: Army Electronic Warfare and Intelligence Board
AEWICS e: Airborne Early Warning and Interceptor Control System
AEWIS e: Army Electronic Warfare Information System
AEWL e: Association of Employers of Waterside Labour
AEWP e: Aerospace Education Workshop Project
AEWPC e: Army Electronic Warfare Policy Committee
AEWR e: Adverse Effect Wage Rate
AEWRADAR e: Airborne Early Warning Radar
AEWRON e: Airborne Early Warning Squadron
AEWS e: Advanced Earth Satellite Weapon System
- e: Advanced Electronic Warfare System
- e: Aircraft Early Warning System
AEWSPS e: Aircraft Electronic[s] Warfare Self-Protection System
AEWTF e: Aircrew Electronic Warfare Tactics Facility
AEWTS e: Advanced Electronic Warfare Test Set
AEWTU e: Airborne Early Warning Training Unit
AEWW e: Airborne Early Warning Wing
aex e: Automatic Electronic Exchange
AExO e: Assistant Experimental Officer
AEXP e: Applications Experience
AEZ d: Armbindeerkennungszeichen
AF d: Ablenk[ungs]faktor
- d: Absorberfläche
- d: Ackerfläche
- d: alle Fahrt
- d: alternative Frequenz[en]
- d: Anfangsfolge[n]
- d: angewandte Forschung
- d: Anglistische Forschungen (journ.)
- d: Anilin-Formaldehydplast
- d: Anlagenfaktor
- d: Anordnungsfreiheit
- d: Anschlußfeld
- d: Anwendungsfeld
- d: Arbeitsfähigkeit
- d: Atemfrequenz
- d: audiofrequent
- d: Audiofrequenz
- d: Auffänger
- d: Ausführen
- d: Ausgangsfilter
- d: Ausgleichsforderung
- d: Auslandsfernamt
- e: Abnormal Frequency
- e: Absolute Function
- e: Absorbing Field
- e: Absorbing Filter
- e: Absorption Fading
- e: Absorption Filter
- e: Accounting Form
- e: Accumulation Factor
- e: Accuracy Factor
- e: Accuracy Figure
- e: Acid-Fast
- e: Acre-Foot
- e: Across Flats
- e: Activating Factor
- e: Activity File
- e: Addition Formula
- e: Address[ed] Field
- e: Address File
- e: Address Format
- e: Address Frame
- e: Addressing Format
- e: Administrative Function
- e: Admiral of the Fleet
- e: Advance[d] Freight[s]
- e: Advanced Function
- e: Advance Freight
- e: Aeronautically Fixed
- e: Affirmative Flag
- e: Affix
- e: Afghanistan
- e: Africa[n]
- e: Aft Fuselage
- e: Aggregation Factor
- e: Agricultural Farm
- e: Airfile
- e: Air Filter
- e: Air Fitter
- e: Air Flight
- e: Air Forager
- e: Air Force
- e: Air Foundation
- e: Air Frame
- e: Air France
- e: Air Freight
- e: Albumin-Free
- e: Aldehyde-Fuchsin
- e: Algebraic Function
- e: Alphabetic Field
- e: Alphabetic File
- e: Alternating Field
- e: Alternating Flow
- e: Alternative Fertility
- e: Alternative Frequencies
- e: Alternative Frequency
- e: Aluminium Federation
- e: American Fabrics (journ.)
- e: Aminofluorene
- e: Amniotic Fluid
- e: Amphiphilic Flavin
- e: Amplification Factor
- e: Analog[ue] Feedback
- e: Anchoring Fibril
- e: Angiogenesis Factor
- e: Angle Frame
- e: Anterior Foot
- e: Antiferromagnetic
- e: Aortic Flow
- e: Apparent Fault
- e: Apply Force
- e: Arbitrary Function
- e: Architectural Forum (journ.)
- e: Arctic Front
- e: Arithmetical Function
- e: Arithmetic Flag
- e: Armed Forces
- e: Army Force
- e: Army Form
- e: Array Factor
- e: Arthritis Foundation
- e: Artificially Fed
- e: Asiatic Fleet
- e: Assembly and Fabrication
- e: Assembly Fixture
- e: Associate Fellow
- e: Associational Fluency
- e: Asymmetry Factor
- e: Atmospheric Flight
- e: Atomic Fluorescence
- e: Atomic Forum
- e: Atrial Fibrillation
- e: Atrial Flutter
- e: Attenuation Factor
- e: Audio Fidelity
- e: Auricular Fibrillation
- e: Austrian Forum
- e: Auto-Fiche
- e: Autocorrelation Function
- e: Autofocus
- e: Automatic Filter
- e: Automatic Focusing
- e: Automatic Follower
- e: Automatic Following
- e: Automation Foundation
- e: Auxiliary Feed
- e: Auxiliary Feedwater
- e: Auxiliary Field
- e: Availability Factor
- e: Aviation Facilities
- e: Aviation Forum
- e: Award Fee
- e: Axial Flow
- e: Axle Fluid
af a: afgang
- d: aschefrei
- e: air filter
- e: ale firkin
- e: anti-foam
- e: audiofrequency
- i: anno futuro
Af d: Abfrage
- d: Arbeitseffektivität
- e: African[s]
- e: Afrikaans
- f: Académie française
a. F. d: alte Fassung
- d: alte Folge
- d: alte Form
Af. d: Artilleriefahrzeug
aF e: Abfarad
- e: Attofarad
Af. e: Africa[n]
a. f. e: ash-free
- l: adfinem
- l: anni futuri
A & F e: Accounting and Finance
- e: Agriculture and Forestry
- e: Aircraft and Facilities
- e: Arming and Fusing
A/F e: Air/Fuel
- e: Air Field
A-F e: Anglo-French
a-f e: audio-frequency
ÄF d: Ämter des Fernmeldewesens
AF35m e: Auto-Focus 35mm-camera
AFA d: Accumulatoren-Fabrik AG
- d: Akkumulatorenfabrik
- d: Arbeitsgemeinschaft für Abfallbeseitigung
- d: Arbeitskreis für Arbeitsstudien
- d: Auslandsfernamt
- d: automatische Feinabstimmung
- d: automatische Frequenzabstimmung
- e: Abstract Family of Acceptors
- e: Access File Attribute
- e: Acute Focal Appendicitis
- e: Aerial Field Artillery
- e: AIDS Follow-up Assessment Questionnaire
- e: Air Force Academy
- e: Air Force Act
- e: Air Force Advisory
- e: Air Force Association
- e: Air Force Auxiliary
- e: Air Frame Assembly
- e: Alien Firearms Act
- e: Allergy Foundation of America
- e: American Federation of Astrologers
- e: American Federation of Aviculture
- e: American Firearm Association
- e: American Forensic Association

- e: American Forestry Association
- e: American Foundrymens Association
- e: Amfonelic Acid
- e: Analog[ue] Filter Assembly
- e: Application Fit Analysis
- e: Application for Federal Assistance
- e: Armed Forces Act
- e: Armo[u]red Field Artillery
- e: Army Flight Activity
- e: Aspirin Foundation of America
- e: Assembly Fixture Accessory
- e: Assistant Freight Agent
- e: Associate in Fine Arts
- e: Associate of the Faculty of Actuaries
- e: Associate of the Faculty of Auditors
- e: Association of Federal Architects
- e: Association of Flight Attendants
- e: Audio-Frequency Amplifier
- e: Audio-Frequency Apparatus
- e: Australian Field Artillery
- e: Automated Frequency Assignment
- e: Automatic False Alarm
- e: Automatic Field Assistant
- e: Automatic Fund Allocation
- e: Azimuth Follow-up Amplifier

AfA d: Abschreibung für Abnutzung
- d: Absetzung für Abnutzung
- d: Amt für Agrarordnung
- d: Amt für Agrarstruktur
- d: Arbeitsgemeinschaft freier Angestelltenverbände

afa e: dry-type forced-air cooled [transformer]

AFAA e: Automatic Fire Alarm Association

AFAADS e: Advanced Forward Area Air Defense [System]

AFAADW e: Advanced Forward Area Air Defense Weapon

AFAAEC e: Air Force Academy and Aircrew Examining Center

AFAAR e: American Fund for Alternatives to Animal Research
- e: Automated Forward Area Alerting Radar

AFA Art e: Associate in Fine Arts in Art

AFAB e: Air Force Academy Board
- e: Air Force Audit Branch

AFABS e: Air Force Arctic Broadcasting Squadron

Afa-Bund d: Allgemeiner freier Angestelltenbund

AFAC e: Air-Forced-Air Cooling
- e: Airborne Forward Air Controller
- e: Air Force Acquisition Circular
- e: Air Force Armament Center
- e: American Fisheries Advisory Committee
- e: Automatic Field Analog[ue] Computer

AFACAL e: Associate Fellow of American College of Allergists

AFACE e: Army Field Artillery Combat Effectiveness

AFACG e: Army Air Forces Commanding General

AFAC & IC e: Air Force Aeronautical Chart and Information Center

AFA Cooling e: Air-Forced-Air Cooling

AFAD e: Air Force Acquisition Document
- e: Air Force Ammunition Depot
- e: Air Force Authorization Document
- e: Armed Forces Acquisition Document

AFADGRU e: Air Force Advisory Group

AFADO e: Association of Food and Drug Officials

AFADS e: Advanced Forward Air Defense System

AFADTC e: Air Force Armament Development and Test Center

AFADVMC e: Air Force Advanced Management Class

AFAE e: Air Force Acquisition Executive
- e: American Foundation on Automation and Employment

AFAF e: Air Force Assistance Fund
- e: Air Force Auxiliary Field
- e: Air Force Director of Accounting and Financing
- e: Atlantic Fleet Amphibious Force

AfÄF d: Akademie für Ärztliche Fortbildung

AFAFC e: Air Force Accounting and Finance Center

AFAFFO e: Air Force Aerospace Fuels Field Office

AFAFPSO e: Air Force Aerospace Fuel Petroleum Supply Office

AFAFS e: Air Force Advanced Flying School

AFAG e: Air Force Advisory Group

AFAI e: Air Force Agent Installation

AFAIAA e: Associate Fellow of the American Institute of Aeronautics and Astronautics

AFAIDSR e: American Foundation for AIDS Research

AFAIM e: Associate Fellow of the Australian Institute of Management

AFAITC e: Armed Forces Air Intelligence Training Center

AFAL e: Air Force Avionics Laboratory

AFALC e: Air Force Acquisition Logistics Center

AFALD e: Air Force Acquisition Logistics Division

AFALLS e: Air Force Alaskan Long Line System

AFAM e: Air Field Attack Munition
- e: Air Force Armament Museum
- e: Automated Frequency Assignment Model

afam e: airfeld attack ammunition
- e: automatic frequency-assignment model

AFAMA e: Air Force Air Materiel Area

AFAMIC e: Air Force Aerospace Materials Information Center

AFAMM e: Aerial Field Artillery Multi-Mode

AFAMMOD e: Air Force Ammunition Ordnance Depot

AFAMRL e: Air Force Aerospace Medical Research Laboratory

Af Am[s] e: African American[s]
- e: Afro-American[s]

AFA Mus e: Associate in Fine Arts in Music

AFAO e: Air Force Administration and Organization
- e: Air Force Administrative Order
- e: Approved Force Acquisition Objective

AFAP e: Airfield Ammunition Park
- e: Australian Federation of Airline Pilots
- e: Aviation Fuel and Ammunition Park

afap e: artillery-fired atomic projectile

AFAPL e: Air Force Aero-Propulsion Laboratory

AFAPO e: Air Force Accountable Property Officer

AFAR e: Advanced Field Array Radar
- e: Airborne Fixed Array Radar
- e: American Foundation for AIDS Research
- e: Automatic False Alarm Rate
- e: Azores Fixed Acoustic Range

AFARS e: Army Federal Acquisition Regulations Supplement

AFARV e: Armored Forward Area Rearm Vehicle
- e: Armored Forward Area Resupply Vehicle

AFAS e: Advanced Field Artillery System
- e: Air Flow Actuated Switch
- e: Area Fire Armor System
- e: Associate of the Faculty of Architects and Surveyors
- e: Associate of the Faculty of Astrological Studies
- e: Automated Frequency-Assignment System
- e: Auxiliary Feedwater Actuating System
- f: Association Française pour l'Avancement des Sciences

AFAS-C e: Advanced Field Artillery System - Cannon

AFASC e: Air Force Aeronautical Systems Command

AFASD e: Air Force Aeronautical Systems Division

AFASE e: Association for Applied Solar Energy

AFASED e: African Association of Education for Development

AFA-SEF e: Air Force Association - Space Education Foundation

AFASPO e: Air Force Automated Systems Project Office

AF/ASPR e: Air Force / Armed Service Procurement Regulation

AFAST e: Army Field Assistance and Technology

AFATDS e: Advanced Field Artillery Tactical Data System

AFATL e: Air Force Armament Technology Laboratory

AFAUD e: Air Force Auditor General

AFAUX e: Air Force Auxiliary [Field]

AFAVC e: Atlantic Fleet Audio-Visual Center

AFaxA e: American Facsimile Association

AFB d: Abraumförderbrücke
- d: Allgemeine Feuerversicherungs-bedingungen
- d: Amtliches Fernsprechbuch
- d: Arbeitsausschuß für Berufsausbildung
- d: Arbeitsgemeinschaft Fachärztlicher Berufsverbände
- d: Ausführungsbestimmung[en]
- d: Automatische Fahr- und Bremssteuerung
- d: Automatisierte Fernbedienung
- e: Acid-Fast-Bacillus
- e: Acid-Fast-Bacteria
- e: Active Feedback Block
- e: Aeromedical Field Laboratory
- e: Aflatoxin B
- e: Aircooled Fluidized Bed
- e: Air Force Base
- e: Air Force Bulletin
- e: Airframe Bulletin
- e: Air Freight Bill
- e: American Farm Bureau
- e: Anal Fin Base
- e: Aorto-Femoral Bypass
- e: Atmospheric Fluid Bed
- e: Automatic Fall-Back
- e: Auxiliary Function Bit

AfB d: Arbeitsgemeinschaft für Bildungsfragen

afb e: anti-friction bearing
- n: afbeelding

AFB_1 e: Aflatoxin B_1

AFBA e: Armed Forces Broadcasters Association

AFBC e: Atmospheric Fluidized-Bed Combustion
- e: Atmospheric Fluidized Bed Coal
AFB Combustion e: Atmospheric Fluid Bed Combustion
AFBDA e: Anti-Friction Bearing Distributors Association
AFBDC e: Air Force Base Development Center
AFBF e: American Farm Bureau Federation
AFBFAR e: Air Force Bent Fin Artillery Rocket
AFBFS e: Air Force Basic Flying School
AFBG e: Arteriofemoral Bypass Graft
AFBIS e: Associate Fellow of the British Interplanetary Society
AFBITS e: Air Force Base Information Transfer System
AFBM e: Air Force Ballistic Missile
AFBMA e: Air Force Ballistic Missile Arsenal
- e: Anti-Friction Bearing Manufacturers Association
AFBMC e: Air Force Ballistic Missile Center
- e: Air Force Ballistic Missile Committee
AFBMD e: Air Force Ballistic Missile Division
AFBMIR e: Air Force Ballistic Missile Installation Regulation
AFBMTC e: Air Force Ballistic Missile Training Center
AFBNM e: Agate Fossil Beds National Monument
AFBO e: Approved Force Budget Objective
AFBP e: Air Force Bailment Property
AFBRMC e: Air Force Business Research Management Center
AFBS e: Air Force Board Structure
AFBSD e: Air Force Ballistic Systems Division
AFB-Steuerung d: automatische Fahr-Brems-Steuerung
AFBTR e: Association for Brain Tumor Research
AFBU e: Air Force Base Unit
AFBW e: Analog[ue] Fly by Wire
AFC e: Adjustable Focus Control
- e: Advanced Fighter Capability
- e: Advanced Fire Control
- e: Advanced Function of Communications
- e: Aerodynamic Flight Control
- e: Affiliation Code
- e: African Farmers Committee
- e: African Forestry Commission
- e: Airfield Construction
- e: Air Flow Control
- e: Air Force Circulars
- e: Air Force Component
- e: Air Force Comptroller
- e: Air Force Council
- e: Air Force Cross
- e: Airframe Change
- e: Airworthiness and Flight Characteristics
- e: Alkaline Fuel Cell
- e: Alumin[i]um Field Coil
- e: American Forest Council
- e: Amplitude-Frequency Characteristic
- e: Analog[ue]-to-Frequency Conversion
- e: Analog[ue]-to-Frequency Converter
- e: Antenna for Communications
- e: Antibody-Forming Cell[s]
- e: Apollo Flight Control
- e: Aquatic Federation of Canada
- e: Area Forecast Center
- e: Area Frequency Coordinator
- e: Armament and Fuel Coordinator
- e: Armed Forces Comptroller
- e: Armed Forces Council
- e: Armo[u]red Flexible Conveyor
- e: Army Field Code
- e: Army Field Command
- e: Army Finance Center
- e: Army Flying Corps
- e: Assimilation and Fractional Crystallization
- e: Atomic Fluid Cell
- e: Atomic Fuel Corporation
- e: Audio Freight Claims
- e: Audio Frequency Change[r]
- e: Audio Frequency Choke
- e: Audio Frequency Coder
- e: Audio Frequency Coding
- e: Australian Fluorine Chemicals
- e: Automated Fare Collection
- e: Automated Filing Cabinet[s]
- e: Automatic Fare Collection
- e: Automatic Feedback Control
- e: Automatic Fidelity Control
- e: Automatic Fire Control
- e: Automatic Flatness Control
- e: Automatic Flight Control
- e: Automatic Flow Control
- e: Automatic Flushing Cistern
- e: Automatic Following Control
- e: Average Fixed Cost
- e: Aviation Fire Controlman
- e: Axial Flow Compressor
afc e: anti-fading compensation
A & FC e: Armament and Fire Control
a.f.c. e: audio frequency change
- e: automatic frequency control
AFCA e: American Fastener and Closure Association
- e: Area Fuel Consumption Allocation
- e: Armed Forces Chemical Association
- e: Armed Forces Communications Association
- e: Assistant Freight Claim Agent
- e: Atom Fluorescence for Chemical Analysis
AFCAC e: Air Force Computer Acquisition Center
AFCAD e: Automatic File Control and Documentation
AFCAI e: Associate Fellow of the Canadian Aeronautical Institute
AFCAL f: Association Française de Calcul
AFCALT f: Association Française de Calcul et de Traitement de l'Information
AFCAM e: Air Force Coated Aluminum Metal
AFCAN e: Analog[ue] Factor Calibration Network
- f: Association Française de Calcul Numérique
AFCAO e: Air Force Computer Acquisition Office
AFCAP e: Air Force Capability Assessment Program[me]
AFCAPS e: Air Force Civilian Automated Pay System
AFCAS e: Advanced Flight Control Actuation System
AFCAS-AE e: Advanced Flight Control Actuation System-All Electric
AFCAT e: Alert Force Capability Test
AFCB e: Armed Forces Combat Bulletin
AFCC e: Air Force Combat Command
- e: Air Force Communications Center
- e: Air Force Communications Command
- e: Air Force Component Commander
- e: Air Freight Control Centre
- e: Assault Fire Command Post
- e: Australian Federation of Construction Contractors
- e: Avaible Frame Capacity Count
AFCCB e: Air Force Configuration Control Board
AFCCDC e: Air Force Command and Control Development Center
AFCCDD e: Air Force Command and Control Development Division
AFCCE e: Air Force Cost Center
- e: Association of Federal Communications Consulting Engineers
AFCCG e: Atlantic Fleet Combat Camera Group
AFCCP e: Air Force Command and Control Post
- e: Air Force Component Command Post
AFCCPC e: Air Force Communications Computer Programming Center
AFCCS e: Air Force Command and Control System
AFCD e: Advanced Fighter Capability Demonstrator
- e: Air Force Cryptologic Depot
AFCE e: Air Force Civil Engineer (journ.)
- e: Air Force Civil Engineering Unit
- e: Allied Forces Central Europe
- e: American Foundation for Continuing Education
- e: Armed Forces Chemical Association
- e: Associate in Fuel Technology and Chemical Engineering
- e: Automatic Flight Control Equipment
AFCEA e: Armed Forces Communications and Electronics Association
AFCEC e: Air Force Civil Engineering Center
AFCEL e: Air Forces Contractor Experience List
AFCENT e: Allied Forces Center
AFCF e: Anthropology Film Center Foundation
AFCFP e: Arab Federation of Chemical Fertilizer Producers
afcfs e: advanced fighter-control-flight simulator
AFCFS e: Air Force; Chief of Flight Safety
AFCGIC e: Air Force Ceramics and Graphite Information Center
AFCGM e: Air Force Center; Guided Missiles
AFCH e: Air Force Component Headquarters
AF Ch Sch e: Air Force Chaplain School
AFCI e: American Foot Care Institute
- e: Associate of the Faculty of Commerce and Industry
AFCIA e: Armed Forces Civilian Instructors Association
AFCIE e: Automatic Frequency Control Intermediate Frequency
AFCIF e: Automatic Frequency Control Intermediate Frequency
AFCIG e: Air Force Inspector General
AFCIP e: Air Force Center for International Programs
AFCIS-L e: Armor Full Crew Research Simulator Center - Laboratory
AFCK e: Aerial Field Charge Kit
- e: Antenna Field Charge Kit
AFCL e: Africa Container Lines
- e: Associate of the Farriers Company of London
AFCLC e: Air Force Contract Law Center
AFCM e: Air Force Central Museum
- e: Air Force Commendation Medal

AFCMA e: Aluminum Foil Container Manufacturers Association
- e: Australian Fibreboard Container Manufacturers Association
AFC-Man e: Automatic Frequency Control - Manual
AFC-MAN e: Automatic Frequency Control-Manual
AFCMC e: Air Force Contract Maintenance Center
AFCMR e: Armed Forces Central Medical Registry
AFCO e: Admiralty Fleet Confidential Order
- e: Air Force Contracting Officer
- e: Air Force Control Office
- e: Automatic Fuel Cut-Off
AFCOA e: Air Force Chief of Operations Analysis
- e: Air Force Contracting Office Approval
AFCOLR e: Air Force Coordinating Office for Logistics Research
AFCOM e: Air Force Communications
- e: Association for Computer Operations Managers
AFCOMAC e: Air Force Combat Ammunitions Center
AFCOM Center e: Aeronautical Fixed Telecommunications Network Communications Center
AFCOMMSTA e: Air Force Communications Station
AF Comp e: Air Force Composite
AFCOMPMET e: Air Force Comptroller Management Engineering Team
AF Compt e: Air Force Comptroller
AFCOMS e: Air Force Commissary Service
AFCOMSEC e: Air Force Communications Security
AFCOMSECCEN e: Air Force Communications Security Center
AFCOMSECM e: Air Force Communications Security Manual
AFCON e: Air Force Controlled [Unit]
AFCORS e: Advanced Fire Control Radar System
AFCOS e: Air Force Combat Operations Staff
- e: Armed Forces Courier Service
AFCP e: Advanced Flight Control Program[me]
- e: Air Force Civilian Personnel
- e: Air Force Command Post
- e: Air Force Computer Program[me]
- e: Automatic Flight Control Panel
AFCPL e: Air Force Computer Program[me] Library
AFCPM e: Air Force Civilian Personnel Manual
AFCPMC e: Air Force Civilian Personnel Management Center
AFCPR e: Air Force Cost Reduction Programm
AFCR e: American Federation for Clinical Research
AFCRB e: Air Force Central Review Board
AFCRC e: Air Force Cambridge Research Center
- e: Air Force Cambridge Research Computer
AFCRL e: Air Force Cambridge Research Laboratories
- e: Air Force Cambridge Research Library
AFCRL-PSRP e: Air Force Cambridge Research Laboratories-Physical Sciences Research Paper
AFCRP e: Air Force Cost Reduction Program[me]
AFCRS e: Airborne Fire Control Radar Set
AFCS e: Adaptive Flight Control System
- e: Advanced Flight Control System
- e: Air Force Communications Service
- e: Air Force Communications Squadron
- e: Air Force Communications System
- e: Armament Flight Control System
- e: Army Facilities Components System
- e: Army Functional Component System
- e: Automatic Fare Collection System
- e: Automatic Fire Control System
- e: Automatic Flight Control System
- e: Auxiliary Flight Control System
- e: Avionic Flight Control System
AFCSA e: Air Force Center for Studies and Analyses
AFCSC e: Air Force Command and Staff College
- e: Air Force Cryptologic Support Center
AFCSCP e: Automatic Flight Control System Control Panel
AFCSDG e: Air Force Container System Development Group
AFCSDO e: Air Force Communications-computer Systems Doctrine Office
AFCS E&I e: Air Force Communications Service-Engineering and Installation
AFCSL e: Air Force Communications Security Letter
AFCSM e: Air Force Communications Security Manual
- e: Air Force Communications Service Manual
AFCSP e: Air Force Communications Security Pamphlet
AFCSS e: Air Force Communications Support System
- e: Army in the Field Containers System Study
AFCT e: Affect
- e: Alert Force Capability Test
AFCTCP e: Air Force Combat Theater Communications Program[me]
AFC & TO e: Air Force Clothing and Textile Office
AFCTS e: Air Force Combined Tomography System
AFCTV e: Action for Childrens Television
AFCU e: American Friends of Cambridge University
- e: Association of Federal Computer Users
AFCWF e: Air Force Civilian Welfare Fund
AFD d: Archiv für Diplomatik (journ.)
- d: automatisierte Fernbedienung
- d: Avalanche-Fotodiode
- e: Accelerated Freeze Drying
- e: Account Feed[ing] Device
- e: Acid Fractionator Distillate
- e: Acoustic Firing Device
- e: Acoustic Flaw Detector
- e: Active Filter Design
- e: Admiralty Floating Dock
- e: Aft Flight Deck
- e: Airborne Frequency Doubler
- e: Air Force Depot
- e: Air Force Detachment
- e: Air Force Directive
- e: Airport / Facility Directory
- e: Alternating Field Demagnetization
- e: Amplitude-Frequency Distortion
- e: Amplitude-Frequency Distribution
- e: Appointed Factory Doctor
- e: April Fools Day
- e: Arithmetic Function Designator
- e: Arm / Fire Device
- e: Armed Forces Day
- e: Arming and Fusing Device
- e: Army Field Depot
- e: Assistant Field Director
- e: Assistant Flight Director
- e: Automated Flaw Detector
- e: Automatic Fast Demagnetization
- e: Automatic Feeding Device
- e: Automatic Fire Detection
- e: Automatic Fire Detector
- e: Auxiliary Floating Dock
- e: Axial Flux Density
- e: Axial Flux Difference
afd a: afdeling
AfD d: Amt für Datenverarbeitung
AFDA e: Abstract Family of Deterministic Acceptors
- e: Air Force Data Automation
- e: Air Force Distribution Agency
- e: Axial Flux Difference Alarm
AFDAA e: Air Force Data Automation Agency
AFDAMET e: Air Force Data Automation Management Engineering Team
AFDAP e: Air Force Data Automation Planning
- e: Air Force Designated Acquisition Program[me]
- e: Air Force Directorate of Advanced Technology
AFDAR e: Air Force Defense Acquisition Regulations
AFDAS e: Aircraft Fatigue Data Analysis System
- e: Airframe Fatigue Data Analysis System
AFDASTA e: Air Force Data Station
AFDAT e: Air Force Directorate of Advanced Technology
AFDATACOM e: Air Force Data Communications
- e: Air Force Data Communications System
AFDATACOM System e: Air Force Data Communications System
AFDATASTA e: Air Force Data Station
AFDB e: African Development Bank
- e: Alternative Fuel Data Bank
- e: Armed Forces Development Board
- e: Auxiliary Floating Dock; Big
AFDC e: Agriculture and Fishery Development Corporation
- e: Aid to Families with Dependent Children
- e: Air Force Department Constabulary
- e: Automatic Formation Drone Control
AFDCB e: Armed Forces Disciplinary Control Board
AFDCCO e: Air Force Departmental Catalog Coordinating Office
AFDCMI e: Air Force Policy on Disclosure of Classified Military Information
AFDCO e: Air Force Distribution Control Office
AFDCOM System e: Air Force Data Communications System
AFDCP e: Aft Flight Deck Control Panel
- e: Air Force Decision Coordinating Paper
AFDCS e: Automatic Film Data Collection System
AFDCUF e: Aid to Families with Dependent Children of Unemployed Fathers
AFDDA e: Air Force Director of Data Automation
AFDE e: American Fund for Dental Education
- e: Arctic Fuel Dispensing Equipment
AFDEC e: Association of Franchised Distributors of Electronic Components
AFDET e: Automatic Fault Detection
- e: Automatic Fault Detector

AFDFD e: Air Force Deputy for Foreign Development
AFDFR e: Air Force Development Field Representative
AFDFS e: Air Force Department Fire Service
AFDH e: American Fund for Dental Health
AFDI e: Australian Forest Development Institute
AFDIERSS e: Air Force Departmental Industrial Equipment Reserve Storage Site
AFDIS e: Air Force Director of Inspection Services
AFDK e: After Dark
AFDL e: Abstract Family of Deterministic Languages
AFDM e: Air Force Driver Magazine
- e: Ash-Free Dry Mass
- e: Auxiliary Floating Dock; Medium
AFDMA e: Air Force Director of Management Analysis
AFDMIC e: Air Force Defense Metals Information Center
AFDMP e: Air Force Directorate of Materials and Processes
AFDMS e: Airborne Flight Detection Measurement System
AFDO e: Aft Flight Deck Operator
- e: Air Force Duty Officer
- e: Assistant Fighter Director Office
- e: Assistant Flight Dynamics Officer
- e: Association of Food and Drug Officials
AFDOA e: Armed Forces Dental Officers Association
AFDOUS e: Association of Food and Drug Officials of the United States
AFDP e: Air Force Departure Point
- e: Army Force Development Plan
AFDPDB e: Aft Flight Deck Power Distribution Box
AFDPRC e: Air Force Disaster Preparedness Resource Center
AFD Process e: Accelerated Freeze-Drying Process
AFDRD e: Air Force Director[ate] of Research and Development
AFDRIF e: Air Force Directory of Resident Inspection Facilities
AFDRQ e: Air Force Director of Requirements
AFDS e: Advanced Fighter Diagnostic System
- e: Air Foil Design System
- e: Amphibious Flagship Data System
- e: Automatic Flight Director System
- e: Autopilot Flight Director System
- e: Auxiliary Fighter Director Ship
AFDSC e: Air Force Data Services Center
AFDSDC e: Air Force Data System Design Center
AFDSEC e: Air Force Data Systems Evaluation Center
AFDT e: Air Freight Decision Tool
AFDTL e: Air Force Drug Testing Laboratory
AFDU e: Air Fighting Development Unit
- e: Air Force Development Unit
- e: Alternative Fuels Development Unit
AFDW e: Air Force District of Washington
AFE d: Arbeitsgemeinschaft zur Förderung der Elektrowirtschaft
- d: Auftrags[fern]eingabe
- d: automatische Frequenzentlastung
- d: automatischer Fehleremulator
- e: Accredited Financial Examiner
- e: Aerospace Facilities Engineer
- e: Air Force Experiment
- e: Air Force in Europe
- e: Allowed Failure Effect
- e: Alternate Fighter Engine
- e: Amniotic Fluid Embolism
- e: Anti-Ferro-Electric
- e: Apple File Exchange
- e: Authority for Expenditure
- e: Automatic Fire Extinguisher
AfE d: Archiv für Elektrotechnik
AFEA d: Ausfalleffektanalyse
- e: American Farm Economic Association
- e: Automobile Fuel Efficiency Act
- e: Aviation Facilities Energy Association
AFEAS e: Alternative Fluorocarbon Environmental Acceptability Study
AFeb d: amtliches Fernsprechbuch
AFEB e: Armed Forces Epidemiological Board
AFEBS e: Air Force European Broadcasting Squadron
AFEDPC e: Air Force Electronic Data Processing Center
AFEE e: Airborne Forces Experimental Establishment
AFEES e: Armed Forces Examining and Entrance Station[s]
AFEI e: Americans for Energy Independence
- e: Association of Finnish Electric Industries
AFEIS e: Armed Forces Examining and Induction Stations
AfELF d: Ausschuß für Ernährung, Landwirtschaft und Forsten
AFELIS e: Air Force Engineering and Logistics Information System
AFEM e: Armed Forces Expeditionary Medal
AFEMMIS e: Air Force Equipment Maintenance Management Information System
AFEMS e: Air Force Equipment Management System
AFEMST e: Air Force Equipment Management Survey Team
AFEMT e: Air Force Equipment Management Team
AFEOAR e: Air Force European Office of Aerospace Research
AFEOS e: Air Force Electro-Optical Site
AfEP d: Amt für Erfindungs- und Patentwesen (DDR)
AFEP e: Army Facilities Energy Program[me]
AFEPI e: Air Force Equipment Procurement Instruction
AFEPIC e: Air Force Electronic Properties Information Center
AFER e: Air Force Engineering Responsibility
AFERB e: Air Force Educational Requirements Board
AFERC e: Air Force Edwards Research Center
AFERSS e: Air Force Environmental Rocket-Sounding System
AFES e: Admiralty Fuel Experimental Station
- e: Aggregate Field Expense Study
- e: Air Force Examining Station[s]
- e: Air Force Exchange Service
- e: Armed Forces Examining Station
- e: Armed Forces Exchange Service
- e: Automatic Feature Extraction System
AFESA e: Air Force Engineering and Services Agency
AFESC e: Air Force Electronic Security Command
- e: Air Force Engineering and Services Center
AFESC/ESL e: Air Force Engineering and Services Center / Engineering and Services Laboratory
AFESD e: Air Force Electronic Systems Division
AFESMET e: Air Force Engineering and Services Management Engineering Team
AFESP e: Air Force English Syntax Project
AFETAC e: Air Force Environmental Technical Applications Center
AFETO e: Air Force Engineering Technology Office
AFETR e: Air Force Eastern Test Range
AFETRM e: Air Force Eastern Test Range Manual
AFETS e: Air Force Engineering and Technical Service
AFEWC e: Air Force Electronic Warfare Center
AFEWES e: Air Force Electronic Warfare Evaluation Simulator
AFEX e: Air Forces Europe Exchange
- e: Ammonia Freeze Explosion
- e: Armed Forces Exchange
AFF d: Ansteuerungsfunkfeuer
- d: automatische Fahrzeugführung
- e: Above Finished Floor
- e: Accelerator Free Fall
- e: Acceptance and Ferry Flight
- e: Advanced Function Feature
- e: Affair[s]
- e: Affecting
- e: Afferent
- e: Affiliated
- e: Affinity
- e: Affirmative
- e: Affirming
- e: A Flip-Flop
- e: Agriculture, Forestry, Fishing
- e: American Farm Foundation
- e: Army Field Forces
- e: Associated Fresh Foods
- e: Atrial Filling Fraction
- e: Auditory Flutter Fusion
- e: Automated Field Fire
- e: Automatic Fast Feed
- e: Automatic Feeding Form
- e: Automatic Frequency Follower
- e: Automatic Frequency Following
- e: Axisymmetrical Flow Field
aff. l: affinis
AFFAS e: Armed Forces Financial Advisory Services
AFFB e: Army Field Forces Board
AFFC e: Air Force Ferry[ing] Command
- e: Alumin[i]um Foil Field Coil
AFFCO e: Alaska Forest Fire Council
AFFCS e: Advanced Fuze Function Control System
AFFC System e: Advanced Fire/Flight Control System
AFFD e: Affirmed
aff'd e: affirmed
affd e: affixed
- e: afford[able]
- e: affordability
AFFDL e: Air Force Flight Dynamics Laboratory
AFFE e: Airborne Fire Fighting Equipment
- e: Air Force Far East
- e: Army Forces Far East
affet i: affettuoso

AFFF e: American Fish Farmers Federation
- e: Aqueous Film-Forming Foam
- e: Aqueous Film Forming Foam
AFFFA e: American Forged Fitting and Flange Association
AFFFT e: Academy of Family Films and Family Television
AFFG e: Affirming
AFFI e: American Frozen Food Institute
- e: Arab Federation for Food Industries
affiec e: affect[at]ion
- e: affective
affil. e: affilate[d]
AFFIL e: Affiliate
affil e: affiliated
Affil RSH e: Affiliate of the Royal Society of Health
Affin e: Affinity
AffInstSM e: Affiliate of the Institute of Sales Management
AffIP e: Affiliate of the Institute of Plumbing
affirm e: affirmative
AFFIRM e: Analyzer for Formula Translation Incremental Reengineering Methodology
- e: Association for Federal Information Resources Management
AffIWHTE e: Affiliate of the Institution of Works and Highways Technician Engineers
affl e: affiliate
Affl e: Affluent
afflat l: afflatus
AFFLC e: Air Force Film Library Center
Afflt. e: Affluent
AFFM e: Armo[u]red Force Field Manual
AFFOM e: Air Force Field Office Manager
AFFOR e: Air Force Forces
AFFORD e: Analysis for Forces Objectives and Resources Determination
Affores. e: Afforestation
AFFPC e: Air Force Financial Postal Clerk
affret i: affrettando
AFFRMN e: Affirmation
AFFS d: automatisiertes Feldführungssystem
- e: Air Force Flying Safety
- e: Army Field Feeding System
- e: Audio-Frequency Frequency Shift[ing]
AFFSE e: Armed Forces Food Science Establishment
AFFSIM e: Airborne Formation Flight Simulator
afft e: affdavit
AFFT e: Association of Federal Fiscal Technicians
- e: Auditory Flutter Fusion Threshold
AFFTC e: Air Force Field Technical Center
- e: Air Force Flight Test Center
- e: Air Force Flying Training Command
AFFTD e: Air Force Foreign Technology Division
AFFTIS e: Air Force Flight Test Instrumentation System
AFFTU e: Augmentor Fuel Flow Test Unit
AFG d: Arbeitsförderungsgesetz
- e: Aerial Field Gain
- e: Aflatoxin G
- e: Allied Freighter Guard
- e: Alternative Force Generator
- e: Amniotic Fluid Glucose
- e: Analog[ue] Frequency Generator
- e: Analog[ue] Function Generator
- e: Analytical Function Generator
- e: Antenna Field Gain
- e: Arbitrary Function Generator
- e: Argyrophil, Fluorescent, Granulated
- e: Army Force Guidance
- e: Audio Function Generator
- e: Automatic Function Generator
afg a: afgang
Afg. e: Afghan[s]
Afg e: Afghani
- e: Afghanistan
AFGC e: American Forage and Grassland Council
AFG Cells e: Agryrophil, Fluorescent, Granulated Cells
AFGCM e: Air Force Good Conduct Medal
AFGE e: American Federation of Government Employees
- e: American Forum for Global Education
AFGEN e: Arbitrary Function Generation
AFGF e: Acidic Fibroblast Growth Factor
Afgh. d: Afghanistan
Afghanistan e: Republic of Afghanistan [Doulat i Jamhouri ye Afghánistán]
AFGIS e: Aerial Free Gunnery Instruction School
AFGK d: Archiv für Frankfurts Geschichte und Kunst (journ.)
AFGL e: Air Force Geophysics Laboratory
AFGM e: American Federation of Grain Millers
AFGN d: Antiformantglied; Nasal
AFGO e: Air Force General Orders
AFGP e: Antifreeze Glycoprotein
AFGR e: Approved Force Cross Requirement
AFGS e: Air Force Guide Specification
- e: American-French Genealogical Society
AFGSL d: Antiformantglied; Stimmlos
AFGU e: Aerial Free Gunnery Unit
AFGW e: American Flint Glass Workers Union of North America
AFGWC e: Air Force Global Weather Center
- e: Air Force Global Weather Central
AFGWRP e: Air Force Global Weather Reconnaissance Program[me]
AFGWU e: American Flint Glass Workers Union
AFH e: Aerial Feed Horn
- e: Air Force Hospital
- e: American Foundation for Health
- e: American Foundation for Homeopathy
- e: Antenna Feed Horn
- e: Anterior Facial Height
- e: Army Family Housing
- e: Automatic Feed Hopper
AFHC e: Air Force Headquarters Command
AFHF e: Air Force Historical Foundation
- e: American Foot Health Foundation
AFHG e: Additive-Free Hard Gold
AFHLO e: Air Force Hurricane Liaison Office
AFHPSP e: Air Force Health Professions Scholarship Program[me]
AFHQ e: African Force Headquarters
- e: Air Force Headquarters
- e: Allied Forces Headquaters
- e: Armed Forces Headquarters
- e: Army Field Headquarters
AFHQ(CIC) e: Allied Forces Headquarters (Counter Intelligence Corps)
AFHQPS e: Allied Forces Headquarters Petroleum Section
AFHRL e: Air Force Human Relations Laboratory
- e: Air Force Human Resources Laboratory
A-Fhrzg. d: Aufklärungsfahrzeug
AFHS e: Average Flying Hours per Sortie
AFHV e: Americas Funniest Home Videos
AFI d: Arbeitsgemeinschaft für Fachinformation
- d: automatische Fahrplaninformation
- d: automatische Fehleridentifikation
- e: Adjusted Family Income
- e: African/Indian Ocean
- e: Air Filter Institute
- e: Air Flow Indicator
- e: Air Forces; Iceland
- e: Amaurotic Familial Idiocy
- e: American Fiber Institute
- e: American Filter Institute
- e: American Firearms Industry
- e: American Forest Institute
- e: American Fur Industry
- e: Armed Forces Institute
- e: Associate of the Faculty of Insurance
- e: Association of Food Industries
- e: Automatic Fault Isolation
AfI d: Amt für Information
AFIA d: Arbeitsgemeinschaft für industrielle und gewerbliche Abfallfragen
- e: American Feed Industry Association
- e: American Footwear Industries Association
- e: Associate of the Federal Institute of Accountants
AFIAeS e: Associate Fellow of the Institute of Aeronautical Sciences
AFIAS e: Associate Fellow of the Institute of Aeronautical Sciences
- e: Associate Fellow of the Institute of the Aerospace Sciences
AFIB e: Air Force Test Base
- e: Atrial Fibrillation
AFIC e: Air Force Intelligence Center
afic s: aficinnado
AFICCS e: Air Force Integrated Command and Control System
- e: Air Force Interim Command and Control System
AFICD e: Associate Fellow of the Institute of Civil Defence
AFICE e: Air Forces; Iceland
AFID e: Alkali Flame Ionization Detector
- e: Arithmetic Function Identifier
- f: Agence Française d'Information et de Documentation
AFIDA e: Agricultural Foreign Investment Disclosure Act
AFIDS e: Automatic Firearms Identification System
AFIE e: Abnormal Fluctuation in the Economy
- e: Armed Forces Information and Education
AFIED e: Armed Forces Information and Education Division
AFIF e: Air Force Industrial Fund
- e: Armed Forces Information Film
- e: Associated Fraternity of Iron Forgers
- e: Association of Foremen Iron Founders
AFIGAC e: Air Force Inspector General Activities Center
AFII e: American Federation of International Institutes
AFIIM e: Associate Fellow of the Institute of Industrial Managers
A-FILE e: Adolescent-Family Inventory of Life Events and Changes
AFILR e: Approved Force Investment Level Requirement
AFIM e: After Image
- e: Air Force Inventory Manager
AFIMA e: Associate Fellow of the Institute of Mathematics and its Applications
AFIM-Datei e: After Image-Datei

AFIMS e: Air Force Information Management Study
AFIMTS e: Air Force Instruction and Military Training School
AFINE f: Association Française pour l'Industrie Nucleaire d'Equipement
AFINS e: Airways Flight Inspector
AFINSPATH e: Armed Forces Institute of Pathology
AF Inst Pet e: Associate Fellow of the Institute of Petroleum
AFINTELMET e: Air Force Intelligence Management Engineering Team
AFIO e: Agreement for Fighter Interceptor Operations
- e: Approved Force Inventory Objective
- e: Association of Former Intelligence Officers
- e: Authorization for Interceptor Operations
AFIP e: Air Force Information Program[me]
- e: Air Force Intelligence Publication
- e: American Federation of Information Processing
- e: Armed Forces Information Program[me]
- e: Armed Forces Institute of Pathology
- e: Automated Financial Improvement Program[me]
AFIPS e: American Federation of Information Processing Societies
AFIPS Conf Proc Fall Spring Jt e: American Federation of Information Processing Societies. Conference Proceedings. (journ.)
AFIR d: Ausschuß für industrielles Rechnungswesen
- e: Air Force Installation Representative
AFI-RAN e: Africa, India Ocean Region Air Navigation
AFIRB e: Armed Forces Identification Review Board
AFIRE e: Association of Fundamental Institutions of Religious Education
afirm e: Affirmative
AFIRO e: Air Force Installations Representative Officer
AFIRSS e: Automatic Fact Information Retrieval and Storage System[s]
A-FIRST e: Advanced-Far Infrared Search Track
AFIS d: angewandte Fertigungs- und Informationssysteme
- d: automatisches Fehleridentifikations-system
- e: Aerodrome Flight Information Service
- e: Air Force Intelligence Service[s]
- e: Air Force Intelligence Study
- e: American Forces Information Service
- e: Armed Forces Induction Station
- e: Armed Forces Information School
- e: Armed Forces Information Service
- e: Armed Forces Information System
- e: Army Force Integration Study
- e: Automated Field Interview System
- e: Automated Financial Information System
- e: Automated Fingerprint Indentification System
AFISC e: Air Force Inspection and Safety Center
AFISM e: Alumin[i]um-Free Inorganic Suspended Material
AFISOL e: Aerodrome Flight Information Service Officers Licence
AFISR e: Air Force Industrial Security Regulations
AFIT e: Airblast Fuel Injection Tube
- e: Air Force Institute of Technology
- e: American Fabricating Institute of Technology
- e: Armed Forces Institute of Technology
- e: Automatic Fault-Isolation Test[er]
AFITAE f: Association Française d'Ingénieurs et Techniciens de l'Aéronautique et de l'Espace
AFITC e: Armed Forces Intelligence Training Center
AFIT/ST e: Air Force Institute of Technology School of Systems and Logistics
AFJ e: Air Force Jet
- e: Armed Force[s] Journal
AFJAG e: Air Force Judge Advocate General
AFJAGS e: Air Force Judge Advocate General School
AFJKT e: Air Force Job Knowledge Test
AFJMSNS e: Air Force Justification for Major System New Start
AFJROTC e: Air Force Junior Reserve Officers Training Corps
AFK d: Aachener Fluidtechnisches Kolloquium
- d: Akademie für Führung und Kommunikation
- d: Amplituden / Frequenz-Kennlinie
- d: Amplituden/Frequenz-Kennlinie
- d: aramidfaserverstärkte[r] Kunststoff[e]
- d: Arbeitsgemeinschaft für Friedens- und Konflikt-Forschung
AfK d: Arbeitsgemeinschaft für Kerntechnik
- d: Arbeitsgemeinschaft für Kommunikationsforschung
- d: Arbeitsgemeinschaft für Korrosionsfragen
- d: Ausschuß für Klassifikation
afk n: afkorren, afkorting
AFKAG e: Air Force Cryptographic Aid; General
AFKG d: Arbeitsförderungskonsolidie-rungsgesetz
AFKN e: American Forces Korea Network
AFL d: Abfanglinie
- d: Antifibrinolysin
- e: Above Field Level
- e: Abstract Family of Languages
- e: Active Fuel Length
- e: Advanced Flow Laser
- e: Aflatoxicol
- e: Air Force Letter
- e: Air Force Liaison
- e: Air Force List
- e: Air Freight List
- e: American Federation of Labor
- e: Animated Film Language
- e: Anti-Fatty-Liver
- e: Antifibrilysine
- e: Antifibrinolysine
- e: Antifibrolysin
- e: Antisymmetric Filter
- e: Applied Fisheries Laboratory
- e: Automatic Fault Location
- R: Aeroflot
Afl. d: Artillerieflieger
afl e: atrial flutter
- n: aflevering
AFLA e: Adolescent Family Life Act
- e: Air Lubricated Free Axis Trainer
- e: Asian Federation of Library Associations
- e: Associate of Fire Loss Adjusters
AFLANT e: Air Forces; Atlantic
AFLAS e: Aviation Fuels Logistical Area Summary
AFLAT e: Air Force Language Aptitude Test
AFLC e: Air Force Logistics Center
- e: Air Force Logistics Command
- e: Australian Federal Libraries Committee
AFLCF e: Air Force Logistics Command Form
AFLCG e: Air Force Logistics Control Group
AFL-CIO e: American Federation of Labor-Congress of Industrial Organizations
AFLCL e: Air Force Logistics Command Letter
AFLCM e: Air Force Logistic Command Manual
AFLCN e: Air Force Logistics Communi-cations Network
AFLCON e: Air Force Logistics Command Operations Network
- e: Air Force Logistics Communications Network
AFLCP e: Air Force Logistics Command Pamphlets
AFLCPs e: Air Force Logistics Command Pamphlets
AFLCR e: Air Force Logistics Command Regulations
AFLD e: Airborne Frauenhofer Line Discriminator
afld e: airfield
AFLETS e: Air Force Law Enforcement Terminal System
AFLI e: Air Force Legislative Item
- e: Association for Library Information
AFLIR e: Advanced Forward Looking Infrared
AFLL e: Army Fuels and Lubricant Laboratory
AFLMC e: Air Force Logistics Management Center
AFLOGMET e: Air Force Logistics Management Engineering Team
AFLOSH e: Automatic Dynamic Digital Test Fault Logic Simulation Hybrid
AFLP e: Acute Fatty Liver of Pregnancy
- e: Armed Forces Language Program[me]
AFLR e: Armed Forces Liaison Representative
AFLRL e: Army Fuels and Lubricants Research Laboratory
AFLRS e: Allied Forces Local Resources Section
AFLS e: Active Flight Load System
- e: Air Force Library Service
- e: Approach Flash-Lighting System
- e: Armed Forces Librarians Section
AFLSA e: Air Force Longevity Service Award
aflt e: afloat
AFLT e: Audio Frequency Line Transformer
AFLU e: Available for Local Use
AfLuT d: Amt für Landwirtschaft und Tierzucht
AFM d: Analog-Meßwerterfassung, -Fernmessung und -Meßwertverarbeitung
- d: Anschlußsteuerung Fernschreibmaschine
- d: Arbeitsgemeinschaft freier Mineralöl-importeure
- d: Arbeitsgemeinschaft Freier Mitarbeiter
- d: Arbeitsgemeinschaft für fertigungs-technisches Meßwesen
- d: Außenhandelsverband für Mineralöl
- d: automatische Fräsmaschine
- e: Abrasive Flow Machining
- e: Absorption Frequency Meter

- e: Academy of Family Mediators
- e: Accredited Farm Manager
- e: Acoustic Fast Motion
- e: Adhesive Film Mechanism
- e: Aflatoxin M
- e: Air Flow Meter
- e: Air Force Manual
- e: Air Force Museum
- e: Airplane Flight Manual
- e: Amplitude-Frequency Modulation
- e: Analysis and Forecasting Mode
- e: Annular Fire Missile
- e: Application Functions Module
- e: Approved Flight Manual
- e: Armed Forces Management
- e: Armed Forces Movement
- e: Associated Foam Manufacturers
- e: Associated Fur Manufacturers
- e: Associative Film Memory
- e: Atomic Force Microscope
- e: Audio Frequency Modulation
- e: Automatic Flight Management
- e: Aviation Fleet Maintenance

AfM d: Archiv für Musikwissenschaft (journ.)
A&FM e: Aerodynamics and Flight Mechanics
afm e: Antifriction Metal
A/FM e: Arm / Firing Mechanism
AFMA e: Access Floor Manufacturing Association
- e: Air Force Manual of Abbreviations
- e: American Feed Manufacturers Association
- e: American Fiber Manufacturers Association
- e: American Fur Merchants Association
- e: American Furniture Manufacturers Association
- e: Armed Forces Management Association
- e: Autobody Filler Manufacturers Association

AFMAB e: Atmospheric Forcings for the Mid-Atlantic Bight
AFMAG e: Air Force Management Analysis Group
AFMAINMET e: Air Force Maintenance Management Engineering Team
AFM/AP e: A File Management Auxiliary Processor
AFMBE e: Australian Federation for Medical and Biological Engineering
AFMC e: Armed Forces Mail Call
- e: Asian Fluid Mechanics Committee
- e: Automotive Filter Manufacturers Council

AFMCC e: Air Freight Motor Carriers Conference
AFMCO e: Army Force Modernization Coordination Office
AFMDC e: Air Force Machinability Data Center
- e: Air Force Missile Development Center

AFME e: Airframe
AFMEA e: Air Force Management Engineering Agency
AFMED e: Allied Forces Mediterranean
AFMEDMET e: Air Force Medical Management Engineering Team
AFMETC e: Air Force Meteorological Center
AFMF e: Air Fleet Marine Force
AFMFIC e: Associated Factory Mutual Fire Insurance Companies
AFMFP e: Aircraft, Fleet Marine Force; Pacific
AFMH e: American Foundation for Mental Hygiene
AFMI e: Association of French Mechanical Industries
AFMIC e: Air Force Materials Information Center
- e: Armed Forces Medical Intelligence Center

AFMIDPAC e: Army Forces; Middle Pacific
AF-MIPR e: Air Force-Military Interdepartmental Purchase Requests
AFMIS e: Army Food Management Information System
A/F Mixture e: Air / Fuel Mixture
AFML e: Air Force Materials Laboratory
- e: Armed Forces Medical Library

AFMLO e: Air Force Medical Logistics Office
AFMMFO e: Air Force Medical Materiel Field Office
AFMML e: Air Force Medical Materiel Letter
afmo s: afectisimo
AFMPA e: Air Force Medical Publications Agency
- e: Armed Forces Medical Procurement Agency
- e: Armed Forces Medical Publication Agency

AFMPC e: Air Force Military Personnel Center
- e: Assistant for Material Program[me] Control

AFMPDC e: Air Force Mechanical Properties Data Center
AFMR d: anti-ferromagnetische Resonanz
- e: American Foundation for Management Research
- e: Anti-Ferromagnetic Resonance
- e: Anti-Ferromagnetic Resonator
- e: Armed Forces Master Records
- e: Armed Forces Military Report
- e: Assistant Firemaster

AFMRB e: Air Force Material Review Board
AFMRS e: Anti-Submarine Warfare Formatted Message Reporting System
AFMS d: Artilleriefunkmeßstation
- e: Airborne Frequency Multiplexing System
- e: Air Force Manpower Standards
- e: Air Force Medical Service
- e: Airlift Field Maintenance Section
- e: American Federation of Mineralogical Societies

AFMSC e: Air Force Medical Service Center
- e: Air Force Medical Specialist Corps

AFMSL e: Air Force Measurement Standards Laboratories
AFMSMET e: Air Force Maintenance and Supply Management Engineering Team
AFMSP e: Air Force Meteorological Satellite Program[me]
AFMT e: Air Force Manufacturing Technology
AFMTC e: Air Force Military Training Center
- e: Air Force Missile Test Center

AFMVOP e: Air Force Motor Vehicle Operator Test
AFMW e: Australian Federation of Medical Women
AFN d: Anzeigegerät für Funknavigation
- d: automatische Frequenznachsteuerung
- d: automatische Frequenznachstimmung
- d: automatische Frequenzsteuerung
- e: Active Filter Network
- e: Air Force[s] Network
- e: All Figure Number
- e: American Forces Network
- e: Armed Forces Network
- e: Automatic Feature Negotiation
- e: Average Failure Number
- f: Afrique du Nord

AfN d: Arbeitsausschuß für Netzplantechnik
AF/N e: Air Force/Navy
AFNA e: Air-Force-Navy
AF/NA e: Air-Force / Navy Aeronautical
AFNA e: Air Force with Navy
AFNAB e: Air-Force-Navy Aeronautical Bulletin
AFNAS e: Air-Force-Navy Aeronautical Standard
AFNB e: Armed Forces News Bureau
AFNC e: Air Force Nurse Corps
AFNCOAR e: Air Force Noncommissioned Officer Academy Ribbon
AFNE e: Allied Forces Northern Europe
- e: American Forces Network; Europe
- e: Americans for Nuclear Energy
- e: Armed Forces Network; Europe

AFNEA e: Air Force Notice to Airman Exchange Area
AFNETF e: Air Force Nuclear Engineering Test Facility
AFNETR e: Air Force Nuclear Engineering Test Reactor
AFNETSTA e: Air Force[s] Network Station
AFNFICM e: Atlantic Fleet Naval Forces Intelligence Collection Manual
AFNI e: Armed Forces Network; Island
AFNIL f: Agence Francophone pour la Numérotation Internationale du Livre
AFNON e: Allied Forces North Norway
AFNOR e: Armed Forces Press Service
- f: Association Française de Normalisation

AFNORTH e: Allied Forces Northern Europe
AFNRC e: Armed Forces National Research Council
AFNS e: Air Force News Service
AFO d: Arbeitsfolge
- d: Archiv für Orientforschung (journ.)
- e: Accounting and Finance Office[r]
- e: Admiralty Fleet Order
- e: Advanced File Organization
- e: Air Force Order
- e: Airports Field Office
- e: Ankle-Foot Orthosis
- e: Announced Flight Opportunity
- e: Announcement of Flight Opportunity
- e: Army Forwarding Officer
- e: Artillery Forward Observer
- e: Association of Field Ornithologists
- e: Average Fractional Overflow
- e: Axial Flux Offset

A Fo e: Army Form
AFÖ d: Akademische Föderation Österreichs
AFOAE e: Air Force Office of Atomic Energy
AFOAR e: Air Force Office for Aerospace Research
AFOAS e: Air Force Office of Aerospace Sciences
AFOAT e: Air Force Office for Atomic Energy
- e: Assistant for Atomic Energy

AFOB e: Air Force Operations Base
AFÖB d: Arbeitsgemeinschaft zur Förderung der österreichischen Bauwirtschaft

AFOC e: Air Force Comptroller (journ.)
- e: Air Force Operations Center
- e: Auditor Freight Overcharge Claim
- e: Automatic Flight Operation Center
AFOE e: Amphibious Follow-on- Echelon
- e: Assault Follow-on Echelon
AFOECP e: Air Force Officer Education and Commissioning Program[me]
AFOEHL e: Air Force Occupational and Environmental Health Laboratory
AFOEP e: Air Force Officer Education Program[me]
AF of L e: American Federation of Labor
AFOG e: Asian Federation of Obstetrics and Gynaecology
AFOIC e: Air Force Officer in Charge
AFOLDS e: Air Force On-Line Data System
AFOMJ e: Association of Fats and Oil Manufacturers of Japan
AFOMS e: Air Force Office of Medical Support
AFOPA e: Air Force Office of Public Affairs
AFOQT e: Air Force Officer Qualifying Test
AFOR e: Air Force Operations Room
- e: Air Force Overseas Replacement
AFORA e: Air Force Office of Research Analysis
AFORD e: Air Force Overseas Replacement Depot
AFOREP e: Air Force Operational Report
AFORMS e: Air Force Operations Resource Management System
AFORS e: Air Force Office of Scientific Research
a fort i: a fortiori
AFOS e: Advanced Field Operating System
- e: Advanced Field Operations System
- e: Air Force Objective Series
- e: Anglo-French Offshore Surveys
- e: Armed Forces Optometric Society
- e: Automation of Field Observations and Services
- e: Automation of Field Operations and Services
AFOSCR e: Air Force Organization Status Change Report
AFOSH e: Air Force Occupational Safety and Health
AFOSR e: Air Force Office of Scientific Research
AFOTC e: Air Force Operational Test Center
AFOTEC e: Air Force Operational Test and Evaluation Center
AFOUA e: Air Force Outstanding Unit Award
AFOUAR e: Air Force Outstanding Unit Award Ribbon
AFP d: alternierendes Flankenpulsverfahren
- d: Arbeitsflugplatz
- d: Arbeitsgemeinschaft Flughafenplanung
- d: Artilleriefährprahm
- e: Abstract Family of Processors
- e: Add Fixed Point
- e: Adiabatic Fast Passage
- e: Advanced Fileable Processor
- e: Advanced Flexiple Processor
- e: Advanced Function Printing
- e: Aflatoxin P
- e: Air Force Pamphlet
- e: Air Force Plan
- e: Air Force Plant
- e: Air Force Police
- e: Alpha-Feto Protein
- e: Alternate Flight Plan
- e: Alternative Fertility Proportion
- e: American Standard Code for Information Interchange Front Panel
- e: Amniotic Alphafetoprotein
- e: Annual Financial Plan
- e: Annual Funding Program[me]
- e: Antifreeze Polypeptide
- e: Antifreeze Protein
- e: Aperture File Protocol
- e: Apple Talk Filing Protocol
- e: Armed Forces Police
- e: Army Force Program[me]
- e: Army Fuze Program[me]
- e: Associated Fantasy Publishers
- e: Association for Finishing Processes of Society of Manufacturing Engineers
- e: Association of Federal Photographers
- e: Associative File Processor
- e: Attached Formula Translation Processor
- e: Audio Flat Panel
- e: Authority for Purchase
- e: Automatic Feed Punch
- e: Automatic Floating Point
- f: Agence France Press
afp e: anterior faucial pillar
- e: automated filter photometer
AFPA e: Agricultural Fair Practices Act
- e: Airborne Flate Plate Array
- e: American Fisheries Protection Act
- e: American Folklife Preservation Act
- e: Aquarama and Fairmount Park Aquarium
- e: Automatic Flow Process Analysis
AFPAM e: Air Force Pamphlet
- e: Automatic Flight Planning and Monitoring
AFPAO e: Air Force Property Accountable Officer
AFPAV e: Airfield Pavement
AFPBS e: Air Force Pacific Broadcasting Squadron
AFPC e: Air Force Postal Clerk
- e: Air Force Procurement Circular
AF/PC e: Automatic Frequency / Phase-Controlled
AFPC e: Automatic Frequency and Phase Control
AFPCB e: Armed Forces Pest Control Board
AFPCH e: Army Force Planning Cost Handbook
AFPC Loop e: Automatic Frequency Phase Controlled Loop
AFPCP e: Air Force Potential Contractor Program[me]
AFPCS e: Automatic Flight Path Control System
AFPD e: Air Force Procurement District
- e: Armed Forces Police Department
- e: Authorization for Program[me] Development
AFPDA e: Army Force Planning Data and Assumptions
AFPDAB e: Air Force Physical Disability Appeal Board
AF-PDC e: Air Force Publications and Distribution Center
AFPDS e: Armed Forces Production Distribution Service
AFPE e: Air Force Planning Element
- e: Air Force Preliminary Evaluation
- e: American Foundation for Pharmaceutical Education
- e: Association for Progressive Education
AFPEA e: Air Force Packaging Evaluation Agency
AFPEC e: Armed Forces Product Evaluation Committee
AFPG e: Air Force Planning Guide
AFPH e: American Federation of the Physically Handicapped
AFPI e: Air Force Procurement Instructions
- e: American Forest Products Industries
AFPID e: Air Force Purchase Item Description
AFPL d: Ausweichflugplatz
- e: Air Force Packaging Laboratory
AFPM e: Aircraft Force Projection Model
AFPMO e: Air Force Polaris Material Office
AFPMPMS e: Air Force Professional Manpower and Personnel Management School
AFPO e: Air Force Procurement Officer
- e: Air Force Property Officer
- e: Air Force Purchasing Office
- e: Arithmetical Fixed Point Operation
AFPOM e: Air Force Program[me] Objectives Memorandum
AFPP e: Air Force Procurement Procedure[s]
AFPPA e: American Federation of Poultry Producers Associations
AFPPF e: Automatic Fluorescent Penetrant Processing Facility
AFPPS e: American Forces Press and Publications Service
AFPR e: Air Force Plant Representative
- e: Air Force Procurement Regulation
- e: Air Force Procurement Representative
- e: Air Force Project Representative
- e: Armed Forces Procurement Regulations
AFPRDS e: Air Force Petroleum Retail Distribution Station
AFPRL e: Air Force Personnel Research Laboratory
AFPRP e: Air Force Production Reserve Policy
AFPS e: Adjustable Frequency Power Supply
- e: American Forces Press Service
- e: Armed Forces Press Service
- e: Army Film and Photographic Section
- e: Aseptic Food Processing System
AFPS's e: Adjustable Frequency Power Supplies
AFPT e: Air Force Personnel Test
- e: Auxiliary Feed Pump Turbine
AFPTC e: Agricultural Farm Produce Trade Corporation
AFPTRC e: Air Force Personnel and Training Research Center
AfPU e: African Postal Union
AFPU e: Air Force Postal Unit
- e: Army Film and Photographic Unit
AFQ e: Aflatoxin Q
- e: Airframe Flight Qualification
- f: Association Forestière Québeçoise
AFQA e: Air Force Quality Assurance
AFQAR e: Air Force Quality Assurance Representative
AFQC e: Air Force Quality Control
AFQCR e: Air Force Quality Control Representative
AFQPL e: Air Force Qualified Products List
AFQT e: Armed Forces Qualification Test
AFQTVA e: Armed Forces Qualification Test; Verbal Arithmetic
AFR d: Abfühlregister
- d: Alkohol-Fällung-Rückstand
- d: Antifibrinolysinreaktion
- d: Archiv für Reformationsgeschichte (journ.)
- d: Automatische Frequenzregelung

- e: Absolute Filtration Rating
- e: Abstract Family of Relations
- e: Acceptable Failure Rate
- e: Acceptance Failure Rate
- e: Access Function Register
- e: Acid Fractionator Recycle
- e: Admiralty Fuel Research
- e: Advanced Fault Recognition
- e: Advanced Fault Recognizor
- e: Advanced Fault Resolution
- e: Africa
- e: After Flame Ratio
- e: Air-Fuel Ratio
- e: Aircraft Flight Report
- e: Air Flow Relay
- e: Air Force Regulation[s]
- e: Air Force Reserve
- e: Airframe
- e: Alternating Frequency Rejection
- e: Amplitude-Frequency Response
- e: Application Function Routine
- e: Aqueous Flare Response
- e: Arithmetic Factor Register
- e: Armed Forces Radio
- e: Artillery Flash Ranging
- e: Ascorbic Free Radical
- e: Atrial Filling Rate
- e: Audio-Frequency Range
- e: Audio-Frequency Reactor
- e: Auditor Freight Receipts
- e: Automated Format Recognition
- e: Automatic Field Recognition
- e: Automatic Format Recognition
- e: Available for Release
- e: Awaiting Forward Release
- e: Away from Reactor
- e: Axial Flow Reactor

afr a: afrikansk
- e: automatic feld/format recognition

Afr. d: Afrika
- d: Afrikaner

afr. d: afrikanisch
- d: altfranzösisch

AfR d: Arbeitsgruppe für Informationsrecht und Politik
- d: Arbeitskreis für Rechtschreibregelung

Afr e: African
- e: Afrikaans

A-Fr e: Anglo-French

AFRA e: American Farm Research Association
- e: Armed Forces Reserve Act
- e: Average Freight Rate Assessement

AFRADBIORSCHINST e: Armed Forces Radiobiology Research Institute

AF/RADC e: Air Force Rome Air Development Center

AFR Ae S e: Associate Fellow of the Royal Aeronautical Society

AFRAIDS e: Acute Fear Regarding AIDS

AFRAM e: Afro-American

AFRAMS e: Air Force Recoverable Assembly Management System

AF Range e: Audio Frequency Range

afranz. d: altfranzösisch

AFRAP e: Air Force Recruiter Assistance Program[me]
- e: American Foundation of Religion and Psychiatry

AFRAPT e: Air Force Research in Aircraft Propulsion Technology

AFRAS e: Associate Fellow of the Royal Aeronautical Society

AFRASEC e: Afro-Asian Organization for Economic Cooperation

AFRASIA e: Africa and Asia

A/F Ratio e: Air-to-Fuel Ratio

AFRB e: Australian Family Research Bulletin

AFRC e: Agricultural and Food Research Council
- e: Air-Fuel Ratio Control
- e: Air Force Records Center
- e: Air Force Regional Civil Engineer
- e: Area Frequency Response Characteristic
- e: Armed Forces Reserve Center
- e: Automatic Frequency Ratio Control

AFRCD e: Air-Fuel Ratio Control Device

AFRCE e: Air Force Regional Civil Engineer

AFRCSTC e: Air Force Reserve Combat Support Training Center

AFRCTC e: Air Force Reserve Combat Training Center

AFRD e: Acute Febrile Respiratory Disease
- e: Air Force Research Directorate
- e: Air Force Research Division

AFRDB e: Air Force Research and Development Branch

AFRDE e: Air Force Regional Civil Engineer[s]

AFREDCOM e: Armed Forces Readiness Command

AFREIC e: Air Force Radiation Effects Information Center

Afr Eng e: African English

AFREP e: Air Force Representative

AFREQ e: Army Air Forces Requirements Division

AFRES e: Air Force Reserve

AFRESBSGP e: Air Force Reserve Regions Base Support Group

AFRESM e: Armed Forces Reserve Medal

AFRESNAVSQ e: Air Force Reserve Navigation Squadron

AFRESR e: Air Force Reserve Regions

AFRESRGP e: Air Force Reserve Regions Group

AFRESS e: Air Force Reserve Sectors

AFRFTC e: Air Force Reserve Flying Training Center

AFRI e: Acute Febrile Respiratory Illness
- e: American Fur Resources Institute
- e: Applied Forest Research Institute

African languages e: Hausa in Central and West Africa
- e: Rwanda in southern Central Africa
- e: Somali in East Africa
- e: Swahili in parts of East Africa
- e: Xhosa and Zulu in South Africa
- e: Yoniba and Ibo in West Africa

AFRID e: Automatic Fuze Radiograph Inspection Device

afrik a: afrikansk

afrik. d: afrikanisch

Afrik e: Afrikaans
- n: Afrikaans

Afrik Eng e: Afrikaans English

AFRIS e: Australian Financial Review Information Service

AfrLS e: African Language Studies (journ.)

AFRM e: Air Frame

afrm e: airframe

AFRM e: Armed Forces Reserve Medal

AFRO e: Air Force Research Objectives
- e: Air Force Reserve Order[s]
- e: Air Force Routine Order

Afro-Amer e: Afro-American

Afro-American e: African-American

AFROASI e: Authority for Removal of Accepted Spacecraft Installations

Afro-Bras e: Afro-Brasiliero

AFROC e: Association of Freestanding Radiation Oncology Centers

Afro-Carib e: Afro-Caribbean
- s: Afro-Caribeño

AFROIC e: Air Force Resident Officer in Charge

Afro[s] e: Afro-American[s] Black[s], Negro[es]

AFRP e: Air Force Recurring Publication

AFRPL e: Air Force Rocket Propulsion Laboratory

AFRR e: Air Force Reserve Region
- e: Air Force Resident Representative

Afr Res Doc e: African Research and Documentation

AFRRGp e: Air Force Reserve Recovery Group

AFRRI e: Armed Forces Radiobiology Research Institute

AFRS e: Advanced Fighter Radar System
- e: Advanced Financial Reporting System
- e: Agricultural and Food Research Service
- e: Air Force Rescue Service
- e: Air Force Reserve Sector
- e: American Forces Radio Station
- e: Approved Force Retention Stock
- e: Armed Forces Radio Service
- e: Armed Forces Recruiting Stations
- e: Automatic Flight Reference System
- e: Auxiliary Flight Reference System

AfrS e: African Studies (journ.)

AFRSF e: Air Force Range Support Facility
- e: Atlantic Fleet Range Support Facility

AFRSI e: Advanced Flexible Reusable Surface Insulation

AFRST e: Automatic Focusing Random Scene Tracker

AFRSTC e: Air Force Reserve Specialist Training Center

AFRT e: Air Freight
- e: Armed (or American) Forces Radio and Television
- e: Armed Forces Radio and Television

AFRT-B e: American Forces Radio and Television-Bentwaters

AFRTC e: Air Force Research Training Center
- e: Air Force Reserve Training Center

AFRTD e: Air Force Research and Technology Division

AFRTN e: Armed (or American) Forces Radio and Television Network

AFRTS e: Armed (or American) Forces Radio and Television Service
- e: Armed Forces Radio and Telegraph Service
- e: Armed Forces RadioTelevision Service

AFRTS-BC e: Armed Forces Radio and Television Service-Broadcast Center

AFRTS-W e: American Forces Radio and Television Service-Worldwide

AFRTV e: Armed (or American) Forces Radio and Television

AFRTVS e: Armed (or American) Forces Radio and Television Service

AFRU e: Armed Forces Reporting Unit

afrz. d: altfranzösisch

AFS d: Anschlußsteuerung für Folienspeicher
- d: Atomfluoreszenzspektroskop[ie]
- d: automatische Fernschreib-Speicher-vermittlung
- d: automatische Fertigungssteuerung
- d: automatisches Führungssystem
- d: automatisierte Fertigungssteuerung
- d: automatisiertes Führungssystem
- e: Active Fuzing System
- e: Administrative Fact Sheet
- e: Advanced Fermentation System

- e: Advanced Figure Sensor
- e: Advanced Firing System
- e: Advanced Flying School
- e: Aerial Feed System
- e: Aerial Film Speed
- e: Aerial Fire Support
- e: Aerodrome Fire Service
- e: Aerometric Information Retrieval System Facility Subsystem
- e: Aeronautical Fixed Service
- e: Air Flow Sensor
- e: Air Force School
- e: Air Force Speciality
- e: Air Force Standard
- e: Air Force Station
- e: Air Force Stock
- e: Air Force Supply
- e: Airline Feed System
- e: Airway Facilities Service
- e: Airways Facilities Sector
- e: Airways Facilities Shop
- e: American Federation of Scientists
- e: American Finance System
- e: American Fisheries Society
- e: American Foundrymens Society
- e: Animal Feeding Stuffs
- e: Antenna Feed System
- e: Applicant File Search
- e: Arming and Fusing System
- e: Army Fire Service
- e: Army Food Supply
- e: Asian Folklore Studies (journ.)
- e: Associative Film Store
- e: Atlantic Ferry Service
- e: Atomic Fluorescence Spectroscopy
- e: Atomic Frequency Standard
- e: Audio-Frequency Shift
- e: Audio Frequency Shift
- e: Automatic Fault Simulator
- e: Automatic Firing Sequencer
- e: Automatic Flight System
- e: Automatic Frequency Stabilization
- e: Auxiliary Feedwater System
- e: Auxiliary Fire Service
- e: Aviation Facilities Service
- e: Azimuth Fellowup System
- e: Azimuth Follow-up System

afs a: afsender
AfS d: Absetzung für Substanzverringerung
- d: Amt für Standardisierung (DDR)
- d: Arbeitsausschuß für Sensitometrie

afs. d: Aforesaid
AFSA e: Air Force Senior Advisory
- e: Air Force Sergeants Association
- e: American Federation of School Administrators
- e: American Fire Sprinkler Association
- e: Application for Federal Student Aid
- e: Automated Fire Support Artillery
- e: Aviation Force Structure for the Army

AFSAB e: Air Force Science Advisory Board
- e: Air Force Scientific Advisory Board

AFSAM e: Air Force School of Aviation Medicine
AFSAMSO e: Air Force Space and Missile Systems Organization
AFSARI e: Automation for Storage and Retrieval of Information
AFSAS e: Academy for Friends of Secretarial Arts and Sciences
- e: Advanced Fire Support Avionics System
- e: American Federation of School Administrators and Supervisors

AFSAT e: Air Force School of Applied Tactics
AFSATCOM e: Air Force Satellite Communications
AFSATCOMS e: Air Force Satellite Communications System
AFSATCOM System e: Air Force Satellite Communications System
AFSATLCF e: Air Force Satellite Control Facility
AFSAW e: Air Force Special Activities Wing
AFSAWC e: Air Force Special Air Warfare Center
Af Sax e: Afro-Saxon
AFSB e: Air Force Specification Bulletin
- e: Australian Flying Saucer Bureau

AFSC e: Air Force Service Center
- e: Air Force Service Command
- e: Air Force Skill Code
- e: Air Force Specialty Code
- e: Air Force Supply Catalog
- e: Air Force Supply Code
- e: Air Force Systems Command
- e: Armed Forces Staff College
- e: Assessment of Fluency in School-Age Children

AFSCA e: Amalgamated Flying Saucers Clubs of America
AFSCC e: Air Force Satellite Control Center
- e: Air Force Security Communications Center
- e: Air Force Special Communications Center
- e: Armed Forces Supply Control Center

AFSC-DH e: Air Force Systems Command Design Handbook
AFSC/DL e: Air Force Systems Command Director of Laboratories
AFSCE e: American Fertility Society Classification of Endometriosis
AFSCF e: Air Force Satellite Control Facilities
- e: Air Force Systems Command Form

AFSCI e: American Foundation for the Science of Creative Intelligence
AFSCIC e: Air Force Systems Command Inspection Center
AFSCL e: Air Force Systems Command Letter
AFSCM e: Air Force Systems Command Manual
AFSCN e: Air Force Satellite Control Network
AFSCOORD e: Assistant Fire Support Coordinator
AFSCP e: Air Force Systems Command Pamphlet
- e: Air Force Systems Concept Paper

AFSCPP e: Air Force Systems Command Procurement Production
AFSCR e: Air Force Systems Command Regulation
AFSCS e: Air Force Satellite Communications System
- e: Army Field Stock Control System

AFSC/SSD e: Air Force Systems Command / Space Systems Division
AFSC/STLO e: Air Force Systems Command / Scientific Technical Liaison Office
AFSD d: automatische Fernschreib-Speichervermittlung beim Deutschen Wetterdienst
- e: Air Force Space Division
- e: Air Force Supply Date
- e: Air Force Supply Depot
- e: Air Force Supply Directive[s]

afsd e: aforesaid
AF/SD e: Air Force and Space Digest
AFSE e: Allied Forces Southern Europe
AFSEM e: Army Food Service Energy Management
AFSERT e: Associate Fellow of the Society of Electronic and Radio Technicians
AFSF e: Advanced Field Site Facility
- e: Air Force Stock Fund
- e: Air Force Supply Force

AFSFO e: Airways Facilities Sector Field Office
AFSI e: Architectural Fabric Structures Institute
AFSIC e: Air Force Surveys in Geophysics
AFSIE e: Air Force Standard Items and Equipment
AFSIG e: Ascent Flight Systems Integration Group
AF SIG GEN e: Audio-Frequency Signal Generator
AFSINC e: Air Force Service Information and News Center
AFSIP e: Air Force Standard Intelligence Publication
afsk e: audio-frequency shift keying
AFSK e: Audio Frequency Shift Keyed
- e: Audio Frequency Shift Keying

AFSL e: Approved Fastener Substitution List
AFSLAET e: Associate Fellow of the Society of Licensed Aircraft Engineers and Technologists
AFSM e: Air Force Screen Magazine
- e: Artillery Forces Simulation Model
- e: Association for Food Service Management
- e: Augmented Finite State Machine

AFSN e: Air Force Serial Number
- e: Air Force Service Number
- e: Air Force Stock Number

AFSO e: Aerial Fire Support Officer
- e: Air Force Service Office
- e: Airways Facilities Sector Office

AFSOB e: Air Force Special Operation Base
AFSOON e: Air Force Solar Observing Optical Network
AFSOUTH e: Allied Forces Southern Europe
AFSP e: Acute Fibrinoserous Pneumonia
- e: Air Force Space Plan
- e: Air Force Space Program[me]
- e: Air Force Standard Practice

AFSPCOMMCEN e: Air Force Special Communications Center
AFSPD e: Airfield Spare Parts Depot
- e: Air Force Systems Project Division

AFSR e: Advanced Foreign System Requirements
- e: Argonne Fast Source Reactor
- e: Armed Forces Screen Reports

AFSRAN e: Air Force Stock Record Account Number
AFSS e: Air Force Security Service
- e: Air Force Service Statement
- e: Automated Flight Service Station

AFSSA e: Army Financial Stock Summary Analysis
AFSSC e: Armed Forces Supply Support Center
AFSSD e: Air Force Space Systems Division
AFSSMET e: Air Force Special Staff Management Engineering Team
AFSSO e: Air Force Special Security Officer

AFSSOP e: Air Force Security Service Office of Production
AFSS System e: Air Force Supply Services System
AFST e: Assured Field Shop Task
- e: Auxiliary Feedwater Storage Tank
AFSTC e: Air Force Satellite Test Center
- e: Air Force Space Technology Center
- e: Air Force Space Test Center
- e: Army Foreign Science and Technology Center
- e: Army Foreign Science Technical Center
AFSTRIKE e: Air Force Strike Command
AFSTS e: Air Force Space Technology Satellite
AFSU e: Auxiliary Ferry Service Unit
AFSUB e: [Army] Air Forces Anti-Submarine Command
AFSV d: automatische Fernmelde- und Speichervermittlung
AF Sv Comd e: Air Force Service Command
AFSWA e: Armed Forces Special Weapons Agency
AFSWC e: Air Force Special Weapons Center
AFSWP e: Air Force Special Weapons Project
- e: Armed Forces Special Weapons Project
AFT d: Antifibrinolysintest
- e: Acceptance Functional Test
- e: Acetate Film Tape
- e: Active File Table
- e: Adaptive Ferroelectric Transformer
- e: Adiabatic Flame Temperature
- e: Advance Feed Tape
- e: Aerodynamic Flight Test
- e: Aeronautical Fixed Telecommunication[s]
- e: Aflatoxin
- e: After[noon]
- e: Air Freight Terminal
- e: American Federation of Teachers
- e: Analog[ue] Facility Terminal
- e: Animal-Facilitated Therapy
- e: Annual Field Training
- e: Annual Financial Target
- e: Anterior Fold from Typhlosole
- e: Assembly Facility Tool
- e: Asynchronous Framing Technique
- e: Atmospheric Flight Test
- e: Audio-Frequency Transformer
- e: Audio-Frequency Transmitter
- e: Auditor Freight Traffic
- e: Autogenic Feedback Training
- e: Automated Funds Transfer
- e: Automatic Fine Tuning
- e: Automatic Flight Termination
- e: Automatic Frequency Tuner
- e: Automatic Fund Transfer
AFTA e: Acoustic Fatigue Test Article
- e: Advanced First-Term Avionics
- e: Aft Frame Tilt Actuator
- e: Atlantic Free Trade Area
- e: Automated Fault Tree Analyzer
- e: Avionics Fault Tree Analyzer
AFTAAS e: Advanced Fast Time Acoustic Analysis System
AFTAC e: Air Force Tactical Air Command
- e: Air Force Tactical Center
- e: Air Force Technical Applications Center
AFTAT e: Air Force Technical Approval Team
AFTB e: Afterburner
- e: Air Force Test Base
AFTBC e: Air Flow Thermal Balance Calorimeter
AFTC e: Airborne Flight Training Command
- e: Apparent Free Testosterone Concentration
AFTCC e: Air Force Troop Carrier Command
AFTCLR e: After Cooler
AFTCom e: Associate of the Faculty of Teachers in Commerce
AFTD e: Air Force Test Director
- e: American Foreign Trade Definitions
AFTDS e: Automated Flight Test Data System
AFTE e: American Federation of Technical Engineers
- e: Arab Federation for Technical Education
- e: Authority for Tooling Expenditures
AFTEC e: Air Force Flight Test Center
- e: Air Force Test and Evaluation Center
AFTER e: Automatic Functional Test and Evaluation Routine
AFTEX e: Australia France Technical Exchange
AFTF e: Air Force Task Force
AFTFWC e: Air Force Tactical Fighter Weapons Center
AFTH e: Air Force Technical Headquarters
AFTI e: Advanced Fighter Technology Integration
- e: American Film Technologies, Incorporated
AFTIA e: Armed Forces Technical Information Agency
AFTIC e: Air Force Technical Intelligence Center
AFTJ e: Airborne Fuze Test Jammer
AFTM e: Additive Full-Time Manning
- e: American Foundation for Tropical Medicine
- e: Assistant Freight Traffic Manager
AFTMA e: American Fishing Tackle Manufacturers Association
AFTN e: Aeronautical Fixed Telecommunication[s] Network
aftn e: afternoon
AFTN e: Autonomously Functioning Thyroid Nodule
AFTN-Programm d: Auffüllung der technologischen Normale-Programm
AFTO e: Air Force Technical Order
afto s: afecto
AFTOD e: Air Force Technical Objectives Documents
AFTOSB e: Air Force Technical Order Standardization Board
A/FTP e: Acceptance Functional Test Procedure
AFTP e: Additional Flight Training Period
- e: Advanced Fault Tree Analysis Program[me]
- e: Aircrew Flight Training Period
AFTPS e: Air Force Test Pilot School
AFTR e: Air Force Technical Report
- e: American Federal Tax Reports (journ.)
- e: Army Flying Time Report
AFTRA e: American Federation of Television and Radio Artists
AF Transmitter e: Audio Frequency Transmitter
AFTRC e: Air Force Technical Training Command
- e: Air Force Training Center
- e: Air Force Training Command
AFTRCC e: Aerospace Flight Test Radio Coordinating Council
AFTRRC e: Animal Feed and Tissue Residue Research Center
AFTRS e: Australian Film, Television and Radio School
AFTS e: Adaptive Flight Training System
- e: Aeronautical Fixed Telecommunications Service
- e: Airborne Flight Test System
- e: Air Force Tactical Shelter
- e: Aseptic Fluid Transfer System
- e: Australian Film and Television School
- e: Australian Flying Training School
- e: Automated Funds Transfer System
- e: Automatic Flexible Test Station
- e: Automatic Frequency Tone Shift
AFTSC e: Air Force Technical Service Command
AFTTC e: Air Force Technical Training Center
AFTTH e: Air Force Technical Training Headquarters
Aft Tot e: After Total
AFTU d: Archiv für Reformationsgeschichte. Texte und Untersuchungen (journ.)
- e: Air Force Test Unit
- e: Arizona Federation of Teacher Unions
AFTV e: American Forces Television
- e: Armed Forces Television
AFTWDS e: Afterwards
AFU d: Analog/Frequenz-Umsetzer
- e: Active Filter Unit
- e: Advanced Flying Unit
- e: Assault Fire Unit
- e: Autonomous Functional Unit
- e: Auxiliary Functional Unit
A.Fu. d: Agentenfunk
AfU d: Arbeitsgemeinschaft für Unfallverhütung
- d: Ausführungsbehörde für Unfallversicherung
AFUD e: American Foundation for Urologic Diseases
AFUE e: Annual Fuel Utilization Efficiency
AFuG d: Amateurfunkgesetz
- d: Gesetz über den Amateurfunk
AFULE e: Australian Federated Union of Locomotive Enginemen
AFUNTS e: Air Force Undergraduate Navigator Training System
AFUPO e: Air Force Unit Post Office
AFUR e: Amplified Failure or Unsatisfactory Report
AFUS e: Air Force of the United States
- e: Armed Forces of the United States
AFuVK d: Außerordentliche Funkverwaltungskonferenz
AFUW e: Australian Federation of University Women
AFV d: Auslandsfleischbeschauverordnung
- e: Aerospace Flight Vehicle
- e: Afluidal Variant
- e: Anti-Flood Valve
- e: Armo[u]red Fighting Vehicle
- e: Armo[u]red Force Vehicle
- e: Armored Family of Vehicles
AfV d: Amt für Verteidigungslasten
- d: Anzeigenblatt für Verkehr
- d: Ausschuß für Verkehr
AFVA e: Air Force Visual Aids
- e: American Film and Video Association
AFVB d: Allgemeine Feuerversicherungsbedingungen
AFVCS e: Automatic Fingerprint Verification Computer System
AFVG e: Anglo-French Variable Geometry
AFVG Aircraft e: Anglo-French Variable Geometry Aircraft

AFVG Fighter e: Anglo-French Variable Geometry Fighter
AFVN e: American Forces Vietnam Network
AFVPA e: Advertising Film and Videotape Producers Association
AFW d: Akademie für Führungskräfte der Wirtschaft
- d: Arbeitskreis zur Förderung dekorativer Wirtschaftswerbung
- e: Air Foirce With Army
- e: Air Force Weapon
- e: Air Force Wing
- e: Army Field Workshop
- e: Auxiliary Feedwater
- e: Auxiliary Fresh Water
- e: Axial Flow Wheel
AfW d: Amt für Wasserwirtschaft
- d: Arbeitsgemeinschaft für Weltraumforschung
AFWA e: Air Force with Army
AFWAB e: Army Fixed Wing Aptitude Battery
AFWAL e: Air Force Wright Aeronautical Laboratories
AFWAL/ML e: Air Force Wright Aeronautical Laboratories Materials Laboratory
AFWAR e: Air Force with Army
AFWAS e: Auxiliary Feedwater Actuating System
AFWC e: Auxiliary Feedwater Control
AFWE e: Air Forces; Western Europe
AFWESPAC e: Army Forces; Western Pacific
AFWET e: Air Force Weapons Effectiveness Testing
AFWETS e: Air Forces Weapons Effectiveness Testing System
AFWFA e: Australian Fresh Water Fishermens Assembly
AFWL e: Air Force Weapon[s] Laboratory
AFWMAA e: Air Force Wide Mission Area Analysis
AFWN e: Air Force With Navy
AFWOFS e: Air Force Weather Observing and Forecasting System
AFWR e: Approved Force War Reserve
- e: Atlantic Fleet Weapons Range
AFWRT e: Air Force Western Test Range
AFwS d: Auslandsfernwählsystem
AFWS e: Advanced Filament Wound Structure
- e: Auxiliary Feedwater System
AFWST e: Armed Forces Womens Selection Test
AFWTF e: Atlantic Fleet Weapons Training Facility
AFWTR e: Air Force Western Test Range
AFWTRM e: Air Force Western Test Range Manual
AFWWMCCS e: Air Force World Wide Military Command and Control System
AFXF e: Advanced Flash X-ray Facility
AFY e: Air Facility
AFYDP e: Armys Five-Year Defense Program[me]
AFZ d: Allgemeine Fischwirtschafts-Zeitung
- d: Allgemeines Flugfunksprechzeugnis
- d: Arbeits- und Forschungsgemeinschaft "Graf Zeppelin"
- d: Art des Fernmeldezeugs
- d: Auffassungszone
- d: Aufzeichnung
- e: Anatolian Fault Zone
A.F.Z. e: Aircraft Flighting Zone
AFZIN e: Air Force Zone of Interior
AG d: Aargau (Kfz)
- d: Abgas
- d: Abwehrgruppe
- d: Aktiengesellschaft
- d: Aktionsgemeinschaft
- d: Allgemeine Genehmigung
- d: Altersgeld
- d: Amt gehend belegen
- d: Amtsgehilfe
- d: Amtsgericht
- d: Anatomische Gesellschaft
- d: Anschlußgerät
- d: Anstellungsgrundsätze
- d: Anti-Globulin
- d: Anwaltsgericht
- d: Appellationsgericht
- d: Arbeitgeber
- d: Arbeitsgang
- d: Arbeitsgebiet
- d: Arbeitsgegenstand
- d: Arbeitsgemeinschaft
- d: Arbeitsgericht
- d: Arbeitsgrundlage
- d: Arbeitsgruppe
- d: Aschegehalt
- d: Astronomische Gesellschaft
- d: Atomgewicht
- d: Aufbaugesetz
- d: Aufklärungsgruppe
- d: Auftraggeber
- d: Aufwertungsgesetz
- d: Ausfuhrgenehmigung
- d: Ausführungsgesetz
- d: Ausgabe[gerät]
- d: Automatengraph
- d: Automatisierungsgeräte
- d: autonomes Gebiet
- d: Axialglas
- d: Vereinigung Volkseigener Betriebe Automatisierungsgeräte (DDR)
- e: Acceptance Gauge
- e: Accessory Gland
- e: Accountant General
- e: Acoustic Generator
- e: Acting
- e: Adder Gate
- e: Address Generator
- e: Address Group
- e: Adjutant General
- e: Advanced Guard
- e: Advisory Group
- e: Aerographers Mate
- e: Aeronautical Standards Group
- e: Aerospace Group
- e: Again[st]
- e: Against Grain
- e: Agency
- e: Agent General
- e: Aggressive Growth
- e: Agiate
- e: Air Ga[u]ge
- e: Air Gap
- e: Air Group
- e: Air Gunner
- e: Alignment Group
- e: Allegheny Ludlum Steel
- e: Alternate Gradient
- e: Alternating Gradient
- e: American [Wire] Gage
- e: Aminoguanosine
- e: Analytical Grade
- e: Anastomosis Group
- e: And Gate
- e: Annual Goal
- e: Anti-Gas
- e: Anti-Gentilism
- e: Antigen
- e: Antigravity
- e: Armo[u]r Grating
- e: Army Group
- e: Army Guidance
- e: Arresting Gear
- e: Artificial Gravity
- e: Assault Gun
- e: Association Graph
- e: Atlas Gemini
- e: Atrial Gallop
- e: Attitude Gyro
- e: Attorney-General
- e: Attribute Grammer
- e: Auditor General
- e: Automatic Gauge
- e: Availability Guarantee
- e: Axiogingival
- e: sonar research ship
a.G. d: ab Grenze
Ag. d: Agent
- d: Agentschaft
- d: Agentur
- d: Antigen[e]
Ag d: Alfagras
- d: Aufzuchtergebnis
- e: Agreement
- e: Agricultural
- e: Argentum
- l: argentum
A.G. d: als Gast
- d: Arbeitsgeschütz
- d: Artilleriegeschoß
a. G. d: altes Gesetz
a. g. d: anderweitig genannt
aG d: auf Gegenseitigkeit
ag. d: ausgelaufen
- e: Agreed
ag e: agitate
- e: agribusiness
- e: agriculture
- e: agrobiology
- e: agroindustrial
- e: agrology
- e: agronomy
- e: alternate geologies
A/G e: Airgraph
- e: Albumin / Globulin
a/g e: air-to-ground
a-g e: air-to-ground
- e: anti-gas
a g f: à gauche
Äg. d: Ägypten
- d: Ägypter
äg. d: ägyptisch
ÄG d: Änderungsgesetz
AG1 e: Aerographers Mate, First Class
AG2 e: Aerographers Mate, Second Class
AG3 e: Aerographers Mate, Third Class
AGA d: Allgemeine Geschäftsbedingungen für das Anzeigenwesen
- d: Arbeitgeberanteil
- d: Arbeitgeberverband Groß- und Außenhandel
- e: Accredited Gemologists Association
- e: Aceglutamide Alumin[i]um
- e: Adjutant General of the Army
- e: Aerodromes and Ground Aids
- e: Agricultural Administration
- e: Air-to-Ground-to-Air
- e: Alabama Gas
- e: Alterable Gate Array
- e: Amalgamated Gas Accumulation
- e: American Galvanizers Association
- e: American Gas Association
- e: American Gastroenterological Association

- e: American Gastroscopic Association
- e: American Genetic Association
- e: American Glassware Association
- e: American Goiter Association
- e: American Gold Association
- e: American Graniteware Association
- e: American Guppy Association
- e: Antigliadin Antibodies
- e: Apple Growers Association
- e: Appropriate for Gestational Age
- e: Associated Geographers of America
- e: Attitude Gyro Assembly
- e: Australian Gas Association
- e: Automated Genetic Analyzer
- e: Automatic Gas Analyzer
- e: Average Global Automobile
- e: Azimuth Gimbal Assembly

AgA d: Ausschuß für gefährliche Arbeitsstoffe
aga e: accelerated growth area
a-g-a e: air-to-ground-to air
AGAA e: Attitude Gyro Accelerometer Assembly
- e: Automatic Gain Adjusting Amplifier

Ag-AB e: Antigen-Antibody
AG-AbfG d: Ausführungsgesetz zum Abfallbeseitigungsgesetz
AGACS e: Automatic Ground-Air[craft] Communication[s] System
AGA-CS e: Air-Ground-Air Communications System
AGADS e: Advanced Graphics Avionics Display System
AGAF d: Arbeitsgemeinschaft für Ausbildung und Fortbildung
A-G AF e: Adjutant General; Air Force
AG-AFE d: Arbeitsgemeinschaft der Aus- und Fortbildungseinrichtungen im Informations- und Dokumentationswesen
AGAG e: Acidic Glycoaminoglycan
AGAL e: American Gas Association Laboratories
AGA Mon e: American Gas Association. Monthly
AGAMP e: Automatic Gain Adjusting Amplifier
AGANI e: Apollo Guidance and Navigation Information
AGAP e: Attitude Gyro Accelerometer Package
- e: Automated Graphics Application Program[me]

AGARD e: Advisory Group for Aeronautical Research and Development
- e: Advisory Group for Aerospace Research and Development

AGAS e: Aviation Gasoline
AGASS e: Automated Geomagnetic Airborne Survey System
AGATE e: Air-to-Ground Acquisition and Tracking Equipment
AGA Timber e: Agarwood Timber
AgaV d: Arbeitsgemeinschaft alternativer Verlage und Autoren
AGAVE e: Automatic Gimballed- Antenna Vectoring Equipment
AGB d: Allgemeine Geschäftsbedingungen
- d: Amerika-Gedenkbibliothek
- d: Amtsgerichtsbezirk
- d: Anhaltische Geschichtsblätter (journ.)
- d: Arbeitsgesetzbuch (DDR)
- e: Accessory Gear Box
- e: Advanced Geometry Blade
- e: American Glycerin Bomb
- e: Angora Goat Breeders
- e: Assault Gun Battalion
- e: Association of German Broadcasters
- e: Association of Governing Boards of Universities and Colleges
- e: Asymptotic Giant Branch
- e: Audits of Great Britain

agb e: any good brand
a.g.b. e: armo[u]red gunboat
AGBAD e: Alexander Graham Bell Association for the Deaf
AGBGB d: Ausführungsgesetz zum Bürgerlichen Gesetzbuch
agbio e: agrobiology
AGBM e: Antiglomerular Basement Antibody Test
agbm e: antiglomerular basement membrane antibody
AGBS e: Artillery Ground Burst Simulator
AGBUC e: Association of Governing Boards of Universities and Colleges
AGBUS e: Analog[ue] Ground Bus
agbus e: agribusiness
AGBz d: Amtsgerichtsbezirk
AGC d: Anhydroglucochloral
- e: Adaptive Gain Computer
- e: Advanced Gas Centrifuge
- e: Advanced Graduate Certificate
- e: Aerojet-General Corporation
- e: African Groundnut Council
- e: Air-Ground Chart
- e: Air-Ground Communications
- e: Air-to-Ground Code
- e: Air-to-Ground Communication[s]
- e: Air Ground Chart
- e: American Golden Cross
- e: American Grassland Council
- e: Amphibious Group Command
- e: Amplitude Gain Control
- e: Ancient Gneiss Complex
- e: Apollo Guidance Computer
- e: Armed Guard Center
- e: Asrtonomical Great Circle
- e: Assessment Guidance Centre
- e: Astronomical Great Circle
- e: Automatic Ga[u]ge Control
- e: Automatic Gain Control
- e: Automatic Gain Correction
- e: Automatic Generation Control
- e: Avocado Growers Council

A/GC e: Acting Group Captain
AGCA e: Altitude Gyroscope Control Assembly
- e: Automatic Ground-Control[led] Approach

AGCAP e: Automated Generic Case Analysis Program[me]
AGCC e: Air-Ground Communications Channel
- e: Airborne and Ground Communications Central
- e: American Guernsey Cattle Club
- e: Automatic Gain Control Calibration

AGC/CFAR e: Automatic Gain Control/ Constant False Alarm Rate
AGCD e: Association of Green Crop Driers
AGCE e: Atmosphere General Circulation Experiment
AGCF e: Air-Ground Correlation Factor
AGCG e: Associated Granite Craftsmens Guild
AG Chem-Dok. d: Arbeitsgemeinschaft Chemie-Dokumentation
AGCI e: Australian Genealogical Computer Index
- e: Automatic Ground-Controlled Intercept[ion]

AGCIC e: Association of German Chambers of Industry and Commerce
AGC-ING e: Automatic Gain Controlling
agcl e: automatic ground-controlled landing
AGCM e: Army Good Conduct Medal
- e: Atmospheric General Circulation Model

AgCN e: Silver Cyanide
AGCO e: Air-Ground Cooperation Officer
agcol e: agricultural college
AGCP e: Automatic Gain Calibration Program[me]
AGCRSP e: Army Gas-Cooled Reactor Systems Program[me]
AGCS e: Advanced Guidance and Control System
- e: Automatic Ground Checkout System
- e: Automatic Ground Computer System
- e: Automatic Ground Control Station

AGCSC e: Automatic Ground Control System Computer
AGCT e: Antiglobin Consumption Test
- e: Army General Classification Test

AGCTS e: Armed Guard Center Training School
AGCU e: Air-Ground Cooling Unit
- e: Attitude Gyro Coupling Unit
- e: Autopilot Ground Control Unit

AGCW e: Autonomous Guidance for Conventional Weapons
AGCY e: Agency
AGD d: Amtsgerichtsdirektor
- e: Academy of General Dentistry
- e: Adjutant Generals Department
- e: Agar Gel Diffusion
- e: Aircraft Gunfire Detector
- e: Alaskan General Depot
- e: American Gage Design
- e: Associated Gas Distributors
- e: Atlanta General Depot
- e: Attack Geometry Display
- e: Australian Geodetic Datum
- e: Axial Gear Differential

agd e: agreed
AGDA e: American Gasoline Dealers Association
- e: American Gun Dealers Association

AGDBT d: Arbeitsgemeinschaft der Datenbankteilnehmer
AGDC e: American Gauge Design Committee
AGD/CSD e: Axial Gear Differential / Constant-Speed Drive
AgDDC e: Silver Diethyldithiocarbonate
Ag. Dei l: Agnus Dei
Ag Dept e: Agriculture Department
AGDFV d: Arbeitsgruppe Datenfernverarbeitung
AGDIC e: Astro Guidance Digital Computer
Ag Digest e: Agricultural Digest
AGDir d: Amtsgerichtsdirektor
AGDL d: Arbeitsgemeinschaft Deutscher Lehrerverbände
A-GDL e: Army Gas Dynamic Laser
AGDLS e: Air-to-Ground Data Link System
AGDMilNW d: Arbeitsgruppe zur Herstellung und Aufrechterhaltung der Datenverarbeitungskompatibilität
AGDOC e: Agricultural Documentation
AG/DR e: Assistant Gunner / Driver
AGDS e: American Gage Design Standard
- e: deep submergence support ship

AGDT e: Advisory Group on Data Transmission
- e: Agar Gel Immunodiffusion Test

AGE d: Arbeitsgemeinschaft Energie
- d: Arbeitsgemeinschaft Entwicklungsländer
- d: Ausgleichseinrichtung
- e: Admiralty Gunnery Establishment
- e: Adult Growth Examination
- e: Advanced Glycosylated End-product
- e: Advisory Group on Energy
- e: Aerospace Ground-Support Equipment
- e: Affiliated Gas Equipment
- e: Agarose Gel-zone Electrophoresis
- e: Air-Ground Equipment
- e: Allyl Glycidyl Ether
- e: Amarillo Grain Exchange
- e: Angle of Greatest Extension
- e: Apollo Guidance Equipment
- e: Asian Geotechnical Engineering
- e: Asian Information Centre for Geotechnical Engineering
- e: Assistant Garrison Engineer
- e: Associated Ground Equipment
- e: Associate in General Education
- e: Auditory Gross Error
- e: Automatic Ground Equipment
- e: Automatic Guidance Electronics
- e: Auxiliary Ground Equipment

age. e: aerospace ground equipment
Ag.E. e: Agricultural Engineer
AgE e: Antigen E
AGEAS e: Automatic Ground Effect Augmentation System
AGEC e: Army General Equipment Command
- e: Automatic Ground Environment Computer

AGED e: Advisory Group on Electronic Devices
- e: Aerospace Ground Equipment Department
- e: Army Group Effects Department

AGEd e: Associate in General Education
AGED e: Automated General Experimental Device
AGEH d: Arbeitsgemeinschaft für Entwicklungshilfe
AGEI e: Aerospace Ground Equipment Installation
AGEN d: Arbeitsgemeinschaft für elektrische Nachrichtentechnik
AGEn d: Ausgleichseinrichtungen
Ag Eng e: Agricultural Engineering
AGENTEFO d: automatisches Generieren von Tests und Fehler-Ortungshilfen
AGEOCP e: Aerospace Ground Equipment out of Commission for Parts
AGEP e: Advisory Group on Electronic Parts
AGEPC e: Acetyl-Glyceryl-Ether Phosphorylcholine
AGERD e: Aerospace Ground Equipment Recommendation Data
- e: Aerospace Ground Equipment Requirements Data

agerm. d: altgermanisch
AGERS e: Auxiliary General Electronics Research Ship[s]
AGES d: Aktionsgemeinschaft Greifvogel- und Eulenschutz
- e: Air-Ground Engagement Simulation
- e: Air-to-Ground Engagement System
- e: Aircrew Gliding Escape System

AG & ES e: American Gas & Electric System
AGES e: American Gas and Electric Services
AGE/SE e: Aerospace Ground Equipment / Support Equipment
AGET e: Advisory Group on Electronic Tubes
AGETS e: Automated Ground Engine Test System
A.-Gew. d: Atomgewicht
AGF d: Arbeitsgemeinschaft der Großforschungseinrichtungen
- d: Arbeitsgemeinschaft Feuerschutz
- d: Arbeitsgemeinschaft für Forschung des Landes Nordrhein-Westfalen
- d: Arbeitsgemeinschaft Futter und Fütterung
- d: Arbeitsgemeinschaft Getreideforschung
- d: Arbeitsgemeinschaft Güterfernverkehr im Bundesgebiet
- e: Adjutant General to the Forces
- e: Adrenal Growth Factor
- e: Alternating Gradient Focusing
- e: Angle of Greatest Flexion
- e: Army Ground Forces
- e: Automatic Guided Flight
- e: Aviation Guided Flight

AGFA d: Aktiengesellschaft für Anilinfabrikation
- e: Assistant General Freight Agent

AGFC e: Automatic Gain and Frequency Control
AGFCS e: Automatic Gunfire Control System
AGFF d: Arbeitsgemeinschaft zur Förderung des Futterbaues
AGFIZ d: Arbeitsgemeinschaft der Fachinformationszentren
AGFL e: Airborne Ground Fire Locator
AGFLS e: Airborne Ground Fire Location System
AGFR e: Automatic Gain and Frequency Response
AGFRTS e: Air and Ground Forces Resources and Technical Staff
AGFSR e: Aircraft Ground Fire Suppression and Rescue
AGFSRS e: Aircraft Ground Fire Suppression and Rescue Systems
AGFW e: Arbeitsgemeinschaft Fernwärme e.V.
AGG d: Arbeitsgerichtsgesetz
- e: Agammaglobulinemia

agg e: agammaglobulinaemia[c]
agg e: agglutination[ed]
- e: aggravate[ed]
- e: aggregate[d]
- e: aggregation

AGGAFT e: American Grape Growers Alliance for Fair Trade
AGGD e: Apollo Guidance Ground Display
aggie e: agriculture
aggies e: agate playing marbles students of agricultural colleges or schools
aggl e: agglutinated
AGGL e: Agglutination
agglut e: agglutinated
AGGLUT e: Agglutination
AGGMR e: Advanced Gas-cooled Graphite Moderated Reactor
AGGR e: Air-to-Ground Gunnery Range
Aggr. d: Aggregat
aggr e: aggregate
aggred. feb. l: aggrediente febre
AGGREG e: Aggregation
AGGS e: Antigas Gangrene Serum
AGH e: Addison Gilbert Hospital, Gloucester, Mass.
- e: American Guild of Hypnotherapists
- e: Arc Gas Heater
- e: Army Group Headquarters
- e: Australian General Hospital

AGHA d: Archiv für Geschichte des Hochstifts Augsburg (journ.)
AGHDEA e: Association of General Heating and Domestic Engineer Assistants
AGHE e: Association for Gerontology in Higher Education
- e: Association for Gerontology in Higher Education agi adjusted gross income

AGI d: Adolf-Grimme-Institut
- e: Adjusted Gross Income
- e: Adjutant General Inspection
- e: Agreement Item
- e: Air Gunnery Instructor
- e: American Gas Institute
- e: American Geographical Institute
- e: American Geological Institute
- e: Annual General Inspection
- e: Associate of the Greek Institute
- e: Australian Grain Institute

A/GI e: Anti-Gas Instructor
AGIC e: Air-Ground Information Center
- e: Andrus Gerontological Information Center
- e: Auto Glass Industry Council
- e: Automatically-Generated Integrated Circuit

AGICHS e: Auto Glass Industry Committee for Highway Safety
AGID e: Association of Geoscientists for International Development
AGIFORS e: Airline[s] Group of International Federation of Operational Research Societies
AG i. L. d: Aktiengesellschaft in Liquidation
AGIL e: Airborne General Illumination Light
- e: American Gum Importers Laboratories
- e: Analytic Geometry Interpretative Language

AGILE e: Analytic Geometry Interpretive Language
- e: Auto-Graphics Interactive Library Exchange
- e: Autonetics General Information Learning Equipment

ag imps hnd e: agricultural implements hand
ag imps ot hand e: agricultural implements other than hand
agind e: agroindustrial
Aging e: American Association of Hospital Accountants
AGIO e: Armed Guard Inspection Officer
AGIP i: Azienda Generale Italiana Petroli
AGIPA e: Adaptive Ground- Implemented Phased Array
AGIPAC e: Agie (A.G. für Industrielle Elektronik) Interactive Programming and Calculation[
- e: Agie Interactive Programming and Calculation

AGIS d: automatisches Gefechts-informationssystem
- d: automatisiertes Gefechts- und Informationssystem
- d: automatisiertes Gefechts- und Informationssystem [auf] Schnellbooten
- e: Acoustographic Imaging System
- e: Agricultural Information System
- e: Agricultural Statistical Information System
- e: Air-Ground Integration System
- e: Armed Guard Inspection System

agisra d: Arbeitsgemeinschaft gegen internationale sexuelle und rassistische Ausbeutung
agit e: agitatum
agit. l: agita
agit. a. us. l: agita ante usum

agit. bene l: agita bene
agit. vas. l: agitato vase
AGK d: Arbeitsgemeinschaft Kälteindustrie
AGKO e: Argenine, Glutamate a-Ketoglutorate Oxalacetate
AGKT d: Antiglobin-Konsumptionstest
- d: Arbeitsgemeinschaft Kritische Tiermedizin
AGL d: Arbeitsgemeinschaft Luftfahrtausrüstung
- d: Arbeitsgutlage
- d: automatisierte Gruppen-Lerndynamik
- e: Above Ground Level
- e: Absolute Ground Level
- e: Acute Granulocytic Leukemia
- e: Agricultural
- e: Argon Gas Laser
- e: Argon Glow Lamp
- e: Australian Gas Light
- e: Automated Group Learning
- e: Automatic Gun-Laying
- e: Automatic Gun Layer
a.g.L. d: akademisch geprüfter Landwirt
agl e: airborne gun laying
- e: aminoglutethimide
Agla d: Ausgleichsamt
AGLC e: Air-to-Ground Liaison Code
aglfrz. d: anglofranzösisch
AGLINE e: Agricultural Online
AGLINET e: Agricultural Libraries Information Network
- e: Agricultural Library and Information Network
- e: Agricultural Library Networks
aglm e: agglomerate
AGLMB d: Arbeitsgemeinschaft der Leitenden Medizinalbeamten
AGLN e: Asian Grain Legumes Network
a.gl.O. d: am gleichen Ort
AGLO e: Air Ground Liaison Officer
AGLR e: Airborne Gun Laying Radar
AGLS e: Association for General and Liberal Studies
AGLSP e: Association of Graduate Liberal Studies Programs
AGLT e: Acidified Glycerol Lysis Test
- e: Airborne Gun-Laying [for] Turrets
- e: Automatic Gun-Laying Turrets
AGL Timber e: Aglaia Timber
A-Glue e: Airplane Glue
AGM d: Ablenkgeschwindigkeits-modulation
- d: angeborener gestaltbildender Mechanismus
- e: Admiralty General Message
- e: Advanced Glass Melter
- e: Air-to-Ground Missile
- e: Alternative Generator Model
- e: American Green Movement
- e: Annual General Meeting
- e: Assistant General Manager
- e: Auxiliary General Missile
Agm. d: Arbeitsgemeinschaft
AGM-53A e: North AmericanRockwell Condor air-to-surface missile
AGMA e: American Gear Manufacturers Association
- e: Assistant General Manager for Administration
- e: Athletic Goods Manufacturers Association
AGMC e: Aerospace Guidance and Meteorology Center
AGMEPS e: Advisory Group on Management of Electronic Parts Specifications
AGMF e: Agents Master File
AGMIA e: Assistant General Manager for International Activities
AGMIS e: Adjutant General Management Information System
AGMIV e: African Green Monkey Immunodeficiency Virus
AGMK e: African Green Monkey Kidney
AGMM d: Arbeitsgemeinschaft für Musikpflege und Musikerziehung
AGMO e: Assistant General Manager for Operations
AGMPP e: Assistant General Manager for Plans and Production
AGMR e: Advanced Gas-Cooled Maritime Reactor
AGMRD e: Assistant General Manager for Research and Development
AGMS e: Aircraft Ground Mobility System
AGMT e: Agreement
- e: Augment
AGMTI e: Air-to-Ground Moving Target Indicator
AGμC d: Arbeitsgemeinschaft Mikrocomputer
AGMV e: Agropyron Mosaic Virus
AGN d: Anzeiger des Germanischen Nationalmuseums (journ.)
- d: Arbeitsgemeinschaft Güternahverkehr im Bundesgebiet
- d: Arbeitsgemeinschaft Normenpraxis
- e: Active Galactic Nucleus
- e: Acute Glomerulonephritis
- e: Additive Gaussian Noise
- e: Aerojet-General Nucleonics
- e: Aerojet General Nucleonics
- e: Again
- e: Agnosia
- e: Apollo Guidance and Navigation
- e: Applied Genetics News
- e: Augmentation
agn e: active galactic nuclei
- e: agnomen
Agñ e: Agaña, Guam
AGNBC e: Advisory Group on National Bibliographic Control
AGNCY e: Agency
AGNIS e: Apollo Guidance and Navigation Industrial Support
AGNM d: Anzeiger des Germanischen Nationalmuseums (journ.)
AGNOS e: Agnostic
agnos e: agnosticism
- g: agnostos
AGNPP e: Above-Ground Net Primary Production
AGNQ e: Apollo / Ground Operations Support System Navigation Qualification
AGNS e: Allied General Nuclear Services
- e: Allied Gulf Nuclear Services
AGNST e: Against
AGNVH e: Association of Growers of the New Varieties of Hops
AGO d: Arbeitsgemeinschaft des Omnibusgewerbes
- d: Ausschuß für die Gebührenordnung
- e: Air Gunnery Officer
- e: Antigas Officer
- e: Army General Orders
- e: Arresting Gear Officer
- e: Atmospheric Gas Oil
a.g.O. d: am gegebenen Ort
AGODDS e: Auxiliary General Oceanographic Research Oceanographic Digital Data Syste
AGOE e: Advisory Group for Ocean Engineering
AGOES e: Advanced Geosynchronous Observation Environment Satellite
AGÖF d: Arbeitsgemeinschaft Ökologischer Forschungsinstitute
AGOP e: Australian Government Offsets Program[me]
AGOR e: Auxiliary General Oceanographic Research
AGOR Vessel e: Auxiliary General Oceanographic Research Vessel
AGOS e: Air-Ground Operations School
- e: Air-Ground Operations Section
- e: Air-Ground Operations System
- e: American Gynecological and Obstetrical Society
- e: Aviation Gunnery Officers School
AGOSP e: Advisory Group on Science Program[me]
AGOSS e: Automated Ground Operations Scheduling System
AGP d: Arbeitsgemeinschaft der Produktionsgenossenschaften
- d: Arbeitsgemeinschaft Personenverkehr
- d: Archiv für Geschichte der Philosophie (journ.)
- d: Armeegerätepark
- e: Acid Glycoprotein
- e: Adjutant Generals Pool
- e: Advanced Guided Projectile
- e: Aircraft Grounded for Parts
- e: Aircraft Gun Pod
- e: Algal Growth Potential
- e: Antisymmetrized Geminal Power
- e: Archtves of General Psychiatry
- e: Arctic Gas Profile
- e: Army Ground Pool
- e: Army Group
- e: Automatic Gang Punch
- e: Automatic Guidance Programming
- e: Auxiliary Generating Plant
- e: Aviation General Policy
agp e: above-ground pool
AGPA e: a-Aminoguanidinopropionic Acid
- e: American Group Practice Association
- e: American Group Psychotherapy Association
AGPB e: Advanced General Purpose Bomb
AGPC e: Adjutant General Publications Center
AGPCH e: Association of General Practitioner Community Hospitals
AGPDC - e: Aeronutronic General Perturbations Differential Correction Program[me]
agpe e: angle plate
AGPh d: Archiv für Geschichte der Philosophie (journ.)
AGPhD d: Allgemeine Gesellschaft für Philosophie in Deutschland
AGPI e: Automatic Ground Position Indicator
AGPL p: Administração-Geral do Porto de Lisboa
AGPLAN d: Arbeitsgemeinschaft für Planungsrechnung
- d: Arbeitsgemeinschaft Planung
AGPM e: Associated Glass and Pottery Manufacturers
AGPMR e: Agricultural Property Management Regulations
AGPPI e: American Grain Products Processing Institute
AGPR e: Agricultural Procurement Regulations
AGPrÄs. d: Amtsgerichtspräsident

AGPS e: Automatic Gun Positioning System
AGPT e: Agar-Gel Precipitation Test
AGP Test e: Algal Growth Potential Test
AGPTT e: Aerial Gunnery Part Task Trainer
AGPU e: Aviation Ground Power Unit
AGPUBLDEP e: Adjutant Generals Publications Depot
AG & QMG e: Adjutant-General and Quartermaster-General
AGR d: Alarmgruppenregister
- d: Amtsgerichtsrat
- e: Advanced Gas-Cooled Reactor
- e: Advanced Graphite Reactor
- e: Agree
- e: Air-to-Ground Ranging
- e: Air-to-Ground Rocket
- e: Alarm Group Register
- e: Albumin / Globulin Ratio
- e: American-German Review (journ.)
- e: Annual Growth Rate
- e: Anticipatory Goal Response
- e: Autonetics Generalized Reset
a.Gr. d: ab Grenze
AGr d: Ackergrünland
Agr d: Agrarier
- e: Agricultural[ist]
- e: Agriculture
agr. d: agrarisch
- d: agronomisch
- d: altgriechisch
- e: agronomic[al]
Agr. d: Agrikultur
- d: Agronom[ie]
- d: Agronomist
- e: Agronomy
AGr. d: Anstellungsgrundsätze
- d: Arbeitsgruppe
agr e: advanced gascooled graphite-moderated reactor
- e: agree[ment]
- e: agricultural
AGRA e: Army Group Royal Artillery
- e: Automatic Gain Ranging Amplifier
agrar e: agrarian[ism]
AgrarRefG d: Agrarreformgesetz
A/G Ratio e: Albumin / Globulin Ratio
Agra U e: Agra University
AgrBauR d: Agrarbaurat
AGRC e: American Graves Registration Command
AGRE e: Atlantic Gas Research Exchange
AGREE e: Advisory Group on [the] Reliability of Electronic Equipment
Agr Eng e: Agricultural Engineering
Ag Rep e: Agricultural Representative
AGREP e: Agricultural Research Project[s]
Ag Res e: Agricultural Research
AGRF e: American Geriatric Research Foundation
Agr Hist e: Agricultural History
agri e: agricultural[ist[
- e: agriculture
AGRI e: American Genealogical Research Institute
agribusiness e: agricultural business
agric e: agricultural
- e: agriculture
Agric Adm e: Agricultural Administration
Agric Asia e: Agriculture Asia
Agric Bull e: Agriculture Bulletin
Agric C e: Agricultural Code
AGRICC e: Automatic Generation and Retrieval of Information on Chemical Components
Agric. Chem. e: Agricultural Chemicals (journ.)
Agric. E. e: Agricultural Engineer
Agric Econ e: Agricultural Economist
Agric Educ e: Agricultural Education
Agric Eng e: Agricultural Engineering
Agric. Eng. e: Agricultured Engineering (journ.)
Agric. Engr. e: Agricultural Engineer
Agric Hist e: Agricultural History
Agric Index e: Agricultural Index
Agric. Mark. e: Agricultural Marketing (journ.)
Agric Met e: Agricultural Meteorology
AGRICOLA e: Agricultural On-Line Access
Agric Prog e: Agricultural Progress
Agric Rec e: Agricultural Record
Agric Res e: Agricultural Research
Agric Sci Dig e: Agricultural Science Digest
Agric Sci Rev e: Agricultural Science Review
Agric Syst e: Agricultural Systems
Agric Tech e: Agricultural Technologist
Agri Ind e: Agriculture Index
agri-indus e: agricultural-industrial [complex]
agrimech e: agriculture mechanized
agripower e: agricultural power
AGRIS d: Agrar-Informationssystem
- e: Agricultural [Research] Information System
AgRISTARS e: Agriculture and Resources Inventory Survey through Aerospace Remote Sensing
AGRL e: Agricultural
Agrl. E. e: Agricultural Engineer
AGRM e: Adjutant General, Royal Marines
- e: Agreement
AGRMT e: Agreement
agro a: agronomiae
- e: aggravation
- e: agrobiological
- e: agrobiology
- e: agrologic[al]
- e: agronomical
- e: agronomist
Agro. e: Agrobiologist
- e: Agronomy
Agro e: Agronomics
agrobio e: agrobiologic[al][ly]
- e: agrobiologist
- e: agrobiology
agrogeol e: agrogeology
agroind e: agroindustrial[ly]
- e: agroindustrialization
- e: agroindustrialize[r]
- e: agroindustry
AGROINFORM e: Agricultural Information
Agrom. e: Agronomy
AGRON e: Agronomy
Agron. J. e: Agronomy Journal
agros e: agrostology
agrost e: agrostology
Agr Progr e: Agricultural Progress
Agr Res e: Agricultural Research
AGRs e: Advanced Gas-cooled Reactors
AGRT e: Agreement
- e: Automatic Guard Receiver Terminal
AGRU d: Ansage geänderter Rufnummern
AGS d: Amt für Gewässerschutz
- d: Amtsgerichtssekretär
- d: Arbeitsgangstufe
- d: Arbeitsgemeinschaft der Selbständigen
- d: Arbeitsgemeinschaft Schallplatte
- d: Ausgangs-Gruppensucher
- d: Ausgangsgruppensucher
- d: automatische Grauwertstabilisierung
- e: Abort Guidance Section
- e: Abort Guidance System
- e: Abort Guidance System[s]
- e: Academic Guidance Service
- e: Acoustic Guidance Sonar
- e: Adreno-Genital-Syndrome
- e: Advanced Guidance System
- e: Aero Gun Sight
- e: Airborne[-radar] Gunsight
- e: Aircraft General Standards
- e: Aircraft General Stores
- e: Air Ground Section
- e: Air Gunnery School
- e: Allied Geographic Section
- e: Alternating Gradient Synchrotron
- e: Alternation Gradient Synchrotron
- e: American Geographical Society
- e: American Geriatrics Society
- e: American Goat Society
- e: American Gynecological Society
- e: Amoco Graphics System
- e: Appalachian Geological Society
- e: Applicon.Graphic-System
- e: Applied Graphics System[s]
- e: Armed Guard School
- e: Armo[u]red Gun System
- e: Army General Staff
- e: Army Guard School
- e: Army Guard Ship
- e: Artificial Gravity Structure
- e: Ascent Guidance-and-Control System
- e: Associate in General Studies
- e: Australian Geographic Society
- e: Australian Geomechanics Society
- e: Automated Graphic System
- e: Automatic Gain Stabilization
- e: Automatic Gain Stabilizer
ags. d: angelsächsisch
Ags e: aguascalientes
ags e: antigens
A.G.S. e: Associate in General Studies
Agsb d: Augsburg
AGSCC e: Army General Supplies Commodity Center
Ag Sci J e: Agricultural Science Journal
Ag Sci R e: Agricultural Science Review
AGSD e: Association for Glycogen Storage Disease
AGSE e: Aerospace Ground Support Equipment
- e: Aircraft Ground Support Equipment
AGSGG d: Ausführungsgesetz zum Sozialgerichtsgesetz
AGSI e: Advanced Genetic Sciences, Incorporated
- e: Automatic Government Source Inspection
AGSIDC e: Arab Gulf States Information Documentation Center
AGSIM e: American Graduate School of International Management
AGSM d: Ausgangsgruppensucher-Markierer
- e: Associate of the Guilhall School of Music
AGSP e: Alignment Group Sensing Platform
- e: Atlas General Survey Program[me]
AGSq e: Aircraft Generation Squadron
AGSR e: Advanced Ground Surveillance Radar
- e: All-Weather Ground Surveillance Radar
AGSRO e: Association of Government Supervisors and Radio Officers
AGS-RTO e: Automatic Ground Spoiler - Rejected Takeoff

AGSS e: American Geographical and Statistical Society
- e: Attitude Ground Support System
- e: Auxiliary Research Submarine
AGST e: Against
- e: Applied General System Theory
AGStPO d: Ausführungsgesetz zur Strafprozeßordnung
AGSW e: Air-to-Ground Stand-off Weapon
AGT d: Anschlußgerät Telegrafie
- d: Antriebs- und Getriebe-Technik (journ.)
- d: Ausrüstungen-Geräte-Typen
- e: Above Ground Test
- e: Additional Gunner Training
- e: Advanced Gas Turbine
- e: Advanced Ground Transport
- e: Advanced Guidance Technology
- e: Against
- e: Aircraft-derivative Gas Turbine
- e: Aircraft Gas & Turbine
- e: Alberta Government Telephone
- e: Alberta Government Telephones
- e: Allison Gas Turbine
- e: Antiglobulin Test
- e: Army Gunner Training
- e: Arresting Gear Tester
- e: Associate in Glass Technology
- e: Association of Geology Teachers
- e: Audiographic Teleconference
- e: Automated Graphic Technology
- e: Automated Guideway Transit
- e: Automatic Guideway Transit
- e: Aviation Gas Turbine
Agt. d: Agent
- d: Agentur
Agt d: Anschaltgerät
AGt d: Anschlußgerät
- d: Ausschuß / Gebrauchstauglichkeit
agt e: agreement
A/GT e: Antigas Training
AGTA e: Australian Geography Teachers Association
agtand[s] e: agricultural land[s]
AGTC e: Airport Ground Traffic Control
AGTELIS e: Army Ground Transportable Emitter Location Identification System
- e: Automated Ground Transportable Emitter Location and Identiftcation System
- e: Automatic Ground Transportable Emitter Location and Identification System
AGTELS e: Automated Ground Tactical Emitter Location System
AGTF e: Alternate Geology Test Facility
AGTH e: Adrenoglomerulotropic Hormone
AGTL e: Alberta Gas Trunkline Co. Ltd.; Calgary
agto s: agosto
AGTP e: Automatically Generated Test Analysis and Programs
Agtr. d: Agentur
AGTr e: Adrenoglomerulotrophin
AGTS e: Advanced Gunnery Target Systems
- e: Aerial Gunnery Target System
- e: Association for Gifted and Talented Students
- e: Automated Guideway Transit System
- e: Automated Gyro Test Set
AGTT e: Abnormal Glucose Tolerance Test
agtv e: advanced ground transportation vehicle
AGU d: Arbeitsgemeinschaft für Umweltfragen
- e: Address Generation Unit
- e: Adenine-Guanine-Uracil
- e: Aerospace Ground Unit
- e: All Got Up
- e: American Geophysical Union
- e: Anhydroglucose Unit
- e: Automatic Ground Unit
- e: Aviation Ground Unit
Aguacates e: Aguacate Mountains, Costa Rica
Aguas e: Aguascalientes
Agu Cur e: Agulhas Current
AGUW d: Ausbildungsgerät Unterwasserwaffen
AGUX e: Adenine-Guanine-Uracil-Xanthine
AGV d: Akademischer Gesangverein
- d: Angestelltenversicherungsgesetz
- d: Arbeitgeberverband
- d: Arbeitgebervereinigung
- d: Arbeitsgemeinschaft der Graphischen Verbände
- d: Arbeitsgemeinschaft Verpackung
- d: Ausgabeverstärker
- d: Ausgabeverteiler
- d: Auslandsfleischbeschaugebühren-Verordnung
- e: Alkali-Gravity-Viscosity
- e: Aniline Gentian Violet
- e: Automated Guided Vehicle
- e: Automatic Guided Vehicle
AGVC e: Automatic Governing Valve Control
AGvga e: Anglo-German variable-geometry aircraft
AGVS e: Air-to-Ground Voice System
- e: Automatic Guided Vehicle System[s]
AGVT Project e: Advanced Ground Vehicle Technology Project
AGW d: Amtsgruppenwähler
- d: Anthropologische Gesellschaft in Wien
- d: Arbeitsgemeinschaft der Verbände der Gas- und Wasserwerke
- d: Arbeitsgemeinschaft der Werbefachverbände
- d: Atemgrenzwert
- d: Auslandsgruppenwähler
- d: Aussonderungsgrenzwert[e]
- e: Acoustic-Gravity Wave
- e: Actual Gross Weight
- e: Adjusted Gross Weight
- e: Advanced Graphics Workstation
- e: Air Gap Width
- e: Allowable Gross Weight
- e: Alternate Gross Weight
AGWAC e: Australian Guided Weapons [and] Analog[ue] Computer
AGWG d: Abhandlungen der Gesellschaft der Wissenschaft zu Göttingen (journ.)
AGWI e: American Gulf and West Indies
AGWN e: Additive Gaussian White Noise
AGWS e: Advanced Gun Weapon System
AGWSE e: Association of Ground Water Scientists and Engineers
AGWt e: Actual Gross Weight
AGWTR d: Auslandsgruppenwähler für den Transitverkehr
AGZ e: Actual Ground Zero
- e: Actual Ground Zone
agz. d: angezeigt
AGZM d: Auslandsgesprächszeitmeß-einrichtung
AGZPO d: Ausführungsgesetz zur Zivilprozeßordnung
AGZPOKO d: Ausführungsgesetz zur Zivilprozeßordnung und zur Konkursordnung
AGZR d: Auslandsgesprächszeit-registereinrichtung
AH d: Ahorn
- d: Amtshilfe
- d: Antennenhalter
- d: Arbeitshöhe
- d: Außenhandel
- e: Abdominal Hysterectomy
- e: Absolute Humidity
- e: Academy of Homiletics
- e: Accelerated Hypertension
- e: Acceptor Handshake
- e: Access Hole
- e: Accidental Hypothermia
- e: Accumulator High
- e: Acid Hematine
- e: Active Homing
- e: Actual Hours
- e: Add Halfword
- e: Adenomatous Hyperplasia
- e: Adult Heart
- e: After-Hyperpolarization
- e: After Hatch
- e: After Hours
- e: Agricultural History
- e: Agriculture Handbook
- e: Aircraft Handler
- e: Airfield Heliport
- e: Air Headquarters
- e: Air History
- e: Airline Hostess
- e: Alarm Horn
- e: Alcoholic Hepatitis
- e: Allen Head
- e: Allowance Holder
- e: Alter Heading
- e: Alternate Headquarters
- e: Ampere Hour
- e: Analog[ue] Hybrid
- e: Anhydrous Hydrazine
- e: Animal Husbandry
- e: Annunciator Horn
- e: Anti-Halation
- e: Antihalation
- e: Antihunt
- e: Antihyaluronidase
- e: Apache Helicopter
- e: Army Helicopter
- e: Army Hospital
- e: Arterial Hypertension
- e: Artificial Heart
- e: Artificial Horizon
- e: Ascites Hepatoma
- e: Astigmatism; Hypermetropic
- e: Atrial His-Bundle
- e: Attack Helicopter
- e: Attention Handling
- e: Attitude Hold
- e: Auger Hole
- e: Autonomic Hyperreflexia
- e: Available Hours
ah d: Amperestunde[n]
- e: acetohexamide
- e: agricultural homestead
- e: air-hardened
- e: amenorrhea and hirsutism
- e: aminohippurate
- e: anhydrite
Ah. d: Anhang
- d: Aushang
Ah d: Anhänger
- d: Aushilfe
- e: Ampere-Hour
- e: hyperopic astigmatism
aH e: Attohenry
A. H. l: Anno Hebraico

A.H. e: Antwerp/Hamburg
- l: Anno Hebraico
- l: Anno Hegirae

A & H e: Accident and Health
- e: Agricultural and Horticultural
- e: Alive and Healthy

A/H e: Air Handling
- e: Air over Hydraulic
- e: Amenorrhea / Hyperprolactinemia

A-H e: Alaska-Hawaii Time
- e: American-Hawaiian Line
- e: Arrow-Hart & Hegeman Electric Company

a-h e: ampere-hour

AH-1 e: military aircraft carrying machinegun

AH-64 e: Attack Helicopter

AHA d: Abhörapparat
- d: Außenhandelsausschuß
- e: Acetohydroxamic Acid
- e: Additive Histologic Assessment
- e: Adirondack Historical Association
- e: Agricultural and Horticultural Engineering Abstracts
- e: Alberta Hospital Association
- e: American Hardboard Association
- e: American Health Association
- e: American Heart Association
- e: American Hepatitis Association
- e: American Historical Association
- e: American Hospital Association
- e: American Hovercraft Association
- e: American Hypnotherapy Association
- e: Anterior Hypothalamic Area
- e: Area Health Authority
- e: Association for Humane Abortion
- e: Australian Hospital Association

aha e: acquired hemolytic anemia
- e: alpha hydroxy acid
- e: autoimmune hemolytic anemia

AHAA e: American Hearing Aid Association

AHAB e: Australian Historical Association Bulletin

ahab e: attacking hardened air bases

AHAC e: Automated (or Automatic) High Accuracy Comparator

AHAF e: American Handwriting Analysis Foundation
- e: American Health Assistance Foundation

AHAK d: Antiherz-Antikörper

AHAM e: Association of Home Appliance Manufactures

AHAMS e: Advanced Heavy Antitank Missile System

AHAP e: Apartment House Addressing Program[me]

ahas e: acetohydroxy acid synthase

AHAS e: Automatic Helicopter Approach System

AHA(T) e: Area Health Authority (Teaching)

AHAT e: Arylhydroxamicacyltransferase

AHAU e: Anti-Hemagglutinating Unit

AHAW d: Abhandlungen der Heidelberger Akademie der Wissenschaft (journ.)

AHAWS e: Advanced Heavy Antitank Weapon System

AHB d: Anschlußheilbehandlung
- e: Africanized Honey Bee
- e: Air Heater Blower
- e: Assault Helicopter Battalion
- e: Attack Helicopter Battalion
- e: Australian Honey Board

AHBAI e: American Health and Beauty Aids Institute

AHBD e: Alpha-Hydroxybutyric Dehydrogenase

AHB-ETEI d: Außenhandelsbetrieb Elektrotechnik Export-Import (DDR)

AHB-ME d: Außenhandelsbetrieb Maschinenexport (DDR)

AH Bty e: Active Hostile Battery

AHC e: Academy of Hospital Counselors

AHC e: Acute Haemorrhagic Conjunctivitis
- e: Acute Haemorrhagic Cystitis
- e: Airport Handling Committee
- e: Allan Hancock College
- e: Allied High Commission
- e: American Hardware Corporation
- e: American Health Consultants
- e: American Helicopter Company
- e: American Horse Council
- e: American Horticultural Council
- e: American Hospital Corps
- e: Ampere-Hour Capacity
- e: Animal Health Committee
- e: Anthropogenic Hydrocarbons
- e: Army Hospital Corps
- e: Assault Helicopter Company
- e: Association for the History of Chiropractic
- e: Attack Helicopter Company
- e: Australian Horticultural Corporation
- e: Automatic Headway Control

AHCA e: Afghan Hound Club of America
- e: American Health Care Association

AHCAA e: American Health Care Advisory Association

AHCC e: Academy of Health Care Consultants

AHCEI e: Ad Hoc Committee on Equipment Interoperability
- e: American Histadrut Cultural Exchange Institute

AHCF e: Alternate Headquarters Command Facility
- e: Aqueous Homogeneous Critical Facility

AHCIMA e: Associate of the Hotel, Catering and Institutional Management Association

AHCIS e: Ambulatory Health Care Information System

AHCM e: Academy of Hazard Control Management

AHCo e: Assault Helicopter Company

AHCRA e: Arabian Horse Club Registry of America

AHCS e: Advanced Hybrid Computer System
- e: Advanced Hybrid Computing System
- e: Alternative Health Care Systems, Inc.

AHCT e: Ascending Horizon Crossing Time

AHCTL e: Acetylhomocysteinethiolactone

AHD d: Antihyaluronidase
- e: Advanced Helicopter Development
- e: Ahead
- e: Airborne and Helicopter Division
- e: Airhead
- e: Airport Hotel Directory
- e: American Health Decisions
- e: American Heritage Dictionary
- e: Anti-Helicopter Device
- e: Antihypertensive Drug
- e: Arc Heating Device
- e: Audio High Density
- e: Auto, Head and Drive
- e: Autoimmune Haemolytic Disease

ahd. d: althochdeutsch

ahd e: acetaldehyde dehydrogenase
- e: arteriosclerotic heart disease

AhD e: Arrowhead

AHDC e: Atari Hard-Disk Controller

AH Device e: Anti-Hunt Device

AHDGA e: American Hot Dip Galvanizers Association

AHDME e: Association of Hospital Directors of Medical Education

AHDP e: Advanced Helicopter Development Program[me]
- e: Azacycloheptane Diphosphonate

AHDR e: Air Header

AHDT e: Alaska-Hawaii Daylight Time

A-H DT e: Alaska-Hawaii Daylight Time

AHE d: Antihyaluronidase-Einheit
- e: Acute Haemorragic Encephalomyelitis
- e: Air to Heat Exchanger
- e: Alternatives in Higher Education
- e: Ammunition Handling Equipment
- e: Armament Handling Equipment
- e: Associate in Home Economics
- e: Association for Higher Education

AHEAD e: Army Help for Education and Development
- e: Australian Health Education Advisory Digest

AHEC e: American Hardwood Export Council
- e: Appropriate Home Energy Cooperative
- e: Area Health Education Center
- e: Arrowhead Energy Corporation

AH Ec e: Associate in Home Economics

AHEL e: Army Human Engineering Laboratory

AHEM e: Association of Hydraulic Equipment Manufacturers

AHEO e: Area Health Education Officer

AHEPA e: American Hellenic Educational Progressive Association

AHERA e: Asbestos Hazard Emergency Response Act

AHES e: American Humane Education Society
- e: Artificial Heart Energy System

AHESC e: Airport Handling Equipment Sub-Committee

AHF d: Antihämophilie-Faktor
- e: Active History File
- e: Acute Heart Failure
- e: American Health Foundation
- e: American Hospital Formulary (journ.)
- e: American Hull Form
- e: Anhydrous Hydrofluoric Acid
- e: Anhydrous Hydrogen Fluoride
- e: Antihaemolytic Factor
- e: Antihaemophilic Factor
- e: Antihaemophytic Factor
- e: Antihemophilic Factor
- e: Army Historical Foundation
- e: Asian Hospital Federation
- e: Associated Health Foundation
- e: Auto Hold Fire
- e: Automatic Hold Fire

Ahf e: Argentinian hemorrhagic fever

AHFCR e: Anderson Hospital for Cancer Research

AHFLCD e: Association of Health Facility Licensure and Certification Directors

AHFMR e: Alberta Heritage Foundation for Medical Research

AHFR e: Argonne High Flux Reactor

AHFRAC e: Army Human Factors Research Advisory Committee

AHFRDC e: Army Human Factors Research and Development Committee

AHFS e: American Hospital Formulary Service

AHG d: antihämophiles Globulin
- e: American High-Density Gradient
- e: Anhydroglucose
- e: Antihemolytic Globulin
- e: Anti-Human Globulin
- e: Antihaemophilic Globulin
- e: Australian Historical Geography
Ahg. d: Anhang
- d: Anhänger
- d: Aushang
AHGAK d: Archiv für Hessische Geschichte und Altertumskunde (journ.)
AHGB d: Allgemeines Handelsgesetzbuch
AHGC e: Advanced Hardened Guidance Computer
- e: Australian Horticultural Growers Council
AHGMRF e: Ad Hoc Group for Medical Research Funding
AHGS e: Acute Herpetic Gingival Stomatitis
- e: Advanced Harpoon Guidance System
- e: Attitude Heading Gyroscope System
AHH e: Acryl Hydrocarbon Hydroxylase
- e: Arc Heater Housing
- e: Aromatic Hydrocarbon Hydroxylase
- e: Aryl Hydrocarbon Hydroxylase
ahh e: alpha-hydrazine analog of histidine
AHHS e: Alexander Hamilton High School
- e: American Hackney Horse Society
AHHSA e: American Holistic Health Sciences Association
AHI d: Archiv für Handel und Industrie
- e: Active Hostility Index
- e: Aerodynamic Heating Indicator
- e: Afro-Hispanic Institute
- e: American Healthcare Institute
- e: American Health Institute
- e: American Honey Institute
- e: American Hospital Institute
- e: Animal Health Institute
- e: Animal Health Insurance
AHI e: Approach; Horizon Indicator
- e: Artificial Horizon Indicator
- e: Augmented Human Intellect
AHIC e: Art Hazards Information Center
AHID e: Association of High Tech Distributors
AHII e: American Health Industries Institute
AHIL e: Association of Hospital and Institution Libraries
AHIL Q. e: Association of Hospital and Institution Libraries Quarterly (journ.)
AHIP e: Advanced Helicopter Improvement Program[me]
- e: Army Helicopter Improvement Program[me]
- e: Assisted Health Insurance Plan
- e: Australian Health Insurance Program[me]
AHIRC e: Aroostook Health Information and Resource Consortium
AHIRS e: Australian Health Information and Research Service
AHIS e: Association of Heads of Independent Schools
- e: Autonetics General Information Learning Equipment Homing Interceptor Simulation
- e: Automated Hospital Information System
AHI System e: Augmented Human Intellect System
AHIT e: Attack Helicopter Instrument Test
AHITI e: Animal Health and Industries Training Institute
AHK d: Alliierte Hohe Kommission
- d: Anhängerkupplung
AHKABl. d: Amtsblatt der Alliierten Hohen Kommission
AHKG d: Gesetz der Alliierten Hohen Kommission
AHL e: A Hardware-programming Language
- e: Alaska Historical Library
- e: Auroral Hydrogen Line
- e: Average Hearing Level
- l: Ad Hunc Locum
ahl e: alcohol-induced hyperlipidemia
a.h.l. l: ad hunc locum
ahle e: acute hemorrhagic leukoencephalitis
AHLE e: Auroral Hydrogen Line Emission
AHLF e: Additional High Layer Functions
AHLG d: Anti-Human-Lymphozyten-Globulin
A-HLH e: Amphipathic Helix-Loop- Helix
AHLI e: American Home Lighting Institute
AHLMA e: American Home Laundry Manufacturers Association
ahls e: antihuman-lymphocyte serum
AHLV e: American Hop Latent Virus
AHM e: Acutely Hazardous Material
- e: Airport Handling Manual
- e: Allowance Holder Monthly
- e: American Heatth Magazine
- e: Ammonium Heptamolybdate
- e: Ampere-Hour-Meter
- e: Anterior Hyaloid Membrane
- e: Auxiliary Handling Machine
AHMA e: Advanced Hypersonic Manned Aircraft
- e: Allied Hat Manufacturers Association
- e: American Hardware Manufacturers Association
- e: American Hemisphere Marine Agencies
AHMC e: American Horticultural Marketing Council
- e: Australian Health Ministers Conference
AHMD e: Airborne Helmet Mounted Display
AHMH e: Association of High Medicare Hospitals
AHMI e: American Holistic Medical Institute
- e: Appalachian Hardwood Manufacturers, Incorporated
AHMJ e: Arc-Heated Materials Jet
AHMS e: Abstracts of Hospital Management Studies
AHMV d: Arbeitskreis Heizungs- und Maschinenwesen staatlicher und kommunaler Verwaltunge
AHN e: Adventist Health Network of North America
- e: Army Health Nurse
- e: Assistant Head Nurse
AhndG d: Ahndungsgesetz
AHNO e: Association of Head and Neck Oncologists of Great Britain
AHO d: Allgemeine Hafenordnung
- d: Ausschuß für die Honorarordnung der Ingenieure
- d: Außenhandelsorganisation
- e: Anharmonic Oscillator
- e: Applicant Holding Office
- e: Attack Helicopter Operations
A/H/O e: Abort / Hold / Orbit
AHOAG e: Attack Helicopter Operations and Analysis Group
AHOD e: Areal Hypolimnetic Oxygen Deficit
AHOI d: Allgemeines Hochschul-Orientierungs- und Informationssystem
AHOP e: Assisted Home-Ownership Program[me]
AHOP-Syndr. d: Adipositas-Hypertermie-Oligomenorrhoe-Parotis-Syndrom
A-horiz e: A-horizon
AHOS e: Automatic Hydrologic Observing System
AHP d: Abhörpunkt
- e: Absorption Heat Pumping
- e: Accelerator Heel Point
- e: Acute Hemorrhagic Pancreatitis
- e: Adaptive Hopping Permit
- e: Affordable Housing Program[me]
- e: Air Horsepower
- e: Allied Health Professionals
- e: Allied Hydrographic Publication
- e: Alternative Health Plans
- e: American Health Professionals
- e: American Homeopathic Pharmacopoeia
- e: Analytic Hierarchy Process
- e: Aniline Hydrogen Phthalate
- e: Army Heliport
- e: Assistant Head Postmaster
- e: Assistant House Physician
- e: Association for Humanistic Psychology
- e: Attitude Hold Pitch
AhP d: Anhaltspunkt
Ahp d: Aushilfsposten
ahp e: acute hemorrhagic pancreatitis
- e: air at high pressure
- e: aviation horsepower
AHPA - e: Accumulator High-Pressure Air
- e: Accumulator High Pressure Air
- e: Adeninylhydroxypropanoic Acid
- e: American Health Planning Association
- e: American Herbal Products Association
- e: American Honey Producers Association
- e: American Horse Protection Association
- e: Arthritis Health Professions Association
AHPAT e: Allied Health Professions Admissions Test
AHPB e: Association for Humanistic Psychology in Britain
AHPC e: Aging Health Policy Center
AHPl d: Arbeits- und Haushaltsplan
AHPL e: A Hardware Programming Language
AHPNAS e: Advanced High Performance Nuclear Attack Submarine
AHPOA e: Anterior Hypothalamus; Preoptic Area
AH/PO Area e: Anterior Hypothalamus/ Preoptic Area
AHPP e: Association of Humanistic Psychology Practitioners
AHPR e: Academy of Hospital Public Relations
AHPS e: Auxiliary Hydraulic Power Supply
AHPSC e: Airport Handling Procedures Sub-Committee
AHQ e: Advanced Headquarters
- e: Air Headquarters
- e: Allied Headquarters
- e: Anthrahydroquinone
- e: Area Headquarters
- e: Arkansas Historical Quarterly (journ.)
- e: Army Headquarters
- e: Association for Healthcare Quality
AHqM e: Air Headquarters; Malta
AHQS e: Asymmetric Hindquarters Syndrome
AHR d: Adressenhilfsregister
- d: Agglutinationshemmungsreaktion
- d: Antihyaluronidasereaktion
- d: Antihyaluronidasreaktion
- d: Außenhandelsrundschreiben
- e: Ablative Heat Rate
- e: Academy of Human Rights
- e: Acceptable Hatard Rate
- e: Acceptable Hazard Rate
- e: Active High Resolution

- e: Adsorptive Heat Recovery
- e: American Historical Review (journ.)
- e: Anchor
- e: Andalusian Horse Registry
- e: Annual History Review
- e: Annual Hospital Report
- e: Aqueous Homogeneous Reactor
- e: Arabian Horse Registry of America
- e: Association for Health Records
- e: Attitude Hold Roll
- e: Automatic Handwriting Reader

A/HR e: Ampere / Hour
AHRA e: Advanced Helmet-sight Reticle Assembly
- e: American Healthcare Radiology Administrators

AHRC e: Alister Hardy Research Centre
- e: Animal Health Research Center

AHRF e: American Hearing Research Foundation
AHRL e: Arctic Health Research Laboratory
AHRP e: Academy on Human Rights and Peace
AHRRN e: Automatic Hydrologic Radio Reporting Network
AHRS e: Attitude and Heading Reference System
- e: Automatic Heading Reference System

AHRTAG e: Appropriate Health Resources and Technologies Action Group
AHRU e: Armo[u]r Human Research Unit
- e: Attitude Heading Reference Unit
- e: Aviation Human Research Unit

AHRW e: Alcohol Health and Research World
AHS d: Arbeitshöhe; Soll-
- e: Ablative Heat Shield
- e: Academy of Health Sciences
- e: Aerial Homing System
- e: Aerospace High School
- e: African Horsesickness
- e: Agricultural History Society
- e: American Hearing Society
- e: American Helicopter Society
- e: American Horticultural Society
- e: American Hospital Society
- e: American Hospital Supply
- e: American Housing Survey
- e: American Hypnodontic Society
- e: Ammunition Handling System
- e: Amtrak Historical Society
- e: Annual Housing Survey
- e: Annual Hull Survey
- e: Antenna Homing System
- e: Arlington Hall Station
- e: Assistant Head of Section
- e: Assistant House Surgeon
- e: Associated Heat Services
- e: Attack Helicopter Support
- e: Australian Historical Statistics
- e: Australian Historical Studies
- e: Automatic Highway System
- e: Aviation High School
- e: Aviation Historical Society
- e: Azores Hot Spot

A. H. S. l: anno humanae salutatis
AHSA e: African Heritage Studies Association
- e: American Hampshire Sheep Association
- e: American Home Satellite Association
- e: American Horse Shows Association
- e: Art, Historical and Scientific Association
- e: Association of Hospital Security Administrators
- e: Aviation Historical Society of Australia

AHSC e: American Home Sewing Council
AHSCA e: American Home Sewing and Craft Association
A-H Scale e: Anti-Hispanic Scale
AH-Schaltung d: Abtast- und Halte-Schaltung
AHSCo e: Assault Helicopter Support Company
AHSCP e: Army High School Completion Program[me]
AHSE e: Assembly, Handling and Shipping Equipment
AHSF e: American Hungarian Studies Foundation
AHSI e: Advanced Human Systems Institute
AHSIC e: Alaskan Health Sciences Information Center
AHSM e: Academy for Health Services Marketing
- e: Alpha Hand and Shoe Monitor (Radiation Detection)

AHSME e: Annual High School Mathematics Examination
AHS-MS e: American Housing Survey-Metropolitan Sample
AHSN e: Assembly of Hospital Schools of Nursing
AHSNZ e: Aviation Historical Society of New Zealand
AHSP e: Automatic High Speed Plot
AHSPI e: Actions Having Significant Personnel Implications
AHSR e: Airborne Height-Surveillance Radar
- e: Airborne Height Surveillance Radar
- e: Air Height Surveillance Radar
- e: Association for Health Services Research

AHSRC e: American High-Speed Railway Corporation
- e: Arctic Health Services Research Center

AHSS e: Association of Home Study Schools
AHSSPPE e: Association of Handicapped Student Service Programs in Postsecondary Education
AHST e: Alaska-Hawaii Standard Time
- e: Anchor Handling Salvage Tug
- e: Association of Highway Steel Transporters
- e: Attack Helicopter Self Test

AHStat d: Außenhandelsstatistik
AHStatG d: Gesetz über die Außenhandelsstatistik
AHStG d: Allgemeines Hochschulstudiengesetz
AHSV e: African Horsesickness Virus
AHT d: Agrar- und Hydrotechnik
- d: Antihyaluronidasetest
- d: Arbeitsgemeinschaft Härtereitechnik
- d: Arbeitsgemeinschaft Heizungs- und Lüftungstechnik
- d: Arbeitshöhe; tatsächliche
- d: Austenitformhärten bei hoher Temperatur
- e: Acoustic Homing Torpedo
- e: Anchor Handling Tug
- e: Animal Health Technician
- e: Animal Health Trust
- e: Attack Helicopter Team
- e: Augmented Histamine Test
- e: Average Holding Time

aht e: antihyaluronidase titer
AHTC e: Advances in High Temperature Chemistry
AHTGG e: Antihuman Thymocyte Gamma Globulin
AHTN e: Association of Hospital Television Networks
AHTP e: Aerodynamic Heat Test Plans
AHTR e: Acute Hemolytic Transfusion Reaction
AHTS e: American Health and Temperance Society
AHU d: Arbeitshygienische Untersuchungsstelle
- d: Außenhandelsunternehmen
- e: Accumulated Heat Unit
- e: Air-Handling Unit
- e: Anti-Halation Undercoat

ahu e: after hangup
AHV d: Ausfuhrhändlervergütung
- e: Acceleration; High Voltage
- e: Accelerator High Voltage
- e: Aircraft Handling Vehicle

ahv e: alternative fuel vehicle
a.h.v. l: adhunc vocem
AHVB d: Allgemeine Haftpflichtversicherungsbedingungen
AHVMA e: American Holistic Veterinary Medical Association
AHVMF d: Archiv des Historischen Vereins von Mainfranken (journ.)
AHVN e: American Hospital Video Network
AHVNRh d: Annalen des Historischen Vereins für den Niederrhein (journ.)
AHVO d: Allgemeine Hafenverordnung
AHVR e: Automatic High-Voltage Regulator
AHVUA d: Archiv des Historischen Vereins von Unterfranken und Aschaffenburg (journ.)
AHW e: Atomic Hydrogen Welding
AHWA e: American Hazardous Waste Association
AHWC e: Associate of Heriot-Watt College
AHWG e: Ad Hoc Working Group
AHX e: Azahypoxanthine
AHYD e: Aromatics Hydrogenation
AHZB d: anzusetzender Hinzurechnungsbetrag
AHZV d: Anhängerzugvorrichtung
AI d: Aerodynamisches Institut
- d: Agenturinkasso
- d: Amtsinspektor
- d: Anschlußliste
- d: Anthropologisches Institut
- d: Appenzell-Innerrhoden (Kfz;CH)
- d: applikative Information
- d: Arbeitsintensität
- d: Archäologisches Institut
- d: Artikulationsindex
- d: Ausgangsinformation
- d: Aussteuerinstrument
- d: Automatik intern
- d: automatisierter Index
- d: Automobilindustrie
- e: Aaland Islands
- e: Access / Information
- e: Access Instruction
- e: Accidental Injuries
- e: Accounting Information
- e: Accumulated Interest
- e: Accured Interest
- e: Acoustic Impedance
- e: Acoustic Intensity
- e: Acquisition Institute
- e: Action Identifier
- e: Action Item
- e: Active Ingredient
- e: Activity Index
- e: Activity Interrupt
- e: Actual Instruction
- e: Actuator / Indicator
- e: Add Impulse

- e: Add Instruction
- e: Additional Instruction
- e: Address Incomplete
- e: Address Index
- e: Address Information
- e: Address Instruction
- e: Addressless Instruction
- e: Adjustment Inventory
- e: Administrative Instructions
- e: Admiralty Instructions
- e: Admiralty Islands
- e: Adrenergic ß-Inhibitors
- e: Advocacy Institute
- e: Aerial Impedance
- e: Aerial Input
- e: Aerospace Intelligence
- e: After Image
- e: Aged Individual
- e: Agricultural Index
- e: Agricultural Institution
- e: Airborne Intercept[ion]
- e: Aircraft Identification
- e: Aircraft Industry
- e: Aircraft Intercept[ion]
- e: Airfield Index
- e: Air India
- e: Air Injection
- e: Air Inspector
- e: Air Installation[s]
- e: Air Intelligence
- e: Air Intercept
- e: Air Interdiction
- e: Airship Industries Ltd.
- e: Airspeed Indicator
- e: Airways Inspector
- e: Allergy and Immunology
- e: All Iron
- e: Alphabetic Information
- e: Alphabetic Interpreter
- e: Alphabetic Item
- e: Alphanumeric Information
- e: Alphanumeric Instruction
- e: Alphanumeric Item
- e: Alternate Instruction
- e: Altimeter Indicator
- e: Altitude Indicator
- e: Altitude Indoctrination
- e: Alumin[i]um Ion
- e: American Institute
- e: Amide-Imide
- e: Amorphous Inclusion
- e: Amplifier Input
- e: Analog[ue] Information
- e: Analog[ue] Input
- e: Andaman Islands
- e: AND Inverter
- e: Angiogenesis Inhibitor
- e: Angle Iron
- e: Anglo-Indian
- e: Anglo-Irish
- e: Anglo-Israelism
- e: Annals of Iowa (journ.)
- e: Annoyance Index
- e: Antecedent Index
- e: Antenna Impedance
- e: Antenna Input
- e: Anthracite Institute
- e: Anthropological Institute
- e: Anti-Icing
- e: Anxiety Index
- e: Aortic Incompetence
- e: Aortic Insufficiency
- e: Apical Impulse
- e: Apparent Impedance
- e: Application and Industry
- e: Appraisal Institute
- e: Appreciation Index
- e: Aptitude Index
- e: Arctic Institute
- e: Area Inspector
- e: Arithmetic[al] Instruction
- e: Arithmetic[al] Item
- e: Army Intelligence
- e: Array Identifier
- e: Articulation Index
- e: Artificial Insemination
- e: Artificial Intelligence
- e: Aspen Institute
- e: Asphalt Institute
- e: Assembler Instruction
- e: Assembly Instruction
- e: Assignment Instructions
- e: Assistance and Instructions
- e: Assistant Inspector
- e: Assistant Instructor
- e: Astrologers International
- e: Atherogenic Index
- e: Atomic[s] International
- e: Atrial Insufficiency
- e: Attention Interruption
- e: Attenuation Index
- e: Attitude Indicator
- e: Auditory Induction
- e: Austrian Institute
- e: Authorized Inspector
- e: Autographics International
- e: Automated Index
- e: Automated Instruction
- e: Automatic Input
- e: Automatic Interrupter
- e: Automatic Interruption
- e: Automation Institute
- e: Automotive Industries
- e: Avery Island
- e: Avionic Instrument
- e: Avionics Integration
- e: Awaiting Instruction
- e: Axiocisal
- e: Azimuth Indicator
- e: Inherent Availability
- f: Altesse impériale

ai. d: altindisch

ai e: accidentallyincurred
- e: achievement viaindependence
- e: airborneintercept
- e: axioincisal

a.i. e: after inspection
- l: anno inventionis

A.I. e: Airborne Interceptor
- e: Aircraft Interceptor

a. i. l: ab initio
- l: ad interim

A & I e: Abstracting and Indexing
- e: Afars and Issas, Republic of Djibouti
- e: Agricultural and Industrial
- e: Agriculture and Industrial
- e: Alteration and Improvement
- e: Alteration and Inspection
- e: Arts and Industry

a & i e: accident and indemnity

A/I e: Accident / Incident
- e: Aptitude Index

a^I e: Anode of Section One

AIA d: Aerodynamisches Institut [der Technischen Hochschule] Aachen
- d: Anti-Intrusion Alarm
- e: Action Item Assignment
- e: Activity Interrupt Acknowledgement
- e: Acupuncture International Association
- e: Advise if Able
- e: Aerialift Industries Association
- e: Aerospace Industrial Association
- e: Aerospace Industries Association
- e: Aestheticians International Association
- e: Airborne Integration Area
- e: Airborne Interception of Aircraft
- e: Allergy Information Association
- e: Allred Interaction Analysis
- e: Allyl-Isopropyl-Acetamide
- e: American Importers Association
- e: American Institute of Accountants
- e: American Institute of Aeronautics
- e: American Institute of Arch[a]eology
- e: American Institute of Architects
- e: American International Academy
- e: American Inventors Association
- e: Amylase Inhibitor Activity
- e: Anthracite Industry Association
- e: Anti-Icing Additive
- e: Anti Intrusion Alarm
- e: Antivenin Institute of America
- e: Anything Invented Anywhere
- e: Appliance Industry Association
- e: Arch[a]eological Institute of America
- e: Arctic Institute of America
- e: Argentine Interplanetary Association
- e: Army Information Architecture
- e: Army Institute of Administration
- e: Army Intelligence Agency
- e: Artificially Induced Aurora
- e: Asbestos International Association
- e: Associate of the Institute of Actuaries
- e: Associate of the Institute of Architects
- e: Association of Industrial Advertisers
- e: Association of Industrial Archaeology
- e: Association of International Accountants
- e: Authors Institute of America
- e: Auto International Association
- e: Automated Image Analysis
- e: Automated Imaging Association
- e: Automatic Information Acquisition
- e: Automation One Association
- e: Automotive Industry Authority
- f: Association Internationale d'Allergologie

AiA d: Angestellter im Außendienst
- e: Accuracy in Academia

AIa e: Annals of Iowa (journ.)

AIAA e: Aerospace Industries Association of America
- e: American Industrial Arts Association
- e: American Institute of Aeronautics and Astronautics
- e: American Institute of Aeronautics and Astronautics Journal
- e: Associate of the Institute of Affiliate Accountants

AIAA Bull e: American Institute of Aeronautics and Astronautics Bulletin

AIAA J e: American Institute of Aeronautics and Astronautics Journal

AIAAM e: Advanced Interceptor Air-to-Air Missile

AIAA-TIS e: American Institute of Aeronautics and Astronautics-Technical Information Service

AIAB e: Associate of the International Association of Bookkeepers

AIAC e: Air Industries Association of Canada
- e: Associate of the Institute of Company Accountants
- e: Automotive Industries Association of Canada

AIAD e: Acronyms, Initialisms and Abbreviations Dictionary

AIAE e: Associate of the Institute of Automobile Engineers
- e: Australian Institute of Art Education

AIAESD e: American International Association for Economic and Social Development

AIAF e: Absolutely Integral Algebraic Function
- e: American Institute of Architects Foundation
AIAG d: Aluminium-Industrie AG
- e: Automotive Industry Action Group
AIAgrE e: Associate of the Institution of Agricultural Engineers
AIAI e: American Indian Archaeological Institute
AIAIS e: American In-Vitro Allergy/ Immunology Society
AIA J e: American Institute of Architects Journal
AIAL e: Associate of the Institute of Arts and Letters
AIAM e: Association of International Automobile Manufacturers
AIA/NA e: Asbestos Information Association-North America
AI and Q e: Animal Inspection and Quarantine
AI and S e: Army Intelligence and Security
AIANZ e: Associate of the Institute of Accountants; New Zealand
AIAOS e: Academic Instructor and Allied Officer School
AIAP e: Ardmore Industrial Air Park
- e: Armys Incentive Awards Program[me]
AIAPN e: Australian Institute of Air Pilots and Navigators
AIAR e: American Institute for Archaeological Research
- e: American Institute of Aerological Research
AIArb e: Associate of the Institute of Arbitrators
AIARD e: Association for International Agriculture and Rural Development
AIAS e: American Institute of Architecture Students
- e: Anti-Intrusion Alarm Set
- e: Army Institute of Advanced Studies
- e: Associate of the Incorporated Association of Architects and Surveyours
- e: Australian Institute of Aboriginal Studies
- e: Australian Institute of Agricultural Science
- e: Australian Institute of Agriculture and Science
AIASA e: American Industrial Arts Student Association
AIAT e: Associate of the Institute of Animal Technicians
- e: Association of the Institute of Asphalt Technology
- e: Attitude-Interest Analysis Test
- e: Auditory Integrative Abilities Test
AIAVA e: Aminoiodoacetamidovaleric Acid
AIB d: Abdichtung von Ingenieurbauwerken
- d: Aerodynamisches Institut der Technischen Hochschule Braunschweig
- d: Aminoisobuttersäure
- e: Accidents Investigation Branch
- e: Address and Instruction Bus
- e: Advances in Inorganic Biochemistry
- e: Agency Investigation Board
- e: Agricultural Information Bulletin
- e: Aircraft Instrument Bulletin
- e: Allied Intelligence Bureau
- e: American Institute of Baking
- e: American Institute of Banking
- e: Amino-Isobutyric [acid]
- e: Analog[ue] Input-output Board
- e: Analog[ue] Input Base
- e: Analysis and Information Branch
- e: Anthracite Information Bureau
- e: Aptitude Index Battery
- e: Armored Infantry Battalion
- e: Army Infantry Board
- e: Associate of the Institute of Bankers
- e: Associate of the Institute of Building
- e: Australian Institute of Biology
- e: Automatic Intercept Bureau
- e: Avionics Integration Bench
- f: Association des Industries de Belgique
AIBA e: Agricultural Information Bank for Asia
- e: Air Intercept Battle Analysis
- e: Alpha-Aminoisobutyric Acid
- e: Amino-Isobutyric Acid
- e: Associate of the Institute of Business Administration
AIB Acid e: Amino-Isobutyric Acid
AIBC e: Architectural Institute of British Columbia
A-IBCM e: Anit-Intercontinental Ballistic Missile
AIBCS e: American Intersociety Board of Certification of Sanitarians
AIBD e: American Institute of Building Design
- e: Asia-Pacific Institute for Broadcasting Development
- e: Associate of the Institute of British Decorators
AIBE e: Associate of the Institute of Building Estimators
aibf e: advanced internally blown jet flap
AIBGA e: All-Island Banana Growers Association [of Jamaica]
AIBICC e: Associate of the Incorporated British Institute of Certified Carpenters
aibm e: anti-intercontinental ballistic missile
AIBM e: Associate of the Institute of Baths Management
- f: Association Internationale des Bibliothèques Musicales
AIBN d: Azoisisobytronitril[e]
- e: Azoisobutyronitrile
AIBNRM e: American Institute of Bolt, Nut and Rivet Manufacturers
AIBP e: Associate of the Institute of British Photographers
AIBS e: Active Isolation / Balance System
- e: American Institute of Biological Sciences
- e: Associate of the Institute of Building Societies
AIBS Bull e: American Institute of Biological Sciences Bulletin
AIBT in e: Active Item Balance Tape in
AIBT out e: Active Item Balance Tape out
AIC d: Arbeitsgruppe Informationswissenschaft in der Chemie
- e: Accelerator Information Center
- e: Accretion-Induced Collapse
- e: Acoustic Isolation Chamber
- e: Acting Inspector of Constabulary
- e: Activity Identification Code
- e: Adaptive Inferential Control
- e: Address Information Center
- e: Advanced Image Compression
- e: Advanced Intelligence Center
- e: Aerodynamic-Influence Coefficient
- e: Aeronautical Information Circular
- e: Agricultural and Industrial Chemistry
- e: Agricultural Improvement Council
- e: Agricultural Institute of Canada
- e: Airborne Information Correlation
- e: Aircraft In Commission
- e: Aircraft Industries Center
- e: Aircraft Industry Conference
- e: Aircraft Information Correlator
- e: Aircraft Intercommunication
- e: Air Information Center
- e: Air Information Codification
- e: Air Inlet Controller
- e: Air Intercept Controller
- e: Air Interception and Control[ler]
- e: Air Interception Committee
- e: Air Interception Control
- e: Air Interceptor; Centimeter
- e: Allied Intelligence Center
- e: Allowance Item Code
- e: American Institute of Chemists
- e: American Institute of Constructors
- e: American Institute of Cooperation
- e: American Institute of Criminology
- e: American International College
- e: American International Communications Corp.
- e: Aminoimidazolecarboxamide
- e: Ammunition Identification Code
- e: Analysts International Corporation
- e: Analytical Instruction Code
- e: Apollo Intermediate Chart
- e: Apprenticeship Information Center
- e: Aquaculture Information Center
- e: Arab Information Center
- e: Army Industrial College
- e: Army Intelligence Center
- e: Arthritis Information Clearinghouse
- e: Artificial Illumination Center
- e: Artificial Intelligence Corporation
- e: Art Information Center
- e: Art Institute of Chicago
- e: Asbestos Information Centre
- e: Asian Information Centre
- e: Associate of the Institute of Chemistry
- e: Asymmetric Illumination Contrast
- e: Atlantic Intelligence Center
- e: Attack Information Center
- e: Australian Institute of Cartographers
- e: Automatic Information Center
- e: Automatic Initiation Circuit
- e: Automatic Intercept Center
- e: Automatic Iris Control
- e: Automatic Item Correction
- e: Automotive Information Council
- e: Awaiting Incoming Continuity
- f: Association Internationale de Cybernétique
- i: Associazione Italiana per il Calco Automatico
AICA e: Alliance of Independent Colleges of Art
- e: American Institute of Commemorative Art
- e: Amino-Imidazole-Carboxylic Acid[e]
- e: Anterior Inferior Cerebellar Artery
- e: Anterior Inferior Communicating Artery
- e: Associate of the Institute of Company Accountants
- e: Australasian Institute of Cost Accountants
- e: Automobile Importers Compliance Association
- f: Association Internationale pour le Calcul Analogique
- i: Associazione Italiana per il Calcolo Automatico
AICAE e: American Indian Council of Architects and Engineers
AICAP e: American Institute of Computerized Accounting Professionals
AICAR d: 5 (4) Aminoimidazol-4(5)-carboxamidribotid
- e: Amino-Imidazole-Caboxamide Ribonucleotide
AICB e: African Institute for Economic Development and Planning
- f: Association Internationale Contre le Bruit

AICBM e: Anti-Intercontinental Ballistic Missile
A-ICBM e: Army Intercontinental Ballistic Missile
AICC e: Action Item Control Card
- e: Automatic Interactive Computer Control
aicc e: antibody-induced cell-mediated cytoxicity
AICCC e: Alarm Industry Committee for Combating Crime
- e: American Institute of Child Care Centers
AICCF f: Association Internationale du Congrès des Chemins de Fermé
AICCP e: Association of the Institute for Certification of Computer Professionals
AICCT e: Association for the Improvement of Community College Teaching
AICCU e: Association of Independent California Colleges and Universities
AICD e: Accelerated Individual and Company Development
- e: Activation-Induced Cell Death
- e: Authorized Industrial Control Distributor
- e: Automatic Implantable Cardioverter-Defibrillator
AICE e: Agency for Information and Cultural Exchange
- e: American Institute of Chemical Engineers
- e: American Institute of Consulting Engineers
- e: American Institute of Crop Ecology
- e: Associate of the Institute of Civil Engineers
- e: Associate of the Institution of Commercial Engineers
AI-CE e: Atomic International-Combusting Engineering
- e: Atomic International-Combustion Engineering
AI Centre e: Artificial Insemination Centre
AICeram e: Associate of the Institute of Ceramics
AICerE e: American Institute of Ceramic Engineers
AICF e: Auto-Immune Complement Fixation
aich e: automatic integrated container handling
AI Character e: Attention Identification Character
AIChE e: American Institute of Chemical Engineering
- e: American Institute of Chemical Engineers
AIC-INT e: Aerodynamic Influence Coefficients with Interference
AICK e: Associated Independent Colleges of Kansas
AICL e: American Institute of Child Life
AICMA f: Association Internationale des Construcreurs de Matériel Aéronautique
AICMIP e: All-India Coordinated Millet Improvement Programme
AICNY e: American Institute of the City of New York
AICO e: Action Information Control Office[r]
- e: American Insulator Corporation
AICOM e: Ai-Electronics Computer
AIConsE e: American Institute of Consulting Engineers
AICP e: American Institute of Certified Planners
- e: Army Internal Control Program[me]
- e: Army Inventory Control Point
- e: Artificially Intelligent Computer Performer
- e: Atomic Incident Control Plan
AICPA e: American Institute of Certified Public Accountants
AICPOA e: Advanced Intelligence Center Pacific Ocean Area
AICQ i: Associnzione Italiana per il Controllo della Qualità
AICR e: American Institute for Cancer Research
- e: Association for International Cancer Research
AICRIP e: All India Coordinated Rice Improvement Programme
AICS e: Action Item Closeout Sheet
- e: Adaptive Interference Cancellation System
- e: Advanced Imaging Communications System
- e: Advanced Interior Communication System
- e: Air Induction Control System
- e: Air Inlet Control System
- e: Air Intercept Controller Supervisor
- e: American Institute of Ceylonese Studies
- e: Army Intelligence Center and School
- e: Assistance for Isolated Children Scheme
- e: Associate of the Institute of Chartered Shipbrokers
- e: Association of Independent Colleges and Schools
- e: Association of Independent Computer Specialists
- e: Association of Industrial Colleges and Schools
- e: Automated Industrial Control System
- e: Automatic Inlet Control System
- e: Automatic Intersection Control System
- f: Association Internationale du Cinéma Scientifque
AICSC e: Aircraft Integrated Crew Station Concepts
AICSTS e: Air Inlet Control System Test Set
AICT e: Atlantic Information Center for Teachers
- e: Automatic Integrated Circuit Tester
AICTA e: Associate of the Imperial College of Tropical Agriculture
AICU e: Association of International Colleges and Universities
AICUM e: Association of Independent Colleges and Universities in Massachusetts
- e: Association of Independent Colleges and Universities of Michigan
AICUN e: Association of Independent Colleges and Universities of Nebraska
AICUO e: Association of Independent Colleges and Universities of Ohio
AICUZ e: Air Installation Compatible Use Zone
AICUZ Program[me] e: Air Installation Compatible Use Zones Program[me]
AICV e: Armored Infantry Combat Vehicle
AICW e: Advanced Individual Combat Weapon
- e: Associate of the Institute of Clerks of Works
AID d: Aerodynamisches Institut der Technischen Hochschule Darmstadt
- d: Aktiver Informations-Dienst
- d: Allgemeiner Informationsdienst
- d: Anmeldeidentifikation[skennzeichen]
- d: Auftragsabwicklung im Dialog
- d: Auswertungs- und Informationsdienst für Ernährung, Landwirtschaft und Forsten
- d: Autointeractive-Interactive Design
- e: Abbreviated Item Description
- e: Abortion Information Data-Bank
- e: Acceptable Intake Daily
- e: Accident / Injury / Damages
- e: Accident, Incident, Deficiencies
- e: Acquired Immunodeficiency
- e: Acronyms and Initialisms Dictionary
- e: Action Item Directive
- e: Active Integral Defense
- e: Acute Infectious Disease
- e: Adaptive Intelligent Dialog[ue]
- e: Adult Information on Drugs
- e: Advanced Integrated Diagnostics
- e: Advanced Interactive Draughting
- e: Advanced Ionization Detector
- e: Advanced Ionization Development
- e: Advance Information Document
- e: Aeronautical Inspection Directorate
- e: Aerospace Information Digest
- e: Aerospace Information Division
- e: Agency for International Development
- e: Agricultural Industrial Development
- e: Agricultural Information Division
- e: Airborne Intelligent Display
- e: Air Information Device
- e: Air Information Division
- e: Air Inlet Damper
- e: Air Inspection Directorate
- e: Air Intake Duct
- e: Airline Interline Development
- e: Airport Information Desk
- e: Algebraic Interpretive Dialog[ue]
- e: American Institute of Decorators
- e: American Instructors of the Deaf
- e: Analog[ue] Input Description
- e: Analog[ue] Input Device
- e: Analog[ue] Input Differential
- e: Analog[ue] Interface Device
- e: Analytical Instrument Development
- e: Applied Information and Documentation
- e: Applied Information Development
- e: Area Imaging Device
- e: Argon Ionization Detector
- e: Argonne Interactive Display
- e: Army Information Digest
- e: Artificial Insemination by Donor
- e: Assembly Instruction Device
- e: Assistance and Independence for the Disabled
- e: Association for International Development
- e: Associative Interactive Dictionary
- e: Attached Inflatable Decelerator
- e: Attached Inflatable Detector
- e: Attention Identification
- e: Attention Identifier
- e: Audit Item Disposition
- e: Auditory Information Display
- e: Augmented Index and Digest
- e: Authorized Industrial Distributor
- e: Auto-Instructional Device[s]
- e: Autoimmune Deficiency
- e: Automatic Implantable Defibrillator
- e: Automatic Incident Detector
- e: Automatic Indexing of Documents
- e: Automatic Industrial Drilling
- e: Automatic Information Distribution
- e: Automatic Initial Distribution
- e: Automatic Interaction Detection
- e: Automatic Interaction Detector[program]
- e: Automatic Internal Diagnosis
- e: Automatic Interrogation Distortion
- e: Automotive Industry Data
- e: Avalanche Injection Diode
A & ID e: Acquisition and Improvement District
aid. e: applications interface device

AIDA e: Advanced Integrated-circuit Design Aids
- e: American Indicator Digest Average
- e: Associated Independent Dairies of America
- e: Attention, Interest, Desire, Action
- e: Australian Industries Development Association
- e: Automatic Inspection Data Accumulator
- e: Automatic Inspection of Data
- e: Automatic Instrumented Diving Assembly
- e: Automatic Intruder Detector Alarm
- e: Automobile Information Data Advertising
- e: Automobile Information Disclosure Act

AIDAC e: Assistance Information and Data Acquisition Center
AIDAP e: Automatic Inspection, Diagnostic and Prognostic
AIDAPS e: Automatic Inspection, Diagnostic and Prognostic Systems
AIDAP System e: Automatic Inspection Diagnostic and Prognostic System
AIDAS e: Advanced Instrumentation and Data Analysis System
- e: Advanced Integrated Digital Avionics System

AIDAT e: Automated Integrated Dynamic Avionics Tester
- e: Automatic Integrated Dynamic Avionics Tester

AIDATS e: Army In-Flight Data Transmission System
AIDC e: American Industrial Development Council
- e: Arkansas Industrial Development Commission
- e: Asian Industrial Development Council
- e: Association of Information and Dissemination Center[s]
- e: Australian Industries Development Corporation
- e: Automatic Image Density Control

AIDD e: American Institute of Design and Drafting
- e: Association of Insulin-Dependent Diabetics

AIDE e: Accountability in Data Entry
- e: Adapted Identification Decision Equipment
- e: Adaptive and Integrated Decision Expeditor
- e: Administration Information Display Equipment
- e: Aerospace Installation Diagnostic Equipment
- e: Airborne Insertion Display Equipment
- e: Aircraft Installation Diagnostic Equipment
- e: Automated Image Device Evaluator
- e: Automated Integrated Design and Engineering
- e: Automated Integrated Director Equipment
- e: Automatic Inline Device Evaluator
- e: Automatic Instrumented Decision Equipment
- f: Association Internationale des Distributions d'Eau

AIDEC e: Association for Industrial Development and Economic Cooperation
AIDECS e: Automatic Inspection Device for Explosive Charge Shell
AIDER e: Analog[ue] Implemented Differential Equation-solving Routine
AIDES e: Airborne Infrared Decoy Evaluation System
- e: American Independent Designers and Engineers Society
- e: Analyst Information Display and Exploitation System
- e: Analyst Intelligence Display and Exploitation System
- e: Automated Image Data Extraction System
- e: Automated Interactive Design and Evaluation System

AIDE/TPS e: Advanced Interactive Data Entry/Transaction Processing System
AIDI i: Associazione Italiana per la Documentazione e l'Informazione
AIDIA e: Associate of the Industrial Design Institute of Australia
AID-IFP e: Automatic Indexing of Documentation of Integrated File Processor
AIDIS s: Asociación Interamericana de Ingeneria Saniraria
AIDJEX e: Arctic Ice Dynamics Joint Experiment
AIDL e: Auckland Industrial Development Laboratory
- e: Australian Industrial Development Laboratories

AIDMS e: Applied Information and Data Management System
AIDO e: Air Intelligence Duty Officer
- e: Arab Industrial Development Organization

AIDOS d: automatisiertes Informations- und Dokumentationssystem
- e: Archaeological Information and Documentation Services

AIDOS-FRS d: automatisiertes Informations- und Dokumentationssystem-Faktenrecherchesystem
AIDP e: Advanced Institutional Development Program[me]
- e: Advances in Disease Prevention
- e: Associate of the Institute of Data Processing

AIDPM e: Associate of the Institute of Data Processing Management
AIDR e: Aerospace Internal Data Report
- e: Army Institute of Dental Research

AIDS e: Abort Inertial Digital System
- e: Abstract Information Digest Service
- e: Abstracts Information Dissemination System[s]
- e: Academy of International Dental Studies
- e: Acceptor Idle State
- e: Accident / Incident Data System
- e: Account Identification and Description Services
- e: Acoustic Intelligence Data System
- e: Acquired Immune Deficiency Syndrome
- e: Action Information Display System
- e: Activity Information Data System
- e: Adaptive Intrusion Data System
- e: Administrative Information Data System
- e: Advanced Impact Drilling System
- e: Advanced Integrated Data System
- e: Advanced Integrated Display System
- e: Advanced Interactive Debugging System
- e: Advanced Interactive Display System
- e: Advanced Interconnection Development System
- e: Aerospace Intelligence Data System
- e: Agricultural Information and Documentation Section
- e: Agricultural Information Development Scheme
- e: Airborne Inertial Data System
- e: Airborne Information Display System
- e: Airborne Initial Data System
- e: Airborne Integrated Display System
- e: Aircraft Integrated Data System
- e: Aircraft Interface Data Summaries
- e: Aircraft Intrusion Detection System
- e: Air Force Intelligence Data-handling System
- e: Air Integrated Data System
- e: Alabama Information Development System
- e: All-Purpose Interactive Debugging System
- e: All Individuals Deserve Support
- e: Almost Ideal Demand System
- e: Amdahl Internally Developed Software
- e: American Institute for Decision Science
- e: Analyst Intelligence Data System
- e: Architectural Interaction Design System
- e: Area Intrusion Detection System
- e: Army Information and Data Systems
- e: Army Inventory Data Systems
- e: Artwork-Interactive Design System
- e: Association of Information and Dissemination Centres
- e: Attitudinal Information Data System
- e: Augmented Ignition Delay Sensor
- e: Automated Identification Division System
- e: Automated Information and Documentation System
- e: Automated Information Data (or Display) System
- e: Automated Information Directory System
- e: Automated Information Dissemination System
- e: Automated Installation and Diagnostic Services
- e: Automated Integrated Data System
- e: Automated Integrated Debugging System
- e: Automated Intelligence Data System
- e: Automated Inventory Distribution System
- e: Automatic Illustrated Document[ation] System
- e: Automatic In-Flight Data System
- e: Automatic Information Display System
- e: Automatic Integrated Debugging System
- e: Automatic Integrating Debugging System
- e: Automatic Interactive Debugging System
- e: Automatic Inventory Dispatching System
- e: Automation Instrument Data Service

aids e: acute infectious deseases
AI & DSC e: Army Information and Data System Command
AIDSCOM e: Army Information and Data Systems Command
AID System e: All-purpose Interactive Debugging System
AIDTAC e: Analog[ue] Integrator with Digital Tachometer
AIDUS e: Automated Information Directory Update System
- e: Automated Input and Document Updating System
- e: Automated Input und Document Update System

AIE d: Adsorptions-Isotopie-Effekt
- e: Acceptance Inspection Equipment
- e: Add Instruction Execution
- e: Airborne Interceptor Equipment
- e: American Institute of Esthetics
- e: Analog[ue] Input Expander
- e: Analytical Instruction Element
- e: Army Information Engineering
- e: Associate of the Institute of Education

AIEA e: Australian Institute of Engineering Associates
- f: Agence Internationale de l'Energie Atomique
- s: Agencia Internacional de Energia Atómica

AIEAL e: Australian Institute of Engineering Associates Ltd

AIEC e: Advanced Ion Exchange Cellulose
- e: All-Industry Electronics Conference
- e: Assoctation of Iron Exporting Countries
AIECA e: Associated Independent Electrical Contractors of America
AIECF e: American Indian and Eskimo Cultural Foundation
AIED e: American Institute for Economic Development
AI Ed e: Associate in Industrial Education
AIEd e: Associate of the Institute of Education
AIEDP e: Asian Institute for Economic Development and Planning
AIEE e: American Institute of Electrical Engineers
- e: Associate of Institution of Electrical Engineers
- e: Australias International Engineering Exhibition
AIEE Proc e: American Institute of Electrical Engineers Proceedings
AIEI e: Association of Indian Engineering Industry
AIEKF e: Adaptive Iterated Extended Kalman Filtering
AIEM e: Associate of the Institute of Executives and Managers
AIENDF e: Atomics International Evaluated Nuclear Data Files
- e: Atomics International Evaluation Nuclear Data File[s]
aiep e: amount of insulin extracted from the pancreas
AIER e: American Institute for Economic Research
AIERE e: Associate of the Institution of Electronic and Radio Engineers
AIERI f: Association Internationale des Etudes et Recherches sur l'Information
AIES e: Artificial Intelligence Expert System
- e: Australian Institute of Emergency Services
AIESEC f: Association Internationale des Etudiants en Sciences Economiques et Commerciales
AIET e: Average Instruction Execution Time
- e: Average Instruction Executive Time
AIETA e: Airborne Infrared Equipment for Target Analysis
AIEU e: Armament and Instrument Experiment Unit
AIEWROC e: Army Intelligence / Electronic Warfare Reorganization Overwatch Committee
AIExpE e: Associate of the Institute of Explosives Engineers
AIF d: Amt für industrielle Formgestaltung
- d: Anweisung für den internationalen Fernsprechdienst
- d: Arbeitsgemeinschaft Industrieller Forschungsvereinigungen
- e: Addressless Instruction Format
- e: Aerospace Intelligence File
- e: Affiliated Inventors Foundation
- e: Air Interceptor Fuze
- e: American Industry Foundation
- e: American Institute of France
- e: Amphibian Imperial Forces
- e: Animal Industry Foundation
- e: Annual Improvement Factor
- e: Anti-Invasion Factor
- e: Army Industrial Fund
- e: Artificial Interstitial Fluid
- e: Assembly Inactive File
- e: Atomic Industrial Forum
- e: Atomic International Forum
- e: Audience Interest Factor
- e: Australian Institute of Fuel
- e: Automated Installation File
- e: Automated Intelligence File
AIFA e: Agricultural Insecticide and Fungicide Association
- e: Associate of the International Faculty of Arts
AIFB e: American Institute of Financial Brokers
AIFCS e: Airborne Interception Fire Control System
AIFD e: Acute Intrapartum Fetal Distress
- e: Alaska Institute for Fisheries Development
- e: American Institute of Food Distribution
AIFE e: American Institute for Exploration
- e: Associate of the Institution of Fire Engineers
AIFEE e: All-India Federation of Electricity Employees
AIFG e: American Institute of Professional Geologists
AIFI e: Automatic Inflight Insertion
AI Fire E e: Associate of the Institute of Fire Engineers
AIFLD e: American Institute for Free Labor Development
AIFM e: Associate of the Institute of Factory Managers
- e: Automatic Integrating Fluctation Meter
- f: Association Internationale des Femmes Médecins
AIFOS e: Academic Instructor and Foreign Officer School
AIFRB e: American Institute of Fishery Research Biologists
AIFS e: Advanced Indirect Fire System
- e: Advanced Instruction Flying School
- e: Advanced Integrated Flight System
- e: American Institute for Foreign Study
- e: Australian Institute of Forensic Sciences
AIFSA e: American Institutions Food Service Association
AIFSSF e: American Institute for Foreign Study Scholarship Foundation
AIFST e: Associate of the Institute of Food Science and Technology
- e: Australian Institute of Food Science and Technology
AIFT d: Aluminium-Formol-Toxoid
- e: American Institute for Foreign Trade
- e: Americans for Indian Future and Traditions
- e: Audio Input Frequency Tolerance
AIFTA e: Anglo-Irish Free Trade Area
- e: Associate of the Institute of Freight Trades Association
AIFTDS e: Airborne Integrated Flight Test Data System
AIFU e: Automated Instruction Fetch Unit
AIFUC e: Assignment Instructions were Furnished to Your Command
AIFUCTO e: All-India Federation of University and College Teachers Organization
AIFURC e: Assignment Instructions were Furnished to Your Command
AIFV e: Armored Infantry Fighting Vehicle
AIG d: Arbeitskreis Internationaler Gemeinschaftsdienste
- d: Auslandsinvestitionsgesetz
- e: Accident Investigation
- e: Address-Indicating Group
- e: Adjutant Inspector General
- e: Air Inspector General
- e: Air Intelligence Group
- e: Alcohol-Insoluble Glutamine
- e: Alcohol-Insoluble Glutelin
- e: All-Inertial Guidance
- e: Angle of Inner Gimbal
- e: Anti-Immunoglobulin
- e: Architectural Inventory Group
- e: Army Inspector General
- e: Artificial Intelligence Group
- e: Assistant Inspector General
- e: Assistant Instructor in Gunnery
- e: Australian Institute of Geoscientists
- f: Association Internationale de Géodesia
- f: Association pour l'Informatique de Gestion
Aig f: Aiguille
AIGA e: American Institute of Graphic Arts
AIGBI e: Archaeological Institute of Great Britain and Ireland
AIGC e: American Indian Graduate Center
- e: American Institute of Group Counseling
AIGCM e: Associate of the Incorporated Guild of Church Musicians
AIGD e: Airborne Interception and Gun Direction
AIGE e: Astro Inertial Guidance Equipment
AIGM e: Associate of the Institute of General Managers
AIGS e: Acoustic Intelligence Gathering System
- e: Agricultural Investment Grant Scheme
- e: All Inertial Guidance System
- e: Auxiliary Inerting Gas Subsystem
AIGSS e: Annual Inert Gas System Survey
AIGT e: Association for the Improvement of Geometrical Teaching
AIH e: All in Hand
- e: American Institute of Homeopathy
- e: American Institute of Hydrology
- e: Artificial Insemination by Husband
- e: Aspen Institute of the Humanities
- e: Australian Institute of Health
AIHA e: American Industrial Hygiene Association
- e: Associate of the Institute of Hospital Almoners
- e: Autoimmune Hemolytic Anemia
AIHC e: American Industrial Health Conference
- e: American Industrial Health Council
AIHE e: Association for Innovation in Higher Education
AIHEC e: American Indian Higher Education Consortium
AIHED e: American Institute for Human Engineering and Development
AIHEX e: Asian International Hardware Exposition
AIHP e: American Institute of the History of Pharmacy
AIHR e: American Indian Horse Registry
- e: Association of International Health Researchers
AIHS e: American Indian Historical Society
- e: American Irish Historical Society
- e: Aspen Institute of Humanistic Studies
- e: Aspen Institute of Human Studies
- e: Australian Institute of Health Surveyors
- f: Association Internationale d'Hydrologie Scientifique
AIHSA e: American Insurers Highway Safety Alliance
AIHSC e: Auto Industry Highway Safety Committee
AIHsg e: Associate of the Institute of Housing

AIHVE e: Associate of the Institution of Heating and Ventilating Engineers
AIHX e: Auxiliary Intermediate Heat Exchanger
AII e: Acceptance Inspection Instruction
- e: Acquired Intelligence, Incorporated
- e: Aerial Inspection Instrument
- e: Air India International
- e: American Institute, Incorporated
- e: American Interprofessional Institute
- e: Apollo Implementing Instructions
- e: Army Intelligence Interpreter
- e: Australian Insurance Institute
- e: Automatic Imagery Interpretation
aII e: Anode of Section Two
AIIA e: Associate of the Indian Institute of Architects
- e: Associate of the Institute of Industrial Administration
- e: Associate of the Insurance Institute of America
- e: Atlantic Institute for International Affairs
- e: Australian Information Industry Association
- e: Australian Institute of International Affairs
AIIAL e: Associate of the International Institute of Arts and Letters
AIIB e: Allied Irish Investment Bank
AIIC e: American Institute of Intermediate Coinage
- e: Army Imagery Intelligence Corps
- e: Associate of the Insurance Institute of Canada
AIID e: American Institute of Interior Design
AIIDC e: Authorized Item Identification Data Collaborator
AIIDC Code e: Authorized Item Identification Data Collaborator Code
AIIDR e: Authorized Item Identification Data Receiver
AIIDR Code e: Authorized Item Identification Data Receiver Code
AIIDS e: Authorized Item Identification Data Submitter
AIIDS Code e: Authorized Item Identification Data Submitter Code
AIIE e: American Institute of Industrial Engineers
AIIEA e: All-India Insurance Employees Association
AIIF e: Automated Installation Intelligence File
AIIHPH e: All India Institute of Hygiene and Public Health
AIIL e: American Institute of International Law
AIIM e: Associate of the Institution of Industrial Managers
- e: Association for Information and Image Management
AIIMS e: All-India Institute of Medical Sciences
AIInfSc e: Associate of the Institute of Information Scientists
AIIP e: Advanced Instrumentation for Industry Project
- e: Associate of the Institute of Incorporated Photographers
- e: Association of Independent Information Professionals
- e: Australian Institute of Industrial Psychology
AIIPC e: American International Institute for the Protection of Childhood
AIIPMR e: All India Institute of Physical Medicine and Rehabilitation
AIIR e: Air Intelligence Information Report
AIIS e: American Institute for Imported Steel
- e: American Institute for International Steel
- e: American Institute of Indian Studies
- e: American Institute of Iranian Studies
- e: American Institute of Islamic Studies
- e: American Intraocular Implant Society
- e: Anterior Inferior Iliac Spine
- e: Automated Import Inspection System
AIISL e: Associate of the International Institute of Arts and Letters
AIITech e: Associate of the Institute of Incorporated Technologists
AIJ e: Ampullary-Isthmic Junction
AIJA e: Australian Institute of Judicial Administration
AIJPA e: Artificial Intelligence Job Performance Aid
AIJWF e: All-India Jute Textile Workers Federation
AIK d: Agrarinvestitionskredit
- d: allgemeiner Immittanz-Konverter
AIKD e: American Institute of Kitchen Dealers
AIKR e: Artificial Intelligence Knowledge Representation
AIL e: Academy of International Law
- e: Action Item List
- e: Administrative / Intelligence / Logistics
- e: Advance Information Letter
- e: Aeronautical Instrument[s] Laboratory
- e: Aerospace Instrumentation Laboratory
- e: Aileron
- e: Airborne Instruments Laboratory
- e: Aircraft Instrument Laboratory
- e: Air Intelligence Liaison
- e: Amber Indicating Lamp
- e: Amber Indicating Light
- e: American Institute of Laundering
- e: Angioimmunoblastic Lymphadenopathy
- e: Argon Iron Laser
- e: Arithmetic Input Left
- e: Array Interconnection Logic
- e: Artificial Intelligence Laboratory
- e: Art Institute of Light
- e: Associate of the Institute of Linguists
- e: Association of International Libraries
- e: Audio Input Level
- e: Australian Institute of Librarians
- e: Average Inventory Level
- e: Aviation Instrument Laboratory
- e: Avionics Integration Laboratory
A.I.L. e: Associate of the Institute of Linguistics
AILA e: Airborne Instruments Laboratory Approach
- e: American Institute of Landscape Architects
- e: Associate of the Institute of Land Agents
- e: Associate of the Institute of Landscape Architects
- e: Australian Institute of Landscape Architects
- f: Association Internationale de Linguistique Applique
AILACT e: Association for Informal Logic and Critical Thinking
AILAM e: Associate of the Institute of Leisure and Amenity Management
AILAS e: Airborne Integrated Light Avionics System
- e: Automatic Instrument Landing Approach System
- e: Automatic Instrument Low-Approach
AILD e: Angioimmunoblastic Lymphadenopathy with Dysproteinemia
AILE e: Arterial Insufficiency of the Lower Extremities
AILGA e: Associate of the Institute of Local Government Administrators
AIL/NA e: Association of International Libraries / North America
AILO e: Air Intelligence Liaison Officer
AI Loco E e: Associate of the Institution of Locomotive Engineers
AILS d: Automatisiertes Informationslenkungssystem der Standardisierung (DDR)
- e: Advanced Integrated Landing System
- e: Airborne Infrared Live Scanner
- e: Angular Intensity Light Scattering
- e: Automatic Instrument Landing System
AILSA e: Aerospace Industrial Life Sciences Association
AILSS e: Advanced Integrated Life Support System
AILV e: Artichoke Italian Latent Virus
AIM d: Arbeitsgemeinschaft Information Meeresforschung und Meerestechnik
- d: Arbeitskreis industrielles Messen
- d: assoziative Index-Methode
- d: Automatikimpulszeitmesser
- d: automatisches Informationsmanagement
- e: Abridged Index Medicus (journ.)
- e: Abstracts of Instructional Material[s]
- e: Abstracts of Instructional Materials in Vocational and Technical Education (journ.)
- e: Academy for Interscience Methodology
- e: Academy Introduction Mission
- e: Access Isolation Mechanism
- e: Access to Information for Medicine
- e: Achievement Identification Measure
- e: Action for Independent Maturity
- e: Active Inert Missile
- e: Active Integrated Module
- e: Adaptive Injection Moulding
- e: Add, Initial, Multiprecision
- e: Adhesive Insulation Material
- e: Advanced Information Management
- e: Advanced Information Manager
- e: Advanced Intercept Missile
- e: Advanced Interface Module
- e: Advance Information Memory
- e: Aerial Independent Model
- e: Aerial Intercept Missile
- e: Aerosol Inhalation Measurement
- e: Aerospace Industrial Modernization
- e: Aerothermodynamic Integration Model
- e: Aerotriangulation [by Observation] of Indepenent Models
- e: Africa Inland Mission
- e: After Image
- e: Aid of International Medicine
- e: Aid to Improved Marksmanship
- e: Air-Isolated Monolithic [Circuit]
- e: Air-launched Interceptor Missile
- e: Airborne Infrared Mapper
- e: Air Defense Command Intelligence Memorandum
- e: Air Intercept Missile
- e: Air Isolated Monolithic
- e: Airmans Information Manual
- e: Alarm Indicating Monitor
- e: Alarm Indication Monitor
- e: American Innerspring Manufacturers
- e: American Institute for Microminiaturization
- e: American Institute of Maintenance
- e: American Institute of Management
- e: American Institute of Musicology

- e: American Interactive Media, Inc.
- e: Analog[ue] Input Module
- e: Analog[ue] Input Multiplexer
- e: Annals of Internal Medicine (journ.)
- e: Apogee Injection Module
- e: Application Integrated Module
- e: Application Integration Module
- e: Application Interface Module
- e: Application of Integrated Modules
- e: Area Interdiction Mine
- e: Armored-Infantry-Mechanized
- e: Army Installation Management
- e: Artificial Intelligence in Medicine
- e: Asian Institute of Management
- e: Assembly Instruction Mnemonics
- e: Assistant Industrial Manager
- e: Associated Industries of Massachusetts
- e: Associated Information Managers
- e: Associate in Industrial Management
- e: Associate of the Institution of Metallurgists
- e: Association for Information Management
- e: Association for Innovative Marketing
- e: Association of Independent Microdealers
- e: Association of Information Management
- e: Association of Information Managers for Financial Institutions
- e: Associative Index Method
- e: Astrometric Interferometry Mission
- e: Asynchronous Interface Module
- e: Atoms in Molecules
- e: Australian Inland Mission
- e: Australian Institute of Management
- e: Australian Institute of Metals
- e: Authoring of Instructional Materials
- e: Automated Information Management
- e: Automated Integrated Manufacturing
- e: Automated Intelligent Microscope
- e: Automated Inventory Management
- e: Automatic Identification Manufacturers
- e: Automatic Inflation Module
- e: Automatic Invoicing Machine
- e: Automation of Interlending by Microcomputer
- e: Automotive Industrial Motor
- e: Autonomous Infantry Mortar
- e: Avalanche Induced Migration
- e: Awaiting Incoming Message

AIM-47A e: Hughesair-to-air missile
AIM-9 e: Sidewinderair-to-air missile
AIMA e: Acoustical and Insulating Materials Association
- e: American Industrial Music Association
- e: As Interest May Accrue
- e: As Interest May Appear
- e: Associate of the Institute of Municipal Administration
- e: Australian Information Management Association

AIMACC e: Air Material Command Compiling [System]
AIMACC System e: Air Material Command Compiling System
AIMACO e: Air Material Command
- e: Air Material Computer
- e: Air Materiel Command Compiler [Language]

AI Mar E e: Associate of the Institute of Marine Engineers
AIM/ARM e: Abstracts of Instructional Material/Abstract of Research Material
AIMB e: American Institute of Mortgage Brokers
AIMBE f: Association Internationale de Medicine et de Biologie de l'Environnement
AIMBM e: Associate of the Institute of Municipal Building Management
AIMBW e: American Institute of Mens and Boys Wear
AIMC e: Academic Information Management Center
- e: American Indian Medical Clinic
- e: American Institute of Medical Climatology
- e: Army Installation Management Course
- e: Association of Internal Management Consultants
- e: Auto-Initiate Manual-Confirm

AIMCA e: Associate of the Institute of Management Consultants in Australia
AIMCAL e: Association of Industrial Metallizers, Coaters and Laminators
AIMCC e: Audi International Motor Car Club
AIM Circuit e: Air Isolated Monolithic Circuit
AIMD e: Aircraft Intermediate Maintenance Department
- e: American Institute of Marine Development

AIME e: American Institute for Mining Engineers
- e: American Institute of Mechanical Engineers
- e: American Institute of Mining Engineers
- e: American Invitational Mathematics Examination
- e: Associate of the Institute of Marine Engineers
- e: Associate of the Institute of Mechanical Engineers
- e: Associate of the Institute of Mining Engineers
- e: Associate of the Institute of Municipal Entertainment
- e: Association for Informational Media and Equipment
- e: Australian Institute of Mining Engineers
- e: Automatic-Digital [Information] Network Interface in a Multiprogramming
- e: Automatic In-process Microcircuit Evaluation
- e: Average Indexed Monthly Earnings

AIMEA e: Applied Immunoenzymometric Assay
AIMechE e: American Institute of Mechanical Engineers
AI Mech E e: Associate of the Institution of Mechanical Engineers
AIMED e: Association of Independent Mailing Equipment Dealers
AIMEE e: Associate of the Institution of Mechanical Engineers
AIMES e: Association of Independent Medical Equipment Suppliers
- e: Association of Interns and Medical Students

Aimes e: Automated Inventory Management Evaluation System
AIMES e: Avionics Integrated Maintenance Expert System
AIMEVAL e: Airborne Intercept Missile Evaluation
AIMF e: Australian Institute of Metal Finishing
AIMH e: Academy of International Military History
AIMI e: Airborne Infrared Measurement Instrument
- e: Aircraft Intensively Managed Items
- e: Associate of the Institute of the Motor Industry
- e: Aviation Intensive Management Items

AIMILO e: Army/Industrial Material Information Liaison Office[s]
AIMinE e: Associate of the Institution of Mining Engineers
AIMIS e: Advanced Integrated Modular Instrumentation System
AIMIT e: Associate of the Institute of Musical Instrument Technicians
- e: Associate of the Institute of Musical Instrument Technology

AIMLC e: Association of Island Marine Laboratories of the Caribbean
AIMLS e: Associate of the Institute of Medical Laboratory Sciences
AIMLT e: Australian Institute of Medical Laboratory Technology
AIM & M e: American Institute of Mining and Metallurgy
AIMM e: Associate of the Institution of Mining and Metallurgy
- e: Australasian Institute of Mining and Metallurgy

AIMME e: American Institute of Mining and Metallurgical Engineers
AIMMI e: Australasian Institute of Mining and Metallurgy
AIMMPE e: American Institute of Mining, Metallurgical and Petroleum Engineers
AIMO e: Air Mo[u]ld
- e: Audible Instructed Manufacturing Operation[s]

AIMP e: Air Intercept Missile Package
- e: Anchored Interplanetary Monitoring Platform

AIMPA f: Association Internationale de Météorologie et de Physique de l'Atmosphère
AIMPAP e: Asbestos Inspection and Management Plan Assistance Program[me]
AIMPE e: Australian Institute of Marine and Power Engineers
AIMPG e: American Importers Meat Products Group
AIM Process e: Avalanche-Included Migration Process
AIMR e: Association for the Improvement of the Mississippi River
AIMRA e: Agricultural and Industrial Manufacturers Representatives Association
AIM/RDB e: Advanced Information Manager/Relational Data Base
AIMRT e: American Institute for the Medical Research of Trauma
AIMS e: Abnormal Involuntary Movement Scale
- e: Academic Instructional Measurement System
- e: Advanced Image Management Software
- e: Advanced Imagery Manipulation System
- e: Advanced Inertial Measurement System
- e: Advanced Information Management System
- e: Advanced Institutional Management Software, Inc.
- e: Advanced Intercontinental Missile System
- e: Agricultural Information and Marketing Services
- e: Airborne Identification Mobile System
- e: Airborne Integrated Maintenance System
- e: Aircraft Identification Military System
- e: Aircraft Inflight Monitoring System
- e: Aircraft Integrated Munition System
- e: Air Infiltration Measurement Service
- e: Airplane Information Management System
- e: Airport Income Management System

- e: Air Traffic Control Radar Beacon/ Identification Friend or Foe/ Mark XII/System
- e: Airways Integrating and Monitoring System
- e: Altitude Identification Military System
- e: American Institute for Mathematical Statistics
- e: American Institute for Mental Studies
- e: American Institute of Maritime Services
- e: American Institute of Marxist Studies
- e: American Institute of Mathematical Statistics
- e: American Institute of Merchant Shipping
- e: American Institute of Musical Studies
- e: American International Managers Society
- e: Analysis of Internal Management Systems
- e: Annual Improvement Maintenance and Support
- e: Applied Information Management Sciences, Inc.
- e: Applied Information Management System
- e: Arbat Interactive Multi-user System
- e: Architecture to an Interactive Migration System
- e: Arlington Inventory Management System
- e: Army Information Management System
- e: Army Insecticide Measuring System
- e: Army Integrated Meteorological Systems
- e: Army Integrated Microfilm System
- e: Arson Information Management System
- e: Arthritis Impact Measurement Scales
- e: Assessment, Improvement and Monitoring System
- e: Associated Iron Moulders of Scotland
- e: Association for International Medical Study
- e: Audit Information Management System
- e: Auerbach Information Management System
- e: Australian Industries Management Services
- e: Australian Institute of Marine Science
- e: Australian Institute of Marine Studies
- e: Autoadaptive Inventory Management System
- e: Automated Industrial Management System
- e: Automated Information and Management System
- e: Automated Instructional Management System
- e: Automated Instructional Materials-handling System
- e: Automated Instructional Materials Services
- e: Automatic Industrial Management System
- e: Automatic Interference Measurement System
- e: Automotive Information Management System

AIMs e: Advanced Information Memos
AIMS Group e: Audiovisual Instructional Media Services Group
AIMSO e: Aircraft Intermediate Maintenance Support Office
AIM System e: Advanced Information Management System
AIMT e: Association for Integrated Manufacturing Technology
- e: Australian Institute of Medical Technologists

AIMTA e: Associate of the Institute of Municipal Treasures and Accountants
AIM-TWX e: Abridged Index Medicus-Teletypewriter Exchange
AIM-TWX Network e: Abridged Index Medicus via the Teletypewriter Exchange Network
AIMU e: American Institute of Marine Underwriters
AIN e: Aircraft Identification Number
- e: Airframe Integrated Nozzle
- e: Alternative Information Network
- e: American Information Network
- e: American Institute of Nutrition
- e: Anal Intraepithelial Neoplasia
- e: Analog[ue] Input
- e: Approved Item Name
- e: Army Installation
- e: Assembly Identification Number
- e: Assistant in Nursing
- e: Auditory Interneuron
- e: Australian Institute of Navigation

ain e: acute interstitial nephritis
A/in e: Amperes per inch
A/in² e: Amperes per Square inch
AINA e: American Institute of Nautical Archaeology
- e: Arctic Institute of North America
- e: Associate of the Institution of Naval Architects

aina e: automated immunoephelometric assay
AINAI e: African Integrated Network of Administrative Information
AINBN e: Association for the Introduction of New Biological Nomenclature
aind. d: altindisch
AIND d: Aussageindex
A-Ind e: Anglo-Indian
AINDT e: Australian Institute for Non-Destructive Testing
AINDTN e: Air Induction
AINFBESCH d: Ausgangsinformations-beschreibung
A in G Ed e: Associate in General Education
A/INL e: Air Inlet
AINM e: Assistant Inspector; Navy Material
- e: Assistant Inspector of Naval Materiel

AINO e: Assistant Inspector of Naval Ordnance
AINRP e: Approved Item Name Reclassification Program[me]
AINS e: Advanced Inertial Navigation System
- e: Area Inertial Navigation System
- e: Assateage Island National Seashore

AINSE e: Argonne Institute of Nuclear Science and Engineering
- e: Australian Institute of Nuclear Science and Engineering

A Inst AM e: Associate of the Institute of Administrative Management
AInstBB e: Associate of the Institute of British Bakers
AInstBCA e: Associate of the Institute of Burial and Cremation Administration
A.Inst.C.E. e: Associate of the Institution of Civil Engineers
A Inst ExE e: Associate of the Institute of Executive Engineers
AInstFF e: Associate of the Institute of Freight Forwarders
A Inst HE e: Associate of the Institution of Highway Engineers
A Inst M e: Associate of the Institute of Marketing
A Inst MO e: Associate of the Institute of Market Officers
A Inst MSM e: Associate of the Institute of Marketing and Sales Management
A Inst NA e: Associate of the Institution of Naval Architects
A Inst P e: Associate of the Institute of Patentees
- e: Associate of the Institute of Physics

A Inst PA e: Associate of the Institute of Park Administration
AInstPet e: Associate of the Institute of Petroleum
AInstPI e: Associate of the Institute of Patentees and Inventors
A Inst PS e: Associate of the Institute of Purchasing and Supply
AInstSMM e: Associate of the Institute of Sales and Marketing Management
AInstTA e: Associate of the Institute of Transport Administration
AInsuf e: Aortic Insufficiency
A Int e: Air Intelligence
AIN Timber e: Aini Timber
AINTM e: American Institute of Nail and Tack Manufacturers
A Int O e: Air Intelligence Officer
AINWR e: Aleutian Islands National Wildlife Refuge
AINZ e: Advertising Institute of New Zealand
AIO e: Action Information Organization
- e: Activity, Interest and Opinion
- e: Admiralty Inspection Officer
- e: Airborne Infrared Observatory
- e: Airborne Interceptor Officer
- e: Airborne Ionospheric Observatory
- e: Air Installation Office
- e: Air Intelligence Officer
- e: Air Intercept Officer
- e: Aldosterone-Induced Polypeptide
- e: Aldosterone-Induced Protein
- e: American Institute of Organbuilders
- e: American Institute of Physics
- e: Analog[ue] Input-Output
- e: Analog[ue] Input Operation

AIO e: Anti-Inflammatory Protein
- e: Arecibo Ionospheric Observatory
- e: Arithmetic and Logic Unit Input and Output
- e: Army Intelligence Officer
- e: Army Inventary Objective
- e: Artillery Intelligence Officer
- e: Assistant Information Officer

AI & O e: Annual Inspection and Overhaul
AIOA e: American Iron Ore Association
AIOB e: American Institute of Oral Biology
- e: Analog[ue] Input-Output Board
- e: Associate of the Institute of Builders
- e: Associate of the Institute of Building

AIOC e: Anglo-Iranian Oil Company
- e: Assistant Instrumentation Operations Coordination
- e: Associate of the Institute of Carpenters

AIOCC e: Associate Infantry Officer Career Course
AIOD e: Aorto-Iliac Occlusive Disease
- e: Automatic Identification of Outward Dialling
- e: Automatic Identification on Out-Dialled calls
- e: Automatic Identified Outward Dialling

AIOD Calls e: Automatic Identification on Out-Dialled Calls
AIOEC e: Association of Iron Ore Exporting Countries

AI of M&M e: Australasian Institute of Mining and Metallurgy
AION e: Alphabetical Index of Names
AIOP e: Analog[ue] Input / Output Package
AIOPI e: Association of Information Officers in the Pharmaceutical Industry
AIOS e: Auxilary Input/Output Statement
- e: Auxiliary Input/Output Software
AIOTT e: Action Information Operations Tactical Trainer
AIOW e: Association of Independent Optical Wholesalers
AIP e: Ablative Insulative Plastic
- e: Acceptance Inspection Package
- e: Accident Insurance Policy
- e: Acute Intermittent Porphyria
- e: Advanced Information Processing
- e: Advanced Interceptor Propulsion
- e: Aeronautical Information Publication
- e: Aeronautical Information Publication[s]
- e: Airborne Instrumentation Platform
- e: Air Intake Panel
- e: Aldosterone-Induced Polypeptide
- e: Aldosterone-Induced Protein
- e: Allied Intelligence Publication
- e: Alphanumeric Impact Printer
- e: Alternate Inspection Policy
- e: American Institute for Psychoanalysis
- e: American Institute of Physics
- e: American Institute of Planners
- e: American International Pictures, Incorporated
- e: Amplifier Input
- e: Annual Implementation Plan
- e: Annual Improvement Factor
- e: Anti-Inflammatory Protein
- e: Approval in Principle
- e: Army Information Program[me]
- e: Arylene Isopropylidene Polymers
- e: Assault on Illiteracy Program[me]
- e: Associate of the Institute of Physicians
- e: Associate of the Institute of Physics
- e: Associate of the Institute of Plumbing
- e: Association of Independent Producers
- e: Attention and Information Processing
- e: Australian Institute of Petroleum
- e: Australian Institute of Physics
- e: Automated Imagery Processing
- e: Automated Implementation Plan
- e: Automated Information Processing
- e: Automatic Information Processing
- e: Automatic Input Processing
- e: Aviation Indoctrination Program[me]
- e: Avionics Integration Plan
- f: Association Internationale de Papyrulogues
- f: Association Internationale de Pediatrie
A-I-P e: Afghanistan-Iran-Pakistan
aip e: average intravascular pressure
AIPA d: Allgemeines Informations- und Planungsamt
- e: American Ionospheric Propagation Association
- e: Australian Institute of Park Administration
- f: Association Internationale de la Psychologie Adlerienne
AIPB e: American Institute of Professional Bookkeepers
AIPC e: American Institute of Polish Culture
- e: Army Installations Planning Committee
- e: Australian Institute of Pest Control
- f: Association Internationale de Prophylaxie de la Cécité
AIPCEE f: Association des Industries du Poisson de la Communauté Economique Européenne
AIPCN f: Association Internationale Permanente des Congrés Navigation
AIPD e: Army Institute for Professional Development
AIPE e: American Institute for Professional Education
- e: American Institute of Park Executives
- e: American Institute of Plant Engineers
- e: Associate of the Institution of Production Engineers
AIPEDD e: American Institute for the Prevention and Eradication of Dental Disease
AIPet e: Associate of the Institute of Petroleum
- e: Australian Institute of Petroleum
AIPFR e: Atomics International Prototype Fast Reactor
AIPG e: American Institute of Professional Geologists
AIPH e: Agricultural and Industrial Process Heat
AIPHE e: Associate of the Institution of Public Health Engineers
AIPL e: Animal Improvement Programs Laboratory
AIPLU e: American Institute for Property and Liability Underwriters
AIPM e: Associate of the Institute of Personnel Management
AIPMI e: Acute Infero-Posterior Myocardial Infarction
AIPO e: American Institute of Public Opinion
AIPP e: American Institute of Pollution Prevention
- e: Arctic Islands Pipeline Program[me]
- e: Army Industrial Preparedness Program[me]
- e: Australian Institute of Professional Photography
AIPR e: American Institute of Pacific Relations
- e: American Institute of Public Relations
- e: Applied Imagery Pattern Recognition
- e: Australian Institute of Parks and Recreation
- e: Australian Institute of Psychic Research
- e: Automated Information Processing Request
- e: Average Input Pulse Rate
AIPS e: Advanced Integrated Propulsion System
- e: Advanced Interactive Presentation System
- e: AIDS-Induced Panic Syndrome
- e: Applications Information Processing System
- e: Army Information Processing Standards
- e: Astronomical Image Processing System
- e: Australian Institute of Political Science
- e: Automated Information Processing System
- e: Automated Intelligence Processing System
- e: Automatic Indexing and Proofreading System
- e: Average Instructions per Second
- f: Association Internationale pour la Prevention du Suicide
AIPSO e: Auto Insurance Plans Services Office
AIP System e: Automated Immunoprecipitin System
AIPT e: Assistant Inspector of Physical Training
AIPTS e: Active Imaging Pointer-Tracker System
AIPU e: Associative Information Processing Unit
AI & Q e: Animal Inspection and Quarantine
AIQ e: Associate of the Institute of Quarrying
AIQPS e: Associate of the Institute of Qualified Private Secretaries
AIQS e: Associate of the Institute of Quantity Surveyors
AIR d: Amino-Imidazole-Ribonucleotide
- e: Accelerated Item Reduction
- e: Acoustic Intercept Receiver
- e: Action for Industrial Recycling
- e: Action Item Report
- e: Action of Instant Recording
- e: Active Index Register
- e: Acute Insulin Response
- e: Adaptive Intercommunication Requirement
- e: Additional Information Request
- e: Advanced Integration Research
- e: Aerial Input Resistance
- e: Aeronautical Information Report
- e: Aerospace Information Report
- e: After Initial Release
- e: Air-Arming Impact Rocket
- e: Airborne Interceptor Rocket
- e: Airborne Intercept Radar
- e: Aircraft Incident Report
- e: Aircraft Inspections and Repair
- e: Aircraft International Register
- e: Aircraft Inventory Record
- e: Air Inflatable Retarder
- e: Air Injection Reactor
- e: Air Intercept Rocket
- e: Airworthiness
- e: All[-]India Radio
- e: Alphabetical Information Recording
- e: American Institute for Research
- e: American Institute of Refrigeration
- e: American Institute of Research
- e: Antenna Input Resistance
- e: Applied Information Resources
- e: Archives of Interamerican Rheumatology
- e: Arithmetic Input Right
- e: Army Institute of Research
- e: Army Intelligence Reserve
- e: Assembly Inspection Report
- e: Association for Institutional Research
- e: Association Interrogation Register
- e: Attention Interrupt Request
- e: Australian Industrial Refractories
- e: Australian Institute of Radiography
- e: Automated Image Retrieval
- e: Automated Interrogation Routine
- e: Automatic Indexing and Retrieval
- e: Automatic Indication of Range
- e: Automatic Information Reduction
- e: Automatic Interrogation Routine
- e: Average Injection Rate
- e: Aviation Item Reports
air. d: altirisch
a.i.R. d: auch in Reinschrift
AIR-2A e: Douglas air-to-air rocket
AIRA e: Air Attaché
- e: Alberta Information Retrieval Association
- e: Allergies and Intolerant Reactions Association

AIRAC e: Aeronautical Information Regulation and Control
- e: All-Industry Research Advisory Council
- e: Atmospheric Infrared Attenuation Coefficient
AIRACCDT e: Air Accident
AIRACLIS e: Air Activities Logistic Information System
AIRACS e: Aircraft Acquisition and Support
AIRAD e: Airmens Advisory
AI Radar e: Air Interception Radar
AIRAD net e: Air Administration net
AIRAF e: Aircraft; Asiatic Fleet
AIRAH e: Australian Institute of Refrigeration, Air-Conditioning and Heating
AIRAP e: Air Pressure Analysis Program[me]
AIRARMUNIT e: Aircraft; Armament Unit
AIRBAREX e: Air Barrier Exercise
AIRBASECOM e: Air Base Commander
AIRBM e: Anti-Intermediate Range Ballistic Missile
AIRBR f: Association Internationale du Registre des Bateaux du Rhin
AIRBS e: American Institute for Research in the Behavioral Sciences
AIRBUT e: Automatic Resupply and Build-up Time
AIRC e: Association of Independent Radio Contractors
AIRCAL e: Air California
Air Can e: Air Canada
aircat e: automated integrated radar control for air traffic
Air Cav e: Airmobile Cavalry
Air Cdr e: Air Commander
AIRCENT e: Allied Air Forces Central Europe
AIRCEY e: Air Ceylon
Aircft e: Aircraft
AIRCLNR e: Air Cleaner
Air Cmdre e: Air Commodore
AIRCO e: Air Coordinator
- e: Air Reduction Chemical Company
Air Coal e: Airport Coalition
AIRCOM e: Aerospace Communications
- e: Air Command
- e: Air Communications
- e: Air Force Communication Complex
- e: Airways Communications
AIRCOMD net e: Air Command net
AIRCOMNET e: Air Communications Network
- e: Air Force Communications Network
AIRCOMS e: Airways Communications System
AIRCOM System e: Air[ways] Communications System
AIRCON e: Automated Information and Reservation Combined Operations Network
- e: Automated Information and Reservation Computer Operated Network
- e: Automated Information and Reservation Computer Oriented Network
AIRCOND e: Air-Condition[ed]
- e: Air Conditioning
Air Cond. Engr. e: Air Conditioning Engineer
AIRCSC e: Air Command and Staff College
AIRD e: American Indian Research and Development
- e: Australian Industrial Research Directory
AIRDEF e: Air Defense
AIRDEFCOM e: Air Defense Commander
AirDefSch e: Air Defense School
Air Def Sec e: Air Defense Sector
AIRDELOPS e: Air Delivery Operations
AirDevCen e: Air Development Center
AIRDIVDEF e: Air Division Defense
AIRE e: American Institute of Radio Engineers
- e: Australian Institute of Radio Engineers
AIREA e: American Institute of Real Estate Appraisers
AIRELO e: Air Electrical Officer
AIREN e: American Institute for Research and Education in Naturopathy
AIRENGPROPACCOVERHAUL
- e: Airplane Engine, Propeller and Accessory Overhaul
AIREO e: Air Engineer Officer
AIREP e: Aircraft Report
AiRepDn e: Aircraft Repair Division
AIRES e: Advanced Imagery Requirements and Exploitation System
- e: Automated Information Resource System
AIREVAC e: Air Evacuation
AIREW e: Airborne Infrared Early Warning
AIRF e: Aircraft Instrument Repair Facility
- e: All-India Railwaymens Federation
- e: Assignment Instructions Remain Firm
AIRFAM e: Aircraft Familiarization
airfd e: Airfield
airfil e: air filter[s]
AIRGI e: Airmans Guide (journ.)
AIRGLO e: Airborne Infrared Gunfire Locator
AirGndOpSch e: Air-Ground Operations School
AIRGRP e: Air Group
AIRH f: Association Internationale des Recherches Hydrauliques
AIRHC e: Alaska International Rail and Highway Commission
air hp e: air horsepower
AIRHPER e: Alberta Information Retrieval for Health, Physical Education and Recreation
AIRI e: Associate of the Institution of the Rubber Industry
- e: Association of Independent Research Institutes
- e: Atomic Industry Research Institute
AIRIMP e: Air Reservations Interline Message Procedure
AIR INTEL e: Air Intelligence
AIRIS e: Advanced Infrared Imaging Seeker
- e: Air-Store Issuing Ship
Air Jam e: Air Jamaica
AIRL e: Aeronautical Icing Research Laboratory
- e: Automation Industries Research Laboratory
Air LO e: Air Liaison Officer
AIRLOC e: Air Lines of Communication
AIRLORD e: Airlines Load Optimization Recording and Display
AIRLORD PNC e: Airlines Load Optimization Recording and Display Passenger Name Check-in
AIRLORDS e: Airlines Load Optimization Recording and Display System
AIRLS e: Advanced Interrogation, Recording and Location System
AIRM e: Airborne Infrared Mapper
Air M e: Air Marshal
AIRMA e: Air Military Attaché
- e: All India Radio Manufacturers Association
Air Mad e: Air Madagascar
AIRMADA e: Airplane Armada
AIRMAP e: Air Monitoring Analysis and Prediction
AIRME e: Apollo Initiator Resistance Measuring Equipment
AIRMET e: Airmens Meteorological Information
AIRMG e: Aircraft; Machine Gunner
AIRMICS e: Army Institute of Research in Management Information and Computer Science
airmiss e: aircraft-in-flight collision barely missed
air/mmh e: acoustic intercept receiver/ multimode hydrophone
airmnap e: air monitoring, analysis and prediction
AIRMOVE e: Air Movement
AIRNAVAID e: Air Navigational Aid
AIRNAVO e: Air Navigation Office
Air Niu e: Air Niugini
AIRNORTH e: Allied Air Forces Northern Europe
Air NZ e: Air New Zealand
Air Obs e: Air Observer
AIROF e: Anodic Iridium Oxide Film
AIROP e: Airborne Operation
AIROPNET e: Air Operational Network
AIROPNSO e: Air Operations Officer
AIROPS e: Air Operations
AIRPAC e: Air Pacific
AIRPAP e: Air Pressure Analysis Program[me]
AIRPASS e: Airborne Interception Radar and Pilots Attack Sight System
AIRPL e: Airplane
airpl e: airplane[s]
Air Plt e: Airplane Pilot
Air Pollut Assoc J e: Air Pollution Control Association Journal
Air Pollut News e: Air Pollution News
Air Port e: Air Portugal
Air Prog e: Air Progress
airpt art e: airport art
Air Qual Instrum e: Air Quality Instrumentation
airr e: adjusted internal rate of return
AIRRES e: Air Rescue
AirResCen e: Air Reserve Center
Air Res Dist e: Air Reserve District
AIRS d: automatisierte[s] Informations- und Rechensystem[e]
- e: Ablator Insulated Ramjet Study
- e: Accident / Incident Reporting System
- e: Accident Information Retrieval System
- e: Accounting Incomplete Records System
- e: Advanced Inertial Reference Sphere
- e: Advanced Instrumentation for Reflood Studies
- e: Advanced Integrated Reconnaissance System
- e: Aerometric Information Retrieval System
- e: Airborne Infrared Radiometer System
- e: Airborne Integrated Reconnaissance System
- e: Aircraft Inventory Reporting System
- e: Airline Interline Reservations System
- e: Airport Information Retrieval System
- e: Alliance of Information and Referral Systems
- e: American Information Retrieval Service
- e: Appalachia Improved Reference Services (Library)
- e: Army Information Radio Service
- e: Audit Integrated Reporting System
- e: Australian Information Retrieval Services

- e: Automated Information Reference Systems
- e: Automated Insurance Rating Services, Inc.
- e: Automatic Image Retrieval System
- e: Automatic Information Retrieval System

AIRSHTR e: Air Shutter
AIRSKEDELFLT e: Aircraft Schedule for Delivery to Fleet
AIRSOLS Command e: Air Solomons Command
AIRSOUTH e: Allied Air Forces Southern Europe
AIRSS e: Advanced Ballistic Reentry System Instrumentation Range Safety Systems
AIRSTORDEP e: Air Stores Depot
airsurance e: Air Insurance
AIRSYSCOM e: Air Systems Command
AIRTASS e: Airborne Towed Array Sonar System
AIRTE e: Associate of the Institute of Road Transport Engineers
AIRTRANS e: Airport Transportation System
AIRTRANSEX e: Air Transportation Exercise
Air U e: Air University
AIRV e: Air Injection Relief Valve
AIRVAN e: Air-Mobile Van
AIRWC e: Air War College
AIRWEANET e: Air Weather Network
AIRWORK e: Airwork Atlantic Limited
AIRX e: American Industrial Radium and X-Ray Society
AIRXRS e: American Industrial Radium and X-Ray Society
AIRY-Sh. e: Airy-Shaw
AIS d: Alarmindikationssignal

- d: Aortenisthmusstenose
- d: automatisiertes Informations-[verarbeitungs]system
- e: Abbreviated Injury Scale
- e: Ablating Inner Surface
- e: Academic Instructors School
- e: Academy of Independent Scholars
- e: Accelerated Inspection System
- e: Accounting Information System
- e: Accumulator Injection System
- e: Acorn Interactive System
- e: Acoustic Imaging System
- e: Action Item Sheet
- e: Adaptive Information Selector
- e: Administration Information System
- e: Advanced Indications Structure
- e: Advanced Indications System
- e: Advanced Information System[s] [Company]
- e: Advanced Instructional System
- e: Advanced Ionospheric Sounder
- e: Advanced Isotope Separation
- e: Advance In Schedule
- e: Aeronautical Information Service
- e: Aeronautical Information Specialist
- e: Airborne Imaging Spectrometer
- e: Airborne Infrared Spectrometer
- e: Airborne Initiation System
- e: Airborne Instrumentation Subsystem
- e: Aircraft Inspection System
- e: Aircraft Instrument Subsystem
- e: Air Intelligence Service
- e: Airlock Illumination Subassembly
- e: Air System Interrogator
- e: Alarm Indication Signal
- e: Alarm Inhibit Signal
- e: Alarm Integrated Signal
- e: Alcohol Insoluble Solid
- e: Alphabetical Indication Store
- e: Alphabetical Information Storage
- e: Alternate Interim Successor
- e: Altitude Indicating System
- e: Altitude Indication System
- e: American Indian Studies
- e: American Information Services
- e: American Interplanetary Society
- e: American Iris Society
- e: Analog[ue] Input System
- e: Analog[ue] Instrumentation Subsystem
- e: Androgen Insensitivity Syndrome
- e: Annual Inspection Summary
- e: Antenna Interface Subsystem
- e: Anti-Icing System
- e: Apollo Instrumentation Ship
- e: Arithmetic IF Statement
- e: Army Industrial School
- e: Army Infantry School
- e: Army Information Systems
- e: Army Intelligence and Security
- e: Army Intelligence School
- e: Army Intelligence Survey
- e: Artillery Intelligence School
- e: Associate of the Institute of Statisticians
- e: Association for Integrative Studies
- e: Association of Iron and Steel
- e: Assured Information Systems, Incorporated
- e: Attitude Indicating System
- e: Australian Information Service
- e: Australian Iron and Steel
- e: Automated Identification System
- e: Automated Indicator System
- e: Automated Information System
- e: Automated Instrumentation System
- e: Automated Insurance Service
- e: Automatic Idle Speed
- e: Automatic Image Screening
- e: Automatic Inercept System
- e: Automatic Input of Subroutines
- e: Automatic Intercept Switch
- e: Automatic Intercept System
- e: Automatic Intercity Station
- e: Automatic Intermediate Station
- e: Award Information System
- f: Association Internationale de la Savonnerie et de la détergence
- f: Association Internationale de Sociologie

ais e: agreed industry standard

- e: Automatic Interplanetary Station

AI & S e: Army Intelligence and Security
AISA e: Agricultural Information Society for Asia

- e: American Institute for Shippers Associations
- e: American Institute of Supply Associations
- e: Analytical Isoelectrofocusing Scanning Apparatus
- e: Associate of Incorporated Secretaries Association
- e: Australian Institute of Systems Analysis
- f: Association Internationale pour la Sécurité Aérienne

AI San E e: Associate of the Institution of Sanitary Engineers
AISAR e: Accidental Incident Sabotage Assistance Request
AISB e: Artificial Intelligence and Simulation [of] Behavio[u]r

- f: Association Internationale de Standardisation Biologique

AISC e: Advanced Information Systems Company

- e: American Indian Studies Center
- e: American Institute of Steel Construction
- e: Army Information Systems Command
- e: Assessment and Information Services Center
- e: Association of Independent Software Companies
- e: Association of Informed Senior Citizens
- e: Australian Institute of Steel Construction

AISCR e: Automated Interpretation System of Chest Roentgenograms
AISD e: Abstracting and Indexing Services Directory

- e: Analog[ue] Information Storing Device

AISE e: American Intercultural Student Exchange

- e: Association of Iron and Steel Engineers
- e: Average Integral Square Error
- f: Association Internutionale des Sciences Économiques

AISES e: American Indian Science and Engineering Society
AISF e: Airlift Industrial Services Flight

- e: Avionic Integration Support Facility

AISG e: Accountants International Study Group
AISHWC e: Australian Industrial Safety, Health and Welfare Cases
AISI e: Advanced International Studies Institute

- e: Airborne Instrumentation Subsystem Internal
- e: American Iron and Steel Institute
- e: Associate of Iron and Steel Institute

AI Signal e: Alarm Inhibit Signal
AISIS e: Advanced Icing Severity Indication System
AIS ISGI d: Automatisiertes Informationssystem des Internationalen Systems für gesellschaftschaftswissenschaftliche Information
AISJ f: Association Internationale des Sciences Juridiques
AISL e: Aviation Information Services Limited
AISM f: Association Internationale des Sociétés de Microbiologie
AISÖ d: Arbeitsgemeinschaft Internationaler Straßenverkehrsunternehmer Österreichs
AISOB e: Associate of the Incorporated Society of Organ Builders
ais (or Lunik III) e: automatic interplanetary station
AISP e: Alternative Information Sources Project

- e: Association of Information Systems Professionals

AISR e: Army Institute of Surgical Research
AISRCM e: Australian Institute of Swimming and Recreation Centre Management
AISS e: Airborne Infrared Surveillance Set

- e: Australian Innovation Sourcing Service
- e: Automatic Intercom[munication] Switching System
- f: Association Internationale de la Science du Sol

AIST e: Agency of Industrial Science and Technology

- e: Association for Intelligent Systems Technology
- e: Australian Institute of Science Technology
- e: Automatic Informational Station

AISTM e: Associate of the Institute of Sales Technology and Management
AI Struct E e: Associate of the Institution of Structural Engineers

AISV e: Amphibious Infantry Support Vehicle
AISW e: Aircraft Interception Special Weapon
AIT d: analytischer Intelligenztest
- e: Academy for Implants and Transplants
- e: Acoustic Impact Technique
- e: Acoustic Impact Test
- e: Acute Intensive Treatment
- e: Adelaide Institute of Technology
- e: Advanced Identification Techniques
- e: Advanced Individual Training
- e: Advanced Infantry Training
- e: Advanced Information Technology
- e: Advanced Interceptor Technology
- e: Afghan Institute of Technology
- e: Agency for Instructional Technology
- e: Agency for Instructional Television
- e: Agglutination-Inhibition Test
- e: Air Injection Tube
- e: Air Intake Temperature
- e: Aiton
- e: All India Television
- e: American Industrial Transport, Inc.
- e: American Institute of Technology
- e: Analytic Intelligence Test
- e: Architect-in-Training
- e: Army Individual Test
- e: Army Intelligence Translator
- e: Asian Institute of Technology
- e: Assured Intermediate Task
- e: Attribute Information Table
- e: Auto-Ignition Temperature
- e: Automatic Information Test
AiT af: Anjuman-i-Tarikh [Historical Society of Afghanistan)
AI & T e: Assembly Integration and Test
AITA e: Advanced Individual Training Available
- e: Air Industries and Transport[s] Association
- e: Australian Industrial Truck Association
AITAA e: Advanced Individual Training Attrition Analysis
AITC e: Action Information Training Center
- e: Air Intelligence Training Center
- e: Alabama International Trade Center
- e: Allyl Isothiocyanate
- e: American Institute of Timber Construction
- e: Australian Industry and Technology Council
- e: Australian Information Technology Council
- f: Association Internationale des Traducteurs de Conference
AITD e: Autoimmune Thyroid Disease
AITE e: Advanced Indication Technology Experiment
- e: Aircraft Integrated Test Equipment
- e: Australian International Technology Exhibition
- e: Automatic Intercity Telephone Exchange
AITEP e: Association for International Technical Promotion
AITF e: Ammunition Initiatives Task Force
AITI e: Aero Industries Technical Institute
- e: American Institute of Technical Illustrators
AITIA e: American Institute of Technical Illustrators Association
AITIC e: Applied Information Technologies Corporation
AITM e: American Institute of Tack Manufacturers
AITP e: Allergy, Immunology and Transplantation Program[me]
AITRC e: Applied Information Technologies Research Center
AITS e: Action Item Tracking System
- e: Automated Information Transfer System
- e: Automatic Integrated Telephone System
AITSA e: Associate of the Institute of Trading Standards Administration
AITT e: Augmented Insulin Tolerance Test
AITU e: Alliance of Independent Telephone Unions
AITVS e: Association of Independent Television Stations
AIU e: Abort Interface Unit
- e: Absolute Iodine Uptake
- e: Abstract Information Unit
- e: Advanced Instrumentation Unit
- e: Alarm Interface Unit
- e: Alarm Interference Unit
- e: Array Interface Unit
- e: Avionics Interface Unit
AIUM e: American Institute of Ultrasound in Medicine
AIURA e: American Institute of Urban and Regional Affairs
AIUS e: Australian Institute of Urban Studies
AIV d: Akademischer Ingenieur- Verband
- d: Architekten- und Ingenieurverein
- d: Außenhandelsausschuß industrieller Verbände
- d: automatische Informationsverarbeitung
- d: Institut für Automatische Informationsverarbeitung
- e: Accelerated Inverse Voltage
- e: Advanced Interactive Video
- e: Armo[u]red Infantry Vehicle
- e: Artillery Meteorological System Instrumentation Vehicle
AIV f: Association Internationale de Volcanologie
A.I.V. e: Automatic Inboard Vent-tank
AIVF e: Association of Independent Video and Filmmakers
aiv-Institut d: Institut für Automatische Informationsverarbeitung
AIVP e: American Institute of Vocal Pedagogy
AIVR e: Accelerated Idioventricular Rhythm
AIVS d: automatisiertes Informations-verarbeitungssystem
AIW d: Arbeitskreis Fachagenturen und Berater für Industriewerbung
- d: Arbeitskreis für Internationale Wissenschaftskommunikation
- e: Analytical Instruction Word
- e: Asbestos Insulated Wire
- e: Atlantic Intracoastal Waterway
- e: Auroral Infrasonic Wave
AIWE e: Associate of the Institution of Water Engineers
AIWEC e: Advanced / Innovative Wind Energy Concept
AIWF e: American Institute of Wine and Food
AIWHTE e: Associate of the Institution of Works and Highways Technician Engineers
AIWI e: American Industrial Writing Institute
AIWM e: American Institute of Weights and Measures
AIWMA e: Associate of the Institute of Weights and Measures Administration
AIWPHSA e: American Institute of Wholesale Plumbing and Heating Supply Associations
AIWRS e: Arctic International Wildlife Range Society
AIWS e: Advanced Interdiction Weapon System
Aix e: Aix-en-Provence
AIX e: A version of unIX
AIZ d: Arzneimittel-Informationszentrum
AiZ d: Automation im Zug
AJ e: Absolute Judgement
- e: Absolute Jump
- e: Acting Judge
- e: Adding Jump
- e: Adjustment
- e: Air Jordan
- e: Alloy Junction
- e: American Jurisprudence (journ.)
- e: Analog[ue] Junction
- e: Ankle Jerk
- e: Anti-Jam[ming]
- e: Archaeological Journal (journ.)
- e: Architects Journal (journ.)
- e: Arc Jet
- e: Area Junction
- e: Art Journal
- e: Assembly Jig
- e: Associate in Journalism
- e: Associate Justice
- e: Astronomic Journal (journ.)
- e: Attack Jet
- e: Auxiliary Junctor
AJ-37 e: Swedish multimission combat aircraft
AJA e: Adjacent
- e: American Jewish Archives (journ.)
- e: American Journal of Anatomy (journ.)
- e: American Journal of Archaeology (journ.)
AJAC e: American Journal of Alzheimers Care (journ.)
- e: Automatic Jamming Avoidance Circuitry
AJAG e: Assistant Judge Advocate General
AJAI e: Anti-Jamming, Anti-Interference
AJAO e: American Juvenile Arthritis Organization
AJAQ e: Army Job Activities Questionnaire
AJAr e: American Journal of Archaeology (journ.)
AJAR e: Automatic Junction Analyzer and Recorder
AJAS e: Australian Journal of Applied Science (journ.)
Ajax e: Douglas Nike-Ajax surface-to- air missile
AJAZ e: Australian Journal of Botany (journ.)
AJB e: Adjustable Bed
- e: Audio Junction Box
- e: Australian Journal of Botany
AJBO e: Anti-Jam[ming] Black-Out
AJB Press e: Adjustable Bed Press
AJC e: Alvin Junior College
- e: Austin Junior College
aJc s: antes de Jesucristo
AJCC e: Alternate Joint Command Center
- e: Alternate Joint Communications Center
- e: American Jersey Cattle Club
- e: American Joint Committee on Cancer
AJCL e: American Journal of Comparative Law
- e: Australia-Japan Container Line
AJCM e: American Journal of Chinese Medicine (journ.)
AJC Method e: Alternative Justifiable Cost Method
AJCN e: American Journal of Clinical Nutrition (journ.)

AJCO e: American Journal of Clinical Oncology (journ.)
AJCOR e: Australian Joint Council for Operational Research
AJCP e: American Journal of Clinical Pathology (journ.)
- e: Australian Joint Copying Project
AJCU e: Air Jet Control Unit
AJD e: All-Japanese Districts
- e: Anti-Jam[ming] Display
AJDC e: American Journal of Diseases of Children (journ.)
AJDDA e: American Journal of Digestive Diseases (journ.)
AJDE e: Aerospace Installation Diagnostic Equipment
AJDG e: Air Jet Distortion Generator
AJDT e: Australian Journal of Dairy Technology (journ.)
AJE e: Adjusting Journal Entry
- e: Adult Jewish Education
- e: American Journal of Education (journ.)
- e: American Journal of Epidemiology (journ.)
- e: Anti-Jam[ming] Equipment
- e: Australian Journal of Education (journ.)
- e: Autostart Job Entry
AJEB d: Arbeitsgemeinschaft für Jugend- und Erwachsenenbildung
AJ Ecol e: Australian Journal of Ecology (journ.)
AJ Ed e: Australian Journal of Education (journ.)
AJER e: Alberta Journal of Educational Research (journ.)
AJES e: American Journal of Economics and Sociology (journ.)
AJET e: Australian Journal of Educational Technology (journ.)
AJF e: Anti-Jam[ming] Frequency
AJFH e: Anti-Jam[ming] Frequency Hopper
- e: Australian Journal of French Studies (journ.)
AJG e: American Journal of Gastroenterology (journ.)
- e: Automatic Job-stream Generator
AJH e: American Journal of Hematology (journ.)
- e: American Journal of Hygiene (journ.)
- e: Anti-Jam[ming] Hopper
AJHE e: Australian Journal of Higher Education (journ.)
AJHG e: American Journal of Human Genetics (journ.)
AJHP e: American Journal of Hospital Pharmacy (journ.)
AJHPER e: Australian Journal for Health, Physical Education and Recreation (journ.)
AjHQ e: American Jewish Historical Quarterly (journ.)
AJHS e: American Jewish Historical Society
- e: Andrew Jackson High School
AJHSNME e: American Junior High School National Mathematics Exam[ination]
AJI e: American Jewish Institute
- e: American Journal of Insanity (journ.)
- e: American Justice Institute
- e: Anti-Jam[ming] Improvements
AJIC e: American Journal of Infection Control (journ.)
AJIL e: American Journal of International Law (journ.)
AJIM e: American Journal of Industrial Medicine (journ.)
AJIS e: Automated Jail Information System
AjLE d: Ausschuß junger Lehrer und Erzieher
AJLL e: Australian Journal of Labour Law (journ.)
AJLM e: American Journal of Law and Medicine (journ.)
AJLO e: Avalanching Junction Light Output
AJLS e: Australian Journal of Law and Society (journ.)
AJM e: Abrasive-Jet Machining
- e: American Journal of Mathematics (journ.)
- e: American Journal of Medicine (journ.)
- e: Analog[ue] Junction Module
- e: Australian Journal of Management (journ.)
- e: Australian Journal of Mining (journ.)
AJMA e: Anti-Jam Manpack Antenna
AJMD e: American Journal of Mental Deficiency (journ.)
AJME e: Australian Journal of Music Education (journ.)
AJMG e: American Journal of Medical Genetics (journ.)
AJMM e: Asian Journal of Modern Medicine (journ.)
AJMS e: American Journal of Mental Sciences (journ.)
- e: American Journal of the Medical Sciences (journ.)
AJMs e: Analog[ue] Junction Modules
AJMT e: American Journal of Medical Technology
AJN e: American Journal of Numismatics (journ.)
- e: American Journal of Nursing (journ.)
AJNHS e: Andrew Johnson National Historic Site
AJNR e: American Journal of Neuroradiology (journ.)
AJO e: American Journal of Optometry (journ.)
- e: American Journal of Orthodontics (journ.)
- e: Anti-Jam[ming] Operator
- e: Australian Journal of Ophthalmology (journ.)
AJOC e: Assistant Instrumentation Operations Coordination
AJOG e: American Journal of Obstetrics and Gynecology (journ.)
AJOPs e: American Journal of Orthopsychiatry (journ.)
AJOT e: American Journal of Occupational Therapy (journ.)
AJP e: American Journal of Pathology (journ.)
- e: American Journal of Pharmacy (journ.)
- e: American Journal of Philology (journ.)
- e: American Journal of Psychiatry (journ.)
- e: American Journal of Psychoanalysis (journ.)
- e: American Journal of Psychology
- e: Australasian Journal of Philosophy (journ.)
- e: Australian Journal of Psychology (journ.)
AJPA e: American Journal of Physical Anthropology (journ.)
AJPC e: American Jewish Physicians Committee
AJPh e: American Journal of Philology (journ.)
A.J.Ph. e: American Journal of Physics (journ.)
AJPH e: American Journal of Public Health
AJPM e: American Journal of Preventive Medicine (journ.)
AJPP e: Australian Journal of Plant Physiology (journ.)
AJPs e: American Journal of Psychology
AJPS e: Australian Journal of Pharmaceutical Sciences (journ.)
AJPSA e: American Journal of Psychiatry (journ.)
AJPst e: American Journal of Psychotherapy
AJPsy e: American Journal of Psychiatry (journ.)
AJQ e: Army Job Questionnaire
AJR e: American Journal of Roentgenology (journ.)
- e: Australian Journal of Reading (journ.)
- e: Automatic Job Recovery
AJRC e: American Junior Red Cross
- e: Australia-Japan Research Centre
AJRI e: American Journal of Reproductive Immunology (journ.)
AJRL e: Automation Industries Research Laboratory
AJS e: American Journal of Science (journ.)
- e: American Journal of Sociology (journ.)
- e: American Journal of Surgery (journ.)
- e: Angle-Jam[ming] System
- e: Anti-Jam[ming] Synthesizer
- e: Axisymmetric Jet Stretcher
AJSI e: Australian Journal of Social Issues (journ.)
AJSJ e: American Justinian Society of Jurists
AJSL e: American Journal of Semitic Language and Literatures (journ.)
AJSM e: American Journal of Sports Medicine (journ.)
AJSO e: American Journal of Sociology (journ.)
AJ Soc e: American Journal of Sociology (journ.)
AJ Soc Is e: Australian Journal of Social Issues (journ.)
AJSP e: American Journal of Surgical Pathology (journ.)
AJSR e: Australian Journal of Scientific Research (journ.)
AJT e: Advanced Jet Trainer
- e: American Journal of Theology (journ.)
- e: Anti-Jam[ming] Technique
AJTh e: American Journal of Theology
AJTR e: Allowance Prescribed in Joint Travel Regulation
AJTWC e: Alternate Joint Typhoon Warning Center
AJUS e: Antarctic Journal of the United States
AJVD e: Abrupt Junction Varactor Doubler
AJVR e: American Journal of Veterinary Research (journ.)
AJW e: Anti-Jam[ming] Waveform
AJWR e: Alternate Joint War Room
AJZO e: Australian Journal of Zoology (journ.)
AK d: Abbruchkosten
- d: Absorptionskoeffizient
- d: Adressenkonstante
- d: Adreßkode
- d: Akkumulator
- d: Aktienkapital
- d: Aktivkohle
- d: Alaska
- d: Ansatzkuppe
- d: Anschlußkasten
- d: Anschlußkennung
- d: Anschlußkosten
- d: Antike Kunst (journ.)
- d: Äquivalenzklasse

- d: Arbeitskarte
- d: Arbeitskontakt
- d: Arbeitskopie
- d: Arbeitskraft
- d: Arbeitskräfte
- d: Arbeitskreis
- d: Art der Kurzwahl
- d: Aufnahmekopf
- d: Aufräumungskosten
- d: Ausfuhrkontrolle
- d: Ausgabekonverter
- d: Ausgabe Kopplung
- d: Auskunft
- d: Außenkante
- d: äußerste Kraft
- d: Auswahlkippstufe
- d: Auszahlungskontrolle
- d: Autokarte
- d: Autokassettenrecorder
- d: automatische Koppelung
- d: Autorenkollektiv
- d: Autorenkorrektur
- d: Azbel-Kaner
- d: ß-Aspartokinase
- e: Above Knee
- e: Accessory Kit
- e: Adapter Kit
- e: Adaption Kit
- e: Addition Key
- e: Address Kind
- e: Adenosine Kinase
- e: Adenylate Kinase
- e: Advanced Cruiser
- e: Advancing Key
- e: Alaska
- e: Alaska Coastal-Ellis Airlines
- e: Allotment Key
- e: Alphanumeric Keyboard
- e: Amount Keyboard
- e: Amplitude Keyed
- e: Anterior Kidney
- e: Apogee Kick
- e: Armature Knockoff
- e: Attention Key
- e: Automatic Clock
- e: Automatic Key
- R: Avtomat Kalashnikov

ak d: Actinolite
- d: altkatholisch

Ak d: Akademie
- d: Angorakanin[wolle]
- d: Antikörper

Ak. d: Akademiker
- d: Akustik[er]

ak. d: akademisch
- d: akustisch
- d: akut

A.K. d: Aktivitätskoeffizient

a. K. d: auf Kohle
- d: auf Kommissionsbasis
- d: auf Kredit
- d: außer Konkurrenz

a.K. d: auf Kohlebasis
- d: auf Kriegsdauer
- d: außer Kraft

A-K d: Anode-Kat[h]ode

AK 47 e: Automatic rifle

AKA d: Abschnittsknotenamt
- d: Absorptionskälteapparat
- d: Akademische Auslandsstelle
- d: Ausfuhrkreditanstalt
- d: Autokorrelationsanalyse
- d: automatische Kohlenaufbereitung
- e: Above Knee Amputation

AkA d: Arbeitsgemeinschaft für kommunale Abfallwirtschaft

aka e: also known as

Akad a: Akademi
- e: Akademie

Akad. d: Akademie
- d: Akademiker

akad. d: akademisch

Akad. d. K. d: Akademie der Künste

Akad. d. Wiss. d: Akademie der Wissenschaft[en]

Akad.-Fam. d: Akademikerfamilie

AkadGFG d: Gesetz über die Führung akademischer Grade

Akad Nauk R: Akademiya Nauk

AK Amp e: Above Knee Amputation

Akan e: Akan National Park on Hokkaido Island, Japan

akath. d: altkatholisch

AKAW d: Anzeiger der Kaiserlichen Akademie der Wissenschaften (journ.)

AKB d: allgemeine Kraftverkehrs-versicherungsbedingungen
- d: allgemeine Kundendienstbedingungen
- d: Alliierte Kommandantur[a] Berlin
- d: Anschlußsteuerung für Kassettenmagnetbandeinheit
- d: Atomkraftwerk Bayern GmbH
- d: autonome komplette Blöcke

a.K.B. d: außer Kriegsbereitschaft

AKBS d: aufgabenklassenbezogenes Betriebssystem
- e: Advanced Kinematic Bombing System

AKC e: Associate of Kings College

AKCL e: Associate of Kings College London

AKD d: Arbeitsgemeinschaft kommunale Datenverarbeitung
- e: Automatic Key Distribution

AKdÄ d: Arzneimittelkommission der deutschen Ärzteschaft

AKDB d: Anstalt für kommunale Datenverarbeitung in Bayern

Ak. d. K. d: Akademie der Künste

AkDR d: Akademie des Deutschen Rechts

AKDStV d: Arbeitsgemeinschaft katholischer deutscher Studentenverbindungen

AKDV d: Anstalt für kommunale Datenverarbeitung

Ak. d. Wiss. d: Akademie der Wissenschaft[en]

AKE d: Aktionskreis Energie eV
- d: Arbeitskräfteerhebung

AKES e: Automatic Kinetic Enzyme System

AKEW d: Arbeitsgemeinschaft Kernkraftwerk [der österreichischen Elektrizitätswirtschaft]

AK-F d: Arbeitskräfte-Fachbericht

AKF d: Autokorrelationsfunktion
- e: American Kidney Fund
- e: Australian Koala Foundation

AKFn d: Autokorrelationsfunktionen

AKG d: Allgemeines Kriegsfolgengesetz
- d: Archiv für Kulturgeschichte (journ.)
- d: Astronomische Gesellschaft Katalog
- d: automatisches Kontrollgerät
- e: Auxiliary Killing Ground

AKGD e: Acknowledged

AKH d: Allgemeines Krankenhaus

AKh d: Arbeitskräftestunde[n]
- d: Arbeitskraftstunde[n]

AKHB d: allgemeine Kraftfahr-Haftpflichtversicherungsbedingungen

AKh/ha d: Arbeitskraftstunde[n] pro Hektar

AKh/Kuh d: Arbeitskraftstunde[n] pro Kuh

AKI d: Arbeitsgemeinschaft Deutsche Kunststoffindustrie
- d: Arbeitsgemeinschaft Keramische Industrie
- d: Arbeitskreis für Information
- e: Aircraft Kill Indicator
- e: Alfred Kinsey Institute
- e: Anti-Knock Index
- e: Automix Keyboards, Incorporated

AKIS d: alphabetisches Kundeninformationssystem
- d: Arbeitskreis Informations-System
- d: Auftragskontroll- und Informationssystem

AKJ d: Arbeitskräftejahr[e]

AKK d: Amt für Kernforschung und Kerntechnik (DDR)
- d: Anstalt für Kabelkommunikation

akk a: akkusativ

Akk. d: Akkord[eon]
- d: Akkumulator
- d: Akkusativ

Akk e: Akkadian

Akkr. d: Akkreditiv

Akku d: Akkumulator

Akku-E-lok d: Akkumulatorenelektro-lokomotive

Akkulok d: Akkumulatorenlokomotive

AKL d: Abspannklemme
- d: Anrufkontrollampe
- d: Arbeitskreisleiter
- e: Auckland, New Zealand (airport)

AKLD e: Auckland

AKliAnS d: Auslandsklinkenanschaltesatz

AKM d: Autoren, Komponisten und Musikverleger
- d: Axialkolbenmaschine
- e: Apogee Kick Motor
- e: Automatic Key Management
- e: Soviet military weapon

A-K-Maschine d: Arbenz-Kammerer-Maschine

AKML d: Abhandlungen zur Kunst-. Musik- und Literatur -Wissenschaft (journ.)

AKN d: automatisches Kommunikationsnetz

AKNF d: Adair-Koshland-Nemethy- Filmer

AKO d: automatisierte Korrespondenz

A-Kohle d: Adsorptionskohle
- d: Aktivkohle

AKommG d: Kommanditgesellschaft auf Aktien

AKONA d: automatischer Kontenausgleich

AKOTECH d: Arbeitsgemeinschaft für Auslands- und Kolonialtechnik

AKP d: Afrika, Karibik und Pazifischer Raum
- d: Anschluß für Kassettenplattenspeicher
- d: Antennenkeulenposition

A.K.P. d: Armeekraftwagenpark

AKR e: Auroral Kilometric Radiation
- e: Address Key Register

Akr. d: Arbeitskreis

a. Kr. d: auf Kredit
- d: auf Kriegsdauer

Akr e: Akron
- g: Akmtirion
- g: Akra

AKRA d: Arbeitsgemeinschaft Kraftwagenspedition

AKrB d: Allgemeine Krankenversicherungsbedingungen

AKrFB d: Arbeitskräfte im Fernmeldebau

AKriegsfG d: Allgemeines Kriegsfolgengesetz

AKRO e: Acknowledge Receipt Of

AkrV d: Arbeitskreis Vermessungswesen

AKS d: Allgemeine Kurbelschwingungen
- d: Anerkanntes Saatgut
- d: Arbeitsgemeinschaft der freiberuflich tätigenr öffentlich bestellten und vereidigten Kraftfahrzeug-Sachverständigen
- d: Arbeitsgemeinschaft Kernkraftwerk Stuttgart
- d: Arbeitsgemeinschaft zum Studium der Errichtung eines Kernkraftwerkes Stuttgart
- d: Ausgang für Kontrollsignal
- e: Advanced Keyboarding System
- e: Amplitude Key Shift
- e: Amplitude Key Shifting

AKSB d: Allgemeine Bedingungen für die Kraftverkehrs-Strafrechtsschutzversicherung
AKSESS e: Algorithmic Key Selection, Entry and Summarization System
AKSR e: American Karakul Sheep Registry
AKT d: Analogkonzentrator
- d: Arbeitsplatz für Konstrukteure und Technologen
- e: Akita Television [Japan]

Akt. d: Akten
- d: Aktion
- d: Aktionär

akt. d: aktiv
Akt. d: Aktiva
- d: Aktivität
- d: Aktualität
- d: Aktuar

AktG d: Aktiengesetz
Akt-Ges d: Aktiengesellschaft
Akt.-Ges. d: Aktiengesetz
AKTIND d: Aktienindex[berechnung]
Aktivier. d: Aktivierung
Akt.-Nr. d: Aktennummer
AktO d: Aktenordnung
AktOBGH d: Aktenordnung Bundesgerichtshof
AktR d: Aktienrecht
- d: Aktionsradius

aktTGl d: aktives T-Glied
AKTU d: Aktualisierungsprogramm
Akt.-Verz. d: Aktenverzeichnis
Akt.Z. d: Aktenzeichen
AKU d: Arbeitskreis für Umweltschutz
- e: Automatic Keying Unit

Aku d: Akkumulator
Akü d: Abkürzung
Akülex. d: Abkürzungslexikon
AKURON e: Autonetics Kalman Utilization of Reference for Optimal Navigation
Aküspr. d: Abkürzungssprache
Akust. d: Akustik
Akust Zh R: Akusticheshkii Zhurnal (journ.)
Aküv d: Abkürzungsverzeichnis
AKV d: Allgemeine Krankenversicherungsbedingungen
- d: Allgemeiner Knappschaftsverein
- d: Anzeige-Kat[h]odenverstärker
- d: äußerste Kraft voraus
- d: automatische kundenwunschabhängige Vorbereitung [der Produktion]

AKW d: Allgäuer Kraftwerke
- d: Amt für Kontrolle des Warenverkehrs
- d: Arbeitskreis Werbefernsehen
- d: Atomkraftwerk

AKWAS e: Author and Keyword[s] in Alphabetical Sequence
AKWIC e: Altered Keyword in Context
- e: Author and Keyword[s] in Contex

AKZ d: Akkumulator-Nahtstellenzusatz
- d: Alte Kennzeichen
- d: Anmeldeaktenzeichen
- d: Antriebskennzahl

Akz. d: Aktenzeichen
- d: Akzeleration
- d: Akzent
- d: Akzept[ant]
- d: Akzeptation
- d: Akzeß
- d: Akzession
- d: Akzidenz
- d: Akzise

akz. d: akzeptiert
- d: akzessorisch

Akz.-Dr. d: Akzidenzdruck[erei]
Akz.-Kr. d: Akzeptkredit
AL d: Adressenleitung
- d: Adressenlesedraht
- d: Alabama
- d: Alaska
- d: Albanien (Kfz)
- d: Alphanumerischer Leser
- d: Alphanumerischer Locher
- d: Alumin[i]um
- d: Ammoniumlaktat
- d: Amtsleitung
- d: Anfangslader
- d: Anfangsladung
- d: Anlage[n]
- d: Anruflampe
- d: Anzeigelampe
- d: Assimilationsleistung
- d: Assimilationsleitung
- d: Auftragsliste
- d: Ausfuhrliste
- d: Ausgang Löschen
- d: Ausgangslogik
- d: Außenlautsprecher
- e: Abnormal Lungs
- e: Absolute Level
- e: Absolute Limen
- e: Absolute Loader
- e: Acceptor Level
- e: Accession List[s]
- e: Accidental Loss
- e: Accommodation Ladder
- e: Accrued Liability
- e: Accumulator Low
- e: Accuracy Landing
- e: Acoustics Laboratory
- e: Acquisition and Logistics
- e: Acquisition Logistician
- e: Acute Leukemia
- e: Adaption Level
- e: Adaptive Logic
- e: Additional Listing
- e: Add Logical
- e: Addressable Location
- e: Address Language
- e: Address Latch
- e: Addressless
- e: Address Level
- e: Address Line
- e: Address List
- e: Admiralty Letter
- e: Advisory Leaflet
- e: Aerial Laboratory
- e: Aerodynamics Laboratory
- e: Aeromechanics Laboratory
- e: Aeronautical Laboratory
- e: Aeronomy Laboratory
- e: Aerophysics Laboratory
- e: Agricultural Labourer
- e: Aircraft Laboratory
- e: Aircraft Logistics
- e: Air Letter
- e: Air Liaison
- e: Air Lift
- e: Air Lime[s]
- e: Alarm
- e: Alarm Relay
- e: Albania
- e: Algorithmic Oriented Language
- e: Algorithmic Language
- e: Allegheny Airlines
- e: Allowance List
- e: Alphabetic[al]
- e: Alphabetic Letters
- e: Alphabetic Listing
- e: Aluminum Limited
- e: Amber [glow] Lamp
- e: Amber Light
- e: Amendment List
- e: American Literature (journ.)
- e: Amplitude Limiter
- e: Analog[ue] Link
- e: Analog[ue] Loop-back
- e: Analytical Laboratory
- e: Analytical Limits
- e: Angle Lock
- e: Anglo-Latin
- e: Annual Leave
- e: Anodic Layer
- e: Antenna Laboratory
- e: Architectural League
- e: Arm Length
- e: Army List
- e: Array List
- e: Arrival Locator
- e: Articular Layer
- e: Artificial Language
- e: Artificial Line
- e: Artificial Luminance
- e: Assembler Language
- e: Assembly Language
- e: Assembly List
- e: Assembly Listing
- e: Assets and Liabilities
- e: Assistance Line
- e: Associated Laboratories
- e: Associative Law
- e: Astronomical League
- e: Astronuclear Laboratory
- e: Astropower Laboratory
- e: Autograph Letter
- e: Automatic Logging
- e: Automobile Liability
- e: Aviation Electronicsman
- e: Avionics Laboratory
- e: Axiolingual
- l: Anno Lucis
- s: América Latina

al. d: alemannisch
- l: alias

Al. d: Alemannische
- d: Alimente
- d: Alkohol
- d: Alumne
- d: Anlasser

Al d: Alginat
- d: Alkalilöslichkeit
- d: Alpakawolle
- d: Alumnat
- d: Anschlußleitung
- e: Allied
- e: All lengths
- e: Aluminum

al d: Almandine
- e: action limit
- e: air lock
- e: albumin
- e: alcohol
- e: alias
- s: alumnos

a. L. d: an der Lahn
- d: an Land

a.L. d: angesaugte Luft
- d: auf Lebenszeit
- d: auf Lieferung

A. L. d: Arabische Liga
- e: Arab League

aL e: assumed latitude
a l f: après livraison
a. l. l: ad libitum
A & L e: Administration and Logistics
- e: Approach and Landing

A-L e: Allegheny-Ludlum
- e: Anglo-Latin

A/L e: Assemble / Load
ä. L. d: ältere Linie
AL0B e: Alphabetic 0 (Zero) Bit
AL 1 B e: Alphabetic 1 Bit
AL-60 e: transport aircraft
ALA d: Alabama
- d: Alanin[e]
- d: Allgemeines Luftwaffenamt
- d: Arbeitsgemeinschaft der schweizerischen Aluminiumverarbeiter, -walzwerke und -hütten
- e: Actual Leaf Area
- e: Aerial Lightning Arrester
- e: Agriculture Labor
- e: Amalgamated Lithographers of America
- e: American Laminators Association
- e: American Landscape Architects
- e: American Laryngological Association
- e: American Library Association
- e: American Lighting Association
- e: American Literary Anthology
- e: American Lung Association
- e: Aminolevulinic Acid
- e: Antenna Lightning Arrester
- e: Arc Lamp Assembly
- e: Area Letter of Acceptance
- e: Arkansas Library Association
- e: Army Launch Area
- e: Assault Landing Area
- e: Assembly of the Librarians of the Americas
- e: Associated Locksmiths of America
- e: Associate in Liberal Arts
- e: Associate[ship] of the Library Association
- e: Authorized Landing Area
- e: Authors League of America
- e: Axiolabial
- e: ë-Aminolevulinic Acid

Ala. e: Alabamian
ala e: alanine
- e: alighting area

A.L.A. e: Associate of the [British] Library Association
ALAA e: Associate of the Library Association of Australia
- e: Associate of the London Association of [Certified and Corporate] Accountants

Ala Agric Exp Stn e: Alabama Agricultural Experiment Station
alaar e: air-launched air-recoverable rocket
ALAAR e: Air Launched Air-to-Air Rocket
A-lab e: acid laboratory
alabaster e: calcite
ALABEL e: American Library Association Board of Education for Librarianship
ALABM e: Air Launched Anti-Ballistic Missile
ALABOL e: Algorithmic and Business Oriented Language
ALA Bull e: American Library Association Bulletin (journ.)
ALAC e: Alaska Air Command
- e: Artificial Limb and Appliance Centre

ALAD e: Aminolaevulinate Dehydratase
- e: Automatic Liquid Agent Detector

ALA-D e: ë-Aminolevulinic Acid Dehydratase
alad e: abnormal left axis deviation
ALADDIN [Rocket] e: Atmospheric Layer and Density Distribution of Ions and Neutrons [Rocket]
ALADIN d: Anlagen-Dialog Informationssystem
- e: A Language for Decision Information
- e: Algebraic Automated Digital Iterative Network
- e: Automated Laboratory Diagnostic Instrument

ALADIN Testing e: Advanced Laser-Aided Defect Inspection in Nondestructive Testing
ALAE e: Association of Licensed Aircraft Engineers
ALAEA e: Australian Licensed Aircraft Engineers Association
alag e: axiolabiogingival
Ala Geol Surv e: Alabama Geological Survey
AlaHQ e: Alabama Historical Quarterly (journ.)
ALAIRC e: Alaskan Air Command
ALAIRS e: Advanced Low Altitude Infrared-Reconnaissance Sensor
ALA-ISAD e: American Library Associations Information Science and Automation Division
Ala.J.Med. Sci. e: Alabama Journal of Medical Sciences (journ.)
ALAL d: Axiolabiolingual
ALA [I] e: Amalgamated Lithographers of America
ALALY e: Alumin[i]um Alloy
ALAM e: Associate of the London Academy of Music
- e: Association of Licensed Automobile Manufacturers
- e: Atlas List Processing Algebraic Manipulator

Ala Mar Resour e: Alabama Marine Resources
Alands e: Aland Islands
ALANO e: All Accident Notice Offices
ALAP e: As Late as Possible
- e: As Low as Possible
- e: Associative Linear Array Processor

ALAPCO e: Association of Local Air Pollution Control Officials
AlaR e: Alabama Review (journ.)
ALARA e: As Low as Reasonably Achievable
ALARACT e: All Army Activities
ALARM e: Advanced Low-Altitude Radar Model
- e: Air-Launched Advanced Ramjet Missile
- e: Air-Launched Anti-Radiation Missile
- e: Airborne Laser Receiver Module
- e: Alerting Long-Range Airborne Radar for Moving Target Indicator
- e: Automatic Light Aircraft Readiness Monitor
- e: Automatic Log and Restart Mechanism

ALARR e: Air-Launched Air-Recoverable Rocket
- e: Air-Launched Rocket Recovery

ALART e: Army Low-Speed Air Research Tasks
Alas. d: Alaska
- e: Alaska

ALAS e: Alaska[n]
- e: Aminolevulinic Acid Synthetase
- e: Approach Landing Autopilot System
- e: Army Library Automated System[s]
- e: Array Logic and Storage
- e: Artillery Location Acoustic System
- e: Associate in Letters, Arts and Sciences
- e: Associate of the Land Agents Society
- e: Asynchronous Look-Ahead Simulator
- e: Automated Library Acquisitions System
- e: Automated Literature Alerting System
- e: Automatic Load Alleviation System

A.L.A.S. e: Associate in Letters, Arts and Sciences
ALASC e: Aircraft Launching Accessory Service Change
ALASCOM e: Alaska Communications
Alas DST e: Alaskan Daylight Saving Time
Alasia e: Australasia
Alas ST e: Alaskan Standard Time
ALAST e: Advanced Laser Spot Tracker
Ala St Found e: Alabama State Foundation
ALAT e: Active Look-Aside Table
- e: Aircraft Latitude
- e: Alanine Transaminase
- e: Army Language Aptitude Test

alat. d: altlateinisch
Al-At e: Alma-Ata
ALATB e: Airlanding Antitank Battery
ALAW e: Advanced Light Antitank Weapon
ALB d: Albumin
- d: allgemeine Lagerbedingungen
- d: allgemeine Liefer- und Leistungsbedingungen
- d: allgemeine Lieferbedingungen (des RGW)
- d: Allgemeines Literaturblatt (journ.)
- d: Alumin[i]um Bronze
- d: Anschlußsteuerung für Lochbandeinheit
- d: Arbeitsgemeinschaft Landwirtschaftliches Bauwesen
- d: Armeelazarettbasis
- d: automatische, lastabhängige Bremskraftregelung
- d: automatische lastabhängige Bremse
- d: automatisiertes Liegenschaftsbuch
- e: Academic Libraries of Brooklyn
- e: Air-Launched Booster
- e: Aircraft Launching Bulletin
- e: Alphabetic[al]
- e: Antilock Brake
- e: Arithmetic and Logic Box
- e: Assumed Leading Bit
- e: Automatic Light Block
- e: Automatic Loc-Bottom

alb d: albanisch
- e: albumin
- e: aluminum-bronze

Alb. d: Albatros
- d: Album

Alb e: Albania
- e: Albion

ALBA d: allgemeine Bauabrechnung
- e: Aluminium Bahrain

Albania e: Peoples Republic of Albania
Albatross e: Grumman amphibian transport aircraft
ALBC e: Anti-Laser Beam Coating
Al-BD d: Aluminiumband
ALBEN e: Aerodynamic Load Balanced Elliptical Nozzle
Albert e: Albert Canal connecting Antwerp and Liege
- e: Albert National Park in Zaire
- e: Albert Nyasa (Albert Lake)

ALBERT e: A Low Budget Easy Retrieval Technique
ALBI e: Air-Launched Ballistic Intercept
- e: Air-Launched Boost[er] Intercept

ALBIS e: Air-Launched Ballistic Intercept System
- e: Atmosphere Launched Boost[er] Intercept System
- e: Australian Library Based Information System
Alb LS e: Albany Law School
ALBM e: Air-Land Battle Management
- e: Air-Launched Ballistic Missile
ALBO e: Automatic Line Build-Out
- e: Automatic Line Buildout
Albq e: Albuquerque
ALBS e: Air-Launched Balloon System
- e: Anti-Lock Braking System
ALBSU e: Adult Literacy and Basic Skills Unit
ALBUM d: anwendungsvariables, literaturauswertendes, bibliographiengenerierendes Uniterm
Albuq e: Albuquerque
ALBUS e: All Bureaus
ALBz d: Aluminiumbronze
ALC - e: Accredited Land Consultant
- e: Acquisition Life Cycle
- e: Adaptive Linear Combiner
- e: Adaptive Logic Circuit
- e: Administrative / Logistics Center
- e: Advanced Library Concepts, Inc.
- e: Aerial Loading Coil
- e: Aft Load Controller
- e: Agricultural Land Commission
- e: Air Launchable Concept
- e: Air Logistics Center
- e: Air Logistics Command
- e: Alberta Livestock Co-operative
- e: Alkali Light Chain
- e: Allocate
- e: Alumin[i]um Casting
- e: American Language Course
- e: American Locomotive Company
- e: Analytical Liquid Chromatograph[y]
- e: Antenna Loading Coil
- e: Arab League Council
- e: Area Logistics Center
- e: Area Logistics Command
- e: Armament Logistics Center
- e: Army Logistics Center
- e: Artificial Luminous Cloud
- e: Assembly Language Coding
- e: Assembly Language Compiler
- e: Associate of Loughborough College [of Advanced Technology]
- e: Astro Launch Circuit
- e: Audio Load Compensator
- e: Automatic Landing Control
- e: Automatic Level[ling] Controll[er]
- e: Automatic Light Control
- e: Automatic Load[ing] Control
- e: Automatic Locking Circuit
- e: Axiolinguocervical
alc e: alcohol
- e: approximate lethal concentration
- e: avian leukosis complex
ALCA e: Aft Load Control Assembly
- e: American Leather Chemists Association
- e: Automatic Level Control Assembly
ALCAC e: Airlines Communications Administrative Council
ALCAD e: Alumin[i]um-Clad
ALCAL e: Algebraic Calculator Language
- e: Alloy Coated Alumin[i]um
AlCan e: Alaska-Canada
ALCAN e: Alaskan-Canadian [Highway]
- e: Aluminium Company of Canada
Alcan eclipse d: Alaska-Canada eclipse
ALCANUS e: Alaska, Canada, United States
ALCAPP e: Automatic List Classification and Profile Production
ALCARS e: Airborne Launch Control and Recovery System
ALCATS e: Automated Lines of Communications and Target System
ALCC e: Airborne Launch-Control Center
- e: Airlift Control Center
- e: Army Logistical Control Center
- e: Association of London Computer Clubs
ALCCAM e: Army Life Cycle Cost Analysis Model
ALCCM e: Army Life Cycle Cost Model
ALCD e: Alumin[i]um-Clad
- e: Associate of the London College of Divinity
ALCE e: Advanced Laser Communication Experiment
- e: Airlift Control Element
ALCENT e: Airlift Center
ALCH e: Alcohol
- e: Approach Light Contact Height
Alch e: Alchemy
alch e: approach-light contact height
alchem e: alchemy
alcid e: alcohol and acid
alcism e: alcoholism
ALCJ e: Army Logistics Command Japan
ALCL e: Assembly Line Communications Link
- e: Association of London Chief Librarians
ALCLAN e: Air Logistics Command Local Area Network
- e: Air Launched Cruise Missile
- e: Associate of the London College of Music
ALCMGS e: Air-Launched Cruise Missile Guidance Set
ALCMs e: Air-Launched Cruise Missiles
- e: Air Launched Cruise Missiles
ALCO e: Airlift Coordinating Office[r]
- e: Airlift Launch Control Officer
- e: Airlift Liaison Coordination Officer
- e: Algorithmic Language Compiler
- e: Aluminum Company of Colorado
- e: American Lava Corporation
- e: American Locomotive Company
ALCOA e: Aluminium Company of America
Al Coax PL e: Alumin[i]um Coaxial Plug
Alco Bev e: Alcoholic Beverage
ALCOH e: Alcohol
alcoh e: alcohol[ic]
alcohol e: ethyl alcohol
alcol e: alcohol
alcolic e: alcoholic
ALCOM e: Alaskan Command
- e: Algebraic Compiler
- e: Algebraic Computer
- e: ALGOL Compiler
- e: All Commands
AL/COM e: Applied Logics Comprehensive Computing
ALCOMLANT e: All Commands; Atlantic [Fleet]
ALCOMPAC e: All Commands; Pacific [Fleet]
ALCON e: All Concerned
ALCOP e: Alternate Command Post
ALCOR e: Algorithmic Language Compiler
- e: Algorithmic Language Conversion
- e: Algorithmic Language Converter
ALCOR e: Advanced Research Projects Agency-Lincoln Coherent Observable Radar
- e: Advanced Research Projects Agency Lincoln C-band Observables Radar
ALCORCEN e: Air Logistics Coordination Center
ALCOR Code e: Algorithmic Language Converter Code
ALCORSS e: Australian Liaison Committee on Remote Sensing by Satellite
alcotrician e: alcohol and nutrition
ALCP e: Alternate Command Post
- e: Area Local Control Panel
ALCPT e: American Language College Placement Test
ALc R e: Alcohol Rub
ALCr e: Alumin[i]um Crown
ALCS e: Airborne Launch-Control System
- e: Analog[ue] Large Capacity Store
- e: Automatic Launch Control System
ALCS/T e: All Air Traffic Service Combined Station/Towers
ALCt e: Assault Landing Craft
ALCT e: Attempt to Locate
ALCTP e: Academic Library Consultant Training Program[me]
ALCTS e: Association for Library Collections and Technical Services
ALCU d: Algorithmierung von Lehrprogrammen für den computergesteuerten Unterricht
- e: Arithmetic Logic and Control Unit
- e: Asynchronous Line Control Unit
alcu l: Altocummulus
AL-Cu e: Alumin[i]um-to-Copper
ALcup d: Aluminium mit Kupferschicht
ALD d: Alarmeingabe; dynamische
- e: Acceptable Limit for Dispersion
- e: Acoustic Locating Device
- e: Activity Level Dependent
- e: Adrenoleukodystrophy
- e: Advanced Laser Designator
- e: Advanced Logistics Development
- e: Agricultural Lime Department
- e: Airborne Line Discriminator
- e: A Later Date
- e: Alcoholic Liver Disease
- e: Aldolase
- e: American Library Directory
- e: American Lobbyists Directory
- e: Analog[ue] Line Driver
- e: Anterior Lateral Dendrites
- e: Anterior Latissimus Dorsi
- e: Appraisal of Language Disturbance
- e: Approximate Lethal Dose
- e: Asbestos Lung Disease
- e: Assistant Laboratory Director
- e: Assistant Local Director
- e: Asynchronous Limited Distance
- e: Asynchronous Line Driver
- e: At a Later Date
- e: Automated Logic Design
- e: Automated Logic Diagram
- e: Automatic Locking Differential
- e: Automatic Logic Diagram
- e: Available-to-Load Date
- e: Average Length of Detection
Ald. d: Aluminiumdose
ald. e: aldehyde
AL & D e: Army List and Directory
ALDA e: Analytic Learning Disability Assessment
- e: Apollo Launch Data System
ALDA System e: Apollo Launch Data System
ALDBAS e: Army Logistics Data Base and Access System
ALDC e: Acetolactate Decarboxylase
- e: Army Logistics Data Center
- e: Automatic Load and Drive Control

ALDCS e: Active Lift Distribution Control System
ALDE d: Alarm- /Digital-Eingabe
- d: Alarm-/Digital-Eingabe
- d: alarmbildende digitale Eingänge
Aldehyd l: Alcoholus dehydrogenatus
aldehyde e: al[cohol] dehy[drogenated]- dehydrogenated [oxidized] alcohol
ALDEP e: Automated Layout Design Program[me]
ALDEV e: African Land Development
ALDG e: Automatic Logic Design Generator
ALDH d: Aldehyddehydrogenase
ALDI Message e: Associated Long Distance Interstate Message
ALDIS e: Australian Laser Disk Information Services
ALDO e: Activity Level Dependent Operations
aldo d: Aldosterone
- e: aldosterone
ALDOC e: Arab League Documentation and Information Centre
ALDOS d: automatisches Lager-, Dispositions- und Optimierungssystem
ALDP e: Academic Library Development Program[me]
- e: Automatic Language Data Processing
ALDPS e: Automated Logistics Data Processing System
- e: Automatic Low Data Processing System
ALDRI e: Automatic Low Data Rate Input
ALDS e: Apollo Launch-Data System
- e: Automatic Lightning Detection System
- e: Automatic Line Device Store
- e: Automatic Literature Distribution System
ALDT e: Argon Laser Discharge Tube
ALD Timber e: Alder Timber
ALE d: Alarmeinrichtung
- d: Arbeitsgemeinschaft der Leiter deutscher nichtbundeseigener Eisenbahnen
- d: arithmetische logische Einheit
- e: Actuarial Life Expectancy
- e: Adaptive Line Enhancement
- e: Adaptive Line Enhancer
- e: Additional Living Expense
- e: Address Latch Element
- e: Address Latch Enable
- e: Airborne Laser Experiment
- e: Airport Landing Equipment
- e: Airport Lighting Equipment
- e: Alternate Low Energy
- e: Arid Land Ecology
- e: Army Liaison Element
- e: Association for Liberal Education
- e: Atmospheric Lifetime Experiment
- e: Atomic Layer Epitaxy
- e: Automated Large Experiment
- e: Automatic Laser Encoder
- e: Automatic Line Equalization
Ale s: Alemania
ALEA d: allgemeine Lieferbedingungen für den Export von Anlagegütern
ALEC e: Algebraic Exponents and Coefficients
- e: Analysis of Linear Electronic Circuits
ALECS e: Air Force - Los Alamos Electromagnetic Pulse Calibration Simulator
- e: Automated Law Enforcement Communications System
ALECSO e: Arab League Educational, Cultural and Scientific Organization
al. ed. l: alia editione
ALEDC e: Associate Logistics Executive Development Course
ALEM e: Apollo Lunar Exploration Mission
alem. d: alemannisch
ALEMS e: Apollo Lunar Excursion Module Sensor[s]
ALEP e: Audio Lingual Education Press
ALE Project e: Arid Land Ecology Project
Ale RD s: República Democrática Alemana
Ale RF s: República Federal Alemana
alerfa e: alert phase
alerg. d: allergisch
ALERT e: Acute Launch Emergency Reliability Tip
- e: Affiliated League of Emergency Radio Teams
- e: Alcohol Level Evaluation Roadside Test
- e: Alcohol Level Evaluation Road Tester
- e: Alerting
- e: American Library for Education, Research and Training
- e: Architecture to Logic Equation Realization Technique
- e: Automated Law Enforcement Reporting Technique
- e: Automated Linguistic Extraction and Retrieval Technique
- e: Automated Local Evaluations in Real Time
- e: Automatic Linguistic Extraction and Retrieval Technique
- e: Automatic Logging Electronic Reporting and Telemetering
- e: Automatic Logical Equipment Readiness Tester
- e: Automatic Logician Evaluation Readiness Tester
ALERTCONS e: Alert Conditions
ALERTE f: Association pour la Liberté d'Expression à la Radio et à la Télévison
ALERT II e: Automatic Law Enforcement Response Time
ALERTS e: Airborne Laser Equipment Real Time Surveillance
ALESCO e: American Library and Educational Services Company
ALESEP e: Airfoil Leading Edge Separation
Aleut e: Aleutian [Islands]
Aleutians e: Aleutian Islands
- e: Aleutian Mountains
A Level e: Advanced Level
Alex d: Alexandria
ALEX e: Alert Exercise
Alexanders e: Alexander Archipelago
- e: Alexander Islands
Alex City e: Alexander City, Alabama
ALF d: Auftragserfassung, Lagerwirtschaft, Fakturierung
- d: Ausschuß Luftfahrtforschungsprogramm
- e: Absorption Limiting Frequency
- e: Accelerated Loading Facility
- e: Accuracy Limit Factor
- e: Aloft
- e: Alphanumeric
- e: Application Library File
- e: Assisted-Living Facility
- e: Audiographic Learning Facility
- e: Automatic Lead Former
- e: Automatic Letter Facer
- e: Automatic Letter Facing
- e: Automatic Line Feed
- e: Auxiliary Landing Field
- e: Average Load Factor
ALFA d: automatisiertes Luftfracht-Abwicklungsverfahren
- e: Advanced Laser Flow Analysis
- e: Advanced Liaison Forward Area
- e: Air Lubricated Free Attitude
- e: A Language for Automation
- e: A Language for Financial Analysis
- e: Anonima Lombarda Fabbrica Automobili
- e: Automated Loading of Features and Assemblies
- e: Automatic Ledger Feet Attachment
- e: Automatic Line Fault Analysis
Alfa e: letter A radio code
ALFAD e: Acoustic Low-Flying- Aircraft Detector
ALFC e: Automatic Load Frequency Control
ALFCENT e: Allied Land Forces Central Europe
ALFd e: Advanced Landing Field
ALFGL Circuit e: Automatic Low-Frequency Gain-Limiting Circuit
ALFI e: Air-Land Forces Integration
- e: Air-Land Forces Interface
alflt e: airlift
ALFMED e: Apollo Light Flash Moving Emulsion Detector
ALFORD e: Appalachian Laboratory for Occupational Respiratory Diseases
ALFRED e: Associate Learning from Relative Environmental Data
- e: Automatic Listing and Flowcharting Reorganized into an Elementary Dictionary
ALFS e: Airborne Low-Frequency Sonar
- e: U.S. Navys Airborne Low-Frequency Sonar Program[me]
ALFs e: Absorption Limiting Frequencies
- e: Audiographic Learning Facilities
ALFSS e: All Flight Service Station
ALFT d: Aluminium-Formol-Toxoid
- e: Airlift
- e: Approach and Landing Flight Test
ALFTRAN e: ALGOL-to-FORTRAN-Translator
ALG d: Algier (Flughafen)
- d: Antilymphozyten-Globulin
- e: Advanced Landing Ground
- e: Air Algérie
- e: Aircraft Landing Gear
- e: Air Logistics
- e: Antilymphocyte Globulin
- e: Antilymphocytic Globulin
- e: Asynchronous Line Group
- e: Axiolinguogingival
Alg. d: Algae
- d: Algerien
- d: Algier
- d: Algorithmus
alg d: Algal
- d: Algebra
- e: algae
- e: algebra
- e: algebraic[al]
- e: allergic[al]
- e: allergy
- e: along[side]
- n: algerneen
alg. d: algebraisch
Alg e: Algeria
- e: Algiers
- e: Algorithm
ALGA e: Associate in Local Government Administration
AlGaAs e: Alumin[i]um Gallium Arsenide
ALGADDR e: Algebraic Adder
ALGEB e: Algebra
ALGEC e: Algorithmic Language for Economic Calculations
- e: Algorithmic Language for Economic Problems
ALGENDEP e: Alaskan General Depot
Alger e: Algernon

Algeria e: Democratic and Popular Republic of Algeria
- f: République Algérienne Démocratique et Populaire
algett i: allegretto
ALGLYN e: Alumin[i]um Glycinate
ALGM e: Air-Launched Guided Missile[s]
- e: Algorithm
ALGN e: Alignment
ALGO e: Algorithm
ALGOL e: Algebraic Oriented Language
- e: Algorithmic Language
- e: Algorithmic Oriented Language
ALGOL Compiler e: Algorithmic Language Compiler
ALGOL Program[me] e: Algorithmic Language Program[me]
ALGOL Symbol e: Algorithmic Language Symbol
ALGOL System e: Algorithmic Language System
Algonquin e: Algonquin Peak
- e: Algonquin Provincial Park, Ontario
ALGOPOL e: Algorithmic Problem-Oriented Language
ALGS e: Approach and Landing Guidance System
ALGTG e: Alighting
algtto i: allegretto
ALGW d: Aluminiumgleichwert
ALH e: Anterior Lobe Hormone
- e: Anterior Lobe of the Hypophysis
- e: Atypical Lymphoid Hyperplasia
ALHARD e: Air-Launched High- Altitude Reconnaissance Drone
ALHFAM e: Association of Living Historical Farms and Agricultural Museum
ALHHS e: Association of Librarians in the History of the Health Sciences
ALHS e: Abraham Lincoln High School
ALHT e: Apollo Lunar Hand Tool[s]
ALHTC e: Apollo Lunar Hand Tool Carrier
ALI d: Autofahrer-Leit- und Informations-system
- d: Auto[fahrer]-Lenkungs- und Informationssystem
- e: Activity Level Independent
- e: Acton Laboratory, Inc.
- e: Agricultural Limestone Institute
- e: Airborne Laser Illuminator
- e: Air Launched Interceptor
- e: Alberta Legislation Information
- e: American Ladder Institute
- e: American Law Institute
- e: American Library Institute
- e: Apron Lighting
- e: Arc Lamp Igniter
- e: Arm Length Index
- e: Associate of the Landscape Institute
- e: Asynchronous Line Interface
- e: Australian Laser Institute
- e: Automated Logic Implementation
- e: Automatic Language Identification
- e: Automatic Line Integration
- e: Automatic Location Identification
- e: Automatic Logic Implementation
- e: Automobile Leading & Information
- e: Automotive Lift Institute
- e: Autonomous Learner Index
- e: Awaiting Laboratory Input
ALIA e: Royal Jordanian Airlines
ALIADS e: Alaskan Integrated Air Defense System
ALIAS e: Algebraic Logic Investigation of Apollo Systems
- e: Australias Library, Information and Archives Services
ALIATCS e: All-International Air Traffic Communications Station
ALIATSC e: All International Aeronautical Telecommunications Switching Center[s]
ALIB e: Army Library
ALIC e: Arid Lands Information Center
- e: Australian Libraries and Information Council
ALICE e: Adaptive Line Canceller and Enhancer
- e: Adiabatic Low-Energy Injection and Capture Experiment
- e: Alaskan Integrated Communications Exchange
- e: All-purpose Lightweight Individual Carrying Equipment
- e: Application Language Interface Conversion and Extension
- e: Applicative Language Idealized Computing Engine
- e: Artillery Line Communications Equipment
- e: Automated Library; Information Control and Exchange
- e: Automated Location of Isolation and Continuity Errors
- e: Automatic Landing Instrument Control Equipment
- e: Automatic Laundering Instrument Control Equipment
Alice e: The Alice-Alice Springs, Northern Terntory, Australia
ALICS e: Advanced Logistics Information and Control System
- e: Alaska Integrated Communications System
- e: Assembly Language by Information Control System
ALID e: Automated Library Issue Document
ALIDO e: Amplitude and Latency Instrument with Digital Output
alien e: alienist
ALIFSS e: All International Flight Service Stations
ALIGN e: Alignment
ALII e: Advanced Libraries & Information, Incorporated
ALIM e: Air-Launched Intercept[or] Missile
ALIMD e: Association of Life Insurance Medical Directors
ALIMDA e: Association of Life Insurance Medical Directors of America
Al Imp Reps e: Alert Implementation Reports
ALIMREP e: Alert Implementation Report
ALIMS e: Automatic Laser Instrumentation Measuring System
ALIN e: Agricultural Libraries Information Network
- e: Alignment
ALIO e: Activity Level Independent Operations
ALIP e: Abnormal Localization of Immature Precursors
- e: Annular Linear Induction Pump
ALIR e: Adder Left Input Register
- e: Advanced Laser Intercept Receiver
- e: Australian Library and Information Research
ALIRATS e: Airborne Laser Illuminator Ranging and Tracking System
ALIRT e: Adaptive Long-range Infrared Tracker
- e: Analytic Library Information Retrieval and Transfer
ALIS d: Autofahrer-Leit- und Informationssystem
- d: Auto[fahrer]-Lenkungs- und Informationssystem
- e: Advanced Life Information System
- e: Arid Lands Information System
- e: Automated Library [and] Information System
- e: Automated Library Information Service
- e: Automated Library Information System
- e: Automatic Link Setup
ALISE e: Association for Library and Information Science Education
Al. Isls e: Aleutian Islands
ALISS d: automatisiertes Leitungs- und Informations-System Standardwerk (DDR)
ALISSE e: Agricultural Libraries and Information Services in the South-East
ALIST e: Attributes List
ALI-System d: Auto-Lenkungs- und Informationssystem
A Lit e: Associate in Literature
ALIT e: Automatic Line Insulation Test[er]
ALITALIA e: Italian [Inter]national Airline
ALIVE e: Air-Launched Instrumented Vehicle Evaluation
ALJ e: Australian Library Journal (journ.)
aljak e: Alumin[i]um-Jacketed [coaxial cable]
ALJD e: Administrative Law Judge of the Department
ALJH e: Association of Libraries of Judaica and Hebraica
ALJR e: Australian Law Journal Reports
ALK d: Alarmerfassungs- und Schleifensteuerungs-Konzentrator
- d: Alkyl
- d: Amt für Landeskultur
- d: Ausgabelochkarte
- d: Auslaufkammer
- d: automatisierte Liegenschaftskarte
Alk. d: Alkali
- d: Alkohol
alk d: Alkali[ne]
- e: alkali
alk. d: alkalisch
- d: alkoholisch
alkal. d: alkalisch
Alkali e: Soviet air-to-air radar-guided homing missile
alki e: alcohol
ALKM d: Archiv für Literatur und Kirchengeschichte des Mittelalters (journ.)
alkoh. d: alkoholisch
alk phos e: alkaline phosphatase
alk p'tase d: alkaline Phosphatase
alkums e: air-launched cruise missiles
ALKY e: Alkalinity
Alkylier. d: Alkylierung
ALL d: Antilärmliga
- e: Above Lower Limit
- e: Accelerated Learning of Logic
- e: Acute Lymphatic Leukemia
- e: Acute Lymphoblastic Leukemia
- e: Acute Lymphocytic Leukemia
- e: Address Locator Logic
- e: Admiralty List of Lights
- e: Airborne Laser Laboratory
- e: Aircraft Landing Lamp
- e: All-Logic Level
- e: Allergy
- e: Allowance Load List
- e: Application Language Liberator
- e: Application Load List
- e: Arc Laser Light
- e: Argon Laser Lining

all. d: alleinig
- d: alliiert
- e: airborne laser laboratory
- e: allergy
All. d: Alliierte
- d: Alliteration
all e: allegoric
- e: allegorical
- e: allegory
ALL. e: Allioni
al.l. l: alia lectio
ALLA d: allgemeine Anweisung
- e: Allied Long Lines Agency
Allagash e: Allagash River and Allagash Wilderness Waterway, Maine
ALLC e: Association for Literary and Linguistic Computing
ALLCAT e: All Critical Atmospheric Turbulence
ALLCAT Program[me] e: All Critical Atmosphere Turbulence Program[me]
ALLCB e: Association for Literary and Linguistic Computing Bulletin
ALLCE e: Allowance
ALLCJ e: Association for Literary and Linguistic Computing Journal
ALLD e: Airborne Laser Locator Designator
- e: Airborne Locator Laser Designator
- e: Allowed
ALLDEPHI d: Allgemeine Deutsche Philips Industries GmbH
alleg. d: allegorisch
AL-Leg. d: Aluminiumlegierung
Alleg. e: Allegory
Alleghenies e: Allegheny Mountains of Pennsylvania, Maryland, Virginia and West Virginia
Alleg. Pow. Syst. e: Allegheny Power System
Allem f: Allemagne
Allergol. e: Allergologic[al]
allerh. d: allerheiligst
- d: allerhöchst
ALLF e: Additional Low Layer Functions
ALLG d: Archiv für Lateinische Lexikographie und Grammatik (journ.)
Allg. d: Allgäu
allg. d: allgemein
Allg. d: Allgemeinheit
AllgDSeeVersBed. d: Allgemeine Deutsche Seeversicherungsbedingungen
AllgDTranspVersBed. d: Allgemeine Deutsche Binnentransportversicherungsbedingungen
allgem d: allgemein
AllGeschBed. d: Allgemeine Geschäftsbedingungen
AllgFeuerversBed. d: Allgemeine Feuerversicherungsbedingungen
Allg Fisch-Ztg d: Allgemeine Fischerei-Zeitung (journ.)
Allg Forstztg d: Allgemeine Forstzeitung (journ.)
allg. Gen. d: allgemeine Genehmigung
AllgHaftpflVersBed. d: Allgemeine Haftpflichtversicherungsbedingungen
AllgKrankVersBed. d: Allgemeine Krankenversicherungsbedingungen
AllgKVerkVersBed. d: Allgemeine Kraftverkehrsversicherungsbedingungen
AllgLagBed. d: allgemeine Lagerbedingungen
AllGO d: Allgemeine Gebührenordnung
Allg. Prakt. Chem. d: Allgemeine und Praktische Chemie (journ.)
AllHKomm. d: Alliierte Hohe Kommission
ALLI e: Australian Legal Literature Index
Alligators e: Alligator Rivers of Australias Northern Territory
Allis-Chalmers Eng. Rev. d: Allis- Chalmers Engineering Review (journ.)
allit e: alliteration
- e: alliterative
allj. d: alljährlich
All J Chem d: Allgemeines Journal der Chemie (journ.)
ALL-MBE e: Atomic Layer-by-Layer Molecular Beam Epitaxy
ALLMIS e: Army Lessons Learned Management Information System
ALLN e: Anterior Lateral Line Nerve
ALLNAVSTAS e: All Naval Stations
ALLO e: Atypical Legionella-Like Organism
allo e: allonym
Allo i: allegro
ALLOC e: Allocate
alloc e: Allocate[d]
- e: allocation
Al Loco E e: Associate of the Institution of Locomotive Engineers
allop e: allophone
ALLOT e: Alloted
ALLOW e: Allowable
- e: Allowance
ALLP e: Arc Laser Light Pump
- e: Audio-Lingual Language Programming
Allr. d: Allrad[antrieb]
ALLS e: Apollo Lunar Logistic Support
ALLSTNS e: All Stations
Alltto i: allegretto
alluv e: alluvial
- e: alluvium
AllVr d: Allverstärker
ALLVRJ e: Air-Launched Low-Volume Ramjet
ALL Wea Tra U e: All-Weather Training Unit
AllWeaTraULant e: All-Weather Training Unit; Atlantic Fleet
ALM d: Alarm
- d: asynchroner Linearmotor
- e: Adding Listing Machine
- e: Advanced List of Materials
- e: Aerophysics Laboratories Memorandum
- e: Air-Launched Missile
- e: Aircraft Limited Model
- e: Airfield Landing Mat
- e: Airlift Loading Model
- e: Alarm Master
- e: Alerting Message
- e: American Leprosy Missions
- e: Apollo Lunar Module
- e: Applied Laboratory Method
- e: Arm Lock Magnet
- e: Asynchronous Line Module
- e: Asynchronous Line Multiplexer
- e: Audio Level Meter
- e: Audio Lingual Method
- e: Augmented Lunar Module
- n: Antilliaanse Luchtvaart Maatschappij
alm a: almindelig
Alm. d: Almanach
alm. e: alarm
ALM. e: Almandine
A & LM e: Arkansas & Louisiana-Missouri
A.L.M. l: Artium Liberalium Magister
ALMA e: Alphanumeric Language for Music Analysis
- e: American Loudspeaker Manufacturers Association
- e: Analytical Laboratory Managers Association
ALMAJCOM e: All Major Commands
ALMB e: Air-Launched Missile Ballistics
ALMBz d: Aluminiummehrstoffbronze
ALMC e: Air-Launched Missile Change
- e: Army Logistics Management Center
- e: Army Logistics Management College
ALMDA e: Airlock Multiple Docking Adapter
ALME e: Acetyllysine Methyl Ester
AL-Methode d: Ammoniumlaktatessigsäure-Methode
ALM Frequency e: Absorption Limiting Frequency
ALMI e: Alpha Microsystems
- e: Anterior Lateral Myocardial Infarct
ALMICS e: Automated Logistics Management and Inventory Control System
ALMIDO e: Amplitude and Latency Instrument with Digital Output
ALMIDS e: Army Logistic[s] Management Integrated Data System
ALMIDSY e: Army Logistics Management Integrated Data System
ALMILACT e: All Military Activities
ALMIMSIP e: Air-Launched Missile Intermediate Maintenance System Program[me]
ALMIOS e: Air-Launched Missile Inventory Objectives Study
ALMIRBM e: Air-Launched Medium-Intermediate Range Ballistic Missile
ALMM d: Algorithmus und Lösung für MM (mathematisches Modell)
ALMPT e: Air-Launched Missile Propulsion Technology
ALMRS e: Automated Land and Minerals Records System
ALMS e: Air-Launched Missile System
- e: Aircraft Landing Measurement System
- e: Air Lift Management System
- e: A Library Management System
- e: Analytic Language Manipulation System
- e: Atomic Line Molecular Spectroscopy
- e: Automated Library Management System
- e: Automated Logic Mapping System
- e: Automated Logic Matrix System
- e: Auxiliary Liquid Metal System
ALMSA e: Automated Logistics Management Systems Activity
- e: Automated Logistics Management Systems Agency
ALMT e: Allotment
ALM Timber e: Almondwood Timber
ALMV e: Air-Launched Miniature Vehicle
- e: Anterior Leaflet of Mitral Valve
ALN d: allgemeines Lichtnetz
- e: Accounting Line Number
- e: Adaptive Learning Network
- e: Advanced Land Navigation
- e: Align
- e: Ammunition Lot Number
- e: Anterior Lateral Nerve
- e: Attribute Level Number
- e: Australian Library News (journ.)
aln e: anterior lymph node
ALNA e: Algebraic Logic Network Analysis
- e: Automatic Logic Network Analyzer
ALNASTA[s] e: All Naval Station[s]
ALNI e: Alloyed Nickel

ALNICO d: Aluminium-Nickel-Cobalt
- e: Alloyed Nickel Cobalt Steel
alnmt e: alignment
ALNNO e: Air-Launched Non-Nuclear Ordnance
ALNOT e: Alert Notice
ALNP e: Abraham Lincoln National Park
ALNTS e: Automatic Liquid Nitrogen Transfer System
ALNW e: Air-Launched Nuclear Weapon
ALNZ e: Air League of New Zealand
ALO e: Administrative Liaison Officer
- e: Admiralty Liaison Officer
- e: Advanced Lunar Operation
- e: Air Liaison Officer
- e: Allied Liaison Office[r]
- e: Alternate Launch Officer
- e: Alternate Liaison Officer
- e: American Liaison Office[r]
- e: Apollo Lunar Orbit
- e: Arm Length Order
- e: Army Liaison Officer
- e: Authorized Level of Organization
- e: Automatic Lock-On
- e: Axiolinguo-Occlusal
ALOA e: Amalgamated Lithographers of America
- e: Assembly of Librarians of the Americas
ALOAL e: Autonomous Lock-On After Launch
ALOB e: Alphabetic O Bit
ALOC e: Administrative and Logistics Operations Center
- e: Air Line[s] of Communication
- e: Air Logistics Chain
- e: Allocation
- e: Alternate Launch Officer Console
- e: Apollo Launch Operations Committee
- e: Army Logistic Operations Center
aloc e: allocate
ALOFT e: Airborne Light Optical Fiber Technology
- e: A Language Oriented to Flight-engineering and Testing
ALOG d: Anti-Logarithmus
- e: Administration-Logistics
- e: Army Logistician (journ.)
ALOHA e: Aboriginal Lands of Hawaiian Ancestry
- e: Aloha Airlines
ALOHA System e: Additive Links Online Hawaii Area System
ALOM e: Air-Land Operations Manual
ALON e: Aircraft Longitude
- e: Air Liaison Officer Net
ALOO e: Albuquerque Operations Office
ALOP e: Apollo Launch Operation Panel
- e: Army Logistics Objectives Program[me]
ALOPE e: Airborne Light Detection and Ranging Oceanographic Probing Experiment
ALOR e: Adder Left Output Register
- e: Advanced Lunar Orbital Rendezvous
ALOREP e: Airlift Operational Report
ALORS e: Advanced Large Object Recovery System
ALOS e: Apollo Lunar Orbital Science
- e: Average Length of Stay
ALOSH e: Appalachian Laboratory for Occupational Safety and Health
ALOT e: Adaptive Laser Optics Techniques
- e: Airborne Lightweight Optical Tracker
- e: Airborne Lightweight Optical Tracking
- e: Allotment
- e: Allotted
ALOTM e: Allotment
ALOTMT e: Allotment
ALOTS e: Airborne Lightweight Optical Tracking System[s]
Alouette e: Aernspatiale armed helicopter
ALOX d: Aluminiumoxid
ALP d: Agrarleitplan
- d: Alveolarproteinose
- d: automatischer Lernprozeß
- e: Acute Lupus Pericarditis
- e: Adam Language Processor
- e: Advanced Language Processor
- e: Advanced Language Program[me]
- e: Advanced Lunar Project[s]
- e: Advisory Light Panel
- e: Air-Launched Platform
- e: Air; Low Pressure
- e: Airborne Line Printer
- e: Air Liaison Post
- e: Airport Layout Plan
- e: Airport Location Point
- e: Alkaline Phosphatase
- e: Allied Liaison and Protocol
- e: Allied Logistic Publication
- e: Alphalytic Protease
- e: Alternate Program[me]
- e: Ambulance Loading Post
- e: Anterior Lobe of Pituitary
- e: Arithmetic [and] Logic Processor
- e: Assembly Language Preprocessor
- e: Assembly Language Processing
- e: Assembly Language Program[me]
- e: Assembly Language Programming
- e: Assistant for Logistics Planning
- e: Asynchronous Line Pairs
- e: Autocode List Processing
- e: Automated Language Processing
- e: Automated Learning Process
- e: Automated Library Program[me]
- e: Auxiliary Language Processor
Alp e: Alphen
- e: Alpine
ALPA e: Air Line Pilots Association
- e: Alaskan Long-Period Array
ALPAC e: Automatic Language Processing Advisory Committee
ALPAK e: Algebra Package
ALPB e: Alphabetic[al]
ALPC e: Adaptive Linear Predictive Coding
ALPEC e: Ammunition Loading Production Engineering Center
ALPERSCOM e: All Personnel Communication
ALPETH e: Alumin[i]um and Polyethylene
ALPH e: Alpha[nu]meric
Alph. d: Alphabet
alph. d: alphabetisch
ALPHA e: Action League for Physically Handicapped Advancement
- e: Alkali Plasma Hall Accelerator
- e: Alphabetic[al]
- e: Alphanumeric[al]
- e: Automated Literature Processing, Handling and Analysis
- e: Automatic Literature Processing, Handling and Analysis
Alpha e: letter A radio code
ALPHA Machine e: Automatic Lector Position Handwriting Machine
alphameric e: alphabetic numeric
- e: alphanumeric and alphabetic-numeric
alphametic e: alphabet arithmetic
ALPHANUM e: Alphanumeric
alphanumeric e: alphabetical-numerical
- e: alphabetical numerical
alpha order e: alphabetical order
ALPHAS e: Automated Literature Processing Handling and Analysis System
ALPHA System e: Automated Literature Processing, Handling and Analysis System
ALPL e: Advanced Lunar Projects Laboratory
ALPM e: Alphanumeric
alpo e: Apollo lunar polar orbiter
ALPO e: Association of Lunar and Planetary Observers
ALPR e: Argonne Low-Power Reactor
ALPS e: Advanced Linear Programming System
- e: Advanced Liquid Propulsion System
- e: Advanced Logic Processing System
- e: Approach and Landing Procedures Simulator
- e: Assembly Line Planning System
- e: Associated Logic Parallel System
- e: Associative Logic Parallel System
- e: Associative Logic Processor System
- e: Automated Language Processing System
- e: Automated Letter Preparation System
- e: Automated Library Processing Services
- e: Automated Library Processing System
- e: Automatic Landing Positioning System
- e: Automatic License Plate Scanning
- e: Automatic Linear Positioning System
Alps e: Alpine Mountains
ALPSP e: Association of Learned and Professional Society Publishers
ALPTA e: American Low Power Television Association
Alpujarras e: Alpujarras Mountains of Almeria and Granada in Spain
AL-PUR-AL d: Aluminium-Polyurethan-Aluminium
AL PUR COM e: All-Purpose Communication[s]
ALPURCOMS e: All-Purpose Communication[s] System
ALQ e: Abraham Lincoln Quarterly (journ.)
- e: Almost Letter Quality
ALQAS e: Aircraft-Landing Quality-Assessment Scheme
ALR d: Allgemeines Landrecht für die preußischen Staaten
- d: Alternatives Lokalradio
- d: automatische Lautstärkeregelung
- d: automatischer Lüftungsregler
- d: Reinstaluminium
- e: Active Line Rotation
- e: Aden Law Reports
- e: Airborne Laser Rangefinder
- e: Alarm Reset
- e: Alert
- e: American Law Reports
- e: Artillery Locating Radar
- e: Automated Legal Research
- e: Automatic Level Recorder
- e: Automatic Load Regulator
Alr. d: Aluminiumröhre
alr l: aliter
ALRAFAC e: All-Radar Facilities
ALRC e: Anti-Locust Research Centre
- e: Area Learning Resource Center
ALRCP e: Army Long Range Capabilities Plan
ALRD e: Army Logistics Research & Development
ALREMP e: Aircraft Launch and Recovery Equipment Maintenance Program[me]
ALREP e: Air Launched Report
ALRH e: Apollo Lunar Radioisotope Heater
ALRI e: Advanced Long-Range Interceptor
- e: Airborne Long-Range [Radar] Input
- e: Airborne Longe-Range Intercept
ALRIE e: Airborne Long-Range Input Electronic

AL RL e: Alumin[i]um Research Laboratories [of the Aluminium Company of America]
ALROS e: American Laryngological, Rhinological and Otological Society
ALRP e: All Line Repeat Printing
ALRR e: Armes Laboratory Research Reactor
ALRS e: Admiralty List of Radio Signals
- e: Arithmetic Logic Register Stack
ALRT d: Arbeitsgemeinschaft Luftfahrt- und Raumfahrttechnik
ALRTF e: Amyotrophic Lateral Sclerosis
- e: Antilymphocytic Serum
- e: Army-Longe-Range Technological Forecast
ALRU e: Automatic Line Record Update
ALS d: Ablaufsteuerung
- d: amyotrophe Lateralsklerose
- d: Arbeitsgemeinschaft ländlicher Siedlungsträger
- d: Arbeitsgemeinschaft lebensmittel-chemischer Sachverständiger
- d: Arbeitsgemeinschaft lichttechnischer Spezialfabriken
- d: Arbeitskreis Lichttechnischer Spezialfabriken
- d: automatisches Leitungssystem
- e: Activated Low Structure
- e: Advanced Laser System
- e: Advanced Library Systems, Inc.
- e: Advanced Life Support
- e: Advanced Logistics Spacecraft
- e: Advanced Low-power Schottky
- e: Advanced Lunar Studies
- e: Agricultural Land Service
- e: Airborne Laser System
- e: Airborne Live Scanner
- e: Aircraft Landing System
- e: Alignment Level Signal
- e: Alphanumeric Logging System
- e: Alumin[i]um-Sheat
- e: American Lumber Standards
- e: Anti-Lymphocyte Stimulator
- e: Anti-mouse Lymphocyte Serum
- e: Anticipated Life Scan
- e: Anticipated Life Span
- e: Antilymphocyte Serum
- e: Antilymphocytic Serum
- e: Antilyphocyte Serum
- e: Approach and Landing Simulator
- e: Approach Light System
- e: Arithmetic Logic Section
- e: Assembler Language Statement
- e: Associate List Searcher
- e: Associate List Selection
- e: Associate List Selector
- e: Associate of the Linean Society
- e: Associative List Selection
- e: Associative List Selector
- e: Augmented Logistics Support
- e: Australian Literary Studies (journ.)
- e: Autograph Letter Signed
- e: Automated Library System[s]
- e: Automatic Landing System
- e: Automatic Level Setting
- e: Automatic Line Selection
- e: Automatic Line Supervision
- e: Auxiliary Lighter
- e: Azimuth Laying Set
als e: aldolase
- e: amyotrophic lateral sclerosis
A.L.S. e: Associate of the Linnean Society
ALSA e: American Land Sailing Association
- e: Astronaut Life Support Assembly
ALSAC e: Aiding Leukemia Stricken American Children
ALSAFECOM e: All Safety Commands
ALSAM e: Air-Launched Ship-Attack Missile
- e: Air Launched Surface Attack Missile
ALSC e: American Lumber Standards Committee
- e: Association for Library Service to Children
- e: Automatic Level and Slope Control
ALSCC e: Apollo Lunar Surface Close-up Camera
ALSCP e: Appalachian Land Stabilization and Conservation Program[me]
ALSD e: Apollo Lunar Surface Drill
ALSE e: Apollo Lunar Sounder Experiment
- e: Apollo Lunar Surface Experiment
AL SEA FRON e: Alaskan Sea Frontier
AL SEC e: Alaskan Sector
- e: Aleutian Sector
Al seg i: al segno
ALSEP e: Apollo Lunar Surface Experiments Package
ALSER e: Axle-Load Survey Recorder
ALSI e: Alumin[i]um Silicon
- e: Analog[ue] Large Scale Integration
Alsk e: Alaska[n]
ALSK e: Alaska [Railroad]
ALSL e: Assembly List Shortage Log
ALSOR e: Air-Launched Sounding Rocket
ALSP e: Army Logistics Study Program[me]
ALSRC e: Apollo Lunar Sample Return Container
ALSS e: Advanced Laser System Study
- e: Advanced Location Strike System
- e: Airline System Simulator
- e: Apollo Logistic[s] Support System
ALST e: Alaska Standard Time
- e: Altostratus
ALSTA e: All Stations
ALSTAR e: Altitude Layer Surveillance Terminal Area Radar
ALS Technology e: Advanced Low-power Schottky Technology
ALSTG e: Altimeter Setting
ALS-TTL e: Advanced Low-power Schottky Transistor-Transistor Logic
ALSU e: Autonomous Line Scanning Unit
ALT d: Arbeitsgemeinschaft Luftfahrttechnik
- e: Accelerated Life Testing
- e: Actuarial Life Table
- e: Algebraic Language Translator
- e: Alternate
- e: Alternative Mode
- e: Alternator
- e: Altitude
- e: Amber Light
- e: Approach and Landing Test
- e: Australian Law Times
- e: Automatic Language Translation
- e: Automatic Line Testing
- e: Automatic Line Tracking
- e: Automatic Logic Tester
alt. d: altaisch
- d: alternativ
- d: Alternative
Alt. d: Altona
A-L-T. d: Arzneimittel-Liste und -Tarif
alt e: academic learning time
- e: alanine aminotransferase
- e: alter[ation]
- e: altered
- e: altering
- e: alternating
- e: alternation[s]
- e: alternative
- e: altimeter
Alt e: alternating [light]
ält. d: ältere[r] [s]
- d: älteste[r] [s]
Alta e: Alberta
ALTA e: American Library Trustee Association
- e: Association of Local Transport Airlines
ALTAC e: Algebraic Transistorized Automatic Computer
- e: Algebraic Translator and Compiler
- e: Algebraic Translator Assembler Compiler
Alta Freq. i: Alta Frequenza (journ.)
ALTAIR e: Advanced Research Projects Agency Longe-Range Tracking and Instrumentation Radar
- e: Automatic Logical Translation and Information Retrieval
Altais e: Altai Mountains
ALTAN e: Alternate Alerting Network
ALTAPE e: Automatic Line Tracer and Programming Equipment
- e: Automatic Line Tracing and Programming Equipment
ALTARE e: Automatic Logic Testing and Recording Equipment
ALTA System e: Automatic Line Test and Administrative System
ält. Ausg. d: ältere Ausgabe
AltbG d: Altbankengesetz
ALTCENTRE e: Advanced Level Telecommunications Center
alt-ch e: alternate-channel
ALTCOM e: Alternate Command[er]
ALTCOMCEN e: Alternate Command Center
ALTD e: Altered
altd. d: altdeutsch
altd e: alloted
ALTDEV e: Alternative Device
alt dieb. l: alternis diebus
ALTDS e: Apollo Launch Trajectory [Data] System
altdt. d: altdeutsch
ALTE e: Altitude Error
ALTEL e: Association of Long Distance Telephone Companies
Al Tel Ca e: Allens Telegraph Cases
ALTER e: Alteration
- e: Alternate
ALTER System e: Automated Legal Text Entry and Revision System
ALTF e: Alternate Field
Alt. F. Fl. e: Alternating Fixed and Flashing
Alt. F. Gp. Fl. e: Alternating Fixed and Group Flashing
Alt. Fl. e: Alternating Flashing
Alt Got e: Alternate Gothic
Alt. Gp. Fl. e: Alternating Group Flashing
altgr. d: altgriechisch
Alt. Gr. Fl. e: Alternating Group Flashing
alt. hor. l: alternis horis
ALTHQ e: Alternate Headquarters
ALTID e: Algorithmic Language for Teaching and Information Dialog[ue]s
- e: Alteration Identification
ALTIHP e: Avionics Laboratory Technical Information Handling Profile
altim e: altimeter
ALTINSAR e: Alternate Instruction Address Register
altk. d: altkatholisch
altl. d: altlateinisch
ALTLIB e: Alternate Library

Alt.Lt. e: Alternating Light
altm e: altimeter
ALTN e: Alternate
altn e: alternation
alt noc. l: alternis noctibus
ALT NOT e: Alternation Note
ALTNTR e: Alternator
alto e: alternator
alt of inst e: Alteration of Instruments
AltölG d: Altölgesetz
ALTP e: Automatic Linear Temperature Programmer
ALTPM d: Algorithmus für technisch-physikalisches Modell
ALTR e: Alteration
- e: Alternate
- e: Alternator
ALTRAN e: Algebraic Translator
- e: Assembly Language Translator
ALTRD e: Altered
ALTREC e: Automatic Life Testing and Recording of Electronic Components
ALTRN e: Alteration
altru e: altruism
- e: altruist
- e: altruistic
ALTRV e: Altitude Reservation
alts. d: altsächsisch
ALTS e: Advanced Lunar Transportation System
- e: Analog[ue] Line Termination Subsystem
- e: Automated Library Technical Services
- e: Automatic Line Test Set
alt.set. e: altimeter setting
altspr. d: altsprachlich
Alt T e: Alternate Target
alttest. d: alttestamentlich
ALTU e: Adder, Logical and Transfer Unit
alt udk a: alt udkomne
alt xyl e: alto xylophone
ALU e: Advanced Logical Utility
- e: Arithmetic and Logic[al] Unit
- e: Arithmetic Logic[al] Unit
- e: Asynchronous Line Unit
Alu d: Aluminium
Alucon e: Aluminium Conductors
alue e: admissible linear unbiased estimator
ALUM e: Alumin[i]um
ALUMN e: Aluminium
ALUP d: automatische lochstreifengesteuer-te Universalprüfeinrichtung
- d: automatische lochstreifengesteuerte universelle Prüfeinrichtung
ALUSLO e: American Legation; United States [Naval] Liaison Officer
ALUSNA e: American Legation; United States Naval Attaché
ALUSNOB e: American Legation; United States Naval Observer
ALUTROL e: Arithmetic Logic Unit for Control
ALV d: Archiv für Literatur und Volksdichtung (journ.)
- e: Air Launched Vehicle
- e: Annual Lease Value
- e: Apple Leafspot Virus
AlV e: Alarm Valve
alv e: alveolar
- e: avian leukemia virus[es]
- e: avian leukosis virus
ALVAO f: Association des Langues Vilvantes pour l'Afrique Occidentale
alv. deject. l: alvi dejecriones
ALVIN e: Antenna Lobe for Variable Ionospheric Nimbus
ALVRJ e: Air-Launched Low Volume Ramjet
ALVRT e: Air Launcher Low Volume Ramjet
ALV Vent d: alveoläre Ventilation
alvx e: alveolectomy
ALW d: Archiv für Liturgiewissenschaft (journ.)
- e: Admiralty List of Wireless Signals
- e: Air Launched Weapon
- e: Allowance
- e: Arch-Loop-Whorl
ALWAC e: Axel-Wennergren Automatic Computer
ALW-ACC e: Always Accessible
ALW-AFL e: Always Afloat
Alweg e: Axel Lennard Wenner Gren
ALWIN e: Algorithmic Wiswesser Notation
ALWL e: Army Limited War Laboratory
alws e: Allowances
alwt e: advanced lightweight torpedo
Alx e: Alexandria
ALY e: Alloy
ALY STL e: Alloy Steel
ALZ d: Alarmzentrale
- d: Aluminium-Zentrale
- e: Assault Landing Zone
alz i: alzamento
ALZUDI d: algorithmische Zuordnungsdidaktik
AM d: Abdichtungsmittel
- d: Abrechnungsmaschine
- d: Abschnittsmarke
- d: Abschreckmittel
- d: Addressograph Multigraph GmbH
- d: Adressenmarke
- d: Adressenmodifikation
- d: Aktionsmodul
- d: Amplitudenmodulation
- d: Anfangsmarke
- d: Anlagenmodul
- d: Anschlußmöglichkeit[en]
- d: Anti-Materiel
- d: Arbeitsmittel
- d: Arbeitsmodul
- d: arithmetischer Mittelwert
- d: arithmetisches Mittel
- d: astabiler Multivibrator
- d: Asynchronmaschine
- d: Auslösemagnet
- d: Aussteuerungsmesser
- e: Abrasive Machining
- e: Absorption Modulation
- e: Abstract Mathematics
- e: Academy of Management
- e: Access Manager
- e: Access Matrix
- e: Access Mechanism
- e: Access Method
- e: Access Module
- e: Accounting Machine
- e: Accounting Method
- e: Accumulative Multiplication
- e: Acoustic Memory
- e: Acquisition Manager
- e: Active Memory
- e: Actomyosin
- e: Actual Measurement Weight
- e: Adaptive Multiplexer
- e: Adding Machine
- e: Adding Mechanism
- e: Addition Matrix
- e: Addressable Memory
- e: Addressed Memory
- e: Addressing Machine
- e: Addressing Matrix
- e: Addressing Method
- e: Address[ing] Mode
- e: Address Mark[er]
- e: Address Match
- e: Address Memory
- e: Address Modification
- e: Address Modifier
- e: Addressograph Multigraph
- e: Adjoint Matrix
- e: Administrative Manual
- e: Administrative Message
- e: Air-Cooled Motor
- e: Airlock Module
- e: Air Mail
- e: Air Marshal
- e: Air Material
- e: Air[plane] Mechanic
- e: Alarm Message
- e: Alignment Mark
- e: Alternate Mark
- e: Alternate Mode
- e: Alveolar Macrophages
- e: Ambient
- e: America[n]
- e: American Machinist (journ.)
- e: Ammeter
- e: Amplifier
- e: Amplitude Modulated
- e: Analog[ue] Memory
- e: Analog[ue] Module
- e: AND Module
- e: Angle Meter
- e: Angular Momentum
- e: Annular Magnet
- e: Anti-Mines
- e: Anticipatory Modification
- e: Applied Microbiology (journ.)
- e: Area Multiplexer
- e: Arithmetic Mean
- e: Arms Material
- e: Army Manual
- e: Army Map
- e: Assignment Memorandum
- e: Assistant Manager
- e: Associate[d] Member
- e: Associative Memory
- e: Astable Multivibrator
- e: Asynchronous MODEM
- e: Atlantic Monthly (journ.)
- e: Atomic Migration
- e: Auto-Mechanic
- e: Automatic Manual
- e: Automatic Memory
- e: Automatic Monitoring
- e: Auxiliary Marker
- e: Auxiliary Memory
- e: Available Memory
- e: Aviation [Structural] Mechanic
- e: Aviation Medicine
- e: Aviation Metalsmith
- e: Axiomesial
- e: Master of Arts
- f: Alpes Maritimes
- f: Association Mondiale Vérérinaire
- s: Almacenes Maritimos
- s: Almanaque Mundial
am a: amerikansk
- d: amerikanisch
- d: Amphibole
- d: Attometer (= 10^{-18}m)
- e: amphibian
- e: amplitude
am. d: amorph
- e: air-locked module
- e: aircooled motor
- e: ammeter
- e: amplitude modulation
- e: asymmetric multiplier
- e: auditory [sequential] memory

Am. d: Amateur
- d: Amerika[ner]
- d: Amerikanismus
- d: Ametropia

Am d: Amerizium
- d: Ammonium
- d: Ammunition
- d: Amyl
- d: Anmeldestelle
- e: Aeromexico
- e: Air Medal
- e: Air Ministry
- e: Amatol
- e: Amazonas
- e: America
- e: Americum
- e: Armenia
- e: large minesweeper
- e: metric angle
- s: Aeronaves de México

a.m. e: above-mentioned
- l: ante meridien

a. m. l: anno mundi
- l: ante mortem

A. M. l: Artium Magister

A.M. e: Access Manway
- e: Air Mail

a. M. d: als Mitglied
- d: amtliche Mitteilung
- d: anderer Meinung

a.M. d: am Main
- d: angewandte Mathematik

A & M e: Agricultural and Mechanical College of Texas
- e: Architectural and Mechanical
- e: Archives and Manuscripts (journ.)
- e: Assembly and Maintenance

a & m e: ancient and modern

A-M e: Addressograph, Multigraph
- e: Alpes-Maritimes

a-m e: Ampere-Minute[s]
- e: Amplitude-Modulated
- e: Amplitude Modulation

A-m e: Ampere[s]-Minute[s]

A/M e: Auto / Manual
- e: Automatic/Manual
- e: Aviation Medicine
- e: Command Module

A/m d: Ampere/Meter
- e: Ampere[s] per Meter

AM1c e: Aviation Metalsmith First Class

AM 1 Sb d: Einseitenband-Amplitudenmodulation

A/m² d: Ampere / Quadratmeter
- d: Ampere/Quadratmeter
- e: Ampere[s] per square meter

Am² e: Amperemeter Squared

a/m² e: amperes per square meter

AM2c e: Aviation Metalsmith Second Class

AM 2 Sb d: Zweiseitenband-Amplitudenmodulation

AM-3C e: Aeritalia-Aermacchi single-engine three-place armed-trainer aircraft

AM3c e: Aviation Metalsmith Third Class

AMA d: Antimalarial Agent
- e: Academy of Model Aeronautics
- e: Acoustical Materials Association
- e: Acoustic Materials Association
- e: Actual Mechanical Advantage
- e: Advanced Minuteman Accelerometer
- e: Aerospace Medical Association
- e: Against Medical Advice
- e: Agricultural Marketing Administration
- e: Air Material[s] Area
- e: American Machinery Association
- e: American Management Association
- e: American Maritime Association
- e: American Medical Association
- e: American Meteor Association
- e: American Microfilm Association
- e: American Military Attaché
- e: American Motorcycle Association
- e: Amplitron Amplifier
- e: Analog[ue] Major Alarm
- e: Antimitochondrial Antibody
- e: Associate of the Mathematical Association
- e: Associate of the Museums Association
- e: Associative Memory Address
- e: Associative Memory Array
- e: Asynchronous Multiplexer Adapter
- e: Australian Medical Association
- e: Australian Meteorological Association
- e: Automated Modification Analyzer
- e: Automatic Malfunction Analysis
- e: Automatic Memory Allocation
- e: Automatic Message Accounting
- e: Automobile Manufacturers Association

AmA e: American Annual (journ.)
- e: American Anthropologist (journ.)

AMAA e: Association of Medical Advertising Agencies

AMAB d: Allyltrimethyl-Ammoniumbromid[e]
- e: Allyltrimethyl-Ammoniumbromide
- e: Army Medical Advisory Board

AMAC e: Aircraft Monitor and Control
- e: Aldermaston Multi-Access Configuration
- e: Automated Multiparameter Analyzer for Cells

Am Acad Arts and Sci Proc e: American Academy of Arts and Sciences; Proceedings (journ.)

Am Acad Pol Soc Sci e: American Academy of Political and Social Science

Am Acad Rel e: American Academy of Religion

AMA/CMA e: American Medical Association/California Medical Association

AMACO e: American Medical Assurance Company

AMACS e: Automatic Message Accounting Collecting System

AMACUS e: Automated Microfilm Aperture Card Updating System

AMAD e: Activity Median Aerodynamic Diameters
- e: Aircraft-Mounted Accessory Drive

AMA-DE e: American Medical Association-Drug Evaluation[s]

AMAE e: American Museum of Atomic Energy

AMAERF e: American Medical Association Education and Research Foundation

AMAFA e: Air Mass and Front Analysis

amagn. d: amagnetisch

AMAIMM e: Associate Member of the Australasian Institute of Mining and Metallurgy

AMAIS e: Agricultural Materials Analysis Information Service

AMAL e: Aero-Medical Acceleration Laboratory
- e: American Medical Acceleration Laboratory

amal d: Amalgam
- e: amalgam[ate]
- e: amalgamate[d]
- e: amalgamation

am. alc. e: Amyl Alcohol

AMALG e: Amalgamated

amalgam e: mercury and silver mixture

AMAmIEE e: Associate Member of the American Institute of Electrical Engineers

AMA-MTR e: Automatic Message Accounting-Magnetic Tape Recording

AMANDA e: Automated Management of Document Access
- e: Automatic Measurement and Analysis
- e: Automized Medical Anamnesis Dialog[ue] Assistant

AMANDA Computer e: Automized Medical Anamnesis Dialog[ue] Assistant Computer

AM and M e: Applied Mathematics and Mechanics

Am Anthropol e: American Anthropologist (journ.)

Am Antiq e: American Antiquarian (journ.)

Am Antiq Soc Proc e: American Antiquarian Society; Proceedings (journ.)

AMAP e: Advanced Multiprogramming Analysis Procedure
- e: Advanced Multiprogramming Analysis Program[me]

AMAPS e: Advanced Manufacturing, Accounting and Production System
- e: Apogee Motor Assembly with Paired Satellites

AMAR e: Analog[ue] Multiplexer Address Register
- e: Annual Major Additions Rate
- e: Anti-Missile Array Radar

AMARC e: Automatic Message Accounting Recording Center

Am Arch e: American Architect (journ.)

AMARS e: Advanced Micrographic Access and Retrieval System
- e: Air Mobile Aircraft Refueling System
- e: Automatic Message Accounting Recording System
- e: Automatic Message Address Routing System
- e: Autonetics Modular Airborne Radar System

AMAS e: Advanced Midcourse Active System
- e: Automated Maneuvering Attack System
- e: Automatic Message Accounting System

AMASCP e: Air Material Area Stock Control Point

Am Assn Blood e: American Association of Blood Banks

Am Assn Coll Pharm e: American Association of Colleges of Pharmacy

AMAT e: American Air Transport

AMATA e: Audible Manufacturing Training Aid

AMATC e: Air Materiel Armament Test Center

Amateur FG d: Amateurfunkgesetz

AMA Timber e: Amari Timber

AMATIS e: Automated Meteorological Terminal Information Service

amatol e: ammonia & toluene (explosive)

A Matter e: Advance Matter

AMATYC e: American Mathematical Association of Two-Year Colleges

AMAUS e: Aero Medical Association of the United States

AMAWA e: American Medical Association Womens Auxiliary

AMAX e: American Metal Climax

Amaz e: Amazonas

AMB d: Allgemeine Maschinenversicherungsbedingungen
- d: allgemeine Montagebedingungen
- d: ambulante medizinische Betreuung
- d: Auftrags- und Maschinenbestand
- d: Ausgabe Magnetband

- e: Address Mark Bit
- e: Aerospace Medicine and Biology
- e: Air-launched Missile Bulletin
- e: Aircraft Maintenance Base
- e: Air Material Base
- e: Air Ministry Bulletin
- e: Airways Modernization Board
- e: Amber
- e: Ambulance
- e: Ambulatory
- e: Antarctic Meteorite Bibliography
- e: Armament Material Bulletin
- e: Asbestos Millboard
- e: Auto-Manual Bridge
- e: Autolyzed Mouse Brain
- f: Association Maritime Belge
- p: Associação Médica Brasileira

amb. d: ambulant
Amb. d: Ambulanz
amb e: ambient
AMB. e: Anti-Motorboat
AMBA d: Ausgabe Magnetbanddatei
AMBAC e: Associate Member of the British Association of Chemists
Amb Brdg e: Ambassador Bridge
Amb Col e: Ambassador College
amb. dvr. e: Ambulance Driver
AMBEG d: Automatische Maschinen zur Bearbeitung des Glases
AMBEL e: Ambiguity Eliminator
AMBER d: Analyse der Montage-Bereitstellung
- e: Air Movements; British Empire Routes

ambg e: ambigous
- e: ambiguity

Am Bibl e: American Bibliographic Center-Clio Press
AMBIG e: Ambiguity
ambig e: ambiguous
AMBIM e: Associate Member of the British Institute of Management
AMBIT e: Acronym May Be Ignored Totally
- e: Algebraic Manipulation by Identify Translation

ambiv e: ambivalence
- e: ambivalent

Am Bk e: American Book Company
Am Bk Prices e: American Book Prices Current
AMBL e: Airmobile
- e: Ambulatory

AMBLADS e: Advise Method; Bill of Lading and Data Shipped
Amb Lib e: Ambrosian Library
Am Booksellers e: American Booksellers Association
Ambos e: Amboseli National Park
Amboys e: New Jerseys Perth Amboy and South Amboy
Ambridge e: American Bridge [company]
AM Brit IRE e: Associate Member of the British Institution of Radio Engineers
AMBRL e: Army Medical Biomechanical Research Laboratory
ambros e: ambrosia
Ambrosian e: Ambrosian Library
Amb. Serv. e: Ambulance Service
Amb. Serv. Pk. e: Ambulance Service Park
ambt d: ambulant
- e: ambulant

AMBT e: Ambulatory
Amb. Temp. e: Ambient Temperature
ambu e: ambulance
ambul d: ambulant
- e: ambulance
- e: ambulate[d]
- e: ambulating
- e: ambulation
- e: ambulatorily
- e: ambulatory
- l: ambulacrum

AMBUSH e: Advanced Model Builder Shell
AMC d: Acetylmethylcarbinol
- d: Auto- und Motorrad-Club
- e: Acceptable Means of Compliance
- e: Account Manager Code
- e: Addressable Memory Clock
- e: Advanced Minuteman Computer
- e: Advance Message Center
- e: Aeromedical Monitor Console
- e: Aerospace Medical Command
- e: Air-Launched Missile Change
- e: Airborne Mode Control
- e: Aircraft Motion Compensation
- e: Air Mail Center
- e: Air Materiel Command
- e: Air Monitoring Center
- e: Alarm Monitor Computer
- e: Albany Medical Center
- e: Albany Medical College
- e: Alberta Microelectronic Centre
- e: Alternate Media Center
- e: Ambulatory Medical Care
- e: American Maritime Cases
- e: American Medical Center
- e: American Motors Corporation
- e: American Music Center
- e: Angular Motion Compensator
- e: Animal Medical Center
- e: Anti-Malaria Campaign
- e: Armament Material Change
- e: Armed Merchand Cruiser
- e: Armo[u]red Car
- e: Army Maintenance Command
- e: Army Materiel Command
- e: Army Medical Center
- e: Army Medical Corps
- e: Army Missile Command
- e: Army Mobility Command
- e: Army Munitions Command
- e: Arthrogryposis Multiplex Congenita
- e: Art Masters Certificate
- e: Associative Memory Computer
- e: Attribute Mark Count
- e: Australian Micrographic Congress
- e: Auto-Manual Center
- e: Automated Measurement Corporation
- e: Automatic Message Counting
- e: Automatic Mission Control
- e: Automatic Mixture Control
- e: Automatic Modulation Control
- e: Autonomous Multiplexer Channel
- e: Axiomesiocervical

AMc e: coastal minesweeper
a. m. c. l: a mundo condito
AMCA e: Air Movement and Control Association
AMCALMSA e: Army Materiel Command Automated Logistics Management Systems Agency
AMCAP e: Advanced Microwave Circuit Analysis Program[me]
AMCAR e: Automatic Magnetic Card Reader
AMCAS e: American Medical College Application Service
AMC-ASC e: Air Materiel Command-Aeronautical Systems Center
AMCAT e: Addressograph Multigraph Computer Access Terminal
AMCAWS e: Advanced Medium Caliber Aircraft Weapon System
AMCB e: Access Method Control Block
- e: American Medical Center for Burma

AMCBH e: Auxiliary Machine Casing Bulkhead
AMCC e: American Medical Computer Center
AMCCOMNET e: Air Material Command Communications Network
AMCD e: Addressograph Multigraph Copier Duplicator
- e: American Medical Center at Denver

AMCE e: Aldershot Military Corrective Establishment
AMCEC e: Allied Military Communications Electronics Committee
AMCEE e: Association for Media-based Continuing Education for Engineers
AMCF e: Air Materiel Command Forms
AMCFASC e: Army Materiel Command Facilities and Services Center
AMCHA e: Amino-Methyl-Cyclohexane Carboxylic Acid
Am Chem e: American Chemical Society
Am Child e: American Child (journ.)
Am Childhood e: American Childhood (journ.)
AMCHQ e: Air Materiel Command Headquarters
AMCI e: Ameritech Mobile Communications, Incorporated
AMCIA e: Associate Member of the Association of Cost and Industrial Accountants
AMCIB e: Associate Member of the Corporation of Insurance Brokers
Am City e: American City (journ.)
Am City (C ed) e: American City (City Edition) (journ.)
Am City (T & C ed) e: American City (Town and Country Edition) (journ.)
AMCL e: African Metals Corporation Limited
- e: Air Materiel Command Letter
- e: Amended Clearance
- e: Association of Metropolitan Chief Librarians

AMCLDC e: Army Materiel Command Logistic Data Center
am. clk. e: ammunition clerk
AMCLO e: Air Materiel Command Logistics Officer
AMCLSSA e: Army Materiel Command Logistics Systems Support Agency
AMCM e: Airborne Mine Countermeasures
- e: Air Materiel Command Manual

AMCMFO e: Air Materiel Command Missile Field Office
A.M.C.O. e: Air Ministry Command Order
- e: Air Ministry Confidential Order

AMCO e: American Command
AMCODE System e: American Stock Exchange Computerized Order Display and Execution System
AMCOM e: American Stock Exchange Communications
AMCONREPO e: American Consular Reporting Officer
Am Cont e: American Continent
AMCOS e: Aldermaston Mechanized Cataloguing and Order[ing] System
- e: Australian Mechanical Copyright Owners Society

AMCP e: Army Materiel Command Pamphlet
- e: Associated Medical Care Plans

AMCPI e: Army Materiel Command Procurement Instruction[s]

AMCPSCC e: Army Materiel Command Packaging, Storage and Containerization Center
AMCR e: Air Materiel Command Regulation[s]
- e: Army Materiel Command Regulation[s]
- e: Automatic Magnetic Card Reader
AMCRD e: Air Materiel Command Research and Development
- e: Army Materiel Command Research and Development
AMCRUS e: Adaptable Microcomputer Realtime User System
AMCS e: Adaptive Microprogrammed Control System
- e: Advanced Manpower Control System
- e: Airborne Missile Control System
- e: Aircraft Maintenance Control System
- e: Aircraft Mounted Control System
- e: Army Mobilization Capabilities Study
- e: Automatic Measurement and Compensation System
AMCSA e: Army Materiel Command Support Activity
AMCSS e: Airborne Missile Control Subsystem
AMCST e: Associate of the Manchester College of Science and Technology
AM & CT e: Antibiotic Medicine and Clinical Therapy
AMCU e: Australian Malaria Control Unit
am. cur. l: amicus curiae
AMCW e: Amplitude Modulation Continuous Wave
Am Cyc e: American Cyclopedia
AMD d: Actinomicin D
- d: Arbeitskreis makromolekulare Dokumentation
- d: Arbeitskreis mittlere Datentechnik
- e: Accident Model Document
- e: Acid Mine Drainage
- e: Actinomcin D
- e: Address Mark Detection
- e: Administrative and Miscellaneous Duties
- e: Admiralty Machinery Depot
- e: Advanced Memory Development
- e: Advanced Micro Devices
- e: Advance Manufacturing Directive
- e: Aero-Mechanics Department
- e: Aerospace Material Document
- e: Aerospace Medical Division
- e: Air Movement Data
- e: Air Movement Designator
- e: Alpha-Methyldopa
- e: Amendment
- e: Amplitude Modulation Distortion
- e: Antimorphine Dose
- e: Army Medical Department
- e: Associative Memory Data
- e: Associative Memory Device
- e: Atomic and Molecular Data
- e: Automated Maintenance Depot
- e: Automatic Map Display
- e: Auxiliary Memory Drum
- e: Axiomesiodistal
amd. d: altmitteldeutsch
- d: altmodisch
AMDA e: Advanced Maneuvering Demonstration Aircraft
- e: Airlines Medical Directors Association
- e: American Microcomputer Dealers Association
- e: Associated Minicomputer Dealers Association
AMDAC e: Amdahl Diagnostics Assistance Center
AMdAK d: Apparatur zur Messung und diskreten Abnahme der Koordinaten
AMDAPS e: Automatic Meteorological Data Acquisition and Processing System
AMDAS e: Advanced Micro Devices Assembler
AMDB e: Agricultural Machinery Development Board
AMDC e: Army Missile Development Center
- e: Assistant Marshal of the Diplomatic Corps
AMDEA e: Associated Manufacturers of Domestic Electrical Appliances
AM-Dec d: Amplitudendecodierung
AMDEC e: Agricultural Marketing Development Executive Committee
- e: Associated Manufacturers of Domestic Electric Cookers
AMDEL e: Australian Mineral Development Laboratories
Am Dent e: American Dental Association
AMDES e: Automatic Masking Data-generation for Electron-beam-exposure System
AMDF e: Army Master Data File
- e: Average Magnitude Difference Function
a.m.d.F.b. d: augenblicklich mit der Führung beauftragt
AMDFRMS e: Army Master Data File Reader Microfilm System
a.m. D.g. l: ad majorem Dei gloriam
AMDI e: Automatic Miss Distance Indicator
AMDL e: Abstract Machine Description Language
AMDLEVAC e: Aeromedical Evacuation
AMD-MOS e: Advanced Micro Devices-Metal Oxide Semiconductor
Amd. No. e: Amendment Number
Amdoc e: American Doctors
Am Dot Inst e: American Documentation Institute
AMDP e: Aircraft Maintenance Delayed for Parts
- e: Air Member for Development and Production
- e: Ammunition Distributing Point
AMDR e: Advance Missile Deviation Report
AMDRS e: Automatic Mobile Director and Reporting System
AMDS e: Advanced Microcomputer Development System
- e: Agri-Markets Data Service
- e: Association of Military Dental Surgeons
- e: Automatic Message Distribution System
AM-DSB e: Amplitude Modulation Double Sideband
AM-DSB/SC e: Amplitude Modulation Double Sideband; Suppressed Carrier
AMDT e: Active Maintenance Down Time
amdt e: amendment
a.m.d.Tit. d: auch mit dem Titel
AME d: Arbeitskreis Musikelektronik
- d: Ausgabeeinheit Meldungen
- e: Access Method Executor
- e: Acoustomagnetoelectric
- e: Admiralty Mining Establishment
- e: Advanced Manufacturing Engineering
- e: Advanced Master of Education
- e: Aero-Medical Evacuation
- e: Aeromedical Evacuation
- e: Amplitude Modulation Equivalent
- e: Angle Measuring Equipment
- e: Anti-Multipath Equipment
- e: Associated Memory Equipment
- e: Automatic Message Exchange
- e: Automatic Microfiche Editor
- e: Average Magnitude of Error
- e: Aviation Medical Examiner
- e: Aviation Medical Examiners
AmE e: American English
AMEA e: Association of Machinery and Equipment Appraisers
AMEAMS e: Adaptive Multibeam Experiment for Aeronautical and Maritime Services
AMEC e: Allied Middle East Command
- e: Australian Minerals and Energy Council
AMECD e: Antimechanized
amech e: account mechanical
Am Econ Assn Bul e: American Economic Association Bulletin (journ.)
Am Econ Assoc Pub e: American Economic Association Publications (journ.)
Am Econ R e: American Economic Review (journ.)
Am Econ R Papers & Proc e: American Economic Review; Papers and Proceedings (journ.)
AMECOS e: Automatic Measuring Computing and Sorting
AME COTAR e: Angle Measuring Equipment-Correlation Tracking and Ranging
amecz e: antimechanized
AMED e: American Education (journ.)
AMEDA e: Automatic Microscope Electronic Data Accumulator
AMEDAS e: Automated Meteorological Data Acquisition System
AMEDA System e: Automated Meteorological Data Acquisition System
AMedD e: Army Medical Department
AMEDDPAS e: Army Medical Department Property Accounting System
A Med Lab e: Army Medical Laboratory
AMedP e: Army Medical Publication[s]
AMEDPC e: Automotive Manufacturers Electronic Data Processing Council
AMedS e: Army Medical Service
AMEE e: Admiralty Marine Engineering Establishment
AMEI e: American Macaroni Export Institute
AMEIC e: Associate Member of the Engineering Institute of Canada
AMEL e: Aero Medical Equipment Laboratory
amelior e: Amelioration
AMEM e: Association of Marine Engine Manufacturers
AMEME e: Association of Mining, Electrical and Mechanical Engineers
AMEND[T] e: Amendment[s]
Am Engr e: American Engineer
amer a: amerikansk
amer. d: amerikanisch
Amer e: America
AMER e: American Economic Review (journ.)
amér f: américain
Amer Acad Arts & Sci Mem e: American Academy of Arts and Sciences Memoirs (journ.)
Amer Acad of Arts and Sciences
- e: American Academy of Arts and Sciences Proceedings (journ.)
AMERADC e: Army Mobility Equipment Research and Development Center

Amer Anthropol e: American Anthropologist (journ.)
Amerasians e: American-Asians
Amer. Assoc. Pet. Geol. Bull. e: American Association of Petroleum Geologists Bulletin
Amer. Bee J. e: American Bee Journal
Amer Blk e: American Black
Amer. Ceram. Soc. Bull. e: American Ceramic Society Bulletin
Amer. Dyestuff Rep. e: American Dyestuff Reporter (journ.)
Amer Eng e: American-English
Amer F e: American French
Amer. Fam. Physician e: American Family Physician (journ.)
Amer. Heart J. e: American Heart Journal
American Virgins e: U.S. Virgin Islands
Americas e: Western Hemisphere
America's Tropical Islands e: Florida Keys
- e: Guam
- e: Hawaiian Islands
- e: Padre Islands, Texas
- e: Puerto Rico
- e: U.S. Virgin Islands
Americo-Libs e: Americo-Liberians
AMERICOM e: American Communications
Ameridish e: American Yiddish
amerik. d: amerikanisch
Amerind e: American & Indian
Amer Ind e: American Indian
Amer. Ind. Hyg. Assoc. J. e: American Industrial Hygiene Association Journal
Amerindians e: American Indians
Ameringlish e: American English
Amer. Inst Ind. Eng.; Proc. An
- e: American Institute of Industrial Engineering; Proceedings of the Annual National
AMERITECH e: American Information Technologies
Amer. J. Anat. e: American Journal of Anatomy (journ.)
Amer. J. Bot. e: American Journal of Botany (journ.)
Amer. J. Cardiol. e: American Journal of Cardiology (journ.)
Amer. J. Clin. Nutr. e: American Journal of Clinical Nutrition (journ.)
Amer. J. Clin. Pathol. e: American Journal of Clinical Pathology (journ.)
Amer. J. Dig. Dis e: American Journal of Digestive Diseases (journ.)
Amer. J. Dis. Child. e: American Journal of Diseases of Children (journ.)
Amer. J. Epidemiol. e: American Journal of Epidemiology (journ.)
Amer. J. Gastroenterol. e: American Journal of Gastroenterology (journ.)
Amer. J. Hosp. Pharm. e: American Journal of Hospital Pharmacy (journ.)
Amer. J. Hum. Genet. e: American Journal of Human Genetics (journ.)
Amer. J. Int. Law e: American Journal of International Law (journ.)
Amer. J. Math. e: American Journal of Mathematics (journ.)
Amer. J. Med. e: American Journal of Medicine (journ.)
Amer. J. Med. Sci. e: American Journal of Mental Deficiency (journ.)
- e: American Journal of the Medical Science (journ.)
Amer. J. Med. Technol. e: American Journal of Medical Technology (journ.)
Amer. J. Obstet. Gynecol. e: American Journal of Obstetrics and Gynecology (journ.)
Amer. J. Opthalmol. e: American Journal of Opthalmology (journ.)
Amer. J. Optom. e: American Journal of Optometry (journ.)
Amer. J. Orthodont. e: American Journal of Orthodontics (journ.)
Amer. J. Orthop. e: American Journal of Orthopedics (journ.)
Amer. J. Orthopsychiat. e: American Journal of Orthopsychiatry (journ.)
Amer Jour Psych e: American Journal of Psychology (journ.)
Amer. J. Pathol. e: American Journal of Pathology (journ.)
Amer. J. Pharm. e: American Journal of Pharmacy and Sciences Supporting Public Health (journ.)
Amer. J. Pharm. Educ. e: American Journal of Pharmaceutical Education (journ.)
Amer. J. Phys. e: American Journal of Physics (journ.)
Amer. J. Phys. Anthropol. e: American Journal of Physical Anthropology (journ.)
Amer. J. Physiol e: American Journal of Physiology (journ.)
Amer. J. Phys. Med. e: American Journal of Physical Medicine (journ.)
Amer. J. Psychiat. e: American Journal of Psychiatry (journ.)
Amer. J. Psychol. e: American Journal of Psychology (journ.)
Amer. J. Public Health e: American Journal of Public Health and the Nations Health (journ.)
Amer. J. Roentgenol. Radium Th
- e: American Journal of Roentgenology, Radium Therapie and Nuclear Medicine (journ.)
Amer. J. Sci. e: American Journal of Science (journ.)
Amer. J. Surg. e: American Journal of Surgery (journ.)
Amer. J. Trop. Med. Hyg. e: American Journal of Tropical Medicine and Hygiene (journ.)
Amer. J. Vet. Res. e: American Journal of Veterinary Research (journ.)
Amer. Mach./Metalwork Mfg. e: American Machinist/Metalworking Manufacturing (journ.)
Amer. Math. Mon. e: American Mathematical Monthly (journ.)
Amer Men Sci e: American Men of Science (journ.)
Amer. Midland Nat. e: American Midland Naturalist (journ.)
Amer. Mineral. e: American Mineralogist (journ.)
Amer. Nat. e: American Naturalist (journ.)
AMER NATL STD e: American National Standard
Amer Oriental Soc Jour e: American Oriental Society Journal (journ.)
Amer. Potato J. e: American Potato Journal (journ.)
Amer. Production and Inventory
- e: American Production and Inventory Control Society; Annual Conference Proceedings
Amer. Rev. Respirat. Dis. e: American Review of Respiratory Diseases (journ.)
Amer. Sch. e: American Scholar (journ.)
Amer. Sci. e: American Scientist (journ.)
AmerSp e: American Spanish
Amer Sp e: American Speech (journ.)
Amer Std e: American Standard
Amer. Surg. e: American Surgeon (journ.)
Amer. Zool. e: American Zoologist (journ.)
AMES e: Aeromedical Evacuation System
- e: Air Ministry Experimental Station
AMeS e: American Meteorological Society
AMES e: American Mosquito Extermination Society
- e: Association of Marine Engineering Schools
- e: Automated Medical Examination System
- e: Automatic Message Entry System
AMeßE d: automatische Meßeinrichtung für Leitungen
AMeßSD d: automatischer Meßsatz für Leitungen für Durchgangsbetrieb
AMeßSE d: automatischer Meßsatz für Leitungen für Endvermittlungsstelle
AMeßSEv d: Automatischer Meßsatz für Leitungen für Endvermittlungsstelle; vereinfacht
AMeßUe d: automatische Meßübertragung
AMet e: Associate of Metallurgy
AMETA e: Army Management Engineering Training Agency
AMETS e: Artillery Meteorological System
AMEU e: American Energy Union
AMEWA e: Associated Manufacturers of Electric Wiring Accessories
AMEX e: American Express Company
AMEXCO e: American Express Company
AmEz d: Amtsleitungsentzerrer
AMF e: Acoustic Match Filter
- e: AIDS Medical Foundation
- e: Air Mail Facility
- e: Air Mail Field
- e: Allied Mobile Force
- e: American Machine and Foundry
- e: Analog[ue] Matched Filter
- e: Annual Material Forecast
- e: Arc Melting Furnace
AMFAR e: American Foundation for Aids Research
AMFF e: Advanced Materials Fabrication Facility
AMFGC e: Association of Midwest Fish and Game Commissioners
AMFIC e: Automatic Microfilm Information System
AMFIS e: American Microfilm Information Society
- e: Automatic Microfilm Information System
AM-FM e: Amplitude Modulation- Frequency Modulation
AM-FM-MPX e: Amplitude Modulation-Frequency Modulation- Multiplex[ing]
Am For e: American Forests (journ.)
AMFR d: automatische Maschinenfließreihe
AmFr e: American French
AmFuG d: Amateurfunkgesetz
AMFUR e: Amplified Failure or Unsatisfactory Report
AMG d: Arbeitsgemeinschaft Moderne Getränkeverpackung
- d: Arzneimittelgesetz
- e: Address Mark Gate
- e: Aircraft Machine Gunner
- e: Allied Military Government
- e: American Military Government
- e: Amplitude Modulation Generator
- e: Angle of Middle Gimbal
- e: Assistant Major-General
- e: Automatic Magnetic Guidance
- e: Automatic Model Generator
- e: Axiomesiogingival

amg. e: among
AMGA e: American Gas Association
Am Geog Soc Bull e: American Geographical Society Bulletin (journ.)
Am Geol e: American Geological Institute
Am Geophysical e: American Geophysical Union
AMGFC e: Association of Midwest Game and Fish Commissioners
AMGO e: Assistant Master-General of Ordnance
AMGOT e: Allied Military Government in Occupied Territories
AMGRA e: American Milk Goat Record Association
AMGS e: Acceleration Monitoring Guidance System
AMH e: Alaska Military Highway
- e: Application Message Handler
- e: Australian Military Hospital
- e: Automated Medical History
- e: Aviation Mechanic; Hydraulics
Amh e: Amharic
amh e: astigmatism with myopia predominating
AMHA e: Anchorage Museum of History and Art
AMHCI e: Associate Member of the Hotel and Catering Institute
Am Hist Assn Rept e: American Historical Association Reports (journ.)
Am Hist R e: American Historical Review (journ.)
Am Hist Rec e: American Historical Record (journ.)
Am Hist Reg e: American Historical Register (journ.)
Am Hist Res e: American History Research Associates
AMHS e: Alaska Marine Highway Authority
- e: Australian Maritime Historical Society
- e: Automated Materials Handling System
- e: Automated Message Handling System
AMHT e: Automated Multi-phase Health Testing
- e: Automated Multiphasic Health Testing
- e: Automated Multiphasing Health Testing
AMHTS e: Automated Multi-phase Health Testing and Services
- e: Automated Multiphasic Health Testing and Services
- e: Automated Multiphasing Health Testing and Services
AMI e: Acute Myocardial Infarction
- e: Acute Myocardial Insufficiency
- e: Address Modification Instructions
- e: Advanced Manned Interceptor
- e: Aerospace Materials Information
- e: Aircraft Multiplex Intercommunications
- e: Air Mileage Indicator
- e: Alternate Mark Insertion
- e: Alumin[i]um Measuring Instrument
- e: American Maritime Institute
- e: American Meat Institute
- e: American Medical International
- e: American Medical International (journ.)
- e: American Medical International, Inc.
- e: American Microsystems, Incorporated
- e: American Military Institute
- e: American Movers Institute
- e: American Mushroom Institute
- e: Amitriptyline
- e: Anterior Myocardial Infarction
- e: Apogee Motor Igniter
- e: Association of Medical Illustrators
- e: Australian Motor Industries
- e: Austrian Microsystems, Incorporated
- e: Automated Media Information
- e: Average Mutual Information
- e: Axiomesioincisal
- i: Aeronautica Militare Italiana
aMI d: akuter Myokardinfarkt
a.m.i. e: Alternate Mark Inversion
AMIA e: American Metal Importers Association
- e: Angular Magnetic-hydrodynamic Integrating Accelerometer
- e: Associate Member of the Institute of Almoners
AMIADB e: Army Member; Inter-American Defense Board
AMIAE e: Associate Member of the Institute of Automobile Engineers
- e: Associate Member of the Institute of Automotive Engineers
AMI Ae E e: Associate Member of the Institution of Aeronautical Engineers
AMIAMA e: Associate Member of the Incorporated Advertising Managers Association
AMIAP e: Associate Member of the Institution of Analysts & Programmers
AMIBritF e: Associate Member of the Institute of British Foundrymen
AMIC e: Aerospace Matenals Information Center
- e: Air Movement Information Center
- e: Australian Microcomputer Industry Clearinghouse
- e: Australian Mining Industry Council
AMICA e: Automatic Module for Industrial Control Analysis
AMICE e: Associate Member of the Institution of Civil Engineers
AMICEI e: Associate Member of the Institution of Civil Engineers of Ireland
AMI Chem E e: Associate Member of the Institution of Chemical Engineers
AMICO e: American Measuring Instrument Company
AMI-Code e: Alternate Mark Inversion-Code
AMICOM e: Army Missile Command
AMICS e: Aircraft Maintenance Irregularity Control System
AMIDI i: Associazione Medici Dentisti Italiani
AMIDS e: Advanced Multispectral Image Description System
- e: Advanced Multispectral Image Descriptor System
Amids e: Amidships
AMIEA e: Associate Member of the Institution of Engineers; Australia
AMIE Aust e: Associate Member of the Australian Institution of Engineers
AMIED e: Associate Member of the Institution of Engineering Designers
AMIEE e: Associate Member of the Institution of Electrical Engineers
AMIEI e: Associate Member of the Institution of Engineering Inspection
AMIE Ind e: Associate Member of the Institution of Engineers; India
AMIERE e: Associate Member of the Institution of Electronic and Radio Engineers
AMIF e: American Meat Institute Foundation
- e: Automated Map Information File
AMI Fire E e: Associate Member of the Institution of Fire Engineers
AMI Gas E e: Associate Member of the Institution of Gas Engineers
AMIGO d: allgemeine Misch- und Gruppenbearbeitungsorganisation
AMIGOS e: Access Method for Indexed Data Generalized for Operating System
AMIH e: Associate Member of the Institute of Housing
AMIHVE e: Associate Member of the Institution of Heating and Ventilating Engineers
AMII e: Association of Musical Instrument Industries
AMiK d: Aufnahmemikrofon
AMIL e: A Microprogramming Language
AMI Loco E e: Associate Member of the Institution of Locomotive Engineers
AMILPRI e: Amilprilose Hydrocholoride
AMI Mar E e: Associate Member of the Institute of Marine Engineers
AMIME e: Associate Member of the Institution of Minning Engineers
AMI Mech E e: Associate Member of the Institution of Mechanical Engineers
AMIMI e: Associate Member of the Institute of the Motor Industry
AMI Min E e: Associate Member of the Institution of Minning Engineers
AMIMM e: Associate Member of the Institution of Mining and Metallurgy
AMI/MOS e: American Micro-Systems, Incorporated/Metal Oxide Semiconductor
AMIMunE e: Associate Member of the Institution of Municipal Engineers
A/min d: Anschläge pro Minute
Amin d: Arbeiterminute
AMIN e: Advertising and Marketing International Network
AMINA e: Associate Member of the Institution of Naval Architects
- f: Association Mondiale des Inventeurs
Am Ind e: American Indian
Am Indus Arts e: American Industrial Arts Association
AMINE e: Associate Member of the Institute of Naval Engineers
AMinisTech e: Associate in Minerals Technology
aminos e: amino acids
AM Inst BE e: Associate Member of the Institution of British Engineers
AM Inst CE e: Associate Member of the Institution of Civil Engineers
AM Inst Chem E e: Associate Member of the Institution of Chemical Engineers
AM Inst EE e: Associate Member of the Institution of Electrical Engineers
AM Inst Gas E e: Associate Member of the Institution of Gas Engineers
AMInstHE e: Associate Member of the Institution of Highway Engineers
AM Inst M & Cy E e: Associate Member of the Institution of Municipal and County Engineers
AMInstPC e: Associate Member of the Institute of Public Cleansing
AMInstR e: Associate Member of the Institute of Refrigeration
AM Inst T e: Associate Member of the Institute of Transport
A.M. Inst. T. e: Associate Member of the Institute of Transport
AM Inst TE e: Associate Member of the Institution of Transport Engineers

AMINTAPHIL e: American Section, International Association for Philosophy of Law and Social Philosophy
AM International e: Addressograph-Multigraph-International
AMINucE e: Associate Member of the Institution of Nuclear Engineers
AMIO e: Arab Military Industrialization Organization
AMIOB e: Associate Member of the Institute of Building
AMIOP e: Associate Member of the Institute of Printing
AMIP e: Analysis of Management Inventory Policy
- e: Army Management Information Program[me]
AMIPA e: Associate Member of the Institute of Park Administration
- e: Associate Member of the Institute of Practioners in Advertising
AMIPE e: Associate Member of the Institution of Production Engineers
AMIPHE e: Associate Member of the Institution of Public Health Engineers
AMIPlantE e: Associate Member of the Institution of Plant Engineers
AMIProdE e: Associate Member of the Institution of Production Engineers
AMIPtgM e: Associate Member of the Institute of Printing Management
Amirantes e: Amirante Islands
AMIRE e: Associate Member of the Institution of Radio Engineers
AMIR System e: Audit Management Information Reporting System
AMIS e: Accounting Management Information System
- e: Acquisition Management Information System
- e: Advanced Management Information System
- e: Agribusiness Management Information System
- e: Aircraft Multiplex Intercommunications System
- e: Air Movements Identification System
- e: Air Movements Information Section
- e: Airport Management Information System
- e: Alternate Mark Inversion Signal
- e: American Musical Instrument Society
- e: Army Information System
- e: Army Management Information System
- e: Atari Message and Information System
- e: Australian Municipal Information System
- e: Automated Management Information System
- e: Automated Minerals Information System
AMISanE e: Associate Member of the Institution of Sanitary Engineers
AMISE e: Associate Member of the Institution of Sanitary Engineers
AMI Service e: Advertising and Marketing Intelligence Service
AMISI e: Associate Member of the Iron and Steel Institute
AMI Signal e: Alternate Mark Inversion Signal
Amistad · e: Amistad National Recreation Area near Del Rio, Texas
AMI Struct E e: Associate Member of the Institute of Structural Engineers
AMIT e: Ampex to IBM Tape
- e: Associate Member of the Institute of Transport
AMITA e: Associate Member of the Industrial Transport Association
AMI Tec E e: Associate Member of the Institute of Technical Engineers
AMIWE e: Associate Member of the Institution of Water Engineers
AMIWM e: Associate Member of the Institution of Works Managers
AMJ e: Automation Journal of Japan
Am J Archaeol e: American Journal of Archaeology (journ.)
Am J Dis Child e: American Journal of Diseases of Children (journ.)
AM J Econ e: American Journal of Economics & Sociology (journ.)
Am J Educ e: American Journal of Education (journ.)
AM J Nurs e: American Journal of Nursing (journ.)
Am Jour Sci e: American Journal of Science (journ.)
AM Jour Soc e: American Journal of Sociology (journ.)
AM J Phil e: American Journal of Philology (journ.)
AmJ Phys e: American Journal of Physics (journ.)
Am J Phys Anthrop ns e: American Journal of Physical Anthropology; New Series (journ.)
Am J Phys Anthropol e: American Journal of Physical Anthropology
AmJ Psy e: American Journal of Psychiatry (journ.)
Am J Psychol e: American Journal of Psychology (journ.)
Am J Pub Health e: American Journal of Public Health (journ.)
AmJS e: American Journal of Sociology (journ.)
Am J School Hygiene e: American Journal of School Hygiene (journ.)
Am J Sci e: American Journal of Science (journ.)
Am J Sem Lang e: American Journal of Semitic Languages (journ.)
Am J Soc e: American Journal of Sociology (journ.)
Am J Soc Sci e: American Journal of Social Science (journ.)
AM J Theol e: American Journal of Theology (journ.)
AMK Arbeitsmittelkarte
- d: Arzneimittelkommission der deutschen Ärzteschaft
- d: Aufnahmemagnetkopf
amkal. d: ammoniakalisch
AMKL d: Asynchronmaschine mit Käfigläufer
AMkr d: Auslandsmarkierer
AMKTU e: Army Markmanship Training Unit
AML d: Atemmittellage
- e: Aberdeen Marine Laboratory
- e: Actual Measured Loss
- e: Acute Monocytic Leuk[a]emia
- e: Acute Myeloblastic Leuk[a]emia
- e: Acute Myelocytic Leuk[a]emia
- e: Acute Myelogenous Leuk[a]emia
- e: Acute Myoblastic Leuk[a]emia
- e: Admiralty Materials Laboratory
- e: Advanced Manual Line
- e: Aeromedical Laboratory
- e: Aeronautical Materials Laboratory
- e: Air Mail Letter
- e: Algebraic Manipulation Language
- e: Allied Military Liaison
- e: A Manufacturing Language
- e: American Meteorite Laboratory
- e: Amplitude-Modulated Link
- e: Amplitude Modulation with Limiter
- e: Animated Movie Language
- e: Anterior Mitral Leaflet
- e: Application Macro Library
- e: Application Module Library
- e: Applied Mathematics Laboratory
- e: Applied Microwave Laboratory
- e: Approved Materials List
- e: Army Medical Library
- e: Array Machine Language
- e: Assembly Micro Library
- e: Automatic Machine Loading
- e: Automatic Magazine Loader
- e: Automatic Magazine Loading
- e: Automatic Modulation Limiter
- e: Automatic Modulation Limiting
- e: Auxiliary Mine Layer
- e: Aviation Material Laboratories
AMLA e: Airplane Model List of America
Am Labor Leg R e: American Lobor Legislation Review (journ.)
AMLC e: Aerospace Medical Laboratory; Clinical
- e: Asynchronous Multi-Line Controller
AMLCC e: Asynchronous Multi-Line Communications Coupler
AMLG e: Allied Military Headquarters; Greece
- e: Amalgam
Am Lib Dir e: American Library Directory
Am.Libr. e: American Libraries (journ.)
Am Librarians e: American Librarians Agency
Am-Lib[s] e: Americo-Liberian[s]
AmLit e: American Literature (journ.)
Am Lit M e: American Literary Magazine (journ.)
AMLLV e: Advanced Multipurpose Large Launch Vehicle[s]
AMLO e: Aeromedical Liaison Officer
- e: Assistant Military Landing Officer
AMLP e: Amplitude Modulation Link Program[me]
AMLS e: American Medico-Legal Society
AMLS e: Antimouse Lymphocyte Serum
- e: Master of Arts in Library Science
AML Timber e: Amla Timber
AMM e: Adaptive Mathematical Model
- e: Additional Memory Module
- e: Agnogenic Myeloid Meta-plasia
- e: Aircraft Maintenance Manual
- e: Air Mining Mission
- e: Amalgam
- e: American Merchant Marine
- e: American Military Mission
- e: Ammamonium
- e: Amman (airport)
- e: Ammeter
- e: Ammunition
- e: Analog[ue] Monitor Module
- e: Anomalous Magnetic Moment
- e: Anti-Missile Missile
- e: Associated Maintenance Module
- e: Auxiliary Magnetic Memory
- e: Aviation Machinists Mate
- f: Association Medicale Mondiale
Amm d: Ammoniak
AM/M e: Air Ministry Message
amm e: ammonia
amm. e: Ammoniacal
AM & M e: Applied Mathematics and Mechanics
- e: Automatic Maintenance Monitor
AMM1c e: Aviation Machinists Mate First Class

AMM2c e: Aviation Machinists Mate Second Class
AMM3c e: Aviation Machinists Mate Third Class
AMMA d: Acrylnitril-Methylmethacrylat[e]
- e: American Museum of Marine Archeology
- e: Automated Media Management System
Am Mach e: American Machinist
Am Malacologists e: American Malacologists
Am Map e: American Map Company
AM M Art e: American Magazine of Art (journ.)
Am Math Mo e: American Mathematical Monthly (journ.)
Am Math Soc e: American Mathematical Society
AMMC e: Aviation Machinists Mate; Carburetor Mechanic
AM/MDA e: Airlock Module and Multiple Docking Adapter
AMME e: Automated Multi-Media Exchange
Am Meteorite e: American Meteorite Laboratory
Ammeter d: Amperemeter
ammeter e: amperemeter
AMMF e: Aviation Machinists Mate; Flight Engineer
AMM forum e: American Metal Market forum
AMMH e: Aviation Machinists Mate; Hydraulic Mechanic
AMMI e: American Merchant Marine Institute
- e: Aviation Machinists Mate; Instrument Mechanic
AMMIC e: Armament Maintenance Management Information Center
Am Midl Nat e: American Midland Naturalist (journ.)
AMMINET e: Automated Mortgage Management Information Network
AMMIP e: Aviation Material Management Improvement Program[me]
AMMIS e: Aircraft Maintenance Manpower Information System
AMML e: Acute Myelo-Monocytic Leuk[a]emia
- e: Automated Microbial Metabolism Laboratory
Ammn e: Ammunition
Ammn Sub Pk e: Ammunition Subpark
AMMO e: Air Ministrys Meteorological Office
- e: American Mathematical Monthly (journ.)
- e: Army Mobile Missile Operation
Ammo e: American Motors
ammobr e: ammunition bearer
ammo hand e: ammunition handler
ammon e: ammonia
Ammon e: Ammonite
ammonia water e: ammonium hydroxide
ammo rcd clk e: Ammunition Records Clerk
ammo renov e: Ammunition Renovator
AMMORK e: Ammunition Rack
ammo sec ch e: Ammunition Section Chief
ammo sp e: Ammunition Specialist
ammo tn e: Ammunition Train
Am Motors e: American Motors
AMMP e: Advanced Manned Missions Program[me]
- e: Apollo Master Measurements Program[me]
- e: Approved Modernization Maintenance Program[me]
- e: Aviation Machinists Mate; Propeller Mechanic
AMMPE e: American Mining, Metallurgical and Petroleum Engineers
AMMR e: Advanced Multimission Radar
- e: Advanced Multivision Radar
- e: Aircraft Maintenance Manpower Requirement
AMMRA d: Arbeitsgemeinschaft mittelständischer Mineralölraffinerien
AMMRC e: Army Materials and Mechanics Research Center
AMMRE d: Anstalt für mechanisch-musikalische Rechte
Am MRG d: Gesetz der amerikanischen Militärregierung
AMMRL e: Aircraft Maintenance Material Readiness List
Am MRVO d: Verordnung der amerikanischen Militärregierung
AMMS e: Aerospace, Maritime and Military Systems
- e: Automated Multi-Media Switch
- e: Automatic Multimode Mass Spectrometry
- e: Auxiliary Memory Management System
AMMSS e: Automatic Message and Mail Sorting System
AMMT e: Automated Multi-Media Terminal
Amm Tn e: Ammunition Train
AMMTR d: Ammeter
- d: Amperemeter
- e: Anti-Missile Test Range
ammu e: Ammunition
AMN e: Alloxazine Mononucleotide
- e: Alpha Motoneuron
- e: Ammunition
- e: Anterior Median Nerve
- e: Atomic Mass Number
Amn d: Anschlußmöglichkeiten
- d: astabile Multivibratoren
- e: Airman
AMNA e: Ammonia
Am Na Att e: American Naval Attaché
Am Nat e: American Naturalist (journ.)
Am Natural e: American Naturalist (journ.)
AMNCS e: Advanced Multiplatform Navy Computer System
AMND e: Ammend[ment]
amnes e: amnesia[c][al][ly]
AMNH e: American Museum of Natural History
AMN Inst E e: Associate Member of the National Institute of Engineering
AMNIP e: Adaptive Man-machine Non-arithmetic Information Processing
AMNIPS e: Adaptive Man-machine Non-numerical Information Processing System
AMNIP System e: Adaptive Man-machine Non-numerical Information Processing System
AM NIT e: Ammonium Nitrate
AMNL e: Amplitude-Modulation Noise Level
- e: Army Medical Nutrition Laboratory
AmnM e: Airmans Medal
AM Noise e: Amplitude Modulation Noise
AMNSWP e: Acoustic Mine Sweeping
AMNZIE e: Associate Member of the New Zealand Institution of Engineers
AMO d: Analogeingabe für Momentanwerte
- e: Adder for Multiple Operands
- e: Adder of Multiple Operands
- e: Administrative Medical Officer
- e: Advance Material Order
- e: Aircraft Materiel Officer
- e: Air Mail Only
- e: Air Mass Zero
- e: Air Ministry Order
- e: Allied Meteorological Office
-e: Alternate Molecular Orbit[al]
- e: American Medical Optics
- e: American Motors
- e: Answering Machine Owner
- e: Area Medical Officer
- e: Area Milk Officer
- e: Army Medical Officer
- e: Assistant Medical Officer
- e: Aviation Maintenance Officer
- e: Aviation Medical Officer
- e: Axiomesioocclusal
AMÖ d: Arbeitsgemeinschaft Möbeltransport
AMOAC e: Automatic Multiloop Optimal Approach Controller
AMOB e: Automatic Meteorological Oceanographic Buoy
Am Obs BN e: American Observer; British Navy
AMOCOM e: Army Mobility Command
AMOE d: Analogeingabe für Momentanwerte; Erweiterungsbaustein
AMOF e: Advanced Material Order Form
AMOHST e: Ammunition Hoist
AMOHSTDR e: Ammunition Hoist Drive
amol acute e: monocytic leukemia
AMONO e: Agricultural Machinery Operation and Management Office
AMOR d: Anwendersoftware-Monitor
- e: Associative Memory with Ordered Retrieval
amor e: Amorphous
Am Oriental Soc J e: American Oriental Society Journal (journ.)
AMORPH e: Amorphous
amort. e: amortizable
- e: amortization
- e: amortize[d]
- e: amortizement
- e: amortizing
AMOS e: Acoustic, Meteorological and Oceanographic Survey
- e: Additionally-awared Military Occupational Specialty
- e: Adjustable Multi-class Organizing System
- e: Advanced Metal Oxide Semiconductor
- e: Advanced Research Projects Agency Midcourse Observatory Site
- e: Aerospace Maintenance and Operational Status
- e: Alpha Microsystems Operating System
- e: Alternate Military Occupational Speciality
- e: A Multi-access Online System
- e: Anodically-formed Metal Oxide Semiconductor
- e: Anodized Metal Oxide Semiconductor
- e: Antireflection-coated Metal-Oxide Semiconductor
- e: Assembly Management Operating System
- e: Associative Memory Organizing System
- e: Automated Materials Ordering System
- e: Automatic Meteorological Observation Station
- e: Automatic Meteorological Observation System
AMOSFET[s] e: Anodized Metal Oxide Semiconductor Field Effect Transistor[s]
AMOSS e: Adaptive Mission-Oriented Software System

AMOST e: Anodized Metal Oxide Semiconductor Transistor
AMP d: 2-Amino-2-Methyl-1-Propanol
- d: Adenosinmonophosphat
- d: Adenosinmonophosphorsäure
- d: Aminomethylpropanol
- d: Arzneimittelprüfung
- d: Ausgabe Magnetplatte
- e: Accelerometer Monitoring Program[me]
- e: Accounting Machine Program[me]
- e: Acid Mucopolysaccharide[s]
- e: Acquisition Management Plan
- e: Adaption Mathematical Processor
- e: Add Multi-Precision
- e: Additional Military Production
- e: Adenosine Cyclic Monophosphate
- e: Adenosine Monophosphate
- e: Advanced Management Program[me]
- e: Advanced Manned Penetrator
- e: Advanced Minuteman Platform
- e: Aerospace Medical Panel
- e: Air-Mail Pioneer
- e: Aircraft-Missile Project
- e: Air Member for Personnel
- e: American Melting Point
- e: American Military Police
- e: American Museum of Photography
- e: Ammonium Molybdophosphate
- e: Amplidyne
- e: Amplifier
- e: Amplitude
- e: Apollo Mission Programs
- e: Applied Mathematical Panel
- e: Area Mail Processing
- e: Argonne Microprocessor
- e: Army Marine Products
- e: Army Materiel Plan
- e: Army Mine Planter
- e: Aseptic Maintenance by Pressurization
- e: Associate Memory Processor
- e: Association of Media Producers
- e: Associative Memory Processor
- e: Asymmetric Multiprocessing System
- e: Atmospheric Meteorological Probe
- e: Aurora Memorial Park
- e: Automated Maintenance Program[me]
- e: Automated Manufacturing Planning
- e: Automated Mathematics Program[me]
- e: Automatic Multi-Program[me]
- e: Automatic Multipattern
- e: Automatic Multiplying Punch
- e: Avalanche Mode Photodiode
- e: Average Mean Pressure
- e: Aviation Modernization Program[me]
- e: Axially Magnetized Plasma
Amp. d: Ampere
- d: Ampulle
- d: Amputation
amp. d: amputiert
amp e: amperage
- e: ampere[s]
- e: amphetamine
- e: ampicillin
- e: amplification
- e: ampule
- e: amputation
- e: anmpule
AMPA e: American Medical Publishers Association
- e: Azimuth Mark Pulse Amplifier
AMP AC e: Ampere[s] Alternating Current
AMPAS e: Academy of Motion Picture Arts and Sciences
AMPC e: Air Mail Processing Center
- e: Area Mail Processing Center
- e: Automatic Message Processing Center
Amp Cal e: Amplifier Calibration
AMPCO e: Associated Missile Products Corporation
AMPCR e: Alternate Microprogram[me] Count Register
AMPD d: 2-Amino-2-Methyl- 1.3-Propane-Diol
- e: Army Mobilization Program[me] Directive
AMP DC e: Ampere[s] Direct Current
amp det e: amplitude detector
AMPDS e: Advanced Missile Propulsion Definition Study
AMPEP e: Amplitude-Modulated Peak Envelope Power
AMPERES e: Automated Maintenance Performance and Engineering Reliability Evaluation System
AMPEX e: Automatic Programming-System Extended
AMP FAC e: Amplification Factor
AMPFUR e: Amplifying Failure; Unsatisfactory or Removal Report
AMPG e: Air Material Proving Ground
ampg e: Airmile[s] per Gallon
Amph. d: Amphibienflugzeug
AMPH e: Amphibia[n]
- e: Amphoric
amp.-h. e: Ampere-Hour[s]
amph e: amphibious
- e: amphimict
Amph Comd e: Amphibious Command
Amphet. d: Amphetamine
amphetamine e: alphamethyl-penethyl-amine
AMPHIB e: Amphibia[n]
amphib e: amphibious
amphibex e: amphibious exercise
Amphib Recon e: Amphibious Reconnaissance
amphig e: amphigoric[al]
- e: amphigorist
- e: amphigory
Am Philos Soc e: American Philosophical Society
AMP-HR e: Ampere-Hour[s]
amph tk e: Amphibious Tank
amph tk crm e: Amphibious Tank Crewman
Amph Trk e: Amphibious Track
Am Phys Educ R e: American Physical Education Review (journ.)
AMPIC e: Atomic [and] Molecular Processes Information Center
AMPL e: Advanced Microprocessor Laboratory
- e: Advanced Microprocessor Programming Language
- e: Advanced Microprocessor Prototyping Laboratory
- e: Advanced Microprocessor Prototyping Language
- e: A Macro Programming Language
- e: A Manuscript Preparation Language
- e: Amplifier
- e: Amplitude
ampl - i: ampliata
- l: amplus
AMPLAs e: Apparatus Mounted in Plastic
AMPLDN e: Amplidyne
AMPLE e: A Multi-Programming Language-Extended
AMPLFD e: Amplified
AMPLG e: Amplidyne Generator
AMPLIDYNE e: Amplifier Dyne
AMPLIF e: Amplifier
AMPLMG e: Amplidyne Motor Generator
AM-PM d: Amplitudenmodulation mit Phasenmodulation
AM/PM e: Amplitude Modulation/Pulse Modulation
AMPM e: Auto[matic]-Manual Probe Multiplexer
AMP MIN d: Ampere Minute[s]
AMPOL e: American Petroleum
Am Pol Sci R e: American Political Science Review (journ.)
AM-Post d: Allierte Militärpost
AMPP e: Advanced Microprogrammable Processor
- e: Advanced Multi-Programming Processor
- e: Arctic Meteorology Photographic Probe
AMPPD e: Army Mobilization Planning and Programming Directive
AMPPGD e: Army Mobilization Planning and Programming Guidance Document
AMPPL e: Associative Memory and Parallel Processing Language
AMPR e: Advanced Multi-Purpose Radar
- e: Aeronautical Manufactures Planning Report
- e: Aircraft Manufacturers Progress Report
- e: Airframe Manufacturers Planning Report
- e: Amplifier
- e: Application Module Processing Routine
- e: Assembly Manufacturing Planning Record
- e: Assembly Manufacturing Process Record
- e: Automatic Manifold Pressure Regulator
AMPS e: Accured Military Pay System
- e: Acid Mucopoly Saccharide
- e: Adaptive Mode Planning System
- e: Adenosine Monophosphate Succinate
- e: Advanced Maneuvering Propulsion System
- e: Advanced Manned Penetrator System
- e: Advanced Mobile Phone Service
- e: Advanced Mobile Phone System
- e: Assembly Manufacturing Payrolls System
- e: Assignment Management Planning System
- e: Atmospheric, Magnetospheric and Plasmas in Space
- e: Automated Material Processing System
- e: Automated Merchandise Processing System
- e: Automatic Message Processing System
amps e: amperes
- e: ampules
AMPS AC e: Amperes Alternating Current
AMPS Acid e: 2-Acrylamido-2-Methylpropane-Sulfonic Acid
AMPS DC e: Amperes Direct Current
AMP/SEC e: Ampere[s]/Second
AMPSIN e: Adaptive Mode Planning System Input
AMPSLIN e: Adaptive Mode Planning System Input
AMPSR e: American Political Science Review
AMPSS e: Advanced Manned Penetractor Strike System
- e: Advanced Manned Precision Strike System
AMPS/SEC e: Amperes/Second
Amp. St. d: Amperestunde[n]
Am Psychologist e: American Psychologist (journ.)
AMP System e: Automatic Message Processing System
AMPT e: Advanced Maneuvering Propulsion Technology
- e: Alpha-Methylparatyrosine
- e: Automatic Module Production Tester

AMPTD e: Amplitude
AMPTD-MODUL e: Amplitude Modulation
AMPTE e: Active Magnetospheric Particle Tracer Experiment
AMPTEs e: Active Magnetosphere Particle Explorers
AMPTP e: Association of Motion Pictures and Television Producers
AMPTS e: Automatic Mobile Public Telephone System
amp-turn[s] e: ampere-turn[s]
ampul. l: ampulla
ampus[s] e: amputee[s]
AMQ e: American Medical Qualification
- e: Analog[ue] Multiplexer Quantitizer
- e: Aparent Molar Quantity
am + r d: angewandte meß- und regeltechnik (journ.)
AMR d: Anweisung über Mindestreserven
- e: Arbeitsgemeinschaft meerestechnisch-gewinnbarer Rohstoffe
- e: Abnormal Mission Routine
- e: Activity Metabolic Rate
- e: Activity Metabolic Request
- e: Address Modification Register
- e: Advanced Management Research
- e: Advanced Research Projects Agency Measurement Radar
- e: Advance Material Request
- e: Aerospace Medical Research
- e: Airborne Magnetic Recorder
- e: Airman Military Record
- e: Air Movement Recorder
- e: Air Musketry Range
- e: Alternating Motion Rate
- e: Altitude Marking Radar
- e: Altitude Marking Range
- e: Applied Mechanics Reviews (journ.)
- e: Arithmetic Mask Register
- e: Atlantic Missile Range
- e: Automated Management Report[s]
- e: Automated Message Recording
- e: Automatic Map Reader
- e: Automatic Map Reading
- e: Automatic Message-Registering
- e: Automatic Message Recording
- e: Automatic Message Registering
- e: Automatic Message Routing
- e: Automatic Meter Reading
- e: Aviation Medical Report[s]
Am R e: Ammunition Reserve
AMRA e: Agricultural Machinery Research Association
- e: American Medical Records Association
AMRAAM e: Advanced Medium-Range Air-to-Air Missile
AMRAC e: Anti-Missile Research Advisory Council
AMRAD e: Advanced Research Projects Agency Measurement Radar
Am Radio e: American Radio Relay League
AMRC e: Advanced Metals Research Corporation
AM RC e: Amplitude Modulation-Reduced Carrier
AMRC e: Army Mathematics Research Center
- e: Army Mobility Research Center
A.M.R.D. e: Air Member for Research and Development
AM RD e: Air Ministry Radio
AMR & DL e: Air Mobility Research & Development Laboratory
Am Rec G e: American Record Guide (journ.)
Am Res e: American Research Council
AM & RF e: African Medical and Research Foundation
AMRF e: Ammended Route of Flight
AMRINA e: Associate Member of the Royal Institution of Naval Architects
AMRIP e: Avionics Module Repair Improvement Program[me]
AMRL e: Aerospace Medical Research Laboratories
- e: Applied Mechanics Research Laboratory
- e: Army Medical Research Laboratory
AMRNL e: Army Medical Research and Nutrition Laboratory
AMRO e: Association of Medical Record Officers
- e: Atlantic Missile Range Operations
AMRPD e: Applied Manufacturing Research and Process Development
AMRR e: Army Materials Research Reactor
AMRRDC e: Army Manpower Resources Research and Development Center
AMRS e: Air Maintenance and Repair Station
- e: Air Ministry Radio Station
- e: Automatic Message Registering System
AMR System e: Automatic Message Registering System
AMR Timber e: Amra Timber
AMRTS e: Associate Member of the Royal Television Society
- e: Atlantic Missile Range Telemetry Submodule
AMRV e: Astronaut Maneuvering Research Vehicle
AMS d: Antikörpermangel-Syndrom
- d: Äthylmethansulfonat
- d: automatisiertes Materialflußsystem
- e: Access Method Services
- e: Accounting Management System
- e: Acoustic Measurement System
- e: Acoustic Monitor System
- e: Acta Medica Scandinavica
- e: Actual Military Service
- e: Actual Minus Sign
- e: Administrative Management Society
- e: Admiral of Mine Sweepers
- e: Advanced Machining System
- e: Advanced Manned Spacecraft
- e: Advanced Marksman Series
- e: Advanced Memory Systems
- e: Advanced Meteorological System
- e: Advanced Microcomputer System
- e: Advanced Minuteman System
- e: Advanced Missile System
- e: Advanced Mission Studies
- e: Aerial Measuring System
- e: Aeronautical Material Specification[s]
- e: Aerospace Material Specification
- e: Agricultural Marketing Service
- e: Airborne Maintenance System
- e: Air Management Station
- e: Air Mass
- e: Air Missile System
- e: Air Movements Section
- e: Alarm Monitoring System
- e: Altitude Measurement System
- e: American Management Systems
- e: American Marketing Service
- e: American Market Selection
- e: American Mathematical Society
- e: American Meteorological Society
- e: American Meteor Society
- e: American Microchemical Society
- e: American Microscopical Society
- e: American Military Standard
- e: American Mineral Spirits
- e: American Museum of Safety
- e: Ammonium Sulfamate
- e: Amplifier Subsystem
- e: Analog[ue] Master Slice
- e: Antenna Measuring System
- e: Apollo Mission Simulator
- e: Application Management Services
- e: Application Management System
- e: Application Mapping Service
- e: Applied Mathematic Series
- e: Area Music Shop
- e: Arista Manufacturing Systems
- e: Army Management School
- e: Army Management Structure
- e: Army Map Service
- e: Army Medical Service[s]
- e: Army Medical Staff
- e: Articulated Mirror System
- e: Artillery and Missile School
- e: Assistant Military Secretary
- e: Associative Memory System
- e: Asymmetric Multiprocessing System
- e: Atmospheric Monitor System
- e: Auditory Memory Span
- e: Australian Medical Services
- e: Australian Motor Sports & Automobiles (journ.)
- e: Authority for Material Substitution
- e: Automated Multiphasic Screening
- e: Automatic/Manual Station
- e: Automatic Message Switching
- e: Automatic Mode Status
- e: Automatic Monitoring System
- e: Automatic Music Scan
- e: Automatic Music Sensor
- e: Autopilot Mode Selector
- e: Auxiliary Memory Set
- e: Auxiliary Motormine-Sweeper
- e: Aviation Structural Mechanic
- e: Avionics Maintenance Shop
- f: Assemblée Mondiale de la Santé
AMs e: Air Movements
Am S e: American Studies (journ.)
ÄMS d: Äthylmethansulfonat
AMSA e: Advanced Manned Strategic Aircraft
- e: American Metal Stamping Association
- e: American Museum of Social Anthropology
- e: Australian Marine Sciences Association
- e: Australian Medical Students Association
AM Sam e: American Samoa
AMSAM e: Anti-Missile Surface-to-Air Missile
AMSANZ e: Aviation Medicine Society of Australia and New Zealand
AMSAS e: Advanced Strategic Aircraft
AMSAT d: Amateur[funk]-Satellit
AM-SC e: Amplitude Modulation-Suppressed Carrier
AMSC e: Army Materiel Supply Command
- e: Army Medical Specialist Corps
- e: Army Mobility Support Center
AMSCD e: Amount Subject to Cash Discount
Am Sch Athens e: American School of Classical Studies at Athens
Am Sch Bd J e: American School Board Journal (journ.)
AM Schneiden d: anodenmechanisches Schneiden
Am Sch Orient Res Bul e: American Schools of Oriental Research Bulletin (journ.)
Am Sci & Eng e: American Science and Engineering, Inc.
AMS/CO e: Access Method Services Cryptographic Option

AMSCO e: American Sterilizer Company
- e: Army Medical Supply Control Officer
AMSDL e: Acquisition Management System and Data List
AMSE e: Aircraft Maintenance Support Equipment
- e: Associate Member of the Society of Engineers
Am Sec e: Ammunition Section
AMSEC e: Analytical Method for System Evaluation and Control
AMSEF e: Anti-Mine Sweeping Explosive Float
AMSF e: Area Maintenance Supply Facility
AMSGS e: Army Medical Service Graduate School
am sgt e: Ammunition Sergeant
AMSI e: Atlantic Merchant Shipping Instructions
AMSIR e: Agriculture, Marine, Scientific and Industrial Research Ministry
AMSL d: Asynchronmotor mit Schleifringläufer
- e: Above Mean Sea Level
- e: Applied Mathematics and Statistics Laboratory
- e: Approved Material Substitution List
AMS-M e: Advanced Microcomputer System-Multibus-Architecture
AMSMH e: Association of Medical Superintendents of Mental Hospitals
AMSO e: Air Member for Supply and Organization
- e: Ammunition Shipment Order
AMSOC e: American Miscellaneous Society
Am Soc Afr Cult e: American Society of African Culture
Am Soc HRAC Eng e: American Society of Heating, Refrigerating and Air-Conditioning Engineers
Am Soc Indxrs e: American Society of Indexers
Am Sociol R e: American Sociological Review (journ.)
Am Soc Metals e: American Society for Metals
Am Soc Soc e: American Sociological Society
Am Soc Tool and Mfg Eng e: American Society of Tool and Manufacturing Engineers
AMSP e: Allied Military Security Publication
Am Sp e: American Speach (journ.)
AMSP e: Army Master Study Program[me]
- e: Array Machine Simulation Program[me]
Am Sp e: American Spanish
AMSPAR e: Association of Medical Secretaries, Practice Administrators and Receptionists
AMSR e: Air Member for Supply and Research
- e: Automated Microform Storage and Retrieval
- e: Autonomous Missile Site Radar
AMSRDC e: Army Medical Service Research and Development Command
AMSS e: Advanced Manned Space Simulator
- e: Advanced Meteorological Sounding System
- e: Aeronautical Mobile Satellite Service
- e: American Milking Shorthorn Society
- e: Army Medical Service School
- e: Automatic Master Sequence Selector
AM-SSB e: Amplitude Modulation-Single Sideband
AM-SSB/RC e: Amplitude Modulation-Single Sideband / Reduced Carrier
AM-SSB/SC e: Amplitude Modulation-Single Sideband/Suppressed Carrier
AMSSS e: Actron Microprocessor Software Support System
Amst d: Amsterdam
- e: Amsterdam
AMST e: Advanced Medium Short Take-off and Landing Transport
- e: Almost Minimal Spanning Tree
- e: Association of Maximum Service Telecasters
- e: Automated Maintenance Support Tool
A/M Station e: Automatic/Manual Station
AM STD e: American Standard
AMSTs e: Almost Minimal Spanning Trees
AMSTS e: Automatic Multiparameter Semiconductor Test Set
AMSU e: Auto-Manual Switching Unit
Am-Sup d: Ammoniumsuperphosphat
AM-Sup-Ka d: Ammoniak-Superphosphat-Kalisalz
Am-Sup-Ka d: Ammoniaksuperphosphatkali
AMSWIE e: Associate Member of the South Wales Institute of Engineers
AM Switch e: Automatic Manual Switch
AMSY e: American Management Systems
AMT d: Alpha-Methyl-p-Tyrosin
- d: Ausrüstungen, Maschinen und Transportmittel
- e: a-Methyl-p-Tyrosine
- e: Academy of Medicine Toronto, Canada
- e: Active Maintenance Time
- e: Advanced Manufacturing Technology
- e: Aerial Mail Terminal
- e: Air Mail Transfer
- e: Air Mail Transmission
- e: Air Member for Training
- e: Alkali-Metal Turbine
- e: Alpha-Methyltyrosine
- e: American Medical Technologists
- e: American Telecommunications Corp.
- e: Amplitude-Modulated Transmitter
- e: Analog[ue] Magnetic Tape
- e: Angular Mapping Transformation
- e: Apogee Motor Timer
- e: Associate in Mechanical Technology
- e: Associate in Medical Technology
- e: Astrograph Mean Time
- e: Audio-Magnetotelluric
- e: AUTODIN Multimedia Terminal
- e: Automated Microfiche Terminal
- e: Automatic Moon Tracking
- e: Automatic Motor Tester
- e: Available Machine Time
- e: Avalanche Memory Triode
amt e: amethopterin
- e: amount
- e: amphetamine
AMTA e: Airborne Moving Target Attack
- e: American Massage and Therapy Association
amtamk e: amphibious tank
AMTAS e: Automatic Modal Tuning and Analysis System
AMTB e: Anti-Motor-Torpedo-Boat
AMTC e: Academic Member of Trinity College
- e: Air Material Armament Test Center
- e: American Manchester Terrier Club
- e: Army Missile Test Center
- e: Art Masters Teaching Certificate
AMTCL e: Association for Machine Translation and Computational Linguistics
AMTD e: Advanced Mechanical Transport Depot
- e: Automatic Magnetic Tape Dissemination
AMTDA e: Agricultural Machinery and Tractor Dealers Association
AMTD Service e: Automatic Magnetic Tape Dissemination Service
AMTEC e: American Metalworking Technology for the European Community
- e: Ampex-Time-Element-Compensator
- e: Automatic Time Element Compensator
AM Tech I e: Associate Member of the Technology Institute
Am Tech Soc e: American Technical Society
Am Tel & Tel e: American Telephone and Telegraph
AMTEX e: Air-Mass Transformation Experiment
amti e: Airborne Moving Target Indicator
AMTI e: Area Moving Target Indicator
- e: Automatic Moving Target Indicator
AMTIDE e: Aircraft Multipurpose Test Inspection and Diagnostic Equipment
AMTIR e: Advanced Moving Target Indicator [Radar]
AMTIS e: Airborne Moving Target Indicator System
AMTK e: Amphibious Tank
amtl. d: amtlich[erseits]
amtl. Begr. d: amtliche Begründung
amtl. Bek. d: amtliche Bekanntmachung
Am Tn e: Ammunition Train
AMTPI e: Associate Member of the Town Planning Institute
AMTR d: Ammeter
- e: Atlantic Missile Test Range
AMTRAC e: Ammunition Track
- e: Amphibious Tractor
AMTRAN e: Automatic Mathematical Translation
- e: Automatic Mathematical Translator
AMTRANS e: Army Missile Transport System
AMTRC e: Amount Received
AMTREX e: Amphibious Training Exercise
Am Trk e: Amphibious Truck
AMTS e: Advanced Mobile Telephone System
- e: Air Member for Technical Services
amt[s] e: amphetamine[s]
AMTS e: Analog[ue] Magnetic Tape Storage
- e: Associate Member of the Television Society
Amtsbl. d: Amtsblatt
Amtsbl. Eur. Gem. d: Amtsblatt der Europäischen Gemeinschaft
Amtsbl.-Vfg. d: Amtsblattverfügung
Amtsdt. d: Amtsdeutsch
Amtsh. d: Amtshandlung
- d: Amtshauptmann
- d: Amtshauptmannschaft
Amtsspr. d: Amtssprache
AMT Timber e: Amaltas Timber
AMTU e: Automatic Margin Test Unit
AMU e: Adult Male Unit
- e: Aerial Matching Unit
- e: Air Mileage Unit
- e: Air Mission Unit
- e: Alarm Monitor Unit
- e: Aligarh Muslim University
- e: American Malacological Union
- e: Analog[ue] Multipler Unit
- e: Antenna Matching Unit

- e: Army Medical Unit
- e: Associated Midwestern Universities
- e: Association of Minicomputer Users
- e: Astronaut Maneuvering Unit
- e: Atomic Mass Unit
- e: Auxiliary Memory Unit
- e: Average Monthly Usage

AMUA e: Associate of Music; University of Adelaide

AMUCOM e: Army Munition Command

AMUN e: Air Munition

AMUNC e: Army Munitions Command

AMUP e: Averaged Motor Unit Potential

A Mus e: Associate of Music

AMUX d: Analog-Multiplexer
- e: Avionics Multiplex

AMUX System e: Avionics Multiplex System

AMV d: Arabis-Mosaikvirus
- d: astabiler Multivibrator
- d: Atemminutenvolumen
- e: Actual Measurement Weight
- e: Alfalfa Mosaic Virus
- e: Amphibious Warfare
- e: Arithmetic Mean Value
- e: Astable Multivibrator
- e: Avian Myeloblastitis Virus
- f: Association Mondiale Veterinaire

Am V e: Armo[u]red Vehicle

AMVAP e: Associate Manufacturers of Veterinary and Agricultural Products

AMVER e: Atlantic Merchant Vessel Report
- e: Automated Merchant Vessel Emergency Rescue
- e: Automated Merchant Vessel Report[ing]
- e: Automated Mutual-Assistance Vessel Rescue System

AMVER Network e: Automated Merchant Vessel Emergency Rescue Network

AMVERS e: Atlantic Merchant Vessel Reporting System
- e: Automated Merchant Vessel Reporting System

AMVF e: Amplitude Modulated Voice Frequency

AMVFT e: Amplitude Modulated Voice Frequency Telegraph

AM-VFT System e: Amplitude Modulated Voice Frequency Telegraphy System

AMVM e: Administrative Motor Vehicle Management

AMVSB e: Amplitude Modulation with Vestigial Side-Band
- e: Amplitude Modulation with Vestigial Sideband

AmVSt d: Anmeldevermittlungsstelle

AMW d: Auslandsmischwähler
- e: Active Microwave Workshop
- e: Actual Measurement Weight
- e: Angular Momentum Wheel
- e: Antimaterial Warhead
- e: Antimissile Warfare
- e: Average Molecular Weight

AMWA e: American Medical Womens Association
- e: American Medical Writers Association
- e: Area Microwave Assembly

AMWD e: Address Mark Write Data
- e: Advanced Millimeter Wave Device

AMW Detector e: Analog Moving Window Detector

Amwelsh Coal Charter e: Americanized Welsh Coal Charter

AMWG e: Academy of Master Wine Growers

AMWH e: Antimaterial Warhead

AMWR e: Air Ministry War Room

AMWS e: American Men and Women of Science

AM-WT d: amplitudenmodulierte Wechselstromtelegrafie

AM-WT-System d: amplitudenmoduliertes Wechselstromtelegrafie-System

AMX d: Ausgabematrix
- e: Automatic Message Exchange

AMX-105 f: French self-propelled 105mm howitzer

AMX-13 f: French light tank

AMX-155 f: French self-propelled 155mm howitzer

AMX-30 f: French medium tank

AMX-VTT f: French armored personnel carrier

Amy. d: Amytal

amy e: amytal

AMY e: Automatic Monitor of Yarn

amyl e: amyl nitrate

AMZ d: abhängiger Maximalstromzeitschutz
- d: Absorptionsmeßzange
- d: Auto, Motor und Zubehör (journ.)
- e: Air Mass Zero

am. Z. d: amerikanische Zone

AN d: Ammoniumnitrat[e]
- d: Anfangszeichen
- d: Anlage
- d: Anpassungsnetzwähler
- d: Anpassungsnetzwerk
- d: Anschaltung
- d: Antenne
- d: Anzeige
- d: Arbeitnehmer
- d: Arbeitsnachbereitung
- d: Arbeitsnachweis
- d: Arbeitsnorm
- d: Artikelnumerierung
- d: Artikelnummer
- d: Asparagin
- d: Auftragnehmer
- d: Ausrüstungsnachweis
- e: Above-Named
- e: Absolute Name
- e: Abstract Number
- e: Accession Number
- e: Account Number[ing]
- e: Acid Number
- e: Acrylonitrile
- e: Add Normalized
- e: Advanced Navigator
- e: Aerodynamics Note
- e: Aids to Navigation
- e: Aircraft Number
- e: Air Force-Navy
- e: Airmail Notice
- e: Airman
- e: Air Natural (Cooling)
- e: Air Navigation
- e: Air Navigator
- e: Air Reduction
- e: Alaska Network
- e: Alpha[nu]meric
- e: American Navy
- e: Analog[ue] Network
- e: Analyzer
- e: AND
- e: And (Algol)
- e: AND NOT
- e: Anglo-Norman
- e: Anorexia Nervosa
- e: Antenna
- e: Antinode
- e: Arithmetic Notation
- e: Army-Navy
- e: Army Number
- e: Arrival Note
- e: Arrival Notice
- e: Art News (journ.)
- e: Associate in Nursing
- e: Astronautics Notice
- e: Atomic Number
- e: Autonetics
- e: Netherlands Antilles

an. d: altnordisch
- d: analog
- d: anomal
- d: anorganisch
- d: anormal
- e: annual
- e: anode
- e: anonymous
- l: anno
- l: ante

an e: annual
- e: anorthite
- e: atmosphere; normal

a. n. l: absque nota
- l: ad notam

a. N. d: am Neckar

An. d: Analyse
- d: Anamnese
- d: Anode
- d: Anschrift

An d: Anisotropie
- d: Anströmung
- d: normale Atmosphäre
- e: Akon
- e: Aktinon
- e: Annam
- e: Annamese

A/N d: alphanumerisch
- e: Acid[ic] and Neutral
- e: Allied/Neutral
- e: Alphanumeric[al]
- e: Alphanumerics

A_n e: normal atmosphere

AN-12 R: Soviet Antonov 100-passenger cargo plane

AN-14 R: Soviet Antonov 6-seat transport aircraft

AN-14M R: Soviet Antonov 15-passenger turboprop plane

AN-2 R: Soviet Antonov 14-passenger biplane

AN-22 R: Soviet Antonov 22

AN-26 R: Soviet Antonov 50-passenger transport plane

ANA d: atomgesteuerte Navigationsanlage
- d: automatischer Netzwerk-Analysator
- e: Adaptive Null Aerial
- e: Adaptive Null Antenna
- e: Adminstration for Native Americans
- e: Aerojet Network Analyzer
- e: Air Force-Navy Aeronautical
- e: Air Navigation Act
- e: All Nippon Airways
- e: Alpha-Naphtalenic Acid
- e: Alpha-Naphthalenic Acid
- e: American Nature Association
- e: American Neurological Association
- e: AND to Accumulator
- e: Antibodies to Nuclear Antigen
- e: Antinuclear Antibody
- e: Army-Navy Aeronautical [Specifications]
- e: Assigned Night Answer
- e: Associate National Academician
- e: Associate of the National Academy
- e: Automated Naval Architecture
- e: Automatic Network Analyzer
- e: Automatic Number Analysis
- e: Automatic Number Announcer

- s: Asociación Nacional Auromovilista
ana e: anesthesia[c]
- l: ana partes aequales
Aña e: Agaña, Guam
ANAAS e: Australian and New Zealand Association for the Advancement of Science
anac e: anachronism
- e: anachronistic
ANACDUTRA e: Annual Active Duty for Training
ANACHEM e: Association of Analytical Chemists
anacol e: anacoluthon
ANACOM e: Analog[ue] Computer
ANACONDA e: Analytical Control and Data
anacreon e: anacreontic[s]
- e: anacreontist
anacru e: anacrusis
ANAD e: Anniston Army Depot
ANADIS e: Australian [National] Animal Disease Information System
anaes e: anaestetic
- e: anaesthesia
anaesth d: Anaesthesiologist
- e: anaesthesia
- e: anaesthesiologist
- e: anaesthesiology
- e: anaesthetic[s]
ANAF e: Army-Navy-Air Force
ANAFJ e: Army-Navy-Air Force Journal
ANAG e: Abstracts of North American Geology (journ.)
anag e: anagramist
- e: anagrammatic[al] [ly]
- e: anagram[s]
ANAGOL d: Analogrechner Algorithmic Language
ANAHL e: Australian National Animal Health Laboratory
ANAI e: African Network of Administrative Information
ANAKON d: Analytikerkonferenz
ANAL d: Analysis
- e: Analog[ue]
- e: Analyst
- e: Analytic
- e: Analyze[r]
Anal. d: Analogie
- d: Analyse
- d: Analytik
anal. d: analytisch
- e: analog
- e: analogous
- e: analogy
anal - e: Analgesic
- e: analysis
- e: analyze
- e: analytic[al]
Anal. Biochem. e: Analytical Biochemistry (journ.)
Anal. Chem. e: Analytical Chemistry (journ.)
analg e: analgesic
ANALIT e: Analysis of Automatic Line Insulation Tests
Anal. Lett. e: Analytical Letters (journ.)
ANALOG d: Analogrechner
- d: Analogrechner Algol (Algorithmic Language)
anal psychol e: analytical psychology
analy d: analyze
analys e: analysis
analyst e: psychoanalyst
analyt. d: analytisch
- e: analytical
Analyt. Engr. e: Analytical Engineer
ANAN Cooling e: Air-Natural Cooling
AnAp d: Anschalteapparat
ANAP e: Agglutination Negative; Absorption Positive
ANAPAC e: Analysis Package
AN/ARC e: Army Navy / Airborne Radio Communications
ANARC e: Association of North American Radio Clubs
ANARE e: Australian National Antarctic Research Expedition[s]
AnÄsth. d: Anästhesie
- d: Anästhesist
- d: Anästhetikum
ANAT e: Anatomy
Anat. d: Anatolien
- d: Anatomie
anat. d: anatolisch
- d: anatomisch
anat d: Anatomist
- e: anatomic[al]
- e: anatomist
ANATC e: Air Navigation and Traffic Control
anath e: anathema[tize]
ANATRAN e: Analog[ue] Translator
Anat. Rec. e: Anatomical Record (journ.)
ANATRON e: Analog[ue] Translator
ANAU d: Analogausgabe
- d: Analogausgabegerät
ANAUe d: Analogausgabegeräte
ANAV e: Air Navigator
- e: Area Navigation
ANB d: Anschlußbedingung
- e: Air Navigation Bureau
- e: Ambient Noise Buoy
- e: Army-Navy British
- e: Australian National Bibliography
aNb d: anerkannter Nachbau
ANBA d: Amtliche Nachrichten der Bundesanstalt für Arbeit
ANBFM Modem e: Adaptive Narrow-Band Frequency Modulation Modem
ANB-HP d: Anschlußbaugruppe für Hochpegel
ANBN e: Alpha-Nitrose-Beta-Naphthol
ANBS e: Armed Nuclear Bombardment Satellite
- e: Army-Navy British Standard
ANB Standard e: Army-Navy-British Standard
ANBU d: Anlagenbuchhaltung-[Programmpaket]
ANBw d: Amt für Nachrichtenwesen der Bundeswehr
ANBW d: Arbeitsgemeinschaft Norddeutscher Bildungswerke der Wirtschaft
ANC d: Anchorage (Flughafen)
- e: Adaptive Noise Canceling
- e: Aerojet Nuclear Company
- e: African National Congress
- e: Airborne Navigation Computer
- e: Air Force-Navy-Civil
- e: All Number[s] Calling
- e: Ammonium-Nitrate-Carbon
- e: Answer-Charge
- e: Answer signal; Charge
- e: Arlington National Cemetery
- e: Army-Navy-Civil
- e: Army-Navy-Commercial
- e: Assistant Network Controller
AnC e: Analog[ue] Computer
anc e: ancient
Anc e: Ancona
a. n. C. l: ante nativitatem Christi
ANCA e: Allied Naval Communications Agency
ANCAR e: Australian National Committee for Antarctic Research
ANCC e: Anodal Closure Contraction
ANCCAC e: Australian National Committee on Computation and Automatic Control
ANCEA i: Associazione Nazionale Commercianti Elettrodomestici Affini
ANCFIL e: Anchored Filament
ANCFX e: Allied Naval Commander; Expeditionary Force
ANCG e: Announcing
ANCH e: Anchorage
anchor. e: alphanumeric character generator
a. n. Chr. l: ante nativitatem Christi
ANCIC e: Australian National Copyright Information Centre
ancienn f: anciennement
ANCIF e: Automated Nautical Chart Index File
ANC-in-C e: Allied Naval Commander-in-Chief
ANCIRS e: Automated News Clipping, Indexing and Retrieval System
ANCIR/TS e: Analog[ue] Circuit-Analysis by Time-share System
ANCLAV e: Automatic Navigation Computer for Land and Amphibious Vehicles
ANCON s: Asociación Nacional para la Conservactón de la Naturaleza
ancova e: analysis of covariance
ANCOVS e: Active Night Covert Viewing System
ANCR e: Aitcraft Not Combat Ready
ANCS e: Airborne Night Classification System
- e: Alphanumeric Channel Set
- e: American Numerical Control Society
- e: Autonomous Navigation System Concept
anct. e: ancient
ANCXF e: Allied Naval Commander; Expeditionary Force
AND d: Andorra (Kfz)
- d: Andromeda
- d: Arbeitsgemeinschaft numerische Drehautomaten
- e: Air Force-Navy Design
- e: Air Navigation Device
- e: Air Navigation Direction[s]
- e: Air Navigation Directive[s]
- e: Airplane Nose Down
- e: All Numeral Dial[ling]
- e: Alphanumeric Display
- e: Army-Navy Design
- e: Automatic Network Dialling
and. d: altniederländisch
- d: andauernd
- d: andere
- d: anders
And. d: Andacht
- d: Andenken
and e: altnordisch
- i: andante
And e: Andalucia
- e: Andaman Islands
- e: Andorra
- e: Andromeda
- l: Andromeda
Änd. d: Änderung
ÄndA d: Änderungsanordnung[en]
andalusite e: aluminum silicate
Andamans e: Andaman Islands

Andamans and Nicobars e: Andaman and Nicobar Islands of Burma in the Bay of Bengal
ÄndAO d: Änderungsanordnung[en]
ANDAS e: Automatic Navigation and Data Acquisition System
Änd B d: Änderungsbestimmung[en]
Änd.-Bek. d: Änderungsbekanntmachung[en]
ÄndBest. d: Änderungsbestimmung[en]
ANDE e: Alphanumeric Display Equipment
- s: Administración Nacional Electricidad
ANDEP d: alphanumerischer Datenerfassungsplatz
ANDEPe d: alphanumerische Datenerfassungsplätze
Andes s: Cordillera de los Andes
ANDEW CHEM INT ED d: Angewandte Chemie; International Edition (journ.)
and. g. d: anderweitig genannt
ÄndG. d: Änderungsgesetz
ÄndG AnVNG d: Gesetz für Änderung des Angestelltenversicherungs-Neuregelungsgesetzes
and. gen. d: anderweitig genannt
Änd.-Ges. d: Änderungsgestz
ÄndG zum PatG d: Änderungsgesetz zum Patentgesetz
ANDI d: analoges Diffusionsmodell
- d: Anmeldedienst
And. I. e: Andaman Islands
AnDig e: Analog[ue]-Digital
ANDIP e: American National Dictionary for Information Processing
ANDIPS e: American National Dictionary for Information Processing Systems
AN-DK-Sprengstoffe d: Ammoniumnitrat-Dieselkraftstoff-Sprengstoffe
ANDL e: Army Nuclear Defense Laboratory
ANDMS e: Advanced Network Design and Management System
Andno i: andantino
Ando e: Andorra[n]
andorr. d: andorranisch
Andover e: Hawker-Siddeley troop transport
AndR d: Abfluß nach der Regelung
Andr e: Andromeda
ANDRO e: Androsterone
- e: Army Non-Destructive Readout
androg e: androgyn
ANDRZ. e: Andrzjowsky
ANDS e: Advanced Navy Display System
- e: Alphanumeric Displays
- e: Automated Newspaper Delivery System
AnDTe e: Anodal Duration Tetanus
Andte i: andante
ANDUS e: Anglo-Dutch-United States
ÄndV d: Änderungsverordnung
Änd.-V d: Änderung von Vorschriften
Änd.-Verf. d: Änderungsverfügung
Änd.-Verord. d: Änderungsverordnung
Änd.-Vfg. d: Änderungsverfügung
ÄndVFO d: Änderungsverfügung zur Fernmeldeordnung
- d: Verordnung zur Änderung der Fernmeldeordnung
Änd.-Vorschl. d: Änderungsvorschlag
- d: Änderungsvorschläge
Änd.-Vorschr. d: Änderungsvorschrift
ANDVT e: Advanced Narrow-band Digital Voice Terminal
Änd.-Vtg. d: Änderungsvertrag
Änd.-Vtrg. d: Änderungsvertrag
ANDZ e: Anodize
AnE d: Antriebselement[e]
ANE e: Acoustic Noise Environment
- e: Aeronautical and Navigation Electronics
- e: Aerospace and Nagigational Electronics
ANEAC e: Australian National Energy Advisory Committee
ANEBA e: Alphanumeric Electronic Bit Analyzer
ANEC e: Analysis Evaluation and Computation
ANEDA f: Association Nationale d'Etudes pour la Documentation Automatique
ANEPA e: Army-Navy Electronics Production Agency
ANERAC e: Annual North Eastern Regional Antipollution Conference
anerk. d: anerkannt
Anerk. d: Anerkennung
anes d: An[a]esthesia
- d: An[a]esthesiologist
- e: an[a]esthesia
- e: an[a]esthesiologist
- e: an[a]esthesiology
- e: an[a]esthetician
- e: an[a]esthetic[s]
anesth e: anesthetic
Anesth. Analg. e: Anesthesia and Analgesia (journ.)
anesthesiol e: An[a]esthesiology
ANEVAC e: Analysis Evaluation and Computation
ANEX e: Anode Excitation
ANF d: Anfang
- d: Anforderung
- d: antinuklearer Faktor
- e: Air Navigation Facilities
- e: Air Navigation Facility
- e: American National Fine
- e: American National Form
- e: Anchored Filament
- e: Anti-Nuclear Factor[s]
- e: Army-Navy Fuel
- e: Army News Features
- e: Associate Normal Form
- e: Atlantic Nuclear Force
- e: Atomic Nuclear Forum
- e: Audio Notch Filter
- e: Automatic Noise-Figure Indicator
- e: Aviation News Features
anf. d: anfahren
- d: anfallen
- d: anfangen
- d: anfangs
- d: anfassen
- d: anfechten
- d: anfordern
- d: anführen
Anf. d: Anfahrt
- d: Anfall
- d: Anfänger
- d: Anfechtung
- d: Anführer
ANFADR d: Anfangsadresse
AnfÄrb. d: Anfärben
Anf.-Bil. d: Anfangsbilanz
AnFd d: Anschaltefeld
ANFE e: Aircraft not Fully Equipped
AnfG d: Anfechtungsgesetz
Anf.-Geh. d: Anfangsgehalt
- d: Anfangsgehälter
anfgl. d: anfänglich
ANFI e: Automatic Noise Figure Indicator
ANFIA i: Associazione Nazionale fra le Industrie Automobilistiche
ANFO e: Ammonium Nitrate and Fuel Oil [explosive]
Anford. d: Anforderung[en]
anfr. d: altniederfränkisch
- d: anfragen
Anfr d: Anfrage
Anfr. d: Anfrager
ANFRA e: American National Fur Breeders Association
anfrÄnk. d: altniederfränkisch
anfrk. d: altniederfränkisch
ANG d: Angola
- d: Ausschuß, Nacharbeit, Garantie
- d: automatischer Nummerngeber
- e: Acoustic Noise Generator
- e: Air-Force Navy-Army Guided-Missiles
- e: American Natural Gas
- e: Australian New Guinea
an.g d: anderweitig genannt
a.n.g. d: anderweitig nicht genannt
Ang. d: Angabe[n]
- d: Angebot[e]
- d: Angehörige[r]
- d: Angelegenheit[en]
- d: Angestellte[r]
- d: Angriff
ang. d: angeblich
- d: angeboten
- d: angehend
- d: angehörend
- d: angehörig
- d: angekündigt
- d: angelegt
- d: angemeldet
- d: angenommen
- d: angestellt
- d: angewandt
- d: angular
Ang e: Anchorage
- e: Anglo-
- e: Angola
ang e: angiogram
- e: angle
ANG/ANL/R-O e: Angle Analog[ue] Read-Out
angeb. d: angeboren
- d: angeboten
- d: angebunden
Angeb. d: Angebot[e]
angef. d: angefangen
- d: angefordert
- d: angeführt
angegl. d: angeglichen
- d: angegliedert
angeh. d: angehend
- d: angehoben
- d: angehörend
- d: angehörig
Angeh. d: Angehörige[r]
angek. d: angekommen
- d: angekündigt
Angel. d: Angelegenheit[en]
angel. d: angelegt
- d: angelernt
- d: angeliefert
ANGELL e: Associated Nursery Guides Emphatically Lacking in Leisure
angem. d: angemahnt
- d: angemeldet
- d: angemessen
- d: angemietet
angen. d: angenommen
anger. d: angereichert
- d: angerissen
- d: angerostet
angesch. d: angeschafft
- d: angeschossen
- d: angeschrieben
angeschl. d: angeschlagen
- d: angeschlossen

angeschr. d: angeschrieben
angeschw. d: angeschweißt
- d: angeschwemmt
- d: angeschwollen
angest. d: angestellt
- d: angestoßen
Angest. d: Angestellte[r]
angew. d: angewachsen
- d: angewandt
- d: angewendet
- d: angewiesen
Angew. Bot. d: Angewandte Botanik (journ.)
ANGEW CHEM d: Angewandte Chemie (journ.)
ANGEW CHEM INT ED d: Angewandte Chemie; International Edition (journ.)
Angew. Inf. d: Angewandte Informatik (journ.)
angiol e: angiology
Angkor e: Angkor Temple
ANG-Kosten d: Ausschuß, Nacharbeit, Garantie-Kosten
angl. d: angleichen
- d: anglikanisch
Angl. d: Angleich[ung]
- d: Angliederung
- d: Anglikaner
- e: Anglican
ANGLFN e: Angle Function
Anglo-Afg e: Anglo-Afghan
Anglo-Afr e: Anglo-African
Anglo-Amer e: Anglo-American
Anglo-Ant e: Anglo-Antarctic[an]
- e: Anglo-Antillean
Anglo-Arab e: Anglo-Arabian
Anglo-Arg e: Anglo-Argentine
Anglo-Art e: Anglo-Arctic
Anglo-Aus e: Anglo-Australian
- e: Anglo-Austrian
Anglo-Bah e: Anglo-Bahaman
Anglo-Barb e: Anglo-Barbadian
Anglo-Bas e: Anglo-Basque
Anglo-Bei e: Anglo-Belizian
Anglo-Belg e: Anglo-Belgian
Anglo-Bhu e: Anglo-Bhutanese
Anglo-Bol e: Anglo-Bolivian
Anglo-Bots e: Anglo-Botswana
Anglo-Braz e: Anglo-Brazilian
Anglo-Bul e: Anglo-Bulgarian
Anglo-Bur e: Anglo-Burman
- e: Anglo-Buru-Indian
Anglo-CA e: Anglo-Central American
Anglo-Cam e: Anglo-Cameroonian
Anglo-Can[ad] e: Anglo-Canadian
Anglo-Cat e: Anglo-Catalan
Anglo-cath e: Anglo-Catholic
Anglo-Cey e: Anglo-Ceylonese
Anglo-Chi e: Anglo-Chinese
Anglo-Chil e: Anglo-Chilean
Anglo-Col e: Anglo-Colombian
Anglo-Cub e: Anglo-Cuban
Anglo-Cyp e: Anglo-Cypriot
Anglo-Czech e: Anglo-Czechoslovak[ian]
Anglo-Dah e: Anglo-Dahomean
Anglo-Dan e: Anglo-Danish
Anglo-Du e: Anglo-Dutch
Anglo-Ecu e: Anglo-Ecuadorean
Anglo-Egypt e: Anglo-Egyptian
Anglo-Eng e: Anglo-English
Anglo-Ethio e: Anglo-Ethiopian
Anglo-Fin e: Anglo-Finnish
Anglo-Fr e: Anglo-French
Anglo-Gael e: Anglo-Gaelic
Anglo-Gam e: Anglo-Gambian
Anglo-Ger e: Anglo-German
Anglo-Ger. Med.Rev. e: Anglo-German Medical Review (journ.)
Anglo-Gr e: Anglo-Greek
Anglo-Guy e: Anglo-Guyanese
Anglo-Hond e: Anglo-Honduran
Anglo-Hung e: Anglo-Hungarian
Anglo-Ice e: Anglo-Icelandic
Anglo-Ind e: Anglo-Indian
Anglo-Indo e: Anglo-Indonesian
Anglo-Ir e: Anglo-Iranian
- e: Anglo-Iraqi
- e: Anglo-Irish
Anglo-Isr e: Anglo-Israeli
Anglo-Ital e: Anglo-Italian
Anglo-Jam e: Anglo-Jamaican
Anglo-Jap e: Anglo-Japanese
Anglo-Jew e: Anglo-Jewish
Anglo-Jor e: Anglo-Jordanian
Anglo-Ken e: Anglo-Kenyan
Anglo-Kuw e: Anglo-Kuwaiti
Anglo-Lat e: Anglo-Latin
Anglo-Mal e: Anglo-Malawian
- e: Anglo-Malaysian
- e: Anglo-Maltese
Anglo-Mald e: Anglo-Maldivian
Anglo-Mex e: Anglo-Mexican
Anglo-N e: Anglo-Norse
Anglo-Nep e: Anglo-Nepalese
Anglo-Nig e: Anglo-Nigerian
Anglo-Nor e: Anglo Norwegian
Anglo-Norm e: Anglo Norman
Anglo-NZ e: Anglo-New Zealand
Anglo-Pak e: Anglo-Pakistani
Anglo-Para e: Anglo-Paraguayan
Anglo-Per e: Anglo-Persian
- e: Anglo-Peruian
Anglo-Pol e: Anglo-Polish
Anglo-Port e: Anglo-Portuguese
Anglo-Rho e: Anglo-Rhodesian
Anglo-Rom e: Anglo-Romanian
Anglo-Rus[s] e: Anglo-Russian
Anglo[s] e: Anglo-Saxon[s]
Anglo-SA e: Anglo-South African
- e: Anglo-South American
Anglo-Sam e: Anglo-Samoan
Anglo-Sax e: Anglo-Saxon
Anglo-Scot e: Anglo-Scottish
Anglo-SL e: Anglo-Sierra Leonean
Anglo-Som e: Anglo-Somali
Anglo-Sov e: Anglo-Soviet
Anglo-Span e: Anglo-Spanish
Anglo-Sud e: Anglo-Sudanese
Anglo-Swe e: Anglo-Swedish
Anglo-Swi e: Anglo-Swiss
Anglo-Tanz e: Anglo-Tanzanian
Anglo-Tob e: Anglo-Tobagan
Anglo-Togo e: Anglo-Togolese
Anglo-Ton e: Anglo-Tongan
Anglo-Trin e: Anglo-Trinidadian
Anglo-Turk e: Anglo-Turkish
Anglo-Ugan e: Anglo-Ugandan
Anglo-Uru e: Anglo-Uruguayan
Anglo-Ven e: Anglo-Venezuelan
Anglo-W e: Anglo-Welsh
Anglo-Yem e: Anglo-Yemini
Anglo-Yugo e: Anglo-Yugoslav[ian]
Anglo-Zamb e: Anglo-Zambian
AnglVO d: Angleichungsverordnung
Angl.-Vtrg. d: Angliederungsvertrag
Angl-Yingl e: Anglish-Yinglish
ANGM e: Air National Guard Manual
ANGMOM e: Angular Moment
Angola e: Peoples Republic of Angola
ang pec l: angina pectoris
AnGR d: Anschaltegestellrahmen
AN/GRA e: Air Force/Navy Ground Radar
Ang-Sax Amer e: Anglo-Saxon America
Angst. d: Angestellte[r]
ANGTS e: Alaska Natural Gas Transportation System
ANGUS e: Acoustically-Navigated Geophysical Underwater Survey
- e: Air National Guard of the United States
ANGUS A e: Navigable General-Purpose Unterwater Surveyor
AngV d: Angestelltenversicherung
AngVG d: Angestelltenversicherungsgesetz
AngVPfl. d: Angestelltenversicherungs-pflicht
angw. d: angewandt
Anh. d: Anhalt
- d: Anhang
- d: Anhänger
- d: Anhörung
- d: Anhydrid
anh. d: anhalten
- d: anhängend
- d: anhängig
- d: anheften
- d: anhören
- d: Anhydride
- d: anhydrisch
- e: anhydrite
anh e: anhydrous
ANH e: Australian National Highway
ANHAL e: Australian National Humanities and Arts Library
anhed. d: anhedral
ANHF e: Australian National Historic Site
anhic. e: Anhydritic
ANHS e: Adams National Historic Site
anhy d: Anhydride
- e: anhydrous
anhyd d: Anhydride
ANHYD e: Anhydrous
anhydr. d: anhydrisch
ANI e: Advanced Network Integration
- e: Animal
- e: Army-Navy-Industry
- e: Australian National Industries
- e: Automatic Number Identification
- e: Automatic Number Identifier
- f: Association Nationale pour l'Infographie
- p: Agéncia Nacional de Informaçao
- s: Agencia Nacional de Informaciones
A & NI e: Andaman and Nicobar Islands
ani f: atmosphère normale internationale
ANICA i: Associazione Nazionale Industrie Cinematografiche e Affini
ANIF e: Automatic Number Identification Failure
anil. e: aniline
aniline e: phenyl amine
anim e: animal
- i: animato
anim. e: animate
- e: animated
Anim. e: Animism
ANIM e: Association of Nuclear Instrument Manufacturers
animad e: animadversion
ANInstCE e: Associate Member of the Institution of Civil Engineers
ANIP e: Army-Navy Instrumentation Program[me]
ANIPLA i: Associazione Nazionale Italiana per l'Automazione
ANIRC e: Annual National Information Retrieval Colloquium
ANIS d: allgemeines nichtnumerisches Informationssystem

ANIT d: Alpha-Naphthylisothiocyanat[e]
- e: Automatic Network for Indirect Transmission
ANITA d: Analyse des Auftragsbestandes und Integration der Auftragsbearbeitung
aniv e: anniversary
- s: anniversario
ANJ Timber e: Anjan Timber
ANK d: Arbeitsgruppe Neue Kommunikationstechniken
- e: Alphanumeric Keyboard
- e: Ankara (airport)
ank a: ankomst
- e: ankle
- n: ankomen
Ank. d: Ankauf
- d: Ankäufe
- d: Anker
- d: Ankündigung
- d: Ankunft
ank. d: ankommen[d]
Anka d: Antennenkappe
ANKB e: Alphanumeric Keyboard
Ankf d: Anschaltekoffer
Ankl. d: Anklage
- d: Ankläger
- d: Anklang
Ankl.-Vertr. d: Anklagevertreter
- d: Anklagevertretung
ANKOR e: Autonetics Kalman Optimum Reset
Ank.-St. d: Ankündigungssteuer
AnkVSt d: Ankunftsvermittlungsstelle
ANL e: Add Normalized Long
- e: Analog[ue]
- e: Anneal[ed]
- e: Annoyance Level
- e: Annual
- e: Argonne National Laboratory
- e: Army Natick Laboratories
- e: Australian National Library
- e: Automatic New Line
- e: Automatic Noise Levelling
- e: Automatic Noise Limited
- e: Automatic Noise Limiter
- e: Automatic Noise Limiting
Anl. d: Analyse
- d: Anlage
- d: Anlasser
- d: Anlauf
- d: Anlaut
- d: Anleihe
- d: Anleitung
- d: Anlieferung
- d: Anlieger
anl. d: anläßlich
- d: anlautend
- d: anlegen
- d: anleiten
- d: anlernen
- d: anliefern
- d: anliegend
anl e: animal
- e: anneal
AnL e: Anthropological Linguistics (journ.)
ANLAD e: Argonne National Laboratory; Accelerator Division
ANLAS d: Anlagenabrechnungssystem
ANLA-System d: Anlagenabrechnungssystem
Anl.-Bl. d: Anlageblatt
ANLCA e: Alaska Native Land Claims Act
anld e: annealed
AnlfdB d: Anleitung für die Bedienung
Anl. f. d. Bed. d: Anleitung für die Bedienung
AnlG d: Anlagegesetz
- d: Anleihegesetz
anlg e: analog[ue]
ÄNLG d: Ärzteniederlassungsgesetz
ANLG Function e: Antilogarithmic Function
ANLID e: Argonne National Laboratory; Idaho Division
Anl.-K. d: Anlagekapital
- d: Anlagekosten
- d: Anlegekosten
- d: Anliegerkosten
ANLO e: Assistant Naval Liaison Officer
ANLOR e: Angle Order
Anl.-P. d: Anleihepapier
ANLP e: Alphanumeric Logic Package
Anl.-Pl. d: Anlegeplatz
ANLR e: Angular
- e: Annular
ANLSC e: Additive Noise Linear Sequential Circuit
Anl.-St. d: Anlegestelle
ANLT e: Anchor Light
an. lt e: anchor light
anlys e: analysis
ANLYST e: Analyst
Anm. d: Anmelder
- d: Anmeldung
Anm. d: Anmerkung[en]
ANM e: Additive Noise Machine
- e: Admiralty Notice to Mariners
Anm. Best. Pat. d: Anmeldebestimmungen für Patente
ANMC e: Alaska Native Medical Center
- e: American National Metric Council
- e: Assistant Navy Mail Clerk
ANMCC e: Alternate National Military Command Center
Anm. d. Red. d: Anmerkung der Redaktion
Anm.-Form. d: Anmeldeformular
ANMI e: Air Navigation Multiple Indicator
ANMP e: Account Network Management Program[me]
Anm.-Pfl. d: Anmeldepflicht
ANMR e: Acoustic Nuclear Magnetic Resonance
ANMRC e: Australian Numerical Meteorology Research Center
Anm.-St. d: Anmeldestelle
anmt e: announcement
ANN d: Abteilung Nichtnumerik
- d: Arbeitsgruppe für Nichtnumerik
- e: Annealed
- e: Annex
- e: Announcement
- e: Annunciator
- e: Answer-No-charge
- e: Average Nearest Neighbour
Ann. d: Annahme
- d: Annonce
- d: Annullierung
- e: Annal[s]
Ann d: Annalen
- f: Annales
- i: Annali
ann. d: annoncieren
- d: annonciert
- d: annullieren
- d: annulliert
- l: anni
- l: anno
AnN d: Anschaltenetz
ann e: annotated
- e: announce[ment]
- e: announcer
- e: annual[ly]
- e: annuity
ANNA e: Air Force-Navy-NASA-Army
- e: Army-Navy-NASA; Air-Force
Ann. Alergy e: Annals of Allergy (journ.)
Ann Am Acad Pol Sci e: Annals for the American Academy of Political and Social Science (journ.)
Ann Chim Phys f: Annales de Chimie et de Physique (journ.)
ANNECS e: Automated Nikkei News Editing and Composing System
Ann. Entomol. Soc. Amer. e: Annals of the Entomological Society of America (journ.)
ANNET d: Analyse von Netzwerken
Ann.-Exp. d: Annoncenexpedition
Ann Fluiddyn e: Annals of Fluid Dynamics (journ.)
an.n.g. d: anderweitig nicht genannt
Ann Geophys f: Annales de Geophysique (journ.)
Ann Inst Henn Poincaré f: Annales de Institut Henri Poincaré (journ.)
Ann. Intern. Med. e: Annals of Internal Medicine (journ.)
anniv e: anniversary
anniv. l: anniversarium
Ann. Libr. Sci. Docum. e: Annals of Library Science and Documentation (journ.)
annly e: annually
annm a: anmaerkning
Ann. Math e: Annals of Mathematics (journ.)
Ann. Math. Atatist. e: Annals of Mathematical Statistics (journ.)
Ann. N.Y. Acad. Sci e: Annals of the New York Academy of Sciences (journ.)
Ann. Occup. Hyg. e: Annals of Occupational Hygiene (journ.)
ANNOD e: A Navigator of Natural Language Organized Data
ANNOT e: Annotated
- e: Annotation
Ann. Otol.; Rhinol.; Laryngol. e: Annals of Otology; Rhinology; Laryngology (journ.)
Ann Phys d: Annalen der Physik (journ.)
- f: Annules de Physique (journ.)
Ann. Phys. e: Annals of Physics (journ.)
Ann. Rep. e: Annual Report
Ann Rev Nucl Sci e: Annual Review of Nuclear Science (journ.)
Ann. Rheum. Dis. e: Annals of the Rheumatic Diseases (journ.)
Ann.-St. d: Annahmestelle
Ann. Surg. e: Annals of Surgery (journ.)
Ann. Thorac. Surg. e: Annals of Thoracic Surgery (journ.)
Ann. Trop. Med. Parasitol. e: Annals of Tropical Medicine and Parasitology (journ.)
annu e: annual[e]
- i: annuario
annuit e: annuitant
Annuit Coeptis l: Annuit coeptis novus ordo seculorum
annul. e: annulment
Annunc e: Annunciation
Annu. Rep. Progr. Chem.; Sect. e: Annual Reports on the Progress of Chemistry; Section A. General (journ.)
- e: Annual Reports on the Progress of Chemistry; Section B. Organic Chemistry (journ.)
Annu. Rev. Astron. Astrophys. e: Annual Review of Astronomy and Astrophysics (journ.)

Annu. Rev. Biochem. e: Annual Review of Biochemistry (journ.)
Annu. Rev. Entomol. e: Annual Review of Entomology (journ.)
Annu. Rev. Fluid Mech. e: Annual Review of Fluid Mechanics (journ.)
Annu. Rev. Genet. e: Annual Review of Genetics (journ.)
Annu. Rev. Med. e: Annual Review of Medicine (journ.)
Annu. Rev. Microbiol. e: Annual Review of Microbiology (journ.)
Annu. Rev. Nucl. Sci. e: Annual Review of Nuclear Science (journ.)
Annu. Rev. Pharmacol. e: Annual Review of Pharmacology (journ.)
Annu. Rev. Phys. Chem. e: Annual Review of Physical Chemistry (journ.)
Annu. Rev. Physiol. e: Annual Review of Physiology (journ.)
Annu. Rev. Phytopathol. e: Annual Review of Phytopathology (journ.)
Annu. Rev. Plant. Physiol. e: Annual Review of Plant Physiology (journ.)
Annu. Rev. Psychol. e: Annual Review of Psychology (journ.)
an. n. v. d: anderweitig nicht vergeben
- d: anderweitig nicht vorgesehen
Anny. e: Annuity
ANO d: Anordnung
- e: Above-Named Officer
- e: Air Navigation Order[s]
- e: Alpha-Numeric Output
- e: AND NOT Operation
- e: Anti-Narcotics Office
AnOC e: Anodal Opening Contraction
ANOCOVA e: Analysis of Covariance
anod e: anodize[d]
anolyte e: Anodic Electrolyte
ANOM e: Analysis of Means
- e: Assistant Network Operations Manager
anon. d: anonym
anon e: anonymous[ly]
Anon. l: anonymous
anop e: assembly no operation
anord. d: altnordisch
- d: anordnen
Anord. d: Anordnung[en]
anordn. d: anordnen
Anordn. d: Anordnung[en]
anorex e: anorexia nervosa
anorg. d: Anorganisch
anorg. Chem. d: anorganische Chemie
Anorm e: Aircraft Not Operationally Ready-Maintenance
anor nerv l: anorexia newosa
ANORS e: Aircraft Not Operationally Ready-Supplies
- e: Anticipated Not Operationally Ready-Supply
ANOT e: 3-Amino-5-Nitro-0-Tolumanide
- e: Annotate
An Oto Rhiao Laryagol e: Annals of Otology, Rhinology and Laryngology (journ.)
ANOV e: Analysis of Variance
anova e: analysis of variance
ANP d: Anpassung
- d: Anpassungsgerät
- d: Arbeitsausschuß Normenpraxis
- e: Arbeitsgemeinschaft Nukleare Prozeßwärme
- d: Armeenachrichtenpark
- d: Ausschuß für Normenpraxis
- e: A-Norprogesterone
- e: Aberdare National Park
- e: Acadia National Park
- e: Adaptive Noise Processor
- e: Aircraft Nuclear-Power[ed] [Program]
- e: Aircraft Nuclear-Propulsion [Program[me]]
- e: Akan National Park
- e: Albert National Park
- e: Amsha National Park
- e: Angkor National Park
- e: Archives of Neurology and Psychatry (journ.)
- e: Arusha National Park
- e: Awash National Park
anp. d: anpassen
- d: anpeilen
ANPA e: Appalachian National Park Association
ANPAM e: African Network for Public Administration and Management
ANPA-RI e: American Newspaper Publishers Association-Research Institute
ANPAT e: American Newspapers Publishers Abstracting Technique
ANPC e: Air-Nitrogen Pressurization Control
- e: Archives of Neurology and Psychatry-Chicago (journ.)
ANPD e: Aircraft Nuclear Propulsion Department
ANPDGE e: Aircraft Nuclear Propulsion Department; General Electric [Company]
anp.-f. d: anpassungsfähig
AnpG d: Anpassungsgesetz
ANPL e: Archives of Neurology and Psychatry-London (journ.)
ANPN e: Army Navy Performance Number
ANPO e: Aircraft Nuclear Propulsion Office
ANPOD e: Aerial Positioning Device
- e: Antenna Positioning Device
ANPP e: Aircraft Nuclear Power Plant
- e: Aircraft Nuclear Propulsion Program[me]
- e: Aircraft Nuclear Propulsion Project
- e: Army Nuclear Power Program[me]
ANPPC e: Army-Navy Petroleum Pool; Pacific Coast
ANPPF e: Aircraft Nuclear Power Plant Facility
ANP Program[me] e: Aircraft Nuclear Propulsion Program[me]
ANPSF e: Active Neighbourhood Pattern Sensitive Fault
ANPT e: Aeronautical National Taper Pipe Thread[s]
AnPt d: Anschalteplatte
an.pt. e: aniline point
AnpUe d: Anpassungsübertragung
Anp VO d: Anpassungsverordnung
ANPWS e: Australian National Parks and Wildlife Service
ANR d: Advisornummer
- d: Arbeitskreis Norddeutscher Rechenzentren
- d: Armeenachrichtenregiment
- d: Auftragsnummer
- d: automatischer Neunfach-Repeater
- e: Alphanumeric Replacement
- e: American Natural Resources
- e: Association of Neutron Radiographers
- e: Australian National Railways
- e: Awaiting Number Received
ANr d: Aktennummer
Anr. d: Anrechnung
- d: Anruf[er]
AnR d: Anschalterahmen
ANRA e: Air Navigation Radio Aids
- e: Amistad National Recreation Area
- e: Arbuckle National Recreation Area
ANRAC e: Aids to Navigation Radio Control
ANRAO e: Australian National Radio Astronomy Observatory
ANRC e: American National Red Cross
- e: Animal Nutrition Research Council
- e: Australian National Railways Commission
- e: Australian National Research Council
Anreg. d: Anreger
- d: Anregung
ANRF e: Area Normalization with Response Factors
ANROA e: Australian National Radio Astronomy Observatory
ANRS e: Acceptor not Ready State
- e: Automatic Noise Reduction System
AnrSchr d: Anrechnungsschreiben
ANRT f: Association Nationale de la Recherche Technique
ANS d: Arbeitskreis zur Nutzbarmachung von Siedlungsabfällen
- d: Atemnot-Syndrom
- d: autonomes Nervensystem
- e: 1,8-Anilinonaphtalenesulfonate
- e: Active Network Synthesis
- e: Actual Noise Silencer
- e: Add Normalized Short
- e: Airborne Navigation Sensor
- e: Air Naval School
- e: Air Navigation School
- e: American National Standard
- e: American Nuclear Society
- e: American Nutrition Society
- e: AND to Storage
- e: Answering
- e: Apollo Network Simulations
- e: Army and Navy Stores
- e: Army Nursing Service
- e: Associated Nuclear Services
- e: Astronomical Netherlands Satellite
- e: Astronomical Satellite
- e: Autograph Note Signal
- e: Autograph Note Signed
- e: Automatic Navigation System
- e: Autonomic Nervous System
Ans. d: Ansage
- d: Ansager[in]
- d: Ansammlung
- d: Ansehen
- d: Ansicht
- d: Ansiedelung
Ans d: Ansatz
AnS d: Anschalt[e]satz
- d: Anschalteschiene
- d: Anschaltsatz
ans e: answer
- e: answered
ANSA e: Advanced Network System Architecture
- e: Aminonaphtholsulfonic Acid
- e: Automatic New Structure Alert
A[N]SA e: American Standards Association
ANSAM e: Anti-Missile Surface-to-Air Missile
ANSBR e: American National Standard for Bibliographic References
ANSC e: American National Standards Committee
- e: American Nuclear Science Corporation
A-N Scale e: Anti-Negro Scale [measuring negative attitudes]
ANSCH e: American National Standard Code for Information Interchange
ansch. d: anschaffen
- d: anschuldigen

Ansch. d: Anschaffung
- d: Anschuldigung
AnSch d: Anschalteschiene
Ansch.-K. d: Anschaffungskosten
Ansch.-Kr. d: Anschaffungskredit
Anschl. d: Anschlag
- d: Anschluß
anschl. d: anschließen[d]
Anschr. d: Anschreiben
- d: Anschrift[en]
AnschrSt d: Anschriftenstelle
ANSCHS d: Anschaltesatz
Ansch-Us d: Anschalteumsetzer
ANSCII e: American National Standard Code for Information Interchange
ANSCOBOL e: American National Standard Institute Common Business Oriented Language
ANSCR e: Alphanumeric System for Classification [of] Recordings
ANSCSOCR e: American National Standard Character Set for Optical Character Recognition
ANSER e: Agricultural Network Serving Extension and Research
Anseri e: Anseriformes
ANSETT e: Ansett Airways
ANSETT ANA e: Ansett Australian National Airways
AnsGt d: Ansagegerät
ANS HD d: Anschaltsätze für den Hinweisdienst
ANSI e: American National Standards Institute
- e: Australian National Standards Institute
ANSIC e: Aerospace Nuclear Safety Information Center
ANSICOBOL e: American National Standard Institute Common Business Oriented Language
ANSIM e: Analog[ue] Simulator
ANSL e: Australian National Standards Laboratory
ANSLICS e: Aberdeen and North of Scotland Library and Information Cooperative Service
ANSMusN e: American Numismatic Society Museum Notes (journ.)
ANSO e: Assistant Naval Stores Officer
ANSORC e: Analog[ue] System for Optical Reading of Characters
ANSP e: Academy of Natural Sciences of Philadelphia
ANSPA e: Astronomical Netherlands Satellite Programme
Anspr. d: Ansprache
- d: Anspruch
- d: Ansprüche
anspr. d: ansprechen
- d: ansprechend
- d: anspringen
ANSR e: Add-on Non-Stop Reliability
- e: Advanced Naval System Requirements
ANSS e: American Nature Study Society
- e: Associate of the Normal School of Science
ANSSIR e: A Network of Social Security Information Resources
ANSSL e: Australian National Social Sciences Library
ANSSMFE e: Australian National Society of Soil Mechanics and Foundation Engineering
ANSSR e: Aerodynamically Neutral Spin Stabilized Rocket
Anst d: Ansagetaste
Anst. d: Anstalt
- d: Anstellung
- d: Anstich
- d: Anstoß
anst. d: anstechen
- d: anstecken[d]
- d: anstellen
- d: ansteuern
- d: anstoßen
Anst.-Bed. d: Anstellungsbedingungen
AnSteKf d: Anschaltesteuerkoffer
ANSTEL e: Australian National Scientific and Technological Library
Anst.-Gr. d: Anstellungsgrundsätze
ANSTI e: African Network for Scientific and Technological Institutions
Anst.-L. d: Anstaltsleiter
Anst.-Ltr. d: Anstaltsleiter
AnstO d: Anstaltsordnung
Anst.-Schr. d: Anstellungsschreiben
Anst.-Vertr. d: Anstellungsvertrag
AnsUe d: Ansageübertragung
ANSVIP e: American National Standard Vocabulary for Information Processing
AnsVr d: Ansageverstärker
AnSw d: Anwender-Software
ANSW e: Anti-Nuclear Submarine Warfare
ANSWER e: Algorithm for Non-Synchronized Waveform Error Reduction
- e: Automated Network Schedule with Evaluation of Resources
ANT d: 2-Aminonitrothiazol
- d: A.N. Tupolev
- d: Antenne
- d: Anthanthrene
- d: Ausschuß Normungstechnik
- d: Austenitformhärten bei niedriger Temperatur
- e: Acoustic Noise Test
- e: Aerial Noise Temperature
- e: Antenna
- e: Antenna Noise Temperature
- e: Antigua [Station]
- e: Antiquity
- e: Antonym
- e: Assessment of New Techniques
- e: Australian National Times
- e: Australian Northern Territory
- e: Autonomous Navigation Technology
- R: AN Tupolev
AnT d: Anlaßtaste
- d: Anschaltetaste
Ant d: Anteil
- e: Antillean
- e: Antillea West Indian Federation
- e: Antilles
- e: Antwerp
Ant. d: Anteil[e]
- d: Antiklinale
- d: Antiphon
- d: Antiqua
- d: Antiquar[iat]
- d: Antiquitäten
- e: Antarctica
- e: Antigua
- e: Antiquarian
ant. d: anteilig
- d: antik
- d: Antilog[arithm]
- d: antiquarisch
- e: antenna[s]
- e: antilog
- e: antilogarithm
- e: antique
- e: antiquities
- e: antonym
ant d: Antriebstechnik (journ.)
- e: anterior
- e: anticipate[d]
- e: antique
- i: antico
- i: antiporta
- s: anterior
- s: antologia
Ant A e: Anti-Aircraft
ANTAARES e: Antenna Tracking Altitude, Azimuth and Range by Electronic Scan
ANTAC e: Air Navigation and Tactical Control
ANTACCS e: Advanced Navy Tactical Command and Control System
antag e: Antagonistic
Antarc e: Antarctic
- e: Antarctica
Antarc. Circ. e: Antarctic Circle
Antarc O e: Antarctic Ocean
Antarctic J.U.S. e: Antarctic Journal of the United States
ant ax line e: anterior axial line
Ant & Bar e: Antigua and Barbuda
ANTC e: Air Navigation Traffic Control
Ant-C e: Antennapedia Complex
ANTC e: Antichaff Circuit
- e: Australian National Television Council
Ant Chil s: Antarctica Chilena
ANTCOMDUSARCARIB e: Antilles Command; United States Army Caribbean
ANTCP e: Anticipate
Ant Cur e: Antilles Current
ant. d e: anterior diameter
antec e: annual technical conference
ANTELCO s: Administración Nacional de Telecomunicaciones
antennafier e: antenna and [radiofrequency] amplifier
antennamitter e: antenna [and] transmitter
antennaverter e: antenna and converter
ANTEX e: Alphanumeric Terminal Executive
Antf e: Antofagasta
ant. frt. e: Anticipated Freight
Anth. d: Anthologie
Anthol. d: Anthologie
anthol e: anthological[ly]
- e: anthologist
- e: anthologize
- e: anthology
ANTHOL e: Anthologist
- e: Anthologize
- e: Anthology
AnthQ e: Anthropological Quarterly (journ.)
Anthr. d: Anthropolog[i]e
anthr. d: anthropologisch
Anthrnposophic e: Anthroposphic Press
anthro d: Anthropologist
- e: Anthropogeography
- e: anthropological
- e: anthropologist
- e: anthropology
- e: anthropometry
- e: anthropomorphism
- e: anthropophagy
anthroco e: anthrocosis
Anthrop. d: Anthropolog[i]e
anthrop. d: anthropologisch
anthrop e: anthropological
- e: anthropology
Anthrop J e: Anthropological Institute Journal (journ.)
anthropol. d: anthropologisch
anthropol e: anthropologist
- e: anthropology

Anthropolog. d: Anthropolog[i]e
anthropom e: Anthropometry
Anthrop R e: Anthropological Review (journ.)
antiallerg. d: antiallergisch
antibio[s] e: antibiotic[s]
antichlor e: antichlorine
- e: antichloristic
anticli e: anticlimactic[al][ly]
- e: anticlimax
- e: anticlinal
- e: anticline
- e: anticlinorium
Anti-Corros.Methods Mater. e: Anti-Corrosion Methods and Materials
antidis e: antidisestablishmentarianism
antid[s] e: antidote[s]
antifreeze e: grain or methyl alcohol mixture
Antig e: Antigua
Antillas Mayores s: Cuba, Hispaniola, Jamaica, Puerto Rico
Antillas Menores s: Leeward and Winward Islands
Antillas Neerlandesas s: Aruba, Bonaire, Curacao, Saba, Sint Eustatius and half of Sint Maarten
Antilles e: West India Islands excluding Bahamas
antilog e: antilogarithm
antimag e: antimagnetic
antimat e: antimatter
Antimicrob. Agents Chemother. e: Antimicrobial Agents and Chemotherapy (journ.)
ant. ind. e: antenna inductance
ant. ind. sw. e: antenna inductance switch
ant. ind. tuning e: antenna inductance tuning
antinuke e: anti nuclear
antip e: anti-poverty
- e: antiparasitic
- e: antiparticle
- e: antipathetic
- e: antiperiodic
- e: antIpodal
- e: antipode
- e: antipollution
- e: antiproton
- e: antipsychotic
- e: antipyretic
- e: antipyrine
antiphon e: antiphonal[ly]
Antipodes e: Australia and New Zealand
ANTIPOL e: Antipollutant
antipol e: antipollution
Antiq d: Antiquar[iat]
- e: Antiquity
antiq. d: antiquarisch
Antiq. d: Antiquitäten
- e: Antiquarian
ANTIQ e: Antiquarian
- e: Antique
antiq égypt f: antiquité égyptienne
antiq gr f: antiquité grecque
Antiq.-Hdl. d: Antiquitätenhandel
- d: Antiquitätenhändler
antiq hébr f: antiquité hébraique
Antiq.-Kat. d: Antiquariatskatalog
antiq rom f: antiquité româine
Antiquar. e: Antiquarian
Antiques e: Antiques Publications
antis. d: antiseptisch
anti-Sem e: anti-Semitic
antisub e: anti-submarine
AntiSubDef e: Anti-Submarine Defense
ANTISUBM e: Anti-Submarine
Anti-TR-Switch e: Anti-Transmit/Receive Switch
- e: Anti-Transmitting/Receiving Switch
antivox e: Antivoice-Operated Transmission
Antl e: Antlia
Ant Lat e: Antique Latin
Anto e: Antofagasta
anton e: antonym
Ant Ops e: Antarctic Operations
Ant.-P. d: Anteilprovision
Ant Pen e: Antarctic Peninsula
ant pit e: Anterior Pituitary
Antr. d: Antrag
- d: Anträge
- d: Antrieb
- d: Antritt
antr. d: antreiben
- d: antreten
Antr e: Antrim
ANTRAS d: Analyse tragender Systeme
Antrims e: Antrim Mountains of Northern Ireland
ANTS e: Advanced Naval Training School
- e: Advanced Research Projects Agency Network Terminal System
- e: Airborne Night Television System
- e: American National Taper Shank
- e: Andover Newton Theological School
- e: Any Tape Search [Computer Program[me]]
- e: Assessment of New Techniques Scheme
- e: Automatic Nitrogen Transfer System
ant[s]. e: antonym[s]
ant sup spine e: anterior superior spine
ANTU e: Alphanaphthylthiourea
Antuer p: Antuérpia
Antw d: Antwerpen
Antw. d: Antwort[en]
Antw.-K. d: Antwortkarte
Antw.-Pk. d: Antwortpostkarte
Antw.-Schr. d: Antwortschreiben
ANU d: Antigua (Flughafen)
- e: Airframe Nose Up
- e: Australia[n] National University
ANUDS e: Army Nuclear Data System
AnUe d: Anschalteübertragung
An Ue d: Anschlußübertragung
anug e: acute necrotizing ulcerative gingivitis
Anuga d: Allgemeine Nahrungs- und Genußmittel-Ausstellung
ANULAS e: Australian National University Library Automated Systems
ANUNET e: Australian National Universitys Communications Network
ANUP e: Australian National University Press
AnUs d: Anschalteumsetzer
anv a: anverdelse
a. n. v. d: anderweitig nicht vergeben
- d: anderweitig nicht vorgesehen
an. v. d: anderweitig vergeben
- d: anderweitig vorgesehen
AnV d: Angestellten[renten]versicherung
- d: Arbeitnehmerverband
- d: Arbeitnehmervertreter
- d: Arbeitnehmervertretung
An V d: Anlaßventil
ANV d: Arbeitsnormenvertrag
- e: Air; Non-Ventilated
an. verg. d: anderweitig vergeben
AnVersPfl. d: Angestelltenversicherungspflicht
AnVG d: Angestelltenversicherungsgesetz
AnVM d: Angestelltenversicherungsmarke
AnVNG d: Angestelltenversicherungs-Neuregelungsgesetz
ANVO e: Accept No Verbal Orders
an. vorg. d: anderweitig vorgesehen
AnVPfl. d: Angestelltenversicherungspflicht
ANW d: Anweisung
- e: Apollo Network
ANw d: Anpassungsnetzwerk
AnW d: Anschaltewähler
Anw. d: Anwalt
- d: Anwaltschaft
- d: Anwärter
- d: Anwartschaft
- d: Anwendung
- d: Anwesen
anw. d: anwesend
Anw. d: Anwohner
ANWA e: Abstracts of New World Archaeology (journ.)
AnwBVW d: Anweisung für das betriebliche Vorschlagwesen
ANWCG e: Army Nuclear Weapon Coordination Group
AnwDA d: Anwärterdienstalter
ANWDOK d: Anwendungsdokumentation
AnwDZ d: Anwärterdienstzeit
Anwend. d: Anwendung
AnwFZw d: Anweisung für das Fernmeldezeugwesen
ANWG e: Apollo Navigation Working Group
ANWIHO d: Anlauf; Wiederhochlauf (z.B. Asynchronmotor)
Anw.-K. d: Anwaltskammer
- d: Anwaltskosten
Anw.-L. d: Anwesenheitsliste
ANWR e: Agassiz National Wildlife Refuge
- e: Arctic Natural Wildlife Refuge in Alaska
- e: Arkansas National Wildlife Refuge
- e: Arrowhead National Wildlife Refuge
- e: Audubon National Wildlife Refuge
a.n.wt. e: actual net weight
Anw.-Tech. d: Anwendungstechnik[er]
anx e: annex
ANZ d: Agrarische Nachrichtenzentrale
- d: Anzapf-
- d: Anzeige[n]
- e: Air New Zealand
- e: Australia and New Zealand
Anz. d: Anzahl
- d: Anzahlung
- d: Anzeiger
- d: Anzug
ANZAAS e: Australian and New Zealand Association for the Advancement of Science
Anz.-Abt. d: Anzeigenabteilung
ANZAM e: Australia, New Zealand and Malaysia
ANZAMRS e: Australia and New Zealand Association for Medieval and Renaissance Studies
ANZAR e: Australian and New Zealand Association of Radiology
AnzBedFd d: Anzeige- und Bedienfeld
Anz.-Betr. d: Anzahlungsbetrag
ANZCAN e: Australia-New Zealand-Canada (cable system)
ANZCP e: Australian and New Zealand College of Psychiatrists
Anz f d Alert d: Anzeiger für Deutsches Altertum (journ.)
ANZGAM e: Australian and New Zealand Graduates Association of Malaysia
ANZHES e: Australian and New Zealand History of Education Society

ANZIA e: Associate of the New Zealand Institute of Architects
ANZIC e: Associate of the New Zealand Institute of Chemists
ANZJS e: Australian and New Zealand Journal of Sociology
ANZLA e: Associate of the New Zealand Library Association
Anz.-Pfl. d: Anzahlungspflicht
- d: Anzeigepflicht
anzpfl. d: anzeigepflichtig
Anz.-Schl. d: Anzeigenschluß
ANZSNM e: Australian and New Zealand Society of Nuclear Medicine
ANZSOM e: Australian and New Zealand Society of Occupational Medicine
ANZSOS e: Australian and New Zealand Society of Oral Surgeons
ANZSTS e: Australia and New Zealand Society for Theologyical Study
ANZUS e: Australia, New Zealand, United States
Anz.-Vertr. d: Anzeigenvertreter
- d: Anzeigenvertretung
AO d: Abfrageoperation
- d: Abgabenordnung
- d: Advokatenordnung
- d: Akridinorange
- d: Anordnung
- d: Anrufordner
- d: Anti-Oxidant
- d: Apothekenordnung
- d: Arbeitsordnung
- d: Arbeitsort
- d: Argon-Oxygen
- d: Artillerieoffizier
- d: Atomorbitale
- d: Ausfuhrordnung
- d: Auslandsorganisation
- e: Abort Output
- e: Absolute Output
- e: Access Opening
- e: Accountant Officer
- e: Accounts Office
- e: Accumulator Overflow
- e: Acridine Orange
- e: Action Officer
- e: Actual Operation
- e: Add-On
- e: Adder Output
- e: Adding Operator
- e: Address[ing] Operation
- e: Address Order
- e: Address Out
- e: Adjusted Output
- e: Administrative Operation[s]
- e: Administrative Order
- e: Aerial Observer
- e: Aeronautical Order
- e: Airdrome Officer
- e: Air Officer
- e: Air Ordnance
- e: Air Over
- e: All Over
- e: Alphabetic Order
- e: Alphanumeric Order
- e: Alphanumeric Output
- e: American Optical
- e: Amplifier Output
- e: Analog[ue] Output
- e: AND Operation
- e: AND Operator
- e: AND OR
- e: Angola
- e: Anodal Opening
- e: Answer Only
- e: Area Officer
- e: Area of Operation[s]
- e: Arithmetic Operation
- e: Arithmetic Operator
- e: Arithmetic Organ
- e: Army Order
- e: Army Ordnance
- e: Artillery Officer
- e: Assembly Order
- e: Assembly Outline
- e: Astronomical Observatory
- e: Asynchronous Operation
- e: Atomic Orbital
- e: Atrioventicular Opening
- e: Audio Operator
- e: Audio Oscillator
- e: Authorized Order
- e: Automated Operator
- e: Automatic Observer
- e: Automatic Operation
- e: Autonomous Oblast
- e: Auxiliary Operation
- e: Auxiliary Oscillator
- e: Aviation Ordnanceman
- e: Axio-Occlusal
- e: Outward Operator
a. O. d: alte[r] Ordnung
a O d: an der Oder
ao. d: außerordentlich
ao e: anti-oxidant
- e: aorta
- e: aortic opening
- e: axioocclusal
A.O. e: A-Operator
a. o. e: account of
- l: ab ovo
A/O e: Answer/Originate
A.ö. d: Auslaßventil öffnet
ÄO d: Äthylenoxyd
AOA d: Allgemeiner Organisationsausschuß
- e: Accessible Operand Affiliation
- e: Air Officer [in charge] of Administration
- e: American Oceanology Association
- e: American Optometric Association
- e: American Orthopedic Association
- e: American Orthopsychiatric Association
- e: American Osteopathic Association
- e: American Overseas Airlines
- e: American Overseas Association
- e: Amphibious Objective Area
- e: Angle of Attack
- e: Ascending Order Arrangement
- e: Atlantic Ocean Area
- e: At Or Above
- e: Australian Optometrical Association
- e: Australian Orthopaedic Association
AÖA d: Abteilung für Öffentlichkeitsarbeit
AOAA e: Aminooxyacetic Acid
AOAC e: Army Ordnance Ammunition Command
- e: Association of Official Agricultural Chemists
- e: Association of Official Analytical Chemists
AOAD e: Army Ordnance Arsenal District
AOAG e: Automated Office AG
AOAI e: Angle-of-Attack Indicator
A-O-AMPL e: AND-OR Amplifier
AOAO e: Advanced Orbiting Astronautical Observatory
AO/Arb d: Arbeitsordnung für die Arbeiter (der Deutschen Bundesbahn)
AOAT e: Allowed Off Aircraft Time
AÖAW d: Anzeiger der Österreichischen Akademie der Wissenschaften (journ.)
AOB d: Anordnungsbeziehung[en]
- e: Adder Output Buffer
- e: Advanced Operational Base
- e: Airborne Optical Beacon
- e: Air[craft] Observer
- e: Air Order of Battle
- e: Alcohol on Breath
- e: Angle of Bank
- e: Angle of Beam
- e: Angle on the Bow
- e: Approved Operating Budget
- e: At or Below
AÖB d: allgemeine öffentliche Bücherei (DDR)
AOBD e: Acousto-Optic Beam Deflector
AOBMO e: Army Ordnance Ballistic Missile Office
AOBR e: Analog[ue] Output Buffer Register
A-O-B satellite e: Amsat-Oscar-B Satellite
AOBSR e: Air Observer
AO Buffer e: Address Out Buffer
AOC e: Advanced Office Computer
- e: Air Officer Commanding
- e: Air Oil Cooler
- e: Air Operation Central
- e: Air Operations Center
- e: All Ones Counter
- e: American Opthalmological Color
- e: American Optical Company
- e: American Orthoptic Council
- e: Anodal Opening Clonus
- e: Anodal Opening Contraction
- e: Arabian Oil Company
- e: Architect of the Capitol
- e: Army Operations Center
- e: Array Out Counter
- e: Attached to Other Correspondence
- e: Attitude and Orbit Control
- e: Auditor Overcharge Claims
- e: Automatic Operation Control
- e: Automatic Output Control
- e: Automatic Overload Circuit
- e: Automatic Overload Control[ler]
- e: Average Operating Cost
- e: Aviation Officer Candidate
- e: Awaiting Outgoing Continuity
A o C e: Academy on Computers
a. o. c. l: anno orbis conditi
AOCA e: American Osteopathic College of Anesthesiologists
AOCBAF e: Air Officer Commanding; Base Air Forces
AOCCA e: Air Officer Commanding; Coastal Area
AOCCC e: Air Officer Commanding-in-Chief; Coastal Command
AOCE e: Attitude and Orbit Control Electronics
AO-CHIDis e: Army Ordnance-Chicago District
AOCI e: Anodal Opening Clonus
AOC-in-C e: Air Officer Commanding-in-Chief
AOCL e: Average Outgoing Count Limit
AO-CLEVDis e: Army Ordnance-Cleveland District
AOCM e: Advanced Optical Counter Measures
- e: Aircraft out of Commission for Maintenance
AOCP e: Airborne Operational Computer Program[me]
- e: Aircraft out of Commission for [lack of] Parts
AOCR e: Advanced Optical Character Reader
- e: Advanced Optical Character Recognition
- e: Aircraft Operating Cost Report

AOCS e: Alpha Omega Computer System
- e: American Oil Chemists Society
- e: Atlantic Outer Continental Shelf
AOCT e: Analog[ue] Output Conversion Time
AOCU e: Arithmetic Output Control Unit
- e: Associative Output Control Unit
AOD d: aktionsorientierte Datenverarbeitung
- d: Argon-Oxygen Decarburization
- e: Academy of Oral Dynamics
- e: Acknowledgement of Debt
- e: Advanced Ordnance Depot
- e: Airfield Operations Designator
- e: Airlift Operations Directive
- e: Air Officer of the Day
- e: Analog[ue] Output Device
- e: Analog[ue] Output Differential
- e: AND-OR Delay
- e: Angle of Descent
- e: Apollo Operations Director
- e: Arithmetic Output Data
- e: Army Ordnance Depot
- e: Arterial Occlusive Disease
- e: Auriculo-Osteodysplasia
AODA e: Analog[ue] Output Driver Amplifier
ao diag e: acridine-orange diagnosis
AODICV e: American Optical Differential Interference Contrast Microscope
AODP e: Acquisition Orbit Determination Program[me]
AODS e: An Ordnance Destruct System
- e: Atlas Operational Data Summary
AODV d: aktionsorientierte Datenverarbeitung
AOE e: Add Order Execution
- e: Aerodrome of Entry
- e: Airborne Operational Equipment
- e: Auditing Order Error
AOEC e: Airways Operations Evaluation Center
AOEL e: Advanced Ocean Engineering Laboratory
AOES e: Advanced Orbit/Ephemeris Subsystem
- e: Arctic Ocean Environment Simulator
AOF d: Archiv für Orientforschung (journ.)
- e: Active Optical Fuze
- e: Advanced Operating Facility
- f: Afrique Occidentale Francaise
A of L e: Algebra of Logic
A of M e: Academy of Music
AOF/OPF e: Advanced Operating Facility/Operating Procedure Facility
A of P e: Arms of Precision
AOFS e: Active Optical Fuzzing System
A of S e: Academy of Science
- e: Angle of Sight
AÖFS d: Arbeitsgemeinschaft Öffentlicher Fotosammlungen
AOG e: Aircraft on Ground
- e: AND/OR Graph
- e: Arrival of Goods
- e: Atlantic Oceanographic Group
- e: Australian Oil and Gas
- e: gasoline tanker
AÖG d: Archiv für Östereichische Geschichte (journ.)
AOGB e: American Outpost in Great Britain
AOGeh d: Anordnung über Geheimpatente
AOGM e: Army of Occupation of Germany Medal
AOGMS e: Army Ordnance Guided Missile School
AOGO e: Advanced Orbiting Geophysical Observatory
ao. GV d: außerordentliche Generalversammlung
ao. H. d: außerordentlicher Haushalt
AOH e: Acid Open Hearth
- e: Add-On Header
- e: Apollo Operations Handbook
- e: Aviators Oxygen Helmet
- e: Awaiting Office Hours
- e: Awaiting Overhaul
ao. HV d: außerordentliche Hauptversammlung
AOI d: Arabische Organisation für Industrialisierung
- e: Accent on Information
- e: Acousto-Optical Imaging
- e: Advance Ordering Information
- e: AND/OR Invert[er]
- e: Angle of Incidence
- e: Automated Operator Interface
- e: Automated Outgoing Interface
A o I e: Aims of Industry
AOIC e: Assistant Officer in Charge
AOI-Gate e: AND-OR-Invert Gate
A Oil e: Aviation Oil
AOIPS e: Atmospheric and Oceanic Information Processing System
AOIT e: Automatic Outgoing Intertoll Trunk-test
AOIV e: Automatically Operated Inlet Valve
AOJ e: Acquire on Jam[ming]
ao. J.C. f: aprés Jésus-Christ
AOK d: Allgemeine Ortskrankenkasse
- d: Armeeoberkommando
- e: All OK
AOL e: Absence over Leave
- e: Admiralty Oil Laboratory
- e: Application Oriented Language
- e: Atlantic Oceanographic Laboratories
- e: Atlantic Oceanography Laboratories
AÖL d: Arbeitsgemeinschaft Österreichischer Lochkartenfachleute
äol d: äolisch
AOLM e: Apollo Orbiting Laboratory Module
AOLO e: Advanced Orbital Launch Operations
AOLR e: Amplifier Open Loop Response
AOLUG e: Arizona On-Line Users Group
AOLW/D e: American Optical Laser Welder/Driller
AOM e: Acoustic-Optic Modulator
- e: Active Oxygen Method
- e: Add-on Memory
- e: Aviation Ordnance Man
aoM d: außerordentliches Mitglied
A.O.M. e: Air over Motor
AOM1c e: Aviation Ordnance Man First Class
AOM2c e: Aviation Ordnance Man Second Class
AOM3c e: Aviation Ordnance Man Third Class
AOMA e: American Occupational Medical Association
AOMB e: Aviation Ordnanceman; Bombsight Mechanic
AOMC e: Army Ordnance Missile Center
- e: Army Ordnance Missile Command
AOME e: Assistant Ordnance Mechanical Engineer
ao. Mitgl. d: außerordentliches Mitglied
AOML e: Atlantic Oceanographic and Meteorological Laboratories
AOMM e: Add-on Main Memory
AOMPS e: Automatic Outgoing Message Processor System
AOMT e: Aviation Ordnanceman; Turret Mechanic
aö Münz d: anrufbarer öffentlicher Münzfernsprecher
ao. MV d: außerordentliche Mitgliederversammlung
AONB e: Area[s] of Outstanding Natural Beauty
AONET e: American Osteopathic Network
AoN-Prinzip d: Alles-oder-Nichts-Prinzip
AOO e: American Oceanic Organization
- e: Anodal Opening Odor
- e: Aviation Ordnance Officer
AOOSY e: Automatic Orbiting Operations System
AOP d: Gesellschaft für Analyse, Organisation und Programmierung
- e: Advanced Onboard Processor
- e: Aerospace Observation Platform
- e: Air Observation Plane
- e: Air Observation Post
- e: Allied Ordnance Publication
- e: Anodal Opening Picture
- e: Aortic-Pressure Pulse
- e: Apprenticeship Outreach Program[me]
- e: Armo[u]red Observation Post
- e: Assembly and Operation Plan
- e: Association of Optical Practitioners
- e: Association of Osteopathic Publications
- e: Astrophysical Observatory; Potsdam
- e: Automatic Operations Panel
a.o.P. d: außerordentlicher Professor
AoP e: Aortic Pressure
AOPA e: American Orthotic and Prosthetic Association
AOPC e: Australian Overseas Projects Corporation
A-operator e: Outward operator
AO-PHILDis e: Army Ordnance-Philadelphia District
AOPI e: AND OR Power Inverter
AOPL e: Association of Oil Pipe Lines
AO Pressure e: Aortic Pressure
ao.Prof. d: außerordentlicher Professor
AOPSA e: Advanced Optical Power Spectrum Analyzer
AOPV e: Air-Operated Plastic Valve
AOQ e: Average Outgoing Quality
AOQL e: Average Outgoing Quality Level
- e: Average Outgoing Quality Limit
AOQS e: Acousto-Optic Q-Switch
AOR e: Acknowledgement of Receipt
- e: Add One Right
- e: Aerial Ohmic Resistance
- e: Air / Oil Ratio
- e: Airborne Overland Radar
- e: Air Operations Room
- e: Alphanumeric Optical Reader
- e: Anchor Order
- e: Angle of Reflection
- e: Antenna Ohmic Resistance
- e: Argon-Oxygen Refining
- e: Army Operational Research
- e: Assembly Operations REcord
- e: Atlantic Ocean Region
- e: Auxiliary Oil Replenishment
AöR d: Anstalt des öffentlichen Rechts
- d: Archiv des öffentlichen Rechts
AORA e: Atlantic Ocean Recovery Area
Aorangi e: Mount Cook, New Zealand
AORC e: Association of Official Racing Chemists
AORD e: Astronaut Operations Requirement Document
AORF e: Amplifier Oscillator; Radiofrequency

AORL e: Apollo Orbital Research Laboratory
AOrPA e: American Orthopsychiatric Association
AORT e: Association of Operating Room Techniques
aort regurg e: aortic regurgitation
aort sten e: aortic stenosis
Ao-RV e: Aorto-Right Ventricle
AOS d: Aufgabe der optimalen Standardisierung
- e: Acquisition of Sight
- e: Acquisition of Signal[s]
- e: Activity Operating Schedule
- e: Add or Substract
- e: Advanced Office System
- e: Advanced Operating System
- e: Agricultural Organization Society
- e: Air Observer School
- e: Air Oil Separator
- e: Air Operations Specialist
- e: Algebraic Operating System
- e: Algebraic Operation System
- e: American Opthalmological Society
- e: American Oriental Society
- e: American Otological Society
- e: Amphibious Objective Studies
- e: Amplifier Output Stage
- e: Analog[ue] Output Submodule
- e: Angle of Sight
- e: Anodal Opening Sound
- e: Application-Oriented Subprogram[me]
- e: Astronomical Observatory Satellite
- e: Asynchronous Overlay Supervisor
- e: Atlantic Ocean Ship
- e: Azimuth Orientation System
AOSAP e: Airway Operations Specialist; Airport
AOSC e: Association of Oilwell Servicing Contractors
A-O Scale e: Anti-Oriental Scale
AOSE e: American Order of Stationary Engineers
AOSG e: Airways Operations Specialist; General
AOSL e: Authorized Organizational Stockage List
AOSM e: Associate of the Otago School of Mines
AOSML e: Army Ordnance Submarine Mine Laboratory
AOSO e: Advanced Orbiting Solar Observatory
AOSP e: Active Optics Simulation Program[me]
- e: Automatic Operating and Scheduling Program[me]
AÖSp d: Allgemeine Österreichische Spediteurbedingungen
AOSPS e: American Otorhinologic Society for Plastic Surgery
AOSPV e: Airways Operations Supervisor
AOS/RT e: Advanced Operating System/ Real Time
AOSS e: Active Optics Simulation System
- e: Advanced Operations Support System
- e: Airborne Oil Surveillance System
- e: Airways Operations Specialist
AOS/VS e: Advanced Operating System/Virtual Storage
AOT e: Acquisition on Target
- e: Active on Target
- e: Alameda-Oakland Tunnel
- e: Alignment Optical Telescope
- e: Analog[ue] Output Timer
- e: AND-OR-Trigger
- e: Angle on Target
- e: Anodal Opening Tetanus
- e: Antarctic Observation Team
- e: Askania Optical Tracker
- e: Association of Occupational Therapists
- e: Auxiliary Output Tester
- e: Average Operation Time
AOTA e: American Occupational Therapy Association
- e: Australian Overseas Trading Corporation
AOTC e: Accountant Officers Technical Course
AOTE e: Amphibious Operational Training Element
AOTe e: Anodal Opening Tetanus
AO Technique e: Acridine-Orange Technique
AOTOP e: Advent Orbital Test and Operation Plan
AOtS e: American Otological Society
AOTS e: Aviation Ordnance Test Station
AOTT e: All-Ordnance Thrust Termination
- e: Automatic Outgoing Trunk Test
AOU e: Air Operated Unit
- e: American Ornithologists Union
- e: Apparent Oxygen Utilization
- e: Arithmetic Output Unit
- e: Associative Output Unit
- e: Australian Ornithologists Union
- e: Azimuth Orientation Unit
AOV d: Arbeitsgemeinschaft Ostwestfälischer Versorgungsbetriebe; Gütersloh
- e: Amplified Output Voltage
- e: Analysis of Variance
AÖV d: Arbeitsgemeinschaft Österreichischer Verkehrsflughäfen
AOW d: akustische Oberflächenwellen
- e: Alabama Ordnance Works
- e: Army Ordnance Workshop
- e: Articles of War
AOW-Bauelement d: akustisches Oberflächenwellen-Bauelement
AOWC e: Army Ordnance Weapons Command
AOWS e: Aircraft Overhaul Work Stoppage
- e: Army Ordnance Workshop Services
AOX e: AND-OR Extender
AOZ d: Adresse ohne Zusatzinformation
- d: Arm-Ohr-Zeit
AÖZ d: Anodenöffnungszuckung
AP d: Abonnentenpunkt
- d: Abrufprogramm
- d: Abschlußprovision
- d: Abschreibung; proportional
- d: Adreßpuffer
- d: Aktionspotential
- d: algorithmisch-physikalisch
- d: Alkaline Phosphatase
- d: alkalische Phosphatase
- d: Allpaß
- d: Altpapier
- d: Ammonium Perchlorat[e]
- d: Ammonium Persulphate
- d: Amsterdamer Pegel
- d: Amtsperiode
- d: Analogpunkt
- d: Analysenprobe
- d: Anfangspunkt
- d: Anforderungsprofil
- d: Anhaltspunkt
- d: Anilinpunkt
- d: Anlegepunkt
- d: Anpassungsschaltung
- d: Anschaltepunkt
- d: Anschlußplatte
- d: Anschlußprozessor
- d: Anschlußpunkt
- d: Anwenderprogramm
- d: Anwendungsprogramm
- d: Anwendungsprojekt[e]
- d: Anwendungsprozessor
- d: Applikationspatent
- d: Applikationsprogramm
- d: Arbeitsplanung
- d: Arbeitsplatzdatei
- d: Arbeitsproduktivität
- d: Arbeitsprogramm
- d: Arbeitspunkt
- d: Archiv für Psychiatrie und Nervenkrankheiten (journ.)
- d: Asbestpappe
- d: Aufnahmepunkt
- d: Auftragspuffer
- d: Ausgabeprogramm
- d: Ausgabepuffer
- d: Ausschließungspatent
- d: automatische Programmierung
- e: A-Pulse
- e: Above Proof
- e: Absolute Peak
- e: Absolute Permeability
- e: Absolute Permitivity
- e: Absolute Program[me]
- e: Absolute Programming
- e: Absorbed Power
- e: Absorbing Power
- e: Absorptive Power
- e: Accelerometer Package
- e: Acceptable Program[me]
- e: Access Panel
- e: Access Permit[tee]
- e: Access Point
- e: Access Prefix
- e: Accidental Printing
- e: Accounting Period
- e: Accounting Process
- e: Accounts Payable
- e: Accurate Programming
- e: Acid-Proof
- e: Acid Phosphatase
- e: Acquisition Plan
- e: Action Post
- e: Action Potential
- e: Active Program[me]
- e: Acyclic Process
- e: Adapter Panel
- e: Adding Punch
- e: Additional Premium
- e: Addition Position
- e: Additive Process
- e: Add Packed
- e: Addressing Program[me]
- e: Address Part
- e: Address Portion
- e: Address Printing
- e: Address Program[me]
- e: Adjustable Point
- e: Administrative Processor
- e: Administrative Program[me]
- e: Admiralty Pattern
- e: Advanced Park
- e: Advanced Post
- e: Advanced Processor
- e: Advanced Procurement
- e: Advanced Programming
- e: Advance Payment
- e: Advance Preparation
- e: Aerial Pattern
- e: Aerial Port
- e: Aft[er] Perpendicular
- e: After Peak
- e: Aiming Point
- e: Aircraft Park
- e: Air Passage

- e: Airplane[s]
- e: Airplane Pilot
- e: Air Pollution (journ.)
- e: Airport
- e: Air Position
- e: Air Publication
- e: Algorithmic Procedure
- e: Alignment Periscope
- e: Alkaline Phosphatase
- e: All-Purpose
- e: Allied Papers
- e: Alphabetic Punch
- e: Alphanumeric Punching
- e: Alpha Particle [Spectrometer]
- e: Alteration Program[me]
- e: Alternate Print
- e: Alternate Program[me]
- e: Alternative Patch
- e: Alternative Program[me]
- e: Alum Precipated
- e: Alum Precipitated
- e: American Patent
- e: American Pharmacopoeia
- e: American Plan
- e: American Pyschologist (journ.)
- e: Amino-Purine
- e: Amino-Peptidase
- e: Ammonium Perchlorate
- e: Ammunition Point
- e: Amplitude Phase
- e: Anal Pore
- e: Analytical Plotter
- e: Andra Pradesh
- e: Angina Pectoris
- e: Angle Point
- e: Anomalous Propagation
- e: Answer Print
- e: Antenna Pattern
- e: Anterior Pituitary
- e: Anteroposterior
- e: Antipersonnel
- e: Aortic Pressure
- e: Aperture
- e: Apollo Program[me]
- e: Appendectomy
- e: Application Process[or]
- e: Application Program[me]
- e: Application Programming
- e: Applications Processor
- e: Applied Physics (journ.)
- e: Apply Pressure
- e: A Pulse
- e: Arbitrary Parameter
- e: Argument and Programming
- e: Argument Pointer
- e: Arithmetic[al] Progression
- e: Arithmetic Problem
- e: Arithmetic Processor
- e: Arithmetic Progression
- e: Armo[u]r Piercing
- e: Army Panel
- e: Aromatic Polyamide
- e: Array Processor
- e: Arrest Point
- e: Arterial Pressure
- e: Artificial Pneumothorax
- e: Ascent Phase
- e: As Prescribed
- e: As Purchased
- e: Assembled Program[me]
- e: Assembler Program[me]
- e: Assembly of Parties
- e: Assembly Program[me]
- e: Assignment Phase
- e: Assignment Problem
- e: Assignment Program[me]
- e: Assistance Program[me]
- e: Assistant Paymaster
- e: Associated Press
- e: Associative Processor
- e: Assumed Position
- e: Atmospheric Pressure
- e: Atomic Power[ed]
- e: Attached Processor
- e: Attenuator Pad
- e: Aurora Pump
- e: Authors Proof
- e: Automatic Pistol
- e: Automatic Program[me]
- e: Automatic Programming
- e: Automatic Punch[er]
- e: Automatic Punching
- e: Autopilot
- e: Auxiliary Point
- e: Auxiliary Program[me]
- e: Average Payable
- e: Aviation Pilot
- e: Awaiting Parts
- e: Axial Pitch
- e: Axiopulpal
- e: Azimuth; Predicted
- e: L-Aspartyl-L-Phenylalanine
- R: Aviapolk

Ap. d: Abpraller
- e: Apocalypse
- e: April
- l: Apostolus

a. P. d: als Prüfer

A.P. d: Amerikanisches Patent
- d: Anlaßpumpe
- e: Armo[u]r Plated

ap d: angewandte planung (journ.)
- e: apatite
- e: apothecaries
- e: apothecary
- e: appendices
- e: appendix
- e: approach
- i: anno passato

Ap d: Anschlußplan
- d: Apotheke[r]
- d: Apparat
- d: Aufputz
- d: Aufstellungsplan
- d: Auftragspost
- d: Auftreffpunkt
- e: Apogam
- e: Apparat[e]
- l: Apus

ap. e: apex
- e: apparent
- l: apud

a.P. d: auf Probe

a.p. d: außerplanmäßig
- e: almost periodical
- l: ante prandium

a. p. l: ad publicandum
- l: anni praesentis
- l: anni praeteriti
- l: ante partum

A & P e: Airframe and Powerplant
- e: Apogee and Perigee
- e: Atlantic and Pacific

A/P e: Aerials and Propagation
- e: Antennas and Propagation
- e: Automatic Pilot

a/p e: angle point
- e: authority to pay
- e: autopilot

A-P e: Arithmetic and Program[me]
- e: Auscultation and Percussion

a.-p. l: anterior-posterior

a_p e: geomagnetic index

AP1c e: Aviation Pilot First Class

AP2c e: Aviation Pilot Second Class

AP3c e: Aviation Pilot Third Class

APA d: alle Punkte adressierbar
- d: Arzneimittel-Prüfungsanstalt
- d: Austria Presse-Agentur
- e: Abort Programmer Assembly
- e: Advanced Program[me] Authorization
- e: Aerial Pattern Analyzer
- e: Agricultural Publishers Association
- e: Aircraft Plume Analysis
- e: Aldosterone-Producing Adenoma
- e: Allowance for Project Adjustment
- e: All Points Addressable
- e: American Pharmaceutical Association
- e: American Philological Association
- e: American Philosophical Association
- e: American Physiotherapy Association
- e: American Pilots Association
- e: American Podiatry Association
- e: American Polygraph Association
- e: American Poultry Association
- e: American Psychiatric Association
- e: American Psycho-analytical Association
- e: American Psychological Association
- e: American Psychosomatic Association
- e: American Psychotherapy Assoc iation
- e: American Pyrotechnics Association
- e: Amino-Penicillanic Acid
- e: Animation Producers Association
- e: Antenna Pattern Analyzer
- e: Anthracite Producers Association
- e: Antipernicious Anemia Factor
- e: Apache Railway
- e: Architectural Photographers Association
- e: Architectural Precast Association
- e: Associate in Public Administration
- e: Associate Principle for Addition
- e: Association for the Prevention of Addiction
- e: Association of Physicians of Australasia
- e: Augmented Predictive Analyzer
- e: Automatic Photographic Analysis
- e: Auxiliary Personnel Attack
- e: Axial Pressure Angle
- s: Aerovias Panamá

APAAC e: Asian-Pacific-American Advocates of California

APA Algorithm e: Arbitrary Precision Approximating Algorithm

APAC e: Aerial Photographic Analysis Center
- e: Aerial Pointing Angle Change
- e: Airborne Parabolic Arc Computer
- e: Airsearch Parabolic Analog[ue] Computer
- e: Airsearch Pneumatic Analog[ue] Computer
- e: Antenna Pointing Angle Change
- e: Asphalt-Plastic-Asphalt-Chip
- e: Automatic Positioning and Control

APACE e: Aldermaston Project for the Application of Computers to Engineering

APACHE e: Accelerated Project to Automate Critical Hardware Systems
- e: Accelerator for Physics and Chemistry of Heavy Elements
- e: Accelerator for Physics and Chemistry of Heavy Metals
- e: Analog[ue] Programming and Checking
- e: Application Package for Chemical Engineers

AP-Achse d: Außenplaneten-Achse

APACI s: Asociación Peruana para el Avance de la Ciencia

APACS e: Adaptive Planning and Control Sequence
- e: Airborne Position and Attitude Camera System

APAD d: 3-Acetylpyridin-Adenindinukleotid
- e: 3-Acetylpyridine-Adenine Dinucleotide

APADAS e: Automated Phase and Amplitude Data Acquisition System
APADS e: Automatic Programmer and Data System
- e: Automatic Programming and Data System

APAD System e: Application-Adaptable Database System
APAF e: Antipernicious Anemia Factor
APAG d: Aluminium-Präzisions-Armaturen-Guß (VEB; DDR)
A-Pakt d: Atlantikpakt
APAL e: Array Processor Assembly Language
AP & AM e: Adler Planetarium and Astronomical Museum
APAM e: Array Processor Access Method
APANZ e: Associate of the Public Accountants of New Zealand
APAP d: Acetyl-P-Amino-Phenol
- e: Apollo Propulsion Analysis Program[me]

AP/APC e: Associate Processor/Associative Processor Controller
APAR e: Adaptive Phase Array Radar
- e: Approved Program[me] Analysis Request
- e: Authorized Program[me] Analysis Report
- e: Automatic Placement and Routing
- e: Automatic Processing and Recording
- e: Automatic Program[me] Analysis Report
- e: Automatic Programming and Recording

apar e: Apparatus
APAREL e: A Parse Request Language
APARS e: Arab Passenger Airline Reservation System
a-part e: alpha particle[s]
Apart. e: Apartment
Aparts. e: Apartments
APAS e: Automated Programmable Assembly System
- e: Automatic Performance Analysis System
- e: Automatic Programmable Assembly System
- e: Automatic Protocol Analysis System

APAs e: American Polygraph Association members
APase d: alkalische Phosphatase
APAT e: Acetyl-Phenyl-Alanyl-Tyrosine
- e: Armo[u]r-Piercing Anti-Tank

APATS e: Aerial Pattern Test System
- e: Antenna Pattern Test System
- e: Automatic Programming [or Programmer] and Test System

APAVC e: Almost Periodic Amplitude Variation Coding
APAW d: Abhandlungen der Preussischen Akademie der Wissenschaften (journ.)
APB e: Active Position Backward
- e: Adrema Pitney Bowes
- e: Air Portable Bridge
- e: All-Points Bulletin
- e: Anti-Personal Bomb
- e: Anti-Phase Boundary
- e: Application Program[me] Block
- e: Associated Press Broadcasters
- e: Atrial Premature Beat

A.P.B. e: Absolute Premissive Block[ing]
APBA e: Associated Press Broadcasters Association
APBE d: Arbeitsplatzbelegung[sdaten]
APBH e: After Peak Bulkhead
APBHR e: American Part-Blooded Horse Registry
APBI e: Advanced Planning Briefing for Industry
APBIC e: Action for Prevention of Bum Injuries to Children
APBM d: Außenkabel, Papierisolation, Bleimantel
APBO d: Apotheken-Betriebsordnung
APBS e: Accredited Poultry Breeding Station
- e: Advanced Post Boost System

APB Scheme e: Absolute Permissive Block Scheme
APB Signalling e: Absolute Permissive Block Signalling
APBS Scheme e: Accredited Poultry Breeding Stations Scheme
APB System e: Absolute Permissive Block System
APBX e: Automatic Private Branch Exchange
APC d: Arbeitsplatzcomputer
- d: Aspirin-Phenacetin-Coffein
- e: Abbreviated Performance Campaign
- e: Accounting Processing Code
- e: Acoustical Phase Constant
- e: Acoustical Plaster Ceiling
- e: Acoustical Propagation Constant
- e: Adaptive Plate Cabinet
- e: Adaptive Predict[ive] Coding
- e: Adaptive Processing Control
- e: Additional Program[me] Cycle
- e: Address Plate Cabinet
- e: Address Punched Card
- e: Adenoidal-Pharyngeal-Conjunctival
- e: Adjustable Pressure Conveyor
- e: Advanced Performance Computer
- e: Advanced Personal Computer
- e: Advanced Profession Computer
- e: Advanced Programming Course
- e: Advanced Propulsion Cooling
- e: Aeronautical Planning Chart
- e: Agricultural Prices Commission
- e: Air-Pollution Control
- e: All-Purpose Capsule
- e: All Purpose Computer
- e: Alphabetic Print Control
- e: American Power Conference
- e: Ammonium Perchlorate
- e: Analog[ue]-to-Pressure Converter
- e: Angular Position Counter
- e: Annular Primary Combustor
- e: Antenna Pattern Correction
- e: Antiphlogistic Corticoid
- e: Aperture Current
- e: Appalachian Power Company
- e: Approach Control
- e: Approach Power Compensator
- e: Area Positive Control
- e: Arkansas Polytechnic College
- e: Armo[u]r-Piercing-Capped
- e: Armo[u]red Personnel Carrier
- e: Army Petroleum Center
- e: Army Pictorial Center
- e: Army Postal Clerk
- e: Aspirin, Phenacetin and Caffeine
- e: Assembled Program[me] Count
- e: Assembly Program[me] Code
- e: Assistant Principal Chaplain
- e: Associative Processor Control[ler]
- e: Atomic Power Constructions [Limited]
- e: Atrial Premature Contractions
- e: Augmented Parity Check
- e: Australian Personal Computer
- e: Auto-Pilot Controller
- e: Autographed Presentation Copy
- e: Automated Packaging Code
- e: Automated Production Control
- e: Automatically Prepunched Card
- e: Automatic Performance Control
- e: Automatic Peripheral Control
- e: Automatic Phase Control
- e: Automatic Picture Control
- e: Automatic Pitch Control
- e: Automatic Position Control
- e: Automatic Potential Control
- e: Automatic Pressure Conveyor
- e: Automatic Process Control
- e: Automatic Program[me] Control
- e: Automatic Punching of Cards
- e: Average Power Control

A/PC e: Auto-Pilot Capsule
apc e: virus adenoidal, pharyngeal, conjuctival virus
a. p. C. l: anno post Christum
APCA e: Air Pollution Control Association
- e: Anti-Parietal Cell Antibody
- e: Audio Peak Clipping Amplifier
- e: Australian Physical Culture Association
- e: Automatic Phono-Cardiac Analyzer

APCAPS e: Automated Payroll, Cost and Personnel System
APCB e: Air Pollution Control Board
APCBC e: Armo[u]r-Piercing-Capped; Ballistic Cap
APCC e: American Potash & Chemical Corporation
- e: Apollo Program[me] Control Center
- e: Association of Professional Computer Consultants

apc-c e: aspirin, phenacetin, caf feine-with codeine
APC Code e: Augmented Parity Check Code
APCD e: Air Pollution Control Department
- e: Air Pollution Control District

APCE e: Air-Pollution Control Equipment
APCEC e: Army Precommission Extension Course
APCEF e: Advanced Power Conversion Experimental Facility
APCG e: Aperture Plate Character Generator
- e: Association of Pacific Coast Geographers

APCH e: Approach
APCHE e: Armo[u]r-Piercing-Capped; High Explosive
- e: Automatic Programmed Checkout Equipment
- e: Automatic Programming Checkout Equipment

a. p. Chr. l: anno post Christum
a. p. Chr. n. l: anno post Christum natum
APCI e: Apollo Computer, Incorporated
A-PCI e: Armo[u]r-Piercing-Capped Incendiary
APCI-T e: Armo[u]r-Piercing-Capped-Incendiary with Tracer
APCL e: Atmospheric Physics and Chemistry Laboratory
- e: Atomic Power Construction, Ltd.

APCLC e: Armo[u]r-Piercing-Capped; Long Case
APCM e: Adaptive Pulse Code Modulation
- e: Adaptive Pulse Code Modulator
- e: Authorized Protective Connecting Module

APC-MLQ d: adaptive prädikative Codierung mit Maximum-Likelihood Quantisierung
- e: Adaptive Predictive Coding with Maximum-Likelihood Quantization

APCN e: Active Pulse Compression Network
a. p. C. n. l: anno post Christum natum

APCO e: Air Pollution Control Office
- e: Application Configurator
- e: Associated Public-Safety Communications Officer[s]
APCON e: Approach Control
APCP e: Activation Project Control Plan
- e: Aquatic Plants Control Program[me]
APC/QC e: Armo[u]red Personnel Carrier/Qualification Course
APCR e: Alabama Power Company Reactor
- e: Apollo Program[me] Control Room
- e: Automatic Punching Card Recording
APC Regulations e: Air Pollution Control Regulations
APCS e: Approach Power Compensator System
- e: Associative Processor Computer System
- e: Attitude and Pointing Control System
- e: Automated Process Control System
APCSE e: Advanced Power Conversion Skid Experiment
APCT e: Armo[u]r-Piercing-Capped with Tracer
APCTL e: Application Control
- e: Autopilot Control
APCV e: Air Piloted Control Valve
APC-Viren e: Adenoidal-Pharyngeal-Conjunctival-Viren
APD d: Arbeitsplatzdatei
- d: Avalanchephotodiode
- e: Above Platen Device
- e: Acquisition Processor Display
- e: Action Potential Duration
- e: Active Position Down
- e: Additional Program[me] Drum
- e: Adjustable Pitch Device
- e: Advanced Planning Document
- e: Advanced Program[me] Development
- e: Aiming Point Determination
- e: Air Pollution Division
- e: Airport Directory
- e: Air Procurement Directive
- e: Air Procurement District
- e: Air to Pneumatic Distribution
- e: Albany Port District [Railroad]
- e: Amplitude Probability Distribution
- e: Analog[ue]-to-Pulse Duration
- e: Angular Position Digitizer
- e: Anteroposterior Dianmeter
- e: Apollo Program[me] Directive
- e: Apollo Project Directive[s]
- e: Approach Progress Display
- e: Approximate Pore Diameter
- e: Area Postal Directory
- e: Atomic Products Division
- e: Authors and Printers Dictionary
- e: Automated Powder Diffractometer
- e: Auxiliary Power Distribution
- e: Avalanche Photodiode
- e: Average of Percent Differences
- e: Average Particle Diameter
- e: Average Pore Diameter
- e: Average Power Dissipation
AP & D e: Advanced Planning and Design
APDA e: American Parkinsons Disease Association
- e: Army Physical Disability Activity
- e: Auxiliary Pump-Drive Assembly
APDAR e: Advanced Polarization Diversity Autotrack Receiver
APDC e: Agricultural Pest Destniction Council
- e: Air Procurement District Commander
- e: Ammonium Pyrrolidine-Dithio Carbonate
- e: Ammonium Pyrrolidine Dithiocarbamate
APDE e: Atomic Power Development Association
APDF e: Apollo Propulsion Development Facility
- e: Asian-Pacific Dental Federation
APDL e: Algorithmic Processor Description Language
APDM e: Associative Push Down Memory
APDP e: Apollo Program[me] Definition Phase
APdrugs e: Antiparkinsonian Drugs
APDS e: Advanced Personal Data System
- e: Advanced Personnel Data System
- e: Advanced Planning Data Sheet
- e: American Pediatric Society
- e: Armo[u]r-Piercing Discarding Sabot
APDs e: Amplitude Probability Distributions
- e: Avalanche-Photodiodes
APDSMS e: Advanced Point Defense Surface Missile System
APDU e: Application Protocol Data Unit
- e: Association of Public Data Users
APE d: Anpassungseinheit
- d: Arbeitsplatzerfassungsbeleg
- d: automatische Programmladeeinrichtung
- d: automatische Prüfeinrichtung
- e: Adapted Physical Educator
- e: Advanced Production Engineering
- e: Aerial Port of Embarkation
- e: Air Pollution Engineer
- e: Alfven Propulsion Engine
- e: Amalgamated Power Engineering
- e: Aminophylline, Phenobarbital, Ephedrine
- e: Ammunition Peculiar Equipment
- e: Anchor Placement Equipment
- e: Anterior Pituitary Extract
- e: Army Premilinary Evaluation
- e: Assistant Project Engineer
- e: Atomic Photoelectric Effect
- e: Automatic Photomapping Equipment
- e: Automatic Positioning Equipment
- e: Autonomous Processing Element
- e: Available Potential Energy
- e: Available Power Efficiency
APe d: Anwende[r]programme
APEA e: Aerial Pattern Error Analysis
- e: Antenna Pattern Error Analysis
- e: Association of Professional Engineers
- e: Australian Petroleum Exploration Association
APEC e: All-Purpose Electronic Computer
- e: Automated Procedures for Engineering Consultants
Apeco e: American Photocopy Equipment Company
APE Computer e: Arizona Pedagogical Computer
AP Ed e: Associate in Physical Education
APED e: Atomic Power Equipment Department
APEI e: Associated Poultry and Egg Industries
APEL e: Aeronautical Photographic Experimental Laboratory
APEn d: Anpassungseinheiten
APEO d: Alkylphenolethoxylat[e]
APEQS e: Aerial Photography of the Eclipse of the Quiet Sun
APER e: Air Pollutant Emissions Report
apers. d: altpersisch
apers e: antipersonnel
APERT e: Aperture
AP-ESSQ e: All-Pass Error Spectrum Shaping Quantizer
APET e: Application Program[me] Evaluator Tool
APEX e: Additive System of Photographic Exposure
- e: Advance-Purchase Excursion
- e: Air Pollution Exercise
- e: All-Purpose Electronic X-ray-computer
- e: All-Purpose Rocket for Collecting Atmospheric Soundings Piggyback Emulsion
- e: Amplitude and Phase Extraction
- e: Amplitude Phase Extractor
- e: Analysis of Performance and Expense
- e: Assembler and Process Executive
- e: Assembler and Program[me] Executive
- e: Automated Planning and Execution
APEXC e: All Purpose Electronic X-ray Computer
APEX Computer e: All-Purpose Electronic X-ray Computer
APEX Plan e: Advance Purchase Excursion Plan
APF d: Alkohol-Probefrühstück
- d: Anti-Perniziosa-Faktor
- d: Archiv für das Post- und Fernmeldewesen
- d: Archiv für Papyrus-Forschung und verwandte Gebiete (journ.)
- e: Accurate Position Finder
- e: Acid Proof Floor
- e: Active Position Forward
- e: Adjustable Pawl Fastener
- e: Advanced Printer Function
- e: Advanced Program[me] Feature
- e: Animal-Protein Factor
- e: Anterior Pericardial Fluid
- e: Area Protect Feature
- e: Artillery Preparation Fire
- e: Association of Pacific Fisheries
- e: Atomic Packing Factor
- e: Australian Pharmaceutical Formulary
- e: Authorized Program[me] Facilities
- e: Automatic Press Feed
- e: Automatic Program[me] Finder
- e: Automatic Program[me] Finding
APFA e: Accelerator-Pulsed Fast Assembly
- e: American Pipe Fittings Association
- e: Associated Poultry Farmers of Australia
- f: Association des Professeurs Franco-Americains
APFC e: Air Publications and Forms Centre
- e: American Plant Food Council
- e: Asia-Pacific Forestry Commission
- e: Assembly Process Flow Chart
- e: Automatic Phase and Frequency Control
APFCS e: Automatic Power Factor Control System
APFD e: Autopilot and Flight Director (Marconi-Elliot)
APFM e: All Purpose Fast Memory
- e: Australian Postgraduate Federation in Medicine
AP FORTRAN e: Array Processor Formula Translation
APFR e: Automatic Programmable Film Reader
APFRI e: American Physical Fitness Research Institute
APFS e: Auto Program[me] Find System
APFSD Sabot e: Armo[u]r Piercing, Fin-Stabilized, Discarding Sabot
AP fuel e: Ammonium Perchlorate Fuel
APFX e: Apply Fixture
APG d: Algopupillogramm
- d: allgemeiner Programmgenerator
- d: Anregungspulsgenerator
- d: Apothekengesetz
- d: Arbeitsplatzgruppe
- d: Äthyl-Polyglykol
- d: Auswerte-Programmgenerator

- e: Aberdeen Proving Ground
- e: Accessory Pedal Ganglia
- e: Aerial Power Gain
- e: Air Proving Ground
- e: American Programmers Guild
- e: Antenna Power Gain
- e: Application Program[me] Generator
- e: Army Proving Ground
- e: Australian Proving Ground
- e: Automated Process Generator
- e: Automatic Priority Group[ing]
- e: Automatic Program[me] Generator
- e: Azimuth Pulse Generator

APG/BRL e: Aberdeen Proving Ground / Ballistic Research Laboratory
APGC e: Air Proving Ground Center
- e: Air Proving Ground Command

APGCE e: Air Proving Ground Center-Eglin Air Force Base
APGCU e: Autopilot Ground Control Unit
APGES e: Application Program[me] Generating and Executive System
APG-HEL e: Aberdeen Proving Ground-Human Engineering Laboratory
APG/HEL e: Aberdeen Proving Ground-Human Engineering Laboratory
APG-OBDC e: Aberdeen Proving Ground - Ordnance Bomb Disposal Center
APGOEF e: Air Proving Ground-Eglin; Florida
APG-OTC e: Aberdeen Proving Ground-Ordnance Testing Command
APGr d: Arbeitsplatzgruppe
APGrn d: Arbeitsplatzgruppen
APGS e: Apollo Propellant Gaging System
APH e: Access Permit Holder
- e: Acetylphenylhydrazine
- e: Active Position Home
- e: Antepartum Hemorrhage
- e: Anterior Pituitary Hormone
- e: Apethic
- e: Aphasia
- e: Aphelion
- e: Automatic Parts Handler
- e: Aviators Protective Helmet

A Ph e: Associate in Philosophy
APHA e: American Printing History Association
- e: Australian Pneumatic and Hydraulic Association

APhA e: American Pharmaceutical Association
A-PhÄnomen d: Aubert-Phänomen
A-Phase d: Ausführungsphase
AP-HC e: Armo[u]r Piercing-Hard Core
a.ph.c. e: Automatic Phase Control
APHE e: Armo[u]r-Piercing; High Explosive
APHEBC e: Armo[u]r-Piercing; High Explosive; Ballistic Cap
aphet e: aphetic
APHF e: American Poultry and Hatchery Federation
APHFFF e: Ames Prototype Hypersonic Free Flight Facility
AphG d: Antennensystem mit phasiertem Gitter
APHIS e: Animal and Plant Health Inspection Service
APhO e: Area Pharmaceutical Officer
aphp e: antipseudomas human plasma
aphro[s] e: aphrodisiac[s]
APHS e: Australian Psychology and Hypnotherapy Association
APHV e: Armo[u]r-Piercing; Hyper Velocity
API d: Anwendersoftware-Paket ISI (industrielles Steuerungs- und Informationssystem)
- d: Anwendungsprogramm-Interface
- d: außerplanmäßige Instandhaltung
- e: Academic Press, Inc.
- e: Acceptable Periodic Inspection
- e: Advance Performance Interceptor
- e: Aerial Photointerpretation
- e: Aerial Position Indicator
- e: Air Pollution Index
- e: Air Position Indicator
- e: Alabama Polytechnic Institute
- e: Alignment Progress Indicator
- e: American Paper Institute
- e: American Paramedical Institute
- e: American Petroleum Institute
- e: American Pipe Institute
- e: American Potash Institute
- e: American Press Institute
- e: AND Power Inverter
- e: Angle Position Indicator
- e: Antecedent Precipitation Index
- e: Antenna Position Indicator
- e: Application Program[me] Interface
- e: Architectural Periodicals Index
- e: Armo[u]r-Piercing-Incendiary
- e: Assembly Program[me] Instruction
- e: Associate of the Plastics Institute
- e: Australian Pharmaceutical Industries
- e: Australian Planning Institute
- e: Automated Pronunciation Instructor
- e: Automatic Priority Interrupt
- e: Automatic Program[me] Interrupt
- e: Automatic Programming Instruction
- f: Association Phonetique Internatinnale

APIA e: Animal Protection Institute of America
- e: Arkansas Poultry Improvement Association

APIC e: Allied Press Information Center
- e: American Petroleum Industries Committee
- e: Apollo Parts Information Center
- e: Army Photo Interpretation Center
- e: Asian Packaging Information Centre
- e: Association for Practitioners in Infection Control
- e: Automatic Power Input Control
- e: Automatic Power Input Controller
- e: Automatic Programming Information Center

APICON e: Aircraft Position Information Converter
APICS e: Air Pollution Information and Computer System
- e: American Production and Inventory Control Society

API/CS e: Application Program[me] Interface/Communication Services
APICSC e: Atlantic-Pacific Interoceanic Canal Study Commission
apicult e: apiculture
APID e: Army Photo Interpretation Detachment
API Degree e: American Petroleum Institute Degree
APIE e: American Psycho / Info Exchange
- e: Antioch Program[me] for Interracial Education

APIF e: Automated Process Information File
APILIT e: American Petroleum Institute Literature
A Pill e: Abortion Pill
APIM f: Association Professionelle Internutionaledes Médecins
APIN e: Atlas Propulsion Information Notice
APIP e: Apollo Personnel Identification Program[me]
- e: Apollo Personnel Investigation Program[me]

APIPOCC e: Appropriating Property in Possession of Common Carrier
APIR e: American Petroleum Institute Research
APIS e: Air Position Indicating Station
- e: Alternative Press Information Service
- e: Architectural and Planning Information Service
- e: Array Processing Instruction Set
- e: Automated Process Interrupt System

APISC e: Atlantic-Pacific Interoceanic-Canal Study Commission
API-T e: Armo[u]r-Piercing-Incendiary with Tracer
APIU e: Air Photographic Interpretation Unit
- e: Army Photo Interpretation Unit

APIVR e: Artificial Pacemaker Ventricular Rhythm
APJ e: American Power Jet
- e: Angle Panel Jack
- e: Astrophysical Journal

APJE e: Association of Philosophical Journals Editors
APK d: Äthylen-Propylen-Kautschuk
- d: Auslandspostkarte
- e: Accelerometer Package
- e: Amplitude and Phase Keying (or Keyed)

APk e: Aircraft Park
APK System e: Amplitude Phase-Keyed System
APL d: Ablaufplanungsverfahren
- e: A Program[me] Language
- e: A. Programming Language
- e: Absolute Power Level
- e: Absolute Program[me] Loader
- e: Actual Parameter List
- e: Actuarial Programming Language
- e: Additional Programming Language
- e: Adult Performance Level
- e: Advanced Product Line
- e: Advanced Programming Language
- e: Advance Procurement List
- e: Aero Propulsion Laboratory
- e: Air Pacific Limited
- e: Airplane
- e: Air Provost Marshal
- e: Akron Public Library
- e: Albany Public Library
- e: Albuquerque Public Library
- e: Algorithmic Procedural Language
- e: Algorithmic Programming Language
- e: All-Pass Loop
- e: All-Purpose Linotype
- e: Allowance Parts List
- e: All Purpose Linotype
- e: Aluminum-Polythene Laminate
- e: Anterior Pituitary Like Hormone
- e: Aperture Lip
- e: Applications Programming Language
- e: Applied Physics Laboratory
- e: Approval
- e: Approved Parts List
- e: Armo[u]r Plate
- e: Army Personnel Letter
- e: Army Promotion List
- e: Artificial Programming Language
- e: Asian Program[me] Library
- e: Assembly Parts List
- e: Assembly Program[me] Language
- e: Assembly Programming Language

- e: Association of Programmed Learning
- e: Associative Programming Language
- e: Astronomical Position Line
- e: Augusta Public Library
- e: Automated Programming Language
- e: Automatic Phase Lock
- e: Automatic Production Line
- e: Automatic Program[me] Load[ing]
- e: Automatic Programming Language
- e: Automotive Pigeon Loft
- e: Average Picture-signal Level
- e: Average Picture Level

Apl. d: Absatzplan[ung]
- d: Arbeitsplan[ung]
- d: Ausbildungsplan
- d: Ausbildungsplatz
- e: April

A-Pl d: Ausbauplanung
apl. d: außerplanmäßig
APLA d: Arbeitsplanausführung
APLA Bull. e: Atlantic Provinces Library Association Bulletin (journ.)
APLAC e: Analysis Program[me] Linear Active Circuits
APLADA d: Arbeitsplandatei
- d: Arbeitsplatzdatei

APL (Appl. Phys. Lab.) Tech. Di
- e: Applied Physics Laboratory Technical Digest (journ.)

AP & Lat e: Anterior, Posterior & Lateral
APLC e: Army Propulsion Laboratory and Center
- e: Automated Parking Lot Control

APLD e: Automatic Program[me] Locate Device
APLE e: Association of Public Lighting Engineers
APLET e: Association for Programmed Learning and Educational Technology
APlFG d: Ausbildungsplatzförderungsgesetz
APL Hormone e: Anterior Pituitary Like Hormone
APL/JHU e: Applied Physics Laboratory / Johns Hopkins University
APLL e: Analog[ue] Phase-Locked Loop
- e: Automatic Phase-Locked Loop

APlLW d: Auslands-Platzleitungswähler
APLM e: Asynchronous Pulse-Length Modulation
APLN e: Apollo Program[me] Logic Network
- e: Armo[u]r-Piercing; Long Nose

aplns e: applications
APLO e: Aerial Port Liaison Officer
- e: Automation Planning and Liaison Office

apl. Prof. d: außerplanmäßiger Professor
APLQ e: Applique
APLS e: American Plant Life Society
- e: American Private Line Services, Inc.
- e: American Psychology Law Society
- e: Apparel Performance Level Standards
- e: A Programming Language-Structured
- e: Automated Process Line System
- e: Automatic Printing Line Selection
- e: Automatic Process Line System

APLSTATPACK e: Advanced Programming Language Statistical Package
APL Substance e: Anterior Pituitary Like Substance
APLSV e: A Programming Language with Shared Variables
aplt e: applicant
APLT e: Automatic Programming Language Translation
APLU e: Army Pay Ledger Unit
- e: Automatic Program[me] Loading Unit

APLUM e: A Programming Language/University of Massachusetts
ApM d: Anschläge pro Minute
APM e: Absorption Power Meter
- e: Academy of Physical Medicine
- e: Academy of Psychosomatic Medicine
- e: Accurate Piston Motion
- e: Acid Precipitable Material
- e: Add Punch Machine
- e: Advanced Program[me] Monitor
- e: Aerial Pattern Measurement
- e: Aerial Positioning Mechanism
- e: Aim-Point-Miss
- e: Air Permeability Meter
- e: Air Power Museum
- e: Air Provost Marshal
- e: Alarm Panel Monitor
- e: Alumin[i]um Powder Metallurgy
- e: Aluminium Powder Material
- e: Amplitude Phase Modulation
- e: Analog[ue] Panel Meter
- e: Antenna Pattern Measurement
- e: Antenna Positioning Mechanism
- e: Anti Personnel Missile
- e: Apomict
- e: Applied Magnetics
- e: Applied Mechanics
- e: Arbitrary Precision Multiplication
- e: Arithmetic Processing Module
- e: Army Procost Marshal
- e: Assistant Paymaster
- e: Assistant Provost Marshal
- e: Association for Psychoanalytic Medicine
- e: Associative Principle for Multiplication
- e: Automated Processing Method
- e: Automatic Plate Modulation
- e: Automatic Plugging Meter
- e: Automatic Printing Machine
- e: Automatic Programming Machine
- e: Auxiliary Processing Machine
- e: Average Potential Model
- e: L-Aspartyl-L-Phenylalamine Methylester

APMA e: Advance Payment of Monetary Allowance in Lieu of Transportation Authorized
- e: Automatic Phonograph Manufacturers Association

APMB e: Armo[u]r-Piercing; Mono Block
APMBA e: Assistant Project Manager for Business Administration
APMC e: Academy of Physchologists in Marital Counseling
- e: Automatic Permanent Magneticfield Compensator

A/P MCU e: Autopilot Monitor and Control Unit
APME d: automatische Prüf- und Meßeinrichtung
- e: Area Precipitation Measurement Equipment
- e: Automated Processing of Medical English

APMG e: Assistant Postmaster General
APMI e: American Powder Metallurgy Institute
- e: Area Precipitation Measurement Indicator

APMIS e: Automated Project Management Information System
APMMRI e: Automatic Point Marking, Measuring and Recording Instrument
A/P MON e: Autopilot Monitor
APMR e: Association for Physical and Mental Rehabilitation
APMS e: Advanced Power Management System
- e: Automatic Pulse Measurement System

Ap Msgr e: Airplane Messenger

APMT e: Advanced Planetary Mission Technology
- e: Aerial Pattern Measurement Test
- e: Antenna Pattern Measurement Test
- e: Antenna Pattern Measuring Test

APn d: Anschlußplatten
APN e: All Pass Network
- e: American Practical Navigator
- e: Apron
- e: Artificial Pneumothorax
- e: Augmented Proportional Navigation
- e: Authorized Part Number
- e: Average Peak Noise

Apn e: Apocryphon
APNA e: American Power Net Association
APNEU e: Auxiliary Pneumatic
APNIC e: Automatic Programming National Information Center
APNL e: Army Personnel Newsletter
APNP e: Arthurs Pass National Park
APNSS e: Analog[ue] Private Network Signalling System
ApO d: Apotheken-Ordnung
APO d: Allgemeine Prüfungsordnung
- d: Ausbildungs- und Prüfungsordnung
- e: Accountable Property Officer
- e: Acting Pilot Officer
- e: Action Print Only
- e: Advanced Post Office
- e: Advanced Programming Option
- e: Airforce Post Office
- e: Air Pilot Officer
- e: American Potash & Chemical
- e: Animal Procurement Officer
- e: Annual Program[me] Objectives
- e: Airport
- e: Apogee
- e: Apolipoprotein
- e: Apollo Project Office
- e: Apomorphine
- e: Area Petroleum Officer
- e: Army Pay Officer
- e: Army Post Office
- e: Asian Productivity Organization
- e: Assembly Production Order
- e: Association of Physical Oceanographers
- e: Association of Physical Oceanography
- e: Australian Post Office
- e: Automatic Paging Optimizer
- e: Automatic Power Off
- e: Auxiliary Patrol Officer

APOAF e: All Present Or Acconted For
APOB e: Airplane Observation
APOC e: Anglo-Persian Oil Company
- e: Army Point of Contact

Apoc e: Apocalypse
APOD e: Aerial Port of Debarkation
APOE e: Aerial Port of Embarkation
ApoG d: Apothekengesetz
apog e: Apogee
APOL e: Advanced Program[me]-Oriented Language
- e: Aerospace Program[me]-Oriented Language

apol e: apologete
- e: apologetic[al]
- e: apologetics
- e: apologia
- e: apologize
- e: apology

APOLLO e: Article Procurement with On-Line Local Ordering
APON e: Association of Pediatric Oncology Nurses
APOP e: Apollo Preflight Operations Procedures

APOS d: allgemeine polytechnische Oberschule (DDR)
- e: Adaptive Problem-Oriented Software
- e: Advanced Polar Orbiting Satellite
A-Position e: Outgoing Position
APOSS e: Automated Point of Sale System
- e: Automatic Point of Sale System
a post l: a posteriori
ÄPost d: Ämter des Postwesens
Apostles e: Apostle Islands, Lake Superior
- e: New Testament disciples
APOT e: Annulus-Pressure-Operated Tester
APOTA e: Automatic Positioning of Telemetering Aerial
- e: Automatic Positioning of Telemetering Antenna
apotek a: apoteket
apoth e: apothecaries
- e: apothecary
ApothBetrO d: Apothekenbetriebsordnung
A-Powered e: Atomic-Powered
APP d: Aneurin-Pyrophosphat
- d: Anwenderprogrammpaket[e]
- d: argininreiches Polypeptid
- e: A Posteriori Probabilities
- e: Accelerated Pacification Program[me]
- e: Access Point Pace
- e: Actual Parameter Part
- e: Advance[d] Planetary Probe
- e: Advanced Parts Procurement
- e: Advanced Project Planning
- e: Advance Procurement Plan
- e: Aerial Position Programmer
- e: Aggregate Production Planning
- e: Air Parcel Post
- e: Air Pollution Potential
- e: Algonquin Provincial Park
- e: Alphabetic Printing Punch
- e: Antenna Position Programmer
- e: Antipodal Propagation Phenomena
- e: Appearance
- e: Apprehend
- e: Approach
- e: Armo[u]r Piercing Proof
- e: Army Procurement Procedure
- e: Association of Professional Photogrammetrists
- e: Association Parallel Processor
- e: Associative Parallel Processor
- e: Atomic Power Plant
- e: Automatical Prepunching
- e: Automatic Plate Processor
- e: Automatic Position Planning
- e: Automatic Power Protection
- e: Automatic Priority Processing
- e: Auxiliary Pneumatics Panel
- e: Auxiliary Power Plant
App d: Apparat[ur]
- e: Appendage
- e: Appendix
- e: Apperception
- e: Appetite
App. d: Appartement
- d: Apposition
- l: Apostoli
app e: apparatus
- e: apparel
- e: apparent[ly]
- e: append[ed]
- e: appetizer
- e: applicant
- e: applied
- e: appoint[ed]
- e: appropriate
- e: appropriation
- e: approval
- e: approve[d]
- e: approximate[ly]
- e: approximation
app. e: apprentice
APPA e: Advise Present Position and Altitude
- e: American Public Power Association
- e: Artificial Pilot Phased Array
appar e: apparatus
appar. e: apparent[ly]
APPARAT e: Archive Preservation Program[me] and Retrieval by Automated Techniques
APPA Relief Valve e: Axial Pneumatic Pilot Assisted Relief Valve
APPAUC e: Association of Physical Plant Administrators of Universities and Colleges
APPC e: Advanced Program[me]-to-Program[me] Communication[s]
- e: Automatic Power Plant Checker
App. Chem. e: Applied Chemistry
APPCON e: Approach Control
APPC/PC e: Advanced Program[me]-to-Program[me] Communication[s] for Personal Computer
APPD e: Approved
- e: Aviation Personnel Planning Data
APPECS e: Adaptive Pattern-Perceiving Electronic Computer System
append e: Arithmetic and Controls
APPF e: Additional Print Position Feature
App. Fabr. d: Apparatefabrik
APPI e: Advance[d] Planning Procurement Information
APPIS e: Arkansas Public Policy Information System
Appl. d: Applikant
- d: Applikation
APPL e: Advance Procurement Planning List[s]
- e: Appellant
- e: Applicable
- e: Appliance
- e: Applicant
- e: Application
- e: Applied
Appl. Biochem. Microbiol. e: Applied Biochemistry and Microbiology (journ.)
APPLE e: Advanced Propulsion Packaged Liquid Engine
- e: Analog[ue] Phased Processing Loop Equipment
- e: Apollo Payload Exploration
- e: Ariane Passenger Payload Experiment (India)
- e: Associative Processor Programming Language Evaluation
Appl. Electr. Phenom. e: Appl. Electrica Phenomena (journ.)
APPLIC e: Application
applican l: applicandus
Applic. Engr. e: Application Engineer
Appl. Microbiol. e: Applied Microbiology (journ.)
appln e: application
Appl. Opt. e: Applied Optics (journ.)
Appl. Phys. e: Applied Physicist
Appl. Phys. Lett. e: Applied Physics Letters (journ.)
Appl. Polym. Symp. e: Applied Polymer Symposia (journ.)
Applsch. d: Applikationsschule
Appl. Spectrosc. e: Applied Spectroscopy (journ.)
Appl. Ther. e: Applied Therapeutics (journ.)
App ME e: Applied Mechanics Engineer
Appmt e: Appointment
APPN e: Advanced Peer-to-Peer Network
- e: Appropriation
App^{no} i: Appennino
AP & PO e: Advance Programming and Proposal Operations
A/P POI e: Autopilot Positioning Indicator
APPON e: Application
Appos. d: Apposition
APPP d: Abhandlungen der Philosophie, Psychologie und Pädagogik (journ.)
- e: Association of Planned Parenthood Physicians
APPR e: Appreciation
- e: Apprehend
- e: Apprehension
- e: Apprentice[ship]
- e: Approval
- e: Approve
- e: Approve[d]
- e: Approximate[ly]
- e: Approximation
- e: Army Package Power Reactor
appr. d: appretieren
- d: appretiert
- d: approbieren
- d: approbiert
- e: approximativ
Appr. d: Appretur
- e: Approbation
appren e: apprentice
APPRES e: Acoustic Projector Pressure Relief and Equalization System
appro e: approbation
- e: approval
approb. d: approbieren
- d: approbiert
Approb. e: Approbation
approp e: appropriation
APPROX e: Approximate[d]
- e: Approximate[ly]
- e: Approximating
- e: Approximation
- e: Approximative
Approx. d: Approximation
approx. d: approximativ
APPROXN e: Approximation
apprs e: approaches
APPS e: Australian Physiological and Pharmacological Society
- e: Australian Plant Pathology Society
- e: Automated Packaging Planning System
apps e: appendices
- e: appendixes
APPT e: Alberta Pilot Projects for Television
- e: Appoint[ment]
apptd e: appointed
apptmt e: appointment
APPTU e: All-Pakistan Post and Telegraph Union
APPU e: Air Photo Production Unit
- e: Application Program[me] Preparation Unit
APPV e: Approve
APPVL e: Approval
APPX e: Appendix
appy e: appendectomy
APQ e: American Philosophical Quarterly (journ.)
- e: Arithmetic Processor Queue
- e: Available Page Queue

APR d: Abwehrproteinase-Reaktion
- d: Allgemeine Papier-Rundschau (journ.)
- d: April
- d: Arbeitsplatzrechner
- d: Arithmetikprozessor
- d: Ausgabe-Pufferregister
- d: automatische Pegelregelung
- d: automatisches Produktions-Registriersystem
- e: Absolute Position Register
- e: Accelerator-Pulsed Reactor
- e: Acoustic Paramagnetic Resonance
- e: Active Page Register
- e: Active Position Return
- e: Active Prominence Region
- e: Additional Platen Roller
- e: Advanced Production Release
- e: Aerial Pattern Recorder
- e: Aerial Photographic Reconnaissance
- e: Aerial Position Recorder
- e: Agency Progress Report
- e: Airborne Profile Recorder
- e: Airman Performance Report
- e: Air Priority
- e: Air Priority Rating
- e: All Purpose Room
- e: Alphanumeric Page Reader
- e: Alternate Path Reentry
- e: Alternate Path Retry
- e: Amoebic Prevalence Rate
- e: Analog[ue] Parameter Record
- e: Annual Percentage Rate
- e: Annual Progress Report
- e: Antenna Aperture
- e: Antenna Pattern Recorder
- e: Antenna Position Recorder
- e: Anterior Pituitary Reaction
- e: Anti-Plugging Relay
- e: Apollo Program[me] Requirements
- e: Apprentice
- e: Assembly Program[me] Run
- e: Assigned Procurement Responsibility
- e: Association of Petroleum Refiners
- e: Automated Punch Requisition
- e: Automatic Passbook Reader
- e: Automatic Passbook Recording
- e: Automatic Pattern Recognition
- e: Automatic Production Recording
- e: Automatic Production Registering
- e: Automatic Programming and Recording
- e: Available Power Response

apr. d: apriorisch
APr d: Asbestpapier
- d: automatisches Prüfen

Apr e: April
- R: Aprel

apr i: aprile
a. pr. l: anni praesentis
- l: anni praeteriti

a pr. l: a principio
- l: a priori

APRA d: Auswahlprogrammrahmen
- e: Army Pulsed Reactor Assembly

APRAC e: Air Pollution Research Advisory Committee
APRANET e: Advanced Research Projects Agency Network
APRC e: Army Physical Review Council
a. p. R. c. l: anno post Romam conditam
APRCAS e: All-Purpose Rocket for Collecting Atmospheric Soundings
APRCH e: Approach
APRDC e: Army Polar Research and Development Center
APrE d: automatische Prüfeinrichtung
APRE e: Aerospace Photographic Reconnaissance Equipment
- e: Air Procurement Region; Europe
- e: Army Personnel Research Establishment

APrEARkpl d: automatische Prüfeinrichtung für Auslandsrichtungskoppler
APrEF d: automatische Prüfeinrichtung für Einrichtungen der Fernwahl
APrE/KN d: automatische Prüfeinrichtung für Koppelnetze
APrEL d: automatische Prüfeinrichtung für Leitungen
APREQS e: Approval Requests
APrERg d: automatische Prüfeinrichtung für Register
APrE/S d: automatische Prüfeinrichtung für Sätze
après JC f: après Jesus Christ
APrETW d: automatische Prüfeinrichtung für Telegrafenwählnetze
APrEUmw d: automatische Prüfeinrichtung für Umwerter
APrEZIG d: automatische Prüfeinrichtung für Zählimpulsgeber
APrF d: Altpreussische Forschungen (journ.)
APRF e: Army Pulsed Reactor Facility
- e: Army Pulse Radiation Facility

APRFE e: Air Procurement Region; Far East
APRG e: Aerial Phenomena Research Group
APRI e: Air Priority
a pri. l: a priori
APRIL e: Aquaplanning Risk Indicator for Landings
- e: Associative Processor with Integrated Logic
- e: Automatically Programmed Remote Indication Logged
- e: Automatically Programmed Remote Indication Logging

A/Prin e: Assistant Principal
APRIS e: Aluminium Company of America Picture-phone Remote Information System
Apr J-C f: Après Jésus-Christ
APRL e: American (or Army) Prosthetic Research Laboratory
APrLW d: Auslandsprüfleitungswähler
APRM e: Application Programmers Reference Manual
- e: Automatic Position Reference Monitor
- e: Average Power Range Monitor

APRN e: Armo[u]r-Piercing; Round Nose
APRNT e: Apparent
APrO d: Allgemeine Prüfungsordnung
APRO e: Aerial Phenomena Research Organization
APROC e: Adaptive Processor
aprop e: appropriate
APrPl d: Automatischer Prüfplatz
APRR e: Association for Planning and Regional Reconstruction
APRRN e: Advance Personnel Requirements Research Note
APRS e: Affirmative Poll Response State
AprS e: American Proctologic Society
APRS e: Applied Physics Research Section
- e: Association for the Preservation of Rural Scotland
- e: Association of Professional Recording Studios
- e: Automatic Position Reference System
- e: Automatic Production Recording System

APrSLD d: automatischer Prüfsatz für Leitungen-Durchgangsverkehr
- d: automatischer Prüfsatz für Leitungen für Durchgangsbetrieb

APrSLE d: automatische Prüfsätze für Leitungen-Endverkehr
- d: automatischer Prüfsatz für Leitungen für Endvermittlungsstellen

APRST e: Automatic Progressive Register Sender Tester
APR System e: Automatic Production Recording System
APRT e: Adenine Phosphoribosyltransferase
- e: Airport

APRTA e: Associated Press Radio [and] Television Association
APRU e: Asia Pacific Research Unit
APrUe d: Auslandsprüfübertragung
APrUE d: automatische Prüfübertragungen
APRVL e: Approval
APRX e: Approximate[ly]
APRXLY e: Approximately
APS d: Aminopenizillansäure
- d: Analyse- und Prognosesystem
- d: anlagenbezogenes Programmsystem
- d: Anwenderprogrammiersprache
- d: Anwenderprogrammsystem
- d: anwendungsorienterte Programmiersprache
- d: Applikationsprogrammsystem
- d: Arbeitsplatzsystem
- d: automatischer Probenspeicher
- d: automatisches Prüfsystem
- d: automatisierte Produktionssteuerung
- e: Aboriginal Protection Society
- e: Absolute Pressure Sensor
- e: Academy of Political Science
- e: Accelerated Photosynthetic System
- e: Accelerometer Parameter Shift
- e: Accessible Program[me] Sequence
- e: Accessory Power Supply
- e: Accessory Power System
- e: Actual Parade State
- e: Actual Plus Sign
- e: Additional Program[me] Step
- e: Addressed Packet System
- e: Adenosine Phosphosulfate
- e: Advanced Personnel System
- e: Advanced Photosynthetic System
- e: Advanced Propellant System
- e: Aerial Pointing Subsystem
- e: Aerodynamic Particle Size
- e: Airborne Power Supply
- e: Airborne Power System
- e: Airborne Pulse Search
- e: Air Pollution Syndrome
- e: Air Pressure Switch
- e: Airraft Prepared for Service
- e: Allegheny Power System
- e: Alphanumeric Photocomposer System
- e: American Metal Products
- e: American Pediatric Society
- e: American Pheasant Society
- e: American Philosophical Society
- e: American Physical Society
- e: American Physics Society
- e: American Physiological Society
- e: American Phytopathological Society
- e: American Plant Selections
- e: American Polar Society
- e: American Postal Service
- e: American Primrose Society
- e: American Proctologic Society
- e: American Prosthodontic Society
- e: American Psychosomatic Society
- e: Ammonium Persulfate

- e: Amplifier Power Supply
- e: Angstrom Pyrheliometric Scale
- e: Angular Position Sensor
- e: Anode Potential Stabilized
- e: Antenna Pointing Subsystem
- e: Apollo Program[me] Specifications
- e: Appearance Potential Spectroscopy
- e: Application[s] Program[me] Sampler
- e: Application[s] Program[me] System
- e: Application Process Subsystem
- e: Applied Peripheral Systems, Incorporated
- e: Arc Plasma Spray[ing]
- e: Armo[u]r-Piercing Shell
- e: Army Pilot School
- e: Army Postal Service
- e: Array Processing Subroutine
- e: Array Processor Software
- e: Array Processor System
- e: Ascent Propulsion [Sub]system
- e: Assembly Programming System
- e: Assimilations per Second
- e: Assistant Private Secretary
- e: Associate of the Pharmaceutical Society
- e: Associate of the Philosophical Society
- e: Atmospheric Pollution Sensor
- e: Atmospheric Protection System
- e: Atomic Power Station
- e: Attached Processor for Speech
- e: Attached Processor System
- e: Attended Pay Station
- e: Attitude Propulsion Subsystem
- e: Australian Psychological Society
- e: Autograph Postcard Signed
- e: Automated Patching System
- e: Automated Production System
- e: Automatically rotatable Phase Shifter
- e: Automatic Page Search
- e: Automatic Patching System
- e: Automatic Picture Setting
- e: Automatic Pilot System
- e: Automatic Piot System
- e: Automatic Program[me] Search
- e: Automatic Programming System
- e: Auxiliary Power Supplies
- e: Auxiliary Power Supply
- e: Auxiliary Power System
- e: Auxiliary Program[me] Storage
- e: Auxiliary Propulsion System
- e: Avionics Processing System

APs e: American Psychologist (journ.)
A & PSA e: Aden and Protectorate of South Arabia
APSA e: Advanced Polysilicon Self-Aligned
- e: American Political Science Association
- e: Associate of the Photographic Society of America
- e: Association of Professional Scientists of Australia
- e: Austitralian Pharmaceutical Sciences Association
- e: Australian Political Studies Association
- s: Aerolíneas Peruanas SA
- s: Azufrera Panamericana SA

APsaA e: American Psychoanalytic Association
AP & SC e: Army Port and Service Command
APSDs e: Auto-Power Spectral Densities
APSE e: Abstracts of Photographic Science and Engineering
- e: Abstracts of Photographic Sciences and Engineering Literature
- e: Ada Programming and Support Environment
- e: Automatic Programming and Scaling of Equations

AP-SE e: Armo[u]r Piercing-Special Effects
APSFSL e: Assistant Private Secretary to the First Sea Lord
APSG e: Area Program[me] Support Group
APSGB e: Associate of the Philosophical Society of Great Britain
APSGD e: Army Procurement-Sharpe General Depot
APSI e: Advanced Propulsion Subsystem Integration
- e: Aircraft Propulsion Subsystem Integration
- e: Aircraft Propulsion System Integration
- e: Allstates-Programming and Systems, Incorporated
- e: Ampere[s] per Square Inch

APS-INC e: Alphanumeric Publishing System, Incorporated
APSK d: Arbeitsplanstammkarte
- e: Amplitude and Phase Shift Keyed
- e: Amplitude and Phase Shift Keying

APSKn d: Arbeitsplanstammkarten
APSL e: Access Path Specification Language
- e: Acting Paymaster Sub-Lieutnant

APSM d: autonomes Programmentwicklungssystem für Mikrocomputer
- e: Auxiliary Power Supply Module

APSM Letters e: Additional Program[me] Support Materials Letters
APSN e: Association Package Sequence Number
APSO e: A Priori Stochastic-Optimal
APSP e: Array Processing Subroutine Package
- e: Array Processor Subroutine Package

APSQ e: Advance Payment of Subsistence and Quarters
APSR e: American Political Science Review (journ.)
- e: Axial Power Shaping Rod

APS Radar e: Airborne Pulse Search Radar
APSRS e: Aerial Photography Summary Record System
APSS d: Arbeitsplanstammsätze
- d: automatisches Programmsuchsystem
- e: Academy of Political and Social Science
- e: Advanced Planetary Spacecraft System
- e: American Polled Shorthorn Society
- e: Analysis and Programming for Space Systems
- e: Association for the Psychophysiological Study of Sleep
- e: Automated Program[me] Support System
- e: Automatic Program[me] Search System

APSSNM e: Advisory Panel on Safeguarding Special Nuclear Material
A.P.S.T. e: Associate in Public Service Technology
APS Technique e: Arc Plasma Spray Technique
APSTLGRU e: Air Postal Group
APSTRAT e: Aptitude Strategies
APSTRON e: Air Postal Squadron
A/P STSPG e: Autopilot Test Programmer
APSU e: Air Port Signal Unit
APSW d: altes Programmstatuswort
APsych e: Acta Psychologica (journ.)
APsychoA e: American Psychoanalytic Association
APsychosomS e: American Psychosomatic Society
APsychpthA e: American Psychopathological Association
AP-System d: Applikationsprogrammsystem
APT e: Actual Parameter Area
- e: Adaptive Programming Technology
- e: Advanced Passenger Train
- e: Advanced Patent Technique
- e: Advanced Pilot Training
- e: African Petroleum Terminals Ltd.
- e: After Peak Tank
- e: Airborne Pointer-Tracker
- e: Airman Proficiency Test
- e: Airport [Control] Tower
- e: All-Purpose Terminal
- e: All Picture Transfer
- e: Alum-Precipitated Toxoid
- e: Aminopropylisothiourea
- e: Analog[ue] Pressure Transducer
- e: Analog[ue] Program[me] Tape
- e: Angular Position Transducer
- e: Apartment
- e: Apollo Pad Test
- e: Appoint[ed]
- e: Appointment
- e: Aptitude
- e: Arizona Photopolarimeter Telescope
- e: Armo[u]r-Piercing with Tracer
- e: Army Parachute Team
- e: Army Pay Tables
- e: Asia Pacific Telecommunications
- e: Augmented Programming Training
- e: Automate Picture Taking
- e: Automatically Programmed Tool[s]
- e: Automatically Programmed Tooling
- e: Automatic Parts Testing
- e: Automatic Photoelectric Telescope
- e: Automatic Photometric Telescope
- e: Automatic Picture Taking
- e: Automatic Picture Transmission
- e: Automatic Planning and Technology
- e: Automatic Position Telemetering
- e: Automatic Position Telemetry
- e: Automatic Programmed Tooling
- e: Automatic Programmed Tools
- e: Automatic Programming for Tools
- e: Automatic Programming of Tools
- e: Automatic Programming Technique
- e: Automatic Programming Tool
- e: Automatic Progression Testing
- e: Automation Planning and Technology
- e: Automotive Professional Training

APTA e: American Physical Therapy Association
- e: American Public Transit Association
- e: American Public Transport Association
- e: Aptitude Area

APT-AC e: Automatically Programmed Tool-Advanced Contouring
APT Ammo e: Armo[u]r-Piercing and Tracer Ammunition
APT-BP e: Automatically Programmed Tool-Basic Positioning
APTC e: Airport Traffic Controller
- e: Automatically Programmed Tool Compiler

APTD e: Aid to the Permanently and Totally Disabled
Aptdo s: apartado
APT-E e: Advanced Passenger Train-Experimental
APTE e: Automatic Production Test Equipment
APTF e: Automated Program[me] Testing Facility
APTG e: Adult Porcine Tyorid Gland
apti e: actions per time interval
APTI e: Association of Principals of Technical Institutions
- e: Automatic Print Transfer Instrument

APTIC e: Air Pollution Technical Information Center
APT-IC e: Automatic Programmed Tool-Intermediate Contouring
APTIF e: Average Process Time Inverted File
APT-IGS e: Automatically Programmed Tool-Integrative Graphics System
ÄPTK d: Äthylen-Propylen-Terpolymer-Kautschuk
APT-Kautschuk e: Äthylen-Propylen-Terpolymer-Kautschuk
APT Language e: Automatically Programmed Tool Language
APTO e: Alumin[i]um Plastic Tearoff
a. pto. l: anno passato
APTP e: Arithmetic Proficiency Training Program[me]
APTPDA e: Advance Payment of Travel per Diem Authorized
APTR e: Advanced Pressure Tube Reactor
- e: Association Pointer
- e: Automatic Picture Transmission Receiver
AP-Trockner d: Allphasen-Trockner
APTS e: Advanced Passenger Trains
- e: Apartments
- e: Application Program[me] Test System
- e: Automatically Programmed Tool System
- e: Automatic Picture Transmission Subsystem
- e: Automatic Picture Transmission System
- e: Automatic Programmer and Test System
- e: Automatic Programming Testing System
APTS System e: Anti-Pollution Transfer and Storage System
A/P TSTMN d: Autopilot Test Monitor
A/P TSTPG e: Autopilot Test Programmer
APT System e: Automatically Programmed Tool System
- e: Automatic Picture Transmission System
A-PTT e: Activated Partial Thromboplastin Time
APTT e: Apollo Parts Task Trainer
APTU e: African Postal and Telecommunication[s] Union
- e: Air Transport Pressurizing Unit
- e: Australian Postal and Telecommunications Union
APTUS e: Apparatus
APTW e: Asiatic-Pacific Theater of War
APU d: Ausgabepuffer
- d: automatische Programmunterstützung
- e: Accessory Power Unit
- e: Acoustica Propellant Utilization
- e: Active Position Up
- e: Additional Peripheral Unit
- e: After Power Unit
- e: Airborne Power Unit
- e: Airborne Processing Unit
- e: Analog[ue] Processing Unit
- e: Analytical Processing Unit
- e: Applied Psychology Unit
- e: Arab Postal Union
- e: Arithmetic Processing Unit
- e: Arithmetic Processor Unit
- e: Army Postal Unit
- e: Asynchronous Processing Unit
- e: Audio Playback Unit
- e: Auxiliary Power Unit
- e: Auxiliary Processing Unit
- e: Auxiliary Propulsion Unit
APUB e: American Public Utilities Bureau
APUD e: Amine Precursor Uptake Decarboxylation
APUHS e: Automatic Program[me] Unit; High-Speed
APULS e: Automatic Program[me] Unit; Low-Speed
APUPA e: Alien, Penumbral, Umbral, Penumbral, Alien
APUS e: Analysis and Programming for Unmanned Satellites
- e: Auxiliary Power Unit Subsystem
- e: Auxiliary Power Unit System
A-Put e: Associate Pulmonary Technologist
APV d: Alarm- und Prüfsteuersignal-Verarbeitung
- e: Accidents, Poisonings and Violence
- e: Air Piloted Valve
- e: Approve
- e: Armo[u]red Personnel Vehicle
- e: Automatic Patching Verification
- e: Average Particle Volume
APVD e: Approved
APVDC e: Association of Parents of Vaccine-Damaged Children
APVL e: Approval
APV-Measurement e: Auto-Photovoltaic Measurement
APVO d: Ausbildungs- und Prüfungsverordnung
APW d: Arbeitsgruppe für physikalische Weltraumforschung
- e: Abnormally Polarized Waves
- e: Absorption Peak Wavelength
- e: Accelerated Public Works
- e: American Prisoner of War
- e: Anti-Personnel Weapon
- e: Apia, Western Samoa
- e: Architectural Projected Window
- e: Artifical Pond Water
- e: Artificial Pond Water
- e: Augmented Plane Wave
APWA e: American Public Works Association
A/PW Converter e: Analog[ue]-to-Pulsewidth Converter
APWD e: Aircraft Proximity Warning Device
APWI e: Air Prisoner of War Interrogation
APWL e: Automatically Processed Wire List
APWM e: Augmented Plane Wave Method
APWO e: Assistant Public Works Officer
APWP e: Accelerated Public Works Program[me]
APWR e: American Polish War Relief
APWS e: Aircraft Proximity Warning System
- e: Artificial Pond Water with additional Saccharose
APWSS e: Asian-Pacific Weed Science Society
APX e: Ampex
- e: Ampex [Corporation]
- e: Appendix
- e: Approximate[ly]
- e: Automatic Programming-System Extended
APZ d: Anwendungs-Programm-Zentrum
- d: automatischer Peilzusatz
- e: Assiniboine Park Zoo
APZF d: Arbeitsplatzzeitfond
AQ d: adaptive Quantisierung
- d: Ausführungsqualität
- d: Ausgabequittung
- e: Accomplishment Quotient
- e: Achievement Quotient
- e: Acquisition
- e: Acquisition Message
- e: Acquittance
- e: Adaptive Quantization
- e: Address Queue
- e: Aircraft Quality
- e: Air Quenched
- e: American Quarterly (journ.)
- e: Analog[ue] Quantity
- e: Antarctica
- e: Any Quantity
- e: Apollo Qualification
- e: Aqueous
- e: Assimulatory Quotient
- e: Asynchronous Quenching
- e: Attainment Quotient
- e: Australian Quarterly (journ.)
Aq. d: Aquarium
- e: Aqueduct
aq. l: aqua
Äq d: Äquator
äq. d: äquatorial
Äq. d: Äquivalent
- d: Äquivalenz
AQA e: Air Quality Act
AQAB e: Air Quality Advisory Board
AQAP e: Allied Quality Assurance Publication[s]
AQAPs e: Allied Quality Assurance Publications
Aqar e: Aquarius
aq. astr. l: aqua adstricta
AQB of Tests e: Army Qualification Battery of Tests
aq. bull. l: aqua bulliens
AQC e: Airmans Qualification Card
- e: Air Quality Criteria
- e: Associate of Queens College
- e: Automatic Quench Calibration
- e: Automatic Quench Correction
aq. cal. l: aqua calida
AQCL e: Analytical Quality Control Laboratory
aq. corn. l: aqua communis
AQCR e: Air Quality Control Region
AQD e: Analysis of Quantitative Data
aq. dest. l: aqua destillata
AQDHS e: Air Quality Data Handling System
AQDM e: Air Quality Dispersion Model
- e: Air Quality Display Model
AQE e: Airmen Qualifying Examination
- e: Allocated Queue Element
AQF e: Air Quality Forecast
- e: Automatic Query Formulation
aq. ferv. l: aqua fervens
aq. fluv. l: aqua fluvi
aq. font. l: aqua fontana
aq. fort. l: aqua fortis
aq. frig. e: aqua frigida
AQGV e: Azimuth Quantized Gated Video
AQl d: Auslandsquer[verbindungs]leitung
AQL e: Acceptable Qualifying Level
- e: Acceptable Quality Level
- e: Advanced Query Language
- e: Approved Quality Level
- e: A Query Language
- e: Average Quality Level
- e: Average Quality Limit
- e: Average Quality of [the] Lot
Aql. l: Aquila
AQL-g d: Auslandsquerverbindungsleitung-gehend
AQL-k d: Auslandsquerverbindungsleitung-kommend
AQM e: Air Quartermaster
- e: Assistant Quartermaster
- e: Atmospheric Quality and Modification
- e: Atmospheric Quarterly and Modification
aq. mar. l: aqua marina
aq. ment pip. l: aqua menthae

AQ Message e: Acquisition Message
AQMG e: Assistant Quartermaster General
AQN e: Armor Qualification Number
- e: Azimuthal Quantum Number
aq. niv. l: aqua nivalis
AQP d: Akquisitionsphase
- e: A Data Base Management System Query Processor
aq. pluv. l: aqua pluvialis
Äq. Pr. e: Äquator[ial] Projection
aq. pur. l: aqua pura
AQR d: Ausgabequittungsregister
- e: Air Quality Region
- e: Assembly Quality Record
Aqr e: Aquarius
Aqr. l: Aquarius
AQREC e: Army Quartermaster Research and Engineering Command
aq. regia l: aqua regia
AQS e: Air Quality Standards
- e: Approximate Quadratic Search
- e: Automated Quotation System[s]
AQ SOL e: Aqueous Solution
AQT e: Acceptable Quality Test
- e: Airman Qualifying Test
- e: Applicant Qualification Test
aq. tep. l: aqua tepidis
aqu e: aqueous
Aqu l: Aquila
Äqu. d: Äquivalent
aqua. e: aquaria
- e: aquarium
- e: aquatic
Aqua e: Aquarium
- l: Aquarius
aqua e: Aquatic
aquacult e: aquaculture
aquar e: aquarium
Aquar e: Aquarius
AQUARIUS e: A Query and Retrieval Interactive Utility System
aque e: aqueduct
Aqued e: Aqueduct
Äqu.-Gew. d: Äquivalentgewicht
AQUIS e: Acquisition
- e: Aquaculture Information System
Äquiv. d: Äquivalent
- d: Äquivalenz
Äquiv.-Gew. d: Äquivalentgewicht
AR d: Abderhaldensche Reaktion
- d: Abfrageregister
- d: Abgasrückführung
- d: Abraum
- d: Abrufroutine
- d: Abschreibungsrate
- d: Abstimmanzeigeröhre
- d: Adaptivregler
- d: Adreß-Rechnung
- d: Adressenregister
- d: Adreßregister
- d: Ahnenreihe
- d: akkumulatives Register
- d: Aktienrecht
- d: Aktionsradius
- d: Alkalireserve
- d: allgemeines Register
- d: Amateur Radio
- d: Amtsrichter
- d: Analogrechner
- d: Anheizrelais
- d: Antireflektion
- d: Anzeigeröhre
- d: Arbeit-Ruhe-Kontakt
- d: Arbeitsrechner
- d: Arbeitsrecht
- d: Arbeitsregister
- d: Archivrat
- d: A Register
- d: Artillerieregiment
- d: Ausgaberegister
- d: Ausgangsrechnung
- d: Ausgleichsrechnung
- d: ausregelbar
- d: Außenhandelsrecht
- d: automatische Regelung
- d: automatischer Regler
- d: autonome Republik
- e: Accelerator Research
- e: Access Request
- e: Accomplishment Ratio
- e: Accounting Routine
- e: Accounts Receivable
- e: Accounts Register
- e: Accumulating Reproducer
- e: Accumulative Register
- e: Accumulator Register
- e: Achievement Ratio
- e: Acid-Resisting
- e: Acid Rutile
- e: Acknowledgement of Receipt
- e: Acoustic Reflex
- e: Acquisition Radar
- e: Acrylic Rubber
- e: Action Register
- e: Active Rate
- e: Active Resistance
- e: Active Routine
- e: Activity Resume
- e: Actual Range
- e: Addend Register
- e: Addition Register
- e: Add Register
- e: Address Range
- e: Address Register
- e: Administrative Rule
- e: Administrative Ruling
- e: Admiralty Representative
- e: Advanced Reactor
- e: Advice of Receipt
- e: Advisory Report
- e: Aerial Reconnoiterer
- e: Aerial Refueling
- e: Aerial Resistance
- e: Aerodynamic Report
- e: Aerolineas Argentinas
- e: Aeronautical Radio [Navigation]
- e: Aeronautical Research
- e: Age Replacement
- e: Aircraft Rocket
- e: Airman Records
- e: Airman Recruit
- e: Air Raid
- e: Air Receive
- e: Air Reconnaissance
- e: Air Refueling
- e: Air Regulator
- e: Air Requirement
- e: Air Resistance
- e: Airship Rigger
- e: Alarm Reaction
- e: Allocated Reserve
- e: Allocation Register
- e: Allocator Routine
- e: Allowance Regulations
- e: All Rail
- e: All Risks
- e: Alphabetic Register
- e: Alphanumerical Reader
- e: Alphanumeric Representation
- e: Alternate Route
- e: Ambiguity Resolver
- e: Amendment Request
- e: American Rationalist (journ.)
- e: Ammunition Reserve
- e: Amphibian Reconnaissance
- e: Amplifier
- e: Amplifier Research
- e: Analog[ue] Representation
- e: Analytical Reagent
- e: Analyzed Reagent
- e: Annual Register
- e: Annual Report
- e: Annual Return
- e: Annual Review
- e: Anode Reaction
- e: Anodic Reaction
- e: Antenna Resistance
- e: Anti-Racketeering
- e: Anti-Reflection
- e: Anti-Reflective
- e: Anticipatory Response
- e: Aortic Regurgitation
- e: Apical-Radial
- e: Application Routine
- e: Applied Research
- e: Approved for Release
- e: Archaeological Reports (journ.)
- e: Architectural Review (journ.)
- e: Area of Resolution
- e: Argentina
- e: Arithmetic Register
- e: Arkansas
- e: Armed Reconnaissance
- e: Armed Robbery
- e: Armo[u]red Reconnaissance
- e: Army Receiving-Valve
- e: Army Regulation[s]
-e: Array
- e: Arrester
- e: Arrival [and Return]
-e: Arrive[d]
- e: Arsenophyrite
- e: Artificial Respiration
- e: Aspect Ratio
- e: As Required
- e: Assembly and Repair
- e: Assembly Routine
- e: Assigning Range
- e: Assignment Routine
- e: Assistant Resident
- e: Associate in Retailing
- e: Associative Register
- e: Asynchronous Repeater
- e: Atlantic Records
- e: Attack and Release
- e: Attendance Record
- e: Attention Routine
- e: Attenuation Reaction
- e: Audio Response
- e: Augmentation Reaction
- e: Automatic Radio
- e: Automatic Reader
- e: Automatic Restart
- e: Automatic Rifle[man]
- e: Automatic Routine
- e: Autoregressive
- e: Auxiliary Register
- e: Auxiliary Routine
- e: Aviation Radionavigation
- e: Avionic[s] Requirement[s]
- e: Awaiting Reply
- e: Axial Ratio
- e: Azimuth Ridge

a. R. d: alte[r] Regel[ung]
- d: altes Recht
- d: am Rande
- d: am Rennsteig
- d: am Rhein
- d: auf Rechnung
- d: auf Reise[n]
a.R. d: außer Reichweite
ar. d: altrussisch
- d: arabisch
Ar. d: Araber
- d: Arabien
- d: Aroma
Ar d: Argon
- e: Argon
- e: Aryl
ar d: aromatisch
- e: area
- e: aromatic
A & R e: Analysts and Researchers
- e: Approved and Removed
- e: Arithmetic and Register
- e: Assault and Robbery
A.R. e: Antwerp-Rotterdam
A. R. l: Academiae Rector
a. r. l: anno regni
Ar e: Arab
- e: Arabia[n]
- e: Arabic
- e: Aragon
A/R N: Aksjeredert
a/r e: at the rate [of]
ARA d: Abrechnungsautomat
- d: Abwasserreinigungsanlage
- d: Amsterdam-Rotterdam-Antwerpen
- d: Arbeitskreis Rundfunk e.V. Aachen
- e: Abbreviated Registered Address
- e: Abnormal Return Address
- e: Academy of Rehabilitative Audiology
- e: Accelerated Readiness Analysis
- e: Accelerated Relaxation Algorithm
- e: Accredited Rural Appraiser
- e: Actual Range Angle
- e: Adapter; Right Angle
- e: Aerial Rocket Artillery
- e: Aerospace Research Associates
- e: Aerospace Research Association
- e: Agricultural Research Administration
- e: Airborne Radar Approach
- e: Airborne Radar Attachement
- e: Airborne Receiving Aerial
- e: Airborne Receiving Antenna
- e: Airborne Rocket Artillery
- e: Aircraft Replaceable Assemblies
- e: American Radio Association
- e: American Railway Association
- e: American Rationalist Association
- e: American Recreational Activities
- e: American Republics Area
- e: American Rheumatism Association
- e: Analog[ue] Radar Absorber
- e: Antireceptor Antibody
- e: Antwerp-Rotterdam-Amsterdam
- e: Area Redevelopment Act
- e: Associate of the Royal Academy
- e: Association of Radiologists of Australasia
- e: Asynchronous Ripple Adder
- e: Automatic Route Advance
- e: Auxiliary Recovery Aerial
- e: Auxiliary Recovery Antenna
- e: Average Response Amplitude
- s: Armada República Argentiná
Ara e: Arabic
- e: Arabinose
- e: Argentina
Ara-A e: 9-ß-D-Arabinofuranosyladenine
ARAA e: Aerodrome Radar Approach Aid[s]
- e: Aerodrome Radio Approach Aid[s]
- e: Airdrome Radio Approach Aid[s]
- e: American Registry of Architectural Antiquities
ARAACOM e: Army Anti-Aircraft Command
ARAAV e: Armo[u]red Reconnaisance Airborne Vehicle
ARAB e: A-Register A-Bit
Arab. d: Araber
- d: Arabien
- e: Arabia[n]
- e: Arabic
arab. d: arabisch
Arab Emirates e: United Arab Emirates
Arabsat e: Arab Communications Satellite
ARABSAT e: Arab Satellite Communications Organization
Ara-C e: 1-ß-D-Arabinofuranosylcytosine
ARAC e: Aerospace Research Applications Center
- e: Area Airports Checked
- e: Army Radar Approach Control
- e: Array Reduction Analysis Circuit
- e: Associate of the Royal Agricultural College
Arach. e: Arachnology
ARACI e: Associate of the Royal Australian Chemical Institute
ARAD e: Airborne Radar and Doppler
- e: Alpha Research and Development
- e: Automated Requirements Allocation Data
- e: Average Response Amplitude Data
ARADCOM e: Army Air Defense Command
ARADMAC e: Army Aeronautical Depot Maintenance Center
ARADSCH e: Army Air Defense School
AR Ae S e: Associate of the Royal Aeronautical Society
ARAeSI e: Associate of the Royal Aeronautical Society and Institution
Arafura e: Arafura Sea between Australia and New Guinea
AR AGT e: Army Agent
ARAIA e: Associate of the Royal Australian Institute of Architects
Arak d: Atomrakete
Aral e: Aral Sea
ARAL e: Aromate and Aliphate
- e: Associate of the Royal Academy of Literature
- e: Automatic Record Analysis Language
ARAM e: Analog[ue] Random Access Memory
- e: Associate of the Royal Academy of Music
ARAMCO e: Arabian-American Oil Company
ARAMIS e: American Rheumatism Association Medical Information System
ARAND e: Analysis of Random Data
AR and D e: Advanced Research and Development
- e: Aeronautical Research and Development
- e: Air Research and Development
ArAnS d: Anrufanschaltesatz
Aransas e: Aransas National Wildlife Refuge, Texas
ARAP e: Annual Review in Automatic Programming
- e: Astronaut Rescue Air Pack
ARAPT e: Atlantic Research Automatic Position Telemetering
ARARAT e: Alternative Research in Architecture, Resources, Art and Technology
Ararat e: Mount Ararat
ARAS d: aufsteigendes (oder aszendierendes) retikuläres Aktivierungssystem
- e: Ascending Reticular Activating System
- e: Associate of the Royal Aeronautical Society
- e: Associate of the Royal Agricultural Society
- e: Associate of the Royal Astronomical Society
ARAT e: Automatic Random Access Transport
A-Raum d: Aufnahmeraum
ARAV e: Army Aviator
Arava e: Israeli light transport plane
AR Av Bad e: Army Aviator Badge
AR Av MO Bad e: Army Aviation Medical Officers Badge
ARAZ e: Antwerp-Rotterdam-Amsterdam-Zeebrugge
ARB d: Allgemeine Bedingungen für die Rechtsschutzversicherung
ARB e: Accounting Research Bulletin (journ.)
- e: Administrative Research Bulletin
- e: Aircraft Reactor Branch
- e: Aircraft Recovery Bulletin
- e: Air Research Bureau
- e: Air Resources Board
- e: All Routes Busy
- e: Arbitration
- e: Arbitrator
- e: Armed Services Technical Information Agency Report Bibliography
- e: Army Rearming Base
- e: Automatic Program Checkout Equipment Relay Box
- e: Automatic Radar Beacon
Arb. d: Arbeit
Arb d: Arbeiten
arb e: Arbitrary
- e: Arbor
ArbA d: Arbeitsamt
- d: Arbeitsanweisung
ARBA e: American Reference Books Annual
- e: Associate of the Royal Society of British Artists
Arb. Anw. d: Arbeitsanweisung
arb & aw e: Arbitration and Award
Arb.-B. d: Arbeitsbeginn
- d: Arbeitsbuch
ARBB e: A-Register B-Bit
ARBBA e: American Railway Bridge and Building Association
Arb.-Bl. d: Arbeitsblatt
- d: Arbeitsblätter
ARBD d: Allgemeiner Radiobund Deutschlands
Arb.-Eins. d: Arbeitseinsatz
Arbeitssch. d: Arbeitsschutz
Arbeitsstd. d: Arbeitsstunde[n]
Arbf. d: Arbeitsfeld
- d: Arbeitsförderung
ArbFördG d: Arbeitsförderungsgesetz
ArbG d: Arbeitgeber
- d: Arbeitsgericht
- d: Arbeitsgesetz
Arb.-Geb. d: Arbeitsgebiet
Arb.-Gem. d: Arbeitsgemeinschaft
ArbGEntsch. d: Arbeitsgerichtsentscheidung
ArbGG d: Arbeitsgerichtsgesetz
ArbGR d: Arbeitsgerichtsrat

Arb.-Gr. d: Arbeitsgröße
- d: Arbeitsgruppe
ARBIT d: Arbeitsgemeinschaft der Bitumen-Industrie
ARBITER e: Access Refusal and Barrier Interface Terminal
ARBITS e: Army Base Information Transfer System
Arb.-Kr. d: Arbeitskreis
ArbM d: Arbeitsminister[ium]
ARBM e: Association of Radio Battery Manufacturers
Arb.-Min. d: Arbeitsminister[ium]
ArbN d: Arbeitnehmer
Arb.-Nachw. d: Arbeitsnachweis
ArbnErfG d: Gesetz über Arbeitnehmererfindungen
ArbNV d: Arbeitnehmervertreter
- d: Arbeitnehmervertretung
ArbO d: Arbeitsordnug
arbo e: arthropod-borne
arbor e: arboriculture
ARBOR e: Arthropod-Borne
arbor. e: virus arthropod-borne virus
Arb.-Ord. d: Arbeitsordnung
ARBOR Experiment e: Argonne Boiling Reactor Experiment
ARBOR Facility e: Argonne Boiling Reactor Facility
Arb.-Pl. d: Arbeitsplan[ung]
- d: Arbeitsplatz
- d: Arbeitsplätze
ArbplSchG d: Arbeitsplatzschutzgesetz
ARB Press e: Arbor Press
ArbR d: Arbeitsrecht
Arb.-Richtl. d: Arbeitsrichtlinie[n]
ARBRSI e: Annual Report of the Board of Regents of the Smithsonian Institution (journ.)
ARBs e: Air Resources Boards
ARBS e: Angle Rate Bombing System
- e: Associate of the Royal Society of British Sculptors
- e: Automatic Radar Beacon Sequencer
Arb.-St. d: Arbeitsstoff[e]
- d: Arbeitsstudie[n]
- d: Arbeitsstunde[n]
ArbStÄttV d: Arbeitsstättenverordnung
ArbStV d: Arbeitsstoffverordnung
Arb.-Tg. d: Arbeitstag[e]
- d: Arbeitstagung
arbtrn e: arbitration
arbtror e: arbitrator
ArbUSt d: Arbeitsuntersuchungsstelle
ArbZ d: Arbeitszeit
ArbZO d: Arbeitszeitordnung
ArbZVO d: Arbeitszeitverordnung
ARC e: Acoustic Research Center
- e: Addiction Research Center
- e: Add Row and Constant
- e: Adjusted Range Correction
- e: Administrative Radio Conference
- e: Advanced Reentry Concepts
- e: Agricultural Relations Council
- e: Agricultural Research Center
- e: Agricultural Research Council
- e: Aids-Related Complex
- e: Aiken Relay Calculator
- e: Airborne Radio Communicating
- e: Airborne Radio Communication
- e: Airborne Research Capsule
- e: Aircleaning
- e: Aircraft Radio Corporation
- e: Air Reduction Center
- e: Air Rescue Center
- e: Air Reserve Center
- e: Alternate Route Cancel
- e: Altitude Rate Command
- e: American Railway Cases
- e: Ames Research Center
- e: Amphibious Research Craft
- e: Amplitude and Risetime Compensation
- e: Amplitude and Risetime Compensator
- e: Analog[ue] Recursive Computer
- e: Analog[ue] Response Conditioner
- e: Analyzer-Recorder-Controller
- e: Anomalous Retinal Correspondence
- e: Antigen-Reactive Cell
- e: Applied Research Corporation
- e: Approach Radar Controller
- e: Arcade
- e: Area of Responsibility Centre
- e: Argonne Reactor Computation
- e: Arthritis and Rheumatism Council
- e: Asian Research Center
- e: Askew Redundancy Check
- e: Association of Rehabilitation Centers
- e: Atlantic Research Center
- e: Atlantic Research Corporation
- e: Atomedic Research Center
- e: Attached Resource Computer
- e: Attendant Recall
- e: Attended Resource Computer
- e: Audio Response Control
- e: Augmentation Research Center
- e: Auto-Refrigerated Cascade
- e: Automated Rent Cellections
- e: Automatically Repaired Computer
- e: Automatic Radio Control
- e: Automatic Ram Control
- e: Automatic Range Control
- e: Automatic Rate Changer
- e: Automatic Ratio Control
- e: Automatic Reading of Characters
- e: Automatic Reception Control
- e: Automatic Recovery Computer
- e: Automatic Relay Calculator
-e: Automatic Relay Computer
-e: Automatic Remote Control
- e: Automatic Reset Counter
- e: Automatic Resolution Control
- e: Automatic Revenue Collection
- e: Automatic Routing Character
- e: Autopilot Rate Control
- e: Auxiliary Roll Control
- e: Average Response Computer
Arc e: Arch
- e: Archaic
- e: Arctic
arc e: arcus
Ar C e: Armo[u]red Car
ARCA e: Adaptive Reliability Control System
- e: American Rice Growers Cooperative Association
- e: Associate of the Royal Cambrian Academy
- e: Associate of the Royal Canadian Academy
- e: Associate of the Royal College of Arts
ARCAD e: Association of Renault and Chrysler for Automotive Development
ARCADE e: Argonne Computer-Aided Diffraction Equipment
- e: Automatic Radar Control and Data Equipment
ARCAIC e: Archive and Record Cataloguing and Indexing by Computer
ARCAN e: Aeronautical Radio of Canada
Arc Arch e: Arctic Archipelago
ARCAS e: All-Purpose Rocket for Collecting Atmospheric Samples
- e: All-Purpose Rocket for Collecting Atmospheric Soundings
- e: Automatic Radar Chain Acquisition System
ARCAT e: Army Radio Code Aptitude Test
ARCB e: A-Register C-Bit
ARCC e: Askew Redundancy Check-Character
Arc. Cir. e: Arctic Circle
arc cos d: Arkuskosinus
- e: inverse cosine (arc cosine)
arc cosec e: inverse cosecant
arc cosech e: inverse hyperbolic cosecant
arc cosh e: inverse hyperbolic cosine
arc cot d: Arkuskotangens
- e: inverse cotangent (arc cotangent)
arc coth e: inverse hyperbolic cotangent
arccsc e: arccosecant
arc csc e: inverse cosecant
arc ctg e: inverse cotangent (arc cotangent)
Arc Cur e: Arctic Current
ARCE e: Amphibious River Crossing Equipment
Arc.E. e: Architectural Engineer
ARCen e: Air Reserve Center
arcft e: aircraft
ARCG e: American Research Committee on Grounding
arch. d: archaisch
- d: archäologisch
- d: architektonisch
- d: archivieren
- e: archaic
- e: archipela
- e: architect[s]
- e: architectural
- e: architecture
Arch. d: Archäolog[i]e
- d: Archipel
- d: Architekt
- d: Architektur
- d: Archiv[ar]
arch d: Archive
Arch s: archipiélago
ARCH e: Advanced Reconnaissance Helicopter
- e: Architectures
- e: Archive[s]
- e: Articulated Computer Hierarchy
- e: Articulated Computing Hierarchy
- e: Automated Reports Control Handling
- e: Automated Resolution from Century to Honeywell
archaeol. e: archaeological
- e: archaeologist
archaeol e: archaeology
Archäol. d: Archäolog[i]e
Arch.-Ass. d: Archivassessor
ArchBibl d: Archäologische Bibliographie (journ.)
Arch. Biochem. Biophys. e: Archives of Biochemistry and Biophysics (journ.)
archcrit e: architectural critic[ism]
Arch de Cln e: Archipelago de Colón
Arch.Dermatol. e: Archives of Dermatology (journ.)
Arch.Dermatol.Forsch. d: Archiv für Dermatologische Forschung (journ.)
Arch Dig e: Architectural Digest
Arch.-Dir. d: Archivdirektor
Arch E e: Architectural Engineer

ARCHEDDA e: Architectures for Heterogeneous European Distributed Databases
Arch. Eisenhüttenw. d: Archiv des Eisenhüttenwesens (journ.)
Arch. Elektron. Übertragungste d: Archiv für Elektronik und Übertragungstechnik (journ.)
Arch. Elektrotech. (Berlin) d: Archiv für Elektrotechnik (Berlin) (journ.)
Arch. Energiewirtsch. d: Archiv für Energiewirtschaft (journ.)
Arch Entwicklungsmech. Organ. d: Archiv für Entwicklungsmechanik der Organismen; Wilhelm Roux (journ.)
Arch. Environ. Health e: Archives of Environmental Health (journ.)
archeo e: archeological
- e: archeologist
- e: archeology
archeol e: archeology
Archeoz e: Archeozoic
Arch Forum d: Architektengesetz
- e: Architectural Forum (journ.)
ArchG d: Architektengesetz
Arch. Geflügelk. d: Archiv für Geflügelkunde (journ.)
Arch. Gen. Psychat. e: Archives of General Psychatry (journ.)
Arch.Gesmte Virusforsch. d: Archiv für die Gesamte Virusforschung (journ.)
Arch. Hist. Exact Sci. e: Archive for History of Exact Sciences (journ.)
Arch. Hydrobiol. d: Archiv für Hydrobiologie (journ.)
Arch. Hyg. Bakteriol. d: Archiv für Hygiene und Bakteriologie
ARCHI d: Archivierungsprogramm
archi d: Archive
- d: Archivist
- e: archival
- e: archive
- e: archivist
Arch. Immunol. Exp. Ther e: Archives of Immunology and Experimental Therapy (journ.)
Arch. Intern. Med. e: Archives of Internal Medicine (journ.)
archip e: archipelago
archipel n: archipelago
ARCHISYST e: Architectural Information System
archit. e: Architect[ure]
- e: Architectural
Archit R e: Architectural Review (journ.)
Archit Rec e: Architectural Record (journ.)
Archive e: Archive Press
Archives Soc Hist e: Archives of Social History
Archiv PF d: Archiv für das Post- und Fernmeldewesen
Archiv Philos d: Archiv für Philosophie (journ.)
Arch. Kinderheilk. d: Archiv für Kinderheilkunde (journ.)
Arch. klin. Chir. d: Archiv für klinische Chirurgie; Langenbecks (journ.)
Arch. klin. Exp.-Ohren-Nasen-K d: Archiv für klinische und experimentalle Ohren-Nasen- und Kehlkopfheilkunde (journ.)
Arch. klin. Exp. Opthalmol. d: Archiv für klinische und experimentelle Opthalmologie; Albrecht von Grafes (journ.)
Arch. Lebensmittelhyg. d: Archiv für Lebensmittelhygiene (journ.)
Arch. Meteorol. Geophys. Biokl d: Archiv für Meteorologie, Geophysik und Bioklimatologie; Serie A (journ.)
- d: Archiv für Meteorologie, Geophysik und Bioklimatologie; Serie B (journ.)
Arch. Mikrobiol. d: Archiv für Mikrobiologie (journ.)
Arch. Neurol. e: Archives of Neurology (journ.)
archoe e: archeologist
- e: archeology
ARCHON System e: Archives Online System
Arch. Opthalmol. e: Archives of Opthalmology (journ.)
Arch. Oral Biol. e: Archives of Oral Biology (journ.)
Arch. OrthopÄd. Unfallchir. d: Archiv für Orthopädische und Unfallchirurgie (journ.)
Arch. Otolaryngol. e: Archives of Otolaryngology (journ.)
Arch. Pathol. e: Archives of Pathology (journ.)
Arch. Pharm. d: Archiv der Pharmazie (journ.)
Arch PuT d: Archiv für das Post- und Telegrafenwesen
Arch.-R. d: Archivrat
Arch. Ration. Mech. Anal. e: Archive for rational Mechanics and Analysis (journ.)
Arch Rec e: Architectural Record (journ.)
Arch Rec Bks e: Architectural Record Books
Arch. Surg. e: Archives of Surgery (journ.)
archt e: Architect
Arch. Tech. Messen d: Archiv für Technisches Messen und Industrielle Messtechnik (journ.)
archtl e: Architectural
Arch. Toxikol. d: Archiv für Toxikologie (journ.)
archv e: archive
Archvl. d: Archivalien
ARCI e: American Railway Car Institute
- e: Associate of the Royal Colonial Institute
ARCIS e: African Regional Centre for Information Science
- e: Agricultural Research Council Computerized Information System
- e: Agricultural Research Current Information System
ARCLOS e: Army Close Support
ARCM e: Associate of the Royal College of Music
ARCN e: Agricultural Research Council of Norway
ARCNET e: Attached Resources Computer Network
ARCO e: Agricultural Research Center Operations
- e: Alitalia Reservation and Communication
Arc O e: Arctic Ocean
ARCO e: Automatic Reservations and Communications
arco i: arcata
ARCOM e: Army Research Command
ARCOMSAT e: Arabian Communication Satellite
- e: Arctic Communication Satellite
ARCOM Satellite e: Arabian Communication Satellite
- e: Arctic Communication Satellite
ARCOMSAT System e: Arab Communications Satellite System
ARCON e: Automatic Rudder Control
ARCOPS e: Arctic Operations
ARCOS d: Arzt-Computersystem
arcos e: arc cosine
ar cosh d: Areakosinus
ARCOS System e: Automatic Reservations and Communications System
ARCO System e: Alitalia Reservation and Communication System
ar coth d: Areakotangens
ARCP e: Air Refueling Control Point
ARCR e: Arthritis and Rheumatism Council for Research
ARCRA e: Arabian Horse Club Registry of America
ARCRAN e: Agricultural Research Council of Rhodesia and Nyasaland
ARCRL e: Agricultural Research Council Radiobiological Laboratory
ARCS d: automatisiertes Reproduktions-, Copier- und Sortiersystem
- e: Achievement Reqards for College Scientists
- e: Adaptive Reliability Control System
- e: Advanced Reconfigurable Computer System
- e: Air Resupply and Communications Service
- e: Associate of the Royal College of Science
- e: Associate of the Royal College of Surgeons
- e: Attached Resource Computer System
- e: Automated Reproduction and Collating System
- e: Automated Revenue Collection System
- e: Automated Ring Code Search
- e: Automated Ring Code System
- e: Automatic Recognition of Continuous Speech
- e: Automatic Route Control System
ARCSA e: Aviation Requirements for the Combat Structure of the Army
ARCSc e: Associate of the Royal College of Science
ARCSEC e: Inverse Secant (Arc Secant)
arc sech e: inverse hyperbolic secant
arc sin d: Arkussinus
- e: inverse sine (arc sine)
arc sinh e: inverse hyperbolic sine
ARCST e: Associate of the Royal College of Science and Technology
ARC System e: Attached Resource Computer System
ARCT e: Air Refueling Control Time
Arct e: Arctic
ARCT e: Army Radio Code Aptitude Test
- e: Army Radio Code Test
- e: Askew Redundancy Check-Track
arctan d: Arkustangens
arc tan e: inverse tangent (arc tangent)
arc tanh e: inverse hyperbolic tangent
Arct. Circ. e: Arctic Circle
arc tg e: inverse tangent (arc tangent)
arctgh e: inverse hyperbolic tangent
ARCTS e: Automated Radar Controlled Terminal System
ARC Unit e: Automatic Remote Control Unit
ARCUP e: Atlantic Region Canadian University Press
ARCVS e: Associate of the Royal College of Veterinary Surgeons
ARC/W e: Arc Weld[ing]
ARD d: Antennenrichtdiagramm
- d: Arbeitsgemeinschaft der [öffentlich-rechtlichen] Rundfunkanstalten der Bundesrepublik Deutschland
- e: Abnormal Return Address
- e: Accelerated Rural Development
- e: Acute Respiratory Disease

- e: Address Register Decoding
- e: Air Reserve District
- e: Answering, Recording and Dialling
- e: Armament Research Development
- e: Army Research and Development
- e: Automatic Recorder of Deformation
- e: Automatic Release Date
- e: Auxiliary Repair Dock
- e: Average Response Data

AR & D e: Advanced Research and Development
- e: Air Research and Development

ar & d e: aeronautical research and development

ARd e: Arterial Road

ARDA e: Advanced Reactor Development Associates
- e: American Rabbit Dealers Association
- e: Analog[ue] Recording Dynamic Analysis
- e: Analog[ue] Recording dynamic Analyzer
- e: Atomic Research and Development Authority
- e: Automated Rotary Dipping Apparatus

ARD Abstracts e: Arthritis and Rheumatic Diseases Abstracts

ARDC e: Aberdeen Research and Development Center
- e: Air Research and Development Center
- e: Air Research and Development Command
- e: American Research and Development Corporation
- e: Applied Research and Design Center
- e: Army Research and Development Center

ARDC-A e: Air Research and Development Command - Andrews Air Force Base

ARDC-BMD e: Air Research and Development Command, Ballistic Missile Department

ARDCF e: Air Research and Development Command Forms

ARDCM e: Air Research and Development Command Manual

ARDCR e: Air Research and Development Command Regulations

ARD Dock e: Auxiliary Repair Dry Dock

ARDE e: Armament Research and Development Establishment

ARDEC e: Armament Research and Development Establishment, Canada

ARDECO e: Aluminium Resources Development Company

AR Def e: All-Round Defense

ARDF e: Airborne Radio Direction Finding

ARDI e: Automatic Relay of Digital Information

ARDIS d: automatisiertes Recherche- und Dokumentations-Informationssystem
- e: Army Research and Development Information System

ARDISO e: Army Research and Development Information Systems Office

ARDM e: Auxiliary Repair Dock; Medium

ARDME e: Automatic Radar Data Measuring Equipment
- e: Automatic Range Detector and Measuring Equipment

ARDP e: Army Requirements Development Plan

ARDS e: Advanced Remote Display Station
- e: Advance[d] Remote Display System

ard's e: analog recording dynamic analyzers

A-Rdschr. d: Ausfuhrrundschreiben

ARDT e: Automatic Remote Data Terminal

ARDXC e: Australian Radio DX Club

ARE d: Arbeitsgemeinschaft der regionalen Elektrizitätsversorgungsunternehmen
- d: Ausgabeeinheit Rechneranschluß
- e: Acoustic Radiation Element
- e: Advanced Real-time Executive
- e: Aerial Range Equipment
- e: Aeronautical Research Establishment
- e: Aircraft Reactor Experiment
- e: Aircraft Recovery Equipment
- e: Air Reactor Experiment
- e: Algebraic Riccati Equation
- e: American Railway Express
- e: Anion-Responsive Electrode
- e: Antenna Range Equipment
- e: Appollo Reliability Engineering
- e: Arab Republic of Egypt
- e: Associate in Religious Education
- e: Asymptotic Relative Efficiency
- e: Automated Responsive Environment
- e: Automatic Record Evaluation
- e: Auxiliary Rocket Engine

Ar E e: Architectural Engineer

ArE e: Arecibo

AREA e: Amateur Radio Emergency Association
- e: American Railway Engineering Association
- e: Army Reactor Experimental Area
- e: Army Recorder Experimental Area
- s: Aerovias Ecuatorianas

AREA Proc. e: American Railway Engineering Association Proceedings

AREAPWOFC e: Area Public Works Officer

AREBA e: Accelerated Reeducation of Emotions, Behaviour and Attitudes

AREC e: Agricultural Research and Educational Center
- e: Amateur Radio Emergency Corps

ARED e: Acoustic Reflex Ear Defender
- e: Aperture Relay Experiment Definition

AREE e: Admiralty Regional Electrical Engineer
- e: Apollo Reliability Engineering Electronics

A-REG d: Akkumulator-Register

AREG e: Accumulator Register
- e: Address Register
- e: Apparatus Repair-strategy Evaluation Guidelines
- e: A Register

A Reg 1 e: A-Register 1-Bit

A Reg 1 B e: A-Register One Bit

A Reg 2 B e: A-Register Two Bit

A Reg 4 B e: A-Register Four Bit

A Reg 8 e: A-Register 8-Bit

A Reg 8 B e: A-Register Eight Bit

A Reg A e: A-Register A-Bit

A Reg B e: A-Register B-Bit

A Reg BB e: A-Register B-Bit

A Reg CB e: A-Register C-Bit

A-Register d: Akkumulatorregister

A Reg Wd Mk e: A-Register Word Mark

A Rel e: Associate in Religion

ARELEM e: Arithmetic Element

ARELEM Program[me] e: Arithmetic Element Program[me]

ARENA e: Adoption Resource Exchange of North America

ARENTS e: Advanced Research Environmental Test Satellite

ARENTS ARPA e: Environmental Test Satellite

AREP e: Air Refueling Egress Point

AREPS e: Advanced Reconnaissance Electrically Propelled Spacecraft

ARES e: Advanced Radiation Effects Simulation
- e: Advanced Research Electromagnetic Simulation
- e: Advanced Research Electromagnetic [Pulse] Simulator
- e: Advanced Rocket Engine Storable
- e: Agricultural Research
- e: Amateur Radio Emergency Service
- e: Antireticulo-Endothelial Serum
- e: Automatic Record Evaluation System

AREs e: Algebraic Riccati Equations

ARESTEM e: A Recording Stray Energy Monitor

ARETO e: Arab Republic of Egypt Telecommunications Organization
- e: Arab Republic of Egypt Telephone Organization

ARETS e: Arizona Regional Ecological Test Site

ARF d: Arbeitsunterlagen für Radio- und Fernsehtechniker
- e: Acute Respiratory Failure
- e: Acute Rheumatic Fever
- e: Alkali Resistant Factor
- e: American Radio Forum
- e: American Rationalist Federation
- e: American Rehabilitation Foundation
- e: Animal Research Facilities
- e: Application Replacement Factor
- e: Area Resource File
- e: Arthritis and Rheumatism Foundation
- e: Automatic Report Feature
- e: Automatic Return Fire

ARFA e: Allied Radio Frequency Agency

ARFC e: Air Reserve Flying Center
- e: Average Rectified Forward Current

ARFCOSTA e: Armed Forces Courier Station

ARFDC e: Atomic Reactor and Fuel Development Corporation

ARFF d: Allrichtungsfunkfeuer

ARFMSR e: Air Reserve Forces Meritorious Service Ribbon

Arfo d: Arbeitsgemeinschaft für Forstschutz und Naturkunde

ARFOR e: Area Forecast

ARFORSTAT e: Army Force Status Reporting system

ARFPDS e: Air Reserve Forces Personnel Data System

ARG d: Agrarreformgesetz
- d: Argument
- e: American Record Guide
- e: Arginin[e]
- e: Argument[ation]
- e: Argus
- e: Arresting
- s: Aerolineas Argentinas

Arg. d: Argentinien
- d: Argonaut[en]
- d: Argument[e]

arg. d: argentinisch

ARg d: Auslandsregister

Arg e: Argentina
- e: Argumentator
- s: Argelia

ARGA e: Appliance, Range, Adjust [data processing]

arg. ad bac. l: argumentum ad baculum

arg. ad hom. l: argumentum ad hominem

arg. ad ignor. l: argumentum ad ignorantum

arg. ad miser. l: argumentum ad misericordiam

arg. ad veri. l: argumentum ad vericundiam

ARGCA e: American Rice Growers Cooperative Association

Arge d: Arbeitsgemeinschaft
ArGeChemDOK d: Arbeitsgemeinschaft für Chemie-Dokumentation
Argel p: Argelt
- s: Argelia
Argen e: Argentin
- e: Argentine
- e: Argentinian
argent. d: argentinisch
Argentina s: República Argentina
ARGE Rhein d: Arbeitsgemeinschaft der Länder zur Reinhaltung des Rheins
argmt e: arrangement
ARGO e: Advanced Research Geophysical Observatory
ARGONAUT e: Argonne Nuclear Assembly for University Training
Argonne Rev. e: Argonne Reviews (journ.)
Argosy e: Argosy-Antiquarian Limited
- e: Hawker-Siddeley four-engine turboprop
ARGP e: Aeronautical Radionavigation Glide Path
ARGR e: Association for Research in Growth Relationship
ARGS e: American Rock Garden Society
- e: Anti-Radar Guidance Sensor
ARGUS d: Abonnement-Recherche zur gezielten Unterrichtung über Schrifttum
- e: Advanced Research on Groups Under Stress
- e: Analytical Reports Gathering and Updating System
- e: Associative Registers for Generalized User Switching
- e: Automatic Routine Generating and Updating System
Argus e: Canadair long-range reconnaissance plane
a. Rh. d: am Rhein
ARh d: Anreizhilferelais
ARH e: Active Radar Homing
- e: Aerial Radome Heater
- e: Ammunition Railhead
- e: Antenna Radome Heater
- e: Anti-Radiation Homer
- e: Anti-Radiation Homing
ARHA e: Associate of the Royal Hibernian Academy
ARHIR e: Anti-Radiation Homing Infrared
Ari d: Artillerie
ARI d: Astronomisches Recheninstitut
- d: Autofahrer-Rundfunk-Information[ssystem]
- e: Acne Research Institute
- e: Aerial Radiac Instrument
- e: Aeronautical Radio, Incorporated
- e: Agricultural Research Institute
- e: Air-Conditioning and Refrigeration Institute
- e: Airborne Radio Installation
- e: Airborne Radio Instrument
- e: Aircraft Radio Instrument
- e: Allied Research Institute
- e: Aluminum Research Institute
- e: Aluminum Roofing Institute
- e: American Rayon Institute
- e: American Refractories Institute
- e: American Russian Institute
- e: Animal Research Institute
- e: Applications Reference Index
- e: Aricraft Radio Instrument
- e: Army Research Institute
- e: Automated Readability Index
- e: Automatic Radio Information
- e: Automobile-drivers Radio Information
- n: Amsterdamsch Radio Institut
ARIA e: Advanced Range Instrumentation Aircraft
- e: American Radio Importers Association
- e: Apollo Range Instrumentation Aircraft
ARIANA e: Ariana Afghan Airlines
ARIAS e: Augmented Relational Intelligence Analysis System
ARIBA e: Associate of the Royal Institute of British Architects
ARIC e: Agricultural Research Information Centre
- e: Associate of the Royal Institute of Chemistry
ARICN e: Asian Regional Irrigation Communication Network
ARICRSU e: American Russian Institute for Cultural Relations with the Soviet Union
ARICS e: Associate of the Royal Institution of Chartered Surveyors
ARIEL e: Automated Realtime Investment Exchange Limited
- e: Automatic Retrieval of Information Electronically
ARIEM e: Army Research Institute of Environmental Medicine
ARIES e: Advanced Radar Information Evaluation System
- e: Astronomical Radio Interferometric Earth Surveying
- e: Atmospheric Research Information Exchange Study
- e: Authentic Reproduction of an Independent Earth Satellite
- e: Automated Reliability Estimation Program[me]
- e: Automated Reliability Interactive Estimation System
ARIMA e: Autoregressive Integrated Moving Average
ARIMA Process e: Autoregressive Integrated Moving Average Process
ARIMS e: Airborne Radar Inflight Monitoring System
ARINA e: Associate of the Royal Institution of Naval Architects
ARINC e: Aeronautical Radio Incorporated
ARIP e: Air Refueling Ingress Point
- e: Air Refueling Initial Point
- e: Automatic Rocket Impact Predictor
ARIPHH e: Associate of the Royal Institute of Public Health and Hygiene
ARIS d: automatisches Retrieval Informationssystem
- e: Academic Research Information System, Incorporated
- e: Acid Rain Information System
- e: Activity-Reporting Information System
- e: Advanced Range Instrumentation Ship[s]
- e: Advanced Range Instrumentation System[s]
- e: Advanced Remote Imaging Systems
- e: Advanced Research Instrument System
- e: Aerial Radiac Instrument System
- e: Aerial Radio Instrument System
- e: Aeronautical Research Institute of Sweden
- e: Ag[e]ing Research Information System
- e: Airborne Range Instrumentation Station
- e: Aircraft Recording Instrumentation System
- e: Aircraft Research Instrumentation System
- e: Alabama Resources Information System
- e: A Relational Information System
- e: Atlantic Range Instrumentation Ship
- e: Atomic Reactor in Space
- e: Attitude and Rate Indicating System
- e: Automated Reactor Inspection System
- e: Automatic-Recording Infrared Spectrometer
ARISE e: Adult Referral and Information Service in Education
- e: Appointment, Registration, Information System and Evaluation
ARIST e: Annual Review of Information Science and Technology (journ.)
ARISTOTLE e: Annual Review and Information Symposium on the Technology of Training, Learning and Education
- e: Automated Retrival Information System to Track and Optimize the Training and Learning
ARI-System d: Autofahrer-Rundfunk-Informationssystem
ARIT e: American Registered Inhalation Therapists
- e: American Research Institute in Turkey
arit s: aritmética
Arith. d: Arithmetik
arith. d: arithmetisch
Arith e: Arithmetic
arith e: arithmetic[al][ly]
- e: arithmetician
Arith Elem e: Arithmetic Element
Arith Log U e: Arithmetic Logic Unit
Arith Org e: Arithmetic Organ
Arith U e: Arithmetic Unit
ARIÜ d: automatisches Registriergerät für innere Überspannungen
Ariz. d: Arizona
Ariz e: Arizona
Ariz H e: Arizona Highways (journ.)
Ariz Hist Found e: Arizona Historical Foundation
ARJ e: Acquisition Radar Jamming
ARJIS e: Automated Regional Justice Information System
ARJ Timber e: Arjun Timber
Ar/K d: Abraum-Kohle
ARK e: Air Rescue Kit
Ark e: Arkansas
Ark f Fys a: Arkivfor Fysik
ar Kontakt d: Arbeit-Ruhe-Kontakt
ARkpl d: Auslandsrichtungskoppler
ArKr d: Arbeitskreis
arkt. d: arktisch
ARL e: Abortion Rights League
- e: Acceptable Reliability Level
- e: Admiralty Research Laboratory
- e: Aerial Reconnaissance Laboratory
- e: Aeromedical Research Laboratory
- e: Aeronautical Research Laboratory
- e: Aerospace Research Laboratory
- e: Aircraft Radio Laboratory
- e: Aircraft Rocket Launcher
- e: Air Resources Laboratory
- e: Air Run Landing
- e: Algorithmic Language Realtime Language
- e: Amateur Radio League
- e: Anesthesia Research Laboratories
- e: Applied Research Laboratory
- e: Arctic Research Laboratory
- e: Army Radiation Laboratory
- e: Atlantic Regional Laboratory
- e: Automatic Record Level
- e: Average Run Length
arl e: aerial
arl° s: articum
Arlanda e: Stockholm, Swedens airport
ARLB e: Automatic Radio Location Beacon
Arlis e: Arctic Research Laboratory Ice Station
ARLIS e: Arctic Research Laboratory Island
ARLIS/NA e: Art Libraries Society/North America

ARLO e: Air Reconnaissance Liaison Officer
- e: Art Reference Libraries of Ohio
arlob e: aviation lubricant
ARLS e: Automatic Resupply Logistics System
ARIST e: Anual Review of Information Science and Technology
ARM d: Anreizmelder
- d: Armatur
- e: Abstracts of Research and Related Materials in Vocational and Technical Education
- e: Accelerated Relaxation Method
- e: Access Regulation Mechanism
- e: Address Register Modification
- e: Agricultural Requisites and Mechanization
- e: Aircraft Radio Mechanic
- e: Algorithmic Remote Manipulation
- e: Allergy Relief Medicine
- e: Amateur Radio Monitor
- e: Ampex Replacement Memories
- e: Ampex Replacement Memory
- e: Anhysteric Remanence
- e: Anhysteric Remanent Magnetization
- e: Anti-Radar Missile
- e: Anti-Radiation Missile
-e: Anti-Radio Missile
- e: Apollo Requirements Manual
- e: Applied Research Management
- e: Area Radiation Monitor[s]
- e: Armament
- e: Arming
- e: Army Ready Material
- e: Artificial Rupture of the Membranes
- e: Assistant Regional Manager
- e: Asynchronous Response Mode
- e: Atmosphere Radiation Monitor
- e: Automated Route Management
- e: Automated Route Manager
- e: Availability, Reliability [and] Maintainability
- e: Aviation Radioman
Arm. d: Armee
- d: Armenien
- d: Armierung
- e: Armature
- e: Armenia[n]
arm. d: armenisch
- e: armiert
- e: armo[u]r[ed]
Ar.M. e: Master of Architecture
- l: Architecturae Magister
ARM 1 c e: Aviation Radioman First Class
ARM 2 c e: Aviation Radioman Second Class
ARM 3 c e: Aviation Radioman Third Class
ARMA e: Accumulator Reservoir Manifold Assembly
- e: American Records Management Association
- e: American Registry of Medical Assistants
- e: Armature
- e: Association of Records Managers and Administrators
- e: Auto-Regressive-Moving Average
ARMAC e: Aviation Radioman; Combat Aircrewman
ARM ACCEL e: Armature Acceleration
ARMACS e: Aviation Resources Management and Control System
ARMADA e: Aircraft Reliability, Maintainability, Availability Design Analysis
ARMAN e: Artificial Methods Analyst
ARMC e: Automatic Repeat Request Mode Counter
ARM CA e: Armo[u]red Cable
ARMCM e: Anti-Radiation Missile Countermeasures
- e: Associate of the Royal Manchester College of Music
ARMCOMSAT e: Arab Communications Satellite System
ARMD e: Armo[u]red
ARMDAS e: Army Damage Assessment System
ARMDC e: Armo[u]red Car
Armd CC e: Armo[u]red Combat Command
Armd Div Amu Pk e: Armo[u]red Division Ammunition Park
Armd Div Pet Sub-Pk e: Armo[u]red Division Petrol Subpark
Armd Div Sup Pk e: Armo[u]red Division Supply Park
Armd Sch e: Armo[u]red School
Armd Sig e: Armo[u]red Signals
ARMEL e: Armament and Electronics
Armen e: Armenia[n]
Armeter e: Automatic-Remote-Meter
ARMF e: Advanced Reactivity Measurement Facility
ARMI e: American Rack Merchandisers Institute
- e: American Research Merchandising Institute
- e: Army Resources Management Institute
ARMIC e: Army Industrial Center
Armier. d: Armierung
ARMIP e: Accounting and Recording Management Improvement Program[me]
ARMIS e: Agricultural Research Management Information System
- e: Association of Retail Management Information Systems
ARMIT e: Associate of the Royal Melbourne Institute of Technology
ARML e: Airmail
ARMM e: Analysis and Research of Methods for Management
ARMMS e: Automated Reliability and Maintainability Measurement System
- e: Automatic Reconfigurable Modular Multicompressor System
- e: Automatic Reconfigurable Modular Multiprocessor System
armn. e: airman
AR-MONP e: Ayers Rock-Mount Olga National Park
arm pl e: Armo[u]r Plate
arm.pltd. e: Armo[u]r Plated
armr e: Armorer
Armr Sergt e: Armorer Sergeant
ARMS e: Accounts Receivable Management System
- e: Advanced Receiver Model System
- e: Aerial Radiological Measurements and Survey
- e: Aerial Radiological Measuring Survey
- e: Aerial Radiological Measuring System
- e: Aerial Radiological Monitoring System
- e: Airborne Ruggedized Memory System
- e: Aircraft Reliability and Maintainability Simulation
- e: Aircraft Resources Management System
- e: Amateur Radio Mobile Society
- e: Application of Remote Manipulators in Space
- e: Army Reader Microfilm System
- e: Associate of the Royal Society of Miniature Painters, Sculptors and Gravers
- e: Automated Records Management System
- e: Automatic Data Processing Equipment Resources Management System
- e: Automatic Radiation Monitoring System
- e: Automatic Radio Meteorological Station
ARM SH e: Armature Shunt
ARMSPAC e: Aircraft Resources Management System; Pacific
ARMSPAN e: Argonne Multi-channel Stored Program[me] Analyzer
ARMS Robot e: Application of Remote Manipulators in Space Robot
ARM System e: Archival Records Management System
ARMT e: Armament
- e: Armature
ARMTE e: Army Missile Test and Evaluation
ARMU e: Addressable Remote Multiplexer Unit
- e: Associated Rocky Mountain Universities
ARMY AEL e: Army Aerobiology and Evaluation Laboratory
ARMY CERC e: Army Coastal Engineering Research Center
ARMY ECOM e: Army Electronics Command
ARMY NDL e: Army Nuclear Defense Laboratory
ARMY TRCO e: Army Transportation Research Command
a. Rn. d: am Rhein
ARN e: Address Reference Number
- e: Air Request Net
- e: Associated Radio Navigation
- e: Atmospheric Radio Noise
arn s: ácido ribonucleico
ARNA e: Association of Radio-Television News Analysts
ARND e: Around
ARNG e: Arcing
- e: Arrangement
ARNM e: Aztec Ruins National Monument
ARNMD e: Association for Research in Nervous and Mental Disease
ARNO e: Air Indicator Not Operating
- e: Armo[u]r in Night Operations
ARNR d: Artikelnummer
AR(NSW) e: Arbitration Reports (New South Wales)
ARO e: After Receipt of Order
- e: Airborne Range Only
- e: Air Radio Officer
- e: Algonquin Radio Observatory
- e: Analysis of Records Obtained
- e: Applied Research Objective
- e: Army Routine Order
- e: Assistant Research Officer
- e: Association for Research in Ophthalmology
- e: Association of Roentgenological Organizations
- e: Automatic Range Only
- e: Auxiliary Read-Out
AROC e: Air Rescue Operations Center
AROCC e: Association for Research of Childhood Cancer
AROD e: Airborne Radar Orbital Determination
- e: Airborne Range and Orbit Determination
- e: Airborne Ranging and Orbit Determination
ARODS e: Airborne Radar Orbital Determination System
AROD System e: Airborne Ranging and Orbit Determination System

AROM e: Alterable Read Only Memory
- e: Aromatic
- e: Artificial Rupture of Membranes
- e: Associative Read-Only Memory
arom. d: aromatisch
- d: aromatisieren
- d: aromatisiert
aromat. d: aromatisch
AROMs e: Associative Read-Only Memories
AR-ONP e: Ayers Rock-Olga National Park
AROP d: arithmetische Operation[en]
ARO Radar e: Airborne Range Only Radar
AROS e: Alterable Read-only Operating System
- e: Annual Report on Stress
AROU e: Aviation Repair and Overhaul Unit
A Route e: Alternate Route
ARp d: apothekenrezeptpflichtig
ARP d: arbeitsspeicherresidentes Programm
- d: Arithmetikprozessor
- e: Active Recording Program[me]
- e: Active Rest Point
- e: Activity Request Packet
- e: Address Register Position
- e: Advanced Reentry Program[me]
- e: Advanced Research Projects
- e: Aerial Radiation Pattern
- e: Aerodrome Reference Point
- e: Aeronautical Recommened Practice
- e: Aerospace Recommened Practice
- e: Aerospace Reference Project
- e: Airborne Radar Platform
- e: Aircraft Reference Point
- e: Aircrew Respiratory Protection
- e: Airport Reference Point
- e: Air Raid Precautions
- e: Air Raid Protection
- e: Air Refueling Probe
- e: Alternator Research Package
- e: American Registry of Pathologists
- e: Ammunition Refilling Point
- e: Analogous Random Process
- e: Analytical Rework Program[me]
- e: Antarctic Research Program[me]
- e: Antenna Radiation Pattern
- e: Antilles Research Program[me]
- e: Area Redevelopment Program[me]
- e: Army Research Plan
- e: Association for Realistic Philosophy
- e: Atmospheric Research Program[me]
- e: Australian Reptile Park
- e: Automatic Receiver Program[me]
- e: Automatic Recovery Program[me]
- e: Automatic Relative Plotter
- e: Avalanche Resonance Pumped
- e: Azimuth Reference Pulse
- e: Azimuth Relative Pulse
Arpa d: Arbeitsgemeinschaft für Parodontoseforschung
ARPA e: Advanced Research Projects Agency
ARPANET e: Advanced Research Project Agency Network
ARPA-NIC e: Advanced Research Projects Agency-Network Information Center
ARPAP e: Aluminium-Resin-Polyethylene-Aluminium-Polyethylene
ARPAS e: Air Reserve Pay and Allowance System
ARPAT e: Advanced Research Projects Agency Terminal
ARP-Betrieb d: Avalanche Resonance Pumped-Betrieb
ARPC e: Air Raid Precautions Controller
- e: Air Reserve Personnel Center
- e: Australian Radio Propagation Committee
ARPD e: Air Raid Precautions Department
- e: Applied Research Planning Document
ARPDP e: Association of Rehabilitation Program[me]s in Data Processing
ARPEGE e: Air Pollution Episode Game
ARPES e: Angle-Resolved Photoelectron Spectroscopy
ARPESH e: Accurate and Reliable Prototype Earth Sensor Head
ARPIC e: Aerospace Radioisotope Power Information Center
ARPL e: A Retrieval Process Language
ARPM e: Average Revenue per Message
ARPN e: Aircraft and Related Procurement; Navy
ARPO e: Air Raid Precautions Officer
ARPQ e: Asset Record Performance Quota
Arprt e: Airport
ARPs d: Archiv für Religionspsychologie (journ.)
- e: Automatic Relative Plotters
ARPS e: Advanced Radar Processing System
- e: Advanced Real-time Processing System
- e: Aerospace Research Pilot School
- e: Air Raid Precautions Service
- e: Air Reserve Pay System
- e: Arab Physical Soctety
- e: Associate of the Royal Photographic Society [of Great Britain]
- e: Australian Radiation Protection Society
ARPSC e: Amateur Radio Public Service Corps
ARPS-E e: Aerospace Research Pilot School-Edwards Air Force Base
Arpt e: Airport
ARPT e: American Registry of Physical Therapists
- e: Army Registry of Physical Therapists
ARPUD d: Abteilung Raumplanung Universität Dortmund
a-r pulse s: apical-radial pulse
ARPV e: Apricot Ring Pox Virus
ARQ e: Accept Request
- e: Annual Report Questionnaire
- e: Annual Review Questionnaire
- e: Automatic Repeat [on] Request
- e: Automatic Repetition Requested
- e: Automatic Request
- e: Automatic Retransmission of Request
- e: Automatic Retransmission Request
ARQP e: Asset Record Quota Points
Arqto s: Arquitecto
arquo p: arquipélago
Arr. d: Arrangement
arr. d: arrangieren
ARR d: automatische Relaisrechenmaschine
- e: Address Recall Register
- e: Address Reference Register
- e: Add Row and Row
- e: Advanced Restricted Report
- e: Advanced Rocket Ramjet
- e: Advance Release Record
- e: Aerial Radiation Resistance
- e: Aerial Rotation Rate
- e: Aeronautical Radionavigation Radar
- e: Aerospace Rescue and Recovery
- e: Airborne Radio Receiver
- e: Aircraft Radio Regulations
- e: Air Regional Representative
- e: Air Reserve Record[s]
- e: Alaskan Railroad
- e: Alkaline Rust Remover
- e: Altitude Referenced Radiometer
- e: Antenna Radiation Resistance
- e: Antenna Rotation Rate
- e: Anti-Repeat Relay
- e: Apollo Program[me] Requirements
- e: Aricraft Radio Regulation[s]
- e: Armo[u]r Research Reactor
- e: Army Retail Requirements
- e: Arrange[d]
- e: Arrester
- e: Arrival
- e: Arrive[d]
- e: Arrives
- e: Arriving
- e: Association for Radiation Research
- e: Automatic Rerouting
- e: Automatic Retransmission Request
- l: Anno Regni Regis [or Reginae]
ARRA e: Amateur Radio Retailers Association
ARRAPS e: Automatic Repetitive Reset and Program[me] System
ARRC e: Aerospace Rescue and Recovery Center
- e: Air Reserve Records Center
- e: Associate of the Royal Red Cross
ARRCO e: American Resource Recovery Corporation Washington DC
arrd e: arrived
ARRE e: Alarm Receiving and Reporting Equipment
- e: Average Relative Representation Error
ARREP e: Arrival Report
ARREPISIC e: Arrival Report Immediate Superior In Command
ARRES e: Automatic Radar-Reconnaissance Exploitation System
arret. d: arretieren
- d: arretiert
Arret. d: Arretierung
AR-RET-ST e: Arm Retracting Strut
ARRF e: Automatic Recording and Reduction Facility
ArRg d: Anrufregister
arrgt e: arrangement
ARRIVEDREP e: Arrived Report
ARRJ e: Advanced Rocket Ramjet
ARRL e: Aeronautical Radio and Radar Laboratory
- e: American Radio Relay League
arr.n. e: arrival notice
arro s: arroyo
arro seco s: arroyo seco
ARROW e: Aircraft Routeing Right-of-Way
Arr Rad e: Array Radar
ARRS e: Aircraft Refueling and Rearming System
- e: American Roentgen Ray Society
- e: Attitude-Referenced Radiometer Study
ARRT e: American Registry of Radiologic Technologists
Arrt e: Arrangement
ARRTC e: Aerospace Rescue and Recovery Training Center
ARRUNA e: Arrived Unannounced
arrvg e: arriving
arrvl e: arrival
ARS d: aktive Reduktionsschutzeinrichtung
- d: allgemeine und rekursive Strukturierung
- d: Analog-Regelsystem
- d: Anrufrelaissatz
- d: Arbeitsausschuß Regel- und Sicherheitseinrichtungen für Gas und Ölfeuerungen
- d: Arbeitsrechnersystem
- d: Arbeitsrechtssammlung
- d: automatischer Regalstapler
- d: automatisiertes Reservierungssystem

- e: Absolute Radiation Scale
- e: Access Relay Set
- e: Accounts Receivable Statement
- e: Active Repeater Satellite
- e: Advanced Reconnaissance Satellite
- e: Advanced Reconnaissance System
- e: Advanced Record System
- e: Advanced Reentry System
- e: Advanced Rescue System
- e: Aerial Reconnaissance and Security
- e: Aeronautical Research Scientist
- e: Aeroplane Repair Section
- e: Aerospace Research Satellite
- e: Agricultural Research Service
- e: Airborne Ranging System
- e: Airborne Refrigeration System
- e: Aircraft Radiosight
- e: Aircraft Repair Section
- e: Airline Reservations System
- e: Alpha-Ray Spectrometer
- e: American Radium Society
- e: American Recorder Society
- e: American Rhododendron Society
- e: American Rocket Society
- e: American Rose Society
- e: Amplified Response Spectrum
- e: Analog[ue] Recording System
- e: Angular Rate Sensor
- e: Apollo Reentry Ship
- e: Area Re Supply
- e: Army Radio School
- e: Army Radio Station
- e: Arsine
- e: Asbestos Roof Shingles
- e: Atmosphere Revitalization Section
- e: Atmosphere Revitalization System
- e: Attitude Reference System
- e: Audio Response System
- e: Automated Retrieval System
- e: Automatic Recording Spectrometer
- e: Automatic Route Selection
- e: p-Azopenzenearsonate

Ars. d: Arsenal
ARs e: Action Requests
- e: Address Registers
- e: Antarctic Regions
- e: Army Regulations

Ars e: Arsenal
ARSA e: Aeronautical Repair Station Association
- e: Associate of the Royal School of Art
- e: Associate of the Royal Scottish Academy
- e: Associate of the Royal Society of Antiquaries
- e: Associate of the Royal Society of Arts
- e: Austrian Radar Site Analysis

AR San I e: Associate of the Royal Sanitary Institute
ARSAP e: Army Small Arms Program[me]
ARSB e: Anchored Radio Sound Buoy
- e: Automated Repair Service Bureau

ARSC e: Aircraft Repair and Supply Center
- e: Analog[ue] Rotation Speed Control
- e: Association for Recorded Sound Collections

ARSD e: American Refineries Standard Dimension
- e: Aviation Repair Supply Depot

ARS/E d: automatisiertes Reservierungssystem für Eisenbahnen
ARSE e: Alpha-Ray Spectrometric Equipment
ARSECH e: Inverse Hyperbolic Secant
ARSEM e: Army Registry of Special Educational Materials
arsen. d: Arsenal
arsen e: arsenal
Ar. Sgt. e: Armo[u]rer Sergeant
ARSH e: Area Hyperbolic Sine
- e: Associate of the Royal Society for the Promotion of Health

ARSI d: Anlagenüberwachung Regelung, Steuerung, Informationsgewinnung
arsin e: arc sine
arsinh e: Ainverse Hyperbolic Sine
- e: Areasinus

ARSIP e: Arrears in Pay
ARSJ e: American Rocked Society Journal
ARsl e: Arsenal
ARSL e: Associate of the Royal Society of Literature
ARSM d: Auswertung von Richtungs- und Streckenmessungen
- e: Air Rescue Service Manual
- e: Associate of the Royal School of Mines
- e: Associate of the Royal Society of Musicians

ARSN d: automatische Regelung, Speisung, Netz
ARSO e: African Regional Standards Organization
ARSP d: Archiv für Rechts- und Sozialphilosophie (journ.)
- e: Aerospace Research Satellite Program[me]
- e: Aerospace Research Support Program[me]

ARSPA e: Aerial Reconnaissance and Surveillance Penetration Analysis
ARSPH e: Associate of the Royal Society for the Promotion of the Health
ARSPT e: Air Reconnaissance Support
ARSR e: Air Rescue Service Regulation
- e: Air Route Surveillance Radar
- e: Arrester

ARSS e: Atmosphere Reactants Supply System
ARSSA e: Aerial Reconnaissance and Surveillance Survivability Analysis
ARSTN e: Airways Radio Station
AR Stns e: Airways Radio Stations
ARSV e: Apple Ring Spot Virus
- e: Armo[u]red Reconnaissance Scout Vehicle

ART d: additiver Referenzträger
- d: Arbeitsprogramm Rückkehrtechnologie
- e: Accredited Record Technician
- e: Accumulator Read Track and Amsterdam; Rotterdam; Terneuzen
- e: Accumulator Read Truck
- e: Action for Rational Transit
- e: Actual Retention Time
- e: Added Reference Transmission
- e: Additional Reference[-carrier] Transmission
- e: Admissible Rank Test
- e: Advanced Reactor Technology
- e: Advanced Research and Technology
- e: Aerial-Receiver-Transmitter
- e: Airborne Radiation Thermometer
- e: Aircraft Reactor Test
- e: Air Reserve Technician
- e: Alarm Reporting Telephone
- e: Alert Reaction Time
- e: Algebraic Reconstruction Technique
- e: Animated Reconstruction of Telemetry
- e: Antenna-Receiver-Transmitter
- e: Arc-Resistance Tester
- e: Arithmetic Reading Test
- e: Arithmetic Reasoning Test
- e: Arithmetic Reconstruction Technique
- e: Army Research Task
- e: Arterial
- e: Artery
- e: Article
- e: Assisted Reproductive Technologies
- e: Augmented Reentry Test
- e: Authorization and Resource Table
- e: Automated Request Transmission
- e: Automatic Radiating Tester
- e: Automatic Range Tracker
- e: Automatic Range Tracking
- e: Automatic Reporting Telephone
- e: Average Run-Time
- e: Aviation Radio Technician

art a: artikel
- e: artificial
- e: artificer
- p: artigo
- s: articulo

Art. d: Arterie[n]
- d: Artikulation
- d: Artillerie
- e: Artisan

art. d: artifiziell
- d: artikuliert
- e: articulation
- e: artifact
- e: artifilcial
- e: artillery

ARTA e: American Radio Telegraphists Association
- e: Assistant Regional Technical Adviser

Art.-Abt. d: Artillerieabteilung
ARTAC e: Advanced Reconnaissance and Target Acquisition Capabilities
ARTACOM e: Army Requirements for Tactical Communications
ARTADS e: Army Tactical Data System[s]
ARTAD System e: Army Tactical Data System
artanh e: Areatangens
- e: Inverse Hyperbolic Tangent

ArtB e: Art Bulletin (journ.)
ARTC e: Air Route Traffic Control
- e: Associate of the Royal Technical College

ARTCC e: Air Route Traffic Control Center
ARTCCs e: Air Route Traffic Control Centers
ARTCLD e: Articulated
ARTCS e: Advanced Radar Traffic Control System
ARTC System e: Advanced Radar Traffic Control System
ARTDT e: Army Tactical Data Test
ARTE e: Admiralty Reactor Test Establishment
Art. E. e: Artificer Engineer
ARTEC e: Association of Radio and Television Employees of Canada
Art.-Einh. d: Artillerieeinheit
ARTEMIS e: Automated Retrieval of Text from Europes Multinational Information Service
- e: Automatic Retrival of Test through European Multipurpose Information Services

ARTF e: Advanced Radiation Technology Facility
- e: Artificial

artgh e: Inverse Hyperbolic Tangent
arth e: Area Hyperbolic Tangent
Arth e: Arthropoda
Arthritis Rheum. e: Arthritis and Rheumatism (journ.)
ARTI e: Acoustic Ray Trace Indicator
- e: Arab Regional Telecommunication Institute

artic e: articulate[d]
- e: articulation

ARTIC System e: Associometrics Remote Terminal Inquiry Control System
ARTIF e: Artificial[ly]
artif. e: artificer
ARTINS e: Army Terrain Information System
art insem e: artificial insemination
ARTIS d: Artikelinformationssystem
Art J e: Art Journal (journ.)
ARTL e: Awaiting Results of Trial
ARTLY e: Artillery
Art N e: Art News (journ.)
ARTNA e: Association of Radio-Television News Analysts
art nr a: artikelnummer
Art.-Nr. d: Artikelnummer
ARTO e: Advanced Radiation Technology Office
- e: Air Run Take-Off
art° i: articulo
ARTOB e: Amateur Radio Translater on Balloon
- e: Amateur Radio Translation on Balloon
ARTOC e: Army Tactical Operational Control
- e: Army Tactical Operations Center
- e: Army Tactical Operations Central
Art.-Off. d: Artillerieoffizier
ARTP e: Air Reserve Technician Plan
- e: Air Reserve Technician Program[me]
- e: Army Rocket Transportation System
ARTPS e: Army Rocket Transporting Systems
Art Q e: Art Quarterly (journ.)
ART/R e: Airframe Repair Technician-Repairman
ARTRAC e: Advanced Range Testing, Reporting and Control
- e: Advanced Ranging Testing, Reporting and Control
- e: Advanced Real-Time Range Control
- e: Automatic Range Testing, Reporting and Control
Art.-Rgt. d: Artillerieregiment
ART RON e: Artificial Neutron
artron[s] e: artificial neuron[s]
ARTS e: Advanced Radar Terminal System
- e: Advanced Radar Tracking System
- e: Advanced Radar Traffic [Control] System
- e: Adventist Radio Television Services
- e: All-Altitude Air-Bearing Research and Training Simulator
- e: Amateur Radio-Teletype Society
- e: American Radio Telephone System
- e: Annual Report of Transport Statistics
- e: Annual Research Task Summary
- e: Arkansas Research Test Station
- e: Armed-Forces Radio und Television System (or Service)
- e: Army Research Task Summary
- e: Associated Radio and Television Servicemen
- e: Audio Response Time-Shared System
- e: Automated Radar Terminal System
- e: Automated Radar Tracking System
- e: Automatic Radar Traffic Control System
- e: Automatic Resistance Test Set
- e: Automatic Response Tape System
- e: Automatic Response Tuning System
Arts e: Articles
ARTS-DB e: Analysis of Real-Time Systems-Data Base-oriented
ARTS & P e: Australian Radio Technical Sevices and Patents
ARTSS e: Audio Response Time Shared System
ARTS System e: Audio Response Time Shared System
ARTTIS e: Advanced Real Time Total Information System
ARTU e: African Radio and Television Union
- e: Automatic Range Tracking Unit
ARTVS e: Australian Rocket Test Vehicle Simulator
Arty e: Artillery
Arty Am DP e: Artillery Ammunition Distributing Point
Arty engmt e: Artillery Engagement
Arty FC e: Artillery Fire Control
Arty Ln O e: Artillery Liaison Officer
Arty Pk e: Artillery Park
Arty prepn e: Artillery Preparation
arty r e: Artillery Reconnaissance
Arty Sch e: Artillery School
ARU e: Acoustic Resistance Unit
- e: Address Recognition Unit
- e: Aeromedical Research Unit
- e: Aircraft Reception Unit
- e: Air Rescue Unit
- e: Air Reserve Unit
- e: Analog[ue] Remote Unit
- e: Analog[ue] Response Unit
- e: Application Resource Unit
- e: Architecture Research Unit
- e: Arithmetical Unit
- e: Armo[u]red Replacement Unit
- e: Astrophysics Research Unit
- e: Asynchronous Response Mode
- e: Attitude Reference Unit
- e: Audio Response Unit
- e: Automatic Range Unit
ArUe d: Anrufübertragung
ARU-G e: Air Reserve Unit-General Training
ArUmr d: Anrufumrechner
ARUMW d: Anrufumwerter
ARUNK e: Arrival Unknown
aruss. d: altrussisch
ARV d: Arbeitsgemeinschaft Rationeller Verkehr
- d: Auslandsverkehr-Richtungsverbinder
- e: Aeroballistic Re-entry Vehicle
- e: Aerospace Research Vehicle
- e: Afferent Renal Vein
- e: Alternate Record-Voice
- e: American Revised Version (Bible)
- e: Armoured Recovery Vehicle
arv e: arrive
ARVA e: Aircraft Repair Ship for Aircraft
- e: Associate of the Rating and Valuation Association
AR Valve e: Amplifying Relay Valve
ARVE e: Aircraft Repair Ship for Engines
arvg e: arriving
ARVH e: Aircraft Repair Ship for Helicopter
ARVIA e: Associate of the Royal Victorian Institute of Architects
arvl e: Arrival
ARVO e: Association for Research in Vision and Ophthalmology
ARVT d: Anrufverteiler
- d: Anrufverteilung
ARVTE d: Anrufverteileinrichtung
ARW d: Amtsrufweiterschaltung
- d: Anrufweiterschaltung
- d: Arbeitsgemeinschaft der Rundfunkwerbung
- d: Arbeitsgemeinschaft Rheinwasserwerke
- d: Archiv für Religionswissenschaft (journ.)
- d: Auslandsrichtungswähler
- d: Autoreparaturwerkstatt
- e: Air Raid Warden
- e: Air Raid Warning
- e: Anti-Reset Windup
- e: Atmospheric Radio Wave
- e: Attitude Reaction Wheel
ArW d: Anrufwähler
ARWA e: Associate of the Royal West of England Academy
ARWAB e: Army Rotary Wing Aptitude Battery
AR-Wire e: Address Read Wire
ARWM e: A-Register Word Mark
ARWO e: Air Raid Warning Officer
ARWP d: Archiv für Rechts- und Wissenschaftsphilosophie (journ.)
ARWS e: Antiradiation Weapon System
ARX e: Air Regenerative Exhaust
- e: Automatic Retransmission Exchange
ARXT e: American Registry of X-ray Technicians
ARY e: Array
ARZ d: Achillessehnenreflexzeit
- d: Anrufzusatz
- d: Ausbildungsrechenzentrum
- d: Automatisierungs- und Rationalisierungszentrum
- d: Autoreisezug
- e: Apothekenrechenzentrum
Arzneim.-Forsch. d: Arzneimittel-Forschung. Drug Research (journ.)
Arzn G d: Arzneimittelgesetz
ArznMVO d: Arzneimittelverordnung
ArznMWO d: Arzneimittelverordnung
ärztl. d: ärztlich
Ärztl. Forsch. d: Ärztliche Forschung (journ.)
ÄrztO d: Ärzteordnung
AS d: Abfrageschalter
- d: Abfragesignal
- d: Abraumschnitt
- d: Abrechnungssystem[e]
- d: Absatzsystem
- d: Abschirmung
- d: Abschlußschütz
- d: Absetzer
- d: Absperrschieber
- d: Abweichungssignal
- d: Ackerschlepper
- d: Adreßsteuerung
- d: Aktionsstrom
- d: Allstrom
- d: Allylsulfonsäure
- d: Aluminiumsulfat
- d: Aminosäure
- d: Aminosilan
- d: Ammoniumsalpetersprengstoff
- d: Amperesekunde[n]
- d: Amtliche Sammlung
- d: Amtssatz
- d: Anaerobierserum
- d: Anflugführungssender
- d: Anilidsäure
- d: Anodenschalter
- d: Anrufsender
- d: Anrufsucher
- d: Anschluß
- d: Anschlußschnur
- d: Anschlußstecker
- d: Anschlußsteuereinheit
- d: Anschlußsteuerung
- d: Anschlußstufe
- d: Anti-Streptolysin
- d: Antiserum
- d: Antriebseite
- d: Antriebssystem
- d: Anwendersoftware
- d: Anwendersystem

- d: Anwendungsschnittstelle
- d: Anwendungssimulation
- d: Anwendungssystem
- d: Aortenstenose
- d: Apollo Saturn
- d: Arbeitsschritt
- d: Arbeitsschutz
- d: Arbeitssicherheit
- d: Arbeitsspeicher
- d: Arbeitsstudium
- d: Aristolochiasäure
- d: Arsen
- d: Arteriosklerose
- d: Ascorbinsäure
- d: Assemblersprache
- d: Assemblierer
- d: Astigmatismus
- d: Audiosignal
- d: Aufbauschule
- d: Aufschalten
- d: Aufschalteschutz
- d: Aufzeichnungsspalt
- d: Aufzugsachverständiger
- d: Ausfahrtsignal
- d: Ausführungsstruktur
- d: Ausgang setzen
- d: Ausgangsschalter
- d: Ausgangssignal
- d: Auslegeschrift
- d: Ausschalter
- d: Außendienststeuerung
- d: Außenstation
- d: Austast-Synchronsignal
- d: Austauschselektivität
- d: Automatik Sprinkler
- d: Automatische Steuerung
- d: Automatisierungssystem
- d: Autoschutz
- d: Azetylenschweißen
- e: Abilene & Southern
- e: Abnormal Statement
- e: Absolute Sensitivity
- e: Absolute Skip
- e: Absolute Symbol
- e: Absolute System
- e: Absorption Signal
- e: Absorption Switch
- e: Academy of Science[s]
- e: Acceptance Sampling
- e: Accessorial Service
- e: Access Speed
- e: Access Station
- e: Access Switch
- e: Accounting Sequence
- e: Accounting Symbol
- e: Accounting System
- e: Account Sales
- e: Account Sequence
- e: Account Sheet
- e: Account Stated
- e: Accumulating Shift
- e: Accumulating Speed
- e: Accumulator Sign
- e: Accumulator Specification
- e: Acknowledging Switch
- e: Acoustic Storage
- e: Acoustic Store
- e: Acquisition Section
- e: Active Storage
- e: Active Store
- e: Activity Start
- e: Actual Sign
- e: Actuarial Science
- e: Actuarial Statistics
- e: Adapter; Straight
- e: Adapter Section
- e: Add-Substract
- e: Adder Stage
- e: Additional Storage
- e: Addition Slip
- e: Addressable Storage
- e: Addressable Store
- e: Addressing Scheme
- e: Addressing Signal
- e: Addressing System
- e: Address Search
- e: Address Section
- e: Address Selection
- e: Address Selector
- e: Address Space
- e: Address Start
- e: Address Statement
- e: Address Storage
- e: Address Store
- e: Address Substitution
- e: Address Switch
- e: Address Syllable
- e: Address System
- e: Add Statement
- e: Admiral Superintendent
- e: Adrenergic ß-Stimulators
- e: Advanced System[s]
- e: Advance Selector
- e: Aerial Assembly
- e: Aeronautical Standard[s]
- e: Aerospace Standard[s]
- e: Aerospace Studies
- e: After Sight
- e: Aggregate Signal
- e: Agricultural Society
- e: Air-Shuttle
- e: Air-Space[d]
- e: Air-to-Surface Missile
- e: Aircraft Standard[s]
- e: Airless Spraying
- e: Airports Service
- e: Airscoop
- e: Air Screw
- e: Air Seasoned
- e: Air Service
- e: Air Specification
- e: Air Speed
- e: Air Staff
- e: Air Station
- e: Air Support
- e: Air Surveillance
- e: Alaska Airlines
- e: Algebraical Sum
- e: Algebraic Sign
- e: Algorithmic System
- e: Alignment Signal
- e: Alloter Switch
- e: Alloy Steel
- e: Alongside
- e: Alphabetic Sequence
- e: Alphabetic Sort
- e: Alphabetic Storage
- e: Alphabetic String
- e: Alphanumeric Store
- e: Alternating Series
- e: Alter Statement
- e: Alto-Stratus
- e: Alumin[i]um and Steel
- e: American Samoa
- e: American Speech (journ.)
- e: Amicable Settlement
- e: Ammeter Switch
- e: Ammunition Ship
- e: Ammunition Store
- e: Amperemeter Switch
- e: Ampere Second
- e: Amplitudensieb
- e: Analog[ue] Selection
- e: Analog[ue] Signal
- e: Analog[ue] Store
- e: Analog[ue] System
- e: Angle of Sight
- e: Ankylosing Spondylitis
- e: Annals of Science
- e: Antenna Assembly
- e: Anthranilate Synthetase
- e: Anti-Submarine
- e: Antiseptic
- e: Antisurface
- e: Aortic Stenosis
- e: Application Software
- e: Applied Science
- e: Apprentice Seaman
- e: Aqueous Solution
- e: Area Surveillance
- e: Arithmetic Section
- e: Arithmetic Series
- e: Arithmetic Shift
- e: Arithmetic Statement
- e: Arithmetic Subroutine
- e: Armament Supplies
- e: Armature Shunt
- e: Armor School
- e: Army Security
- e: Array-Scaling
- e: Array Segment
- e: Arsenic
- e: Arteriosclerosis
- e: Articulation Score
- e: Artificial Satellite
- e: Ascendance Submission
- e: Ascent Stage
- e: Assembler
- e: Assemble Subprogram[me]
- e: Assembly Statement
- e: Assembly System
- e: Assignment Statement
- e: Assignment Symbol
- e: Assistant Secretary
- e: Assistant Surgeon
- e: Associate in Science
- e: Associative Storage
- e: Associative Store
- e: As Stated
- e: Astronomy
- e: Atlantic Semiconductor
- e: Atmosphere and Space
- e: Atomic Strike
- e: At Sight
- e: Audible Signal
- e: Audio Sensitivity
- e: Audio Station
- e: Auger Spectroscopy
- e: Augmentation System
- e: Augmented Surveyor
- e: Australian Standard[s]
- e: Automatic Skin
- e: Automatic Sprinkler
- e: Automatic Stop
- e: Automation System
- e: Autostore
- e: Auxiliary Software
- e: Auxiliary Storage
- e: Auxiliary Store
- e: Availability Store
- e: Available Store
- e: Aviation Support
- e: Avoidable Shock
- R: Aviaeskadra

as. d: altsächsisch
- d: aseptisch
- d: asiatisch
- d: asymmetrisch
- e: asymmetric
- n: aanstaande

a. S. d: am Samstag
- d: an der Saale
- d: auf Seite
- d: auf Sicht
As d: Asbest
- e: Asbestos
- e: Astigmatism
- e: Asymmetric[al]
- e: Asymmetry
- e: Asymptotic[al]
As e: Asia[n]
- e: Australia[n]
As. d: Aspe
- e: Asiatic
A.s. d: Auslaßventil schließt
a.s. l: auris sinistra
A/s d: Anschläge pro Sekunde
A/S e: Adder/Substracter
- e: Asynchronous/Synchronous
a/s. f: aux soins de
A-S e: Anglo-Saxon
A & S e: Arts and Sciences
- e: Arts and Sciences (journ.)
A_s e: atmosphere standard
AS-11 e: Nord air-launched antitank missile
AS-12 e: Nord automatic-telecommand antitank missile
AS-20 e: Nord air-to-surface radio-controlled missile
AS-30 e: improved version of AS 20
AS-33 e: Nord air-launched inertial-guidance missile
ASA d: Acrylester-Styrol-Acrylnitril
- d: Antriebsschlupf-Anlage
- d: Anzeiger für Schweizerische Altertumskunde (journ.)
- d: Arbeitskreis Sanitär-Ausstattungen
- d: Arbeitsschutzartikel
- d: Arbeitsschutzausschuß
- d: Assoziationsstrukturanalyse
- d: automatische Scharfabstimmung
- d: Automobil-Service-Ausrüstungen
- e: Abort Sensor Assembly
- e: Accelerated Storage Adapter
- e: Access Store Address
- e: Acetylsalicyclic Acid
- e: Acoustical Society of America
- e: Acoustics and Signal Analysis
- e: Acrylonitrile / Styrene / Acrylate
- e: Actuarial Society of America
- e: Adjustable Shock Absorber
- e: Advanced System Avionics
- e: African Studies Association
- e: Alaska Airlines
- e: Aluminum Siding Association
- e: Always Safely Afload
- e: American Scientific Affilliation
- e: American Shellfisheries Association
- e: American Shorthorn Association
- e: American Society for Aesthetics
- e: American Society of Agronomy
- e: American Society of Anesthesiologists
- e: American Sociological Association
- e: American Sociometric Association
- e: American South African Line
- e: American Soybean Association
- e: American Standards Association
- e: American Statistical Association
- e: American Studies Association
- e: Amplifier and Switch Assembly
- e: Anterior Sorting Area
- e: Anthroposophical Society of America
- e: Anti-Static Additive
- e: Area Scanning Alarm
- e: Arginino-Succinic Acid
- e: Army Seal of Approval
- e: Army Service Area
- e: Asparate ß-Semialdehyde
- e: Asynchronous-Synchronous Adapter
- e: Asynchronous Adapter
- e: Atomic Scientists Association
- e: Atomic Security Agency
- e: Auditory Sensation Area
- e: Australian Society of Anaesthetists
- e: Automatic Secondary Allocation
- e: Automatic Separation System
- e: Automatic Spectrum Analyzer
- e: Automatic Steering Antenna
- e: Automatic Systems Analysis
- e: Auxiliary Store Area
- e: Aviation Supply Annex
- e: Azide Styphnate Alumin[i]um
- e: Azimuth Servo Assembly
- s: Aerovias Sud Americana
ASAA e: American Society for the Abandonment of Acronyms
- e: Asian Studies Association of Australia
- e: Associate of the Society of Incorporated Accountants and Auditors
ASAALH e: Association for the Study of Afro-American Life and History
ASAAS e: Asymmetric Stress Analysis of Axisymmetric Solids
ASAB e: Association for the Study of Animal Behavior
ASAC e: Active Satellite Attitude Control
- e: Air Service Area Command
- e: Altostratus and Altocumulus
- e: American Society for African Culture
- e: American Society of Agricultural Consultants
- e: Anti-Submarine Air Control[er]
- e: Asian Standards Advisory Committee
- e: Automatic Selection of Any Channel
asächs. d: altsächsisch
ASACS e: Airborne Surveillance and Control System
ASAD e: Assembly Aid
- e: Authorized Shortages and Discrepancies
ASAE e: Advanced School of Automobile Engineering
- e: American Society of Agricultural Engineers
- e: American Soeiety of Automotive Engineers
ASAES e: Army Small Arms Experimental Station
ASAF e: Assistant Secretary of the Air Force
ASAF(M) e: Assistant Secretary of the Air Force (Management)
ASAF(MA) e: Assistant of the Air Force (Materiel)
ASAG d: Auftragssteuer- und Arbeits-gangdaten
ASAH e: American Society of Association Historians
ASA (I&L) e: Assistant Secretary of the Army (Installations & Logistics)
ASAIO e: American Society for Artificial Internal Organs
ASAK d: Anzeiger für Schweizerische Altertumskunde (journ.)
ASALA e: Associate of the South African Library Association
ASA-Leder d: Arbeitsschutzartikel-Leder
ASALM e: Advance[d] Strategic Air-Launch[ed] [Multimission] Missile
ASAM d: Automatisierungssystem mit autonomen Mikroprozessoren
- e: American Society for Abrasive Methods
- e: Aspen Systems Access Method[s]
- e: Associate of the Society of Art Masters
ASAMP e: Airplane Sizing and Mission Performance
ASA(MP&RF) e: Assistant Secretary of the Army (Manpower and Reserve Forces)
ASA(M&RA) e: Assistant Secretary of the Army (Manpower and Reserve Affairs)
ASAN i: Adriatica Società per Azioni di Navigazione
ASAnes e: American Society of Anesthesiologists
ASAO d: Arbeitsschutzanordnung[en]
- e: Association for Social Anthropology in Oceania asap analog system assembly pack
ASAP d: Asynchronanpassung
- e: Academic and Social Anxiety Program[me]
- e: Advanced Scientific Array Processor
- e: Aircraft Synthesis Analysis Program[me]
- e: Air Service Acceptance Park
- e: Alcohol Safety Action Program[me]
- e: Alcohol Safety Action Project
- e: American Society for Automation in Pharmacy
- e: Analog[ue] System Assembly Pack
- e: Antenna-Scatterer Analysis Program[me]
- e: Anti-Submarine Attack Plotter
- e: Applied Systems and Personnel
- e: Army Scientific Advisory Panel
- e: A Scientific Application Programmer
- e: As Soon As Possible
- e: Australian Society of Animal Production
- e: Automated Statistical Analysis Program[me]
- e: Automated Symbolic Artwork Program[me]
- e: Automatic Spooling with Asynchronous Processing
- e: Auxiliary Spooling Analysis Program[me]
- e: Auxiliary Storage and Playback
ASAP Assembly e: Auxiliary Storage and Playback Assembly
ASAPR e: Accelerated Strike Aircraft Program[me] Requirement
ASAR e: Advanced Surface-to-Air Ramjet
- e: Analog[ue] System Assembly Rack
ASA (R&D) e: Assistant Secretary of the Army (Research & Development)
AS Area e: Anterior Speech Area (brain)
ASARS e: Advanced Synthetic Aperture Radar System
ASAS e: Active Scattering Aerosol Spectrometer
- e: Alternate Stability Augmentation System
- e: American Society of Agricultural Sciences
- e: American Society of Animal Science
- e: Atkins (Computing Services Ltd.) Stress Analysis System
ASA System e: Automatic Separation Assurance System
ASAT d: Arbeitsgemeinschaft Satellitenträger
- e: American Standard Acoustical Terminology
- e: Anti-Satellite
- e: Automatic Spares Analysis Technique
ASATs e: Antisatellite Weapons
ASATT e: Advanced Small-Axial-Turbine Technology
ASAW d: Abhandlungen der Sächsischen Akademie der Wissenschaften zu Leipzig (journ.)
ASAWS e: Advanced Surface-to Air Weapon System
ASB d: Anschlußbereich
- d: Antreiben, Steuern, Bewegen
- d: Arbeitsschutzbestimmung[en]

- d: Arbeitssteuerblock
- d: Aussetzbetrieb
- d: Aussetzschaltbetrieb
- d: Autoschienenbahn
- e: Accumulator Storage Battery
- e: Additional Sense Byte
- e: Administration and Storage Building
- e: African Studies Bulletin (journ.)
- e: Airborne Special Bombing
- e: Aircraft Safety Beacon
- e: Aircraft Services Base
- e: Alternating Sideband
- e: Alternating Sideband Note
- e: American Society of Bacteriologists
- e: American Society of Biochemists
- e: Anencephaly of Spina Bifida
- e: Anti-Shock Body
- e: Anxiety Scale for the Blind
- e: Armed Services Bulletin
- e: Army Supply Base
- e: Asbestos
- e: Associate in Science in Business
- e: Association of Southeastern Biologists
- e: Asymmetric Sideband

asb d: Apostilb
Asb d: Asbest
AS & B e: Aloin, Strychnine and Belladonna
ASBA e: American Shorthorn Breeders Association
- e: American Soy Bean Association

ASBAH e: Association for Spina Bifda and Hydrocephalus
ASBAL e: A Stack Based Abstraction Language
ASBC e: Advanced Standard Buried Collector
- e: American Society of Biological Chemists
- e: American Society of Biophysics and Cosmology
- e: American Society of Brewing Chemists
- e: American Standard Building Code

ASB-C e: Asbestos Covered
ASBC Technology e: Advanced Standard Buried Collector Technology
ASBD e: Advanced Sea-Based Deterrent
ASBD Program[me] e: Advanced Sea-Based Deterrent Program[me]
A.S.B.E. e: Associate in Science in Basic Engineering
ASBH d: Arbeitskreis der Studenten an Berufspädagogischen Hochschulen
ASBHCA e: Average Season Busy Hour Call Attempts
ASBHCC e: Average Season Busy Hour Call Completions
ASB & I e: Aloin, Strychnine, Belladona and Ipecac
ASBIPC e: American Sugar Beet Industry Policy Committee
ASB & I Pills e: Aloin, Strychnine, Belladonna and Ipecac Pills
ASBL e: Assemble[r]
ASBLY e: Assembly
ASBM e: Associate in Business Management
ASB Method e: Alternating Sideband Method
AS-Boot d: Artillerieschnellboot
ASBPA e: American Shore and Beach Preservation Association
AS & B Pills e: Aloin, Strychnine and Belladonna Pills
ASBSTS e: Asbestos
ASBU e: Arab States Broadcasting Union
ASB-Verfahren d: Alternierend-Seitenband-Verfahren
ASC d: Anti-Schlupf-Control
- e: Abbe Sine Condition
- e: Accelerometer Signal Conditioner
- e: Acetylsulfanilyl Chloride
- e: Acquisition and Serials Control
- e: Adaptive Signal Correcting
- e: Adaptive Signal Correction
- e: Adaptive Speed Control
- e: Adaptive Subband Coding
- e: Add-Substract Counter
- e: Additive Source Current
- e: Address Substitution Command
- e: Adelaide Steamship Company
- e: Administrative Service Center
- e: Advanced Scientific (or Science)
- e: Advanced Scientific (or Science) Computer
- e: Advanced Seismic Computer
- e: Advanced Space Computer
- e: Advanced System Concept
- e: Aeronautical Systems Center
- e: Aerospace Static Converter
- e: Agricultural Stabilization and Conservation
- e: Air-Space[d] Cable
- e: Aircraft Service Change
- e: Air Service Command
- e: Air Support Command
- e: Air Support Control
- e: Air Systems Command
- e: Alabama State College
- e: Alaska Steamship Company
- e: All-Sky Camera
- e: All-Steps Control
- e: Allowance Source Code
- e: All Steps Control
- e: Altered State of Consciousness
- e: Alternate Squadron Commander
- e: American Satellite Company
- e: American Satellite Corporation
- e: American Semiconductor Corporation
- e: American Society for Cybernetics
- e: American Society of Cytology
- e: Amount Sign Code
- e: Ampere[s] per Square Centimeter
- e: Analog[ue]-to-Stochastic Converter
- e: Analog[ue] Signal Converter
- e: Analog[ue] Signal Correlator
- e: Analog[ue] Strip Chart
- e: Annapolis Science Center
- e: Anti-Static Compound
- e: Arbitrary Sequence Computer
- e: Arizona State College
- e: Arkansas State College
- e: Army School of Cookery
- e: Army Service Command
- e: Army Signal Corps
- e: Army Subsistence Center
- e: Assigned Service Contractor
- e: Associated Chip Chandlers
- e: Associated Schools of Construction
- e: Associate in Science in Commerce
- e: Associative Structure Computer
- e: Astronautics Support Center
- e: Asynchronous Communication Procedure
- e: Atlantic Seaboard Circuit
- e: Atomic Sciences Committee
- e: Atomic Support Command
- e: Audio-Selectronic
- e: Aughey Spark Chamber
- e: Automated System Character
- e: Automated System Charter
- e: Automatic Digital Network Switching Center
- e: Automatic Scan Counter
- e: Automatic Selective Control
- e: Automatic Selectivity Control
- e: Automatic Sensitivity Control
- e: Automatic Sequence Control
- e: Automatic Switching Center
- e: Automatic Switching Control
- e: Automatic Synchronized Control
- e: Automatic System Control[ler]
- e: Auxiliary Store Capacity
- e: Auxiliary Store Carrier
- e: Auxiliary Switch Closed

as & c e: aerospace surveillance and control
ASc e: Annals of Science (journ.)
- e: Associate in Science

asc e: arteriosclerosis
- e: arteriosclerosistic
- e: ascarid
- e: ascend

ASCA e: American Society of Consulting Arborists
- e: American Standard Chinchilla Association
- e: Association for Science Cooperation in Asia
- e: Automatic Science Citation Alert[ing]
- e: Automatic Subject Citation Alert

AS cable e: Air-Space Cable
ASCAC e: Anti-Submarine Classification Analysis Center
- e: Anti-Submarine Contact Analysis Center

ASCAP e: Artwork and Schematic Analysis Program[me]
- e: At-Sea Calibration Procedure

ASCAS e: All-Service Close Air Support
ASCAT e: Analog[ue] Self-Checking Automatic Tester
- e: Anti-Submarine Contact Analysis Team

ASCATS e: Apollo Simulation Checkout and Training System
ASCB e: Address Space Control Block
- e: American Society for Cell Biology
- e: American Society of Cell Biology

ASCC e: Advanced System Computer Center
- e: Aeronautical Satellite Communications Center
- e: Aeronautical Services Communication Center
- e: Air Standardization Coordinating Committee
- e: American Small Center Cap
- e: American Society for Concrete Construction
- e: American Society for the Control of Cancer
- e: Army Strategic Communications Command
- e: Australian Society of Cosmetic Chemists
- e: Automatic Sequence-Control[led] Calculator
- e: Automatic Sequence-Controlled Computer

ASC Calculator e: Automatic Sequence-Controlled Calculator
ASC Computer e: Automatic Sequence-Controlled Computer
ASCD e: American Society of Computer Dealers
ASCE e: Abrupt Space Charge Edge
- e: Air Standard Cycle Efficiency
- e: Allied Supreme Commander in Europe
- e: American Society of Civil Engineers
- e: Annual Schedule of Circuit Estimates

ASCEA e: American Society of Civil Engineers and Architects
Ascen. e: Ascension

ASCEND e: Advanced System for Communications and Education in National Development
ASCENT e: American Scientist (journ.)
- e: Assembly System for Central Processor ASci
ASCENT Processor e: Assembly System for Central Processor
ASCEP e: Australian Society of Clinical and Experimental Pharmacologists
ASCET e: American Society of Certifed Engineering Technicians
ASCGW e: Action-Study Center for a Governed World
A Sch e: Armament School
ASCHAL e: American Society of Corporate Historians, Archivists and Librarians
ASCHE e: American Society of Chemical Engineers
ASchO d: allgemeine Schulordnung
Aschr. d: Anschrift
ASci e: American Scientist (journ.)
ASCI e: American Society for Clinical Investigation
ASCID e: Altered States of Consciousness Induction Device
ASCIE e: American Standard Code for Information Exchange
ASCII e: American Standard Code for Information Interchange
ASCII Code e: American Standard Code for Information Interchange Code
ASCIIK e: American Standard Code for Information Interchange Keyboard
ASCIS e: Australian School Cataloguing Information Service
ASCL e: Advanced System Concepts Laboratory
- e: American Sugar Cane League
ASCLA e: Association of Specialized and Cooperative Library Agencies
ASCLS e: American Society of Contact Lens Specialists
ASCM e: Antiship Capable Missile
- e: Association of Steel Conduit Manufacturers
As/cm² d: Amperesekunde[n] pro Quadratzentimeter
ASCN e: American Society of Clinical Nutrition
ASCO e: Abort Sensing Control Unit
- e: American Society of Contemporary Ophthalmology
- e: Anti-Submarine Commander
- e: Arab Satellite Communications Organization
- e: Automatic Sustainer Cut-Off
- e: Auxiliary Sustainer Cut-Off
ASC-O e: Access Opening
ASCOM e: Anti-Static Computer
- e: Area Service Command
- e: Army Satellite Communication Project
ASCOM Project e: Army Satellite Communication Project
ASCON e: Automatic Switched Communications Network
ASContl e: Air Support Control
ASCOP e: A Statistical Computing Procedure
ASCORE e: Automatic Shipboard Checkout and Readiness Equipment
ASCOT e: Advanced Storage Control Test
- e: Analog[ue] Simulation of Competitive Operational Tactics
A Scot e: type-A Scottish influenza virus
ASCP e: Air Standardization Coordination Program[me]
- e: American Society of Clinical Pathologists
- e: American Society of Consulting Pharmacists
- e: Army Strategic Capabilities Plan
- e: Automatic System Checkout Program[me]
ASCPC e: American Society of Clinical Pharmacology and Chemotherapy
ASCPT e: American Society for Clinical Pharmacology and Therapeutics
ASCR e: Advanced Sodium Cooled Reactor
- e: Analog[ue] Strip Chart Recorder
- e: Asymmetric[al] Silicon Controlled Rectifier
ascr l: ascriptum
ASCS e: Agricultural Stabilization and Conservation Service
- e: American School of Classical Studies
- e: American Standard Character Set
- e: Area Surveillance Control System
- e: Atmosphere Storage and Control Section
- e: Attitude and Spin Control Subsystem
- e: Attitude Sensing and Control System
- e: Automatic Scan Counter System
AS & CS e: Automatic Stabilization and Control System
A/S Cse e: Anti-Submarine Course
ASCT e: Address Space Control Task
ASCTA e: Association of Short Circuit Testing Authorities
ASCU e: Air Support Control Unit
- e: Armament Station Control Unit
- e: Association of Small Computer Users
- e: Association of State Colleges and Universities
- e: Automatic Scanning Control Unit
ASCUE e: Association of Small Computers in Education
ASCUFRO e: Association of State Colleges and Universities Forestry Research Organizations
ASCVD e: Arteriosclerotic Cardiovascular Disease
A Sc W e: Association of Scientific Workers
ASD d: Auftragssteuerdaten
- d: automatisches Sperrdifferential
- e: Acceleration Spectral Density
- e: Adjustable Speed Drive
- e: Advanced Ship Development
- e: Advanced Supply Depot
- e: Advanced Systems Development
- e: Advanced Systems Division
- e: Adverse State Detector
- e: Advise Shipping Date
- e: Aeronautical System Development
- e: Aeronautical Systems Division
- e: Aldosterone Secretion Defect
- e: Alphabetic Store Device
- e: Alternate Source Development
- e: Ammunition Subdepot
- e: Ammunition Supply Dump
- e: Ampere[s] per Square Decimetre
- e: Apollo Standard Detonator
- e: Army Shipping Document
- e: Artial Septal Defect
- e: Assignment Selection Date
- e: Assign Symbolic Device
- e: Assistant Secretary of Defense
- e: Atomic Solution Diffusion
- e: Atrial Septal Defect
- e: Automated Structural Design
- e: Automatic Synchronized Discriminator
- e: Aviation Supply Depot
- f: Association Suisse de Documentation
ASDA e: Accelerate-Stop Distance Available
- e: Advanced Saturn Data Adapter
- e: American Seafood Distributors Association
ASDA e: Atomic and Space Development Authority
ASDACS e: Acoustic Signal Data Analysis and Conversion System
ASD-ALA e: Adult Services Division-American Library Association
ASDAR e: Aircraft to Satellite Data Relay
ASDA Supermarket e: Associated Dairies Supermarket
ASDC e: Aeronomy and Space Data Center
ASDD e: Advanced Systems Development Division
- e: Apollo Signal Definition Document
ASDE e: Aerial Slave Data Equipment
- e: Aircraft Surface Detection Equipment
- e: Airport Surface Detection Equipment
- e: Antenna Slave Data Equipment
ASDEC e: Applied System Development Evaluation Center
- e: Automatic Selection of Digital Electronic Computers
ASDER e: Airfield Surface Detection Radar
ASDEX d: axialsymmetrisches Divertorexperiment
ASDF e: Aeronautical Systems Division Form
ASDG e: Automated Systems Development Group
ASD/H&M e: Assistant Secretary of Defense / Health and Medical
ASDI e: Automatic Selective Dissemination of Information
ASDIC e: Anti-Submarine Depth Indicating Control
- e: Armed Services Documents Intelligence Center
ASD/I&L e: Assistant Secretary of Defense / Installations & Logistics
ASDIRS e: Army Study Documentation and Information Retrieval System
ASDL e: Aeronautical Satellite Data Link
- e: Automated Ship Data Library
ASDM e: Aeronautical Systems Division Manual
- e: Apollo-Soyuz Docking Module
ASDM e: Automated Systems Design Methodology
ASDNG e: Ascending
ASDP e: Advanced Sensor Development Program[me]
- e: Automatic Shot Dispensing Pump
ASD/P & I) e: Assistant Secretary of Defense / Properties and Installations
ASDPSIM e: Advanced System Data Processing Simulation
ASDR e: Aeronautical Systems Division Regulation
- e: Airborne Sample Data Reduction
- e: Airport Surface Detection Radar
- e: American Society of Dental Radiographers
- e: American Society of Dernmatological Retailers
ASD/R&D e: Assistant Secretary of Defense / Research and Development
ASD/R&E e: Assistant Secretary of Defense Research and Engineering
ASDS d: automatisiertes System der Dispatchersteuerung
- e: Aircraft Sound Description System
- e: American Society of Dental Surgeons

ASD/SA e: Assistant Secretary of Defense / System Analysis
ASD/S&L e: Assistant Secretary of Defense Supply and Logistics
ASDSRS e: Automatic Spectrum Display and Signal Recognition System
ASDSVN e: Army Switched Data and Secure Voice Network
ASDT e: Accumulated Standardized Data Tape
ASD Tape e: Accumulated Standardized Data Tape
ASDTIC e: Analog[ue] Signal to Discrete Time Interval Converter
ASDTP e: Apollo Spacecraft Development Test Plan
ASE d: adaptives Schwellenwertelement
- d: Anschlußsteuereinheit
- d: Ansteuereinheit
- d: Anti-Streptolysin-Einheiten
- d: Arbeitsgemeinschaft Solarenergie
- e: Active Seismic Experiment
- e: Active Switching Element
- e: Admiralty Signal Establishment
- e: Advanced Systems Engineering
- e: Aerospace Support Equipment
- e: Airborne Search Equipment
- e: Airborn Support Equipment
- e: Aircraft Search Equipment
- e: Aircraft Stores Establishment
e: Airport Surface-Detection Equipment
- e: Air Standard Efficiency
- e: Air Store Establishment
- e: Allowable Steering Error
- e: Amalgamated Society of Engineers
- e: American Scientific Engineering
- e: American Society of Engineers
- e: American Society of Enologists
- e: American Steel Equipment
- e: American Stock Exchange
- e: Amplified Spontaneous Emission
- e: Anisotropic Stress Effect
- e: Anti-Submarine Establishment
- e: Application Support Environment
- e: Army School of Education
- e: Army Survey Establishment
- e: Associate in Engineering
- e: Associate of the Society of Engineers
- e: Association of Science Education
- e: Audio Support Equipment
- e: Australian Society of Engineers
- e: Automatic Stabilization Equipment
- e: Automatic Support Equipment
- f: Association Suisse des Electriciens
AS & E e: American Science and Engineering
ASEA e: American Society of Engineers and Architects
- r: Allmänna Svenska Elektriska Aktiebolaget
ASEAG d: Aachener Straßenbahn- und Energieversorgungs-AG
ASEAN e: Association of South-East Asian Nations
ASEB e: Aeronautics and Space Engineering Board
- e: Assam State Electricity Board
Asec d: Amperesekunde[n]
ASEC e: American Standard Elevator Code[s]
- e: Ampere[s]/Second
ASECA e: Association for Education and Cultural Advancement
ASECH d: Acetylselenocholine
ASECS e: American Society for Eighteenth-Century Studies
ASED e: Army School of Education and Depot
ASEE e: American Society for Ecological Education
- e: American Society for Engineering Education
- e: American Society for Environmental Education
- e: American Society of Electrical Engineers
- e: American Society of Electronics Engineers
A/SEE e: Anti-Submarine Experimental Establishment
ASEE e: Associate in Science in Electronic[s] Engineering
- e: Association of Supervising Electrical Engineers
ASEET e: Associate in Science in Electronic Engineering Technology
ASEM e: Anti-Ship Euro-Missile
ASEP e: Advanced Science Education Program[me]
- e: American Society for Experimental Pathology
- e: Array Structure Experiment Package
- e: Automatic Sequence Execution and Processor
ASEPTICS e: Aseptically Packaged Liquids
ASER e: Amplification based on Stimulated Emission of Radiation
- e: Amplification by Stimulated Emission of Radiation
- e: Amplifier Based on Stimulated Emission of Radiation
aserb. d: altserbisch
- d: aserbeidschanisch
A Se S e: Associate in Secretarial Science
ASES e: Automated Software Evaluation System
ASESA e: Armed Services Electro-Standards Agency
A Se Sc e: Associate in Secretarial Science
ASESS e: Aerospace Environment Simulation System
ASET d: Arbeitskommission Sender- und Empfängertechnik
- e: Advanced Sensor Evaluation and Test
- e: Aeronautical Satellite Earth Terminal
- e: American Society for Engineering Technology
- e: Assistant Secretary for Energy Technology
ASETC e: Armed Services Electron Tube Committee
ASETS e: Automatic Storage Evaluator Test System
ASEW e: Airborne and Surface Early Warning
ASEWS e: Airborne and Surface Early Warning System
ASEX e: Asexualization
ASF d: Allgemeine Sicherheitsvorschriften für Fabriken und gewerbliche Anlagen
- d: Anrufsucher für Fernverbindung
- d: automatische Systemfunktion
- e: Active Segment Field
- e: Additional Selection Factor
- e: Additional Storage Feature
- e: Advanced Simulation Facility
- e: Aeromedical Staging Facility
- e: African Swine Fever
- e: Aircraft Services Facility
- e: Air Superiority Fighter
- e: Alaskan Sea Frontier
- e: Alternative Salient Features
- e: American Schizophrenia Foundation
- e: Ammunition Storage Facility
- e: Amorphous Semiconducting Film[s]
- e: Amperes per Square Foot
- e: Amplitude Scale Factor
- e: Aniline Sulphur and Formaldehyde
- e: Anti-Submarine Frigate
- e: Arithmetic Statement Function
- e: Army Stock Fund
- e: Assignable Square Feet
- e: Association of State Foresters
- e: Atomic Scattering Factor
- e: Australian Speleological Federation
- e: Automatic Sheet Feeder
- e: Automatic Signal Filtration
- e: Automatic Signal Finding
- e: Automatic Store and Forward
ASFA e: Aquatic Sciences and Fisheries Abstracts
ASFC e: Atlantic Salt Fish Commission
AsFd d: Anschlußfeld
A/SFDO e: Anti-Submarine Fixed Defenses Officer
ASFE e: Accelerometer Scale Factor Error
- e: American Society For Aesthetics
ASFEC e: Arab States Fundamental Education Centre
ASFFHF e: Association of Science Fiction, Fantasy and Horror Films
ASFG e: Atmospheric Sound-Focusing Gain
ASFIP e: Accelerometer Scale Factor Input Panel
ASFIR e: Active Swept Frequency Interferometer Radar
ASFIS e: Aquatic Science and Fisheries Information Service
- e: Aquatic Science[s] and Fisheries Information System
ASFIS Abstracts e: Aquative Sciences and Fisheries Abstracts
ASFISS e: Advance Simulation Facility Interconnection and Setup System
ASFIT e: Anisotropic Source Flux Iteration Technique
ASFM e: Association of Field Service Managers
ASFMRA e: American Society of Farm Managers and Rural Appraisers
ASFO d: Aufbereitungssystem; festplattenspeicherorientiert
ASFS e: Air Force Stock Fund
ASFTC e: Army Service Forces Training Center
ASFTS e: Airborne Systems Functional Test Stand
ASFTS e: Auxiliary Systems Function[al] Test Stand
ASFW d: automatisiertes System des Finanzwesens
ASFX e: Assembly Fixture
ASG d: Agrarsoziale Gesellschaft
- d: Allwetterschutzgehäuse
AS(g) d: Anrufsucher [für den] Grundverkehr
ASG d: Arbeitsgemeinschaft der Sozialdemokraten im Gesundheitswesen
- d: Arbeitsgemeinschaft Sozialpädagogik und Gesellschaftsbildung
- e: Acid-Saline Giemsa
- e: Acoustic Signal Generator
- e: Acting Secretary General
- e: Aeronautical Standards Group [Air Force and Navy]
- e: Airless Spray Gun
- e: Alcohol-Soluble Glutelin
- e: American Society of Genetics
- e: Antenna Steering Group
- e: Army Surgeon General
- e: Arsenic Silicate Glass

- e: Assign[ed]
- e: Assignment
- e: Automatic Spray Gun

ASGA d: Archiv für Sächsische Geschichte und Altertumskunde (journ.)
ASGAN e: Assistant Secretary of Air Navigation
ASGAT e: Assistant Secretary General for Air Transport
ASGB e: Aeronautical Society of Great Britain
ASGBFR e: Assign Buffer
ASGBI e: Association of Surgeons of Great Britain and Ireland
ASGC e: Advanced Saturn Guidance Computer
ASGD e: Assigned
ASGE e: Acoustic Signal Generator [System]
- e: American Society for Gastrointestinal Endoscopy

ASGLS e: Advanced Space-Ground Link System
- e: Advanced Space Ground Link Subsystem

ASGMT e: Assignment
ASGN e: Assign[ment]
ASGND e: Assigned
ASGp e: Aeronautical Standards Group
ASGP e: Australian Society of General Practitioners
AS/GPD e: Attitude Set and Gimbal Position Display
ASGROW e: Associated Seed Growers
ASGS e: American Scientific Glassblowers Society
ASGW e: Air-to-Surface Guided Weapon
ASH d: Aldosteron-stimulierendes Hormon
- d: Arbeitsgemeinschaft sportwissenschaftlicher Hochschuleinrichtungen
- e: Action on Smoking and Health
- e: Aerial Scout Helicopter
- e: Airship
- e: American Society of Hematology
- e: Antiself Homing
- e: Armature Shunt
- e: Army School of Hygiene
- e: Assault Support Helicopter
- e: Association for the Study of Human Infertility
- e: Australian Society of Herpetologists
- e: Australian Stationary Hospital
- e: Aviation Support; Hydraulic

AsH e: Hyperopic Astigmatism
Ash e: Soviet infrared and radar-homing missile
ASHA e: Aerospace Human Factors Association
- e: American Shire Horse Association
- e: American Suffolk Horse Association

ASHACE e: American Society of Heating and Air-Conditioning Engineers
ASHD e: Arteriosclerotic Heart Disease
ASHE e: American Society of Hospital Engineers
- e: Association for the Study of Higher Education

ASHG e: American Society of Human Genetics
Ash Mus e: Ashmolean Museum
ASHOF e: Air Shutoff
ASHP e: American Society of Hospital Pharmacists
ASHPS e: American Scenic and History Preservation Society
ASHRAE e: American Society of Heating, Refrigerating and Air-Conditioning Engineers
ASHS e: Advanced Study of Human Sexuality
- e: American Society for Horticultural Science

ASH Timber e: Ash Timber
ASHUR e: Apollo Spacecraft Hardware Utilization Request[s]
ASI e: Advanced Scientific Instruments
- e: Advanced Systems Incorporated
- e: Aero-Space Institute
- e: Aerospace Static Inverter
- e: Aerospace Studies Institute
- e: Africa Service Institute
- e: Airsignal International
- e: Air Speed Indicator
- e: Alaska Salmon Industry
- e: Alphabetic[al] Subject Index[ing]
- e: Altimeter Setting Indicator
- e: Amended Shipping Instruction
- e: American Soybean Institute
- e: American Specifications Institute
- e: American Standards Institute
- e: American Statistics Index (journ.)
- e: American Statistics Institute
- e: American Swedish Institute
- e: Amount of Semantic Information
- e: Ampere[s] per Square Inch
- e: Analytic Subject Index
- e: Anti-Saturation Inverter
- e: Apollo Standard Initiator
- e: Application Software, Incorporated
- e: Articulated Subject Index
- e: Asian Statistical Institute
- e: Asynchronous Serial Interface
- e: Augmented Spark Igniter
- e: Augmented System Igniter
- e: Australian Shipbuilding Industries
- e: Automated System Initialization
- e: Aviation Simulations International
- e: Azimuth and Speed Indicator

ASIA e: Army Signal Intelligence Agency
- e: Asian Studies in America
- e: Australian Scientific Industry Association

ASIAC e: Aerospace Structures Information and Analysis Center
Asian-Am[s] e: Asian-American[s]
ASIC d: Anwendersystem-Information Center
- e: Agricultural Science Information Center
- e: Air Service Information Circular (journ.)
- e: Application-Specific Integrated Circuit
- e: Area Security Information Center
- e: Australian Standard Industry Classification

ASICA f: Association Internationale pour la Calcul Analogique
ASID e: Address Space Identification
- e: Address Space Identifier
- e: Alarm Signal Input Device
- e: Australian Society of Implant Dentistry
- e: Automatic Station Identification Device

ASIDIC e: Association of Scientific Information Dissemination Centers
ASIDP e: American Society of Information and Data Processing
ASIDS e: Airborne Surveillance and Intercept Defense System
ASIF e: Avionics Software Integration Facilities
ASiG d: Arbeitssicherheitsgesetz
ASIGCEN e: Area Signal Center
A-Signal d: Austastsignal
A Sigs e: Army Signals
ASI & H e: American Society of Ichthyologists and Herpetologists
ASII e: American Science Information Institute
ASIL e: American Society of International Law
- e: Asymptotic Stability in the Large

ASIM d: Analogsimulation
- e: American Society of Internal Medicine
- e: Analog[ue] Simulation
- e: Associate in Industrial Management

ASIMET s: Asociacion de Industrias Metalurgicas
ASIN e: Agricultural Sciences Information Network
- e: Agricultural Services Information Network

ASINUS d: automatische Schaltungsinterpretation- und -untersuchung
ASIP e: Aircraft Structural Integrity Program[me]
ASIPRE e: Army Snow, Ice and Permafrost Research Establishment
ASIR e: Airport Special Information and Regulations
ASIRC e: Aquatic Sciences Information Retrieval Center
ASIS d: Arbeitsschutzinformationssystem
- e: Abort-Sensing [and] Implementation System
- e: Advanced Scientific Instruments
- e: Alpine Science Information Service
- e: American Society for Industrial Security
- e: American Society for Information Science
- e: Ammunition Stores Issue Ship
- e: Anterior Superior Iliac Spine

AS-ISES e: American Section-International Solar Energy Society
ASIST e: Advanced Scientific Instruments Symbolic Translator
- e: Alzheimer Support, Information and Service Team

ASIT e: Adaptive Surface Interface Terminal
- e: Associate of the Society of Instrument Technology

ASIV d: automatisches System der Informationsverarbeitung
ASIWPCA e: Association of State and Interstate Water Pollution Control Administrators
ASJ d: Arbeitsgemeinschaft Sozialdemokratischer Juristen
- e: Asiatic Society of Japan
- e: Automatic Search Jammer

ASJA e: Assistant Staff Judge Advocate
ASJV d: Allgemeiner Schweizerischer Jagdschutzverein
ASK d: Anionensorptionskapazität
- d: Anschlußkabel
- d: Antistreptokinase
- d: Arbeitsgruppe für Sozial- und Konsumforschung
- d: Arbeitsplan-Stammkarte
- d: Auswahl- und Sachkatalog
- e: Access to Source of Knowledge
- e: Actively Shared Knowledge
- e: Adjustable Stroke Kit
- e: Aircraft Station Keeper
- e: Alphanumeric Standard Keyboard
- e: Amplitude-Shift Keyed
- e: Amplitude-Shift Keying
- e: Analog[ue] Select Keyboard
- e: Applications Software Kit
- e: Askania
- e: Astronaut Survival Kit

ASKA e: Automatic System for Kinematic Analysis
ASKE d: Antistreptokinase-Einheit
AsKf d: Anschlußkoffer
askg e: asking

ASKL d: Anruf- und Schlußkontrollampe
ASKO d: Aufbereitunggssystem; kernspeicherorientiert
ASKS e: Automatic Station Keeping System
ASL d: Aktion Saubere Landschaft e.V.
asl. d: altslawisch
ASL d: Anruf- und Schlußlampe
- d: Anschlußleitung
- d: Antistreptolysin
- d: Atomsicherheitslinie
- d: automatischer Sendersuchlauf
- d: automatisiertes System der Leitung (DDR)
- e: Abbes Sine Law
- e: Above Sea Level
- e: Advanced Simulation Language
- e: Advanced Student in Law
- e: Advanced Systems Laboratory
- e: Aeronautical Structures Laboratory
- e: American Association of State Libraries
- e: American Sign Language
- e: Antenna Systems Laboratory
- e: Applied Science Laboratories
- e: Arithmetic Shift Left
- e: Army Standards Laboratory
- e: Assistant Squad Leader
- e: Association of Symbolic Logic
- e: Astigmatic Spectral Line
- e: Astrosurveillance Science Laboratory
- e: Asymptotic Stability in the Large
- e: Atmospheric Sciences Laboratory
- e: Atomic Safety Line
- e: Australian Society for Limnology
- e: Authorized Stockage List
- e: Automated Soft Lander
- e: Automatic Stereo Level
- e: Auto Stop Latch
- e: Available Space List
- e: Average Service Life
- e: Aviation Systems Laboratory
AsL d: Anschlußlinie
ASLA e: American Society of Landscape Architects
- e: Arizona State Library Association
- e: Australian School Library Association
aslaw. d: altslawisch
ASLB e: Atomic Safety and Licensing Board
ASLBM e: Anti-Submarine Launched Ballistic Missile
ASLBP e: Atomic Safety and Licensing Board Panel
ASLD e: Anti Shop Lifting Device
ASLE e: American Society of Law Enforcement
- e: American Society of Lubrication Engineers
ASLE (Amer. Soc. Lubric. Eng.) e: American Society of Lubrication Engineers Transactions (journ.)
ASLEEP e: Automated Scanning Low Energy Electron Probe
ASLEF e: Associated Society of Locomotive Engineers and Firemen
ASLEP e: Apollo Surface Lunar Experiments Package
ASLH e: American Society for Legal History
ASLI e: American Savings and Loan Institute
- e: Automated System of Legal Information
- e: Automatic Safe Load Indicator
ASLIB e: Association of Special Libraries and Information Bureaus
ASLIB Proc. e: Association of Special Libraries Proceedings (journ.)
ASLIS e: Association of Special Libraries and Information Services
ASLM d: asynchroner/synchroner Leitungsmultiplexer
AsLn d: Anschlußleitungen
ASLO e: American Society of Limnology and Oceanography
- e: Assembly Lay-Out
aslo e: assembly layout
ASLO e: Australian Scientific Liaison Office
ASLP e: Association of Special Libraries in the Philippines
ASLR e: Assembler
ASLT e: Advanced Solid Logic Technology
Aslt e: Assault
Aslt Gn e: Assault Gun
ASL-Titer d: Antistreptolysin-Titer
ASLU e: Antenna Select Logic Unit
ASLV e: Advanced Small Launch Vehicle
ASM d: Arbeitsspeicher-Multiplexer
- e: Adaptive System
- e: Advanced Surface Missile
- e: Advanced System Management
- e: Aerial Switching Matrix
- e: Aerospace Structural Material
-e: Air-to-Surface Missile
- e: Airlift Simulation Model
- e: Air Service Mechanic
- e: Air Superiority Mission
- e: Algorithmic State Machine
- e: Allocation Strategy Module
- e: American Society for Metals
- e: American Society for Microbiology
- e: American Society of Mammalogists
- e: American Society of Microbiology
- e: Ampex Semiconductor Memory
- e: Analog[ue] Storage Module
- e: Analog[ue] Subscriber Module
- e: Angular Second-Moment
- e: Anion Selective Membrane
- e: Antenna Switching Matrix
- e: Apollo Service Module
- e: Apollo Systems Manual
- e: Area Stores Module
- e: Assemble[r]
- e: Assembly
- e: Assistant Staff Meteorologist
- e: Assistant Station Master
- e: Associated Semiconductor Manufacturers
- e: Association for Systems Management
- e: Asynchronous Sequential Machine
- e: Asynchronous State Machine
- e: Audiosonometry
- e: Automated Space Management
- e: Autonomous Sequential Machine
- e: Auxiliary Storage Manager
- e: Available Seat Miles
- e: Aviation School of Medicine
- e: Avionics Shop Maintenance
AsM e: Astigmatism; Myopic
As/m² d: Amperesekunde[n] pro Quadratmeter
ASMA e: Aerospace Medical Associadon
- e: Anti-Smooth Muscle Antibody
- e: Army Signal Material Support Agency
ASMB e: Acoustical Standards Management Board
ASMBD e: Assembled
ASMBL e: Assemble
ASMBLR e: Assembler
AS & MC e: Army Supply and Maintenance Command
ASM/C e: Assemble Complete
ASMC e: Automatic Digital Information Network Station Maintenance Console
- e: Aviation Surface Material Command
ASMD e: American Society of Metal Definitions
- e: Anti-Ship Missile Defense
ASMDDA e: Air Surface Missile Digital Differential Analyzer
ASMD/EW e: Antiship-Missile Defense/ Electronic Warfare
ASME e: American Society of Mechanical Engineers
- e: Association for the Study of Medical Education
ASMEDIT e: Assembler-Editor
AsMem e: Associate Member
ASMER e: Association for the Study of Man-Environment Relations
ASMFC e: Atlantic States Marine Fisheries Commission
ASM/GEN System e: Assembler Generating System
ASMI e: Airfield Surface Movement Indication
ASMM e: American Supply and Machinery Manufacturers
ASMMA e: American Supply and Machinery Manufacturers Association
ASMO d: Aufbereitungssystem; magnetbandorientiert
- e: Arab Standardization and Metrology Organization
ASMOLV e: Afford Service Member Opportunity to Apply for Ordinary Leave
ASMOS d: automatisierte Senderegie Mono oder Stereo
ASMP e: Army Survival Measures Plan
ASMPA e: Armed Services Medical Procurement Agency
ASMPE e: American Society of Motion Picture Engineers
ASM Program[me] e: Aircraft Survival Measures Program[me]
ASMPTE e: American Society of Motion Picture and Television Engineers
ASMR e: Advanced Shortto-Median-Range
- e: Australian Society for Medical Research
ASMRO e: Armed Services Medical Regulating Office
ASMS e: Advanced Surface Missile System
- e: Advanced Synchronous Meteorological Satellite
- e: Associate in Science in Medical Secretarial
- e: Atmosphere Sensing and Maintenance System
ASMT e: Air-Space Multiple-Twin
- e: American Society of Medical Technologists
asmt. e: Assortment
ASMT Cable e: Air-Space Multiple-Twin Cable
ASMU e: Astronaut Stabilizer Maneuvering Unit
ASMUS d: allgemeines Steuerprogramm für Multitask-Systeme
ASMW d: Amt für Standardisierung, Meßwesen und Warenprüfung (DDR)
asm/w e: assemble and weld
ASMWS e: Army Snow and Mountain Warfare School
ASn d: Anschlußsteuereinheiten
ASN d: Asparagin[e]
- d: automatische Sternnavigation
- e: Acknowledged Sequence Number
- e: Allotment Serial Number
- e: American Society for Neurochemistry
- e: Army Serial Number
- e: Army Service Number

- e: Assistant Secretary of the Navy
- e: Associate in Nursing
- e: Asynchronous Sequential Network
- e: Atlantic Satellite Network
- e: Atlantic Steam Navigation
- e: Atomic Scientists News (journ.)
- e: Atomic Strike Net
- e: Audio Signal-to-Noise
- e: Average Sample Number
- e: Axially Symmetric Nozzle

Asn e: Association
ASNA e: Amalgamated Software of North America, Inc.
- e: American Special Mobile Radio Network Association

ASNAP e: Automatic Steerable Null Antenna Processor
ASNDE e: Associate of the Society of Non-Destructive Examination
ASNDT e: American Society for Nondestructive Testing
ASNE e: American Society of Naval Engineers
- e: Assistant Secretary for Nuclear Energy

ASN I/L e: Assistant Secretary of the Navy Installations and Logistics
ASNLH e: Association for the Study of Negro Life and History
ASNOL e: Assistant Senior Naval Officer; Landing
ASNP e: Army Student Nurse Program[me]
ASNPIdentBad e: Army Student Nurse Program[me] Identification Badge
ASNR e: American Society of Neuroradiology
ASN (R&D) e: Assistant Secretary of the Navy; Research and Development
As & Ns e: Andamans and Nicobars
ASNs e: Asynchronous Sequential Networks
ASNSL d: Archiv für das Studium der Neueren Sprachen und Literatur (journ.)
ASNSW e: Anthropological Society of New South Wales
- e: Astronomical Society of New South Wales

ASO d: Arbeitsschutzordnung
- e: Accommodation Sales Order
- e: Advanced Solar Observatory
- e: Aeronautical Supply Office
- e: Air Signal Officer
- e: Air Staff Officer
- e: Air Staff Orientation
- e: Air Supply Officer
- e: Air Surveillance Officer
- e: Alarm System Operation
- e: Allele-Specific Oligonucleotide
- e: American School of Orthodontists
- e: American Society for Oceanography
- e: Ammunition Supply Officer
- e: Anisotropic Spin-Orbit
- e: Antistreptolysin-O
- e: Area of Safe Operation
- e: Area Specialist Officer
- e: Area Supply Officer
- e: Armament and Supply Officer
- e: Army Signal Officer
- e: Arteriosclerosis Obliterans
- e: Assistant Section Officer
- e: Atomic Spin Orbital
- e: Australian Society of Orthodontists
- e: Auxiliary Switch Open
- e: Average Sample Number
- e: Aviation Supply Officer

ASOA e: Avicultural Society of America
ASOAP e: Army Spectrometric Oil Analysis Program[me]
ASOC e: Air Support Operations Center
- e: Assistant Supply Officer-in-Chief

ASocR e: American Sociological Review (journ.)
ASOC(T) e: Assistant Supply Officer-in-Chief (Technical)
ASODDS e: Anti-Submarine Warfare Environmental Prediction Service Submarine Ocean
ASOL d: automatisiertes System der organisatorischen Leitung
ASOP e: Altostratus Opacus
- e: Army Strategic Objectives Plan
- e: Atomic Standing Operation Procedures
- e: Automated Structural Optimization Program[me]
- e: Automatic Scheduling and Operating Program[me]

ASOR e: American School of Oriental Research
ASOS e: American Society of Oral Surgeons
- e: Antimony Sulfide-Oxysulfide
- e: Assistant Supervisor of Shipbuilding
- e: Automated Surface Observing Systems
- e: Automatic Storm Observation Service

ASO Titer e: Antistreptolysin Titer
ASOTS e: Africa South of the Sahara
ASP d: Ablaufsteuer[ungs]programm
- d: Adressenspeicher
- d: afrikanische Schweinepest
- d: Aluminiumsilikat-Pigment
- d: Anwender-Satzprogramm
- d: Anwendungssteuerungsprogramm
- d: Arbeitsgemeinschaft für Sportpsychologie
- d: Arbeitsspeicher
- d: Archiv für slavische Philologie (journ.)
- d: Asparagin[säure]
- d: Asphalt
- e: Abstract Service Primitive
- e: Active Sonar Processor
- e: Activity Scheduling Program[me]
- e: Actual Ship Position
- e: Adaptive Satellite Power[-sharing]
- e: Addressable Storage Position
- e: Advanced Signal Processor
- e: Advanced Software Products, Inc.
- e: Advanced Study Program[me]
- e: Advanced Supply Point
- e: Advanced Support Processor
- e: Advanced System Planning
- e: Aerial Signal Processing
- e: Aerodynamic Source of Power
- e: Aerospace Plane
- e: African Special Project
- e: Airborne Support Platform
- e: Aircraft Servicing Platform
- e: Air Stores Park
- e: Air Support Program[me]
- e: All-Altitude Spin Projected
- e: All South Pole
- e: Alumin[i]um Silicate Pigment
- e: American Selling Price
- e: American Society of Parasitologists
- e: American Society of Pharmacognosy
- e: American Society of Photogrammetry
- e: American Standard Pipe
- e: Ammunition Subpark
- e: Ammunition Supply Point
- e: Amphibious Supply Platform
- e: Analyze Sample Pulse
- e: Anglo-Saxon Protestant
- e: Annual Service Practice
- e: Antenna Signal Processing
- e: Anti-Siphonage Pipe
- e: Anti-Submarine Patrol
- e: Anti-Submarine Projector
- e: Apollo Simple Penetrometer
- e: Apollo Spacecraft Project
- e: Apollo Space Program[me]
- e: Area Specialist Program[me]
- e: Armed Serviced Papers
- e: Army Strategic Plan
- e: Army Supply Program[me]
- e: Assignment Source Point
- e: Association-Storing Processor
- e: Association for Software Protection
- e: Association Storage Processor
- e: Association Storing Processor
- e: Associative-Data Structure Package
- e: Associative Storing Processor
- e: Astronautics Standard Practice
- e: Astronomical Society of the Pacific
- e: A System for Programmers
- e: Atmosphere-Sounding Projectile
- e: Atmospheric Sounding Projectile
- e: Atomic Strike Plan
- e: Attached Support Processor
- e: Attached Support Program[me]
- e: Audio Signal Processor
- e: Augmented Support Period
- e: Augmented Support Plan
- e: Australian Society for Parasitology
- e: Australian Society of Prosthodontists
- e: Automated Sneak Program[me]
- e: Automated Spooling Priority
- e: Automatic Sample Processor
- e: Automatic Schedule Procedure
- e: Automatic Schedule Processor
- e: Automatic Scheduling Procedure
- e: Automatic Service Panel
- e: Automatic Servo Plotter
- e: Automatic Specimen Positioning
- e: Automatic Steel Products
- e: Automatic Switching Panel
- e: Automatic Synthesis Program[me]
- e: Automatic System Procedure
- e: Automtic Synthesis Program[me]
- e: Auxiliary Spacecraft Power
- e: Average Speach Power

Asp d: Ansatzspitze
- d: Aufstellungsplan

ASp d: Arbeitsgemeinschaft Spedition und Lagerei
Asp. d: Aspekt[e]
- d: Aspiration
- e: Asparagine
- e: Aspirant

asp. e: aspartic acid
- e: aspen

a s p f: accepté sous protêt
A-Sp d: Aufnahme-Speicher
ASPA e: Airspace
- e: Aluminosilicate Polyacrylate
- e: Armed Services Procurement Act
- e: Australian Sugar Producers Association
- e: Automated System Performance Analysis
- e: Automatic Systems Pressure Alarm
- e: Auxiliary Storage and Playback Assembly

Asp A e: Aspartic Acid
ASPAC e: Asian Pacific Council
ASP Acid e: Aspartic Acid
A Span e: Anticipation Span
A/S Pat e: Anti-Submarine Patrol
ASPAU e: African Scholarship Program[me] of American Universities
ASpB d: Arbeitsgemeinschaft der Spezialbibliotheken
ASPB e: American Society of Professional Biologists
- e: Assault Support Patrol Boat

ASPC e: Air Space Paper Core
- e: American Sheep Producers Council
- e: American Shetland Pony Club
- e: Analysis of Spare Parts Change

ASPCA e: American Society for the Prevention of Cruelty to Animals
ASPC cable e: Air Space Paper Core Cable
ASPDE e: Automatic Shaft Position Data Encoder
ASPECT e: Acoustic Short-Pulse Echo Classification Technique
- e: Approach to Systematic Planning and Evaluation of Clinical Trials
ASPER e: Assembly System for [the] Peripheral Processor[s]
ASPER Processor e: Assembly System Peripheral Processor
ASPERS e: Armed Services Procurement Regulations
ASPET e: American Society for Pharmacology and Experimental Therapeutics
ASPEX e: Automated Surface Perspectives
ASPh d: Archiv für systematische Philosophie (journ.)
Asph. d: Asphalt
asph. d: asphaltieren
- d: asphaltiert
- e: asphaltic
asph e: asphalt
ASPH e: Australian Society of Professional Hypnotherapists
asphalt e: solid bitumen pitch
asphaltum e: mineral pitch
ASPH C e: Asphalt Concrete
asphic. e: asphaltic
asph mac e: asphalt macadam
ASPHPF e: Asphalt-Plank Floor
ASPHRS e: Asphalt Roof Shingles
asphyx. e: asphyxiant
- e: asphyxiated
ASPI e: Advanced Propulsion Subsystem Integration
- e: Apollo Supplemental Procedural Information
- e: Asynchronous-Synchronous Programmable Interface
ASPIC e: Armed Services Personnel Interrogation Center
ASPIL e: American Standard Practice for Industrial Lighting
ASPIRE e: Access Service for Profitable Information Resource Exchanges
- e: Achieve Successful Performance. Intensify Reliability Effort
- e: Advanced Special Projects in Radiation Effects
ASPJ e: Advanced Self-Protective Jammer
- e: Airborne Self-Protection Jammer
ASPL e: American Society for Law Pharmacy
ASPLP e: American Society for Political and Legal Philosohy
ASPLS e: Any Special Symbol
ASpM d: Aufnahmespulenmotor
ASPM e: American Society of Paramedics
- e: Armed Services Procurement Manual
- e: Automated System for Production Management
ASPMIS e: Apollo Spacecraft Parts and Materials Information Services
aspn e: asparagine
ASPO d: Aufbereitungssystem; plattenspeicherorientiert
- e: Ammunition Supply Point Officer
- e: Apollo Spacecraft Program[me] Office
- e: Apollo Spacecraft Project Office
- e: Avionics System Project Officer
ASPOB d: automatisiertes System der Projektierung und Organisation des Bauwesens
ASPOD e: Autonomous Space Processor for Orbital Debris
ASPOL e: A Simulation Process Oriented Language
AS/POT e: Agena Systems/Power On Test
ASPP e: Alloy-Steel Protective Plating
- e: American Society of Plant Physiologists
- e: Antenna Solar Panel Positioner
- e: Australian Society of Plant Physiologists
ASPPO e: Armed Services Procurement Planning Officer
ASPPR e: Association of Sugar Producers of Puerto Rico
ASPPT e: Association for Suppliers of Printing and Publishing Technologies
ASPR e: Alarm Signal Processing Routine
- e: American Soeiety of Psychical Research
- e: Armed Services Procurement Regulation[s]
- e: Association of South Polar Research
ASPRIN e: Applications Systems and Program[me]s Reference Information Network
ASprK d: Atomsprengkörper
ASPRL e: Armament Systems Personnel Research Laboratory
ASPRM e: Armed Services Procurement Regulation Manual
AS Proj. d: automatisiertes System der Projektierung
ASPRS e: American Society of Plastic and Reconstructive Surgeons
- e: Armed Services Procurement Regulation Supplement
ASPRTR e: Aspirator
ASPS d: Auslands-Schalt- und Prüfstelle
- e: Acoustic Ship-Positioning System
- e: Acoustic Signal Processing System
- e: African Succulent Plant Society
- e: Annular Suspension and Pointing System
- e: Automatic Spacing Pause System
- e: Automatic Specimen Positioning System
Asp-SÄure d: Asparaginsäure
ASPSn d: Auslands-Schalt-und-Prüfstellen
ASPT e: American Society of Plant Taxonomists
- e: Army School of Physical Training
Aspt. e: Aspirant
A Spt Contl e: Air Support Control
ASQ d: Arbeitsgemeinschaft für statistische Qualitätskontrolle
- e: Administrative Science Quarterly
- e: Analytic Solution to Queues
- e: Anthropological Society of Queensland
ASQC e: American Society for Quality Control
ASR d: Ablaufsteuerrechner
- d: Achillessehnenreflex
- d: alphabetisches Sachregister
- d: Amplitudensignalregelung
- d: Ansteuerrechner
- d: Antistreptolysinreaktion
- d: Antriebsschlupfregelung
- d: Arbeitskreis freier Sanitär-Röhrenhändler
- d: Arbeitsstättenrichtlinien
- d: Aufbausystem für die Steuerungs- und Regelungstechnik
- d: automatische Senderegie
- e: Acceptance Summary Report
- e: Accommodation Sales Requisition
- e: Accumulator Shift Right
- e: Active Status Register
- e: Address Shift Register
- e: Address Start Register
- e: Address Switch Register
- e: Admiralty Standard Range
- e: Advanced Surveillance Radar
- e: Agricultural Science Review
- e: Air[borne] Search Radar
- e: Air[borne] Surveillance Radar
- e: Airborne Scanning Radiometer
- e: Airfield Surveillance Radar
- e: Airport Surveillance Radar
- e: Air Search Radar
- e: Air Sea Rescue
- e: Aldosterone Secret in Rate
- e: American Sociological Review (journ.)
- e: American Sugar Refining
- e: Analog[ue] Shift Register
- e: Annual Summary Report
- e: Answer, Send and Receive
- e: Answer Seizure Point
- e: Anti-Streptolysine Reaction
- e: Anti-Submarine Rocket
- e: Apical Sensory Region
- e: Approved System Requirement
- e: Area Surveillance Radar
- e: Arithmetic Shift Right
- e: Assigned Slot Release
- e: Assistant for Service Requirements
- e: Atmospheric Sound Refraction
- e: Atomic Strike Recording
- e: Australasian Software Report
- e: Automatic Send and Receive
- e: Automatic Sending and Receiving
- e: Automatic Speech Recognition
- e: Automatic Speech Recognizer
- e: Automatic Sprinkler Riser
- e: Automatic Step Regulator
- e: Automatic Storage and Retrieval
- e: Automatic Strength Regulation
- e: Auxiliary Submarine Rescue
- e: Available Supply Rate
- e: Aviation Safety Regulation
ASRA e: Automatic Data Processing Systems Resources Analysis
- e: Athwartships Reference Axis
- e: Automatic Stereo[phonic] Recording Amplifier
ASRADI e: Adaptive Surface-signal Recognition and Direction Indicator
ASRAL e: Analog Schematic Translator for Algebraic Language
ASRAPS e: Acoustic Sonar Range Prediction System
ASRC e: Active Stock Record Card
- e: Air-Sea Rescue Craft
- e: Alabama Space and Rocket Center
- e: Atmospheric Sciences Research Center
ASRD e: Aircraft Shipment Readiness Date
ASRDI e: Aerospace Safety Research and Data Institute
ASRDL e: Army Signal Research and Development Laboratory
ASRE e: Admiralty Signal and Radar Establishment
- e: American Society of Refrigerating Engineers
ASRF e: American Seed Research Foundation
ASRI e: Aluminum Smelters Research Institute
ASRL e: Aeroelastic and Structures Research Laboratory
- e: Average Sample Run Length
ASRM e: Antenna System Readiness Monitor
ASRO e: Assistant Superintendant Range Operations
- e: Astronomical Roentgen Observatory

ASROC e: Anti-Submarine Rocket [Computer]
ASRPA e: Army Signal Radio Propagation Agency
ASRPP e: American Society for Research in Psychosomatic Problems
ASRS e: Adjusted Service Rating Score
- e: Advanced Strategic Reconnaissance System
- e: Air-Sea Rescue Service
- e: Airborne Satellite Receiving System
- e: Anglo-Soviet Recognition Signals
- e: Apollo Simulated Remote Site
- e: Automated Shareholder Record System
- e: Automatic Seat Reservation System
- e: Automatic Send [and] Receive Set
- e: Automatic Storage and Retrieval System
- e: Auxiliary Support Reaction System
- e: Aviation Safety Reporting System
AS/RS e: Automated Storage/Retrieval System
ASR Set e: Automatic Send [and] Receive Set
ASRSWS e: Assembly Sequence Record Sheet-Work Sheet
ASR System e: Automatic Storage and Retrieval System
ASRT e: Air Support Radar Team
- e: American Society of Radiologic Technologists
- e: Assort
- e: Atlantic Salmon Research Treaty
ASR Terminal e: Automatic Send-Receive Terminal
ASRU e: Automatic Signal Recognition Unit
ASRV e: Angle Stop Radiator Valve
ASS d: Acetylsalicylsäure
- d: Anwender-System-Software
- d: Anwendersoftwaresystem
- d: Arbeitsgemeinschaft der Schulbuchsortimenter
- d: Arbeitskreis Soziale Sicherheit
- d: Assemblersprache
- d: Außendienst-Steuerungssystem
- d: automatisiertes System der Statistik
- d: Azetylsalizylsäure
- e: Absent Subscribers Service
- e: Accessory Supply System
- e: Acquisition Sun Sensor
- e: Advanced Space Station
- e: Aerosol Sampling System
- e: Aerospace Surveillance System
- e: Airborne Surveillance Set
- e: Air Sampling System
- e: Air Surveillance System
- e: Altitude Sensing System
- e: Analog[ue] Simulation System
- e: Analog[ue] Switching Subsystem
- e: Anglo-Swedish Society
- e: Anterior Superior Spine
- e: Architectural Science Series (Elsevier)
- e: Army Signal School
- e: Army Signal Squadron
- e: Assembler
- e: Assemblies
- e: Assembly
- e: Assistant Secretary of State
- e: Associated Sociologists Society
- e: Associate in Secretarial Science
- e: Associate in Secretarial Studies
- e: Automatic Stabilization System
- e: Automatic Start-up System
- e: Automatic Station Selection
- e: Autopilot Surface Servo
- e: Auxiliary Store System
- e: Aviation Security System
- l: Acta Sanctae Sedis
Ass. d: Assekuranz
- d: Assessor[in]
- d: Assimilation
- d: Assoziation
- d: Assozierung
- d: Assoziierte[r]
- d: Assyrer
- d: Assyrien
ass. d: assistieren
- d: assistiert
- d: assoziert
- d: assoziieren
- d: assyrisch
- e: assortment
Ass e: Assistant
- e: Assurance
- e: Assyrian
ASSA e: American Shetland Sheepdog Association
- e: American Society for the Study of Allergy
- e: American Soil Survey Association
- e: Astronomical Society of South Australia
Ass.A. e: Assembly Area
ASSAM e: Automatic Support Systems Symposium for Advanced Maintainability
Ass Anw d: Assessorenanwärter
- d: Assistentanwärter
ASSArthr e: American Society for the Study of Arthritis
Ass.-Arzt d: Assistenzarzt
ASSASSIN e: Agricultural System for Storage and Subsequent Selection of Information
- e: A System for Storage and Subsequent Selection of Information
Ass.-Ausb. d: Assessorenausbildung
ASSAW e: Aerospace Surveillance and Warning
ASSB e: Assembler
- e: Associative Buffer
- e: Asynchronous Single Sideband
- e: Avionics Subsystem for Strategic Bombers
AS&SB e: Automated Systems and Service Branch
ASSBL e: Assemble
ASSBT e: American Society of Sugar Beet Technologists
ASSC e: Air Service Signal Corps
- e: Army Subsistence Supply Center
Assce e: Assurance
A/S Sch e: Anti-Submarine School
ASSCO e: American Steam Ship Company
assd e: Assessed
- e: assigned
- e: assisted
- e: assorted
- e: assured
AssDA d: Assessorendienstalter
ASSE e: American Society of Safety Engineers
- e: American Society of Sanitary Engineers
- e: Automation System for Scientific Experiments
ASSEA e: Assistant Secretary of State for Economic Affairs
ASSEM e: Assemble[r]
- e: Assembly
ASSERON e: Army Service Squadron
ASSESMT e: Assessment
ASSESS e: Airborne Science Shuttle Experiment System Simulation
- e: Airborne Science Spacelab Experiments-Simulation System
- e: Analytical Studies of Surface Effects of Submerged Submarines
ASSET e: Advanced System and Evaluation Technique
- e: Advanced Systems Synthesis and Evaluation Technique
- e: Aerothermodynamic-Aerothermoelastic Structural Systems Environmental Test
- e: Aerothermodynamic Elastic Structural System Environmental Tests
- e: Air Storage System Energy Transfer
- e: Air Surveillance Sub-System Evaluation and Training
- e: Automated Spares Simulation Estimating Technique
ASSG e: Assignment
ASSGN e: Assign
ASSI e: Address Space Identifier
- e: Assignment
- e: Association for the Study of Snow and Ice
assig e: assignment
assigt e: assignment
assim e: assimilate[d]
- e: assimilation
ASSIST e: A Simple Systematic Integration of Statistical Techniques
- e: Assistant
- e: Automatic System of Secondary Index Selection Technique
- e: Award Scheme for Science, Industry and School-Teaching
AssistO d: Assistentenordnung
Assist.-St. d: Assistentenstelle[n]
ASSK d: Antriebssteuerschaltkreis
ASSKP d: automatisiertes System des Staatlichen Komitees für Preise
AS & SL e: All Ships and Stations Letters
ASSLGE d: Aussagenlänge; einzelne Aussage
ASSLT e: Automatic System Self-Test
asslt wpn e: assault weapons
asslt wpn pl e: assault weapons platoon
ASSM e: Assembler
- e: Assembly
- e: Associative Memory
- e: Authorization for Sale of Salvage Material
ASSMET d: Assembler-Makros für Entscheidungstabellen
assmnt e: assessment
assmt e: assessment
ASSN e: Association
Assn. Brit. Zool. e: Association of British Zoologists
assnce e: assurance
Assn Clin Biochem e: Association of Clinical Biochemists
Assn Comp Mach J e: Association for Computing Machinery Journal
assnd e: Assigned
Assn Ed Comm Tech e: Association for Educational Communications and Technology
Assn Study Anim Behav e: Association for the Study of Animal Behaviour
AssO d: Assistentenordnung
ASSOC e: Associate[d]
- e: Associates
- e: Association
- e: Associative
Assoc Bk e: Associated Booksellers
assocd e: Associated
Assoc. Eng. e: Associate in Engineering
Assoc. Engr. e: Associate Engineer
Assoc IEE e: Associate of the Institution of Electrical [Electronic] Engineers
Assoc. I. Min. E. e: Associate of the Institute of Mining Engineers

Assoc INA e: Associate of the Institute of Naval Architects
- e: Associate of the Institution of Naval Architects
Assoc. Inst. T. e: Associate of the Institute of Transport
Assoc ISI e: Associate of the Iron and Steel Institute
Assoc. Met. e: Associate in Metallurgy
Assoc MICE e: Associatec Member of the Institution of Civil Engineers
Assoc. M. Inst. Gas E. e: Associate Member of the Institute of Gas Engineers
Assoc. S. e: Association Science
Assoc. Sci. e: Associate in Science
assoc w e: associated with
ASSORT e: Automatic System for Selection of Receiver and Transmitter
ASSOTW e: Airfield and Seaplane Stations of the World
ASSP e: Acoustics, Speech and Signal Processing (journ.)
- e: Acoustics, Speech and Signal Processing
- e: Aerospace Systems Security Program[me]
Ass. P e: Assembly Position
ASSPA f: Association Suisse pour l'Automatique
Ass.-Pr. d: Assessorenprüfung
ASSR d: Autonome Sozialistische Sowjetrepublik
- e: Airborne Sea Swell Recorder
- e: American Society for the Study of Religion
- e: Assembler
- e: Autonomous Soviet Socialist Republic
ASSRI e: Arctic and Antarctic Scientific Research Institute
ASSRM e: Ancient and Accepted Scottish Rite Masons
ASSRS e: Adaptive Step-Size Random Search
ASSS e: American Society for the Study of Sterility
- e: American Suffolk Sheep Society
- e: Associate in Science in Secretarial Studies
- e: Australian Society of Soil Science
AssSpr d: Assemblersprache
Ass.-St. d: Assistentenstelle[n]
ASST e: Acetic-acid Salt Spray Test
- e: Advanced Supersonic Transport
- e: Assessment
- e: Assignment
- e: Assist[ant]
- e: Assistance
- e: Assort
- i: Azienda de Stato per i Servizi Telefonici
Asst. Chf. e: Assistant Chief
Asst. Chf. BuAer e: Assistant Chief; Bureau of Aeronautics
Asst. Chf. Engr. e: Assistant Chief Engineer
Asst. Chf. for Acctg. & Disb. e: Assistant Chief for Accounting and Disbursing
Asst. Chf. Nav. Op. e: Assistant Chief of Naval Operations
Asst. Chf. Nav. Op. (Log.) e: Assistant Chief of Naval Operations; Logistics
Asst. Chf. Nav. Op. (Nav. Res.) e: Assistant Chief of Naval Operations; Naval Reserves
ASST COMP[T] e: Assistant Comptroller
Asst C RE e: Assistant Commander; Royal Engineers
Asst CS e: Assistant Chief Surgeon
Asstd e: Assisted
asst'd e: assorted
Asst. Dir. e: Assistant Director
Asst Engr e: Assistant Engineer
ASS Test e: Acetic-acid Salt Spray Test
Asst Flt Liaison Off e: Assistant Fleet Liaison Officer
Asst JAG e: Assistant Judge Advocate General
Asst. Lib. e: Assistant Librarian
Asst Rep e: Assistant Representative
Asst. Sec. War e: Assistant Secretary of War
Asst. Surg. e: Assistant Surgeon
ASSU e: Air Support Signal Unit
assur e: assurance
ASSVD d: automatisiertes System zur Sammlung und Verarbeitung von Daten
ASS-Versuch d: Azetylsäuresalzsprühversuch
ASSVI d: automatisiertes System zur Sammlung und Verarbeitung von Informationen
ASSW e: Antistrategic Submarine Warfare
assy e: assembly
Assyr. d: Assyrer
- d: Assyrien
- d: Assyriolog[i]e
assyr. d: assyriologisch
- d: assyrisch
Assyr e: Assyria[n]
Assyr Babyl e: Assyro-Babylonian
AST d: Ablaufsteuerung
- d: Anfertigungs-Stapelware
- d: Antistreptolysin-Test
- d: Arbeitsfeld-Steuerwerk
- d: asynchroner Teil
- d: asynchrones Steuergerät
- d: Asynchronsteuergerät
- d: Atemstoßtest
- d: Aufgabenstellung
- d: Auftragssteuerung
- d: Ausgabesteuerung
- e: Abort-Scan Table
- e: Absolute Space Time
- e: Abstract Syntax Tree
- e: Acompanying Sound Trap
- e: Action Speed Tactical
- e: Active Segment Table
- e: Add-Subtract Time
- e: Address Selection Track
- e: Address Synchronizing Track
- e: Administrative-Supply Technician
- e: Advanced Simulation Technology
- e: Advanced Supersonic Technology
- e: Advanced Supersonic Transport
- e: Advanced System Technology
- e: Aerospace Technology
- e: Air and Surface Transport
- e: Aircraft Systems Trainer
- e: Air Service Training
- e: Air Supported Threat
- e: Air Surveillance Technican
- e: Air Surveillance Technician
- e: Alaskan Standard Time
- e: All Systems Test
- e: Alternate Sector Table
- e: American Radiator and Standard Sanitary
- e: Anti-Sidetone
- e: Anti-Streptolysine Test
- e: Apollo Systems Test
- e: Apparent Sidereal Time
- e: Apparent Solar Time
- e: Arming System Tester
- e: Army Satellite Tracking
- e: Army SpecializedTraining
- e: Assembly and Structure Test
- e: Astronomical Society of Tasmania
- e: Asynchronous System Trap
- e: Atlantic Standard Time
- e: Atomized Suspension Technique
- e: Audio Station
ASt d: Abfragestelle
- d: Abfragestöpsel
- d: alterungsbeständiger Stahl
- d: Amtsstelle
- d: Anmeldestelle
- d: Antragsteller
- d: Antragstellung
- d: Asiatische Studien (journ.)
- d: Astigmatismus
- d: Ausgabestelle
- d: Außenstelle
- d: Auswertungsstelle
- e: Air Station
a. St. d: alter Stil
- d: am Stück
- d: auf Stapel
Ast e: Assistant
ast d: astigmatisch
ast. e: astrology
- e: astronomer
- e: astronomic[al]
- e: astronomy
A. St. e: Altostratus
AStA d: Allgemeiner Studentenausschuß
ASTA e: Adaptive Search and Track Array
- e: Aerial Surveillance and Target Acquisition
- e: American Seed Trade Association
- e: Army Strategy and Tactics Analysis
- e: Association of Shortcircuit Testing Authorities
- e: Asynchronous-to-Synchronous Transmission Adapter
- e: Australian Science Teachers Association
- e: Automatic System Trouble Analysis
- e: Aviation Systems Test Activity
ASTABA d: Auftragsstammband
ASTALIN d: AIV (Institut für Automatische Informationsverarbeitung) Standard-Listennormierung
A. Stand e: Assistant for Standardization
ASTAP e: Advanced Statistical Analysis Program[me]
- e: Automatic Statistical Analysis Program[me]
ASTAS e: Antiradar Surveillance and Target Acquisition System
ASTC e: Advanced Satellite Tracking Center
- e: Allied Signal Training Center
- e: Army Satellite Tracking Center
- e: Associate of the Sydney Technical College
A.S.T.C. e: Associate of the Sydney Technical College
ASTC e: Australian Science and Technology Council
AST Center e: Army Satellite Tracking Center
ASTD e: Air-Supported Threat Defense
- e: Area Scale Temperature Display
ASTDS e: Air Supported Threat Defense System
ASTe d: asynchrone Steuergeräte
ASTE e: American Society of Tool Engineers
ASTEC e: Advanced Solar Turbo-Electric Concept
- e: Advanced Solar Turbo-Electric Conversion
- e: Advanced Systems Technology
- e: Aerospace System Test and Evaluation Complex

- e: Anti-Submarine Technical Evaluation Center
- e: Australian Science and Technology Council
AS Technology e: Advanced Schottky Technology
ASTEG e: All Systems Test Equipment Group
Asten e: Aortic Stenosis
a sten e: aortic stenosis
ASTEO f: Association Scientifque et Technique pour l'Exploitation des Océans
ASTER Missile e: Anti-Submarine Terrier Missile
ASTF d: automatisches System der Truppenführung
- d: automatisiertes System der Truppenführung
- e: Aeropropulsion System Test Facility
- e: Aerospace Structures Test Facility
AST (G-M) d: asynchrones Steuergerät (geräteseitig-multiplexerseitig)
Asth. d: Asthenie
- d: Astheniker
asth. d: asthenisch
asth e: asthenopia
ASTHE e: Average Straight Time Hourly Earning
astheno e: asthenosphere
ASTHMA e: Aerotherm Axi-Symmetric Transient Heating and Material Ablation
A.S.Thread e: American Standard Thread
ASTI e: American School of Technical Intelligence
- e: Anti-Spasticity Index
- e: Anti-Submarine Training Indicator
- e: Applied Science and Technology Index
- e: Automated System for Transport Intelligence
ASTIA e: Armed Services Technical Information Agency
ASTIC e: Anti-Sidetone Induction Coil
ASTIG e: Astigmatic
- e: Astigmatism
- e: Astigmatizer
- e: Astigmatoscope
- e: Astigmatoscope (or Astigmatoscopy)
- e: Astigmia
- e: Astigmometer
- e: Astigmoscope
ASTIMS e: Agricultural Science and Technology Information Management System
astiodyn e: astrodynamic[al][ly]
- e: astrodynamic[ist]
ASTIP e: Army Scientific and Technical Information Program[me]
ASTIS e: Arctic Science and Technology Information System
ASTL e: Approved Supplier Tab List
Astlo Obsv e: Astrophysical Observatory
ASTIP e: Army Scientifiic and Technical Information Program[me]
AstM d: Aussteuerungsmesser
ASTM e: American Society for Testing and Materials
- e: American Society of Tropical Medicine
- e: American Standard Test Method
ASTM (Amer. Soc. Test Mater.) e: American Society for Testing and Materials Special Technical Publication
ASTM-D e: American Society for Testing and Materials-Draft
ASTME e: American Society of Tool and Manufacturing Engineers
ASTMH e: American Society of Tropical Medicine and Hygiene
ASTM-IP e: American Society for Testing and Materials-Institute of Petroleum
ASTMS e: Association of Scientific, Technical and Managerial Staffs
ASTN e: Astern
- e: Astronic
AST-O d: Antistreptolysin-O
ASTO d: Aufbereitungssystem; trommelorientiert
- e: Assembly Tool
as tol e: as tolerated
ASTOR e: Address Storage Register
- e: Anti-Ship Torpedo
- e: Anti-Submarine Torpedo
ASTOR e: Anti-Submarine Warfare Torpedo Rocket
ASTP d: automatisches System der technologischen Produktionsvorbereitung
- e: Ageing and Surveillance Test Program[me]
- e: Apollo-Soyuz Test Program[me]
- e: Apollo-Soyuz Test Project
- e: Army Special Training Program[me]
ASTR e: Aerospace Systems Test Reactor
- e: American Society for Theater Research
- e: American Society of Therapeutic Radiologists
- e: Astrionics
- e: Astronomical
- e: Astronomy
- e: Asynchronous/Synchronous Transmitter/ Receiver
Astr. d: Astrologe
- d: Astrologie
- d: Astronaut
- d: Astronautik
- d: Astronom
astr. d: astrologisch
- d: astronautisch
- d: astronomisch
Astr d: Astronomie
ASTRA d: Adaptierter Schwimmbecken-Tank-Reaktor; Austria
- d: Ausschreibung im Straßenbau
- e: Adapted Swimming-pool Tank Reactor; Austria
- e: Advanced Static Test Recording Apparatus
- e: Advanced Structural Analyzer
- e: Advanced Structure Analyzer
- e: Advanced System for Radiological Assessment
- e: Air Space Transportation
- e: Air Staff Training
- e: Application of Space Techniques Relating to Aviation
- e: Applied Space Technology-Regional Advancement
- e: Astronomical and Space Techniques for Research on the Atmosphere
- e: Astronomical Space Telescope Research Assembly
- e: Attitude, Steering, Turn, Rate, Azimuth
- e: Automated System-Trust Real-Time Accounting
- e: Automatic Scheduling with Time-Integrated Resource Allocation
-e: Automatic Sorting, Testing, Recording Analysis
- e: Automatic Strobe Tracking
ASTRAC e: Arizona Statistical Repetitive Analog[ue] Computer
astrag e: astragalus
ASTRAL e: Analog[ue] Schematic Translator to Algebra[ic] Language
- e: Assurance and Stabilization Trends for Reliability by Analysis of Lots
As-Trans e: Altostratus Translucidus
ASTRAP e: Application of Space Techniques Relating to Aviation Panel
ASTREC e: Atomic Strike Evaluation Center
ASTREC System e: Atomic Strike Recording System
astr. Einh. d: astronomische Einheit
ASTRG e: Air Starting
ASTRI e: Astrionics
astrion e: astrionic[al][ly]
- e: astrionics
Astrl e: Australia
Astro. d: Astrograph
ASTRO e: Advanced Spacecraft Truck/Trainer/Transport Reusable Orbiter
- e: Army Strategic and Tactical Reorganization Objective
- e: Artificial Satellite Time and Radio Orbit
astro e: astrograph[ic]
- e: astrolabe
- e: astronautics
- e: astronomical
- e: astrophysics
Astro. e: Astrology
- e: Astrometry
- e: Astronomer
- e: Astronomy
astrobio e: astrobiological
- e: astrobiologist
- e: astrobiology
ASTROBIOL e: Astrobiology
ASTROC e: Automatic Stellar Tracking, Recognition and Orientation Computer
astrochem e: astrochemical[ly]
- e: astrochemist[ry]
astrochronics e: astrochronological relatives
Astrodome e: Astrodome Stadium, Houston
astrog. e: astrogeological
astrog e: astrogeologist
- e: astrogeology
astrogen e: astrogenealogy
astrogeo e: astrogeology of celestial bodies
Astrol. d: Astrologe
- d: Astrologie
astrol. d: astrologisch
- e: astrologer
- e: astrological
- e: astrology
ASTROMED e: Astromedicine
astromonk e: astronautical monkey
Astron. d: Astronom
- d: Astronomie
astron. d: astronomisch
- e: astronomical
astron e: astionomy
- e: astronomer
- e: astronomic[al][ly]
Astron e: Astronomy
astronaut e: astronautics
- e: astronautical[ly]
Astronaut. Aeronaut. e: Astronautics and Aeronautics (journ.)
Astron J. e: Astronomical Journal
Astron Nachr d: Astronomische Nachrichten (journ.)
Astron Zl R: Astronomicheskii Zhurnal (journ.)
Astro. Obsv. e: Astrophysical Observatory
Astrophot. d: Astrophotografie
Astrophys e: Astronomy and Astrophysics
astrophys e: Astrophysical
Astrophys e: Astrophysical Journal (journ.)
astrophys e: astrophysics
Astrophys. J. e: Astrophysical Journal

Astrophys. J. Suppl. Ser. e: Astrophysical Journal; Supplement Series
Astrophys. Lett. e: Astrophysical Letters (journ.)
ASTROS e: Advanced Systematic Techniques for Reliable Operation Software
- e: Automated Shell Theory for Rotating Structures
ASTROSPACE e: Astronautics and Space
ASTRP e: Army Specialized Training Reserve Program[me]
ASTS e: Advanced Semi-Automatic Ground Environment Tracking Study
- e: Air-to-Surface Transport System
- e: Airport Surface Traffic Simulator
as'ts e: assists
ASTSECAF e: Assistant Secretary of the Air Force
ASTSECNAV e: Assistant Secretary of the Navy
ASTSECNAV (Air) e: Assistant Secretary of the Navy (Air)
ASTSECNAVMAT e: Assistant Secretary of the Navy for Materiel
ASTT e: Action-Speed Tactical Teacher
- e: Action-Speed Tactical Trainer
- e: American Society of Traffic and Transportation
- e: Associate in Science in Teacher Training
- e: Associate in Science Teacher Training
Ast. T. e: Astronomical Time
ASTTL e: Advanced Schottky Transistor-Transistor Logic
ASTTL Technology e: Advanced Schottky Transistor-Transistor Logic Technology
ASTU e: Air Support Test Unit
- e: Air Support Training Unit
- e: Army Specialized Training Unit
- e: Automatic Systems Test Unit
ASTUTE e: Automated System for Thesaurus Updating, Testing and Editing
ASU d: Abgassonderuntersuchung
- d: Arbeitsgemeinschaft selbständiger Unternehmer
- d: Asynchron-Synchron-Umsetzer
- d: automatische Störunterdrückung
- e: Acknoledgment Signal Unit
- e: Add-Subtract Unit
- e: Administrative Service Unit
- e: Aircraft Scheduling Unit
- e: Aircraft Starting Unit
- e: Altitude Sensing Unit
- e: Analog[ue] Stimulus Unit
- e: Applied Systems Unit
- e: Area Service Unit
- e: Arizona State University
- e: Automatic Steering Unit[s]
- e: Automatic Switching Unit
- e: Autonomous Switch[ing] Unit
- e: Auxiliary Storage Unit
ASU-57 e: Soviet self-propelled
A-Sub. e: Atomic Submarine
ASUBJSCD e: Army Subject Schedule
ASUG e: American Software Users Group
ASUPP e: Air Supply
ASUPT e: Advanced Simulator for Undergraduate Pilot Training
ASUSSR e: Academy of Sciences of the USSR
ASUT e: Adapter Sub-Unit Tester
ASUTS e: American Society of Ultrasound Technical Specialists
ASV d: Allgemeiner Schriftstellerverein
- d: Arbeitsgemeinschaft selbständiger Vermessungsingenieure
- d: Atlas der Schweizerischen Volkskunde
- d: Auslandsfleischbeschaustellen-Verordnung
- d: automatische Sprecher-Verifikation
- e: Acceleration-Switching Valve
- e: Active Service
- e: Aerospace Vehicle
- e: Aerothermodynamic Structural Vehicle
- e: Air-to-Surface Vessel
- e: Airborne Radar for Detecting Surface Vessels
- e: Airborne Search for Surface Vessels
- e: Air Shutoff Valve
- e: Air Solenoid Valve
- e: All Systems Vehicle
- e: American Standard Version (bible)
- e: Ampere-Seconds per Volt
- e: Angle Stop Valve
- e: Anodic Stripping Voltametry
- e: Anthropological Society of Victoria
- e: Anti-Snake Venom
- e: Arithmetic Simple Variable
- e: Arterio/Superficial Venous
- e: Astronomical Society of Victoria
- e: Automatic Self-Verification
- e: Automatic Self Verification
- e: Automatic Shuttle Valve
- e: Automatic Speaker Verification
- e: Avian Sarcoma Virus
ASVA e: Associate of the Society of Valuers and Auctioneers
ASVAB e: Armed Services Vocational Aptitude Battery
ASV American Standard Version e: Arterio/Superficial Venous
ASVC e: Automatic Secure Voice Communications
ASV Detection e: Airborne Surface Vessel Detection
ASVEO e: Advanced Space Vehicle Engineering Operation
ASVIP e: American Standard Vocabulary for Information Processing
ASVO d: Arbeitsschutzverordnung
ASVOL e: Assign Volume
ASVS e: Airborne Stabilized Viewing System
- e: Automatic Signature Verification System
ASVT e: Applications Systems Verification Test
ASW d: Anwendersoftware
- d: Arbeitsgemeinschaft für Sicherheit in der Wirtschaft
- d: außersinnliche Wahrnehmung
- d: Auswahl
- e: Acoustic Surface Wave
- e: Air-to-Surface Weapon
- e: Anti-Aircraft Switch
- e: Anti-Satellite Weapon
- e: Anti-Submarine Warfare
- e: Anti-Submarine Weapon[s]
- e: Anti-Submarine Work
- e: Application Software
- e: Arithmetic Series Weight
- e: Artificial Seawater
- e: Association of Scientific Writers
- e: Australian Standard White
- e: Auxiliary Switch
AS & W e: American Steel and Wire
ASW-A d: Auswahl-Ausgabe
ASWA e: Anthropological Society of Western Australia
- e: Assistant Secretary of War of Air
- e: Audio Switch Assembly
A/SWA e: Aviation/Space Writers Association
ASW/AAW e: Anti-Submarine Warfare and Anti-Air Warfare
ASWACS e: Anti-Submarine Warfare Air Control Ship
ASWB e: Anti-Submarine Warfare Barriers
ASWBPL e: Armed Services Whole Blood Processing Laboratory
ASWBP Laboratory e: Armed Services Whole Blood Processing Laboratory
ASWC e: Anti-Submarine Warfare Center
- e: Army Special Warfare Center
ASWCCS e: Anti-Submarine Warfare Command and Control System
ASWCR e: Airborne Surveillance Warning and Control Radar
ASWD e: Army Special Weapons Depot
ASWE e: Admiralty Surface Weapons Establishment
ASWEPS e: Anti-Submarine Warfare Environmental Prediction System
- e: Anti-Submarine Warfare Environmental Prediction Services
ASWEX e: Anti-Submarine Warfare Exercise
ASWF e: Arithmetic Series Weight Function[s]
ASWG e: American Standard Wire Gage
- e: American Steel and Wire Gage
ASWI e: Anti-Submarine Warfare Installations
ASWILS e: Anti-Submarine Warfare Improved Localization System
ASWIPT e: Anti-Submarine Warfare In-Port Training
ASWL e: Anti-Submarine Warfare Laboratory
ASWLC e: American Shortwave Listeners Club
ASW [LR] e: Anti-Submarine Warning [long-range]
ASWO e: Air Stations Weekly Orders
ASWOT e: Anti-Submarine Warfare Opposed Transit
ASWP 1 e: Asymptotic Stability with Probability One
ASWRC e: Anti-Submarine Warfare Research Center
ASWS e: Aerospace Surveillance and Warning System
- e: Anti-Submarine Warfare Systems
- e: Audubon Shrine and Wildlife Sanctuary
ASWSCCS e: Anti-Submarine Warfare Ship Command and Control System
ASW/SICS e: Anti-Submarine Warfare/Ship Command-and-Control System
ASWSignal e: All-Seems-Well Signal
ASWSOW e: Anti-Submarine Warfare Standoff Weapon
ASW [SR] e: Anti-Submarine Warning [short-range]
ASWTC e: Anti-Submarine Warfare Training Center
ASWTDS e: Anti-Submarine Warfare Tactical Data System
ASWTechCen e: Anti-Submarine Warfare Technical Center
ASWTI d: automatisiertes System der Wissenschaftlich-Technischen Information
ASWTNS e: Anti-Submarine Warfare Tactical Navigation System
ASWTRACEN e: Anti-Submarine Warfare Training Center
ASWTU e: Anti-Submarine Warfare Training Unit
ASWU e: Anti-Submarine Warfare Unit
ASW/UW e: Anti-Submarine Warfare / Underwater Warfare
ASWV e: Arithmetic Series Weight Vector

ASY e: Asymmetric
ASYLCU e: Asynchronous Line Control Unit
ASYLT e: Asynchronous Line Terminator
asym. d: asymmetrisch
ASYM e: Asymetrical
- e: Asymmetric[al]
- e: Asymmetry
asymm. d: asymmetrisch
ASYMP e: Asymptote
asyn. d: asynchron
ASYN d: Asynchronverfahren
- e: Asynchronous
ASYNC e: Asynchronous
ASYNCH e: Asynchronous
asynchr. d: asynchron
ASYSTD e: Advanced System Time Domain
ASZ d: Anodenschließungszuckung
- d: Aufschaltezeichen
- e: American Society of Zoologists
ASZD e: American Society for Zero Defects
AT d: Abfragetaste
- d: Abgasturbine
- d: Ablaufteil
- d: Achtertelegrafie
- d: adjunktive Therapie
- d: Adreßteil
- d: Aktionsteil
- d: Alarmtableau
- d: allgemeiner Teil
- d: alphanumerische Tastatur
- d: Alt-Tuberculin
- d: Altes Testament
- d: aluminiumthermisch
- d: aluminothermisch
- d: Amitriptylin
- d: Anaphylatoxin
- d: Anpaßteil
- d: Anpaßtransformator
- d: Anruftaste
- d: Antititanisch
- d: Anwendungstechnik
- d: Anzeigeteil
- d: Aortenton
- d: Arbeitstag[e]
- d: Arbeitstechnik
- d: Arzneitaxe
- d: Aufbereitungstechnik (journ.)
- d: Aufschaltetaste
- d: Ausgangstext
- d: Auslandstarif
- d: Auslösetaste
- d: Ausnahmetarif
- d: Austausch
- d: autogenes Training
- d: Automatentabelle
- d: Automatisierungstechnik
- d: Autotransformator
- d: Autotransformer
- e: 3-Amino-1,2,4-Triazole
- e: Absolute Term
- e: Absolute Threshold
- e: Absolute Time
- e: Absorption Trap
- e: Acceptance Test[s]
- e: Acceptance Tester
- e: Acceptance Trial[s]
- e: Access Technique
- e: Access Time
- e: Accounting Tabulating
- e: Accounting Technique
- e: Achievement Test
- e: Across Tape
- e: Action Taken
- e: Action Time
- e: Adder Tree
- e: Addition Theorem
- e: Addressing Technique
- e: Address Table
- e: Address Tape
- e: Address Track
- e: Address Transfer
- e: Address Translation
- e: Address Translator
- e: Add Time
- e: Adjacent Tone
- e: Admiralty Trawler
- e: Advanced Technology
- e: Advanced Trainer
- e: Advanced Training
- e: Aerial Tape
- e: Aerial Torpedo
- e: Air-Transportable
- e: Aircraft Tender
- e: Air Technician
- e: Air Temperature
- e: Airtight
- e: Air Traffic
- e: Air Training
- e: Air Transmit
- e: Air Transport
- e: Air Trapping
- e: Air Travel
- e: Algorithm Theory
- e: Algorithm Translation
- e: Alignment Technique
- e: All Told
- e: Alphabetical Tabulator
- e: Alternative Technology
- e: Ambient Temperature
- e: Ambulance Train
- e: American Terms
- e: American Translation
- e: Amino-Transferase
- e: Ampere Turn[s]
- e: Amsterdam Time
- e: Analog[ue] Transistor
- e: Analysis Time
- e: Angle Template
- e: Answer Table
- e: Antenna
- e: Antitank
- e: Antitorpedo
- e: Apparent Time
- e: Approach Time
- e: Appropriate Technology
- e: Areole Trace
- e: Arithmetic Technique
- e: Armament Test
- e: Armour Tenderometer
- e: Arrest Temperature
- e: Articular Tubercle
- e: Artillery Training
- e: Asphalt Tile
- e: Assay Ton
- e: Astronomical Time
- e: Atherton Tablelands
- e: Atlantic Time
- e: Atomatic Typewriter
- e: Atomic Time
- e: Attack Teacher
- e: Attenuator
- e: Audio Tape
- e: Audit Trail
- e: Aural Transducer
- e: Auroral Time
- e: Ausschuß Terminologie
- e: Austria
- e: Autogenic Training
- e: Automatic Telephone
- e: Automatic Test
- e: Automatic Ticketing
- e: Automatic Tracking
- e: Automatic Translation
- e: Automatic Transmission
- e: Automatic Transmitter
- e: Automatic Typewriter
- e: Auxiliary Timer
- e: Availability Table
- e: Awaiting Transportation
At. d: Abtaster
- d: Atelier
- d: Atom
aT d: aktuelle Transpiration
- d: auf Tausend
- d: außer Tarif
At d: Antenne
- d: Astatin
at d: technische Atmosphäre
- e: Alunite
- e: Astatine
at. e: asphalt
- e: asphaltic
- e: atmosphere[s] [technical]
- e: atomic
a t i: a tempo
A/T e: Acceptance and Transfer
A & T e: Agriculture and Technical
- e: Assemble and Test
- e: Assembly and Test
äT d: äußerer Totpunkt
at. % e: atomic percent
AT_1 e: 9 dihydrotachysterol
AT_7 e: hexachlorophene
AT 26 e: Aermachin jet-trainer ground-attack aircraft
ATA d: Anschlußsteuerung Tastatur
- d: Anwendungstechnische Abteilung
- d: Armee-Telegrafenabteilung
- e: Abstracts on Tropical Agriculture (a database)
- e: Academic Travel Abroad
- e: Actual Time of Arrival
- e: Additional Tape Attachment
- e: Administrative Telecommunications Agency
- e: Advertising Typographers Association
- e: Air-Force Training Auxiliary
- e: Air-to-Air
- e: Airborne Target Augmeter
- e: Air Transport; Auxiliary
- e: Air Transport Association
- e: Air Turbine Alternator
- e: Albanian Telegraph Agency
- e: Alimentary Toxic Aleukia
- e: Alternate Track Address
- e: Alternate Track Assignment
- e: Amateur Television Association
- e: American Thyroid Association
- e: American Topical Association
- e: American Transit Association
- e: American Translators Association
- e: American Transport Association
- e: American Tree Association
- e: Anatomical Transplant Association
- e: Antitank Assault
- e: Applied Technology Associates
- e: Appropriation Transfer Account
- e: Assembly and Test Area
- e: Associate Aide
- e: Associated Television Authority
- e: Associate in Technical Affairs
- e: Associate in Technical Arts
- e: Associate Technical Aide
- e: Asynchronous Terminal Adapter
- e: Atlantic Treaty Association
- e: Australian Telecommunications Authority
- e: Automated Testing Analyzer
- e: Automatic Threshold Adjust
- e: Automatic Tracking Aerial

- e: Automatic Tracking Antenna
- e: Automatic Trouble Analysis
- e: Auxiliary Transport Aviation
- e: Average Turnaround
- e: Azimuth Torque Amplifier
- f: Admission Temporair

ata d: Atmosphäre[n] absolut
- e: absolute atmosphere

ATAA e: Advertising Typographers Association of America

ATAAD e: Antitank Assault Air Defense

ATABCS e: Aircraft Tactical Air Battle Control System

ATABE e: Automatic Target Assignment and Battery Evaluation

at abs d: absolute Atmosphäre[n]

At. Absorption Newslett. e: Atomic Absorption Newletter (journ.)

ATAC e: Air Transportable Acoustic Communications
- e: All Tariffs Computerized
- e: Anatomical Transplant Association of California
- e: Army Tank-Automotive Command
- e: Army Tank and Automotive Command
- e: Army Tank Automotive Center

ATACAP e: Antenna-to-Antenna Compatibility Analysis Program[me]

ATACC e: Advanced Tactical Air[borne] Command and Control

ATACCS e: Advanced Tactical Air[borne] Command and Control System
- e: Advanced Tactical Aircraft Command and Control System
- e: Airborne Tactical Air Command and Control System

ATACC System e: Advanced Tactical Air[borne] Command and Control System

ATACS e: Army Tactical Communications System

ATAD e: Absent and Temporary Additional Duty
- e: Absent on Temporary Additional Duty
- e: Atlanta Army Depot
- e: Automatic Target Designation

ATAE e: Associated Telephone Answering Exchanges
- e: Association of Telephone Answering Exchanges

ATAF e: Agricultural Technical Assistance Foundation
- e: Allied Tactical Air Force
- e: A Test Application Facility

ATAL e: Automatic Test Application Language

ATALA f: Association pour l'Etude et de la Linguistique Appliquee

ATAM e: Association for Teaching Aids in Mathematics

atan e: arc tangent

AT-Ang. d: außertarifliche[r] Angestellte[r]

AtAnl d: Antennenanlage

AtAnLV d: Atomanlagenverordnung

AtAp d: Antennenaufstellungsplan

AT Ap e: Advanced Training Airplane

ATAP e: Alternate Track Assignment Program[me]
- e: Apollo Telemetry Aircraft Project

ATAR e: Above Transmitted As Received
- e: Acquisition Tracking and Recognition
- e: Advanced Tactical Avionics Radar
- e: Air-to-Air Recovery
- e: Anti-Tank Aircraft Rocket
- e: Automated Travel Agents Reservation

ATARS e: Anti-Terrain-Avoidance Radar System
- e: Automatic Traffic Advisory and Resolution Service

ATAS e: Academy of Television Arts and Sciences
- e: Advanced Technology Alert System
- e: Air Traffic Area Supervisor
- e: Association of Telephone Answering Services
- e: Automated Telephone Answering System
- e: Automatic Terrain Avoidance System
- e: Automatic Test Analysis System
- e: Automatic Three-Axis Stabilization
- e: Automatic Tracking Antenna System

ATASCII e: Atari-Version American Standard Code for Information Interchange

atav e: atavism
- e: atavist[ic][al][ly]

ATAW e: Advanced Tactical Assault Weapon

ATB d: allgemeine Tarifbestimmungen
- d: Austauschbesatzung
- e: Access Type Bit
- e: Acetyle Tetrabromide
- e: Address Trace Block
- e: Address Translation Buffer
- e: Advanced Technology Bomber
- e: Aeration Test Burner
- e: Airborne Test Bed
- e: Aircraft Technical Bulletin
- e: All Trunks Busy
- e: Amphibious Training Base
- e: Antitank Battery
- e: Apollo Test Box
- e: Arctic Test Branch
- e: Army Telephone Book
- e: Army Transport Boat
- e: Asphalt-Tile Base
- e: Association of Tropical Biology
- e: Stealth Bomber

ATBA e: Absolute Time Base Accuracy
- e: Automatic Test Break & Access

ATBC d: adaptive Teilbandcodierung

ATB Detector e: Ampere Turn Balance Detector

ATBE e: Absolute Time Base Error

ATBM e: Advanced Tactical Ballistic Missile
- e: Advanced Technology Ballistic Missile
- e: Anti-Tactical Ballistic Missile
- e: Average Time Between Maintenance

ATBMC e: Airborne Test Bed Mode Control

ATB Missile e: Anti-Tank Ballistic Missile

ATBT e: Acoustic Telemetry Bathythermometer
- e: Airborne Test Bed Turret

ATC d: adaptive Transformationscodierung
- e: Ablative Thrust Chamber
- e: Acoustical Test Chamber
- e: Acoustic[al] Tile Ceiling
- e: Acoustic Test[ing] Chamber
- e: Active Thermal Control
- e: Active Transfer Command
- e: Adaptive Transform[ation] Coding
- e: Additional Trunk Capacity
- e: Address Translation Chip
- e: Adiabatic Toroidal Compression
- e: Adiabatic Toroidal Compressor
- e: Advanced Terminal Controller
- e: Advanced Thermodynamic Cycle
- e: Advertising Training Center
- e: Aerial Tuning Capacitor
- e: Aerial Tuning Condenser
- e: After Top Centre
- e: Airborne Test Conductor
- e: Aircraft Technical Committee
- e: Air[port] Traffic Control[ler]
- e: Air Training Command
- e: Air Transport Certificate
- e: Air Transport Command
- e: Airway Traffic Control
- e: Alert Transmit Console
- e: Allergic to Combat
- e: Alloy-Tin Couple
- e: American Standard Code for Information Interchange Terminal Controller
- e: Americn Technical Ceramics
- e: Analog Technology Company
- e: Annotated Tax Cases
- e: Annular Turbojet Combustor
- e: Antenna Tuning Capacitor
- e: Apollo Time Conditioner
- e: Approved-Type Certificate
- e: Architecture Technology Corporation
- e: Arctic Test Center
- e: Area Traffic Commissioner
- e: Armament Test Center
- e: Armament Training Camp
- e: Armo[u]red Carrier
- e: Army Topographic Command
- e: Army Training Center
- e: Army Training Circular
- e: Art Teachers Certificate
- e: Assistant Test Chief
- e: Assistant Test Conductor
- e: Asynchronous Terminal Controller
- e: Australian Telecommunications Commission
- e: Autocoder to Common Business-Oriented Language
- e: Automated Technical Control
- e: Automatic Tap-Changing
- e: Automatic Target Counting
- e: Automatic Telephone Call
- e: Automatic Telephone Company
- e: Automatic Temperature Compensation
- e: Automatic Temperature Control
- e: Automatic Threshold Control
- e: Automatic Time Control
- e: Automatic Timing Control
- e: Automatic Tint Control
- e: Automatic Tone Correction
- e: Automatic Tool Changing
- e: Automatic Tracking Control
- e: Automatic Traffic Control
- e: Automatic Train Control
- e: Automatic Truck Control
- e: Automatic Tuning Control
- e: Automation Training Center
- e: Average Total Cost

ATCA e: Advanced Tanker/Cargo Aircraft
- e: Air Traffic Conference of America
- e: Air Traffic Control Association
- e: Allied Tactical Communications Agency
- e: Attitude and Translation Control Assembly
- e: Automatic Transistor Card Analyzer

ATCAC e: Air Traffic Control Advisory Committee

ATCAC Rep e: Air Traffic Control Advisory Committee Report

ATCAP e: Army Telecommunications Automation Program[me]
- e: Automatic Thyristor Circuit Analysis Program[me]

AT Card e: Accounting Tabulating Card

ATCAS e: African Training Centre for Agricultural Statistics
- e: Air-Traffic-Control Automated System
- e: Air-Traffic Control Automation System

ATCASE d: Aspartat-Transcarbamylase
ATCBGS e: Air Traffic Control Beacon Ground Station
ATCBI e: Air Traffic Control Beacon Interrogator
ATCC e: Aerial Target Control Center
- e: Aerospace Traffic Control Center
- e: Air Traffic Control Center
- e: Air Traffic Control Communications
- e: Air Training Corps Cadet
- e: American Type Culture Collection
- e: Automated Telecommunications Center
- e: Automatic Train Control Center
ATCCC e: Advanced Tactical Command and Control Capabilities
- e: Air Traffic Control Coordination Center
ATCCE e: Air Traffic Control Chief Engineer
ATCCS e: Air Traffic Control Communications System
ATCCs e: Automated Telecommunications Centers
ATCD e: Automatic Telephone Call Distribution
ATCE e: Ablative Thrust Chamber Engine
- e: Automatic Test and Checkout Equipment
ATCEU e: Air Traffic Control Evaluation Unit
- e: Air Traffic Control Experimental Unit
ATCF e: Air Traffic Control Facility
atch e: attach[ment]
- e: attaching
atchd e: attached
atchmt e: attachment
ATCK e: Attack
ATCL e: Acceleration-Type Control Law
- e: Air Traffic Control Line
- e: Associate of Trinity College, London
ATCLO e: Amphibious Training Command Liaison Officer
ATCM e: Addressing Toroidal Core Matrix
- e: Air Training Command Manual
- e: American Theater Campaign Medal
ATC Machine Tool e: Automatic Tool Changing Machine Tool
ATCMD e: Atlanta Contract Management District
ATCMS e: Advanced Technology Cruise Missile Study
ATC/NAV e: Air Traffic Control and Navigation
ATCO e: Air Taxi-Commercial Operator
- e: Air Traffic Control Office[r]
- e: Air Traffic Coordinating Office[r]
ATCOG e: Air Traffic Communications Operators Guild
ATCOM e: Air Traffic Compiler
- e: Air Training Command
- e: Atlantic Computer Microfilm Corporation
ATCOR e: Air Traffic Control Operations Representative
- e: Air Traffic Coordinator
ATCOREU e: Air Traffic Coordinator; Europe
ATCORUS e: Air Transport Coordinator; United States
ATCOS e: Atmospheric Compensation Satellite
ATCP e: Air Traffic Control Procedures
- e: Air Training Command Pamphlet
- e: Antarctic Treaty Consultative Parties
ATCP Acid e: 4-Amino-3.5.6-Trichloropicolinic Acid
ATCR e: Air Training Command Regulation
ATCRB e: Air Traffic Control Radar Beacon
ATCRBS e: Air Traffic Control Radar Beacon System
ATCRS e: Air Traffic Control Radar System
ATCRU e: Air Traffic Control Radar Unit
ATCS e: Active Thermal Control System
- e: Air Traffic Communications Service
- e: Air Traffic Communications Station
- e: Air Traffic Control Specialist
- e: Air Traffic Control Station
atc's e: anyterrain motorcycles
ATCSS e: Air Traffic Control Signalling System
ATCT e: Airport Traffic Control Tower
- e: Air Traffic Control Transponder
ATCU e: Air Transportable Communication[s] Unit
- e: Association of Texas Colleges and Universities
- e: Attitude and Translation Control Unit
ATC & W e: Air Traffic Control and [Aircraft] Warning
ATD e: Absent [on] Temporary Duty
- e: Acceptance and Take-over Date
- e: Actual Time of Departure
- e: Aid to the Totally Disabled
- e: Air Transportable Dispensary
- e: Air Turbine Drive
- e: Alternate Tape Drive
- e: Analog[ue] to Time to Digital
- e: Anthropomorphic Test Dummy
- e: Armo[u]red Tank Destroyer
- e: Articulotrochanteric Distance
- e: Art Teachers Diploma
- e: Assistant Test Director
- e: Asynchronous Time Division
- e: Attached
- e: Audio Tone Decoder
- e: Australian Tax Decisions
- e: Automatic Tape Degausser
- e: Automatic Tape Delete
- e: Automatic Tape Dispenser
- e: Automatic Teaching Device
- e: Automobile Traffic Density
ATDA e: Alternate Target Docking Adapter
- e: Augmented Target Docking Adapter
- e: Australian Telecommunications Development Association
ATDC e: After Top Dead Centre
- e: Association of Thalidomide-Damaged Children
ATDD e: Average Total Diametrical Displacement
ATDDL e: Apollo Technical Documentation Distribution List
ATDESA e: Automatic Three-Dimensional Electronics Scanning Array
ATDG e: Automated Test Data Generator
ATDLP e: Apollo Trajectory Decision Logic Prototype
ATDM e: Asynchronous Time-Division Multiplex[er]
- e: Asynchronous Time-Division Multiplexing
ATDMA e: Advanced Time-Division Multiple Access
ATDM Method e: Asynchronous Time-Division Multiplex Method
ATDP e: Air Traffic Data Processor
ATDPS e: Airborne Tactical Data Processing System
ATDS e: Air[borne] Tactical Data System
- e: Airborne Technical Data System
- e: Airways Technical District Supervisor
- e: Automated Data and Telecommunications Service
- e: Automatic Telemetry Decommutation System
- e: Auxiliary Track Data Storage
- e: Aviation Tactical Data System
atdt e: attendant
ATDU e: Alphanumeric Terminal Display Unit
at. E. d: atomische Einheit
ATE d: automatische Tast[en]einrichtung
- e: Aerospace Test Equipment
- e: Air-Turbo Exchanger
- e: Airborne Teletypewriter Equipment
- e: Airplane Test Equipment
- e: Altitude Transmission Equipment
- e: Altitude Transmitting Equipment
- e: Alumin[i]um Triethyl
- e: Anisotrophic Etch
- e: Area Test Equipment
- e: Artificial Traffic Equipment
- e: Assembler-Text Editor
- e: Associated Telephone Exchange[s]
- e: Atlantic City Electric
- e: Atlantic Tropical Experiment
- e: Autographic Theme Extraction
- e: Automatic Telephone and Electric
- e: Automatic Telephone Exchange
- e: Automatic Test[ing] Equipment
ATEA e: American Technical Education Association
- e: American Transit Engineering Association
- e: Australian Telecommunications Employees Association
- e: Automatic Test Equipment Association
ATEC e: Air Transport Electronics Council
- e: Army Test and Evaluation Command
- e: Automated Technical Control
- e: Automated Test Equipment Complex
- e: Automatic Technical Control
- e: Automatic Test Equipment Complex
- e: Aviation Technician Education Council
ATech e: Associate in Technology
A.Tech. e: Associate in Technology
A-Technik d: Aufnahmetechnik
ATECO d: automatisches Telegramm-vermittlungssystem mit Computer[n]
- d: Automatische Telegrammvermittlung mit Computern
- e: Automatic Transmission of Telegrams by (or with) Computer
- e: Automatic Transmission of Telegrams by Computer
ATEE d: N-Acetyl-L-Tyrosinäthylester
ATEGG e: Advanced Turbine Engine Gas Generator
ATEL d: Aare-Tessin-AG für Elektrizität
- e: Audio Techniques and Evaluation Laboratory
ATEM e: Aircaft Test Equipment Modification
ATEMIS e: Automated Traffic Engineering and Management Information System
- e: Automatic Traffic Engineering and Management Information System
A temp i: a tempo
ATEN f: Association Technique pour la production et l'utilisation de l'Energie Nucleaire
ATENE e: Asahi Television News
At Energ R: Atomnaya Energiya (journ.)
At. Energy Aust. e: Atomic Energy in Australia (journ.)
At. Energy Rev. e: Atomic Energy Review (journ.)

ATEP e: Advanced Technical Education Program[me]
- e: Annual Training Equipment Pools
- e: Augment[ed] Thermally Electric Propulsion
ATEPO e: Auto Test in Elementary Programming and Operation
ATER e: Automatic Testing, Evaluation and Reporting
ATERM e: Air Terminal
ATES e: Air Transportable Earth Station
ATESTO e: Automatic Telegraph Equipment for Stockholm Telegraph Office
ATET e: Advanced Technical Experimental Translation
- e: Advanced Technology Experimental Transport
ATEV e: Approximate Theoretical Error Variance
ATEWA e: Automatic Target Evaluator and Weapon Assignor
ATEWS e: Advanced Tactical Electronic Warfare System
ATEX e: Atlantic Trade Experiment
- e: Atlantic Tropical Experiment
- e: Automatic Test Equipment Conference and Exposition
ATF d: Aspartotransferase
- d: Ausschuß für Technik in der Forstwirtschaft
- e: Acceptance Test Facility
- e: Accounting Tabulating Form
- e: Accounting Transfer Function
- e: Actual Time of Fall
- e: Actuating Transfer Function
- e: Advanced Technical Fighter
- e: Advanced Technology Fighter
- e: Aerial Test Facility
- e: Algebraic Technological Function
- e: Altitude Test Facility
- e: Amorphous Thin Film
- e: Antenna Test Facility
- e: Army Training Film
- e: Asphalt-Tile Floor
- e: Automatic Target Finder
- e: Automatic Target Follower
- e: Automatic Terrain Following
- e: Automatic Text Formatter
- e: Automatic Transmission Fluids
- e: Aviation Turbine Fuel
ATFA d: Acetyltrifluoracetate
- e: American Turpentine Farmers Association
ATFAC e: American Turpentine Farmers Association Cooperative
ATF/ATA e: Automatic Terrain Following / Automatic Terrain Avoidance
ATFC e: Automatic Traffic-Flow Control
ATFG e: Analog[ue] Test Function Generator
AtFk d: Aufteilungsfernkabel
ATFM e: Associative Thin-Film Memory
ATFNÜW d: automatische Trägerfrequenz-Netzüberwachung
ATFOS e: Alignment and Test Facility for Optical Systems
ATFP e: Alliance of Television Film Producers
ATFR e: Automatic Terrain-Following Radar
ATFRAM e: Airborne Time-Frequency Range-Altitude Monitor
ATG d: Anmeldetag
- d: Auftastgenerator
- d: Austauschgetriebe
- d: automatisches Testgerät
- e: Adaptive Threshold Gate
- e: Advanced Technology Group
- e: Air-to-Ground
- e: Air Turbine Generator
- e: Anti-Tank Gun
- e: Antithymocyte Globulin
- e: Army Technical Group
- e: Asynchronous Task Group
- e: Automated Test-case Guidance
- e: Automatic Test Generator
- e: Automatic Test Grading
- e: Aviation Turbine Gasoline
Atg. d: Antrag
AtG d: Atomgesetz
At.G. d: Atomgewicht
ATGAF e: Advanced Technology Ground Attack Fighter
ATGAR e: Anti-Tank Guided Air[craft] Rocket
At.-Gew. d: Atomgewicht
ATGF e: Automatic Test Generation Facility
ATGM e: Anti-Tank-Guided Missile
ATGn e: Antitank Gun
ATGS e: Advanced Terminal Guidance System
ATGSB e: Admission Test for Graduate Study in Business
ATGW e: Anti-Tank Guided Weapon[s]
ATH d: Allylthioharnstoff
- d: August-Thyssen-Hütte
- e: Above the Horizon
- e: Air Transportable Hospital
- e: Athwartship
- e: Automatic Attitude Hold
- e: Automatic Tape Handler
Ath d: Athens
- e: Athenium
A Th e: Associate in Theology
ath e: atheism
- e: atheist[ic]
Äth. d: Äther
- d: Äthiopien
- d: Äthyl
äth. d: ätherisch
- d: ätherisieren
- d: äthiopisch
AThChI e: Acetyl-Thiocholine Iodide
athe e: allotetrahydrocortisol
Athen. e: Athenian
atheol e: atheological
- e: atheologist
- e: atheology
äther. d: ätherisch
ATHESA e: Automatic Three-Dimensional Electronic Scanned Array
ATHODYD e: Aero-Thermo-Dynamic-Duct
Ath R e: Arithmetic Register
ATHSC e: Atheroslerosis
at. ht. e: Atomic Heat
athw. e: Athwartship
ATI d: Aluminiumtitanat
- e: Action Tape In
- e: Actual Time of Interception
- e: Aerial Tuning Inductance
- e: Air Technical Index
- e: Air Technical Information
- e: Air Technical Intelligence
- e: American Technology Institute
- e: American Television, Incorporated
- e: American Television Institute
- e: American Toy Institute
- e: American Trousers Institute
- e: Analog Threshold Logic
- e: Antenna Tuning Inductance
- e: Asbestos Technical Institute
- e: Asbestos Textile Institute
- e: Asphalt Tile Institute
- e: Associate of the Textile Institute
- e: Audiometer Telephone Interface
- e: Australian Textile Institute
- e: Automatic Target Identification
- e: Automatic Track Initiation
- e: Automation Techniques, Incorporated
- e: Auxiliary Tape Input
- e: Average Total Inspection
- i: Aero Transporti Italiani
A & TI e: Agricultural and Technical Institute
ATIC e: Aerospace Technical Intelligence Center
- e: Air Technical Intelligence Center
- e: Australian Tin Information Centre
ATIC Aerospace Technical Intel e: Air Technical Intelligence Center
ATID e: American Trade and Industrial Development
ATIG e: Alternative Technology Information Group
ATIGS e: Advanced Tactical Inertial Guidance System
ATII e: Advanced Techniques for Imagery Interpretation
ATIL Program[me] e: Air Target Intelligence Liaison Program[me]
ATIP e: Analog[ue] Tune in Progress
ATIRI e: Ahmedabad Textile Industrys Research Institute
ATIS e: Advanced Thermal Imaging Scanner
- e: Air Technical Intelligence Service
- e: Air Technical Intelligence Study
- e: Aluminium Technical Information Service
- e: Antenna and Transmitter Improvement Study
- e: Asahi Technical Information System
- e: Automatic Terminal Information Service
- e: Automatic Transmitter Identification System
ATISC e: Air Technical Intelligence Services Command
ATJC e: Annular Turbojet Combustor
ATJS e: Advanced Tactical Jamming System
ATK d: Alttuberculin Koch
- d: Anlagentechnischer Katalog
- e: Alphabetic Typewriter Keyboard
- e: Available Ton-Kilometres
- e: Aviation Turbine Kerosene
Atk d: Antennenkabel
- d: Antennenkopf
- d: Aufteilungskabel
- e: Afrikaans
- e: Anti-Tank
atk e: attack
Atk Avn e: Attack Aviation
ATKE d: Atomkernenergie (journ.)
atkm d: angebotene[r] Tonnenkilometer
a/tk. r. e: antitank rifle
ATKSC e: Attack Surveillance Coverage
ATL d: Abgasturbolader
- e: Arbeitsgemeinschaft Technik in der Landwirtschaft
- e: Acoustic Test[ing] Laboratory
- e: Active Task List
- e: Active Time List
- e: Actual Total Loss
- e: Adult T cell leukemia
- e: Adult T cell lymphoma
- e: Advanced Table Look-up
- e: Advanced Technology Laboratories
- e: Aeronautical Turbine Laboratories
- e: Alexander Turnbull Library

- e: Analog[ue] Threshold Logic
- e: Anti-Thrust Law
- e: Appliance Testing Laboratories
- e: Application[s] Terminal Language
- e: Artificial Transmission Line
- e: Associated Testing Laboratories
- e: Automated Tape Library
- e: Automatic Telling
- e: Automatic Test Line
- e: Automatic Turret Lathe
- e: Atlantic [Ocean]

Atl. d: Atlant[en]
- d: Atlantik
- d: Atlas

atl. d: atlantisch
AT/L e: Advanced Technology / Libraries
Atl e: Atlanta
- e: Atlantic
- e: Australia
- f: Atlantique
- i: Atlantico
- s: Atlantico
- s: Australia

ATLA e: Alternatives to Laboratory Animals
- e: American Theological Library Association

Atlantol e: Atlantolgist[ic][al][ly]
- e: Atlantologic[al][ly]
- e: Atlantology

ATLAS e: Abbreviated Test Language for All Systems
- e: Abbreviated Test Language for Avionic[s] Systems
- e: Actual Target Labor Auditing System
- e: Advanced Tactical Lightweight Avionics System
- e: Anti-Tank Laser Assisted System
- e: Atlas Mountains of Algeria and Morocco
- e: A Total Library Automation System
- e: Automated Tape Label Assignment System
- e: Automated Tape Lay-up System
- e: Automated Telephone Line Address System
- e: Automated Testing and Load Analysis System
- e: Automatic, Tabulating, Listing and Sorting System
- e: Automatic Tape Load Audit System
- e: Automatic Thin-Layer Analytical System
- e: Automatic Translation Aiding System

Atlas Agena e: two-stage launch vehicle
Atlas-Centaur e: first American high-energy launch vehicle for space exploration
Atlas-E e: intercontinental ballistic missile
atlas fol e: atlas folio
Atlas icbm e: Atlas Intercontinental Ballistic Misile
ATLAS System e: Automatic Tabulating, Listing and Sorting System
atlatox e: aflatoxin
Atl. C. e: Atlantic City
ATLC e: Automatic Total Last Card
ATLD e: Air-Transportable Loading Dock
ATLE e: Adaptive Threshold Logic Element
ATLEs e: Adaptive Threshold Logic Elements
ATLIS e: Army Technical Libraries and Information Systems
- e: Army Technical Library Improvement Studies
- e: Australian Transport Literature Information System
- e: Automated Technical Library Information System
- e: Automatic Tracking Laser Illumination System

ATLO e: Acceptance Test and Launch Operations
- e: Air Transport Liaison Officer

Atl O e: Atlantic Ocean
ATLP e: Army-Wide Training Literature Program[me]
ATLS e: Air Transport Long Range Navigation System
ATM d: Antriebsmotor
- d: arbeitsschutztechnische Mittel
- d: Archiv für Technisches Messen
- d: Archiv für technisches Messen und industrielle Meßtechnik (journ.)
- d: Austauschmotor
- e: Address Translation Memory
- e: Advanced Ticket Machine
- e: Aerial Test Model
- e: Aerial Turning Motor
- e: Air Terminal
- e: Air Traffic Management
- e: Air Training Manual
- e: Air Training Memorandum
- e: Air Turbine Motor
- e: Alternative Test Method
- e: Alumin[i]um Trimethyl
- e: Amateur Telescope Making
- e: Ampere Turns per Meter
- e: Analog[ue] Trunk Module
- e: Antenna Test Model
- e: Antenna Turning Motor
- e: Antitank Mine
- e: Antitank Missile
- e: Apollo Telescope Mount
- e: Arc Tangent Mechanism
- e: Army Training Memorandum
- e: Assistant Traffic Manager
- e: Association of Teaching Aids in Mathematics
- e: Asynchronous Time Multiplexer
- e: Asynchronous Time Multiplexing
- e: Asynchronous Transfer Mode
- e: Atmosphere[s]
- e: Atmospheric[s]
- e: Atomic Demolition Munition
- e: Authentication Maneuver
- e: Automated Teller Machine
- e: Automatic Teller Machine
- e: Auxiliary Tape Memory
- e: Axial Turbo Machine

Atm. d: Atmosphäre
- d: Atmung

atm. d: atmosphärisch
AtM d: Aufteilungsmuffe
atm e: atmospheric[al]
at. m. e: atomic mass
ATMA e: American Textile Machinery Association
atm. abs. e: atmosphere[s] absolute
ATMAC e: Advanced Technology Microelectronic Array Computer
- e: Air Traffic Management Automated Centers

ATMARS e: Automatic Transmission Measuring and Recording System
ATMAR System e: Automatic Transmission and Measuring Recording System
ATMC e: Army Transportation Materiel Command
atm. chgs. e: atmospheric changes
AtMD d: Aufteilungsmuffendeckel
ATM/D e: Automatic Teller Machine Deposit
ATMDC e: Apollo Telescope Mount Digital Computer
ATME d: automatische Test- und Meßeinrichtung
- e: Atmospheric Transmission Measurement Equipment
- e: Automatic Transmission Measuring Equipment

ATM Equipment e: Automatic Transmission Measuring Equipment
AtMG d: Aufteilungsmuffengestell
ATMG e: Atomizing
ATMI e: American Textile Manufacturers Institute
ATMN e: Amalgamated Tin Mines of Nigeria
ATMOS e: Adjustable Threshold Metal Oxide Semiconductor
- e: Atmosphere
- e: Atmospheric

Atmos. Environ. e: Atmospheric Environment (journ.)
Atmos. Oceanic Phys. e: Atmospheric and Oceanic Physics (journ.)(GB)
- e: Atmospheric and Oceanic Physics (journ.) (US)

ATMOSs e: Adjustable Threshold Metal Oxide Semiconductors
AT-Motor d: Austauschmotor
ATMP e: Air Target Materials Program[me]
atm. pr. e: Atmospheric Pressure
atm. press. e: Atmospheric Pressure
ATMRS e: Automatic Transmission and Measuring Recording
ATMR System e: Automatic Transmission and Measuring Recording System
ATMS e: Advanced Terminal Management System
- e: Advanced Text Management System (IBM)

ATMs e: Advanced Ticket Machines
- e: Automatic Teller Machines

ATMS e: Automatic Teller Machine System[s]
- e: Automatic Transmission Measuring System
- e: Automatic Trunk Measuring System

ATMSS e: Automatic Telegraph Message Switching System
ATMU e: Aircraft Torpedo Maintenance Unit
ATM/W e: Automatic Teller Machine cash with Drawal
ATMX e: Railcar used by the Department of Energy for Shipping Defense Waste
ATN d: automatischer Telefon-Notruf
- e: Actual Test Number
- e: Acute Tubular Necrosis
- e: Aid-to-Navigation
- e: Astrogeophysical Transmission Network
- e: Attention
- e: Augmented Transition Network
- e: Autonomous Thyroid Nodule
- e: Aviation Technician; Navigation

ATNAV e: Acoustic Transponder Navigation
ATN Buoy e: Aid-to-Navigation Buoy
Atndt. e: Attendant
ATNM e: Antitank Nonmetallic
at. no. e: atomic number
ATNP e: Atherton Tablelands National Parks
ATO d: Allgemeine Tarifordnung
- d: Allgemeine Treuhandorganisation
- d: Autotransportordnung
- e: Action Technical Order
- e: Actual Time Over
- e: Address Transfer Operation

- e: Aeronautical Telecommunications Operator
- e: Aircraft Technical Order
- e: Aircraft Transfer Order
- e: Air Targets Officer
- e: Air Terminal Officer
- e: Air Training Officer
- e: Air Travel Order
- e: Allied Travel Office
- e: Allocate to Order
- e: Ammunition Technical Officer
- e: Area Traffic Officer
- e: Assisted Take-Off
- e: Automatic Train Operation

AT & O e: Apollo Test and Operations
A to A e: Air-to-Air
ATOA e: American Tung Oil Association
ATOC e: Air Transport Operations Centre
- e: Average Total Operating Cost

A-to-D e: Analog[ue]-to-Digital
AtoG d: Atomgewicht
ATOG e: Automatic Test Program[me] Generator
ATOG Weight e: Allowable Take-Off Gross Weight
AtOk d: Aufteilungsortskabel
ATOLL e: Acceptance Test or Launch Language
ATOLS e: Automatic Testing On-Line System
ATOM e: Advanced Technology of Management
- e: Analog[ue] Tree-Organized Multiplexer
- e: Arizona Trade-Off Model
- e: Astronomical Telescope Orientation Mount
- e: Automatic Testing Operating and Maintenance
- e: Automatic Transmission of Mail

atomdef e: atomic defense
atomdev e: atomic device
Atomgew. d: Atomgewicht
ATOMIC e: Automatic Train Operation by Mini-Computer
ATOMICS e: Advanced Traffic Observation and Management Information Collection System
Atom-Inform. d: Atom-Information (journ.)
ATOMS e: Automated Technical Order Maintenance Sequence[s]
- e: Automated Technical Order Maintenance Systems

ATOMSTATREP e: Atomic Status Report
Atomwirtsch.; Atomtech. d: Atomwirtschaft; Atomtechnik (journ.)
ATOM WONS e: Atomic Weapons
ATOM WPNS e: Atomic Weapons
ATOM WPNS ELEC ASSR e: Atomic Weapons Electrical Assembler
ATOM WPNS MECH ASSR e: Atomic Weapons Mechanical Assembler
ATOM WPNS NUC ASSR e: Atomic Weapons Nuclear Assembler
ATON e: Aids to Navigation
ATORP e: Antitorpedo
- e: Atomic Torpedo

ATOS e: Assisted Take-Off System
ATOT e: Actual Time over Target
- e: Angle Track on Target

A to Z e: from the beginning to the end
ATP d: Adenosintriphosphat
- d: Adernaufteilungsplan
- d: arbeitsteiliger Prozeß
- d: Auslandtelegrammprüfstelle
- e: Acceptance Test Plan
- e: Acceptance Test Procedure
- e: Accumulated Total Punching
- e: Actual Time of Penetration
- e: Actual Track Pointer
- e: Adenosine Triphosphate
- e: Advanced Technology Program[me]
- e: Advanced Test in Psychology
- e: Advance Test Plant
- e: Air Technical Publication[s]
- e: Air Traffic Procedures
- e: Allied Tactical Publication
- e: Allied Technical Publication[s]
- e: Alternate Target Point
- e: Alternate Term Plan
- e: Anode Tapping Point
- e: Anti-Torque Pedal
- e: Antitank Platoon
- e: Army Training Program[me]
- e: Array Transform Processor
- e: Arsenical Tough Pitch
- e: Assembly Test Program[me]
- e: Astronautics Test Procedures
- e: Asynchronous Transaction Processing
- e: At Time or Place
- e: Authority to Proceed
- e: Automated Testing Procdure
- e: Automated Test Plan
- e: Automated Test Procedure
- e: Automatic Text Processing
- e: Automatic Train Protection

atp f: a tout prix
ATPA e: Auxiliary Turbopump Assembly
ATPase d: Adenosin Triphosphatase
- e: Adenosine Triphosphate

ATPC e: Assistant for Telecommunications Program[me] and Control
ATPCC e: Attitudes Toward Parental Control of Children
ATPD e: Ambient Temperature and Pressure; Dry
ATPF e: Armament Test Preparation Facility
ATP Facility e: Asynchronous Transaction Processing Facility
ATPG e: Automated Test Pattern Generation
- e: Automatic Test Pattern Generation
- e: Automatic Test Pattern Generator
- e: Automatic Test Program[me] Generator

ATPI e: American Textbook Publishers Institute
At Plat e: Antitank Platoon
ATPOS e: Atomic Post-Strike Analysis Report
ATPP e: Aptitude Test for Programmer Personnel
ATPR e: Advanced Test Prototype Reactor
- e: Annual Technical Progress Report

at pres e: at present
ATPS e: Ambient Temperature and Pressure; Saturated [with Water Vapor]
- e: Automatic Type Placement System

AT-PSK e: Adjacent Tone-Reference Phase-Shift Keying
ATR d: Abgastemperaturregelung
- d: Arbeitsausschuß Transportrationalisierung
- d: Auto-Transformer
- e: Access Time of Reading
- e: Access Time to Reading
- e: Accumulator Track Read
- e: Achilles Tendon Reflex
- e: Admiralty Test Rating
- e: Advanced Tactical Radar
- e: Advanced Test Reactor
- e: Advanced Tracker Radar
- e: Aerial Transmit-Receive
- e: Air-launched Trainer Rocket
- e: Airborne Test Reactor
- e: Aircraft Transmitter Receiver
- e: Aircraft Trouble Report
- e: Airline Transport Rating
- e: Air Traffic Regulations
- e: Air Traffic Rules
- e: Air Transport of Radiation
- e: Air Transport Radio
- e: Air Transport Rating
- e: Air Turbo Rocket
- e: All Transistor
- e: Ambient Temperature Range
- e: Analog[ue] Tape Recorder
- e: Angle Tracking System
- e: Answer[ing] Time Recorder
- e: Antenna Transmit-Receive
- e: Anti-Transmit[ter]-Receive[r]
- e: Apollo Test Requirement[s]
- e: Assembly Test Recording
- e: Atrophy
- e: Attenuated Total Reflectance
- e: Attenuated Total Reflection
- e: Attribute
- e: Audiotape Recorder
- e: Australian Telecommunications Research
- e: Automatic Target Recognition
- e: Automatic Traffic Recorder
- e: Automatic Transmitter
- e: Automatic Trunk Routiner
- e: Aviation Technician; Radar
- e: Aviation Training Record

ATRAC e: Angle Tracking Computer
ATRAN e: Automatic Terrain Recognition and Navigation
ATRAX e: Air-Transportable Communications Complex
ATRC e: Air Traffic Regulation Center
- e: Air Training Command
- e: Anti-Tracking Control
- e: Army Transportation Research Command
- e: Atlantic Research Corporation

ATRCE e: Advanced Test Reactor Critical Experiment
ATRD e: Automatic Target Recognition Device
ATREP e: Air Traffic Representative
ATR FIB e: Atrial Fibrillation
ATRI e: Artists Technical Research Institute
ATRIAL e: Atrial Fibrillation
ATRIB e: Average Transfer Rate of Information Bits
ATRID e: Automatic Target Recognition Identification and Detection
ATRIMA e: As Their Respective Interests May Appear
ATRIS e: Air Traffic Regulation Identification System
- e: Air Transportation Research Information Service

ATRL e: Anti-Tank Rocket Launcher
- e: Atrial

ATRLS e: Actual Time of Release
ATRM e: After Torpedo Room
ATRO d: absolut trocken
- d: Ausgabe Trommeldatei
- e: Acting Transporting Officer
- e: Actual Time of Return to Operation

atrop e: atrophy
ATrP e: Allied Training Publication
ATRR e: Advanced Threat-Reactive Receiver
- e: Antitrust & Trade Regulation Report

ATRS e: Advanced Technology Reader Sorter
- e: American Tanker Rate Schedule
- e: American Tanker Rate Scheme
- e: Assembly Test Recording System
- e: Automatic Tape Response System

ATRSO e: Accepts Transfer as Offered
ATR System e: Automatic Tape Response System
ATR T[ube] e: Anti-Transmit-Receive Tube
ATRVAL e: Attribute Value
ATS d: Abtastsystem
- d: Antitetanus-Serum
- d: Anwendungs-Technologiesatellit
- d: Außentonsäule
- d: automatisches Test-System
- d: automatisiertes Transportsystem
- e: Absolute Temperature Scale
- e: Abstract Test Suite
- e: Accelerometer Timer Switch
- e: Acoustic Telemetry Subsystem
- e: Acoustic Transmission System
- e: Acquisition and Tracking Station
- e: Acquisition and Tracking System
- e: Acquisition Target and Search
- e: Active Television System
- e: Address Transfer Sequence
- e: Administrative Terminal System
- e: Advanced Technological Satellite
- e: Advanced Technology Systems, Inc.
- e: Advanced Teleprocessing System
- e: Advanced Terminal System
- e: Advanced Text System
- e: Advanced Time Sharing
- e: Aided Tracking System
- e: Air-to-Ship
- e: Air-to-Surface
- e: Airport-radar Terminal System
- e: Air Tactical School
- e: Air Traffic System
- e: Air Training Scheme
- e: Air Training School
- e: Air Transportable Sound Navigation and Ranging
- e: Air Turbine Starter
- e: Alarm Termination Subsystem
- e: American Television Society
- e: American Theological Society
- e: American Therapeutic Society
- e: American Thesaurus of Slang
- e: Ampere-Turns
- e: Analog[ue] Transmission System
- e: Animal Tub Sized
- e: Anti-Tetanic Serum
- e: Anti-Thymocyte Serum
- e: Anti-Thymus Serum
- e: Anxiety Tension State
- e: Application[s] Technology Satellite
- e: Arbitrating Test-and-Set
- e: Army Transport Service
- e: Associate of Theological Study
- e: Association of Theological Schools
- e: Astronomical Time Switch
- e: Asymptotic Threshold Shift
- e: Atlantic Tracking Ship
- e: Atlantic Trade Study
- e: Attitude Thrustor System
- e: A Tutorial System
- e: Audio Test Set
- e: Automated Telemetry System
- e: Automated Terminal Station
- e: Automated Text System
- e: Automated Trading System
- e: Automatic Telemetry System
- e: Automatic Telephone Set
- e: Automatic Terminal System
- e: Automatic Test[ing] System
- e: Automatic Test Scoring
- e: Automatic Test System
- e: Automatic Train Stop
- e: Automatic Train Stopping
- e: Automatic Train Supervision
- e: Automatic Transfer Service
- e: Automatic Transmission System
- e: Automatic Trim System
- e: Automatic Trunk Synchronizer
- e: Auxiliary Tug Salvage

ATS-A d: automatisches Test-System für Analogtest
ATSA e: Aero Transportes
ATS-A e: Applications Technology Satellite Model A
ATS-B e: Applications Technology Satellite Model B
A.T.S.B.E. e: All Time Saved; Both Ends
ATSC e: Air Technical Service Command
ATS-C e: Applications Technology Satellite Model C
ATSC e: Assistant for Telecommunications System[s] Coordination
AT-Schweißen d: alumionthermisches Schweißen
ATS-D d: automatisches Test-System für digitale Schaltungen
- d: automatisches Test-System für Digitaltest
- e: Applications Technology Satellite Model D

ATSD e: Airborne Traffic and Situation Display
ATSD(AE) e: Assistant of the Secretary of Defense (Atomic Energy)
ATSDR e: Agency for Toxic Substances and Disease Registry
ATSE e: Advanced Throttling Slurry Engine
ATS-E e: Applications Technology Satellite Model E
ATSE e: Assistant for Telecommunications System[s] Engineering
ATSEFI e: Auxiliary Territorial Service Expeditionary Force Institutes
AT Set e: Auxiliary Tank Set
ATS-F e: Applications Technology Satellite Model F
ATSF f: Avion de Transport Supersonique Futur
ATSG e: Acoustic Test Signal Generator
ATSIT e: Automatic Techniques for the Selection and Identification of Targets
ATSJEA e: Automatic Test System for Jet Engine Accessories
ATSL e: A Test Specification Language
ATSM e: Advanced Tactical Stand-off Missile
ATSMC e: Army Transportation Supply and Maintenance Command
ATSOCC e: ApplicationsTechnology Satellite Operations Control Center
ATSP e: Absolute Text and Storage Project
ATSR e: Argonne Thermal Source Reactor
- e: Army Transport Service Regulation

ATSS d: Ausblendung, temperaturgeregelte Stromquelle, Spannungsabschalter
- e: Acquisition and Tracking Sub-System
- e: Advanced Technical Service School
- e: Augmented Target Screener Subsystem
- e: Augmented Target Screening Subsystem
- e: Automatic Telecommunications Switching System
- e: Automatic Telegraph Subsystem
- e: Automatic Telephone Switching System

ATS's e: Advanced Technological Satellites
ATSS-DS e: Automatic Telecommunications Switching System-Data Services
ATSSS e: Air Transportable Sonar Surveillance System
AtST e: Atlantic Standard Time
At. Strom d: Atom und Strom (journ.)
ATSU e: Association of Time-Sharing Users
- e: Automatic Tool Setting Unit

ATT e: Accelerated Test Technology
- e: Address Translation Table
- e: Advanced Technology Transport
- e: Air Traffic Transponder
- e: American Telephone and Telegraph [Company]
- e: Application Transfer Teams
- e: Arginine Tolerance Test
- e: Army Training Test
- e: Attach[ed]
- e: Attachment
- e: Attend[ed]
- e: Attended Public Telephone
- e: Attention
- e: Attenuated
- e: Attenuation
- e: Attenuator
- e: Attic[a]
- e: Attorney
- e: Attribute
- e: Audit Trail Tape
- e: Automatic Toll Ticketing
- e: Avalanche Transit Time
- e: Average Task Time
- e: Average Total Time

Att. d: Attest
- d: Attika
- d: Attitude
- d: Austauschteil

A & TT e: Alcohol and Tobacco Tax
AT & T e: American Telephone & Telegraph
ATTACH e: Attachment
ATTAS e: Advanced Technology Testing Aircraft System
ATTC e: Air Transportability Training Center
- e: Amarillo Technical Training Center
- e: American Telephone and Telegraph Company
- e: Atlantic Transportation Terminal Command
- e: Automatic Transmission Test and Control
- e: Aviation Technical Training Center

ATTCOM e: American Telephone and Telegraphy [Company] Communications
ATTD e: Attached
- e: Attitude
- e: Avalanche Transit-Time Diode

ATT-Diode e: Avalanche Transit-Time Diode
ATTE e: Automatic Transistor Test Equipment
- e: Automatic Transmission Testing Equipment

attech d: Atmosphäre; technische
ATTEN e: Attention
- e: Attenuation
- e: Attenuator

ATTENUAT e: Attenuator
ATTESTN e: Attestation
ATTF e: Advanced Technical Training Facility
ATTG e: Attending
Att-Gen e: Attorney-General
ATTI e: Arizona Transportation and Traffic Institute
- e: Association of Teachers in Technical Institutions

- e: American Telephone and Telegraph [Company] International
ATTIS e: American Telephone and Telegraph [Company] Information Systems
ATTIX e: American Telephone and Telegraph [Company] Interexchange Carrier
ATTM e: Authorization of Transfer Material
ATTM Equipment e: Automatic Trunktesting and Transmission Measuring Equipment
ATTN e: Attention
- e: Attenuation
ATTND e: Attendant
ATTNINQ e: Attention Inquiry
atto e: 10^{-18}
- s: atento
ATTO e: Attorney
- e: Avalanche Transit-Time Oscillator
ATTR e: All Thrust Terminate Relay
- e: Attribute
- e: Average Time to Repair
AtTr d: Antennenträger
Attr. d: Attraktion
- d: Attrappe
- d: Attribut
attr. d: attraktiv
- d: attributiv
- e: attractive
ATTRA e: Automatic Tracking Telemetry Receiving Antenna
attrd e: attributed
ATTREF e: Attitude Reference [Program]
attrib e: attribute[d]
- e: attributive
attrit e: attrition
Atts d: Austauschteilsatz
ATTS e: All-Through Training School
- e: Automatic Telemetry Tracking System
ATTT e: American Telephone and Telegraphy [Company] Technologies
ATTU e: Analog[ue] Trunk Terminating Unit
ATTUN e: Automatic Tuning
Atty. e: Attorney
Atty & C e: Attorney and Client
Atty.-Gen. e: Attorney-General
AT Type e: Adenine and Thymine Type
ATU d: Arbeitstakt des Umsetzers
- e: Address Translation Unit
- e: Advanced Training Unit
- e: Aerial Tuning Unit
- e: Airship Training Unit
- e: Alcohol Tax Unit
- e: Alliance of Telephone Unions
- e: Analysis and Transformation Unit
- e: Anchorage Telephone Utility
- e: Antenna Tuning Unit
- e: Application Terminal Unit
- e: Arab Telecommunications Union
- e: Army Transportation Unit
- e: Atomic Time Unit
- e: Automatic Tracking Unit
- e: Autonomous Transfer Unit
ATu d: Abdampfturbine
atu d: Atmosphäre[n] Unterdruck
atü d: Atmosphäre[n] Überdruck
ATUC e: African Trade Union Confederation
- e: Average Total Unit Cost
ATUG e: Australian Telecommunications Users Group
AtUm d: Antennenumschalter
ATUM e: Antitank Nonmetallic
ATUR e: Apollo Test Unsatisfactory Report
- e: Automatic Telephone Using Radio
ATURS e: Automatic Traffic Usage Recording System
ATUS e: Antitumour Serum
ATV a: Akademiet for de Tekniske Videnskaber
- d: Abwassertechnische Vereinigung
- d: Allgemeine Deutsche Transport-versicherungsbedingungen
- d: Allgemeine Tarifvorschriften
- d: allgemeine technische Vorschriften
- d: Antennenverstärker
- d: Automobil-Treuhandverband
- e: Advanced Test Vehicle
- e: Aerodynamic Test Vehicle
- e: Agena Target Vehicle
- e: Air-to-Vessel
- e: Air Test Vehicle
- e: All-Terrain Vehicle
- e: Amateur Television
- e: Associated Tele Vision
- e: Associated Television
- e: Automatic Threshold Variation
- e: Automatic Ticket Vendor
ATVC e: Automatic Thrust Vector Control
ATVen d: allgemeine technische Vertrags-bedingungen
ATVM d: Attenuator-Thermoelement Voltmeter
AtVO d: Atomverordnung
at. vo. e: atomic volume
AT. VOL d: Atomvolumen
at. vol. e: Atomic Volume
AtVr d: Antennenverstärker
ATVS e: Advanced Television Seeker
ATVSC e: Advanced TV Systems Committee
ATVWS e: Airport Trailing Vortex Warning SYstem
atw d: Atomwirtschaft (journ.)
ATW e: Access Time of Writing
- e: Accumulator Track Write
- e: Aerospace Test Wing
- e: Aircraft Tail-Warning
- e: Air Transport Wing
- e: Anti-Tank Weapon
-e: Atomic Hydrogen Weld
- e: Automatic Tape Winder
AT/WB e: Ampere Turns per Weber
ATWDDS e: Automated Terminal Weather Dissemination Display System
AtWe d: Antennenweiche
ATWE e: Association of Technical Writers and Editors
ATWg e: Air Transport Wing
AT-Winkel d: Antetorisionswinkel
AT W/OD e: Automatic Transmission with Overdrive
ATWS e: Adjustable Thermal Wire Stripper
- e: Anticipated Transient without Scram
- e: Automatic Track While Scan[ning]
ATWT e: Atomic Weight
ATX e: Air Taxi
- e: Automatic Telex
- e: Automatic Telex Exchange
ATXPL e: Atomic Explosion
A TypI f: Association Typographique Internationale
ATYROPT e: At your Option
ATZ d: Abtastzyklus
- d: Aufzeichnungstaktzähler
- d: Automatische Telefonzentrale
- d: Automobiltechnische Zeitschrift
- e: Aerodrome Traffic Zone
AT-Zellulose d: Äthylzellulose
ATZm d: automatische Telefonzentrale; mobil
AU a: Aarhus Universitet
- d: Ankerkreisumschaltung
- d: Anpassungsübertrager
- d: Anwendungsunterstützung
- d: Arbeitsumschaltkontakt
- d: Arbeitsunfähigkeit
- d: Arbeitsunterlage[n]
- d: Arbeitsunterweisung
- d: Audion
- e: Ausführungsunterlage
- d: Ausgangsumsetzer
- e: A-Upper
- e: Aarhus University
- e: Acadia University
- e: Access Unit
- e: Adder Unit
- e: Address Unit
- e: Adjunct Unit
- e: Administrative Audit
- e: Advanced Undersea [Weapon]
- e: Air University
- e: Alabama University
- e: Alfred University
- e: Allahabad University
- e: Allen University
- e: All Up
- e: American University
- c: Analyzer Unit
- e: Andrews University
- e: Angstrom Unit
- e: Answer Unit
- e: Antitoxin Unit
- e: Arbitrary Unit
- e: Arithmetic[al] Unit
- e: Assault Unit
- e: Assembler Unit
- e: Associated Universities
- e: Assumption University
- e: Astromical Unit
- e: Atlanta University
- e: Attached Unassigned
- e: Auburn University
- e: Audio
- e: Augite
- e: Australia
- e: Author
- e: Autocoder
- e: Automatic
- e: Azauridine
- e: gold
au. d: arbeitsunfähig
a. u. l: ab usu
- l: ad usum
- l: anno urbis
a.u. l: aures unitas
- l: au usum
Au l: aurum
AÜ d: Additionsübertrag
- d: Adressenübertrag
- d: Analog[wert]überwachung
- d: Anpassungsübertrager
- d: Aufgabenübertragung
- d: Ausgangsübertrager
- d: Auslandsüberweisung
- d: Außenübertragung
- z: Ankara Üniversitesi
Au'9^{8} e: radioactive gold
AU-23 e: A Fairchild piston-powered stol aircraft
a. u. a. d: auch unter ander[e]n
AUA e: American Urological Association
- e: Associate of the University of Adelaide
- e: Austrian Airlines

AuÄ d: Arbeitsgemeinschaft unabhängiger Ärzte
AUAF e: Association of University Affilliated Facilities
AUAIP e: Aeronautics Upper Atmosphere Impact Program[me]
AUAS e: Academy of Underwater Arts and Sciences
AUB d: Allgemeine Unfallversicherungsbedingungen
- d: Ausfuhr-Unbedenklichkeitsbescheinigung
- d: Automatikbetrieb
- e: American University of Beirut
AUBC e: Association of Universities of the British Commonwealth
AUBER e: Association for University Business and Economic Research
AUBI d: Autofahrerbier (DDR)
aubo. d: außenbord
Aubo-Moto d: Außenbordmotor
AUBV d: Allgemeine Unfallversicherungsbedingungen
AUC d: Ammoniumuranylcarbonat
- e: Aberystwyth University College
- e: Additive Unit Codes
- e: Administrative Unit Commandant
- e: Alternate Unit Codes
- e: American University, Cairo
- e: American University of the Caribbean
- e: Average Unit Cost
a.u.c. l: ab urbe condita
a. u. c. l: anno urbis conditae
AUCAS e: Association of University Clinical Academic Staff
AUCBE e: Advisory Unit for Computer Based Education
AUCC e: American Used Computer Corporation
- e: Association of Universities and Colleges of Canada
AUC Corporation e: American Used Computer Corporation
Aucklands e: Auckland Islands
AUCSRLFRVWAM e: All Union Central Scientific Research Laboratory for the Restoration of Valuable
auct l: auctorum
Aud. d: Auditorium
AUD d: Auftragsdatei
- e: Asynchronous Unit Delay
- e: Audibility
- e: Audible
- e: Audio
- e: Audiogram
- e: Audiometry
- e: Audiovisual
- e: Audition
- e: Audit[or]
- e: Auditorium
AÜD d: automatische Überwachungseinrichtung für Datenwählleitungen
AUDACIOUS e: Automatic Direct Access to Information with On-Line UDC System
AUDAR e: Autodyne Detection and Ranging
AUDB e: Arming Unit Distribution Box
AUDDITS e: Automated Dynamic Digital Test System
AUDDIT System e: Automatic Dynamic Digital Test System
AUDE d: Aufbereitung der Deskriptorenkonkordanzen
AUD.EQUIP. e: Audio Equipment
AUDEV e: Audio Devices
Aud-Gen e: Auditor-General
AUDGENAV e: Auditor General of the Navy
AUDI d: Auskunftsdienst
- e: IEE Audio and Electroaccoustics (journ.)
AUDICS e: Auditable Internal Control Systems
Audimax d: Auditorium Maximum
Audio. e: Audiofrequency
- e: Audiogram
- e: Audiometer
- e: Audiophile
AUDIO e: Audiogenic
- e: Audiology
- e: Audiometry
- e: Audiovisual [Aids]
Audiogr. d: Audiogramm
audiol e: audiologist
Audiol. e: Audiology
Audiom. d: Audiometer
- d: Audiometrie
audiom. d: audiometrisch
AUDIOM e: Audiometer
- e: Audiometric[al][ly]
- e: Audiometrist
AUDIOVIS d: Audiovisual
- e: Audiovisual [Aids]
AUDIT e: Aircraft Unitized Diagnostic Inspection and Test
- e: Army Uniform Data Inquiry Technique
- e: Auditory Input Task
- e: Automatic Unattended Detection Inspection Transmitter
AUDRE e: Audio Response
- e: Augment[ation]
- e: Augmentative
- e: Automatic Digit Recognizer
AUDREC e: Automatic Digit Recognition
- e: Automatic Digit Recognizer
AUDREY e: Audio Reply
- e: Automatic Digit Recognition
- e: Automatic Digit Recognizer
AUDRI e: Automated Drug Identification
AUDS d: Arbeitsgemeinschaft für Unterhaltung der deutschsprachigen Sender
aud.snl. e: audio signal
a. u. d. T. d: auch unter dem Titel
AUDVIS e: Audiovisual
AUE d: allgemeine Überwachungseinrichtung
AUe d: Amtsübertrager
- d: Amtsübertragung
- d: Auslandsübertragung
AUEC e: Association of University Evening Colleges
AUe-g d: Amtsübertragung; gehend
AUe-gk d: Amtsübertragung-gehend/kommend (wechselseitig)
AUe-k d: Amtsübertragung; kommend
AUET e: Armo[u]red. Universal Engine Tractor
AUEW e: Amalgamated Union of Engineering Workers
AUF d: Archiv für Urkundenforschung (journ.)
- d: Auftragsabwicklung
- e: Asset Utilization Factor
Auf. d: Aufenthalt
Aufb. d: Aufbau[ten]
- d: Aufbewahrung
Aufber. d: Aufbereitung
Aufbereit.-Tech. d: Aubereitungs-Technik (journ.)
Aufber.-Techn. d: Aufbereitungstechnik
Aufbew. d: Aufbewahrung
AufbG d: Aufbaugesetz
Aufbr. d: Aufbruch
Aufdr. d: Aufdruck[e]
Auff. d: Auffassung
- d: Auffindung
- d: Aufforderung
- d: Aufführung
- d: Auffüllung
Auffüll. d: Auffüllung
Aufg. d: Aufgabe
- d: Aufgang
- d: Aufgebot
aufgef. d: aufgefordert
- d: aufgeführt
aufh. d: aufheben
- d: aufhören
Aufh. d: Aufhebung
- d: Aufhellung
AufhA d: Anordnung zur Aufhebung
Aufhell. d: Aufhellung
AufhG d: Gesetz zur Aufhebung
Aufhgg. d: Aufghängung
AufhVorschr. d: Vorschrift[en] zur Aufhebung
Aufkl. d: Aufklarung
- d: Aufklärung
Aufl d: Auflage
Aufl. d: Auflockerung
- d: Auflösung
Auflös. d: Auflösen
- d: Auflöser
- d: Auflösung
AuflVO d: Auflassungsverordnung
Aufn. d: Aufnahme
Aufn.-Antr. d: Aufnahmeantrag
Aufn.-Bed. d: Aufnahmebedingung[en]
Aufn.-Best. d: Aufnahmebestimmung[en]
Aufn.-Nr. d: Aufnahmennummer
AUFPRI d: Auftragspriorität
Aufpro d: Aufprojektion
aufr. d: aufrecht
- d: aufreißen
Aufr. d: Aufriß
Aufs. d: Aufsatz
Aufs d: Aufsicht
AUFS e: Absorbance Unit Full Scale
- e: American Universities Field Staff
Aufs.-B. d: Aufsichtsbeamter
- d: Aufsichtsbehörde
Aufs.-Beam. d: Aufsichtsbeamter
Aufs.-Beh. d: Aufsichtsbehörde
Aufsch. d: Aufschub
- d: Aufschüttung
Aufschl. d: Aufschlag
- d: Aufschluß
Aufschr. d: Aufschrift
Aufs.-Pers. d: Aufsichtsperson
- d: Aufsichtspersonal
Aufs.-R. d: Aufsichtsrat
Aufst. d: Aufstellung
- d: Aufstufung
Aufstell. d: Aufstellung
AufT d: Aufschaltetaste
Auft. d: Aufteilung
Auftlg. d: Aufteilung
Aufton d: Aufschaltcton
Auftr. d: Auftrag
- d: Aufträge
- d: Auftritt[e]
Auftr.-Best. d: Auftragsbestätigung
AUFTR-Datei d: Auftragsdatei
Auftr.-Nr. d: Auftragsnummer
Auftr.-Verg. d: Auftragsvergabe
- d: Auftragsvergütung
Aufw. d: Aufwand
aufw. d: aufwärts
- d: aufweisen
- d: aufwenden
Aufw. d: Aufwendung[en]

aufw. d: aufwerten
Aufw. d: Aufwertung
Aufw.-Entsch. d: Aufwandsentschädigung
AufwG d: Aufwertungsgesetz
Aufz. d: Aufzeichnung
- d: Aufzug
Aufz.-Bed. d: Aufzugsbedienung
Aufz.-Nr. d: Aufzugsnummer
AufzV d: Aufzugsverordnung
AUG d: August
- e: Amdahl Users Group
- e: Augment[ed]
- e: Augmentative
Aug d: Augen-
aug e: Augmentation
AUGE d: Arbeitsgruppe Umwelt, Gesellschaft, Energie
augm e: augment[ed]
- f: augmente
augm. e: augmentation
- e: augmentative
AUGU e: Augmenting Unit
AUI e: Associated Universities Incorporated
- e: Attachment Unit Interface
AUJ Box e: Air-to-Umbilical Junction Box
AUJS e: Advanced Universal Jamming System
AUK e: Alphanumeric Universal Keyboard
Auk. e: Auckland
AUKO-T d: Ausgabekonvertierung-tetradische
AUL d: Anrufumleitung
- e: Aberdeen University Library
- e: Above Upper Limit
- e: Add Unnormalized Long
- e: Air University Library
AULC e: American University Language Center
AulKopfVStHand d: Auslandskopf-vermittlungsstelle mit Handbedienung
AULLA e: Australasian Universities Language and Literature Association
AULLAJ e: Australian Universities Language and Literature Association Journal
AULR e: Attrition, Utilization and Loss Rate
AULS e: Audio-visual Library Services
AUM e: Advanced Underwater Missile
- e: Air-to-Underwater Missile
aum s: aumentado
AUMA d: Ausstellungs- und Messe-ausschuß der deutschen Wirtschaft
AU/MIR e: Administrative Unit/Medical Inspection Room
AUMLA e: Australian Universities Modern Language Association
AUM-N e: Air-to-Underwater Missile-Nuclear
AUmr d: Anrufumrechner
AUmw d: Auslandsumwerter
a. u. n. l: abesque ulla nota
AUNT e: Automatic Universal Translator
AUNTIE e: Automatic Unit for National Taxation and Insurance
AU of B e: American University of Beirut
AU of C e: American University of Cairo
AUOS d: automatische universelle Orbitalstation
- e: Automatic, Unified Orbiting Station[s]
AUP d: Ausgabeprogramm
AUPAC Brushes e: Automatically Compacted Brushes
AUPELF e: Association of Wholly or Partially French Language Universities
AUPF e: Australian Uranium Producers Forum
AUPG e: American University Publishers Group
AUPHA e: Association of University Programs in Hospital Administration
AuPO d: Ausbildungs- und Prüfungs-ordnung
AUPO e: Association of University Professors of Ophthalmology
AUPOSTCOM e: Australian Postal Commission
AU-PS d: Australische Patentschrift
AUr d: Auslandsumrechner
AUR e: Aberdeen University Review (journ.)
- e: Aircraft Utilization Report
- e: Association of University Radiologists
- e: Aural
- e: Auricle
- e: Auricular[is]
- e: Aurum
AÜR d: automatischer Überdeckungs-repeater
AURA e: Association of Universities for Research in Astronomy
AURBO e: Aurora Borealis
AUR FIB e: Auricular Fibrillation
AURIC e: Auricular
AURIS e: Albuquerque Urban Regional Information System
AURISA e: Australian Urban and Regional Information Systems Association
aurist. l: auristillae
AURS e: Advanced Unit Record System
- e: Alone Unit Record System
- e: Automatic Unit Record Store
AUrTr d: Auslandsumrechner für den Transitverkehr
AUS d: absolute Unterschiedsschwelle
- d: Arbeiten unter Spannung
- d: Australien (Kfz)
- e: Add Unnormalized Short
- e: Army of the United States
- e: Australia
AUs d: Amtsumsetzer
- d: Anpassungsumsetzer
- d: Auslandsumsetzer
aus d: ausgeschaltet
Aus e: Austin
Aus e: Austria[n]
a. u. s. l: actum ut supra
AUSA e: Aid of the United States of America
A.U.S.(A.) e: Assistant Undersecretary of State (Air Staff)
Ausb. d: Ausbau
- d: Ausbesserung
- d: Ausbeute
- d: Ausbilder
- d: Ausbildung
AUSB d: Ausgabe binär
AUSBCM Program[me] e: Apollo Unified S-Band Circuit Margin Program[me]
ausbez. d: ausbezahlen
- d: ausbezahlt
Ausbild. d: Ausbildung
Ausbild.-Beih. d: Ausbildungsbeihilfe
AusbO d: Ausbildungsordnung
Ausbreit. d: Ausbreitung
AUSC e: Auscultation
AUSCOR e: Automatic Scanning Correlator
AusE e: Australian English
A.U.S.(E.O.) e: Assistant Undersecretary of State (Establishment and Organization)
Ausf. d: Ausfahrt
- d: Ausfall
- d: Ausfällung
- d: Ausfertigung
- d: Ausflockung
- d: Ausfuhr
- d: Ausführung
AusfA d: Ausführungsanordnung[en]
Ausf.-Anord. d: Ausführungsanordnung[en]
Ausf. Anw. d: Ausführungsanweisung[en]
AusfB d: Ausführungsbestimmung[en]
Ausf.-Beh. d: Ausführungsbehörde
Ausf.-Bek. d: Bekanntmachung[en] zur Ausführung
Ausf.Best. d: Ausführungsbestimmung[en]
Ausf.-Bürg. d: Ausfuhrbürgschaft
Ausf.-Erkl. d: Ausfallerklärung
- d: Ausfuhrerklärung
Ausf.-Erl. d: Ausführungserlaß
Ausf.-Fin. d: Ausfuhrfinanzierung
Ausf.-Ford. d: Ausfallforderung
Ausf.-Förd. d: Ausfuhrförderung
AusfFördG d: Ausfuhrförderungsgesetz
AusfG d: Ausführungsgesetz[e]
Ausf.-Gar. d: Ausfuhrgarantie
Ausf.-Geb. d: Ausfertigungsgebühr
Ausf.-Gen. d: Ausfuhrgenehmigung
Ausf.-Ges. d: Ausfuhrgesetz[e]
- d: Ausführungsgesetz[e]
Ausf.-H. d: Ausfuhrhandel
Ausf.-Kontr. d: Ausfuhrkontrolle
Ausflock. d: Ausflockung
Ausf.-Med. d: Ausführungsmeldung
Ausf.-Meld. d: Ausfuhrmeldung
Ausf.-Nr. d: Ausfertigungsnummer
- d: Ausführungsnummer
AusFörd. d: Ausfuhrförderung
Ausf.-Pr. d: Ausfuhrprämie
Ausf.-Send. d: Ausfuhrsendung[en]
Ausfsig d: Ausfuhrsignal
Ausf.-Überw. d: Ausfuhrüberwachung
- d: Ausführungsüberwachung
ausführl. d: ausführlich
AusfV d: Ausführungsverordnung
Ausf.-Verb. d: Ausfuhrverbot
Ausf.-Verf. d: Ausfuhrverfahren
Ausf.-Verord. d: Ausführungsverordnung
Ausf.-Vers. d: Ausfallversicherung
Ausf.-Vertr. d: Ausfuhrvertrag
- d: Ausfuhrverträge
Ausf.-Vol. d: Ausfuhrvolumen
Ausf.-Vorschr. d: Ausführungs-vorschrift[en]
Ausg. d: Ausgabe
- d: Ausgang
ausg. d: ausgeben
- d: ausgegeben
- d: ausgesetzt
Ausg.-Dat. d: Ausgabedatum
- d: Ausgangsdatum
ausgearb. d: ausgearbeitet
ausgebess. d: ausgebessert
ausgeg. d: ausgegeben
ausgegl. d: ausgeglichen
ausgel. d: ausgelagert
- d: ausgeliefert
ausgen. d: ausgenommen
- d: ausgenützt / ausgenutzt
ausges. d: ausgesetzt
ausgesch. d: ausgeschieden
ausgeschl. d: ausgeschlossen
Ausgest. d: Ausgestaltung
ausgest. d: ausgestattet
- d: ausgestellt
- d: ausgestopft
- d: ausgestorben

ausgew. d: ausgewählt
- d: ausgewiesen
- d: ausgewogen
- d: ausgewuchtet
ausgez. d: ausgezahlt
- d: ausgezählt
- d: ausgezeichnet
Ausgl. d: Ausgleich
AusgN d: Ausgabennachweis
Ausg.-Nr. d: Ausgabenummer
- d: Ausgangsnummer
Ausg.-P. d: Ausgangspunkt
Ausg.-Pos. d: Ausgangsposition
Ausg.-St. d: Ausgabestelle
- d: Ausgangsstation
- d: Ausgangsstempel
Aush. d: Aushang
- d: Aushilfe
AUSH e: Australian Serum Hepatitis
AUSH-AG e: Australian Serum Hepatitis Antigen
Aush.-Ang[est]. d: Aushilfsangestellte[r]
Aush.-Pers. d: Aushilfspersonal
- d: Aushilfspersonen
AUSIMM e: Australian Institute of Mining and Metallurgy
AUSINET e: Australian Information Network
Ausk. d: Auskommen
- d: Auskunft
- d: Auskünfte
ausk.-ber. d: auskunftsberechtigt
Ausk.-Ber. d: Auskunftsberechtigung
Auskleid. d: Auskleidung
Ausk.-Pfl. d: Auskunftspflicht
auskpfl. d: auskunftspflichtig
Ausk.-St. e: Auskunftstelle
Ausl. d: Ausladung
- d: Auslage[n]
- d: Auslagerung
- d: Ausland
- d: Ausländer
- d: Ausläufer
- d: Auslaut
- d: Ausleihe
- d: Auslieferung
- d: Auslosung
ausl. d: auslagern
- d: ausländisch
- d: auslegen
- d: ausliefern
- d: auslosen
- d: auslösen
AuslagVO d: Auslagenerstattungsverordnung
AuslAk d: Auslandsauskunft
Ausl.-Auf. d: Auslandsaufenthalt
AusLD d: Auslandsdienst
Ausl.-Ers. d: Auslagenersatz
AuslfG d: Auslieferungsgesetz
AUSLFL e: All-Union State Library of Foreign Literature
AuslG d: Ausländergesetz
AuslKopfVSt d: Auslandskopfvermittlungsstelle
AuslKopfVStHand d: Auslandskopfvermittlungsstelle mit Handbedienung
AuslSdg d: Auslandssendung
Ausl.-Term. d: Auslieferungstermin
AuslVSt d: Auslandsvermittlungsstelle
AuslVStHand d: Auslandsvermittlungsstelle mit Handbedienung
Ausn. d: Ausnahme
- d: Ausnutzung
- d: Ausnützung
Ausn.-Gen. d: Ausnahmegenehmigung
Ausn.-T. d: Ausnahmetarif
Ausn.-Tar. d: Ausnahmetarif
Ausn.-Zust. d: Ausnahmezustand
ausr. d: ausreichend
- d: ausrichten
- d: ausrüsten
Ausr. d: Ausrichter
- d: Ausrichtung
- d: Ausrüstung
Ausrüst. d: Ausrüstung
äuss. d: äußerlich
AUSSAT e: Australian Satellite
Aussch. d: Ausschuß
- d: Ausschüsse
Ausschalt. d: Ausschalter
- d: Ausschaltung
Ausscheid. d: Ausscheider
- d: Ausscheidung
Ausschl. d: Ausschlag
- d: Ausschließung
- d: Ausschluß
ausschl. d: ausschließend
- d: ausschließlich
Ausschn. d: Ausschnitt
Außenst. d: Außenstände
- d: Außenstelle
Außenste. d: Außenstelle
Aussetz. d: Aussetzung
Aussie[s] e: Australian[s]
Ausspr. d: Aussprache
- d: Ausspruch
Ausst. d: Ausstattung
- d: Ausstellung
Ausstatt. d: Ausstattung
Ausstattg. d: Ausstattung
AUST d: Ausgabesteuerung
AusT d: Auslösetaste
Aust. d: Austausch
Aust. e: Australia[n]
AusT e: Australian Terms
Aust Alps e: Australian Alps of New South Wales and Victoria
Aust. Chem. Eng. e: Australian Chemical Engineering (journ.)
Aust. Chem. Process. Eng. e: Australian Chemical Processing and Engineering (journ.)
Aust. Civ. Eng. e: Australian Civil Engineering (journ.)
Aust Comput Bull e: Australian Computer Bulletin (journ.)
Aust Comput J e: Australian Computer Journal (journ.)
Aust. Cur. e: Australian Current (journ.)
Aust. Dent. J. e: Australian Dental Journal (journ.)
Aust. Electron. Eng. e: Australian Electronics Engineering (journ.)
Aust. Electr. World e: Australian Electrical World (journ.)
austen e: austenitic
Austen e: Australian sten gen
Aust Engl e: Australian English
Aust. Geomech. J. e: Australian Geomechanics Journal (journ.)
AUSTIN e: Automatic Substrate Tinner Inline
Aust. J. Agric. Res. e: Australian Journal of Agricultural Research (journ.)
Aust. J. Biol. Sci. e: Australian Journal of Biological Sciences (journ.)
Aust. J. Bot. e: Australian Journal of Botany (journ.)
Aust. J. Chem. e: Australian Journal of Chemistry (journ.)
Aust. J. Dermatol. e: Australian Journal of Dermatology (journ.)
Aust. J. Exp. Agric. Anim. Hus e: Australian Journal of Experimental Agriculture and Animal Husbandry (journ.)
Aust. J. Exp. Biol. Med. Sci. e: Australian Journal of Experimental Biology and Medical Science (journ.)
Aust. J. Instrum. Contr. e: Australian Journal of Instrumentation and Control (journ.)
Aust. J. Med. Technol. e: Australian Journal of Medical Technology (journ.)
Aust. J. Phys. e: Australian Journal of Physics (journ.)
Aust. J. Phys. Astrophys. Supp e: Australian Journal of Physics; Astrophysical Supplement (journ.)
Aust. J. Zool. e: Australian Journal of Zoology (journ.)
Austl. e: Australia[n]
Aust. Libr. J. e: Australian Library Journal (journ.)
Aust. Miner. Ind. e: Australian Mineral Industry (journ.)
Aust. N.Z.J. Med. e: Australian and New Zealand Journal of Medicine (journ.)
Aust. N.Z.J. Obstet. Gynaecol. e: Australian and New Zealand Journal of Obstetrics and Gynaecology (journ.)
Aust. N.Z.J. Surg. e: Australian and New Zealand Journal of Surgery (journ.)
Aust. P. d: australisches Patent
AUST P e: Australian Patent
AUSTPAC Service e: Australian Packet Switching Service
Aust. Phys. e: Australian Physicist (journ.)
Aust. Prior. d: australische Priorität
Austr. d: Australien
- d: Australier
austr. d: australisch
Austr. Acad. Res. Libr. e: Australian Academic and Research Libraries (journ.)
AUSTRAFORD e: Ford Motors of Australia
Austral. d: Australien
- d: Australier
- e: Australasia
- e: Australian
austral. d: australisch
Australas. e: Australasian
Australas. Bull. Med. Phy. Bio e: Australasian Bulletin of Medical Physics and Biophysics (journ.)
Australas. Corros. Eng. e: Australasian Corrosion Engineering (journ.)
Australas. Eng. e: Australasian Engineer (journ.)
Australasian e: Australia, Tasmania, New Zealand and islands of Melanesia (journ.)
Australas. Inst. Mining Met. P e: Australasian Institute of Minning and Metallurgy Proceedings (journ.)
Australas. Radiol. e: Australasian Radiology (journ.)
- e: Australian Road Research (journ.)
Australs e: Austral Islands of Polynesia, also called the Tubuais
Austria d: Republik Österreich
- e: Republic of Austria
AUSTRIATOM e: Austrian Atomic Energy Group
Aust. Road Res. e: Australian Road Research (journ.)
Aust. Road Res. Bd. Proc. e: Australian Road Research Board Proceedings (journ.)

Austro-Hungarian Empire e: Austria, Bohemia, Bosnia, Croatia, Moravia, Bukovina, Transylvania, Galicia, Hungaria
Austronesia e: Islands of South Pacific from Madagascar in Indian Ocean to Hawaiian Islands in the Pacific
Aust. Sci. Teach J. e: Australian Science Teachers Journal (journ.)
Aust. Spec. Libr. News e: Australian Special Libraries News (journ.)
Aust Telecomm Res e: Australian Telecommunication Research
Aust. Transp. e: Australian Transport (journ.)
Aust. Vet. J. e: Australian Veterinary Journal (journ.)
- e: Australian Welding Journal (journ.)
AusV d: Ausstellungsversicherung
Ausv. d: Ausverkauf
Ausverk. d: Ausverkauf
Ausw. d: Auswage
- d: Auswahl
- d: Auswechselung
- d: Ausweis
- d: Ausweitung
- d: Auswertung
ausw. d: auswählen
- d: auswärtig
- d: auswärts
- d: auswechseln
- d: ausweiten
- d: auswerten
AuswA d: Auswechslungsamt
Auswasch. d: Auswaschung
Auswirk. d: Auswirkung[en]
Ausw.-Koll. d: Auswahlkollektion
Ausz. d: Auszahlung
- d: Auszeichnung
Ausz d: Auszählung
- d: Auszug
- d: Auszüge
AuszL d: Auszahlungsliste
Ausz.-Sch. d: Auszahlungsschein
Ausz.-T. d: Auszahlungstermin
- d: Auszeichnungstext
- d: Auszugstermin
AUT d: Automatik
- d: Automobile
- e: Advanced Unit Training
- e: Advanced User Terminal
- e: American United Telecommunications
- e: Amplifier Under Test
- e: Assembly under Test
- e: Association of University Teachers
- e: Author
- e: Automated Unit Test
- e: Automatic
- i: autore
aut. d: autark
- d: automatisch
- d: automatisiert
- d: autonom
- d: autorisiert
- d: autoritär
- e: autogen
Aut. d: Autarkie
- d: Autograph
- d: Automat[ion]
- d: Automobil
- d: Autonomie
- d: Autor
- d: Autorität
AUTD e: Automated
AUTE e: Association of University Teachers of Economics
AUTEC e: Atlantic Underwater (or Undersea) Test Evaluation Center
AUTELCOM e: Australian Telecommunications Commission
AUT.EQ. e: Automatic Equipment
AUTEVO d: automatisierte technische Vorbereitung
AUTEX d: automatische Telex- und Teletexauskunft
AutG d: Gesetz über den Verkauf von Waren aus Automaten
auth. d: authentisch
- d: autorisiert
AUTH e: Authentic
- e: Authenticate
- e: Authenticity
- e: Author[ized]
- e: Authorities
- e: Authority
- e: Authorization
- e: Authorize
AUTHAB e: Authorized Abbreviation
AUTHEN e: Authentic
AUTHGR e: Authority Granted
AUTHN e: Authentication
Auth. Ver. e: Authorized Version
AUTIREM d: automatisiertes Informations- und Recherchesystem der Metallurgie
AUT[l] e: Association of University Teachers
AUTM e: Automatic
AUT MEAS e: Automatic Measurement
AUTMV e: Automotive
AUTMWTRCK e: Automatic Water Check
AUTO d: Automobile
- d: Autotypie
- e: Automatic
- e: Automobile
- e: Automotive
AUTOALARM e: Automatic Alarm
AUTO ATR e: Automatic Start
Autobiogr. d: Autobiographie
AUTOBIRD e: Automatic Bird Feeder
AUTOBSTRACT e: Automatic Abstract
Autobus d: Autoomnibus
AUTOBUS e: Automated Budget System
AUTOC e: Autocorrelation
AUTOCAT e: Automatic Cat Feeder
AUTOCAT e: Automatic Control of Air Transmissions
AUTOCHART e: Automatic Charting
AUTOCLASS e: Automatic Classification
AUTOC Method e: Autocorrelation Method
AUTOCOLOR e: Automatic Color [TV]
AUTOCOM e: Automated Combustor
AUTOCOMPILER e: Autoprogramming Compiler
AUTOCON e: Automatic Contour
AUTOCORREL e: Autocorrelation
AUTO CTDG HDL e: Automatic Cartridge Handler
AUTO CV e: Automatic Check Value
AUTODIDAC e: Autodidact[ic][al][ly]
AUTODIN d: automatisches digitales Netz
- e: Automatic Digital [Information] Network
AUTODIN Network e: Automatic Digital Information Network
AUTODOC e: Automated Documentation
- e: Automatic Documentation
AUTODOG e: Automatic Dog Feeder
Autog. d: Autograph
AUTOG e: Autograph[ed]
Autogr. d: Autogramm
- d: Autographie
- e: Autograph
autogr. d: autographiert
- d: autographisch
AUTOGROM e: Autoprompter
AUTOHAND e: Automatic Hand
Autoklavier. d: Autoklavierung
AUTOLABS e: Automatic Low Altitude Bombing System
AUTOLAND e: Automatic Landing
AUTO LEAN e: Automatic Lean
AUTOLIGN e: Automatic Alignment
AUTOLING e: Automated Linguistics
AUTOLOAD e: Automatic Loading
Autom. d: Automatik
- d: Automatisation
- d: Automatisierung
- e: Automation
- e: Automobile
autom. d: automatisch
autom e: automatic
-e: automation
- e: automotive
- i: automobile
- p: automóvel
- s: automóvil
AUTOMAD e: Automatic Adaption Data
Auto Mag e: Automobile Magazine (journ.)
AUTOMAP e: Automatic Machining Program[me]
AUTOMAST e: Automatic Mathematical Analysis and Symbolic Translation
- e: Automatic Mathematical Analysis and Symbolic Translator
AUTOMAT e: Automatic Methods and Times
automát i: automatica
- s: automático
AutomatG d: Gesetz über den Verkauf von Waren aus Automaten
AUTOMATIC e: Automatic Revolver
AUTOMATION e: Automatic Action
- e: Automatic Operation
- e: Automation Action
- e: Automatization
Automat. Remote Contr. e: Automation and Remote Control (journ.)
Automat. Weld. e: Automatic Welding (journ.)
Autom. Doc. Math. Linguist. e: Automatic Documentation and Mathematical Linguistics (journ.)
AUTOMET Correction e: Automatic Meteorological Correction
AUTOMEX e: Automatic Message Exchange
AUTOMEX Service e: Automatic Message Exchange Service
AUTOMN e: Automation
Automob.-techn. Z. d: Automobil-technische Zeitschrift (journ.)
Automot. Eng. e: Automotive Engineering (journ.)
AUTOMTN e: Automation
AUTOMV e: Automotive
auton. d: autonom
- e: autonomous
Auton. d: Autonomie
- e: Autonomy
AUTON e: Automation
AUTONEST Program[m] e: Automatic Nesting Program[me]
AUTONET d: automatisches Zeichnen von Netzplänen
- e: Automatic Network
AUTONET Display e: Automatic Network Display
AUTO OVLD e: Automatic Overload

AUTOP d: Autoren-Operator
- e: Automatic Pistol
- e: Autopsy
AUTOPAP Machine e: Automatic Position[ing] and Pierce Machine
AUTOPASS e: Autopass
AUTOPET e: Automatic Pet Feeder
AUTOPIC e: Automatic Personnel Identification Code
AUTOPILOT e: Automatic Pilot
AUTOPISTOL e: Automatic Pistol
AUTOPIT d: automatisches Programmieren inklusive Technologie
- e: Automatic Programming Including Technology
AUTOPLOT e: Automatic Plotter
AUTOPOL e: Automated Programming of Lathes
- e: Automatic Programming of Lathes
AUTOPOL Indication e: Automatic Polarity Indication
AUTOPOLL e: Automatic Poll[ing]
AUTOPOS e: Automatic Trains, Protection, Operation and Supervision
- e: Automatic Transportation, Protection, Operation and Supervision
AUTOPROD e: Automated Projective Drawing
AUTOPROG e: Automatic Programming
AUTOPROMPT e: Automated Programming of Machine Tools
- e: Automatic Programming of Machine Tools
AUTOPROPS e: Automatic Programming for Positioning Systems
AUTOPROS e: Automated Process Planning System
AUTOPSY e: Automatic Operating System
AUTO PTS e: Automobile Parts
AUTOQEST e: Automatic generation of Requests
autor. d: autorisieren
- d: autorisiert
AUTO RECL e: Automatic Reclosing
AUTO RICH e: Automatic Rich
AUTORIF e: Automatic Rifleman
autoris. d: autorisiert
AUTOROTIC[S] e: Automobile Neurotic[s]
AUTOS e: Automatics
- e: Automobiles
AUTOSATE e: Automated [Data] System[s] Analysis Technique
AUTOSCAN e: Automatic Satellite / Computer Aid to Navigation
- e: Automatic Scanner
- e: Automatic Stereo Broadcast Scanner
AUTOSCRIPT e: Automated System for Composing, Revising, Illustrating and Phototypesetting
AUTO S&CV e: Automatic Stop and Check Valve
AUTO SEQ e: Automatic Sequence
AUTOSERVCen e: Automated Service Center
AUTOSEVC e: Automatic Secure Voice Communication[s]
AUTOSEVCOM e: Automatic Secure Voice Communications
AUTOSEVOCOM e: Automatic Secure Voice Communication[s]
AUTOSEVOCOM Network e: Automatic Secure Voice Communication[s] Network
AUTOSPEC e: Automated Specification[s]
AUTOSPOT e: Automated System for Positioning [of] Tools
- e: Automatic System for Positioning [of] Tools
AUTO S & SV e: Automatic Stop and Check Valve
AUTOSTARTER e: Autotransformer Starter
AUTO-STATIS e: Automatic Statewide Auto Theft Inquiry System
AUTOSTOP e: Automatic Stop
AUTO STR e: Automatic Start
AUTOSTRAD e: Automated System for Transportated Data
- e: Automated System for Transportation [of] Data
- e: Automated System for Transporting Data
AUTOSTRT e: Automatic Starter
AUTOSTRTG e: Automatic Starting
AUTOSYN e: Automatically Synchronous
AUTO TR d: Autotransformer
AUTOTR e: Autotransformer
AUTOTRACK e: Automatic Tracking
AUTOTRAN e: Automatic Translation
AUTO TRANS e: Automatic Transmission
AUTO-TRIP e: Automatic Transportation Research Investigation Programme
AUTOVERIFIER e: Automatic Verifier
AUTOVON e: Automatic Voice [Switching] Network
AUTO W/OD e: Automatic [Transmission] with Overdrive
AUTO XFMR d: Autotransformer
AUTPOL Indication e: Automatic Polarity Indication
AU TR e: Aural Training
AUTRA d: automatische Tragwerkberechnung
- d: Automatisierung und Rationalisierung
AUTRAN e: Automatic Target Recognition Analysis
- e: Automatic Translation
- e: Automatic Utility Translator
AUTRANAVS e: Automated Transponder Navigation System
AUTRANs e: Automatic Utility Translators
AUTRAX e: Automatic Traffic Recording and Analysis Complex
AUT[S] e: Association of University Teachers
AUTSWITCH e: Automatic Switching
AutTln d: automatischer Teilnehmer
AUT TR e: Automatic Transmitter
AUTTUN e: Automatic Tuning
AutV d: Automatenversicherung
AUT[W] e: Association of University Teachers
AUU e: Association of Urban Universities
AUUA e: American Univac Users Association
AUV d: Auslandsumzugskostenverordnung
- e: Administrative Use Vehicle
- e: Air University
- e: Armo[u]red Utility Vehicle
AUVB d: Allgemeine Versicherungsbedingungen für die Unfallversicherung
Au virus e: Australian antigen
AUVL e: Airborne Ultraviolet Laser
AUVMIS e: Administrative Use Vehicle Management Information System
AUVS e: Association of Unmanned Vehicle Systems
AUW e: Advanced Undersea Weapon[s]
- e: Advanced Underwater Weapon[s]
- e: Airframe Unit Weight
- e: All-Up-Weight
- e: Antiunderwater Warfare
AU/W e: Automatic Winder
AÜW d: Allgäuer Überlandwerk
AUWE e: Admiralty Underwater Weapons Establishment
AUWS e: Automatic Unmanned Weather Station
AUX e: Auxiliary
- e: Auximones
- e: Auxocardia
AUXBOX e: Auxiliary Box
AUX FS e: Auxiliary Fire Station
AUXFUN e: Auxiliary Function
AUX GCS e: Auxiliary Ground Control Station
AUXIL e: Auxiliary
AUXILIARY e: Recording Control
AUXM e: Auxiliary Machinery
AUXOSC e: Auxiliary Oscillator
Aux. Pat. e: Auxiliary Patrol
AUX R e: Auxiliary Register
AUXRC e: Auxiliary Recording Circuit
- e: Auxiliary Recording Control
AUXS e: Auxiliaries
AuxS e: Auxiliary Switch
AUXY e: Auxiliary
AUZ d: Arbeitsunfähigkeits-Zusatzversicherung
- d: Austreibungszeit
- e: Authorized
a. u. Z. d: auf unbestimmte Zeit
AV d: Abhörverstärker
- d: Ablaufverfolger
- d: Abnormal Voltage
- d: Abschlußvermittler
- d: Absolute Voltmeter
- d: Addierverstärker
- d: alle Fahrt voraus
- d: allgemeiner Vertrag
- d: allgemeine Verfügung
- d: Allgemeine Verwaltungsvorschrift
- d: Allgemeine Vorschrift
- d: Altersversicherung
- d: Altersversorgung
- d: Ameisenverbreitung
- d: Amtsübertragungsvorsatz
- d: Amtsvorstand
- d: Amtsvorsteher
- d: Angestelltenversicherung
- d: Anlagevermögen
- d: Anschlußverteiler
- d: Antennenverstärker
- d: Anwendungsverfahren
- d: Arbeitsgemeinschaft Volksernährung
- d: Arbeitsvertrag
- d: Arbeitsvorbereitung
- d: Arbeitsvorlage
- d: Architektenverein
- d: Atem[zug]volumen
- d: Atrioventrikulär
- d: Audiovision
- d: audiovisuell
- d: Aufnahmeverstärker
- d: Aufzeichnungsverstärker
- d: Aufzugsverordnung
- d: Ausbildungsverordnung[en]
- d: Ausbildungsvorschrift[en]
- d: Ausführungsverordnung[en]
- d: Ausführungsvorschrift[en]
- d: Auslaßventil
- d: Außenhandelsverband
- d: Außenvolumen
- d: Die Arbeitsvorbereitung (journ.)
- e: Absolute Value
- e: Absolute Vector
- e: Absorbed Voltage
- e: Abvolt
- e: Acceptable Variation

- e: Accumulated Value
- e: Accumulator Volume
- e: Acid Value
- e: Actual Value
- e: Actual Velocity
- e: Address Validity
- e: Address Value
- e: Address Verification
- e: Advanced Voyager
- e: Aerospace Vehicle
- e: Afferent Vessel
- e: Aggregate Value
- e: Air Vent
- e: Algorithmic Variable
- e: Allocation Voucher
- e: Alphabetic Verifier
- e: Alternating Voltage
- e: American Viewpoint
- e: Analysis of Variance
- e: Angular Velocity
- e: Anion Vacancies
- e: Anti-Vehicle
- e: Aortic Valve
- e: Apical Vacuoles
- e: Arbitrary Value
- e: Arithmetic Verb
- e: Armo[u]red Vehicle
- e: Array/Vector
- e: Array Variable
- e: Arteriovenous
- e: Association Value
- e: Atomic Volume
- e: Atrioventricular
- e: Attribute Value
- e: Audio-Visual
- e: Auriculo Ventricular
- e: Auriculoventricular
- e: Authorized Version
- e: Availability
- e: Available
- e: Average Variability
- e: Aviation
- e: Aviator
- e: Avionics
- e: Avoirdupois
- f: avril
- i: alto voltaggio
- i: a vista
- p: alta voltagem
- R: Avtomat Kalashnikov
- s: alto voltaje

Av d: Adern; verseilt
av d: arbeitsverwendungsfähig
aV e: attovolt
Av. e: Avenue
av. e: average
a. v. l: ad valorem
A. V. l: ad vivum
A. V l: Anno Vixit
A/V d: Ampere/Volt
- e: Audiovisual

A & V d: Ankauf und Verkauf
A-V e: Atrionventricular
AV-8B e: U.S. Marine Corps fighter-bomber jump jet
AVA d: Aerodynamische Versuchsanstalt
- d: Arbeitsgemeinschaft zur Verbesserung der Agrarstruktur
- d: arteric-venöse Anastomose
- d: Auftrags-Vergabe-Abrechnung
- d: Automobil- und Warenkredit-Verkehrsanstalt GmbH
- e: Absolute Value Amplifier
- e: Absolute Virtual Address
- e: Activity Vector Analysis
- e: Address-Value Attribute
- e: Alternate Volume Allocation
- e: American Video Association
- e: Arteriovenous Anastomosis
- e: Audio/Visual Aid[s]
- e: Audio/Visual Annunciator
- e: Australian Veterinary Association
- e: Automatic Voice Advice
- e: Automatic Voice Alarm
- e: Average Alarm
- e: Azimuth Versus Amplitude

AVA Array e: Active Van Atta Array
AVAC e: Automated Vacuum
- s: Asociación Venezolana para la Avance de la Ciencia

AVADS e: Autotrack Vulcan Air Defense System
av/af e: anteverted/antetlexed
AVAIL e: Availability
- e: Available

AVAL e: Availability
- e: Available

AVAM d: audiovisuelle Arbeitsmittel
AVASI e: Abbreviated Visual Approach Slope Indicator
AVASIS e: Abbreviated Visual Approach Slope Indicator System
AVATI e: Asphalt and Vinyl Asbestos Tile Institute
AV-AWOS e: Aviation Automated Weather-Observation System
AVB d: Allgemeine Verfrachtungs-bedingungen
- d: Allgemeine Versicherungsbedingungen
- d: audiovisuelle Bildungsmittel
- e: Allgemeine Versorgungsbedingungen
- e: Analog[ue] Video Bandwidth
- e: Avionics Bulletin

avb r: avbeta
Av Bad e: Army Aviation (or Aviator) Badge
AVBL e: Armo[u]red Vehicle Bridge Launcher
- e: Armo[u]red Vehicle Launch Bridge
- e: Available

AV-Block d: atnoventrikulärer Block
AVBM d: Arbeitgeberverband der Berliner Metallindustrie
Avbr d: automatische Vakuumbremse
- d: automatische Vakuumbremse

AVBS e: Absolute Value Bit Synchronizer
AVBSchG d: Ausführungsverordnung zum Berufsschulgesetz
AVBSG d: Verordnung zur Ausführung des Bundesseuchengesetzes
AVBTÄO d: Verordnung zur Ausführung der Bundestierärzteordnung
AVBTL d: Allgemeine Versicherungs-bedingungen für Tierlebensversicherung
AVC e: Absolute Value Computer
- e: Abdominal Vena Cava
- e: Acceleration Vector Control
- e: Allantoid Vaginal Cream
- e: Altitude Velocity Chart
- e: American Video Channels, Inc.
- e: Arc Vacuum Cast
- e: Association of Virginia Colleges
- e: Association of Visual Communications
- e: Association of Vitamin Chemists
- e: Audio / Visual Center
- e: Aural and Visual Code
- e: Automatic Vibration Control
- e: Automatic Voltage Clamp
- e: Automatic Voltage Control[ler]
- e: Average Variable Costs
- e: Aviation Cadet
- e: Avionics Change

av C i: avanti Cristo
AVCA e: Acceleration Vector Control
- e: Allantoid Vaginal Cream
- e: Automatic Volume Control

AVCAD e: Aviation Cadet
AVCAL e: Aviation Consolidated Allowance List
- e: Aviation Coordinated Allowance List

AVCAT e: Aviation High-Flash Turbine Fuel
AVCC e: Altitude Vertical Velocity Computer
AVCCOA e: Audiovisual, Computer and Communication Office Automation
Av. Cert. e: Aviators Certificate
AVCIR e: Aviation Crush Injury Research
AVCM e: Associate of Victory College of Music
AVCN e: Antero-Ventral Cochlear Nucleus
AVCOM e: Aviation Command
AVCS e: Advanced Videcon Camera System[s]
- e: Atrioventricular Conduction System
- e: Avionics Components and Subsystems

AVCSA e: Assistant Vice Chief of Staff Army
avd a: avdeling
AvD d: Automobilclub von Deutschland
AVD e: Absolute Value Device
- e: Air Velocity Detector
- e: Alternate Voice [and] Data
- e: Alternative Voice [and] Data
- e: Alternative Voice Data
- e: Anode Voltage Drop
- e: Army Veterinary Department
- e: Automatic Vertical Door
- e: Automatic Voice Data
- e: Automatic Voltage Digitizer

AVDA d: Ausgabe Verarbeitungsdatei
- e: American Venereal Disease Association
- e: American Veterinary Distributors Association

AVD Controller e: Advanced Video Display Controller
A-V Difference e: Alteriovenous Concentration Difference
AVDO e: Aerospace Vehicle Distribution Officer
AVDP e: Aided Visual Development Program[me]
- e: Avoirdupois

AvdR d: Abfluß von der Regelung
AVDS e: Articulated Vehicle Dynamic Simulator
- e: Automated Vehicle Diagnostic System
- e: Automatic Vacuum Deposition System
- e: Automatic Voice Data Switching

AVD Service e: Alternate Voice and Data Service
AVDS Model e: Articulated Vehicle Dynamic Simulation Model
AVDT e: Angular Variable Differential Transformer
AVDU e: Audio-Visual Display Unit
AVE d: Allgemeine Verbindlichkeits-erklärung
- d: Alternativen zur Verstädterung Europas
- e: Aerospace Vehicle Electronics
- e: Aerospace Vehicle Equipment
- e: Airborne Vehicle Equipment
- e: Atmospheric Variability Experiment
- e: Automatic Volume Expander
- e: Automatic Volume Expansion
- e: Avenue
- e: Average
- e: Averaging Enhancement

AVEA e: American Veterinary Exhibitors Association

AVEBD e: Average Blank Data
AVEC e: Amplitude Vibration Exciter Control
- e: Automatic Vibration Exciter Control
AVEDAY e: Average Day
av.eff. e: average efficiency
AVEM e: Association of Vacuum Equipment Manufacturers
AVENS e: Audiovisual Education in Neurosurgery
AVENSA s: Aerovias Venezolanas
AVEP e: Average Visual Evoked Potential
AVEPS e: Alderbourne Valleay Environmental Protection Society
AVER d: Abkommen über die Verkehrsteilung zwischen der Eisenbahn und der Rheinschiffahrt
- d: Art der Verletzung
- d: Auftragsverwaltung
- e: Average
AVERDISROP e: Avert Disruption of Operation
AVers. d: Altersversicherung
- d: Altersversorgung
- d: Angestelltenversicherung
- d: Arbeitslosenversicherung
AVERT e: Association of Voluntary Emergency Radio Teams
- e: Automatic Verification, Evaluation and Readiness Tester
AVerzDX d: Amtliches Verzeichnis der Datexteilnehmer
AVerzTX d: Amtliches Verzeichnis der Telexteilnehmer
AVerzTxTtx d: Amtliches Verzeichnis der Teletex- und Teletextteilnehmer
AVF d: Ausbildungsvorschrift für die Feuerwehr
- d: äußerer Verzögerungsfaktor
- d: Availability Factor
- e: Arteriovenous Filstula
- e: Azimuthally Varying Field
AVF Accelerator e: Azimuthally Varying Field Accelerator
AVfg. d: allgemeine Verfügung
AVFR e: Available For Reassignment
AVFUEL e: Aviation Fuel
AVG d: Abfallverbrennungsgesellschaft mbH; Hamburg
- e: Air Ventilation Garment
- e: Audio-Visual Group
- e: Average
Avg R: Avgust
AVGAS e: Average Diameter
- e: Aviation Gasoline
AVGE e: Average
AV-GerÄt d: audiovisuelles Gerät
AVGrdstVG d: Verordnung zur Ausführung des Grundstücksverkehrsgesetzes
AVH d: Aderverbindungshülse[n]
- d: Arbeitsrechtliche Vereinigung Hamburg
AVH e: Acute Viral Hepatitis
- e: Average Heading
AVHRR e: Advanced Very High Resolution Radiometer
AvHSt d: Alexander-von-Humboldt-Stiftung
AVI d: Arbeitsgemeinschaft der eisen- und metallverarbeitenden Industrie
- e: Adaptive User Interface
- e: Adjustable Voltage Inverter
- e: Airborne Vehicle Identification
- e: Air Velocity Index
- e: American Virgin Islands
- e: Anti-Vivisection League
- e: Audio-Visual Institute
- e: Automatic Vehicle Identification
- e: Aviation
- f: Association Universelle d'Aviculture Scientifque
AVIA e: Aviation
Aviaco s: Aviación y Comercio
AVIANCA e: Aerovias Nacionates de Colombia
Avian Dis. e: Avian Diseases (journ.)
AVIAT e: Aviation
AVIATECA s: Empresa Guatemalteca de Aviacion
Aviat. Week Space Technol. e: Aviation Week and Space Technology (journ.)
AVID e: Advanced Visual Information Display
- e: Airborne Vehicle Identification
- e: American Veterinary Identification Device
- e: Audio-Visual Instruction Department
AVIDAC e: Argonne Version of Institute Digital Automatic Computer
Avigation e: Aircraft Navigation
AVIOB e: Aviation Observation
AVIONICS e: Aviation And Astronautics Electronics
- e: Aviation Electronics
AVIP e: Association of View-data Information Providers
AVIS e: Active Vibration Isolation System
- e: Advanced-Visual Information System
- e: Audio-Visual Information System
- e: Automatic Visual Inspection System
AVISAM e: Average Index[ed] Sequential Access Method
AVISCO e: American Viscose Corporation
AVISPA s: Aerovias Interamericanas de Panamá
AVISTA e: Audio-Visual Tracking of Aircraft
AVIT e: Audio-Visual Instruction[al] Technology
AVJ e: Anti-Vibration Joint
av JC f: avant Jésus Christ
AVK d: Allgemeine Verwaltungskosten
- d: Arbeitsgemeinschaft verstärkte Kunststoffe
AV-Knoten d: Atrioventrikularknoten
AVKrpflG d: Verordnung zur Ausführung des Krankenpflegesetzes
AVKÜ d: automatische Verkehrsüberwachung
AVL d: Amt für Verteidigungslasten
- d: Amtsverbindungsleitung
- d: automatische Vakuumbremsleitung
- e: Angle Versus Length
- e: Anti-Vivisection League
- e: Approved Vendors List
- e: Asheville (airport)
- e: Associated Veterinary Laboratories
- e: Automatic Vehicle Locating
- e: Automatic Vehicle Location
- e: Available
- e: Average Length
AVLA e: Audio-Visual Language Association
Av Labs e: Aviation Laboratories
AVLABS e: Aviation Laboratories
AVL-Baum e: Adelson-Velskii; Landis-Baum
AVLBL e: Availability
AVLF e: Airborne Very Low Frequency
AVLINE e: Audiovisual[s] On-Line (Computer Retrieval System)
AVLM e: Anti-vehicle Land Mine
AVLOC e: Airborne Visible Laser Optical Communication
- e: Audio-Visual Learning Objectives Catalog
AVLS d: automatisches Verkehrserfassungs- und Lenkungs-System
- e: Audio-Visual Library Service
- e: Automatic Vehicle Location System
AVLSI e: Advanced Very-Large-Scale Integration
AVLStVG d: Verordnung zur Ausführung des Landesstraf- und Verordnungsgesetzes
AVLUB e: Aviation Lubricant
AVM d: audiovisuelle Medien
- d: Auslandsvermittlungssystem
- e: Advanced Virtual Machine
- e: Airborne Vibration Monitor
- e: Air Velocity Meter
- e: Air Vice Marshal
- e: Analysis Virtual Machine
- e: Animal Vegetable and Mineral
- e: Antivehicle Mine
- e: Arteriovenous Malformadon
- e: Audio-Visual Modulator
- e: Automatic Vehicle Monitoring
- e: Automatic Voltage Margin
- e: Automatic Voting Machine
- e: Aviation Medical
AVMA e: American Veterinary Medical Association
- e: Audio-Visual Management Association
AVMARC e: Audio-Visual Machine Readable Cataloguing
Av Mech e: Aviation Mechanic
AVM/EF e: Advanced Virtual Machine / Extended Facility
AVMF R: Aviarsiya Voenno Morskikh Flota
AVMH e: Available Manhours
AV Munition e: Anti-Vehicle Munition
AVMZ d: audiovisuelles Medienzentrum
AVN d: Allgemeine Vermessungs-Nachrichten (journ.)
- d: Angehörige[r] der Vereinten Nationen
- e: Air Vietnam
- e: Atrioventricular Node
- e: Aviation
- e: Aviation News (journ.)
- e: Avonmouth
AVNAG d: Annalen des Vereins für Nassauische Altertumskunde und Geschichtsforschung (journ.)
AVNAKGF e: Annalen des Vereins für Nassauische Altertumskunde und Geschichtsforschung (journ.)
Avn Eng e: Aviation Engineer
AVNET e: Aviation Network
AVNG e: Afferent Vein to Nephridial Gland
AVNL e: Automatic Video Noise Levelling
- e: Automatic Video Noise Limiter
- e: Automatic Video Noise Limiting
AVNMED e: Aviation Medicine
Av. No. e: Average Number
AV Node e: Artenoventricular Node
- e: Atrioventricular Node
AVNot. d: Allgemeine Verfügung betreffend Angelegenheiten der Notare
Avn Sch e: Aviation School
AVNU e: Aviation Unit
AVO d: allgemeine Vollzugsordnung
- d: Ampere, Volt, Ohm
- d: Anerkennungsverordnung
- d: Arbeitsverordnung
- d: Arbeitsvertragsordnung
- d: Ausfallvergütungsordnung
- d: Ausführungsverordnung
- e: Administrative Veterinary Officer

- e: Ampere-Volt-Ohm
- e: Avocado
- e: Avoid Verbal Orders
ÄVO d: Änderungsverordnung
AVOCON e: Automated Vocabulary Control
Avog. d: Avogadro
AVOID e: Accelerated View of Input Data
AVOID e: Airfield Vehicle Obstacle Indicating Device
- e: Airfield Vehicle Obstacle Indication Device
- e: Avionic Observation of Intruder Danger
AVOIDS e: Avionic Observation of Intruder Danger System[s]
AVOIL e: Aviation Oil
Avoir e: Avoirdupois
AVOLO e: Automatic Voice Link Observation
AVO Meter d: Ampere, Volt, Ohm Meter
AVON d: Amtliches Verzeichnis der Ortz[netz]kennzahlen
AVOR d: Arbeitsvorbereitung
AVOS e: Acoustic Valve Operating System
AVOSGG d: Verordnung zur Ausführung des Sozialgerichtsgesetzes
AVP d: Arbeitsgemeinschaft Versuchsreaktor
- d: Arge Versuchsreaktor; Jülich
- d: Armeeversorgungspunkt
- d: Arterienvolumenpuls
- d: Atomversuchsreaktor
- e: Address Verification Pulse
- e: Antiviral Protein
- e: Arginine Vasopressin
- e: Army Validation Program[me]
- e: Attached Virtual Processor
- e: Automatic Variable Perforating
- e: Automtic Variable Perforating
- f: Avoirdupois
AVPBefG d: Verordnung zur Ausführung des Personenbeförderungsgesetzes
AVPC e: Attached Virtual Processor Control
AVR d: Allgemeiner Vliesstoff-Report (journ.)
- d: Allgemeine Versicherungsbedingungen für Rindviehversicherung
- d: Arbeitsgemeinschaft Versuchsreaktor
- d: Atomversuchsreaktor
- d: Ausgangsverstärker
- d: automatische Verstärkungsregelung
- e: Absolute Value Representation
- e: Adjustable Voltage Rectifier
- e: Aortic Valve Replacement
- e: Auto-Voltage Regulator
- e: Automatic Voice Recognition
- e: Automatic Voice Relay
- e: Automatic Voltage Regulation
- e: Automatic Voltage Regulator
- e: Automatic Volume Recognition
- e: Automatic Volume Regulation
AVRA e: Audio-Visual Research Association
AVRAC e: Agricultural and Veterinary Research Advisory Committee
AVRAD e: Altitude Variation Rate and Displacement
AVRC e: Askew and Vertical Redundancy Check
AVRI e: Animal Virus Research Institute
AVRO n: Algemeene Vereniging Radio Omroep
AVRP e: Advanced Vehicle Reception Park
- e: Atrioventricular Refractory Period
- e: Audiovisual Recording And Presentation
AVRS d: allgemeiner Vertrags-Rechtsschutz
- e: Audio/Video Recording System
- e: Audiovisual Recording System
AVS d: Amtsverbindungssatz
- d: Arbeitsvorbereitungssystem
- e: Absolute Value Sign
- e: Absolute Vectors
- e: Adjustable Voltage Screw-down
- e: Advance in Schedule
- e: Aerospace Vehicle Simulation
- e: Aided Visual System
- e: Airborne Viewing System
- e: Air Valve Silencer
- e: Altitude Vertical Scale
- e: American Vacuum Society
- e: American Videotext Services, Inc.
- e: Anti-Vivisection Society
- e: Applied Viewdata Systems
- e: Association for Voluntary Sterilization
- e: Audio-Visual Section
- e: Automated Verification System
- e: Automatic Volume Sensing
- e: aviation supply ship
A/VSA e: Altimeter/Velocity Sensor Antenna
AVSBR d: automatische Vakuumschnellbremse
AVSC e: Audio-Visual Support Center
AVSCOM e: Army Aviation & Surface Command
- e: Army Aviation Systems Command
- e: Aviation and Surface-materiel Command
AVSEP e: Audio Visual Superimposed Electrocardiogram Presentation
AVSER e: Aviation Safety Engineering and Research
AVSL e: Association of Visual Science Librarians
- e: Available Space List
AVSM e: Auxiliary Video Switching Matrix
AVSS e: Aided Visual Sensor System
- e: Apollo Vehicle Systems Section
AVSt d: A-Vermittlungsstelle
AVST d: Auslandsvermittlungsstelle
- e: Advanced Vehicle Simulation Technique
AVST e: Automated Visual Sensitivity Test[er]
AVSYCOM e: Aviation Systems Command
AVSYN e: Air Vehicle Synthesis
AVT d: Audiovisionsgesellschaft Tageszeitung
- d: automatischer Vielfachsondentaster
- e: Added Value Tax
- e: Adult Vocational Training
- e: Air Velocity Transducer
- e: Air Vibration Table
- e: Allen Vision Test
- e: Apollo Validation Test
- e: Applications Vertical Test
- e: Arginine Vascotocin
- e: Attribute Value Time
- e: Audiovisual Tutorial
- e: Automatic Vision Testing
- e: Automatic Vocal Transaction Analyzer
- e: Available Time
- e: Aviation Medicine Technician
AVTA e: Automatic Vocal Transaction Analysis
- e: Automatic Vocal Transaction Analyzer
- e: Automatic Vocal Translation Analyzer
AVTAG e: Aviation Wide-Cut Turbine Fuel
AVTC e: American Video Teleconferencing Corporation
- e: Army Vocational Training Centre
AVTE e: Adult, Vocational and Technical Education
AV TMP e: Average Temperature
AVTP e: Adult Vocational Training Program[me]
AVT Program[me] e: Applications Vertical Test Program[me]
AVTR e: Airborne Videocassette Tape Recorder
- e: Analog[ue] Video Tape Recorder
AVTRW e: Association of Veterinary Teachers and Research Workers
AVTUR e: Aviation Turbine [Fuel]
avu d: arbeitsverwendungsunfähig
AVUS d: Automobil- Verkehrs- und Übungsstraße
- d: Automobile Versuchs and Untersuchungs Strecke
AVV d: Aachener Verkehrsverbund
- d: Adreßbuchverlegerverband
- d: Allgemeiner Verband der Versicherungsangestellten
- d: Allgemeine Verwaltungsvorschrift[en]
- d: Antennen-Verteiler-Verstärker
- d: Arbeitsgemeinschaft der Vereinigung öffentlicher Verkehrsbetriebe
Av. Val. e: Average Value
AV VALES e: Atrioventricular Valves
AVVC e: Altitude Vertical Velocity Computer
AVVFStr. d: Allgemeine Verwaltungsvorschrift für die Auftragsverwaltung der Bundesfernstraße
AVVI e: Altitude-Vertical Velocity Indicator
AV W e: Average Width
AVwGebO d: Allgemeine Verwaltungsgebührenordnung
AvWk e: Aviation Week (journ.)
AVWRRE e: American Cargo War Risk Reinsurance Exchange
AV WT e: Avoirdupois Weight
AVwV d: allgemeine Verwaltungsvorschriften
AVZ d: Arbeits- und Verfügungszentrum
- d: Audiovisionszentrale
- d: Aufbau- und Verfügungszentrum
AW d: Abtastwert
- d: Abwasser
- d: Addierwerk
- d: Amperewindung[en]
- d: Amt für Wasserwirtschaft
- d: Amtswähler
- d: Anfangswort
- d: Anfertigungsware
- d: Anpassungsnetzwerk
- d: Anschaffungswert
- d: Antiklinalwände
- d: Anwartschaftsversicherung
- d: Anwendung
- d: Atemwiderstand
- d: Aufwertung
- d: Ausbesserungswerk
- d: Ausbildungswesen
- d: Ausgabewandler
- d: Ausgangswert
- d: Ausgewählte Werke
- d: Ausgleichswandler
- d: Außenwirtschaft
- d: Ausweisleser
- d: Automobilwirtschaft (journ.)
- e: A-Wire
- e: Absorption Wavemeter
- e: Accidentally Wounded
- e: Acid Waste
- e: Actual Weight
- e: Adding Wheel
- e: Additional Word
- e: Addressable Word
- e: Address[ed] Word

- e: Add Word
- e: Advanced Weapon[s]
- e: Advance Winding
- e: Air-to-Water
- e: Aircraft Warning
- e: Air Warning
- e: Air Washer
- e: Airway[s]
- e: Alignment Window
- e: All Water
- e: All Weather
- e: All Widths
- e: Alphabetic Word
- e: Alphanumeric[al] Word
- e: Alternate Weapon
- e: Amba
- e: American Welding
- e: Anti-Wear
- e: Apparent Watt
- e: Arming Wire
- e: Armstrong-Withworth
- e: Arm Width
- e: Army Wide
- e: Articles of War
- e: Atomic Warfare
- e: Atomic Weight
- e: Attowatt
- e: Augmentor Wing
- e: Automatic Weapon[s]
- e: Automatic Word
- e: Auxiliary Winding
- e: Aviation Week (journ.)
- e: Axial Width

a.W. d: ab Werk
- d: auf Wartegeld
- d: auf Widerruf
- d: auf Wunsch

Aw d: Anwärter
- d: Auswerter
- e: AIDS-related virus
- n: Antwerpen

Aw. d: Anwartschaft
- d: Anwesende
- d: Anwesenheit
- d: Ausweis
- d: Auswertung

a.w. e: above water
- e: all wool

A/W d: Aufnahme/Wiedergabe
- e: Against Wind
- e: Atual Weight

a/w e: Airworthy

A & W e: Alive and Well

A-W e: Audible Warning

ä.W. d: äußere Weite

AWA d: Anstalt zur Wahrung der Aufführungsrechte
- d: Arbeitsgemeinschaft Wasser
- d: Ausführungsanweisung zur Außenwirtschaftsverordnung
- e: Acoustic Wave Analysis
- e: Advise When Able
- e: Air Warfare Analysis
- e: All-Weather Attack
- e: All the Worlds Aircraft
- e: All Wave Aerial
- e: All Wave Antenna
- e: Aluminum Wares Association
- e: American Watch Association
- e: American Waterfowl Association
- e: American Wine Association
- e: American Wire Association
- e: Anomalous Winter Absorption
- e: Antique Wireless Association
- e: Assist Work Authorization
- e: Association of Women in Architecture
- e: Atmospheric Winds Aloft
- e: Audio Warning Amplifier
- e: Audio Wave Analyzer
- e: Automatic Warning Aid

AWAC e: Airborne Warning and Control
- e: Airborne Weapon and Control

AWACS e: Advanced Warning and Control System
- e: Airborne Warning and Command System
- e: Airborne Warning and Control System[s]
- e: Amalgamated Wireless Australasia Computer Services

AWAC System e: Airborne Warning and Command System
- e: Airborne Warning and Control System

AWADS e: Adverse Weather Aerial Delivery System
- e: All-Weather Aerial Delivery System

A-Waffen d: Atomwaffen

AWAG d: automatisches Wähl- und Ansagegerät

AWAL e: American-West African Line

AWANAS e: Aviation Weather and Notice to Airmen System

AWAR e: Area Weighted Average Resolution

AWARE e: Advance Warning Equipment
- e: Airborne Warning and Recording Equipment

AWARS e: Airborne Weather [and] Reconnaissance System

AWAS e: Acoustic Wave Analysis System

AWASP e: Advance Weapon Ammunition Support Point

AWAT e: Area Weighted Average T-Number

AWB d: Atomwirkungsberater
- e: Agricultural Wages Board
- e: Airway Bill
- e: Australian Wine Board
- e: Australian Wool Board

AWBAbh d: (Königlich-Preussische) Akademie der Wissenschaften zu Berlin; Abhandlungen (journ.)

AWB/CN e: Air Waybill/Consignment Note

AWBE e: Automatic Weather Broadcast Equipment

AWBR d: Arbeitsgemeinschaft Wasserwerke Bodensee-Rhein

AWBSb d: (Königlich-Preussische) Akademie der Wissenschaften zu Berlin; Sitzungsberichte (journ.)

AWC e: Acting Wing Commander
- e: Agricultural Wages Committee
- e: Airborne Weapons Control
- e: American Watershed Council
- e: American Wood Council
- e: American Wool Council
- e: Anaconda Wire & Cable [Company]
- e: Army Weapons Command
- e: Assisting Work Center
- e: Association for Woman in Computing
- e: Australian Wool Commission

AWCAP e: Air War College Associate Program[me]

AWCI e: Aircraft and Weapons Control Interceptor

AWCIS e: Aircraft and Weapons Control Interceptor System

AWCLS e: All Weather Carrier Landing System

Aw/cm d: Amperewindungen pro Zentimeter

AW Co. e: Air World Company

AWCS e: Agency-Wide Coding Structure
- e: Air[borne] Weapons Control System
- e: Automatic Warning and Control System
- e: Automatic Weapons Control System

AWC System e: Automatic Warning and Control System

AWCTA e: Atomic-Weapon-Carrying Tactical Aircraft

AWCU e: Association of World Colleges and Universities

AWCWP e: Absolute Wet Center Manifold Pressure

AWD d: automatischer Wähldienst
- d: automatische Wähleinrichtung[en] für Datenübertragung
- d: automatische Wähl[er]einrichtung[en] für Datenverbindungen
- d: automatische Wähleinrichtung[en] für Datenverkehr
- e: Astrogeodetic World Datum
- e: Average Working Depth
- e: Award[ed]

AWDATS e: Artillery Weapon[s] Data [Transmission] System

AWDAT System e: Artillery Weapon[s] Data Transmission System

AWDI e: Automated Worthless Document Index

AwDm d: Auswerte-Demodulator

AWDR e: Advanced Weapon Delivery Radar

AWDREY e: Atomic Weapon[s] Detection, Recognition and Yield

AWE d: Allwellenempfänger
- d: Arbeitsgruppe Werbung Elektroindustrie
- d: automatische Wähleinrichtung
- d: automatische Wiedereinschaltung
- e: Accepted Weight/Estimate
- e: Advise When Established
- e: All-Weather Electronics
- e: Average Weekly Earnings

AWEA e: American Wind Energy Association
- e: Australian Wind Energy Association

AWEASVC e: Air Weather Service

AWECOM e: Army Weapons Command

AWES e: Army Waterways Experimental Station

AWF d: Ausschuß für wirtschaftliche Fertigung
- e: Acceptable Workload Factor
- e: Adjoint Wave Function
- e: Adrenal Weight Factor
- e: African Wildlife Foundation
- e: All-Weather Fighter
- e: All Weather Flare
- e: American Wildlife Foundation
- e: Aviation Weather Facility

AWFI e: American Wood Fabric Institute

AWG d: Adresswort-Generator
- d: Arabische Wirtschaftsgemeinschaft
- d: Arbeiterwohnungsbaugenossenschaft
- d: Artikelnummer, Weite, Größe
- d: Außenwirtschaftsgesetz
- e: Air Weather Group
- e: American Wire Gage
- e: Art Work Generator

AwG d: Aufwertungsgesetz

AWGN d: Additive White Gaussian Noise

AWHQ e: Amphibious Warfare Headquarters

AWI d: Arbeitsgemeinschaft wirtschaftswissenschaftliche Information
- e: Accommodation Weight Investigation
- e: All-Weather Interceptor
- e: American Watchmakers Institute
- e: Animal Welfare Institute
- e: Architectural Woodwork Institute
- e: Arm Width Index
- e: Australian Welding Institute
- e: Australian Wire Industry
- e: Awake Indication

A & WI e: America and West Indies
- e: Atlantic and West Indies

AWIA e: American Wood Inspection Agency

AWIP e: Assembly Work in Process
- e: Automatic Wave Information Processor

AWIPS e: Advanced Weather Interactive Processing System

AWIS d: automatisches Wartungsinformationssystem
- e: Arizona Water Information System
- e: Association of Women in Science
- e: Automatic Wave Information System

AWIY e: As We Informed You

AWJSRA e: Augmenter Wing Jet Short Takeoff and Landing Research Aircraft

AWK d: Aufnahme-Wiedergabekopf
- d: Aufzeichnungs-Wiedergabekopf
- d: Ausschuß für Wärme- und Kraftwirtschaft

AWKlUe d: Anschlußleitungswählklinkenübertragung

AW-Kopf d: Aufnahme/Wiedergabe (Ton)-Kopf

AWks e: Air Workshop

AWL d: Abwasserlabor
- d: Adria-Wien-Leitung
- d: Arbeitsgemeinschaft Wissenschaftliche Literatur
- d: Atlas-Warten-Leitsystem
- d: Auswahllogik
- d: auswärtiger Leiverkehr
- e: Absent With Leave
- e: Absent Without Leave
- e: Administrative Weight Limitation
- e: Admiralty Wireless-Signals List
- e: All Weather Landing
- e: Animal Welfare League
- e: Artesian Well Lease
- e: Average Word Length
- e: Average Word Load

Awl e: Soviet infrared or radarguidance system

AWLF e: African Wildlife Leadership Foundation

AWLNET e: Area Wide Library Network

AWLOGS e: Army Wholesale Logistic System

AWLS e: All-Weather Landing System

AWL-System d: Ackermann-Wechsel-Ladesystem

AWM d: Aktionsgemeinschaft wirtschaftlicher Mittelstand
- d: Arbeitsgemeinschaft Werbeagenturen und Marketingberatung
- d: Arbeitskreis Werbung und Messewesen
- e: Appliance Wiring Material
- e: Arc Welding Machine
- e: Association of Women Mathematicians
- e: Automatic Washing Machine
- e: Awaiting Maintenance

AWMAbh d: Akademie der Wissenschaften, München, Abhandlungen (journ.)

AWMC e: Advanced Weapons Monitoring System
- e: Army Weapons and Mobility Command

AWMF d: Arbeitsgemeinschaft wissenschaftlich-medizinischer Fachgesellschaften

AWMI e: Anterior Wall Myocardial Infarction

AWMS e: Automatic Water Monitoring Station[s]

Awn d: Anwärter[in]

AWN d: Automaten-Werkzeug-Normen
- e: Activation Work Notice
- e: Aston Whole Number
- e: Automated Weather Network
- e: Awning

AWNCS e: Automated Weather Network Coordinating System

AWNDLS e: Anchor Windlass

AWngSvc e: Air Warning Service

AWNS e: Acceptor Wait for New-cycle State
- e: Acceptor Wait for New cycle State

AWO d: Arbeiterwohlfahrt
- e: Accounting Work Order
- e: Agricultural Workers Organization
- e: Airway Obstruction
- e: Air Weather Officer
- e: American Waterways Operators
- e: Army Welfare Officer
- e: Army Wireless Officer

AWOC e: Agricultural Workers Organizing Committee

AWOI e: American Waterways Operators, Incorporated

AWOL e: Absent Without Official Leave

AWOP e: All-Weather Operations Panel

AWP d: arbeitswissenschaftliche Projektierung
- e: Actual Working Pressure
- e: Addressed Word Position
- e: Adria-Wien-Pipeline
- e: Allied Weather Publication
- e: Anti-Submarine Warfare Program[me]
- e: Awaiting Parts

AWPA e: American Wire Producers Association
- e: American Wood Preservers Association

AWPB e: American Wood Preservers Bureau

AWPI e: American Wood Preservers Institute

AWPL e: Australia-West Pacific Line

A. Wpn. e: Automatic Weapon[s]

AWR d: Absolutwertrechenmaschine
- d: Arbeitsgemeinschaft der Deutschen Werkredakteure
- d: Ausgabewortregister
- d: automatischer Wagenrücklauf
- e: Actual Weight Report
- e: Adaptive Waveform Recognition
- e: Air/Water Ratio
- e: Arctic Wildlife Refuge
- e: Army War Room
- e: Association for the Study of the World Refugee Problems

AWRA e: American Water Resources Association
- e: Australian Welding Research Association

AWRC e: Australian Water Resources Council

AWRE e: Atomic Weapons Research Establishment

AWRIS e: Army War Room Information System

AWRN e: Awareness

AWRNCO e: Aircraft Warning Company

AWRO e: Atomic Weapon Retrofilt Order

AWRS e: Airborne Weather Radar System
- e: Airborne Weather Reconnaissance System
- e: Aircraft Weapons Release Set
- e: All-Weather Reconnaissance System
- e: Aviation Weather Reporting Station

AWRT e: American Women in Radio and Television

AWRTV e: American Women in Radio and Television

AWS d: Antennenwahlschalter
- d: Arbeitsgemeinschaft wissenschaftlicher Sortimenter
- d: Ausgabewarteschlange
- e: Active Work Space
- e: Adjustable Wire Stripper
- e: Air[craft] Warning Service
- e: Aircraft Warning Set
- e: Air[craft] Warning System[s]
- e: Air Weapon System[s]
- e: Air Weather Service
- e: American War Standard[s]
- e: American Weather Service
- e: American Welding Society
- e: Amphibious Warfare School
- e: Anti-Submarine Warfare Systems
- e: Area Working Standard
- e: Astronaut Work Station
- e: Atlas Weapon System
- e: Attack Warning System
- e: Automatic Warning System
- e: Automatic Weather Station
- e: Automatic Welding System
- e: Aviation Weather Service

AWSA e: Airborne Waveguide Slotted Array
- e: Air Warfare Systems Analysis

AWSAA e: Airborne Waveguide Slot Array Aerial
- e: Airborne Waveguide Slot Array Antenna

AWSCOM e: Advanced Weapons Support Command

AWSG e: Armstrong-Withworth and Sperry Gyroscope

AWSM e: Air Weapons Systems Management
- e: Air Weather Service Manual

AWSO e: Assembly Work Schedule Order

AWSP e: Air Weapons Systems Plan

Awst d: Ausbesserungswerkstätte[n]

AW&ST e: Aviation Week and Space Technology (journ.)

AWSTG e: Air Weather Service Training Guide

Awstn d: Ausbesserungswerkstätten

AWSVC e: Air Warning Service

AWT d: Anwendungstechnik
- d: Arbeitsgemeinschaft für Wirkstoffe in der Tierernährung
- e: Actual Work Time
- e: Advanced Waste Treatment
- e: Aerolastic Wind-Tunnel
- e: Anechoic Water Tank
- e: Arc-Jet Wind Tunnel
- e: Associate in Wildlife Technology
- e: Atomic Weight
- e: Average Wait Time
- e: Await

AWTAS e: Automated Weapons Test Analysis System

AWTB e: Aerated Waters Trade Board

AWTBS e: All-Weather Tactical Bombing System

AWTC e: Airway Traffic Controller

AWTF d: Ausschuß für wissenschaftliche und technische Forschung

AWTI e: Air Weapons Training Installation

AWTID d: Ausschuß für wissenschaftlich-technische Information und Dokumentation
AWTMS e: All-Weather Topographic Mapping System
AWTR Process e: Advanced Waste Treatment Process
AWTS e: All Working Time Saved
AWTSC e: Atomic Weapons Tests Safety Committee
AWTT e: Above Water Torpedo Tube
AWTTP e: Apollo Wind-Tunnel Testing Program[me]
AWU e: Agricultural Workers Union
- e: Aluminum Workers Union
- e: Atomic Weight Unit
AWUTC e: Aircraft Warning Unit Training Center
AWV d: Arbeitsgemeinschaft für wirtschaftliche Verwaltung
- d: Arbeitsgemeinschaft wissenschaftlicher Verlage
- d: Ausschuß für wirtschaftliche Verwaltung in Wirtschaft und öffentlicher Hand
- d: Außenwirtschaftsverordnung
- d: automatische Wählvermittlung
- e: Atmospheric Wind Velocity
AWW e: Address Write Wire
AWWA e: American Water Works Association
- e: American Withe-Water Affiliation
- e: Australian Water and Wastewater Association
AWWDs d: (Königliche) Akademie der Wissenschaften zu Wien; Denkschriften (journ.)
AWWF e: All-Weather Wood Foundation[s]
AWWI e: American Wash and Wear Institute
AW Wire e: Address Write Wire
AWWM e: Automatic Wire Wrap Machine
AWWS e: Automated Want and Warrant System
AWWSb d: Akademie der Wissenschaften zu Wien; Sitzungsberichte (journ.)
AWWU e: American Watch Workers Union
AWWV d: Arbeitsgemeinschaft der Wasserwirtschaftsverbände
AWX e: All-Weather Aircraft
AWX(F) e: All-Weather Aircraft (Fighter)
AWX(I) e: All-Weather Aircraft (Intruder)
AWY e: Airway
AWYDC e: All Weather Yaw Damper Computer
AWZ d: Ackerwertzahl
AX e: A-12 fighter plane
- e: AND Extender
- e: Attack Experimental
- e: Axes
- e: Axinite
- e: Axiom[atic]
- e: Axion
- e: Axis
ax. d: axiometrisch
AXAF e: Advanced X-ray Astrophysics Facility
AXBT e: Aircraft Expendable Babythermograph
AXD e: Auxiliary Drum
AxE d: Axeloth-Einheit
AXE e: Automatic Electronic Exchange
- e: Automatic Exchange Equipment
AXF e: Advanced X-ray Facility
AX FL e: Axial Flow
AXFMR e: Automatic Transformer
ax. grad. e: Axial Gradient
axidnt e: accident
AXIO e: Axiogist
- e: Axiological[ly]
- e: Axiology
- e: Axiom
- e: Axiomatic[al][ly]
AXL Timber e: Axlewood Timber
AXM e: Acetocycloheximide
AXM Language e: Axiomatic Language
AXO e: Assistant Experimental Officer
AXP e: Allied Exercise Publication
- e: Associative Crosspoint Processor
- e: Axial Pitch
AXPS e: Air Express
AXR e: Asynchronous Exit Routine
- e: Automatic X-ray Radiograph
AXS e: Auxiliary Storage
- e: Auxiliary Store
Ax Sig Com e: Axis of Signal Communication
AXSIGCOMM e: Axis of Signal Communication
AXTL e: Axial Flow
AY e: Aerology
- e: Anthophyllite
- e: Anyone
AYB e: Accessory Bulletin
AYC e: Accessory Change
- e: Aerodynamic Yaw Coupling
AyCC e: Army Airspace Command and Control
AYCP e: Aerodynamic Yaw Coupling Parameters
Ayers e: Ayers Rock National Park, in Australias Northern Territory
AYF e: Antiyeast Factor
AYI e: Academic Year Institute
- e: Angle of Yaw Indicator
AYLI e: As You Like It
AYP e: Alaska-Yukon Pacific
Ayr. e: Ayrshire
AYT e: Are You There
AYV e: Aster Yellow Virus
AZ d: 1.1'-Azobisformamide
- d: Abstandszünder
- d: Ackerwertzahl
- d: Aktenzeichen
- d: aktive Zone
- d: Alkalitätszahl
- d: Alkoholzahl
- d: allgemeiner Zustand
- d: Allgemeinzustand
- d: Amerikanische Zone
- d: Amtszeichen
- d: Anordnungszwang
- d: Anrufzeichen
- d: Anschlußzentrale
- d: Anschlußzone
- d: Arbeitsausschuß Zeichnungen
- d: Arbeitszeit
- d: Archivalische Zeitschrift (journ.)
- d: Asbestzement
- d: Aufschlagzünder
- d: Ausfallzeit
- d: Austreibungszeit
- d: Azetylzahl
- d: Aziditätsgrad
- d: Aziditätsrad
- e: Active Zone
- e: Adjustable Zero
- e: Air Zero
- e: Alitalia
- e: Alzheimers disease
- e: Arizona
- e: Azerbaijan
- e: Azimuth
- n: Akademisch Ziekenhuis
Az. d: Acetylzahl
- d: Auszahlung
a. Z. d: als Zugabe
- d: auf Zeit
Az e: Azores
- e: Aztec
az. e: Azure
a Z n: aan Zee
AZA d: Arbeitszeitaufwand
- d: automatische Zeichenanlage
AZAR e: Adjustable Zero Adjustable Range
AZAS e: Adjustable Zero Adjustable Span
AzB d: Anleitung zur Berechnung
Azb e: Azerbaijan
Azb. e: Azerbaijan[ian]
AZBN e: Azobisisobutyronitrile
AZC e: Automatic Zero Check
AZE d: agrartechnische Zusammenarbeit und Entwicklung
A-Zeit d: Ausgabezeit
Az/EL e: Azimuth and Elevation
azeotrop. d: azeotropisch
Azer e: Azerbaidzhan
AZF d: Abzinsungsfaktor
- d: Allgemeines Zeugnis für den Flugfunk
- d: Arbeitszeitfond
AzFV d: Anhang zu den Fahrdienstvorschriften
AzG d: Abzweiggestell
AZG d: Alarmzeichengeber
- d: Allgemeines Zuständigkeitsgesetz
- d: Anteilzollgesetz
AZGS e: Azusa Ground Station
AZGW d: Auslandszentralgruppenwähler
AZH d: Arbeitsgemeinschaft Zubehör; Heiz-, Koch- und Wärmegeräte
AZI d: Ausgabeziel
- e: American Zinc Institute
- e: Azimuth
Az I e: Azores Islands
AZIG d: Auslandszählimpulsgeber
AZII e: American Zinc Institute, Inc.
Az Is e: Azores Islands
AzK d: Abzweigkasten
- d: Aufzeichnungskopf
Azl d: Abzweigleitung
AZl d: Auslandszentralvermittlungsleitung
AZL d: automatische Zuglenkung
Az-Lok d: Arbeitszuglokomotive
AzM d: Abzweigmuffe
AZM e: Azimuth
AzMPr d: Arbeitszettel für manuelles Prüfen
AZN e: Azurin
AZO d: Allgemeine Zollordung
- d: Arbeitszeitordnung
Azo. e: Azores
AZokpl d: Auslandszonenkoppler
AZON e: Azimuth Only
Azores e: Azores Islands
Azotier. d: Azotierung
AZOV d: Aufschlagzünder ohne Verzögerung
- e: Sea of Azov
AZP d: Abtastzeitpunkt
- d: Assoziativprozessor
- d: Ausbildungszentren-Platzbuchung
- e: Acoustic Zone Plate
AZ-Platte d: Asbestzementplatte
Az-PUR-Az d: Asbestzement-Polyurethan-Asbestzement
AZQ d: Atemzeitquotient

AZR d: Abzweigregler
- d: Adressenzählregister
- d: Anrufzugangsregelung
- d: Anzeigenregister
- d: Aschheim-Zondek Reaktion
- d: Aufzeichnungsregister
- d: Ausgabezielregister
- d: Ausländerzentralregister

Azr. e: Azores
AZ/RAN e: Azimuth and Range
AZRE d: Ausnutzung der Zentralrecheneinheit
AZRI e: Arid Zone Research Institute
AZROCK e: Asbestos Rock
AZ-Rohr d: Asbestzement-Rohr
Azru e: Aztec Ruins National Monument
AZS d: automatische Zugsteuerung
- e: Alloyed Zinc Sheet
- e: Automatic Zero Set

AZSS d: Allzweck-Schnellspeicher
AzStö d: Arbeitszettel für Störbeseitigung
AZT d: Aschheim-Zondek Test
- d: Aufzähltest
- e: Azidothymidine, one of Several Drugs Used in the Treatment of AIDS
- e: Azusa Transponder

Azt e: Aztec
AZTC e: Azusa Transponder Coherent
Aztec Ruins e: Aztec Ruins National Monument, New Mexico
AZ Test d: Ascheim-Zondek Test
AZTG d: Auslandszeittaktgeber
AZTÜE d: Auslandsverkehr-Zeittaktüberwachungseinrichtung
AZTÜEstaff d: Auslandsverkehr-Zeittaktüberwachungseinrichtung mit gestaffelten Zeittakten
Az TÜp d: Arbeitszettel für technische Überprüfung
Azubi d: Auszubildende[r]
Azubis d: Auszubildende
AZUSA e: Azimuth, Speed, Altitude
AZV d: Abtastzeitrasterverschiebung
- d: Anzeigeverstärker
- d: Arbeitszeitverordnung
- d: Autofahrer-Zentralverband
- d: automatische Zielverfolgung

Az Verb d: Arbeitszettel für Verbundarbeiten
AZVO d: Arbeitszeitverordnung
AZVST d: Auslandszentralvermittlungsstelle
AZ WO e: Azure Wove
AZY e: Azyme

B

B d: Bachelor
- d: Ballast
- d: Band
- d: Bandbreite
- d: Bandeisen[bewehrung]
- d: Basis
- d: Basisanschluß
- d: basisch
- d: Basiselektrode
- d: Basiszahl
- d: Batterie
- d: Bauchflosse
- d: Baud
- d: Baumschule
- d: Baumwolle
- d: Bayern
- d: Bazillus
- d: Bedienfunktion
- d: Bedienungsteil
- d: Befehlsregister
- d: Behälter
- d: Beleuchtung
- d: Beleuchtungsanlage
- d: Belgien (Kfz)
- d: Berechtigung
- d: Bereich
- d: Berg
- d: Berichterstattung
- d: Besitzer
- d: Bessemer Stahl
- d: Bestand
- d: Beton
- d: Bild
- d: Billion
- d: Binäranzeige
- d: Binärstelle
- d: Blank
- d: Blindleitwerk
- d: Blindleitwert
- d: Block
- d: Bodenkraft
- d: Bogen
- d: Bohrung
- d: Bombenflugzeug
- d: Bord
- d: Brandbreite
- d: Brauner
- d: Braunerde
- d: Braunhumussäure
- d: Bravo
- d: Breite
- d: Brennstoff
- d: Bronchial
- d: Bronchus
- d: Brückenschaltung
- d: Brückensymmetrie
- d: Buch
- d: Bucht
- d: Bug
- d: Bundesstraße
- d: Byte
- e: Bacillus
- e: Back
- e: Backlash
- e: Bag
- e: Ballistic
- e: Balloon
- e: Baltic
- e: Bandwidth
- e: Barge
- e: Base
- e: Bat
- e: Batch
- e: Bath
- e: Battery
- e: Battle
- e: Baumé[degree]
- e: Baumé scale
- e: Bay
- e: Beach
- e: Beacon
- e: Bearing[angle]
- e: Beecham Product[s]
- e: Behaviour
- e: Belgium
- e: Bell
- e: Belt[ed]
- e: Benoist scale
- e: Benzene
- e: Bias
- e: Bible
- e: Bicuspid
- e: Bill
- e: Binary
- e: Biological
- e: Biplane
- e: Bit
- e: Bitch
- e: Black
- e: Blend
- e: Blood
- e: Blower
- e: Blue
- e: Boat
- e: Boatswain
- e: Body
- e: Boeing
- e: Boilermaker
- e: Boils at
- e: Bomb[er]
- e: Bombardier
- e: Bombardment
- e: Boolean
- e: Booster
- e: Bor
- e: Borderline
- e: Bore
- e: Boron
- e: Bottom
- e: Bound
- e: Bowels
- e: Box
- e: Braid
- e: Branch
- e: Brass
- e: Bravo-code for letter B
- e: Break
- e: Break Contact
- e: Brick
- e: Brigade
- e: Brightness
- e: British
- e: Broken [sea]
- e: Brother
- e: Buccal
- e: Buddhist
- e: Budget
- e: Building
- e: Bulb (camera setting)
- e: Bulk-Carrier
- e: Bulletin
- e: Buoyancy
- e: Bus
- e: Butcher
- e: flux density
- e: Fraunhofer line caused by terrestrial oxygen
- f: Baie
- n: Baai
- p: Baia
- R: Bukhta
- s: Bahlo
- s: Baja
- s: bajar

B. d: Bahn
- d: Beobachter
- d: Biegeprobe
- d: Bildung
- d: Burg
- d: Büro
- e: Bank
- e: Battleship

b d: Bar
- d: Barometer[stand]
- d: Beschleunigung
- e: bale
- e: barn
- e: beam
- e: bel
- e: best
- e: bituminous
- e: boiling
- e: bolt
- e: bond[ed]

- e: book
- e: born
- e: breadth
- e: bridge
- e: broadcast[ing]
- e: bulb (camera exposure device)
- e: wing span

b. l: bis
B- d: B Minus
- e: B Negative

B+ d: B Plus
- e: B-positive
- e: B plus

°B e: degrees Baumé
B0 e: Bit 0
B-1 e: B-negative [battery terminal]
- e: B minus
- e: North American-Rockwell strategic supersonic bomber

b 1 e: booster 1
B-18 e: Rockwell International long-range supersonic sweptwing bomber
B1AC e: Binary One-Address Computer
B1AS e: Binary One-Address System
B1...B8 e: Bit 1...Bit 8
B1c e: Boilermaker First Class
b 1 p e: booster 1 pitch
B1S e: Beaded One Side
b 1 y e: booster 1 yaw
B-2 e: Advanced Technology Fighter aircraft
b 2 e: booster 2
B-25 e: World War II light bomber
B26 e: modernized Douglas B-26 Invader
B2BCONV e: Bits-to-Bytes Conversion
B2c e: Boilermaker Second Class
B2F e: Boeing 720 fan jet airplane
b 2 p e: booster 2 pitch
B2S e: Beaded Two Sides
b 2 y e: booster 2 yaw
B3F e: Boeing 320 aircraft
b4 e: before
B47 e: Stratojet all-weather strategic medium bomber
B-52 e: Stratofortress all-weather intercontinental strategic heavy bomber
B-57 e: Two-place twinengine turbojet all-weather tactical bomber
B-58 e: Strategic allweather supersonic bomber
B-66 e: Destroyer twin-engine turbo jet tactical all-weather light-bombardment aircraft
B6ZS e: Bipolar with Six Zero Substituting
B-707 e: one of a Boeing aircraft series
B-747 e: Boeing jumbo jet aircraft
B 77 e: Bratislava 77 [viral] strain
B7F e: Boeing 707 fan jet airplane
BA d: Basisanschluß
- d: Batterie
- d: Bauabschnitt
- d: Bauamtmann
- d: Beckenausgang
- d: Befehlsaufruf[steuerung]
- d: bemanntes Amt
- d: Benzyladenin
- d: Berufsausbildung
- d: Beschaffungsanweisung
- d: Betriebsanleitung
- d: Betriebsanzeige
- d: Betriebsart
- d: betriebswirtschaftliche Auswertung
- d: bibliothekarische Auslandsstelle
- d: Bildaustastsignal
- d: Bildaustastung
- d: Bildschirmanzeigegerät
- d: Bildsignal mit Austastung
- d: Binärausgang
- d: binäre Addition
- d: blankes Aluminium
- d: Bodenanteil
- d: Bromacetyl
- d: Bronchial Asthma
- d: Buchungsautomat
- d: Buenos Aires
- d: Busadapter
- e: Bachelor
- e: Bachelor of Agriculture
- e: Bachelor of Arts
- e: Backlash Allowance
- e: Bad Attitude
- e: Balloon Astronomy
- e: Bank Angle
- e: Barite
- e: Base
- e: Base Activation
- e: Base Address
- e: Base Adressing
- e: Base Area
- e: Base Assembly
- e: Base Line
- e: Basic
- e: Basic Access
- e: Basic Address
- e: Basic Agreement
- e: Basic Airman
- e: Basic Allowance
- e: Basic Assembler
- e: Basic Authorization
- e: Batch Assembly
- e: Battery
- e: Battery Adjust
- e: Batting Average
- e: Beach Abort
- e: Beginning Address
- e: Bell Aerosystems Company
- e: Bell Alarm
- e: Bending Allowance
- e: Berkshire Athenium
- e: Bettmann Archive
- e: Binary Add
- e: Binary Adder
- e: Binary Addition
- e: Binary Arithmetic
- e: Biological Abstracts (journ.)
- e: Bit Access
- e: Bit Address
- e: Blank Address
- e: Blanking Amplifier
- e: Blind Approach
- e: Block Adress[ing]
- e: Blood Agar
- e: Board of Agriculture
- e: Boeing
- e: Boeing Airplane
- e: Bomb Aimer
- e: Books Abroad (journ.)
- e: Boolean Addition
- e: Boolean Algebra
- e: Boreal Asiatic
- e: Boric Acid
- e: Bottom Ash
- e: Boundary Alignment
- e: Bowker Annual (journ.)
- e: Braking Action
- e: Branch Address
- e: Branch Assignment
- e: Breaking Action
- e: Breaks Above
- e: Breathing Apparatus
- e: Bridge Amplifier
- e: Brigade Area
- e: Bright Annealed
- e: British Academy
- e: British Admiralty
- e: British Airways
- e: British America
- e: British Association [for the Advancement of Science]
- e: Brome-Aceton
- e: Buccoaxial
- e: Budget Activity
- e: Budget Authorization
- e: Buenos Aires
- e: Buffer Address
- e: Buffer Amplification
- e: Building Advisor
- e: Bulb Angle
- e: Bus Available
- e: Business Automation
- e: Busted Aristocrat

Ba d: Bahnarzt
- d: Banane
- d: Blütenachse
- e: Barium
- e: Barye
- p: Baia

B/A e: Backache
b/a e: billed at
B.A. e: British American [oil company]
- e: Buffer Amplifier
- l: Baccalaureus Artium

b.a. e: Bridging Amplifier
B & A e: Bangor & Aroostook
- e: Boston & Albany

b-a e: bare ass[ed]
- e: unclothed

B-A e: Basses-Alpes
BAA d: Betonaufbereitungsanlage
- e: Bachelor of Applied Arts
- e: Bachelor of Arts of Architecture
- e: Benz(a)anthracene
- e: Benzoyl Arginine Amide
- e: Billeting and Acommodations Advisory
- e: Brewers Association of America
- e: British Acetylene Association
- e: British Air Attaché
- e: British Archaeological Association
- e: British Astronomical Association
- e: Broadband Active Analyzer
- e: Budget Activity Account
- e: Buffer Address Array

BA Acc e: Bachelor of Science in Accounting
BAADI d: Bundesanstalt für Agrardokumentation und -information
BAADS e: Bangor Air Defense Sector
BAAF e: Butts Army Airfield
BAAJ e: British Archaeological Association Journal
Baal e: Baalbek
BAAL e: Black Academy of Arts and Letters
- e: British Association for Applied Linguistics

BAAN e: Budget Authorization Account Number
BAAP e: Bilateral Aid to Asia and Pacific [program]
BAAPCD e: Bay Area [Air] Pollution Control District
BAAR e: Board for Aviation Accident Research
- e: Board of Aviation Accident Research

BAAS d: Butylammoniumallylsulfonat
- e: British Association for the Advancement of Science
- e: Broadband Acoustic Array Section

BAASB e: British Association for American Studies Bulletin (journ.)
BAAss. d: Bauamtsassessor
BAAT e: N-Bromoacetyl -mono (p-Azobenzene Arsonic acid)-L-Tryosine

BAATC e: Bay Area Army Terminal Center
BAB d: Betriebsabrechnungsbogen
- d: betriebswirtschaftlicher Auswertungsbogen
- d: bewegliche Armeebasis
- d: Bildschirmanzeigebaugruppe
- d: Bundesautobahn
- e: Blood Agar Base
- e: Blythe Air Base
- e: Booster Assembly Building
- e: Boundary Address Block
- e: British Airways Board
Bab e: Babbitt
- e: Babylon
Bab. e: Babylonia[n]
BABA e: Bachelor of Arts in Business Administration
- e: British Antiquarian Booksellers Association
Babars e: Babar Islands of Indonesia
BA/BASIC e: Business Applications written in Beginners All-purpose Symbolic Instruction Code
BABB e: Babbit Metal
BABBITT e: Babbitt Metal
Babbs e: Computer
Babbage's e: Computer (the first computer)
BABF d: Bezirksaufsichtsbeamter für das Fernmeldewesen
BAbfG d: Bundesabfallgesetz
BABI e: Brooke Army Burn Institute
Babines e: Babine Mountains of British Columbia
BABL d: Bundes-Arbeitsblatt
BAB MET e: Babbitt Metal
BABO e: Boolean Approach for Bivalent Optimization
BABS d: Bildausgabe über Bildschirm
- e: Batch Automated Balancing System
- e: Beam Approach Beacon System
- e: Blind Approach Beacon System
- e: Blind Approach Beam System
BAB Timber e: Babul Timber
Baby e: Babylon[ia]
Baby. e: Babylonia[n]
BAC d: Benzalkonium Chloride
- e: Bacilli
- e: Background Analysis Center
- e: Bacteria
- e: Bacterial Adherent Colonies
- e: Bacterial Antigen Complex
- e: Bacteriological
- e: Bacteriologist
- e: Bacteriology
- e: Bankers Automated Clearing Service
- e: Barometric Altitude Control
- e: Base Area Commandant
- e: Basic Address Calculation
- e: Battleship Aircraft Carrier
- e: Below All Clouds
- e: Bendix Aviation Corporation
- e: Billing and Accounting Computer
- e: Binary-Analog[ue] Conversion
- e: Binary Asymmetric Channel
- e: Biomedical Applications of Computers
- e: Bipolar Active-plastic Cell
- e: Block-Acknowledged Counter
- e: Blood Alcohol Concentration
- e: Blower Access Cover
- e: Boating Anti-Pollution Council
- e: Boeing Airplane Company
- e: Brigade Ammunition Column
- e: British Aircraft Corporation
- e: British Association of Chemists
- e: British Automation Company
- e: Bromacetyl Cellulose
- e: Buccoaxiocervical
- e: Buffer Address Counter
- e: Bunker Adjustment Charge
- e: Bus Adaptor Control
BAc e: Bachelor of Accounts
Bac. e: Bacillus
- l: Baccalaureus
bac. e: Bacterial
bac f: bachot
BAC-145 e: British Jet Provost trainer aircraft
BACA e: Berlin Airways Control Approach
- e: British Advisory Committee for Aeronautics
- e: British Aerospace Commercial Aircraft
- e: British Agricultural Contractors Association
BACAH e: British Association of Consultants in Agriculture and Horticulture
BACAIC e: Boeing Airplane Company Algebraic Interpretive Computing
BACAICS e: Boeing Airplane Company Algebraic Interpreter Coding System
BACAIC system e: Boeing Airplane Company Algebraic Interpretive Computing System
BACAL e: Butter and Cheese Association Limited
BACAN e: British Association for the Control of Aircraft Noise
BA Cap e: Bayonet Automobile Cap
BACAT e: Barge Aboard Catamaran
BAC BAG e: Bactine Bag
B.Acc. e: Bachelor of Accountancy
BACC e: British-American Collectors Club
B Acc's e: Bachelor of Accountsy
BACD e: Boeing Airplane Company Design
BACE e: Bachelor of Air Conditioning Engineering
- e: Basic Aircraft Checkout Equipment
- e: Basic Automatic Checkout Equipment
- e: British Association of Consulting Engineers
- e: Bureau of Agricultural Chemistry and Engineering
BAC Eng e: Bachelor of Air Conditioning Engineering
BACGA e: British-Australian Cotton Growing Association
Bach e: Bachelor
BACH e: Bachelor of Arts in Chemistry
BACHAC e: Bachelor of Arts in Chemistry
BA Chem e: Bachelor of Applied Chemistry
B.A. Chem. e: Bachelor of Arts in Chemistry
Bach of Arts e: Bachelor of Arts
BACI e: Brazilian-American Cultural Institute
BACIE e: British Association for Commercial and Industrial Education
BACIS e: Budget Accounting Information System
back. e: backwardation
BACLIN e: Baroclinic
BACO e: Base Activation Chance Order
- e: British Aluminum Company
BACS d: Business Automation & Computer Services GmbH
- e: Backup Acquisition System
- e: Banks Automated Clearing Services
BACS e: Boeing Applied Computing Service
BACSEB e: Bureau of Naval Weapons Aviation Clothing and Survival Equipment Bullet
BACT e: Bacteria
- e: Bacteriological
- e: Bacteriologist
- e: Bacteriology
- e: Bacterium
- e: Best Available Control Technology
BACTER e: Bacteriological
- e: Bacteriologist
- e: Bacteriology
BACTERIOL e: Bacteriologic[al][ly]
- e: Bacteriologist
- e: Bacteriology
Bacteriol; Rev e: Bacteriological Reviews (journ.)
BACTRIAN e: Bactrian Camel [twohump Camel of Asia]
BACU e: Battle Area Control Unit
BACUS e: Booz, Allen and Hamilton Computer Utilization System
BACUSA e: British Aircraft Corporation USA
BAD d: Busadapter
- e: Base Air Depot
- e: Berlin Airlift Device
- e: Bit Anomaly Detector
BÄD d: Bahnärztlicher Dienst
BADA e: Base Air Depot Area
BADAS e: Binary Automatic Data Annotation System
BADC e: Binary Asymmetric Dependent Channel
BADCT e: Best Available Demonstrated Control Technology
BADGE e: Base Air Defense Ground Environment
BADGES e: Base Air Defense Ground Environment System
BADIC e: Biological Analysis Detection Instrumentation and Control
BADir. d: Bauamtsdirektor
BADL d: Bonner Arbeiten zur Deutschen Literatur (journ.)
- e: N-Bromoacetyldinitrophenyl-Lysine
B Adm e: Bachelor of Administration
B Adm Eng e: Bachelor of Administrative Engineering
B.Admin. e: Bachelor of Administration
BADS e: Bangor Air Defense Sector
- e: British Association of Dermatology and Syphilology
BAdv Art & Des e: Bachelor of Advertising Arts and Design
BAE d: betriebswirtschaftliche Analysen; Einzelhandel
- d: VEB Berliner Akkumulatoren- und Elementefabrik (DDR)
B Ae e: Bachelor of Aeronautical Engineering
- e: Bachelor of Aeronautics
B.A.E. e: Bachelor of Aeronautical Engineering
- e: Bachelor of Agricultural Engineering
- e: Bachelor of Architectural Engineering
- e: Bachelor of Art Education
- e: Bachelor of Arts in Education
BAE e: Barium Enema
- e: Beacon Aerial Equipment
- e: Beacon Antenna Equipment
- e: Board of Architectural Examiners
- e: Bureau of Agricultural Economics
- e: Bureau of Agricultural Engineering
- e: Bureau of American Ethnology

BAe e: British Aerospace
B.A.E.A. e: British Atomic Energy Authority
BAEC e: British Agricultural Export Council
- e: British Amateur Electronics Club
- e: British Atomic Energy Corporation
B.A.Econ. e: Bachelor of Arts in Economics
BA Ed e: Bachelor of Art Education
B.A.Ed. e: Bachelor of Arts in Education
BA Ed e: Bachelor of Arts in Education
BAED e: British Airways European Division
BA Educa e: Bachelor of Arts in Education
B.Ae.E. e: Bachelor of Aeronautical Engineering
BAEE e: Benzoylarginine Ethyl Ester
BA ENEM e: Barium Enema
BAEng. e: Bureau of Agricultural Engineering
BAEO d: Bundesärzteordnung
BAEQS e: Bid-Asked Electronic Quotation System
BAER e: Brainstem Auditory-Evoked Response
BAERE e: British Atomic Energy Research Establishment
B Aero E e: Bachelor of Aeronautical Engineering
B Ae S e: Bachelor of Aeronautical Science
B Ae Sc e: Bachelor of Aeronautical Science
BAE Timber e: Baen Timber
BAF d: Bamberger Abhandlungen und Forschungen (journ.)
- d: Bundesanstalt für Fleischforschung
- e: Baffle
- e: Board of Agriculture and Fisheries
- e: Britains Atomic Factories
- e: British Air Ferries
- e: British Air Force
- e: Bunker Adjustment Factor
- e: Burma Air Force
- e: Burundi Air Force
BAFCOM e: Basic Armed Forces Communication
BAFCOM Plan e: Basic Armed Forces Communication Plan
BAFESA d: Bahnfernschreib[selbstanschluß]anlage
BAFF d: Bedienungsanforderungs-Flipflop
BAFM e: British Association of Forensic Medicine
BAFOC e: British Air Force[s] of Occupation Command
BAFÖG d: Bundesausbildungsförderungsgesetz
BAFS e: British Academy of Forensic Science
BAFSVS e: British Armed Forces Special Vouchers
BAFT d: Beratender Ausschuß für Forschung und Technologie
BAFVC e: Bids Accepted for the Following Vacancies
BAG d: Badenwerk AG
- d: Beiträge zur alten Geschichte (journ.)
- d: Bildaufzeichnungsgerät
- d: Bundesanstalt für den Güterfernverkehr
- e: Bag All Garbage
- e: Baggage
- e: Ballistic Attack Game
- e: Beta Absorption Gauge
- e: Buccoaxiogingival
Bag. d: Baghdad
B.Ag. e: Bachelor of Agriculture
BÄG d: Bayerisches Ärztegesetz
BA-gauge e: Bayard-Alpert gauge
BAGe d: Bildaufzeichnungsgeräte
B.Ag.E. e: Bachelor of Agricultural Engineering
BAGG e: Buffered Acide Glucose Glycerol
B.Agr. e: Bachelor of Agriculture
BAGR e: Bureau of Aeronautics General Representative
B Agr E e: Bachelor of Agricultural Engineering
B.Agr.Eco. e: Bachelor of Agricultural Economics
B Agr (Hort) e: Bachelor of Agriculture (Horticulture)
B.Agric. e: Bachelor of Agriculture
B Agr S e: Bachelor of Agricultural Science
B Agr Sc e: Bachelor of Agricultural Science
BAGS e: Bullpup All-Weather Guidance System
B.Ag.Sc. e: Bachelor of Agricultural Science
BAGü d: Bundesanstalt für den Güterfernverkehr
BAH d: Bahrein (Flughafen)
- d: Biologische Anstalt Helgoland
- e: Basic Adaptive Hardware
- e: Basic Attention Handling
- e: Booz-Allen-Hamilton
- e: British Airways Helicopters
Bah d: Bergahorn
- e: Bahama Islands
- e: Bahanmas
- e: Bahia
- e: Bahrain
B-A H e: British-American Hospital
Bahams e: Bahamas
Bahngel. d: Bahngelände
Bahnw. d: Bahnwärter
- d: Bahnwesen
Bahr e: Bahrain
Bahrain e: Bahrain
BAH Re e: Bachelor of Arts in Human Relations
BAHS e: British Agricultural History Association
- e: British Agricultural History Society
BAH Timber e: Bahera Timber
B-A-I d: Beschlüsse-Anweisungen-Informationen
BAI d: Bundesverband der Agraringenieure
- d: Büro für angewandte Informatik GmbH
- e: Base Activation Instruction
- e: Boeing Airborne Instrumentation
- e: Buffer Alignment Information
- e: Bureau of Animal Industry
Ba. I. e: Bahama Islands
BAIB e: Beta-Amino-Isobutyric [Acid]
BAIC e: Binary Asymmetric Independent Channel
- e: Bureau of Agricultural and Industrial Chemistry
BAID e: Boolean Array Identifier
BAIE e: British Association of Industrial Editors
BA in A & Sci e: Bachelor of Arts in Arts and Science
BA in B & E e: Bachelor of Arts in Business and Economics
BA in Cer A e: Bachelor of Arts in Ceramic Art
BA in E & B e: Bachelor of Arts in Economics and Business
BA in E ed e: Bachelor of Arts in Elementary Education
BA in J e: Bachelor of Arts in Journalism
BA in M Ed e: Bachelor of Arts in Music Education
BA in Rel Ed e: Bachelor of Arts in Religious Education
BAINS e: Basic Advanced Integrated Navigation System
BAInsp. d: Bauamtsinspector
BA in Sp e: Bachelor of Arts in Speech
Ba. Is. e: Bahama Islands
BAIS e: Blind Approach Instrument System
- e: Bulletin Articles Information Subsystem
BAIT e: Bacterial Automated Identification Technique
BA J e: Bachelor of Arts in Journalism
Baj. e: Bajonett
BA Jour e: Bachelor of Arts in Journalism
BAK d: Berliner Arbeitskreis für Information
- d: Blutalkoholkonzentration
- e: Backup
- e: Barrier Arresting
- e: Broadband Aerial Kit
- e: Broadband Antenna Kit
Bak e: Baku
baka e: African pygmy
Bakt. d: Bakterie[n]
- d: Bakteriolog[i]e
bakt. d: bakteriell
- d: bakteriologisch
BAKT d: Basisanschluß-Konzentrator
BAK Timber e: Bakota Timber
BAL d: Baltimore (Flughafen)
- d: Baustellenausstattungs- und Werkzeugliste
- d: Betriebsanalysenlaboratorium
- e: Balance[d]
- e: Balancing
- e: Ballistic[s]
- e: Balsam
- e: Base Authorization List
- e: Basic Assembler Language
- e: Basic Assembly Language [computer programming]
- e: Biological Assessment Laboratory
- e: Blood Alcohol Level
- e: Branch and Link
- e: British Anti-Lewisite
- e: Broad Absorption Line
- e: Business Application Language
Bal. e: Balcony
- e: Baleares
- e: Ballast
B.A.I. l: Baccalaureus in Arte Ingeniaria
BALAD e: Bleachable Absorber Laser Amplifier and Detector
BALANCE e: Basic and Logically Applied Norms-Civil Engineering
BALARE e: Buoyancy Actuated Launch and Retrieval Elevator
bal. arenae l: balneum arenae
BALAST e: Balloon Astronomy
balc e: balconette
Balc. e: Balcony
BAL CL e: Balancing Coil
BALCOL e: Burroughs Algebraic Compiler
BALDICER e: Balanced Diet Certificates
BALDNY e: Ballistic Density
Baldria e: Baltic-Alexandria
BALGOL e: Burroughs Algebraic Compiler
- e: Burroughs Algorithmic Language
BALH e: British Association for Local History
BALID e: Ballisti[s]c Identification
BALINVERTER e: Balanced Input Converter
BALIS d: Bayerisches Landwirtschaftliches Online Informationssystem
- e: Bay Area Library and Information System

BALITAC e: Basic Literal Automatic Coding
BALL e: Ballast
- e: Ballistic
- e: Balloon
Ball e: Ballerup
ball. d: ballistisch
Ball. d: Ballon
- e: Ballismus
BALLAD e: Ballistic Loran Assist Device
Ball Coll e: Balliol College Oxford
Ballist d: Ballistik
BALL LP e: Ballast Lamp
BALLOTS e: Bibliographic Automation of Large Library Operations using Time Sharing
BALLOTS System e: Bibliographic Automation of Large Library Operations on a Time Sharing System
BALLT e: Ballast Tube
BALLUTE e: Balloon [and] Parachute
BALLWIN e: Ballistic Wind
BALM e: Block and List Manipulator
BALMI e: Ballistic Missile
Bal-Mol e: Ballester-Molina (Argentine semi-automatic pistol)
BAL NET e: Balancing Network
BALOC e: Base Logistical Command
BALOP e: Balopticon [projector]
BALP d: Buenos Aires - La Plata
B Alp e: Basses-Alpes
BALPA e: Balance of Payments
- e: British Air Line Pilots Association
B-alpes e: Basses-Alpes
BAL RHEO e: Balancing Rheostat
BALS d: Balsam
- e: Bachelor of Arts in Library Science
Bals e: Bahama Islands
BALS e: Balancing Set
- e: Balearic Islands
- e: Balsam
- e: Blind Approach Landing Sysem
bals. l: balsamum
BALSPACON e: Balance of Space to Space Control Agencies
Bal T e: Balance Test
Balt. e: Baltic
BALTA d: ballistische Tageseinflüsse
BALTAP e: Baltic Approaches
BALTHUM e: Balloon Temperature and Humidity
Baltic[s] e: Estonia, Latvia, Lithuania
BAL Timber e: Ballagi Timber
BALTPULP e: Baltic Pulp and Paper
BAL TR e: Balancing Transformer
Baltwood e: Baltic Wood
Baluch d: Baluchistan
BALUN e: Balanced-to-Unbalanced
- e: Balancing Unit
BALUN Network e: Balance[d]-to-Unbalance[d] Network
BALUTE e: Balloon Parachute
BAL V e: Balanced Voltage
- e: Balance Voltage
BAL VOLT e: Balanced Voltage
BALWND e: Ballistic Wind
BAM d: Belegartmerkmal
- e: Bachelor of Applied Mathematics
- e: Bachelor of Arts in Music
- e: Baikal-Amur-Magistral
- e: Ballistic Advanced Missile
- e: BankAmerica Corporation
- e: Base Automotive Maintenance
- e: Basic Access Method
- e: Bio-Activated Mineral Lotion
- e: Bit Access Memory
- e: Block Access Method
- e: Block Allocation Map
- e: Bonded Atom Model
- e: Bradley Aberration Method
- e: British Air Ministry
- e: Broadcasting Amplitude Modulation
- e: Brooklyn Academy of Music
- e: Buenes Aires Musical (journ.)
- e: Buffered Access Method
- e: Business Analysis Model
BAMBI e: Ballistic Anti-ballistic Missile Boost Intercept
- e: Ballistic Missile Bombardment Interceptor
- e: Ballistic Missile Burning Intercept
- e: Bayesian; Analysis Modified by Inspection
BAMBP e: Butyl-Alpha-Methyl-Benzylphenol
BAMBSR e: British Admiralty Maintenance and Supply Representative
BAMC e: Brooke Army Medical Center
B. Am. D. e: Base Ammunition Depot
BAME e: Benzoyl Arginine Methyl Ester
BAMG e: Browning Aircraft Machine Gun
BAM hrs e: Branch-office Average Maintenance Hours
BAMIRAC e: Ballistic Missile Radiation Analysis Center
BAMMEX e: British Agricultural Machinery Manufacturers Export Group
BAMO e: Bureau of Aeronautics Material Officer
BAMOS e: Batchprocessing Multi-Language Operating System
BAMP e: Basis Analysis and Mapping Program[me]
BAMR e: Bureau of Aeronautics Maintenance Representative
BAMRO e: Bureau of Aeronautics Maintenance Repair Officer
- e: Bachelor of Ayurvedic Medicine and Surgery
- e: Broadcast to Allied Merchant Ship[s]
BAmtm d: Bundesbahnamtmann
BAMTM e: British Association of Machine Tool Merchants
BA Mus Ed e: Bachelor of Arts in Music Education
BAMX d: Basisanschluß-Multiplexer
BAN d: bundeseinheitliche Artikel-Nummer[ierung]
- e: Bantu
- e: Base Activation Notice
- e: Best Asymptotically Normal
- e: Bionics Adaptive Network
- e: Bond Anticipation Not
- e: British Association of Neurologists
- e: Bulletin of the Astronomical Institutes of the Netherlands (journ.)
- R: Biblioteka Akademii Nauk
BANCS e: Bell Administrative Network Communications System
band. d: banbdagiert
- d: bandagieren
Band. d: Bandage
- d: Banderole
- d: Bandung
B and ARP e: Bare and Acid-Resisting Paint
B and E e: Beginning and Ending
Bandeirante e: Brazilian 12-passenger transport
B and F e: Bell and Flange
B and I e: Base and Increment
B and L e: Ball and Lever
- e: Business and Loan
B and O e: Belladonna and Opium
B and S e: Beams and Stringers
- e: Bell and Spigot
- e: Booster and Sustainer
BANEC e: Balanced Nuclear Economy Code
BANF d: Blockanfang[sadresse]
Banffs. e: Banffshire
BAng d: Bundesbahnangestellter
Bangla e: Bangladesh
Bangladesh e: Peoples Republic of Bangladesh
BANG[S] e: Bombing[s]
- e: Explosion[s]
BANG TORP e: Bangalore Torpedo Banir Bombing and Navigation Inertial
BANIR e: Bantam Inertial Reference
- e: Bombing & Navigation Inertial Reference
Bank. e: Bangkok
BANL d: Bundesanstalt für Naturschutz und Landschaftspflege
BANMP e: Basic Analysis And Mapping Program[me]
BANS e: Basic Air Navigation School
- e: Bond Anticipation Notes
- e: Bright Alphanumeric Subsystem
- e: British Association of Numismatic Societies
BANSDOC e: Bangladesh National Scientific and Technical Documentation Centre
BANSHEE e: Balloon and Nike Scaled High Explosive Experiment
BAN-System d: bundeseinheitliches Artikel-Numerierungssystem
Bantam e: Bantam Books
- e: Swedish antitank guided missile
BAN Timber e: Banati Timber
B.A. Nurs. e: Bachelor of Arts in Nursing
BANWR e: Bosque Apache National Wildlife Refuge
Banz d: Bundesanzeiger
BANZARE e: British-Australian-New Zealand Antarctic Research Expedition
BAO e: Bachelor of Art of Oratory
- e: Bachelor of Obstetrics
- e: Basal Acid Output
- e: Base Account Officer
- e: Base of Air Operations
- e: Basic Arithmetic Order
- e: Binary Arithmetical Operation
- e: British American Oil
- e: British Association of Otolaryngologists
B.A.O. e: Bachelor of Arts in Oratory
- e: Bachelor of the Art of Obstetrics
BÄO d: Bunesärzteordnung
B.A.-ohm e: British Association Ohm
BAOInsp. d: Bauamtsoberinspektor
BAO-MAO e: Basal Acid Output to Maximal Acid Output
BAOSekr. d: Bauamtsobersekretär
BAP d: Basis-Assemblerprogramm
- d: Basisanschlußpunkt
- d: bedingte Anforderung an die Programmsteuerung
- d: Benzyl-[p]-Aminophenol
- d: Bildschirmarbeitsplatz
- d: Branchenanwendungsprogramm
- e: Band Amplitude Product
- e: Baptism
- e: Base Auxiliary Power
- e: Basic Assembler Program[me]
- e: Basic Assembly Program[me]
- e: Battery Aiming Point
- e: Beacon Aircraft Position
- e: Beginning at a Point

- e: Benzo[a]pyrene
- e: Best Adaptive Path
- e: Binary Array Processor
- e: Biological Abstracts Previews
- e: Blood-Agar Plate
- e: Brachial Artery Pressure
- e: Bus Available Pulse

BAP & C e: British Association of Print & Copyshops
BAPCO e: Bahrain Petroleum Company
BAPCT e: Bachelor of Arts in Practical Christian Training
BAPD e: Base Air Post Depot
BAPE e: Bachelor of Arts in Physical Education
- e: Balloon Atmospheric Propagation Experiment
- e: Base Plate

BA Phys Med e: British Association of Physical Medicine
BAPL e: Base Assembly Parts List
- e: Bettis Atomic Power Laboratory

BAPM e: British Association of Physical Medicine
BAPN d: Beta-Amino-Propionitrite
BAPNA e: Benzoyl-Arginine-p-Nitroanilide
BAPP e: British Association of Pig Producers
BAppArts e: Bachelor of Applied Arts
BAppSc e: Bachelor of Applied Science
BAP REPT e: Beds and Patients Report
BAPS e: Beacon-Automated Processing System
- e: Beam and Plate System
- e: British Association of Pediatric Surgeons
- e: British Association of Plastic Surgeons
- e: Bureau of Air Pollution Sciences

BAPs e: Binary Array Processor[s]
BAPSA e: Broadcast Advertising Producers Society of America
BAPSI d: Bildschirm-Arbeitsplatz-Simulation
BAPT d: Bundesamt Post und Telekommunikation
Bapt. e: Baptist
BAPT e: Basic Applications Programmer Training
- e: British Association of Physical Training

BAPTA e: Bearing and Power Transfer Assembly
BAQ d: Barranquilla (Flughafen)
- e: Bachelor Airmens Quarters
- e: Basic Allowance for Quarters

Bar. d: Baracke
- d: Barbados
- d: Barometerstand

BAR d: Barometer
- d: Basisadreßregister
- d: Befehlsadreßregister
- d: Bereitadressenregister
- d: Berliner Außenring
- e: Babinet Absorption Rule
- e: B Adress Register
- e: Barometer
- e: Barometric
- e: Barometric[al]
- e: Baroque
- e: Barque
- e: Barrage
- e: Barrel
- e: Barrister
- e: Base Adder Register
- e: Base [Address] Register (computer)
- e: Basic Address Register
- e: Basic Army Regulations
- e: Battery Acquisition Radar
- e: Board of Anthropological Research
- e: Book Auction Records
- e: Bookkeeping Address Register
- e: Broadcast Advertisers Reports (journ.)
- e: Browning Automatic Rifle
- e: Budget Adjustment Request
- e: Buffer Address Register
- e: Bureau of Aeronautics Representative
- e: Bureau of Automotive Repair
- e: Butylacetyl Ricinoleate
- e: Byte Address Register

B.Ar. e: Bachelor of Architecture
BARA f: Bureau d'Analyse et de Recherche Appliquées
Barak f: Israeli version of the French Mirage military aircraft
Barastu-Verfahren d: Balcke-Rapid-Stufenverfahren [der Wasserenthärtung]
BARB d: Barbiturat
- d: Barbitursäure
- e: Barbados
- e: Barbarian
- e: Barbary Ape
- e: Barbary Horse
- e: Barbecue
- e: Barbiturate
- e: British Angular Rate Bombsight

BArb. d: Bundesbahnarbeiter
BArbBl. d: Bundesarbeitsblatt
BArbGG d: Bundesarbeitsgerichtsgesetz
BARBI e: Baseband Radar Bag Initiator
- e: Baseline Radar Bag Initiator

BARBIES e: Barbiturates
BAR BLN e: Barrage Balloon
Barbs. e: Barbiturates
barbus s: barbudos
Barc. d: Barcelona
Barc e: Barcelona
BARC e: Barge, Amphibious, Resupply, Cargo
- e: Bhaba Atomic Research Centre
- e: Bhaba Atomic Research Committee
- e: British Aeronautical Research Committee

BARCAP e: Barrier Combat Air Patrol
B Arc E e: Bachelor of Architectural Engineering
B Arch e: Bachelor of Architecture
B Arch (Arch) e: Bachelor of Architecture in Architecture
B Arch (Arch E) e: Bachelor of Architecture in Architectural Engineering
B Arch Des e: Bachelor of Architectural Design
B Arch E e: Bachelor of Architectural Engineering
B Arch in City Pl e: Bachelor of Architecture in City Planning
B Arch & TP e: Bachelor of Architecture and Town Planning
BARCS e: Battlefield Area Reconnaissance System
BARD e: British Database on Research for Aids for the Disabled
BARDS e: Bucknell Automated Retrieval and Display System
B Ar E e: Bachelor of Architectural Engineering
BA Reg e: Buffer A Register
Barents e: Barents Sea
Barg. e: Bargain
Barisans e: Barisan Mountains
BARITE e: Barium Sulfate
BARITT e: Barrier-Injection and Transit-Time
BARITT-Diode e: Barrier-Injection [and] Transit-Time Diode
Bark. d: Barkasse
BARN e: Biological Agricultural Reactor of the Netherlands
- e: Body Awareness Resource Network
- e: Bombing and Reconnaissance Navigation

BARNS e: Bombing and Reconnaissance Navigation System
BARO d: Barometer
- d: Barostat

BARON e: Business-Accounts Reporting Operating Network
BARONS e: Business-Accounts Reporting Operating Network System
BAROPS e: Barrier Operations
BAROSWITCH e: Barometric Switch
B & ARP e: Bare and Acid-Resisting Paint
BARP e: British Association of Radiology and Physiotherapy
Barpr. d: Barpreis
BARR e: Barrier
- e: Barrister
- e: British Association of Rheumatology and Rehabilitation

BARS e: Backup Attitude-Reference System
- e: Ballistic Analysis Research System
- e: Barrels
- e: Bell Audit Relate System
- e: British Association of Residential Settlements
- e: Budget Analysis Reporting System

BARSA e: Billing, Accounts Receivable, Sales Analysis
B Ar Sc e: Bachelor of Arts and Sciences
BARSR ru: Biblioteca Academiei Republicii Socialiste România
BARSTUR e: Barding Sands Underwater Test Range
- e: Barking Sands Tactical Underwater Range

BART e: Bay Area Rapid Transit
- e: Brooklyn Army Terminal

BAR TD e: Bay Area Rapid Transit District
BARTOB e: Bavarian Amateur Radio Translator on Balloon
Barts e: Saint Bartholomews Hospital in London
- e: Saint-Barthelemy in the French West Indies

BART System e: Bay Area Rapid Transit System
BARV e: Beach Armo[u]red Recovery Vehicle
BARZREX e: Bartoc Archives Z-symbol Rhythm Extraction
BAS d: Basalt
- d: Basis
- d: basisch
- d: Basophil
- d: Berichts- und Auskunftssystem
- d: Berufsaufbauschule
- d: Besetztanrufsucher
- d: Bibliotheks-Automatisierungs-System
- d: Bild-, Austast-, Synchronsignal
- d: Bild-Austast-Synchron
- d: Bildaustastsynchron-[Signal]gemisch
- d: Bildsignal mit Austastung und Synchronisierung
- d: blockorientiertes Automatisierungssystem
- e: Bachelor of Agricultural Science
- e: Bachelor of Applied Science
- e: Bachelor of Arts in Speech
- e: Basic Activity Subset
- e: Basic Air Speed
- e: Basic Allowance for Subsistence
- e: Basic Angle System
- e: Basil
- e: Basilica
- e: Basilicata

- e: Basophil[s]
- e: Basset
- e: Bass Strait
- e: Bastogne
- e: Bastrop
- e: Basuto
- e: Basutoland
- e: Beacon Airborne S-band
- e: Beam-Alignment Sensor
- e: Behavioral Approach Scale
- e: Bell Audit System
- e: Benzyl Analog[ue] of Serotonin
- e: Berkshire Archaeological Society
- e: Bessemer Acid Steel
- e: Bit Access Storage
- e: Bit Access Store
- e: Block Automation System
- e: Block of Assignment Statements
- e: Boatswain; Antisubmarine Service
- e: Bomb Alarm System
- e: Bomb Assembly Spares
- e: Books-Across-the-Sea
- e: Boolean Assignment Statement
- e: Boric Acid Solution
- e: Brazilian American Society
- e: Brewers Association of Scotland
- e: British Acoustical Society
- e: British American Standard
- e: British Antarctic Survey
- e: British Arachnological Society
- e: British Association Standard
- e: Buenos Aires
- e: Bulletin of the Atomic Scientists (journ.)
- e: Business Accounting Systems

BASA d: Bahn-Selbst[wähl]anschlußanlage
- e: British Atomic Scientists Association

BAS ADV TRA e: Basis Advance Training
Basa-Ue d: Bahnselbstanschlußanlage-Übertragung
BASB e: Body Acceleration Given Synchronously with the Heartard
- e: Bulimia Anorexia Self Help

BASBOL e: Basic Business-Oriented Language
BA Sc e: Bachelor of Agricultural Science
- e: Bachelor of Applied Science

BASc e: Bachelor of Arts and Science
BASC e: Base Activation Statistical Control
- e: Berlin Air Safety Center
- e: Booth American Shipping Corporation

BASCOM e: Beginners All-purpose Symbolic Instruction Code Compiler
BASD e: Basic Active Service Date
BASDU e: Barr and Stroud Dentrometer User
BASE e: Bank Americard Authorization System Exchange
- e: Basic Army Strategic Estimate
- e: Basic Semantic Element
- e: Brokerage Accounting System Elements
- e: Buildings, Antennas, Spans, Earth Formations

Base 16 e: Hexadecimal
Base 2 e: Binary
Base B e: Octal 10 Decimal
Ba Sec e: Base Section
BASEC e: British Approvals Service for Electric Cables
BASEDAC e: Beginners All-purpose Symbolic Instruction Code Data Acquisition
BASEEFA e: British Approvals Service for Electrical Equipment in Flammable Atmospheres
BASekr. d: Bauamtssekretär
BASEOPS e: Base Operations
BASEX e: Beginners All-purpose Symbolic Instruction Code Extension
BASEX Language e: Basic Experimental Language
BASF d: Badische Anilin- und Sodafabrik AG
BASH e: Body Acceleration Synchronous with Heart
B.A.S.I. e: Brown Animal Sanatory Institution
BASIC e: Basic Algebraic Symbolic Interpretive Compiler
- e: Basic Appraisal System for Incoming Components
- e: Basic Automatic Stored Instruction Computer
- e: Battle Area Surveillance and Integrated Communications
- e: Beginners Algebraic Symbolic Interpretive Compiler
- e: Beginners All-purpose Symbolic Instruction Code
- e: Biological Abstracts Subjects in Context
- e: British-American Scientific International Commercial
- e: British-American Scientific International Commercial [English]

Basic [English] e: British-American Scientific International Commercial [English]
BASICO e: Behavior Science Corporation
BASICPAC e: Basic Package Computer
- e: Basic Processor and Computer
- e: Battle Area Surveillance and Integrated Communications Processor and Computer

BASIC/PTS e: Beginner`s All-purpose Symbolic Instruction Code Paper Tape System
BASICTNG e: Basic Training
BASIC-X e: Beginners All-purpose Symbolic Instruction Code-Extended
BASIE e: British Association for Commercial and Industrial Education
BA-Signal d: Bild- [und] Austastsignal
BASIL e: Barclays (Bank) Advanced Staff Information Language [Basís]
basil. e: basilect[ic][al][ly]
BASIS d: Banken-Service-Informations-system
- d: Bausteine für Archivsystematiken und Informationssysteme
- d: Bausteinsystem für integrierte Steuerung
- d: bibliothekarisch[es]-analytisches System zur Informationsspeicherung
- e: Bank Automated Service Information System
- e: Battelle Automated Search Information System
- e: Bay Area Spatial Information System
- e: Bibliographic Author or Subject Interactive Search
- e: British Airways Staff Information System
- e: Budgetary and Scheduling Information System
- e: Bulletin of the American Society for Information Science
- e: Burroughs Advanced Statistical Inquiry System
- e: Burroughs Applied Statistical Inquiry System

BASIS-H e: BASIS-History
BA-SIS-H e: Bibliographic Author or Subject Interactive Search-History
BASIS-L d: bibliothekarisches-analytisches System zur Informationsspeicherung - Leihverkeh
BASIS-P e: BASIS Political Science, Public Administration, Urban Studies and International
BA-SIS-P e: Bibliographic Author or Subject Interactive Search-Political Science
BA-SIS-S e: Bibliographic Author or Subject Interactive Search-Sociology
BASIS-S BASIS e: Sociology
BASJE e: Bolivian Air Shower Joint Experiment
BASN e: Block-Acknowledged Sequence Number
BAS NET e: Basic Network
BASO d: Bandsortierprogramm
- e: Base Accountable Supply Officer
- e: Brigade Air Support Officer
- e: Bureau of Aeronautics Shipping Order

BASOPS e: Base Operating Information System
- e: Base Operations

BASOR e: Bulletin of the American Schools of Oriental Research
baso[s] e: basophile[s]
BASP d: Bild-, Austast-, Synchronsignal und Prüfzeile
- d: Bild-Austast-Synchron-Pegel

BA Sp e: Bachelor of Arts in Speech
BASP e: Backspace
- e: Biomedical Analog[ue] Signal Processor

BAS project e: Books-Across-the- Sea-project
B.A.S.R. e: Business Automation Specification Reports
BASRA e: British Amateur Scientific Research Association
B/ASRE e: British Admiralty Signal Radar Establishment
BAss. d: Bauassessor
- d: Bundesbahnassistent

BASS e: Bachelor of Arts in Social Studies
- e: Basic Analog[ue] Simulation System
- e: Behavioral and Social Sciences
- e: Best Available Shelter Survey

B As S[c] e: Bachelor of Association Science
BAssess. d: Bundesbahnassessor
BAS-Signal d: Bild-, Austast- [und] Synchronsignal
BAssist. d: Bauassistent
- d: Bundesbahnassistent

BASSR e: Bashkirian Autonomous Soviet Socialist Repubviet Socialists Republic
BASt d: Bundesanstalt für Straßenwesen
BAST e: Boric Acid Storage Tank
BASTEI d: Bankspeicherung technischer Informationen
BAS Timber e: Balasu Timber
B/A-Stücke d: Bekleidungs- und Ausrüstungsstücke
BASW e: Bell Alarm Switch
BAS. WT. e: Basic Weight
BASYS e: Basic System
- e: Building Aid System

BAT d: Bedingungsanzeigeteil
- d: Bundesangestelltentarif
- e: Basic Assurance Test
- e: Batch
- e: Battery
- e: Battle[ship]
- e: Behavioral Avoidance Test
- e: Best Available Technology
- e: Biological Abstract on Tape
- e: Biomedical Application Team[s]

- e: Blind Approach Training
- e: Boeing Air Transport
- e: Bolshoi All-azimuth Telescope
- e: Brown Adipose Tissue
Bat d: Batterie
Bat. e: Batavia
BATAB System e: Baker and Taylors Automated Buying System (for Libraries)
BATC e: Base Air Traffic Control
- e: Boing Atlantic Test Center
- e: British Amateur Television Club
- e: Burnside-Ott Aviation Training Center
BATCH e: Block Allocated Transfer Channel
- e: Block Allocator Transfer Channel
bat. chg. e: Battery Charger
- e: Battery Charging
BAT CO e: Battery Cut-Off
BATE e: Base Activation Test Equipment
- e: Base Assembly and Test Equipment
- e: Baseband Adaptive Transversal Equalizer
BATEA e: Best Available Technology Economically Achievable
- e: Best Available Technology Economically Available
BATES e: Ballistic Test and Evaluation Systems
- e: Ballistic Test Evaluation and Sealing
B-A Test e: Blood-Alcohol Test
BATEX e: Batch Executive
BAT FU e: Battery Fuse
BATH e: Best Alignment True Heading
- e: Best Available True Heading
- e: Boron-carbide, Alumin[i]um, Titanium Hybride
BA Theol e: Bachelor of Arts in Theology
Batho. d: Bathometer
BAT HOLD e: Battery Holder
BA thread e: British Association thread
Bathy. e: Bathymeter
- e: Bathyscaphe
- e: Bathysphere
BATM e: British Admiralty Technical Mission
BATO e: Balloon-Assisted Takeoff
BATP e: Bachelor of Arts in Town [and Country] Planning
- e: Boric Acid Transfer Pump
BATRECON e: Battle Reconnaissance
BAT rifle e: Battalion Anti-Tank rifle
BATROP e: Baratropic
BATRS e: British Amateur Tape Recording Society
BATRY e: Battery
BATS e: Ballistic Aerial Target System
- e: Basic Additional Teleprocessing Support
- e: Basic Additional Teleprocessing System
- e: Bit Access Test System
- e: British Additional Teleprocessing Support
BATSC e: Bay Area Transportation Study Commission
BATSHIP e: Battleship
BAT SUP e: Battery Supply
BATT d: Batterie[n]
- e: Batten
- e: Batteries
- e: Battery
BATT OPER e: Battery Operated
BATTR d: Batterie[n]
BATT.-Rel. d: Batterie-Relais
BATT VOLT e: Battery Voltage
BATTY e: Battery
BAU e: Bangladesh Agricultural Union
- e: Baseband Assembly Unit
- e: Basic Assembly Unit
- e: British Absolute Unit
- e: British Association Unit
Bau e: Bauhaus
Bau Ass[ess]. d: Bauassessor
BauaufsG d: Bauaufsichtsgesetz
Bauausf. d: Bauausführende[r]
- d: Bauausführung
Baubeschr. d: Baubeschränkung
- d: Baubeschreibung
BAUD e: Telecommunication Unit Measunng Speed of Signalling and Equal to one Code Element
- f: Baudot
BAUD Code e: Baudot Code
BauDir. d: Baudirektor
BauDVO d: Baudurchführungsverordnung
BauE d: Baumann-Einheit
B Au E[ng] e: Bachelor of Automobile Engineering
bauf. d: baufällig
Bauf. d: Bauführer
- d: Bauführung
Baufinzg. d: Baufinanzierung
Baufs. d: Betriebsaufseher
Baufzg. d: Baufinanzierung
BauG d: Baugesetz
Baugel. d: Baugelände
Baugen. d: Baugenehmigung
Baugest. d: Baugestaltung
BauGO d: Baugebührenordnung
Bauing. d: Bauingenieur
Bauj. d: Baujahr
Bauk. d: Baukasten
- d: Baukosten
- d: Baukunst
Baul. d: Bauland
- d: Bauleiter
BaulB[esch]G d: Baulandbeschaffungsgesetz
BaulG d: Baulandgesetz
- d: Baulenkungsgesetz
Baumasch. Bautech. d: Baumaschine und Bautechnik (journ.)
Bau NVO d: Baunutzungsverordnung
BauO d: Bauordnung
Baupfl. d: Baupflege
- d: Baupflicht
Baupl. d: Bauplan
- d: Baupläne
- d: Bauplatz
- d: Bauplätze
baupol. d: baupolizeilich
Baupr. d: Baupreis[e]
Bau PrüfV d: Bautechnische Prüfungsverordnung
Bau Prüf V[O] d: Bautechnische Prüfungsverordnung
Bau-R. d: Baurat
BAUS e: British Association of Urological Surgeons
BauStatG d: Baustatistikgesetz
Bautech[n]. d: Bautechnik[er]
bautechn d: bautechnisch
BAUTIS d: Bau-Technologie-Informationssystem
Bautr. d: Bauträger
Bau/UNL d: Baustahl; unlegiert
BauVorlV[O] d: Bauvorlagenverordnung
Bauw. d: Bauweise
- d: Bauwesen
BAV d: Bau- und Ausrüstungsverordnung
- d: Breitbandanschlußvermittlung
- d: Bürstenabhebevorrichtung
Bav e: Bavaria
- e: Bavarian
BAV l: Biblioteca Apostolica Vaticana
BAVA e: Bureau of Audio-Visual Aids
BAVE e: Bureau of Audio-Visual Educational
BAVIP e: British Association of Viewdata Information Providers
BAVTE e: Bureau of Adult, Vocational and Technical Education
Baw d: Bahnwärter
BAW d: Bundesanstalt für Wasserbau
- e: Back Artwork
- e: Bare Alumin[i]um Wire
- e: Beach Armored Recovery Vehicle
- e: Bulk Acousic Wave
- e: [n-]Butanol Acetic-acid Water
BAWAG d: Bayerische Wasserkraftwerke AG
BAWI e: Balance Agriculture with Industry Program[me]
BAWTR e: Babcock and Wilcox Test Reactor
BAX e: Beacon Airborne X-band
Bay d: Bayern
bay e: bayonet
Bay B[au]O d: Bayerische Bauordnung
BAYC e: Baxonet Candelabra
- e: Bayonet Candelabra
BAY CAND DC e: Bayonet Candelabra Double Contact
BAY CAND SC e: Bayonet Candelabra Single Contact
BayEBG d: Bayerisches Eisenbahn- und Bergbahngesetz
Bayer. Akad. d. Wiss. Philos-P d: Bayerische Akademie der Wissenschaften. Philosophisch-Philologische und Historische Abteilung
Bayer Akad Wiss Math-Naturw Ab d: Bayerische Akademie der Wissenschaften. Mathematisch-Naturwissenschaftliche Abteilung
Bayer Akad Wiss Philos-Hist Ab d: Bayerische Akademie der Wissenschaften. Philosophisch-Historische Abteilung
Bayr d: Bayreuth
BAYS e: British Association of Young Scientists
BAYSK e: Bayonet Skirted
Bay Str WG d: Bayerisches Straßen- und Wegegesetz
Bay WG d: Bayerisches Wassergesetz
BAZ d: Bandmarke aufzeichnen
- d: Bandmarkenaufzeichnung
- d: Batteriezusatz
- d: Bauaufsichtliche Zulassungen
- d: Bauzement
- d: Bearbeitungszentrum
- d: Bearbeitungszyklus
Baz. d: Bazillen
- d: Bazillus
BB d: Bandblock
- d: Basisband
- d: Baukastenbohreinheit
- d: Beobachtung
- d: Berufsberater
- d: Berufsberatung
- d: Beschleunigungsbetrieb
- d: Bestellbuch
- d: Betriebsberater
- d: Betriebsbereitschaft
- d: Beugungsbild
- d: Block-Basis-Schaltung
- d: Blutbild
- d: Braunkohlebrikett
- d: Breitbahn
- d: Breitband
- d: Brennelement-Bündel
- d: Buchungsbeleg

- d: Buchungsbescheinigung
- d: Bundesbahn
- d: Bundesbahnbetriebsinspektor
- d: Butan-Butylen-Gemisch
- d: Butane-Butylene
- e: Bachelor of Bacteriology
- e: Bachmanns Bundle
- e: Back-to-Back
- e: Back Bearing
- e: Backboard
- e: Back to back
- e: Bail Bond
- e: Bailey Bridge
- e: Balloon Barrage
- e: Band Block
- e: Bang-Bang
- e: Bank Burglar[y]
- e: Barbados
- e: Bare Base
- e: Base on Balls
- e: Basic Block
- e: Bayonet Base
- e: Beacon Buoy
- e: Bead Butt
- e: Beam Bolster
- e: Bearer Bond
- e: Beer Barrel
- e: Begin Block
- e: Begin Bracket
- e: Best of Breed
- e: Bill Book
- e: Binary Base
- e: Bit/Byte
- e: Black Body
- e: Black Bordered
- e: Black Brant
- e: Block Basis
- e: Blocking Back
- e: Blood Bank
- e: Blood Buffer [Base]
- e: Blow Back
- e: Blue Bloaters
- e: Blue Book
- e: Bomb
- e: Bomb Bay
- e: Boole & Babbage Company
- e: Both [Bones]
- e: Bottom Bounce
- e: Branch Bill
- e: Breadboard
- e: Bread Bulk
- e: Breaker Block
- e: Breaks Below
- e: Breakthrough Bleeding
- e: Breast Biopsy
- e: British Blue
- e: Broadband
- e: Bronchial Bars
- e: Brush Border
- e: Buffer Base
- e: Bulletin Board
- e: Bunch[ing] Block
- e: Bungling Bureaucrat
- e: Bureau of Biologics
- e: Burton and Bitter
- e: Bus-Bar
- e: Busy Bit
- e: Double Black

B.b. d: Backbord
Bb. d: Baubeschreibung
- d: Buchbinder[ei]

b.b. e: ball bearing
- e: baseband
- e: below bridges
- e: best best

B-B d: Boden-Boden
- e: Bogie-Bogie (single-unit locomotive)

B/B e: Bead / Bead
- e: Both to Blame

b/b e: bottled in bond
B & B e: Bell and Bell
BBA d: Bauelemente- und Bauzubehöranzeiger (journ.)
- d: Baugerät- und Baumaschinenanzeiger (journ.)
- d: Baumaterial- und Baufertigteileanzeiger (journ.)
- d: Behälter-Behandlungsanweisung
- d: Betriebsbedarfs- und Ausrüstungsanzeiger (journ.)
- d: Biologische Bundesanstalt
- d: Bürotechnik- und Büroeinrichtungsanzeiger (journ.)
- e: Bachelor of Business Administration
- e: Baseband Amplifier
- e: Born Before Arrival
- e: British Beekeepers Association
- e: Broad Band Aerial
- e: Broad Band Antenna

BBAB e: N-p-Butoxybenzylidene-p-Aminobenzonitrile
B Bac e: Bachelor of Bacteriology
BBAC e: British Balloon and Airship Club
- e: Bus-to-Bus Access Circuit

BB Ad[m] e: Bachelor of Business Administration
Bbados e: Barbados
BBahnG d: Bundesbahngesetz
BbAmtm d: Bundesbahnamtmann
BBass d: Bundesbahnbetriebsassistent
B-battery e: Plate battery
BBauBl d: Bundesbaublatt
BBauG d: Bundesbaugesetz
BBauref d: Bundesbahnbaureferendar
BBB d: Baggern, Bugsieren, Bergen
- d: Bezirksberatungsbeamter
- d: Börsenblatt für den Deutschen Buchhandel (journ.)
- d: Bundesstelle für Büroorganisation und Bürotechnik
- e: Bags, Barrels or Boxes
- e: Baseband Breadboard
- e: Basic Boxed Base
- e: Before Breaking Bulk
- e: Best Berlin Broadcast
- e: Best British Briar
- e: Bis-Benzimidazo Benzophenanthroline
- e: Blood Brain Barrier
- e: Bundle Branch Block

BB/B e: Body-Bound Bolts
BBBA e: British Bird Breeders Association
BBBl. d: Bundesbaublatt
BBBS e: Bang-Bang-Bang Surfaces
BBC d: Brombenzylzyanid
- e: Bachelor of Beauty Culture
- e: Bachelor of Building Construction
- e: Barrels, Boxes or Crates
- e: Basic Bibliographic Citation
- e: Battery Booster Cable
- e: Before Bottom Center
- e: Bermuda Base Command
- e: Binary Block Coding
- e: Black Body Cavity
- e: Blockhouse Battery Charger
- e: British Broadcasting Corporation
- e: Broadband Conducted
- e: Broadband Control
- e: Bromobenzyl Cyanide [Gas]
- e: Brown, Boveri und Cie
- e: Building Block Concept

BBC-1 e: British Broadcasting Corporations first television network
BBC-2 e: British Broadcasting Corporations second television network
BBC Cable e: Broadband Coaxial Cable
B.B.C.D. e: Before Bottom Dead Center
BBC Dissociation e: Braid-Berheim-Charcot Dissociation
BBCL e: Bermuda Broadcasting Company Limited
BBC-Nachr. d: BBC-Nachrichten, Mannheim
BBCRD e: British Broadcasting Corporation Research Department
BBCW e: Bare Beryllium Copper Wire
Bbd d: Breitband
BBD e: Baby Born Dead
- e: Back Bias Detector
- e: Beta-Binomial Distribution
- e: Blackboard
- e: Blackbord
- e: Bomb Bay Door
- e: Bucket Brigade Device[s]

BBDC e: Before Bottom Dead Center
BbDir. d: Bundesbahndirektor
BBDL e: Bucket Brigade Delay Line
BBDs e: Bucket Brigade Devices
BbdVr d: Breitbandverstärker
BBE d: Braunkohlebrikketteinheit
- e: Binary Block Encoding

BBed. d: Bahnbediensteter
B.-Bed. d: Bürobedarf
BB Ed e: Bachelor of Business Education
BBES e: Bang-Bang-Erection System
BBev. d: Bahnbevollmächtiger
BBF d: Becken-Bein-Fußgipsverband
- d: Berufsbildungsforschung

Bbf d: Betriebsbahnhof
BBF e: Balloon-Borne Filter
- e: Benzo[a]fluoranthene
- e: Biblioteca Benjamin Franklin
- e: Boron Based Fuel

BBFO e: Bardsey Bird and Field Observatory
BBFR e: Balloon-Borne Filter Radiometer
BBG d: Berufsbildungsgesetz
- d: Blätter für das Bayerische Gymnasial-Schulwesen (journ.)
- d: Bodenbearbeitungsgerät
- d: Bundesbahngesetz
- d: Bundesbaugesetz
- e: Berlin Border Guard
- e: Bermuda Botanical Gardens
- e: Board of Broadcast Governors
- e: Bouncing-Ball Generator
- e: Brooklyn Botanic Garden
- e: Bundle Branch Block

BBGA e: Belgian Begonia Growers Association
BBGKY e: Born-Bogolyubov-Green-Kirkwood-Yvon
BB/GMB d: VEB Bergmann-Borsig / Görlitzer Maschinenbau (DDR)
BBH d: Bergbauernhilfe
- e: Battalion Beachhead

BBHC e: Buffalo Bill Historical Center
B.B.H.S. e: Birmingham Botanical and Horticultural Society
BBHW e: Baseboard Hot Water
BBI e: Barbecue Briquet Institute
- e: Biscuit Bakers Institute
- e: British Bottlers Institute
- e: Builders Benevolent Institution

BBIA e: Billiard and Bowling Institute of America
B Bib Art e: Bachelor of Biblical Arts
B Bibl f: Bachelier en Bibliothèconomie
B Bi Ch[em] e: Bachelor of Biological Chemistry
B Bi E[ng] e: Bachelor of Biological Engineering
BBII e: Brass and Bronze Ingot Institute
BBIM e: Buoyant Ballistic Inertial Missile
BBIP e: British Books in Print
B Bi Phy e: Bachelor of Biological Physics
B Bi Sc e: Bachelor of Biological Sciences
B. Bisc. e: Bay of Biscay
BBJ e: Ball Bearing Joint
BBK d: bibliothekarisch-bibliographische Klassifikation
- d: Breitbandkommunikation
- d: Brown Boverie Krupp-Reaktorbau
- d: Buchführung-Bilanz-Kostenrechnung
- d: Burstbeginnkennzeichen
- e: Bread-Board Kit
BBKA e: British Bee Keepers Association
BBKG d: Beiträge zur Bayerischen Kirchengeschichte (journ.)
BBKK d: Bundesbahnbetriebskrankenkasse
BBK-Kabel d: Breitbandkommunikations-Kabel
BBK-Netz d: Breitbandkommunikations-Netz
BB/KR d: Brown Boveri / Krupp Reaktorbau
BBL d: Beiblatt
- d: Bibliothek
- e: Barrel
- e: Basic Business Language
- e: Beacons and Blind Landing
- e: Branch Back and Link
- e: British Barrel
- e: Buried Bit Line
BB Lamp e: Bayonet Base Lamp
BBldg e: Bachelor of Building
BBldgSc e: Bachelor of Building Science
BBLDir d: Bibliotheksdirektor
BBLI d: Bibliotheksinspektor
BBLOI d: Bibliotheksoberinspektor
BBLOR e: Bibliotheksoberrat
BBLR d: Bibliotheksrat
BBLS e: British Barrels
BBLS/D e: British Barrels per Day
BB Lumber e: B and Better Lumber
BBM d: Beförderungsbedingungen des deutschen Möbeltransports
- e: Bachelor of Business Management
- e: Books by Mail
- e: Break-Before-Make
- e: Building Block Monochromator
BBN d: Baubedarfsnachweis[ung]
- e: Balloon-Borne Nephelometer
BBNNR e: Braunton Burrows National Nature Reserve
BB No e: Bromide-Bromate Number
BBNP e: Big Bend National Park
BBNR e: Back Bay National Refuge
BBO d: Bugsierboot
- d: Büromaschinen, Büromöbel, Organisationsmittel e.V.
- e: Binary Boolean Operation
- e: Booster Burn-Out
B-Body e: Bingham Body
BBOI d: Bundesbahnoberinspektor
BBOL e: Building Block Oriented Language
B-Bomb e: Benzedrine Bomb (Slang-Benzedrine Inhaler)
BBOR d: Bundesbahnoberrat
BBOT e: 1,5-Bis[2-(5-tert.-Butylbenzoxazolyl)] Thiophene
BBP d: Benzylbutylphthalat
- d: Briefbahnpost
- d: Butylbenzylphthalat
- d: Butyl Benzyl Phthalate
- e: Batch Bulk Processing
- e: Bavarian Border Police
- e: Beam Bolster with Plate
- e: Boxes, Barrels or Packages
- e: Building Block Principle
- p: Boletim de Bibliografa Portuguesa (journ.)
BBPN e: Balloon-Borne Polar Nephelometer
BBPS e: Behaviour-Based Personnel Systems
- e: Bureau of Biological and Physical Sciences
BBR d: Babcock Brown-Boveri Reaktorbau GmbH
- d: Bänder Bleche Rohre (journ.)
- d: Bedarfs- und Bestellrechnung
- d: Bentbromarone
- d: Bundesbahnrat
- e: Balloon-Borne Radio
- e: Benzbromarone
- e: Black Body Radiator
- e: Broadband Radiated
BB Reg e: Buffer B Register
BBRG e: Ball Bearing
BBRL e: British Branch of the Radiation Laboratory
BBRP d: Berliner Beiträge zur Romanischen Philologie (journ.)
BBRR e: Brookhaven Beam Research Reactor
BBRS e: Balloon-Borne Radio System
BBRT d: Betriebsbruttotonne
BBS d: Band-Betriebssystem
- d: Bang-Bang Servo
- d: Basis-Betriebssystem
- d: Bedienungsblattschreiber
- d: Betriebsberufsschule (DDR)
- d: Betriebsblattschreiber
- e: Bachelor of Business Science
- e: Ball Bearings
- e: Barber Suggestibility Scale
- e: Bare Base Set
- e: Barrels of Basic Sediment
- e: Bermuda Biological Station
- e: Box Bark Strips
- e: British Bryological Society
- e: Brunei Broadcasting Service
- e: Bulletin Board System
- e: Bus-Bars
- e: Business Batch System
BB Sc e: Bachelor of Business Science
BB-Schaltung d: Block-Basis-Schaltung
BBSE e: Board of Behavioral Science Examiners
BBSI e: Beauty and Barber Supply Institute
- e: British Boot and Shoe Institute
BBSJ e: Ball Bearing Swivel Joint
BB Socket e: Bayonet Base Socket
BBSP e: Balloon-Borne Solar Pointer
- e: Building Block Signal Processor
BBSR e: Bermuda Biological Station for Research
- e: Bucket Brigade Shift Register
BBSRs e: Bucket-Brigade Shift Registers
BBS & W e: Barrels of Basic Sediment and Water
BBT d: Bundesverband Bürotechnik
- e: Ball Bearing Torque
- e: Baombardement
- e: Basal Body Temperature
- e: Bombardment
- e: Buck-Boost Transformer
BB & TC e: Bahamas Broadcasting and Television Commission
BB Technique e: Branch and Bound Technique
BBtm d: Bundesbahnbetriebsmeister
B Btry e: B Battery
- e: Counterbattery
BBU e: Bam Bolster Upper
- e: Baseband Unit
- e: Battery Backup
- e: Beefmaster Breeders Universal
- e: Blue-Brick University
BBUG e: Bode & Babbage Users Group
bb.-unb. d: bedingt beständig bis unbeständig
B Bus Ad e: Bachelor of Business Administration
Bbv d: Bahnbevollmächtigter
BBV d: Breitbandvermittlungsstelle
BBW d: Berliner Bauwirtschaft (journ.)
- d: Bundesbahnbetriebswart
- d: VEB Berliner Bremsenwerk (DDR)
- e: Bare Brass Wire
BBWC e: Broad-Band Waveguide Circulator
BBWG e: Broad Band Waveguide Circulator
BBY e: Brush Back Yearly
BBYCY e: Bow Buoyancy
BBz d: Baubezirk
- e: Bearing Bronze
BBZ d: Berbaubedarf-Beschaffungszentrale
- d: Beregnungsbedürftigkeitswertzahl
BC d: Bedienungscode
- d: Befehlscode
- d: Binärcode
- d: Bürocomputer
- e: Bachelor of Chemistry
- e: Bachelor of Classic
- e: Bachelor of Commerce
- e: Back Connected
- e: Back Course
- e: Bacone College
- e: Bad Check
- e: Balance Card
- e: Balance Column
- e: Balance Computer
- e: Balance Control
- e: Balance Counter
- e: Balanced Current
- e: Balliol College [Oxford]
- e: Ballistic Camera
- e: Ballistic Coefficient
- e: Balloon Command
- e: Bank Clearing
- e: Bar Chair
- e: Bard College
- e: Bare Copper
- e: Barge, Cargo
- e: Barium Crown
- e: Barnard College
- e: Barrel Coating
- e: Barrington College
- e: Barry College
- e: Baruch College
- e: Base Collector
- e: Base Command
- e: Base Connection
- e: Base Count
- e: Basic Code
- e: Basic Component

- e: Basic Computer
- e: Basic Control[ler]
- e: Batched Compilation
- e: Bates College
- e: Bathyconductograph
- e: Battery Charger
- e: Battery Commander
- e: Battery Control
- e: Battle Casualities
- e: Battle Commander
- e: Battle Cruiser
- e: Bayonet Cap
- e: Bayonet Catch
- e: Beam Catcher
- e: Beam Collimator
- e: Bearer Capability
- e: Beaver College
- e: Beckley College
- e: Before Christ
- e: Before Computer
- e: Begin Column
- e: Beginning Character
- e: Beginning Climb
- e: Belgian Congo
- e: Belhaven College
- e: Bellarmine College
- e: Bell Cord
- e: Belmont College
- e: Beloit College
- e: Benedict College
- e: Bennett College
- e: Bennington College
- e: Berea College
- e: Berry College
- e: Bethany College
- e: Bethel College
- e: Between Centers
- e: Bias Cell
- e: Bibliographic Classification
- e: Bidecimal Code
- e: Bidirectional Counter
- e: Bin[ary] Card
- e: Binary Cell
- e: Binary Channel
- e: Binary Character
- e: Binary Code[d]
- e: Binary Coding
- e: Binary Command
- e: Binary Conversion
- e: Binary Count[er]
- e: Bio-Conversion
- e: Biological Conservation
- e: Biotin-Carboxylase
- e: Biquinary Code
- e: Birth Certificate
- e: Birth Control
- e: Bishop College
- e: Bit Capacity
- e: Bit Chain
- e: Bit Check
- e: Bit Code
- e: Bit Combination
- e: Bit Configuration
- e: Blackburn College
- e: Blank Card
- e: Blank Character
- e: Blank Column
- e: Blank Command
- e: Blanking Circuit
- e: Blind Copy
- e: Blinn College
- e: Bliss College
- e: Block Check
- e: Block Command
- e: Block Count[er]
- e: Bloomfield College
- e: Bluefeld College
- e: Bluffton College
- e: Body-Centered
- e: Bogus Check
- e: Bolometric Correction
- e: Bolt Circle
- e: Bomber Command
- e: Bonded [single] Cotton
- e: Bone Conduction
- e: Bone Connection
- e: Bookcase
- e: Boolean Calculation
- e: Boolean Coding
- e: Boolean Complementation
- e: Boresight Camera
- e: Boron Carbon
- e: Boston College
- e: Bottom Center
- e: Bottom Chord
- e: Bottom Contour
- e: Boundary Condition
- e: Bourget College
- e: Bowdoin College
- e: Boxes or Crates
- e: Branch and Condition
- e: Branch Conditional
- e: Branch on Condition
- e: Brandon College
- e: Brasenose College
- e: Breaking Capacity
- e: Brenau College
- e: Brentwood College
- e: Brescia College
- e: Brevard College
- e: Briarcliff College
- e: Bridgewater College
- e: Brightness Contrast
- e: Bristol Channel
- e: British Candle
- e: British Columbia
- e: British Commonwealth
- e: British Council
- e: British Currency
- e: Broadcast [Control]
- e: Broadcasting
- e: Bronchus-Carcinom
- e: Brooklyn College
- e: Bruyere College
- e: Bryant College
- e: Bubble Chamber
- e: Buccal Cavity
- e: Buccocervical
- e: Bucking Coil
- e: Budgeted Cost
- e: Buffer[ed] Computer
- e: Buffer Cell
- e: Buffer Cycle
- e: Building Center
- e: Building Code
- e: Bulk Cargo
- e: Bulk Cement
- e: Bulk Core
- e: Bulkhead Connector
- e: Burden Center
- e: Burdett College
- e: Bursa Copulatrix
- e: Bursting Charge
- e: Bus Compatible
- e: Bus Coupler
- e: Business Computer
- e: Butler College
- e: Byte Counter

B/C e: Bales of Cotton
- e: Bill[s] for Collection

BCA d: Blätter für Christliche Archäologie und Kunst (journ.)
- e: Bachelor of Commercial Arts
- e: Basic Channel Adapter
- e: Battery Commander Assistant
- e: Battery Control Area
- e: Best Cruise Altitude
- e: Bilderberg Continuum Atmosphere
- e: Binary Coded Address
- e: Bisynchronous Communications Adapter
- e: Blood Colour Analyzer
- e: Bovine Carbonic Anhydrase
- e: British Caledonian Airways
- e: Buffered Communications Adapter
- s: Biblioteca

BCAA e: Branched-Chain Amino Acids
- e: British Columbia Automobile Association

BCAB e: Birth Control Advisory Bureau
- e: British Computer Association for the Blind

BCAC e: Biology Classroom Activity Checklist
- e: British Conference on Automation and Computation

BCAL e: British Caledonian Airways

BCALA e: Black Caucus of the American Library Association

BCAM e: Basic Communication Access Method
- e: Breast Cancer Awareness Month (journ.)

BC AMPL e: Broadcast Amplifier

B Can L e: Bachelor of Canon Law

BCAP e: Beacon CollisionAvoidance Program[me]

BCAR e: British Civil Aircraft Requirements
- e: British Civil Airworthiness Requirement[s]

BC-ARS e: Birth Control-Abortion Referral Service

BCARs e: British Civil Aircraft Requirements

BCAS e: Beacon[-compatible] Collision-Avoidance System
- e: British Compressed Air Society

B. Casters e: Broadcasters

BCAT e: Birmingham College of Advanced Technology

BCAVM e: British Catalogue of Audiovisual Materials

BCB e: Base Current Branch
- e: Battery Control Building
- e: Binary Code Box
- e: Bit Control Block
- e: Block Control Byte
- e: Broadcast Band
- e: Button Cell Battery

BCBH e: Boiler Casing Bulkhead

BCC e: Basis-Computer-Center
- e: Ballistic Camera Control
- e: Battery Control Center
- e: Battery Control Central
- e: Beacon Control Console
- e: Beam Coupling Coefficient
- e: Behavio[u]r Classification List
- e: Benard Convection Cell
- e: Berkshire Community College
- e: Bi-Cylindrical-Conical
- e: Binary Code Conversion
- e: Binary Coded Character
- e: Binary Convolutional Code
- e: Binary Convolution Code
- e: Birth Control Campaign
- e: Birth Control Clinic
- e: Blind Carbon Copy
- e: Blind Courtesy Copy
- e: Block-Completed Counter
- e: Block Character Check
- e: Block Check[ing] Character
- e: Block Coded Decimal
- e: Body-Centered-Cubic

- e: Body-Centered Cubic
- e: British Communications Corporation
- e: British Crown Colony
- e: Broadcast Control Center
- e: Broadcasting Corporation of China
- e: Bronx Community College
- e: Burlington Community College
- e: Business Computer Center

BCCBP s: Biblioteca de Cataluña y Central de Biblioteca Populares
BCCC e: Ballistic Compressor Computer Code
BCCCU e: Biomedical Computing Council of California Universities
BCCD e: Bulk-channel Charge-Coupled Device
- e: Bulk Charge Coupled Device
- e: Buried-channel Charge-Coupled Device

BCCDs e: Bulk-channel Charge Coupled Devices
BCCG e: British Cooperative Clinical Group
BCCO e: Base Consolidation Control Officer
BCCP e: Biotin-Carboxyl-Carrier-Protein
- e: Boeing Contract Change Proposal

BCCS e: Body-Centered Cubic System
BCCW e: Bare Copper Clad Wire
BCD d: Binärcode für Dezimalziffern
- d: binär codiert dezimal
- d: binär codierte Dezimale
- d: binär codierte Dezimalform
- d: binär codierte Dezimalziffer
- e: Bad Conduct Discharge
- e: Battle Correlator Display
- e: Behind Completion Date
- e: Binary Coded Data
- e: Binary Coded Decimal
- e: Binary Coded Digit
- e: Binary Counted Decimal
- e: Blank Column Detection
- e: Blank Column Detector
- e: Borderline Between Comfort and Discomfort glare
- e: Burst Can Detector
- e: Burst Cartridge Detection
- e: Business Conditions Digest
- e: Business Cycle Developments

BcD e: Barrels per Calendar Day
BCD/B e: Binary Coded Decimal / Binary
BCDC e: Bay Conservation and Development Commission
- e: Binary-Coded Decade Counter
- e: Binary Coded Decimal Counter

BCD Code e: Binary Coded Decimal Code
BCDCs e: Binary Coded Decade Counters
BCDD e: Binary Coded Decimal Digit
- e: British Chemical Defence Department

BCD Device e: Blank Column Detection Device
BCD Digit e: Binary Coded Decimal Digit
BCDDP e: Breast Cancer Detection Demonstration Project
BCDF e: B-Cell Differentiation Factor
BCDIC e: Binary Coded Decimal Information Code
- e: Binary Coded Decimal Interchange Code

BC Dis e: Bad Conduct Discharge
BCDNAF e: Binary Coded Decimal Non-Adjacent Form
BCD Notation e: Binary Coded Decimal Notation
BCD Number e: Binary Coded Decimal Number
BCDP e: Battery Control Data Processor
- e: Bubble Chamber Data Processing

BCD/Q e: Binary Coded Decimal / Quarternary
BCDS e: Board Command Data Sort
BCD System e: Binary Coded Decimal System
BCDTA e: British Chemical and Dyestuffs Traders Association Ltd.
BCD Translation e: Binary Coded Decimal Translation
BCE e: Bachelor of Chemical Engineering
- e: Bachelor of Christian Education
- e: Bachelor of Civil Engineering
- e: Barium Cloud Experiment
- e: Basal Cell Epithelioma
- e: Base Checkout Equipment
- e: Base Civil Engineer
- e: Basic Comparison Element
- e: Beam Collimation Error
- e: Before the Christian era
- e: Bench Checkout Equipment
- e: Boolean-Controlled Element[s]
- e: Branch Command Execution
- e: British Columbia Electric
- e: Bubble Chamber Experiment
- e: Bundle Controlled Expansion

Bce s: Belice
BC Ed e: Bachelor of Commercial Education
BCED e: Bibliography of the Computer in Environmental Design
BCEE e: Belgian Committee for Electroheat and Electrochemistry
- e: Bis (2-Chloroethyl) Ether

B Ce Eng e: Bachelor of Cement Engineering
BCEIUA e: Building and Civil Engineering Industries Uniformity Agreement
BCEM e: Bureau of Community Environmental Management
B Cer E[ng] e: Bachelor of Ceramic Engineering
BCEX e: Bundle Controlled Expansion
BCF e: Bachelor of City Forestry
- e: Bandpass Crystal Filter
- e: Basic Control Frequency
- e: Battle Cruiser Fleet
- e: Battle Cruiser Flotilla
- e: Battle Cruiser Force
- e: Beam Correction Factor
- e: Billion Conductor Feet
- e: Bromochloride-Fluoromethane
- e: Bromochlorodifluoromethane
- e: Bulked Continous Fibre
- e: Bureau of Commercial Fisheries

BCF Extinguisher e: Bromchloride Fluormethane Fire Extinguisher
BCFGA e: British Columbia Fruit Growers Association
BCFS e: British Columbia Forestry Society
BCFSK e: Binary-Code[d] Frequency-Shift Keying
BCFT e: Beechcraft
BCFTE e: British Commonwealth Forest Translation Exchange
BCG d: Bacillus von Calmette und Guérin
- d: Ballistocardiogramm
- d: Binet-Cauchy-Gesetz
- e: Bacillus Calmette-Guérin
- e: Ballistocardiogram
- e: Bidirectional Categorical Grammar
- e: British Gas Council
- e: Bromcresol Green
- e: Bucking Current Generator

BC generation e: Back-Cross generation
B-CGF e: B-cell Growth Factor
BCGNM e: Black Canyon of the Gunnison National Monument
BCG test e: Bicolo[u]r Guaiac test
BCH e: Bachelor of Chemistry
- e: Basal Cell Hyperplasia
- e: Bicycloheptadien
- e: Binary Coded Hollerith
- e: Bits per Circuit per Hour
- e: Block Control Header
- e: Bose, Chandhuri and Hocquenghem
- e: Bunch

BCh e: Bristol Channel
BCHA e: British Columbia Health Association
B Channel e: Right Stereo Channel
BCH Code e: Bose, Chandhuri and Hocquenghem Code
B Chem e: Bachelor of Chemistry
B Ch[em] E[ng] e: Bachelor of Chemical Engineering
B Ch Eng e: Bachelor of Science in Chemical Engineering
BCHG e: Bunching
Bchl e: Bacteriochlorophyll
Bchl b e: Bacteriochlorophyll b
BCHNR e: Batch Number
BCHPA e: British Columbia Hydro and Power Authority
BCHPH e: British Columbia Hydro and Power Authority
BCHP Program[me] e: Board Command History Purge Program[me]
B Chr Ed e: Bachelor of Christian Education
B Chrom e: Bachelor of Chromatic
Bch[s]t. d: Buchstabe
BCH Timber e: Black Chuglam Timber
BCI d: binär codierte Information
- e: Balance Computer Index
- e: Barro Colorado Island
- e: Basic Command Interpreter
- e: Battery Condition Indicator
- e: Battery Council International
- e: Binary Coded Information
- e: Biomedical Communications Inventory
- e: Birth Control Institute
- e: Bit Count Integrity
- e: Bituminous Coal Institute
- e: Boxcarintegrator
- e: Brazilian Coffee Institute
- e: Broadcast Interference

B.C.I. e: Butchers Charitable Institution
BCIA e: Bounded Carry Inspection Adder
- e: British Columbia Institute of Agrologists

BCIE e: Bis (2-Chloroisopropyl) Ether
BCIP e: Belgium Centre for Information Processing
BCIS e: Binary Constitution Information Service
BCIs e: Boxcarintegrators
BCISC e: British Chemical Industrial Safety Council
BCIU e: Bus Control Interface Unit
BCJ e: Broadcasting Corporation of Japan
BCK d: Block
BCKT e: Becket
BCL e: Bachelor of Cannon Law
- e: Bachelor of Civil Law
- e: Bachelor of Commercial Law
- e: Base Coupled Logic
- e: Basic Command Language
- e: Basis Contour Line
- e: Basis Counter Line
- e: Behavio[u]ral Checklist
- e: Belfast City Library
- e: Bicycle
- e: Biological Computer Laboratory
- e: Books for College Libraries (journ.)
- e: British Council Library

- e: Broadcast Listener
- e: Broom Closet
- e: Burroughs Common Language
- e: Bus Clear
- e: Business Computer Limited

BCLA e: Birth Control League of America
BCLDD-MOSFET e: Buried-Channel Lightly Doped Drain-Metal Oxide Semiconductor Field Effect Transistor
b/cm d: Bit pro Zentimeter
BCM e: Bachelor of Church Music
- e: Balance Calibration Machine
- e: Balanced Crystal Mixer
- e: Ballistic Correction of the Moment
- e: Basic Control Memory
- e: Basic Control Monitor
- e: Battery Control and Monitor
- e: Best Cruise Mach Number
- e: Beyond Capability of Maintenance
- e: Binary Choice Model
- e: Binary Coded Matrix
- e: Binary Code Memory
- e: Binary Conversion Matrix
- e: Blunt Conical Model
- e: Body Cell Mass
- e: Bougainville Copper Mine
- e: Brightest Cluster Member
- e: British Catalogue of Music
- e: British Commercial Monomark
- e: British Consular Mail
- e: Business Calculating Machine
- e: Business Center Map

BCME e: Bi-Chlormethyl Ether
- e: Butylcarbamic Acid; Methyl Ester
- e: Chlormethyl Ether

BCML e: Burroughs Current Mode Logic
BCMN f: Bureau central de measures nucléaires
BC Mode e: Basic Control Mode
BCMOS e: Bipolar Complementary [-symmetry] Metal Oxide Semiconductor
- e: Buried Channel Metal Oxide Semiconductor

BCMOSs e: Buried Channel Metal Oxide Semiconductors
BCMR e: Bidirectional Current Monitoring Relay
BCN e: Beacon
- e: Binary Coded Notation
- e: Binary Coded Number
- e: Biomedical Communications Network
- e: British Commonwealth of Nations

BCNP e: Bryce Canyon National Park
BCNRA e: Bighorn Canyon National Recreation Area
BCN System e: Binary Coded Number System
BCNU e: Bis-Chlorethyl-Nitroso-Urea
BCNY e: Brazilian Center of New York
BCNZ e: Broadcasting Corporation of New Zealand
BCO d: Binärcode für Oktalziffern
- e: Basic Calculating Operation
- e: Battery Control Officer
- e: Battery Cut-Off
- e: Binary Coded Octal
- e: Booster [engine] Cut-Off
- e: Bridge Cut-Off

BCOA e: Bituminous Coal Operators Association
BCOB e: Booster Cut-Off Backup
BCOE e: Bench Checkout Equipment
B Coeff e: Block Coefficient
B/C of A e: British College of Aeronautics
BCol e: Book Collector (journ.)
BCOM e: Burroughs Computer Output [to] Microfilm
B Com Adm e: Bachelor of Commercial Administration
B Com[m] e: Bachelor of Commerce
B Com Sc e: Bachelor of Commercial Science
B-Contact e: Back Contact
- e: Break Contact (of the Contactor)

BCP d: Benzylchlorophenol
- e: Bachelor of City Planning
- e: Basic Control Program[me]
- e: Basket Coupled Pliers (slang)
- e: Batch Communications Program[me]
- e: Battery Command Post
- e: Beam Candle Power
- e: Blue Cross Plan
- e: Bootstrap Commissioning Program[me]
- e: Bromocresyl Purple
- e: Budget Change Proposal
- e: Byte Control Protocol

BCPA e: British Copyright Protection Association
BCPE e: Board of Certification in Professional Ergonomics
BC-Piston e: Boost Control Piston
BCPL e: Basic Combined Programming Language
- e: Basic Computer Programming Language
- e: Bootstrap Combined Programming Language

BCPM e: British Columbia Provincial Museum
BCPMA e: British Chemical Plant Manufacturers Association
BCPN e: N-Butyl-N-(3-Carboxypropyl) Nitrosamine
BCPS e: Beam Candle Power Second[s]
BCPSM e: Board of Certifed Product Safety Management
BCQ e: Book-Collectors Quarterly (journ.)
B Cr e: Bachelor of Criminology
BCR e: Badge Card Reader
- e: Battery Charge Regulator
- e: Battery Control Radar
- e: Best Critical Region
- e: Billing-Collecting-Remitting
- e: Bituminous Coal Research
- e: Boron-Carbon Resistor
- e: Branch and Count Register
- e: British Columbia Railway
- e: Byte Count Register
- f: Bureau communautaire de références

BCRI e: Bituminous Coal Research Institute
- e: British Ceramic Research Institute

B Crk e: Bell Crank
BCRLRA e: Beverage Container Recycling and Litter Reduction Act
BCROS e: Balanced Capacitor Read Only Storage
BCRS e: Bar-Code Reader / Sorter
BCRST e: Bibliography on Cold Regions Science and Technology
BCRT e: Binary Coded Range Time
BCRTS e: Binary Coded Range Time Signal
BCRU e: British Committee on Radiological Units
BCRUM e: British Committee on Radiation Units and Measurement[s]
BCRV e: Blunt Conical Reentry Vehicle
BCS e: Bachelor of Chemical Science
- e: Bachelor of Commercial Science
- e: Bank Control System
- e: Banking Communication System
- e: Bardeen-Cooper-Schrieffer
- e: Basic Combined Subset
- e: Basic Communications Support
- e: Basic Computer System
- e: Basic Contract Specification
- e: Basic Control System
- e: Basic Court System
- e: Battery Computer System
- e: Beam Communication Set
- e: Bidirectional Category System
- e: Binary Synchronous Communications
- e: Biocomputer Statistics
- e: Biomedical Computing Society
- e: Blip Counter System
- e: Block Check Sequence
- e: Block Control Signal
- e: Body-Centered Tetragonal
- e: Boeing Computer Services
- e: Bombing Computer Set
- e: Bounded Cellular Space
- e: Branch and Condition Status
- e: Bridge Control System
- e: British Calibration Service
- e: British Computer Society
- e: Broadcast Communications System
- e: Buffer Cycle Select
- e: Business Communications Sciences, Incorporated
- e: Business Communication System
- e: Business Computer System
- e: Business Control System

BCs e: Battery Chargers
BCSAA e: British Computer Society Algorithmic Language Association
BC Sc e: Bachelor of Commercial Science
BCSC e: British Columbia Steamship Company
BC Se e: Bachelor of Commercial Service
BCSN e: Block-Completed Sequence Number
BCSOC e: Binary Convolutional Self-Orthogonal Code
BCSP e: Board of Certified Safety Professionals
BCS Program[me] e: Board Command Sort Program[me]
BCST e: Broadcast
BCS-Theory e: Bardeen-Cooper-Schrieffer Theory
BCSTN e: Broadcasting Station
BCT e: Babesiasis Card-agglutination Test
- e: Bachelor of Christian Training
- e: Bandwidth Compression Technique
- e: Basic Combat Training
- e: Between Commands Testing
- e: Binary Coded Tape
- e: Body-Centered Tetragonal
- e: Branch on Count
- e: Brown-Curtis Turbine

BCT e: Bushing Current Transformer
BCTIC e: Biomedical Computing Technology Information Center
BCTS e: Byte Channel Transfer Switch
BCU e: Basic Counter Unit
- e: Bay Cabinet Unit
- e: Big Close-Up
- e: Binary Counting Unit
- e: Block Control Unit
- e: Branch Control Unit
- e: Buffer Control Unit
- e: Bus Control Unit

BCUA e: Business Computers Users Association
BCUG e: Bilateral Closed User Group
BCUGO e: Bilateral Closed User Group with Outgoing Access
BCURA e: British Coal Utilization Research Association
BCUSA e: Buddhist Center of the United States of America

BCV e: Ball Check Valve
- e: Basal Cerebral Vigilance
- e: Black Cancer Virus
- e: Brightness Contrast Value
BCVX e: Biconvex
BCW e: Bare Copper Wire
- e: Binary Code Word
- e: Bobbin Coil Winder
- e: Brush Clean and Withe
- e: Buffer Control Word
- e: Bureau of Child Welfare
BCWA e: British Cotton Waste Association
BCWD e: Biological and Chemical Warfare Division
BCWL e: Basic Carbonate White Lead
- e: Binary Code Word Length
BCWMA e: British Chemical Ware Manufacturers Association
BCWP e: Budgeted Cost for Work Performed
BCWS e: Budgeted Cost for Work Scheduled
BD d: Band
- d: Banddurchhang
- d: Basis-Datenverarbeitung
- d: Begrenzer / Demodulator
- d: Belegdrucker
- d: Bergungsdampfer
- d: Besatzdichte
- d: Bescheiddienst
- d: Bewegungsdetektor
- d: Bezugsdämpfung
- d: binärer Dekoder
- d: Bitdichte
- d: Blattdrucker
- d: Blitzductor
- d: Blitzschutzdiode
- d: Blockdiagramm
- d: Blutdruck
- d: Brenndauer
- d: Bulldozer
- d: Bundesrepublik Deutschland
- e: Bachelor of Divinity
- e: Backward Definition
- e: Bangladesh
- e: Bar Draft
- e: Bar Draught
- e: Barrels per Day
- e: Base Deflection
- e: Base Depot
- e: Base Detonating
- e: Base Diameter
- e: Base Down
- e: Basic Data
- e: Battle Dress[ing]
- e: Baud[s]
- e: Bead[ed]
- e: Beam Degrader
- e: Beginning Descent
- e: Behavio[u]ral Differential
- e: Belladonna
- e: Below Day
- e: Bend Down
- e: Bernoulli Disk
- e: Bibliographic Data
- e: Bill[s] Discounted
- e: Binary Decode[r]
- e: Binary Digit
- e: Binary Display
- e: Binary Divide[r]
- e: Bit Density
- e: Blank Display
- e: Blastomyces Dermititis
- e: Block Data
- e: Block Delete
- e: Block Diagram[me]
- e: Blowing Dust
- e: Board
- e: Bold
- e: Bomb Disposal
- e: Bond[ed]
- e: Booking Data
- e: Boolean Dependency
- e: Boom Defense
- e: Booster Development
- e: Brass Disc
- e: Broad
- e: Bronchidesmus
- e: Brought Down
- e: Buccodistal
- e: Bud Duct
- e: Bulk Density
- e: Bundle
- e: Bursal Dependent
- e: Bus Driver
- e: Bus Duct
- e: Business Data
- e: Butadiene
- e: Byte Device
bd a: bind
- d: baumaschinendienst (journ.)
Bd d: Bad
- d: Bandgerät
- d: Baud
- d: Breitendurchmesser
- e: Bound[ary]
Bd. d: Blechdose
- d: Bund
- d: Bündel
- e: Boulevard
B/D e: Bank Draft
- e: Binary to Decimal
- e: Blood Drainage [Ratio]
- e: Braided
B & D e: Black & Decker
BDA e: Bachelor of Domestic Arts
- e: Bachelor of Dramatic Arts
- e: Battle Damage Assessment
- e: Bermuda [Tracking Station]
- e: Binary Digit Adder
- e: Block Decoder Assembly
- e: Bomb Damage Assessment
- e: Booster-Distribution Amplifier
- e: Booster Distribution Amplifier
- e: Breakdown Acid
- e: Business Data Analysis
BDAC e: Bureau of Drug Abuse Control
B & DAF e: Bounded and Described as Follows
BDAM e: Basic Data Access Method
- e: Basic Direct Access Method
B Day e: Barbarossa Day
BDB d: Börsenblatt für den Deutschen Buchhandel (journ.)
BdB d: Bund deutscher Baumschulen
BDB e: Big Dumb Booster
- e: Bis-Diazotized-Benzidine
BdBVM d: Beauftragter des Bundesverkehrsministeriums
BDC e: Back-up Digital Computer
- e: Berlin Documents Center
- e: Binary-to-Decimal Conversion
- e: Binary[-to-]Decimal Converter
- e: Binary[-to-]Decimal Counter
- e: Binary Differential Counter
- e: Bin Data Cell
- e: Boeing Development Center
- e: Bomb Data Center
- e: Bonded Double-Cotton
- e: Boolean Differential Calculus
- e: Bottom Dead Center
- e: Broadcast
- e: Building Data Council
BDCB e: Buffered Data and Control Bus
BDCE e: Butanediol Diglycidyl Ether
Bdch. d: Bändchen
BDCT e: Broadcast
B-D cu e: Boron Deoxidized Copper
BDD e: Bedded
- e: Binary-to-Decimal Decoder
- e: Binary-to-Digital Decoder
- e: Binary Digital Data
BDDA e: Basic Direct Data Access
BDDA Method e: Basic Direct Data Access Method
BDDI e: Bending Die
BDDRE e: Brigadier and Deputy Director of Royal Engineers
BDDV e: Beading Device
- e: Biocular Display Drivers Viewer
BDE d: Bedieneinheit
- d: Betriebsdatenerfassung
- e: Beta Disintegration Energy
- e: Board of Dental Examiners
- e: Bomber Development Establishment
- e: Bright Display Equipment
- e: British Destroyer Escort
- e: Butadiene Diepoxide
Bde d: Bände
Bde. d: Binde
BDEAP d: Bedienungs-Ein-/Ausgabe-Programm
BDEC d: Befehlsdecodierung
BDED e: Bounded
BDEF-Jb. d: Bundesverband Deutscher Eisenbahnfreunde-Jahrbuch
BD ELIM e: Band Elimination
BDent Sci e: Bachelor of Dental Science
B Des e: Bachelor of Design
BDES e: Bibliographic Data Exchange System
BDE-Station d: Betriebsdatenerfassungs-station
BDE-System d: Betriebsdatenerfassungs-system
BDE-T[erminal] d: Betriebsdaten-erfassungsterminal
BDF d: Bund Deutscher Forstmänner
- e: Backwards Differentiation Formulas
- e: Base Detonating Fuse
- e: Basic Display File
- e: Beam Deviation Factor
- e: Birth Defects Foundation
- e: Boxcar Doppler Filter
- e: Bus Differential
B.D.F.A. e: British Dairy Farmers Association
BDFGA e: Bio-Dynamic Farming and Gardening Association
BD/FKD-Us d: Bescheiddienst / Fernsprechkundendienst-Umsetzer
BDFS e: Basic Disk Filing System
BD FT e: Board-Foot
BdFt e: Board Feet
BDG d: Bediengerät
- d: Bindegewebe
- d: Bund der Diplomgärtner
- e: Beacon Data Generation
- e: Binding
- e: Boarding
- e: Bridge
- e: Bridging
- e: Buffered Desoxycholate Glucose
- e: Building
Bdg. d: Bindung
BDGA d: Bund Deutscher Garten- und Landschaftsarchitekten
BDGH e: Base Depot and General Hospital
- e: Binding Head
BDG Pl e: Bedding Plane

BDGT e: Budget
BDH d: Bund Deutscher Holzwirte
- e: Bearing, Distance and Heading
- e: Binding Head
BDHI e: Bearing, Distance and Heading Indicator
BDHSA e: Bomb Director High Speed Aircraft
BDI d: Bundesverband der Deutschen Industrie
B Di e: Bachelor of Didactics
BDI e: Base-Diffusion Isolation
- e: Bearing Deviation Indicator
- e: Beck Depression Inventory
- e: Binary Discrete Input
- e: Biological Damage Indicator
- e: Bombardier Data Indicator
- e: Both Dates Included (or Inclusive)
- e: Buffered Direct Injection
- e: Bureau of Dairy Industry
- e: Burroughs Direct Interface
BDIA e: Base Diameter
BDIAC e: Batelle Defense Information Analysis Center
BDIC e: Batelle Defense Information Center
- e: Binary-coded Decimal Interchange Code
B Did e: Bachelor of Didactics
B Di E e: Bachelor of Diesel Engineering
B Di Eng e: Bachelor of Diesel Engineering
BDIHIST Program[me] e: Board Interface History Program[me]
BD in E e: Bachelor of Divinity in Education
B Dipl e: Bachelor of Diplomacy
BDI Process e: Base-Diffusion Isolation Process
BDir. d: Baudirektor
BDir d: Bundesbahndirektor
BDIR e: Bus Direction
BDI Technique e: Base-Diffusion Isolation Technique
BDI Technology e: Base-Diffusion Isolation Technology
BDK d: Bedienungselement; kombiniert
BDK d: Beiträge zur Deutschen Klassik (journ.)
B.d.K. d: Bevollmächtigter des Kraftfahrwesens
BDk e: Bridge Deck
BDKA d: Beiträge zur Deutschen Klassik; Abhandlungen (journ.)
Bdl d: Bahnhofsdispatcherleitung
B.d.L. d: Befehlshaber der Linienschiffe
- d: Befehlshaber der Luftstreitkräfte
BDL d: Bund der Landwirte
- e: Bachelor of Devine Literature
- e: Bad Data Lister
- e: Base Deflection Left
- e: Baseline Demonstration Laser
- e: Base of Dorsal Left
- e: Basic Design Language
- e: Battery Data Link
- e: Beach Discharge Lighter
- e: Best Dressed List
- e: Billeting Distribution List
- e: Build Definition Language
- e: Bundle
- e: Business Definition Language
BDL BK S e: Bundled Bark Strips
BDLC e: Burroughs Data Link Control
bdle[s] e: bundle[s]
BDLg d: Blätter für deutsche Landesgeschichte (journ.)
BDLS e: Battery Data Link System
BDLS e: Bundles
BD LT e: Bow Designation Light
BDM e: Basic Data Management
- e: Binary Delta Modulation
- e: Binary Digital Multiplier
- e: Births, Deaths and Marriages
- e: Bomber Defense Missile
- e: Bubble Domain Memory
BDMA e: Benzyl Dimethyl Amine
- e: British Disinfectant Manufacturers Association
BDMAA e: British Direct Mail Advertising Association
BDMN e: Backend-image Database Management Network
BDMR e: Bandmaster
BDMS e: Base Depot of Medical Stores
- e: Berkeley Database Management System
- e: Burroughs Data Management System
BDN d: Bausteine zum Deutschen Nationaltheater (journ.)
- e: Bank Draft Number
- e: Bell Data Network
- e: Bend Down
- e: Blue Danube Network
BDNG e: Bedding
BDO e: Bow Door
- e: Bus for Data Output
B/DOE e: Barrels per Day Oil Equivalent
BDOS e: Basic Disk Operating System
- e: Basic Disk Operation System
- e: Batch Disk Operating System
BDP d: Beiträge zur Deutschen Philologie (journ.)
- d: Bezirksdirektion der Deutschen Post (DDR)
- e: Base Development Plan
- e: Base Diffusion Process
- e: Basic Data Processing
- e: Biparietal Diameter
- e: Bonded Double Paper
- e: Bottom Dead Point
- e: Breakdown Pressure
- e: Brookes Deflection Potentiometer
- e: Building Design Partnership
- e: Bulk Data Processing
- e: Business Data Processing
BDPA e: Black Data Processing Associates
BDPEC e: Bureau of Disease Prevention and Environmental Control
BDPI e: Base Data Processing Installation
BDPN e: Bedpan
B Dpo e: Base Depot
BDPO e: Business Data Processing Operation
BDPS e: Brigade Data Processing System
BDPSC e: Basic Data Processing Systems Center
BDPSK e: Binary Differential Phase-Shift Keying
BDR d: Baudeckenreflex[e]
- d: Binder
- d: Bombardier
- e: Base Deflection Right
- e: Basic Data Records
- e: Bell Does not Ring
- e: Best Depth Range
- e: Bi-Duplexed Redundancy
- e: Binary Dump Routine
- e: Bomb Damage Repair
- e: Border
- e: Brigadier
- e: Bus Data Register
Bdr. d: Buchdruck
Bdr d: Bundesdruckerei
B Dr Art e: Bachelor of Dramatic Art
Bdr-Gnr e: Bombardier-Gunner
BDRI e: Bright Display Radar Indicator
BDRY e: Boundary
BDS d: Barbados (Kfz)
BDS d: Betriebsdatenspeicher
- d: Betriebsdatenverarbeitungsstation
- e: Bachelor of Dental Surgery
- e: Base Data System
- e: Base Design Section
- e: Base Divider Strip
- e: Bias Discrete Spot
- e: Binary Decade Scaler
- e: Binary Digit System
- e: Biological Defense System
- e: Biological Detection System
- e: Block Data Statement
- e: Block Data Subprogram[me]
- e: Bomb Damage Survey
- e: Bomb Director Set
- e: Bomb Director System
- e: Bounded Double Silk
- e: Bound in Boards
- e: Brass Divider Strip
- e: Bulk Data Switching
- e: Bundles
- e: Business Definition System
- e: Buttres Double Seal
BDs e: Backward Definitions
Bds e: Boards
- e: Bonds
BDSA e: Business and Defense Services Administration
BD Sc e: Bachelor of Dental Science
BDSD e: Base Detonating; Self-Destroying
BDSG d: Bundesdatenschutzgesetz
BDSI e: Bangladesh Standards Institution
- e: Basic Direct Shipping Instructions
BDSM e: Base Depot Stockage Model
BDS-OFI e: Base Design Section - Operational Facilities Installations
BDSP d: Bandsperre
- e: Basic Data Set Project
BDSS e: Bayesian Decision-Support System
BDS-SFI e: Base Design Section - Support Facilities Installations
BDST e: British Double Summer Time
- e: Broadcast
BDT d: Bedientastatur
- d: Büro- und Datentechnik GmbH
- d: Bürodatentechnik
- e: Beam Deflection Tube
- e: Binary Deck-to-[binary] Tape
- e: Bone Dry Ton[s]
- e: Burst Delay Timer
Bdt. d: Bidet
BDTD e: Balanced Digital Transmission Device
BDTH e: Breadth
BDTR e: Basic Data Transmission Routine
BDTS d: Bescheid-Teilnehmerschaltung
- e: Buffered Data Transmission Simulator
- e: Bulk Data Transfer Subsystem
BdU d: Bund der Umweltschützer
BDU e: Bargraph Display Unit
- e: Basic Device Unit
- e: Basic Display Unit
- e: Battery Display Unit
- e: Biomedical Display Unit
- e: Bomb; Dummy Unit
- e: Bomb Disposal Unit
- e: Bromodeoxyuridine
BDV d: Basis-Datenverarbeitung
- d: Breitbanddurchgangsvermittlung
- e: Breakdown Voltage
- e: Bud Drop Virus
BDVA d: Beiträge zur Deutschen Volks- und Altertumskunde (journ.)
BD/VEG d: Bezirksdirektion Volkseigner Güter (DDR)
BDVG e: Bow Diving

BDW e: Binary Data Word
- e: Block Descriptor Word
- e: Blunted Delkta Wing
- e: Buffered Distilled Water
- e: Buried Distribution Wire
BDWEL e: Board Discrete Wire Equivalent List
BDWL Program[me] e: Board Discrete Wire List Program[me]
BDX e: Bar Double-X
BDXAV e: Bendix Aviation
BDXR e: Block Demultiplexer
BDY e: Body
Bdy e: Boundary
Bd. Z. d: Bodenzünder
Bd. Zdr. d: Bodenzünder
BDZR e: Bulldozer
BE d: Basiseinheit
- d: Basiseinrichtung
- d: Batterieentladung
- d: Bauelement
- d: Bauelemente der Elektrotechnik (journ.)
- d: Bazillenemulsion
- d: Bearbeitungseinheit
- d: Beckeneingang
- d: Beckenendlage
- d: Bedienereingriff
- d: Befehlsempfänger
- d: Behandlungseinrichtung
- d: Behandlungserfolg
- d: Behringeinheit
- d: Benzoat-Einheit
- d: Benzylamphetamin
- d: Beschallungseinheit
- d: Beschaltungseinheit
- d: besondere Einflüsse
- d: Betriebseinheit
- d: Betriebselektronik
- d: Betriebserde
- d: Bewerten
- d: Bezugsebene
- d: Bezugselektrode
- d: Bildschirmeinheit
- d: Bildsenderendstufe
- d: Binäreingang
- d: biologische Einheit[en]
- d: Bohreinheit
- d: Bose-Einstein
- d: Brockhaus-Enzyklopädie
- d: Broteinheit
- e: Bachelor of Economics
- e: Bachelor of Education
- e: Bachelor of Elocution
- e: Bachelor of Engineering
- e: Bachelor of English
- e: Bachelor of Expression
- e: Bachelor of the Elements
- e: Bacillary Emulsion
- e: Back End
- e: Band Elimination
- e: Base Ejection
- e: Base Equation
- e: Base Excess
- e: Basic Encyclopedia
- e: Basic Equation
- e: Beacon Explorer
- e: Beam Expander
- e: Begin
- e: Beginning Experience
- e: Bell End
- e: Below Elbow
- e: Beryl
- e: Beveled Edge
- e: Biennial
- e: Binding Energy
- e: Block End
- e: Boolean Encoding
- e: Boolean Expression
- e: Booster Engine
- e: Borough Engineer
- e: Both Ends
- e: Breaker End
- e: British Element
- e: British Empire
- e: British English
- e: British Experimental
- e: Buddhist Era
- e: Bus Enable
be. d: bezüglich
- e: bilingual education
Be. e: Belgian
- e: Belgium
Be e: Berkeley
- e: Beryllium
Bé e: Baumé
B/E e: Bill of Entry
- e: Bill of Exchange
B & E e: Breaking and Entering
b & e e: Beginning and Ending
BEA d: Bedienungseinrichtungsanschaltung
- e: Background Equivalent Activity
- e: Basic Electric Arc
- e: Blue Etch-Anodize
- e: Blue Etch-Anodizing
- e: British East Africa
- e: British Electricity Authority
- e: Broadcast Education Association
- e: Bureau of Economic Analysis
- e: Burroughs Extended Algorithmic Language
BEAB e: British Electricity Approvals Board
BEAC e: Boeing Electronic Analog Computer
- e: Boeing Engineering Analog Computer
BEACON e: British European Airways Computerized Office Network
- e: British European Airways Computer Network
BEAIRA e: British Electrical and Allied Industries Research Association
BEAM e: Bidders Early Alert Message
- e: Brain Electrical-Activity Mapping
- e: Building Equipment Accessories and Materials
- e: Burroughs Electronic Accounting Machine
BEAMA e: British Electrical and Allied Manufacturers Association
BEAMOS e: Beam-Accessable Metal Oxide Semiconductor
- e: Beam-Addressed Metal Oxide Semiconductor
BE Amp e: Below Elbow Amputation
BEAMS e: Base Engineer Automated Management System
BEAP d: Bedienungs-Ein-/Ausgabe-Programm
- d: Bibliotheks-Eingabe-Ausgabeprogramm
BEAR e: Bearing
- e: Biological Effects of Atomic Radiation
- e: Bombardment Excited Amplification in a Semiconductor
- e: Bulk Easy Access Readout
- e: Bureau of Electronic and Appliance Repair
BEARA e: British Electronic & Applied Research Association
Bearb. d: Bearbeiter
bearb. d: bearbeitet
Bearb. d: Bearbeitung
BEAR HUG e: Basic Extended Acronym Human Users Guide
BEART e: Beaver Army Terminal
BE-A Satellite e: Beacon Explorer A Satellite
BEAST e: Basic Experimental Automatic Syntactic Translator
- e: Brookings Economics and Statistical Translator
BEASY e: Boundary Element Analysis System
BEAT e: Breaking and Entering and Auto Theft
B.E.A.T.S. e: Both Ends; All Time Saved
Beau e: Beaufort
BEB e: Beam Entry Buffer
- e: Best Ever Bottled
- e: Binary Exponential Backoff
- e: Bridge Erection Boat
- e: British Electricity Board
Bebavioral Res e: Behavioral Research Laboratories
BEBC e: Big European Bubble Chamber
BEBO e: Bond-Energy Bond-Order
BEBS d: Bestellblattschreiber
BE-B Satellite e: Beacon Explorer B Satellite
BEBU d: Betriebsbuchhaltung
BEc e: Bachelor of Economics
BEC e: Bachelor of Engineering Construction
- e: Back End of Chest
- e: Base Engineering Course
- e: Base Equipment Container
- e: Base Examination Course
- e: Base Extension Course
- e: Basic Education Centre
- e: Binary Erasure Channel
- e: Bio-Electro-Chemistry
- e: Boron Electron Center
- e: British Electric Council
- e: British Electrotechnical Committee
- e: Broad Economic Categories
- e: Burst-Error Channel
- e: Business Electronics Computer
bec. e: because
BECC e: Boeing Engineering & Consruction Company
- e: British Empire Cancer Campaign
BECCR e: British Empire Cancer Campaign for Research
BECE e: Bachelor of Electro-Chemical Engineering
Bech. e: Bechuanaland
BEChem e: Bachelor of Chemical Engineering
BECKTRAN e: Beckman Translation
BECO e: Booster Engine Cut-Off
BEcon e: Bachelor of Economics
BEconIA e: Bachelor of Economics in Industrial Administration
BEconPA e: Bachelor of Economics in Public Administration
BE COUP e: Band-Edge Coupling
BECR e: Bit Error Correction Rate
BECS e: Basic Error Control System
BECTIS e: Bell College of Technology Information Service
- e: Bell College Technology Information System
BECTO e: British Electric Cable Testing Organization
BECUN e: Battelle Educational Computer Users Network
Bed. d: Bedarf
- d: Bedeutung
- d: Bedingung[en]
- d: Bedürfnis
bed. d: bedeutend
- d: bedeutet

bed d: bedienen
Bed d: Bedienung
B Ed e: Bachelor of Education
BED e: Bachelor of Elementary Didactics
- e: Base Engineer Depot
- e: Box External Data
- e: Bridge-Element Delay
BEDA d: Bearbeitungsdauer
- d: Bedarfsauslösung
- d: Bedarfsermittlung
- e: British Electrical Development Association
BEDAC e: Burst Error Detection and Correlation
Bedampf. d: Bedampfung
BEDAS d: Betriebs-Daten-System
bed. best. d: bedingt beständig
Bedeut. d: Bedeutung
BedFd d: Bedienungsfeld
BedGt d: Bedienungsgerät
B Ed in Phys Ed e: Bachelor of Education in Physical Education
BEdN e: Bachelor of Education in Nursing
BEDOC e: Beds Occupied
BEdPE e: Bachelor of Education in Physical Education
BEdSc e: Bachelor of Educational Science
BEDT e: Brooklyn Eastern District Terminal
BEDT Railroad e: Brooklyn Eastern District Terminal Railroad
B Educ e: Bachelor of Education
BEDU-System d: Bodenheizung, Bodendämpfung und Unterflurbewässerung-System
BEE e: Bachelor of Electrical Engineering
- e: Band Edge Energy
- e: Basic Earthquake Education
- e: Basic Economic Education
- e: Books for Equal Education
- e: British Employers Executive
- e: Bulletin of Environmental Education
BE-Ebene d: Beckeneingangsebene
BEEC e: Binary Error-Erasure Channel
- e: Binary Error Erasure Channel
Beech 99 e: Beechcraft seventeenseat aircraft
Beech Queen e: Air Beechcraft Seminole transport aircraft
BEEF e: Business and Engineering Enriched Formular Translation
BEEF Translator e: Business and Engineering Enriched Formula Translator
Beeh F-33C e: five-place training airplane
BE ELIM e: Band Elimination
BE Eng e: Bachelor of Electrical Engineering
BEEP e: Battalion Equipment Evaluation Program[me]
BEER e: Binary-Element Error Ratio
- e: Bombardment Enhanced Fetch Rate
BEET e: Best-estimated Evaluation Trajectory
BEF d: Band-Elimination Filter
- d: Bauelementefunktion
- e: Basic Energy Foods
- e: Basic External Function
- e: Bayes Empirical-Distribution Function
- e: Blunt End First
- e: Blunt End Forward
- e: Buffered Emitter Follower
bef. d: befähigt
- d: befördert
- d: befugt
- e: before
Bef. d: Befähigung
- d: Befeuerung
- d: Beförderung[sbuch]
- d: Befugnis
- d: Befund
Bef d: Befehl
Befa d: Beobachtungsfahndung
BEFA d: Bildeingabe vom Fernsehabtaster
BE FAP e: Before
- e: Bell Laboratories Formula Translation Assembly Program[me]
Befb. d: Beförderungsbuch
Bef.-Bed. d: Beförderungsbedingungen
Bef.-Beschr. d: Beförderungsbeschränkung[en]
Bef. B. Mö. d: Beförderungsbestimmungen für den Möbelverkehr
BefC d: Befehlscode
Befeucht. d: Befeuchtung
Bef.-Fr. d: Beförderungsfrist
BefG d: Beförderungsgesetz
BEFM e: Bending Form
Bef.-Nachw. d: Befähigungsnachweis
Bef.-Pfl. d: Beförderungspflicht
Bef.-Pl. d: Beförderungsplan
BefPz d: Befehlspanzer
Befr. d: Befragung
- d: Befristung
befr. d: befristet
BEFR d: Betriebs-/Ersatz-Fixierregister
BefReg d: Befehlsregister
Befr.-Erg. d: Befragungsergebnisse
Bef.-Sp. d: Beförderungssperre
Befstw d: Befehlsstellwerk
Bef.-Vertr. d: Beförderungsvertrag
Bef.-Vorschr. d: Beförderungsvorschrift[en]
BefW d: Befehlswort
Bef.-Weg d: Beförderungsweg
BefZ d: Befehlszähler
Bef.-Zeugn. d: Befähigungszeugnis
Bef.-Zgn. d: Befähigungszeugnis
BefZReg d: Befehlszählregister
BefZyk d: Befehlszyklus
Beg. d: Beginn
BEG d: Belgrad (Flughafen)
- d: Bodeneffektgerät
- e: Belgrade (airport)
- e: Bureau of Economic Geology
beg. e: begin[ning]
Begl. d: Beglaubigung
- d: Begleitung
- e: Begleiter
begl. d: begleitet
Begl.-Sch. d: Begleitschein
Begl.-Schr. d: Begleitschreiben
BEGNR e: Beginning Number
Beg o Rec e: Beginning of Record
begr. d: begradigen
- d: begrenzen
- d: begrenzt
- d: Begrenzung
- d: begründen
- d: begründet
Begr. d: Begradigung
- d: Begriff
- d: Begründer
- d: Begründung
BEGSR e: Begin Subroutine
BEGUM d: Belegverarbeitung für Girozentrale unter Monitorsteuerung
Beg.-Z. d: Beginnzeichen
Beh. d: Behälter
- d: Behandeln
- d: Behelf
beh. d: behandelt
- d: behelfsmäßig
BEH e: Behaviour
- e: Bulk Erase Head
Behandl. d: Behandlung
Behav. e: Behavior
behav e: behavioral
- e: behaviorist[ic]
Behav. Sci. e: Behavioral Science (journ.)
BEHC e: Bio-Environmental Health Center
Beheiz d: Beheizung
BEHP e: Bisethylhexyl Phthalate
BEHVL e: Behavioral
BEI e: Bridgeport Engineering Institute
- e: Budget Executives Institute
- e: Butanol-Extractable Iodine
Beibl. d: Beiblatt
- d: Beiblätter
BEIC e: Battelle Energy Information Center
BEID e: Behavioral Effects of Infections Diseases
B-Eier d: Boden-Eier
beif. d: beifolgend
- d: beifügen
beiflgd d: beifolgend
beig. d: beigeheftet
beigeb. d: beigebunden
beigedr. d: beigedruckt
beigef. d: beigefügt
beigeh. d: beigeheftet
Beih. d: Beiheft[e]
beih. d: beiheften
Beil. d: Beilage[n]
beil d: beiliegend
BEIR e: Biological Effects Of Ionizing Radiation
Beis. d: Beisatz
- d: Beisitzer
BEIS e: British Egg Information Service
Beisp. d: Beispiel[e]
beisph. d: beispielhaft
beispw. d: beispielsweise
Beitr. d: Beitrag
- d: Beiträge
Beitr. Biol. Pflanz. d: Beiträge zur Biologie der Pflanzen (journ.)
Beitr. Pathol. d: Beiträge zur Pathologie (journ.)
Beitr. Phys. Atmos. d: Beiträge zur Physik der Atmosphäre (journ.)
Beiw. d: Beiwagen
BEJ d: mit Butanol extrahierbares Jod
- e: Base-Emitter Junction
B/E Junction e: Base/Emitter Junction
BEK d: Bauelementekonstruktion
- d: Betriebsmitteleingabekarte
- e: Bovine Embryonic Kidney
- e: Butyl Ethyl Ketene
Bek[Ämpf]. d: Bekämpfung
BEKBC DIC e: Binary-Coded Decimal Interchange Code
BEKR d: Betriebs-/Ersatz-Kippstufenregister
BEL d: Bauelement
- d: Bestückungselement[e]
- d: Betriebserdungsleiter
- e: 10 decibels
- e: Bachelor of English Literature
- e: Basic Equipment List
- e: Battery and Earth Loop
- e: Belgium
- e: Belize
- e: Belgian
- e: Bell
- e: Belorussia
- e: Bus Exchange Logic
- p: Bacharel

bel. d: beladen
- d: belasten
- d: belastet
- d: belegen
- d: belegt
- d: beleuchten
- d: beleuchtet
Bel. d: Belag
- d: Beläge
- d: Belastung[en]
- d: Beleg[e]
- d: Belegung[en]
Bel. d: Beleuchtung
- d: Belichtung
bel. d: belichten
- d: belichtet
- e: below
B EL e: Bachelor of Elocution
BELADO d: Belastungsmodell Dortmund
BEL CRK e: Bell Crank
BELe d: Bestückungselemente
B Ele e: Bachelor of Elements
B.Elec. & Tel. Eng. e: Bachelor of Electronics and Telecommunication Engineering
Bel.-EKG d: Belastungs-Elektrokardiogramm
B EL Eng e: Bachelor of Electrical Engineering
Beleucht. d: Beleuchtung
bel ex f: bel example
BELF d: Bundesministerium für Ernährung, Landwirtschaft und Forsten
BELFA d: VEB Berliner Elemente Fabrik (DDR)
BELFAST d: Bezugsregulierung, Lieferscheine, Fakturierung, Statistiken
belg a: belgisk
BELG d: Bayerische Electricitäts-Lieferungs-Gesellschaft AG; Bayreuth
Belg e: Belgian
- i: Belgio
- s: Belgica
Belg. e: Belgium
Bel.-Gew. d: Belastungsgewicht
BELGFE d: Beleglesefehlerprogramm für Datensichtgeräte
Belgolux e: Belgium and Luxembourg
Belg. P[at.] e: Belgian Patent
Bel./h d: Belege pro Stunde
BELIMATIC e: Bell Laboratories Machine-Aided Technical Information Center
BELINDIS e: Belgian Information and Dissemination Service
Belkaw d: Bergische Licht-, Kraft- und Wasserwerke GmbH
BELL d: Bildeingabe vom Lochbandleser
- e: Bell Aircraft
- e: Bell System
- e: British Electric Lamp, Limited
Bell 204 e: gunship helicopter
Bell 206 e: five-seat turbo-powered helicopter
BELLA e: Belladonna
BELLDEX e: Bell Indexing
BELLE e: Bellewe New York City Hospital
Bell Lab. Rec. e: Bell Laboratories Record (journ.)
BELLMATIC e: Bell .Laboratories Machine-Aided Technical Information Center
BELLREL System e: Bell Laboratories Library Real-time Loan System
Bell Syst Tech. J. e: Bell Systems Technical Journal (journ.)
Bel.-M. d: Belichtungsmesser
Bel./min d: Belege pro Minute
Bel.-Nr. d: Belegnummer
Bel.-Pr. d: Belastungsprobe
BeLR e: Bell Laboratories Record (journ.)
Bel.-St. d: Belastungsstärke
- d: Beleuchtungsstärke
- d: Belichtungsstärke
Bel.-Str. d: Belichtungsstärke
Belt 47 e: Sioux utility helicopter
BEL Timber e: Bael Timber
Belüft. d: Belüftung
BELUM-Programm d: Belegumschreibung[s]-Programm
Belvac Société f: Belge de Vacuologie et de Vacuotechnique
Bel.-Zeit d: Belastungszeit
- d: Belegungszeit
- d: Belichtungszeit
Bem. d: Bemerkung
- d: Bemessung
BEm d: Bildempfänger
BEM e: Bachelor of Engineering of Mines
- e: Baggage Express Messenger
- e: Basic Editor Monitor
- e: Behavior Engineering Model
- e: Biax Element Memory
- e: Boundary Element Method
- e: British Empire Medal
BEMA d: Bewertungsmaßstab
- e: Business Equipment Manufacturers Association
BEMAR e: Backlog of Essential Maintenance and Repair
BE-Matrix d: Bindungselektronen-Matrix
BE-ME e: Bachelor of Engineering in Mechanical Engineering
BEME e: Brigade Electrical and Mechanical Engineer
BEMF e: Back Electromotive Force
Bem.-Gr. d: Bemessungsgrenze
Bem.-Grundl. d: Bemessungsgrundlage
Bem Richtl d: Bemessungsrichtlinien
BEMS e: Bioelectromagnetics Society
- e: Building Energy Management Systems
Bemss. d: Bemessung
BEN d: Benghasi (Flughafen)
- e: Bench
- e: Bus Enable
- f: Bureaue of d'Études nucleaires
Ben. d: Benachrichtigung
- d: Benotung
- d: Benutzung
ben. d: benutzen
- d: benutzt
B En e: Bachelor of English
BENCOM e: Benefical Communications
BENDEX e: Beneficiary Data Exchange
Bendix Tech.J. e: Bendix Technical Journal (journ.)
B en Dr f: Bachelier en Droit
bene e: benzine
Benelux e: Belgium, Netherlands and Luxemburg
Benetz. d: Benetzung
benev. e: benevolent
B Eng e: Bachelor of Engineering
Beng. e: Bengal[i]
B Eng A e: Bachelor of Engineering in Agriculture
B Eng Physics e: Bachelor of Engineering Physics
B Eng. Sc[i] e: Bachelor of Engineering Science
B.Eng.Tech. e: Bachelor of Engineering Technology
Benin e: Peoples Republic of Benin
BENNIES e: Benzedrine Stimulants Benny Benzedrine
BEN Press e: Bench Press
BENS e: Bounded Error Navigation System
B Ent e: Bachelor of Entomology
BENT e: Beginning Evening Nautical Twilight
- e: Bent of Tau Beta Pi (journ.)
BEN Timber e: Benteak Timber
BENZ e: Benzedrine
- e: Benzidine
- e: Benzine
BeO e: Beryllium Oxide[s]
BEO e: Borough Education Officer
beob. d: beobachten
- d: beobachtet
Beob. d: Beobachter
- d: Beobachtung[en]
BEOC e: Battery Echolon Operating Control
BEOG e: Basic Educational Opportunity Grants
BEOP e: Best Estimate of Orbital Parameters
BEP d: Bedieneinheit Prozeßrechner
- d: Beiträge zur Englischen Philologie (journ.)
- e: Bachelor of Engineering Physics
- e: Benzo[e]pyrene
- e: Best Efficiency Point
- e: Bit Error Probability
- e: Budget Execution Plan
- e: Bureau of Engraving and Printing
- e: Burst-Error Processor
- f: Brevet d'Etudes Professionelles
BEPA e: Binding Energy per Atom
B EpA e: British Epilepsy Associatlon
BEPC e: Beijins Electron Positron Collider
- e: British Electrical Power Convention
BE Phy e: Bachelor of Engineering Physics
BEPI e: Budget Estimates Presentation Instructions
BEPO e: British Experimental Pile Operation
- e: British Experimental Pile Zero
BEPOC e: Burroughs Electrographic Printer-plotter for Ordnance Computing
BEPP e: Binding Energy per Particle
BEPQ e: Bureau of Entomology and Plant Quarantine
BEP's e: Break Even Points
BEPT e: Boiler Feed Pump Turbine
BEPTL e: Bionomics, Environment, Plasmodium, Treatment, Immunity
BEQ e: Bachelor Enlisted Quarters
ber d: beraten
- d: berechnen
- d: berechnet
- d: berufen
Ber. d: Berater
- d: Beratung
- d: Berechnung
- d: Bereich
- d: Bericht
- d: Bericht[e]
- d: Berichtigung
- d: Berufung
BER d: Berlin (Flughafen)
- e: Basic Electrical Rhythm
- e: Beyond Economic Repair
- e: Binary Error Rate
- e: Bit Error Rate
- e: Block Error Rate
- e: Boiling Experimental Reactor
- e: British Experimental Rocket
- e: Budget Execution Review
- e: Buffer

- e: Bureau of Economic Regulation
BeR e: Berkeley Review (journ.)
BerADev e: Berlin Airlift Device
BE Railway e: Boston Elevated Railway
berat. d: beraten[d]
Berat. d: Beratung
berat. Ing. d: beratender Ingenieur
BE-Raum d: Beckeneingangsraum
Berb. d: Berber
Berb e: Berber
BerBl d: Berichtigungsblatt
Ber Bunsenge Phys Chem d: Berichte der Bunsengesellschaft für Physikalische Chemie (journ.)
BERC e: Biomedical Engineering Research Corporation
- e: Black Economic Research Center
BERCO e: British Electric Resistance Company
Ber. Dtsch. Bot. Ges. d: Berichte der Deutschen Botanischen Gesellschaft (journ.)
Ber. Dtsch. Keram. Ges. d: Berichte der Deutschen Keramischen Gesellschaft (journ.)
Berechn. d: Berechnung
berg e: iceberg
Bergass. d: Bergassessor
Bergb. d: Bergbau
Bergd d: Bergungsdampfer
Bergdir. d: Bergdirektor
BERGE d: Berichtgenerator
BergG d: Berggesetz
Berg- HüttenmÄnn. Monatsh. d: Berg- und Hüttenmännische Monatshefte (journ.)
Berging. d: Bergingenieur
Berginst. d: Berginstitut
BergR d: Bergrat
Ber.-Grundl. d: Berechnungsgrunglage
Bergs-M d: Bergungsmotorschiff
Bergw d: Bergwerk
Berg.-Wiss. d: Bergbauwissenschaften und Verfahrenstechnik (journ. bis 1971)
BERH e: Board of Engineers for Rivers and Harbors
Ber.-Ing. d: beratender Ingenieur
Ber. Is e: Bermuda Islands
Berk. e: Berkeley
Berlin. München. TierÄrztl. Wo d: Berliner und Münchener Tierärztliche Wochenschrift
Berm e: Bermuda
- e: Bermuda Islands
BERM e: Bit Error Rate Monitor
BERMUDA d: Benutzerservice für Maskenunterstützung an Datenstationen
BERNET d: Berliner Rechnernetz
BERPEX d: Berlin Packet Exchange
Ber. Phys.-Med. Ges. Würzburg d: Berichte der Physikalisch-Medizinischen Gesellschaft zu Würzburg (journ.)
BERPM e: Basic Exchange Rate Planning Model
BERs e: Binary Error Rates
- e: Bit Error Rates
B & ERS e: Boat and Engine Repair Shop
BERS e: British Empire Receiving Station
BERSAFE e: Berkeley Stress Analysis Finite Elements
BerSchG d: Berufsschulgesetz
BERSIGAN-System d: Berichts-, Signal- und Analyse-System (der Firma Siemens)
BERT e: Basis Energy Reduction Technology
- e: Bit Error Rate Tester
BERTA e: British Empire Radio Transmission Award
BERTH e: Berthing
BER TRI e: Bermuda Triangle
BERTS e: Bit Error Rate Test System
BERU e: Building Economics Research Unit
Berufsausb. d: Berufsausbildung
Berufsber. d: Berufsberatung
Ber. u. Inf. EG d: Berichte und Informationen der Europäischen Gemeinschaft (journ.)
BERYL e: Beryllium Aluminum Silicate
BES d: Behandlungssystem
- d: besetzt
- d: Besetztzeichen
- d: besonders
- d: Bilderkennungssystem
- d: Bilderzeugungssystem
- d: Breitband-Einheitssystem
- e: Bachelor of Engineering Science
- e: Balanced Electrolyte Solution
- e: Basic Executive System
- e: Bass Equalizer System
- e: Bessels Functions
- e: Biax Element Store
- e: Biological Engineering Society
- e: Black Enamel Slate
- e: Block Error Status
- e: Blue-Enhanced Silicon
- e: Booster Exhaust Stream
- e: Bose-Einstein Statistics
- e: British Ecological Society
- e: Buildings and Equipment Section
- e: Bureau of Employment Security
- e: Bureau of Environmental Statistics
Bes. d: Besatzung
- d: Besichtigung[en]
bes. d: besichtigen
BESA e: British Engineering Standards Association
Bes A[usf]B[est] d: besondere Ausführungsbestimmungen
Bé Scale e: Baumé Scale
besch. d: beschädigen
- d: beschädigt
Besch. d: Beschädigung
- d: Bescheid
- d: Bescheinigung
- d: Beschichtung
- d: Beschuß
Beschaff. d: Beschaffung
Beschaffenh. d: Beschaffenheit
Beschick. d: Beschickung
beschl. d: beschleunigen
Beschl[eunig]. d: Beschleunigung
beschn. d: beschnitten
beschr. d: beschränken
- d: beschrankt
Beschr. d: Beschränkung
- d: Beschränkung[en]
- d: Beschreibung
BeschwF d: Beschwerdeführer
BESD e: Base Engineer Store Depot
- e: Basic Enlisted Service Date
Beseitig. d: Beseitigung
BESEP e: Base Electronics System Engineering Plan
BESEX e: Bering Sea Expedition
- e: Bering Sea Experiment
Bes Function e: Bessels Function
BESI e: Bus Electronic Scanning Indicator
BesL d: Besetztlampe
BESL d: Betriebserdungs- und Schutzleiter
besp. d: bespannen
- d: bespannt
BESP e: Blue Enhanced Silicon Photodiode
BES Photodiode e: Blue-Enhanced Silicon Photodiode
Bespr. d: Besprechung[en]
bespr. d: besprühen
BESR d: Betriebs-/Ersatz-Steuerregister
BESRL e: Behavioral Science Research Laboratory
BESS e: Bank of England Statistical Summary
- e: Bessemer
- e: Binary Electromagnetic Signal Signature
- e: Bottom Environmental Sensing System
B es S f: Bachelier es Sciences
BESSY d: Berliner Elektronenspeicherring für Synchrotronstrahlung
- d: Betriebsdaten-Erfassungs- und Steuerungs-System
Best. d: Bestand
- d: Bestandteil[e]
- d: Bestellung[en]
- d: Bestimmung
BEST e: Ballastable Earthmoving Sectionalized Tractor
- e: Base-Emitter Self-aligned Technology
- e: Basic Executive Scheduler and Timekeeper
- e: Better Electronic Service Technicians
- e: Bias, Equalization, Sensitivity of the Tape
- e: Bicycle Ergometer Stress Test
- e: Bit Efficient Storage Technique
- e: Blockhaus Equipment Switching Test
- e: Boeing Extractive Sludge Treatment
- e: British Expertise in Science & Technology
- e: Business Electronic Data Processing Systems Technique[s]
- e: Business Electronic System Technique
- e: Business Equipment Software Technique[s]
- e: Business Executive System for Time-sharing
BestA d: Bestimmungsamt
beständ. d: beständig
BE-Statistics e: Bose-Einstein Statistics
Bestäub. d: Bestäubung
Best.-A[ufn]. d: Bestandsaufnahme
best.-bb. d: beständig bis bedingt beständig
Best Betr Fu d: Bestimmungen über Funkanlagen des beweglichen Betriebsfunks
Best.-H. d: Bestimmungshafen
Best.-Nr. d: Bestellnummer
Bestpr Dr Anl d: Bestimmungen über private Drahtfernmeldeanlagen
BEST Program[me] e: Balanced, Effective, State-related and Training-oriented Program[me]
Bestrahl. d: Bestrahlung
Best.-Sch. d: Bestellschein
BET d: Be- und Entladetechnik
- d: Berufseignungstest
- d: Beschichten, Entwickeln, Tempern
- d: Blockeigenbedarfstransformator
- e: Balanced-Emitter Technology
- e: Balanced-Emitter Transistor
- e: Balanced Emitter Transistor
- e: Best Estimated Trajectory
- e: Best Estimate of Trajectory
- e: Between
- e: Between and Brunauer-Emmett-Teller
- e: Block Equivalence Tape
- e: British Electric Traction
- e: Business Experience Training

BETA e: Basic English for Testing Applications
- e: Battlefield Exploitation and Target Acquisition
- e: Betamax (video-cassette system)
- e: Billion-Channel ExtraTerrestrial Assay
- e: British Equipment Trade Association
- e: Business Equipment Trade Association
Betätspg. d: Betätigungsspannung
Betäub. d: Betäubung
BE Te-Com e: Bachelor of Telecommunication Engineering
BET Equation e: Brunauer-Emmett-Teller Equation
BET-Gleichung d: Brunauer-Emmett-Teller-Gleichung
Bethlehem e: Bethlehem Steel Corporation
Beth Steel e: Bethlehem Steel
BETINA d: Bausteine für ein technisches Informationssystem mit Netzanalyse
BET Method e: Brunauer-Emmett-Teller Method
BET-Methode d: Brunauer, Emmett und Teller-Methode
BetMVVO d: Betäubungsmittel-Verschreibungsverordnung
betn e: between
Betonstein Ztg. d: Betonstein Zeitung (journ.)
Betr. d: Betrag
- d: Betrieb
betr. d: betreffend
- d: betrifft
- e: better
Betra d: Betriebs- und Bauanweisung
Betr.-Ber. d: Betriebsberater
- d: Betriebsberatung
Betr.-Ord. d: Betriebsordnung
Betr.-Sp[g]. d: Betriebsspannung
Betr.-Spgn. d: Betriebsspannungen
BetrSt d: Betriebsstelle
Betr. St. d: Betriebsstoff
betr.-techn. d: betriebstechnisch
BetrV d: Betriebsvermittlung
BetrVfG d: Betriebs-Verfassungsgesetz
BETSIE e: Bets Settlement Investment Engine
Better F e: Better Farming (journ.)
BET-Theorie d: Brunauer-Emmett-Teller-Theorie
BETTS e: Bolt Extrusion Thrust Termination System
betw e: between
BEU e: Base Equipment Unit
- e: Basic Encoding Unit
- e: Binary Elementary Unit
Beug. d: Beugung
Beurt. d: Beurteilung
BEV d: Bahnenergieversorgung
- d: Bergbau-Elektrizitäts-Verbundgemeinschaft
- d: besondere Ersatzstromversorgung
- d: betrieblicher Eigenverbrauch
- d: Böschungsentfernungsverhältnis
- e: Bevatron
- e: Bevel[led]
- e: Beverage
- e: Billion Electron[en] Volt[s]
- e: Bovine Enterovirus
BeV d: Billion Elektronenvolt
BEVR e: Broadcast Electronic Video Recording
BEW d: Bayerische Elektrizitäts-Werke GmbH
bew. d: bewaffnet
- d: beweglich
- d: bewegt
- d: beweisen
- d: bewerten
- d: bewertet
- d: bewurzelt
Bew. d: Bewaffnung
- d: Bewässerung
- d: Bewegung
- d: Beweis
- d: Bewertung
BEW e: 1-Butanol-Ethanol-Water
BEWAG d: Berliner Elektrizitätswerke AG
Bew.-Anl. d: Bewässerungsanlage
Bewässer. d: Bewässerung
bewegl. d: beweglich
Bewert. d: Bewertung
Bewi d: Betriebswirtschaft
bewLaFuSt d: bewegliche Landfunkstelle
Bew.-Richtl. d: Bewertungsrichtlinien
BewTln d: beweglicher Teilnehmer
B Ex e: Bachelor of Expression
BEX e: Broadband Exchange
- e: Broadcast Exchange
BEXEC e: Budget Execution
BEXs e: Broadband Exchanges
BEXTREME e: Beam Extreme
BEY d: Beirut (Flughafen)
bez. d: bezahlt
- d: bezogen
- d: bezüglich
Bez. d: Bezeichnung
- d: Bezirk
- d: Bezug
- e: Bezirk
Bez.-B[au]m d: Bezirksbaumeister
Bezeich. d: Bezeichnung
Bezieh. d: Beziehung[en]
BEZNR d: Bezirk-Nummer
bezw. d: beziehungsweise
BF d: Bachforelle
- d: Bahnhof
- d: Bandfilter
- d: Basisbandfrequenz
- d: Basisfrequenz
- d: Basisfrequenzband
- d: Bastfaser
- d: Batterieschaltfeld
- d: Bauelementefunktion
- d: Bedien[ungs]feld
- d: Befehlsfolge[register]
- d: Befehlszähler
- d: Beförderung
- d: benutzt für
- d: Beschreibungsfunktion
- d: Betriebsfläche
- d: Bewertungsfaktor
- d: Biegefestigkeit
- d: Blasfolienanlage
- d: Blockfaktor
- d: Block Format
- d: Blockungsfaktor
- d: Brikettfabrik
- e: Bachelor of Finance
- e: Bachelor of Forestry
- e: Back-Feed
- e: Back-Folded
- e: Backface
- e: Back Feed
- e: Back Focal
- e: Barrage Fire
- e: Basal Fold
- e: Base Formula
- e: Base Frequency
- e: Base Funded
- e: Base Fuse
- e: Base Fuze
- e: Basic Format
- e: Batch Fabrication
- e: Battery Fire
- e: Bayonet Fighting
- e: Bead Flush
- e: Beam-Foil
- e: Beam Forming
- e: Beat Frequency
- e: Best Fit
- e: Binary Function
- e: Blank Flange
- e: Blind Flange
- e: Block Factor
- e: Blocking Factor
- e: Blue Flag
- e: Boiler Feed
- e: Bold Face
- e: Boolean Factor
- e: Boolean Format
- e: Boolean Function
- e: Both Faces
- e: Bottom Face
- e: Branching Filter
- e: Breaker Failure
- e: Break Field
- e: Brief[ly]
- e: British Frisian
- e: Buffer[ed]
- e: Buffer Fat
- e: Buffer Free
- e: Butter Fat
- e: Byte Field
B/F e: Background/Foreground
- e: Black Female
- e: Brought Forward
- e: Brown Female
- e: Buffer/Formatter
bf i: bassa frequenza
BF 3-MEA d: Borfluorid-3-Monoäthylamin
BFA d: Betriebs-Fernsprechanlage[n]
- d: Betriebsfernsprechanlage
- d: Bildfilmaufzeichnung
- d: Braunkohlenfilterasche
- d: Buchungs- und Fakturierautomat
- d: Bundesforschungsamt
- d: Bundesforschungsanstalt
- e: Bachelor of Fine Arts
- e: Basal Forebrain Area
- e: Broadcasting Foundation of America
BFAC e: Bulletin of the Faculty of Arts; Cairo (journ.)
B-FACT e: Booster Flight-Acceptance Composite Test
BFA in D[r] A[rt] e: Bachelor of Fine Arts in Dramatic Art
BFA [in] Ed e: Bachelor of Fine Arts in Education
BFA [in] Mus e: Bachelor of Fine Arts in Music
BFA in PS e: Bachelor of Fine Arts in Painting and Sculpture
BFA [in] Sp e: Bachelor of Fine Arts in Speech
BFALA e: Bachelor of Fine Arts in Landscape Architecture
BF Algorithm e: Best Fit Algorithm
BFALN e: Buffer Boundary Alignment
BFAn d: Betriebs-Fernsprechanlagen
BFAP e: Binary Fault Analysis Program[me]
BFAS e: Basic File Access System
BFB d: Bibliotheksforum Bayern (journ.)
- d: Branchenfernsprechbuch
- d: Bundesanstalt für Bodenforschung
- e: Broad-Flanged Beam
- e: Bureau of Forensic Ballistics
BFBC e: British Forces Broadcasting Service
BF-Bd d: basisfrequentes Band

BFBS e: British Forces Broadcasting Service
- e: British Forces Broadcasting Station
BFBS-TV e: British Forces Broadcasting Service-Television
BFBU e: British Forces Broadcasting Unit
BFC d: benutzte Facette
- e: Bale Feet; Cubic
- e: Basic/Four Corporation
- e: Basic File Channel
- e: Bending Feedback Control
- e: Benign Febrile Convulsion
- e: Blast Furnace Cement
- e: Brake Force Coefficient
- e: Braking Force Coefficient
- e: Budget and Forecast Calendarization
BFCO e: Band Filter Cut-Off
BFCS e: British Friesian Cattle Society
BFCT e: Boiler Feed Compound Tank
BfD d: Beauftragter für Datenschutz
BFD e: Back Focal Distance
- e: Basic Field Descriptor
- e: Basic Floppy Disk
- e: Beat-Frequency Detection
- e: Bill Feed Device
- e: Book Form Drawing
- e: Boolean Function Designator
BFDC e: Battalion Fire Distribution Center
BF Distance e: Back Focal Distance
Bfe d: Büro für Erfindungs- und Vorschlagwesen (DDR)
BFE e: Bachelor of Forest Engineering
- e: Beam Forming Electrode
- e: Bell Telephone Laboratories Furnished Equipment
- e: Bromotrifluoroethylene
BFeB d: Branchenfernsprechbuch
BF Eng e: Bachelor of Forest Engineering
BFER e: Base Field-Effect Resistor
BFET e: Bipolar-mode Field Effect Transistor
BFET Logic e: Buffered Field Effect Transistor Logic
BFF d: Bundesfachgruppe Bühne, Film, Fernsehen
- e: British Fishing Federation
- e: Buffered Flip-Flop
BFG d: Bearbeitungsfolgegraf
- d: Briefing
- e: Beat Frequency Generator
- e: Binary Frequency Generator
- e: Brute Force Gyro
- e: Buffing
BFH d: Binärzeichenfehlerhäufigkeit
- d: Bitfehlerhäufigkeit
BFHA e: Bulletin of the Friends Historical Association (journ.)
BFHS e: Benjamin Franklin High School
BFI d: Betriebsforschungsinstitut
- e: Batch Freeform Input
- e: Battlefield Interdiction
- e: Beam-Forming Interfact
- e: Bismuth Formic Iodide
- e: British Film Industries
- e: British Film Institute
- e: Business Forms Institute
BFIA e: British Flour Industry Association
BFIC e: Binary Fault Isolation Chart
BFID e: Boolean Function Identifier
BFJ e: Booster Fuel Jacket
BfK d: Bahnhofskasse
BFK d: benutzte Facettenkombination
BfK d: Briefkasten
Bfk d: Brikettfabrik
BFL d: Bandreduktion durch Frequenzverschachtelung und mit Hilfe eines Laufzeitgliedes
bFl d: beweglicher Flugfunkdienst
BFL e: Bachelor of Family Life
- e: Back Focal Length
- e: Beginning File Label
- e: Buffered Field Effect Transistor Logic
BFLF i: Biblioteca della Facoltà di Lettere e Filosofia
BFLs e: Buffered Field Effect Transistor Logics
BFLW e: British Federation of University Women
BFM e: Barium Ferrite Magnet
- e: Basic Field Manual
BFMA e: British Forces Maintenance Area
- e: Business Forms Management Association
B-FMDCPSK e: Binary-Frequency- Modulated Differentially Coherent Phase-Shift-Keyed
- e: Binary-Frequency-Modulated Differentially Coherent Phase-Shift-Keying
BFMDS e: Base Flight Management Data System
BFMIRA e: British Food Manufacturing Industries Research Association
BFMO e: Base Fuels Management Officer
BFMRA e: British Food Manufacturers Research Association
BF Muscles e: Biceps Femoris Muscles
BFn d: Beschreibungsfunktion
Bfn e: Bloemfontein
BFN e: British Forces Network
Bfo d: Bahnhofsfahrordnung
BFO e: Balance Forearm Orthesis
- e: Beat Frequency Oscillator
- e: Blood Forming Organ[s]
BFor e: Bachelor of Forestry
BFORM e: Budget Formulation
BForSc[i] e: Bachelor of Forestry Science
BFOZP e: Best-Fit Optic Z-plane
BFP e: Battery Fuse Panel
- e: Bill Feed Printer
- e: Biological False Positive [Reactions]
- e: Boiler Feed Pump
- e: Bottom Finding Pinger
- e: Buffered Fast Printer
BFPDDA e: Binary Floating Point Digital Differential Analyzer
BFPHC e: Basic Filter Power Handling Capacity
BFpl d: Buchfahrplan
BFPPS e: Bureau of Foods, Pesticides and Product Safety
BFPR e: Binary Floating Point Resistor
BFP Reaction e: Biological False Positive Reaction
BFR d: Befehlsregister
- d: Betonfenster
- d: Betriebsführungsrechner
- d: Bitfehlerrate
- e: Barrier Film Rectifier
- e: Before
- e: Biologic False Reactor
- e: Block Format Recording
- e: Blood Flow Rate
- e: Bone Formation Rate
- e: Bridged Frequency Ringing
- e: Brush Front Regularly
- e: Buffer
Bfrd e: Buffered
BFRL e: Basic Facility Requirements List
BFRP e: Boron Fibre (or Fiber) Reinforced Plastics
BFRS e: Bio-Feedback Research Society
BFR SOL e: Buffered Ringers Solution
BFRST e: Buffered Start
BFS d: Bundesanstalt für Flugsicherung
- e: Bachelor of Foreign Service
- e: Band Filter Set
- e: Basic File System
- e: Battlefield Surveillance
- e: Beam-Foil Spectroscopy
- e: Blast Furnace Slag
- e: Blood Fasting Sugar
- e: Business File System
BfS d: Büro für Standardisierung (DDR)
BFSO e: Base Fuel Supply Officer
BFStrG d: Bundesfernstraßengesetz
BFt d: Beamter im mittleren fernmeldetechnischen Dienst
BFT e: Bachelor of Foreign Trade
- e: Batch Fabrication Technique
- e: Bentonite Flocculation Test
- e: Betonite Flocculation Test
- e: Biofeedback Training
BFTC e: Boeing Flight Test Center
BFTP e: Bailment Flight Test Program[me]
BFTSC e: Brassboard Fault Tolerant Spaceborne Computer
BFTU e: Boeing Field Test Unit
BFU e: Benjamin Franklin University
BFV d: Bayerischer Forstverband
BFVMTL e: Baseline Flight Vehicle Mission Time Line
BfVS d: Bund für Vogelschutz
BFW d: Büro für Wirtschaftsinformatik GmbH
- e: Boiler Feedwater
BFZ e: Branch of Fall Zero
BG d: Bandgerät
- d: Bandgeschwindigkeit
- d: Basisgerät
- d: Baugruppe
- d: Baugruppen und Geräte
- d: Baukastengewindebohreinheit
- d: Bedienungsgerät
- d: Bedienungsgröße
- d: bedingt gestattet
- d: Befehlsgeber
- d: Berg
- d: Biegung
- d: Bildschirmgerät
- d: Bindegarn
- d: Bindegewebe
- d: Bitgruppe
- d: Blechschraubengewinde
- d: Bogen
- d: Buchungsstelle für Fernmeldegebühren
- d: Bulgarien (Kfz)
- e: Back Gear
- e: Bag
- e: Bale / Grain
- e: Basal Groove
- e: Basic Group
- e: Bearing
- e: Bengal
- e: Bevel Gear
- e: Billing Group
- e: Birmingham Gauge
- e: Bit Group
- e: Block Gap
- e: Blood Group
- e: Bluish Green
- e: Bonded Goods
- e: British Gauge
- e: British Guyana
- e: Broad Gauge
- e: Buccal Ganglion
- e: Buccogingival
- e: Bulgaria

- e: Bus Grant
- f: Bibliothèque publique et universitaire de Genève
- i: Buccogingival
BGA d: Betriebsgeräuschspannungsabstand
- e: Beet Growers Association
- e: Blue-Green Algae [Virus]
BGAM e: Basic Graphic Access Method
Bg.-Ass[ess]. d: Bergassessor
Bgb. d: Bergbau
BGB d: Bundesbahn-Güterkursbuch
- d: Bürgerliches Gesetzbuch
- e: Booksellers of Great Britain
B-G B[azillus] d: Bordet-Gengou Bazillus
BGBL d: Bundesgesetzblatt
BGC e: Binary Group Code
- e: Blood Group Class
- e: British Gas Corporation
- e: British Gas Council
BGCG e: British Guiana Consolidated Goldfields Ltd.
BGCOB e: Background-compiler Common Business Oriented Language
BGD e: Billion[s] Gallon[s] per Day
BGDA e: Blue Grass Depot Activity
BGDD e: Board Gardner Denver Data
BGDSL d: Beiträge zur Geschichte der Deutschen Sprache und Literatur (journ.)
BGDSLH d: Beiträge zur Geschichte der Deutschen Sprache und Literatur; Halle (journ.)
BGDSLT d: Beiträge zur Geschichte der Deutschen Sprache und Literatur; Tübingen (journ.)
BGE e: Bachelor of Geological Engineering
- e: Bull-General Electric
- e: Butyl Glycidyl Ether
BG & E e: Baltimore Gas and Electric
B Ge E[ng] e: Bachelor of Geological Engineering
BGeh d: Bundesbahngehilfe
BGEN e: Biogen Incorporated
B G[en] Ed e: Bachelor of General Education
BGF d: Blutgerinnungsfaktoren
- e: Basic Group Formation
BGFE e: Boston Grain and Flour Exchange
bgft d: bergfeucht
BGFZ d: beschränkt gültiges (Flug-) Funksprechzeugnis
BGG e: Black Granite Gage
- e: Booster Gas Generator
- e: Bovine Gamma Globulin
BGGRA e: British Gelatine and Glue Research Association
BGH e: Bovine Growth Hormone
- e: Bovine Pituitary Growth Hormone
- e: Brooke General Hospital
Bg.-H[au]ptm. d: Berghauptmann
BGIRA e: British Glass Industry Research Association
BGK d: Bundesbahn-Güterkursbuch
Bgk e: Bangkok [airport]
BGK-Equation e: Bhatnagar-Gross-Krook Equation
BGKL d: Brandgefahrenklasse
B.G.L. e: Bachelor of General Laws
BGL e: Below Ground Level
BGLB e: Brilliant Green Lactose Broth
BGLR e: Bugler
BGM e: Basegram
- e: Bethnal Green Museum
- e: Bulgemaster
BGMA e: British Gear Manufacturers Association
Bgm[str]. d: Bergmeister
Bgmstr 1c e: Buglemaster First Class
Bgmstr 2c e: Buglemaster Second Class
BGn d: Baugruppen
BGN e: Board of Geographic Names
- e: Board on Geographic Names
BG-NR d: Baugruppennummer
BGNR e: Barren Grounds Nature Reserve
BGP d: Betriebsgüteprüfgerät
- e: B-Glycerophosphase
- e: B-Glycerophosphate
- e: Background Program[me]
- e: Barrier; Breaseproof
- e: Below-Ground Pool
BGPDC Program[me] e: Brouwer General Pertubations Differential Correction Program[me]
BG PROGRAM e: Background Program[me]
BGPW e: Bare Gold Plated Wire
Bg Pz d: Bergpanzer
BGr d: Baugruppe
BGR d: Baugruppenrahmen
- d: Begrenzungsregelung
- e: Back-Gear[ed]
- e: Basal Granule
- e: Bombing and Gunnery Range
bgr. d: begradigen
Bgr. d: Begradigung
BgR d: Bergrat
BGRR e: Brookhavens Graphite Research Reactor
BGRV e: Boost Glide Reentry Vehicle
Bgrz. d: Begrenzung
bgrzt. d: begrenzt
BGS d: Baugruppensystem
- d: Bildschirmgruppensteuergerät
- d: Blockgruppensteuerung
- e: Bachelor of General Studies
- e: Backup Gimbal Servo
- e: Bags
- e: Basic Graphic System
- e: Beacon Ground S-Band
- e: Bntish Geriatrics Society
- e: Boeing Ground Support
- e: Bombing and Gunnery School
- e: British Goat Society
- e: British Grassland Society
BGSA e: Blood-Granulocyte Specific Activity
BGSC e: Belfer Graduate School of Science
BGSIA e: Beginning Standard Instrument Approach
BGSM e: Bowman Gray School of Medicine
BGSS e: Battalion Ground Surveillance Section
BGSTA e: Beginning Straight-in Approach
BGSU e: Bowling Green State University
BGT d: Baugruppenträger
- e: Basic Group Translator
- e: British Guinea Time
BGTT e: Borderline Glucose Tolerance Test
BGU e: Ben Gurion University
- e: Bowling Green University
- e: British Guiana
- p: Biblioteco General da Universidade
BgUL f: Bibliothèque générale de l'Université de Liège
BGU Timber e: Blue Gum Timber
BGV d: Bezirksgasverteilung
BGW d: Bereichsgruppenwähler
Bgw. d: Bergwerk
BGW d: Bezirksgruppenwähler
- d: Bundesverband der deutschen Gas- und Wasserwirtschaft
- d: VEB Berliner Glühlampenwerk (DDR)
- e: Back Gear Winder
- e: Battlefeld Guided Weapon
Bgw.-Dir. d: Bergwerksdirektor
Bg-Wz. d: Bogen-Wasserzeichen
BgZ d: Beginnzeichen
BGZ d: Berufsbildungs- und Gewerbeförderungszentrum
- d: Blutgerinnungszeit
BH d: Bahnhelfer
- d: Bauherr
- d: Bauhöhe
- d: Befehl Holen
- d: behelfsmäßig
- d: Berghauptmann
- d: Brammen mit Haube
- d: Brest/Hamburg
- d: British-Honduras (Kfz)
- d: Buchhaltung
- d: Buchhandlung
- e: Bachelor of Hebrew
- e: Bachelor of Humanics
- e: Bahrain
- e: Base Hospital
- e: Basic Hardware
- e: Basic Heterostructure
- e: Beachhead
- e: Betriebstunde[n]
- e: Bill of Health
- e: Binary-to-Hexadecimal
- e: Blasthole
- e: Block Handler
- e: Block Head
- e: Boiler House
- e: Borehole
- e: Bottomhung
- e: Brain Hormone
- e: Brake Horsepower
- e: Branch Highway
- e: Breast Height
- e: Bridewell Hospital
- e: Brinell Hardness
- e: British Honduras
- e: Bulkhead
- e: Buried Heterostructure
- e: Burlington House
- e: Busy Hour
B.-H. e: Bordeaux to Hamburg
BHA d: Butylhydroxyanisol[e]
- e: Bachelor of Hospital Administration
- e: Base Helix Angle
- e: Benzene Hexachloride
- e: Bleed Hose Assembly
- e: British Homeopathic Association
- e: British Honduras Airways
- e: British Horse Association
- e: Butylated Hydroxyanisole
BHAD e: Broach Adapter
BH Adm e: Bachelor of Hospital Administration
BHAS e: Burroughs Hospital Accounting System
- e: Burroughs Hospital Administration System
Bhaufs d: Betriebshauptaufseher
BHB acid e: D(-)-ß-Hydroxybutyric acid
BHBNM e: Big Hole Battlefield National Monument
BHBR e: Boat Harbor
BHbv d: Bundesbahn-Hausbrandversorgung
BHC d: Benzolhexachlorid
- e: Ballistic Height Correction
- e: Beaching Cradle
- e: Beam Heated Cathode
- e: Benzene Hexachloride
- e: Bishydroxycoumarin
- e: Block Hardware Code
- e: Blockhouse Computer
- e: Bore Hole Capsule
- e: British Hovercraft

- e: Busy Hour Call[s]
BHCA e: Bassett Hound Club of America
- e: Busy Hour Call[s] Attempt[s]
BHCC e: Better Heating-Cooling Council
BHD e: Beachhead
- e: Billhead
- e: Binary Homing Device
- e: Bulkhead
BHDF e: British Hospital Doctors Federation
BHDM e: British Home Defense Measures
BHE e: Bachelor of Home Economics
- e: Bi-Harmonic Equation
BH Ec e: Bachelor of Home Economics
BHEc e: Bachelor of Household Economics
BHET e: Bis (ß-Hydroxyethyl) Terephthalate
BHE Timber e: Bhendi Timber
BHEW e: Benton Harbor Engineering Works
Bhf d: Bahnhof
BHF d: Bonner Historische Forschungen (journ.)
- e: Background Heat Flux
BHFR e: Brookhaven High Flux Reactor
BHFX e: Broach[ing] Fixture
BHG d: Bäuerliche Handelsgenossenschaft
- d: Bauhauptgruppe[n]
- d: Bundeshochschulgesetz
BHG Equation e: Booth-Henry Gorin Equation
BHHI e: British Home and Hospital for Incurables
BHI d: Bank für Handel und Industrie
- e: Brain Heart Infusion
- e: British Horological Institute
- e: British Humanities Index
- e: Burst-Height Indicator
- f: Bureau Hydrographique Internationale
BHIB e: Beef Hearth Infusion Broth
BHIF e: Better Highways Information Foundation
BHIPS e: Burroughs Hospital Information Processing System
BHIS e: Burroughs Hospital Information System
BHJ e: Bulkhead Jack
BHK e: Baby Hamster Kidney
- e: type-B Hong Kong influenza virus
BHL e: Bachelor of Hebrew Letters
- e: Bachelor of Hebrew Literature
- e: Bachelor of Humane Letters
- e: Biological Half Life
- e: Busy Hour Load
BHLF e: Basic High Layer Functions
BHL Syndrome e: Bilateral Hilar Lymphoma Syndrome
bhm. d: behelfsmäßig
BHM d: Berg- und Hüttenmännische Monatshefte (journ.)
- e: Beef Heart Mitochondria
- e: Birmingham
- e: Busy Hour Mode
BHMA e: British Hard Metal Association
BH Macro e: Block Handling Macro
BHMO e: Blue Hill Meteorological Observatory
BHMRA e: British Hydromechanics Research Association
BHN d: Babyhamsterniere[n]
BHN e: Bephenium Hydroxynaphthoate
Bhn e: Brinell Hardness Number
BHNWR e: Bombay Hook National Wildlife Refuge
BHO e: Branch Hydrographic Office
B Ho Ec e: Bachelor of Household Economy
BHOLD e: Background Hold
B Hond e: British Honduras
B&hor e: B-horizon
B Hor e: Bachelor of Horticulture
B Ho Sc e: Bachelor of Household Science
BHP e: Biological Hazard Power
- e: Boiler Horsepower
- e: Bottom Hole Pressure
- e: Brake Horsepower
- e: Brashear-Hastings Prism
BHP-HR e: Brake Horsepower-Hour
BHR d: Bauchhautreflex[e]
Bhr e: Bahrain
BHR e: Basal Heart Rate
- e: Biotechnology & Human Research
- e: Block Handling Routine
- e: Bulkhead Receptable
BHRA e: British Hydromechanical Research Association
- e: British Hydromechanics Research Association
BHRBASAL e: Biotechnology And Human Research
BHRF d: Bohrfett
Bhrg. d: Bohrung
BHRI e: Brewers Hop Research Institute
B & HRO e: Biotechnology and Human Research Office
BHS d: Batteriehauptschalter
- d: Bluthirnschranke
- d: Bundesbahnhauptsekretär
- e: Bachelor of Home Science
- e: Balboa High School
- e: Base Heat Shield
- e: Basic Hole System
- e: Betahemolytic Streptococcus
- e: Binding Head Steel
- e: Bore Hole Seismometer
- e: Boys High School
- e: Bulletin of Hispanic Studies
- e: Burlesque Historical Society
- e: Bushwick High School
BHSASSC e: Block Handler Set Association
BHSc e: Bachelor of Household Science
BHsch d: Bundesbahnhauptschaffner
BH Set e: Block Handler Set
BHSL e: Basic Hytran Simulation Language
BHSMCo e: Bulletin of the Historical Society of Montgomery County (journ.)
BHSS e: Bronx High School of Science
BHST e: Bottom-Hole Static Temperature
BHT- d: Braunkohlenhochtemperatur-
BHT d: Butylhydroxytoluen
- d: Butylhydroxytoluol
- d: butyliertes Hydroxytoluol
- e: Bureau of Highway Traffic
- e: Butylated Hydroxytoluene
BHTD e: Bureau of Hygiene and Tropical Diseases
BHT-Koks d: Braunkohlenhochtemperatur-koks
B Hu e: Bachelor of Humanities
BHU e: Banaras Hindu University
Bhu. e: Bhutan
BHU e: Busy Hour Usage
Bhutan e: Kingdom of Bhutan
Bh V d: Beihilfevorschriften
BH/VH e: Body Hematocrit/Venous Hemocrat [Ratio]
BHW e: Boiling Heavy Water
- e: Boston Hospital for Women
BHWR e: Boiling Heavy Water Reactor
B Hy[g] e: Bachelor of Hygiene
BI d: Bauingenieur
- d: Bauinspektor
- d: Bipolar
- d: Birke
- d: Bundesbahninspektor
- d: Wismut
- e: Babson Institute
- e: Background Investigation
- e: Backward Indicator
- e: Bacteriological Index
- e: Bahama Islands
- e: Balearic Islands
- e: Base Identifier
- e: Base Ignition
- e: Base Infantry
- e: Base Injection
- e: Base of Prism in
- e: Basic Information
- e: Basic Instruction
- e: Batch Input
- e: Battery Inverter
- e: Battlefield Illumination
- e: Bearing Inch
- e: Beer Institute
- e: Bermuda Islands
- e: Bicycle
- e: Billing Instructions
- e: Binary
- e: Binary Information
- e: Binary Input
- e: Binary Item
- e: Biopsy
- e: Biot[ite]
- e: Bismuth
- e: Bismuthum
- e: Bits
- e: Black Iron
- e: Blanking Input
- e: Blank Instruction
- e: Block Instruction
- e: Body Injury
- e: Branch[ing] Instruction
- e: Breakpoint Instruction
- e: British India
- e: British Isles
- e: Brookings Institution
- e: Buffer Index
- e: Bulk Issue
- e: Burn Index
- e: Burundi
- e: Bus Interface
B & I e: Base and Increment
BIA e: Bachelor of Industrial Arts
- e: Barack Inventory Accounting
- e: Bee Industries Association
- e: Bicycle Institute of America
- e: Binding Industries of America
- e: Boost, Insertion and Abort
- e: Booster Insertion and Abort
- e: Booster Interstage Assembly
- e: Braille Institute of America
- e: Brazilian International Airlines
- e: Brick Institute of America
- e: Bristle Institute of America
- e: British Institute of Acupuncture
- e: Building Industry Association
- e: Buyers Information Advisory
BIAA e: Bee Industries Association of America
BIAE e: British Institute of Adult Education
BIanw d: Bundesbahninspektoranwärter
BI-APS e: Battery Inverter Accessory Power Supply
BIAR e: Base Installation Action Requirements
- e: Base Interrupt Address Register
BI Arch e: Bachelor of Interior Architecture

BI Arch E[ng] e: Bachelor of Interior Architectural Engineering
BIAS d: Bibliotheks-Ausleihesystem
- e: Battlefield Illumination Airborne System
- e: Broadcast Industry Automation System
- e: Brooklyn Institute of Arts and Sciences
BIAT e: Burn-in/Ag[e]ing Tester
- e: Burn-In/Aging Tester
BIAX e: Biaxial
BIB a: bibliotek
- d: Berliner Institut für Betriebsführung
- d: Bibtiothek
- d: Bildungszentrum für informationsverarbeitende Berufe
- e: Baby Incendiary Bomb
- e: Backward Indicator Bit
- e: Balanced Incomplete Block
- e: Biennale of Illustrations Bratislava
- e: Biographical Information Blank
- e: Biographical Inventory Blank
- e: Bottled in Bond
- f: bibliothèque
- g: bibliotheke
- i: biblioteca
- l: biblioteca
- l: bibliotheka
- n: bibliotheek
- p: biblioteca
- P: biblioteka
- r: bibliotek
- R: biblioteka
- s: biblioteca
Bib e: Bible
- l: Biblica
BIBA e: Babson Institute of Business Administration
Bib Amb i: Biblioteca Ambrosiana
Bib Apo i: Vat Biblioteca Apostolica Vaticana
bib b a: biblioteksbind
Bib Bod l: Bibliotheca Bodmeriana
Bib Cen s: Biblioteca Central / Mexico Citys Ciudad Universitari
BIBD e: Balanced Incomplete Block Designs
BIBDES e: Bibliographic Data Entry System
BIBEAS d: Bibliotheks-Eingabe-Ausgabe-System
Bib Ecu s: Biblioteca Ecuatoriana / Quito
Bib Esc s: Biblioteca de San Lorenzo el Real de El Escorial
BIBL d: Bibliothek
- e: Biblical
- e: Bibliographer
- e: Bibliographical
- e: Bibliography
- e: Bibliotec[-]
- e: Bibliothèque
BIBLICOS e: Bibliographic Computer System
BIBLIO d: Bibliotheksautomatisierung
- e: Biblioclasm
- e: Biblioclast
- e: Bibliogenesis
- e: Bibliognost
- e: Bibliogony
- e: Bibliograph
- e: Bibliographer
- e: Bibliographic[al]
- e: Bibliography
- e: Bibliographical imprint or note
BIBLIOC e: Biblioclasm
- e: Biblioclast
BIBLIO-DAT d: Datenbank der Deutschen Bibliothek
BIBLIOG e: Bibliographee
- e: Bibliographer
- e: Bibliographic[al]
- e: Bibliography
Bibliogr. d: Bibliographie
BIBLIOGRAPH e: Bibliographer
- e: Bibliography
Bibliogr. Dokum. Terminol. d: Bibliographie, Dokumentation, Terminologie
BIBLIOKLEPT e: Biblikleptomania[c]
BIBLIOL e: Bibliolater
- e: Bibliolatrous
- e: Bibliolatry
- e: Bibliological
- e: Bibliologist
- e: Bibliology
BIBLIOM e: Bibliomancy
- e: Bibliomane
- e: Bibliomania[c]
- e: Bibliomanist
BIBLIOP e: Bibliopegic
- e: Bibliopegist
- e: Bibliopegy
- e: Bibliophagist
- e: Bibliophile
- e: Bibliophilia
- e: Bibliophobe
- e: Bibliophobia
- e: Bibliopole
BIBLIOPSY e: Bibliopsychology
BIBLIOS e: Book Inventory Building and Library Information Oriented System
BIBLIOTBER e: Bibliotherapeutic
- e: Bibliotherapist
- e: Bibliotherapy Bibliotrain Railroad Car
BIBLIOTHEC e: Bibliotheca[l]
- e: Bibliothecar[y]
BIBL MUN f: Bibliothèque Municipale
BiblZ d: Biblische Zeitschrift (journ.)
Bib Mus i: Biblioteca Musicale
Bib Nac s: Biblioteca Nacional
Bib Nar P: Biblioteka Narodowa
Bib Nat f: Bibtiothèque Nationale
Bib Naz Bra i: Biblioteca Nazionale Braidense
Bib Naz Cen i: Biblioteca Nazionale Centrale
BIBNET e: Bibliographic Network
Bi/Bo d: Benzin/Benzol
BIBO e: Bounded Input Bounded Output
BIBOM e: Binary Input-Binary Output Machine
BIBOMM d: Binary Input-Binary Output Moore Maschine
Bib Pal s: Biblioteca de Palacio
BIBR e: Backward Indicator Bit Received
BibR e: Biblical Review (journ.)
BIBRA e: British Industrial Biological Research Association
BIBS e: Bibliographies
- e: Built-in Breathing System
Bib Soc Am e: Bibliographical Society of America
Bib Soc Can e: Bibliographical Society of Canada
Bib Sor f: Bibliothèque de la Sorbonne
BIBT e: Backward Indicator Bit to be Transmitted
Bib Uni s: Biblioteca Universitaria
BibWorld e: Biblical World (journ.)
BIBX e: Backward Indicator Bit Expected
BIC d: 4-(3,3-Bis-(2-chloräthyl)-1-triazeno)-imidazol-5-carboxamid
- e: Barium Ion Cloud
- e: Battlefield Information Center
- e: Bear[er] Identification Code
- e: Biodeterioration Information Centre
- e: Biographical Inventory Creativity
- e: Biomedical Instrumentation Consultant
- e: Bipolar Integrated Circuit
- e: Bombardment-Induced Conductivity
- e: British Information Centre
- e: British Institute of Cinematographers
- e: British Insulated Cables
- e: Broadband Interface Controller
- e: Buffer Interlace Controller
- e: Buried Isolation Capacitor
- e: Business Information Center
- e: Byte Input Control
- f: Bureau International des Containers
BICAP e: Binary Capacitor
BICAPs e: Binary Capacitors
BICARB e: [Sodium] Bicarbonate
BICARSA e: Billing, Inventory Control, Accounts Receivable, Sales Analysis
BICC e: Battlefield Information Control Center
- e: British Insulated Callenders Cables
BICE e: Base for Information and Communication in Europe
BICEMA e: British Internal Combustion Engine Manufacturers Association
BICEP e: British Industrial Collaborative Exponential
BICEPS e: Basic Industrial Control Engineering Programming System
BICEPT e: Book Indexing with Context and Entry Points from Text
BICERA e: British Internal Combustion Engine Research Association
BICERI e: British Internal Combustion Engine Research Institute
BICFET e: Bipolar Inversion Channel Field Effect Transistor
BICHLOME e: Sodium Bichromate
BICHS e: Biconditional
- e: Bulletin of the International Committee of Historical Sciences (journ.)
BICMOS e: Bipolar Complementary Metal Oxide Semiconductor
BICP e: Biomedical Interdisciplinary Curriculum Project
BICS e: British Institute of Cleaning Science
- e: Burroughs Inventory Control System
BICSc e: British Institute of Cleaning Science
BICT e: Beijing Institute of Computer Technology
BICV e: Biconcave
BICX e: Biconvex
BICYPLANE e: Bicycle-Powered Airplane
BID d: Beta-Ionisationsdetektor
- e: Bachelor of Industrial Design
- e: Bachelor of Interior Design
- e: Block Interaction Diagram
- e: Blow in Door
- e: Brought in Dead
BId e: Bureau of Identification
BIDAP e: Bibliographic Data Processing [Program[me]]
BIDAP Program[me] e: Bibliographic Data Processing Program[me]
BIDEC e: Binary-to-Decimal [Converter]
BIDEMATRON e: Beam-Injected Distributed-Emission Magnetron
bidest. d: bidestilliert
BIDOPS e: Bi-Doppler Scoring
BIDP e: Basic Institutional Development Program[me]
BIDS e: Bendix Integrated Data System
- e: Burroughs Input and Display System
BIDT e: Burroughs Input and Display Terminal

BIE e: Bachelor of Industrial Engineering
- e: Binural Intensity Effect
- e: Branch Instruction Execution
- e: British Institute of Engineers
- e: Bureau of Industrial Economics
- e: Business-Industry-Education
- f: Bureau Internationale d'Electrothermie
BIEE e: British Institute of Electrical Engineers
Bieg. d: Biegung
BIEN e: Basic Income European Network
BI Eng e: Bachelor of Industrial Engineering
BIET e: British Institute of Engineering Technology
BIETA s: Biblioteca Interamericana de Estadistica Teórica y Aplicada
BiF d: Biegeform
BIF d: Bildfernsprechen
- d: Bildfernsprecher
- e: Banded Iron Formation
- e: Bombardiers Information File
- e: British Industries Federation
BIFA d: Bundesinstitut für Arbeitsschutz
BIFAC e: British Isles Federation of Agricultural Co-operatives
BIFET d: bipolarer Feldeffekttransistor
- e: Bi-Field Effect Transistor
- e: Bipolar[-junction] Field Effect Transistor
BIFET-Technik d: bipolare Feldeffekt-transistor-Technik
BIFET Technology e: Bipolar[-junction] Field Effect Transistor Technology
BIFID Program[me] e: Bifurcation Identification Program[me]
BIFN e: Bifurcation
BIFOA d: Betriebswirtschaftliches Institut für Organisation und Automation
BIFORE e: Binary Fourier Representation
BIFS e: Bell Information Flow System
BIFUS e: British, Italian, French, United States
BIG d: Bund der Ingenieure des Gartenbaues
- e: Best in Group
- e: Biological Isolation Garment
- e: Business Investment Game
Big Belts e: Big Belt Mountains of Montana
BIGENA e: Bibliography and Index of Geology Exclusive of North America
BIGFERN d: breitbandiges integriertes Glasfaser-Fernnetz
BIG-FET e: Bipolar Insulated-Gate Field-Effect Transistor
BIGFON d: breitbandiges integriertes Glasfaser-Fernmelde-Ortsnetz
BIGIT e: Binary Digit
BIGOT e: British Intelligence General Orders Transferred
BIGS e: Biological Isolation Garment
BIGS e: Boeing Interactive Graphics System
- e: Booster Inertial Guidance System
BIH e: Benign Intracranial Hypertension
- e: Beth Israel Hospital
- e: Built-in Hold
BIHFS e: British Institute of Hardwood Flooring Specialists
bihor. l: bihorium
BIHP e: Bulletin of the Institute of History and Philology (journ.)
BIHU-Gas d: biologisches Humusgas
BII e: Background Illumination Intensity
- e: Beckman Instruments, Inc.
- e: Biosophical Institute, Inc.
- e: Butanol Insoluble Iodine
BIIA e: British Institute of Industrial Art
BIIL e: Basic Issue Item List
BI-IPS e: Battery Inverter Instrument [action] Power Supply
- e: Battery Inverter Instrument[ation] Power Supply
BIIT e: Blood Incubation Infectivity Test
- e: Blood Inoculation Infectivity Test
- f: Bureau International d`Information sur les Télécommunications
bijb n: bijbelse term
BIJ Timber e: Bijasal Timber
bijv n: bijvoorbeeld
BIKAS d: Bielefelder Katalogsystem
BIKE e: Bicycle
- e: Biotechnology Information Knot for Europe
BIKERS e: Motorcyclists
BiKM d: biologische Kampfmittel
BIKO d: Bildschirmkommandoprogramm
BIL d: Bundesverband der Ingenieure für Landbau
- e: Base Isolation Level
- e: Basic [Impulse] Insulation Level
- e: Belgian Institute in London
- e: Billet
- e: Billion
- e: Block Input Length
- e: Blue Indicating Lamp
- e: Blue Indicating Light
- e: Blue Indicator Light
- e: Braille Institute Library
- e: Brother-in-Law
- e: Bulk Items List
- e: Buried Injector Logic
BILA d: Bibliothek Laden
- d: Bilanzierung
- e: Battelle Institute Learning Automation
- e: Bible Institute of Los Angeles
BILAN d: Bilanzierung
BILAT e: Bilateral
BILB e: Built-in Light Beacon
BILBO e: Built-in Logic Block Observation
- e: Built-in Logic Block Observer
Bild. d: Bildung
BILD d: bistabile Laserdiode
- e: Bistable Laser Diode
Bildg. d: Bildung
bildl. d: bildlich
Bild Wiss. d: Bild der Wissenschaft (journ.)
BILE e: Balanced-Inductor Logical Element
BILED d: bistabile lichtemittierende Diode
- e: Bistable Light Emitting Diode
BILG e: Building Industry Libraries Group
BIL GAL e: Billion Gallons
BILGE e: Beating Inflation in Local Government Establishment
- e: Binary Load Generation
BILI e: Basic Issue List Items
BILI e: Bilirubin
BILING e: Bilingual[ism]
- e: Bilingualist[ic][al][ly]
BilK e: Bilge Keel
BILLI e: Billionth
BILLION e: 10^{12} (GB)
- e: 10^{9} (US)
BIL[S] e: Billion[s]
BILS e: British International Law Society
- e: Butterworth Industrial Laws Service
BIM e: Bachelor of Indian Medicine
- e: Bachelor of Industrial Management
- e: Basic Industrial Materials
- e: Beginning of Information Mark[er]
- e: Best In Match
- e: Biographical Inventory for Medicine
- e: Blade Inspection Method
- e: British Institute of Management
- e: British Institute of Metals
- e: Bus[link-]Interface Module
BI-M e: Bimonthly
bim i: bimensile
- i: bimestrale
- i: bimestre
BIMA e: Business and Industry Management Abstracts
BIMAC e: Bistable Magnetic Core
BIMACS e: Blood Bank Information and Management Control System
BIMAG e: Bistable Magnetic
BIMAG Core e: Bistable Magnetic Core
BIMAP e: Bill of Material Processor
BIMATRON e: Beam Injected Magnetron
BIMC e: Baltic and International Maritime Conference
BIMCAM e: British Industrial Measuring and Control Apparatus Manufacturers Association
BIMCAM Association e: British Industrial Measuring and Control Apparatus Manufacturers Association
BIMCO e: Baltic and International Maritime Conference
BI-MM e: Base Installation Minuteman
BIMOS e: Bipolar combined with Metal Oxide Semiconductor Field-Effect Transistor
- e: Bipolar Metal Oxide Semiconductor
BIMOS Technique e: Bipolar and Metal Oxide Semiconductor Technique
BIMOT e: Bimotor
BIMS e: Battlefield Integration Management System
- e: Business Information Management System
BimSchG d: Bundesimmissionsschutzgesetz
BImSchV d: Bundesimmissionsschutz-verordnung
BImSchVwV d: Bundesimmissionsschutz-Verwaltungsverordnung
BIN d: binär
- e: Bank Identification Number
- e: Basic Identification Number
- e: Basic Information Network
- e: Bell Information Network
- e: Billion Instructions
- e: Binary
- e: Boeing-Michigan Air Research Center Interception
- e: Bulletin of International News
- e: Business Information Network
Bin. d: Binnenschiff
BINA f: Bureau International des Normes de l'Automobile
BINAC e: Binary Automatic Computer
- e: High-Speed Electronic Digital Computer
Bin Cde Dec S e: Binary Coded Decimal System
Bin Cl e: Binary Cell
BINCODE e: Binary Code
BINCOS e: Binder Control Subsystem
Bind. d: Bindung
B Ind e: Bachelor of Industry
BIND e: Binding
B Ind E e: Bachelor of Industrial Engineering
B Ind Ed e: Bachelor of Industrial Education
bind H e: Binding Head
Bin Dig e: Binary Digit
B Ind Mgt e: Bachelor of Industrial Management

BINDT e: British Institute of Non Destructive Testing
BINET e: Bicentennial Information Network
- e: Binet-Simon [psychological test]
Bing d: Betriebsingenieur
BINIT e: Binary Digit
BINM e: Buck Island National Monument
BIN NO SYS e: Binary Number System
BinnSchG d: Binnenschiffahrtsgesetz
Binn Sch SO d: Binnenschiffahrtsstraßen-ordnung
Bino d: Baukasteninformationssystem nach genormten Operationskarten
BINOCAM e: Binocular and Camera
BINOCS e: Binoculars
BINOM e: Binomi[n]al
BINOMEXP e: Binomi[n]al Expansion
BINOVC e: Break in Overcast
Bin Pt e: Binary Point
BINR e: Baseband Intrinsic Noise Ratio
BINS e: Barclays Integrated Network System
BInsp. d: Bauinspektor
B. Intel. O. e: Base Intelligence Officer
B Int L e: Bachelor of International Law
Bin Tr e: Binary Trigger
BINV d: Binärsignalverarbeitung
BINWR e: Blackbeard Island National Wildlife Refuge
BIO d: Bildinterpretation-Online
- e: Base Installation Officer
- e: Basic Input/Output
- e: Bedford Institute of Oceanography
- e: Biographical
- e: Biography
- e: Biologic[al]
- e: Biological Information-Processing Organization
- e: Biology
- e: Biomedical Information-Processing Organization
- e: Biophysics
- e: Branch Intelligence Officer
- e: Brigade Intelligence Officer
- e: Broadcasting Information Office
BI/O e: Buffered Input/Output
BIOACT e: Bioactive
- e: Bioactivity
BIOASTRO e: Bioastronaut[ic][al][ly]
BIOAUTO e: Bioautograph[ic][al][ly]
BIOCAM e: Binocular Camera
BIOCHEM e: Biochemical
- e: Biochemist
BIOCH[EM] e: Biochemistry
Biochem. Biophys. Res. Commun. e: Biochemical and Biophysical Research Communications (journ.)
Biochem. Genet. e: Biochemical Genetics (journ.)
Biochem. J. e: Biochemical Journal (journ.)
Biochem. Med. e: Biochemical Medicine (journ.)
Biochem. Pharmacol. e: Biochemical Pharmacology (journ.)
BIOCHRON e: Biochronometry
BIOCID e: Biocidal
- e: Biocide
BIOCLEAN e: Biologically Clean
BIOCON e: Biocontamination
BIOCYB e: Biocybernetics
BIOD e: Bell Integrated Optical Device
BIODEF e: Biological Defense
BIODEG e: Biodegradability
- e: Biodegradable
- e: biodegradation
- e: Biodegrade[rs]
- e: Biodegrading
BIODES e: Biodestructible
BIODET e: Biodeterioration
BIOELECTROG e: Bioelectrogenesis
BIOELECTRON e: Bioelectron[ic][al][ly]
- e: Bioelectronics
BIOENG e: Bioengineer[ing]
- e: Biological Engineer[ing]
BIOENV e: Bioenvironment[al][y]
- e: Bioenvironmentalist
BIOEX e: Bioexperiment[ation]
BIOG e: Biographer
- e: Biographical[ly]
- e: Biography
BIOGEO e: Biogeology
BIOGEOG e: Biogeographer
- e: Biogeographic[al]
- e: Biogeography
BIOI e: Block Input/Output Input
BIO-INSTR e: Biological-Instrumentation
BIOINSTRU e: Bioinstrument[al][ly]
- e: Bioinstrumentation
BIO IN UNS ZEIT d: Biologie in unserer Zeit (journ.)
biol a: biologi or biologisk
BIOL d: Biologe
- d: Biologie
- d: Biologin
- d: biologisch
- d: Biologist
- e: Biologic[al]
- e: Biologist
- e: Biology
BIOLAB e: Biological Laboratories
Biol Abstr e: Biological Abstracts (journ.)
Biol. Bull. e: Biological Bulletin (journ.)
BIOLDEF e: Biological Defense
BIOLOP e: Biological Operation[s]
BIOLREPT e: Biological Report
Biol. Rev. e: Biological Review (journ.)
BIOLWPN e: Biological Weapons
BIOLWPNSYS e: Biological Weapon[s] System
BIOM e: Buffer Input-Output Memory
BIOMASS e: Biological Investigadons of Marine Antarctic Systems and Stocks
- e: Biological Mass Source of Ethanol and Methanol From Crops And Trees
- e: Mass of Biological Material
BIOMATH e: Biomathematician
- e: Biomathematics
BIOMED d: biomedizinische Datenbank
- e: Biomedical
- e: Biomedicine
Bio-Med Eng. e: Bio-Medical Engineering (journ.)
BIOMOD e: Biochemical Modelling
- e: Biological Modelling
BIONICS e: Biological Electronics
- e: Biology-Electronics
Bionik d: Biologie und Technik
BIOO e: Block Input/Output Output
BIO-ORG e: Bio-Organic[al][y
BIOPACK e: Biological Pack
BIOPHYS e: Biophysical
- e: Biophysicist
- e: Biophysics
Biophys. J. e: Biophysical Journal (journ.)
BIOPOL[S] e: Biopolymer[s]
BIOR e: Business Input/Output Rerun
BIORE e: Bioresearch[er]
BIORED e: Biological Resources Development
BIOREP e: Biological Attack Report (journ.)
- e: Biological Report (journ.)
BIOS e: Basic Input/Output Subsystem
- e: Basic Input/Output Supervisor
- e: Basic Input/Output System
- e: Biological Investigations of Space
- e: Biological Satellite
BIOSAT e: Biosatellite
BIOSCI e: Bioscience
- e: Bioscientific
- e: Bioscientist
BIOSEN[S] e: Biosensor[s]
BIOSIS e: Bio-Science Information Service [of Biological Abstracts]
BIOSOMA e: Biological, Social, Machine
BIOSPEL e: Biospeleologist[ic][al][ly]
- e: Biospeliology
BIOSTAT e: Biostatistic[s]
BIOT e: British Indian Ocean Territories
- e: British Indian Ocean Territory
BIOTA e: Biological Institute of Tropical America
BIOTEC[CH] e: Biotechnical[ly]
- e: Biotechnological[ly]
- e: Biotechnologist
- e: Biotechnology
BIOTEL e: Biotelemetric
- e: Biotelemetry
BIOTRANS e: Bio-transforrner
- e: Biotransformation
BIO Unit e: Bottom Input/Output Unit
BIOWAR e: Biological Warfare
BIP d: Brutto-Inlandsprodukt
- e: Background Interference Procedure[s]
- e: Bacterial Intravenous Protein
- e: Balanced in Plane
- e: Balloon Interrogation Package
- e: Beginners All-purpose Symbolic Instruction Code Interpreter Package
- e: Binary Image Processor
- e: Bipartite
- e: Bipropellant
- e: Bismuth Iodoform Paraffin
- e: Bitumen Product
- e: Board for International Broadcasting
- e: Books in Print
- e: Bootstrap Input Program[me]
- e: Border Industrial Program[me]
- e: Borough-Interborough Problem[s]
- e: British Institute in Paris
- e: British Institute of Philosophy
- e: British Institute of Physics
- e: Bulk Information Processing
- e: image processor
BIP 8 Code e: Bit-Interleaved Parity 8 Code
BIPAD e: Binary Pattern Detector
BIPCO e: Built-in-Place Component
BIPD e: Biparting Doors
BIPFA e: Bipartite Food, Agriculture and Forestry Group
Bi-Phase-L e: Bi-Phase-Level
Bi-Phase-M e: Bi-Phase-Mark
Bi-Phase-S e: Bi-Phase-Space
BIPHET e: Biphetamine
Bi-ϕ-L e: Bi-Phase-Level
Bi-ϕ-M e: Bi-Phase-Mark
Bi-ϕ-S e: Bi-phase Space
BIPLEX e: Binary Pattern Laser Extraction
- e: Binary Pattern Logic Extraction
BIPM f: Bureau International des Poids et Mesures
BIPO d: Bipolar
- e: British Institute of Public Opinion
BIPP e: Bismuth Iodoform Paraffin Paste
- e: British Institute of Practical Psychology

BIPS e: Bibliographic Information Publication System
- e: Billion[s] Instructions per Second
- e: British Integrated Programmed Suite

bi-py[r] e: Bipyramidal
Bi-Quin e: Bi-Quinary
BIR e: Basic Incidence Rate
- e: Bell-Independent Relations
- e: Break-in Relay
- e: British Institute of Radiology
- e: Bureau of Intelligence and Research
- e: Bus Interface Register
- p: Birmânia
- s: Birmania

BIRC e: Bio-Integral Resource Center
BIRD e: Bibliographic Information Retrieval and Dissemination
BIRDCAP e: Bird Combat Air Patrol
BIRDIE e: Battery Integration and Radar Display Equipment
B Ir E e: Bachelor of Irrigation Engineering
BIRE e: British Institute of Radio Engineers
- e: British Institution of Radio Engineers

B Ir Eng e: Bachelor of Irrigation Engineering
BIRF e: Biochemical Research Foundation
BIRI e: Brewing Industries Research Institute
BIRMA e: British Infra-Red Manufacturers Association
BIRMPDIS e: Birmingham Procurement District
BIRMS e: Battelle Interactive Resources Management System
BIROS e: Bibliographic Information Retrieval Online Service
BIRP e: Beverage Industry Recycling Progranm
B.I.R.-Programm d: Büro-Inseln-im-(Groß-) Raum-Programm
BIRS e: Basic Indexing and Retrieval System
- e: Basic Information Retrieval System
- e: British Institute of Recorded Sound
- f: Bureau international de recherche sur la sauvagine

BIRT e: Bolt Installation and Removal Tool
BIRTHQUAKE e: Population Explosion
BIR Timber e: Birch Timber
BIS d: Bausteininterpretationssystem
- d: Beispiele integrierter Schaltungen
- d: Betriebliches Informations-System
- d: betriebliches Informationssystem
- d: betriebswirtschaftliches Informationssystem
- d: Bibliothekssatz
- d: Büro-Informations-Systeme
- e: Basic Instruction Set
- e: Basic Interchange System
- e: Battlefield Illumination System
- e: Binary Information Signal
- e: Binary Input Signal
- e: Bisector
- e: Boliden Salt
- e: Brain Information Service
- e: British Information Service
- e: British Interplanetary Society
- e: Broadband Information System
- e: Brought into Service
- e: Burn-in Screening
- e: Business Information System
- e: Business Instruction Set
- f: Bureau d`Information Software

BISAD e: Business Information Systems Analysis and Design
BISAM d: Bibliothekssystem für Ausleihe und Medienerfassung
- e: Basic Index[ed] Sequential Access Method
- e: Basic Indexing Sequential Access Method

Bis. Arch. e: Bismarck Archipelago
Bisc e: Biscayan
BISCLANT e: Bay of Biscay; Atlantic [Subarea]
BISCO e: British Iron and Steel Corporation
BISCOM e: Business Information Communications
BISCOM System e: Business Information Systems-Communications System
BISCUS e: Business Information System/Customer Service
BISDAM e: Basic Index[ed] Sequential Data Access Method
B-ISDN e: Broadband-Integrated Services Digital Network
bisett i: bisettimanale
BISEX e: Bisexual
BISF e: British Iron and Steel Federation
BISFA e: British Industrial and Scientific Film Association
Bish Mus e: Bishop Museum
Bishop e: Bishop Museum
- e: Bishop Museum Press

BISI e: British Iron & Steel Institute
BISL e: British Information Service Library
BISMAC e: Business Machine Computer
BISMAPS e: Business Information System Modelling and Planning System
BISNET e: Bank Information System Network
BISP e: Between Ischial Spines
- e: Bispinous (Interspinous Diameter)
- e: Business Information System Program[me]

BISPA e: British Independent Steel Producers Association
BISPE e: Board of Inspection and Survey; Prelimenary Evaluation
BISR e: Bureau of Information Science Research
BISRA e: British Iron and Steel Research Associates
BISS e: Base and Installation System[s]
- e: Bio-Isolator Suit System

BISSG e: Business Information Systems Specialist Group
BiST d: Biegestanze
BIST e: Built-in Self Test
- e: Bureau of Information for Science and Technology

BISTA e: Bureau of International Scientific and Technological Affairs
BISTSS e: Business Information System/Trunks and Special Services
BISU e: Base Interface Surveillance Unit
- e: Boeing Interface Surveillance Unit

bisw. d: bisweilen
BISY d: Bibliothekssystem
BISYNC e: Binary Synchronous
- e: Binary Synchronous Communication
- e: Binary Synchronous Computer

BIT d: Basisinseltransistor
- d: Berufsinteressentest
- d: Bitumen
- d: Büro- und Informations-Technik
- d: Burroughs integriertes Terminal
- e: Bachelor of Industrial Technology
- e: Band Ignitor Tube
- e: Base Island Transistor
- e: Basic Indissoluble Information
- e: Binary Digit
- e: Binary Digit
- e: Binary Information Transfer
- e: Bituminous
- e: Black Incorporation Time
- e: Boron Injection Tank
- e: Bradford Institute of Technology
- e: Built-in Test[ing]

BITD e: Bibliographic and Information Technology Department
BITE e: Backward Interworking Telephone Event[s]
- e: Backward Interworking Telephony Event[s]
- e: Base Installation Test Equipment
- e: Benthic Inflatable Toolstore Enclosure
- e: Built-in Test Equipment

BiTEG d: Bildschirmtexterprobungsgesetz
Bitel d: Bildschirmtelefon
BITES e: Benthic Inflatable Toolstore Enclosures
BIT-FACS e: Book, Insurance, Tax-Fixed Asset Control System
BITM e: Birla Industrial and Technological Museum
- e: Bituminous

BITN d: Bitumen
- e: Bilateral Iterative Network

BITO e: Benzyl-Thiourea
- e: British Institution of Training Officers
- e: Burnishing Tool

BITS d: Bit pro Sekunde
- e: Backplane Integrated Test System
- e: Binary Digit[s]
- e: Binary Information Transfer System
- e: Binary Intersystem Transmission Standard
- e: Birla Institute of Technology and Science (India)
- e: Bits per Second

Bitsy d: Bildtextsystem
BITTC e: Building Industry Technicians TrainIng Council
BITU e: Benzyl-Thiourea
BITUM d: Bitumen
- e: Bituminous

BITUMD e: Bituminized
BIU e: Basic Information Unit
- e: Bus Interface Unit
- f: Bureau International des Universités

BIUG e: Background Information and User Guide
BIV d: Bezirksinstitut für Veterinärwesen
BiV d: Bildverstärker
BIV e: Bivouac
BIVAR e: Bivariant
BIVMOS d: bipolarer Vertikal-Metall-Oxid-Semiconductor
- e: Bipolar Vertical Metal Oxide Semiconductor

BIW d: Bund der Ingenieure des Weinbaues
- e: Battle Injury and/or Wound
- e: Biweekly
- e: Boston Insulated Wire [and Cable Company]
- e: Business Information Wire

BIWA d: Bibliotheks-Wartungsprogramm
- d: Bibliothekswartung

Biwa d: Bildwandler
BIX e: Binary Information Exchange
BIZ d: Betriebs- und Instandhaltungszentrum
BiZ d: Biblische Zeitschrift (journ.)
BIZJET e: Business-Typejet Airplane
BIZMAC e: Business Machine Computer
BIZNET e: American Business Network (Data base)

BJ d: Baujahr
- e: Bachelor of Journalism
- e: Barrage Jammer[s]
- e: Barrage Jamming
- e: Biceps Jerk
- e: Brass Jacket
- e: Bulkhead Jack

BJA e: Basic Journal Abstracts
- e: British Journal of Aesthetics (journ.)

BJC e: Babinet Jamin Compensator
- e: Baltimore Junior College
- e: Bismarck Junior College
- e: Boise Junior College
- e: Brevard Junior College

BJCEB e: British Joint Communications [and] Electronics Board
BJ Ed e: Bachelor of Jewish Education
BJF e: Ball Joint Fitting
- e: Batch[ed] Job[s] [in] Foreground
- e: Batch Job Foreground
- e: Batch Job Format
- e: Benzo[j]fluoranthene

BJG e: Basic Jumbogroup
BJM e: Bluejackets Manual
BJMPs e: British Journal of Medical Psychology (journ.)
BJP e: Bachelor of Jewish Pedagogy
BJP[s] e: British Journal of Psychology (journ.)
BJRLM e: Bulletin of the John Rylands Library. Manchester (journ.)
BJs e: Bell Jar System
BJT e: Base Junction Transistor
- e: Bipolar Junction Transistor

BJU e: Beach Jumper Unit
- e: Bob Jones University

BJur[is] e: Bachelor of Jurisprudence
BK d: Ballistokardiograf
- d: Bank
- d: Basiskomponente
- d: Baukombinat (DDR)
- d: Baukosten
- d: Bazillus Koch
- d: Belegkassette
- d: Bereichskennung
- d: Berkelium
- d: Berufskrankheit
- d: Beton-Kalender
- d: Betriebskanal
- d: Betriebskontrolle
- d: Betriebskontrollschaltung
- d: bewegter Kopf
- d: Bewertungskennzeichen
- d: biologische Kampfmittel
- d: Blank
- d: Blendkörper
- d: Block[stelle]
- d: Bornaische Krankheit
- d: Braunkohle
- d: Breitbandkabel
- d: Breitbandkommunikation
- d: Brennkammer
- d: Bürokommunikation
- d: Buskoppler
- e: Back
- e: Barkhausen-Kurz[oscillation]
- e: Below Knee
- e: Berkelium
- e: Bilge Keel
- e: Black
- e: Block Keyboard
- e: Book
- e: Brake
- e: Break[-in Keying]
- e: Broken
- e: Brook

BKA d: Bereichsknotenamt
- d: Breitbandkabelanlage
- e: Broadband Klystron Amplifier

BK Amp e: Below Knee Amputation
B-Kanal d: Basiskanal
BK-Anlage d: Breitbandkabelanlage
BKB d: Bildungskursbuch
- d: Blockkontrollbyte
- d: Braunkohlenbergwerk (DDR)
- d: Braunschweigische Kohlen-Bergwerke

BKBLE e: Bookable
- e: Bookmobile

BKBNDG e: Bookbinding
BKBNDR e: Bookbinder
BKC e: Benzalkonium Chloride
BKCA e: Bank Cable
BKD e: Bilateral Kinaesthetic Differences
- e: Blackboard

BKDI e: Brake Die
BKDN e: Breakdown
BKDNDIO e: Breakdown Diode
BKE d: Betriebskontrolleinrichtung
- d: Bildkontrollempfänger
- d: Brechkrafteinheit
- d: Buskoppeleinheit

BKER e: Block Error Rate
BKERs e: Block Error Rates
BKEX e: Book Exchange
BKF d: Berufskraftfahrer[ausbildung]
- e: Benzo[k]fluoranthene
- e: Blocking Factor

BKG d: Ballistokardiografie
- d: Ballistokardiogramm
- d: Bürokopiergerät[e]
- e: Bookkeeping
- e: Breakage
- e: Breaking

BKGD e: Background
BKGE e: Brokerage
BK-Gestell d: Betriebskontrollgestell
BKH d: Bikarbonathärte
BKI e: Better Kitchens Institute
- e: Break-in

BKK d: Bangkok (Flughafen)
- d: Betriebskrankenkasse
- d: Braunkohlenkombinat

BKL d: Bekleidung
- d: Beleg[ungs]kontrollampe

BKLR e: Black Letter
BKLT e: Booklet
BKL Timber e: Bakul Timber
Bklyn Brdg e: Brooklyn Bridge
Bklyn Mus e: Brooklyn Museum
BKM d: Buskoppelmodul
BKN d: Bau-Kasten-Netzplan
- d: Breitbandkoppelnetz
- e: Baken
- e: Barkentine
- e: Birkenhead
- e: Broken [clouds]

BKN Clouds e: Broken Clouds
BK-Netz d: Betriebskanal-Netz
- d: Breitbandkommunikations-Netz

BKO d: Barkhausen-Kurz Oszillator
- d: Betriebskosten
- e: Barkhausen-Kurz Oscillator

B/K Oscillator e: Barkhausen/Kurz Oscillator
BKP d: Baukostenplan
- e: Bookplate

BKPG e: Bookkeeping
BKPR e: Bookkeeper
BKR d: Baukontenrahmen
- d: Bogenkreuzung
- e: Breaker

BKR1c e: Baker First Class
BKS d: Basis-Kalkulationssystem
- d: Baukastensystem
- d: Beckenkelchsystem
- d: Blutkörperchensenkung[sreaktion]
- d: Bundesverband Kabel und Satellit
- e: Backstrip
- e: Barracks
- e: Biotech Knowledge Sources
- e: Books
- e: Brakes

BK-Schwingung d: Barkhausen-Kurz-Schwingung
Bks for Libs e: Books for Libraries (journ.)
Bk. Sh. e: Bookshelf
BKSLF e: Back Shelf
BKSP e: Backspace
BKST d: Betriebskostenstelle[nummer]
BKS Theory e: Bohr-Kramers-Slater Theory
BKSTSJ e: British Kinematograph, Sound and Television Society Journal
BKT d: Bakterie[n]
- e: Basket
- e: Bracket
- e: Bucket

Bkt. Pl. e: Bracket Plate
Bktr d: Betriebskontrolleur
BKTR d: Bibliothekstransformation
BKTT e: Below Knee to Toe
BKUP e: Backup
BKV d: Bahnkreuzungsvorschriften
- d: Bürstenabhebe- und Kurzschließvorrichtung

BkVA d: Blindkilovoltampere
BKVl d: Breitbandkommunikation-Verbindungsleitung
BKVO d: Berufskrankheitenverordnung
BKVrP d: Breitbandkommunikation-Verstärkerpunkt
BKVrSt d: Breitbandkommunikation-Verstärkerstelle
BKW d: Bahnkraftwagen
- d: Bahnkraftwerk
- d: Beobachtungskraftwagen
- d: Bernische Kraftwerke AG (CH)
- d: Blindkilowatt
- d: Bogenkreuzungsweiche
- d: Braunkohlenwerk
- e: Breakwater

BKWD e: Backward
BKWP e: Below Kne Walking Plaster
BKWR e: Breakwater
BKWW d: beabsichtigter Kompaßweg-winkel
BKZ d: Benutzerkennzeichen
- d: Bodenklimazahl
- d: Brandwarnkleinzentrale

BKZn d: Brandwarnkleinzentralen
BL d: Basutoland (Kfz)
- d: Bauschaltliste
- d: Befehlslänge
- d: Belegleser
- d: Belegleser de Belegtlampe
- d: Belegtlampe
- d: Beleuchtung
- d: Betriebsleiter
- d: Betriebsstofflager
- d: Bildleitung
- d: Bitleitung
- d: blank
- d: Blase
- d: Blatt
- d: Blau
- d: Blech
- d: Blei
- d: Blende
- d: Blesse

- d: Blind
- d: Blindgänger
- d: Blinker
- d: Blinklicht
- d: Block
- d: Blockade
- d: Blocklänge
- d: Blut
- d: bürstenlos
- d: Bühnenlaser
- d: Bunte Listen
- e: Bachelor of Law
- e: Bachelor of Letters
- e: Bachelot of Literature
- e: Background Listening
- e: Bale
- e: Barrel
- e: Basal Lamina
- e: Base Line
- e: Basic Letter
- e: Basic Linkage
- e: Basutoland
- e: Battery Leader
- e: Battery Left
- e: Beak Line
- e: Bell
- e: Bend Line
- e: Bilateral
- e: Billet
- e: Bill Lodged
- e: Bill of Landing
- e: Binary Logic
- e: Biological Laboratory
- e: Bit Length
- e: Bit Location
- e: Black
- e: Black Letter
- e: Blade
- e: Blanking
- e: Bleed[ing]
- e: Blind Loaded
- e: Block Label
- e: Block Length
- e: Block Loading
- e: Blood [Loss]
- e: Blue [Lamp]
- e: Blue Light
- e: Blue Line
- e: Bodleian Library
- e: Body Length
- e: Bolt Lock
- e: Bomb Line
- e: Boolean Logic
- e: Bottom Layer
- e: Boundary Layer
- e: Boundary Line
- e: Breath Length
- e: Breech Loader
- e: Breech Loading
- e: Briggian Logarithm
- e: British Library
- e: Buccolingual
- e: Building Line
- e: Bulletin
- e: Bus Link
- e: Bus Load
- e: Busy Lamp
- e: Butt Line
- e: Buttock Line

bl. d: blutig
B-L d: Boden-Luft
- e: Belgium-Luxembourg

B & L e: Ball & Lever
BL²MOS e: Buried Load Logic Metal Oxide Semiconductor
BLA d: Befehlsausgang
- d: Boden-Luft-Abwehr
- d: Bruttolohnabrechnung
- e: Bachelor of Landscape Architecture
- e: Bachelor of Liberal Arts
- e: Bangladesh Library Association
- e: Base Loaded Aerial
- e: Base Loaded Antenna
- e: Bilateral Agreement
- e: Bills of Lading Act
- e: Binary Logical Association
- e: Blocking Acknowledgement
- e: Bombay Library Association
- e: Bracket and Linkage Assembly
- e: British Library Association

BlA e: Black Art (journ.)
BL-AA s: Biblioteca Luis-Angel Arango
BLAB d: Betriebslaboratorium
BLAC e: British Light Aviation Centre
BLADE e: Base Level Automation of Data through Electronics
- e: Basic Level Automation of Data through Electronics
- e: Bell Laboratories Automatic Device
- e: Binary Level Access Dosemeter Issue and Record Keeping Editor

BLADES e: Bell Laboratories Automatic Design System
BLADS e: Bell Laboratories Automatic Design System
BLAH e: British Life, Art and History
BLAISE e: British Library Automated Information Service
B La L e: Bachelor of Latin Letters
BLAM e: Ballistically Launched Aerodynamic Missile
BL[and]Arch e: Bachelor of Landscape Architecture
BLAP d: Basislochbandaufbereitungs-programm
BLARD e: Boat Launching and Recovery Device
BLA Signal e: Blocking Acknowledgement Signal
Blast e: Blastoidea
BLAST e: Blocked Asynchronous Transmission
- e: Bolean Logic and State Transfer

BLAST g: blastos
Blast Furn. Steel Plant e: Blast Furnace Steel Plant (journ.)
BlÄtt. d: Blättchen
BLB d: Bauleitungsbuch
BLB d: Blutbild
- d: Bundesbahnliegenschaftsbuch

BLBD e: Binary Light Beam Deflector
BLBG d: Baulandbeschaffungs-Gesetz
- e: Biological Laboratory, Brunswick, Georgia

BLC d: Befehlslänge[n]code
- e: Balance
- e: Barrier Layer Cell
- e: Base Line Configuration
- e: Blood Culture
- e: Blue Line Copy
- e: Board Level Computer
- e: Boundary Layer Control
- e: Breech Loader Carriage
- e: British Lighting Council
- e: Broadband Latching Circulator
- e: Buffer Location Counter

BLCE[s] e: Base Line Calibration Equipment[s]
BLCK e: Blockade
BLCKF d: Blockfett
BLCMP e: Birmingham Libraries Cooperative Mechanization Project
BLCS e: Boundary Layer Control System
BLCS Program[me] e: Board Line Command Select Program[me]
Bl. Cult. e: Blood Culture
Bl.-D d: Blechdose
BLD d: Blind[er]
- e: Bachelor of Industrial Design
- e: Beam Lead Device
- e: Below Limit of Detection
- e: Binary Load/Dump
- e: Blood [and Lymphatic System]
- e: Bold
- e: Boulevard
- e: Build

BL Des e: Bachelor of Landscape Design
BLDG d: Beladung
- d: Bildung
- e: Building

Bldg E[ngr] e: Building Engineer
BLDGS e: Buildings
BLDI e: Blank Die
BLDR e: Bleeder
- e: Builder

BLDT e: Balloon-Launched Decelerator Test
BLDTV e: Balloon Launch Drop Test Vehicle
BLE d: Betriebsleitebene
- d: Betriebsmittel der Leistungselektronik
- d: Bodenlockerungseinheit
- e: Bachelor of Industrial Engineering
- e: Bachelor of Library Economics
- e: Beam-induced Light Emission
- e: Block Length Error
- e: Blunt Leading Edge
- e: Brotherhood of Locomotive Engineers

BLEACH e: Babel Language Editing and Checking
BLEND e: Birmingham and Loughborough Electronic Network Device
BL Eng e: Bachelor of Landscape Engineering
BLER e: Block Error Rate
BLERT e: Block Error Rate Test
- e: Block Error Ratio Test

BLESP e: Bell Labs Engineering Scholarship Program[me]
BLESSED e: Bell Little Electrodata Symbolic System for the Electrodata
BLEU e: Blind Landing Experimental Unit
BLEVE e: Boiling Liquid-Expanding Vapour [or Vapor] Explosion
BLF d: blätterförmig
- d: Blockungsfaktor
- e: Ballast Lumen Factor
- e: Band-Limiting Filter
- e: Baryta Light Fling
- e: Blocking Factor
- e: Boundary Layer Flow
- e: Branch Loss Factor
- e: Busy Lamp Field

BlFNV f: Banque Internationale pour le Financement de l'Energie Nucléaire
BLG d: Bahnlagernd
- d: bargeldlose Lohn- und Gehaltszahlungen
- d: Baulandgesetz
- d: Belüftungsgerät
- d: Blitzgefährdungsgrad
- e: Beta-Lactoglobulin
- e: Blooming Gate
- e: Breech Loading Gun
- e: Building
- e: Bulgarian

Blg. P. e: Belgian Patent
Bl.-Gr. d: Blutgruppe

BLH e: Baldwin-Lima - Hamilton
- e: Blade Loading Harmonics
- e: Borderline Hypertension
- e: British Legion Headquarters
BLHS e: Ballistic Laser Holographic System
BLI d: Beiträge zur Linguistik und Informationsverarbeitung
- d: Belastungsinkremente
- d: Bodenlockerungsindex
- e: Bachelor of Literary Interpretation
- e: Basic Learning Institute
BLib e: Bachelor of Librarianship
BLibS[ci] e: Bachelor of Library Science
Blickw. d: Blickwinkel
BLIM e: Buffer Location-counter Limit
BLIMP e: Boundary Layer Integral Matrix Procedure (or Program[me])
Blind Sdg d: Blindenschriftsendung
B-LINE e: Bell Line [Drawing Language]
BLINK e: Backward Linkage
BLIP e: Background-Limited Infrared Photoconductor
- e: Background-Limited Infrared Photography
- e: Background-Limited Photodetector
- e: Base Language Input Program[me]
- e: Beam-Lead Interconnect Packaging
- e: Big Look Improvement Program[me]
- e: Bit Line Inductance Program[me]
- e: Block-Diagram Interpreter Program[me]
- e: Boundary Layer Instrument Package
- e: Brookhaven Linear Accelerator Isotope Producer
BLIPs e: Background Limited Infrared Photoconductors
BLIS e: Baseline Information System
- e: Bell Laboratories Interpratative System
BLISS e: Basic Language for Implementation of System Software
- e: Basic List-oriented Information Structures System
B Lit e: Bachelor of Literature
BliWaG d: Blindwarengesetz
BLIZ e: Blizzard[ly]
- e: Blizzardous
BLJ e: Bachelor of Letters in Journalism
- e: Bumper Lift Jack
BLK d: Belastungskapazität[en]
- d: Blank
- d: Blinkfeuer
- d: Block
- d: Blutkörperchen
- d: VEB Betonleichtbaukombinat (DDR)
- e: Black
- e: Blanking
- e: Blocking
- e: Bulk
BLKC e: Bulk Carrier
BLKCNT e: Block Count
BLKD e: Blanked
- e: Blanking
- e: Blocked
- e: Blocking
- e: Bulkhead
BLKG e: Blanking
- e: Blocking
BLKGD e: Blanking Die
BLKH[D] e: Bulkhead
BLK LT e: Black Light
BLKN e: Blacken
BLKNG e: Blackening
BLK Press e: Blanking Press
BLKR e: Block Rule
- e: Bulker
BLKS e: Blanking Signal
- e: Blocking Signal
BLKST e: Ballast
BLKT e: Blanket
BLL d: Ballen
- d: Bedienungslochstreifenleser
- d: blaue Leitlampe
- e: Bachelor of Latin Letters
- e: Base Locator Linkage
- e: Base of Lateral Ship
- e: Below Lower Limit
- e: Biosophical Institute Incorporated
- e: Blue Lamp
BLLD e: British Library Lending Division
BLLE e: Balanced-Line Logic[al] Element
BLLF e: Basic Low Layer Functions
BLLG d: biologische Lösung
BLLRCS e: Bureau of Library and Learning Resources and Community Services
BLL Rev. e: British Lending Library Review (journ.)
BLM d: Blindleistungsmaschine
- d: Blindmuffe
- e: Bachelor of Landscape Management
- e: Ball Limiting Memory
- e: Basic Language Machine
- e: Bilayer Lipid Membrane
- e: Blinking Light Monitor
- e: Book League Monthly (journ.)
- e: Brushless Motor
- e: bürstenloser Motor
- e: Bus and Lining Module
BLMB d: Blocklänge Magnetbanddateien
BLMC e: British Leyland Motor Corporation
BLM Number e: Best Loiter Mach Number
BLMP d: Blocklänge Magnetplattendateien
BLMPS e: Base Level Military Personnel System
BLMPX e: Block Multiplex[er]
BLMR e: Bureau of Labor-Management Reports
BLMs e: Bilayer Lipid Membranes
- e: Bimolecular Lipid Membranes
BLMS e: Book Library Management System
BlmSchG d: Bundeslärmschutzgesetz
BLM System e: Book Library Management System
BLMUX d: Blockmultiplex[kanal]
- e: Block Multiplexer
BLMUX Channel e: Block Multiplex Channel
BLN d: Bank für Landwirtschaft und Nahrungsgüterwirtschaft
BLN e: Balloon
- e: Bronchial Lymph Nodes
BLND e: Blend
BLNG e: Belling
BLNK e: Blank
BLNKT e: Blanket
BLNR e: Benton Lake National Refuge
BLNWR e: Big Lake National Wildlife Refuge
- e: Bitter Lake National Wildlife Refuge
- e: Buffalo Lake National Wildlife Refuge
BLO e: Below
- e: Blocking
- e: Blower
- e: Bombardment Liaison Officer
- e: Butt Line Zero
BLOBS e: Bladder Observation
BLOC d: Blockade
- e: Block Oriented Compiler
- e: Booth Library On-Line Circulation
BLOCK e: Blockade
BLOCS e: Buckwell (University) Online Circulation System
BLODI e: Block Diagram [Compiler]
BLODIC e: Block Diagram Compiler
BLODICB e: Block Diagram Compiler B
BLODI-G e: Block Diagram Graphics
BLOGIC e: Block Logic Computer
BLOM e: Booster Lift of Mass
BLO Signal e: Block[ing] Signal
BLOSSOM e: Basic Liberation of Smokers and Sympathizers of Marijuana
BLOT e: Biotron[ic][al][ly]
- e: British Library of Tape Recordings
BLO Timber e: Black Locust Timber
BLOW e: Booster Lift Off Weight
BLOWS e: British Library of Wildlife Sounds
BLP d: Basisleiterplatte
- d: Belegungsplan
- d: Blättchenpulver
- e: Backloading Point
- e: Ball Lock Pin
- e: Bar Line Printer
- e: Basic Launch Plan
- e: Basic Logistics Plan
- e: Blind Loaded and Plugged
- e: Blue Line Print
- e: Bombing Landplane
- e: Buffer Load Point
- e: Bypass Label Processing
BLPA e: Burroughs Line Printer Adapter
BLPC e: Backward-Limit Photocell
BLPES e: British Library of Political and Economic Science
BLPR e: Blood Pressure
BLPZZ e: Bent Logarithmically Periodic Zig-Zags
BLQ Equation e: Balescu-Lenard-Quernsey Equation
BLR e: Ballistic Research Laboratories
- e: Barrier Layer Rectifier
- e: Baseline Restorer
- e: Beyond Local Repair
- e: Blocking-signal Reception
- e: Boiler
- e: Breech-Loading Rifle
- e: Broad-Line System
BLRC e: Bell Laboratories Record (journ.)
BLRD e: British Library Reference Division
BL-Region e: Black-Light Region
BLRG e: Breech-Loading Rifled Gun[s]
BLRMKR e: Boilermaker
BLROA e: British Laryngological, Rhinological and Otological Association
BLRP e: Boilerplate
BLS d: Belegtlampenschalter
- d: Blattfernschreibapparat
- d: Blattschreiber
- e: Bachelor of Library Science
- e: Bachelor of Library Service
- e: Balanced Line System
- e: Bales
- e: Balloon Launching Station
- e: Band Limited Signal
- e: Barrels
- e: Basic Life Support
- e: Bell Log System
- e: Binary Light Switch
- e: Bit Length Specification
- e: Blocking-signal Sending
- e: Blood and Lymphatic System
- e: Blood Sugar
- e: Boundary Layer Separation
- e: Broadband Latching Switch
- e: Brooklyn Law School
- e: Buccal-Lingual Syndrome
- e: Buffered Line Selector
BLSB e: Ball-Lock Separation Bolt
BLSc e: Bachelor of Library Science
BLS-CGear e: Brush Lifting and Short-Circuiting Gear

BLSGMA e: British Lampblown Scientific Glassware Manufacturers Association
BLSJ e: Beam Lead Sealed Junction
BLSJICP e: Beam Lead Sealed Junction Integrated Circuit Package
BLSN e: Blowing Snow
BLSS e: Base Level Self-Sufficiency
Bl.ST d: Blockstation
- d: Blockstelle
- e: Ballast
BLSTGPWD e: Blasting Powder
BL STL e: Billet Steel
BLSV d: Bundesluftschutzverband
BLSW e: Barrels of Load Salt Water
BLSWD e: Barrels of Load Salt Water per Day
BLT d: Blatt
- d: Blättchen
- e: Bachelor of Literature
- e: Basic Language Translator
- e: Belt
- e: Blanket
- e: Blind Loaded and Traced
- e: Blind Loaded and Tracer
- e: Blind Loaded with Tracer
- e: Blood Type
- e: Blue Light
- e: Bolt
- e: Borrowed Light
- e: Built
BLT BT e: Bolted Bonnet
BLTC e: Baltic
- e: Bottom Loading Transfer Cask
BLTI e: Better Lawn and Turf Institute
BLT-IN e: Built-In
BLU e: Basic Link Unit
- e: Basic Logic Unit
- e: Bessie-Lowry Unit[s]
- e: Bipolar Line Unit
- e: Blue
- e: Bomb, Live Unit
BLUCB e: Bancroft Library of the University of California
BLUE e: Best Linear Unbiased Estimator
BLUN e: Blood Urea Nitrogen
Bl Us d: Blockadeumsetzer
BLV d: Beam-Lead-Verfahren
- d: Bereichslastverteilung[en]
- d: Bundeslaufbahnverordnung
- e: British Legion Village
Bl.-Verschl. d: Bleiverschluß
BLVS d: Bibliothek des Literarischen Vereins; Stuttgart (journ.)
BLW d: Bundesbahn-Landwirtschaft
- e: Below
- e: Black Watch
BLWA e: British Laboratory Ware Association
BLWDN e: Blowdown
BLWG e: Blowing
BLWR e: Blower
BLWRS e: Barriered Landscape Water Renovation System
BLWS e: Bellows
BLWT e: Blowout
- e: Blowtorch
BLZ d: Bankleitzahl
- d: Belegungszähler
- d: Betriebsleitzentrale
- d: Blattzahl
- d: Blitzfeuer
- e: Boundary Layer Zone
BLZn d: Bankleitzahlen
BM d: Bahnmeisterei
- d: Basalmembran
- d: Basilarmembran
- d: Baumeister
- d: Baumuster
- d: Beckenmitte
- d: Befehls- und Meldebus
- d: Belichtungsmesser
- d: Bestimmungsmensur
- d: Betäubungsmittel
- d: Betriebsmeßtechnik
- d: Bildmischeinrichtung
- d: Binärmultiplikation
- d: Bindemittel
- d: bistabiler Multivibrator
- d: Blockmarke
- d: Bordmechaniker
- d: Bremsmotor
- d: Bytemaschine
- d: bytemultiplex
- e: Babbit-Metal
- e: Bachelor of Mathematics
- e: Bachelor of Medicine
- e: Bachelor of Music
- e: Backing Memory
- e: Back Marker
- e: Balancing Machine
- e: Ballistic Missile
- e: Basal Metabolism
- e: Base Machine
- e: Base Maintenance
- e: Basement Membrane
- e: Basic Microprogram[me]
- e: Basilar Membrane
- e: Beachmaster
- e: Beam [Monitor]
- e: Bearing Magnetic
- e: Beginning Mark
- e: Bell Metal
- e: Bench Maintenance
- e: Bench Mark
- e: Bending Moment
- e: Bermuda
- e: Billing Machine
- e: Bill of Materials
- e: Bimonthly
- e: Binary Multiplication
- e: Binary Multiply
- e: Binding Margin
- e: Birthmark
- e: Bistable Multivibrator
- e: Bitumen Mortar
- e: Block Mark
- e: Block Marker
- e: Board Measure
- e: Body Mass
- e: Bone Marker
- e: Bookkeeping Machine
- e: Bootstrap Memory
- e: Boundary Marker
- e: Bowel Movement
- e: Braided Memory
- e: Branch Material
- e: Branch Memorandum
- e: Brass Mounted
- e: Breadth-Mouled
- e: Breech Mechanism
- e: Brightness Merit
- e: British Manufactured
- e: British Monomark
- e: British Museum
- e: Broad Measure
- e: Bromine-Methanol
- e: Bronze Metal
- e: Brooklyn Museum
- e: Brown Male
- e: Bubble Memory
- e: Buccal Mass
- e: Buccomesial
- e: Buffer/Multiplexer
- e: Buffer Mark
- e: Buffer Memory
- e: Buffer Module[s]
- e: Buffer Multiplexer
- e: Bulk Memory
- e: Bureau of Medicine
- e: Bureau of Mines
- e: Burma
- e: Business Machine
- e: Byte Machine
- e: Byte Mask
- e: Byte Multiplex
BM1 e: Boatswains Mate 1st Class
BM2 e: Boatswains Mate 2nd Class
BMA d: Benutzermaschine
- d: Betriebsmonitor-Alarm
- d: Brandmeldeanlage
- e: Bachelor of Municipal Administration
- e: Balanced Magnetic Amplifier
- e: Baltimore Museum of Art
- e: Beach Maintenance Area
- e: Bicycle Manufacturers Association
- e: Body Mounted Accelerometer
- e: British Marine Aircraft
- e: British Medical Association
- e: British Military Attaché
- e: Butyl Methacrylate
B Ma E[ng] e: Bachelor of Marine Engineering
BMAG e: Body-Mounted Attitude Gyro
BMANT e: Boom Antenna
BMAPS e: Buckbee-Mears Automated Plotting System
BMAR e: Ballistic Missile Acquisition Radar
B Mar E[ng] e: Bachelor of Marine Engineering
BMASR e: Bureau of Military Application of Scientific Research
BMAT d: Bundesminister[ium] für Atomforschung
- e: Ballistic Missile Analyst Technician
- e: Beginning of Morning Astronomical Twilight
B.Math. e: Bachelor of Mathematics
BMAT/S e: Ballistic Missile Analyst Technician-Specialist
BMB d: Büromaschinenbörse
- e: Ballistic Missile Branch
- e: Barium Metaborate
- e: Bombardment
- e: British Medical Board
- e: British Medical Bulletin (journ.)
- e: British Metrication Board
- e: Broadcast Measurement Bureau
BMBDR e: Bombardier
B-M-B-Gelenk e: Beat-Müller-Balance-Gelenk
BMBR e: Bomber
BMBW d: Bundesminister[ium] für Bildung und Wissenschaft
BMBW-FB d: Bundesminister[ium] für Bildung und Wissenschaft-Forschungsbericht
Bmb Wg e: Bombardment Wing
Bmb Wg (H) e: Bombardment Wing (Heavy)
Bmb Wg (L) e: Bombardment Wing (Light)
Bmb Wg (M) e: Bombardment Wing (Medium)
BMC e: Ballistic Missile[s] Center
- e: Ballistic Missile Checker
- e: Ballistic Missiles Committee
- e: Base-machine Microinstruction Cycle

- e: Basic Machine Cycle
- e: Battelle Monte Carlo
- e: Bearing Mounted Clutch
- e: Biax Memory Core
- e: Binary Magnetic Core
- e: Binary Memory Cell
- e: Blind Mating Connector
- e: Blockhouse Monitor Console
- e: Block Multiplexer Channel
- e: British Museum Catalogue
- e: Bubble Memory Controller
- e: Buckbee-Mears Company
- e: Bulk Mail Center
- e: Bulk Media Conversion
- e: Bulk Moulding Compound
- e: Burst-mode Multiplexer Channel
- e: Burst Multiplexer Channel
- e: Business Microcomputer

BMC/AMC e: Ballistic Missile Center; Air Materiel Command
BM Car e: Baggage and Mail Car
BMCat e: British Museum Catalogue
BMCBB e: Boatswains Mate; Construction Battalion; Boatswain
BMCBS e: Boatswains Mate; Construction Battalion; Stevedore
BMC Division e: Business Microcomputer Division
BMCET/S e: Ballistic Missile Checkout Equipment Technician-Specialist
BMCI e: British Mosquito Control Institute
BMC Institute e: British Mosquito Control Institute
BMCP e: Biomedical Measurement and Control Panel
BMCS e: Ballistic Missile Cost Study
- e: Bulk Moulding Compounds
- e: Bureau of Motor Carrier Safety
- e: Business Management Control System

BMCSRP e: Business Management Control System Research Project
BMCT e: Beginning of Morning Civil Twilight
BMD e: Ballistic Missile Defense
- e: Baseband Modulator-Demodulator
- e: Becker Muscular Dystrophy
- e: Benchmark Monitor Display
- e: Biomedical Computer Program[me]
- e: Biomedical Data
- e: Births, Marriages and Deaths
- e: Bone Marrow Depression
- e: Bubble Memory Device
- e: Buffer Memory Device
- e: Bureau of Medical Devices

BMDATC e: Ballistic Missile Defense Advanced Technology Center
BMDC e: Ballistic Missile Defense Center
BMDCA e: Ballistic Missile Defense Communications Agency
BMD Car e: Baggage, Mail and Dormitory Car
BMDCP e: Battalion Mortar and Davy Crockett Platoon
BMDEAR e: Ballistic Missile Defense Emergency Action Report
BMDF e: Black Mesa Defense Fund
BMDITP e: Ballistic Missile Defense Integrated Training Plan
BMDM e: British Museum Department of Manuscripts
BMDMP e: Ballistic Missile Defense Master Plan
BMD-NEAT Committee e: Ballistic Missile Defense-Nuclear Effects and Threat Committee
BMDNS e: Basic Mission Design Number and Series
BMDO e: Ballistic Missile Defense Operations
BMDOA e: Ballistic Missile Defense Operations Activity
BMDP e: Biomedical Data Processing
BMDPM e: Ballistic Missile Defense Program[me] Manager
BMDPO e: Ballistic Missíle Defense Program[me] Office[r]
BMDR e: Bombardier
BMDS e: Ballistic Missile Defense System
- e: Base Mail Distribution Scheme

BMDSCOM e: Ballistic Missile Defense System Command
BMD System e: Ballistic Missile Defense System
BME d: Betrieb mit Enddämpfung
- e: Bachelor of Mechanical Engineering
- e: Bachelor of Mining Engineering
- e: Bachelor of Music Education
- e: Basic Medium Edition (of the Universal Decimal Classification)
- e: Bench Maintenance Equipment
- e: Biax Memory Element
- e: Binary Memory Element
- e: Bio-Medical Electronics
- e: Bio-Medical Engineering
- e: Birmingham Electric Company
- e: Born-Mayer Equation
- e: Brotherhood of Marine Engineers

BM/E e: Broadcast Management/ Engineering (journ.)
BME (Aero Option) e: Bachelor of Mechanical Engineering (Aeronautical Option)
BME Car e: Baggage, Mail and Express Car
B Mech E. e: Bachelor of Mechanical Engineering
B Med e: Bachelor of Medicine
BM Ed e: Bachelor of Music Education
BMedBiol. e: Bachelor of Medical Biology
BMEDS e: Base Management Engineering Data System
B Med Sc e: Bachelor of Medical Science
BMEF e: British Mechanical Engineering Federation
BMELF d: Bundesministerium für Ernährung, Landwirtschaft und Forsten
BM Eng e: Bachelor of Mechanical Engineering
BMEP e: Brake Mean Effective Pressure
B Met e: Bachelor of Metallurgy
BMET e: Biomedical Equipment Technician
- e: Biomedical Equipment Technology
- e: Biomet Incorporated

B Met E[ng] e: Bachelor of Metallurgical Engineering
BMETs e: Biomedical Equipment Technicians
BMEWS e: Ballistic Missile Early Warning Station
- e: Ballistic Missile Early Warning System

BMF d: Betriebsmittelfond[s]
- d: Betriebsmittelfond[s]
- e: Boron Metal Fiber

BMFA e: Boston Museum of Fine Arts
BMFT d: Bundesminister[ium] für Forschung und Technologie
BMG d: Betäubungsmittel-Gesetz
- d: Bordmaschinengewehr
- d: Breitenmeßgerät
- e: Browing Machine Gun
- e: Business Machines Group
- e: Business Management Game

BMGes d: Bundesminister für Gesundheitswesen
B Mgt E[ng] e: Bachelor of Management Engineering
BMH e: British Military Hospital
BMHW d: VEB Berliner Metallhütten und Halbzeugwerke
BMI d: Bayerische Mineral-Industrie AG
- e: Ballistic Missile Interceptor
- e: Barley and Malt Institute
- e: Battelle Memorial Institute
- e: Birmingham and Midlands Institute
- e: Body Mass Index
- e: Book Manufacturers Institute
- e: Branch of Minus
- e: Bravais-Miller Indices
- e: British Military Intelligence
- e: Broadcast Music, Incorporated
- e: Broadway Memorial Institute

B Mic e: Bachelor of Microbiology
BMIC e: Bureau of Mines Information Circular
B Mi E[ng] e: Bachelor of Mining Engineering
BMIGT e: Ballistic Missile Inertial Guidance Technician
BMIGT/M e: Ballistic Missile Inertial Guidance Technician-Mechanic
BMILS e: Bottom-Mounted Impact Locations System
B Min E[ng] e: Bachelor of Mining Engineering
BMIP e: Basic Medical Insurance Plan
BMIR e: Below-Market Interest Rate[s]
BMIS e: Back-gate Metal Insulator Semiconductor Field Effect Transistor
- e: Bank Management Information System
- e: Basic Management Information System
- e: Business Management Information System

BMISFET e: Bulk Metal Insulator Semiconductor Field Effect Transistor
- e: Bulk Metal Isolator Semiconductor Field Effect Transistor

BMIS Structure e: Bulk Metal Insulator Semiconductor Structure
BMJ e: British Medical Journal (journ.)
BMK d: Bau- und Montagekombinat (DDR)
- d: Betriebsmittelkonstruktion
- d: Betriebsmotorkompensator
- d: Bordmaschinenkanone
- d: VEB Baumaschinenkombinat (DDR)
- e: Birthmark
- e: Bookkeeping Machine Keyboard
- e: Bookmark[er]

BMKA d: Bytemultiplexkanal-Anschluß
BMKE d: Betriebsmotorkompensator-Einfach-Linienschreiber
BMKR e: Boilermaker
BMKTR d: Betriebsmaschinenkontrolleur
BMKV d: Betriebsmotorkompensator-Vielfachpunktschreiber
BML e: Bachelor of Modern Languages
- e: Base-machine Microroutine Length
- e: Basic Machine Language
- e: Basic Material Laboratory
- e: Belfast & Moosehead Lake
- e: Bodega Marine Laboratory
- e: British Museum Library
- e: Bulk Material Length

BMLC e: Basic Multiline Controller
BMLD e: Binaural Masking Level Difference
- e: Breadth Moulded

BMLET/R e: Ballistic Missile Launch Equipment Technician-Repairman

BMM e: Bachelor of Mining and Metallurgy
- e: Ballistic Missile Manager
- e: Basic Modular Memory
- e: Benthic Metabolism Measurement

BMMA e: Brigade Major; Medium Artillery
- p: Biblioteca Municipal Mário de Andrade

BMMCDC e: Binary Metal and Metalloid Constitution Data Center

BMMD e: Body Mass Measurements Device

BMMG e: British Microcomputer Manufacturers Group

BMMP e: Benign Mucous Membrane Pemphigoid

BMMRG Program[me] e: Board Multiply Merge Program[me]

BMN d: Bundesmarine-Norm
- e: Base Managers Notice
- e: Bone Marrow Necrosis
- e: British Merchant Navy

BMNH e: British Museum of Natural History

BMNP e: Bale Mountains National Park
- e: Blue Mountains National Park

BMNT e: Beginning of Morning Nautical Twilight

BMO e: Ballistic Missile Office[r]
- e: Base-machine Microroutine Overhead
- e: Basic Machine Operation[s]
- e: Beach Modulator Oscillator
- e: Biological Material Oxidizer
- e: Bond Molecular Orbital[s]
- e: Branch Microoperation
- e: Brigade Medical Officer
- e: British Meteorological Office
- e: Business Machine Operator

BMOC e: Ballistic Missile Orientation Course
- e: Big Machine on Campus

BMOM e: Base Maintenance and Operational Model
- e: Base Maintenance and Operations Model

BMOO e: Butyl Mesityl Oxide Oxalate

BMOS e: Back-gate[-input] Metal Oxide Semiconductor

BMOs e: Branch Microoperations

BMOSFET e: Back-gate[input] Metal Oxide Semiconductor Field-Effect Transistor

BMOSs e: Back-gate[-input] Metal Oxide Semiconductors

BMOTR e: Ballistic Missile Operational Training Readiness

B Moulded e: Breadth Moulded

BMP d: Bataillons-Medizinalpunkt
- d: Bundesminister[ium] für das Post- und Fernmeldewesen
- e: Batch Message Processing
- e: Beefy-Meaty-Peptide
- e: Benchmark Program[me]
- e: Berlin Military Post
- e: Bill of Materials Processor
- e: Bone Morphogenetic Protein
- e: Brake Mean Pressure

BMP-76P8 e: Soviet amphibious annored-infantry combat vehicle

BMPA e: Broadband Microwave Power Amplifier

BMPP e: Benign Mucous-Membrane Pemphigus

BMPR e: Bimonthly Progress Report

BMPS e: British Medical Protection Society

BMPSM e: Bachelor of Music in Public School Music

BMPX Program[me] e: Board Multiplex Program[me]

BMPY Program[me] e: Board Multiply Program[me]

BMQ e: British Museum Quarterly (journ.)

BMR d: Bildröhrenmeßregenerator
- d: Bordmikrorechner
- d: Bremsmotor mit Rippenkühlung
- e: Basal Metabolic Rate
- e: Basal Metabolism Rate
- e: Basal Metabolism Reading
- e: Basic Military Requirement[s]
- e: Bibliography of Medical Reviews
- e: Bomber
- e: British Marine Radar
- e: Brookhavens Medical Reactor
- e: Bureau of Mineral Resources

BMRA e: Basic Multi-Role Avionics
- e: Brigade Major; Royal Artillery

BMR Karte d: Bestellmengenrechenkarte

BMRL e: Building Materials Research Laboratories

BMRMO e: Balance Mobilization Reserve Material Objective

BMRR e: British Museum Reading Room

BMRS[YS] e: Ballistic Missile Recovery System
- e: Ballistic Missile Reentry System

BMS d: Buchungsmaschinenstation
- e: Bachelor of Marine Science
- e: Bachelor of Mechanical Science
- e: Bachelor of Medical Science
- e: Background Mapping Sensor
- e: Balanced Magnetic Switch
- e: Ballistic Missile Ship
- e: Ballistic Missile Specification
- e: Basic Mapping Support
- e: Basic Meteorological Services
- e: Behaviour Monitor System
- e: Biomedical Monitoring System
- e: Bit Mark Sequence
- e: Bit Mark Sequencing
- e: Blow Moulding System
- e: Boston Museum of Science
- e: Breathing Metabolic Simulator
- e: British Malachological Society
- e: British Measurement System
- e: British Mycological Society
- e: Buffalo Museum of Science
- e: Building Management System
- e: Bureau of Medical Services
- e: Bureau of Medicine and Surgery
- e: Burroughs Management System
- e: Burst Measuring System
- e: Business Management System
- e: Bypass Monochrome Signal

BM Sc e: Bachelor of Mechanical Science

BMSE e: Baltic Mercantile and Shipping Exchange

BMSJ e: British Medical Students Journal

BMSM e: Bachelor of Music in School Music

BMSO e: Blue Mountain Seismological Observatory

BMSP e: British Modal Speaking Position

BMSR d: Betriebsmeß-, Steuerungs- und Regelungstechnik
- d: Betriebsmessung, Steuerung und Regelung

BMSRC e: Boatswains Mate; Ship Repair; Canvasman
- e: Boatswains Mate; Ship Repair; Crane Operator

BMSRR e: Boatswains Mate, Ship Repair, Rigger

BMSR-Technik d: Betriebsmeß-, Steuerungs- und Regelungstechnik

BMSS e: British and Midlands Scientific Society
- e: Buoy Messenger

B Ms Sc e: Bachelor of Mechanical Sciences

BMSTR d: Baumeister
- d: Baumuster

BMT d: Bau- und Montagetechnologie
- d: Baumaschine + Bautechnik (journ.)
- d: biomedizinische Technik
- d: Büro, Mensch, Technik
- e: Bachelor of Medical Technology
- e: Basic Motion Time
- e: Bibliography of Medical Translations
- e: Bipolar Memory Technology
- e: Block Marker Track
- e: British Mean Time
- e: Brooklyn-Manhattan Transit
- e: Burroughs Modular Terminal

Bmt-Anlagen d: betriebsmaschinentechnische Anlagen

BMTD e: Ballistic Missile Terminal Defense
- e: Base Mechanical Transport Depot

BMTFA e: British Malleable Tube Fittings Association

BMTI e: Block Mode Terminal Interface

BMTLC e: Bimetallic

BMTP e: Bureau of Mines Technical Paper

BMTR d: Barometer

BMTS e: Ballistic Missile Target System

BMTV e: Ballistic Missile Test Vessel

BMU e: Backing Memory Unit
- e: Basic Measurement Unit
- e: Beachmaster Unit
- e: Beta Methyl Umbelliferone
- e: British Medical Union
- e: Bus Monitor Unit

BMus d: Berliner Museen (journ.)

B Mu[s] e: Bachelor of Music

B.Mu[s]. e: British Museum

B Mus E[d] e: Bachelor of Music Education
- e: Bachelor of Music in Education

B Mus (PSM) e: Bachelor of Music in Public School Music

BMUX e: Block Multiplexer

BMUX Channel e: Block Multiplexer Channel

BMV d: bistabiler Multivibrator
- d: Bundesminister[ium] für Verkehr
- e: Base Mount Valve
- e: Basic Minute Values
- e: Bistable Multivibrator
- e: Bromegrass Mosaic Virus

BMVg d: Bundesminister[ium] für Verteidigung

BMVP e: Barrier; Moisture-Vaporproof

BMVR d: Bundesminister des Verteidigungsrates

BMVtdg d: Bundesminister[ium] für Verteidigung

BMVVG d: Betäubungsmittelverschreibungsverordnungsgesetz

BMVVO d: Betäubungsmittelverschreibungsverordnung

BMW d: Bayerische Motoren Werke AG
- e: Bare Molybdenum Wire
- e: Beam Width

BMwF d: Bundesminister[ium] für wissenschaftliche Forschung

BMWi d: Bundesminister[ium] für Wirtschaft

BMWo d: Bundesminister[ium] für Wohnungswesen und Städtebau

BMWS e: Ballistic Missile Weapon System

BMX e: Base-machine Microroutine Execution
- e: Baseband Matrix
- e: Bicycle Motorcross
BMXR e: Batch Mixer
- e: Block Multiplexer
BMZ d: begrenzt abhängiger Maximalstromzeitschutz
- e: Basement Membrane Zone
BN d: Banknote
- d: Baumuster-Nummer
- d: Beiträge zur Namensforschung (journ.)
- d: Belastungsnetzwerk
- d: Besetzungsnormen
- d: Betriebsnorm[en] (DDR)
- d: Betriebsnummer
- d: Bundesbahn-Norm
- e: Bachelor of Nursing
- e: Bahnen
- e: Bank Note
- e: Base Notation
- e: Base Number
- e: Batch Numbering
- e: Beacon
- e: Bearing
- e: Between
- e: Bibliothèque National
- e: Billion
- e: Binary Notation
- e: Binary Number [System]
- e: Biquinary Notation
- e: Biquinary Number
- e: Block Name
- e: Block Number
- e: Bloody Nuisance
- e: Bolt and Nut
- e: Bombing/Navigation
- e: Booking Note
- e: Born
- e: Boron Nitride
- e: Branchial Neuritis
- e: Brown [Norway]
- e: Bundle
- e: Bureau of Narcotics
- e: Bushing
- i: Biblioteca Nazionale [Italian-National Library]
- s: Biblioteca Nacional
BNA d: Baseler Nomenklatur der Anatomie
- e: Bachelor of Navigation
- e: Basle Nomina Anatomica
- e: Beta-Naphthylamine
- e: Brazil Nut Association
- e: British Naval Attaché
- e: British North America
- e: Burroughs Network Architecture
BNAF e: Brazil Nut Advertising Fund
BN Arch e: Bachelor of Naval Architecture
BNAU e: Bulgarian National Agrarian Union
B Nav e: Bachelor of Navigation
BNB e: British National Bibliography
- e: British North Borneo
BNBC e: British National Book Centre
BNC e: Bayonet Nut Connector
- e: Brasenose College
- e: Bulk Negative Conductance
- s: Biblioteca Nacional de Chile
- s: Biblioteca Nacional de Colombia
BNCC e: British National Committee for Chemistry
BNCDST e: British National Committee on Data for Science and Technology
BNCF i: Biblioteca Nazionale Centrale Firenze
BNCGIP e: British National Committee for General Information Programme
BNCH e: Bench
BNCHBD e: Benchboard
BNCIEC e: British National Committee of the International Electrotechnical Commission
BNCL e: Binnacle
BNCLR e: Binocular
BNCM f: Bibliothèque Nationale du Conservatorie de Musique
BNCNDT e: British National Committee for Non-Destructive Testing
BNCOE e: British National Committee for Ocean Engineering
BNCOQ e: Bachelor Noncommissioned Officers Quartiers
BNCOR e: British National Committee for Oceanographic Research
BNCOS e: British Noncommissioned Officers
BNC PL e: Bayonet Nut Connector Plug
BNCS e: British Numerical Control Society
BNC SKT e: Bayonet Nut Connector Socket
BNCSR e: British National Committee for Space Research
BNC STR ADAPT e: Bayonet Nut Connector Straight Adapter
BNC T ADAPT e: Bayonet Nut Connector Tee Adapter
BNC-U ADAPT e: Bayonet Nut Connector to Ultra High Frequency Adapter
BNCVE i: Biblioteca Nazionale Centrale Vittorio Emanuele II
BNCW e: Bare Nickel Chrome Wire
BND e: Band
- e: Bend
- e: Benzoylated-Naphthoylated
- e: Bond[ed]
- e: Bound
- e: British National Discography
BNDAFF e: Best Null Detector and Flip Flop
BNDC e: British Nuclear Design and Construction Limited
- e: Bulk Negative Differential Conductivity
BNDD e: Bureau of Narcotics and Dangerous Drugs
BNDDIS e: Band Display
BNDE d: Bandeisen
BNDG e: Binding
- e: Bonding
BNDL e: Bundle
BNDO f: Bureau National de Données Océaniques
BNDR d: Binder
BNDRY e: Boundary
BNE d: Brisbane (Flughafen)
- e: Bachelor of Naval Engineering
- e: Bathymetric Navigation Equipment
- e: Buffalo Niagara Electric Corporation
- e: But Not Exeeding
BNEC e: British Nuclear Energy Committee
- e: British Nuclear Energy Conference
BN Ed e: Bachelor of Nursing Education
BNEL e: British National Engineering Laboratory
BN Eng e: Bachelor of Naval Engineering
BNEP e: Basic Naval Establishment Plan
BNES e: British Nuclear Energy Society
BNF d: Backus-Normal-Form
- d: Beiträge zur Namenforschung (journ.)
- e: Backus-Naur-Form
- e: Best Noise Figure
- e: Bomb Nose Fuze
- e: Boolean Normal Form
- e: Boron Nitride Fiber
- e: British National Formulary (journ.)
- e: British Nuclear Forum
BNFC e: British National Film Catalogue
BNFL e: British Nuclear Fuels Limited
BNFMF e: British Non-Ferrous Metals Federation
BNFMRA e: British Non-Ferrous Metals Research Association
BNFMTC e: British Non-Ferrous Metals Technology Centre
BN Form e: Backus-Naur Form
BNFP e: Barnwell Nuclear Fuel Plant
BNFSA e: British Non-Ferrous Smelters Association
BNG e: Bending
- e: Branch No Group
- e: Broadband Noise Generator
- e: Bureau of Natural Gas
BnG-DL f: Bibliothèque Nationale du Grand-Duche de Luxembourg
BNGM e: British Naval Gunnery Mission
BNGS e: Bomb Navigation Guidance System
BNH e: Burnish
BNHA e: Badlands Natural History Society
BNHAS e: Burton-on-Trent Natural History and Archaeological Society
BNHS e: British National Health Service
BNI e: Black Nation of Islam
BNIST f: Bureau National de l'Information Scientifique et Technique
BNJM s: Biblioteca Nacional José Martí
BNKNO e: Bank Number
BNL e: Battelle Northwest Laboratories
- e: Berkeley Nuclear Laboratories
- e: Brookhaven National Laboratory
- f: Bibliothèque Nationale du Liban
- p: Biblioteca Nacional de Lisboa
BNLO e: British Naval Liaison Officer
BNM e: Badlands National Monument
- i: Biblioteca Nazionale Martiana
- s: Bajo el Nivel del Mar
- s: Biblioteca Nacional de México
BN-Metall d: Bahnmetall
BNML e: Bitter National Magnet Laboratory
Bn MO e: Battalion Medical Officer
BNN e: Biquinary Number Notation
BNO e: Barrels of New Oil
- e: Bladder Neck Obstruction
- e: British Naval Officer
BNOC e: British National Oil Corporation
BNOV e: But Not Over
BNP d: Betriebsnetzplan
- e: Background Noise Power
- e: Bahamas National Park
- e: Bako National Park
- e: Banff National Park
- e: Belair National Park
- e: Birmingham Nitro Proof
- e: Bontebok National Park
BNPF e: Beginning, Negative, Positive, Finish
BNR e: Bell-Northern Research
- e: Bonded Negative Resistance
- e: British Naval Representative
- e: Burner
- e: (Messenger) Boy Not Returned
BNRC e: Bulletin of the National Research Council
BNRS e: British National Radio School
BNRY e: Binary
BNS d: Benzolschweißen
- d: bevorrechtigte Nebenstelle
- e: Bachelor of Natural Science
- e: Bachelor of Naval Science
- e: Bachelor of Nursing Science
- e: Bathymetric Navigation System

- e: Bill Number Screening
- e: Binary Number System
- e: Bombing-Navigation System

BNSA d: Bereich mit Nullinhalt und Symbol am Anfang
BN Sc e: Bachelor of Nursing Science
BNSH e: Burnish
BNSI e: Barbados National Standards Institution
BNS-KrÄmpfe d: Blitz-Nick-Salaam-Krämpfe
BNSY e: Boston Naval Shipyard
BNT e: Bent
- e: Beta-Naphthoflavone
- e: Bonnet

BNTC e: Bainbridge Naval Training Center
BNTM e: Baltic and North Sea Telecommunication Meeting
BNTO e: Bonding Tool
BNU d: Bund Natur- und Umweltschutz
- e: Basic Notch Unit

B Nurs e: Bachelor of Nursing
B-Nut e: B-shaped Nut
BNV e: Bark Necrosis Virus
- s: Biblioteca Nacional de Venezuela

BNVE i: Biblioteca Nazionale Vittorio Emunuele III
BNW e: Bureau of Naval Weapons
BNWL e: Battelle North-West Laboratory
- e: Blackwater National Wildlife Refuge
- e: Bowdoin National Wildlife Refuge
- e: Brigantine National Wildlife Refuge

BNWR e: Blackwater National Wildlife Refuge
- e: Bowdoin National Wildlife Refuge
- e: Brigantine National Wildlife Refuge

BNX e: British Nuclear Export Executive
BNYD e: Boston Navy Yard
BNZ d: Bezirks-Neurerzentrale (DDR)
- d: Bezirks-Neurerzentrum (DDR)
- e: Byte Not Zero

BNZN e: Benzoin
BO d: Bauordnung
- d: Benzol
- d: Berufsordnung
- d: Betriebsordnung
- d: Body Odour
- e: B-Operator
- e: Bachelor of Oratory
- e: Bachelor of Osteopathy
- e: Back Order
- e: Bad Order
- e: Bail Out
- e: Base [of Prism] out
- e: Base Order
- e: Beat [Frequency] Oscillator
- e: Beat[ing] Oscillator
- e: Bench Order
- e: Binary-to-Octal
- e: Binary Operation
- e: Biological Origin
- e: Bit O (Zero)
- e: Black Oil
- e: Blackout
- e: Blanking Oscillator
- e: Blockhouse Operation
- e: Blocking Order
- e: Blocking Oscillator
- e: Block Operation
- e: Block Order
- e: Blowoff
- e: Blowout
- e: Body Odor
- e: Boil-Off
- e: Boiled Oil
- e: Boiler Oil
- e: Bolivia[n]
- e: Boolean [Operation]
- e: Boolean Operator
- e: Boston [Airport]
- e: Botanical Origin
- e: Boundary Light
- e: Bowel [Obstruction]
- e: Bowels Open
- e: Branch Office
- e: Branch Order
- e: Brought-over
- e: Bubble Oxygenator
- e: Bucco-Occlusal
- e: Burn-Out
- e: Byte Out

B & O e: Belladonna & Opium
Bö d: Betriebsökonomik
BO-5 e: Messerschmidt-Bölkow-Blohm Helicopter
BOA d: Bau- und Betriebsordnung für Anschlußbahnen
- e: Bare of Armatures
- e: Basic Ordering Agreement
- e: Benzyloctyladipat
- e: Bibliography of Agriculture
- e: Bipolar Operational Amplifier
- e: Born on Arrival
- e: Break-off Altitude
- e: British Optical Association
- e: British Orthopedic Association
- e: British Osteopathic Association
- e: British Overseas Airways
- e: Broad Ocean Area

BOAC e: British Overseas African Command
- e: British Overseas Air Charter
- e: British Overseas Airways Corporation

BOADICEA e: British Overseas Airways Digital Information Computer for Electronic Automation
B o A [Disp] e: British Optical Association [Dispensing Certificate]
BOADS e: Boston Air Defense Sector
BOAE e: Bureau of Agricultural Economics
BOAM e: Bell Owned and Maintained
BOA-MILS e: Broad Ocean Area-Missile Impact Locating System
BOAmtm d: Bundesbahnoberamtmann
BOAR d: Bundesbahnoberamtsrat
BOARD e: Battelles Office Automation Research and Development
BOAT e: Better Occupational Awareness Training
BOATS e: Boatswain
BOATU e: Boat Unit
Boaufs d: Betriebsoberaufseher
BOB d: Betrieb ohne Beaufsichtigung
- e: Barges-on-Board
- e: Bay of Bengal
- e: Beginning of Business
- e: Bipolar Offset Binary
- e: Bobbin
- e: Boeing-Michigan Air Research Center B Program[me]
- e: Business Opportunity Bank

BOB CODE e: Bipolar Offset Binary Code
BOBHIP e: Boeing-Michigan Air Research Center B History Printout
BO-BO e: Bo-Bo-Type Locomotive
BOBR e: Boring Bar
BOBS e: Beacon Only Bombing System
- e: Bolted on Base

BOC e: Back Outled Central
- e: Beginning of Cycle
- e: Bell Operating Company
- e: Best Output and Colour
- e: Bevitron Orbit Code
- e: Bingham Oceanographic Collection
- e: Bingham Oceanography Collection
- e: Block-Oriented Computer
- e: Blow Out Coil
- e: Body on Chassis
- e: Bottom of Conduit
- e: British Ornithologists Club
- e: British Oxygen Company
- e: Brittany Oceanological Center
- e: Broadcast Operations Control
- e: Build Out Capacitor
- e: Bus Out Check
- e: Butyloxycarbonyl
- e: Byte Output Control
- e: t-Butyloxycarbonyl

BOccuTher. e: Bachelor of Occupational Therapy
BOCD e: Barrels of Oil per Calendar Day
BOCF e: Bureau of Commercial Fisheries
BOCOL e: Basic Operating Consumer-Oriented Language
BOCS e: Bell Operating Companies
- e: Bendix Optimum Configuration Satellite
- e: Bisynchronous Oriented Communications System

BOD d: Beratungsgesellschaft für Organisation und Datenverarbeitung
- d: Bordeaux (Flughafen)
- e: Base Ordnance Depot
- e: Beneficial Occupancy Data
- e: Biochemical Oxygen Demand
- e: Biological Oxygen Demand
- e: Bistable Optical Device
- e: Black-out Door
- e: Boeing on Dock
- e: Boston Ordnance District
- e: Break Over Diode
- e: Buy-off Date

BOD-DO e: Biochemical Oxygen Demand-Dissolved Oxygen
BODEPE e: Boiler Design and Performance
BODL e: Bodleian Library
BODS e: British Oceanographic Data Service
BOD_5 e: Biological Oxygen Demand after five days
BOD unit e: Bodansky unit
BOE d: Betrieb ohne Enddämpfung
- e: Bachelor of Oral English
- e: Back Outlet Eccentric
- e: Beginning of Extent
- e: Board of Osteopathic Examiners
- e: Branch Order Execution
- e: Buffered Oxide Etchant

BOEC e: British Oil Equipment Credits
Boeing 707 e: four-engine longrange jet transport
BOF e: Barium Oxide Ferrite
- e: Basic Oxygen Furnace
- e: Beginning of File
- e: Bias Oscillator Frequency
- e: Binary Oxide Film
- e: Books-on-File

B of AA e: Bachelor of Aeronautical Administration
BOFADS e: Business Office Force Administration Data System
- e: Business Office Forms Administration Data System

B of Adv Art & Des e: Bachelor of Advertising Arts and Design
B of AE e: Bachelor of Aeronautical Engineering
B of B e: Back of Book
B of BC e: Bachelor of Building Construction

B of EE e: Bachelor of Electrical Engineering
B of EE (Com Opt) e: Bachelor of Electrical Engineering; Communications Option
BOFEEP e: Boulder Optimized Field Engineering Estimating Procedure
B of EE (Power Opt) e: Bachelor of Electrical Engineering; Power Option
B of EP e: Bachelor of Engineering Physics
Boffer e: Best Offer
B of H e: Board of Health
B of HE e: Board of Higher Education
B of I e: Block of Information
- e: Block of Instructions
B of ID e: Bachelor of Interior Design
B of IM e: Bachelor of Industrial Management
B of L cl. e: Bill of Lading; clean
B of L t. e: Bill of Lading; through
B of N e: Bureau of Narcotics
BOFS e: Black Oil Finish Slate
B of TC e: Bachelor of Textile Chemistry
B of TE e: Bachelor of Textile Engineering
B of TM e: Bachelor of Textile Management
BO Furnace e: Basic Oxygen Furnace
BOFX e: Boring Fixture
BOG d: Bogen
- d: Bogota
- d: Büro-Organisations-Gesellschaft
- e: Beginning of Group
- e: Boiling
- e: Buried Optical Guide
BOGIE e: Unidentifiled Aircraft
BOG's e: Basic Oppertunity Grants
BOH e: Basic Open Hearth
- e: Bill of Health
- e: Bottom of Heart
- e: Break-off Height
- e: Bureau of Ordnance and Hydrography
BOHC e: Boron-Oxygen Hole Centre
BoHP & RD e: Bureau of Health Planning and Resources Development
bohr. d: bohren
Bohr[g]. d: Bohrung
BOHS e: British Occupational Hygiene Society
BOI d: Bundesbahnoberinspektor
- e: Basis of Issue
- e: Blind Operator Interface
- e: Branch Output Interrupt
- e: Break of Inspection
BOIE d: baumorientiertes interaktives Entwurfswerkzeug
BOIESA e: Bureau of Oceans and International Environmental and Scientific Affairs
BOIL e: Boiling
BOILPT e: Boiling Point
BOIMARS e: Basis of Issue Monitoring and Recording System
BOIP e: Basic of Issue Plan
BOIS e: Branch Office Information System
BOJ e: Booster Jettison
BOL e: Bachelor of Oriental Languages
- e: Bachelor of Oriental Learning
- e: Beginning of Life
- e: Begin of Life
- e: Bill of Lading
- e: Biotechnology Orbital Laboratory
- e: Block Output Length
- e: Bolivia[n]
- e: Bollard
- e: Bottom of Lung
- e: Boundary Light
- e: Build Out Lattice
BOL-148 e: d-2Bromolysergic Acid Tartrate (Hallucinogen)
BOLD e: Bibliographic On-Line[Library] Display
- e: Bit Oriented Line Discipline
BOLDS e: Burroughs Optical Lens Docking System
BOLIC e: Branch Office Library Inventory Control
BOLO e: Be on the Lookout
BOLOVAC e: Bolometric Voltage and Current
BOLS e: Bolster
BOLT e: Basic Occupational Language Training
- e: Beam-of-Light Transistor
BOL Timber e: Bola Timber
Bolz d: Bolzen
BOM d: Bombay (Flughafen)
- d: Bomber
- d: Bombing
- e: Base Operations Manager
- e: Basic Operating Memory
- e: Basic Operating Monitor
- e: Basic Operation[al] Memory
- e: Beginning of Message
- e: Beginning of the Month
- e: Bill of Material[s]
- e: Binary Order of Magnitude
- e: Bit Organized Memory
- e: Bombardier
- e: Bombardment
- e: Born-Oppenheimer Method
- e: Builders Old Measurement
- e: Byte[s] per Operating Memory
BOMAG e: Bill[s] of Material Analyzer and Generator
BOMAP e: Barbados Oceanographic and Meteorological Analysis Project
BOMARC e: Boeing Long-Range Surface-to-Air Missile Bearing Nuclear Warhead
- e: Boeing Michigan [Aeronautical] Research Center
BOMB d: Bomber
- e: Bombardier
- e: Bombardment
Bombay Hook e: Bombay Hook National Wildlife Refuge
BOMBR e: Bombardier
BOM Chamber e: Bureau of Mines Chamber
BOMCOM e: Bomber Command
BOMD e: Bureau of Medical Devices
BOMEX e: Barbados Oceanographic and Meteorological Experiment
- e: Bomber Exercise
BOMIC e: Beryllium Oxide Microwave Integrated Circuit
BOMOL e: Black Box Model Description Language
BOMOP e: Basic Officer Military Orientation Program[me]
BOMOS e: Buried Oxide Metal Oxide Semiconductor
BOMP e: Base Organizition and Maintenance Processor
- e: Basic Organization and Maintenance Program[me]
- e: Bill-of-Material Processor
BOMREP e: Bombing Report
BOMROC e: Bombardment Rocket
BOMRON e: Bombing Squadron
BOMS e: Bureau of Medical Services
BOMST e: Bombsight
BON e: Beta Oxynaphthoic
BOND e: Bonding
Bon. I. e: Bonin Islands
BONIS e: Bibliography of Old Norse-Icelandic Studies
BONT d: benutzerorientiert
BON Timber e: Bonsum Timber
BONUS d: Banken-Online-Universal-System
- e: Boiling Nuclear Superheat [Reactor]
BOOK e: Bibliographic Online Organized Knowledge
- e: Bound, Ordered, Organized Knowledge
- e: Built-in Orderly Organized Knowledge
BöOK d: Böschungsoberkante
Book R e: Book Reviews (journ.)
BOOL e: Boolean
BOOM e: boomerang
BOOST e: Bettering Oregons Opportunity for Saving Talent
- e: Booster
BOOSTRKT e: Booster Rocket
BOOT e: Bootstrap
BOP e: Balance of Payments
- e: Balance of Power
- e: Base of Preference Program[me]
- e: Basic Operation Plan
- e: Basic Overall Polarity
- e: Basic Oxygen Process[ing]
- e: Bathyscaphe Oceanographic Program[me]
- e: Battery Observation Post
- e: Behavior Observation Program[me]
- e: Best Operating Procedure
- e: Binary Output Program[me]
- e: Bipolar Operational Power [supply]
- e: Bit Oriented Protocol
- e: Blowout Preventer
- e: Board of Pharmacy
- e: Break-off Portion
- e: Buffalo Orphan Prototype [Virus]
- e: Burn-Out Proff
BOPA e: Basic Operating Programming Aid
BOPACE e: Boeing Plastic Analysis Capability for Engines
BOPAT e: Border Patrol
BOPD e: Barrels of Oil per Day
B-Operator e: Inward Operator
B o PM e: Board of Podiatric Medicine
BOPRESS e: Boiler Pressure
BOPS e: Bipolar Operational Power Supplies
- e: Blowout Preventer Stack
BOP Supply e: Bipolar Operational Power Supply
BOP Test e: Basic Overall Polarity Test
BOpt[om] e: Bachelor of Optometry
BOQ e: Bachelor Officers Quarters
BOR d: Bundesbahnoberrat
- e: Bachelor of Oratory
- e: Bachelor of Orientation
- e: Battalion Orderly Room
- e: Beginning of Record
- e: Boring
- e: Bowels Open Regularly
- e: British Other Ranks
- e: Bus Out Register
BORA d: bildschirmorientiertes Reiseveranstalter-Anwendersystem
BORAD e: British Oxygen Research and Development
BORAL d: Bor und Aluminium
BORAM e: Block-Oriented Random-Access Memory
BORAMs e: Block-Oriented Random-Access Memories
BORat d: Bundesbahnoberrat
BORAX e: Boiling-Water Reactor Experiment
- e: Sodium Tetraborate

BORAZON e: Boron Nitride Heated and Pressed with a Catalyst
- e: Boron Nitrogen Compound Harder than Diamond
B or B e: Brass or Bronze
BORD e: Book Order and Record Document
BORE e: Beryllium Oxide Reactor Experiment
BOREP e: Bombing Report
BORI e: Brass or Iron
B Orient e: Bachelor of Oriental Studies
BORIS e: Blackley Order Reception and Invoicing System
BORL e: Borland International, Inc.
BORM e: Bureau of Raw Materials [for American Vegetable Oils and Fats Industries]
BO-Ro Carrier e: Bulk-Oil-Roll-on Carrier
BORR e: Bales or Rolls
BOR's e: British Other Ranks
B Or Sc e: Bachelor of Oratory Science
BORSCHT e: Battery Power Feed; Overvoltage Protection; Ringing; Supervision (or Signalling)
BORSENIC e: Boron and Arsenic
BORSHT e: Battery; Overvoltage; Ring; Supervision; Hybrid Test
BORU e: Boat Operating and Repair Unit
BOS d: Bau- und Betriebsordnung für Schmalspurbahnen
- d: Bohrschablone
- d: Bundesbahnobersekretär
- e: Background Operating System
- e: Base Operating Support
- e: Basic Oblate Speriod
- e: Basic Operating System
- e: Basic Oxygen Steel
- e: Basic Oxygen Steelmaking
- e: Basis-Software-Program
- e: Batch Operating System
- e: Battalion Orderly Sergeant
- e: Bell Operating System
- e: Bit Organized Store
- e: Book Order and Selection
- e: Boston [airport]
- e: Brigade Orders
- e: Building-Out Section
- e: Business Office Supervisor
BOsch d: Bundesbahnoberschaffner
BOSFW e: Bureau of Sport Fisheries and Wildlife
BOSH e: Bottom Oriented Shrimp Harvester
BOSIC d: Bosch symbolischer Instruktions-Code
BOSN e: Boatswain
BOSP e: Bioastronautics Orbital Space Program[me]
BOSR e: Base of Stack Register
- e: Bottom of Stack-Register
BOSS e: Ballistic Missile Early Warning System Operational Simulation System
- e: Basic Operating System Software
- e: Batch Operating Software System
- e: Beehive Office Supervisory System
- e: Betts Open Shop System
- e: Biased Optimal Steering Selector
- e: Bioastronautical Orbiting Space System
- e: Bioastronautic Orbital Space System
- e: Biological Orbiting Space Station
- e: Bistable Optically-controlled Semiconductor Switch
- e: Boeing Operational Supervisory System
- e: Bomb Orbital Strategic System
- e: Broad Ocean Scoring System
- e: Building Research Establishment Online Search System
- e: Burroughs Operational System Simulator
- e: Business Oriented Software System[s]
BOSS-Wedge e: Bomb Orbital Strategic System - Weapon Development Glide Entry
Bost. e: Boston [Airport]
Boston Soc of Nat Hist Proc e: Boston Society of Natural History; Proceedings (journ.)
Boston Spa e: British Library Lending Division in Boston Spa
Boston Tech e: Boston Technical Publishers (journ.)
BOSTPDIS e: Boston Procurement District
BostQ e: Boston Quarterly Review (journ.)
BostR e: Boston Review (journ.)
BO-Strab d: Straßenbahn-Bau- und Betriebsordnung
BOSU e: Bioastronautic Operational Support Unit
BOSWASH e: Boston to Washington
BOT a: botanik or botanisk
- d: Botanik[er]
- d: botanisch
- d: Botanist
- e: Bachelor of Occupational Therapy
- e: Balance of Time
- e: Beat-frequency Oscillator Transformer
- e: Begin[ning] of Tape
- e: Blackout Time
- e: Books On Tape
- e: Botanic[al]
- e: Botanist
- e: Botany
- e: Botswana
- e: Bottle
- e: Bottom
- e: Bought
- e: Build-Operate-Transfer
BOTAN e: Botanic[al][ly]
- e: Botanist
- e: Botany
BOTEL e: Boat Hotel
Bot. Gaz. e: Botanical Gazette
BOTH e: Bombing over the Horizon
BOTLO e: Bureau of Ordnance Technical Liaison Officer
BOTM e: Beginning of Tape Marker
BOTMA e: British Office Technology Manufacturers Alliance
BOT Marker e: Beginning-of-Tape Marker
BOTMG e: Bottoming
BOT-Ohm e: Board-of-Trade-Ohm
BOTOSS e: Bottom Topography Survey System
BOTOX e: Botulinium Toxin
Bots e: Bottles
BOTTS e: Busy Tone Trunks
BOTU e: Board of Trade Unit
BOTU(FW) e: Basic Operational Training Unit (Fixed Wing)
BOT-Unit e: Board-of-Trade-Unit
BOU e: Bit Operation Unit
- e: British Ornithologists Union
BOUMAC e: Boulder Laboratory Macrosystem
BOUND e: Boundary
BOUT e: Bus Out
BOV e: Blending Octane Valve
- e: Brown Oil of Vitriol
- e: Burn-out Velocity
- e: Bus Operation Valid
BOVC e: Base of Overcast
BOVINOL e: Bovinologic[al][ly]
- e: Bovinologist
- e: Bovinology
BOW e: Bag of Water
- e: Barrels of Oil and Water
- e: Beryllium Oxide Washer
- e: Black-out Window
- e: Born out of Wedlock
BOWO e: Brigade Ordnance Warrant Officer
BOWR e: Boiler Water
BOX Isolation Technique e: Buried Oxide Isolation Technique
BOX Timber e: Boxwood Timber
BOYC e: Boyne City [Railroad]
BOZO e: Brawny Intellectual Light-Weight
BP d: Bahnpost
- d: Bandpaß[filter]
- d: Bauprinzip
- d: Bauschaltplan
- d: Bausteinprojektierung
- d: Bedienungsprozessor
- d: Bedien[ungs]pult
- d: Belegungsplan
- d: Benutzerprozeß
- d: Benzin- und Petroleum AG
- d: Berechnungsprotokoll
- d: beschleunigter Personenzug
- d: Betriebs[abgabe]preis
- d: Betriebspädagogik
- d: Bibliotheksprogramm
- d: Bildpunkt
- d: bisexuelle Potenz
- d: Bundespost
- e: Bachelor of Painting
- e: Bachelor of Pedagogy
- e: Bachelor of Pharmacy
- e: Bachelor of Philosophy
- e: Background Processing
- e: Background Program[me]
- e: Back Order
- e: Back Pressure
- e: Back Projection
- e: Backspace Pawl
- e: Bad Parity
- e: Baking Powder
- e: Bandpass
- e: Barber Pole
- e: Barometric Pressure
- e: Base Pay
- e: Base Percussion
- e: Base Pitch
- e: Base Plate
- e: Base Point
- e: BaseProcured
- e: Basic Pay
- e: Basic Protein
- e: Batch Processing
- e: Batch Proessing
- e: Battery Package
- e: Battle Practice
- e: Bayonet Pin
- e: Beacon Point
- e: Beam[-forming] Plate[s]
- e: Beam Power
- e: Bearing Point
- e: Bed Pan
- e: Bed Plate
- e: Before Present
- e: Behaviour Pattern
- e: Below Proof
- e: Benz[o][a]pyrene
- e: Between Perpendiculars
- e: Bill of Parcels
- e: Bills Payable
- e: Binary Place
- e: Binary Point
- e: Binary Punch
- e: Binding Post
- e: Biopack

- e: Bioscience Program[me]
- e: Biotic Potential
- e: Biparletal
- e: Bipolar
- e: Birthplace
- e: Bit Pair
- e: Bit Position
- e: Black Powder
- e: Blast Propagation
- e: Blind Post
- e: Block Packaging
- e: Block Parity
- e: Block Punch
- e: Blood Pressure
- e: Blood Program[me]
- e: Blueprint
- e: Boiler Plate
- e: Boiler Pressure
- e: Boiling Point
- e: Bolted Plate
- e: Bombing Post
- e: Bonded [Single] Paper
- e: Bonded Part
- e: Boolean Part
- e: Bootstrap Program[me]
- e: Border Patrol
- e: Boron Plastic
- e: B Pulse
- e: Branching Program[me]
- e: Branch Program[me]
- e: Brass Plug
- e: Break Point
- e: Bridge Bypass
- e: British Patent
- e: British Petroleum [Company]
- e: British Pharmacopoeia (journ.)
- e: Broadcast Pioneer
- e: Broncho-Plural
- e: Bubble Pulse
- e: Buccopulpal
- e: Buffered Printing
- e: Bullet-Proof
- e: Bullous Pemphigoid
- e: Bureau of Power
- e: Butt Plane
- e: By-Pass
- e: Bypass
- e: Lib Broadcast Pioneers Library
- i: Biblioteca Pubblica
- s: Biblioteca Pública

BPA d: Bahnpostamt
- d: Beginn der Paraffin-Abscheidung
- e: Bachelor of Painting
- e: Bachelor of Professional Arts
- e: Bachelor of Public Administration
- e: Background Program[me] Area
- e: Balloon Platoon of America
- e: Bandpass Amplifier
- e: Beam Plasma Amplification
- e: Biological Photographers Association
- e: Bioshield Power Assembly
- e: Birds Protection Act
- e: Bisphenol A
- e: Blanket Purchase Agreement
- e: Blind Persons Act
- e: Blood Pressure Assembly
- e: Bonneville Power Administration
- e: Boronated Phenylalanine
- e: Bovine Plasma Albumin
- e: Break-Point Address-Register
- e: British Patents Abstracts
- e: British Pediatric Association
- e: Broadband Power Amplifier
- e: Broadcasters Promotion Association
- e: Budget Project Account
- e: Business Publication Audit [of Circulation]

B.P.a. d: Bundespatent angemeldet

BPAC e: Budget Program[me] Activity Code

BP Adm e: Bachelor of Public Administration

BPaed e: Bachelor of Paediatrics

BPAM e: Basic Partitioned Access Method
- e: Basic Program[me] Access Method

BPA Register e: Break-Point Address Register

BPA-SDI e: Bonneville Power Administration Selective Dissemination of Information

BPatGer. d: Bundespatentgericht

BpB d: Bahnpolizeibeamter

BPB d: Benutzer-Programmbibliothek
- e: Bachelor of Physical Biology
- e: Bank Port Bill
- e: Bromphenol Blue

BPBS d: Band-Platte-Betriebssystem

BPBW e: Bare Phosphor Bronze Wire

BPC e: Back Pressure Control
- e: Bandpass Crystal
- e: Base Patrol Center
- e: Base Point Configuration
- e: Basic Peripheral Channel
- e: Basic Programming for Computers
- e: Basic Punched Card
- e: Bilateral Private Circuit
- e: Binary Punched Card
- e: Binding Post Chamber
- e: Blank Punched Card
- e: Blast-furnace Portland Cement
- e: Block Priority Control
- e: Book Prices Current
- e: Boost Protective Cover
- e: Boston Programming Center
- e: British Petroleum Company Ltd.
- e: British Pharmaceutical Codex
- e: British Phosphate Commission
- e: Buffer Punch Card
- e: Bulk Petrol Company
- e: Business and Professional Code

BPCA e: Bioshield Pyrotechnic Control Assembly

B-P Cartridge e: Barncade-Penetrating Cartridge

BPCC e: British Printing and Communication Corporation

BPCD e: Barrels per Calendar Day

BPCDI Unit e: Brookhaven Portable Cesium Developmental Irradiator Unit

BPCF e: Bandpass Crystal Filter

BPCI e: Bulk Packaging and Containerization Institute

BPCM e: Band Predictive Code Modulation
- e: Basic Punched Card Machine

BP/CP e: Base Procured/Central Procured

BPCR e: Brakes on Pedal Cycle Regulations

BPCT e: Best Practical Control Technology

BPCTCA e: Best Practicable Control Technology Currently Available

BPD e: Bachelor of Pedagogics
- e: Bandpass Detector
- e: Barrels per Day
- e: Basic Planning Document
- e: Bio-Psychology Doctor
- e: Blood Program[me] Directives
- e: Boxes per Day
- e: Bronchopulmonary Dysplasia
- e: Bushing Potential Device

BPDA e: Basic Planning Data and Assumption
- e: Bibliographic Pattern Discovery Algorithm

BPDA Method e: Basic Partitioned Data Access Method

BPDC e: Berkeley Particle Data Center

BPDMS e: Basic Point-Defense Missile System

BPDSMS e: Basic Point Defense Surface Missile

BPE e: Bachelor of Pedagogy
- e: Bachelor of Physical Education
- e: Back Porch Effect
- e: Basic Programming Extensions
- e: Best Preliminary Estimate
- e: Binural Phase Effect
- e: Bit-Plane Encoding
- e: Black Powder Express
- e: Boiling Point Elevation

B Ped e: Bachelor of Pedagogics
- e: Bachelor of Pedagogy

BP Ed e: Bachelor of Physical Education

BPED e: Basic Pay Entry Date

B Pe E[ng] e: Bachelor of Petroleum Engineering

B Pet E e: Bachelor of Petroleum Engineering

BPEW Program[me] e: Board Pre-Edit Wires Program[me]

BPF d: Bezirksdirektion für das Post- und Fernmeldewesen (DDR)
- e: Bandpass Filter
- e: Blade Passing Frequency
- e: Blue Print Files
- e: Bottom Pressure Fluctuation
- e: Bradykinin Potentiating Factor
- e: British Pacific Fleet
- e: British Polio Fellowship
- e: Bromine Pentafluoride

BPFA Method e: Basic Partitioned File Access Method

BPFILO e: British Pacific Fleet Intelligence Liaison Officer

BPFLO e: British Pacific Fleet Liaison Officer

BPF's e: Bandpass Filters

BPFS e: Bulk Petroleum Facilities and Systems

BPG d: Benzathin-Penicillin-G
- d: Bundespatentgericht
- e: Blood Pressure Gauge
- e: Boron Pyrolytic Graphite
- e: Break Pulse Generator

BPGC e: Bearing per Gyro Compass

B Ph e: Bachelor of Philosophy

BPH e: Bachelor of Public Health
- e: Barrels per Hour
- e: Benign Prostatic Hyperplasia
- e: Benign Prostatic Hypertrophy
- e: British Pharmacopoeia

BPharm e: Bachelor of Pharmacy

BPhC e: Bachelor of Pharmaceutical Chemistry

BPHE e: Bachelor of Physical and Health Education
- e: Bachelor of Public Health Engineering

BPH Ed e: Bachelor of Public Health Education

BPH Eng e: Bachelor of Public Health Engineering

BPHI e: Boost Phase Intercept

B Phil e: Bachelor of Philosophy

BP & HL e: Brown Picton and Hornby Libraries

BPHN e: Bachelor of Public Health Nursing

B Pho e: Bachelor of Photography

B Ph S e: Bachelor of Physical Science

BPhty e: Bachelor of Physiotherapy

BPhW d: Berliner Philologische Wochenschrift (journ.)

B Phy e: Bachelor of Physics

B Phys Ed. e: Bachelor of Physical Education

B Phys Thy. e: Bachelor of Physical Therapy
BPI e: Bernard Price Institute [of Geophysical Research; Johannesburg]
- e: Bit[s] per Inch
- e: Bituminous Pipe Institute
- e: Bookman Price Index
- e: Book Production Industry (journ.)
- e: Booksellers Provident Institution
- e: Boost Phase Intercept
- e: Bradley Polytechnic Institute
- e: Break-Point Instruction
- e: British Pacific Islands
- e: Broadcasting Publications, Incorporated
- e: Brooklyn Polytechnic Institute
- e: Building Products Institute
- e: Bureau of Plant Industry[, Soils and Agricultural Engineering]
- e: Buying Power Index
- e: Byte[s] per Inch
BPICA f: Bureau Permanent Internationale des Constructeurs d'Automobiles
BPID e: Book Physical Inventory Difference
BPIF e: British Printing Industries Federation
BPISAE e: Bureau of Plant Industry, Soils and Agricultural Engineering
BPIT e: Basic Parameter Input Tape
BPJ e: Balanced Pressure Joint
BPKFL e: Block Packaging Flag
BPKT d: Bestellpunkt
- e: Basic Program[me] Knowledge Test
- e: Basic Programming Knowledge Test
BPL d: Bedienungsplatz
- e: B-Propriolactone
- e: Bachelor of Patent Law
- e: Band-Pass Limiter
- e: Band Pressure Level
- e: Basic Program[me] Loader
- e: Beam Packing Loss
- e: Belfast Public Library
- e: Bell Propulsion Laboratory
- e: Beta-Propriolactone
- e: Binghamton Public Library
- e: Birmingham Public Library
- e: Birthplace
- e: Bleach Powder Lime
- e: Bone Phosphate of Lime
- e: Book-Plate
- e: Boston Public Library
- e: Bridgeport Public Library
- e: British Physical Laboratories
- e: Brooklyn Public Library
- e: Buffalo Public Library
- e: Burroughs Program[me] Language
- e: Burroughs Programming Language
- e: Burst Position Locator
- e: Business Planning Language
- e: Business Programming Language
BPLA e: Bow Plane
BPL/K d: Benutzerprozeß lokal/kommunizierend
BPLQ e: Boston Public Library Quarterly (journ.)
BPLS e: Boston Public Latin School
- e: Brief Psychiatric Rating Scale
BPM d: biologische Prozeßsteuerungsmittel
- d: Breitbandpegelmesser
- d: Buchstaben pro Minute
- d: Bundesministerium für das Post- und Fernmeldewesen
- d: Bundespostminister[ium]
- e: Bank Proof Machine
- e: Barrels per Minute
- e: Batch Processing Monitor
- e: Beam Positioning Monitor
- e: Beats per Minute
- e: Best Practical Means
- e: Bi-Phase Modulation
- e: Blows per Minute
- e: Breaths per Minute
BPM e: Bulletin of Paleomalacology (journ.)
BPMA e: British Machinery Pump Manufacturers
- e: British Printing Association
BPME e: Branching Process with Markovian Environments
BPMEL e: Base Precision Measurement Equipment Laboratory
BpMF e: British Postgraduate Medical Federation
BPMin d: Bundesminister für das Post- und Fernmeldewesen
- d: Bundespostminister[ium]
BPMM e: Bits per Millimeter
BPMS e: Blood Pressure Measuring System
BPMVfg d: Bundespostministerium-Verfügung
BPN e: Balloon-borne Polar Nephelometer
- e: Bandpass Network
- e: Breakdown Pulse Noise
- e: Budget Project Number
BPNHM e: Banff Park Natural History Museum
BPO d: Betriebsprüfungsordnung
- e: Base Pay Office[r]
- e: Base Post Office
- e: Benzoyl Peroxide
- e: Boeing-Michigan Aeronautical research Center Prelaunch Output
- e: British Post Office
BpObl d: Bahnpost-Oberbetriebsleitung
BPOC e: Before Processing on Course
B Pol Sc e: Bachelor of Political Science
BPON e: Broadband Passive Optical Network
BPOS e: Batch Processing Operating System
- e: Block-print Position
BPOST e: Binding Post
BPP e: Beacon Portable Packset
- e: Border Patrol Police
- e: Boundary Phase Plasticity
- e: British Polished Plate
- e: Bubble Pulse Period
- e: Buffer Punch Program[me]
- e: Bursting Pacemaker Potential
BPPF e: Base Program[me] Preparation Facility
BPPG e: Bureau Planned Procurement Guide
BPPSJ e: Balanced Pressure Plane Swivel Joint
BPP-Theory e: Bloembergen-Purcell-Pound Theory
BPQ e: Boston Public Quarterly (journ.)
BPR e: Banana Plug Resistor
- e: Bar-Pattern Response
- e: Battery Powered Recorder
- e: Beryllium Physics Reactor
- e: Bimonthly Progress Report
- e: Biweekly Progress Report
- e: Blood Pressure Recorder
- e: Bonded Particle Rolling
- e: Book Publishing Record
- e: Bridge Plotting Room
- e: Bubble Position Register
- e: Building Products Register
- e: Bulletin of Prosthesis Research
- e: Bureau of Public Roads
- e: Bypass Ratio
BPRA e: British Pattern Recognition Association
BPRE e: Branching Process with Random Environments
BPRF e: Bullet-Proof
BPRS e: Basic Parts Records Summary
- e: Brief Psychatric Rating Scale
BPrSt d: Bundesprüfstelle
BPRT d: Bundesverband privater Rundfunk und Telekommunikation
BPRTHM e: Bureau of Public Transport Highway Mobilization
BPS d: Basisprogrammiersprache
- d: Bit pro Sekunde
- d: Brems-Pferdestärke
- d: Bullenprüfstation
- d: Bundesprüfstelle
- d: Byte pro Sekunde
- e: Bachelor of Psychology
- e: Back Panels
- e: Balanced-Pressure System
- e: Base Postal Section
- e: Basic Programming Suggest
- e: Basic Programming Support
- e: Basic Programming System
- e: Batch Processing System
- e: Baud[s] per Second
- e: Bayou Preservation Society
- e: Beacon Portable Set
- e: Beacon Processing System
- e: Benchmark Portability System
- e: Best Paint Solid
- e: Bill Processor System
- e: Biophysical Society
- e: Bit[s] per Second
- e: Blanked Picture Signal
- e: Blowout Pipe system
- e: Border Patrol Sector (or Station)
- e: British Plain Spirits
- e: British Psychological Society
- e: Buffer Punch Service
- e: Burroughs Poll/Select
- e: Byte[s] per Second
BPSA e: Bachelor of Public School Art
- e: Business Products Standards Association
BPSC e: Bearing per Standard Compass
- e: Bulk Petrol Storage Center
BPSD e: Barrels per Stream Day
- e: Bulk Petrol Supply Depot
BPSH e: Border Patrol Sector Headquarters
BPSI e: Bits per Square Inch
BPSJ e: Balanced Pressure Swivel Joint
BPSK e: Bi-Phase Shift Keying (or Keyed)
- e: Binary Phase-Shift-Keyed
- e: Binary Phase Shift Keying
BPSM e: Bachelor of Public School Music
BPSN e: Budget Project Symbol Number
BPSS e: Base Perimeter Security System
- e: Base Procurement Service Stores
- e: Basic Packet Switched System
- e: Basic Production Scheduling System
- e: Bio-Pack Sub-System
- e: British Psychological Society
B Ps Sc e: Bachelor of Psychic Sciences
BPStgC e: Bearing per Stearing Compass
BpStGC e: Bearing per Stearing Gyro Compass
B Ps Th e: Bachelor of Psychotherapy
BPSV d: Baustein-Programmsystem zur statistischen Versuchsauswertung
B Psych e: Bachelor of Psychology
BPT e: Bachelor of Physiotherapy
BPT e: Bandpass Transformer
- e: Base Point
- e: Battle Practice Target
- e: Beginning Procedure Turn
- e: Best Practical Treatment
- e: Bipolar Transistor
- e: Bi Post

- e: Boiling Point
- e: Bound Plasma Tryptophan
- e: Braille Printing Terminal
- e: Breakpoint
- e: British Petroleum Tanker
- e: Buffer Punch Transfer
- e: Bulk Petrol Transport

BPTI e: Bovine Pancreatic Trypsin Inhibitor
BPTO e: Ballistic Missile Early Warning System Performance Test Outline
BPTS e: Basic Principles Training Simulator
BPTV e: Battleship Propulsion Test Vehicle
BPU d: Basis-Programm[ier]unterstützung
- e: Base Production Unit
- e: Basic Pole Unit
- e: Basic Processing Unit
- e: Boeing-Michigan Air Research Center Pick-Up

BPUNP s: Biblioteca Publica de la Universidad de la Plata
BPV d: Baupreisverordnung
- d: Behälter- und Palettenvorschriften
- e: Back Pressure Valve
- e: Bordetella Pertussis Vaccine
- e: Bovine Papilloma Virus

BPW d: Bahnpostwagen
- d: Berliner Philologische Wochenschrift (journ.)
- e: Bare Platinum Wire
- e: Board of Public Works

BPWR e: Burnable Poison Water Reactor
B Py e: Bachelor of Pedagogy
BPz d: Bergungspanzer
BPZ d: blockweise Prioritätszuteilung
BPZoll d: Betriebsprüfung Zoll
BQ d: Becquerel
- e: Bachelors Quarters
- e: Barque
- e: Basically Qualified
- e: Basic Qualification
- e: Beauty Quotient
- e: Becquerel
- e: Boiler Quality
- f: Bibliotheque nationale du Quebec

BQA e: Benzoquinone Acetic Acid
BQAP e: Bilevel Quality Assurance Program[me]
BQEP e: Brief Qualification Evaluation Program[me]
BQL e: Basic Query Language
- e: Batch Query Language

BQLI e: Brooklyn, Queens, Lond Island
BQM e: Base Quartermaster
BQMS e: Battery Quartermaster Sergeant
BQS d: benutztes Quasisynonym
- e: Basic Query System

BQUE e: Barque
BR d: B-Register
- d: Basic Register
- d: Basisregister
- d: Baurat
- d: Baureihe
- d: Bayerischer Rundfunk
- d: Bedien[ungs]rechner
- d: Befehlsregister
- d: Belastungsregler
- d: bemannter Raumflug
- d: Bereichsrechner
- d: Berliner Rundfunk
- d: Betriebsrechner
- d: Betriebsrechnung
- d: Bibliotheksrechner
- d: Bimetallrelais
- d: Binary Representation
- d: Bomber
- d: Brand
- d: Brasilien (Kfz)
- d: Braun
- d: Breit
- d: Brennschneiden
- d: Brennstoff
- d: Brom
- d: Bronze
- d: Broschur
- d: Brückenhaus
- d: Bundesbahnrat
- e: Bacteriological Reviews
- e: Badge Reader
- e: Band Rate
- e: Band Reject
- e: Bangladesh Railway
- e: Bank Rate
- e: Base Reclamation
- e: Base Record
- e: Base Register
- e: Basic Requirements
- e: Basic Research
- e: Basic Routine
- e: Battery Right
- e: Beam Rider
- e: Bearing [Ratio]
- e: Bedrock
- e: Bend Radius
- e: Bill of Rights
- e: Bills Receivable
- e: Binary Record[ing]
- e: Biosphere Reserve
- e: Birmingham Revision
- e: Birth Rate
- e: Bit Rate
- e: Bit Resolution
- e: Blank Record
- e: Block Reading
- e: Block Reception
- e: Block Register
- e: Block Replacement
- e: Boiler Room
- e: Boiling Range
- e: Boil Resistant
- e: Bombardier
- e: Bomber Reconnaissance
- e: Booking Register
- e: Book of Reference
- e: Book Rack
- e: Booster-Regulator
- e: Bootstrap Routine
- e: Brake [Relay]
- e: Branch
- e: Branching Ratio
- e: Branch Report
- e: Branch Routine
- e: Brass
- e: Brazil
- e: Break [Relay]
- e: Breath
- e: Breeder Reactor
- e: Bridge
- e: Bridging
- e: Briefing Room
- e: Brig
- e: Brigadearzt
- e: British Rail
- e: British Railways
- e: British Revision
- e: Broken and Repaired
- e: Bromine
- e: Bronchial Rings
- e: Brown
- e: Brush
- e: Buffer Register
- e: Builders Risk
- e: Building and Repair
- e: Building Risk
- e: Bulkhead Receptacle
- e: Bulk Resistance
- e: Bunker Ramo
- e: Bus Request
- e: Butadiene Rubber
- e: Byte Rate

Br. d: Breite
- d: Bremse
- d: Brenner
- d: Bruch
- d: Brücke
- d: Brustflosse
- e: British
- e: Brother

br. d: bremsen
- d: broschiert
- d: brutto

BR-1150 e: Breguet maritime-patrol aircraft
BRA d: Bedienungsrechneranschluß
- d: Betriebsregelungsanweisung (DDR)
- e: Base Rate Area
- e: Basic Record Audit
- e: Beam Ride Actuator
- e: Beef Recording Association
- e: Bee Research Association
- e: Bench Replaceable Assembly
- e: Beta Resorcyclic Acid
- e: Brigadier; Royal Artillery
- e: British Robot Association
- f: Bibliothèque royale Albert I
- f: Brassière
- s: Brasil [Spanish-Brazil]

BRA (AA) e: Brigadier; Royal Artillery (Antiaircraft)
BRAB e: Building Research Advisory Board
BRAC e: Britannica Reading Achievement Center
BRACBYCEPHS e: Brachycephalics
BRAD e: Biotechnic Research and Development
- e: British Rate and Data
- e: Brookhaven Raster Display

B Ra E[ng] e: Bachelor of Radio Engineering
BRAF e: Braking Action Fair
BRAFA d: VEB Brandsohlenfabrik (DDR)
BRAG d: Baseler Rheinschiffahrts AG
- e: Braking Action Good

BRAID e: Bidirectional Reference Array; Internally Derived
BRAILLE e: Balanced Resource Allocation Information for Logical Lucid Evaluation
BRAINS e: Behavior Replication by Analog[ue] Instruction of the Nervous System
Br. Air. At. e: British Air Attaché
Br. Am. e: British America
BRAMATEC e: Brain Mapping Technique
BRAN e: Braking Action Nil
BRANCAN Cable e: Brazil-Canary Islands Cable
BRANE e: Bombing Radar Navigation Equipment
BRAP e: Braking Action Poor
BRAS e: Ballitic Rocket Air Suppression
- e: Brasenose College
- f: Brassière
- i: Brasile
- s: Brasil

BRASILSAT e: Brazils satellite
BRASS e: Base Engineer Emergency Forces Reporting, Analysis and Status System
- e: Bottom Reflecting Active Sonar System
- e: Bottom Reflection Active Sonar System

BRASTACS e: Bradford Scientific, Technical and Commercial Service
BRat d: Bundesbahnrat
Br. atm. e: British Atmosphere

BRAVE e: Boeing Robotic Air Vehicle
Bravo letter e: B radio code
BRAXP e: Braking Action Extremely Poor
BRAZ e: Bazil[ian]
- e: Brazier
- e: Brazing
BRAZH e: Brazier head
BRB e: Ballistic Recoverable Booster
- e: Ballistic Reentry Body
- e: Base Rate Boundary
BRBB e: B Register B Bit
BRBC e: Bovine Red Blood Cells
- e: Burro Red Blood Cells
BRBDS e: Barbados
BRBZC e: Brass, Bronze or Copper
BRC e: Balcones Research Center
- e: Base Recovery Course
- e: Base Register Contents
- e: Base Residence Course
- e: Basic Real Constant
- e: Biological Radio Communications
- e: Biological Records Centre
- e: Biological Research Center
- e: Biomedical Recovery Capsule
- e: Bit Reversion Circuit
- e: Block Representation Code
- e: Bounded Right Context
- e: Breeder Reactor Corporation
- e: Bridge Remote Control
- e: British Columbia
- e: British Radio Communication
- e: British Reinforced Concrete
- e: Buffer Read Card
- e: Bunker-Ramo Corporation
- e: Burlingame Research Center
- e: Business Reply Card
BRCA e: British Roller Canary Association
BRCB e: B-Register C Bit
BRCC e: British Roller Canary Club
BRCD e: Binary information with a Residue Check Digest
- e: Braced
BRCDE e: Branch Code
BRCH e: Branch
- e: Broach
Br. Chem. Eng. Process Technol e: British Chemical Engineering and Process Technology (journ.)
BRCMA e: British Radio Cabinet Manufacturers Association
B Rcn e: Battle Reconnaissance
Br. Col. e: British Columbia
Br. Corros. J. e: British Corrosion Journal (journ.)
BRCS e: Ballistic Missile Early Warning SystemRearward Communications System
BRCT e: Broadband Rectangular-to-Circular Transition
Br Cwlth e: British Commonwealth
BRD d: Barbados (Kfz)
- d: Bergrettungsdienst
- d: Bergungs- und Rettungs-Dienst
- d: Bundesrepublik Deutschland
- e: Back Receive Data
- e: Balance Reading Device
- e: Ball Reduction Drive
- e: Base Register Designator
- e: Basic Retirement Date
- e: Binary Rate Divider
- e: Blank Recording Disc
- e: Board
- e: Bomb Release Distance
- e: Booster Requirements Document
- e: Braid[ed]
- e: Broad
BRDAT d: Betriebsregisterdatei
BRDC e: Bare Refractory; Double Containment (Boiler)
- e: Bureau of Research and Development Center
BRDCST e: Broadcast
BRDCT e: Broadcast
BRDF e: Basic Read Data Flow
- e: Bidirectional Reflectance Distribution Function
Brd. Fwd. e: Brought Forward
BRDG e: Bituminous Roads Development Group
- e: Bridge
- e: Bridging
BRDM e: Soviet amphibious reconnaissance vehicle
BRDT e: Bayesian Reliability Demonstration Test[s]
- e: Breadth
BRDU e: Bromodeoxyuridine
BRE d: Bremen (Flughafen)
B Re e: Bachelor of Religion
BRE e: Bachelor of Religious Education
- e: Building Research Establishment
BREAD e: Research Institute of Australia
BRec e: Bills Receivable
BREC e: Brecciated
BRECCIA e: Pyroclastic Volcanic Rock
Brech. d: Brechung
BRECOM e: Broadcast Radio Emergency Communication
BRECSU e: Building Research Energy Conservation Unit
BR Ed e: Bachelor of Religious Education
B Re E[ng] e: Bachelor of Refrigeration Engineering
Bref d: Baureferendar
B Reg d: B-Register
B Reg 1 d: B-Register 1 Bit
B Reg 1 B d: B-Register 1 Bit
B Reg 2 B d: B-Register 2 Bit
B Reg 4 B d: B-Register 4 Bit
B Reg 8 d: B-Register 8 Bit
B Reg 8 B d: B-Register 8 Bit
B Reg A d: B-Register A Bit
B Reg AB d: B-Register A Bit
B Reg B d: B-Register B Bit
B Reg BB d: B-Register 8 Bit
B Reg CB d: B-Register C Bit
B-Register d: Basisregister
- e: Base Register
Breguet 1150 e: Atlantique maritime-patrol aircraft
Breguet 765 e: Sahara flying transport
B Reg Wd MK d: B-Register Word Mark
Breisg d: Breisgau
BREMA e: British Radio Equipment Manufacturers Association
BREN e: Bare Reactor Experiment at Nevada
Brennst. d: Brennstoff[e]
Brennst.-V[erbr]. d: Brennstoffverbrauch
Brennst.-Wärme-Kraft d: Brennstoff-Wärme-Kraft (journ.)
BRER e: Basic Radiation Effects Reactor
BRESTCHAN e: Brest Sub-Area Channel
bret d: bretonisch
BREV e: Brevet
- e: Breviary
- e: Breviate
- e: Brevier
- f: Breveté
- i: Brevetto
brev. l: breviarium
BRF d: Baltische Rotbannerflotte
- d: Band-Rejection Filter
- d: Baurichtlinien für Fallschirme
- d: Bereich Rundfunk und Fernsehen
- d: Betonrohrfertiger
- d: Betriebslaboratorium für Rundfunk und Fernsehen (DDR)
- e: Bell Rings Faintly
- e: Branchial Filament
- e: Brief[ing]
- e: British Road Federation
BRFFI e: Biochemical Research Foundation of the Franklin Institute
BRFG d: Briefing
BRFl d: Breitflachprofil
BRFP e: Baseline Reference Flight Plan
BRG d: Bruttogewicht
- d: Guayana (Kfz)
- e: Barge
- e: Baud Rate Generator
- e: Bearing
- e: Bridge
- e: Bridging
- e: Brig
- e: British Guyana
- n: Bibliotheek van de Rijksuniversiteit te Gent
BRGBLN e: Barrage Balloon
Br.-Gew. d: Bruttogewicht
BRGHD e: Bridgehead
BRGM f: Bureau de Recherches Géologiques et Miniers
BRH e: Blackburn Rivet Head
- e: Bridgehead
BRH e: British Honduras
- e: Brush Holder
- e: Bureau of Radiological Health
BRHLR e: Brush Holder
Br. Hon. e: British Honduras
BRHS e: Bay Ridge High School
- e: Betsy Ross High School
BRI e: Base Recirculation Insulation
- e: Bearing Range Indicator
- e: Behavior Research Institute
- e: Biological Research Institute
- e: Book Review Index
- e: Brain Research Institute
- e: Brand Rating Index
- e: Brightness
- e: British Isles
- e: Building Research Institute
- e: Business Research Institute
BRIA e: Biological Research Institute of America
BRICS e: Bureau of Research Information Control System
BRICSHST e: British Rail Inter-City Service High-Speed Train
BRIDGEX e: Bridge Construction Exercise
BRIG e: Ballistic Missile Early Warning System Raid Input Generator
BRIGATPAD e: British Rail Interactive Graphical Aid to Production and Design
Brigham Young Univ. Sci. Bull. e: Brigham Young University Science Bulletin; Biological Series (journ.)
Brikettier. d: Brikettierung
BRIL e: Brilliance
BRIMEC e: British Mechanical Engineering Federation
BRIMILHOSP e: British Military Hospital
BRIOLIS e: British Institute On-Line Information Service
BRIP e: Biomedical Research in Progress
BRISC e: Biological Recording in Scotland Committee

BRISOL e: British Solvent Oils [Company]
BRIT a: britisk
- d: britisch
- e: Britain
- e: Britannia
- e: Britannica
- e: British
BRITAEA e: British Atomic Energy Agency
Brit Book C[en]tr e: British Book Centre
Brit Britannica e: Encyclopaedia Britannica
BRITE e: Bright Radar-Indicator Tower Equipment
BRITHOSP e: British Hospital
BRITINFO e: British Information Services
BRITIRE e: British Institute of Radio Engineers
British Am Bks e: British American Books
Brit. J. Appl. Phys. e: British Journal of Applied Physics (journ.)
Brit. J Psychiat e: British Journal of Psychiatry (journ.)
Brit. J Surg e: British Journal of Surgery (journ.)
BRITLOG e: British Logistics
BRITMET e: Britannia Metal
Brit Mus e: British Museum
Brit Pat e: British Patent
Brit Pharm e: British Pharmacopoeia
Brit Phos Comm e: British Phosphate Commission
BritRail e: British Railways
Brit. Sam. e: British Samoa
BRITSHIPS e: British Shipbuilding Integrated Production System
Britt. l: Britannorum
Brit Telecom e: British Telecommunications
BRJ e: Berko Rotary Joint
- e: Blind Riveted Joint
Br J App Phys e: British Journal of Applied Physics (journ.)
Br. J. Cancer e: British Journal of Cancer
Br. J. Dermatol. e: British Journal of Dermatology
BRJE e: Burroughs Remote Job Entry
Br. J. Exp. Pathol. e: British Journal of Experimental Pathology
Br. J. Haematol. e: British Journal of Haematology
Br. J. Ind. Med. e: British Journal of Industrial Medicine
Br. J. Non-Destr. Test. e: British Journal of Non-Destructive Testing
Br. J. Opthalmol. e: British Journal of Opthalmology
Br. J. Radiology e: British Journal of Radiology
Br J Surg e: British Journal of Surgery
BRK d: Braunkohle
- d: Brücke
- e: Break
- e: Broken
BRKBD e: Brake Band
Brk.-Bergb. d: Braunkohlebergbau
Brk.-Bergw. d: Braunkohlebergwerk
BRKG e: Breaking
BRK PT e: Break Point
BRKR e: Breaker
BRKT e: Bracket
BRKWTR e: Breakwater
BRL d: Beschaffungsrichtlinie[n]
- d: Blockierringleitung
- d: Busringleitung
- e: Balance Return Loss
- e: Ballistic Research Laboratories
- e: Ballistic Research Laboratory
- e: Barge, Refrigerated, Large
- e: Barrel
- e: Beecham Research Laboratories
- e: Beginning Reel Label
- e: Behavioral Research Laboratories
- e: Bible Research Library
- e: Binary Reloctable Loader
- e: Bionetics Research Laboratory
- e: Birmingham Reference Library
- e: Bit Rate Low
- e: Bomb Release Line
- e: Branch and Link
- e: Breechloader
- e: British Research Library
- e: brown line positive
- e: Butterwick Research Laboratories
- n: Bibliotheek der Rijksuniversiteit te Leiden
BRL 1241 e: Beecham Research Laboratories formula 1241 (methicillan)
BRL 1341 e: Beecham Research Laboratories formula 1341 (penbritin)
Br LC e: Branch on Load Card
BRLESC e: Ballistic Research Laboratory Electronic Scientific Computer
BRLG d: Brennlänge
- e: Bomb, Radio, Longitudinal, Generator-powered
BRLROLL e: Barrel Roller
BRLS e: Barrels
- e: Barrier Ready Light System
BRLSYS e: Barrier Ready Light System
BRLT e: Basic Reference Lottery Ticket
BRM d: Barometer
- d: Brückenmeister[ei]
- e: Baseline Reference Mission
- e: Binary Rate Multiplier
- e: Bit Rate Multiplier
- e: British Racing Motor[s]
BRMA e: Board of Registration of Medical Auxiliaries
BRMAVOFI e: Bureau of Raw Materials for American Vegetable Oils and Fats Industries
BRMBR e: Bear River Migratory Bird Refuge
BRMC e: Barometric[al]
Br. Med. Bull. e: British Medical Bulletin (journ.)
Br. Med. J. e: British Medical Journal (journ.)
BRMGB e: Branch Manager
BRMILHOSP e: British Military Hospital
Br Mn e: Branch on Minus
BRMs e: Binary Rate Multipliers
BrMstr d: Brandmeister
BRN d: Bahrein (Kfz)
- e: Birth Rate Normal
- e: Brown
- e: Brunei
- e: Buccal Retractor Nerve
BRNG e: Burning
BRNNR e: Braunton Burrows National Nature Reserve
BRNO e: Bromine Number
BRNP e: Blue Ridge National Parkway
BRNR e: Burner
BRN Rechner d: Brutto-Tara-Netto-Rechner
BRNSHR e: Burnisher
BRNZ e: Branch on Non-Zero
- e: Bronze
- e: Bronzing
BRO d: Bronze
- e: Badge Readout
- e: Broach
- e: Broadband Remote Oculometer
- e: Bronchoscopy
Br OF e: Branch on Overflow
BROFICON e: Broadcast Fighter Control
BROI e: Branch Operating Instructions
BROILER e: Biopedagogical Research Organization on Intensive Learning Environment Reactions
BROK e: Broker[age]
BROM e: Bipolar Read-Only Memory
- e: Bromide
- e: Bromidic
- e: Bromo
- e: Bromo-Seltzer
BROMAT e: Bromatology
BROME e: Bromeliad
BROMIDROSIS e: Bromohydrosis
Bromier d: Bromierung
BROMO e: Bromidrosis
- e: Bromo-Seltzer
- e: Bromoform
BROMO-SELTZER e: Bromide and Seltzer
BROMOT e: Bromotology
BROMs e: Bipolar Read-Only Memories
BRONC s: Bronco
BRONCH e: Bronchial
- e: Bronchitis
- e: Bronchoscopic
- e: Bronchoscopist
- e: Bronchoscopy
BRONCHO g: Bronchus
Bronx HTF e: Bronx Homicide Task Force
Bronx Zoo e: New York Zoological Gardens
BRONZE e: 92% copper, 6% tin, 2% zinc
Brooklyn HTF e: Brooklyn Homicide Task Force
BROOM e: Ballistic Recovery of Orbiting Man
BROP d: Bundesraumordnungsprogramm
BROS e: Basic Read Only Storage
brosch. d: broschiert
Brosch. d: Broschüre
BROV e: Branch Overflow
BROWNULATED e: Granulated Brown Sugar
Brown U Pr e: Brown University Press
BROWSER e: Browsing On-line with Selective Retrieval
BRP d: Beiträge zur Romanischen Philologie (journ.)
- d: Bereichsplan
- e: Beacon Ranging Pulse
- e: Beacon Reply Processor
- e: Breeder Reactor Program[me]
- e: British Patent
- e: Buffer Read Program[me]
BRPh d: Beiträge zur Romanischen Philologie (journ.)
BRPH e: Bronchophony
BRPK e: Brownish Pink
BRPL e: Baton Rouge Public Library
BRPP e: Basic Radio Propagation Prediction[s]
BR-Pufferlösung d: Britton-Robinson-Pufferlösung
BRR e: Battelle Research Reactor
- e: Bridge Receiving Room
- e: Brookhaven Research Reactor
BRRD e: Barred
Br.-Reg.-T. d: Bruttoregistertonne
BRRG e: Barring

BRRI e: Building and Road Research Institute
BR/RL e: Bomb Rack / Rocket Launcher
BRRL e: British Radio Relay League
- e: British Road Research Laboratory
BRRS e: Banana River Repeater Station
BRR Signal e: Before Record Signal
Br Rys e: British Railways
BRS d: Baurichtlinien Schiffbau
- d: Block rücksetzen
- d: Brandschiefer
- e: B-mode Receiving Station
- e: Ballistic Recording System
- e: Balloon Radio System
- e: Bertrand Russell Society
- e: Bibliographic Retrieval Service[s]
- e: Binary Ring Sequence
- e: Block Received Signal
- e: Bomber-Recorder System
- e: Boron Reinforced Structure
- e: Brass
- e: Break Request Signal
- e: Bridges
- e: Bristol Record Society
- e: British Receiving Station
- e: British Record Society, Limited
- e: British Road Service
- e: British Roentgen Society
- e: Buffer Read Service
- e: Building Research Station
- e: Business Radio Service
BRSF d: Bremsenfett
BRSIT e: Bore Sight
BRSL e: British Road Services Limited
BR sound e: Breath sound
BRST d: Broadcast
- e: Burst[er]
BRSTD e: British Standard
Brstf. d: Brennstoff
BRT a: Bruttoregisterton
- b: Belgische Radio en Televisie
- d: Bauteile- und Relaistabelle
- d: bereit
- d: breit
- d: Brutto-Register-Tonne[n]
- d: Bruttoregistertonnage
- e: Base Resistance Transistor
- e: Before Recorded Time
- e: Binary Run Tape
- e: Bio-Research Titles
- e: Bolt Removal Tool
- e: Bridged Tap[ping]
- e: Bright[ness]
- e: British Time
- e: Brooklyn Rapid Transit [Company]
- e: Buffer Read Transfer
- e: Byte Result Trigger
BRTA e: British Regional Television Association
BR & TC e: Bermuda Radio and Television Company
BRTD e: Bright Radar Tube Display
BRTDC e: British Recorded Tape Development Committee
BRT E[ng] e: Bachelor of Radio and Television Engineering
BRTG e: Bomb, Radio, Transverse, Generator-powered
BRTH e: Breath
BRTHR e: Breather
BRTKM d: Bruttotonnenkilometer
BRTM e: Basic Real-Time Monitor
Br Tn (hv) e: Bridge Train; Heavy
Br Tn (l) e: Bridge Train; Light
BRTS e: Back Request to Send
BRU d: bahnamtlicher Rollfuhrunternehmer
- d: Brunei (Kfz)
- d: Brüssel (Flughafen)
- e: Bearing Repeater Unit
- e: Boat Repair Unit
- e: Brayton Rotating Unit
- e: British Recovery Unit
- e: Brussels (airport)
- n: Bibtiotheek der Rijksuniversiteit te Utrecht
Brüa d: Breite über alles
BRUCE e: Buffer Register under Computer Edict
B Ru E[ng] e: Bachelor of Rural Engineering
BRUIN e: Brown University Interpreter
BRUIN Language e: Brown University Interactive Language
Brun. e: Brunei
Bruns Beitr. klin. Chir. d: Bruns Beiträge zur klinischen Chirurgie (journ.)
Brun U e: Brunel University
BrüPz d: Brückenlegepanzer
BRurSc[i] e: Bachelor of Rural Science
Brus d: Brüssel
- f: Bruxelles
- i: Bruselle
- n: Brussel
- s: Bruselas
BRUSA e: British-United States Agreement
BRU Timber e: Bruguiera Timber
Brux p: Bruxelas
BrV d: Brennstoffventil
BRV e: Ballistic Reentry Vehicle
- e: Basic Research Vehicle
Br. Vet. J. e: British Veterinary Journal (journ.)
BRVMA e: British Radio Valve Manufacturers Association
BRVS d: Bahnrollfuhr-Versicherungsschein
Brw d: Brückenwerkstatt
BRW e: British Relay Wireless
BRWAB e: Branching Random Walk with Absorbing Barriers
BRWg e: Bridge Wing
BRWM e: B-Register Word Mark
BRY e: Bryology
Bryce e: Bryce Canyon National Park
BRYOL e: Bryology
BRZ d: Bezirksrechenzentrum
- d: Bronze
- d: Bruttoraumzahl
- e: Branch on Zero
- e: Braze
- e: bronze
BRZFI d: Branchenzentrum Fertigungsindustrie
BRZG e: Brazing
BRZN d: Bibliotheksrechenzentrum für Niedersachsen
BS d: Bahama Inseln (Kfz)
- d: Balance Selector
- d: Band-Stop[p]
- d: Bandsperre
- d: Bandsystem
- d: Barsortiment
- d: Basisstation
- d: Baustein[e]
- d: Bayerische Staatsbibliothek
- d: Bearbeitungssystem
- d: Bedienungssatz
- d: Bedienungsschlüssel
- d: Befindlichkeitsskala
- d: Behältersachverständiger
- d: Besatzstärke
- d: Betonschwelle
- d: Betriebsstoffsatz
- d: Betriebssystem
- d: Bildschirm[einheit]
- d: Bildschirmsteuereinheit
- d: Bildschirmsteuergerät
- d: Bildsender
- d: Bimetallspirale
- d: Binderschicht
- d: Bit pro Sekunde
- d: Blattschreiber
- d: Bleisammler
- d: Block Scheme
- d: Blockstruktur
- d: Blutsenkung
- d: Bremsstrecke
- d: Brennschluß
- d: Brennschwindung
- d: Brennstrecke
- d: Bromsulfateintest
- d: Brückenschutz
- d: Buchungsstation
- d: Buckelschweißen
- d: Bussteuerung
- d: Byte pro Sekunde
- e: Bachelor of Science
- e: Bachelor of Surgery
- e: Backing Storage
- e: Backing Store
- e: Back Space (Data-Processing Character)
- e: Back Stamp
- e: Backward Signalling
- e: Bags
- e: Bahamas
- e: Balance Selection
- e: Balance Sheet
- e: Bales
- e: Band-Stop
- e: Band Setting
- e: Baroswitch
- e: Base
- e: Based Store
- e: Base Salvage
- e: Base Section
- e: Base Shell
- e: Base Shield
- e: Base Supply
- e: Basic Sediment
- e: Basic Symbol
- e: Basic System
- e: Basophils
- e: Battery Simulator
- e: Battle Ship
- e: Battle Star
- e: Beam Splitter
- e: Beam Stearing
- e: Beam Stop
- e: Bearings
- e: Beaufort Scale
- e: Behind Schedule
- e: Bessemer Steel
- e: Beta Spectrometer
- e: Big System
- e: Bill of Sale
- e: Bill of Sight
- e: Bill of Store
- e: Binary Scale
- e: Binary Scaler
- e: Binary Search
- e: Binary Signal
- e: Binary Subtract[ion]
- e: Binary Symbol
- e: Binary System
- e: Biochemical Society
- e: Biological Society
- e: Biometric Society
- e: Biophysical Society
- e: Biquinary System

- e: Bismuthinite
- e: Bit[s] per Second
- e: Bit Space
- e: Bit String
- e: Bit Sync[hronization]
- e: Bituminous Surface
- e: Black Scale
- e: Blank Space
- e: Block Selection
- e: Block Size
- e: Block Sort
- e: Block Storage
- e: Block Structure
- e: Blood Sugar
- e: Blue Steel
- e: Body Shell
- e: Boiler Space
- e: Boiler Survey
- e: Bomb Service
- e: Bombsight
- e: Bomb Store
- e: Bonded Single-Silk [Insulation]
- e: Booking Store
- e: Borescope
- e: Boresight
- e: Botanical Society [of the British Isles]
- e: Both Sides
- e: Bottom Sediment
- e: Bottom Settlings
- e: Bow Shock
- e: Braided Store
- e: Break Jaw
- e: Breakpoint Symbol
- e: British Standard
- e: British Survey
- e: Broadcast[ing] Satellite
- e: Broadcast[ing] Service
- e: Broadcast[ing] Station
- e: Buffer Storage
- e: Buffer Store
- e: Buffer Survey
- e: Building Science (journ.)
- e: Bulk Storage
- e: Bulk Store
- e: Bump Storage
- e: Bunker Surcharge
- e: Bureau of Ships
- e: Bureau of Standards
- e: Business System
- e: Byte Size

B & S e: Beams and Stringers
- e: Bell and Spigot
- e: Booster and Sustainer

bs l: bancus superior

BSA d: Bereich mit Symbol am Anfang
- d: Betriebsrichtungsschalter-Abfrageseite
- d: Bis-(trimethylsilyl)-acetamid
- d: Bohr-Sommerfeld Atom
- e: Bachelor of Science [in] Agriculture
- e: Bachelor of Scientific Agriculture
- e: Barrack Store Accountant
- e: Basic Standardization Agreement
- e: Basic Stock Allowance
- e: Benzenesulfonic Acid
- e: Bibliographical Society of America
- e: Binary Switching Algebra
- e: Binary Synchronous Adapter
- e: Biofeedback Society of America
- e: Bismuth-Sulphite Agar
- e: Bistrimethylsilylacetamide
- e: Bistrymethyllacetamide
- e: Bit Sync[hronization] Acquisition
- e: Body Surface Area
- e: Botanical Society of America
- e: Bovine Serum Albumin
- e: Bowel Sound Active
- e: British School at Athens
- e: British School of Archaeology
- e: British South Africa

BSAA e: Bachelor of Science in Applied Arts
- e: British School of Archaeology at Athens
- e: British South American Airways

BSA Adm e: Bachelor of Science in Agricultural Administration

BS Ad e: Bachelor of Science in Business Administration

BS Adv e: Bachelor of Science in Advertising

BSAE e: Bachelor of Science in Administrative Engineering
- e: Bachelor of Science in Aeronautical Engineering
- e: Bachelor of Science in Agricultural Engineering
- e: Bachelor of Science in Architectural Engineering
- e: British School of Archaeology in Egypt

BSAE-E e: Bachelor of Science in Aeronautical Engineering-Electronics Major

BS(AE Elec) e: Bachelor of Science with Aeronautical Engineering Electives

BS Aero Adm e: Bachelor of Science in Aeronautical Administration

BS Aero E e: Bachelor of Science in Aeronautical Education
- e: Bachelor of Science in Aeronautical Engineering

BSAF e: British Sulphate of Ammonia Federation

BS Ag E e: Bachelor of Science in Agricultural Engineering

BS Ag[r] e: Bachelor of Science in Agriculture

BS Agr Ed e: Bachelor of Science in Agricultural Education

BS Agr Eng e: Bachelor of Science in Agricultural Engineering

BSAI e: British School of Archaeology in Iraq

BSAJ e: British School of Archaeology in Jerusalem

BSAL e: Base Spares Allowance List
- e: Basic Stock Allowance List
- e: Block Structured Assembly Language

BSALS e: British Society of Agricultural Labour Science

BSAM e: Bachelor of Suddha Ayurvedic Medicine
- e: Basic Sequential Access Method
- e: Binary Sequential Access Method

BS A Math e: Bachelor of Science in Applied Mathematics

BSAME e: Bachelor of Science in Aircraft Maintenance Engineering

BS and W e: Bottom Settlings and Water

BSAO d: Brandschutzanordnung

BS App Arts e: Bachelor of Science in Applied Arts

BSAR e: British School of Archaeology at Rome

BS Arch e: Bachelor of Science in Architecture

BS Arch (ARchE) e: Bachelor of Science in Architecture in Architectural Engineering

BS Arch E e: Bachelor of Science in Architectural Engineering

BS Arch E[ng] e: Bachelor of Science in Architectural Engineering

BS Art Ed e: Bachelor of Science in Art Education

BSAs e: Body Surface Areas

B Sa Sc e: Bachelor of Sacred Sciences

BSAT e: Bachelor of Science in Air Transportation

BSAU d: Blattschreiberausgabe
- e: British Standard for Automobile

BSAVA e: British Small Animals Veterinary Association

BSB d: Bedienungsblattschreiber
- d: biochemischer Sauerstoffbedarf
- d: biologischer Sauerstoffbedarf
- e: Bachelor of Science in Business
- e: Backspace Block
- e: Baseband
- e: Both Sidebands
- e: British Satellite Broadcasting Limited
- e: British Standard Beam
- e: British Standard Brass
- e: Buffer Store Block

BSBA e: Bachelor of Science in Business Administration
- e: British Starter Battery Association

BSBC e: British Social Biology Council

BSB Ed e: Bachelor of Science in Business Education

BSBG e: Burst and Synchronization Bit Generator
- e: Burst and Synchronous Bit Generator

BSBI e: Botanical Society of the British Isles

BS Biol e: Bachelor of Science in Biology

BSB_5 d: biologischer Sauerstoffbedarf in fünf Tagen

BSB_n d: biologischer Sauerstoffbedarf in n Tagen

BS Bus e: Bachelor of Science in Business

BS Bus Ad[m] e: Bachelor of Science in Business Administration

BS Bus and Public Adm e: Bachelor of Science in Business and Public Administration

BS Bus Ed e: Bachelor of Science in Business Education

BS Bus-MR e: Bachelor of Science in Business-Medical Records

BSB-Zahl d: biochemische Sauerstoff-Bedarfszahl

BSC d: binär synchroner Code
- e: Bachelor of Christian Science
- e: Bachelor of Science
- e: Bachelor of Science in Commerce
- e: Back Space Character
- e: Back Space Contact
- e: Back Space Control
- e: Back Spacing Control
- e: Balkan Supply Center
- e: Bank Street College
- e: Base-message Switching Center
- e: Base Site Control
- e: Basic
- e: Basic-message Switching Center
- e: Basic Synchronous Communication[s]
- e: Beltsville Space Center
- e: Bemidji State College
- e: Bethlehem Steel Corporation
- e: Bibliographical Society of Canada
- e: Billet Sequence Code
- e: Binary Store Circuit
- e: Binary Symmetric[al] Channel
- e: Binary Synchronous Communication[s]
- e: Binary Synchronous Control
- e: Biological Stain Commission
- e: Biomedical Sciences Corporation
- e: Biomedical Sciences Corps
- e: Blip-Scan Counter
- e: Bloomsburg State College
- e: Bluefeld State College
- e: Body Support Cradle
- e: Bolted Separable Connector

- e: Booking Search Control
- e: Borosilicate Crown
- e: British Standard Channel [Steel]
- e: British Standard Cycle
- e: British Steel Corporation
- e: Bubble String Comparator
- e: Buffer Store Circuit
- e: Business Service Center
- e: Bus System Control
- e: Byte Size Control

BSCA e: Binary Synchronous Communications Adapter
Bsc A Eco e: Bachelor of Science in Agricultural Economics
BScAg & AH e: Bachelor of Science in Agriculture and Animal Husbandry
B Sc Ag[r] e: Bachelor of Science in Agriculture
BScAgrBio e: Bachelor of Science ín Agricultural Biology
BScAgrEng e: Bachelor of Science in Agricultural Engineering
B scan e: B scanner
BS C and E e: Bachelor of Science in Commerce and Economics
BScApp e: Bachelor of Applied Science
BScArch e: Bachelor of Science in Architecture
BSCBA e: Brown Swiss Cattle Breeders Association of America
BSCC e: Binary Synchronous Communications Controller
- e: British Society for Clinical Cytology

BScCE e: Bachelor of Science in Civil Engineering
BScChemE e: Bachelor of Science in Chemical Engineering
BScD e: Bachelor of Science in Dentistry
BScDent e: Bachelor of Science in Dentistry
B Sc Dn e: Bachelor of Science in Dianoetics
BScDomSc e: Bachelor of Science in Domestic Science
BSCE e: Bachelor of Science in Civil Engineering
- e: Bank Street College of Education

B Sc Econ e: Bachelor of Science in Economics
B Sc EE e: Bachelor of Science in Electrical Engineering
BScElEd e: Bachelor of Science in Elementary Education
BSc Eng e: Bachelor of Science in Engineering
B Sc Eng Elect e: Bachelor of Science in Electrical Engineering
B Sc Eng Phys e: Bachelor of Science in Engineering Physics
BS Cer E e: Bachelor of Science in Ceramic Engineering
B Sc F[or] e: Bachelor of Science in Forestry
BSCH d: Boden-Schiff
- d: Brandschiefer
- d: Braunschimmel
- d: Brennschiefer
- d: Bundesbahnschule
- e: Bachelor of Science in Chemistry
- e: Binary Symmetric Channel

BS Ch E Elect e: Bachelor of Science with Chemical Engineering Electives
BS Chem e: Bachelor of Science in Chemistry
BS Ch[em] E[ng] e: Bachelor of Science in Chemical Engineering
BSchG d: Binnenschiffahrts-Gesetz
BS Chm e: Bachelor of Science in Chemistry
B Sch Mus e: Bachelor of School Music
BSc Hort e: Bachelor of Science in Horticulture
BSchR d: Bezirksschulrat
BSchSO d: Binnenschiffahrtsstraßen-ordnung
BSchUO d: Binnenschiffahrtsuntersu-chungsordnung
B Sc in Agr Engr e: Bachelor of Science in Agricultural Engineering
B Sc in Bact e: Bachelor of Science in Bacteriology
B Sc in CE e: Bachelor of Science in Civil Engineering
B Sc in EE e: Bachelor of Science in Electrical Engineering
B Sc in HE e: Bachelor of Science in Home Economics
B Sc in ME e: Bachelor of Science in Mechanical Engineering
B Sc in Med e: Bachelor of Science in Medicine
B Sc in Med Techn e: Bachelor of Science in Medical Technology
B Sc in Nurs e: Bachelor of Science in Nursing
B Sc in Occ Ther e: Bachelor of Science in Occupational Therapy
B Sc in Opt e: Bachelor of Science in Optometry
B Sc in Phar e: Bachelor of Science in Pharmacy
B Sc in Phys e: Bachelor of Science in Physics
B Sc in Phys Ther e: Bachelor of Science in Physical Therapy
B Sc in Rest Mgt e: Bachelor of Science in Restaurant Management
B Sc in Soc Adm e: Bachelor of Science in Social Administration
B Sc L e: Bachelor of the Science of Law
BSCL e: Bell System Common Language
- e: Business Service Checklist

BSc LA e: Bachelor of Science Laboratory Assistant
BSCM e: Binary Synchronous Communications Macro
BSc ME e: Bachelor of Science Mechanical Engineering
BScMed e: Bachelor of Science in Medicine
BScMet e: Bachelor of Science in Metallurgy
BScMin e: Bachelor of Science in Mining
BSc MLS e: Bachelor of Science Medical Laboratory Science
B Scn e: Bachelor of Scientology
BSCN e: Bit Scan
BSc N[urs] e: Bachelor of Science in Nursing
BSc Nutr e: Bachelor of Science Nutrition
B Sc O e: Bachelor of the Science of Oratory
BSCO e: Brake Specific Carbon Oxide
BSCOM e: Bachelor of Science in Communications
- e: Binary Synchronous Communication

BS Com Ed e: Bachelor of Science in Commercial Education
BS Comm e: Bachelor of Science in Commerce
BS Com[m] Eng e: Bachelor of Science in Commercial Engineering
BSCorp e: British Steel Corporation
BSCP e: Biological Sciences Communication Project
- e: British Standard Code of Practice

BSc P and OT e: Bachelor of Science Physical and Occupational Therapy
BSc PEd e: Bachelor of Science Physical Education
BSc Pharm e: Bachelor of Science Pharmacy
BSC Process e: Binary Synchronous Communication Process
BSc QS e: Bachelor of Science Quantity Surveying
BSCRA e: British Steel Castings Research Association
BS Cr E e: Bachelor of Science in Ceramic Engineering
BSc RS e: Bachelor of Science Rural Science
BSCRT e: Bachelor of Science in Radiologic Technology
BSCS e: Bachelor of Science in Computer Science
- e: Binary Synchronous Communication System
- e: Biological Sciences Curriculum Study
- e: Blip-Scan Counter System

BSc Soc e: Bachelor of Science Sociology
BScSS e: Bachelor of Science in Secretarial Studies
BSC/SS e: Binary Synchronous Communications/Start-Stop
BScSur e: Bachelor of Science in Surveying
BSC System e: Binary Synchronous Communication System
BSCTE e: Bachelor of Science Textile Engineering
- e: Bell System Center for Technical Education

BSc Tech e: Bachelor of Technical Science
BSc Text e: Bachelor of Science Textiles
BSc Vet e: Bachelor of Science Veterinary
BScVetSc e: Bachelor of Science in Veterinary Science
BSD d: bipolarer Schalter und Dekodierung
- e: Bachelor of Science in Dentistry
- e: Bachelor of Science in Design
- e: Bachelor of Science in Didactics
- e: Ballistic Systems Division
- e: Barrels per Steaming Day
- e: Base Supply Depot
- e: Beam Steering Device
- e: Berkeley Software Distribution
- e: Bit Storage Density
- e: Bit String Data
- e: Blast Suppression Device
- e: Block-Surface-Distance
- e: British Space Development
- e: British Standard Dimensions
- e: Buffer Store Device
- e: Building Systems Design (journ.)
- e: Bulk Storage Device
- e: Burst Slug Detection

BSDAM e: Basic Sequential Data Access Method
BSDC e: Binary Symmetric Dependent Channel
- e: British Space Development Company
- e: British Standard Data Code

BS Dent e: Bachelor of Science in Dentistry
BS Des e: Bachelor of Science in Design
BS Des Dec Des e: Bachelor of Science in Design in Decorative Design
BSDF e: Beet Sugar Development Foundation
BSD H[yg] e: Bachelor of Science in Dental Hygiene

BS Di e: Bachelor of Scientific Didactics
BSDL e: Boresight Datum Line
BSDP e: Bell Systems Data Processing
BSD's e: Binary Storage Devices
BSE d: Basis[steuer]einheit
- d: Bereich mit Symbol am Ende
- d: Betriebs- und Schutzerde
- d: Betriebs- und Schutzerdung
- d: Betriebsschutzerde
- d: Betriebssteuereinrichtung
- d: Betriebssysteme
- d: Bildschirmeinheit
- d: Bourquin-Sherman-Einheit BSE
- d: Bussteuereinheit
- e: Bachelor of Sanitary Engineering
- e: Bachelor of Science [in] Education
- e: Bachelor of Science [in] Engineering
- e: Backscatter Electron
- e: Band Splitting Effect
- e: Base Support Equipment
- e: Binary Store Element
- e: Boiler System Efficiency
- e: Booster Systems Engineer
- e: Boston Stock Exchange
- e: Botanical Society of Edinburgh
- e: Botony School of Edinburgh
- e: Breast Self-Examination (Cancer Control)
- e: Broadcast[ing] Satellite-Experimental
- e: Broadcasting Satellite for Experimental Purposes
- e: Bureau of Steam Engineering
B Se A e: Bachelor of Secretarial Arts
BSEA e: Baltic Sea
- e: British School of Egyptian Archaeology
- e: British Standard Equal Angle
BSEAeE e: Bachelor of Science in Engineering in Aeronautical Engineering
BS Ec and Bus Adm e: Bachelor of Science in Economics and Business Administration
BSECBI e: Botanical Society and Exchange Club of British Isles
BSECE e: Bachelor of Science in Engineering and Civil Engineering
BSECh E e: Bachelor of Science in Engineering in Chemical Engineering
BS Ec[on] e: Bachelor of Science in Economics
B Sec Sc e: Bachelor of Secretarial Science
BS Ed e: Bachelor of Science in Education
BS Ed Music Super e: Bachelor of Science in Education; Music Supervision
BS Ed-Phys Ed e: Bachelor of Science in Education-Physical Education
BSEE e: Bachelor of Science in Electrical Engineering
- e: Bachelor of Science in Electronic Engineering
- e: Bachelor of Science in Elementary Education
BSE & E e: Bachelor of Science in Engineering and Economics
BSE EE e: Bachelor of Science in Engineering in Electrical Engineering
BS E E Elect e: Bachelor of Science with Electrical Engineering Electives
BSE EM e: Bachelor of Science in Engineering in Engineering Mechanics
BSEE-ME e: Bachelor of Science in Electrical Engineering and Mechanical Engineering
BS E Engr e: Bachelor of Science in Electrical Engineering
BSE Geod & Surv e: Bachelor of Science in Engineering in Geodesy and Surveying
BSE In E e: Bachelor of Science in Engineering in Industrial Engineering
BSEL e: Bachelor of Science and English Literature
BSELCH e: Buffered Selector Channel
BS EL E e: Bachelor of Science in Electronic Engineering
BS Elect Engr e: Bachelor of Science in Electrical Engineering
BS El[em] Ed e: Bachelor of Science in Elementary Education
BSEM e: Bachelor of Science in Engineering of Mines
BSE Mat E e: Bachelor of Science in Engineering in Materials Engineering
BSE ME e: Bachelor of Science in Engineering in Mechanical Engineering
BSE Met E e: Bachelor of Science in Engineering in Metallurgical Engineering
BSE M & Ind E e: Bachelor of Science in Engineering in Mechanical and Industrial Engineering
BS-EN e: British Standard-European (British Edition of an European Standard)
BSE Nav Arch & Mar E e: Bachelor of Science in Engineering in Naval Architecture and Marine Engineering
BS Eng e: Bachelor of Sanitary Engineering
BS Eng[r] e: Bachelor of Science in Engineering
BS Engr Ad e: Bachelor of Science in Engineering Administration
BS E[ngr] P[hys] e: Bachelor of Science in Engineering Physics
BS Eng[r] Sci e: Bachelor of Science in Engineering Sciences
BSEP e: Basic Skill Education Program[me]
BSE S[c] e: Bachelor of Science in Engineering Sciences
B Se St e: Bachelor of Secretarial Studies
BSF d: Bedien- und Schaltfeld
- d: Betonsteinfertiger
- e: Bachelor of Science in Forestry
- e: Back Scatter Factor
- e: Back[ward] Space File
- e: Ball Spinning Friction
- e: Bandwidth Shape Factor
- e: Basic Skill Film
- e: Benzene Soluble Fraction
- e: Beta-S-Fetoprotein
- e: Blade Slap Factor
- e: Brief Stop for Fuel
- e: British Standard Fine [screw-thread]
- e: Bulk Shielding Facility
BSFA e: Building Science Forum of Australia
BSFC d: benutzte Schlußfacette
- e: Brake Specific Fuel Consumption
BS Fin e: Bachelor of Science in Finance
BSFM e: Bachelor of Science in Forest Management
BSF Mgt e: Bachelor of Science in Fisheries Management
BS For e: Bachelor of Science in Forestry
BSFS e: Bachelor of Science in Foreign Service
BSFs e: Benzene Soluble Fractions
BSF Screw Thread e: British Standard Fine Screw Thread
BS Fsty e: Bachelor of Science in Forestry
BSFT e: Bachelor of Science in Fuel Technology
- e: Basalt Spent Fuel Test[ing]
BSF Thread e: British Standard Fine Thread
BSG d: Ballistospinografic
- d: Betriebssteuergerät
- d: Bildschirmgerät
- d: Blutkörper[chen]senkungsgeschwindigkeit
- d: Bromsulfglutation
- d: Bundesseuchengesetz
- e: Bit Sync[hronization] Generator
- e: Bootstrap Gyro
- e: British Standard Gauge
BS G1 E e: Bachelor of Science in Geological Engineering
BSGe d: Betriebssteuergeräte
BSGE e: Bachelor of Science in General Engineering
- e: Bachelor of Science in Geological Engineering
BS Gen Bus e: Bachelor of Science in General Business
BS Gen Ed e: Bachelor of Science in General Education
BS Gen Eng e: Bachelor of Science in General Engineering
BS Gen Home Econ e: Bachelor of Science in General Home Economics
BS Gen Nur e: Bachelor of Science in General Nursing
B.S.Gen. Nur. e: Bachelor of Science in General Nursing
BS Geog e: Bachelor of Science in Geography
BS Geol e: Bachelor of Science in Geology
BS Geol E[ng] e: Bachelor of Science in Geological Engineering
BS Ggr e: Bachelor of Science in Geography
BS GI e: Bachelor of Science in Geology
BS GI E e: Bachelor of Science in Geological Engineering
BSGL e: Branch Systems General Licence
BSG Mgt e: Bachelor of Science in Game Management
BSGP e: Bachelor of Science in Geology and Physics
BS Gph e: Bachelor of Science in Geophysics
BS Group Work Admin e: Bachelor of Science in Group Work Administration
BS Group Work Ed e: Bchelor of Science in Group Work Education
BSGS e: Base Support Group System
BSH d: Benzolsulfohydrazid
- e: Benzenesulfonyl Histamine
- e: British Society of Hypnotherapists
- e: British Standard of Hardness
- e: Bushel
BSHA e: Bachelor of Science in Hospital Administration
BSHC e: Brake Specific Hydrocarbon[s]
BS Health and Phy Ed e: Bachelor of Science in Health and Physical Education
BS Health Ed e: Bachelor of Science in Health Education
BSHE[d] e: Bachelor of Science in Health Education
BSHG e: Bushing
BS H[ome] E[con] e: Bachelor of Science in Home Economics
BS Home Econ Ed e: Bachelor of Science in Home Economics Education
BSHP e: Beginning Standard Holding Procedure
BSHS e: British Society for the History of Science
BS Hyg e: Bachelor of Science in Hygiene
BSI d: Bildsignal
- d: Bitschiebeimpuls
- e: Basic Shipping Instructions
- e: Bit Sequence Independence
- e: Booster Situation Indicator

- e: Bound Serum Iron
- e: Branch and Store Instruction
- e: British Sailors Institute
- e: British Solomon Islands
- e: British Standard Interface
- e: British Standards Institute
- e: British Standards Institution
- e: Building Societies Institute
- e: Building Stone Institute

BSIA e: Better Speech Institute of America
BSIAP e: Beginning Straight-In Approach
BSIB e: British Society for International Bibliography
BSIC e: Binary Symmetric Independent Channel
BSIE e: Bachelor of Science in Industrial Education
- e: Bachelor of Science in Industrial Engineering
- e: Banking Systems Information Exchange
- e: Bio-Sciences Information Exchange

BSIHE e: British Society for International Health Education
BSIL e: Basic Switching Impulse Insulation Level
BSIM e: Bachelor of Science in Industrial Management
BS in Acc e: Bachelor of Science in Accounting
BS in AE e: Bachelor of Science in Administrative Engineering
- e: Bachelor of Science in Aeronautical Engineering
- e: Bachelor of Science in Agricultural Education
- e: Bachelor of Science in Architectural Engineering

BS in Aero Adm e: Bachelor of Science in Aeronautical Administration
BS in Ae[ro] E e: Bachelor of Science in Aeronautical Engineering
BS in Ag e: Bachelor of Science in Agriculture
BS in Ag DM e: Bachelor of Science in Agriculture in Dairy Manufacturing
BS in Ag & Ed e: Bachelor of Science in Agriculture and Education
BS in Agr & Ch[em] e: Bachelor of Science in Agriculture and Chemistry
BS in Agr Ed e: Bachelor of Science in Agricultural Education
BS in Ag[r] E[ng] e: Bachelor of Science in Agricultural Engineering
BS in Am e: Bachelor of Science in Agricultural Administration
BS in Arch e: Bachelor of Science in Architecture
BS in BMS e: Bachelor of Science in Basic Medical Sciences
BS in Bus e: Bachelor of Science in Business
BS in B[us] A[d] e: Bachelor of Science in Business Administration
BS in B[us] Ed e: Bachelor of Science in Business Education
BS in C e: Bachelor of Science in Commerce
BS in Cart e: Bachelor of Science in Cartography
BS in C & BA e: Bachelor of Science in Commercial and Business Administration
BS in CE e: Bachelor of Science in Chemical Engineering
- e: Bachelor of Science in Civil Engineering

BS in C & Ec e: Bachelor of Science in Commerce and Economics
BS in CE-Music e: Bachelor of Science in Christian Education-Music
BS in Cer e: Bachelor of Science in Ceramics
BS in Cer E e: Bachelor of Science in Ceramic Engineering
BS in Cer Tech e: Bachelor of Science in Ceramic Technology
BS in Ch[em] E[ng] e: Bachelor of Science in Chemical Engineering
BS in Chem Tech e: Bachelor of Science in Chemical Technology
BS in Ch[m] e: Bachelor of Science in Chemistry
BS in Chm E e: Bachelor of Science in Chemical Engineering
BS in Com & Bus e: Bachelor of Science in Commerce and Business
BS in Com Ed e: Bachelor of Science in Commercial Education
BS in Com[m] e: Bachelor of Science in Commerce
BS in Comm Rec e: Bachelor of Science in Community Recreation
BS Ind Art e: Bachelor of Science in Industrial Art
BS Ind Chem e: Bachelor of Science in Industrial Chemistry
BS Ind Ed e: Bachelor of Science in Industrial Education
BS Ind Eng[r] e: Bachelor of Science in Industrial Engineering
BS in Dent e: Bachelor of Science in Dentistry
BS in DH e: Bachelor of Science in Dental Hygiene
BS in E e: Bachelor of Science in Education
- e: Bachelor of Science in Engineering

BS in Ec e: Bachelor of Science in Economics
BS in Ed e: Bachelor of Science in Education
BS in E[lec]E[ngr] e: Bachelor of Science in Electrical Engineering
BS in Elect Eng e: Bachelor of Science in Electronic Engineering
BS in EM e: Bachelor of Science in Engineering of Mines
BS in E Math e: Bachelor of Science in Engineering Mathematics
BS in Eng e: Bachelor of Science in Engineering
BS in E[ngr]P[hys] e: Bachelor of Science in Engineering Physics
BS in ES e: Bachelor of Science in Engineering Sciences
BS in Fin e: Bachelor of Science in Finance
BS in For e: Bachelor of Science in Forestry
BS in FS e: Bachelor of Science in Foreign Service
BS in Fy e: Bachelor of Science in Forestry
BS in GE e: Bachelor of Science in General Engineering
BS in Ge E e: Bachelor of Science in Geological Engineering
BS in Gen Bus e: Bachelor of Science in General Business
BS in Gen Eng e: Bachelor of Science in General Engineering
BS in Gen Nurs e: Bachelor of Science in General Nursing
BS in Gen Sci e: Bachelor of Science in General Science
BS in Gen Std e: Bachelor of Science in General Studies
BS in Geod & Surv e: Bachelor of Science in Geodesy and Surveying
BS in Geol E e: Bachelor of Science in Geological Engineering
BS in Gph E e: Bachelor of Science in Geophysical Engineering
BS in GS e: Bachelor of Science in General Studies
BS in GSM e: Bachelor of Science in General Science and Mathematics
BS in GWE e: Bachelor of Science in Group Work Education
BS in HE[con] e: Bachelor of Science in Home Economics
BS in H Ed e: Bachelor of Science in Health Education
BS in H & PE e: Bachelor of Science in Health and Physical Education
BS in H & RA e: Bachelor of Science in Hotel and Restaurant Administration
BS in IA e: Bachelor of Science in Industrial Arts
BS in IE e: Bachelor of Science in Industrial Engineering
BS in IE & M e: Bachelor of Science in Industrial Engineering and Management
BS in IM e: Bachelor of Science in Industrial Management
BS in Ind Art e: Bachelor of Science in Industrial Art
BS in Ind Ch e: Bachelor of Science in Industrial Chemistry
BS in Ind Ed e: Bachelor of Science in Industrial Education
BS in Ind E[ng] e: Bachelor of Science in Industrial Engineering
BS in Ind Mgt e: Bachelor of Science in Industrial Management
BS in J e: Bachelor of Science in Journalism
BS in Lat e: Bachelor of Science in Latin
BS in LP e: Bachelor of Science in Land Planning
BS in L & S e: Bachelor of Science in Letters and Science
BS in LS e: Bachelor of Science in Library Science
BS in LT e: Bachelor of Science in Laboratory Technology
BS in MA e: Bachelor of Science in Mechanical Arts
BS in Math e: Bachelor of Science in Applied Mathematics
BS in Math Stat e: Bachelor of Science in Mathematical Statistics
BS in MdT e: Bachelor of Science in Medical Technology
BS in M[ech]E[ng] e: Bachelor of Science in Mechanical Engineering
BS in Mech Ind e: Bachelor of Science in Mechanical Industries
BS in Med e: Bachelor of Science in Medicine
BS in Med Rec e: Bachelor of Science in Medical Records
BS in Med Rec Lib e: Bachelor of Science in Medical Records Librarianship
BS in Med S e: Bachelor of Science in Basic Medical Science
BS in Med Sc e: Bachelor of Science in Medical Secretarial Science
BS in Med Tech e: Bachelor of Science in Medical Technology
BS in M Educ e: Bachelor of Science in Music Education
BS in M Engr e: Bachelor of Science in Mechanical Engineering

BS in Met e: Bachelor of Science in Metallurgy
- e: Bachelor of Science in Meteorology
BS in Met E[ngin] e: Bachelor of Science in Metallurgical Engineering
BS in Mgt Engr e: Bachelor of Science in Management Engineering
BS in Mgt Sc e: Bachelor of Science in Management Science
BS in Min e: Bachelor of Science in Mining
BS in Min E[ng] e: Bachelor of Science in Mining Engineering
BS in MRL Sc e: Bachelor of Science in Medical Record Library Science
BS in MS e: Bachelor of Science in Military Science
BS in MT e: Bachelor of Science in Medical Technology
BS in Mu Ed e: Bachelor of Science in Music Education
BS in Mus Ed e: Bachelor of Science in Musical Education
BS in Nat G Engin e: Bachelor of Science in Natural-Gas Engineering
BS in Nat Hist e: Bachelor of Science in Natural History
BS in NS[c] e: Bachelor of Science in Natural Science
BS in N[urs] e: Bachelor of Science in Nursing
BS in N[urs]E[d] e: Bachelor of Science in Nursing Education
BS in O[cc]T[her] e: Bachelor of Science in Occupational Therapy
BS in Ocean e: Bachelor of Science in Oceanography
BS in OH e: Bachelor of Science in Ornamental Horticulture
BS in Opt e: Bachelor of Science in Optics
- e: Bachelor of Science in Optometry
BS in Ortho e: Bachelor of Science in Orthoptics
BS in PA e: Bachelor of Science in Practical Arts
- e: Bachelor of Science in Public Administration
BS in PAL e: Bachelor of Science in Practical Arts and Letters
BS in Pet e: Bachelor of Science in Petroleum
BS in P[et]E[ngin] e: Bachelor of Science in Petroleum Engineering
BS in Ph[ar] e: Bachelor of Science in Pharmacy
BS in PHPM e: Bachelor of Science in Public Health and Preventative Medicine
BS in P[hys] Ed e: Bachelor of Science in Physical Education
BS in Phys Th[er] e: Bachelor of Science in Physical Therapy
BS in Prac Arts e: Bachelor of Science in Practical Arts
BS in Pr Ge e: Bachelor of Science in Professional Geology
BS in Pr Met e: Bachelor of Science in Professional Meteorology
BS in PSM e: Bachelor of Science in Public School Music
BS in R[ad]T[ech] e: Bachelor of Science in Radiological Technology
BS in Rad & TV e: Bachelor of Science in Radio and Television
BS in RAH e: Bachelor of Science in Range Animal Husbandry
BS in RE e: Bachelor of Science in Religious Education
BS in Rec e: Bachelor of Science in Recreation
BS in Rec Lead e: Bachelor of Science in Recreation Leadership
BS in San E e: Bachelor of Science in Sanitary Engineering
BS in San Sci e: Bachelor of Science in Sanitary Science
BS in Sec Ed e: Bachelor of Science in Secondary Education
BS in Sec Sc e: Bachelor of Science in Secretarial Science
BS in Soc St e: Bachelor of Science in Social Studies
BS in Spec Flds e: Bachelor of Science in Special Fields
BS in SS[c] e: Bachelor of Science in Social Science
BS in Stat e: Bachelor of Science in Statistics
BS in Struc E e: Bahelor of Science in Structural Engineering
BS in TE e: Bachelor of Science in Textile Engineering
BS in Text e: Bachelor of Science in Textiles
BS in Th e: Bachelor of Science in Physical and Occupational Therapy
BS in Voc Ag e: Bachelor of Science in Vocational Agriculture
BS in Voc Ed e: Bachelor of Science in Vocational Education
BS in ZS e: Bachelor of Science in Zoological Sciences
BSIR e: Bachelor of Science in Industrial Relations
BSIRA e: British Scientific Instrument Research Association
BSIS e: Bio-Science Information Services
- e: British Society for Information Science
BSIT e: Bachelor of Science in Industrial Technology
- e: Bipolar-mode Static Induction Transistor
BSITs e: Bipolar-mode Static Induction Transistors
BSIU e: British Society for Internationmal Understanding
BSJ e: Bachelor of Science in Journalism
- e: Balanced Swivel Joint
- e: Ball and Socket Joint
BS J[rnl] e: Bachelor of Science in Journalism
BSK d: Braunkohlenschwelkoks
- e: Back Shunt Keying
- e: Back Space Key
- e: Back Spacing Key
BSL d: Basel (Flughafen)
- d: Betonschwellenlager
- d: Blocksteuerung für Linienverkehr
- e: Bachelor of Sacred Literature
- e: Bachelor of Science in Languages
- e: Bachelor of Science in Law
- e: Bachelor of Science in Linguistics
- e: Base Line
- e: Basic Switching-surge Level
- e: Basic Systems Language
- e: Beam Shape Loss
- e: Behavioral Sciences Laboratory
- e: Berkeley Scientific Laboratories
- e: Billet Split Lens
- e: Bills of Lading
- e: Biochemical Society of London
- e: Biofeedback Systems Ltd.
- e: Biot-Savart Law
- e: Bit Scanning Line
- e: Bit Serial Link
- e: Bomb Savety Line
- e: Botanical Society of London
- e: Bulk Semiconductor Limiter
BSLA e: Bachelor of Science in Landscape Architecture
BSLA and Med e: Bachelor of Science in Liberal Arts and Medicine
BSLA and Nurs[e] e: Bachelor of Science in Liberal Arts and Nursing
BS Lab Rel e: Bachelor of Science in Labor Relations
BS Lab Tech[n] e: Bachelor of Science in Laboratory Technology
BSL Arch e: Bachelor of Science in Landscape Architecture
BS Lib Arts e: Bachelor of Science in Liberal Arts
BS Lib Service e: Bachelor of Science in Library Service
BS Light Bldg Industry e: Bachelor of Science in Light Building Industry
BSLM e: Bachelor of Science in Landscape Management
BSLR e: Bus Selector
BSLS e: Bachelor of Science in Library Science
- e: Bachelor of Science in Library Service
- e: Bradford Small Livestock Society
BSLSS e: Buddy Secondary Life Support System
BSM d: Betriebs- und Steuermeßplatz
- d: Betriebssystem München
- d: Bi-Stable Multivibrator
- e: Bachelor of Scared Music
- e: Bachelor of School Music
- e: Bachelor of Science in Medicine
- e: Bachelor of Science in Music
- e: Back Space Mechanism
- e: Backward Set-up Message
- e: Basic Storage Module
- e: Basic Subsystem Module
- e: Basic Systems Memory
- e: Batch Spool Monitor
- e: Battery Shop Maintenance
- e: Bi-Stable Multivibrator
- e: Birmingham School of Music
- e: Bit-Sliced Microprocessor
- e: Both Sides Measured
- e: Bottom Sonar Marker
- e: Braked Servomotor
- e: Bronze Star Medal
BS MA e: British Standard: Marine Series
BS Mar Eng e: Bachelor of Science in Marine Engineering
BSMC e: Bachelor of Science in Mathematics and Chemistry
BSMCP e: Blue Shield Medical Care Plans
BSME e: Bachelor of Science in Mechanical Engineering
- e: Bachelor of Science in Mining Engineering
- e: Bachelor of Science in Music Education
BS Mech E e: Bachelor of Sience in Mechanical Engineering
BS Mech Ind e: Bachelor of Science in Mechanical Industries
BS Mech Techn e: Bachelor of Science in Medical Technology
BS Med e: Bachelor of Science in Medicine
BSM Ed e: Bachelor of Science in Music Education
BS Med Rec e: Bachelor of Science in Medical Records
BS Med Rec Lib e: Bachelor of Science in Medical Records Librarianship
BS Med Sc e: Bachelor of Science in Medical Science

BS Med T[ech] e: Bachelor of Science in Medical Technology
BS ME Elect e: Bachelor of Science with Mechanical Engineering Electives
BS Met e: Bachelor of Science in Metallurgy
- e: Bachelor of Science in Meteorology
BS Met E e: Bachelor of Science in Metllurgical Engineering
BS Met Eng e: Bachelor of Science in Metallurgical Engineering
BSMF e: Bit Sync[hronization] Matched Filter
BSMG e: Basic Super Mastergroup
BS Mg E e: Bachelor of Science in Mining Engineering
- e: Bit Sync[hronization] Matched Filter
BSMG Signal e: Basic Super Mastergroup Signal
BS Mgt Sc e: Bachelor of Science in Management Science
BS Min e: Bachelor of Science in Mineralogy
BS Min e: Bachelor of Science in Mining
BS Min E[ng] e: Bachelor of Science in Mining Engineering
B.S.Min.Eng. e: Bachelor of Science in Mining Engineering
BS Mng E e: Bachelor of Science in Mining Engineering
BSMP e: Brussels Sprouts Marketing Program[me]
BSMPY Program[me] e: Board Sort Multiply Program[me]
BSMS e: Bachelor of Science in Military Science
- e: Basic Storage Modules
BSMT e: Bachelor of Science in Medical Technology
- e: Basement
BS Mt E e: Bachelor of Science in Metallurgical Engineering
BS Mu e: Bachelor of Sacred Music
BS Mus e: Bachelor of School Music
- e: Bachelor of Science in Music
BS Mus Ed e: Bachelor of Science in Music Education
BSMV e: Barley Stripe Mosaic Virus
- e: Bistable Multivibrator
BSN d: Barium-Strontium-Niobat
- e: Bachelor of Science in Nursing
- e: Back-end Storage Network
- e: Backward Sequence Number
- e: Baker School of Navigation
- e: Barium Sodium Niobate
- e: Bowel Sound Normal
- e: Bundle Signal Name
BSNA e: Bachelor of Science in Nursing Administration
- e: Bureau of Salesmens National Associations
BS Nat Hist e: Bachelor of Science in Natural History
BSNAVSHIPYD e: Boston Naval Shipyard
BSNDT e: British Society for Non-Destructive Testing
BSNE[d] e: Bachelor of Science in Nursing Education
BSNEX e: Bundle Signal Name Expanded
BSNH e: Boston Society of Natural History
- e: Buffalo Society of Natural History
BSNIT e: Bachelor of Science in Nautical Industrial Technology
BSNO e: Brake Specific Nitric Oxide
BSNR e: Backward Sequence Number Received
BSNS e: Buffalo Society of Natural Sciences
BSNT e: Backward Sequence Number of next to be Transmitted
BS Nurs e: Bachelor of Science in Nursing
BS Nurs Ed e: Bachelor of Science in Nursing Education
BSO d: Bernsteinsäure-Oxydase
- d: Blindschachtordnung
- e: Bachelor of Sociology
- e: Bachelor of the Science of Oratory
- e: Base Signal Officer
- e: Beach Signal Office
- e: Bit String Operator
- e: Blue Stellar Object
- e: Bomb Savety Officer
- e: Bomb Store Officer
- e: Broader System of Ordering
- e: Broad System of Ordering
BS Occ Ther e: Bachelor of Science in Occupational Therapy
BSocSci e: Bachelor of Social Science
BSocSt e: Bachelor of Social Studies
BSocWk e: Bachelor of Social Work
BSOpt e: Bachelor of Science in Optometry
BS or EOM e: Backspace or End of Message
BS Orn Hort e: Bachelor of Science in Ornamental Horticulture
B So Se e: Bachelor of Social Service
BSOT e: Bachelor of Science in Occupational Therapy
BSP d: Basis-Software Programm
- d: Bedien- und Service-Prozessor
- d: Beispiel
- d: Bruttosozialprodukt
- e: Bachelor of Science in Pharmacy
- e: Bachelor of Speech
- e: Back Space Pawl
- e: Ball Stop
- e: Ballistic Missile Early Warning System Specification
- e: Bell System Practice
- e: Bell Systems Practice
- e: Bering Sea Patrol
- e: Bibliographical Society Publications (journ.)
- e: Birmingham School of Printing
- e: Botanical Society of Pennsylvania
- e: British Society for Parasitology
- e: British Standard Pipe
- e: Bromosulfophthalein
- e: Burroughs Scientific Processor
- e: Business Systems Planning
BSPA e: Bachelor of Science in Practical Arts
- e: Bachelor of Science in Public Administration
BSPAL e: Bachelor of Science in Practical Arts and Letters
BSPE e: Bachelor of Science in Petroleum Engineering
BSPE[duc] e: Bachelor of Science in Physical Education
BSPE Program[me] e: Board Sort Pre-Edited Program[me]
BS Per & In Rel e: Bachelor of Science in Personnel and Industrial Relations
BS Per & Pub Rel e: Bachelor of Science in Personnel and Public Relations
BS Pet e: Bachelor of Science in Petroleum
BS Pet Eng e: Bachelor of Science in Petroleum Engineering
BS Ph e: Bachelor of Science in Pharmacy
BSPH e: Bachelor of Science in Public Health
BS Phar[m] e: Bachelor of Science in Pharmacy
BSPHN e: Bachelor of Science in Public Health Nursing
BS Ph Th e: Bachelor of Science in Physical Therapy
BS Phys e: Bachelor of Science in Physics
BS Phys Ed[u] e: Bachelor of Science in Physical Education
BS Phys Ther e: Bachelor of Science in Physical Therapy
BSPL e: Behavioral Science Programming Language
BSPM e: Battlefield Systems Project Management
- e: Benzene-Soluble Particulate Matter
BSPMA e: British Sewage Plant Manufacturers Association
BSPSM e: Bachelor of Science in Public School Music
BSPT e: Bachelor of Science in Physical Therapy
BSPTEST e: Bromsulphalein Test
BSP Thread e: British Standard Pipe Thread
BS Public Adm e: Bachelor of Science in Public Administration
BSPW e: Bare Silver Plated Wire
BSQ e: Bachelor Sergeant Quarters
- e: Basque
BSR d: Berliner Stadtreinigung[s-Betriebe]
- d: Betriebssystem für reale Adressierung (Siemens)
- d: Bizepssehnenreflex
- d: Blutkörperchensenkungsreaktion
- e: Bachelor of Science in Recreation
- e: Bachelor of Science in Rehabilitation
- e: Backspace Recorder
- e: Backspace Register
- e: Bacteria Survival Ratio
- e: Balloon-Supported Rocket[s]
- e: Bank Sorter-Reader
- e: Basal Skin Resistance
- e: Basic Service Rate
- e: Battle Short Relay
- e: Better Sound Reproduction
- e: Bit Shift[ing] Register
- e: Bit Slippage Rate
- e: Blip-Scan Radar
- e: Blip-Scan Ratio
- e: Blood Sedimentation Rate
- e: Blue Streak Request
- e: Board of Standards Review
- e: Boilermaker; Ship Repair
- e: Bore Sight Restricted
- e: Brain Stimulation Reinforcement
- e: British School at Rome
- e: British Society of Rheology
- e: Buffered Sampler Reader
- e: Buffered Send/Receive
- e: Bulk-Shielding Reactor
BSRA e: British Sound Recording Association
BS Radio-TV e: Bachelor of Science in Radio and Television
BS Rad Tech e: Bachelor of Science in Radiological Technology
BSRAP e: Beginning Standard Range Approach
BSRC e: Battelle Seattle Research Center
- e: Biological Serial Record Center
BSRD e: Behavioral Sciences Research Division
BSRE e: Bachelor of Science in Religious Education
BS Rec e: Bachelor of Science in Recreation
BS Ret e: Bachelor of Science in Retailing

BSRF e: Bell System Reference System
- e: Brain Stem Reticular Formation
BSRFS e: Bell System Reference Frequency Standard
BSRIA e: Building Services Research and Information Association
BSRL e: Boeing Scientific Research Laboratories
BSRRO e: Begin Standard Radar Refuel Orbit
BSRR Program[me] e: Board Sort Routed Records Program[me]
BSRS e: Bell System Repair Specification
BSR System e: Basic Service Rate System
BSRT e: Bachelor of Science in Radiological Technology
BS Ry ME e: Bachelor of Science in Railway and Mechanical Engineering
BSS d: Bausteinsystem
- d: Bestellbestand
- d: Betriebsstufenschalter
- d: Bildschirmsteuergerät
- d: Bildschirmsystem
- e: Bachelor of Sanitary Science
- e: Bachelor of Science in Science
- e: Bachelor of Secretarial Science
- e: Bachelor of Social Science[s]
- e: Backing Store System
- e: Back Space Statement
- e: Balanced Salt Solution
- e: Base Station System
- e: Basic Shaft System
- e: Basic Synchronized Subset
- e: Beach Signal Station
- e: Beam Steering System
- e: Bibliography of Soil Science (journ.)
- e: Bibliothèque Saint-Sulpice / Montreal
- e: Biological and Social Sciences
- e: Bit Storage and Sense
- e: Black-Silk Suture
- e: Boeing-Michigan Aeronautical Research Center Squadron Simulator
- e: British Standard Section
- e: British Standard Sizes
- e: British Standard Specification[s]
- e: Broadcast[ing] Satellite Service
- e: Bronze Service Star
- e: Buffered Saline Solution
- e: Buffered Salt Solution
- e: Bulk Storage System
BSSA e: Bachelor of Science in Secretarial Administration
BSSanE e: Bachelor of Science in Sanitary Engineering
BSSB e: British Society of Soil Science
BS Sc e: Bachelor of Sanitary Science
- e: Bachelor of Science in Science
BSSC e: Bit Synchronizer-Signal Conditioner
BS Sc E[ng] e: Bachelor of Science in Science Engineering
BS School LS e: Bachelor of Science in School Library Science
BS School Mus e: Bachelor of Science in School Music
BS School Supv e: Bachelor of Science in School Supervision
BSSE e: Bachelor of Science in Sanitary Engineering
- e: Bachelor of Science in Secondary Education
BSSE Algorithm e: Binary Shift/Substract Euclidean Algorithm
BS Sec Adm e: Bachelor of Science in Secretarial Administration
BS Sec Ed e: Bachelor of Science in Secondary Education
BS Sec Sc[i] e: Bachelor of Science in Secretarial Science
BS Sec St e: Bachelor of Science in Secretarial Studies
BSSG e: Biomedical Science[s] Support Grant
BSSO e: British Society for the Study of Orthodontics
BS Soc e: Bachelor of Science in Sociology
BS Soc Serv e: Bachelor of Science in Social Service
BS Soc St e: Bachelor of Science in Social Studies
BS Soc WK e: Bachelor of Science in Social Work
BS Sp e: Bachelor of Science in Speech
BSSP e: Battelle Seminars and Studies Program[me]
BSSP e: Broadband Solid State Preamplifier
BSSR e: Bureau of Social Science Research
BSSRS e: British Society for Social Responsibility in Science
- e: Bureau of Safety and Supply Radio Services
Bs-SS d: Bildschirm-Steuersystem
BSSS e: Bachelor of Science in Secretarial Studies
BSSS e: Bachelor of Science in Social Science
- e: British Society of Soil Science
BSS Sc e: Bachelor of Science in Social Science
BSS Sci e: Bachelor of Science in Secretarial Science
BS Struc Eng e: Bachelor of Science in Structural Engineering
BSSW e: Bachelor of Science in Social Work
BSSW e: Bare Stainless Steel Wire
BST d: Basaltemperatur
- d: Baustein
- d: Befehlssteuerwerk
- d: Bildschirmtastatur
- d: Blockschachtelungstiefe
- d: Blockschatelungstiefe
- e: Bachelor of Sacred Theology
- e: Bachelor of Science in Teaching
- e: Bachelor of Statistics
- e: Back Spacing Time
- e: Ballast
- e: Bare Sight Tunnel
- e: Base Stop Tester
- e: Beam-Steering Transducer
- e: Beam-Switching Tube
- e: Bering Standard Time
- e: Bill of Sight
- e: Binary Search Tree
- e: Binary Synchronous Transmission
- e: Bioscript
- e: Blood Serological Test[ing]
- e: Booster
- e: Bore Sight Tracker
- e: Bovine Somatotropin
- e: Breast Stimulation Test (Fetal Test)
- e: Brief Stimulus Therapy
- e: Brief Systems Test
- e: British Standard Time
- e: British Summer Time
- e: Burst
- e: Business Systems Technology, Incorporated
BSTD e: Base Supplies and Transport Depot
- e: Bastard
B-Stelle d: Beobachtungsstelle
B St E[ng] e: Bachelor of Structural Engineering
BS Text e: Bachelor of Science in Textiles
BS Textile Eng e: Bachelor of Science in Textile Engineering
BSTFA e: Bis(trimethylsilyl)trifluoro-acetamide
BSTG e: Bilateral Stellate Ganglionectomy
BST & IE e: Bachelor of Science in Trade and Industrial Engineering
BSTJ e: Bell System Technical Journal
BSTL e: Bistaple
B ST LT e: Blue Stern Light
BSTM e: Biaxial Shock-Test Machine
BST Memory e: Bit Strip Tape Memory
Bston d: Besetztton
BSTR d: Befehlsstruktur
- e: Booster
BS Trans e: Bachelor of Science in Transportation
BS Trk e: Bomb Service Truck
BST RKT e: Boost Rocket
BStrVermG d: Bundesstraßenvermögens-gesetz
BSTU e: Basic Storage Unit
BSU e: Backing Store Unit
- e: Base Selection Unit
- e: Base Service Unit
- e: Basic Selection Unit
- e: Basic Sounding Unit
- e: Basic Storage Unit
- e: Bearer Switchover Unit
- e: Bistable Storage Unit
- e: Bistable Store Unit
- e: British Standard Unit[s]
- e: Business Service Unit
BSUA e: British Standard Unequal Angle
BSUB e: Ball and Socket Upper Bearing
B Sup Airfd. e: Base Supply Airfield
B Sur e: Bachelor of Surgery
BSURE e: Barking Sands Underwater Range Expansion
BSurv e: Bachelor of Surveying
BSUSMA e: Bachelor of Science; United States Military Academy
BSUSNA e: Bachelor of Science; United States Naval Academy
BSUT e: Beam Steering Ultrasonic Transducer
BSV d: Bayerische Saatbau-Vereinigung
- d: Betriebsrichtungsschalter-Verbindungs-seite
- d: Betriebssystem für virtuelle Adressierung (Siemens)
- d: Betriebssystem für virtuelle Speicherung
- e: Bark Split Virus
- e: Batten-Spielmeyer-Vogt [Syndrome]
- e: Boiler Stop Valve
- e: Boolean Simple Variable
BS Voc Ag e: Bachelor of Science in Vocational Agriculture
BS Voc Ed e: Bachelor of Science in Vocational Education
BSW d: Bahnstromwerk
- d: Betonschwellenwerk
- d: Blattschreiber für Wartung
- e: Bachelor of Social Work
- e: Bare Steel Wire
- e: Barrel[s] of Salt Water
- e: Barrel Switch
- e: Biological Society of Washington
- e: Botanical Society of Washington
- e: British Standard Whitworth
BS & W e: Basic Sediment and Water
BS + W e: Bottom Settlings plus Water
BSWCC e: British Short Wave Correspondence Club
BSWD e: Barrel[s] of Salt Water per Day
BSWG e: British Standard Wire Gage
- e: British Standard Wire Gauge

B & SWG e: Brown and Sharpe Wire Gauge
BSWIA e: British Steel Wire Industries Association
BSWM e: Bureau of Solid Waste Management
BSW Thread e: British Standard Whitworth Thread
BSY e: Binary Synchronous
- e: Busy
B Sy Th e: Bachelor of Systematic Theology
BSZ d: Basisband-Symmetrierzusatz
- d: Besetztzeichen
- d: Betriebssteuerzentrale
- d: Betriebsstundenzähler
- d: Bezirkssteuerzentrale
- d: Bildschirmzeitung
- d: Buttersäurezahl
- e: Ballistic Systems Zeus
- e: Block Store Zeros
BT d: Basaltemperatur
- d: Basisteil[e]
- d: Bathythermograf
- d: Bathythermometer
- d: Bauteil
- d: Bauteiletabelle
- d: Bedarfsträger
- d: Bedientisch
- d: Bedingungsteil
- d: Behandlungstage
- d: Belegungstaste
- d: Beleuchtungstransformator
- d: Besetztton
- d: Betriebs-Technik (journ.)
- d: bidirektionales Tor
- d: Bildschirmterminal
- d: Bildschirmtext
- d: Bildtelegrafie
- d: Bildträger
- d: Bittakt
- d: Bit Test
- d: Blinktaktgeber
- d: Block Tag
- d: Block Transfer
- d: Blocktransformator
- d: Block Type
- d: Brennstofftechnik
- d: Brusttiefe
- d: Butanthiol
- e: Babylonian Talmud
- e: Bachelor of Teaching
- e: Bachelor of Theology
- e: Bacillus Thuringiensis
- e: Back Transfer
- e: Balanced Ternary
- e: Bandwidth Time
- e: Barge; Training
- e: Basic Trainer
- e: Basic Training
- e: Basic Typing
- e: Bathythermal Traces
- e: Bathythermograph
- e: Battery Target
- e: Beat
- e: Before Total
- e: Beginning of Tape
- e: Bellini-Tosi
- e: Bent
- e: Benzotriazole
- e: Bering Time
- e: Berth Terms
- e: Bhutan
- e: Bias Temperature
- e: Binary Fourier Representation Transform
- e: Binary Translation
- e: Binary Trigger
- e: Bistable Trigger
- e: Bit
- e: Bitemporal
- e: Bladder Tumor
- e: Blind Toss
- e: Block Template
- e: Blue Tetrazolium
- e: Boat [Tail]
- e: Body Temperature
- e: Bomber-Trooper
- e: Bomber Transport
- e: Bombing Table
- e: Boolean Term
- e: Bootstrap Technique
- e: Bought
- e: Bovine Testicle
- e: Brain Tumor
- e: British Tariff[s]
- e: British Telecom[/Telecommunications]
- e: Broader Term
- e: Builders Trials
- e: Buried Tape
- e: Burn Time
- e: Burst Trapping
- e: Bus Tie
- e: Busy Tone
- e: Byte[s]
- f: Basse Tension
- f: Brevet Technique
B/T d: Breiten-/Tiefengrad
B & T e: Bridges and Tunnels
B_1 e: Thiamine Vitamin
BT 13 e: Vultee two-place basictrainer
B_2 e: Ribotlavin Vitamin
BTA d: Betriebssicherungstaste
- d: betriebstechnische Anlage
- d: biologisch-technische[r] Angestellte[r]
- d: Bürotechnik und Automation (journ.)
- e: Ballistic Track Assignor
- e: Best Times Available
- e: Better than Average
- e: Billable Time Authorization
- e: Blood Transfusion Association
- e: British Transport Advertising
- e: British Tuberculosis Association
BTÄ d: Bleitetraäthyl
BTA + bto d: Bürotechnik Automation + Organisation (journ.)
BTAM e: Basic Tape Access Method
- e: Basic Telecommunications Access Method
- e: Basic Teleprocessing Access Method
- e: Basic Terminal Access Method
BTAO e: Bureau of Technical Assistance Operations
BTÄO d: Bundes-Tierärzteordnung
BTAP e: Bond Table Analysis Program[me]
- e: Bond Trade Analysis Program[me]
BTB d: Branchentelefonbuch
- e: Basic Test Battery
- e: Boric-Acid, Tar-oil and Bentonite
- e: Braided Tube Bundle
- e: Breakage-Fusion-Bridge
- e: Brom[o]thymol Blue
- e: Burst-Time Bias
- e: Bus Tie Breaker
BTBA e: Blood Transfusion Betterment Association
BTBC e: Boehm Test of Basic Concepts (Psychology)
BTBCAP e: Box-to-Box Compatibility Analysis Program[me]
BTC e: Bachelor of Textile Chemistry
- e: Bachelor of Textile Chemistry
- e: Basic Training Center
- e: Batch Terminal Controller
- e: Before Top Center
- e: Bell-Telephone Company
- e: Below Threshold Change
- e: Bench Test Console
- e: Benzene Tetrachloride
- e: Beryllium Thrust Chamber
- e: Bi-purpose Triple Coil
- e: Bilateral Tape Card
- e: Binary Digit Time Counter
- e: Binary Twos Complement
- e: Block Terminating Character
- e: Block Transfer Control[ler]
- e: Business and Technology Center
- e: Butt-Treated Cedar
BTCA e: Basic Tables of Commissioning Allowances
- s: biblioteca
BTCB e: Bumped Tape Carrier Bonding
BTCC e: Basic Traffic Control Center
- e: Broome Technical Community College
BTC Code e: Bipolar Two-Complement Code
BT Ch e: Bachelor of Textile Chemistry
BTCP e: Bachelor of Town and Country Planning
BTCU e: Buffered Transmission Control Unit
BTD e: Bachelor of Textile Dyeing
- e: Back Transmit Data
- e: Balanced Tape Drive
- e: Base Transit Depot
- e: Bias Telegraph Distortion
- e: Binary-to-Decimal
- e: Bomb Testing Device
- e: Brief Task Description
- e: British Transport Docks
- e: Bulk Tape Degausser
- e: Bulk Transmission of Data
- e: Burn to Depletion
BTDA e: Benzophenone Tetracarboxylic [Acid] Dianhydride
BTDAM e: Basic Telecommunications Data Access Method
BTDC e: Before Top Dead Centre
BT Des e: Bachelor of Textile Design
BTDF-L-C e: Butt-Treated Douglas-Fire, Larch and Cedar
BTDL e: Back Transient Diode Logic
- e: Basic Transient Diode Logic
BT-Drucksache d: Bundestag-Drucksache
BTDS e: Benzoylthiamine Disulfide
BTE d: Benzilsäure-Tropinester
- d: Bustreibereinheit
- e: Bachelor of Textile Engineering
- e: Battery Terminal Equipment
- e: Battery Timing Equipment
- e: Bidirectional Transceiver Element
- e: Blunt Trailing Edge
- e: Boltzmann Transport Equation
- e: Bourdon Tube Element
- e: Brayton Turboelectric Engine
- e: Bulk Tape Eraser
- e: Business Terminal Equipment
- f: breveté
BTEC e: Business and Technican Education Council
B Tech e: Bachelor of Technology
B.-Techn. d: Bautechnik[er]
btechn. d: bautechnisch
BTEE d: Bustreibereinheit im Erweiterungsschrank
- e: Brayton Turbo-Electric Engine
B Tel E e: Bachelor of Telecommunication Engineering
B-Tell e: Back Telling
BTEng e: Bachelor of Textile Engineering
BText e: Bachelor of Textiles
BTEZ d: Bustreibereinheit im Zentralschrank

BTF e: Balance to Follow
- e: Balance Transferred
- e: Ballistic Test Facility
- e: Barrels of Total Fluid
- e: Bench-Test Fixture
- e: Benzotrifluoride
- e: Binary Transversal Filer
- e: Binary Transversal Filter
- e: Bomb Tail Fuze
- e: British Trawlers Federation
- e: Bulk Transfer Facility

BTFL[Y] e: Butterfly
BTFLY VLV e: Butterfly Valve
BTFS e: Breast Tumor Frozen Section
BTFWD e: Brought Forward
BTG d: Brennstofftechnische Gesellschaft
- e: Ball Tooth Gear
- e: Battery Timing Group
- e: Beacon Trigger Generator
- e: Boiler Turbine Generator
- e: British Technology Group
- e: Burst Transmission Group

BTGJ e: Bal Tooth Gear Joint
BTH d: Bibliothek der Technischen Hochschule Hannover
- e: Bachelor of Theology
- e: Beyond the Horizon
- e: British Thomson Houston
- e: Bulk Transfer Hose

BTHEFF. e: Brake Thermal Efficiency
BTHS e: Brooklyn Technical High School
BThU e: British Thermal Unit
BTI d: Bedientisch
- d: Bezirkstierzuchtinspektion
- e: Bacillus Thuringiensis Israelensis
- e: Bandung Technical Institute
- e: Bank and Turn Indicator
- e: Basic Timesharing, Incorporated
- e: Boston Theological Institute
- e: Bridge[d] Tape Isolator
- e: British Technology Index
- e: British Technology Index B-Tisch
- e: British Telecom International
- e: Broaching Tool Institute
- e: Byte Transfer In

BTIPR e: Boyce Thompson Institute for Plant Research
BTJ e: Ball Tooth Joint
- e: Board of Trade Journal

Btkm d: Bruttotonnenkilometer
BTL e: Back Focal Length
- e: Backtell
- e: Balanced Transformer-Less
- e: Beacon Tracking Level
- e: Beginning Tape Label
- e: Behind the Lens
- e: Behind the Line
- e: Bell Telephone Laboratories
- e: Between Layers
- e: Board Test Language
- e: Business Translation Language
- e: Bus Transceiver Logic

BTLL e: Bottom Lead Left
BTLP e: Butt-Treated Lodgepole Pine
BTLR e: Bottom Lead Right
BTLS e: Bell Telephone Laboratories System
- e: Breadboard Terminal Landing System

BTLV e: Biological Threshold Limit Value
BTM d: Betäubungsmittel
- d: Betriebsmeister
- d: Bittertechnikmethode
- d: Bleitetramethyl
- e: Back Transfer Mechanism
- e: Balance Totalizing Mechanism
- e: Ballast Tank Meter
- e: Basic Transport Mechanism
- e: Batch Time-sharing Monitor
- e: Beginning of Tape Mark[er]
- e: Bell Telephone Manufacturing
- e: Blast Test Missile
- e: Bottom
- e: British Tabulating Machines Company
- e: British Theatre Museum
- e: Bromotrifluoromethane
- e: Buffered Terminal Multiplexer

BTM 7 Program[me] e: Board Tester M 7 Program[me]
BTMA e: Busy Tone Multiple-Access
BTMC e: British Telecom Mobile Communications
BTME e: Babcock Test of Mental Efficiency
BTMG d: Blätter der Thomas Mann Gesellschaft (journ.)
BTMS e: Body Temperature Measuring System
BTMVVO d: Betäubungsmittel-Verschreibungsverordnung
BTN e: Between
- e: Billing Telephone Number
- e: Brussels Tariff Nomenclature
- e: Button

BTNHD e: Button Head
BTN-Nr. d: Brüsseler Tarif-Nomenklatur-Nummer
BTNR e: British Telecom Network Requirement[s]
BTNS e: Basic Terminal/Network Support
BTnWe d: Bild-Tonweiche
BTNW Program[me] e: Board Tester Not Wired Program[me]
BTO a: Brutto
- d: Brutto
- d: Bürotechnik und Organisation (journ.)
- e: Big-Time Operator
- e: Blocking Tube Oscillator
- e: Block Transfer Order
- e: Bombing Through Overcast
- e: Branch Transportation Officer
- e: Brief Task Outline
- e: Brigade Transport Officer
- e: British Trust for Ornithology
- e: Byte Transfer Out
- s: bruto
- s: bulto

B TO B e: Back to Back
BTO[S] e: Big Time Operator[s]
BTOW e: Boiler Technician of the Watch
BTP d: Baukastensystem der technologischen Projektierung
- e: Bachelor of Town Planning
- e: Bailment Test Program[me]
- e: Ballistic Missile early Warning System Test Procedure
- e: Batch Transfer Program[me]
- e: Body Temperature and Pressure
- e: Branch Technical Position

BT product e: Bandwidth-Time product
BTPS e: Body Temperature, Pressure [and] Saturation
- e: Body Temperature, Pressure, Saturated
- e: Body Temperature, Pressure, Soft
- e: Body Temperature and Pressure

B_{psc} e: Bearing per Standard Compass
B_{pstgc} e: Bearing per Steering Compass
BTPTS e: Building and Town-Planning Technical Staff
BTR d: Beitrag
- d: Betrieb
- d: Bildträger
- d: Brammentemperaturregelung
- e: Ballast Tube Resistor
- e: Ballistic Missile Early Warning System Test Report
- e: Barrel Tile Roof
- e: Basic Transportation Vehicle
- e: Behind-the-Reader
- e: Behind-the-Tape-Reader
- e: Behind Tape Reader
- e: Billable Time Report
- e: Block Tape Recorder
- e: Blood Transfusion Reactions
- e: British Telecommunications Research
- e: Broadcast and Television Receivers (journ.)
- e: Bus Transfer

BTR-40 e: Soviet armored personnel carrier and scout car
BTR-50 e: Soviet amphibious personnel carrier
BTR-60P e: Soviet amphibious armored personnel carrier
BTrA d: Bildträgerabstand
BTRA e: Bachelor of Town and Regional Planning
Btr Aufs d: Betriebsaufseher
BTR Connection e: Behind-the-Tape-Reader Connection
BTrf d: Bautruppführer
BtrI d: Betriebsinspektor
Btr.-Ing. d: Betriebsingenieur
BTRL e: British Telecom Research Laboratories
BTRM e: Soviet armored-infantry combat vehicle
BtrMstr d: Betriebsmeister
BtrOAufs d: Betriebsoberaufseher
BTRP e: Bachelor of Town and Regional Planning
Btry e: Battery
BTS d: Benztraubensäure
- d: biologische Trockensubstanz
- d: Bürotechnische Sammlung
- e: Balloon Transport System
- e: Barrier Terminal Strip
- e: Base of Terminal Service
- e: Basic Tape System
- e: Basic Training School
- e: Batch Terminal Simulator
- e: Beacon Tracking System
- e: Bench Test Specification
- e: Bio Telemetry System
- e: Blood Transfusion Service
- e: Blue Tool Steel
- e: Boeing Test Support
- e: Boolean Time Sequence
- e: Branch Terminal System
- e: Brazilian Thyroid Sludge
- e: British Tanzania Society
- e: British Telecommunications Systems, Ltd
- e: British Television Society
- e: British Textile Society
- e: Broadcast Television Systems

BTSA e: British Tensional Strapping Association
BTSD e: Basic Training for Skill Development
BTSH e: Bovine Thyroid Stimulating Hormone
BTSS e: Basic Time-Sharing System
- e: Braille Time-Sharing System

BT ST e: Billet Steel
BTST e: Busy-Tone Start
BTSTR e: Burster
BTST STR e: Bursting Strength
BT System e: Bellini-Tosi System

BTT d: Bildschirmtextterminal
- d: Braunkohlentieftemperatur-Koks
- e: Bachelor of Textile Technology
- e: Bank Teller Terminal
- e: Beginning to Tape Test
- e: Beginning to Tape Test BTT-Koks
BTTA e: British Thoracic and Tuberculosis Association
BTT-Koks d: Braunkohlentieftemperatur-Koks
BTTNS e: Battens
btto. d: brutto
Bttr. d: Batterie
BTU d: Bildschirmtext-Telex-Umsetzer
- d: Biotechnische Umschau (journ.)
- e: Balloon Training Unit
- e: Base Transfusion Unit
- e: Basic Training Unit
- e: Basic Transmission Unit
- e: Block Transfer Unit
- e: Board-of-Trade Unit
- e: British Thermal Unit
BTU/CU FT e: British Thermal Unit per Cubic Foot
BTU/CU FT DEGF e: British Thermal Unit per Cubic Foot times Degree Fahrenheit
BTU/DEGF e: British Thermal Unit per Degree Fahrenheit
BTU/FT²H e: British Thermal Unit per Square Foot times Hour
BTU/FT³ e: British Thermal Unit per Cubic Foot
BTU/FT³ DEGF e: British Thermal Unit per Cubic Foot times Degree Fahrenheit
BTU/FT H DEG F e: British Thermal Unit per Foot times Hour times Degree Fahrenheit
BTU/H e: British Thermal Unit per Hour
BTU/H°R e: British Thermal Unit per Degree Rankine
BTU/IN² S e: British Thermal Unit per Square Inch times Second
BTU/LB e: British Thermal Unit per Pound
BTU/LB DEG F e: British Thermal Unit per Pound times Degree Fahrenheit
BTU/LB°R e: British Thermal Unit per Pound times Degree Rankine
BTU/S e: British Thermal Unit per Second
BTU/S FT² DEGF e: British Thermal Unit per Second times Square Foot times Degree Fahrenheit
BTU/S FT DEGF e: British Thermal Unit per Second times Foot times Degree Fahrenheit
BTU/SQFT H e: British Thermal Unit per Square Foot times Hour
BTU/SQUIN S e: British Thermal Unit per Square Inch times Second
BTU/S SQFT DEGF e: British Thermal Unit per Second times Square Foot times Degree Fahrenheit
BTU/°R e: British Thermal Unit per Degree Rankine
BTV d: bautechnologische Versorgung
- d: Bild-Ton-Vorstufe
- d: Büro für Textverarbeitung
- d: Bürotextverarbeitung
- e: Basic Transportation Vehicle
- e: Battleship Test Vehicle
- e: Bi-modal Transit Vehicle
- e: Blast Test Vehicle
- e: Bluetongue Virus
- e: Buoyancy Transport Vehicle
- e: Business Television
BTVOR e: Broadcast Terminal Very-High Frequency Omni-Range
BTW e: Bare Tungsten Wire
- e: Between
- e: Bimetal Turbine Wheel
- e: Boat Wave
- e: By the Way
BT-Wagen d: Behältertragwagen
BTWHS e: Booker T. Washington High School
BTWLD e: Butt Weld
BTWN e: Between
BTW Program[me] e: Board Tester Wired Program[me]
BTX d: Bildschirmtext
- e: Benzene, Toluene, Xylene
Btx-Tx-Dienst d: Bildschirmtext-Telex-Dienst
BtxVst d: Bildschirmtext-Vermittlungsstelle
BTXY Program[me] e: Board X-Y-tester wired Program[me]
Bty e: Battery
B-Type e: Basedow Type
BTYPE e: Block Type
BTZ d: Bildungstechnologisches Zentrum
- e: Benzotriazole
BU d: Bandumsetzer
- d: Bandumsetzung
- d: Bereichsumschaltung
- d: Bergische Universität
- d: Berufsunfähigkeit
- d: Betriebsunterbrechung
- d: Boden-Unterseeboot
- d: Breitenunterschied
- d: Brustumfang
- d: Bruttoumsatz
- d: Buche
- d: Buchse
- d: Buchstabe[numschaltung]
- e: 5-Bromouracil
- e: Backup
- e: Baker University
- e: Base [of Prism] Up
- e: Base Unit
- e: Bath Unit
- e: Baylor University
- e: Bend Up
- e: Binding Unit
- e: Biological Urge
- e: Bishops University
- e: Bit Unit
- e: Bloomsburg University
- e: Blue
- e: Boat Unit
- e: Boston University
- e: Bottom Up
- e: Bradley University
- e: Brandeis University
- e: Bread Unit
- e: Brick Unprotected
- e: Bright Up
- e: Brown University
- e: Brunel University
- e: Bucknell University
- e: Build-Up
- e: Build[er]
- e: Bulgaria[n]
- e: Burma
- e: Bushel[s]
- e: But[h]yl
- e: Butler University
- e: Buzzer
BÜ d: Bahnübergang[e]
- d: Bauteilübersicht
- d: Befehlsübermittlung
- d: Berührungsüberhitzer
- d: Betriebsüberwachung
- d: Binärsignalüberwachung
- d: Brückenübertrager
- d: bündig
BUA e: Booster Umbilical Assembly
- e: Booster Union Assembly
- e: British United Airways
- e: Built-up Addressing
Büa d: Breite über alles
BUAER e: Bureau of Aeronautics
BUALC e: Butyl Alcohol
BUAV e: British Union for the Abolition of Vivisection
BuB d: Bücherei und Bildung (journ.)
BUB d: Buch und Bibliothek
- e: Bubble
BUBE d: Buchungsbearbeitung
BUBS e: Bubbles
BUC e: Ballast Undercutter-Cleaner
- e: Bangor University College
- e: Bus Control
BUCC e: Buccal
BUCCS e: Bath University Comparative Catalogue Study
BUCH d: Bequemes Unproblematisches Chancenreiches Heimstudium
Büch. d: Bücherei
Buchdr. d: Buchdruck
- d: Buchdruckerei
BuChE d: Butyrylcholinesterase
Buchh. d: Buchhandlung
BUCH Program[me] e: Board Update Command History Program[me]
Buchst. d: Buchstabe[n]
Buchverl. d: Buchverlag
BUCK e: Buckram
BuckBl d: Buckelblech
Bucknell U Pr e: Bucknell University Press
BUCO e: Buildings Control Officer
BUCOP e: British Union Catalogue of Periodicals
BUCOP-WLSP e: British Union Catalogue of Periodicals-World List of Scientific Periodicals
BUCU e: Burring Cutter
BUD d: Budapest (Flughafen)
- e: Beneficial Use Date
- e: Buddhist
- e: Buddism
- e: Budget
Buda. e: Budapest
BUDC e: Back-Up Digital Computer
BU-DNA e: 5-Bromo-Uracil-DNA
BUDPLAN d: Budget- und Planungsprogramm
- e: Budget and Plans
BUDR e: Bromodeoxyuridine
- e: Bromouracial Deoxyriboside
BUDS e: Building Utility Design System
B. u. E. d: Berichtigungen und Ergänzungen
BUE d: Buenos Aires (Flughafen)
- e: Built-Up Edge
BUEC e: Back-Up Emergency Communication
- e: Back-up Emergency Communications
BUEG d: Berichte und Informationen der Europäischen Gemeinschaften
BUET e: Bangladesh University of Engineering andTechnology
BUF e: Buffalo (airport)
- e: Buffer[ed]
BÜFA d: Büromaschinenfachausstellung
BUFC e: British Universities Film Council
BUFF e: Big Ugly Fat Fellow (=B 52 bomber)
BUFNOM e: Buffers, Number of
BUFORA e: British Unidentified Flying Object Research Association

BUG e: Basic Update Generator
- e: Boeing-Michigan Air Research Center Unintegrated Guidance
- e: Buccal Ganglion
- e: Bugler
Büg d: Bügel
BU/GHW d: Bergische Universität / Gesamthochschule Wuppertal
BUGS e: Brown University Graphics System
BUH e: Bucharest (airport)
BUIA e: British United Island Airways
BUIC e: Back-Up Intercepter Control
BUILD e: Base for Uniform Language Definition
- e: Building
BUIS e: Barrier Up Indicator System
BUISYS e: Barrier-Up Indicating System
- e: Barrier Up Indicator System
BUIT f: Bureau de l'Union Internationale des Télécommunication
BUK d: Bukarest (Flughafen)
BUKO d: Buskoppler
BUL d: Bundes- und Landstraßen
- e: Below Upper Limit
- e: Bombay University Library
- e: Brown University Library
- e: Bulletin
- f: Bibliothèque de l'Université Laval
BULD e: Base for Uniform Language Definition
BULG d: Bulgarien
- e: Bulgaria[n]
BULKIO e: Bulk Input/Output
Bull e: Bulletin
Bull Acad Sci e: Bulletin of the Academy of Sciences (journ.)
Bull. Acad. Sci. USSR; Div. Ch e: Bulletin of the Academy of Sciences of the USSR; Division of Chemical Science (journ.)
Bull. Acad. Sci. USSR; Geol. S e: Bulletin of the Academy of Sciences of the USSR; Geologic Series (journ.)
Bull. Acad. Sci. USSR; Phys. S e: Bulletin of the Academy of Sciences of the USSR; Physical Sciences (journ.)
Bull. Amer. Math. Soc. e: Bulletin of the American Mathematical Society (journ.)
Bull. Amer. Meteorol. Soc. e: Bulletin of the American Meteorological Society (journ.)
Bull Am Phys Soc e: Bulletin of the American Physical Society (journ.)
Bull Arm Astron Soc e: Bulletin of the American Astronomical Society (journ.)
Bull Astron Inst Neth e: Bulletin of the Astronomical Institutes of the Netherlands (journ.)
Bull. At. Sci. e: Bulletin of the Atomic Scientists (journ.)
Bull. Aust. Math. Soc. e: Bulletin of the Australian Mathematical Society (journ.)
BULLCAT f: Bull Calcul Automatique des Temps
Bull. Chem. Soc. Belgr. e: Bulletin of the Chemical Society; Belgrade (journ.)
Bull Chem Soc Jp e: Bulletin of the Chemical Society of Japan (journ.)
Bull. Environ. Contam. Toxicol e: Bulletin of Environmental Contamination and Toxicology (journ.)
Bull. Exp. Biol. Med. e: Bulletin of Experimental Biology and Medicine (journ.)
Bull. Ga. Acad. Sci. e: Bulletin of the Georgia Academy of Science (journ.)
Bull. Geisinger Med. Cent. e: Bulletin of the Geisinger Medical Center (journ.)
bulli. l: bulliat
Bull. Indian Soc. Soil Sci. e: Bulletin of the Indian Society of Soil Science (journ.)
Bull. Inst. Eng. e: Bulletin of the Institution of Engineers (India) (journ.)
Bull. Jap. Soc. Mech. Engrs. e: Bulletin of the Japan Society of Mechanical Engineers (journ.)
Bull. Los Angeles Neurol. Soc. e: Bulletin of the Los Angeles Neurological Society (journ.)
Bull. Math. Biophys. e: Bulletin of Mathematical Biophysics (journ.)
Bull. Med. Libr. Ass. e: Bulletin of the Medical Library Association (journ.)
Bull. N.Y. Acad. Med. e: Bulletin of the New York Academy of Medicine (journ.)
Bull NYZS e: Bulletin of the New York Zoological Society (journ.)
Bull. Parenter. Drug Assoc. e: Bulletin of the Parenteral Drug Association (journ.)
BULLPRF e: Bulletproof
BULLRAC e: Bull Random Access
BULLRAC Memory e: Bull Random Access Memory
Bulls e: Bulletins
Bull Seismol Soc Am e: Bulletin of the Seismological Society of America (journ.)
Bull. Torrey Bot. Club e: Bulletin of the Torrey Botanical Club (journ.)
Bull. Wld. Hlth. Org. e: Bulletin of the World Health Organization (journ.)
BULOGA e: Business Logistics Game
BUL Timber e: Bulletwood Timber
BUM e: Break-Up Missile
BuMa d: Buchungsmaschine
BüMa d: Büromaschine
BuMed e: Bureau of Medicine and Surgery
BUMP e: Back-Up Module Program[me]
- e: Boston University Marine Program[me]
- e: Bottom-up Modular Programming
BUMS e: Bachelor of Urani Medicine and Surgery
Bu M & S e: Bureau of Medicine and Surgery
BUN e: Blood Urea Nitrogen
BUNA d: Butadien und Natrium
- e: Butadiene and Natrium
BuNr d: Buchungsnummer
BUNSENITE e: Nickel Oxide
BUOC e: British Union Oil Company
BUP e: Back-Up Plate
- e: Back-up Processor
- e: Bend Up
- e: Boston University Press
BUPP e: Back-Up Plate; Perforated
BUPS e: Beacon; Ultra Portable S-Band
BUPX e: Beacon; Ultra-Portable X-Band
BUR d: Burma (Kfz)
- e: 5-Bromouridine
- e: Backup Register
- e: Basic Utility Routine
- e: Burbank; Calif
- e: Burried
Buran e: first Soviet space shuttle
B Urb Pl e: Bachelor of Urban Planning
BURD e: Biplane Ultralight Research Device
Burk Fas e: Burkina Faso
Burm. e: Burma
BURP e: Backup Rate of Pitch
BURS e: Basic Unformatted Real System
BURSTDS e: Bureau of Standards
BUS d: Batterieumschalter
- d: Omnibus
- e: Back-up Store
- e: Bank of the United States
- e: Basic Utility System
- e: Bushel[s]
BÜS d: Bedienerübungssystem
- d: Blutübertragungsstation
- d: Brückenübersetzstelle
BUSAC e: Bureau of Ships Analog Computer
BUSAK e: Bus Acknowledgement
BUSCI e: British-United States Convoy Instructions
BUSE e: Boston University Studies in English (journ.)
BUSEN e: Bus Enable
BUS ERR CY e: Bus Error Cycle
BUSH e: Bushing
BUSHIPS e: Bureau of Ships
BUSM e: Boston University School of Medicine
BUSRA e: British-United States Routing Agreement
BUSR[E]Q e: Bus Request
BUSS e: Biomedical Urine Sampling System
- e: Bristol University Spelaeological Society
- e: Buoy Underwater Sound Signal
- e: Bus Unit Scan Station
BU STD e: Bureau of Standards
BUSWREC e: Ban Unsafe Schoolbuses which Regularly Endanger Children
BÜSYS d: Büchereisystem
BUT e: Basic Unit Training
- e: Butter
- e: Button
BUTE e: Butazolidin (Phenylbutazone)
BUTEC e: British Underwater Test and Evaluation Centre
BUTN e: Butane
BUTSUP. e: Battery Supply
BUV e: Backscatter Ultraviolet
BUVO d: Betriebsunfallvorschrift
BUVS e: Backscatter Ultraviolet Spectrometer
BUV Spectrometer e: Backscatter Ultraviolet Spectrometer
BUW P: Biblioteka Uniwersyrecka w Warszawie
BUWEPS e: Bureau of Weapons
BUWEPSFLEREADREP e: Bureau of Naval Weapons Fleet Readiness Representative
BUWEPSFLEREADREPCEN e: Bureau of Naval Weapons Fleet Readiness Representative; Central
BUWEPSFLEREADREPLANT e: Bureau of Naval Weapons Fleet Readiness Repesentative; Atlantic
BUWEPSFLEREADREPPAC e: Bureau of Naval Weapons Fleet Readiness Representative; Pacific
BUWEPS FR e: Bureau of Naval Weapons; Fleet Readiness
BUWEPSINST e: Bureau of Naval Weapons Instruction
BUWEPSREP e: Bureau of Naval Weapons Representative
BUWEPSRESREP e: Bureau of Naval Weapons Resident Representative
BUZ d: Buchner Zahl
- d: Buchungszettel
- e: Buzzer
BÜZ d: Betriebsführungs- und -überwachungszentrale
BUZZ e: Buzzer

BV d: Bandverstärker
- d: Bauverbot
- d: Bauvorschrift[en]
- d: Bedienungsvorschrift
- d: Belegverarbeitung
- d: Beschaltungsverhältnis
- d: Bestandsveränderung
- d: Besteckversetzung
- d: Betonverflüssiger
- d: Bildsendervorstufe
- d: Bildverstärker
- d: Brenngasventil
- d: Buchveröffentlichung
- e: Back View
- e: Balanced Voltage
- e: Bellows Valve
- e: Bible Version
- e: Binary Value
- e: Biologic[al] Value
- e: Blood Vessel
- e: Blood Volume
- e: Blow Valve
- e: Bonnet Valve
- e: Book Value
- e: Boolean Variable
- e: Boom Vessel
- e: Breakdown Voltage
- e: Breviary
- e: Brick Veneer
- e: Bronchovesicular
- e: Busy Verification
- e: Butterfly Valve
- e: By Volume
- f: Bureau Veritas
- n: bijvoorbeeld
B + V e: Blohm und Voss
b.v. l: balneum vaporis
BV 202 N: Norwegian Army armored personnel carrier
BVA d: bemanntes Verstärkeramt
- e: Bachelor of Vocational Adjustment
- e: Bachelor of Vocational Agriculture
- e: British Veterinary Association
- e: British Videogram Assocation
BVB d: Bundesverband der Büromaschinen-Import- und Vertriebsunternehmen
- d: VEB Kombinat Berliner Verkehrsbetriebe (DDR)
BVBRF e: Blood Vessel of Bronchial Filament
BVC e: Black Varnish Cambric
- e: British Veterinary Codex
- e: Buena Vista College
BVD d: Brandverhütungsdienst
- e: Beacon Video Digitizer
- e: Bonus Vacation Days
- e: Bovine Viral Diarrhea
BVDT e: Brief Vestibular Disorientation Test
BVE d: Befehlsvorbereitungseinheit
- d: Bohrvorschubeinheit
- d: Busverstärkereinheit
- e: Bachelor of Vocational Education
- e: Butyl Vinyl Ether
BVE didtribution e: Bivariate Exponential distribution
BVerfGE d: Bundesverfassungs-gerichtsentscheid
BVerwG d: Bundesverwaltungsgericht
- d: Bundesverwaltungsgesetz
BVES e: British Voluntaly Euthanasia Society
BVetMed e: Bachelor of Veterinary Medicine
BVetSci e: Bachelor of Veterinary Science
BVetSur e: Bachelor of Veterinary Surgery
BVF d: Bauvorschriften für Flugzeuge
BVFK d: Bildverstärkerfernsehkette
BVG d: Berliner Verkehrs-Betriebe
- d: Berliner Verkehrsgesellschaft
- d: Bonner Verkehrsgesellschaft
- d: Bremer Verkehrsgesellschaft
BVGE e: Beverage
BVH d: Bauvorhaben
- e: Biventricular Hypertrophy
BVI d: Bundesversuchsinstitut
- e: Better Vision Institute
- e: Blood Vessel Invasion
- e: British Virgin Islands
BVJ e: British Veterinary Journal (journ.)
BVK d: B-Vitaminkomplex
- d: Bestands- und Verfügbarkeitskontrolle
- d: VEB Baustoffversorgungskombinat (DDR)
BVLS e: Battery-Voltage Limit System
BVM d: Bodenverbesserungsmittel
- d: Bundesverkehrsminister[ium]
- e: Bachelor of Veterinary Medicine
- e: Blessed Virgin Mary
- e: Broncho Vascular Markings
BVMA e: British Valve Manufacturers Association
BVMS e: Bachelor of Veterinary Medicine and Surgery
BVN d: Breitbandverteil[er]netz
- d: Breitbandvorläufernetz
BVNP e: Bolusan Volcano National Park
BVO d: Berechnungsverordnung
- d: betriebliches Vorschlagswesen
- e: Brominated Vegetable Oil
BVOR e: Broadcast Very-High Frequency Omni-Range
BVP d: Bundesverband für Personenbeförderung e.V.
- e: Beacon Video Processor
- e: Blood Vessel of Palp
- e: Blood Vessel of Pinnule
- e: Blood Vessel Prostheses
- e: Blood Volume Pulses
- e: Booster Vacuum Pump
- e: Boundary Value Problem
- e: British Volunteer Programme
- n: Bonhoffer-Van der Pol
BVP Equation e: Bonhoffer-Van der Pol Equation
BVPS e: Beacon Video Processing System
- e: Booster Vacuum Pump System
BVQ d: Betriebsstoffverbrauchsquote
BVR d: Begrenzerverstärker
- e: Balance[d] Valve Regulator
- e: Black Void Reactor
- e: British Vehicle Registration
- e: Broadcasting Video Recording
BVRO e: Base Vehicle Reporting Officer
BVRS e: Breadboard Visual Reference System
BVS d: Bandverwaltungssystem
- d: Bauvorschriften für Segelflugzeuge
- d: Bergwerkschaftliche Versuchsstrecke
- d: Bibliotheks-Verbund-System
- d: Bildverarbeitungssystem
- d: Blockverteilersteuerung
- d: Block vorsetzen
- e: Bachelor of Veterinary Science
- e: Bachelor of Veterinary Surgery
- e: Buoyant Venus Station
BVSAW d: Berichte über die Verhandlungen der Sächsischen Akademie der Wissenschaften zu Leipzig
BVSc e: Bachelor of Veterinary Science
BVSc & AH e: Bachelor of Veterinary Science and Animal Husbandry
BVSt d: B-Vermittlungsstelle
BVT e: Brevet[ted]
- e: British Videotext and Teletext
BV-Telegrafierapparat d: Baudot-Verdan-Telegrafierapparat
BVU d: betriebliche Verfahrensunter-suchung
BVV d: Behandlungsverfahren-Verordnung
BVW d: Behördenvorwähler
- d: Betriebliches Vorschlagwesen
- d: Bundesverkehrswacht
- e: Backward Volume Wave
- e: Binary Voltage Weigher
BVWP d: Bundesverkehrswegeplanung
BW d: Bahnbetriebswerk
- d: Bandwagen
- d: Batteriewiderstand
- d: Baumwolle
- d: Baumwollwachsdraht
- d: Bauwagen
- d: Bauweise
- d: Bauwesen
- d: Bauwirtschaft (journ.)
- d: Befehlswerk
- d: Befehlswort
- d: Bergwacht
- d: Bezugswert
- d: Bild der Wissenschaft (journ.)
- d: Bildwandler
- d: biologische Wertigkeit
- d: Blindwatt
- d: Bogenweiche
- d: Brustwirbel
- e: Backward Wave
- e: Bacteriological Warfare
- e: Bacteriological Warhead
- e: Bandwidth
- e: Basal Web
- e: Basic Weapons
- e: Beam Width
- e: Bell Wire
- e: Below Water
- e: Bendix-Westinghouse
- e: Best Welsh (coal)
- e: Biggs Wall
- e: Biological Warfare
- e: Biological Weapon
- e: Birmingham Wire
- e: Birth Weight
- e: Bits per Word
- e: Bit Word
- e: Biweekly
- e: Black Watch
- e: Blood Wasserman
- e: Body Water
- e: Body Weight
- e: Body Wing
- e: Bonded Warehouse
- e: Bonded Winery
- e: Both Ways
- e: Botswana
- e: Bovine Vaginitis Virus
- e: Braided Wire
- e: Brainware
- e: Brass Washer
- e: British Waterways
- e: Buried Wire
- e: Business Week (journ.)
- e: Butt-Weld[ed]
- e: By Weight
- n: bijwoord
BWA d: betriebwirtschaftliche Auswertung
- d: Brustwandabteilung
- e: Backward-Wave Amplifier
- e: Bent Wire Antenna
- e: British Waterworks Association

BWAR e: Budget Workload Analysis Report
BWAR Program[me] e: Board Wire Assignment and Routing Program[me]
BwAst d: Bahnbetriebswerk-Außenstelle
BWaStG d: Bundes-Wasserstraßengesetz
BWB d: Bundesamt für Wehrtechnik und Beschaffung
- e: British Waterways Board
BWBR e: Bureau of Naval Weapon Branch Representative
BWC e: Backward Wave Converter
- e: Basic Weight Calculator
- e: Basic Weight Controller
- e: Biological Weapons Convention
- e: Bonded Wine Cellar
- e: British War Cabinet
- e: Broadband Waveguide Circulator
- e: Buffer Word Counter
B/WCA e: Bomb to Warhead Conversion Components
BWCA e: Boundary Water Canoe Area
BWCC e: British Weed Control Conference
BW Coal e: Best Welsh Coal
BWCP e: Base Wire Communications Program[me]
- e: Bench Welder Control Panel
BW/CW e: Biological Warfare / Chemical Warfare
BWD d: Brustwirbeldornfortsatz
- e: Bacillary White Diarrhoea
- e: Backward
- e: Baldwin Wallace College
- e: Barrel[s] of Water per Day
- e: Bulk Wet Density
BWE d: Brustwandelektro[kardio]gramm
- e: Bachelor of Welding Engineering
BWEA d: Bundeswehrersatzamt
BWF e: Bailett Weighting Function
- e: Breakwater Fort
- e: Burst Wave Form
- e: Butt Welded Filter
BWFachS d: Bundeswehrfachschule
BWG e: Birmingham Wire Gage
- e: Birmingham Wire Gauge
- e: British Wire Gauge
bwgl. d: beweglich
BWH d: biologischer Wirkungsgrad
- d: biologischer Wirkungsherd
- e: Barrel[s] of Water per Hour
- e: Blower Wheel Housing
BWHB e: Black and White Horizontal Bands
BWHS e: Black and White Horizontal Stripes
BWI e: Branch Workload Index
- e: British West Indies
- e: British Workmens Institute
BWIA e: Better Walk if Able
BWIP e: Basalt Waste Isolation Project
BWJ e: Butt Weled Joint
BWK d: Brennstoff-Wärme-Kraft (journ.)
- d: Brustwirbelkörper
- d: Bund der Wasser- und Kulturbauingenieure
- e: Batch Weighting Kit
- e: Brillouin-Wentzel-Kramers
- e: Bulwark
BWKw d: Bezirkswerkstatt der Postkraftwagen
BWL d: Betriebswirtschaftslehre
- e: Belt Work Line
- e: Biological War Laboratory
BW LT e: Bow Light
BWM d: Bundeswirtschaftsminister[ium]
- e: Backward Wave Magnetron
- e: Barrel[s] of Water per Minute
BWMS e: British Wireless Marine Service
BWN e: Basic Weather Network
- e: Brown
BWO e: Backward Wave Oscillator
- e: Baser Work Order
- e: Broadband Waveguide Oscillator
BWOC e: Big Woman on Campus
BWOS e: Backward Wave Oscillator Synchronizer
BWOT e: Backward Wave Oscillator Tube
BWP e: Ballistic Wind Plotter
- e: Barrier Waterproof
- e: Basic War Plan
- e: Boiling Water Proof
BWPA e: Backward Wave Parametric Amplifier
- e: Backward Wave Power Amplifier
- e: British Waste Paper Association
- e: British Wood Preserving Association
BWPD e: Barrel[s] [of] Water per Day
BWPH e: Barrel[s] [of] Water per Hour
BWPUC e: British Waste Paper Utilization Council
BWR d: Boden-Wetter-Radaranlage
- d: Bremswegrechner
- d: Brückenwechselrichter
- e: Bandwidth Ratio
- e: Biweekly Report
- e: Boiling Water Reactor
- e: Boiling Water Resistant
BWRA e: British Water Research Association
- e: British Welding Research Association
BWRC e: Biological Warfare Research Center
BWRE e: Biological Warfare Research Establishment
BW Reaktion d: Bordet-Wassermann Reaktion
BWR-Equation e: Benedict-Webb-Rubin Equation
BWR-FLECHT e: Boiling Water Reactor - Full Length Emergency Cooling Heat Transfer Test
BWRS e: Biological Warfare Rapid Warning System
BWRWS e: Biological Warfare Rapid Warning System
BWS d: Bausteinwarteschlange
- d: Bayerischer Wohnungs- und Siedlungsbau
- d: Bewertungsschema
- d: Bildwiederholspeicher
- d: Brustwirbelsäule
- d: VEB Büromaschinenwerk Sömmerda (DDR)
- e: Bandipur Wildlife Sanctuary
- e: Batch Weighting System
- e: Battlefield Weapons System
- e: Beaufort Wind Scale
- e: Bevelled Wood Siding
- e: Biological Weapon[s] System
- e: British Watercolour Society
BWSL e: Battlefield Weapons Systems Laboratory
BW & SM e: Bread Wrapping and Sealing Machine
BWSO e: Backward Wave Sweep Oscillator
B.W.Sp. d: Betriebswasserspiegel
BWSTRN e: Bowstring
BWSV d: Bundes-Wasser- und Schiffahrtsverwaltung
BWSW d: Bundeswehrsozialwerk
BWT e: Backward-Wave Tube
- e: Boeing Wind Tunnel
- e: Bohr-Wheeler Theory
- e: Both Way Trunk
- e: British Winter Time
BWTF e: Bank Wire Transfer of Funds
BWTR e: Babcock and Wilcox Test Reactor
BWTSDS e: Bare Wire and Telephone System Development Schedule
BW TV e: Black-and-White Television
B & W TVs e: Black-and-White Television Set
BWU e: Blue Whale Unit
BWV d: Bundesverband Werkverkehr
- e: Backwater Valve
BWVO d: Binnenwasserstraßen-verkehrsordnung
BWVS d: Black and White Vertical Stripes
BWW d: Bahnbetriebswagen[werk]
- d: beabsichtigter Wegwinkel
- e: Bad Weather Watch
BWWs e: Biological Warfare Weapons
BWWV e: Bayerischer Wasserwirtschaftsverband
BWY e: Blast Wave Yield
BWYV e: Beet Western Yellow Virus
BWZ d: Bodenwertzahl
- d: Brandwarnzentrale
- d: Bruttowertzuwachs
- d: Bundesminister[ium] für wirtschaftliche Zusammenarbeit
- e: Blank When Zero
BWZSt d: Bodenwertzuwachssteuer
BX e: Base Exchange
- e: Base Indexed
- e: Box
- e: Branch Exchange
- e: British Xylonite
- e: Electrical Cable Contained In Flexible Tubing
BX-C e: Bithorax Complex
BX Cable e: Insulated Wires Within Flexible Tubing
BxDC d: Butoxyethyl Diglycol
BXF e: Box Fin
BxK e: Box Keel
BXK e: Broadband X-Band Klystron
BXL e: Bakelite Xylonite Limited
BXM e: Box Express Messenger
BXS e: Boxes
BY e: Battery
- e: Billion Years
- e: Biopsy
- e: Budget Year
- e: Buoy
- e: Busy
BY C e: Battery Commander
BYDA e: Busy Don't Answer
BYDV e: Barley Yellow Dwarf Virus
BYE e: Byelorussia[n]
BYF e: Battery Fire
BYK e: Broadband Y-Band Klystron
BYMS e: British Yard Motor Minesweepers
BYMUX e: Bytemultiplex[er]
BYMUX Channel e: Byte Multiplex Channel
BYNT e: Buoyant
BYP d: Bypass
- e: Bypass
BYPCAP e: Bypass Capacitor
BYPCOND e: Bypass Condenser
BYPRO[S] e: By-Product[s]
BYPS d: Bytes pro Sekunde
BYTE e: Eight Adjacent Binary Digits that a Computer Processs as a Unit
BYTECY e: Byte Cycle
BYU e: Brigham Young University
BYZ e: Byzantine

BZ d: Bearbeitungszentrum
- d: Bedienungszeit
- d: Befehlszähler
- d: Befehlszählregister
- d: Begriffszusammenhang
- d: Belegungszeichen
- d: Benzene
- d: Benzol
- d: Besetztzeichen
- d: besondere Zusatzbestimmungen
- d: Betriebsstundenzähler
- d: Betriebszentrale
- d: Bezeichnung
- d: Biblische Zeitschrift (journ.)
- d: Binärzähler
- d: Blechschraube mit Zapfen
- d: Blutungszeit
- d: Blutzucker
- d: Bodenzahl
- d: Brennstoffzelle
- d: Brennzünder
- d: Bronze
- e: Beaten Zone
- e: Benzaldehyde
- e: Benzene
- e: Benzodiazepine
- e: Benzoyl
- e: Benzyl
- e: Blank when Zero
- e: Brazil[ian]
- e: Brillouin Zone
- e: Buzzer
- e: Buzzing

BZA d: Bandzugautomatik
- d: Biologische Zentralanstalt
- d: Bombenzielapparat
- d: Bundesbahn-Zentralamt

BZA e: Benzoic Acid
BZAC e: Brazing Accessory
BZALC e: Benzyl Alcohol
BZB d: Bord-zu-Bord
- d: Bundesamt für zivilen Bevölkerungsschutz
- d: Burghagens Zeitschrift für Bürotechnik und Informatik (journ.)

BzBf d: Bezirksbauführer
BzBlG d: Benzinbleigesetz
BzBT d: Bezirksbeamter für Telegrafentechnik
BzBTn/TV d: Bezirksbeamter für Ton- und Fernsehübertragungen (TV = Television)
BzBÜ d: Bezirksbeamter für Übertragungstechnik
BZD d: Benzidine
BzDA Acid e: Benzene-Dioxane-Acetic Acid
BZE d: Bord-zu-Erde
- e: Belize

BZF d: Bremszylinderfett
BZFX e: Brazing Fixture
BZG d: Bombenzielgerät
- d: Bundeszentralgenossenschaft Straßenverkehr

BZGA[K] d: Baseler Zeitschrift für Geschichte und Altertumskunde (journ.)
- e: Benzaldehyde

BZH e: Benzaldehyde
BZK d: Bandzugkraft
BZK d: Bezirkskabel
BZL d: Benzol
- d: Bezirkslinie
- e: Benzyl

BzMA Acid e: Benzene-Methanol-Acetic Acid
Bzn. d: Benzin
BzPrüfpl Fu d: Bezirksprüfplatz für Funkgeräte
BZR d: Befehlszählregister
- d: Bytezähler

BZRG d: Bundeszentralregistergesetz
B & ZS e: Binary & Zero Substitution
BZ Sc e: Bachelor of Zoological Science
BZT d: Bündelzieltaste
- e: Benzethonium Chloride

BZTS d: Bonner Zeitschrift für Theologie und Seelsorge (journ.)
BZV d: Brazzaville (Flughafen)
bzw. d: beziehungsweise
BZW e: Bare Zirconium Wire
BZ-Zellulose d: Benzylzellulose

C

C d: Carbon
- e: Cellatron (DDR)
- d: Charakteristik
- d: Chemisches Zentralblatt (journ.)
- d: Cuba (Kfz)
- d: Cytosin
- d: [elektrotechnische] Kapazität
- d: Karbonylnickel
- d: Kohlenstoff[zahl]
- d: Kollektor[anschluß]
- d: Kompression
- d: Konzentration
- e: Cable
- e: Caesar
- e: Calculate[d]
- e: Calculated Weight
- e: Call
- e: Calm
- e: Calorie
- e: Camp
- e: Can
- e: Canadian
- e: Cancel[ed]
- e: Candella
- e: Candle
- e: Canon
- e: Capacitance
- e: Capacitor
- e: Capacity
- e: Capillary Blood
- e: Captain
- e: Car
- e: Carat
- e: Carbohydrate[s]
- e: Carbon[eum]
- e: Cargo [or Transport Airplane]
- e: Cargo Vessel
- e: Carriage
- e: Carribean
- e: Carrier
- e: Carry
- e: Carton
- e: Case
- e: Cash
- e: Cast
- e: Castle
- e: Catalog[ue]
- e: Catechism
- e: Cathode
- e: Catholic
- e: Caudal
- e: Caught
- e: Cause
- e: Celestial
- e: Cell
- e: Celsius
- e: Celtic
- e: Cent
- e: Center
- e: Centi[-]
- e: Centigrade
- e: Centigram[me]
- e: Centimeter
- e: Centimetre
- e: Central[ized]
- e: Centre
- e: Century
- e: Certified
- e: Cervical
- e: Cervix
- e: Chairman
- e: Chamaephyt
- e: Chancellor
- e: Change
- e: Channel
- e: Chapter
- e: Character
- e: Charge
- e: Chemistry
- e: Chest
- e: Chief
- e: Child
- e: Chipping
- e: Chirp
- e: Chord Length
- e: Christ
- e: Chrominance
- e: Chronometer
- e: Church
- e: Cipher
- e: Circa
- e: Circle
- e: Circling
- e: Circuit
- e: Circular
- e: Circumference
- e: Cirrus
- e: Citation
- e: City
- e: Civil
- e: Class
- e: Clear[ance]
- e: Clock[wise]
- e: Clonus
- e: Closed
- e: Closure
- e: Cloud[s]
- e: Cloudy
- e: Coarse
- e: Coast[er]
- e: Cobalt
- e: Cocaine
- e: Code
- e: Coefficient
- e: Coil
- e: Cold
- e: Collected
- e: Collector
- e: College
- e: Colo[u]r[ed]
- e: Colo[u]rless
- e: Colt
- e: Column
- e: Combat [Aircraft]
- e: Command[ant]
- e: Commander
- e: Commanding
- e: Common
- e: Compare
- e: Compass
- e: Complement
- e: Complex
- e: Compliance
- e: Compound[ed]
- e: Compression
- e: Comptroller
- e: Compute
- e: Compute[r]
- e: Computer Programming Language
- e: Computing
- e: Concentrate[d]
- e: Concentration
- e: Concentration of a Gas in Blood
- e: Condenser
- e: Conductance
- e: Conductivity
- e: Conductor
- e: Confidential
- e: Congregational
- e: Connect
- e: Conservative
- e: Constable
- e: Constant
- e: Contact
- e: Container
- e: Continent[al]
- e: Contra
- e: Contract[ion]
- e: Contrast
- e: Control[ler]
- e: Controlling
- e: Convection
- e: Convergence
- e: Copper
- e: Copy[right]
- e: Core
- e: Correct[ion]
- e: Corrosive
- e: Cortex
- e: Cost
- e: Costa Line
- e: Cotton
- e: Coulomb
- e: Council

- e: Count[er]
- e: County
- e: Coupon
- e: Course
- e: Court
- e: Cranial
- e: Created
- e: Creation
- e: Creative
- e: Creosote
- e: Cross
- e: Crown
- e: Cruiser
- e: Crystal[line]
- e: Cube
- e: Cubic[al]
- e: Curies constant
- e: Curie[s]
- e: Currency
- e: Current
- e: Curve
- e: Cutter
- e: Cycle[s]
- e: Cylinder[s]
- e: Cylindric[al]
- e: Cytidine
- e: Cytisine
- e: Cytochrome
- e: Cytolysine
- e: Fraunhofer line characteristic of hydrogen
- e: Heat Capacity per Mole
- e: Hundredweight
- e: Molecular Heat
- e: Single-Cotton[covered]

c d: chromatiert
- d: Karat
- d: Lichtgeschwindigkeit
- d: Zenti (10^{-2})
- e: Speed of Light
- f: Cap
- f: Col
- l: centum
- l: congius
- s: Cabo

C' e: Bacteriologic Complement

C- e: C-minus
- e: C negative

C+ e: C-plus
- e: C positive

C ++ e: Computer Programming Language

c/ s: cargo
- s: contra

C_1 e: First Class

C-10 e: Decamethonium

C-118 e: Passenger Transport Aircraft

C-123 e: Provider Twin-Engine Assault Transport

C-124 e: Transport Airplane

C-130 e: Military Transport Airliners

C-131 e: Military Transport Airliners

C-133 e: Transport Airplane

C-135 e: Boeing Stratofreighter

C-140 e: Transport Airplane

C-141 e: Transport Airplane

C 19ster e: Steroidscontaining 19 Carbon Atoms

C l, C 2, C 3, etc. e: Cervical Nerves or Vertebrae 1, 2, 3, etc.

C^2 e: Command and Control

C_2 e: Second Class

C 21 ster e: Steroids Containing 21 Carbon Atoms

C^2D^2 e: Combat Development Command Division

C^2L e: Closed Complementary [Symmetry] Metal Oxide Semiconductor Logic

C^2MOS e: Clocked Complementary [-symmetry] Metal Oxide Semiconductor

C^2R e: Charge Control Ring
- e: Charge Coupled Ring

C^2RAM e: Charge-Coupled Random Access Memory

C^2S e: Collector-Coupled Structure

C^3 e: Command, Control and Communication
- e: Computer and Communications in Compact Size

C3 e: Curriculum Committee on Computer

C-3 e: Mentally or Physically Defective

C_3 e: Command, Control, Communications
- e: Load Capacitance
- e: third class

C.3.3. e: cell 3, 3rd landing, gallery

C^3AMS e: Computer-Controlled Container Automatic Marking System

C^3/CM e: Command Control and Communications / Counter-Measures

C3D e: Cascade Charge Coupled Device

C3EM e: Curriculum Commitee on Computer Education for Management

C^3L e: Complementary Constant Current Logic

C^3 Laser e: Cleaved Coupled Cavity Laser

C^3Ls e: Complementary Constant Current Logics

C-3PL e: Cytocomputer Parallel Picture Processing Language

C^3RAM e: Charge-Coupled Random Access Memory
- e: Continuously Charge-Coupled Random Access Memory

C^3RA Memory e: Continuously Charge-Coupled Random Access Memory

C^3RAMs e: Continuously Charge-Coupled Random Access Memories

C^3S e: College Chemistry Consultants Service

C_4 e: Command and Control, Communications and Computers

C-4 e: Controlled Collapse Chip Connection

C4 e: Convair 440 airplane
- e: Crown Quarto [7-½ x 10 inches]

C-4 e: Military Explosive

C-42 e: Utility Aircraft

C-45 e: Transport Plane

C-46 e: Curtiss-Wright World War II Commando 36-passenger transport

C-47 e: Air Transport

C4D e: Conductively-Connected Charge-Coupled Device

C4I e: Command, Control, Communication, Computers and Intelligence

C 4 S d: Chondroitin-4-Sulfat

C-54 e: Passenger Transport Aircraft

C-5A e: Lockheed Military Cargo Transport Airplane

C-6 e: Hexamethonium

C 6 S d: Chondroitin-6-Sulfat

C-9 e: Twinengine Jetliner Designed For Medical Evacuation

C-95 e: Passenger Transport Aircraft

CA d: Capital Account
- d: Cellulose Acetate
- e: Calcareous
- e: Calibrated Altitude
- e: Call Accepted
- e: Call Address
- e: Cambium
- e: Canada
- e: Cancel
- e: Cancelled
- e: Cancer
- e: Cancrinite
- e: Candida Albicans
- e: Candle
- e: Capacitor
- e: Capillary Agglutination
- e: Carbonic Anhydrase
- e: Carcinoma
- e: Card Alignment
- e: Card Application
- e: Card Assembler
- e: Cardiac Arrest
- e: Cariolis Absorber
- e: Carribean Area
- e: Cart Away
- e: Carthographic Assistant
- e: Cascade Amplifier
- e: Case-Annealed
- e: Case[s]
- e: Catecholamine[s]
- e: Cathode
- e: Cellular Automation
- e: Celluloseacetat
- e: Center Angle
- e: Center for Astrophysics
- e: Central America
- e: Central Angle
- e: Cephacetril
- e: Cephalic Artery
- e: Certificate of Airworthiness
- e: Certification Authority
- e: Certified Acupuncturist
- e: Cervioaxial
- e: Chain Address
- e: Chained Addressing
- e: Chaining Address
- e: Chamber of Agriculture
- e: Channel Adapter
- e: Channel Address
- e: Character Alignment
- e: Character Assignment
- e: Check Addition
- e: Chemical Abstracts (journ.)
- e: Chief Advisor
- e: Chief Auditor
- e: Chlormadinone Acetate
- e: Choline Adrenalin
- e: Chronological Age
- e: Circa
- e: Circuit Algebra
- e: Circuit Analysis
- e: Circuitry Adapter
- e: Civil Aviation
- e: Cleaning Arm
- e: Clear and Add
- e: Clear Aperture
- e: Clerical Aptitude
- e: Clipping Amplifier
- e: Close Annealed
- e: Coast Artillery
- e: Coated Abrasive
- e: Coaxial
- e: Cocaine Anonymous
- e: Coded Address
- e: Code of Arms
- e: Cold Acclimated
- e: Cold Adapted
- e: Cold Agglutinin
- e: Cold Air
- e: Combat Aircrew(man)
- e: Command Accountant
- e: Command Address
- e: Commercial Air
- e: Commercial Aviation
- e: Common Antigen
- e: Communications Adapter
- e: Commutator Assembly
- e: Companies Act
- e: Compensation Act
- e: Complement Add

- e: Complex Angle
- e: Compressed Air
- e: Computer Access
- e: Computer Analysis
- e: Computer and Automation
- e: Computer Application
- e: Computer Assembly
- e: Computer Associates
- e: Computer Automation [Limited]
- e: Computing Amplifier
- e: Computing Automaton
- e: Concrete Additive
- e: Concrete Admixture
- e: Concrete Aggregate
- e: Conditional Addition
- e: Conditional Assembly
- e: Configuration Assignment
- e: Connect Acknowledge
- e: Consonant Amplification
- e: Constant Attenuation
- e: Construction Authorization
- e: Contact Ammeter
- e: Content Addressable
- e: Content Addressable [Memory]
- e: Continuation Address
- e: Continued Addition
- e: Control Accelerometer
- e: Control Area
- e: Control Assembly
- e: Controlled Approach
- e: Controlled Atmosphere
- e: Controller of Aircraft
- e: Conversion Angle
- e: Coronary Artery
- e: Correct Algorithm
- e: Corrective Action
- e: Correlation Analysis
- e: Cortisone Acetate
- e: Council Accepted
- e: Council Agenda
- e: Count Attribute
- e: Counterpoise Aerial
- e: Counterpoise Antenna
- e: County Attorney
- e: Cover Aft
- e: Coverage Analysis
- e: Cranial Academy
- e: Crank Angle
- e: Criticality Analysis
- e: Croup Associated
- e: Cruise Altitude
- e: Crushed Aggregate
- e: Current Address
- e: Current Assets
- e: Curvilinear Accelerator
- e: Cylinder Address
- f: Centre Agricole
- p: Comganhia de Navegação Carregudores Açoreanos

Ca d: Calcium
- d: Cauchy-Zahl
- e: Cable
- e: Calcium

ca d: cirka
- i: coll'arco
- l: circa
- p: corrente alternada
- s: corriente alterna

C & A e: Classification and Audit
- e: Computers and Automation (journ.)

C/A e: Command/Address

C-A e: C to A Gate

CAA e: 7-Cyanacetyl- Aminocephalosporanic Acid
- e: Caging Amplifier Assembly
- e: Cambridge Computer Associates
- e: Canadian Automobile Association
- e: Caribbean Atlantic Airlines
- e: Central African Airways
- e: Chief of Army Aviation
- e: Circular Aperture Aerial
- e: Circular Aperture Antenna
- e: Civil Aeronautics Administration
- e: Clean Air Act
- e: Computer-Aided Administration
- e: Computer-Aided Assembling
- e: Computer-Aided Assembly
- e: Computer-Aided Automation
- e: Computer-Assisted Accounting
- e: Computer Amplifier Alarm
- e: Computers and Automation
- e: Consecutive Access Addresses
- e: Cooperative Alumni Association
- e: Crime Aboard Aircraft
- e: Cyanoacrylate Adhesive

CAAA e: California Agricultural Aircraft Association
- e: Coast Artillery Anti-Aircraft

CAAAL e: Classified Abstract Archive of the Alcohol Literature

CAAB e: California Asparagus Advisory Board

CAAC e: China Aeronautics Airline Corporation
- e: College Admissions Assistance Center
- e: Combat Alert Aircrew

CAAD e: Computer-Aided Application Design
- e: Computer-Aided Architectural Design

CAADRP e: Civil Aircraft Airworthiness Data Recording Program[me]

CAAF e: Campbell Army Airfield

CAAFS e: Chinese Academy of Agricultural and Forestry Sciences

CAAIS e: Computer-Aided Action Information System[s]

CAAM e: Civil Aeronautics Administration Manual
- e: Civil Aeronautics Authority Manual

CAAN e: Continental Advertising Agency Network

CAAOR e: Combined Antiaircraft Operations Room

CAAR e: Compressed Air Accumulator Rocket

CAARC e: Commonwealth Advisory Aeronautical Research Council

CAAS e: Ceylon Association for the Advancement of Science
- e: Combined Arms and Support
- e: Computer-Aided Approach Sequencing
- e: Computer-Aided Approach Spacing
- e: Computer-Assisted Acquisition System
- e: Computer-Assisted Approach Sequencing
- e: Computer-Assisted Approach System
- e: Computerized Alarm Analysis System
- e: Connecticut Academy of Arts and Sciences

CAASE e: Computer-Assisted Area Source Emissions

CAAS Scheme e: Computer-Assisted Approach Sequencing Scheme

CAAT e: Canadian Academic Aptitude Test

CAATC e: Civil Aeronautics Administration-Type-Certificate

C & AATC e: Coastal and Anti-Aircraft Training Centre

CA Att e: Civil Air Attaché

CAAV e: Central Association of Agricultural Valuers

CAAW e: Customer Authorization for Additional Work

CAAWEX e: Canned Anti-Air Warfare Exercise

CAB d: Celluloseacetobutyrat
- e: Cabbage
- e: Cabin[et]
- e: Cable
- e: Cabletelevision Advertising Bureau
- e: Cabochon
- e: Cabriolet
- e: Calibration
- e: Canadian Association of Broadcasters
- e: Captive Air Bubble
- e: Captured Air Bubble
- e: Cellulose Acetate Butyrate
- e: Cement Asbestos Board
- e: Change Analysis Board
- e: Channel Address Buffer
- e: Channel Address Bus
- e: Circulation Audit Board
- e: Civil Aeronautics Board
- e: Civil Aeronautics Bulletin
- e: Collating and Binding
- e: Commonwealth Agricultural Bureau
- e: Computer-Aided Building
- e: Computer Address Bus
- e: Controlled-Atmosphere Brazing
- e: Controlled Amortization Bond
- e: Cooperative Analysis of Broadcasting
- e: Coronary Artery Bypass
- e: taxicab
- f: Calculator Arithmétique Binaire

CABA e: Connecticut Artificial Breeding Association

CABAS e: City and Borough Architects Society
- e: Computerized Automated Blood Analysis System

CABB e: Captured Air-Bubble Boat

CABD e: Computer-Aided Building Design

CABEI e: Central American Bank for Economic Integration

CABEX e: Computer Automatic Branch Exchange

CABG e: Coronary Artery Bypass Grafting

CABI e: Commonwealth Agricultural Bureau International

CABLE e: Computer-Assisted Bay-Area Law Enforcement

CABLECAST e: Broadcast by Cable Television
- e: Cablecaster
- e: Cablecasting

CABLECAST[ING] e: Cable Television Telecast[ing]

CABLESE e: Cablegram Language

CABLE TV e: Community-antenna Television

CABLEVISION e: Cable Television

CABM e: Commonwealth of Australia Bureau of Meteorology

CABM[S] e: Chinese-Oriented Antiballistic Missile System

CABP e: Calcium-Binding Protein

Cabr. e: Cabriolet

CABR e: Central Address Buffer Register

CABRA e: Copper and Brass Research Association

CABS e: Cabbages

CABS e: Childrens Adaptive Behavior Scale
- e: Computer-Aided Batch Scheduling
- e: Computer-Augmented Block System
- e: Computer-Automated Block System
- e: Computerized Annotated Bibliographic System
- e: Computerized Annotated Bibliography System
- e: Current Awareness in Biological Sciences

CABT e: Cesium Atomic Beam Tube
CABTMKR e: Cabinetmaker
CABWK e: Cabinetwork
CAC e: Cable Access Cover
- e: Calibration and Certification
- e: Calibration and Checkout
- e: California Aeronautics Commission
- e: Canterbury Agricultural College
- e: Carbon-Arc Cutting
- e: Cardiac-Accelerator Center
- e: Caribbean Air Command
- e: Center for Advanced Computation
- e: Changing to Approach Control
- e: Chemical Abstracts Condensates (journ.)
- e: Chief of Air Corps
- e: Clear All Channels
- e: College Admission Center
- e: Combat Air Crew
- e: Command Address Change
- e: Command and Control
- e: Commander Air Center
- e: Committee on the Application of Computers (in the Construction Industry)
- e: Communication Accessories Company
- e: Communication and Control
- e: Complete Address Constant
- e: Computer-Aided Classification
- e: Computer-Aided Crime
- e: Computer-Assisted Counseling
- e: Computer Acceleration Control
- e: Computer Adaptive Control
- e: Computers in Aerospace Conference
- e: Congressional Agricultural Committee
- e: Constant Alert Cycle
- e: Contact Approach Control
- e: Control and Coordination
- e: Cooperation and Coordination
- e: County Agricultural Committee
- e: Currency Adjustment Charge
- e: Current Actions Center
- e: Cyclic Address Change
CACA e: Canadian Agricultural Chemicals Association
- e: Central After-Care Association
- e: Central and Associated Chambers of Agriculture
- e: Citizens Association for the Care of Animals
- e: Computer-Aided Circuit Analysis
- e: Continuous Accumulation of Coriolis Acceleration
CA-Cancer J. Clin. e: CA-Cancer Journal for Clinicians (journ.)
CACB e: Compressed-Air Circuit Breaker
CACC e: Cathodal Closure Contraction
- e: Civil Aviation Communications Center
- e: Clutter Automatic Gain Control
- e: Coded Automatic Gain Control
CACCI e: Committee in the Application of Computers in the Construction Industry
CACD e: Compromise Approach to Compiler Design
- e: Computer-Aided Circuit [Analysis and] Design
- e: Computer-Assisted Circuit Design
-e: Computer Automated Cargo Documentation
CACDP e: California Association of County Data Processors
CACE e: Chicago Association of Consulting Engineers
- e: Computer-Aided Chemical Engineering
CACEE e: Computer Aid[s] for Chemical Engineering Education
CACH e: Communications Adapter Channel
CACHE e: Computer-controlled Automated Cargo Handling Envelope
- e: Computer Aid[s] for Chemical Education
CACI e: Consolidated Analysis Centers, Incorporated
CACIC e: Chicago Area Computer Information Centers
CACIP e: Central American Cooperative Corn Improvement Project
CACM e: Communications of the Association for Computing Machinery (journ.)
CAC Net e: Control and Coordination Net
CACO e: Casualty Assistance Call Officer
C/A Code e: Course/Acquisition Code
CACOM e: Chief Aircraft Communicator
CACON e: Cargo Container
- e: Chemical Abstracts Condensates
$CaCO_3$ e: Calcium Carbonate
CACP e: Cartridge-Actuated Compaction Press
- e: Corrosion and Cathodic Protection
CACR e: Clean Air Car Race
- e: Council for Agricultural and Chemurgic Research
CACS e: California Aqueduct Control System
- e: Centralized Alarm and Control System
- e: Comprehensive Airport Communications System
- e: Computer-Aided Communication[s] System
- e: Computer-Assisted Communication[s] System
- e: Content Addressable Computing System
- e: Continental Airways and Communications Service
- e: Controller Active State
- e: Core Auxiliary Cooling System
CACSD e: Computer-Aided Control System Design
CACSP e: Coated Aluminium Coated Steel; Polyethylene
CACT e: Civil Air Carrier Turbojet
- e: Command Automatic Card Tester
CACTIS e: Computer Assistant to a Community Telephone Information Service
CACTOS e: Computation and Communication Trade-Off Studies
CACTUS f: Capteur Accelerometrique Capacirif Triaxial UItra-Sensible
CACU e: Coast Artillery Cooperation Unit
CACUL e: Canadian Association of College and University Libraries
CACVE e: California Advisory Council on Vocational Education
CAC & W e: Continental Aircraft Control and Warning
CACW e: Continental Aircraft Control and Warning
CACWS e: Core Auxiliary Cooling Water System
CAD e: Cabling Diagram
- e: Cadaver
- e: Cadence
- e: Cadet
- e: Cadmium
- e: Calculator-Aided Design
- e: Call Acceptance Delay
- e: Cartridge-Activated Device
- e: Cartridge-Actuated Device
- e: Center Area Discrete
- e: Central Aircraft Dispatch
- e: Central Ammunition Depot
- e: Channel Area Difference
- e: Character Assemble/Disassemble
- e: Character Assembler and Distributor
- e: Circulatory Assist Device
- e: Clear Add
- e: Command Address
- e: Communications Access Device
- e: Communications and Data
- e: Compensated Avalanche Diode
- e: Computer-Aided Design
- e: Computer-Aided Detection
- e: Computer-Aided Diagnosis
- e: Computer-Aided Disaster
- e: Computer-Aided Dispatch
- e: Computer-Aided Drafting
- e: Computer-Assisted Design
- e: Computer-Assisted Dialog[ue]
- e: Computer-Automated Design
- e: Computer Access Device
- e: Computer Analysis Department
- e: Computer Application Digest
- e: Continuous Acceleration Device
- e: Contract Award Date
- e: Control and Display
- e: Controlled Avalanche Diode
- e: Coronary Artery Disease
- e: Corrective Action Directive
- e: Corrective Action Disposition
- e: Counter and Decoder
- f: Conversion Analogique-Digitale
CADA e: Clear Air Dot Angle
- e: Computer-Aided Design or Analysis
- e: Computer-Assisted Distribution and Assignment
- f: Centre d'Analyse Documentaire pour l'Archélogie
CADAC e: Clean Arithmetic with Decimal-base and Controlled-precision
CADACS e: Computer-Aided Design and Construction System
CADAD e: Computer-Aided Design and Application of Devices
CADAE e: Computer-Aided Design and Engineering
CADAFE s: Compañía Anónima de Administración y Fomento Electrico
CADAI e: Center Apollo Documentation Administration Instructions
CADAM - e: Computer-Aided Design and Manufacturing
- e: Computer-Augmented Design and Manufacturing
- e: Computer-graphic[s] Augmented Design and Manufacturing
CADAM System e: Computer-graphics Augmented Design and Manufacturing System
CADAN f: Centre d'Analyse Documentaire pour Afrique Noir
CADANCE[s] e: Computer-Aided Design and Numerical Control Effort[s]
CADAPSO e: Canadian Association of Data Processing Service Organizations
CADAR e: Computer-Aided Design, Analysis and Realizability
- e: Computer-Aided Design, Analysis and Reliability
CADAS e: Computerized Automatic Data Acquisition System
CADAVRS e: Computer-Assisted Dial Access Video Retrieval System
CADC e: Cambridge Automatic Digital Computer
- e: Central Air Data Computer
- e: Computer-Aided Design Center
- e: Corrective Action Data Center
CAD/CAM e: Computer-Aided Design and Computer-Aided Manufacture Manufacturing)

CAD/CAT d: Computer-Aided Design and Computer-Aided Test
CADC/CC e: Central Air Data Computer/Central Computer
CAD Centre e: Computer-Aided Design Centre
CADCO e: Core and Drum Corrector
CADCOM e: Computer-Aided Design for Communications
CADCON e: Computer-Aided Design Conference
CADD e: Computer-Aided Database Design
- e: Computer-Aided Design [and] Drafting
- e: Computer-Aided Drafting and Design
CADDA e: Combined Analog[ue]-Digital Differential Analyzer
CADDAC e: Central Analog[ue] Data Distributing and Computing [System]
- e: Central Analog[ue] Data Distributing and Controlling [System]
- e: Central Analog[ue] Data Distribution and Computing
- e: Central Analog[ue] Data Distribution and Controlling
CADDAC System e: Central Analog[ue] Data Distribution and Computing System
- e: Central Analog[ue] Data Distribution and Controlling System
CAD/D-CAM e: Computer-Aided Design, Drafting and Computer-Aided Manufacturing
CADDE e: Central Automatic Digital Data Encoder
CADDS e: Center Apollo Document Description Standards
- e: Computervision Automatic Design and Draughting System
CADE e: Computer-Aided Data Entry
- e: Computer-Aided Design
- e: Computer-Aided Design [and] Engineering
- e: Computer-Aided Design Evaluation
- e: Computer-Assisted Data Entry
- e: Computer-Assisted Data Evaluation
- e: Computer-Assisted Data Engineering
CADEM e: Computer-Aided Design Engineering and Manufacturing
- e: Computer-graphics Augmented Design and Manufacturing
CADEP e: Computer-Aided Design of Electronic Products
- e: Computer-Assisted Description of Patterns
CADES e: Computer-Aided Development and Evaluation System
- e: Computer-Assisted Data Entry System
CADE System e: Computer-Assisted Data Entry System
CADET e: Computer-Aided Design and Electrical Test
- e: Computer-Aided Design Engineering Technique
- e: Computer-Aided Design Experimental Translation
- e: Computer-Aided Design Experimental Translator
- e: Covariance Analysis Describing-function Technique
CADETS System e: Classroom-Aided Dynamic Educational Time-Sharing System
CADEX e: Computer-Aided Design Exhibition
CADF e: Cathode-Ray Tube Automatic Direction Finding (or Finder)
- e: Commutated Aerial Direction Finder
- e: Commutated Antenna Direction Finder
- e: Contract Administrative Data File
CADFISS e: Computation and Data Flow Integrated Subsystem[s] [Test]
CADFISS Test e: Computation and Data Flow Integrated Subsystem Test
CADI e: Central Apollo Data Index
CAdI e: Computer-Administred Instruction
CAD-I e: Computer Access Device-Input
CADIAP e: Comment and Data Integration and Printing
CADIC e: Chemical Analysis Detection Instrumentation Control
- e: Computer-Aided Design of Integrated Circuits
CADICAS e: Computer-Aided Design of Integrated Circuits and Systems
CADICS e: Computer-Aided Design of Industrial Cabling System
CADIS e: Computer-Aided Design Interactive System
- e: Computer-Aided Design of Information System[s]
- e: Coronary Artery Disease
CADISIM e: Computer-Assisted Disposal Simulation
CA Dist e: Coast Artillery District
CADI System e: Computer-Aided Design Interactive System
CADIZ e: Canadian Air Defence Identification Zone
- e: Coastal Air Defense Identification Zone
CADL e: Communications and Data-Link
CADLIC e: Computer-Aided Design of Linear Integrated Circuits
CADM e: Center Apollo Data Manager
- e: Continental United States Air Defense Modernization
CADMAC e: Computer-Assisted Drawing Machine
CADMAT e: Computer-Aided Design, Manufacture and Testing
CADMINI e: Computer Administrative Instruction
CADMOS d: computerorientiertes Administrations- und Organisationssystem für Schulen
CADMP e: Computer-Aided Data Management Procedure
CADNC Tape e: Computer-Aided Design Numerical Control Tape
CADNIP e: Cambridge Atmospheric Density Numerical Integration Program[me]
CADO e: Computer-Aided Document Origination
CADOCR e: Computer-Aided Design of Optical Character Recognition
CADOCS e: Carrier Aircraft Data Operations Control System
CADOP e: Continental Air Defense Objectives Plan
CADOPCART e: Computer-Aided Design of Printed Circuit Artworks
CADPIN e: Customs Automated Data Processing Intelligence Network
- e: Customs Automatic Data Processing Intelligence Network
CADPL e: Communications/Automatic Data Processing Laboratory
CADPO e: Communications and Data Processing Operation
CADPOS e: Communication[s] and Data Processing Operation System
CADR e: Coast Artillery Drill Regulations
CADRC e: Combined Air Documents Research Center
CADRE e: Canadian Armament Research and Development Establishment
- e: Common Applied Data Research Environment
- e: Current Awareness and Document Retrieval for Engineers
CADS e: Cellular Absorbed Dose Spectrometer
- e: Central Air Data Subsystem
- e: Central Air Data System
- e: Computer-Aided Design System
- e: Computer-Aided Diagnosis System
- e: Computer-Aided Digitizing System
- e: Computer Analysis and Design System
- e: Continental Air Defense System
- e: Control, Acessibility, Disposal, Storage
- e: Conversational Analyzer and Drafting System
CADSI e: Communications and Data Systems Integration
CADSS e: Combined Analog[ue]-Digital System[s] Simulator
- e: Communications and Data Subsystem
CADSYS e: Computer-Aided Design System
CADTe e: Cathodal Duration Tetanus
CAD Unit e: Control and Display Unit
CADW e: Civil Air Defense Warning
CADW System e: Civil Air Defense Warning System
CAE e: Caelum
- e: Canadian Aviation Electronics
- e: Carribean Air Express
- e: Carrier Aircraft Equipment
- e: Chartered Automobile Engineer
- e: Chief Activation Engineer
- e: Chief Administrative Engineer
- e: Chief of Aeronautical Engineering
- e: Circular Average Error
- e: Commission for Aerology
- e: Communications and Electronics
- e: Computer-Aided Education
- e: Computer-Aided Engineering
- e: Computer-Assisted Education
- e: Computer-Assisted Electrocardiography
- e: Computer-Assisted Engineering
- e: Computer-Assisted Enrollment
- e: Computer-Assisted Entry
- e: Computer-Assisted Estimating
- e: Computer-Augmented Education
- e: Council on Anthropology and Education
- f: Calculatrice Arithmétique Electronique
- f: Compagnie d`Automatisme Européenne
CAEA e: California Aviation Education Association
CA-Earl e: Computer Associates Easy Access Report Language
CAEC e: Canadian Atomic Energy Commission
- e: County Agricultural Executive Committee
CAEDM e: Community/Airport Economic Development Model
CAEDS e: Computer-Aided Engineering Design System
CAED System e: Computer-Aided Engineering Design System
CAEDTA e: Calcium Disodium Ethylene Diamine Tetra-Acetate
CAEE e: Coast and Antiaircraft Experimental Establishment
CAEI e: Canadian Aeronautical Institute
CAEL e: Caelum
- e: Consolidated Aerospace Equipment List
- e: Council for the Advancement of Experimental Learning

CAEM e: Cargo Airline Evaluation Model
- e: Certified Assistant Export Manager
- e: Commission for Aeronautical Meteorology

CAE/MIS e: Computer-Assisted Estimating and Management Information
CAEN e: Chemical and Engineering News (journ.)
CAENEX e: Complex Atmospheric Energetics Experiment
CAER e: Chief Aerographer
CAERM e: Chief Aerographers Mate
Caes. e: Caesar
CAES e: Canadian Agricultural Economics Society
- e: Compressed-Air Energy Storage
- e: Connecticut Agricultural Experiment Station

CAESAR e: Computerized Automation by Electronic System with Automated Reservations
CAESQ e: Communications and Electronics Squadron
CAEU e: Casualty Air Evacuation Unit
CAF d: Celluloseacetatfilm
- d: Celluloseacetatfolie
- e: Caffeine
- e: Cardiac Assessment Factor
- e: Centralized Authorization File
- e: Ceylon Air Force
- e: Channel Address Field
- e: Clean Assembly Facility
- e: Cloth Assistance Factor
- e: Complete Assembly for Ferry
- e: Computer-Assisted Fraud
- e: Conjunctive Alterations File
- e: Cost, Assurance and Freight
- e: Cost and Freight
- e: Critical Area Flag
- e: Currency Adjustment Factor
- e: Currency Adjustment Fee
- e: Current Approved File

CAFAF e: Commander; Amphibious Force Atlantic Fleet
CAFB e: Charleston Air Force Base
- e: Chemically Active Fluidized Bed
- e: Clark Air Force Base
- e: Columbus Air Force Base
- e: Cooke Air Force Base

CAFC e: Computed Automatic Frequency Control
CAFD e: Contact Analog[ue] Flight Display
CAFE d: Celluloseacetatfolien- Elektrophorese
- e: Computer-Aided Film Editor
- e: Corporate Average Fuel Economy

CAFEE e: Critical Assembly Fuel Element Exchange
CAFEI e: Central American Fund for Economic Integration
CAF-Elektrophorese d: Celluloseacetatfolien-Elektrophorese
CAFF e: Caffeine
- e: Computer-Aided Fault Finding

CAFGA e: Computer Applications for the Graphic Arts
CAFGU e: Citizens Armed Forces Geographical Unit
CAFIC e: Combined Allied Forces Information Center
CAFIT e: Computer-Assisted Fault Isolation Test[ing]
CAFM e: Coast Artillery Field Manual
- e: Commercial Air Freight Movement

CAFMNA e: Compound Animal Feedingstuffs Manufacturers National Association
CAFS e: Cardiac Assessment Factors
- e: Cartridge-Actuated Flame System
- e: Content Addressable File Store
- e: Content Addressed File System
- e: Content Addressed Film System

CAFSC e: Control Air Force Specialty Code
CAFSU e: Catapult and Arresting Gear Field Service Unit
CAFTA e: Central American Free Trade Area
- e: Council of Australian Food Technology Associations

CAFTN e: Commonwealth Air Forces Telecommunications Network
CAG e: Cancer Assessment Group
- e: Capacitance Anode/Grid
- e: Chronic Atrophic Gastritis
- e: Computer-Assisted Graphics
- e: Computer-Assisted Guidance
- e: Computer Analysis Group
- e: Computer Applications Group
- e: Constant Aerial Glide
- e: Constant Altitude Glide
- e: Controller and Auditor General
- e: Cytosine, Adenine And Guanine

CAGA e: California Asparagus Growers Association
CAGC e: Coded Automatic Gain Control
- e: Continuous Access Guided Communication

CAGD[s] e: Computer-Aided Geometric Design[s]
CAGE e: California Almond Growers Exchange
- e: Canadian Air-Ground Environment
- e: Compiler and Assembler by General Electric
- e: Computerized Aerospace Ground Equipment
- e: Convicts Association for a Good Environment

CAGEL e: Consolidated Aerospace Ground Equipment List
CAGI e: Compressed Air and Gas Institute
CAGM e: Commission for Agricultural Meteorology
CAGMS e: Computer-Aided Geometry Modeling Symposium
CAGP e: Canadian Arctic Gas Pipeline
CAG RA e: Commander; Army Group; Royal Artillery
CAG RE e: Commander; Army Group; Royal Engineers
CAGS e: Canadian Arctic Gas Study
- e: Certificate of Advanced Graduate Study
- e: Civil Anti-Gas School

CAG Sigs e: Commander; Army Group; Signals
CAH d: Cyanacethydrazid
- e: Congential Adrenal Hyperplasia
- e: Cyanacetic Hydrazide

CAHD e: Coronary Atherosclerotic Heart Disease
CAHE e: Core Auxiliary Heat Exchanger
CAHI e: Central Aero-Hydrodynamical Institute
Ca-Horizont d: Kalkanreicherungshorizont
CAH Range e: Continent Antwerp Hamburg Range
CAHS e: Comprehensive Automation of the Hydrometeorological Service[s]
CAHT e: Computer Aids for Human Translations
CAI d: computerassistierte Industrie
- e: Cairo (Airport)
- e: Canadian Aeronautical Institute
- e: Canadian Airlines International
- e: Channel Available Interruption
- e: Close Approach Indicator
- e: Coded Acoustic Interrogator
- e: Compressed Air Institute
- e: Computer-Administered Instruction
- e: Computer-Aided Industry
- e: Computer-Aided Information
- e: Computer-Aided Instruction
- e: Computer-Assisted Instruction [Incorporated]
- e: Computer-Augmented Instruction
- e: Computer Analog[ue] Input
- e: Computer Applications, Incorporated
- e: Computer Automation, Incorporated
- e: Conditional Assembler (or Assembly) Instruction
- e: Configuration Acceptance Inspection
- e: Configuration Audit Inspection
- e: Confused Artificial Insemination
- e: Conjunctive Alteration Indicator
- e: Control and Acquisition Interface
- e: Control and Indication
- e: Control and Instrumentation
- e: Current Awareness Information
- e: Gonville and Caius College, Cambridge

CAIB d: Computer-Anwendung Industrie-Beratung GmbH
- e: Certified Associate of the Institute of Bankers

CAIC e: Computer-Aided Indexing and Categorizing
- e: Computer-Aided Indexing and Classification
- e: Computer-Assisted Indexing and Categorizing
- e: Computer-Assisted Indexing and Classification
- e: Computer-Assisted Instruction Center

CAI/CMI e: Computer-Assisted Instruction and Computer-Man Instruction
CAI Col e: Gonville and Caius College-Cambridge
CAIDO e: Chief Advisor; International District Office
CAIFO e: Chief Advisor; International Field Office
CAIL e: Coal and Allied Industries Limited
- e: Computer-Aided Information Logistics

CAIM e: Computer-Aided Instruction Management
- e: Computer-Assisted Instruction Management

CAIMAW e: Canadian Association of Industrial, Mechanical and Allied Workers
CAIMS e: Conventional Ammunition Integrated Management System
CAIN e: Calculation of Interia
- e: Cataloging and Indexing [System]
- e: Cleveland Area Interlibrary Network

CAINS e: Carrier Aircraft Inertial Navigation System
- e: Computer-Aided Instruction System
- e: Computer-Assisted Instruction System

CAIN System e: Cataloging and Indexing System
CAINT e: Computer-Aided Interrogation
- e: Computer-Assisted Interrogation
- e: Counter-Air and Interdiction

CAI/OP e: Computer Analog[ue] Input/Output
CAIP e: Computer-Aided Index[ing] Program[me]
- e: Computer-Assisted Index[ing] Program[me]

CAIR e: Cost Analysis Information Report
- e: Countermeasures Airborne Infra-Red

CAIRA e: Central Automated Inventory and Referral Activity
- e: Central Automated Inventory and Referral System
CAIRC e: Caribbean Air Command
CAIRS e: Central Automated Inventory and Referral System
- e: Computer-Aided Information Retrieval System
- e: Computer-Assisted Information Retrieval System
- e: Computer-Assisted Interactive Resources Scheduling [System]
CAIRS System e: Computer-Assisted Interactive Resources Scheduling System
CAIS e: Canadian Association for Information Science
- e: Central Abstracting and Indexing Service
- e: Communications and Information Systems
- e: Computer-Aided Instructions
- e: Computer-Aided Insurance System
- e: Computer-Assisted Information Service
CaiSe e: Cairo Studies in English (journ.)
CAISIM e: Computer-Aided Industrial Simulation
- e: Computer-Assisted Industrial Simulation
CAISMS e: Computer-Assisted Instruction Study Management System
CAISP e: Computer-Assisted Individualized Study Package
CAISYS e: Computer-Aided Instruction System
CAITR e: Computer Assistance for International Team Research
CAITS e: Chemical Agent Identification Training Set
CAI/VR e: Computer-Assisted Instruction with Voice Response
CAJ e: Calked Joint
- e: Central Asiatic Journal
- e: College Art Journal
CA-JASPER e: Computer Associates Job Accounting System for Performance Evaluation and Report
CAJE e: Consolidated Antijam Equipment
CAK e: Ciphertext Auto Key
- e: Command Acknowledge
- e: Conical Alignment Kit
- e: Cube Alignment Kit
CAKE e: Computer-Assisted Keyboard Evaluator
- e: Computer-Assisted Key Entry
cal d: Kalorie
CAL d: VEB Chemieanlagenbau Leipzig (DDR)
- e: Calcareous
- e: Calculate[d]
- e: Calculated Average Life
- e: Calculator
- e: Calendar
- e: Caliber
- e: Calibrate[d]
- e: Calibration
- e: Calibre
- e: Calorie
- e: Center for Applied Linguistics
- e: Checking Automation Language
- e: China Airlines
- e: Command Authorization List
- e: Common Assembler Language
- e: Common Assembly Language
- e: Commonwealth Acoustic Laboratories
- e: Component Action List
- e: Composite Aeronautical Load List[ing]
- e: Compressed Air Loudspeaker
- e: Computer-Aided Laboratory
- e: Computer-Aided Layout
- e: Computer-Aided Learning
- e: Computer-Assisted Learning
- e: Computer Accounting Limited
- e: Computer Animation Language
- e: Concentrated Ammonia Liquor
- e: Conservation Analytical Laboratory
- e: Continuity Accept Limit
- e: Continuous Address List
- e: Contractor Attention List
- e: Conversational Algebraic Language
- e: Cornell Aeronautic[al] Laboratory
- e: Course Author Language
- e: Cycle Allocation
- e: Cyprus Airways
CALA e: Chinese-American Librarians Association
- e: Civil Aviation Licensing Act
- e: Computer-Aided Loads Analysis
- s: Cruabozo
CALAS e: Calibrated Air Speed
- e: Computer-Aided Laboratory Automation System
- e: Computer-Aided Language Analysis System
- e: Computer-Assisted Language Analysis System
CALB e: Computer-Aided Line Balancing
- e: Computer-Assisted Line Balancing
CALBR e: Calibration
CALC e: Calcareous
- e: Calculate[d]
- e: Calculating
- e: Calculation
- e: Calculator
- e: Calculus
- e: Customer Access Line Charge
CALCALT e: Calculated Altitude
CALCD e: Calculated
CAL/CERT e: Calibration/Certification
CALCG e: Calculating
CALCHEM e: Computer-Assisted Learning in Chemistry
Calcif. Tissue Res. e: Calcified Tissue Reerch (journ.)
CALCN e: Calculation
CALCOFI e: California Cooperative Oceanic Fisheries Investigations
CALCOMP e: California Computer Products
CalComp e: California Computer Products, Incorporated
Calc Univ e: Calcutta University
CALCUS e: Centre for Advanced Land Use Studies
CALD e: Calculated
- e: Computer-Assisted Logic Design
Cal Data Procs e: California Data Processors
CALDIC e: California Digital Computer
CALE e: Canadian Army Liaison Executive
Caled. Can. e: Caledonian Canal
CALEN e: Calendar
- e: Calender
CALEV s: Compañía Anómina Luz Electrica de Venezuela
CALI e: Computer-Assisted Library Index
CALIB e: Calibrate
- e: Calibration
CALIBN e: Calibration
CALICHE e: Calcium Carbonate Coust [or] Dust-$CaCO_3$}
Calif. Agric. e: California Agriculture (journ.)
Calif Hist e: California Historical Society
Calif. Med. e: California Medicine (journ.)
CALIME e: Computer-Assisted Library Instruction for Mechanical Engineers
CALIMS e: Computer-Assisted Library and Information Management
CALINET e: California Information Network
CALIPS e: Calibrated Pressure Switch
CALIT e: California Institute of Technology
CALL e: Canadian Association of Law Libraries
- e: Carat Assemble[d] Logical Loader
- e: Community Access Library Line
- e: Composite Aeronautical Load List
- e: Computer-Aided Loft Lines
- e: Computer-Assisted Learning Laboratory
- e: Computerized Ambulance Location Logic
- e: Counselling at the Local Level
- e: Current Awareness - Library Literature
CALLA e: Common Acute Lymphoblastic Leukemia Antigen
CALLAS d: computerangeschlossenes Lernleitungs-Analysesystem
- e: Calla Lillies
Cal-ID e: California Identification System
CALLIG e: Calligrapher
- e: Calligraphic
- e: Calligraphy
CALLISTO e: Computer-Aided Learning in Library and Information Science Teaching Organization
CALM e: Catapult Arresting-gear and Landing-gear Maintenance
- e: Child Abuse Listening Meditation
- e: Collected Algorithms for Learning Machines
- e: Common Business Oriented Language Automatic Language Modifier
- e: Collected Algorithms for Learning Machines
- e: Computer-Aided Layout of Masks
- e: Computer-Aided Learning Model
- e: Computer-Aided Library Mechanization
- e: Computer-Assisted Library Mechanization
- e: Computer Archive of Language Materials
- e: Continuously Advancing Longwall Mining
CALMACQ e: Computer-Assisted Library Mechanization: Acquisition; Acquisition
CALMS e: Computer Automatic Line Monitoring System
- e: Continuous Automatic Line Monitoring System
CALN e: Computer-Assisted Learning Network
CALOGSIM e: Computer-Assisted Logistics Simulation
CALOMEL e: Mercurous Chloride (Hg_2Cl_2)
CALOUT e: Calibration Output
CAL OUT CF e: Calibrator Output Cathode Follower
CALP e: Computer Analysis of Library Postcards
CalPAW e: California Parks and Wildlife Initiative
CALPs e: Computer-Assisted Learning Packages
CALR e: Calculator
- e: Computer-Assisted Legal Research System
CALRI e: Central Artificial Leather Research Institute
CALRS e: Centralized Automatic Loop Reporting System
- e: Comprehensive Automated Learning Resources System

CALS e: Canadian Association of Library Schools
- e: Communications Area Local Station
- e: Comprehensive Automated Learning System
CalSpace e: California Space Institute
CALT e: Calculated Altitude
CALT e: Cleared Altitude
CALTEC e: California Institute of Technology
Cal Tech e: California Institute of Technology
CALTRANS e: California Department of Transportation
cal_{th} e: calorie (thermochemical calorie)
CALUTRON e: California University Cyclotron
Calv. e: Calvin[ism]
CALVDAC e: California Venereal Disease Advisory Council
CALYPSO e: Computer-Aided Layout of Precision Microelectronics
CAM e: Calculated Access Method
- e: Call Accepted Message
- e: Camber
- e: Cambodia[n]
- e: Camera
- e: Cameroon[s]
- e: Camouflage
- e: Camphor
- e: Care and Maintenance
- e: Cartographic Automatic Mapping
- e: Cascade Access Method
- e: Catapult[ed] Aircraft Merchantship
- e: Cell Adhesive Molecule
- e: Cellulose Acetate Membrane
- e: Central Address Memory
- e: Central America
- e: Character Address Module
- e: Checkout and Automatic Monitoring
- e: Checkout and Maintenance
- e: Chinese Access Method
- e: Chloramphenical
- e: Chorio-Allantoic Membrane
- e: Circular Area Method
- e: Civil Aeronautics Manual
- e: Civil Air Manual
- e: Civil Air Movement
- e: Civil Aviation Medicine
- e: Clear and Add Magnitude
- e: Coast Artillery Memorandum
- e: Cockpit Area Microphone
- e: Commercial Air Movement
- e: Committee for Aquatic Microbiology
- e: Communication[s] Access Manager
- e: Communication[s] Access Method
- e: Comprehensive Achievement Monitoring
- e: Computer-Addressed Memory
- e: Computer-Aided Management
- e: Computer-Aided Manufacture
- e: Computer-Aided Manufacturing
- e: Computer-Assisted Mailing
- e: Computer-Assisted Management
- e: Computer-Assisted Manufacture
- e: Computer-Assisted Manufacturing
- e: Computer Address Matrix
- e: Computer Applications to Manufacturing
- e: Computer Assessment of Media
- e: Computing Accounting Machine
- e: Computing Attachment Machine
- e: Consolidated Aircraft Maintenance
- e: Content-Addressable Memory
- e: Content-Addressed Memory
- e: Contents Addressable Memory
- e: Contingency Analysis Model
- e: Continuous Air Monitor
- e: Contract Audit Manual
- e: Contractor-Acquired Materiel
- e: Conventional Addressing Machine
- e: Correlatable Access Memory
- e: Crassulacean Acid Metabolism
- e: Cruise and Maintain
- e: Cybernetic Anthropomorphous Machine
C''_{am} e: Amylase Clearance
CAMA e: Central Air Materiel Area
- e: Centralized Automatic Message Accounting
- e: Civil Aerospace Medical Association
- e: Computer-Aided Mathematical Analysis
- e: Control and Automation Manufacturers Association
CAMA-C e: Centralized Automatic Message Accounting-Computerized
CAMAC e: Channel Allocation Monitor and Controller
- e: Computer-Aided Measurement and Control
- e: Computer-Aided Monitoring and Control
- e: Computer-Automated Measurement and Control
- e: Computer Application to Measurement and Control
CAM Accounting e: Centralized Automatic Message Accounting
CAMAE e: Central Air Materiel Area; Europe
CAMAL e: Cambridge Algebra [System]
- e: Continuous Airborne Missile Alert
CAMALS e: Cambridge Algebra System
CAMAL System e: Continuous Airborne Missile-launcher and Low-level System
CAMA-ONI e: Centralized Automatic Message Accounting-Operator Number Identification
CAMAR e: Common-Aperture Multifunction Array Radar
Camb. e: Cambrian
Camb J e: Cambridge Journal
Camb. Obs. e: Cambridge Observatory
Cambod. e: Cambodia[n]
Cambridge UP e: Cambridge University Press
CAMCORDER e: Camera [and] Recorder
- e: Camera and Tape Recorder Device
CAMCOS e: Computer-Assisted Maintenance Planning and Control System
CAMD e: Computer-Aided Mechanical Design
- e: Computer-Aided Molecular Design
- e: Computer-Assisted Molecular Design
CAMDA e: Car and Motorcycle Drivers Association
CAMDAP e: Computers and Medieval Data Processing
CAMDI e: Confidential Admiralty Merchant Ship Defense Instructions
CAMDS e: Chemical Agent Munitions Dispersal System
- e: Computer-Aided Medical Diagnostics System
CAMEL e: Collapsible Airborne Military Equipment Lifter
- e: Common Automatic Manifest Language
- e: Component and Material Evaluation Loop
- e: Computer-Aided Machine Loading
- e: Computer-Assisted Education Language
- e: Critical Aeronautical Material and Equipment List
CAMELOT e: Computerization and Mechanization of Local Office Tasks
CAMERA e: Cooperating Agency Method for Event Reporting and Analysis
CAMESA e: Canadian Military Electronics Standards Agency
CAMET e: Computer Archive of Modern English Text
CAMFLG e: Camouflage
CAMFORD e: Cambridge and Oxford
Cam High e: Camden High School
CAMI e: Civil Aeromedical Institute
- e: Coated Abrasive Manufacturers Institute
- e: Computer-Aided Manufacturing-International [Incorporated]
- e: Continuing Action Maintenance Instruction
CAMIL e: Computer-Assisted/Management Instructional Language
CAMIS e: Computer-Assisted Makeup and Imaging System
CAML e: Computer-Accessed Microfiche Library
CAMM d: Chlor-Alkali-Market Model
- e: Canadian Association of Medical Microbiologists
- e: Computer-Aided Manufacturing Management
CAMMIS e: Command Aircraft Maintenance Manpower Information System
CAMMS e: Computer-Assisted Map Manoeuvre System
CAMNET e: Computer-Aided Manufacturing Network
CAMOF e: Camouflage
CAMOL e: Computer-Assisted Management of Learning
CAMOS e: Computer-Aided Abrasive Machining Oscillation Studies
CAMP d: Computergraphics Anwendungen für Management und Produktivität
- e: Adenosine 3'5' Cyclic Monophosphate
- e: Cabin Air Manifold Pressure
- e: Central Access Monitor Program[me]
- e: Coastal Artillery Mine Planter
- e: Command and Management Presentation
- e: Communications Administrative Message Program[me]
- e: Community Air Monitoring Program[me]
- e: Compiler for Automatic Machine Programming
- e: Computer-Aided Maintenance Project
- e: Computer-Aided Mask Preparation
- e: Computer-Aided Mask Production
- e: Computer-Aided Mathematics Program[me]
- e: Computer-Aided Menu Planning
- e: Computer-Aided Movie Production
- e: Computer-Assisted Manual Preparation
- e: Computer-Assisted Mathematics Program[me]
- e: Computer-Assisted Menu Planning
- e: Computer-Assisted Movie Production
- e: Computer-graphics Applications for Management and Productivity
- e: Computer Applications for Military Problems
- e: Computer Assisted Menu Planning
- e: Computerized Aircraft Maintenance Program[me]
- e: Continuous Air Monitoring Program[me]
- e: Continuous Air Monitoring Project
- e: Controls and Monitoring Processor
- e: Cooperative Africana Microform Project
- e: Cyclic Adenosine Monophosphate
CAMPCOS e: Computer-Assisted Maintenance Planning and Control System
CAMPRAD e: Computer-Assisted Message Preparation Relay and Distribution

CAMPS e: Centralized Automated Military Pay System
- e: Computer-Assisted Message Processing System
- e: Cooperative Area Manpower Planning System

CAMPSA s: Compañia Arrendataria del Monopolio de Petroleos
CAMPUS e: Comprehensive Analytical Method of Planning in the University Sphere
- e: Comprehensive Analytical Methods for Planning in University Systems
- e: Computerized Analytical Methods in Planning University System

CAMR e: Camera
CAM/RAM e: Content Addressable Memory/Random Access Memory
CAMRAS e: Computer-Assisted Mapping and Records Activities Systems
CAMS e: Chinese Academy of Medical Sciences
- e: Coastal Anti-Missile System
- e: Communication, Advertising and Marketing Studies
- e: Communications Area Master Station
- e: Computer-Aided Methods
- e: Computer Activity Monitor System
- e: Computer Automated Mailing System
- e: Containerized Avionics Maintenance System
- e: Continuous Air-Monitoring System
- e: Control of Aircraft Maintenance and Servicing
- e: Crew and Administrative Management System
- e: Cybernetic Anthropomorphous Machine[s] [System]

CAMSAT e: Camera Satellite
CAMSI e: Canadian; American Merchant Shipping Instructions
- e: Confidential Admiralty Merchant Shipping Instructions

CAMSIM e: Computer-Assisted Maintenance Simulation
CAMSPEC e: Chemical Analysis by Microwave Spectroscopy
CAMSTA e: Camera Station
CAMUS d: Computergesteuertes Arbeits-Markt-Untersuchungs-System
- e: Captive Adjustable Must-Use System

CAN e: Canada
- e: Canal[ization]
- e: Canalize
- e: Canaveral (Cape)
- e: Canberra (airport)
- e: Cancel
- e: Cancel Character
- e: Cancellation
- e: Cancel[l]ed
- e: Cancer
- e: Canister
- e: Canonically

Can. e: Canon[ry]
- e: Citizens Against Noise
- e: Configuration Accounting Number
- e: Correlation Air Navigation
- e: Customs Assigned Number
- f: Calculator Analogique
- f: Convertisseur Analogique-Numérique

CANACOR e: Canadian Agroindustrial Corporation
CAN/ACT e: Canadian / Audio Cassette Tapes
Canad Geog J e: Canadian Geographical Journal (journ.)
Canad J of Ind e: Canadian Journal of Industry (journ.)
CANAIRDEF Command e: Canadian Air Force Defence Command
CanAirHed e: Canadian Air Force Headquarters
CANAIRMAT Command e: Canadian Air Force Material Command
CANAIRTAC Command e: Canadian Air Force Tactical Command
CANAIRTRAIN Command e: Canadian Air Force Training Command
CANAL e: Command Analysis
Canal Zone e: Panama Canal Zone
CANARI e: Communications and Navigation Airborne Radio Instrumentation
CANAS e: Canadian Naval Air Station
CANAVAT e: Canadian Naval Attaché
CANAVSTORES e: Canadian Naval Stores
CANBER e: Canada-Bermuda
CANBER Cable e: Canada Bermuda Cable
CANC e: Cancel[lation]
- e: Cancel[l]ed
- e: Cancelling
- e: Cancer

CANCEE e: Canadian National Committee for Earthquake Engineering
CANCEL e: Cancelling
- c: Connecticut Aircraft Nuclear Engine Laboratory

CANCER e: Computer Analysis of Nonlinear Circuits Excluding Radiation
Cancer Bull. e: Cancer Bulletin (journ.)
Cancer Chemother. Rep. e: Cancer Chemotherapy Reports (journ.)
CANCER LINE e: Cancer Online
CANCERPROJ e: Cancer Project
Can. Chem. Process. e: Canadian Chemical Processing (journ.)
CANCL e: Cancel
- e: Cancellation
- e: Cancelling

CANCON e: Canadian Control System
Can. Contr. Instrum. e: Canadian Controls and Instrumentation (journ.)
CAN/CRS e: Canadian/Computer-based Reference Service
CAND e: Candelabra
- e: Candidate

C and B e: Chemical and Biological
- e: Collating and Binding

C and C e: Coal and Coke
- e: Command and Control

C and D e: Chemist and Druggist
- e: Collected and Delivered
- e: Collection and Delivery

C and DSS e: Communications and Data Subsystem
CANDE e: Command and Edit
- e: Computer-Aided Network Design

C and E e: Commission and Exchange
CANDE Language e: Command and Edit Language
CANDEP e: Canadian Navy Depot
C and F e: Cost and Freight
C and H e: Cocaine and Heroin
- e: Cold and Hot

C and I e: Communications and Instrumentation
C and M e: Care and Maintenance
- e: Coal and Mining
- e: Cocaine and Morphine
- e: Construction and Machinery

C and N e: Communications and Navigation
- e: Control and Navigation

CANDO e: Computer Analysis of Networks with Design Orientation
CANDOC e: Canadian Electronic Document Ordering Service
CANDOFD e: Computer Analysis of Networks with Design Orientation in the Frequency Domain
C and P e: Carriage and Packing
- e: Collated and Perfect
- e: Cut and Pin

C and R e: Construction and Repair
C and R Sounding e: Cord and Rod Sounding
C and S e: Clean and Sober
- e: Common and Standard
- e: Costs and Schedules

CANDSC e: Candelabra Screw
C and T e: Classification and Testing
CANDU e: Canadian Deuterium Uranium
- e: Chelsea Against Nuclear Destruction United

CANDUR e: Canadian Deuterium-Uranium Reactor
CANDU Reactor e: Canadian Deuterium Uranium Reactor
CANE e: Cationic Asphalt-Neoprene Emulsion
- e: Chemical Applications of Nuclear Explosions
- e: Computer-Aided Navigation Equipment

CANEL e: Connecticut Advanced Nuclear Engineering Laboratory
- e: Connecticut Aircraft Nuclear Engine Laboratory

Can. Entomol. e: Canadian Entomologist (journ.)
CANE SUGAR e: Saccharose or Sucrose
CANFARM e: Canadian Farm Management (Data System)
Can. Field-Nat. e: Canadian Field-Naturalist (journ.)
Can-Fr e: Canadian French
CANFSWPAOPPLAN e: Combined Allied Naval Forces Southwest Pacific Ocean Area Operating Plan
Can. Georg. J. e: Canadian Geographic[al] Journal (journ.)
Can. Geotech[n]. J. e: Canadian Geotechnical Journal (journ.)
CANGO e: Committee for Air Navigation and Ground Organization
Can. H[ist]. R. e: Canadian Historical Review (journ.)
Can. Hosp. e: Canadian Hospital (journ.)
CANINE e: Computer Analysis of Networks via Inversion of Network Equations
Can. Inst. Food Technol. J. e: Canadian Institute of Food Technology Journal (journ.)
Can. I[s]. e: Canary Islands
Canis. e: Canister[s]
Can. J. Anim. Sci. e: Canadian Journal of Animal Science (journ.)
Can. J. Biochem. e: Canadian Journal of Biochemistry (journ.)
Can. J. Bot. e: Canadian Journal of Botany (journ.)
Can. J. Chem. e: Canadian Journal of Chemistry (journ.)
Can. J. Chem. Eng. e: Canadian Journal of Chemical Engineering (journ.)
Can. J. Earth Sci. e: Canadian Journal of Earth Sciences (journ.)
Can. J. Econ. e: Canadian Journal of Economics and Political Science (journ.)

Can. J. Forest Res. e: Canadian Journal of Forest Research (journ.)
Can. J. Genet. Cytol. e: Canadian Journal of Genetics and Cytology (journ.)
Can. J. Med. Technol. e: Canadian Journal of Medical Technology (journ.)
Can. J. Microbiol. e: Canadian Journal of Microbiology (journ.)
Can. J. Ophthalmol. e: Canadian Journal of Ophthalmology (journ.)
Can. J. Pharm. Sci. e: Canadian Journal of Pharmaceutical Sciences (journ.)
Can. J. Phys. e: Canadian Journal of Physics (journ.)
Can. J. Physiol. Pharmacol. e: Canadian Journal of Physiology and Pharmacology (journ.)
Can. J. Plant Sci. e: Canadian Journal of Plant Science (journ.)
Can. J. Public Health e: Canadian Journal of Public Health (journ.)
Can. J. Res. e: Canadian Journal of Research (journ.)
Can. J. Soil Sci e: Canadian Journal of Soil Science (journ.)
Can. J. Zoo. e: Canadian Journal of Zoology (journ.)
CANLANT e: Canadian Atlantic Subarea
Can. Lib. Bull. e: Canadian Library Bulletin (journ.)
Can. Libr. J. e: Canadian Library Journal (journ.)
Can. Mach. Metalwork. e: Canadian Machinery and Metalworking (journ.)
Can. maj. l: Canis major-Astronomical Constellation
CAN MARC e: Canadian Machine Readable Cataloguing
Can. Med. Assoc. J. e: Canadian Medical Association Journal (journ.)
CANMET e: Canadas Center for Mineral and Energy Technology
Can. Metall. Quart. e: Canadian Metallurgical Quarterly (journ.)
Can Met Ser e: Canadian Meteorological Service
Can. min. l: Canis minor - Astronomical Constellation
Can. Min. J. e: Canadian Mining Journal (journ.)
CANNA e: Cannabinoid
- e: Cannabinol
CANO e: Catalogue Number
CANOLE e: Canadian On-Line Enquiry
CAN/OLE System e: Canadian/On-Line Enquiry System
CANOPT e: Cascaded Network Optimization Package
CANOPT Package e: Cascaded Network Optimization Package
CANOT e: Computer Analysis of Troubles on Trunks
CANOTT e: Computer Analysis of Troubles on Trunks
Can Pac e: Canadian Pacific
Can. P[at]. e: Canadian Patent
CAnpUe d: Codeanpassungsübertragung
Can. Res. Dev. e: Canadian Research and Development (journ.)
CANS e: Computer-Assisted Network Scheduling
CANSA e: California Association of Nurses in Substance Abuse
CAN/SDI e: Canadian Selective Dissemination of Information
CANSG e: Civil Aviation Navigational Services Group
CANSIM System e: Canadian Socioeconomic Information Management System
Can. Spectrosc[op]. e: Canadian Spectroscopy (journ.)
CAN/SRP e: Canadian Subject Retrieval Program[me]
CANS System e: Computer-Assisted Network Scheduling System
CANTAP e: Canadian Technical Awareness Programme
CANTAT e: Canadian Transatlantic Telephone[s]
CANTB e: Canine Tuberculosis
- e: Cantilever Bridge
Can Telsat e: Canadian Telecommunications Satellite System
CAnthr e: Current Anthropology (journ.)
CANTIL e: Cantilever
CANTOT e: Computer Analysis of Troubles on Trunk-circuits
CANTRAN e: Cancel [in] Transmission
CANTU s: Compañía Anónima Nacional Telefonos de Venezuela
CANUCNET e: Canadian University Computer Network
CANUK e: Canada-United Kingdom
CANUKUS e: Canada-United Kingdom-United States
CANUKUSJCEC e: Canadian-United Kingdown-United States Joint Communications-Electronics Committe
CANUS e: Canada-United States
CANUSE e: Canada United States Eastern [Interconnection]
Can. Vet. J. e: Canadian Veterinary Journal (journ.)
Can. Welder Fabr. e: Canadian Welder and Fabricator (journ.)
Canyon de Chelly e: Canyon de Chelly National Monument
Canyonlands e: Canyonlands National Park
CAO e: Canadian Association of Optometrists
- e: Chronic Airway Obstruction
- e: Circuit Allocation Order
- e: Collective Analysis Only
- e: Commanding Air Officer
- e: Commissioned Armament Officer
- e: Communications Authorization Order
- e: Completed as Ordered
- e: Computer-Aided Administration and Organization
- e: Computer-Aided Office
- e: Computer-Aided Operating
- e: Computer-Aided Organization
- e: Computer Applications Office
- e: Continuous Analog[ue] Output
- e: County Advisory Officer
- e: County Agricultural Officer
- e: Crimean Astrophysical Observatory
- e: Cultural Affairs Officer
CAOC e: Cathodal Opening Contraction
- e: Constant Axial Offset Control
CAOCI e: Cathodal Opening Clonus
CAOCS e: Carrier Aircraft Operational Compatibility System
CAOCS-MOD 2 e: Carrier Aircraft Operational Compatibility System-Model 2
CAOE e: Contractor-Acquired Operational Equipment
CAOF e: Catholic Association of Foresters
CAOps e: Chief of Air Operations
CAOPT e: Council of American Official Poultry Tests
CAORE e: Canadian Army Operational Research Establishment
CAORF e: Computer-Aided Operations Research Facility
CAOS e: Completely Automated Operational System
- e: Completely Automatic Operational System
- e: Computer-Assisted Oscilloscope System
- e: Computer-Augmented Oscilloscope System
- e: Computer Applications Operating System
- e: Computer Assisted Optical Surfacing
CAO SOP e: Coordination of Atomic Operations-Standing Operating Procedures
CAOSOP e: Coordination of Atomic Operations Standard Operating Procedures
CAOT e: Canadian Association of Occupational Therapy
CAP d: Celluloseacetopropionat
- d: Chloramphenicol
- d: Computer-Analyse und Programmierung GmbH
- d: Computerleistung am Arbeitsplatz
- e: Cable Access Point
- e: California Electric Power
- e: Cambridge Accelerator Project
- e: Campaign Against Pollution
- e: Canadian Association of Pathologists
- e: Capacitance
- e: Capacitor
- e: Capacity
- e: Capital
- e: Capitalize[d]
- e: Capsule
- e: Capture[d]
- e: Card Assembly Plant
- e: Card Assembly Program[me]
- e: Carrier Assignment Program[me]
- e: Catabolite Activator Protein
- e: Catapult and Arresting-gear Pool
- e: Cellulose Acetate Propionate
- e: Central Arizona Project
- e: Chief Administrative Pharmaceutical Office[r]
- e: Chief Aviation Pilot
- e: Chlor[o]acetophenone
- e: Circuit Access Point
- e: Civil Air Control
- e: Civil Air Patrol
- e: Civil Air Publication
- e: Classroom Assembly Program[me]
- e: Clean Air Projector
- e: Cleaner Air Package
- e: Client Assessment Package
- e: Closet Approach Point
- e: Code of Advertising Practice
- e: College of American Pathologists
- e: Combat Air Patrol
- e: Command Analysis Pattern
- e: Commercial Automatic Programming
- e: Common Agricultural Policy
- e: Communications-pool Assembler Program[me]
- e: Communications Analysis Package
- e: Community Action Program[me]
- e: Community Alert Patrol
- e: Component Acceptance Procedure
- e: Comprehensive All-Purpose
- e: Comprehensive Assessment Program[me]
- e: Computer-Aided Planning
- e: Computer-Aided Presentation
- e: Computer-Aided Production
- e: Computer-Aided Programming
- e: Computer-Aided Publishing
- e: Computer-Assisted Placement
- e: Computer-Assisted Printing
- e: Computer-Assisted Production
- e: Computer Address Panel

- e: Computer Analysis and Programmer
- e: Computer Analysis and Programming
- e: Computer Analysts and Programmers Ltd
- e: Computer Assisted Placement
- e: Computer Assisted Printing
- e: Computerized Assignment of Personnel
- e: Computers and Automation and People (journ.) (now C & A)
- e: Condenser Absolute Pressure
- e: Configuration Analysis Program[me]
- e: Conflict Analysis Program[me]
- e: Console Action Processor
- e: Continental Africa Project
- e: Continuous Audit Program[me]
- e: Contractor Acquired Property
- e: Control & Authorization Process
- e: Control Anticipation Parameter
- e: Conversion Aid Program[me]
- e: Corporate Account Profitability
- e: Corps Ammunition Park
- e: Cost Analysis Plan
- e: Council of Advanced Programming
- e: Crew Assignment Program[me]
- e: Cropland Adjustment Program[me]
- e: Cryogenic Associative Processor
- e: Cryotron Associative Processor
- e: Current Assessment Plan
- f: Centre d'Analyse et Programming

CAPA e: Central Airborne Performance Analyzer
- e: Commercial Atomic Power Activity

CAPAB e: Capability
CAPAC e: Capacity
- e: Cathodic Protection

CAPAIDM e: Conference on Computer Architecture for Pattern Analysis and Image Database Management
CAPAL e: Computer and Photographic-Assisted Learning
CAPAL System e: Computer and Photographic-Assisted Learning System
CAPARS e: Computer Aided Placement and Routing System
CAPAV e: Committee on Atmospheric Problems of Aerospace Vehicles
CAPC e: Civil Air Patrol Cadet
- e: Computer-Aided Process Engineering

CAPCHE e: Component Automatic Programmed Checkout Equipment
CAPCO e: Central Area Power Coordination
CAPCOM e: Capsule Communications
CAP/COM e: Capsule Communicator
CAP Compressor e: Centrifugal Air Package Compressor
CAPCON e: Capitol Consortium Network
- e: Capsule Control

CAPCON[S] e: Captured Conversation[s]
- e: Recorded Conversation

CAPCP e: Civil Air Patrol Coastal Patrol
CAP-CPP e: Computer Analysts and Programmers Ltd.-Computer Programme Products
CAPCRA f: Cooperative Agricole des Producteurs de Céréales de la Région d'Arras
CAPD e: Continuous Ambulatory Peritoneal Dialysis
CAPDOC e: Captured Documents
CAPE e: Classification and Placement Examination
- e: Communication Automatic Processing Equipment
- e: Computer-Aided Planning and Estimating
- e: Computer-Aided Process Engineering
- e: Computer Applications in Production and Engineering (Conference)
- e: Computerized Accommodated Percentage Evaluation

CAPER e: Computer-Aided Pattern Evaluation and Recognition
- e: Computer-Aided Perspective
- e: Computers Aided Pattern Evaluation and Recognition
- e: Core Analysis and Program[me] Evaluation Recorder

CAPERTSIM e: Computer-Assisted Program[me] Evaluation Review Technique Simulation
CAPF e: Capacitance Factor
- e: Capacity Factor

CAPH e: California Association of Physically Handicapped
CAPICS e: Computer-Aided Processing of Industrial Cabling Systems
CAPITAL e: Computer-Assisted Placing in the Areas of London
CAPL e: Continuously Annealed Processing Length
- e: Controlled Assembly Parts List

CAPLIN e: Computer-Assisted Physics Laboratory Instruction
CAPM e: Central Processing Unit Access Port Monitor
- e: Computer-Aided Patient Management
- e: Computer-Aided Production Management
- e: Computer-Assisted Production Management

CAPMEC e: Computer Application for Planning Manufacturing Engineering Costs
cap. moll. l: capsula mollis
CAPO e: Computer-Aided Platforms Operation
- e: Computer Aid in the Physicians Office

CAPOSS e: Capacity Planning and Operation Sequencing System
CAPP e: Combined Accelerated Program[me]in Psychiatry
- e: Computer-Aided Partitioning Program[me]
- e: Computer-Aided Part Planning
- e: Computer-Aided Process Planning
- e: Content Addressable Parallel Processor

CAPPI e: Constant Altitude Plan-Position Indicator
CAPPP e: Computer-Aided Process and Production Planning
CAPPS e: Computer-Aided Project Planning System
- e: Council for the Advancement of the Psychological Professions and Sciences

CAPR e: Cable Pair
CAPRI e: Canada Agricultural Plant Research Institute
- e: Card and Printer Remote Interface
- e: Coded Address Private Radio Intercommunications
- e: Compact All-Purpose Range Instrument
- e: Computer-Aided Personal Reference Index
- e: Computer-Automated Procurement-Payment and Receipt of Material Inclusive
- e: Computer Aided Personal Reference Index
- e: Computerized Advance Personnel Requirements and Inventory
- e: Computerized Advance Personnel Requirements Information
- e: Computerized Analysis for Programming Investments
- e: Computerized Area Pricing

CAPRISTOR e: Capacitor-Resistor
CAPRI System e: Computer-Aided Personal Reference Index System
CAPS e: Call Attempts per Second
- e: Capacitors
- e: Capsule
- e: Casette Programming System
- e: Cashiers Automatic Processing System
- e: Cassette Programming System
- e: Cell Atmosphere Processing System
- e: Coastal Aerial Photolaser Survey
- e: Collins Adaptive Processing System
- e: Computer-Aided Packaging System
- e: Computer-Aided Pipe Sketching [System]
- e: Computer-Aided Problem Solving
- e: Computer-Aided Programming Software
- e: Computer-Aided Project Study
- e: Computer-Assisted Placement Service
- e: Computer-Assisted Problem Solving
- e: Computer-Based Aid to-Aircraft Project Studies
- e: Computer Aided Pipe Sketching
- e: Computer Aided Programming Software
- e: Computer Application Service
- e: Computer Assisted Placement Service
- e: Computerized Agency Processing System
- e: Control & Auxiliary Power Supply System
- e: Conversion Aids Programming System
- e: Cooperative Awards in Pure Science
- e: Costs and Productivity Scheme
- e: Courtaulds All-Purpose Simulator
- e: Critical Angle Prism Sensor
- e: Customer Annual Progress Summary

caps. l: capsule
CAP SCR e: Cap Screw
CAPSEP e: Capsule Seperation
CAPSHIPFOR e: Capacity Ships Force
CAPSS e: Computer-Assisted Public Safety System
CAPS System e: Computer Aided Pipe Sketching System
CAP ST e: Capacitor Start
CAPSTAR e: Capacitor Start and Run
CAPT e: Caption
- e: Control for the Advanced Passenger Train

CAPTAIN e: Character and Pattern Telephone Access Information Network
- e: Computer-Aided Processing and Terminal Access Information Network

CAPTAINS e: Character and Pattern Telephone Access Information Network System
CAPTAL e: Computer Assisted Placing in the Areas of London
CAPTIN e: Capacitance Pressure Transmitter Indicator
CAPY e: Capacity
CAQA e: Computer-Aided Quality Assurance
CAQC e: Computer-Aided Quality Control
CAR e: California Association of Reactors
- e: Canadian Association of Radiologists
- e: Cara Loudtop Altitude Radiometer
- e: Carat
- e: Card Adapter Register
- e: Cargo
- e: Caribbean Region
- e: Carpenter
- e: Carriage
- e: Carrier
- e: Carry[ing]
- e: Carry Register
- e: Center for Aging Research
- e: Central Address Register
- e: Central African Republic
- e: Central Apparatus Room
- e: Channel Address Register
- e: Chief Airship Rigger
- e: Civil Aeronautical Regulation[s]

- e: Civil Air Regulation[s]
- e: Cloud Altitude Radiometer
- e: Commonwealth Arbitration Reports
- e: Community Antenna Relay
- e: Computer-Aided Recording
- e: Computer-Aided Repair
- e: Computer-Aided Retrieval
- e: Computer-Aided Roboting
- e: Computer-Assisted Research
- e: Computer-Assisted Retrieval
- e: Conditioned Avoidance Response
- e: Configuration Audit Review
- e: Console Address Register
- e: Contract Authorization Request
- e: Control-unit Accumulator Register
- e: Correct Acceptance Rate
- e: Corrective Action Reply
- e: Corrective Action Report
- e: Corrective Action Request
- e: Cut All Round
- e: Cylinder Address Register
- f: Comité Agricole Régional

CARA e: California Association for Research in Astronomy
- e: Cargo and Rescue Aircraft
- e: Check Area Airports
- e: Combat Aircraft Recovery Aircraft
- e: Combat Air Rescue Aircraft
- e: Current Aerospace Research Activities

CARAC e: Civil Aviation Radio Advisory Committee
CARAD e: Computer-Aided Reliability and Design
CARAM e: Content Addressable Random Access Memory
CARB e: California Air Resources Board
- e: Carbonaceous
- e: Carbon[ate]
- e: Carburetor
- e: Carburize

Carb 14 e: Carbon 14
CARBINE e: Computer Automated Realtime Betting [Information] Network
CARBO e: Carbohydrate
CARBOLOY e: Carbon-Cobalt-Tungsten Alloy
CARBON e: Dioxide Carbonic Acid
CARBONADO e: Black or Grayishblack Industrial Diamond
CARBONIF e: Carboniferous
CARBONTET e: Carbon Tetrachloride
CARBOPOL e: Carboxypolymethylene
CARBS e: Computer-Aided Rationalized Building System
- e: Computer-Assisted Rationalized Building System

CARB TET e: Carbon Tetrachloride
CARC e: Canadian Arctic Resources Committee
Car. C. e: Car Carrier
CARC e: Chicago Amateur Radio Club
CARCC e: Chicago Area Radio Club Council
CARCIAC e: Cardboard Illustrative Aid to Computation
CARD e: Card Automated Reproduction and Distribution [System]
- e: Carrier Detector
- e: Channel Allocation and Routing Data
- e: Civil Aviation Research and Development
- e: Coded Automatic Reading Device
- e: Compact Automatic Retrieval Device
- e: Compact Automatic Retrieval Display
- e: Computer-Aided Radar Design
- e: Computer-Assisted Route Development
- e: Computer-Augmented Road Design
- e: Continuous Automatic Remote Display

CARDA e: Computer-Aided Reliability Data Analysis
- e: Continental Airborne Reconnaissance for Damage Assessment

CARDAMAP e: Cardiovascular Data Analysis by Machine Processing
CARDE e: Canadian Armament Research and Development Establishment
CARDI e: Colorado Allergy and Respiratory Disease Institute
CARDIAC e: Cardboard Illustrative Aid to Computation
CARDIOAC e: Cardioacceleration
- e: Cardioaccelerator[y]
- e: Cardioactive
- e: Cardioactivity

CARDIOG e: Cardiogenesis
- e: Cardiogenetic
- e: Cardiograph[ic]
- e: Cardiography

CARDIOL e: Cardiolith[ic]
- e: Cardiologist[ic][al][ly]
- e: Cardiology
- e: Cardiolysin

CARDIOMEG e: Cardiomegaly
CARDIOMYO e: Cardiomyopathy
CARDIOPUL e: Cardiopulmonary
CARDIOV e: Cardiovascular
Cardiovasc. Res. e: Cardiovascular Research (journ.)
CARDIOVER e: Cardioversion
CARDIS e: Cargo Data Interchange System
CARDPACS e: Card Packet System
CARDS e: Card Automated Reproduction and Distribution System
- e: Combat Aircraft Recording and Data System
- e: Computer-Aided Recording of Distribution Systems
- e: Computer-Aided Reliability Data System[s]
- e: Computer-Assisted Radar Display System
- e: Computer-Assisted Recording of Distribution Systems

CARD[S] e: Computer-Aided Research Device[s]
CARD System e: Card Automated Reproduction and Distribution System
CaRe e: Call Reference
CARE e: Clinical and Administrative Research
- e: Computer-Aided Reliability Estimation
- e: Computer-Assisted Renewal Education
- e: Computer Aided Reliability Estimation
- e: Continuous Aircraft Reliability Evaluation
- e: Customer Automatic Relay and Exchange

CARED e: Centre for Applied Research and Engineering Design
CA Register e: Current Address Register
CAREL e: Central Atlantic Regional Educational Laboratory
CARES e: Computation and Research Evaluation System
- e: Computer-Assisted Regional Evaluation System

CARE Scheme e: Convalescent and Additional Recovery Expenses Scheme
CARESIM e: Computer-Assisted Repair Simulation
CARE-SOM e: Coordinated Accident Rescue Endeavor-State of Mississippi
CARE System e: Clinical and Administrative Record System
CARETS e: Central Atlantic Regional Ecological Test Site
CARF e: Canadian Amateur Radio Federation
- e: Central Altitude Reservation Facility

CARGOTAINER e: Cargo Container
CARHSPD e: Carry; High Speed
CARI e: Civil Aeromedical Agricultural Research Institute
- e: Civil Aeromedical Research Institute
- e: Comparative Administration Research Institute

CARIAC e: Cardboard Illustrative Aid to Computation
CARIB e: Caribbean
CARIBSEAFRON e: Carribean Sea Frontier
CARIC e: Contractor All-Risk Incentive Contract
CARIH e: Childrens Asthma Research Institute and Hospital
CARIMAC e: Caribbean Institute of Mass Communications
CARIN e: Car Information and Navigation
CARINET e: Caribbean Information Network
CARIRI e: Caribbean Industrial Research Institute
CARIS e: Computerized Agricultural Research Information System
- e: Computerized Audio Report Information and Status
- e: Constant Angle Reflection Interference Spectroscopy
- e: Current Agricultural Research Information System

CARISMA e: Correction to Applied Research Laboratories Ion-Sputtering Mass Analyzers
CARL e: Calibration Requirements List
- e: Canadian Academic Research Libraries
- e: Chatfield Applied Research Laboratories
- e: Chemical and Radiological Laboratory
- e: Code Analysis Recording by Letters
- e: Colorado Association of Research Libraries
- e: Computer-Assisted Reference Locator
- e: Computer Audio Research Laboratory

CARLA e: Code Actuated Random Load Apparatus
CARLS e: Centralized Automated Loop Reporting System
CARLSSP e: Common Assembly Language Scientific Subroutine Package
CARMRAND e: Civilian Application of the Results of Military Research and Development
CARMS e: Computer-Assisted Records Management System
CARMSIM e: Computer-Assisted Reliability and Maintainability Simulation
CARN e: Conditional Analysis for Random Network[s]
Carnegie Tech e: Carnegie Institute of Technology
Carnics e: Carnic Alps
Carn Inst e: Carnegie Institute [of Technology]
CARO e: Clock Actuated Readout
CAROL e: Caribean Overseas Lines
CAROT e: Centralized Automatic Recording on Trunks
CARP e: Call Accounting Reconcilliation Process
- e: Carpathian
- e: Citizens concerned About Radiation Pollution
- e: Comprehensive Aerial Rainfall Program[me]

- e: Computed Air Release Point
- e: Computer-Aided Robot Programming
- e: Construction of Aircraft and Related Procurement

CARPAC e: Carriers; Pacific Fleet
CARPET e: Calgary and Region Education Television
CARPS e: Calculus Rate Problem Solver
CAR Pulse e: Carotid Pulse
Car Quals e: Carrier Qualifications
CARR e: Carriage
- e: Carried
- e: Carrier
- e: Combination Acknowledgement Reschedule Request
- e: Customer Acceptance Readiness Review

CARR EQ e: Carrier Equipment
CARR EQ OP e: Carrier Equipment Operator
CAR RET e: Carriage Return
CARR-FREQ e: Carrier Frequency
CARRIE MAC e: California Association of Reactors Mortgage Assistance Corporation
CARR-L e: Computer-Assisted Reporting and Research
CARR PLAT e: Carrier Platoon
CARR RET e: Carriage Return
CARRS e: Close-in Automatic Route Restoral System
CARS e: Canadian Arthritis and Rheumatism Society
- e: Central Agricultural Research Station
- e: Coherent Antistokes Raman Scattering
- e: Collision Avoidance Radar Simulator
- e: Combat Arms Regimental System
- e: Community Antenna Relay Service
- e: Community Antenna Relay Station
- e: Community Antenna Relay System
- e: Computer-Aided Reference Services
- e: Computer-Aided Resistance Spotwelding
- e: Computer-Aided Ride Sharing
- e: Computer-Aided Routing System
- e: Computer-Assisted Reference Selection
- e: Computer-Assisted Reference Service
- e: Computer-Assisted Reliability Statistics
- e: Computer-Augmented Resource System
- e: Computer Audit Retrieval System
- e: Computerized Academic Record-keeping System
- e: Computerized Audit and Reporting System
- e: Computerized Automotive Replacement Scheduling
- e: Computerized Automotive Reporting Service
- e: Continental Airlines Reservation System
- e: Continuous Alarm Reporting Service

CARSIMS e: Carry; Simultaneous
CARS System e: Command, Arming, Recording and Timing
- e: Computer-Aided Ride Sharing System
- e: Containerized Automated Rail-Highway Transportation
- e: Continuous Air Resistance Tester

CART e: Cargo Automation Research Team
- e: Cartilage
- e: Cartographer
- e: Cartographic
- e: Cartography
- e: Cartridge
- e: Central Automated Replenishment Technique[s]
- e: Central Automatic Reliability Tester
- e: Centralized Automatic Recorder and Tester
- e: Collision-Avoidance Radar Trainer
- e: Complete Automatic Rating Technique
- e: Complete Automatic Reliability Tester
- e: Complete Automatic Reliability Testing
- e: Complete Automatic Reliable Testing
- e: Computer-Assisted Radiation Therapy
- e: Computer[ized] Automatic Rating Technique
- e: Containerized Automated Rail-Highway Transportation
- e: Continuous Air Resistance Tester

CARTA e: Computer-Aided Recorder Trap Analysis
- e: Contour[ing] Analysis by Random Triangular Algorithm

CARTB e: Canadian Association of Radio and Television Broadcasters
CARTC e: Coast Artillery Replacement Training Center
CARTDRAFT e: Cartographic Draftsman
CARTECH e: Carpenter Technology
Carth. e: Carthage (or Carthagian)
CAR Timber e: Carallia Timber
CARTO e: Cartography
CARTOBIB e: Cartobibliographer
- e: Cartobibliography

CARTOG e: Cartographer
- e: Cartographic
- e: Cartography

CARTRIVISION e: Cartridge and Television
CARTS e: Computer Automated Rapid Transit System
- e: Computer Automated Reserved Track System
- e: Computerized Access and Records Tracking System

CARTU e: Combat Aircrew Refresher Training Unit
C Arty e: Coast Artillery
CAR/USNH e: Caribbeans to United States North of Hatteras
CARVA Power Pool e: Carolina-Virginia Power Pool
Car Z n: Caribische Zee
CAS d: Casablanca (Flughafen)
- d: VEB Chemieanlagen Staßfurt (DDR)
- e: Cable Activity System
- e: Cable Assembly Set
- e: Cable Audit System
- e: Calculated Airspeed
- e: Calgon Adsorption Service
- e: Calibrated Airspeed
- e: California Academy of Sciences
- e: Call Accounting Subsystem
- e: Cartridge Access System
- e: Cascade
- e: Casein
- e: Castellatus
- e: Casual[ties]
- e: Casualty
- e: Catgut Acoustical Society
- e: Cell Address Set
- e: Center for Acoustic Studies
- e: Center for Auto[mobile] Safety
- e: Center of Advanced Studies
- e: Central Africa Standard
- e: Central Alarm System
- e: Centre of Administrative Studies
- e: Certificate of Advanced Study
- e: Chain Acquisition System
- e: Change Analysis Section
- e: Channel-Associated Signalling
- e: Chemical Abstracts Service (journ.)
- e: Chicago Academy of Sciences
- e: Chief of the Air Staff
- e: Child Attitudes Survey
- e: China Association for Standardization
- e: Chinese Academy of Sciences
- e: Chorioallantoic Sac
- e: Circuits and System[s]
- e: Civil Air Surgeon
- e: Clean[er] Air System[s]
- e: Client Accounting System
- e: Close Air Support
- e: Coast Artillery School
- e: Coded Armaments System
- e: Coincidence Adjusting Scale
- e: Collision Avoidance System
- e: Column-Address-Select
- e: Column-Address Strobe
- e: Combined Activities System
- e: Command Augmentation System
- e: Communication Access System
- e: Communication Administration System
- e: Communication Analysis Section
- e: Communication Application Specification
- e: Communications Aerial Sleeve
- e: Communications Antenna Sleeve
- e: Compare Accumulator with Storage
- e: Compensating Air Supply
- e: Complete Assembly for Strike
- e: Compressed Air Spraying
- e: Computer-Aided Approach System
- e: Computer-Aided Selling
- e: Computer-Aided Service
- e: Computer-Aided Simulation
- e: Computer-Aided System[s]
- e: Computer Accounting System
- e: Computer Acquisition System
- e: Computer Application Summary
- e: Computer Arts Society
- e: Computer Automated System
- e: Computerized Accounting System
- e: Computerized Acquisition System
- e: Computers and Systems
- e: Conditional Array Scaling
- e: Content-Addressable Storage (or Store)
- e: Content-Addressed Storage (or Store)
- e: Content Addressable-file Store
- e: Control Augmentation System
- e: Control Automated System
- e: Controlled Access System
- e: Controlled Air Space
- e: Controller Access System
- e: Controls Assembly Set
- e: Control[s] Automation System
- e: Cooperative Applications Satellite
- e: Coordinated Allocation System
- e: Coronary Artery Stenosis
- e: Customer Application Summary

CASA e: Computer and Automated Systems Association
- s: Construcciones Aeronauticas

CAS-A e: Cooperation Application Satellite-A
CASAD e: Computer-Aided Specification Analysis and Documentation
Casa Grande e: Casa Grande National Monument
CAS Aircraft e: Close Air Support Aircraft
CASANZ e: Clean Air Society of Australia and New Zealand
CASAPS e: Computer Applied Systems Accounts Payable System
CAS Association e: Computer and Automated Systems Association
CASB e: Cost Accounting Standards Board
CASBL e: Continuous Automated Single-Base Line
CASBS e: Center for Advanced Study in the Behavioral Sciences
CASC e: Cascade
- e: Computer-Assisted Cartography
- e: Cooperative Applications Satellite C

CASCADE e: Citizens and Scientists Concerned About Dangers to the Environment
- e: Colorado Automatic Single Crystal Analysis Diffraction Equipment
- e: Computer-Aided System for Circuit Analysis and Design
- e: Computer-Aided Systems Construction and Documentation Environment

CASCAID e: Careers Advisory Service Computerized Aid
CASCAN e: Casualty Cancelled
CA Sch e: Coast Artillery School
CASCOMP e: Comprehensive Airship Sizing and [Performance] Computer Program[me]
CASCON e: Chemical Abstracts Service Condensates
- e: Close Air Support Control

CASCOR e: Casualty Corrected
CASCP e: Caribbean Area Small Craft Project
CAS/CPA e: Computer Accounting System/Computer Performance Analysis
CASCS e: Computer-Aided Serials Control System
CASD e: Computer-Aided System Design
CASDAC e: Computer-Aided Ship Design and Construction
CASDO e: Computer Applications Support and Development Office
CASDOS e: Computer-Aided Structural Detailing of Ships
- e: Computer-Assisted Detailing of Ships

CASDS e: Center for Advanced Study in the Developmental Sciences
CASE e: Coalition for Safe Energy
- e: Committee on Academic Science and Engineering
- e: Committee on the Atlantic Salmon Emergency
- e: Common Access Switching Equipment
- e: Common Application Service Element[s]
- e: Computer-Aided Software Engineering
- e: Computer-Aided Structural Engineering
- e: Computer-Aided System[s] Evaluation
- e: Computer-Assisted System Evaluation
- e: Computer-Automated Support Equipment
- e: Computer and System Engineering
- e: Computer and Systems Engineering Limited
- e: Computer Automated Support Equipment
- e: Cooperative Awards in Science and Engineering
- e: Cordinated Aerospace Supplier Evaluation

CASEA e: Center for the Advanced Study of Educational Administration
CASEE e: Canadian Army Signals Engineering Establishment
CASEVAC e: Casualty Evacuation
CASGTC e: California Academy of Sciences Geology Type Collection
CASH e: Call Accounting System for Hotels
- e: Cashier
- e: Catalog of Available and Standard Hardware
- e: Computer-Aided Stock Holdings
- e: Computer Aided System Hardware
- e: Computer Automated System[s] Hardware

CASH Programm e: Computer-Assisted Subject Headings Program[me]
CASI e: Canadian Aeronautics and Space Institute
CASIA e: Chemical Abstracts Subject Index Alert[s] (journ.)
CASIB e: Center for Advanced Studies in International Business
CASIN e: Computer-Aided Subject Index
CASING e: Cross-linking by Activated Species of Inert Gases
CASIN System e: Computer-Aided Subject Index System
CASIS d: computerunterstütztes Arbeitsschutz-Informationssystem
- e: computergestütztes Arbeitsschutz-Informationssystem

CASL e: Control Automation System Logics
CASLIS e: Canadian Association of Special Libraries and Information Services
CASM e: Close Air Support Missile
- e: Command and Service Module
- e: Cyclic Air Sampling Monitor
- e: Cycling Air Sampling Monitor

CASMIT e: Control Automation System Manufacturing Interface Tape
CASMOS [Circuits] e: Computer Analysis and Simulation of Metal-Oxide Semiconductor Circuits
CASMT e: Central Associabon of Science and Mathematics Teachers
CASO e: Computer-Assisted System Operation
CASOC e: California Arabian Standard Oil Company
CASOE e: Computer Accounting System for Office Expenditure
CASOFF e: Control and Surveillance of Friendly Forces
CA-SORT d: Computer Associates GmbH Sortierprogramm
CASOS e: Center for Advanced Study in Organization Science
CASP e: Capability Support Plan
- e: Cape Arago State Park
- e: Computer-Aided Synthesis Planning
- e: Computer-Assisted Search Planning
- e: Computer-Assisted Synthesis Planning
- e: Country Analysis and Strategy Paper
- e: Crew Activities Scheduling Program[me]

CASPAR e: Cambridge Analog[ue] Simulator for Predicting Atomic Reactions
- e: Clarksons Automatic System for Passenger and Agent Reservations
- e: Cushion Air System Parametric Assessment Rig

CASPER e: Computer-Aided Serial Publications Evaluation Routine
CASPERS e: Computer-Automated Speech-Perception System
CASPIR e: Computer-Aided Searching and Patent Image Retrieval
CASPP e: Compressed Air Storage Power Plant
CAS/PWI e: Collision Avoidance System/Pilot Warning Instrument
- e: Collision Avoidance System Proximity Warning Indicator

CAsR e: Central Asian Review (journ.)
CASR e: Cost and Schedule Reporting
CASREP[s] e: Casualty Reports
CASRS e: Countdown and Status Receiving Station
CASS e: Carrier Aircraft Support Study
- e: Cassette
- e: Center for Astrophysics and Space Scientists
- e: Cluster Activator Systems Specialist
- e: Combat Area Surveillance System
- e: Commanded Active Sonobuoy System
- e: Computer-Aided Schematic System
- e: Computer Applications in Shipping and Shipbuilding
- e: Computer Automatic Scheduling System
- e: Computerized Algorithmic Satellite Scheduler
- e: Copper[-Accelerated] Acetic[-Acid] Salt Spray
- e: Crab-Angle Sensing System

CAS/SA e: Chief of the Air Staff/Scientific Advisor
CASSANDRA e: Chromatogram Automatic Soaking, Scanning and Digital Recording Apparatus
CASSARS e: Computer-Assisted Simulation of Supply and Related Systems
CASSEIVER e: Cassette Recorder [and] Receiver
CASSI e: Chemical Abstracts [Service] Source Index (journ.)
CASSIS e: Communication[s] and Social Science Information Service
CASSM e: Content-Addressed Segment Sequential Memory
- e: Context-Addressed Segment Sequential Memory

CASSP e: Central Ammunition Supply Status Point
CASS Test e: Copper-Accelerated Acetic-Acid Salt Spray Test
- e: Copper Acetic Standard Solution Test

CAST e: Cable and Satellite Television
- e: Center for Application of Sciences and Technology
- e: Chemical Automated Search Terminal
- e: Clearinghouse Announcements in Science and Technology
- e: Computer-Aided Structural Technology
- e: Computer-Assisted Scanning Techniques
- e: Computer-Augmented Scanning Technics
- e: Computer Applications And Systems Technic
- e: Computer Applications and Systems Technology
- e: Computerized Automatic Systems Tester
- e: Cost Analysis System
- e: Council for Agricultural Science and Technology

CASTE e: Civil Aviation Signals Training Establishment
- e: Collision Avoidance System Technical Evaluation

CASTER e: Computer-Assisted System for Total Effort Reduction
- e: Conversational and Service Terminal

CASTI e: Centers for the Analysis of Science and Technical Information
CAS Timber e: Casuarina Timber
CASTLE e: Computer-Assisted System for Theater-Level Engineering
- e: Computer Automated Software for the Total Library Environment

CASTOR e: College Applicant Status Report
CASTS e: Canal Safe Transit System
- e: Commercial Automation Support of Technical Services
- e: Countdown and Status Transmission System

CASUM e: Civil Affairs Summary
CASW e: Close Air Support Weapon
- e: Council for the Advancement of Science Writing

CASWO e: Confidential and Secret Weekly Orders
CASWS e: Close Air Support Weapon[s] System
CASY e: Camera-Synchron

CAT d: Cabinentaxi
- e: Cabin Air Temperature
- e: California Achievement Test
- e: Capillary Agglutination Test
- e: Carburettor Air Temperature
- e: Cartridge Assembly Test
- e: Catalog[ue]
- e: Catalogued
- e: Cataloguer
- e: Catalyst
- e: Catamaran
- e: Catapult
- e: Category
- e: Caterpillar [Tractor]
- e: Catholic
- e: Cattle
- e: Caudé Auxiliary Telescope
- e: Celestial Atomic Trajectile
- e: Centralized Automatic Testing
- e: Character Assignment Table
- e: Childs Apperception Test
- e: Choline[-O-]Acetyltransferase
- e: City Air Terminal
- e: Civil Affairs Training
- e: Civil Air Transport
- e: Civilian Actress Technician
- e: Classical Analytic Technique
- e: Clean Air Turbulence
- e: Clear Air Turbulence
- e: Clerical Aptitude Test
- e: Coast Artillery Training
- e: COBOL Automatic Translator
- e: College Ability Test
- e: Colleges of Advanced Technology
- e: Combined Acceptance Trials
- e: Command and Triangulation
- e: Commander; Amphibious Troops
- e: Commercial Airlift Contract
- e: Commission on Agricultural Technology
- e: Community Antenna Television
- e: Compile and Test
- e: Component Acceptance Test
- e: Compressed Air Tunnel
- e: Computer-Aided (or-Attended) Tracking
- e: Computer-Aided Teaching
- e: Computer-Aided Technology
- e: Computer-Aided Testing
- e: Computer-Aided Tomography
- e: Computer-Aided Tool
- e: Computer-Aided Topography
- e: Computer-Aided Training
- e: Computer-Aided Translation
- e: Computer-Aided Tree
- e: Computer-Aided Typesetting
- e: Computer-Assisted Teaching
- e: Computer-Assisted Teleconferencing
- e: Computer-Assisted Test[ing]
- e: Computer-Assisted Topography
- e: Computer-Assisted Training
- e: Computer-Assisted Transcription
- e: Computer-Assisted Translation
- e: Computer-Automated Tester
- e: Computer-Automated Tracer
- e: Computer Access Time
- e: Computer Analysis of Transistors
- e: Computerized Adaptive Test[ing]
- e: Computerized Axial Test[ing]
- e: Computerized Axial Tomography
- e: Computer of Average Transients
- e: Computing Automation Technique
- e: Configuration and Traceability
- e: Consolidated Atomic Time
- e: Continuous Automatic Test[ing]
- e: Controlled Attenuator Timer
- e: Controlled Avalanche Transistor
- e: Converted Aerial Targets
- e: Cooled-Anode Tube
- e: Copper Alloy Tubing
- e: Crack Arrest Temperature
- e: Cummulative Abbreviated Trouble
- e: Current-Adjusting Type
- e: Customer-Activated Terminal

CATA e: Canadian Air Transport Association
- e: Community Antenna Television Association
- e: Cushion Air Tread Articulate

CATAC e: Central Allied Tactical Air Command

CATAL e: Catalog[ue]

Catal. Rev. e: Catalysis Reviews (journ.)

CATALSYST e: Computer-Assisted Teaching and Learning System

CATC e: Circular-Arc-Toothed Cylindrical
- e: Computer-Assisted Test Construction
- e: Continental Atlantic Tidewater and Cities
- e: Continuous Automatic Train Control

CATC e: Crashed Aircraft Transit Centre

CATCALL e: Completely Automated Technique for Cataloging and Acquisition of Literature for Libraries

CATCC e: Canadian Association of Textile Colorists and Chemists
- e: Carrier Air Traffic Control Center

CATCH e: Character Allocated Transfer Channel
- e: Computer Analysis of Thermo-Chemical

CATCO e: Catalytic Construction Company

CATCUSAF e: Commander; Amphibious Training Command United States Atlantic Fleet

CATE e: Comprehensive Automatic Test Equipment
- e: Computer-Aided Testing Equipment
- e: Computer-Assisted Traffic Engineering
- e: Computer-Controlled Automatic Test Equipment
- e: Computer Assisted Traffic Engineering
- e: Convair Automatic Test Equipment
- e: Current Air Research and Development Command Technical Efforts

CATERTRAC e: Caterpillar Tractor

CATEST e: Calcium Test

CATF e: Canadian Achievement Test in French
- e: Commander; Amphibious Task Force

CATG e: Commander; Air Task Group

CATH e: Cathedral
- e: Catheter
- e: Cathode
- e: Catholic

CATHFOL e: Cathode Follower

CathHR e: Catholic Historical Review (journ.)

Cath Lib Assn e: Catholic Library Association

Cath. Libr. Wld. e: Catholic Library World (journ.)

CATHRES e: Cathode Resistor

Cath U Pr e: Catholic University of America Press

CATIA e: Computer-Aided Three-Dimensional Interactive Application

CATIE s: Centro Agmnómico Tempical de Investigación y Enseñanza

Cat. Index e: Catalogue & Index (journ.)

CATIS e: Computer-Aided Tactical Information System

CATLAS e: Centralized Automatic Trouble-Locating and Analysis System

CATLG e: Catalog[ue]
- e: Cataloging

CATLIB Network e: Cable Television Library Network

CATLINE e: Catalog[ue] Online

CATM e: Canadian Achievement Test in Mathematics
- e: Consolidated Air Tour Manual

CATMN e: Consolidated Air Target Material Notices

ca_n e: Calorie (International Table calorie)

CATNIP e: Computer-Assisted Technique for Numerical Indexing Purposes
- e: Computer-Assisted Technique for Numerical Index Preparation

CATO e: Catapult-Assisted Take-Off
- e: Civil Air Traffic Operation[s]
- e: Compiler for Automatic Teaching Operation

CATOR e: Combined Air Transport Operations Room

CATOX e: Catalytic Oxidation

CATP e: Classified Area Term Pass
- e: Computer-Aided Text Processing
- e: Computer-Assisted Typesetting Process

CATPASS e: Computer-Aided Telephony Performance Assessment

CATPROC e: Catalog[ue] Procedure

CATR e: Central Air Transport

CATS e: Canon Auto Tuning System
- e: Category Switch
- e: Centralized Automatic Test System
- e: Centre for Advanced Television Studies
- e: Chicago Area Teleprinter Society
- e: Chicago Area Transportation Study
- e: Civil Affairs Training School
- e: Communications and Tracking System
- e: Comprehensive Analytical Test System
- e: Compute Air-Trans Systems
- e: Computer-Accessed Telemetry System
- e: Computer-Aided Teaching System
- e: Computer-Aided Trading System
- e: Computer-Aided Troubleshooting
- e: Computer-Assisted Test Shop
- e: Computer-Assisted Trading System
- e: Computer-Automated Test System
- e: Computer-Automated Transit System[s]
- e: Computer Analysis of Tape Signals
- e: Computer Automated Test System

CATSA e: Cooperative Air Transport System for Antarctica

CATS-A/P e: Computerware Automated Total Systems-Account[s] Payable

CATSDM e: Computer-Aided Training Systems Development and Management

CATSS e: Cataloguing Support System

CATT e: Centralized Automatic Toll Ticketing
- e: Controlled Avalanche Transit-Time [Triode]
- e: Conveyorized Automatic Tube Tester
- e: Cooled-Anode Transmitting Tube

CATTALO e: Cattle and Buffalohybrid

CATTCM e: Canadian Achievement Test in Technical and Commercial Mathematics

CATT Device e: Controlled Avalanche Transit Time Device

CAT Test e: Computerized Axial Tomography Test

CATTS e: California and Texas Telecommunications System
- e: Computer-Aided Training in Troubleshooting

CATT Triode e: Controlled Avalanche Transit-Time Triode

CATV e: Cabin Air Temperature Valve
- e: Cable Aerial Television
- e: Cable Antenna Television
- e: Cable Telecommunication and Video
- e: Cable Television
- e: Command/Arithmetic[s] Unit
- e: Common Aerial Television
- e: Community Antenna Television
- e: Cooled-Anode Transmitting Valve

CATVA e: Computer-Assisted Total Value Assessment
- Computer-Augmented Total Value Assessment

CATVI e: Computer Analysis of Time-Varying Images
CATVO Project e: Cable Television Outreach Project
CATVS e: Community Antenna Television System
CAU e: Caucasian
- e: Command/Arithmetic[s] Unit
- e: Command and Arithmetic Unit[s]
- e: Complex Arithmetic Unit
- e: Computer Advisory Unit
- e: Configuration Assignment Unit
- e: Crypto Ancillary Unit
- f: Calculatrice Arithmetique Universelle

CAUB e: College and University Business (journ.)
CAUDAR e: Computer-Assisted Unit Data Acquisition and Reduction
CAUFN e: Caution Advised Until Further Notice
CAUL e: Committee of Australian University Librarians
CAUML e: Computers and Automations Universal Mailing List
CAUS e: Causation
- e: Causative
- e: Command/Arithmetic[s] Units

CAUSAT e: Causative
CAUSE e: Because
- e: College and University System[s] Exchange
- e: Comprehensive Assistance to Undergraduate Science Education
- e: Computer-Assisted Underwriting System at Equitable
- e: Counselor Advisor University Summer Education

CAUST e: Caustic
CAUSTIC POTASH e: Potassium Hydroxide (KOH)
CAUSTIC SODA e: Sodium Hydroxide (NaOH)
CAV d: chemie-anlagen + verfahren (journ.)
- e: Cavitation
- e: Cavity
- e: Class Attendance Verification
- e: Composite Analog[ue] Video
- e: Congenital Absence Of Vagina
- e: Congenital Adrenal Virilism
- e: Constant Angular Velocity
- e: Construction Assistance Vehicle
- e: Continuous Airworthiness Visit

CAVALCADE e: Calibrating, Amplitude-Variation and Level-Correcting Analog[ue]-Digital Equipment
CAVD e: Completion, Arithmetic, Vocabulary, Directions
CAVD Test e: Completion, Arithmetic, Vocabulary, Directions Test
CAVE e: California Association of Vocational Educators
- e: Catholic Audio-Visual Educators Association
- e: Computer Augmented Video Education
- e: Computer Automated Vacuum Evaporator
- e: Consolidated Aquanauts Vital Equipment

CAVEA e: Connecticut Audio-Visual Education Association
CAVEAT e: Code and Visual Entry Authorization Technic
Cavendish e: Cavendish Laboratory
CAVI f: Centre Audio-Visuel International
CAVIOL e: Caviology
CA VIRUS e: Croup-Associated Virus
CAVORT e: Coherent Acceleration and Velocity Observations in Real Time
CAVT e: Constant Absolute Vorticity Trajectory
CAVU e: Ceiling and Visibility Unlimited
CAW e: Cable[s] and Wireless
- e: Cam Action Wheel
- e: Carbon-Arc Weld[ing]
- e: Channel Address Word
- e: Chief of Amphibious Warfare
- e: Common Aerial Working
- e: Common Antenna Working
- e: Computer-Aided Wiring
- e: Computer-Aided Writing
- e: Conflict Alert Warning
- e: Control and Warning

CAWC e: Central Advisory Water Committee
- e: Committee on Air and Water Conservation

CAWF e: Carrier All-Weather Flying
CAWG e: Coaxial Adapter Waveguide
CAWM e: College of African Wildlife Management
CAWS e: Central Aural Warning System
- e: Common Aviation Weather Subsystem
- e: Conflict Alert Warning System
- e: Conservation and Wildlife Studies

CAWSPS e: Computer-Aided Weapons Stowage Planning System
- e: Computer-Aided Weapon[s] Storage Planning System

CAX e: Community Automatic Exchange
Cay. e: Cayenne
Caz d: Cetanzahl
CAZ e: Commutating Auto Zero
CAZRI e: Central Arid Zone Research Institute
CB e: Cadmium Bronze
- e: Capacitor Bank
- e: Cape Breton
- e: Carbobenzoxide
- e: Card Batch
- e: Card Box
- e: Carry Byte
- e: Cast Brass
- e: Catapult Bulletin
- e: Catch Basin
- e: Cement Base
- e: Center Board
- e: Center of Buoyancy
- e: Center of Burst
- e: Centibar
- e: Centibel
- e: Central Battery
- e: Centre Board
- e: Centre of Buoyancy
- e: Centre of Burst
- e: Channel Byte
- e: Check Bit[s]
- e: Chemical and Biological
- e: Chest-Back
- e: Chief Boilermaker
- e: Chlorobiphenyl[s]
- e: Chromatid Breaks
- e: Circuit Breaker
- e: Circulating Beam
- e: Citizen[s] Band
- e: Clear Back
- e: Coated on the Back [side]
- e: Cobalt Bomb
- e: Cobaltite
- e: Code Book
- e: Coin Box
- e: Colliding Beams
- e: Columbium
- e: Column Binary
- e: Command Byte
- e: Common Base
- e: Common Battery
- e: Common Bench
- e: Common Bias
- e: Common Bill
- e: Common Block
- e: Common Business Oriented Language
- e: Communciations Buffer
- e: Communications Bus
- e: Compare Binary
- e: Compass Bearing
- e: Computer Backup
- e: Computing Block
- e: Conditional Branch
- e: Conditional Breakpoint
- e: Condition Bit
- e: Conduction Band
- e: Confidential Book
- e: Confidential Bulletin
- e: Configuration Baseline
- e: Connecting Block
- e: Connection Busy
- e: Contact Breaker
- e: Continuous Blowdown
- e: Continuous Breakdown
- e: Contrast Bath
- e: Control Block
- e: Control Board
- e: Control Booth
- e: Control Branch
- e: Control Break
- e: Control Buffer
- e: Control Button
- e: Control Byte
- e: Correction Bit
- e: Coulomb
- e: Count Bus
- e: Counter Battery
- e: Crash Boat
- e: Crew Boat
- e: Cruising Boost
- e: Cubic
- e: Cummulative Bulletin
- e: Cumulonimbus
- e: Currency Bond
- e: Current Bit
- e: Current Branch
- e: Cytochalasin-B

C & B e: Collating and Binding
C.B. l: Chirurgiae Baccalaureus
CB 1 e: Current Bit 1
CB1-3 e: Clear-Back signal No. 1 to No. 3
CB 2 e: Current Bit 2
CBA e: C-Band-Transponder Antenna
- e: Catholic Broadcasters Association
- e: Central Battery Apparatus
- e: Chemical-Biological Activities (journ.)
- e: Chemical Bond Approach
- e: Chronic Bronchitis with Asthma
- e: Chesapeake Bay Annex
- e: Cocoa Beach Apollo
- e: Colliding Beam Accelerator
- e: Communications Network Architecture
- e: Community Broadcasters Association
- e: Cost-Benefit Analysis
- e: Council for British Archaeology

- e: Crushed Brock Aggregate
CBABG e: Commonwealth Bureau of Animal Breeding and Genetics
CBAC e: Chemical-Biological Activities (journ.)
CBAF e: Cobalt Base Alloy Foil
CBAH e: Commonwealth Bureau of Animal Health
CBAIC e: Chemical-Biological Accident and Incident Control
CBAICP e: Chemical and Biological Accident and Incident Control Plan
CBAL e: Counterbalance
CBALO e: Carrier-Borne Air Liaison Officer
CBAM e: Concerns-Based Adoption Model
CBAN e: Commonwealth Bureau of Animal Nutrition
CBAR d: Zentibar
- e: Counterbore Arbor
CBAs e: Canadian Building Abstracts (journ.)
CBAS e: Central Battery Alarm Signalling
- e: Chemical Bond Approach Study
C-BASIC e: Commercial BASIC
CBA Signal e: Change-Back Acknowledgement Signal
CBAT e: Central Bureau of Astronomical Telegrams
CB/ATDS e: Carrier-Based Airborne Tactical Data System
CBB e: Communication in Behavioral Biology
- e: Compensation Booster Battery
- f: Centre Belge du Bois
Cb base e: Capacitance
CBBI e: Cast Bronze Bearing Institute
CBBII e: Council of the Brass and Bronze Ingot Industry
CBBT e: Chesapeake Bay Bridge Tunnel
CBBUAER e: Chief of the Bureau of Aeronautics
CBC e: Cadmium Bronze Connector
- e: California Botanical Club
- e: Canadian Broadcasting Corporation
- e: Cannabichromene
- e: Cannot be Called (Telephony)
- e: Caribbean Broadcasting Company
- e: Cauchy Boundasry Condition
- e: Cerebrobuccal Commissure
- e: Cerebrobuccal Connectives
- e: Ceylon Broadcasting Corporation
- e: Chain Block Controller
- e: Change-Back Control
- e: Chicago Book Clinic
- e: Chinese Binary Code
- e: Cipher Block Chaine
- e: Cipher Block Chaining
- e: Circuit Board Card
- e: Classification by Computer
- e: Columbia Broadcasting Corporation
- e: Column Binary Code
- e: Combined Blood Count
- e: Complete Blood Count
- e: Computer-Based Concentrator
- e: Computer-Based Conferencing
- e: Computer Bar Code
- e: Consecutive Blank Column
- e: Construction Battalion Center
- e: Contraband Control
- e: Crushed Brick Concrete
- e: Cyclical Binary Code
- f: Code Binaire Cyclique
CBCC e: Chemical-Biological Coordination Center
- e: Common Bias; Common Control
CBC Count e: Complete Blood Cell Count
CBCD e: Complementary Binary-Coded Decimal
CB Circuit e: Common-Base Amplifier for Junction Transistors
cbcm e: Cubic Centimeter (US)
- e: Cubic Centimetre (GB)
CBCN e: Cathode Boundary Charge Neutrality
CBCR e: Contention Based Channel Reservation
CBCS e: C-Band Checkout System
- e: Computer-Based Conference System
CBCT e: Circuit Board Card Tester
- e: Customer-Bank Communication Terminal
CBCU e: Counterbore Cutter
CBD e: Call Box Discrimination
- e: Cannabidiol
- e: Carbide
- e: Closed Bladder Drainage
- e: Common Bile Duct
- e: Computer Block Diagram
- e: Configuration Block Diagram
- e: Constant Bit Density
- e: Cyclobutadien
cbdm d: Kubikdezimeter
CBDP e: Communications Based Data Processing
CBDS e: Carcinogenesis Bioassay Data System
- e: Change-Back Declaration Signal
- e: Circuit Board Design System
CBD Signal e: Change-Back Declaration Signal
CBDST e: Commonwealth Bureau of Dairy Science and Technology
CBD System e: Circuit Board Design System
CBDT e: Cannot Break Dial[ling] Tone
- e: Card Board Description Tape
CBE e: Central Battery Exchange
- e: Centralized Branch Exchange
- e: Cesium Bombardment Engine
- e: Cheese Bureau of England
- e: Chemical Binding Effect
- e: Circuit Board Extractor
- e: Competency-Based Education
- e: Compression-Bonded Encapsulation
- e: Computer-Based Education
- e: Conference of Biological Editors
- e: Council for Basic Education
- e: Council of Basic Education
- e: Council of Biology Editors
CBEL e: Cambridge Bibliography of English Literature
CBEM e: Computer-Based Educational Material
CBEMA e: Canadian Business Equipment Manufacturers Association
- e: Computer and Business Equipment Manufacturers Association
CBER e: Citizens Bander
CBE Survey e: Consumer Buying Expectations Survey
CBEUS e: Center for Business Economics and Urban Studies
CBF d: Containerbahnhof
- e: Cerebral Blood Flow
- e: Childrens Blood Foundation
- e: Commander; Battleship Force
- e: Coronary Blood Flow
CBFM e: Constant Bandwidth Frequency Modulation
CBFP e: Chronic Biological False-Positive
CBFS e: Caesium Beam Frequency Standard
CBFT e: Cubic Feet
- e: Cubic Foot
CB-Funk d: Citizen-Band-Funk
CBG e: Corticosteroid-Binding Globulin
- f: Compagnie des Bauxites de Guinée
CBGLO e: Carrier-Borne Ground Liaison Officer
CBH e: Cannot be Heard (Telephony)
- e: Circuit Board Holder
- e: Commonwealth Bureau of Helminthology
CBHA e: Council for Biology in Human Affairs
CBHPC e: Commonwealth Bureau of Horticulture and Plantation Crops
CBI e: Cape Breton Island
- e: Carbonate[d] Beverage Institute
- e: Central Bible Institute
- e: Chesapeake Bay Institute
- e: China-Burma-India
- e: Common Batch Identification
- e: Compare Binary Immediate
- e: Complementary Binary
- e: Complete Background Investigation
- e: Compound Batch Identification
- e: Computer-Based Instruction
- e: Computer-Based Instrumentation
- e: Conditional Breakpoint Instruction
- e: Confederation of British Industry
- e: Control Break Item
- e: Controlled Batch Identification
- e: Cumulative Book Index
CB & I e: Chicago Bridge and Iron (company)
CBI Code e: Complementary Binary Code
CBIC[s] e: Complementary Bipolar Integrated Circuit[s]
CBIE e: Computer-Based Information Exchange
CBIH e: Chemical and Biological Information Handling
CBIN e: Caribbean Basis Information Network
CBIS e: Campus-Based Information System
- e: Computer-Based Information System
- e: Computer-Based Instruction[al] System
CBI System e: Campus-Based Information System
CBKCD e: Customer Backorder Code
cbkm d: Kubikkilometer
CBK Signal e: Clear-Back Signal
CBL e: Cable
- e: Cannabicyclol
- e: Carlyle Barton Laboratory
- e: Caustic Boundary Layer
- e: Cement Bond Log
- e: Chesapeake Biological Laboratories
- e: Compressed Binary Loader
- e: Computer-Based Learning
- e: Configuration Breakdown List
- e: Counterbalance
CBLFA e: Corn Belt Livestock Feeders Association
CBLMN e: Cableman
CBLRD e: Cable Reed
CBM d: Computer Betriebswirtschaft und Management Service GmbH
cbm d: Kubikmeter
CBM e: Ceramic Based Microcircuit
- e: Certified Ballast Manufacturer
- e: Chemical and Biological Munition[s]
- e: Chlorobromomethane
- e: Column Binary Mode
- e: Common Bulk Memory
- e: Compact Balancing Machine
- e: Confidence Building Measure[s]
- e: Constant Boiling Mixture
- e: Continental Ballistic Missile
- e: Controlled Backmix
- e: Conventional Buoy Mooring

- e: Cubic Meter (US)
- e: Cubic Metre (GB)
CBMA e: Canadian Business Manufacturers Association
CBMIS e: Comprehensive Budget and Management Information System
- e: Computer-Based Management Information System
CBMK n: Centraal bureau voor metingen op het gebied van de kernenergie
CBMKR e: Chief Boilermaker
cbmm d: Kubikmillimeter
CBMM e: Council of Building Materials Manufacturers
CBMPE e: Council of British Manufacturers of Petroleum Equipment
CBMPP e: Cargo Bay Module Personnel Provisions
CBMQA e: California Board of Medical Quality Assurance
CBMS e: Computer-Based Management System
- e: Computer-Based Message System
- e: Conference Board of Mathematical Sciences
CBMSRC e: Chief Boatswains Mate, Ship Repair, Crane-Operator
CBMSRR e: Chief Boatswains Mate, Ship Repair, Rigger
CBM System e: Controlled Backmix System
CBMU e: Current Bit Monitor Unit
CBMUA e: Canadian Boiler and Machinery Underwriters Association
CBN e: Cannabinol
- e: Chemical, Bacteriological, Nuclear
- e: Christian Broadcasting Network
- e: Columbia Carbon Company
- e: Commission on Biochemical Nomenclature
- e: Commission on Biological Nomenclature
- e: Computerized Bibliographic Network
- e: Cubic Boron Nitride
CBNE e: California Bureau of Narcotics Enforcement
CBNM e: Central Bureau for Nuclear Measurements
- e: Custer Battlefield National Monument
CBO e: Coding Board Officer
CBO e: Colombo (Airport)
- e: Compensation by Objectives
- e: Computer Burst Order
- e: Conference of Baltic Oceanographers
- e: Counter-Battery Officer
CBOC e: Completion of Bed Occupancy Care
Cbo-Gruppe d: Carbo-Benzoxy-Gruppe
CBOI e: Complete Basis of Issue
C-Bomb e: Cobalt Bomb
Cbo-Schutzgruppe d: Carbo-Benzoxy-Schutzgruppe
CBP e: Calcium Binding Protein
- e: Ceramic Beam Pentode
- e: Conditional Break-Point
- e: Constant Boiling Point
- e: Critical Packing Pressure
- p: Centro de Biologia Piscatoria
- s: Centro de Biologia Piscatoria
CB-Paper e: Coated Back[side] Paper
- e: Covered-Back Paper
CBPBG e: Commonwealth Bureau of Plant Breeding and Genetics
CBPDC e: Canadian Book and Periodical Development Council
CBPFC e: Commonwealth Bureau of Pastures and Field Crops
CB & PGNCS e: Circuit Breaker and Primary Guidance Navigation Control System
CBPI e: Conditional Break-Point Instruction
CBPP e: Contagious Bovine Pleuropneumonia
CBPU e: Central Branch Processing Unit
CBQ e: Civilian Bachelor Quarters (journ.)
CBR e: California Bearing Ratio
- e: Canberra (airport)
- e: Card Buffer Register
- e: Carry Byte Register
- e: Cast Brass
- e: Cavity-Backed Radiator
- e: Center for Brain Research
- e: Centre for Business Research
- e: Change Board Register
- e: Chemical, Biological, Radiological
- e: Circuit Board Rack
- e: Citizens Band Radio
- e: Column Binary Read
- e: Commercial Breeder Reactor
- e: Communication Barge Receiver
- e: Community Bureau of References
- e: Comprehensive Beacon Radar
- e: Counterbore
- e: Crude Birth Rate
- e: Current Balance Record
CBRA e: Chemical, Biological, Radiological Agency
CBRC e: Chemical, Biological, Radiological Center
- e: Crichton Behavioral Rating Scale
CB RCVR e: Citizens Band Receiver
CBRD e: Canadian Base Reinforcement Depot
- e: Construction Battalion Replacement Depot
CBRE e: Chemical, Biological and Radiological Element
CBREG e: Chemical, Biological, Radiological Engineering Group
CBRI e: Central Building Research Institute
- e: Chester Beatty Research Institute
CBRL e: Chemical, Biological and Radiation Laboratories
CBRM e: Charger Battery Relay Module
CBRN e: Caribbean Basin Radar Network
- e: Chemical, Biological, Radiological and Nuclear
CBRO e: Chemical, Biological, Radiological Officer
CBRS e: Canadian Bond Rating Service
- e: Chemical, Biological and Radiological Selection
- e: Child Behavior Rating Scale
- e: Citizens Band Radio Service
- e: Computer-Based Reference Service
CBRU Project e: Computer Based Resource Units Project
CBRW e: Chemical, Biological and Radiological Warfare
CBS e: Call Box Station
- e: Canadian Biochemical Society
- e: Card Buffer Store
- e: Central Battery Signalling
- e: Central Battery Supply
- e: Central Battery System
- e: Central Bibliographic System
- e: Central Bureau of Statistics
- e: Channel Base Section
- e: Check Bits
- e: China Broadcasting Satellite
- e: Chlorobiphenyls
- e: Chronic Brain Syndrome
- e: Cloth Both Sides
- e: Coastal Base Section
- e: Columbia Broadcasting System
- e: Common Base Station
- e: Common Battery Signalling
- e: Common Battery Supply
- e: Commonwealth Bureau of Soils
- e: Compact Buoy System
- e: Complex Behavior Simulator
- e: Comprehensive Business System
- e: Computer Backing Store
- e: Computer Buffer Store
- e: Concrete Block Stucco
- e: Connecticut Botanical Society
- e: Connector Backing Shell
- e: Consolidated Business Systems
- e: Controlled Barrier System
- e: Controlled Blip Scan
- e: Currubin Bird Sanctuary
- e: N-Cyclohexyl-2-Benzthiazylsulphenamide
- n: Central Bureau de Statistiek
CBSC e: Common Bias; Single Control
CBSR e: Chief Boilermaker; Ship Repair
- e: Coupled Breading Superheating Reactor
CBT e: Center for Building Technology
- e: Central Battery Telephone
- e: Cesium Beam Tube
- e: Cipher-Block Chain
- e: Cognitive-Behavioral Therapy
- e: Combat
- e: Communication Barge Transmitter
- e: Complementary Bipolar Transistor
- e: Complex Binary Fourier Representation Transform
- e: Computer-Based Terminal
- e: Computer-Based Testing
- e: Computer-Based Training
- f: Centre Belge de Traductions
CBTI e: Combat Intelligence
- e: Computerized Breast Thermographic Interpreter[s]
CBTRA e: Computer Basic Technology Research Association
CBT[R]Y e: Counterbattery
CBTS e: Cesium Beam Time Standard
CBU e: Cluster[ed] Bomb Unit[s]
- e: Coefficient of Beam Utilization
- e: Collective Bargaining Unit
- e: Container Bomb Unit
- e: Control Beam Unit
C Buoy e: Can Buoy
CBV e: Central Blood Volume
- e: Circulating Blood Volume
- e: Corrected Blood Volume
CBVHS e: Clara Barton Vocational High School
CBW e: Catholic Book Week
- e: Chemical and Biological Warfare
- e: Chemical and Biological Weapons
- e: Column Bandwidth
- e: Constant Bandwidth
CBX e: C-Band Transponder
- e: Computer-Based Exchange
- e: Computer-Controlled Branch Exchange
- e: Computerized Branch Exchange
- e: Computerized Business Exchange[s]
CBZ e: Cambendazole
- e: Carbobenzoxy
CC d: Consolidated Computer GmbH
- e: Cable Card
- e: Cable Connector
- e: Caius College
- e: Calcite
- e: Calculating Capacity
- e: Calculator
- e: Calibration Count
- e: Call Command
- e: Call Confirmation
- e: Call Connected

- e: Call Contract
- e: Call on Carry
- e: Canonical Correlation
- e: Capacity Change
- e: Capsule Communicator
- e: Carbon Copies
- e: Carbon Copy
- e: Card Capacity
- e: Card Channel
- e: Card Check[ing]
- e: Card Code
- e: Card Collating
- e: Card Collator
- e: Card Column
- e: Card Control[led]
- e: Card Count[ing]
- e: Card Counter
- e: Card Cycle
- e: Carges Collect
- e: Carriage Control
- e: Carrier Current
- e: Cast Concrete
- e: Cast Copper
- e: Central Computer
- e: Central Concentrator
- e: Central Console
- e: Central Control
- e: Centrifugal Coating
- e: Centuries
- e: Ceramic [Capacitor]
- e: Cerebral Commissure
- e: Cevacin
- e: Chain[ed] Command
- e: Chaining Check[ing]
- e: Change[d] [of] Course
- e: Channel Coding
- e: Channel Command
- e: Channel Controller
- e: Chapters
- e: Character Change
- e: Character Code
- e: Character Coded
- e: Character Coding
- e: Character Constant
- e: Character Counter
- e: Charge Coupled
- e: Check Channel
- e: Check Character
- e: Check Code
- e: Check Computation
- e: Chemists Club [of New York]
- e: Chest Complaint
- e: Chief Clerk
- e: Chief Complaint
- e: Chief Constructor
- e: Chief Counsel
- e: Chip Carrier
- e: Cholecalciferol
- e: Chronometer Correction
- e: Circuit Card
- e: Circuit Closing
- e: Circulating Copy
- e: Cirrocumulus
- e: Cirrocumulus Cloud[s]
- e: Civil Commotion
- e: Clare College (Cambridge)
- e: Classification of Characteristics
- e: Clear Chill
- e: Clipping Circuit
- e: Clock Card
- e: Clock Control
- e: Close-Coupled
- e: Close Control
- e: Closed Circuit
- e: Closing Coil
- e: Closure Covering
- e: Cloud Cover
- e: Cluster Controller
- e: Coal Commission
- e: Coarse Concrete
- e: Coarse Control
- e: Coastal Command
- e: Code Chain
- e: Code Character
- e: Code Chart
- e: Code Check
- e: Code Combination
- e: Code Control
- e: Code Conversion
- e: Code Converter
- e: Coded Character
- e: Coded Command
- e: Coefficient of Correlation
- e: Cognitive Complexity
- e: Coiled Coil
- e: Coincidence Circuit
- e: Coincident Current
- e: Coin Collect
- e: Coin Completing
- e: Colby College
- e: Collect Call
- e: Collecting Center
- e: Collector Capacitance
- e: Colon Classification
- e: Colo[u]r Code
- e: Colo[u]r Compact
- e: Colo[u]r Compensation
- e: Colo[u]r Contrast
- e: Colo[u]r Correction
- e: Column Capacity
- e: Combat Center
- e: Combat Command
- e: Combat Control
- e: Combinatory Code
- e: Combined (or Combination) Carrier
- e: Combined Carbon
- e: Comditional Command
- e: Command Card
- e: Command Center
- e: Command Chain[ing]
- e: Command Code
- e: Command Computer
- e: Command Console
- e: Command Control
- e: Command Counter
- e: Commanding Chain
- e: Comment Convention
- e: Commercial Computer
- e: Commission Certified
- e: Common Code
- e: Common Collector
- e: Common Control
- e: Communication[s] Center
- e: Communication[s] Central
- e: Communication[s] Centre
- e: Communication[s] Chief
- e: Communication[s] Code
- e: Communication[s] Computer
- e: Communication[s] Control[ler]
- e: Community College
- e: Compact Cassette
- e: Comparator Circuit
- e: Compass Course
- e: Compatibility Class
- e: Complete Carry
- e: Complete Code
- e: Completely Controllable
- e: Complex-Conjugate
- e: Compound Condition
- e: Compulsory Control
- e: Computation Center
- e: Computer Calculator
- e: Computer Capacity
- e: Computer Center
- e: Computer Channel
- e: Computer Check
- e: Computer Code
- e: Computer Community
- e: Computer Consulting
- e: Computer Control
- e: Computer Controlled
- e: Computer Cycle
- e: Computerized Conferencing
- e: Computing Center (US)
- e: Computing Centre (GB)
- e: Concave
- e: Concentric [Cable]
- e: Condition[al] Code
- e: Conditioning Container
- e: Conducting Channel
- e: Conducting Composition
- e: Conductive Channel
- e: Configuration Card
- e: Configuration Control
- e: Congestion Control
- e: Connecting Carrier
- e: Connecting Circuit
- e: Connection Confirm
- e: Connector Circuit
- e: Consecutive Computer
- e: Console Card
- e: Consolidated Computer
- e: Constant Command
- e: Constant Current
- e: Constructing Contractor
- e: Contact Closure
- e: Contemporary Civilization
- e: Continuation Cards
- e: Continuation Clause
- e: Continuity-Check
- e: Continuous Cassette
- e: Continuous Casting
- e: Continuous Clause
- e: Continuous Cooling
- e: Continuous Current
- e: Contrasting Colo[u]r
- e: Control Card
- e: Control Center
- e: Control Central
- e: Control Change
- e: Control Character
- e: Control Check
- e: Control Circuit
- e: Control Clause
- e: Control Code
- e: Control Column
- e: Control Command
- e: Control Computer
- e: Control Connector
- e: Control Console
- e: Control Contactor
- e: Control Counter
- e: Control Cycle
- e: Conversion Code
- e: Cooling Coil
- e: Coordinate Converter
- e: Coordinates Computed
- e: Copper Chromite
- e: Copper Constantan
- e: Core Concrete
- e: Corrugated or Cupped
- e: Corticosterone-Controlled
- e: Cost Center
- e: Counterclockwise
- e: Counting Code
- e: Country Code
- e: County Clerk
- e: Courtesy Copy
- e: Crate Controller
- e: Critical Condition
- e: Cross-Coupling

- e: Cross Channel
- e: Cross Correlation
- e: Cross Couple
- e: Cross Coupling
- e: Crown Clerk
- e: Cruise Control
- e: Crystal Control[led]
- e: Crystal Current
- e: Cubic Capacity
- e: Cubic Centimeter[s] (US)
- e: Cubic Centimetre[s] (GB)
- e: Cubic Contents
- e: Currency Clause
- e: Current Complaints
- e: Cursor Control
- e: Customs Charge[s]
- e: Cutting Cycle
- e: Cycle Check
- e: Cycle Counter
- e: Cyclic Check
- e: Cyclic Correlation
- e: Cytratcyclus
- s: Corriente Conrinua

C & C e: Calibration and Certification
- e: Calibration and Checkout
- e: Coal & Coke
- e: Command and Control
- e: Communication and Control
- e: Computer[s] and Communication[s]

C-C e: Center-to-Center
- e: Centre-to-Centre

c/c f: compte courant
- i: conto corrente
- p: conta corrente
- s: cuenta corriente

c.c. l: corpora cardiaca

CCA e: Cambridge Computer Associates
- e: Canadian Chemical Association
- e: Cancel Corridor Assignment
- e: Card Counting Attachment
- e: Caribbean Conservation Association
- e: Carrier Controlled Approach
- e: Cash Clothing Allowance
- e: Califomia Chiropractic Association
- e: Cellular Cellulose Acetate
- e: Central Chamber of Agriculture
- e: Central Computer Accounting
- e: Central Computer Agency
- e: Centre for Computing and Automation
- e: Cephalin Cholesterol Antigen
- e: Channel-to-Channel Adapter
- e: Chemical Communications Association
- e: Chemical Corps Association
- e: Chick Cell Agglutination
- e: Chimpanzee Coryza Agent
- e: Chromated Copper Arsenate
- e: Circa
- e: City Center Arts (journ.)
- e: Cloud Chamber Analysis
- e: Coarse Crushed Aggregate
- e: College Characteristics Analysis
- e: Combat Command A
- e: Comics Code Authority
- e: Common Communications Adapter
- e: Communication Carrier Assembly
- e: Communications Channel Adapter
- e: Communication[s] Control Area
- e: Compass Control Alarm
- e: Component Checkout Area
- e: Computer & Control Abstracts (journ.)
- e: Computer Corporation of America
- e: Configuration Control Action
- e: Conquest of Cancer Act
- e: Continental Control Area
- e: Contract Change Authorization
- e: Controlled Circulation Audit
- e: Copper Chrome Arsenate
- e: Copper Conductors Association
- e: Crop-Condition Assessment
- e: Cross Country Award (Radio Amateurs)
- e: Current-Controlled Amplifier
- e: Customer Cost Analysis
- e: Cytidyl-Cydityl-Adenyl
- f: Centre de Calcul Analogique

CC & A e: Computer Control and Auditing

CCAA e: Center for Computer Aided Analysis

CCAAFB e: Cape Canaveral Auxiliary Air Force Base

C-C Adapter e: Channel-to-Channel Adapter

CCAE e: Congressional Committee on Atomic Energy

CCAHC e: Central Council for Agricultural and Horticultural Cooperation
- e: Central Council for Agricultural and Horticultural Cooperatives

CCAI e: Chemical Coaters Association International

CCAIA e: California Council of the American Institute of Architects

CCAIC e: Catholic College Admissions and Information Center

CCAID e: Charge-Coupled Area Imaging Device

CCAIS e: Charge-Coupled Area Image Sensor

CCAIT e: Community College Association for Instruction and Technology

CCAM e: Card Controlled Accounting Machine
- e: Colby College Art Museum
- e: Congenital Cystic Adenomatoid Malformation
- e: Conversational Communication Access Method

CC-AMDF e: Cross-Correlation-Average Magnitude Difference Function

CC and S e: Central Computer and Sequencer

CCAP e: Communication Capability Application[s] Program[me]

CCAP e: Communication Control[led] Application[s] Program[me]
- e: Computer Controlled Application[s] Program[me]
- e: Conversational Circuit Analysis Program[me]
- e: Culture Centre of Algae and Protozoa

CCAPS e: Circuit-Card Assembly and Processing System

CCAQ e: Consultative Committee on Administrative Questions

C Car e: Combat Car

CCAS e: Carrier-Controlled Approach System
- e: Central Computer and Sequencing
- e: Communication Control Aid System
- e: Comprehensive Close Air Support
- e: Council of Colleges of Arts and Sciences

CCAT e: Computer-Controlled Automatic Test
- e: Conglutinating Complement Absorption Test
- e: Controlled Avalanche Transit-Time

CCAT E[quipment] e: Computer Controlled Automatic Test Equipment

CCATS e: Communications, Command and Telemetry Systems

CCATs e: Conglutinating Complement Absorption Tests

CCAU e: Cell Over Arming Unit

CCA Unit e: Chick[en] Cell Agglutination Unit

CCB e: Carbon Circuit Breaker
- e: Change Control Board
- e: Channel Command Block
- e: Channel Control Block
- e: Character Control Block
- e: Circuit Concentration Bay
- e: Code Check Byte
- e: Coded Current Block
- e: Coin Collecting Box
- e: Combat Command B
- e: Combined Communications Board
- e: Command and Control Board
- e: Command Communications Boat
- e: Command Control Block
- e: Common Carrier Bureau
- e: Communications Control Batches
- e: Communications Control Block
- e: Configuration Control Board
- e: Console-to-Computer Buffer
- e: Contraband Control Base
- e: Control Center Building
- e: Convertible Circuit Breaker
- e: Cubic Capacity of Bunkers
- e: Cyclic-Check[ing] Bit
- e: Cyclic-Check[ing] Byte

CCBA e: Central Canada Broadcasters Association

CCBD e: Configuration Control Board Directive
- e: Council for Children with Behavioral Disorders

CCBF d: Computerbearbeitung chemischer und biologischer Forschungsergebnisse

CCBI e: Cubic Capacity of Bales

CC Black e: Conductive Channel Black

CCBM e: Card Controlled Billing Machine
- e: Copper Cylinder and Boiler Manufacturers

CCBO e: Cape Clear Bird Observatory

CCBP e: Combined Communications Board Publications

CCBS e: Center for Computer-based Behavioral Studies
- e: Clear Channel Broadcasting Service
- e: Completion of Calls to Busy Subscriber

CCBV e: Central Circulating Blood Volume

CCBVP e: Compound Correlated Bivariate Poison

CCBW e: Commission on Chemical and Biological Warfare

CC Byte e: Cyclic Check Byte

CCC d: Chlor-Cholin-Chlorid
- e: Calcium Cyanamide Citrated
- e: Caldwell Computer Corporation
- e: Cambridge Communication Corporation
- e: Canadian Computer Conference
- e: Card Controlled Calculator
- e: Caribbean Conservation Corporation
- e: Carrier Current Communication
- e: Cathodal Closure Contraction
- e: Cathode Closure Contraction
- e: Central Communications Controller
- e: Central Computer Center
- e: Central Computer Complex
- e: Ceramic Chip Carrier
- e: Change and Configuration Control
- e: Channel Command Call
- e: Channel Command Control
- e: Channel Control Check
- e: Chicago Clinical Chemist (journ.)
- e: Chip-Carrying Card
- e: Chlorcholine Chloride
- e: Chopin Cultural Center
- e: Citrated Calcium Carbimide
- e: Classified Catalog[ue] Code
- e: Classified Control Clerk
- e: Clipper Cross-Correlation

- e: Clipper Cross-Correlator
- e: Closed Cycle Cooler
- e: Columbian Carbon Company
- e: Combat Command C
- e: Command Control Center
- e: Command Control Console
- e: Common Control Circuit
- e: Communications Control Console
- e: Complex Control Center
- e: Component Change Control
- e: Composite Current Conveyor
- e: Computer-Command Control
- e: Computer Circuit Corporation
- e: Computer Communications Console
- e: Computer Communications Converter
- e: Computer Composition Corporation
- e: Computer Control Company
- e: Console Control Circuit
- e: Coordinate Conversion Computer
- e: Copy Control Character
- e: Corrdinate Conversion Computer
- e: Corrugated Capacitor Cell
- e: Crown Cassette Communications Ltd.
- e: Cube Connected Computer
- e: Cube Connected Cycle
- e: Cycle Control Counter
- e: Cyclic Character Check

CC & C e: Command Control and Communications
CCCA e: Commander, Corps, Coast Artillery
CC-CB e: Common Collector-Common Base
CCCB e: Component Change Control Board
CCCC e: Cape Cod Community College
- e: Center City Commuter Connections
- e: Committee for the Conservation and Care of Chimpanzees
- e: Computerized Conferencing and Communications Center
- e: Conference on College Composition and Communication
- e: Consolidated Computer and Control Center

CCCCD e: Conductivity-Connected Charge-Coupled Device
CCCC Program[me] e: Community Coordinated Child Care Program[me]
CCCD e: California Center for Community Development
CCCE e: Closed Cycle Cryogenic Equipment
- e: Computer Communication Centre; Europe
- e: Computer Controlled Checkout Equipment
- e: Consulting Chemists and Chemical Engineers

CCCF e: Central Committee on Communications Facilities
CCC-FID e: Central Classification Committee-Fédération Internationale de Documentation
CCC Highway e: Cleveland-Columbus-Cincinnati Highway
CCCI e: Command Control, Communications and Intelligence
- e: Computer Control Company, Inc.
- e: Copenhagen Conference on Computer Impact

CCCIE e: Chinese Character Code for Information Exchange
CC Circuit e: Common-Collector Amplifier for Junction Transistors
CCCL e: Cathodal Closure Clonus
- e: Complementary Constant Current Logic
- e: Complementary [-symmetry] Metal Oxide Semiconductor Compact Cell

CCCM e: Central Committee for Community Medicine
- e: Coincident Current Core Memory

CCCMMM e: Closed Chest Cardiac Massage and Mouth-to-Mouth
CCCN e: Caribbean Christian Communications Network
CC Computer e: Cube Connected Computer
CCCP e: Carbonylcyanide m-Chloro-phenylhydrazone
CC & CP e: Command Control and Communications Program[me]
CCCP e: Comprehensive Child Care Project
CC/CP e: Current Contents / Clinical Practice
CCCPS e: Chicago College of Chiropody and Pedic Surgery
CCCR e: Committee on the Conservation of Cultural Resources
- e: Communication and Command Control Requirements

CCCRAM e: Continuously Charge Coupled Random Access Memory
CCCRMB e: Coal Committee of the Combined Raw Materials Board
CCCS e: Central Control Computer System
- e: Coincident Current Core Storage
- e: Command, Control and Communications System
- e: Common Core Character Set
- e: Core Component[s] Cleaning System
- e: Core Component[s] Conditioning Station

CCCs e: Cube Connected Cycles
CCCS e: Current-Controlled Current Source
- e: Current Contents Chemical Sciences (journ.)

CC-CT e: Corticosterone and Testosterone-controlled
CCCU e: Crew Compartment Cooling Unit
- e: Crew Conditioning Calibration Unit

CCD d: Century Computer Deutschland GmbH
- e: Carbonate Compensation Depth
- e: Card Code and Date
- e: Cash Concentration and Disbursement
- e: Cell Current Density
- e: Center for Curriculum Development
- e: Central Command Decoder
- e: Central Commissioning Detail
- e: Central Control Desk
- e: Charge-Coupled Device
- e: Circumscribing Circle Diameter
- e: Coarse Control Damper
- e: Cold Cathode Discharge
- e: Collator Counting Device
- e: Combat Center Director
- e: Combat Command D
- e: Command and Control Director
- e: Command Control Destruct
- e: Common Core of Data
- e: Computer-Controlled Display
- e: Computer Control Division
- e: Computer Crime Digest
- e: Condensed Chemical Dictionary
- e: Configuration Change Directive
- e: Construction Completion Date
- e: Contra-Current Distribution
- e: Contract Completion Date
- e: Copolymer Composition Distribution
- e: Core Current Driver
- e: Cost Center Determination
- e: Counter-Current Distribution
- e: Countercurrent Decantation

CCDA e: Cable Cards Design Automation
- e: Circuit Card Design Automation
- e: Commercial Chemical Development Association
- e: Computer-Compatible Data Acquisition

CCDA System e: Computer-Controlled Data Acquisition System
CCDC e: Canadian Communicable Disease Centre
- e: Central Control and Display Console
- e: Computer Compatible Data Collection

CCDCG e: Clinical Chemistry Data Communication Group
CCDF e: Cambridge Crystallographic Data Files
CCDN e: Central Council for District Nursing
- e: Corporate Consolidated Data Network

CCDP e: Central Computer Development Program[me]
- e: Cooperative College Development Program[me]

CCDs e: Charge-Coupled Devices
CCDS e: Computer Communications and Data Service
- e: Control Circuits Design Section

C/CDSB e: Command/Control Disable
CCDW e: Carrying Concealed Deadly Weapon
CCD-Winkel d: Centrum-Collum Diaphysenwinkel
CCDZ e: Continuous Controller with Dead Zone
CCE e: Carbon-Chloroform Extract
- e: Cesium Contact Engine
- e: Channel Command Entry
- e: Channel Control Error
- e: Charge Composition Explorer
- e: Chief; Corps of Engineers
- e: Chief of Communications Electronics
- e: Chief of Construction Engineering
- e: Clapeyron-Clausius Equation
- e: Combat Communications Equipment
- e: Combat Control Elements
- e: Command Chief Engineer
- e: Command Control Equipment
- e: Communications Control Equipment
- e: Complex Control Equipment
- e: Contract Change Estimate
- e: Contractor Change Evaluation
- e: Countercurrent Electrophoresis
- e: Crusade for a Cleaner Environment

CCEA e: Chemical Corps Engineering Agency
CCEI e: Colorado Committee for Environmental Information
- e: Composite Cost Effectiveness Index

CCEP f: Commission Consultative Etudes Postales
CCES e: Common Control Echo Suppressor
CCET e: Carnegie Commission on Educational Television
CCETT f: Centre Commun d'Etudes de Télévision et de Télécommunication
CCEX e: Clad Controlled Expansion
CCF e: Cancer Coagulative Factor
- e: Central Computer Facility
- e: Central Computing Facility
- e: Central Control Facility
- e: Cephalin-Cholesterol Flocculation
- e: Change Control Form
- e: Chronic Cardiac Failure
- e: Circuit Characteristic Function
- e: Circular Crystal Facet
- e: COBOL Communications Facility
- e: Common Chlorofluorocarbon
- e: Common Cold Foundation
- e: Common Communication Format.
- e: Communication Central Facility

- e: Communications Control Field
- e: Commutated Capacitor Filter
- e: Compound Comminuted Fracture
- e: Compressed Citation File
- e: Concentrated Complete Fertilizer
- e: Configuration Control Function
- e: Congestive Cardiac Failure
- e: Continuity-Failure Signal
- e: Continuous Card Feed
- e: Converter Compressor Facility
- e: Cooperate Central File
- e: Corporate Central File
- e: Cross Correlation Function
- e: Crown Competition Factor
- e: Hundred Cubic Feet

CCFC e: Continental Car Ferry Centre
CCFE e: Commercial Customer Furnished Equipment
- e: Corporate Central File-Engineering

CCFF e: Crew-Compartment Fit and Function
- e: Critical Colo[u]r Flicker Frequency

CCFL e: Capacitor-Coupled Field-Effect Transistor Logic
CCFM e: Cryogenic Continuous Film Memory
CCFMC e: Center for the Coordination of Foreign Manuscript Copying
CCFP e: Catalog[ue] Card Format Program[me]
CCFR e: Commonwealth Committee on Fuel Research
- e: Constant Current Flux Reset

CCFT e: Controlled Current Feedback Transformer
CCG e: Central Clock Generator
- e: Communications Change Group
- e: Computer Communications Group
- e: Constant Current Generator
- e: Cubic Capacity of Grain

CCGA e: California Cactus Growers Association
CCGCR e: Closed Cycle Cas-Cooled Reactor
CCGD e: Canadian Centre for Geoscience Data
CCGE e: California Council for Geographic Education
- e: Cold-Cathode-Gage Experiment
- e: Cold Cathode-Gauge Experiment

CCGS e: Corpus Christi Geological Society
CCGT e: Closed-Cycle Gas Turbine
CCH e: Channel Check Handler
CCH e: Computerized Criminal History
- e: Connections per Circuit per Hour
- e: Cube Corner Holder
- e: Cubic Capacity Of Holds

CCHAP e: Circulation Coupled Hover Analysis Program[me]
CCHEP e: Cement-Coated Heavy Epoxy
CCH File e: Computerized Criminal History File
CCH Indicator e: Continuity-Check Indicator
CCHK e: Continuity Check
CCHM e: Cubic Capacity of Holds in Metres
CC/hr e: Cubic Centimeters (or Centimetres) per Hour[s]
CCHS e: Christopher Columbus High School
CCI e: Calcium Chloride Institute
- e: Canadian Credit Institute
- e: Card Computer Interface
- e: Century Computer International
- e: Charge Coupled Imager
- e: Chronic Coronary Insufficiency
- e: Circuit Condition Indicator
- e: College Characteristics Index
- e: Com-Share Communications Interface
- e: Command Control Interface
- e: Component Cost Index
- e: Computer Communications, Inc.
- e: Computer Communications Interface
- e: Computer Composition International
- e: Computer Consoles, Inc.
- e: Computer Controlled Inking
- e: Computer Controlled Instruction
- e: Concentric Coordinate Incident
- e: Concordia Collegiate Institute
- e: Condition Code Indicator
- e: Consolidated Computer, Inc.
- e: Construction Cost Index
- e: Consultative Committee, International
- e: Continuity-Check Incoming
- e: Continuous Card Input
- e: Control Current Impedance
- e: Convert Clock Input
- e: Corrugated, Cupped of Indented
- e: Corrugated Container Institute
- e: Current-Controlled Inductor
- f: Comité Consultatif International

CCIA e: Computer and Communication Industry Association
- e: Console Computer Interface Adapter

CCIAP e: Cooperative Committee on Interstate Air Pollution
CCIB e: Computerized Central Information Bank
CCIC e: Constant Cost Integer Code
- e: Control Card Installation Card

CCICS e: Canadian Conference on Industrial Computer Systems
CCID e: Charge-Coupled Imaging Device
- e: Control Channel Information Demodulator

CCI Device e: Charge-Coupled Imaging Device
CCIG e: Cold Cathode Ion Gage
- e: Cold Cathode Ion Gauge

CCIL e: Commanders Critical Item List
CCIM e: Card Controlled Invoicing Machine
- e: Coronary Care Intensive Medical

CCINC e: Cabinet Committee for International Narcotic Control
CCIP e: Command and Control Information Processing
CCI P[oint] e: Continuously Computed Impact Point
CCIR e: Center for Communication and Information Research
- e: Commited Change Incorporation Record
- e: Communications Change Initiation Request
- e: Consultative Committee on International Radio
- f: Comité Consultatif International des Radiocommunications

CCIRID e: Charge-Coupled Infra-Red Information Device
CCIRRS e: Container and Chassis Identification Reporting and Recording System
CCIS e: Center for Computer and Information Services
- e: Coaxial Cable Information System
- e: Cold Cathode Ion Source[s]
- e: Command [and] Control Information System
- e: Common Channel Interoffice Signalling
- e: Computer-Controlled Interconnect System
- e: Curricular-Career Information Service

CCIS Network e: Common Channel Interoffice Signalling Network
CCISS e: Computerized Coffee Information Service System
- e: Conversational Computer Information and Statistical System

CCIT e: Cable Communications for Information Transfer
- f: Comité Consultatif International Télégraphique
- f: Comité Consultatif International Téléphonique

CCITT e: Consultative Committee on International Telegraph and Telephone
- f: Comité Consultatif International Télégrafique et Téléfonique

CCIU e: Command Channel Interface Unit
CCIW e: Canada Centre for Inland Waters
CCJC e: Chicago City Junior College
- e: Cook County Junior College
- e: Custer County Junior College

CCJD e: Centre for Contemporary Jewish Documentation
CCK d: Cholecystokinin
- e: Channel Check
- e: Coherent Carrier Keying

CCK-PZ d: Cholecystokinin-Pankreozymin
CCKW e: Counterclockwise
CCL e: Capacitor-Coupled Logic
- e: Central Clock
- e: Channel Code Letter
- e: Coating and Chemical Laboratory
- e: Coded Command Line
- e: Combat Command L
- e: Command Control Language
- e: Common Command Language
- e: Common Computer Language
- e: Common Control Language
- e: Communications Change Log
- e: Communications Circular Letter
- e: Communications Control Language
- e: Composite Cell Logic
- e: Computer Control Language
- e: Computer Control Loading
- e: Concise Command Language
- e: Consolidated Computer Limited
- e: Contact Clock
- e: Continuation Clause
- e: Convective Condensation Level
- e: Conversion and Check Limit
- e: Critical Components List
- e: Customized Card List

C-class e: capable of underwater missile launchings
- e: Soviet nuclear-powered submarines

CCLC e: Cooperative College Library Center
CCLCD e: Customized Card List Control Date
CCLID e: Charge-Coupled Line-Imaging Device
CCLKW[S] e: Counterclockwise
CCLN e: Committee for Computerized Library Networks
- e: Council for Computerized Library Networks

CCLS e: Canadian Council of Library Schools
- e: Central Committee on Lumber Standards
- e: Computer-Controlled Launch Set

CCLSAF e: Central Command, Land, Sea and Air Forces
CCL Signal e: Calling-Party Clear Signal
CCl_4 e: Carbon Tetrachloride
ccm d: Kubikzentimeter

CCM e: California College of Medicine
- e: Canadian Committee on Machine Readable Cataloguing
- e: Card Capacity Memory
- e: Center for Communications Management
- e: Central Control Module
- e: Charge-Coupled Memory
- e: Chopper Circuitized Module
- e: Cloud Camera Multiplexer
- e: Combat Cargo Mission
- e: Combined Cipher Machine
- e: Combined Coding Machine
- e: Commodity Class Manager
- e: Communication[s] Controller Multichannel
- e: Communication[s] Control Module
- e: Computer-Controlled Manufacturing
- e: Computer-Controlled Microscope
- e: Computer Control Mode
- e: Confidential Code Message
- e: Constant Current Modulation
- e: Control Computer Module
- e: Controlled Carrier Modulation
- e: Counter-Countermeasure[s]
- e: Crew Cargo Module
- e: Cross-Connect Multiplexer
- e: Cubic Centimeter[s]
- e: Cubic Centimetre[s] (GB)

CCMA e: Civilian Clothing Maintenance Allowance
CCMC e: Coincident-Current Magnetic Core
- e: Commonwealth Cable Management Committee

CCMD e: Continuous Current Monitoring Device
CC/min e: Cubic Centimeters (or Centimetres) per Minute
CCMIS e: Commodity Command Management Information System
CCMLC e: Chief; Chemical Corps
CCMLO e: Chief Chemical Officer
CCMM e: Complete Correlation Matrix Memory
CCMOS e: Clocked Complementary Metal Oxide Semiconductor
CCMP e: Conversion Complete
CCMPTC e: Central Computer Center
CCMR e: Central Contact Management Region
CCMS e: California College of Mortuary Science
- e: Central Cardiac Monitoring System
- e: Committee on the Challenges of Modern Society
- e: Computer Centre Management System

CCMs e: Charge-Coupled Memories
CCMSS e: Computer Controlled Microform Search System
CCMT e: Catechol-0-Methyl Transferase
CCMT e: Computer Compatible Magnetic Tape
- e: Computer Controlled Machine Tool

CCMTA e: Cape Canaveral Missile Test Annex
CCMTC e: Cape Canaveral Missile Test Center
CCMU e: Computer Controller Multiplexer Unit
CCMV e: Cowpea Chlorotic Mottle Virus
CCN e: Campus Computing Network
- e: Channel Code Number
- e: Classification Change Notice
- e: Close Copper Nailing
- e: Cloud Condensation Nuclei
- e: Cluster Controller Node
- e: Command Control Number
- e: Commonwealth Cable Network
- e: Communication Control Number
- e: Configuration Control Number
- e: Contract Change Notice
- e: Contract Change Notification

CCNC e: Charge Controlled Negative Capacitance
- e: Common-Channel Network Controller

CCNDT e: Canadian Council for Non-Destructive Testing
CC-NDT e: Cannot Call-No Dial[ling] Tone
CCNF e: Canonical Conjunctive Normal Form
CCNG e: Computer Communications Networks Group
CCNI s: Cía Chilena de Navegación Interoceanica
CCNL e: Current-Controlled Negative Inductance (L = Inductance)
CCNM e: Chaco Cave National Monument
CCNP e: Callao Cave National Park
- e: Carlsbad Caverns National Park

CCNR e: Central Commission for the Navigation of the Rhine
- e: Citizens Committee on Natural Resources
- e: Consultative Committee for Nuclear Research
- e: Current Controlled Negative Resistance

CCNS e: Cape Cod National Seashore
CCNSC e: Cancer Chemotherapy National Service Center
CCNT e: Chief Controller
CCNWR e: Cross Creeks National Wildlife Refuge
CCNY e: Chemists Club of New York
CCNZ e: Commodore Commanding; New Zealand-Force
CCO e: Cannot Call Out
- e: Certificate of Consignment / Origin
- e: Chicago College of Osteopathy
- e: Chief Chemical Officer
- e: Chief Cipher Officer
- e: Commonwealth Communications Organization
- e: Complete Carry-Over
- e: Composite Checkout
- e: Composite Cut-Off
- e: Comprehensive Certificate[s] of Origin
- e: Computer Controller Operation
- e: Configuration Change Order
- e: Constant Current Operation
- e: Continuity-Check Outgoing
- e: Coordinating Committee on Oceanography
- e: Coronary Care Nurse
- e: Coronary Care Nursing
- e: Crystal-Controlled Oscillator
- e: Current-Controlled Oscillator

CCOC e: Command Control Operations Center
CCOCP e: Corporate Customer Order Control Program[me]
CCOD e: Coalition to Cease Ocean Dumping
CCOH e: Corrosive Contaminants; Oxygen and Humidity
CCONV e: Common Business Oriented Language Conversion
CCOP e: Centralized Customer Order Processing
- e: Committee for the Co-ordination of Joint Prospecting [for Mineral Resources in A
- e: Consolidated Customer Order Processing
- e: Consolidated Customer Order Program[me]

CCO Pressure e: Constant Control Oil Pressure
CCOR e: Continuously Computed Optimum Release
C/COS e: Computer-Controlled Operating System
CCOS e: Current Communications
CC-OTH e: Cannot Call-Other[s]
CCP e: Cable Connector Panel
- e: California Computer Products
- e: Call Control Process[ing]
- e: Carbonless Copying Paper
- e: Caribbean Conservation Program[me]
- e: Casualty Clearing Point
- e: Catalogue Computerization Project
- e: Central Communications Processor
- e: Central Control Position
- e: Certificate in Computer-Progamming
- e: Certification of Computer Professionals
- e: Certified Computer Programmer
- e: Chance-Constrained Programming
- e: Cibachrome Print
- e: Code of Civil Procedure
- e: Command Console Processor
- e: Command Control Panel
- e: Command Control Post
- e: Commission on College Physics
- e: Communication Control Package
- e: Communication Control Processor
- e: Communications Control Panel
- e: Communications Control Program[me]
- e: Computer Central Processing
- e: Computer Circuit Protector
- e: Computer Controlled Polisher
- e: Conditional Command Processor
- e: Configuration Change Plan
- e: Configuration Control Panel
- e: Connecting Card Perforator
- e: Console Command Processor
- e: Consolidated Cryptologic Program[me]
- e: Consolidated Cryptologic Program[me]
- e: Constant Card Processing
- e: Continuous Correlation Processing
- e: Contract Configuration Process
- e: Control Command Processor
- e: Control Computer Program[me]
- e: Core Component Pot
- e: Critical Compression Pressure
- e: Cropland Conversion Program[me]
- e: Cross-Check Procedure
- e: Cross-Connection Point
- e: Cubic Close-Packed
- e: Cyanocarbonyl m-Chlorophenylhydrazone
- e: Cyclic Convolution Property
- f: Calculatrice de Commande à Program[me]

CCPA e: Cloud Chamber Photographic Analysis
CCPC e: Civil Communications Planning Committee
- e: Communications Computer Programming Center
- e: Control Cable Parallel Connector
- e: Control Center Programming Center

CCPD e: Charge-Coupled Photo Diode
- e: Coupling Capacitor Potential Device

CCPDD e: Center for Clinical Psychopharmacology Data Documetation
CCPDS e: Centralized Cancer Patient Data System
CCPE e: Canadian Council of Professional Engineers
CCPG e: Chemical Corps Proving Ground
CC-PILL e: Compound-Cathartic Pill
CCPL e: Corpus Christi Public Library
CCPR e: Coherent Cloud Physics Radar
- e: Connecting Circuit to Push-button-dialling Receiver

CCPS e: Center for Chemical Process Safety
- e: Consultative Committee for Postal Studies
CCPT e: Computer Controlled Positioning Table
CC Pulse e: Compute Clock Pulse
CCQ e: Central Control of Quality
CCR d: Carcinochrom-Reagenz
- e: Central Communication Region
- e: Central Control Room
- e: Chanel Command Register
- e: Channel Check Record[er]
- e: Channel Command Register
- e: Channel Control Reconfiguration
- e: Channel Control Routine
- e: Circulation Control Rotor
- e: Closed-Cycle Refrigerator
- e: Coaxial Cavity Resonator
- e: Combat Crew
- e: Command Chaining Retry
- e: Command Control Receiver
- e: Committment, Concurrency and Recovery
- e: Common Centre of Research
- e: Common Common-channel-signalling Receiver
- e: Communication Control Record
- e: Communications Change Request
- e: Communications Control Room
- e: Compensated Character Recognition
- e: Complex Chemical Reaction
- e: Complex Control Room
- e: Component Catalog[ue] Reviw
- e: Component Characteristics Record
- e: Computer Character Recognition
- e: Computer Check Routine
- e: Computer Controlled Retrieval
- e: Condition Code Register
- e: Configuration Control Register
- e: Console Card Reader
- e: Consumable Case Rocket
- e: Continuity Check Request
- e: Contract Change Release
- e: Contract Change Request
- e: Contractor Cost Reduction
- e: Control-Circuit Resistance
- e: Control Circuit Resistance
- e: Control Contactor
- e: Conversation Control Record
- e: Cooperative College Registry
- e: Credit Card Reader
- e: Credit Card Register
- e: Critical Compression Ratio
- e: Cross-Channel Rejection
- e: Crystal Can Relay
- e: Cube Corner Reflector
- e: Customer Controlled Reconfiguration
- e: Customized Communications Routine
- f: Commission centrale pour la navigation du Rhin
CCRA e: Commander; Corps; Royal Artillery
CCRAM e: Charge Coupled Random Access Memory
CCRAMs e: Charge Coupled Rando Access Memories)
CCRC e: Combat Crew Replacement Center
CCRDC e: Chemical Corps Research and Development Command
CCRE e: Center Current Range Element
- e: Commander; Corps; Royal Engineers
CCRESPAC e: Current Cancer Research Project Analysis Center
CCRH e: Center for Computer Research in the Humanities
CCRI f: Comité Consultatif de Recherche en Informatique
CCRIA f: Comité Consultatif de la Recherche en Informatique et Automatique
CCRIC e: Canadian Communications Research Information Centre
CCRIS e: Carnegie Commission Reports Information System
CCRL e: Cement and Concrete Reference Laboratory
CCRMA e: Center for Computer Research in Music and Acoustics
CCRMO f: Comité Consultatif de la Recherchc Météorologique Océanique
CCRNS e: Card-Capacitor Read-Only Storage
CCRO e: Clock Controlled Readout
CCROS e: Card Capacitor (or Capacitive) Read-Only Storage
- e: Card Capacity Read Only Storage
CC Rotor e: Circulation Controlled Rotor
CCrP e: Code of Criminal Procedure
CCRP e: Continuously Computed Release Point
CCR & R e: Covenants, Conditions, Restrictions and Reservations
CCRRM f: Comité Consultatif de la Recherche sur les Ressources de la Mer
CCRs e: Channel Command Registers
- e: Circulation Control Rotors
CCR Signal e: Continuity-Check-Request Signal
CCR System e: Credit Card Reading System
CCRT e: Cathodochromic Cathode Ray Tube
- e: Colo[u]r Cathode Ray Tube
CCRU e: Common Cold Research Unit
CCS d: Caracas (Flughafen)
- d: computer-codierter Suchlauf
- e: Calculus of Communicating Systems
- e: Camera Control System
- e: Canadian Cancer Society
- e: Canadian Computer Show
- e: Cape Code System
- e: Captital Computer Suites Limited
- e: Card Capacity Store
- e: Card Cornering Station
- e: Card Counting Sorter
- e: Carrier Colo[u]r Signal
- e: Cast Carbon Steel
- e: Casualty Clearing Station
- e: Cent Call Seconds
- e: Center for Chinese Studies (University of California)
- e: Center for Communication Studies
- e: Center for Comparative Sociology
- e: Center for Cuban Studies
- e: Central Certificate System
- e: Central Computer Station
- e: Central Computing Site
- e: Change Control System
- e: Channel Control System
- e: Chemical Corporation of Singapore
- e: Church of Christ, Scientist
- e: Circular Cylindrical Shell
- e: Classification and Compensation Society
- e: Cold Cranking Simulator
- e: Collective Call Sign
- e: Collector Coupled Structure[s]
- e: Colo[u]rvision Constant Speed
- e: Column Code Suppression
- e: Column Code Supression
- e: Command, Control, Support
- e: Command and Communication System[s]
- e: Command [and] Control System
- e: Command Communication Systems
- e: Common Channel Signal[ling] [Sender]
- e: Communication Collection System
- e: Communications Control System
- e: Component Control Section
- e: Computer City System
- e: Computer Communication Software
- e: Computer Consulting Service
- e: Computer Control Station[s]
- e: Computer Control System
- e: Consolidated Cargo Service
- e: Consolidated Container Service
- e: Contaminant Control System
- e: Continuous Colo[u]r Sequence
- e: Contour Control System
- e: Contrast Contour Seeker
- e: Control Command Store
- e: Control Computer Subsystem
- e: Controlled Combustion System
- e: Controller of Communication Services
- e: Conversational Compiling System
- e: Corporate Communications Services
- e: Corporate Consolidation Program[me]
- e: Corporate Consolidation System
- e: Council of Communication Societies
- e: Counter Control System
- e: Coupled Cavity Structure
- e: Custom Computer System
- e: Customer Conversion Statistics
- e: Hundred Call Seconds
CC & S e: Central Computer and Sequencer
- e: Central Computer and Sequencing
CCS² e: Command, Control and Subordinate Systems
CCSA e: Canadian Committee on Sugar Analysis
- e: Cascaded Carry Save Adder
- e: Cascaded Carry Save Addition
- e: Common Control Switching Arrangement[s]
CCSB e: Computer and Communications Standards Board
CCSD e: Command Communication Service Designator
- e: Common Channel Signalling Device
CCSE e: Corporate Communications Switching Equipment
CC/sec e: Cubic Centimeters (or Centimetres) per Second
CCSEM e: Computer Controlled Scanning Electron Microscope
CCSEP e: Cement-Coated Single Epoxy
CCSF e: City College of San Francisco
CCSG e: Computer Components & System Group
CCSI e: Conditional Control Sequence Interruption
CCSK e: Cyclic Shift Keying
CCSL e: Camp Coles Signal Laboratory
- e: Communications and Control Systems Laboratory
- e: Comparative Current Sinking Logic
- e: Compatible Current Sinking Logic
- e: Compatible Current Sourcing Logic
CCSN e: Common Channel Signalling Network
CCSPS e: Central Computer Support Programming System
CCSR e: Copper Cable Steel-Reinforced
CCSS e: Combat Crew Standardization School
- e: Common Channel Signalling System
- e: Conversational Computer Statistical System
- e: Cooperative College-School Science [Program]
CCST e: Center for Computer Sciences and Technology
- e: Chelsea College of Science and Technology

CCSTP e: Cubic Centimeters at Standard Temperature and Pressure
CCSU e: Computer Cross-Select Unit
CCSW e: Copper-Clad Steel Wire
CC System e: Communication Collection System
CCT e: C-Band Communications Transponder
- e: Carriage Control Tape
- e: Cathodal Closing Tetanus
- e: Cathode Closure Tetanus
- e: Centre-Cracked Tension
- e: Centrifugal Cast Tube
- e: Cesium Contact Thruster
- e: Character Code Translation
- e: Chemistry and Chemical Technology
- e: Circuit [Continuity Tester]
- e: Code Checking Time
- e: Combat Crew Training
- e: Communications Control Team
- e: Complete Calls To
- e: Complex Coordination Test
- e: Computer-Compatible Tape
- e: Computer-Controlled Teletext
- e: Computer-Controlled Terminal
- e: Computer-Controlled Typesetting
- e: Computer Checking Time
- e: Computer Concepts Trainer
- e: Computerized Communication Terminal
- e: Conductivity-Controlled Transistor
- e: Connecting Circuit
- e: Conradson Carbon Test
- e: Constant Current Transformer
- e: Continuous Cooling Transformation
- e: Controlled Cord Traction
- e: Correlated Colo[u]r Temperature
- e: Coupler Cut Trough
- e: Creosote Coal Tar
- e: Crystal-Controlled Transmitter
- e: Cubic Capacity Tonnage
- e: Cumberland College of Tennessee
CC/T e: Combined Center/Tower
CCT e: Common Customs Tariff
C" C_{29}, C_{39}, etc. e: Cytochromes 1, 2, 3, etc.
CCTA e: Cape Canaveral Test Annex
- e: Central Computer and Telecommunications Agency
CCTC e: Chemical Corps Technical Command
- e: Chinese Cultural and Trade Center
CCTe e: Cathodal Closure Tetanus
CCTEP e: Cement-Coated Triple Epoxy
CC Test e: Component Check Test
CC-Test e: Contempory Comparison-Test
CCTF e: Complex Contrast Transmission Function
CCTG e: Configuration Control Task Group
CC & TI e: Community College and Technical Institute
CCTI e: Conditional Control Transfer Instruction
CCTk[s] e: Cubic Capacity of Tank[s]
CCTL e: Collector Coupled Transistor Logic
CCTLP e: Counter-Cast Tunnel Lighting Principle
CCTN e: Correction
CCTP e: Card Controlled Tape Punch
- e: Center City Transportation Program[me]
CC TPDU e: Connection Confirm Transport Protocol Data Unit
CCTS e: Canaveral Council of Technical Societies
- e: Configuration Control Test System
- f: Comité de Coordination de Télécommunication par Satellites
CCTSIM e: Circuit Simulation
CCTS System e: Computer Controlled Traffic Signal System
CCTT e: Cornell Critical Thinking Test
CCTV e: Closed-Circuit Television
CCTV/LSD e: Closed-Circuit Television/Large-Screen Display
CCTVS e: Closed-Circuit Television Subsystem
- e: Closed Circuit Television System
CCU e: Cache Control Unit
- e: Calcutta (airport)
- e: Camera Control Unit
- e: Card Counting Unit
- e: Cardiac Care Unit
- e: Central Calculator Unit
- e: Central Capabilities Unit
- e: Central Computer Unit
- e: Central Control[ler] Unit
- e: Channel Control Unit
- e: Chart Comparison Unit
- e: Colo[u]r-Changing Unit[s]
- e: Colo[u]r-Control Unit
- e: Combined Control Unit
- e: Command Chain Unit
- e: Common Control Unit
- e: Communication[s] Control Unit
- e: Communication[s] Coupling Unit
- e: Computer Control Unit
- e: Console Control Unit
- e: Coronary Care Unit
- e: Correlation Control Unit
- e: Coupling Control Unit
CCUAP e: Computer Cable Upkeep Administrative Program[me]
CCUC e: Conference on Computers in the Undergraduate Carricula
CCUF e: Consolidated Component Usage File
- e: Controlled Component Usage File
- e: Corporate Control Usage File
CCURR e: Canadian Council on Urban and Regional Research
CCUS e: Cleared Customs
CCUs e: Coronary Care Units
C/CUT e: Crosscut
CCV e: Closed Circuit Voltage
- e: Code Converter
- e: Colo[u]r Contrast Value
- e: Combat Command V
- e: Common Control Vector
- e: Control Configured Vehicle
- e: Coolant Control Valve
- e: Credit Card Verification
CCVC e: Consonant-Consonant-Vowel-Consonant
CC-Virus e: Common Cold-Virus
CCVS e: Common Business Oriented Language Compiler Validation System
CCVS e: Current-Controlled Voltage Source
CCW e: Carrying [a] Concealed Weapon
- e: Channel Command Word
- e: Channel Control Word
- e: Command Control Word
- e: Component Cooling Water
- e: Copper-Clad Wire
- e: Counterclockwise
CC-Winkel d: Collum-Corpus-Winkel
CCWO e: Commercial Communications Work Order
- e: Cryptocenter Watch Officer
CC WR HDR e: Canvas-Covered Wire Rope Handrail
CCWS e: Chief of Chemical Warfare Service
- e: Component Cooling Water System
- e: Counterclockwise
CCXD e: Computer-Controlled X-ray Diffractometer
CCY e: Card Cycle[s]
- e: Chief Communication Yeoman
CCZA e: Canadian Coastal Zone Atlantic
CCZP e: Canadian Coastal Zone Pacific
CD d: Coder
- d: Cordoba-Durchmusterung
- d: Kadmium
- e: Cable Duct
- e: Cadmium
- e: Caesarian Delivery
- e: Calculation Diagram
- e: Calendar Day
- e: Calibration Device
- e: Call Detector
- e: Call Dispatch
- e: Calls per Day
- e: Camouflage Detection
- e: Canadian Turboprop Airplane
- e: Candela
- e: Candle
- e: Canonically Decoupled
- e: Capability Data
- e: Capacitive Discharge
- e: Capacitor-Diode
- e: Capacitor Discharge
- e: Capacity Downchange
- e: Card [Deck]
- e: Card Duplication
- e: Carried Down
- e: Carrier Detect[ion]
- e: Cataloged Data
- e: Cathode of Diode
- e: Caudal
- e: Celiac Disease
- e: Cell Data
- e: Cell Dendrites
- e: Center Distance
- e: Center Door
- e: Central District
- e: Centration Distance
- e: Centre Distance
- e: Centre Door
- e: Certificate of Deposit
- e: Certificate of Destruction
- e: Cesarean Delivered
- e: Chain Data
- e: Chain[ing] Data
- e: Change Directive
- e: Channel Data
- e: Character Density
- e: Charge Discharge
- e: Chart Datum
- e: Check Digit
- e: Checking Document
- e: Chemical Department
- e: Chlorine Demand
- e: Chronotropic Dose
- e: Circuit Description
- e: Circular-Dichroism[us]
- e: Circular Dispersion
- e: Classification of Defects
- e: Clearance Diving
- e: Clearence Diver
- e: Clock Driver
- e: Clorine Demand
- e: Clothing Depot
- e: Coast[al] Defence (GB)
- e: Coast[al] Defense [Radar]
- e: Code [Digit]
- e: Coding Delay
- e: Coefficient of Drag
- e: Cold-Drawn
- e: Colidyspepsie
- e: College Diploma
- e: College Discovery

- e: Collision Detect[ion]
- e: Colloid Droplet
- e: Combat Development[s]
- e: Command Decoder
- e: Command Delimiter
- e: Command Device
- e: Commercial Data
- e: Commercial Dock
- e: Common Denominator
- e: Common Digitizer
- e: Common Drain
- e: Communicable Disease
- e: Communication Data
- e: Compact-Disc [Player]
- e: Compact-Disc Record[ing]
- e: Compact Digital
- e: Comparative Data
- e: Compare Decimal
- e: Compiling Duration
- e: Completely Denaturated
- e: Component Demand
- e: Compressed Data
- e: Computer-Dependent
- e: Computer Data
- e: Computer Design (journ.)
- e: Computer Diagram
- e: Computer Digest (journ.)
- e: Computer Display
- e: Computer Drum
- e: Computing Device
- e: Condition of Detail
- e: Conductance
- e: Confidential Data
- e: Confidential Document
- e: Conjugate Diameter
- e: Connecting Devices
- e: Conning Director
- e: Consecutive Days
- e: Constructive Dilemma
- e: Contagious Disease
- e: Contextual Dependency
- e: Continental Divide
- e: Continuous Duty
- e: Control Data
- e: Control Desk
- e: Control Device
- e: Control Dictionary
- e: Convalescent Depot
- e: Converging-Diverging
- e: Convulsive Disorder
- e: Convulsive Dose
- e: Cord
- e: Cored
- e: Corrected Depth
- e: Could
- e: Count Data
- e: Countdown
- e: Counter Drive
- e: Critical Difference
- e: Cross Direction
- e: Crystal Diode
- e: Crystal Driver
- e: Curative Dose
- e: Curative Dosis
- e: Current Density
- e: Current Discharge
- e: Customs Declaration
- e: Cycle Delay
- e: Drag Coefficient
- s: Corriente Directa

C/D e: Card to Disk

C & D e: Chemist and Druggist
- e: Collected and Delivered
- e: Collection and Delivery
- e: Communications and Data
- e: Control[s] and Display

CD-4 e: Four Channel Disk

CD4 System e: Compatible Discrete 4[-channel] System

CDA e: California Dental Association
- e: California Dietetic Association
- e: Canadian Dental Association
- e: Canadian Dietetic Association
- e: Capability Data Acknowledgement
- e: Centralized Data Acquisition
- e: Certificate in Data Processing
- e: Chain[ing] Data Address
- e: Chenodeoxycholic Acid
- e: Closet Distance to Approach
- e: Coast Defense Artillery
- e: Coin Detection and Announcement
- e: College Diploma of Agriculture
- e: Command and Data Acquisition
- e: Completely Denaturated Alcohol
- e: Computer Dealers Association
- e: Constant Delta Altitude
- e: Continuity Data Area
- e: Control Data Australia
- e: Control Data Corporation
- e: Cool Dehumidified Air
- e: Copper Development Association
- f: Convertisseur Digital/Analogique
- s: Compañía Dominicana de Aviación

CDAA d: d-Chlor-N.N-diallylacetamid
- e: Central Data Analysis Area
- e: Chloro-Diallylacetamide

CD & AA e: Coast Defense and Antiaircraft

CDAC e: Command and Data Acquisition Center
- e: Communications Dual Access Controller

CDAEG d: Control Data AEG (Allgemeine Elektricitäts-Gesellschaft)

Cd A Eng e: Commissioned Air Engineer

CDAEP e: Civil Defense Adult Education Program[me]

CDA Flag e: Chain[ing] Data Address Flag

CDAP e: Civil Damage Assessment Program[me]
- e: Climatic Data Analysis Program[me]

CDAPSO e: Canadian Data Processing Service Organization

Cd Armn e: Commissioned Airman

CDAS e: Central Data Acquisition System
- e: Command and Data Acquisition Station

CDA Station e: Command and Data Acquisition Station

CD-Audio e: Compact Disk-Audio

CDB e: Caliper Disk Brake
- e: Capacitance Decode Box
- e: Cast Double Base
- e: Cellular Double Bottom
- e: Central Data Bank
- e: Central Data Base
- e: Coastal Defense Base
- e: Collegiate Dictionary of Botany
- e: Command Definition Block[s]
- e: Common Data Bank
- e: Common Data Base
- e: Common Data Bus
- e: Component Data Bank
- e: Corporate Data Base
- e: Current Data Bit
- e: Czapek-Dox Broth

CDBA e: Clearance Divers Breathing Apparatus

CDBCR e: Common Data-Base in Computer Readable-Form

CDBD e: Cardboard
- e: Chemical Data Base Directory

CDBHSS e: Conference on Data Bases in the Humanities and Social Sciences

CDBN e: Column-Digit Binary Network

CD Bty e: Coast Defense Battery

CDC e: Cadaver Disposal Center
- e: Calculated Data of Confinement
- e: Call-Directing Character
- e: Call-Directing Code
- e: Call-Direction Character
- e: Call-Direction Code
- e: Call Directing Character
- e: Capsule Drive Core
- e: Cell Division Cycle
- e: Center for Disease Control
- e: Central Data Control
- e: Central Digital Computer
- e: Central Dispatch Center
- e: Ceramic Disc Capacitor
- e: Cesspool Detergent Chemistry
- e: Channel Data Check
- e: Characteristic[s] Distortion Compensation
- e: Chemical Data Center
- e: Chemical Development Corporation
- e: Chemical Division Council
- e: Child Development Consultant
- e: Chlorodeoxycellulose
- e: Clamped Dielectric Constant
- e: Classified Document Control
- e: Coaxial Directional Coupler
- e: Code-Directing Character
- e: Coder-Decoder Control
- e: Command and Data-Handling Console
- e: Command Destruct Control
- e: Common Data Carrier
- e: Communicable Disease Center
- e: Communications and Data Centre
- e: Company Data Coordinator
- e: Composite Decompression Chamber
- e: Computer Development Center
- e: Computer Diode Corporation
- e: Computer Directions Corporation
- e: Computer Display Channel
- e: Configuration Data Control
- e: Conservation Data Center
- e: Construction Design Criteria
- e: Construction Design Criterium
- e: Control Data Company
- e: Control Data Computer
- e: Control Data Corporation
- e: Copper Data Center (Battelle Memorial Institute)
- e: Countdown Clock
- e: Course and Distance Calculator
- e: Course and Distance Computer
- e: Cryogenics Data Center
- e: Crystallographic Data Centre
- e: Cyclic Decimal Code
- p: Centro de Documentação Cientifica

CDCA e: Chenodeoxycholic Acid

cd-cb e: common drain-common base

CD & CC e: Central Data and Cataloging Center

cd-cc e: common drain-common collector

CDC Cable e: Command Decoder Coaxial Cable

CDCCP e: Control Data Communications Control Procedure
- e: Control Data Communications Control Program[me]

CDCD e: Certificate of Disposition of Classified Documents

CDCDP e: Civil Defense Career Development Program[me]

CDCE e: Central Data-Conversion Equipment

cd-ce e: common drain-common emitter

CDCEC e: Combat Developments Command Experimental Center

CD Circuit e: Common-Drain Circuit for Field-Effect Transistors

CDCL e: Command Document Capability List
CDCM e: Carbon Dioxide Concentrating Module
cd/cm² e: Candela per Square Centimetre
CDCN e: Computer Decisions (journ.)
Cd CO e: Commissioned Communication Officer
Cd Con. e: Commissioned Constructor
CDCP e: Command Display and Control Processor
CDCR e: Center for Documentation and Communication Research
- e: Control Drawing Change Request
CDCS e: Central Data Collection System
- e: Centralized Digital Control System
CDCSC e: Current-Dependent Current Source Capacitor
CDCT e: Centro de Documentación Cientifica y Téchnica
CDCU e: Communications Digital Control Unit
CDCY e: Card Cycle[s]
CDD e: Central Data Display
- e: Certificate of Disability for Discharge
- e: Channel-to-Drain Depletion
- e: Chart Distribution Data
- e: Chlorodibenzo-p-Dioxin
- e: Coded [Decimal Digit]
- e: Colo[u]r Data Display
- e: Combat Data Director
- e: Command Destruct Decoder
- e: Command Document Discard
- e: Compiler Directing Declarative
- e: Comprehensive Dissertation Database
- e: Computer Directed Drawing
- e: Cosmic Dust Detector
- e: Cratering Demolition Device
- e: Current Discontinuity Device
CDDC e: Center Data Descriptions Catalog
CDDDS e: Card Development Design Daily Status
CDDF e: Central Data Distribution Facility
CDDGP e: Commander; Destroyer Development Group; Pacific
CDDI e: Computer-Directed Drawing Instrument
CD Disease e: Curdy Dwarf Disease
CDDP e: Console Digital Display Programmer
CDDS e: Central Data Distribution System
CDDT e: Countdown Demonstration Test
CDE e: Canine Distemper Encephalitis
- e: Carbon Dioxide Economizer
- e: Central Document Exchange
- e: Certificate in Data Education
- e: Certification in Data Education
- e: Channel Data Error
- e: Coal Development Establishment
- e: Code[d]
- e: Command Document End
- e: Complex Data Entry
- e: Complex Date Entry
- e: Consumption Data Exchange
- e: Contamination-Decontamination Experiment
- e: Contents Directory Entry
- e: Controls and Display Equipment
- e: Cornell Dubilier Electronics
CDEC d: 2-Chlorallyl-diäthyldithiocarbamat
- e: Combat Developments Experimentation Center
- e: Combat Developments Experimentation Command
Cded e: Coded
CDEE e: Chemical Defense Experimental Establishment
C de F f: Collège de France
CDEG e: Chicago District Electric Generating Corporation
°C Degree e: Celsius
- e: Degree Centigrade
CDEI e: Control Data Education Institute[s]
CDEINSN e: Code Instruction
CDEIS e: Component Division Engineering Information System
CDEK e: Computer Data Entry Keyboard
Cd ElO e: Commissioned Electrical [Electronic] Officer
Cd Eng e: Commissioned Engineer
CDEOS e: Civil Defense Emergency Operations System
CDEP e: Civil Defense Education Program[me]
CDER e: Center for Death Education and Research
CDERR e: Card Error
CDES e: Comprehensive Data Entry Software
- e: Computek (Incorporated) Data Entry System
- e: Computer Data Entry System
- e: Conversational Data Entry System
CDETC e: Combat Development and Experimentation Center
CDEVC e: Computer Development Center
CDEX e: Civil Defense Exercise
CDF e: California Department of Forestry
- e: California Division of Forestry
- e: Canadian Department of Forestry
- e: Catalog[ue] Data File
- e: Chain Data Flag
- e: Change of Data Field
- e: Change to Data Field
- e: Chart Dynamic Flow
- e: Chlorodibenzofuran
- e: Class Determination and Finding
- e: Colorado Department of Forestry
- e: Command Decoder Filter
- e: Communications Data Field
- e: Component Description File
- e: Components Data File
- e: Confined Detonating Fuse
- e: Connecticut Department of Forestry
- e: Continuous Document Feeding
- e: Control Data Field
- e: Control Detonating Fuse
- e: Controlled Ductility Forming
- e: Critical Demulsification Temperature
- e: Crystal Defect Free
- e: Cumulative Density Function
- e: Cumulative Distributing Function
- e: Cumulative Distribution Function
- e: Cumulative Distribution of Frequency
CDFAB e: California Dried Fig Advisory Board
CDFCHB e: Command Data Format Control Handbook
CDFD e: Card Feed
CDFEA e: California Dried Fruit Export Association
CDFGI e: Charles Darwin Foundation for the Galápagos Islands
CD Film e: Camouflage Detection Film
cdfr e: Commercial Demonstration Fast Reactor
CDFRS e: Charles Darwin Foundation Research Station, Academy Bay, Santa Cruz, Galápagos
CDFSB e: Canadian Dairy Foods Service Bureau
CDFT e: Cord-Foot
cd/ft² e: Candela per Square Foot
CDG e: Capacitor Diode Gate
- e: Circular Diffraction Grating
- e: Code for Dangerous Goods
- e: Coder-Decoder Group
- e: Coding
- e: Converter Display Group
CDGA e: California Date Growers Association
Cd Gunn e: Commissioned Gunner
CDH e: Cable Distribution Head
- e: Choline Dehydrogenase
- e: College Diploma of Horticulture
- e: Command and Data Handling
- e: Constant Delta Height
- e: Constant Differential Height
CDHC e: Command and Data Handling Console
CDHL e: Cardholder
CDH-LOC e: Constructed Double-Heterojunction Large Optical Cavity
CDHP e: Conjugated Diene Hydroperoxide
CDHS e: Comprehensive Data Handling System
CDI e: Capacitive Discharge Ignition
- e: Capacitor Discharge Ignition
- e: Childrens Depression Inventory
- e: Classified Defense Information
- e: Collector-Diffusion-Isolation
- e: Collector Diffused Isolation
- e: Common Data Interface
- e: Compact Disk, Interactive
- e: Comprehensive Dissertation Index
- e: Computer-Directed Instrument
- e: Computer Devices, Incorporated
- e: Computer Direct Input
- e: Control Data Institute
- e: Course Deviation Indicator
- e: Cutting Die Institute
CDIAC e: Carbon Dioxide Information Analysis Center
CDICS e: Centralized Dealer Inventory Control System
CDIF e: Component Development and Integration Facility
- e: Controller/Director Information File
CD-Institute e: Control Data Institute
CDIP e: Correctional Decision Information Project
CDipAF e: Certified Diploma in Accounting and Finance
CDIS e: Community Data Information System
CDI Technique e: Collector Diffusion Isolation Technique
CDI-TTL e: Collector-Diffusion-Isolated Transistor-Transistor Logic
CDIU e: Central Digital Interface Unit
CDL e: Cable Delay Line
- e: Call Description Language
- e: Canal Defense Light
- e: Carbon Dioxide Laser
- e: Card Lever
- e: Cell Description Language
- e: Central Dockyard Laboratory
- e: Coaxial Diode Limiter
- e: Command Definition Language
- e: Common Display Logic
- e: Compiler Description Language
- e: Compiler Design Language
- e: Computer Description Language
- e: Computer Design Language
- e: Computer Development Laboratory
- e: Condor Data Link
- e: Constant Delay Line
- e: Construction, Demolition and Land-Clearing
- e: Contract Data List

- e: Contract Deficiency Listing
- e: Core Diode Logic
- e: Current Discharge Line
- e: Custom Dynamic Logic

CDLC e: Canadian Dental Laboratory Conference
- e: Cellular Data Link Control

CDLev e: Card Lever

CDL-FT e: Candle Foot

CDLRD e: Confirming Design Layout Report Date

CDLS e: Condor Data Link System

cDm d: Kubikdekameter

cdm d: Kubikdezimeter

CDM e: Card Document Machine
- e: Carriage Drive Mechanism
- e: Central Data Management
- e: Circuit Directory Maintenance
- e: Code Division Multiplex
- e: Commercial Database Machine
- e: Companded Delta Modulation
- e: Computer Description Manual
- e: Consolidated Diamond Mines
- e: Contractor Development Material
- e: Control Data Mathematical [Program]
- e: Core Division Multiplexing
- e: Cubic Decimeter
- e: Cubic Decimetre (GB)

Cd/m² d: Candela pro Quadratmeter
- e: Candela per Square Meter
- e: Candela per Square Metre (GB)

CDMA e: Call Division Multiple Access
- e: Cartridge Direct Memory Access
- e: Code Division Multiple Access

Cd. MAA e: Commissioned Master-at-Arms

CDMC e: Cooperative Data Management Committee

CDMLS e: Commutated Doppler Microwave Landing System

CDMM e: Component Demand Maintenance Module

Cd Mod e: Card Module

CDM Program[me] e: Control Data Mathematical Program[me]

CDM Programming e: Control Data Mathematical Programming

cd/ms e: candela per square meter

CDMS e: Central Data Management System
- e: Civil Data Management System
- e: Coherent Doubler Measurement Systems
- e: Commercial Data Management System
- e: Communications Data Management System

CDMSWA e: Consolidated Diamond Mines of South-West Africa

CDN d: Kanada (Kfz)
- e: Canada
- e: Coded Decimal Notation
- e: Coded Decimal Number
- e: Computer Decisions (journ.)
- e: Control-Dependent Noise
- e: Convergent-Divergent Nozzle
- e: Corporate Data Network
- e: Complementary Deoxyribonucleic Acid

CDNB d: 1-Chlor-2,4-Dinitrobenzol
- e: 1-Chlor - 2,4-Dinitrobenzol

CD/NC e: Computer-Aided Design / Numerical Control

CDNF e: Canonical Disjunctive Normal Form

CDNRA e: Coulee Dam National Recreation Area

CDO e: Cadmium Oxide
- e: California Disaster Office
- e: Change Design Order
- e: Chief Decontamination Officer

CDO e: Communications Duty Officer
- e: Community Dial Office

Cd OE e: Commissioned Ordnance Engineer

CDOG e: Combat Developments Objective Guide

CDOH e: Coupling Display Optical Hand

CDOH Controller e: Coupling Display Optical Hand Controller

CDOP e: Component Description Operation Procedure

CDO[S] e: Commando[s]

CDOS e: Controlled Data of Separation

Cdo. Sch. R. M. e: Commando School; Royal Marines

CDP d: Cytidindiphosphat
- e: Celestial Data Processor
- e: Central Data Processing
- e: Central Data Processor
- e: Central Distribution Programmer
- e: Centralized Data Processing
- e: Centralized Data Processor
- e: Certain Data Processing
- e: Certificate in Data Processing
- e: Certification in Data Processing
- e: Certified Data Plan
- e: Certified Data Processor
- e: Check Digit Procedure
- e: Checkout Data Processor
- e: Combat Development Project
- e: Command Data Processor
- e: Commercial Data Processing
- e: Common Depth Point
- e: Common Document Pool
- e: Communications Data Processor
- e: Compound Defraction Projector
- e: Compressor Discharge Pressure
- e: Concept Definition Proposal
- e: Confidence Development Plan
- e: Configuration Data Processor
- e: Constant Deviation Prism
- e: Contract Definition Phase
- e: Coronary Drug Project
- e: Correlated Data Processor
- e: Correlated Date Processor
- e: Cost Data Plan
- e: Cresyl Diphenyl Phosphate
- e: Critical Decision Point
- e: Crosslinked Dextran Polymer
- e: Cytidine [5-]Diphosphate

CDPA e: Certified Data Processing Auditor
- e: Computer Development Personnel Association

CD Pack e: Compact Disc Packaging

CDPB e: Command Document Page Boundary
- e: Computation and Data Processing Branch

CDPC e: Central Data Processing Computer
- e: Commercial Data Processing Center
- e: Computation and Data Processing Center
- e: Cytidine Diphosphate Choline

CDPCA e: Computing and Data Processing Society of Canada

CDPD e: Cellular Digital Packet Data

CDP Exam e: Certificate in Data Processing Examination

CDPI e: Command Data Processing and Instrumentation

CDPIE e: Command Data[s] Processor Interface Equipment

CDPIR e: Crash Data Position Indicator Recorder

CDPIRS e: Crash Data Position Indicator Recorder Subsystem

CDPL e: Cadmium Plate

CD Player e: Compact-Disc Player

CDPM e: Colo[u]red Digital Panel Meter

CDPP d: Cresyldiphenylphosphat
- e: Central Data Processing Program[me]

CDPR e: Customer Dial Pulse Receiver

Cd-Probe e: Cadmium-Probe

CDPS e: Champion Distributed Processing System
- e: Communications Data Processing System
- e: Computing and Data Processing Society

CDPSC e: Computing and Data Processing Society of Canada

CDPU e: Central Data Processing Unit

CDR e: Calcium-Dependent Regulator
- e: Call Data Recording
- e: Call Detail Record[ing]
- e: Card Reader
- e: Central Data Recording
- e: Channel Data Register
- e: Command Destruct Receiver
- e: Command Distribution Rack
- e: Command Document Resynchronization
- e: Command Document Resynchronize
- e: Compact Disk; Recordable
- e: Compare and Difference Right-half-words
- e: Complete Design Release
- e: Components Data Representative
- e: Composite Damage Risk
- e: Constant Dose Range
- e: Controlled Dynamic Range
- e: Countdown Deviation Request
- e: Critical Design Review
- e: Crude Death Rate
- e: Current Dicision Ratio
- e: Current Directional Relay
- e: Current Distribution Ratio

Cd. RadO e: Commissioned Radio Officer

CDRBTE e: Canadian Defence Research Boards Telecommunications Establishment

CDRC e: Conductivity-Recording Controller

CDRD e: Chemical Defence Research Department
- e: Computations and Data Reduction Division

Cd Rdr e: Card Reader

CDRE e: Chemical Defense Research Establishment

CDRF e: Canadian Dental Research Foundation

CDRI e: Central Drug Research Institute

CDRILL e: Center (or Centre) Drill
- e: Counterdrill

CDRILL-O e: Counterdrill Other [Side]

CDRL e: Contract Data Requirements List

Cdrngtn C e: Codrington College

CD-ROM e: Compact Disk-Read-Only Memory

CD-ROM-XA e: Compact Disk-Read-Only Memory-Extended Architecture

CDRR e: Call Detail Recording and Reporting

CDRS e: Charles Darwin Research Station
- e: Computer Data Recording System

CDRS e: Control and Data Retrieval System

CDRSC e: Childrens Depression Rating Scale for Classrooms

CDRU e: Child Development Research Unit

CDRX e: Critical Damping Resistance External

CDS d: computergesteuertes Datendrucksystem
- e: Cadium Sulfide
- e: California Dental Service
- e: Case Data System
- e: Cataloged Data Set
- e: Cataloging Distribution Service
- e: Cathode Dark Space

- e: Census Data System
- e: Center for Degree Studies
- e: Central Digital-Tone Sample-Generator
- e: Central Dynamic Store
- e: Century Data Systems
- e: Certified Documentary Specialist
- e: Charged Droplet Scrubber
- e: Chemical Discriminator System
- e: Chip Design System
- e: Circadian Data System
- e: Climatological Data Sheet
- e: Coastal Defense Ship
- e: Cold Drawn Seamless
- e: Cold Drawn Steel
- e: Colo[u]r Data System
- e: Colo[u]r Difference Signal
- e: Command Destruct System
- e: Command Document Start
- e: Common Diagram System
- e: Communications and Data Subsystem
- e: Compact Depth Sounder
- e: Compact-Disc[s] [Players]
- e: Compatible Duplex System
- e: Compiler Declarative Statement
- e: Compiler Directing Statement
- e: Component Disassembly Station
- e: Comprehensive Display System
- e: Compressed Data Storage
- e: Computer Data Switchboard
- e: Computer Data System
- e: Computer Duplex System
- e: Computerized Documentation Service
- e: Computerized Documentation System
- e: Concentrated Data Set
- e: Conceptual Design Study
- e: Configuration Data Services, Inc.
- e: Configuration Data Set
- e: Construction Differential Subsidy
- e: Consumer Data-Service System
- e: Container Delivery System
- e: Control Data Set
- e: Controlled Delivery System

C & DS e: Controls and Displays Subsystem
CD & SC e: Central Data & Switching Center
- e: Communications Distribution and Switching Center

CDSC e: Coupling Display Scanning Telescope Control
CDSCD e: Cash Discount Code
CDSE e: Computer Driven Simulation Environment
CDSI e: Computer Data Systems, Incorporated
CDS/ISIS e: Computerized Documentation System/Integrated Set of Information Systems
CDSL e: Capsule
- e: Connect Data Set to (or on) Line

CDSM e: Consolidated Defense Supply Material
- f: Comité sur les défis de la société moderne

cd/sq ft e: Candela per Square Foot
cd/sq in e: Candela per Square Inch
CDSR e: Contractural Data Status Reporting
- e: Contractural Data Status Reporting [System]
- e: Controlled Deployment Specular Reflector

CDSR System e: Contractural Data Status Reporting System
CDSS e: Clinical Decision Support System
- e: Compressed Data Storage System
- e: Customer Digital Switching System

C & DSS e: Communications and Data Subsystem[s]
CDST e: Central Daylight Saving Time
CDSU e: Computer and Data Systems Unit
CDSV e: Composite Data Service Vendor
CDT d: chemisch-desinfizierende Trockenreinigung
- d: Cyclohexylendimethyl
- e: Cable Data Tape
- e: Catholic Dictionary of Theology
- e: Central Daylight Time
- e: Certifed Dental Technician
- e: Channel Descriptor Table
- e: Cold-Drawn Tube
- e: Command Destruct Transmitter
- e: Communications Display Terminal
- e: Compressed-Data Tape
- e: Compressor Discharge Temperature
- e: Computer Data Transmission
- e: Computer Data Transmitter
- e: Conductor
- e: Connectionless Data Transmission
- e: Control Data Terminal
- e: Control Digit Test
- e: Coordinate Data Transmission
- e: Coordinate Data Transmitter
- e: Countdown Demonstration Test
- e: Countdown Time
- e: Credit

Cd_{115} e: Radioactive Cadmium
CD_{50} e: [Median] Curative Dose - 50%
CDTA e: Cyclohexanediaminetetraacetic Acid
- e: Cyclohexylenedinitrilotetraacetic Acid

CDT (ADA) e: Council on Dental Therapeutics
CDTCP e: Communication and Data Transmission Control Program[me]
CdTe e: Cadmium Telluride
C-DTE e: Character-Mode Data Terminal Equipment
CDTEC e: Combat Development Technical Evaluation Center
- e: Compact-Disc Technology

CDTHES e: Computerized Documentation Thesaurus
CDTI e: Cockpit Display of Traffic Information
CDTIS e: Components Division Technical Information System
CDTL e: Common Data Translation Language
- e: Complementary Diode Transistor Logic

CDTS e: Computer-Directed Training System
- e: Constant-Depth Temperature Sensor

CDU d: Computer deutschen Ursprungs
- e: Cable Distribution Unit
- e: Cache Decision Unit
- e: Capacitor Discharge Unit
- e: Cartridge Disk Unit
- e: Central Decode Unit
- e: Central Display Unit
- e: Chemical Dependency Unit
- e: Coastal Defense [Radar] Unit
- e: Command Destruct Unit
- e: Computer Display Unit
- e: Control / Display Unit
- e: Control and Diagnostic Unit
- e: Control and Display Unit
- e: Control Data Unit
- e: Cooling Distribution Unit
- e: Coupling Data Unit
- e: Coupling Display Unit
- f: Calculatrice Digitale Universelle

CDUEP e: Civil Defense University Extension Program[me]
CDUI e: Command Document User Information
CDUO e: Coupling Display Unit Optics
CDV e: Check Digit Verification
CDV e: Compact Disk; Video
CDVSC e: Current-Dependent Voltage Source Capacitor
CDW e: Carrying a Dangerous Weapon
- e: Charge Density Wave
- e: Chilled Drinking Water
- e: Computer Data Word

CDWR e: Chilled Drinking Water Return
CDWs e: Charge Density Waves
CDWS e: Civil Defense Warning System
CDX e: Coastal Defense Exercise
- e: Control Differential [Synchro] Transmitter

CDYN d: Zentidyn
CDZ d: Chlordiazepoxide
- d: Chlordiazepoxyd

CDZ e: Collegiate Dictionary of Zoology
CE d: Cer[ium]
- d: Cholesterol Ester
- d: Cholinesterase
- d: Collip-Einheit
- d: Empfangscontainer
- e: Cable Equalizer
- e: Calculation Experiment
- e: Calibration Error
- e: Canada East
- e: Capacitance; External
- e: Captain Engineer
- e: Carbon Equivalent
- e: Card Ejection
- e: Card Error
- e: Cardiac Enlargement
- e: Casualty Evacuation
- e: Cathode Efficiency
- e: Catholic Encyclopedia
- e: Celestial Equator
- e: Center (or Centre) of Efforts
- e: Center Entrance
- e: Central Exchange
- e: Centre Entrance
- e: Cerussite
- e: Ceylon
- e: Chaining Error
- e: Channel End
- e: Chemical Engineer[ing]
- e: Chemical Explosive
- e: Chief [of] Engineers
- e: Chip-Enable
- e: Christian Era
- e: Chronometer Error
- e: Circuit Common-Emitter Amplifier for Junction Transistors
- e: Circular Error
- e: Civil Engineer
- e: Clear Entry
- e: Coal Equivalent
- e: Code Equipment
- e: Coincidence Element
- e: College English (journ.)
- e: Combat Exhaustion
- e: Combustion Engine
- e: Combustion Engineering
- e: Common Emitter
- e: Common Era
- e: Communications Electronics
- e: Communications Equipment
- e: Commutator End
- e: Compacted Earth
- e: Comparing Element
- e: Compass Error
- e: Compensating Errors
- e: Composition Exploding
- e: Composition Explosive
- e: Compound Expression
- e: Compression Engine

- e: Compute Element
- e: Computer Engineer
- e: Computing Element
- e: Computing Error
- e: Concept Exploration
- e: Conditional Expression
- e: Conducted Emission
- e: Conspicuity Enhancement
- e: Constant Error
- e: Constant Expression
- e: Construction Electrician
- e: Consumer Electronics (journ.)
- e: Contractible Element
- e: Control Element
- e: Control Engineering
- e: Control Equipment
- e: Controller Error
- e: Converting Enzyme
- e: Coomutator End
- e: Corps of Engineers
- e: Cost Effectiveness
- e: Count Enable
- e: Countercurrent Electrophoresis
- e: Current Efficiency
- e: Customer Engineer[ing]
- e: Customs Entry
- e: Cytopathic Effect
- f: Calculateur Electronique

C & E e: Communications and Electronics
- e: Construction and Equipment
- e: Coordination and Equipment

C/E e: Compile and Execute

CEA d: carcinoembryonales Antigen
- e: Cambridge Electron Accelerator
- e: Canadian Education Association
- e: Canadian Electrical Association
- e: Carcinoembryonic Antigen
- e: Cement Employers Association
- e: Central Electricity Authority
- e: Childbirth Education Association
- e: Circular Error Average
- e: Clinical Experience Abstracts
- e: Column Exit[s] Address[es]
- e: Commodity Exchange Authority
- e: Common Carcinoembryonic Antigen
- e: Communications Electronics Agency
- e: Congressional Education Associates
- e: Connecticut Educational Association
- e: Conservation Education Association
- e: Constant Extinction Angle
- e: Control Electronics Assembly
- e: Control Element Assembly
- e: Cooperative Education Association
- e: Correctional Education Association
- e: Cost-Effectiveness Analysis
- e: Council for Energy Awareness
- e: Council of Economic Advisers
- e: Council of Engineering Associations
- f: Commissariat à l'Energie Atomique
- f: Confédération Européenne de l'Agriculture
- s: Comisión Económica de Africa

CEAA e: Center for Editions of American Authors
- f: Communauté Européenne de l'Energie Atomique

CEAC e: Commission for European Airspace Coordination
- e: Consulting Engineers Association of California
- e: Control Element Assembly Computer
- e: Cost and Efficiency Adjustment Clause
- f: Commission Européenne de l'Aviation Civile

CEADI e: Colo[u]red Electronic Attitude Director Indicator

CEA-Krebs-Test d: carcino-embryonaler Antigen-Krebs-Test

CEAL e: Cambridge Electron Accelerator Laboratory

CeAN d: Cer(IV)-Ammoniumnitrat
- e: Cer (IV)-Ammoniumnitrat

CEAN e: Community Energy Action Network
- e: Computerized Electroencephalogram Analysis

CEANAR e: Commission on Education in Agriculture and National Resources
- e: Commission on Education in Agriculture and Natural Resources

CEAPD e: Central Air Procurement District

CEARC e: Computer Education and Applied Research Center

CEASD e: Corporate Engineering and Sales Directive

CEASE e: Cost Engineering Automated System Estimate

CEASRS e: Civil Engineer Automated Specification Retrieval System

CEAT e: Canadian English Achievement Test

CEATOMS e: Control Engineering Approach to Management Systems

Ceb e: Cebu

CEB e: Central Electricity Board
- e: Chicken Embryo Bursal
- e: Comité Electrotechnique Belge
- e: Communication and Electronics Board
- e: Continuing Education Books
- e: Cryogenic Expulsive Bladder

CEBAR e: Chemical, Biological [and] Radiological [Warfare]

CEBAR Warfare e: Chemical, Biological and Radiological Warfare

CEBD e: Controller of Experimental Building Development

CEBEMA e: Computer and Business Equipment Manufacturers Association

CeBIT d: Centrum für Büro- und Informationstechnik

CEBJ e: Commission of Editors of Biochemical Journals

CEBM e: Corona, Eddy Current, Beta Ray, Microwave

CEBUS e: Confirmed Exposure But Unconscious

CE Bus e: Consumer Electronics Bus

Ceb-Vis e: Cebu-Visayan

CEC e: California Energy Commission
- e: Canadian Electric[al] Code
- e: Capsule End Cover
- e: Card Extract Counter
- e: Catholic Education Council
- e: Cation-Exchange Capacity
- e: Cation-Exchange Chromatography
- e: Center for Educational Change
- e: Centicycle[s]
- e: Central Economic Committee
- e: Centralized Electronic Control
- e: Ceramic Educational Council
- e: Certification of Equipment Completion
- e: Cholesteryl Erucyl Carbonate
- e: Chuo Electric Computer
- e: Circular Exhaust Cloud
- e: Civil Engineer-in-Chief
- e: Civil Engineer Corps
- e: Coal Experts Committee
- e: Combined Environmental Chamber
- e: Combined Environment Centrifuge
- e: Commonwealth Edison Company
- e: Communications and Electronics Command
- e: Complex Equipment Contract
- e: Computer Equipment Corporation
- e: Consolidated Edison Company
- e: Consolidated Electrodynamics Corporation
- e: Consolidated Electronics Corporation
- e: Consotidated Electrodynamics Corporation
- e: Constant Electric Contact
- e: Consulting Engineers Council
- e: Continental Entry Charts
- e: Controlled Element Computer
- e: Coulson Electrolytic Conductivity
- e: Crop End Control

CECA f: Conmmunauté Européenne du Charbon et de l'Acier
- s: Comunidad Europea del Carbon y del Acero

CE-CB e: Common Emitter-Common Base

CECC e: California Educational Computer [and] Consortium
- e: California Educational Computing Consortium

CECD e: Coulson Electrolytic Conductivity Detector

CECE f: Centre d`Etude et d`Exploitation des Calculateurs Electroniques
- f: Comite pour l'Exploitation de Calculatrices Electroniques

CECIL e: Compact Electronic Components Inspection Laboratory

CECL e: Civil Engineering Computer Laboratory

CECM e: Composite Engineering Change Memo

CECMRL e: Communications-Electronics Consolidated Mobilization Reserve List
- e: Communications-Electronics Materiel Reserve List

CECO e: Center Engine Cut-Off
- e: Commonwealth Edison Company

CECON e: Cleveland Electronic Conference

CECOS e: Civil Engineers Corps Officers School

CECR e: Central European Communication Region

CECRCA e: Commonwealth and Empire Conference on Radio for Civil Aviation

CECRI e: Central Electro-Chemical Research Institute

CECs e: California Ecology Corpsmen

CECS e: Communications Electronics Coordinating Section

CEC Symposium e: Computers, Electronics and Control Symposium

CECTK e: Committee for Electro-Chemical Thermodynamics and Kinetics

CECUA e: Confederation of European Computer Users Associations

CED e: Capacitance Electronic Disk
- e: Captured Enemy Document[s]
- e: Carbon-Equivalent-Difference
- e: Committee for Economic Development
- e: Communications-Electronics Doctrine
- e: Communications-Electronics Document
- e: Communications and Electronics Division
- e: Computer Entry Device
- e: Constant Energy Difference[s]
- e: Copper Ethylenediamine
- e: Cost Estimate Dispersion
- i: Centro Elletronnico di Documentazione

CEDA e: California Economic Development Agency
- e: Canadian Electrical Distributors Association
- s: Comisión Económica de Oeste Astatico

CEDAC e: Central Differential Analyzer Control
- e: Computerized Energy Distribution and [Automated] Control
- e: Cooling Effect Detection and Control
CEDAM e: Conservation, Exploration, Diving, Archeology, Museums
CEDAR e: Computer-Aided Environmental Design Analysis and Realization
- e: Computers in Education as a Resource
- e: Council for Educational Development and Research
CEDDA e: Center for Experiment[al] Design and Data Analysis
CEDI f: Centre Européen de Documentation et d'Information
- s: Centro Europeo de Documentación e Información
CEDM e: Control Element Drive Motor
CEDO e: Centre for Educational Development Overseas
CEDPA e: California Educational Data Processing Association
- e: Certificate in Electronic Data Processing Auditing
CEDPM e: Central Electronic Data Processing Machine
CEDPO e: Communications Electronics Doctrinal Project Office
CEDs e: Captured Enemy Documents
- e: Constant Energy Differences
CEE e: Captured Enemy Equipment
- e: Carbon Electrode Equipment
- e: Center for Environmental Education
- e: Central Engineering Establishment
- e: Central European Encephalitis
- e: Certificate of Extended Education
- e: Channel Equipment Error
- e: Combat Emplacement Excavator
- e: Common Entrance Examination
- e: Communication Electronics Element
- e: Computer-Enhanced Education
- e: Controlled Environment Enclosure
- e: Controller of Engineering and Equipment
CEEA f: Communauté Européenne de l'Energi Atomique
CEECC e: Consolidated Edison Energy Control Center
CEECS e: Computer Environment Energy Control System
CEEFAX e: See-the-Facts
CEEI e: Center for Energy and Environmental Information
CEEP f: Centre Europeáen d'Etudes de Population
CEER e: Chemical Economy & Engineering Review (journ.)
CEERI e: Central Electronics Engineering Research Institute
CEEs e: Controlled Environment Enclosures
CEEV e: Central European Encephalitis Virus
CEF e: Carrier Elimination Filter
- e: Cellular Expansion Factor
- e: Centrifugal Electrostatic Focusing
- e: Centrifugal Electrostatic Forcused
- e: Channeling Effect Factor
- e: Chick[en] Embryo Fibroblasts
- e: Cloth Elongation Factor
- e: Complementary Emitter Follower
- e: Computer Execute Function
- e: Contemporary Evaluation Form
- e: Corporate Express File (or Format)
- e: Critical Experiment Facility
CEFF e: Controlled Energy-Flow Forming
CEFI e: Contractor Engineer-Furnish and Install
CEFOAM e: Checkout Equipment for Onboard Automatic Maintenance
CEFTRI e: Central Food Technological Research Institute
CEG e: Coalition for Economic Growth
- e: Computer Education Group
CEGB e: Central Electricity Generating Board
CEGB Digest e: Central Electricity Generating Board Digest
CEGEP f: College d Enseignement Géneral et Professionnel
CEGL e: Cause Effect Graph Language
CEGS e: Council on Education in the Geological Sciences
CEHHS e: Charles Evans Hughes High School
CEHS e: Civilian Employees Health Service
CEI e: Center for Energy Information
- e: Chemical Engineering Index
- e: Chip Enable Input
- e: Claremont Economics Institute
- e: Cleveland Electric Illuminating [Company]
- e: Committee for Engineering Information
- e: Committee for Environmental Information
- e: Communication[s] Electronic[s] Instruction[s]
- e: Computer Exchange, Incorporated
- e: Computer Extended Instruction
- e: Consolidated Electronics Industries
- e: Correct End Item
- e: Cost Effectiveness Index
- e: Council of Engineering Institutions
- e: Cycle Engineers Institute
- f: Commission Électrotechnique Internationale
CEIAC e: Coastal Engineering Information Analysis Center
CEIB e: Computer Equipment Information Bureau
CEIE e: Council of European Industrial Federations
CEIMS e: Communications Equipment Information and Modelling System
CE-in-C e: Civil Engineer-in-Chief
CEIP e: Communications-Electronics Implementation Plan
CEIPA e: Communications-Electronics Implementation Plan Amendment
CEIR e: Computer Processing Corporation for Economic and Industrial Research
CEIRD e: Confirming Engineering Informatiion Report Date
CEIS e: California Education Information System
- e: Central Economic Information Service
- e: Cost and Economic Information System
CEISD e: Customer Engineering Instruction System Diagram
CEJ e: Cement-Enamel Junction
CEK e: Chicken Embryo Kidney
CEL e: Acet Cellulose Acetate
- e: California Eastern Laboratories
- e: Carbon-Equivalent; Liquid
- e: Celebes
- e: Celebrate[d]
- e: Cellular
- e: Celluloid
- e: Cellulose
- e: Cellulose Celcarbon-Equivalent-Liquid
- e: Celsius
- e: Central Education Library
- e: Civil Engineering Laboratory
- e: Combat Elevation Launch
- e: Computer Economics Limited
- e: Contrast-Enhanced Lithography
- e: Conversational Extensible Language
- e: Corporate Engineering List
- e: Cosine Emission Law
- e: Crowding Effect Laser
- e: Cryogenic[s] Engineering Laboratory
CELADE s: Centro Latinoamericano de Demografía
CELD e: Chicken Embryo Lethal Dose
CELDS e: Computer-Aided Environmental Legislation Data System
- e: Computerized Environmental Legislative Data System
CELEB e: Celebrate
- e: Celebration
- e: Celebrity
CELEBS e: Celebrities
CELESCOPE e: Celestial Telescope
CELIM e: Continuing Education for Library and Information Management
CELINTREP e: Accelerated Intelligence Report
Cell. d: Cellophan
CELL e: Case Existological Laboratories Limited
- e: Celluloid
- e: Continuing Education Learning Laboratory (of Youth Pride, Inc.)
CELLOPH d: Cellophan
CELLPHONE e: Cellular Telephone
CELNAV e: Celestial Navigation
CELNITR e: Cellulose Nitrate
CELO e: Chicken Embryo Lethal Orphan (Virus)
CELOS n: Centrum voor Landbouwkundig Onderzoek in Suriname
CELO Virus e: Chicken-Embryo-Lethal-Orphan Virus
CELPIS e: Colleges of Education Learning Programmes Information Service
C-E-I-R e: Corporation for Economic and Industrial Research
CELS e: Celsius
- e: Continuing Education for Library Staffs
- e: Continuing Education of Librarians in the Southwest
CEL SHEET e: Cellulose [Plastic] Sheet
CELSS e: Closed Ecological Life Support System
CELT e: Classified Entries in Lateral Transposition
- e: Coherent Emitter Location Testbed
CELTIC e: Cell Transport Integral Calculation
- f: Concentrateur Exploitant Les Temps d'Inoccupation des Circuits
CEL Timber e: Celtis Timber
CEM e: Captured Enemy Material
- e: Cement [asbestos]
- e: Cemetery
- e: Central Enhancement and Maintenance
- e: Channel Electron Multiplier
- e: Chief Electricians Mate
- e: Circular Electric Mode
- e: Communication[s]-Electronic[s]-Meteorological
- e: Composite Electro-Chemical Materials
- e: Compromising Emanations
- e: Conventional Electron Microscope
- e: Cost Element Monitor
- e: Cost Evaluation Model
- e: Counter Electromotive
- e: Customer Engineering Memorandum
- f: Compatibilité Électromagnétique
CEMA e: Canadian Electrical Manufacturers Association
- e: Cement Asbestos
- e: Channel Electron Multiplier Array

CEMAB e: Cement Asbestos Board
CEMAC e: Central Monitoring and Control
CEMAD e: Coherent Echo Modulation and Defection
- e: Coherent Echo Modulation and Detection
CEMAP e: Cotton Export Market Acerage Program[me]
- f: Commission européenne des méthodes d'analyse des pesticides
CEM ASB e: Cement Asbestos
CEMAST e: Control of Engineering Material, Acquisition, Storage and Transport
CEMB e: Communications-Electronic-Meteorological Board
CEMBI e: Conference to Explore Machine-Readable Bibliographic Interchange
CEMCB e: Chief Electrians Mate; Construction Battalion
CEMCBC e: Chief Electricians Mate; Construction Battalion; Communications
CEMCBD e: Chief Electricians Mate; Construction Battalion; Draftsman
CEMCBG e: Chief Electricians Mate; Construction Battalion; General
CEMCBL e: Chief Electricians Mate; Construction Battalion; Line-and-Station
CEMCO e: Continental Electronics Manufacturing Company
CEMF e: Counter Electromotive Force
CEMFL e: Cement Floor
CEMIS e: Client-Employee Management Information System
C-E MIX e: Chloroform-Ether Mixture
CeMM d: Ce-Mischmetall
CEMON e: Customer Engineering Monitor[ing]
CEMP e: Cement Paint
CEMPAC e: Communications-Electronics-Meteorology Program[me]Aggregate Code
CEMPIMS e: Communications Electronics Meteorological Program[me] Implementation Man. System
CEMPLAS e: Cement Plaster
CEMR e: Canadian Energy, Mines and Resources
CEMREL e: Central Midwest Regional Educational Laboratory
CEMS e: Central Electronic Management System
- e: Chemical Education Material Study
- e: Civil Engineer Mangement System
CEMs e: Conventional Electron Microscopes
CEMS e: Critical Equipment Monitoring System
CEMSRG e: Chief Electrians Mate; Ship Repair; General-Electrician
CEMSRS e: Chief Electrians Mate; Ship Repair; Shop-Electrician
CEN e: Capacity Exceeding Number
- e: Cenozoic
- e: Centaur
- e: Centaurus (Astronomical Constellation)
- e: Center[ed]
- e: Central[ization]
- e: Centralize
- e: Centre
- e: Century
- f: Comité Européen de Coordination des Normes
C & EN e: Chemical and Engineering News (journ.)
Cen A e: Central Area
Cen. Am. e: Central America
CENCOMMURGN e: Central Communications Region
CENCOMS e: Center for Communication Sciences
CEND e: Combustion Engineering Nuclear Division
CENDIT e: Centre for Development of Instructional Technology
CENELEC f: Comité Européen de Normalisation Electrotechnique
CENEUR e: Central Europe[an]
CENEUR AACS Reg e: Central Europe Airways and Communication Service Region
CENEX e: Complex Energetics Experiment
CENFAM i: Centro Nazionale di Fisica dell'Atmospera e Meteorologia
C Eng e: Chartered Engineer
- e: Chief Engineer
CENG e: Control Engineering (journ.)
CENIM s: Centro Nacional de Investigaciones Metalúrgicas
CENIS e: Center for International Studies
CENL e: Community Equivalent Noise Level
CENOG e: Computerized Electroneuro-Ophthalmograph
CENOUT e: Central Output
CENPAC e: Central Pacific Area
Cen Proc U e: Central Processing Unit
CENSER e: Census Servomechanism [and Tape Handler]
CENSOR e: Centrifugal Solids Recovery
CENT e: Centaur
- e: Centered
- e: Centering
- e: Centigrade[s]
- e: Centimeter[s]
- e: Centimetre[s] (GB)
- e: Central[ized]
- e: Centrifugal
- e: Century
cent l: centum
CENTACS e: Center for Tactical Computer Sciences
Cent Afr e: Central Africa
Cent Am e: Central America
CENTCON e: Centralized Control
CENTEN e: Centennial
CENTF e: Centrifugal
centi e: 10^{-2}
CENTI f: Centre pour le Traitement de l'Information
CENTIG e: Centigrade[s]
CENTLANT e: Central Atlantic
CENTS e: Census Tabulation System
- e: Centuries
CENTY f: Calculatrice Electronique Numérique Type
CEO e: Chemical Engineering Operations
- e: Chick Embryo Origin
- e: Chief Education Officer
- e: Chief Engineer Officer
- e: Chip Enable Output
- e: Command Education Officer
- e: Commissioned Engineer Officer
- e: Communications Electronics Officer
- e: Comprehensive Electronic Office
- e: Corps Education Officer
- e: Customs Enforcement Officer
CEOI e: Communications-Electronics Operations Instruction
CEOINC e: Civil Engineer; Officer-In-Change
CEOT e: Calcifying Epithelial Odontagenic Tumor
CEP e: Capital Expenditure Proposal
- e: Cepstrum
- e: Chain Elongation Proteins
- e: Chemical Engineering Progress (journ.)
- e: Circle of Equal Probability
- e: Circular Error Probability
- e: Circular Error Probable
- e: Colo[u]r Evaluation Program[me]
- e: Column Exit[s] Pulse
- e: Complementary-Element Pairs
- e: Complementary Even Parity
- e: Component Error Propagation
- e: Computed Ephemeris Position
- e: Computer Enhancement Process
- e: Computer Entry Punch
- e: Concentrated Employment Program[me]
- e: Continuous Estimation Program[me]
- e: Corporate Electronic Publishing
- e: Cortical Evoked Potential
- e: Cotton Equalization Program[me]
- e: Counterelectrophoresis
- e: Cylinder Escape Propability
CEPA e: 2-Chloroethylphosphonic Acid
- e: Civil Engineering Program[me] Application[s]
- e: Computer-applications in Engineering, Planning and Architecture
- e: Coupled Electron Pair Approach
CEPACC e: Chemical Education Planning and Coordinating Committee
CEPACS e: Customs Entry Processing and Cargo System
CEPAs e: Civil Engineering Program[me] Applications
CEPE e: Central Experimental and Proving Establishment
- e: Cylindrical Electrostatic Probe Experiment
CEPER e: Combined Engineering Plant Exchange Record
CEPEX e: Controlled Ecosystem Pollution Experiment
CEPH f: Centre d'Etude du Polymorphisme Humain
CEPH FLOC e: Cephalin Flocculation
CEPH-FLOC Test e: Cephalin Flocculation Test
CEPM e: Civil Engineer Preventine Maintenance
CEP Method e: Cepstrum Method
CEPO e: Central European Pipeline Office
CEPPA e: Customer Engineering Product Performance Analysis
CEPPC e: Central European Pipeline Policy Committee
CEPPF e: California Environmental Protection Program[me]Fund
CEPS e: Central European Pipeline System
CEPs e: Civil Engineering Problems
CEPS e: Command-Module Electrical Power System
CEPs e: Complementary-Element Pairs
Cepsa s: Compañía Española de Petróleos
CEPT e: Conference of European Postal and Telecommunications [Administrations]
- f: Conference Européen des Postes et des Télécommunications
CEPTED e: Accepted
- e: Excepted Cepting Accepting
- e: Excepting
CE Pulse e: Chip-Enable Pulse
CEQ e: Committee on Environmental Quality
- e: Council on Environmental Quality
CEQA e: California Environmental Quality Act (1970)
CEQOM e: Combined Electron Quench and Optical Masker

CER e: Ceramic[s]
- e: Channel Error Record
- e: Channel Error Routine
- e: Character Error Rate
- e: Civil Engineering Report
- e: Coarse Element Refinement
- e: Community Educational Resources
- e: Complete Engineering Release
- e: Complete Engine Repair
- e: Conditioned Emotional Response
- e: Cost Estimating Relations
- e: Cost Estimating Relationship

C & ER e: Combustion and Explosives Research
CERAM e: Ceramic[s]
Ceram. Age e: Ceramic Age (journ.)
CERAMAL e: Ceramic and Alloy
CERAMETERM e: Ceramic-Metal Terminal
CERANOX e: Ceramic Non-Oxide
CERATAB Case e: Ceramic Table Case
CERB e: Coastal Engineenng Research Board
CERC e: Central[ized] Engine Room Control
- e: Childrens Evaluation and Rehabilitation Center
- e: Coastal Engineering Research Center
- e: Coastal Engineering Research Council
- e: Computer Entry and Readout Control
- e: Concurrent Engineering Research Center

CERCA e: Commonwealth and Empire Radio for Civil Aviation
CERCDC e: California Energy Resources Conservation and Development Commission
CERD f: Comité Européen de la Recherche et du Développement
CERDIP e: Ceramic Dual-Inline Package[s]
CERD REPORT e: Central Evidence of Research and Development Report
Cer E e: Ceramic Engineer
CERE e: Computer Entry and Readout Equipment
Cereal Chem. e: Cereal Chemistry (journ.)
Cereal Sci. Today e: Cereal Science Today (journ.)
CEREL e: Civil Engineering and Evaluation Laboratory
CERES e: Center for Research and Education in Sexuality
CERES System e: Computer Enhanced Radio Emission Surveillance System
CERF e: Coastal Education and Research Foundation
CERG e: Concrete-Encased Ring Ground
CERI e: Center for Educational Research and Innovation
- e: Centre for Educational Research and Innovation
- e: Clean Energy Research Institute
- f: Centre d`Etudes et de Recherches en Informatique

CERIS e: Chinese Educational Resources Information System
CERL e: Cambridge Electronic Research Laboratory
- e: Central Electricity Research Laborartory
- e: Coastal Engineering Research Laboratory
- e: Commercial Equipment Requirement List
- e: Computer-Based Education Research Laboratory
- e: Construction Engineering Research Laboratory
- e: Cooperative Educational Research Laboratory

CERLAL s: Centro Regional parael Fomento del Libre en America Lutina
CERLI e: Cooperative Educational Research Laboratory, Inc.
CERMET e: Ceramic and Metal
- e: Ceramic-to-Metal
- e: Ceramic-Metallic
- e: Ceramic Metallized

CERN e: Center for Nuclear Research
- f: Commission Européenne pour la Recherche Nucléaire

CERP e: Civil Engineering Report of Performance
- e: Current Economic Reporting Program[me]

CERPAC[K] e: Ceramic Package
CERR e: Channel Error Recording Routine
CERRC e: Complete Engine Repair Requirements Card
CERS i: Comitato Ruropeo della Ricerca e dello Sviluppo
CERT e: Certain
- e: Certificate
- e: Certification
- e: Certified
- e: Certify
- e: Character Error Rate Tester
- e: Character Error Testing
- e: Combined Environmental Reliability Test
- e: Composite Electrical Readiness Test

CERTIF e: Certificate[d]
- e: Certified
- e: Certify

CERTINV e: Certified Invoice
CERTS e: Certificates
CERTs e: Combined Environmental Reliability Test[s]
CERTSUB e: Certain Submarine
CERV e: Cervical
CES e: Candelabra Edison Screw
- e: Central Electronics System
- e: Central Excitatory State
- e: Centre for Environmental Studies
- e: Characteristic Equation System
- e: Closed Ecological System
- e: Coast Earth Station
- e: College of Extended Studies
- e: Commercial Earth Station
- e: Committee on Earth Sciences
- e: Comprehensive Export Schedule
- e: Compressor End Seal
- e: Computer Education in Schools
- e: Computer Education System[s]
- e: Computer Election Systems, Incorporated
- e: Computer Energy Storage
- e: Consolidated Electronic Services
- e: Constant Elasticity Of Substitution
- e: Constructive Error Source
- e: Control Electronics Section
- e: Controlled Environmental System
- e: Coordinated Evaluation System
- e: Corporate Engineering Standard
- e: Cost Effectiveness Study
- e: Cotton Experiment Station
- e: Council for European Studies
- e: County Experimental Station
- e: Critical Experiment Station
- e: Cyanoethyl Sucrose
- f: Comité Electrotechnique Suisse
- s: Consejo Economico y Social

CESA e: Canadian Engineering Standards Association
CESAALA e: Charles E Stevens American Atheist Library and Archives
CESAC e: Communications-Electronics Scheme Accounting and Control
- e: Communications-Electronics Special Account Code

CESAR e: Capsule Escape and Survival Applied Research
- e: CERN Electron Storage and Accumulating Ring

CESB e: Center for Experimental Studies in Business
CESC e: Calcutta Electric Supply Corporation
- e: Communications Equipment Subcommittee

CESD e: Composite External Symbol Dictionary
- e: Composite External Symbol Directory

CE/SD e: Convolutional Encoding/ Sequential Decoding
CESE e: Captured Enemy Signal Equipment
- e: Civil Engineer Support Equipment

CESEMI e: Computer Evaluation of Scanning Electron Microscope Image
CESG e: Communications-Electronics Security Group
- e: Cryogenic Electrically Suspended Gyroscope

CESI e: Centre for Economic and Social Information (of UN)
CESI e: Closed Entry Socket Insulation
- e: Closed Entry Socket Insulator
- e: Communications-Electronics Standing Instruction
- e: Council for Elementary Science International

CESIL e: Computer Education System[s] Instructional Language
CESK e: Cable End Sealing Kit
CESL e: Camp Evans Signal Laboratory
- e: Civil Engineering Systems Laboratory

CESM e: Continuous Electro-Slag Melting
CESMM e: Civil Engineering Standard Method of Measurement
CESO e: Civil Engineer Support Office
CESO-W e: Council of Engineers and Scientists Organizations-West
CESP e: Correlation Echo Sound Processor
CESPM e: Continuous Electroslag Powder Melting
CESR e: Canadian Electronic Sales Representative[s]
- e: Conduction Electron Spin Resonance

CESS e: Council of Engineering Society Secretaries
CESSL e: Cell Space Simulation Language
CEST e: Central European Summer Time
- e: Compacted Earth Sodium Treated

CET d: computerunterstütztes Entscheidungstraining
- e: Capsule Elapsed Time
- e: Center for Educational Technology
- e: Central European Time
- e: Cerebral Electrotherapy
- e: Certified Electrical Technician
- e: Certified Electronics Technician
- e: Common External Tariff
- e: Concept Evaluation Technique
- e: Console Electric Typewriter
- e: Controlled Environmental Test[ing]
- e: Corrected Effective Temperature
- e: Council for Educational Technology
- e: Critical Emulsification Temperature
- e: Critical Experiment Tank
- e: Cummulative Elapsed Time
- f: Collèges d'Enseignement Technique

CETA e: Comprehensive Employment and Training Act
- e: Corrosion Evaluation and Test Area
CETC s: Centro de Estudios Tecnicos
CETEC e: Consolidated Engineering Technology Corporation
CETEDOC f: Centre de Traitement Electronique des Ducments
CETEX e: Committee on Contamination of Extra-Terrestrial Exploration
- e: Contamination by Extraterrestrial Exploration
CETI e: Communication[s] with Extraterrestrial Intelligence
CETIS f: Centre de Traitement de l'Information Scientifique
- f: Centre Européen d'Étude et de Traitement de l'Information Scientifique
CETME s: Centro de Estudios Tecnicos de Materiales Especiales
CETO e: Center for Educational Television Overseas
- e: Centre for Educational Television Overseas
- e: Civil Effects Test Operations
CETOL e: Cetologic[al]
- e: Cetologist
- e: Cetology
CET/OS e: Corporate Express Transfer/ Obsoletion System
CETR e: Consolidated Edison Thorium Reactor
CETRA f: Centre des Etudes Transatlantigues
CETS e: Commission on the Education of Teachers of Science
- e: Conference on European Telecommunications Satellites
- e: Contractor Engineering and Technical Services
- e: Control Element Test Stand
CETUS e: Computerized Exploration and Technical Underwater Surveyor
CET Virus e: Central European Tick Virus
CEU e: Channel Extension Unit
- e: Common Equipment Unit
- e: Constructional Engineering Union
- e: Continuing Education Unit
- e: Control Electronics Unit
- e: Coupler Electronics Unit
CEUCA e: Customs and Economic Union of Central Africa
CEV e: California Encephalitis-Virus
- e: Carbon Equivalent Valve
- e: Combat Engineer Vehicle
- e: Convoy Escort Vessel
- e: Corona Extinction Voltage
- e: Cryogenic Explosive Valve
CEVAR e: Consumable Electrode Vacuum Arc Remelt
CEVAT e: Combined Environmental, Vibration, Acceleration, Temperature
CEVIT e: Contingency Extravehicular Transfer
CEVM e: Consumable Electrode Vacuum Melting
CEW e: Circular Electric Wire
- e: Cold-Drawn Electric-Resistance Welded
CEWA e: Canadian Electronic Wholesalers Association
CEWCC e: Combined Electronic Warfare Control Center
CEWCSC e: Corps of Engineers Waterborne Commerce Statistics Center
CE-Winkel d: Centrum-Ecken-Winkel
CEWR e: Centimetric Early Warning Radar
CEWRM e: Communications Electronics War Readiness Materiel
CEWS e: Contractors Early Warning System
CEX e: Civil Effects Exercise
- e: Civil Effects Experiments
- e: Colo[u]r External
CEY e: Century Electric
Cey[l] e: Ceylon
CeZ d: Zetanzahl
CF d: Carbolfuchsin
- d: Kresolformaldehyd
- e: Cable; Firing
- e: Cable; Functional
- e: Cable; Fuzing
- e: Calcarine Fissure
- e: Calculating Function
- e: Californium
- e: Camer Frequency
- e: Can't Find
- e: Capacity Factor
- e: Carbol Fuchsine
- e: Carbon Film
- e: Card Face
- e: Card Feed[ing]
- e: Card Field
- e: Card File
- e: Car Ferry
- e: Carriage Free
- e: Carried Forward
- e: Carrier-Free
- e: Carrier Frequency
- e: Carry Forward
- e: Case File
- e: Cast Frame
- e: Cataloged File
- e: Cathode Follower
- e: Cease Firing
- e: Cement Floor
- e: Center Focus
- e: Center Forward
- e: Center Frequency
- e: Center Front
- e: Center of Floatation
- e: Central File
- e: Central Filing
- e: Central Fissure (Brain)
- e: Centre Fire
- e: Centre Focus
- e: Centre Forward
- e: Centre Frequency
- e: Centre Front
- e: Centre of Floatation (GB)
- e: Centrifugal Force[s]
- e: Chaining Field
- e: Chaining File
- e: Change Frequency
- e: Change in Formula
- e: Characteristic Frequency
- e: Characteristic Function
- e: Chemists Federation
- e: Chromosomal Fraction
- e: Circuit Finder
- e: Citrovorum Factor
- e: Clock Frequency
- e: Coal-Fired
- e: Code Figure
- e: Coding Field
- e: Coding Form
- e: Coding Format
- e: Coherence Function
- e: Cold Finish[ed]
- e: Colicin Factor
- e: Colorado Fuel & Iron
- e: Colour Framing
- e: Combined File
- e: Command Field
- e: Commercial Formula Translation
- e: Communication[s] Facility
- e: Communication[s] Factor
- e: Compare
- e: Comparing Feature
- e: Complement-Fixation
- e: Complement-Fixing
- e: Complement Form
- e: Complete Fabrication
- e: Concept Former
- e: Concept Formulation
- e: Confirm[ed]
- e: Consecutive Field
- e: Console File
- e: Constant Frequency
- e: Contact; Fixed
- e: Context Free
- e: Continuous Feed[ing]
- e: Continuous Flow
- e: Continuous Focusing
- e: Continuous Form
- e: Continuous Function
- e: Control Field
- e: Control Flag
- e: Controlled Facility
- e: Controller Function
- e: Conversion Facility
- e: Conversion Factor
- e: Correction Factor
- e: Correlation Factor
- e: Corresponding Fellow
- e: Counterflow
- e: Count Field
- e: Count Forward
- e: Counting Fingers
- e: Cover Forward
- e: Cresol Formaldehyde
- e: Crest Factor
- e: Cross Factor
- e: Crown Flint
- e: Crude Fiber
- e: Crystal Filter
- e: Cubic Feet
- e: Cubic Foot
- e: Current Feedback
- e: Current Force
- e: Current Forcing
- e: Cystic Fibrosis
CF2 e: Central Flow Control Facility
CFA e: California Fertilizer Association
- e: California Forestry Association
- e: Canadian Federation of Agriculture
- e: Canadian Forestry Association
- e: Carrier Frequency Alarm
- e: Center for Astrophysics
- e: Chief of Field Artillery
- e: Cleared for Approach
- e: Cognizant Field Activity
- e: Colour Filter Array
- e: Commonwealth Forestry Association
- e: Companions of the Forest of America
- e: Compass Failure Annunciator
- e: Complement-Fixing Antibody
- e: Complete Freuds Adjunct
- e: Component Flow Analysis
- e: Computer Family Architecture
- e: Concept Feasibility Analysis
- e: Configural Frequency Analysis
- e: Control Flow Analyzer
- e: Cowl-Flap Angle
- e: Crossed Field Amplifier
- e: Current Files Area
C Factor e: Cleverness Factor
CFAD e: Commander; Fleet Air Detachment
CFADC e: Controlled Fusion Atomic Data Center

CFAE e: Contract[or]-Furnished Aerospace Equipment
- e: Contract[or]-Furnished Aircraft Equipment
- f: Centre de Formation en Aérodynamique Expérimentale
CFAL e: Current Food Additives Legislation
CF and EA e: Cash Fund and Expense Accounting
CFANS e: Canadian Forces Air Navigation School
CFAP e: Canadian Foundation for the Advancement of Pharmacy
- e: Cleared for Approach
CFA Program[me] e: Computer Family Architecture Program[me]
CFAR e: Constant False Alarm Rate
CFB e: Cipher Feedback
- e: Circulating Fluidized Bed
- e: Coated Front and Back
- e: Commonwealth Forestry Bureau
- e: Corrugated Fibreboard
CFBO e: Calling for Bunkers Only
CFBS e: Canadian Federation of Biological Societies
- e: Card Feed Barrier Strip
CFC e: C-band Frequency Converter
- e: Canadian Forestry Corps
- e: Capillary Filtration Coefficient
- e: Carbon Fiber-reinforced Composite
- e: Card Feed Compare
- e: Central Fire Control
- e: Channel Flow Control
- e: Chief Fire Controlman
- e: Colony-Forming Cell[s]
- e: Complex Facility Console
- e: Complex Facility Control
- e: Configuration Control Register
- e: Consolidated Freight Classification
- e: Control Feed Chart
- e: Controlled Force Circulation
- e: Crossed-Film Cryotron
CFC-113 e: Chlorofluorocarbon
CFCA e: California Fish Canners Association
CFCB e: Card Feed Circuit Breaker
- e: Computer Format Control Buffer
CFCC e: Coal Fired Combined Cycle
CFCF e: Central Flow Control Facility
CFCFG e: Context-Free Coded Fuzzy Grammar
CFCFL e: Context-Free Coded Fuzzy Language
CFCO e: Chief Fire Controlman; Operator
CFCS e: Chief Fire Controlman; Submarines
CFC[S] e: Chlorotluorocarbon
CFCs e: Colony-Forming Cells
CFCS e: Cross Field Closing Switch
CFCU e: Common File Control Unit
CFD e: Card Feed Device
- e: Compact Floppy Disk
- e: Computer Function Diagram
- e: Conforming Flush-Plate Dipole
- e: Continuous Fire Detection
- e: Contractor Functional Demonstration
- e: Control Functional Diagram
- e: Cubic Feet per Day
- e: Cubic Foot per Day
- e: Cumulative Frequency Distribution
CFDA e: Current File Disk Address
CFDD e: Compact Floppy Disk Drive
CFD Program[me] e: Communications Facilities Determination Program[me]
CFDS e: Corps Field Dressing Station
CFDTS e: Cold-Flow Development Test System
CFE d: Chlortrifluoräthylen
- e: Cathode Flicker Effect
- e: Chlorotrifluoroethylene
- e: Climbing Film Evaporator
- e: Complement Field End
- e: Continued Fraction Expansion
- e: Control Flow Expression
- e: Controlled Flash Evaporation
- e: Current Forcing Element
- s: Comisión Federal de Electricidad
CFEI e: Consumer and Food Economics Institute
CFEME e: Canadian Forces Environmental Medicine Establishment
CFER e: Collector Field-Effect Resistor
CFES e: Center for Energy System[s]
- e: Continuous Flow Electrophoresis System
CFE System e: Climbing Film Evaporator System
CFF e: Carry Flip-Flop
- e: Cold Flow Facility
- e: Compressed Citation File
- e: Computer Forms Feeder
- e: Counter Flip-Flop
- e: Critical Flicker Frequency
- e: Critical Flicker Fusion
- e: Critical Fusion Frequency
- e: Current Fault File
CFFA e: Chemical Fabrics and Film Association
CFFC e: Counterflow Film Cooling
CFFP e: Cooperative Forest Fire Prevention
CFFT e: Critical Flicker Fusion Threshold
CF Furnace e: Coal-Fired Furnace
CFG e: Configuration
- e: Context-Free Grammar
CFGD e: Cubic Feet of Gas [per Day]
CFGE e: California Fruit Growers Exchange
CFGG e: Context-Free Graph Grammar
CFGH e: Cubic Feet of Gas per Hour
CFGM e: Cubic Feet of Gas per Minute
CF Grammar e: Context-Free Grammar
CFH e: Card Feed Hopper
- e: Common Business-Oriented Language File Handler
- e: Cubic Feet per Hour
CFHA e: Colorado Forestry and Horticulture Association
CFHO e: Canada-France-Hawaii Observatory
CFHT e: Canada-France-Hawaii Telescope
- e: Continuous-Flow Hypersonic Tunnel
CFI e: Call Failure Indication
- e: Canadian Film Institute
- e: Canadian Forest Inventory
- e: Central Flight Instructor
- e: Certified Flight Instructor
- e: Chief Flying Instructor
- e: Commonwealth Forestry Institute
- e: Complement-Fixing-Inhibiting
- e: Control Format Item
- e: Cost, Freight and Insurance
- e: Crystal Frequency Indicator
CFIB e: Commonwealth Forestry and Timber Bureau
CFIEM e: Canadian Armed Forces Institute of Environmental Medicine
CFIET e: Call Failure Indication; Engaged Tone
CFIM e: Compatibility Feature Initialize Mode
CFINT e: Call Failure Indication; No Tone
CFK d: kohlenstoffverstärkter Kunststoff
- e: Confidence Firing Kit
CFL e: Call Failed
- e: Call Failure
- e: Carnegie Free Library
- e: Cease Firing (or Fire) Line
- e: Central Film Library
- e: Close Focus Lens
- e: Coded Fuzzy Language
- e: Cold Flow Laboratory
- e: Constand Feed Lubricator
- e: Context-Free Language
CFLC e: Compatibility Feature Load Constant
CFL-DL e: Cease Firing (or Fire) Line - Disarm Line
CFLG e: Counter-Flashing
CFLI e: Clay Flue Lining Institute
C-Flip-Flop e: Carry Flip-Flop
CFLOS e: Cloud-Free Line of Sight
CFL Signal e: Call Failure Signal
CFLT e: Carrier Frequency Line Transformer
- e: Courtesy Flight
CFLV e: Compatibility Feature Load Variable
CFM e: Card Feeding Mechanism
- e: Cash-Flow-Method
- e: Cathode Follower Mixer
- e: Center Frequency Modulation
- e: Central Facility Maintenance
- e: Centre Frequency Modulation
- e: Cerebral Function Monitor
- e: Chill-down Flow Meter
- e: Chlorofluoromethane
- e: Circuit Feasibility Model
- e: Clustered File Migration
- e: Collision Force Method
- e: Combat File Maintenance
- e: Companded Frequency Modulation
- e: Companding and Frequency Modulation
- e: Completion of Calls to Busy Subscriber Facility Message
- e: Computer Feasibility Model
- e: Computer Field Maintenance
- e: Continuous Film Memory
- e: Control Function Memory
- e: Cooperative Forest Management
- e: Crystal Frequency Multiplier
- e: Cubic Feet per Minute
- e: Cylindrical Film Memory
CF/M e: Cubic Feet per Month
CFMA e: Conflict-Free Multi-Access
CFMS e: Chained File Management (IBM)
CFMs e: Chlorofluoromethanes
CFMS e: Combined Field Maintenance Shop
- e: Compatibility Feature Mode Set
CFN e: Canadian Forces Network
- e: Chain Field Number
- e: Check Following Numbers
CFNI e: Caribbean Food and Nutrition Institute
CFNP e: Community Food and Nutrition Programs
CFO e: Calling for Order[s]
- e: Cancelling Former Orders
- e: Carrier Frequency Oscillator
- e: Ceramic Fiber Optics
- e: Channel for Orders
- e: Chlorinated Flour Oil
- e: Complex Facility Operator
- e: Connection Fitting Out
- e: Consolidated Function Ordinary (IBM)
- e: Consolidated Function Order
- e: Critical Flash-Over
CFOC e: Contractor Fin Opener Crank

CFOR e: Conversational Formula Translator
CFOT e: Crossed Field Output Tube
CFO Voltage e: Crystal Flashover Voltage
CFP e: Calculation Fixed Point
- e: Carrier Frequency Pulse
- e: Center of Filtering and Plotting
- e: Change Flight Plan
- e: Cold Front[al] Passage
- e: Computer Forms Printer
- e: Concentric Flight Plan
- e: Concept Formulation Package
- e: Contractor Furnished Property
- e: Corneofundal Potential
- e: Creation Facilities Program[me]
- e: Cystic Fibrosis of the Pancreas
CFPA e: Canadian Food Processors Association
CF Paper e: Coated Front Paper
- e: Covered Front Paper
CFPG e: Context-Free Picture Grammar
- e: Context-Free Programmed Grammar
CFPM e: Crossed-Field Photomultiplier
CFPO e: Center for Populations Options
- f: Compagnie Française des Phosphates de l'Océanie
CFPP e: Cold Filter Plugging Point
- e: Concept Formulation Program[me] Plan
CFPS e: Crossed Field Plasma Sheath
CFPSG e: Context-Free Phrase Structure Grammar
CFP/TDP e: Concept Formulation Package-Technical Development Plan
CFR e: Carbon-Film Resistor
- e: Catastrophic Failure Rate
- e: Center for Future Research
- e: C Frame
- e: Chance Failure Rate
- e: Channel Failure Record
- e: Complement Fixation Reaction
- e: Confer
- e: Contact Flight Rules
- e: Cooperative Fuel Research
- e: Coorong Fauna Reserve
- e: Crossfire
- e: Cumulative Failure Rate
CFR1 e: Commercial Fast Reactor 1
- e: First Commercial Fast Reactor
CFRC e: Cooperative Fuel Research Council
CFRD e: Confidential, Formerly Restricted Data
CFRDD e: Colonial Forest Resources Development Department
CFRE e: Circulating Fuel Reactor Experiment
CFR Engine e: Cooperative Fuel Research [Council] Engine
CFRG e: Carbon-Fiber-Reinforced Glass
CFRGC e: Carbon-Fiber-Reinforced Glass Ceramic
CFRI e: Central Fuel Research Institute
CFRM e: Continuous Fiber Reinforcing Mat
CFRMC e: Complement-Free Reed Muller Canonical
CFRP e: Carbon Fiber Reinforced Plastic
- e: Carbon Fibre Reinforced Plastic (GB)
CFRS e: Central Fisheries Research Station
CFRs e: Commercial Fast Reactors
- e: Contact Flight Rules
CFRS e: Customer Float Reporting Service
CFS e: Calcium-Free Salt
- e: Call for Service
- e: Canadian Forestry Service
- e: Card Feed Stop
- e: Carrier Frequency Shift
- e: Cassegrain Feed System
- e: Center Frequency Stabilization
- e: Central Flying School
- e: Central Forecasting Station
- e: Central Frequency Sounding
- e: Centre Frequency Stabilization
- e: Cesium Feed System
- e: Chronic Fatigue Syndrome
- e: Circuit Failure Simulation
- e: Circulation and Finding System
- e: Cold-Finished Steel
- e: Cold Finished Seamless
- e: Combined File Search
- e: Common File System
- e: Completely-Finished Sets
- e: Component Failure Summary
- e: Composite Feed System
- e: Computer Fast Storage
- e: Computer File System
- e: Computer Firmware Systems, Inc.
- e: Computer Function Symbol
- e: Container Freight Station
- e: Continuous Film Store
- e: Counter Filling System
- e: Critical Field Strength
- e: Cryogenic Fluid Storage
- e: Cubic Foot per Second
CFs e: Carrier Frequencies
- e: Characteristic Frequencies
- e: Characteristic Functions
CFSA e: California Flyers School of Aeronautics
- e: College Food Service Association
CFSC e: Compatibility Feature Store Constant
- e: Cryogenic Fluid Storage Container
CFSDA e: Cross-Frequency Statistical Deconvolution Algorithm
CFSLP e: Center for Short-Lived Phenomena (Smithsonian Institution)
CFSPL e: Canadian Forces Special Projects Laboratory
CFSR e: Contract Fund Status Report
CFSS e: Combined File Search Strategy
- e: Combined File Search System
CFS Signal e: Calls for Service Signal
CFST e: Contact-Free Syntactical Translator
- e: Context-Free Syntactical Translator
CFSTI e: Clearinghouse for Federal Scientific and Technical Information
CFSTR e: Continuous Flow Stirred Tank Reactor
CFSV e: Compatibility Feature Store Variable
CFT d: Cardiolipin-Flockungstest
- e: Charge Flow Transistor
- e: Clinical Full Time
- e: Coated Foam Tape
- e: Cold Flow Test
- e: Communication Format Tape
- e: Complement-Fixation Test
- e: Continuous Fourier Transform
- e: Continuous Fourier Transport
- e: Contract Field Technician
- e: Crossed Field Tube
- e: Cubic Feet
- e: Cubic Foot
- f: Compagnie Française de Télévision
CFTA e: Canadian Food Technologists Association
CFTB e: Commonwealth Forestry and Timber Bureau
- e: Control Flight Test Bed
CFTC e: Central Flying Training Command
- e: Commonwealth Fund for Technical Cooperation
CF Test[s] e: Complement-Fixation Test[s]
CFTG e: Context-Free Transduction Grammer
CFTH f: Compagnie Française Thomson-Houston
CFTRI e: Central Food Technological Research Institute
CFTS e: Captive Firing Test Set
CFTs e: Charge Flow Transistors
- e: Complement Fixation Tests
CFTS e: Computerized Flight Test System
CFTSD e: Canadian Forces Technical Services Detachment
CFU e: Colony-Forming Unit
- e: Control Functional Unit
- e: Corn-Equivalent Feed Unit
C-Function e: Controller Function
CFUs e: Colony-Forming Units
CFV e: Chat Fruit Virus
- e: Conditional Forced Value
- e: Conventional Friend Virus
CFVD e: Constant Frequency Variable Dot
CFW d: VEB Chemiefaserwerk (DDR)
CFWIS e: Central Fighter Weapons Instructors School
CFWS e: Conversion of the Federal Wage System
CFY e: Current Fiscal Year
CG d: Zentigramm
- e: Cage
- e: Camera Gun
- e: Capacitance of Grid
- e: Cardiogreen
- e: Cargo Glider
- e: Categorical Grammar
- e: Cathode Grid
- e: Center of Gravity
- e: Centigram[s]
- e: Central Gray
- e: Centre of Gravity
- e: Cerebral Ganglion
- e: Chain Grate
- e: Character Generator
- e: Chocking Gas
- e: Chorionic Gonadotrophin
- e: Chromatid Gaps
- e: Chronic Glomerulonephritis
- e: Clock Generator
- e: Clutter Gate
- e: Code Generator
- e: Code Group
- e: Coincidence Gate
- e: Colloidal Gold
- e: Colo[u]r Graphics
- e: Command Guidance
- e: Compiler Generator
- e: Complement Gate
- e: Comptroller General
- e: Computer Generation
- e: Computer Graphics
- e: Constant Gradient
- e: Control Gate
- e: Control Grid
- e: Course Generator
- e: Crushed or Ground
C & G e: Control and Guidance
CGA e: Canadian Gas Association
- e: Cargo[s proportion of] General Average
- e: Catabolite Gene Activator
- e: Certified Graphoanalyst
- e: Coarse Gravel Aggregate
- e: Coast Guard Academy
- e: Coast Guard Aviation
- e: Colo[u]r Graphics Adapter
- e: Contrast Gate Amplifier
- f: Compagnie Générale d'Automatisme

CG & A e: Computer Graphics and Applications (journ.)
CGAA e: Canadian Graphic Arts Association
- e: Computer Graphic Aerodynamic Analysis
CGAD e: Coast Guard Aviation Depot
CGAE e: Cincinnati Gas and Electric
C Gal e: Cable Gallery
CGAS e: Coast Guard Air Station
CGASC e: Cornell Guggenheim Aviation Safety Center
CGAS Center e: Cornell Guggenheim Aviation Safety Center
CGB e: Canadian Geographic Board
- e: Circuit Group Blocking
- e: Convert Gray to Binary
CGBA e: Circuit Group Blocking Acknowledgement
CGC e: Ceramic Golg Coating
- e: Circuit Group Congestion
- e: Critical Grid Current
- e: Cruise Guidance Control
CGCA e: Controller General of Civil Aviation
CG Circuit e: Common-Gate Amplifier for Field-Effect Transistors
CGCRI e: Central Glass and Ceramic Research Institute
CGC Signal e: Circuit-Group-Congestion Signal
CGD e: Chronic Granulomatous Disease
- e: Comptroller Generals Decision
CGDS e: Cambridge Geological Data System
CGE e: Canadian General Electric
CGE e: Chadwick-Goldhaber Effect
- e: Charge
- e: Chockpit Geometry Evaluation
- e: Commission on Geography in Education
- f: Compagnie Générale d'Electricité
- f: Compagnie Générale des Eaux
CG & E e: Cincinnati Gas and Electric Company
CGEC e: Committee on Global Ecology Concern
CGE FWD e: Carriage Forward
CGEL & PB e: Consolidated Gas, Electric Light and Power Company of Baltimore
CGE PD e: Carriage Paid
CGF e: Carrier Gas Fusion
- e: Center of Gravity Factor
- e: Chemotaxis-Generating Factor
- e: Coarse Glass Frit
- e: Credit Given for
CGFA e: Consolidated Gold Fields Australia Ltd.
CGFP e: Calcined Gross Fission Product
CGFSA e: Consolidated Gold Fields of South Africa
CGG e: Continuous Grinding Gage
CGH e: Cape of Good Hope
- e: Chorionic Gonadotrophic Hormone
- e: Chorionic Gonadotrophin; Human
- e: Computer-Generated Hologram
CGI e: Chief Ground Instructor
- e: City and Guilds Institute
- e: Clinical Global Impression
- e: Compressed Gas Insulation
- e: Computer-Generated Image[ry]
- e: Computer Graphics Interface
- e: Computer-Guided Instruction
- e: Contact Growing Index
- e: Corrugated Galvanized Iron
- e: Cruise Guide Indicator
- f: Compagnie Générale d`Informatique
CGIAR e: Consultative Group on International Agricultural Research
CGIC e: Ceramic and Graphite Information Center
CGIS e: Canadian Geographic Information System
CGIT e: Compressed-Gas-Insulated Tube
CGJ e: Canadian Geographical Journal
C G Jung Foun e: C. G. Jung Foundation for Analytical Psychology
CGK d: Computer Gesellschaft Konstanz mbH
- e: Cathode-Grid Capacitance
CGL e: Center-of-Gravity Locator
- e: Chronic Granulocytic Leukaemia
- e: Computer Generated Letter
- e: Congolomerate
- e: Continuous Gas Laser
- e: Control Group Level
- e: Controlled Ground Landing
- e: Corrected Geomagnetic Latitude
CGLASTA e: Coast Guard Light Attendant Station
CGLBSTA e: Coast Guard Lifeboat Station
CGLI e: City and Guilds of London Institute
CGLkr e: Cleaning Gear Locker
CGL of C e: Commanding General; Lines of Communication
CGLORSTA e: Coast Guard Long-Range Navigation Transmitting Station
CGLS e: Coast Guard Long-Range Navigation Station
CGLSTA e: Coast Guard Light Station
CGLTG e: Cloud-to-Ground Lightning
CGM e: Centigram[me]
- e: Central Gray Matter
- e: Ciliated Greove to Mouth
- e: Circuit Group-supervision Message
- e: Colo[u]r-Graphic Monitor
- e: Computer Graphics Metafile
- f: Compagnie Générale de Micromatique
CGMID e: Character Generation Module Identifier
CGMIF e: Computer Graphics Metafile Interchange Format
CGMO e: Coast and Geodetic Magnetic Observatory
CGMP d: cyclisches Guanosinmonophosphat
- e: Cyclic Guanosine 3',5'-Monophosphate
CGMW e: Commission for the Geological Map of the World
CGN e: Chronic Glomerulonephritis
- e: Cologne (airport)
- e: Computer Group News (journ.)
- e: Nuclear-Powered Guided-Missile Cruiser
CGNM e: Casa Grande National Monument
CGO e: Cargo
- e: Clear-and-Gate Operation
- e: Comptroller General Opinion
CGOPHEOSE e: Consultative Group on Potentially Harmful Effects of Space Experiments
CG/OQ e: Cerebral Glucose Oxygen Quotient
CGOS e: Computervision Graphics Operating System
CGOU e: Coast Guard Oceanographic Unit
CGP e: Capacitance Grid Plate
- e: Choline Glycerophosphatlde
- e: Chorionic Growth-Hormone Prolactin
- e: Circulating Granulocyte Pool
- e: Colo[u]r Graphics Printer
- e: Computer Graphics Processing
- e: Computervision Graphics Processor
- e: Current Geographical Publications
CGPC e: Cellular General Purpose Computer
CGPL e: Conversational Graphical Programming Language
CGP Language e: Chemical Graphics Programming Language
CGPM f: Conférence Générale des Poids et Mesures
CGPN e: Catalog of Galactic Planetary Nebulae
CGPs e: Current Geographical Publications
CGPSq e: Cartographic and Geodetic Processing Squadron
CGR e: Captured Gamma Ray
- e: Chete Game Reserve
- e: Chewore Game Reserve
- e: Chizaria Game Reserve
- e: Code Generating Routine
- e: Crime on Government Reservation
CGRA e: Canadian Good Roads Association
- e: Chinese Government Radio Administration
CGRADSTA e: Coast Guard Radio Station
CGRE e: Gas-Cooled Reactor Experiment
CGRM e: Containment Gaseous Radiation Monitor
CGRP e: Circuit Group
CGRR e: Circuit Group Reset Receipt
CGRS e: Central Gyro Reference System
- e: Circuit Group Reset Sending
CGS e: Canadian Geographical Society
- e: Canadian Geotechnical Society
- e: Centimeter-Gram[me]-Second
- e: Chromatographic Separation
- e: Coast and Geodetic Survey
- e: Conditional Go-to-Statement
- e: Consumer Goods Standards
- e: Council of Graduate Schools
- e: Cyclic Group Signal
C & GS e: Coast and Geodetic Survey
CGSA e: Cellular Geographic Service Area
- e: Computer Graphics Structural Analysis
CGSB e: Centimeter-Gram-Second-Biot
CGSC e: Columbia Gas Service Corporation
CGSE e: Centimeter-Gram[me]-Second Electrostatic
- e: Coast and Geodetic Survey Electrostatic
- e: Counter Group Selector Entry
CGS Em e: Electromagnetic Centimeter-Gram-Second
CGS Es e: Electrostatic Centimeter-Gram-Second
CGSE System e: Coast and Geodetic Survey Electrostatic System
CGSET e: Circuit Group Set
CGS EX e: Counter Group Selector Exit
CGSF e: Centimeter-Gram-Second-Franklin
CGSFU e: Ceramic Glazed Structural Facing Units
CGSM e: Centimeter-Gram[me]-Second [Electro-]Magnetic
- e: Coast and Geodetic Survey [Electro] Magnetic
CGSS e: Cryogenic Gas Storage System
CGS-System d: Zentimeter-Gramm-Sekunden-System
CGSUB e: Ceramic Glazed Structural Unit Base
CGSUS e: Council of Graduate Schools in the United States
CGS-Verfahren d: chemisch-galvanisches Schichtverfahren

CGT d: Chloriongonadotrophin
- e: Chononic Gonadotropin
- e: Combustible Gas Tracer
- e: Corrected Geomagnetic Time
- e: Current Gate Tube
- f: Compagnie Générale de Télégraphie

CGTFL e: California Grape and Tree Fruit League
CGTM e: Command Guided Tactical Missile
CGTS e: Coast and Geodetic Tide Station
- e: Coast Guard Training Station

CGTSF f: Compagnie de Télégraphie San Fils
CGTT e: Cortisone-Glucose Tolerance Test
CGU d: computergestützter Unterricht
- e: Canadian Geophysical Union
- e: Ceramic Glazed Units
- e: Circuit Group Unblocking
- e: Code Generating Unit

CGUA e: Circuit Group Unblocking Acknowledgement
CGUSACE e: Army Corps of Engineers
CGV e: Critical Grid Voltage
CGVH e: Computer-Generated Volume Hologram
CGVS e: Ciliated Grove to Ventral Sac
CGW e: Corning Glass Works
CGWB e: Canadian Government Wheat Borad
CGWL e: Controller of Guided Weapons and Electronics
CH d: Charakteristik
- d: Chemie
- d: China
- d: Chirurg[ie]
- d: Chondroitin
- d: Schweiz (Kfz)
- e: Cable Hut
- e: Camp Hospital
- e: Can't Hear
- e: Candle Hour[s]
- e: Cannot Hear
- e: Card Holder
- e: Cards per Hour
- e: Carnegie Hall
- e: Carrier Haulage
- e: Case Harden[ed]
- e: Center Half
- e: Central Heating
- e: Centre Half
- e: Chain [Home]
- e: Change
- e: Channel [Hold]
- e: Chapter
- e: Character
- e: Charge[d]
- e: Chart
- e: Chassis
- e: Check[ed]
- e: Chemical
- e: Chemistry
- e: Chiffre
- e: Chile[an]
- e: Chinese
- e: Chlorite
- e: Choice
- e: Choke
- e: Choline
- e: Chronic
- e: Clearing House
- e: Clock Hour
- e: Coastal Harbour
- e: Code Holes
- e: Compare Halfword
- e: Compass Heading
- e: Concentration of Hydrogen [Ions in Moles per Liter]
- e: Conductor Head
- e: Control Hole
- e: Core Hole
- e: Crown-Heel
- e: Custom House
- e: Hyperbolic Cosine

C & H e: Cocaine and Heroin
- e: Cold and Hot

Ch. l: Chirurgtae
Ch0 e: Character 0 (Zero)
CH-113 e: Canadian Boeing-Vertol helicopter
Ch 1...2; 3; 4; 5; 6; 7; 8; 9 e: Character 1...2; 3; 4; 5; 6; 7; 8; 9
C^{14} e: Radioactive Carbon
C'^{1},C^{12}z,C^{13}, etc. e: Complements of Complements
ch^2 e: Square Chain
C^2 D^2 (ARDC) e: Command and Control Development Division
C^3 I e: Command, Control, Communications and Intelligence
CH-46 e: Boeing-Vertol twin-rotor helicopter
CH-47 e: Boeing-Vertol helicopter
C^4D e: Conductively-Connected Charge-Coupled Device
CH-53 e: Sikorsky heavy-assault helicopter
CH-54 e: Sikorsky crane helicopter
C_6H_6 e: Benzene
CHA d: Cyclohexylamin
- d: Cyclohexylaminacetat
- e: Cable-Harness Analyzer
- e: Choline Acetylase
- e: Chronic Hemolytic Anemia
- e: Concentric Hemispherical Analyzer
- e: Congenital Hypoplastic Anemia
- e: Coupled Hard Axis
- e: Cyclohexylamine

c^a s: Compañia
CHABA e: Committee on Hearing and Bio-Acoustics
Ch-Abw d: chemische Abwehr
ChAc e: Chloroacetyl
- e: Choline-Acetylase

CHAD e: Code to Handle Angular Data
CHA Device e: Closed Hard Access Device
CHA DRO e: Coupled-Hard-Axis Destructive Readout
CHAER e: Chief Aerographer
CHAFB e: Chanute Air Force Base
CHAG e: Chain Arrester Gear
CHAIN e: Channeled Arizona Information Network
CHAIS e: Consumer Hazards Analysis Information Service
CHAI-Virus e: Cytopathic Human Auto Interfering Virus
CHALE e: Chalcocite
CHALICE e: Compressional Heating and Linear Injection Cusp Experiment
CHALK e: Calcium Carbonate ($CaCO_3$)
CHAM e: Chamfer
Chambers e: Chambers Dictionary of Science and Technology
CHAMP e: Channel Modelling Program[me]
- e: Character Manipulation Procedures
- e: Civilian Health and Medical Program[me]
- e: Civilian Help and Medical Program[me]
- e: Communications Handler for Automatic Multiple Program[me]s
- e: Community Health Air Monitoring Program[me]
- e: Cranfield Hybrid Automatic Maintenance Program[me]

CHAMPION e: Compatible Hardware and Milestone Program[me] for Integrating Organizational Needs
CHAMPS e: Computer History and Maintenance Planning System
CHAMPUS e: Civilian Health and Medical Program[me]of the Uniformed Services
CHAN e: Channel
CHANCE e: Complete Help and Assistance Necessary for College Education
CHANCOM e: Channel Command
CHAN ENT e: Channel Entry
Chan Is e: Channel Islands
CHANNELTRON e: Channel Electron Multiplier
CHANS e: Channels
CHAN SEL e: Channel Selector
CHAN SHFTR e: Channel Shifter
CHANSTOP Diffusion e: Channel Stopping Diffusion
CHAOS e: Concurrent Highspeed Advanced Operating System
CHAOTIC e: Computer-and-Human-Assisted Organization of a Technical Infonmation Center
CHAP e: Certified Hospital Administration Program[me]
CHAP e: Chapter
- e: Charring Ablation Program[me]
- e: Child Health Associate Program[me]
- e: Computer Charring Ablation Program[me]
- e: Contractor Held Air Force Property
- e: Controlled Helium Atmosphere Plant
- e: Convective Heating and Ablation Program[me]
- e: cychophosphamide, hexamethylmelamine, doxonibicin and cisplatin

CHAPS e: Control, Handling, Analysis, Personnel and Sensors
Char. d: Charité (Berlin)
CHAR e: Character (Data Processing)
CHAR e: Character[istic[s]]
- e: Charge

CHARAC e: Characteristic[s]
CHARACT e: Characteristic[s]
charak. d: charakteristisch
CHARC e: Character
- e: Charcoal

CHAR GEN e: Character Generator
CHARGTEL e: Charge Television
CHARGUID e: Character Guidance
CHARIB[I]DIS e: Chalk River Bibliographic Data Information System
CHARL e: Charles University University of Prague
CHARM e: Civil Aeronautics Administration High-Altitude Remote Monitor[s]
- e: Civil Aeronautics Administration High-Altitude Remote Monitoring
- e: Canadian High Altitude Research Missile
- e: Checking, Accounting and Reporting for Memberfirms

CHARs e: Characters
CHARSEC e: Characters per Second
CHART e: Computerized Hierarchy and Relationship Table
CHAS e: Chassis
- e: Cork Historical and Archaeological Society

CHASE e: Check and Analysis System for Errors
- e: Cornell Hotel Administration Simulation Exercise

- e: Cut Holes and Sink[Th]em
CHASM e: Chicago Access Support Module
- e: Coventry Health and Safety Movement
CHAS PL e: Chassis Plug
CHAS SKT e: Chassis Socket
CHAS SKT SW e: Chassis Socket with Switch
CHAST AFB e: Charleston Air Force Base
CHAT e: Cheap Access Terminal
CHA Timber e: Chilla Timber
CHAUD e: Chemical Audit
CHAUD System e: Chemical Audit and Distribution System
Ch B e: Bachelor of Chemistry
CHB e: Chain Home Beamed
- e: Check Brushes
- f: Compagnie Honeywell Bull
Ch.B. l: Chirurgiae Baccalaureus
CH Bit e: Check Bit
CHBP d: chemischer Beobachtungsposten
- e: Center for Housing, Building and Planning
ChBuAer e: Chief of the Bureau of Aeronautics
ChBuDocks e: Chief of the Bureau of Docks
ChBuMed e: Chief of the Bureau of Medicine
ChBuMed[Surg] e: Chief of the Bureau of Medicine and Surgery
ChBuOrd e: Chief of the Bureau of Ordnance
ChBuShips e: Chief of the Bureau of Ships
ChBuWeps e: Chief of the Bureau of Weapons
CHC e: Cargo Handling Charges
- e: Channel Control
- e: Check Coil
- e: Continuous High Chair
- e: Cyclohexylamine Carbonate
CHCF e: Component Handling and Cleaning Facility
CHCIVENGS e: Chief of Civil Engineers
$CHCl_3$ e: Chloroform
CHCM e: Continuous High Chair for Metal-Deck
ChCommSect e: Chief; Communications Section
CHCP e: Continuous High Chair with Plate
CHCS e: Computer Hardware Consultants and Services
CHCU e: Channel Control Unit
- e: Continuous High Chair Upper
CHD d: Computer-Haus; Darmstadt
- e: Check Digit
- e: Chemistry; Doctor of
- e: Childhood Disease
- e: Compensated Heart Disease
- e: Constitutional Hepatic Dysfunction
- e: Coronary Heart Disease
- e: Doctor of Chemistry
CH & D e: Center Hospital and Domiciliary
CHDB e: Compatible High Density Bipolar
CHDB Code e: Compatible High Density Bipolar Code
CHDE e: Compatible High Density Bipolar
Ch Dis e: Character Display
CHDL e: Computer Hardware Description Language
CHDM e: cyclohexanedimethanol
CHE d: Cholesterin-Esterase
- d: Cholinesterase
- e: Cargo Handling Equipment
- e: Channel End
- e: Channel Hot-Electron
- e: Character Early
- e: Check Entry
- e: Chemical Engineer
- e: Chief Engineer
- e: Chip Enable
- e: Cholinesterase
- e: Component Handling Equipment
CHEAP e: Committee for Handling European Automation Programmes
- e: Computerized Health Education Assessment Program[me]
CHEB e: Chebyshev
CHEB Filter e: Chebyshev Filter
CHEBYTOP e: Chebychev Trajectory Optimization Program[me]
CHEC e: Channel Evaluation and Call
- e: Checked
- e: Commonwealth Human Ecology Council
CHE-CIT e: Chemical Engineering-California Institute of Technology
CHECS e: Check Handling Executive Control System
ChED d: Chemie Experiment + Didaktik (journ.)
CHEF e: Contour-Clamped Homogeneous Electric Field
Chefing d: Chefingenieur
Chefp[il] d: Chefpilot
CHEL e: Chain Home Extra Low
CHELEC e: Chief Electrician
CHELONIOL e: Cheloniologic[al][ly]
- e: Cheloniologist
- e: Cheloniology
- e: Chelons
- e: Chelonians
CHEM d: Chemie
- d: Chemiker
- d: Chemisch
- d: Chemist
- e: Chemical[s]
- e: Chemical Education Materials
- e: Chemist
- e: Chemistry
Chem. Abstr. e: Chemical Abstracts (journ.)
Chem. Age India e: Chemical Age of India (journ.)
CHEMANAL e: Chemical Analysis
CHEMAST [System] e: Chemistry Assistant [System]
Chem Ber[ichte] d: Chemische Berichte (journ.)
Chem Br e: Chemistry in Britain (journ.)
Chem Can e: Chemistry in Canada (journ.)
CHEMDEX e: Chemical Index (journ.)
Chem E e: Chemical Engineer[ing]
Chem Econ e: Chemical Economic Services
Chem Ed[uc] e: Chemical Education Publishing Company
Chem Elements Pub e: Chemical Elements Publishing Company
Chem Eng (London) e: Chemical Engineer (London) (journ.)
Chem Eng News e: Chemical Engineering News (journ.)
Chem Engng e: Chemical Eingineering (journ.)
Chem Eng Progr Symp Ser e: Chemical Engineering Progress; Symposium Series (journ.)
Chem Eng Sci e: Chemical Engineering Science (journ.)
Chem Eng World e: Chemical Engineering World (India) (journ.)
Chem Etch e: Chemical Etching
- e: Chemically Etched
Chem. Fa. d: Chemische Fabrik
CHEMFET[s] e: Chemical[ly][-Sensitiv]e Field-Effect Transistor[s]
Chem Flab d: chemisches Feldlaboratorium
CHEMI e: Chemical Engineering Modular Instruction
Chem Ind e: Chemistry and Industry (journ.)
Chem Ind (Düsseldorf) d: Chemische Industrie (Düsseldorf) (journ.)
Chem-Ing-Tech[n] d: Chemie-Ingenieur-Technik (journ.)
Chem Inst d: Chemisches Institut
Chem Instrum e: Chemical Instrumentation (journ.)
CHEM IN UNS ZEIT d: Chemie in unserer Zeit (journ.)
Chem Lab Betr d: Chemie für Labor und Betriebe (journ.)
CHEMLY e: Chemically
Chem & Met Eng e: Chemical and Metallurgical Engineering (journ.)
CHEM MILL e: Chemically Milled
- e: Chemical Milling
Chem O e: Chemical Officer
CHEMO e: Chemotherapist
- e: Chemotherapy
CHEMONUC e: Chemonuclear
CHEMOS e: Chemosphere
- e: Chemospheric[al][ly]
CHEMOSENS e: Chemosensory
CHEMOSTE e: Chemosterilant
- e: Chemosterilization
- e: Chemosterilize[d]
CHEMOSURG e: Chemosurgery
- e: Chemosurgical[ly]
CHEMOTAX e: Chemotaxonomic[al][ly
- e: Chemotaxonomist
- e: Chemotaxonomy
CHEMOTH e: Chemotherapy
Chem Pet Eng e: Chemical and Petroleum Engineering (journ.)
Chem Phys e: Chemical Physics (journ.)
Chem Phys Lett e: Chemical Physics Letters (journ.)
Chem Process e: Chemical Processing (journ.)
Chem Process Eng e: Chemical and Process Engineering (journ.)
- e: Chemical Processing and Engineering (India) (journ.)
Chem Pub e: Chemical Publishing Company
CHEMPURE e: Chemically Pure
chem Rein d: chemische Reinigung
Chem Rev e: Chemical Reviews (journ.)
CHEMS e: Chemical Education Materials Study
CHEMSEARCH e: Chemicals Selected for Equal, Analogous or Related Character[s]
Chem Sec e: Chemical Section
CHEMSOL e: Chemical Solution
CHEMTEC e: Chemical Technicians Curriculum
Chem Techn d: Chemische Technik (journ.)
Chem Technol e: Chemical Technology (journ.)
Chem Technol Fuels Oils e: Chemistry and Technology of Fuels and Oils (journ.)
CHEMTREC e: Chemical Transportation Emergency Center
Chem Unserer Zeit d: Chemie in Unserer Zeit (journ.)
Chem Vers A d: Chemisches Versuchsamt
Chem War[f] e: Chemical Warfare

Chem Zbl d: Chemisches Zentralblatt (journ.)
Chem Ztg d: Chemiker Zeitung
Ch Eng e: Chief Engineer
CHEOPS e: Chemical Engineering Optimization System
- e: Chemical Operations System
Ch ERA e: Chief Engine Room Artificer
CHERT e: Ironstone Sedimentary Rock
CHESS e: Chemical Engineering Simulation System
- e: Community Health and Environmental Surveillance Studies
- e: Community Health and Environmental Surveillance System
CHETA e: Chemical Thermodynamics and Energy Hazard Appraisal
CHETAEH Evaluation e: Chemical Thermodynamics and Energy Hazard Evaluation
CHE Timber e: Indian Chestnut Timber
CHF e: Congestive Heart Failure
- e: Critical Heart Flux
- e: Critical Heat Flux
Chf Chem e: Chief Chemist
Chf Des Engr e: Chief Design Engineer
Chf E[ngr] e: Chief Engineer
Chf Libr e: Chief Librarian
Chf Plan Engr e: Chief Planning Engineer
Chf Publn Engr e: Chief Publications Engineer
CHFS e: Central Hyperfine Structure
Chf Spec Engr e: Chief Specification Engineer
Chf Surg e: Chief Surgeon
Chf Tech e: Chief Technician
Ch/Fwd e: Charges Forward
CHG e: Change
- e: Charge
- e: Charging
- e: Cross Hatch Generator
CHGBL e: Changeable
CHGD e: Changed
- e: Charge[d]
CHGE e: Charge
CHGFA e: Costs Chargeable to Fund Authorization
CHG Message e: Charging Message
CHGOV e: Change-Over
CHGP e: Charging Pump
CHG PL e: Change Plane
CHGR e: Charge[r]
CHGS e: Charges
CH HLD e: Channel Hold
CHI d: chemotherapeutischer Index
- d: chiffriert
- d: Chicago (Flughafen)
- e: Chicago (Airport)
- e: China
- e: Clutch Hitting Index
- e: Computer Human Interaction
- e: Concordia Historical Institute
CHIC e: Cermet (Ceramic-Metal) Hybrid Integrated Circuit
- e: Complex Hybrid Integrated Circuit
Chicago Med Sch Quart e: Chicago Medical School Quarterly (journ.)
CHICO e: Coordination of Hybrid and Integrated Circuit Operations
CHICODER e: Chinese Language Encoder
CHICOM 1 e: First Chinese Space Satellite
CHID e: Community Human and Industrial Development
CHIDIC e: Chinese Dictionary
CHIEF e: Combined Helmholtz Integral Equation Formulation
- e: Controlled Handling of Internal Executive Functions
Chief Nav Op (Air) e: Chief of Naval Operations (Air)
CHIF e: Channel Interface
CHIL e: Current Hogging Injection Logic
- e: Current Hogging Integrated Logic
CHILD e: Cognitive Hybrid Intelligent Learning Device
- e: Computer Having Intelligent Learning and Development
Child Health Bul e: Child Health Bulletin (journ.)
Child Health M e: Child Health Magazine (journ.)
Child Labor Bul e: Child Labor Bulletin (journ.)
CHILL e: CCITT High [Intelligent] Level Language
CHILs e: Current Hogging Injection Logics
CHIMPONAUT e: Chimpanzee Astronaut
CHIMP[S] e: Chimpanzee[s]
CHIN d: China
- e: Characteristic Information
- e: Community Health Information [Network]
CHINA e: Chronic Infectious Neuropathic Agents
CHINFO e: Chief of Information
Ch-Ing d: Chefingenieur
Chin J Phys e: Chinese Journal of Physics (journ.)
ChinL e: Chinese Literature (journ.)
CHIO e: Character-oriented Input/Output
CHIO Processor e: Character-oriented Input/Output Processor
CHIP e: Character-Handling Interface Processor
- e: Chip Hermeticity in Plastic
CHIPS e: Calculator Help in Processing Signals
- e: Chemical Engineering Information Processing System
- e: Chinese Information Processing System
- e: Clearing House Interbank Payments System
- e: Coherent High-Intensity Photon Source
- e: Consumer Health Information Program[me]and Service[s]
CHIR d: Chirurg[ie]
- d: Chirurgisch
- e: Chiropody
- e: Chirurgical
- i: Chirurgia
Chir Doc. l: Chirurgiae Doctor
Chir Doct e: Chirurgiae Doctor (Doctor of Surgery)
Chir Laz d: chirurgisches Lazarett
CHIRO e: Chirography
- e: Chiropractic
- e: Chiropractor
CHIRP e: Community Housing Improvement and Revitalization Program[me]
CHIS e: Clearinghouse Information System
- e: Computerized Hospital Information System
Chi Spr d: chiffrierter Spruch
CHIT e: Computerized Handheld Industrial Terminal
CHI Timber e: Chikrassy Timber
CHIU e: Cache Host Intercommunication Unit
CHJ e: Cambridge Historical Journal
- e: Chief Justice
CHK e: Chalk
- e: Check
CHKB e: Check Bit
CHKD e: Check Digit
CHKDSK e: Check Disk
CHKNO e: Check Number
CHKPROB e: Check Problem
CHKP[T] e: Checkpoint
CHKR e: Checker
chKS d: chemischer Kampfstoff
CHKV e: Check Valve
CHL d: chirurgisches Lazarett
- e: Chain Home Low
- e: Channel
- e: Character Late
- e: Chinese Hamster Lung
- e: Chloroform
- e: Confinement at Hard Labor
- e: Current Hogging Logic
CHLA d: chemisches Laboratorium
- e: Chimpanzee Leucocyte Antigen
- e: Chlorophyll A
CHLAMY e: Chlamydia
CHLB e: Chlorobutanol
Ch Lbr e: Chief Librarian
CHL Cell e: Chinese Hamster Lung Cell
CHLF e: Chloroform
CHLIC e: Coastal Health Library Information Consortium
CHLO e: Chloride
- e: Chloroform
CHLOR e: Chloride
- e: Chlorination
- e: Chlorine
CHLORIC e: Chloricum
Chlorier. d: Chlorierung
CHLORO e: Chloroform
- e: Chlorophyll
- e: Chloroprene
- e: Chlorus
CHLOROFORNM e: Trichloromethane ($CHCl_3$)
CHLOROPRENE e: Synthetic Rubber (C_4H_5Cl)
CHLs e: Current Hogging Logics
CHL Timber e: Chilauni Timber
CHM d: Chemie
- d: Kubikhektometer
- e: Change-over and Change-back Messages
- e: Chemical Milling
- e: Cleveland Health Museum
- e: Compound Handling Machine
- e: Cooper-Hewitt Museum
Ch.M l: Chirurgiae Magister
Ch Mach e: Chief Machinist
CHMAU e: Cache Host Memory Access Unit
CHMC e: Childrens Hospital Medical Center
Ch Med O e: Chief Medical Officer
ChMNH e: Chicago Museum of Natural History
CHMOS e: Complementary High-performance Metal Oxide Semiconductor
CHM Timber e: Champ Timber
CHN e: Carbon, Hydrogen, Nitrogen [Oxygen, Phosphorus, Sulfur]
CHN e: Chain[ing]
- e: Change
CHN Analyzer e: Carbon Hydrogen Nitrogen Analyzer
Ch NCO e: Chemical Noncommissioned Officer
C/Hnd e: Case-Hardened
CHNG e: Change
CHNL e: Channel

CHNL ENT e: Channel Entry
$C_{17}H_{21}NO_4$ e: Cocaine
CHNOPS e: Carbon, Hydrogen, Nitrogen, Oxygen, Phosphorus and Sulfur
CHNS e: Chains
CHNSRA e: Cape Hatteras National Seashore Recreational Area
CHNSY e: Charleston Naval Shipyard
CHN Timber e: Chinar Timber
CHO d: Chinazolinoxyd
- e: Carbohydrate
- e: Character O
- e: Choke Oil
- e: Crop Husbandry Officer
$C_{12}H_{22}O_{21}$ e: Cane Sugar
CHOB e: Cannon House Office Building
CHO Cell[s] e: Chinese Hamster Ovary Cell[s]
CHOKE e: Care How Others Keep the Environment
CHOL e: Cholera
- e: Cholesterol
- e: Common High Order Language
CHOLEST e: Cholesterol Esters
CHOP e: Change of Operational Control
- e: Changeover Point
CHOPAIR e: Change of Operational Control of the Air
CHOPLN e: Change of Operation Plan
CHOPP e: Columbia Homogeneous Parallel Processor
Ch Op[r] e: Chief Operator
CHOPSUM e: Change of Operational Control Summary
CHOPT e: Charterers Option
CHORD e: Change of Operational Order
CHORD-S e: Computer Handling of Reactor Data-Safety
CHORI e: Chief of Office of Research and Inventions
CHO Timber e: Charoli Timber
- e: Chooi Timber
CHO Tumor e: Chinese Hamster Ovarian Tumor
C/hour e: Cards per Hour
CHO/VAC e: Cholera Vaccine
CHOVR e: Change-Over
CHP d: Cumolhydroperoxyd
- d: schweizer Patent
- e: Central Heating Plant
- e: Channel Processor
- e: Child Psychiatry
- e: Combined Heat and Power
- e: Comprehensive Health Planning
- e: Cumulhydroperoxyde
CH & P e: Crew Habitability and Protection
CHPAE e: Critical Human Performance and Evaluation
CHP Generation e: Combined Heat and Power Generation
CHPHOT e: Chief Photographer
CHPM e: Change Priority Mask
CHPP e: Cypress Hills Provincial Park
CH/PPD e: Charges Prepaid
CHPS e: Characters per Second
Ch Pt e: Check Point
CHP Timber e: Chaplash Timber
CHR d: Cercarien-Hüllen-Reaktion
- d: Chrom
- d: Chronometer[zeit]
- e: Canadian Historical Review (journ.)
- e: Candle-Hour[s]
- e: Character [Register]
- e: Chrome
- e: Chrominance
- e: Chromium
- e: Chromobacterium
- e: Chromosomes
- e: Chronic[les]
- e: Coherent Heterodyne Receiver
- e: Constant Hazard Ratio
CHRELE e: Chief Radio Electrician
CHRG e: Charge
CHRIS e: Cancer Hazards Ranking and Information System
- e: Chemical Hazards Response Information System
CHRM e: Chrome
CHRO e: Chrominance
CHROM d: Chromatisch
- d: Chromatisieren
- d: Chromatisiert
- e: Chrominance
- l: chrom
CHROMO e: Chromolithograph
- e: Chromosome[s]
CHROMOLITH e: Chromolithograph[y]
CHROMOS e: Chromosomes
CHRON d: Chronisch
- d: Chronologie
- d: Chronologisch
- e: Chronicle
- e: Chronogram
- e: Chronograph
- e: Chronologic[al]
- e: Chronology
- e: Chronometer
- e: Chronometry
CHRONO e: Chronologic[al][ly]
- e: Chronology
- e: Chronometer
- e: Chronometric[al][ly]
CHRONOGR d: Chronogramm
- d: Chronographie
- e: Chronograph
CHRONOL e: Chronologic[al]
- e: Chronology
CHRO PLTD e: Chrome Plated
CHRST e: Characteristic
CHRTC e: Characteristic
CHR Timber e: Chir Timber
CHRYS e: Chrysler
CHRYSOBERYL e: Beryllium Aluminate
CHRYSOCOLLA e: Hydrous Copper Silicate
CHS d: Computer Hardware Services GmbH
- e: California Horticultural Society
- e: Canadian Hydrographic Service
- e: Case Hardened Steel
- e: Central Heading System
- e: Channels
- e: Chapters
- e: Character Sense
- e: Characters per Second
- e: Checks
- e: Chediak-Higashi Syndrome
- e: Chicago Historical Society
- e: Chloracetophenone Solution
- e: Collimated Holes Structure
- e: Community Health Service
- e: Connecticut Herpetological Society
- e: Connecticut Horticultural Society
- e: Constant Heat Summation
- e: Cristobal High School
- e: Curtis High School
ChSA d: chemische und Strahlenaufklärung
CHSB e: Connecticut Historical Society Bulletin (journ.)
CHSEC e: Characters per Second
CHSL e: Cleveland Health Sciences Library
CHSQ e: California Historical Society Quarterly (journ.)
CHSS e: Childrens Hypnotic Susceptibility Scale
- e: Cooperative Health Statistics System
CHT e: Cable History Tape
- e: Call Hold and Trace
- e: Call Holding Time
- e: Cathode Heating Time
- e: Ceramic Heated Tunnel
- e: Charactron Tube
- e: Chart
- e: Chute
- e: Convective Heat Transfer
- e: Cylinder-Head Temperature
CH_2O e: Formaldehyde
CH_3COOH e: Acetic Acid
Ch Tech e: Chief Technician
CH Temp e: Charge Temperature
CHTG e: Charting
CHTNP e: Chittagong Hill Tracks National Park
CHTORP e: Chief Torpedoman
CHT Timber e: Chatian Timber
CHU e: Centigrade Heat Unit
CHUM e: Computers and the Humanities (journ.)
- e: Computing and the Humanities (journ.)
Chunnel Project e: Channel Tunnel Project
Chur e: Churchill College, Cambridge
CHUT e: Cable Household[s] Using Television
CHUTE e: Parachute
CHV e: Check Valve
CHW e: Chilled Water
- e: Cladding Hull Waste
- e: Cold and Hot Water
- e: Cold Heading Wire
- e: Constant Hot Water
CHX e: Chiro-Xylographic
- e: Chlorhexidine
- e: Condensate Heat Exchanger
CHY e: Chimney
- e: Commission for Hydrology
CI d: Chemischer Informationsdienst
- d: Computer Information
- e: Call Indicator
- e: Call[ing] Instruction
- e: Capability Inspection
- e: Carcinogenic Index
- e: Cardiac Index
- e: Cardiac Insuffciency
- e: Card Index
- e: Card Input
- e: Card Insertion
- e: Card Interpreter
- e: Carnegie Institute
- e: Carriage-to-Interference
- e: Carriage Interlock
- e: Carrier-to-Interference
- e: Carry In
- e: Cartridge Image
- e: Cast Iron
- e: Center of Impact
- e: Central Interval
- e: Centre of Impact
- e: Centromeric Indices
- e: Cephalic Index
- e: Cerebral Infarction
- e: Chain Index
- e: Chaining Instruction
- e: Channel Instruction
- e: Channel Islands
- e: Characteristic Impedance
- e: Check Indicator
- e: Check Information
- e: Chemical Inspectorate
- e: Chemical Ionization
- e: Chemotherapeutic Index

- e: Chief Instructor
- e: Chief of Information
- e: Chlorine Institute
- e: Circuit Interrupter
- e: Cirrus
- e: Civilian Internee
- e: Classification Inventory
- e: Clonus Index
- e: Close-In
- e: Coded Identification
- e: Coded Instruction
- e: Code Information
- e: Coefficient of Intelligence
- e: Cold Iron
- e: Colloidal Iron
- e: Colo[u]r Index
- e: Column Indicator
- e: Combat Ineffective
- e: Combat Information
- e: Combustion Institute
- e: Comfort Index
- e: Command Information
- e: Commonwealth Institute
- e: Communication Information
- e: Communication Interface
- e: Communications and Instrumentation
- e: Community of Interest
- e: Compare Instruction
- e: Complete Instruction
- e: Complex Instruction
- e: Compression Ignition
- e: Computer Independent
- e: Computer Indicator
- e: Computer Industry
- e: Computer Input
- e: Computer Inquiry
- e: Computer Instruction
- e: Computer Interconnection
- e: Computerized Instruction
- e: Computing Instrument
- e: Concentrator-Identifier
- e: Concept Identification
- e: Conditional Instruction
- e: Confidence Interval
- e: Configuration Interaction
- e: Configuration Interface
- e: Configuration Item
- e: Connective Instruction
- e: Constant Instruction
- e: Contamination Index
- e: Continuous Input
- e: Contrast Index
- e: Control Information
- e: Control Instruction
- e: Control Interface
- e: Control Interval
- e: Conversation Impossible
- e: Coordinate Index
- e: Copper Institute
- e: Copy Instruction
- e: Cordage Institute
- e: Core Insulation
- e: Coronary Insufficiency
- e: Corporate Instruction
- e: Cosine Integral
- e: Counter-Intelligence
- e: Course Indicator
- e: Craft Inclination
- e: Cranberry Institute
- e: Critical Influence
- e: Crystal Impedance
- e: Crystalline Insulin
- e: Crystallinity Index
- e: Cubic Inch
- e: Cumulative Index
- e: Curie[s]
- e: Current-Awareness Information
- e: Current Indicator and Integrator
- e: Current Interruption
- e: Curtis Institute
- e: Cut In
- e: Cybernetics International
- e: Input Capacitance
- f: Cerveau Isolé
- f: Circuit Intégré
- f: Compagnie Internationale pour l'Informatique

C & I e: Commerce and Industry
- e: Communication and Instrumentation

CIA d: Cellulose-Ionenaustauscher
- e: California Institute of the Arts
- e: Center for Interreligious Affairs
- e: Centralized Information Acquisition
- e: Channel Interface Adapter
- e: Chemical Industries Association
- e: Chief Inspecting Armourer
- e: Chief Inspector of Armaments
- e: China Institute in America
- e: Communication[s] Interrupt Analysis
- e: Complement and Increment Accumulator
- e: Computer Industry Association
- e: Computer Interface Adapter
- e: Computers Interface Adaptor
- e: Controllers Institute of America
- f: Conseil International des Archives

CIAC e: Career Information and Counseling
- e: Construction Industry Advisory Council

CIAD e: Central Institute for Art and Design

CIAM e: Computer-Integrated and Automated Manufacturing
- e: Computerized, Integrated, Automated, Manufacturing
- s: Congreso Internacional de Arquitectura Moderna

CIAP e: Climatic Impact Assessment Program[me]

CIAPS e: Customer-Integrated Automated Procurement System

CIAS e: California Institute of Asian Studies

CIASE e: Computer Institute for Application in Science and Engineering

CIAT e: Crew Initiated Automatic Test
- s: Centro Internacional de Agricultura Tropical

CIB e: Central Insecticide Board
- e: Centralized Intercept Bureau
- e: Channel Interface Base
- e: Combat Infantrymans Badge
- e: Command Input Buffer
- e: Common Business Oriented Language Information Bulletin
- e: Computer Integrated Business
- e: Convective Instability Base
- e: Core Image Buffer

CIBC e: Commonwealth Institute of Biological Control

CIB COBOL e: COBOL Information Bulletin

CIBHA e: Congenital Inclusion Body Haemolytic Anaemia

CIBHP e: Closed-in-Bottom Hole Pressure

CIC e: Calcium Information Center
- e: Canadian Institute of Chemistry
- e: Card Inventory Control
- e: Cardio-Inhibitor Center
- e: Carrier Identification Code
- e: Center for Instructional Communications
- e: Change Indicator Code
- e: Change Indicator Control
- e: Chemical Industry Council
- e: Chemical Institute of Canada
- e: Chemistry Information Center
- e: Circuit Identification Code
- e: Cloud in Cell
- e: Combat Information Center
- e: Combat Intelligence Center
- e: Combat Intercept Control
- e: Combined Intelligence Course
- e: Command Information Center
- e: Command Input Coupler
- e: Command Interface Control
- e: Common-Impression Cylinder
- e: Common Information Carrier
- e: Commonwealth Information Centre
- e: Communications Instructor Console
- e: Communications Intelligence Channel
- e: Compensated Ionization Chamber
- e: Completely Interconnected Computer
- e: Complex Integrated Circuit
- e: Comprehensive Inorganic Chemistry (journ.)
- e: Computer Information Centre
- e: Computer Instruments Corporation
- e: Conditional Information Content
- e: Content Indication Codes
- e: Control Inquiry Card
- e: Control Instruction Counter
- e: Coordination and Information Center
- e: Coordinator for Industrial Cooperation
- e: Corporate Information Center
- e: Counter-Intelligence Center
- e: Critical Item Code
- e: Curaçao Information Center
- e: Customer Identification Code
- e: Customer Initiated Call
- e: Custom Integrated Circuit
- f: Centre International de Calcul
- f: Conseil International de la Chasse

CICA e: Chicago Industrial Communications Association
- e: Configuration Identification Control and Accounting
- s: Centro de Investigaciones Ciencias Agronómicas

CICAS e: Computer-Integrated Command-and-Attack Systems

CICE e: Computer, Information and Control Engineering

CICGT e: Canadian Institute of Guided Ground Transport

CICHE e: Computer Information Center for Highway Engineering

CICIN e: Conference on Interlibrary Communications and Information Network[s]

CICIREPATO e: Committee for International Cooperation in Information Retrieval among Examining

CICL e: Centre for Industrial Consultancy and Liaison

CICLOPS e: Computers in Chemistry; Logic Oriented Planning of Syntheses

CI CND BOX e: Cast-Iron Conduit Box

CICP e: Capital Investment Computer Program[me]
- e: Committee to Investigate Copyright Problems
- e: Communications Interrupt Control Program[me]

CICS e: Committee for Index Cards for Standards
- e: Customer Information Control System
- e: Customer Inventory Control System

CICS/VS e: Customer Information Control System/Virtual Storage

CICTRAN e: Computer Information Centre for Transportation

CICU e: Cardiology Intensive Care Unit
- e: Cirrocumulus
- e: Commission for Independent Colleges and Universities
- e: Computer-Integrated Converter Unit

CICUNM e: Council of Independent Colleges and Universities of New Mexico
CICUP e: Commission for Independent Colleges and Universities of Pennsylvania
CICV e: Council of Independent Colleges in Virginia
CICWO e: Combat Information Center Watch Officer
CID e: Cable Interconnection Diagram[me]
- e: Cabling Interface Drawings
- e: Center for Industrial Development
- e: Center for Information and Documentation
- e: Central Institute for the Deaf
- e: Centre for Information and Documentation
- e: Change in Design
- e: Charge Image Device
- e: Charge Injection Device
- e: Chemic-Ionization Detector
- e: Chick Infective Dose
- e: Classifitcation of Instructional Disciplines
- e: Cleanliness Identification
- e: Collection of Indexable Data
- e: Collision-Induced Dissocation
- e: Column Indicating Device
- e: Combined Immunodeficiency Disease
- e: Communication Identifier
- e: Communication Implementation Directive
- e: Compatibility Initialization Deck
- e: Component Identification
- e: Condensed Instruction Deck
- e: Connection Identification
- e: Core Image Dictionary
- e: Core Image Dump
- e: Cubic Inch Displacement
- e: Cytomegalic Inclusion Disease
- f: Centre d'Information et de Documentation
CIDA e: Cancer Information Dissemination and Analysis
- e: Channel Indirect Data Addressing
- e: Current-Input Differential Amplifier
- f: Centre d'Informatique et Documentation Automatique
- s: Comité Interamericano de Desarollo Agricola
CIDAC e: Cancer Information Dissemination and Analysis Center
CIDAM e: Compressed Index sequential Access Method
CIDAS e: Centre of Information and Documentation for Agriculture and Silviculture
- e: Conversational Iterative Digital/Analog[ue] Simulator
CIDB d: Chemie-Information und Dokumentation Berlin
- e: Circuit Information Display Board
CIDC e: Cryogenic Information and Data Section
CIDCON e: Civil Disturbance Readiness Conditions
CIDE f: Conseil International du Droit de l'Environnement
CID/ENDS e: Center for Information and Documentation/European Nuclear Documentation System
CIDF e: Control Interval Definition Field
CIDIN e: Common International Civil Aviation Organization Data Interchange Network
CID Lamp e: Compact Indium Discharge Lamp
CID Module e: Charge Injection Device Module
CID NO e: Component Identification Number
CIDNP e: Chemically Induced Dynamic Nuclear Polarization
CIDP e: Computer Industry Development Potential
CIDPS e: Continental Intelligence Data Processing System
CIDS e: Cellular Immunity Deficiency Syndrome
- e: Charge Image Devices
- e: Charge Injection Devices
- e: Chemical Information and Data System
- e: Comprehensive Industry Distribution System
CIDST e: Committee for Information and Documentation on Science and Technology
- f: Comité pour l'Information et la Documentation Scientifique et Technique
CIE e: Catering Industry Employee (journ.)
- e: Cellulose Ion Exchanger
- e: Center for Independent Education
- e: Cesium Ion Emission
- e: Chrysler Institute of Engineering
- e: Cleveland Institute of Electronics
- e: Cleveland Institute of Engineers
- e: Coherent Infrared Energy
- e: Common Ion Effect
- e: Commonwealth Institute of Entomology
- e: Computer Integrated Environment
- e: Computer Interrupt Equipment
- e: Counter Immunoeletrophoresis
- e: Customer-Initiated Entry
- e: Customs Information Exchange
- f: Centre International de l'Environnement
- f: Commission Internationale de l'Eclairage
CIEA s: Centro Internacional de Estudios Agricolas
CIEBM e: Committee on the Interplay of Engineering with Biology and Medicine
CIECC s: Consejo Interamericano para la Educación, la Ciencia, y la Cultura
CIED e: Card Input Editor
CIEE e: Civil Institute of Electrical Engineers
- e: Companion of the Institution of Electrical Engineers
- e: Computer-Integrated Electronic Engineering
- e: Council on International Educational Exchanges
CIEG e: Center for International Economic Growth
CIEI e: Center for International Environment Information
CIEM f: Conseil International pour l'Exploration de la Mer
CIEME e: Chief Inspectorate Electric and Mechanical Equipment
CIEMS e: Catalog for Information Exchange and Message Standards
CIEN s: Comisión Interamericana de Energía Nuclear
CIENT e: Cambridge and Isle of Ely Naturalist Trust
CIEP e: Concentrated Impact Employment Program[me]
- e: Counterimmunoelectrophoresis
CIES e: Comparative and International Education Society
- e: Council for International Exchange of Scholars
CIESM f: Commission International Exploration Scientifique Mediterranean
CIESMM f: Commission International pour l 'Exploration Scientifque de la Mer Méditerranée
CIESS e: Chief Inspector; Engineer and Signal Stores
CIETA f: Centre International d'Etude des Textiles Anciens
CIF e: Cal-Tech Intermediate Format
- e: Canadian Institute of Forestry
- e: Capacitor Input Filter
- e: Captive Installation Function
- e: Carriage, Insurance and Freight
- e: Central Index[ing] File
- e: Central Information Facility
- e: Central Information File
- e: Central Instrumentation Facility
- e: Central Integration Facility
- e: Common Intermediate Format
- e: Customer Information File
CIFAX e: Enciphered Facsimile
CIFC e: Centre for Inter-Firm Comparison
CIFC e: Council for the Investigation of Fertility Control
CIFE e: Central Index File-Europe
CIFI e: Catholic Institute of the Food Industry
CIFP e: Cancel Instrument Flight Rules Flight Plan
CIFRI e: Central Inland Fisheries Research Institute
CIFRR e: Common Instrument Flight Rules Room
CIFST e: Canadian Institute of Food Science and Technology
CIFT e: Central Institute of Fisheries Technology
CIFV e: Chief Inspector of Fighting Vehicles
CIFW e: Cost, Insurance, Freight, War
CIG d: Chemie-Informationsdienst Graz
- e: Chemical Ion Generator
- e: Clause Interconnectivity Graph
- e: Computer Investors Group, Inc.
- e: Cytoplasmic Immunoglobulin
- f: Comité International de Géophysique
CIGFET e: Complementary Insulated Gate Field-Effect Transistor
CIGGT e: Canadian Institute of Guided Ground Transport
CIGR f: Commission Internationale de Genie Rural
CIGTF e: General Inertial Guidance Test Facility
CIH e: Carbohydrate-Induced Hyperglyceridemia
CII e: Call Identity Index
- e: Cancer Immunology and Immunotherapy (journ.)
- e: Chartered Insurance Institute
- e: Current Indicator and Integrator
- f: Compagnie Internationale pour l'Informatique
CIIC e: Chemical Industry Institute of Toxicology
- e: Counter-Intelligence Interrogation Center
CIID f: Commission Internationale des Irrigations et de Drainage
CIIG Bul[l] e: Construction Industry Information Group Bulletin (journ.)
CII-HB f: Compagnie Internationale pour l'Informatique-Honeywell Bull
CIIN e: Computer-Aided Manufacturing Integrated Information Network
CIIR e: Central Institute for Industrial Research
CIIT e: Chemical Industry Institute of Toxicology
CIIUAP e: Commission on Increased Use of Agricultural Products
CIJE e: Current Index to Journals in Education

CIKCU e: Council of Independent Kentucky Colleges and Universities
CIL e: Call Identification Line
- e: Clear Indicating Lamp
- e: Coded Instruction Line
- e: Coefficient of Intensity; Luminous
- e: Commercial Instrument Landing
- e: Computer Independent Language
- e: Computer Instrumentation Limited
- e: Computer Interpreter Language
- e: Control Interpreter Language
- e: Core Image Library
- e: Critical Item List
- e: Current Inhibit Logic
- e: Current Injection Logic
CILA e: Casualty Insurance Logistics Automated
- s: Centro Interamericano de Libros Académicos
CILB f: Commission Internationale de Lutte Biologique Contre les Animaux et les Plantes
CILCA e: Current Interest Late Charge Amount
CILE e: Call Information Logging Equipment
CILES e: Central Information Library and Editorial Section
CILET e: Circular Letter
CILS e: Computer Information Library Service
- e: Cossor Instrument Landing System
CIL Service e: Computer Information Library Service
CILT e: Center for Information on Language and Teaching
- e: Centre for Information of Language Teaching [and Research]
CIM e: Canadian Institute of Mining and Metallurgy
- e: Cavitation Intensity Meter
- e: Central Inventory Management
- e: Cleveland Institute of Music
- e: Co-operative Investigations in the Mediterranean
- e: Coded-Impulse Modulation
- e: Communication[s] Interface Module[s]
- e: Communication[s] Interface Monitor
- e: Communications Improvement Memorandum
- e: Compound Inserting Machine
- e: Computer-Integrated Manufacturing
- e: Computer Input [by] Microfilm[ing]
- e: Computer Input Matrix
- e: Computer Input Media
- e: Computer Input Multiplexer
- e: Computer Integrated Machinery
- e: Computer Integrated Manufacture
- e: Computer Integrated Manufacturing
- e: Conductance Increase Mechanism
- e: Connection Interface Module
- e: Console Interface Module
- e: Continuous Image Microfilm[ing]
- e: Cooperative Investigations in the Mediterranean
- e: Cork Insulation Material
- e: Crystal Impedance Meter
- e: Cumulated Index Medicus (journ.)
- e: Curtis Institute of Music
Ci/m³ e: Curie pro Kubikmeter
CIMA e: Coordinated Investigation of Micronesian Anthropology
- e: Creek Indian Memorial Association
CIMAV e: Communication Improvement Auxiliary Vessel
CIMC e: Communications Intelligent Matrix Control
- s: Colegio Interamericano Médicos y Cirujanos
CIM Card e: Continuous Image Microfilm Card
CIMCO e: Card Image Correction
CIME e: Chartered Institute of Marine Engineers
CI Mech E e: Companion of the Institution of Mechanical Engineers
CIML e: Contract Item Material List
CIMM e: Canadian Institute of Mining and Metallurgy
- e: Communication Interface and Multiplexer Module
CIMM e: Constant Impedance Mechanical Modulation
CIMMS e: Civilian Information Manpower Management System
CIMOS e: Cincinnati Milacron Operating System
CIMP d: Cyclo Inosin-3',5'-Monophosphat
- e: Cyclo Inosin-3.5-Monophosphat
CIMPAC e: Computer Integrated Manufacturing Process Analysis and Control
CIMPLE e: Card Image Manipulator for Large Entities
CIMR e: Commanders Internal Management Review
CIMS e: Chemical Ionization Mass Spectrometry
- e: Civilian Information Management System
- e: Civilian Information Manpower System
- e: Communications Instructions for Merchant Chips
- e: Computer Installation Management System
- e: Computer Integrated Manufacturing System
- e: Computer Integrated Measurement System
CIMT e: Chief Inspector of Mechanical Transport
CIMTP f: Congrés International de Médecine Tropicale et de Paludisme
CIMU e: Compatibility-Integration Mock-Up
CIN e: Carrier Input
- e: Cerebriform Intradermal Nevus
- e: Cervical Intra-Epithelial Neoplasia
- e: Change Identification Number
- e: Chemical Industry Notes
- e: Cinder
- e: Code Identification Number
- e: Colo[u]r Internal
- e: Communication-Identification-Navigation
- e: Component Identification Number
- e: Cooperative Information Network
CINAP e: Cincinnati Numerical Automatic Programming
CINCAC e: Commander-in-Chief; Continental Air Command
CINCAFLANT e: Commander-in-Chief; Air Force Atlantic Command
CINCAFMED e: Commander-in-Chief; Allied Forces; Mediterranean
CINCAFPAC e: Commander-in-Chief; Army Forces; Pacific
CINCAFSTRIKE e: Commander-in-Chief; Air Force Strike Command
CINCAIRCENT e: Commander-in-Chief; Allied Air Forces; Central Europe
CINCAirChan e: Commander-in-Chief; Allied Air Forces; Channel
CINCAIREASTLANT e: Commander-in-Chief; Allied Air Forces; Eastern Atlantic Area
CINCAirWestLant e: Commander-in-Chief; Allied Air Forces; Western Atlantic Area
CINCALAirCentEur e: Commander-in-Chief; Allied Air Forces; Central Europe
CINC Asiatic e: Commander-in-Chief; Asiatic Fleet
CINCATLFLT e: Commander-in-Chief; Atlantic Fleet
CINCBPF e: Commander-in-Chief; British Pacific Fleet
CINCGCY e: Counterinsurgency
CINCH e: Computerized Information Network for Community Health
C-in-C HF e: Commander-in-Chief; Home Fleet
CINCLANTFLT e: Commander-in-Chief; Atlantic Fleet
CINCMAIRCHAN e: Commander-in-Chief; Maritime; Allied Air Forces; Channel
CINCMEAF e: Commander-in-Chief; Middle East Air Forces
CINCNAVEASTLANTMED e: Commander-in-Chief; Naval Forces Eastern Atlantic and Mediterranean
CINCNAVFOREUR e: Commander-in-Chief; Naval Forces; Europe
CINCNELM e: Commander-in-Chief; Naval Forces; Eastern Atlantic and Mediterranean
CINCPACAF e: Commander-in-Chief; Pacific Air Forces
CINCPacFlt e: Commander-in-Chief; Pacific Fleet
CINCPACHEDPEARL e: Commander-in-Chief; Pacific Fleet Headquarters; Pearl Harbor
CINCRDAF e: Commander-in-Chief; Royal Danish Air Force
CINCRDN e: Commander-in-Chief; Royal Danish Navy
CINCRNorN e: Commander-in-Chief; Royal Norwegian Navy
CINCSouth e: Commander-in-Chief; [Allied Forces] Southern Europe
CINCTAC e: Commander-in-Chief; Tactical Air Command
CINCTURNAV e: Commander-in-Chief; Turkish Navy
CINCUS e: Commander-in-Chief; United States Navy
CINCUSAFE e: Commander-in-Chief; United States Air Forces; Europe
CINCUSAFLANT e: Commander-in-Chief; United States Air Force; Atlantic
CINCUSAFSTRIKE e: Commander-in-Chief; United States Air Force Strike [Command]
CINCUSNAVEUR e: Commander-in-Chief; United States Naval Forces; Europe
CINCWIO e: Cooperative Investigation of the Northern Central West Indian Ocean
CIND e: Computer Index of Neutron Data
CINDA e: Center for Information and Numerical Data Analysis
- e: Chrysler Improved Numerical Differencing Analyzer
- e: Computer Index of Neutron Data
CINDA 3G e: Chrysler Improved Numerical Differencing Analyzer for Third-Generation Computers
CINDAS e: Center for Information and Numerical Data Analysis and Synthesis
CINE e: Cinema[tographic]
- e: Cinematography
- e: Cinetheodolite

CINECA e: Cooperative Investigations of the Northern part of the Eastern Central Atlantic
CINFAC e: Counterinsurgency Information Analysis Center
- e: Cultural Information Analysis Center
CINFO e: Chief of Information
CINFR e: Central Institute for Nutrition and Food Research
CINIC e: Current Inverting Negative Impedance Convertor
CINL e: Cumulative Index to the Nursing Literature
CINM e: Channel Islands National Monument
CINNABAR e: Mercuric Sulfide (H_gS)
CINO e: Chief Inspector of Naval Ordnance
CINS e: Central Treaty Organization Institute of Nuclear Science
- e: Cryogenic Inertial Navigating System
CINS CENTO e: Institute of Nuclear Science
CINTC e: Chief; Intelligence Corps
CINTEL e: Computer Interface for Television
- e: Computer Interface to Television
CIN Timber e: Cinnamon Timber
C_{in} e: Insulin Clearance
CINV e: Control Interval
CIO e: Central Information Office
- e: Central Input-Output (Multiplexer)
- e: Chief Information Officer
- e: Combat Intelligence Officer
- e: Command Issuing Officer
- e: Common Item Order
- e: Confidential Interim Order
- e: Confirming Informal Order
- e: Control Input/Output
- f: Commission Internationale d'Optique
CIOA e: Center for Information of America
- e: Center for Information on America
CIO Buffer e: Counter for Input/Output Buffer
CIOCS e: Communications [Adapter] Input-Output Control System
CIOE e: Computer Input/Output Equipment
CIOM e: Communications Input/Output Multiplexer
CIOMS e: Council for [the] International Organization of Medical Sciences
CIOO e: Chief Inspecting Ordnance Officer
CIOP e: Computer-Aided Measurement and Control Input-Output Processor
CIOS f: Conseil International pour l'Organisation Scientifique
CIOSTA f: Commission Internationale pour l'Organisation Scientifique du Travail en Agriculture
CIOU e: Central Input/Output Unit
- e: Custom Input/Output Unit
CIP d: chemisches Ionenplattieren
- e: Call Information Protocol
- e: Calm Inversion Pollution
- e: Capsule Internal Programmer
- e: Cascade Improvement Program[me]
- e: Cast Iron Pipe
- e: Catalog[u]ing in Publication
- e: Cementation-in-Pulp
- e: Central Input Program[me]
- e: Centralized Information Processing
- e: Certain Information Processing
- e: Certified Inhalation Protection
- e: Chief Industrial Property
- e: Cipher
- e: Civilian Improvement Program[me]
- e: Classical Philology (journ.)
- e: Class Improvement Plan
- e: Clean-in-Place
- e: Command Information Program[me]
- e: Commercial Import Program[me]
- e: Commercial Instruction Processor
- e: Common Input Processor
- e: Communications Interface Processor
- e: Communications Interrupt Program[me]
- e: Communications Pool Interpreter Program[me]
- e: Community Improvement Program[me]
- e: Compatible Independent Peripherals
- e: Compensator Improvement Program[me]
- e: Complex Information Processing
- e: Component Improvement Program[me]
- e: Composite Interface Program[me]
- e: Computer Image Processing
- e: Computer Information Processing
- e: Computer Input Program[me]
- e: Computer Interchange Program[me]
- e: Computing Intensive Program[me]
- e: Console Interface Program[me]
- e: Consolidated Instrument Package
- e: Consolidated Intelligence Program[me]
- e: Construction Inspection Procedure
- e: Contract Information Processor
- e: Control Inlet Panel
- e: Conversion in Place
- e: Cost Improvement Program[me]
- e: Cost Improvement Proposal
- e: Count in Process
- e: Critical Innovation Potential
- e: Crystalized Information Processing
- e: Current Injection Probe
- e: Custom Interest Profile
- f: Comité International de Photobiologique
CIPA e: Chartered Institute of Patent Agents
CIPAC e: Collaborative International Pesticides Analytical Council
CIPAP e: Changes in Intinerary to Proceed to Additional Places
CIPASH e: Committee for an International Program[me] in the Atmospheric Sciences and Hydrology
CIPEC f: Conseil Intergouvernmental des Pays Exportateurs de Cuivre
CIPER e: Central Inventory of Production Equipment Records
CIP File e: Critical Inventory Parts File
CIPH e: Cipher
CIPHER e: Calculations of Patient and Hospitals Education Resources
CIPHONY e: Cipher Telephony
- e: Enciphered Telephony
CIPIG e: Conference on Ionization Phenomena in Gases
CIPL e: Commission of Inquiry into Public Libraries (Australian)
- f: Comité International Permanent de Linguistes
CIPN f: Commission Internationale des Parcs Nationaux
CIPO f: Conseil International pour la Préservation des Oiseaux
CIPOM e: Committee on Computers; Information Processing and Office Machines
CIPP e: Cataloging-in-Publication Program[me]
- e: Computing Technology Center Index Printing Program[me]
- e: Copilots Instrument Power Panel
CIPP Model e: Context, Input, Process Product Model
CIPPP e: Cooperative International Pupil-to-Pupil Program[me]
CIPR e: Continuous Inflight Performance Recorder
- f: Commission Internationale de Protection contre les Radiations
CIPRA e: Cast-Iron Pipe Research Association
- f: Commission Internationale pour la Protection des Régions Alpines
CIPS e: Canadian Information Processing Society
- e: Centralized Information Processing System
- e: Cesium Ion Propulsion System
- e: Corporate Information Processing Standards
- e: Current Intake Processing System
CIPSIO f: Comité Intersecrétariats des Problèmes Scientifiques Intéressant l'Océanographie
CIQ e: Computer Input Quantity
CIR e: Canada-India Reactor
- e: Center for Informatics Research
- e: Change Initiation Request
- e: Channel Inboard Record
- e: Characteristic Impedance Ratio
- e: Chief of Industrial Relations
- e: Circle
- e: Circuit
- e: Circular
- e: Circulation
-e: Circumference
- e: Clearinghouse on Information Resources
- e: Colo[u]r Infrared
- e: Common Instrument Flight Rules Room
- e: Computer Integrated Research
- e: Computerized Information Research
- e: Computerized Information Retrieval
- e: Conditional Interrupt Request
- e: Console Interrupt Request
- e: Control and Interrupt Register
- e: Control Instruction Register
- e: Controlled Intact Reentry
- e: Converter Instruction Register
- e: Cost Information Report
- e: Court Industrial Relations
- e: Current Industrial Reports
- e: Current Instruction Register
- f: Comité International de la Radioelectricité
CIRA e: Committee on International Reference Atmosphere
- e: Committee on Space Research International Reference Atmosphere
CIRADS e: Counter-Insurgency Research and Development System
CIRALR e: Coating for Infra-red and Light Reflection
CIRANT e: Circular Antenna
CIRAS e: Census Information Retrieval and Analysis System
C/I Ratio e: Carrier/Interference Ratio
CIRBKR e: Circuit Breaker
CIRC e: Central[ized] Information Reference and Control
- e: Central Intelligence Retrieval Center
- e: Circle
- e: Circuit
- e: Circular
- e: Circulate
- e: Circulation
- e: Circulation Input Recording Center
- e: Circulator
- e: Circulatory
- e: Circumference[s]
- e: Circumferential[ly]
- e: Circumstance[s]
- e: Collection, Inquiry, Reporting and Communication

- e: Collection, Inquiry, Retrieval and Communication
- e: Computing and Information Resources Committee
- f: Centre International de Recherché sur le Cancer

CIRCA e: Computerized Information Retrieval and Current Awareness
CIRCAD e: Circadiamy
- e: Circadian
- e: Circadic

CIRCAL e: Circuit Analysis
CIRCALS e: Circle Analysis System
CIRCAP e: Circular Capacitor
CIRCARD[S] e: Circuit Card[s]
CIRCE e: Circumstance
CIRC G e: Circular Grate
CIRC IN e: Circular Inch
CIR CL e: Circuit Closing
CIRCLE e: Combined Integrated Resource[s] for Community Learning Experiment
CIRCLIP e: Circular Clip
CIRCLTR e: Circular Letter
CIRC MEM e: Circulating Memory
CIRC MIL e: Circular Mil
CIRCO e: Circulation Control
CIRCOL e: Central Information Reference and Control On-Line
CIRCON e: Circulation Control
Circ Res e: Circular Research [An Official Journalof the American Heart Foundation] (journ.)
CIRC System e: Centralized Information Reference and Control System
- e: Circulation Control System
- e: Circulation Input Recording Center System

CIRCUM e: Circumference
CIRCUP e: Circulation Control Updating Program[me]
CIRCUS e: Calculation of Indirect Resources and Conversion to Unit Staff
- e: Circuit Simulator

CIRD e: Chemical Information Resources Directory
Cir Des Eng e: Circuit Design Engineer
CIRDP e: Cost Information Reporting Data Plan
CI Rel e: Certificate in Industrial Relations
CIRES e: Chief Inspector of Royal Engineer Stores
- e: Communication[s] Instructions for Reporting Enemy Sightings

CIRF e: Corn Industries Research Foundation
- e: Customer-Integrated and/or Reference File

CIRF System e: Customer-Integrated and/or Reference File System
CIRG e: Circulating
CIRGA e: Critical Isotope Reactor; General Atomics
CIRIA e: Construction Industry Research and Information Association
CIRIN e: Circular Inch
CIRIS e: Completely Integrated Range-Instrumentation System
- e: Completely Integrated Reference Instrumentation System

CIRK e: Computing Technology Center Information Retrieval from Keywords
CIR LET e: Circular Letter
CIRM e: Celestial Infra-Red Mapper
- e: Circular Mil
- e: Committee International Radio Maritime
- f: Comité International Radio-Maritime
- p: Comissão Interministerial de Recursos do Mar
- s: Centro Internazionale Radio-Médico

CIR Method e: Courant-Isaacson-Rees Method
CIR MIL e: Circular Mil
CIRO e: Card Inductor Read Only
- e: Crash Injury Research Organization

CIRO Memory e: Card Inductor Read Only Memory
CIR OP e: Circuit Opening
CIRP e: College International Research Production
CIRPA e: Canadian Independent Record Production Association
CIR Program[me] e: Central Input Recording Program[me]
CIRS e: Community Information and Referral Service
- e: Computerized Information Retrieval Services
- e: Customer Information Record System
- e: Customer Information Reference System

CIR Signal e: Calling-Line-Identity-Request Signal
CIRSYS e: Circulation System
CIR Unit e: Colonial Insecticide Research Unit
CIRVIS e: Communication Instructions for Reporting Vital (or Visual) Intelligence Sighting
CIS d: compatible[s] Informationssystem[e]
- d: Computer-Informations-Service
- d: computergestütztes Informationssystem
- d: computerorientiertes Informationssystem
- d: Continental Information Systems GmbH
- e: Cable Information Service[s]
- e: Cambridge Interactive System
- e: Canadian Institute of Surveying
- e: Cancer Information Service
- e: Card Input Station
- e: Career Information Service
- e: Career Information System
- e: Cartographic Information Society
- e: Cassette Information Services
- e: Catalog[u]ing in Source
- e: Catholic Information Services
- e: Center for Information Sciences
- e: Center for Information Services
- e: Center for International Studies
- e: Center of International Studies
- e: Central Information Service
- e: Central Information System
- e: Central Input System
- e: Central Installation Supply
- e: Central Instructor School
- e: Centre for Information Services
- e: Cesium Ion Source
- e: Channel and Isolation Supervision
- e: Character Imaging System
- e: Character Instruction[s] Set
- e: Chemical Information Services
- e: Chemical Information System
- e: Cirrostratus
- e: Clinical Information System
- e: College Information System
- e: Commercial Instruction Set
- e: Common Input System
- e: Communication[s] and Instrumentation System
- e: Communication Interfaces
- e: Communications Information System
- e: Community Information System
- e: Compact Interactive Standard
- e: Company Information System
- e: Compatible Information Systems
- e: Compiler Interpreter System
- e: Complementary Information System
- e: Composition Information Services
- e: Computer and Information Systems
- e: Computer Information Service
- e: Computer-Oriented Information System
- e: Computerized Information System
- e: Computing Information Services
- e: Conductor-Insulator-Semiconductor
- e: Configuration Information System
- e: Congressional Information Service
- e: Console Inquiry Station
- e: Continental Information Systems Corporation
- e: Control Indicator Set
- e: Control Input Signal
- e: Convention Information System
- e: Coordinate Indexing System
- e: Cornwall Information Service
- e: Correction Information System
- e: Corrosion Interception Sleeve
- e: Cost Information System
- e: Counter Information Service
- e: Coupled Impedance Synthesis
- e: Cranbrook Institute of Science
- e: Cryogenic Instrumentation System
- e: Cue Indexing System
- e: Current Information Selection
- e: Customer Information System
- e: Custom Integrated System

CISA f: Commission Internationale pour le Sauvetage Alpin
CISAM e: Compressed Index Sequential Access Method
CISC e: Canadian Institute of Steel Construction
- e: Clearinghouse for Innovation in Scientific Communication
- e: Community Information and Services Center
- e: Complex Instruction Set Computer
- e: Complex Instruction Set Computing
- e: Computer and Information Science Center

CISCO e: Compass Integrated System Compiler
CISD e: Corporate Information Service Department
CIS & DB e: Comprehensive Information System and Data Base
CISDC e: Cancer Information Service for Developing Countries
CISE e: Colleges, Institutes and Schools of Education
CISED e: Corporate Information Systems Education Department
CISHEC e: Chemical Industry Safety and Health
CISI e: Centre for Inventions and Scientific Information
- e: Command Inspection System Inspection
- f: Compagnie Internationale de Services en Informatique

CISIC e: California Information Systems Implementation Committee
CI Sign e: Cartridge Illuminating Signs
CISIL e: Cost Information System Indirect Labour
CISIR e: Ceylon Institute of Scientific and Industrial Research
CISK e: Conditional Instability of the Second Kind
CISLUN e: Cislunar[ian]
CISNE e: Computerized Information[s] System for Nursing Educators
CISP e: Computerized Information System for Pathology
- e: Counterintelligence Specialist

CISPF e: Cast Iron Soil Pipe Foundation

CISPI e: Cast Iron Soil Pipe Institute
CISPR f: Comité International Special de Perturbations Radioelectriques
CISR e: Center for International Systems Research
- e: Computer and Information Science Research Center
- e: Configuration Index and Status Report
- e: Corporate Interplant Status Record
CISRC e: Computer and Information Science Research Center
CISS e: Communication and Instrumentation Support Services
- e: Community Information Sharing Service
- e: Computerized Information Storage System
- e: Conference on Information Sciences and Systems
- e: Consolidated Information Storage System
CISSIG e: Cataloging and Indexing Systems Special Interest Group
CIS Structure e: Conductor Insulator Semiconductor Structure
CIST e: Cirrostratus
- e: Co-orbital Intercepter Scoring Technique
CISTC e: Council on International Scientific and Technological Cooperation
CISTI e: Canada Institute for Scientific and Technical Information
- e: Canada Institute of Scientific and Technical Education
- e: Central Institute for Scientific and Technical Information
CISTIP e: Committee on International Scientific and Technical Information Program[me]
CIT d: charakterologischer Intelligenztest
- e: Cable Interface Tape
- e: Cairo Institute of Technology
- e: California Institute of Technology
- e: Call-in-Time
- e: Card Image Tape
- e: Carnegie Institute of Technology
- e: Case Institute of Technology
- e: Center for Information Technology
- e: Central Institute of Technology
- e: Cislunar Interorbital Transportation
- e: Citation
- e: Citrate
- e: Combinational Iterative Tree
- e: Command Interface Test
- e: Command Interpretation Table
- e: Commission on Insurance Technology
- e: Compressor Inlet Temperature
- e: Computer Integrated Telephoning
- e: Computer Integrated Telephony
- e: Computer Interface Technology
- e: Computer Interface Terminal
- e: Configuration Identification Table[s]
- e: Construction Industry Thesaurus
- e: Contact Ion Thruster
- e: Controlled Interceptor Trainer
- e: Convective Instability Top
- e: Crack Initiation Temperature
- e: Cranfield Institute of Technology
- f: Compagnie Industrielle des Télécommunications
- f: Convention Internationale des Télécommunications
CITA e: Citation
- e: Citric Acid
CITAB e: Computer Instruction and Training Assistance for the Blind
CITACID e: Citric Acid
CITAP e: Computerized Interactive Thermal Analysis Program[me]
CITARS e: Crop Identification Technology Assessment for Remote Sensing
CITB e: Construction Industry Training Board
CITC e: Canadian Institute of Timber Construction
- e: Construction Industry Training Center
- e: Construction Industry Training Centre
CITCA e: Commitee of Inquiry into Technological Change in Australia
CIT-CARB e: Caulfield Institute of Technology - Computer Abuse Research Bureau
CITE e: Capsule Integrated Test Equipment
- e: Compression Ignition and Turbine Engine
- e: Computer Integrated Test Equipment
- e: Computerized Information Transfer in English
- e: Consolidated Index of Translations into English
- e: Contractor Independent Technical Effort
- e: Controller Input Test Equipment
- e: Council of the Institute of Telecommunication Engineers
- e: Current Information [on] Tapes for Engineers
- e: Current Information Transfer in English
CITEC f: Compagnie pour l'Informatique et Technique Electronique de Contròle
CITECH e: Cawkell Information and Technology Services
CI Technology e: Collector Implanted Technology
CITEL e: Conference on Interamerican Telecommunications
- f: Compagnie Internationale de Téléinformatique
- s: Comisión Interamericana de Telecomunicaciones
- s: Conferencia Interaméricana de Telecomunicaciones
CITES e: Current Intelligence Traffic Exploitation System
CIT/GAL e: California Institute of Technology / Guggenheim Aeronautical Laboratory
CITI f: Centre d'Information des Techniques Industrielles
CITIS e: Centralized Integrated Technical Information System
- e: Construction Industry Translation and Information Services
CIT/JPL e: California Institute of Technology / Jet Propulsion Laboratory
CITM e: Certified International Traffic Manager
CITP e: Civilian Industrial Technology Program[me]
CITRAC e: Central Integrated Traffic Control
CITRICULT e: Citriculture
CITRINE e: False Topaz
CITS e: Central Integrated Test Set
CITs e: Combinational Iterative Trees
CITS-MU[LTIPLE]X e: Central Integrated Test System Multiplex
CIU e: Central Index Unit
- e: Central Information Unit
- e: Central Interface Unit
- e: Central Interpretation Unit
- e: Channel Interface Unit
- e: Command Interface Unit
- e: Communications Interface Unit
- e: Community Information Utilities
- e: Computer Interface Unit[s]
- e: Console Inquiry Unit
- e: Corporate Information Unit
CIUP e: Country Installed User Program[me]
CIUS e: Computer Interface Units
- f: Conseil International des Unions Scientifiques
CIV e: Civil
- e: Civilian
- e: Civilization
- e: Civilize
- e: Code Inverter Verifier
- e: Corona Inception Voltage
CIVAVN e: Civil Aviation
CIVCLO e: Civilian Clothing
CIVD e: Cold-Induced Vasodilatation
Civ E[ng] e: Civil Engineer[ing]
Civ Eng (NY) e: Civil Engineering (New York) (journ.)
Civ Eng & PWR e: Civil Engineering and Public Works Review (journ.)
CIVIES a: Civilingenier
CIVM e: Collision-Imparted Velocity Method
CIVPERSINS e: Civilian Personnel Information System
CIVSUB e: Civilian Substitution Program[me]
CIW e: Carnegie Institute of Washington
- e: Ceramic Insulated Wire
- e: Computer Instruction Word
- e: Current Instruction Word
CIWS e: Concentrator Isolation Working Subsystem
Ci(x) e: Cosine Integral
CIXA e: Constant Infusion Excretory Program[me]
CJ e: Card Jam
- e: Chief Justice
- e: Clip Joint
- e: Cold Joint
- e: Cold Junction
- e: Computer Journal
- e: Conditional Jump
- e: Construction Joint
C & J e: China and Japan
CJA e: Computer Journal Abstracts (journ.)
CJB e: Cold Junction Box
CJC e: Colby Junior College
- e: Community Junior College
- e: Continuous Job Card
CJCE e: Canadian Journal of Civil Engineering (journ.)
CJD d: Creutzfeldt-Jakob Disease
- e: Card Jam Detector
- e: Criminal Jurisprudence, Doctor of
CJF e: Conversational Job Facility
CJFET e: Complementary Junction Field Effect Transistor
CJI e: Concrete Joint Institute
- e: Conditional Jump Instruction
CJIS e: California Criminal Justice Information System
- e: Canadian Journal of Information Science
- e: Criminal Justice Information System
CJL e: Canadian Journal of Linguistics (journ.)
- e: Conversational Job Language
CJO e: Communications Jamming Operator
- e: Conditional Jump Operation
- e: Conditional Jump Order
CJP e: Casio Jet Printing
- e: Communication Jamming Processor
- e: Cornu-Jellet Prism
CJPs e: Canadian Journal of Psychology (journ.)
CJRL e: Criminal Justice Reference Library
CJS e: Cotton, Jute or Sisal

CJSCSC e: Cost-Schedule Control Systems Criteria
CJTT e: Computer Jobs Through Training
CJZ e: Cable Jacket Zipper
CK e: Cape Kennedy
- e: Cask
- e: Chalk
- e: Check[ed]
- e: Circuit
- e: Classification Keyboard
- e: Clearing Key
- e: Clerk
- e: Clock
- e: Coke
- e: Control Keyboard
- e: Crystal Kit
- e: Cyanogen Chloride
C/K e: Mol Coulombs per Kilmol
CKAFS e: Cape Kennedy Air Force Station
CKB d: VEB Chemisches Kombinat Bitterfeld (DDR)
- e: Caequot Kite Balloon
- e: Centralized Knowledge Base
CK Bits e: Check Bits
CKC e: Check Character
- e: Chicken Kidney Cell
CKD e: Checked
CKDIG e: Check Digit
CKF d: Chrominanz-Kammfilter
- e: Cork Floor
CKFF e: Cape Kennedy Forecast Facility
CKFM e: Check Form
C/kg d: Coulomb je Kilogramm
C/Kg e: Coulombs per Kilogram
CKGA e: Check Gauge
CKIC e: Chemical Kinetics Information Center
ckm d: Kubikkilometer
CKMTA e: Cape Kennedy Missile Test Annex
- e: Cape Kennedy Missile Test Area
CKMTC e: Cape Kennedy Missile Test Center
CKN e: Cape Kennedy
CKO e: Checking Operation
- e: Checking Operator
CKOS e: Countersink Other Side
CKP e: Cayley-Klein Parameter
- e: Computed Kill Point
CKPT e: Check Point
- e: Cockpit
CKRSO e: Cape Kennedy Range Safety Officer
CKS d: chemische Korrosionsschutzstoffe
CKSNI e: Cape Kennedy Space Network, Incorporated
CKT e: Circuit
CKT BD e: Circuit Board
CKT BKR e: Circuit Breaker
CKT CL e: Circuit Closing
CKT ID e: Circuit Identification
CKT OP e: Circuit Opening
CKTP e: Check Template
CKTS e: Circuits
- e: Countersink this Side
CKU d: computerkontrollierter Unterricht
CKVAR e: Capacitor Kilo Volt Ampere[s] Reactive
CKVLV e: Check Valve
CKW d: chlorierte Kohlenwasserstoffe
- e: Clockwise
CKY d: Conakry (Flughafen)
CL d: Chlor
- d: Clarit[isch]
- d: Container-Ladung
-d: Zentiliter
- e: Calamus Length
- e: Calendar Line
- e: Caliber
- e: Calibre (GB)
- e: Call
- e: Calling Line
- e: Call Library
- e: Cambridge Library
- e: Capital Airlines
- e: Card Lever
- e: Card Lifter
- e: Card Line
- e: Card Load
- e: Car Load
- e: Carload Lot
- e: Carnegie Library
- e: Cartilaginous Layer
- e: Cartridge Loader
- e: Cathode Luminescence
- e: Ceiling Level
- e: Ceiling Value
- e: Cellular Logic
- e: Center Line
- e: Center Left
- e: Centiliter
- e: Centilitre (GB)
- e: Central Laboratory
- e: Central Line
- e: Centre Left
- e: Centre Line
- e: Change List
- e: Channel Loading
- e: Chemiluminescence
- e: Chest and Left Arm (Cardiology)
- e: Chile
- e: Chloride
- e: Chlorine [Gas]
- e: Circuit Layout
- e: Circuit Logic
- e: Circular Letter
- e: Civil Law
- e: Clamp Lamp
- e: Clarendon Laboratory
- e: Class[ical]
- e: Classics
- e: Clear[ance]
- e: Clear Lamp
- e: Climatic Laboratory
- e: Climb
- e: Clinic
- e: Clip
- e: Close
- e: Closed Loop
- e: Closing
- e: Closure
- e: Cloud
- e: Code Language
- e: Code Line
- e: Code List
- e: Coding Language
- e: Coding Line
- e: Coefficient of Lift
- e: Coil
- e: Columbite
- e: Command Language
- e: Command List
- e: Commendation List
- e: Comment Line
- e: Common Language
- e: Common Law
- e: Common Logarithm
- e: Communication Lieutenant
- e: Comparative Literature
- e: Compare Logical
- e: Compare Long
- e: Compatibility List
- e: Compiler Language
- e: Complex Loading
- e: Computational Linguistics
- e: Computer Laboratory
- e: Computer Language
- e: Computing Law
- e: Concentration Length
- e: Conceptual Level
- e: Condylar Cartilage
- e: Confidence Limit
- e: Connecting Line
- e: Connectionless
- e: Consolidated Listing
- e: Contact Loss
- e: Contact Lost
- e: Container Load
- e: Continuation Line
- e: Continuous Load
- e: Control Language
- e: Control Leader
- e: Control Limit
- e: Control Line
- e: Control Logic
- e: Conversion Loss
- e: Cooperative Logistics
- e: Correction List
- e: Cost of Living
- e: Council
- e: Counter Logic
- e: Course Line
- e: Craft Loss
- e: Crane Load
- e: Critical List
- e: Crystallographic Laboratory
- e: Current Logic
- e: Cut Length
- e: Cutter Location
- e: Cycle Length
- e: Cyclotron Laboratory
- e: Cylinder
- e: Cytotoxic Lymphocytes
- l: Corpus Luteum
C & L e: Control and Line
CL-1 e: Computer Language One
CLA e: California Library Association
- e: Canadian Library Association
C & LA e: Cargo and Loading Analysis
CLA e: Cargo Landing Adaptability
- e: Carry Look Ahead [Adder]
- e: Catholic Library Association
- e: Center Line Average
- e: Central Listening Ability
- e: Certifed Laboratory Assistant
- e: Chemiluminescent Analyzer
- e: Chinese Librarians Association
- e: Clamp
- e: Clare College
- e: Clear Accumulator
- e: Clear and Add
- e: Coaxial Line Attenuator
- e: Communication[s] Line Adapter
- e: Communication[s] Line Analyzer
- e: Communication[s] Link Analyzer
- e: Computer Law Association
- e: Computer Lessors Association
- e: Connecticut Library Association
- e: Conservative Library Association
- e: Control Logic Assembly
- e: Copyright Licensing Agency
- e: Cover Layer Assembly
- e: Custom Logic Array
CLA-ACB e: Canadian Library Association-l'Association Canadienne des Bibliothéques
CLA Adder e: Carry Lookahead Adder
CLAB d: Computer lösen Aufgaben des Bildungswesens
CLAC e: Closed Loop Approach Control

CLAD e: Collect Adapter
- e: Controllable Liquid Artificial Dielectric
- e: Cover Layer Automated Design (Program)

CLAD Program[me] e: Cover Layed Automated Design Program[me]
- e: Cover Layer Automated Design Program[me]

CLADS e: Clinical Laboratory Data Acquisition System
CLAE e: Council of Library Associations Executives
CLAFB e: Classified List of Army Forms and Books
CLAFIC e: Class Featuring Information Compression
CLAH e: Conference of Latin American History
CLAIM e: Centre for Library and Information Management
CLAIMS e: Class-codes, Assigned, Index, Method, Search
CLAIMS/CHEM e: Class-codes, Assigned, Index, Method, Search/Chemistry
CLAIMS/CLASS e: Class-codes Assigned, Index, Method, Search/Classification
CLAIMS/GEM e: Class-codes, Assigned, Index, Method, Search/General, Electrical, Mechanical
CLAIRA e: Chalk Lime and Allied Industries Research Association
CLAM e: Chemical Low Altitude Missile
- e: Clear Air Mass
- e: Consortium for Library Automation in Mississipi
- e: Control Data Corporation List Processor Algebraic Manipulator

CLAMP e: Chemical Low Altitude Missile Puny
- e: Computer Listing and Analysis of Maintenance Program[me]

CLAMTI e: Clutter-Locked Airborne Moving Target Indicator
CLAN e: California Library Automation Network
CLAO e: Contact Lens Association of Ophthalmologists
CLAR e: Clarification
- e: Clarify

CLARA e: Cornell Learning and Recognizing Automaton
CLARB e: Council of Landscape Architectural Registration Boards
CLARK d: Chlorarsen-Kampfstoff
- e: Combat Launch and Recovery Kit

CLARUS e: Case Law Report Updating Service
CLAS e: Chemiluminescent Analyzers
- e: Classification
- e: Classify
- e: Clinical Laboratory Automated System
- e: Communication Link Analyzer System
- e: Computerized Library Acquisition System
- e: Computer Library Applications Service
- e: Containerized Lighter Aboard Ship
- e: Crowd, Lift, Actuate, Swing

CLASB e: Citizens League Against the Sonic Boom
CLASN e: Classification
CLASP e: Circuit Layout; Automated Scheduling and Production
- e: Clients Lifetime Advisory Service Program[me]
- e: Closed Line Assembly for Single Particles
- e: Composite Launch & Spacecraft Program[me]
- e: Computer Laboratory System Project
- e: Computer Laboratory Systems Project
- e: Computer Language for Aeronautics and Space Programming
- e: Computer Launch and Separation Problem
- e: Computer Liftoff and Staging Program[me]
- e: Cylindrical Laser Plasma

CLASPS e: Coded Label Additional Security and Protection System
CLASS e: Canberra Laboratories Automation Software System
- e: Capacity Loading and Scheduling System
- e: Carrier Landing-Aid Stabilization System
- e: Centralized Library Automation Service System
- e: Class Action Study and Survey
- e: Classic[al]
- e: Classification
- e: Close Air Support System
- e: Computer-based Laboratory for Automated School Systems
- e: Computerized Literature Access Search Service
- e: Containerized Lighter Aboard Ship System
- e: Cooperative Library Agency for Systems and Services
- e: Current Literature Alerting Search Service

CLASS A's e: Class-A Narcotics
CLASS B's e: Class-B Narcotics
CLASSG e: Classifying
CLASSIC e: Classroom Interactive Computer
CLASSIF e: Classification
ClassJ e: Classical Journal (journ.)
CLASSMATE e: Computer Language to Aid and Stimulate Scientific, Mathematical and Technical Education
CLASS M's e: Class-M Narcotics
CLASSN e: Classification
Class Phiol e: Classical Philology (journ.)
ClassQ e: Classical Quarterly (journ.)
ClassR e: Classical Review (journ.)
Class Soc Bull e: Classification Society Bulletin (journ.)
CLASS X's e: Class-X Narcotics
CLAT e: Communication[s] Line Adapter for Teletype
CLAUDE e: Computerized Listing of Abnormal and Unusual Drug Effects
CLAVE e: Autoclave
- e: Steamclave (Sterilizer)

CLAW e: Clustered Atomic Warhead
Clays Clay Miner. e: Clays and Clay Minerals (journ.)
CLB d: Chemie für Labor und Betrieb (journ.)
- e: Central Logic Bus
- e: Clear Both
- e: Climb
- e: Constant Level Balloon
- e: Continuous Line Bucket
- e: Crash Locator Beacon
- e: Current Line Bucket

CLBA e: Current Logical Byte Address
CLBR e: Calibration
CLB System e: Current Line Bucket System
CLBW e: Closed Loop Bandwidth
CLC e: Card Lever Control
- e: Carrier Level Controller
- e: Ceiling Luminance Coefficient
- e: Cellular Logic Computer
- e: Central Logic Complex
- e: Central Logic Control
- e: Channel, Level, Control
- e: Cholesteric Liquid Crystal
- e: Classification Code
- e: Clear Carry
- e: Communication[s] Line Control[ler]
- e: Communication[s] Link Controller
- e: Compare Logical [Characters]
- e: Compressive Load Cell
- e: Computerized Lubrication Control
- e: Computer Leasing Company
- e: Computer Link Corporation
- e: Contrast Light Compensation
- e: Course Line Computer

CLCC e: Ceramic Leaded Chip Carrier
- e: Closed-Loop Continuity Check

CLCD e: Clear Confirmation Delay
- e: Clearinghouse and Laboratory for Census Data

CLCI e: Common Language Circuit Identification
C I, C II, C III, etc. e: Cranial Nerves, I, II, III, etc.
CLCM e: Communication[s] Line Concentrator Module
CLCN e: Conference on Local Computer Networks
CLCP e: Closed-Loop Conjugate Point
CLCr e: Communications Lieutenant Commander
CLCR e: Controlled Letter Contract Reduction
CLCS e: Cable Launch Control System
CLCs e: Cholesteric Liquid Crystals
CLCS e: Closed-Loop Control System
- e: Consequencing Limiting Control System
- e: Current Logic; Current Switching

ClCS/VS e: Customer Information Control System/Virtual Storage
CLCT e: City of Liverpool College of Technology
- e: Collector

CLD e: Called
- e: Central Library and Documentation
- e: Chain Length Distribution
- e: Children with Learning Disabilities
- e: Cleared
- e: Closed
- e: Cloud
- e: Coincidence-Ledge-Dislocation
- e: Colo[u]red
- e: Compression Load Deflection
- e: Computer Logic Demonstrator
- e: Condensed Logic Diagram
- e: Constrained Layer Damping
- e: Control Logic Diagram
- e: Cooled
- e: Could
- e: Current Limiting Device

CLDA e: Control Logic and Drive Assembly
CLDAS e: Clinical [Laboratory] Data Acquisition System
CLDATA e: Cutter Location Data
CLDG e: Composite Language Development Group
CLDIST e: Clearance Distance
CLDS e: Canada Land Data System
- e: Clouds

CLDWN e: Cool Down
CLE e: Chicago Livestock Exchange
- e: Cleveland (Airport)
- e: Communication[s] Line Expander
- e: Conservative Logic Element
- e: Continuing Legal Education
- e: Counter List Entry

CLEA e: Canadian Library Exhibitors Association
- e: Conference and Laser Engineering and Applications

CLEAN e: Committee for Leaving the Environment of America Natural
- e: Commonwealth Law Enforcement Assistance Network
- e: Comprehensive Lake Ecosystem Analyzer
CLEAPSE e: Consortium of Local Education Authorities for the Provision of Science Equipment
CLEAR e: Center for Labor Education and Research
- e: Center for Lake Erie Area Research
- e: Civic Leaders for Ecological Action and Responsibility
- e: Closed Loop Evaluation and Reporting
- e: Compiler, Loader, Executive-Program[me]; Assembler Routines
- e: Components Life Evaluation and Reliability
- e: Controlled Library Environment and Resources
- e: County Law Enforcement Applied Regionally
CLEARP e: Clearance Plane
CLEAR System e: Closed Loop Evaluation and Reporting System
CLEAT e: Computer Language for Engineers and Technologists
CLEC e: Closed-Loop Ecological Cycle
CLEM e: Cargo Lunar Excursion Module
- e: Closed Loop Ex-vessel Machine
- e: Code Learning Machine
- e: Composite Lunar Excursion Vehicle
CLEN e: Continuing Library Education Network
CLENE e: Continuing Library Education Network [and] Exchange
CLEO e: Clear Language for Expressing Orders
- e: Common Language Equipment Order
CLEOPATRA e: Comprehensive Language for Elegant Operating-System and Translator Design
CLEOS e: Conference on Laser and Electro-Optical Systems
CLEP e: College-Level Examination Program[me]
- e: College Level Education Program[me]
CLER d: contralateraler Extensorreflex
CLETS e: California Law Enforcement Telecommuncations System
Cleve Clin Quart e: Cleveland Clinic Quarterly (journ.)
CLF e: Capacitive Loss Factor
- e: Check Label Field
- e: Clear Forward
- e: Commander; Landing Force
- e: Condensed Logic Flow
CLFC e: Condensed Logic Flow Chart
CLFIC e: Center Launch and Flight Instrumentation Center
CLFILE e: Cutter Location File
CLF Line e: Clear Forward Line
CLFM e: Coherent Linear Frequency Modulated
- e: Coherent Linear Frequency Modulation
CLF Signal e: Clear-Forward Signal
CLFT e: Crippled Leap-Frog Test
CLFT Language e: Closed-Loop Flight Test Language
CLG e: Calling
- e: Cancellation
- e: Cancelling
- e: Ceiling
- e: Change to Lower Grade
- e: Clearing
- e: Closed Loop Gain
- e: Closing
- e: Complete Load and Go
- e: Cooling
- e: Cytoplasmic Immunoglobulin
CLGNR e: Catalog[ue] Number
CLGP e: Cannon-launched Laser-Guided Projectile
CLGSFU e: Clear Glazed Structural Facing Units
CLGSUB e: Clear Glased Structural Unit Base
CLHU e: Computation Laboratory of Harvard University
CLI e: Calamus Length Index
- e: Called Line Identification
- e: Called Line Identity
- e: Calling Line Identification
- e: Calling Line Identity
- e: Capacitor Leakage Indicator
- e: Clear Indication
- e: Coach Lace Institute
- e: Coherent Laser Illumination
- e: Coin Level Indicator (Telephone)
- e: Command Language Interpreter
- e: Command Line Interpreter
- e: Compare Logical Immediately
- e: Computer Linguistics, Inc.
- e: Control Level Item
- e: Core Logic Intervalometer
- e: Cost-of-Living Index
- e: Current Labour Invariance
CLIA e: Clinical Laboratory Improvement Act
- e: Collective Linear Ion Accelerator
CLIBOC e: Chinese Linguistics Bibliography on Computer
C-Library e: Circulating Library
CLIC e: Command Language for Interrogating Computers
- e: Communication Linear Integrated Circuit
- e: Community Library Information Center
- e: Company and Literature Information Center
- e: Conversational Language for Interactive Computing
- e: Cooperative Library and Information Centre
- e: Corporate Library Information Centre
CLICMOS e: Cellular-Layout Integrated Circuit-Metal Oxide Semiconductor
CLICS e: Computer Linked Information for Container Shipping
CLID e: Clustering Identification
CLIF e: Command Line Input Filter
CLIFS e: Cost, Life, Interchangeability, Function and Safety
CLIM e: Calling-Line-Identity Message
- e: Cellular Logic-in-Memory
- e: Climate
- e: Climatic
CLIMAP e: Climate Long-Range Investigation, Mapping and Prediction
CLIMAPS e: Climate Longrange Investigation, Mapping and Prediction Study
CLIMAT e: Climatological
- e: Climatologist
- e: Climatology
CLIMATOL e: Climatology
CLI Message e: Calling-Line-Identity Message
CLIN e: Clinic[ally]
- e: Clinical
- e: Clinometer
- e: Contract Line Item Number
Clin Chem e: Clinical Chemistry (journ.)
CLIN/D e: Clinical Death
Clin Exp Immunol e: Clinical and Experimental Immunology (journ.)
CLINFO e: Clinical Information
Clin Med e: Clinical Medicine (journ.)
Clin Obstet Gynecol e: Clinical Obstetrics and Gynecology (journ.)
Clin Orthop Related Res e: Clinical Orthopedics and Related Research (journ.)
CLIN PATH e: Clinical Pathology
Clin Pediat e: Clinical Pediatrics (journ.)
Clin Pharmacol Ther e: Clinical Pharmacology and Therapeutics (journ.)
CLIN PROC e: Clinical Procedures
CLINPROT e: Clinical Protocol
Clin Sci e: Clinical Science (journ.)
Clin Toxicol e: Clinical Toxicology (journ.)
CLIO e: Columbia Libraries Information Online
- e: Conversational Language for Input/Output
CLIP e: Cancel Launch in Progress
- e: Cellular Logic-array Image Processing
- e: Cellular Logic-array Image Processor
- e: College Library Information Packet
- e: Compiler [and] Language for Information Processing
- e: Computer Layout Installation Planner
- e: Confused, Lacerated, Incised and Punctured
- e: Cooperative Library Information Program[me]
- e: Coordinated Library Information Program[me]
- e: Country Logistics Improvement Program[me]
- e: Current Line Pointer
CLIPER e: Climatology and Persistence
CLIPPR e: Consolidated Logistics Information Planning and Programming Requirements
CLIPR e: Computer Laboratory for Instruction in Psychological Research
CLIPS e: Coincident Light Information Photographic Strips
- e: Computer Launch Interference Problems
- e: Coordinate Light Information Photographic Strips
CLIP System e: Compiler [and] Language for Information Processing System
CLIR e: Clearinghouse for Library and Information Resources
CLIRA e: Closed Loop In-Reactor Assembly
CLIRS e: Computerized Legal Information Retrieval System
CLIS e: Central Library and Information Services
- e: Centre for Library and Information Studies
- e: Classification of Library and Information Science
- e: Clearinghouse for Library [and] Information Sciences
- e: College of Library and Information Services
- e: Computer Library Service, Inc.
- e: Criminalistics Laboratory Information System
CLISP e: Conversational List Processor
CLIST d: Command-Liste[n]
- e: Command List[s]
CLIT e: Clitoral
- e: Clitoridectomy
- e: Clitoris
CLIV e: Cold Leg Isolation Valve
- e: Core Logic Intervalometer

CLJ e: Cambridge Law Journal (journ.)
- e: Commander; Land-Forces Jütland
- e: Control Joint
CLJA e: Closed Loop Jumper Assembly
CLJCC Matrix e: Cluster-Limited Journal Cross-Citing Matrix
CLK e: Clock
CLKG e: Calking
CLKG e: Clocking
CLKIN e: Clock in
CLKOUT e: Clock out
CLKW[S] e: Clockwise
CLKWZ e: Clockwise
CLL d: chronische lymphatische Leukämie
- e: Catholic Listener Library
- e: Central Light Loss
- e: Cholesterol Lowering Lipid
- e: Chronic Lymphocytic Leukemia
- e: Circuit Load Logic
- e: Clear Link
- e: Community Language Learning
- e: Computer/Information/ Library Sciences
- e: Computer Leasing Limited
CLLC e: Corporate Low Level Code
CLM e: Calling-Line Matrix
- e: Cancel Last Message
- e: Cement Lime Mortar
- e: Chinese Literature Monthly (journ.)
- e: Circumlunar Mission
- e: Clamp
- e: Clock Module
- e: Column
- e: Communication[s] Line Multiplexer
- e: Contact Lens Mirror
CLMA e: Contact Lens Manufacturers Association
CLMD e: Clamped
CLML e: Chicago Linear Music Language
- e: Current List of Medical Literature
CL Model e: Control Line Model
CLMS e: Clinical Laboratory Management System
- e: Clinical Laboratory Monitoring System
- e: Cluster Mission Simulator
- e: Company Lightweight Mortar System
CLN e: Clean
- e: Clearance
- e: Clipper Negative
- e: Colon
- e: Combinational Logic Network
- e: Corrective Lens
CLNC e: Clearance
CLNP e: Crater Lake National Park
CLNR e: Cleaner
CLNS e: Cape Lockout National Seashore
- e: Connectionless Network Service
CLNT e: Coolant
CLNWR e: Crescent Lake National Wildlife Refuge
CLO e: Cellular Logic Operation
- e: Clothing
- e: Cod Liver Oil
- e: Complex Logical Operation
- e: Computer Lock-on
- e: Computerized Loan Origination
- e: Concentric Line Oscillator
- e: Cornell Laboratory of Ornithology
CLOAX e: Corrugated-Laminated Coaxial Cable
CLOAX Cable e: Corrugated-Laminated Coaxial Cable
CLOB e: Core Load Overlay Builder¼
CLOC e: Computerized Logging and Outage Control
CLOCE e: Contingency Lines of Communication Europe
CLODA e: Closing Date
CLODS e: Computerized Logic Oriented Design System
Cl of D e: Classification of Documents
CLOG e: Change Log
- e: Complex Logarithm
- e: Computer Logic Graphics
CLOPP e: Continuous Level of Production Plan
CLORA e: Closed-Form Ray Analysis
CLOS e: Closure
- e: Command to Line of Sight
- e: Controlled Line of Sight
CLOSSS e: Check List of Social Science Serials
CLOT e: Closed-Loop Operation Test
CLP e: Campbell-Larsen Potentiometer
- e: Carnegie Library of Pittsburgh
- e: Check Localization Program[me]
- e: Clamp
- e: Clip
- e: Clipper Positive
- e: Command Language Processor
- e: Communication[s] Line Processor
- e: Contact Lens Practitioners
- e: Continuous Linear Part
- e: Continuous Line Plotter
- e: Control Line Platform
- e: Cornell List Processor [Language]
- e: Cross-Linked Polyethylene
- e: Current Laboratory Practice (journ.)
- e: Current Line Pointer
CLPE e: Counter List Printing Exit
CLPFF e: Charles Lathrop Pack Forestry Foundation
CLPG e: Clipping
CLP INS e: Clamp Insulation
CL POW e: Climbing Power
CLPR e: Caliper
CLP SCR e: Clamp Screw
CLQ e: Classical Quarterly (journ.)
- e: Cognitive Laterality Quotient
CLR e: Center of Lateral Resistance
- e: Central Logic Rack
- e: Centurion Laser Rangefinder
- e: Circuit Layout Record
- e: Clear[ance]
- e: Clearing
- e: Clear Request
- e: Colo[u]r
- e: Combined Line [and] Recording
- e: Combustible Limit Relay
- e: Common Line Receiver
- e: Computer Language Recorder
- e: Computer Language Research
- e: Constant Load Rupture
- e: Contact Load Resistor
- e: Cooler
- e: Council on Library Research
- e: Council on Library Resources
- e: Current Limiting Resistor
- f: Compagnie Luxembourgeoise de Radiodiffusion
CLRAP e: Cleared as Planned
CLRC e: Circuit Layout Record Card
CLRD e: Clear Request Delay
CLRG e: Collector Ring
CLRI e: Central Leather Research Institute
- e: Council on Library Resources Incorporated
CLRM e: Classroom
CLRMSG e: Clear Message
CLR Operation e: Combined Line [and] Recording Operation
CLRSRF e: Clearance Surface
CLRSTA e: Clearing Station
CLRTEST e: Chloride Test
CLR Traffic e: Combined Line and Recording Traffic
CLRU e: Cambridge Language Research Unit
CLRV e: Canadian Light Rail Vehicle
- e: Cherry Leaf Roll Virus
CLS e: Cable Laying Ship
- e: Cam Limit Switch
- e: Capacity Limiting System
- e: Certificate in Library Science
- e: Cislunar Space
- e: Class
- e: Classification of Library Science
- e: Clear and Subtract
- e: Clear Screen
- e: Close
- e: Closed-Loop Support
- e: Closed Loop System
- e: Close Lunar Satellite
- e: Common Language System
- e: Communication[s] Line Switch
- e: Comparative Systems Laboratory
- e: Compatible Laser System
- e: Computer Load Simulator
- e: Concept Learning System
- e: Connectionless-Mode Service
- e: Constant Level Speech
- e: Constraint Least Squares
- e: Control Languages Services
- e: Country Library Service
CLSA e: Contact Lens Society of America
- e: Contact Lens Society of Australia
- e: Cooperative Logistic Support Arrangement
CLSB e: California Library Services Board
CLSC e: Chautauqua Literary and Scientific Circle
- e: Communications Security Logistic Support Center
CLSD e: Closed
- e: Culham Language for System Development
CLS Filter e: Constrained Least Squares Filter
CLSG e: Closing
CLSI e: Computer Library Services International
- e: Computer Libray Services, Inc.
CLSIM e: Clocked Logic Simulation
CLSL e: Chronic Lymphosarcoma Leukemia
CLSNG e: Closing
CLSR e: Closure
CLSS e: Combat Logistic Support System
- e: Communication Link Subsystem
- e: Computerized Literature Searching Service
CLSTBB e: Cluster Bomb
CLSU e: Communications Security Logistic Support Unit
CLT d: CO_2-Langtunnel
- e: Canadian Law Times
- e: Cargo Left Trailer
- e: Central Limit Theorem
- e: Channel Load Table
- e: Charlotte (Airport)
- e: Cleat
- e: Clinical Laboratory Technician
- e: Clitoral
- e: Clitoridectomy
- e: Clitoris
- e: Clock Track
- e: Closed Loop Telemetry
- e: Cloting Time
- e: Code Language Telegram
- e: Communications Line[s] Terminal

- e: Communications Line[s] Terminator
- e: Communications Line Terminal[s]
- e: Component Library Tape
- e: Computer Language Translator
- e: Cool Light
- e: Corporate Logic Tester
- e: Crippled Leap-frog Test
- f: Compagnie Luxembourgeoise de Télédiffusion

CLTC e: Chief Launch-vehicle Test Conductor
CLTG e: Collecting
CLTGL e: Climatological
CLTGR e: Climatographer
CLTR e: Collator
- e: Computer Law and Tax Report

CLTRS e: Cambridgeshire Low Temperature Research Station
CLTS e: Cryogenic Linear Temperature Sensor
CLTV e: Closed-Loop Television
CLU e: Central Logic Unit
- e: Circuit Line-Up
- e: Cluster
- e: Clutch
- e: Common Logic Unit

CLUB e: Central Library of the University of Baghdad
CLUC e: Central Library Union Catalog
CLUDIA d: Cluster Analyseprogramme mit Dialogteil
CLUE e: Camera Line-Up Equipment
- e: Clinical Literature Untoward Effects
- e: Compiler Language Utility Extension
- e: Computer Learning Under Evaluation

CLUG e: Community Land Use Game
CLUMIS e: Cadastral and Land-Use Mapping Information System
CLUMP e: Communications Pool Lock-Up-Memory Print
CLU Process e: Creusot Loire; Uddelhom Process
CLUS e: Continental Limits of the United States
CLUSA e: Continental Limits; United States of America
CLUSAN e: Cluster Analysis
CLU Signal e: Calling-Line-Identity-Unavailable Signal
CLUT e: Clutter
CLUT e: Computer Logic Unit Tester
CLV e: Constant Linear Velocity
CLVP e: Craft, Landing, Vehicle-Personnel
CLW e: Correction Location Word
CLWG e: Clear Wire Gas
C-L-X e: Continuous Lightweight Exterior
CLYDE e: Computer-Graphics Language for Your Design Equations
CLZ e: Copper, Lead or Zink
CM d: Computer Magazine
- d: Zentimeter
- e: Absolute Coefficient of Pitching Moments
- e: Calculating Machine
- e: Calculating Memory
- e: Calibrated Magnification
- e: Calibration Magnification
- e: Call on Minus
- e: Cameroon
- e: Carboxymethyl
- e: Cardiac Massage
- e: Card Magazine
- e: Card Matching
- e: Card Memory
- e: Carousel Memory
- e: Cell Membrane
- e: Cell Memory
- e: Cellulose Carboxymethyl Cellulose
- e: Cement Mortar
- e: Center-Matched
- e: Center of Mass
- e: Centimeter[s]
- e: Centimetre
- e: Centimorgan
- e: Central Memory
- e: Central Meridian
- e: Centre Matched
- e: Centre of Mass
- e: Certificate of Merit
- e: Certified Master
- e: Chain Memory
- e: Charge Memory
- e: Chief Mechanic
- e: Chief Metalsmith
- e: Chrom-Molybdenum
- e: Cincinnati Museum
- e: Circuit Merit
- e: Circuit Module
- e: Circular Measure
- e: Circular Mil
- e: Circular Muscle
- e: Circulating Memory
- e: Classified Message
- e: Class Mark[s]
- e: Clear Memory
- e: Code Modulation
- e: Coding Memory
- e: Coincidence Memory
- e: Collector Module
- e: Collegiate Microcomputer
- e: Colo[u]r Modulation
- e: Columellar Muscle
- e: Combined Memory
- e: Combined Multiplication
- e: Command Module
- e: Common Market
- e: Common Meter
- e: Common Metre (GB)
- e: Common Mode
- e: Communication
- e: Communication[s] Multiplexer
- e: Comparator[s]
- e: Compiler Method
- e: Complex Mode
- e: Compound Module
- e: Compression Mo[u]lding
- e: Computer Memory
- e: Computer Microtechnology
- e: Computer Module
- e: Computing Machine[ry]
- e: Computing Media
- e: Computing Medium
- e: Computing Memory
- e: Confidential Memorandum
- e: Connecting Machine
- e: Connection Matrix
- e: Constant Memory
- e: Construction and Machinery
- e: Construction Management
- e: Construction Manager
- e: Construction Mechanic
- e: Contact; Moving
- e: Contamination Mode
- e: Continuity Message
- e: Continuous Monitor
- e: Contrast Media
- e: Control-Monitor
- e: Controlled Minefield
- e: Control Mark
- e: Control Memory
- e: Control Monitor
- e: Conversion Matrix
- e: Core Matrix
- e: Core Memory
- e: Corporate Membership
- e: Corrective Maintenance
- e: Correlation Matrix
- e: Corresponding Member
- e: Corrosion Magazine
- e: Coulomb Meter
- e: Countermarked
- e: Countermeasure[s]
- e: Countermortar
- e: Counting Machine
- e: Counts per Minute
- e: Coupling Multiplier
- e: Court Martial
- e: Cross Modulation
- e: Cruiser Minelayer
- e: Cryogenic Memory
- e: Cryotron Memory
- e: Cubic Meter
- e: Cumulative Mortality
- e: Curium
- e: Cyclically Magnetized
- e: Cyclic Memory
- f: Carat Métrique
- i: Correo Maritimo
- s: Camino

C & M e: Coal and Mining
- e: Cocaine and Morphine
- e: Computers and Medicine (journ.)
- e: Control and Monitor[ing]

c.m. l: causa mortis
- l: cras mane

C.M. l: Chirurgiae Magister
- l: Cumulonimbus Mammatus

C/m² d: Coulomb pro Quadratmeter
cm² d: Quadratzentimeter
c/m² e: Candles per Square Meter
- e: Candles per Square Metre (GB)

C/m² e: Coulombs per Square Meter
- e: Coulombs per Square Metre (GB)

cm² e: Square Centimeter
- e: Square Centimetre (GB)

cm³ d: Kubikzentimeter
- e: Cubic Centimeter
- e: Cubic Centimetre (GB)

CMA d: N-(3-Chlor-4-Methylphenyl)-2-Methylpentanamid
- e: Amplitude Cross Modulation
- e: Cable Makers Australia
- e: Calcium Methanearsonate
- e: California Maritime Academy
- e: California Medical Assoclation
- e: Canadian Medical Association
- e: Canis Major (Astronomical Constellation)
- e: Center for Marine Affairs
- e: Certified Medical Assistant
- e: Chemical Manufacturers Association
- e: Circular Mil Area
- e: Civil-Military Affairs
- e: Classified Mail Address
- e: Cleveland Metal Abrasive (Company)
- e: Cleveland Museum of Art
- e: Clothing Maintenance Allowance
- e: Colleges of Mid-America
- e: Colorado Mining Association
- e: Commonwealth Medical Association
- e: Communications Architecture
- e: Communications Managers-Association
- e: Complement Accumulator
- e: Complex Modulus Apparatus
- e: Composite Medium Amplifier
- e: Computerized Management Account
- e: Computer Management Association
- e: Computer Marked Assignment
- e: Computer Monitor Adapter
- e: Conical Monopole Antenna
- e: Contact Making Ammeter
- e: Contraction Mapping Analysis
- e: Contract Maintenance Activity

- e: Control Memory Access
- e: Cylindrical Mirror Analyzer
CMAA e: Chief Master at Arms
CMAAC e: Certifed Medical Assistant Administrative and Clinical
CMAAO e: Confederation of Medical Associations in Asia and Oceania
CMAC e: Cerebellar Model Articulation Control
- e: Colour/Monochrome Attributes Controller
- e: Computer Management and Control
- e: Computer Monitor and Control
CMACL e: Composite Mood Adjective Check List
CMAD e: Computer Manufacture and Design
CMAE e: Contingency Movement After-Effect
CMAK e: Conical Monopole Antenna Kit
CMAL e: Controlled Multiple Address Letter
CMANY e: Communications Managers Association of New York
CMAO e: Court-Martial Appointing Order
C-MAP e: Carnegie-Mellon Action Project
CMAP e: Central Memory Access Priority
- e: Charge Materials Allocation Processor
- e: Co-ordinate Measurement Analysis Program[me]
CMAR e: Control Memory Address Register
CMAS e: Complete Mixing Activated Sludge
- e: Construction Management Accounting System
- f: Confédération Mondiale des Activités Subaquatiques
CMAT e: Canadian Mathematics Achievement Test
C Mat e: Chief of Material
CMAT e: Circuit Mask Translator
CMAV f: Coalition Mondiale pour l'Abolition de la Vivisection
CMAX e: Maximum Capacitance
CMB d: Colombo (Flughafen)
- e: Carbolic Methylene Blue
- e: Chloromercuri Benzoate
- e: Coal Mines Board
- e: Coastal Motor-Boat
- e: Code Matrix Block
- e: Colombo (Airport)
- e: Combat Medical Badge
- e: Core Memory Block
- e: Core Memory Byte
- e: Corrective Maintenance Burden
- f: Compagnie des Machines Bull
CMBHI e: Craft Member British Horological Institute
CMBI e: Caribbean Marine Biological Institute
Cmbt Engr e: Combat Engineer
CMC d: Carboxymethylcellulose
- e: Cable Maintenance Center
- e: Call Modification Complete
- e: Canadian Maritime Commission
- e: Canadian Museum of Civilization
- e: Carboxymethyl Cellulose
- e: Catholic Microfilm Center
- e: Center for Mass Communication
- e: Centimetre Cube
- e: Chemical Materials Catalog[ue]
- e: Clutter Mapper Card
- e: Code[d] [for] Magnetic Character[s]
- e: Coherent Multi-Channel
- e: Color Mixture Curve
- e: Combined Meteorological Committee
- e: Command Module Computer
- e: Commercial Metals Company
- e: Common Market Computer
- e: Communicating Magnetic Card
- e: Communication[s] Magnetic Card
- e: Communication[s] Management Configuration
- e: Communication[s] Mode Control
- e: Communication[s] Multiplexer Controller
- e: Compact Machine Code
- e: Complement Carry
- e: Computer-Mediated Communication
- e: Computer Machinery Corporation
- e: Computer Maintenance Corporation
- e: Computer Mediated Communications
- e: Contact-Making Clock
- e: Control Magnetization Curve
- e: Coordinated Manual Control
- e: Core Memory Cycle
- e: Critical Micelle Concentration
- e: Cubitron Memory Cell
- e: Curved Motion Cutter
- e: Customer Master Card
CMCA e: Character Mode Communications Adapter
CMCB e: Carpenters Mate; Construction Battalion
CMCBB e: Carpenters Mate; Construction Battalion; Builder
CMCC e: Canadian Memorial Chiropractic College
- e: Classified Matter Control Center
- e: Computer Monitor and Control Console
CMCCP e: Cornell Medical Community Computer Project
CMCR e: Colo[u]r Mobile Control Room
- e: Continuous Melting, Casting and Rolling
CMCS e: Canadian Manufacturers of Chemical Specialties
CMCSA e: Canadian Manufacturers of Chemical Specialties Association
CMCT e: Communicate
- e: Communication
CMCTL e: Current Mode Complementary Transistor Logic
CMD e: Carboxymuconolactone Decarboxylase
- e: Cartridge Module Drive
- e: Cataloging Management Data
- e: Central Meridian Distance
- e: Coal Mines Department
- e: Command[ing]
- e: Commendation
- e: Common Measure Double
- e: Common Meter Double
- e: Common Metre Double
- e: Compression Mould Dies
- e: Computer Magnetic Drum
- e: Computer Memory Devices, Inc.
- e: Conductivity Modulated Device
- e: Core Memory Driver
- e: Count Median Diameter
CM & D e: Counter-Measures and Deception
CMDAC e: Current Mode Digital-to-Analog[ue] Converter
CMDB e: Composite-Modified Double Base
CMDB Propellants e: Composite-Modified Double Base Propellants
CMDG e: Commanding
CMDIS e: Computer Management Distributed Information Software
CMDN e: Catalog[ue] Management Data Notification
CMDO e: Consolidated Material Distribution Objectives
CMDP d: 2-Carbomethoxy-1-methylvinyldimethylphosphat
- e: 2-Carbomethoxyl-1-methylvinyldimethylphosphat
CMDR e: Command Reject
- e: Council for Microphotography and Documentary Reproduction
CMDS e: Card Manufacturing Daily Status
- e: Centralized Message Data System
CMDTY e: Commodity
CME e: Central Medical Establishment
- e: Central Memory Extension
- e: Centrifuge Moisture Equivalent
- e: Chemically Modified Electrode
- e: Coherent Memory Filter
- e: Colloid Microthruster Experiment
- e: Command File
- e: Command Message Formulator
- e: Commercial Measuring Equipment
- e: Commission on Man and Environment
- e: Computer Measurement and Evaluation
- e: Continuing Medical Education
- f: Conférence Mondiale de l 'Energie
CM & E e: Chemical Marketing and Economics
- e: Civil and Mining Engineer
- e: Construction Methods & Equipment (journ.)
CMEA e: Central Medical Establishment; Aviation
- e: Council for Mutual Economic Assistance
CMEC e: Chemical Marketing and Economics
- e: Convectron-Microsyn Erection Circuit
CM-ECS e: Central Memory-Extended Core Storage
CMED e: Central Medical Emergency Dispatcher
- e: Central Medical Equipment Depot
CM & EE e: Chief Mechanical and Electrical Engineer
CMERD e: Centre for Medical
- e: Education, Research and Development
CMERI e: Central Mechanical Engineering Research Institute
CMES e: Center for Marine and Environmental Studies
- e: Computer and Management Education Services
CMET e: Coated Metal
CMF e: C-Band Monopulse Feed
- e: Calcium- and Magnesium-Free
- e: California Medical Facility
- e: Camouflage
- e: Cartesian Mapping Function
- e: Central Maintenance Facility
- e: Circular Mil Foot
- e: Coherent Memory Filter
- e: Color Mixture Function
- e: Combat Mission Folder
- e: Command File
- e: Command Message Formulator
- e: Common Mode Failure
- e: Compelled Multi-Frequency
- e: Composite Medical Facility
- e: Comprehensive Management Facility
- e: Condensed Master File
- e: Cyclophosphamide Methotrexate 5-Fluorouracil
- e: Cylindrical Magnetic Film
- e: Cymomotive Force
CM_F e: Frequency-Domain Cross Modulation
CMFA e: Common Mode Failure Analysis
C/MFI e: Conversion, Memory and Fault Indication
CMFI e: Corporate Machine Features Index

CMFLR e: Cam Follower
CM Foam e: Cold Mo[u]ld Foam
CMFRI e: Central Marine Fisheries Research Institute
CMFSW e: Calcium- and Magnesium-Free Synthetic [Sea]-Water
CMFT d: Cardiolipin-Mikro-Flockungstest
CMG d: Computer Management Gruppe
- e: Central Machine Gun
- e: Code Modulus Generator
- e: Commission for Marine Geology
- e: Computer Measurement Group, Incorporated
- e: Control Moment Gyro[scope]
- e: Corning Museum of Glass
- e: Cystometrogram
CMGA e: Control Moment Gyro Assembly
CMGS d: Zentimeter-Gramm-Sekunde
- e: Control Moment Gyro Subsystem
CMH e: Cambridge Medieval History
- e: Centimetre Height Finder
- e: Centimetre Height Finding
- e: Counter-Measures Homing
cm² e: square centimeter
cm³ e: cubic centimeter
CMHA e: California Mental Health Analysis
C-MHA e: Confidential-Modified Handling Authorized
CMHC e: Community Mental Health Center
CMHCA e: Community Mental Health Centers Act
CMHEC d: Carboxymethylhydroxyäthyl-Cellulose
CMHP e: Community Mental Health Program[me]
CMI e: Call Macro-Instruction
- e: Cambridge Memories, Incorporated
- e: Canis Minor (Astronomical Constellation)
- e: Can Manufacturers Institute
- e: Carbohydrate Metabolism Index
- e: Caribbean Meteorological Institute
- e: Cell Mediated Immunity
- e: Cellular-mediated Immune [Response]
- e: Christian Michelson Institute
- e: Code[d] Mark Inversion
- e: Code Maritime International
- e: Collector Module with Inversion
- e: Common Mode Interface
- e: Common Mode Interference
- e: Commonwealth Mining Investments
- e: Commonwealth Mycological Institute
- e: Computer Managed Instruction
- e: Computer Mediated Interaction
- e: Consolidated Metal Industries
- e: Container Master Information
- e: Cornell Medical Index
- e: Cumulative Monthly Issue
- e: Michelson Institute (for Science and Free Thought)
- f: Comité Maritime International
- f: Comité Météorologique International
CMIA e: Coal Mining Institute of America
- e: Cultivated Mushroom Institute of America
CMIC e: Central Management Information Center
CMIC e: Chemical Corps
- e: Commonwealth Management Information Centre
- e: Computer Microfilm International Corporation
CMID e: Cytomegalic Inclusion Disease
CMIL e: Circular Mil
CM/IM e: Cross Modulation/Inter-modulation
C/min e: Cards per Minute
- e: Cycles per Minute
Cmin e: Minimum Capacitance
C-Mine d: Mine mit Chlorameisensäure-ethylester
CMIO e: Countermortar Intelligence Officer
CMIP e: Common Management Information Protocol
- e: Cost Management Improvement Program[me]
CMIQ e: Cornell Medical Index Questionnaire
CMIR e: Common Mode Input Resistance
CMI Response e: Cell-Mediated Immune Response
CMIS d: computerorientiertes Management-Informationssystem
- e: Common Management Information Services
- e: Common Manufacturing Information System
- e: Complementary Metal Insulator Semiconductor
- e: Computer-oriented Management Information System
- e: Computer Managed Instructions
- e: Computer Micro Image Systems
- e: Controls Management (or Manufacturing) Information System
- e: Corporate Management Information System
CMIS Circuit e: Complementary Metal-Insulator-Semiconductor Circuit
CMISE e: Common Management Information Service Element
CMIT e: Current Medical Information and Terminology
CMK e: Chassis Marking Kit
cmkg d: Zentimeter-Kilogramm
CML e: Cell-Mediated Lympholysis
- e: Central Music Library
- e: Chemical
- e: Chicago Midway Laboratory
- e: Chief Moulder
- e: Chronic Myelocytic Leuk[a]emia
- e: Chronic Myelogenous Leuk[a]emia
- e: Chronic Myeloid Leuk[a]emia
- e: Circuit Micrologic
- e: Collimated Monochromatic Light
- e: Common Machine Language
- e: Common Mode Logic
- e: Communication Measurements Laboratory
- e: Complementary Micrologic
- e: Complement Link
- e: Components Managed Laboratory
- e: Computer Managed Laboratory
- e: Computer Management Learning
- e: Condensed Microprogram[me] Logic
- e: Constrained Maximum Likelihood
- e: Conversational Modelling Language
- e: Critical Mass Laboratory
- e: Current Mode Logic
CMLA e: Canadian Music Library Association
- e: Central Medical Library Association
CML AM[MO] e: Chemical Ammunition
CML AMMO STOR SP e: Chemical Ammunition Storage Specialist
CML AM TN e: Chemical Ammunition Train
CML BD e: Chemical Board
CML C e: Chemical Corps
CMLC e: Corporate Machine Level Control
CMLCBL e: Chemical Corps Biological Laboratories
CML C ENGR COM[D] e: Chemical Corps Engineering Command
CML C MAT COMD e: Chemical Corps Materiel Command
CML CO e: Chemical Company
CML CRD COM e: Chemical Corps Research and Development Command
CML C RES & DEV COMD e: Chemical Corps Research and Development Command
CML C SCH e: Chemical Corps School
CML C TNG COM[D] e: Chemical Corps Training Command
CMLDECON e: Chemical Decontamination
CMLDEF e: Chemical Defense
CMLDEP e: Chemical Depot
CMLE e: Carboxymuconate Lactonizing Enzyme
CMLFLAB e: Chemical Field Laboratory
CML GP e: Chemical Group
CMLHO e: Chemical Corps Historical Office
CMLLAB e: Chemical Laboratory
CML LAB CO e: Chemical Laboratory Company
CML LAB MBL e: Chemical Laboratory; Mobile
CML MAINT CO e: Chemical Maintenance Company
CMLMORT e: Chemical Mortar
CML MTR BN e: Chemical Mortar Battalion
CML O e: Chemical Officer
CMLOPS e: Chemical Operations
CML RTC e: Chemical Replacement Training Center
CMLS e: Cleveland Marshall Law School
- e: Computer Multiple Listing Service
CMLSCNP e: Cradle Mountain Lake Saint Clair National Park
CMLSEC e: Chemical Section
CML SMOKE GENR e: Chemical Smoke Generator
CML SMOKE GENR BN e: Chemical Smoke Generator Battalion
CMLSUP e: Chemical Supply
CMLSUPPT e: Chemical Supply Point
CML SV TRC e: Chemical Service Truck
CML TECH INTEL DET e: Chemical Technical Intelligence Detachment
CMLWFARE e: Chemical Warfare
cmm d: Kubikmillimeter
CMM e: Center for Male Medicine
- e: Chemical Milling Machine
- e: Chief Machinists Mate
- e: Coherent Microwave Memory
- e: Collector Module with Memory
- e: Commission for Marine Meteorology
- e: Commission for Maritime Meteorology
- e: Communication
- e: Communication[s] Multiplexer Module
- e: Computerized Modular Monitoring
- e: Computer Main Memory
- e: Concentration Module Main
- e: Coordinate Measuring Machine
- e: Cordax Measuring Machine
- e: Cubic Millimeter[s]
- e: Cubic Millimetre (GB)
- e: Cutaneous Malignant Melanoma
- e: Cutting or Molding Machine
CMMC e: Chemical Micro Milling Company
CMMCBE e: Chief Machinists Mate; Construction Battalion; Equipment-Operator

CMMDA e: Command Module Multiple Docking Assembly
CMME[s] e: Chloromethyl Methyl Ether[s]
CMMI e: Commonwealth Mining and Metallurgical Institution
CMMISP e: California Manpower Management Information System Project
CMML e: Civilian Manpower Management Letter
CMMND e: Command
CMMP e: Computer Multi-Mini-Processor
CMMR e: Chief Machinists Mate; Refrigeration
- e: Confirmed and Made a Matter of Record
CMMS e: Carbon Monoxide Measuring System
- e: Columbia Mental Maturity Scale
CMMs e: Coordinate Measuring Machines
CMMSSA e: Chemical, Metallurgical and Mining Society of South Africa
CMMU e: Cache Memory Management Unit
CMN e: Canadian Merchant Navy
- e: Cerium Magnesium Nitrate
- e: Cerous Magnesium Nitrate
- e: Chemical Milling Machine
- e: Column
- e: Common Market Nations
- e: Common Mode Noise
- e: Communication
- e: Cystic Medial Necrosis
CMN-AA e: Cystic Medial Necrosis of the Ascending Aorta
CMNCE e: Commence
CM-NCS d: (Tetra-S-Carboxymethyl) Neocarzinostatin
CMND e: Command
CMNH e: Cleveland Museum of National History
- e: Cleveland Museum of Natural History
CMNM e: Capulin Mountain National Monument
- e: Craters of the Moon National Monument
CMO e: Calculated Mean Organism
- e: Cardiac Minute Output
- e: Central Meteorological Office
- e: Chief Medical Officer
- e: Comment Output
- e: Common Mode Operation
- e: Computer Microfilm Output
- e: Confidential Monthly Order
- e: Controlled Materials Officer
- e: Countermortar Officer
- e: Court-Material Orders
- e: Crystal Marker Oscillator
CMOMM e: Chief Motor Machinists Mate
CMOMMSRD e: Chief Motor Machinists Mate; Ship Repair; Diesel Engineering Mechanic
CMOMMSRG e: Chief Motor Machinists Mate; Ship Repair; Gasoline Engineering Mechanic
CMOS e: Carbon Molybdenum Steel
- e: Complementary[-Symmetry] Metal Oxide Semiconductor
CMOSM e: Configuration Management Operating System Manual
CMOSRAM e: Complementary [-Symmetry] Metal Oxide Semiconductor-Random Access Memory)
CMOS/SOS e: Complementary-symmetry Metal Oxide Semiconductor / Silicon-On-Sapphire
CMP d: Cytidinmonophosphat
- e: Cast Metal Part
- e: Catoctin Mountain Park
- e: Central Meridian Passage
- e: Central Monitoring Position
- e: Chief Military Physician
- e: Chronic Mental Patient
- e: Civilian Medical Personnel
- e: Command Module Pilot
- e: Common Business-Oriented Language Macro Processor
- e: Compare
- e: Competitive Medical Plan
- e: Compression
- e: Computation[al]
- e: Computed Maximum Pressure
- e: Computer
- e: Computer Master Program[me]
- e: Concrete Mixing and Placing
- e: Configuration Management Plan
- e: Configuration Management Program[me]
- e: Congruent Melting Point
- e: Console Message Processor
- e: Contemporary Music Project
- e: Contract Monitoring Point
- e: Controlled Materiels Plan
- e: Controlled Materiels Production
- e: Corporate Manufacturing Practice
- e: Corrugated Metal Pipe
- e: Cross Modulation Product
- e: Cytidine 3',5-Monophosphate
- e: Cytosine Monophosphate
CMp e: Phase Cross Modulation
CMPC s: Compañia Manufacturera de Papeles y Cartones
CMPD e: Compound[ed]
- e: Compounding
- e: Computer Design (journ.)
CMPF e: Central Meat Processing Facility
- e: Core Maximum Power Fraction
- e: Cumulative Preferred
CMPI e: Civilian Marine Personnel Instructions
CMP Instruction e: Compare Instruction
CMPJ e: [The] Computer Journal
- e: Complement
- e: Complete
CMPL e: Critical Materiels Parts List
CMPLD e: Compiled
CMPLM e: Complement
CMPLTN e: Completion
CMPLX e: Complex
CMPN e: Compilation
CMPN IAP e: Companion of the Institution of Analysts & Programmers
CMPNT e: Component
CMPP d: 4-Chlor-2-Methylphenoxy-Propionsäure
- e: Controlled Materiel Plan Priority
CMPR e: Compare
- e: Compression
CMPRD e: Compared
CMPRG e: Comparing
CMPRN e: Comparison
CMPS e: Centimeter[s] per Second
- e: Centimetre[s] per Second
CMPS e: Command Module Procedures Simulator
CMPS e: Compass
CMPSN e: Composition
CMPST e: Composite
CMPT e: Compartment
- e: Component
- e: Compute
- e: Computer
CMPTG e: Computing
CMPTOP e: Computer Operator
CMPTR e: Computer
CMPUT e: Compute
CMQ e: Coal Mining and Quarrying
CMR e: Call Modification Request
- e: Carbon Magnetic Resonance
- e: Catalog[ue] Management Routine
- e: Center for Materials Research
- e: Centralized Main Remittance
- e: Cerebral Metabolic Rate
- e: Code Matrix Reader
- e: Committee on Manpower Resources [for Science and Technology]
- e: Common Mode Rejection
- e: Communication Moon Relay
- e: Communications Monitoring Report
- e: Communications Moon Relay
- e: Compare Right-Half-Words
- e: Consolidated Mail Room
- e: Continuous Maximum Rate
- e: Continuous Maximum Rating
- e: Contract Management Region
- e: Cossor Marine Radar
- e: Countdown Modification Request
- e: Countermortar Radar
- e: Court-Material Report[s]
- e: Customer Master Record
- e: Customer Material Return
CMR_2 e: Cerebral Metabolic Rate for Oxygen
CMRA e: Chemical Management and Resources Association
- e: Chemical Market Research Association
CMRB e: Chemicals and Minerals Requirements Board
CMRC e: Commercial
CMRCS e: Command Module Reaction Control System
CMRED e: Council on Marine Resources and Engineering Development
CMRE System e: Computer-Managed Review and Examination System
CMRF e: Childrens Medical Research Foundation
CMRG e: Cerebral Metabolic Rate of Glucose
CMRI e: Colonial Microbiological Research Institute
- e: Combined Removal Interval
- e: Command Maintenance Readiness Inspection
CMRL e: Chamber of Mines and Research Laboratories
CMRN e: Cooperative Meteorological Rocket Network
CMRNWR e: Charles M. Russell National Wildlife Range
CMRO e: Cerebral Metabolic Rate of Oxygen
CMRR e: Common Mode Rejection Ratio
CMRS e: Cellular Mobile Radio Service
- e: Central Mining Research Station
- e: Communication Moon Relay System
- e: Computer-Managed Rotation Schedules
- e: Computerized Material Retrieval System
- e: Countermeasures Receiver System
- e: Countermeasures Receiving System
cm/s d: Zentimeter pro Sekunde
CMS e: Cable Marking System
- e: California Malacozoological Society
- e: California Museum of Science
- e: Cambridge Monitor System
- e: Cement-Modified Soil
- e: Center for Maritime Studies
- e: Center for Measurement [of] Science
- e: Center of Mass System
- e: Centimetres per Second
- e: Central Monitoring Stations
- e: Change Management System
- e: Chemical Metallizing System
- e: Chicago Medical School

- e: Christian Medical Society
- e: Circuit Maintenance System
- e: Circuit Multiplication System
- e: Clay Minerals Society
- e: Cleaning Management Station
- e: Close-Medium Shot
- e: Combined Marine Surcharge
- e: Command Module Simulator
- e: Common Manpower Standards
- e: Common Mode Signal
- e: Communications Management System
- e: Communications Monitor System
- e: Compiler Monitor System
- e: Complete Matched Set
- e: Computerized Message Switching
- e: Computer Management System
- e: Condor Missile System
- e: Conservation Materials and Services
- e: Construction Maintenance Supervisor
- e: Construction Management System
- e: Conversational Monitor System
- e: Core Memory Stack
- e: Correctional Medical Systems
- e: County Medical Services
- e: Cross-section Measurement System
- e: Current-Mode Switching
- f: Compagnie Maritime de la Seine

CMSAF e: Chief Master Sergeant of the Air Force
CMS Antenna e: Canonical Minimum Scattering Antenna
CMSC e: Catalina Marine Science Center
CMSE e: Center for Materials Science and Engineering
cm/sec e: Centimeter[s] per Second
cm/sek d: Zentimetersekunde
CMSER e: Commission on Marine Science, Engineering and Resources
CMS & I e: Califonia Museum of Science and Industry
CM & SI e: California Museum of Science and Industry
CM-SM e: Command and Service Modules
CM/SM e: Command Module/service Module
CMSR e: Controller of Merchant Shipbuilding and Repairs
- e: Core Memory Shift Register

CMSRB e: Carpenters Mate; Ship Repair; Boatbuilder-Wood
- e: Chief Metalsmith; Ship Repair; Blacksmith

CMSRC e: Carpenters Mate; Ship Repair; Carpenter
- e: Chief Metalsmith; Ship Repair; Coppersmith

CMSRJ e: Carpenters Mate; Ship Repair; Joiner
CMSRS e: Carpenters Mate; Ship Repair; Shipwright
- e: Chief Metalsmith; Ship Repair; Sheet Metal Worker

CMSS e: Commission on Molecular Structure and Spectroscopy
CMT e: Cadmium-Mercury-Telluride
- e: California Mastitis Test
- e: Camden Marine Terminals
- e: Card Module Tester
- e: Cassette Magnetic Tape
- e: Catalog[ue] Management Table
- e: Center-Mounted Thermistor
- e: Certified Medical Transcriptionist
- e: Chicago Mean Time
- e: Circuit[s] Master Tape
- e: Class Mask Table
- e: Code Matching Technique
- e: Command Module Technician
- e: Comment
- e: Committee on Marine Technology
- e: Computer Magnetic Tape
- e: Computer Managed Training
- e: Computer Mediated Teleconferencing
- e: Computer Memory Tester
- e: Computer Micrographic Technology
- e: Concora Medium Test
- e: Contraction Mapping Theorem
- e: Conversational Mode Terminal
- e: Corrected Mean Temperature
- e: Council on Medical Television
- e: Current Medical Terminology

CMTA e: Corporate Management Technical Assignment
- e: Corporate Manufacturing Technology Assignment

CM-Taschenrechner d: Casio-Mini-Taschenrechner
CMTC e: Communication Module Terminal Controller
- e: Coupled Monostable Trigger Circuit

CMTCS e: Computer Management Transaction Control System
CMTCU e: Communications Message Traffic Control Unit
CMTD e: Cemented
- e: Common Market Telephone Directory

CMT Detector e: Cadmium-Mercury-Telluride Detector
CMTI e: Celestial Moving-Target Indicator
- e: Central Machine Tool Institute

CMTM e: Capsule Mechanical Training Model
- e: Communications and Telemetry

CM-to-CM e: Common Mode-to-Common Mode
CM-to-DM e: Common Mode-to-Differential Mode
CMTOS e: Cassette Magnetic Tape Operating System
CMTP e: Canadian Manpower Training Programme
CMTS e: Centralized Maintenance Test System
CMTT f: Commission Mixte pour les Transmission Télévisuelles
CMU e: Canadian Municipal Utilities
- e: Carnegie-Mellon University
- e: Central Michigan University
- e: Chlorophenyldimethylurea
- e: Compatibility Mock-Up
- e: Concrete Masonry Unit
- e: Constant Multiplier Unit
- e: Control Maintenance Unit

CMU-DA e: Carnegie-Mellon University-Design Automation
CMUS e: Chief Musician
CMU/WA e: Committee on Mans Underwater Activities
CMV d: Computer-Mißbrauch-Versicherung
- e: Common-Mode Voltage
- e: Contact Making Voltmeter
- e: Controlled Mechanical Ventilation
- e: Control of Movement[s]
- e: Cucumber Mosaic Virus
- e: Cytomegalovirus

CMV-CF e: Cytomegalovirus-Complement Fixing
CMVM e: Contact-Making (or Breaking) Voltmeter
CMVPB e: California Motor Vehicles Pollution Board
CMVs e: Cytomegaloviruses
CMVTR e: Colo[u]r Mobile Video Tape Recorder
CMW d: codierter Meßwert
- d: Zentimeterwelle[n]
- e: Circular Magnetic Wave
- e: Cold Molecular Weld
- e: Consecutive Memory Words
- e: Critical Minimum Weight

CMWS d: Zentimeter Wassersäule
CMX e: Cable Multiplexer
- e: Census Method X
- e: Character Multiplexer
- e: Concentration Module Extension
- e: Customer Multiplexer

CMY e: Civilian Man-Years
CMZ e: Concordant Memory Zone
CMZS e: Corresponding Member of the Zoological Society
CN d: Cellulosenitrat
- e: Absolute Coefficient of Yawning Moments
- e: Canal
- e: Cancel
- e: Cannon
- e: Carbon-to-Nitrogen Ratio
- e: Carrier-to-Noise [Ratio]
- e: Case of Need
- e: Cellulose Nitrate
- e: Centurium
- e: Cetane Number
- e: Chalcedony
- e: Change Notice
- e: Charge Nurse
- e: China
- e: Chlor[o]acetophenone
- e: Cholesteryl Nonanoate
- e: Chromosome Number
- e: Cinder
- e: Circular Note
- e: Code Napoléon
- e: Collective Negotiations
- e: Combat; Naval
- e: Commonwealth Nation
- e: Communication[s] Network
- e: Commutated Network[s]
- e: Compass North
- e: Compensator[s]
- e: Computing Nodes
- e: Condensation Nuclei
- e: Condition Name
- e: Connect
- e: Connection Network
- e: Consignment Note
- e: Contract Note
- e: Contract Number
- e: Coordination Number
- e: Cosine of the Amplitude
- e: Country Note
- e: Cover Note
- e: Credit Note
- e: Cryptococcus Neoformans
- e: Cumulo-Nimbus
- e: Cupro Nickel
- e: Cyanogen
- e: Cyclic Neutropenia

C & N e: Communications and Navigation
CNA e: California Nurses Association
- e: Canadian Nuclear Association
- e: Canadian Numismatic Association
- e: Canadian Nurses Association
- e: Center for Naval Analyses
- e: Certified Nurses Aide
- e: Certified Nurses Assistant
- e: Chemical Notation Association
- e: Chief Naval Adviser
- e: Chief of Naval Air (or Aviation)
- e: Code Not Allocated
- e: Combinations Not Allowed
- e: Common Network Architecture

- e: Communication[s] Network Architecture
- e: Computer Network Architecture
- e: Copper Nickel Alloy
- e: Cosmic Noise Absorption
- f: Convertisseur Numérique Analogique

CNAA e: Council for National Academic Awards
CNAAD[V]TRA e: Chief of Naval Air Advanced Training
CNAB e: Commander; Naval Air Bases
CNA B[asic] Tra e: Chief of Naval Air Basic Training
CNAC e: China National Aviation Corporation
CNAEA e: California Narcotic Addict Evaluation Authority
CNAINTERMTRA e: Chief of Naval Air Intermediate Training
CNAOPTRA e: Chief of Naval Air Operational Training
CNAPIMTRA e: Chief of Naval Air Primary Training
CNARESTRA e: Chief of Naval Air Reserve Training
CNAS e: Chief of Naval Air Services
- e: Civil Navigation Aids System

CNAT e: Chief of Naval Air Training
CNATE e: Chief of Naval Airship Training and Experimentation
CNATECHTRA e: Chief of Naval Air Technical Training
CNATEC (LTA) e: Commander; Naval Air Technical Training (Lighter Than Air)
CNAV e: Canadian Navy Auxiliary Vessel
CNB e: Canberra (Airport)
- e: Commander; Naval Base
- e: Cyanocobalamin

CNBr e: Cyanogen Bromide
CNC e: Call on No Carry
- e: Cancer
- e: Central Navigation Computer
- e: Chief of Naval Communications
- e: Communications Network Controller
- e: Computer-aided Numerical Control
- e: Computer-assisted Numerical Control
- e: Computer-integrated Numerical Control
- e: Computer[ized] Numerical Control
- e: Computer Numerical Control[led]
- e: Condensation Nuclei Counter
- e: Consecutive Number Control
- e: Consonant Nucleus Consonant
- e: Counter Narcotics Center
- f: Commande Numérique par Calculateur

CNCC e: Customer Network Control Center
CNCL e: Cancel
- e: Concealed

CNCMH e: Canadian National Committee for Mental Hygiene
CNCP e: Center for New Corporate Priorities
- e: Communications Network Control Processor

CNC Press e: Computer Numerical Control Press
CN/CP Telecommunications e: Canadian-National/Canadian-Pacific Telecommunications
CNCR e: Concurrent
CNCS e: Central Navigation and Control School
CNCT e: Connect
CNCTRC e: Concentric
CNCV e: Concave
CND d: Christlicher Nachrichtendienst
- e: Campaign for Nuclear Disarmament
- e: Canada (Kfz)
- e: Chief of Naval Development
- e: Code Names Dictionary
- e: Condition[ed]
- e: Conduit

CND BOX e: Conduit Box
CNDCT e: Conduct
- e: Conductivity
- e: Conductor

CNDF e: Complex-valued Non-linear Discriminant Function
CNDI e: Combination Die
CNDL e: Conditional
CNDO e: Chief Navy Disbursing Officer
- e: Complete Neglect of Differential Overlap

CNDP e: Communication[s] Network Design Program[me]
- e: Communications Network Design Procedure[s]
- e: Continuing Numerical Data Projects

CNDS e: Condensate
CNE e: Chronic Nervous Exhaustion
- e: Combined Neutral and Earth
- e: Communication[s] Network Emulator

CNEAF e: Commander of Northern European Air Forces
CNEE e: Consignee
CNEL e: Community Noise Equivalent Level
CNEN s: Comisión Nacional de Energía Nuclear
CNEngO e: Chief Naval Engineering Officer
CNEP e: Cable Network Engineering Program[me]
CNES f: Centre National d'Etudes Spatiales
CNESF e: Commander of Northern European Sea Forces
CNET e: Communication[s] Network
- f: Centre Nationale d'Études des Télécommunications

CNEXO e: Centre National pour l'Exploitation des Océans
CNF e: Cannot Find
- e: Caribbean National Forest
- e: Chomsky Normal Form
- e: Cleveland National Forest
- e: Commander; Naval Forces
- e: Commonwealth Nurses Federation
- e: Confine
- e: Conjunctive Normal Formula
- e: Cursor Next Field

CNFS e: California National Fuchsia Society
CNG e: Canadian Naval Gunboat
- e: Chief of Naval Group
- e: Compressed Natural Gas
- e: Connecticut Natural Gas [Corporation]
- e: Consolidated Natural Gas [Company]

CNGA e: California Natural Gas Association
CN-Gas e: Cyanide Gas
CNGCOL e: Change a Column
CNG Tone e: Calling Tone
CNHD e: Congenital Nonspherocytic Hemolytic Disease
CNHM e: Chicago Natural History Museum
CNHS e: Center for Neo-Hellenic Studies
CNHSS e: Croydon Natural History and Scientific Society
CNI e: Called Number Identification
- e: Changed Number Interception
- e: Chief of Naval Information
- e: Chief of Naval Intelligence
- e: Coalition for Networked Information
- e: Committee for Nuclear Information
- e: Communication[s] Navigation [and] Identification
- e: Communications Navigation Instrument
- e: Community Nutrition Institute
- e: Composite Noise Index
- f: Commissariat National à l'Informatique

CNIB e: Canadian National Institute for the Blind
CNIC f: Centre National de l'Information Chimique
CNIF f: Conseil National des Ingénieurs Français
CNIN e: California Narcotic Information Network
CNIPE f: Centre National d'Information pour le Productivé des Entreprises
CNIPTG e: Communications, Networks and Information Processing Theory Group
CNJ e: Copper-Nickel Jacket
CNJA e: Chairman; Naval Job Analysis
CNK e: Common Noun Keyword[s]
CNL e: Canadian National Library
- e: Cancel[lation]
- e: Cardiolipin Natural Lecithin
- e: Carrier Noise Level
- e: Circuit Net Loss
- e: Commonwealth National Library
- e: Constant Net Loss
- e: Control
- e: Cursor New Line

CNLA e: Canadian National Library Association
- e: Council of National Library Associations

C&N LAB e: Communication and Navigation Laboratory
CNLIA e: Council of National Library and Information Associations
CNL Message e: Cancellation Message
CNM e: Cabrillo National Monument
- e: Central Nervous-Tissue Mass
- e: Certified Nurse Midwife
- e: Chief of Naval Material
- e: Chiricahua National Monument
- e: Colombo National Museum
- e: Colorado National Monument
- e: Communication[s] Network Management
- e: Communication[s] Network Manager
- e: Cyclic NOR (Negative OR) Machine

CNMA e: Communications Network for Manufacturing Applications
CNMI e: Communication[s] Network Management Interface
CNMT e: Consignment
CNN e: Cable News Network
- e: Condensed Nearest Neighbour

CNN Design Rule e: Condenset Nearest Neighbour Design Rule
CNNR e: Caerlaverock National Nature Reserve
- e: Cairngorms National Nature Reserve

CNO e: Cannot Observe
- e: Carbon, Nitrogen and Oxygen
- e: Chief of Naval Operations
- e: Computer Non-Operational
- e: Computer Not Operational

CNOB e: Commander; Naval Operating Base
CNOCOM/MIS e: Chief of Naval Operations Command / Management Information System
CNOP e: Conditional Non Operation
- e: Conditional No Operation

CNP e: Canyonlands National Park
- e: Caramoan National Park
- e: Card Network Planning
- e: Cleveland National Park
- e: Colonial National Park
- e: Communication[s] Network Processor
- e: Connection Not Possible
- e: Consecutive Number Printer
- e: Corbett National Park
CN Plat e: Cannon Platoon
CNP Machine e: Count Number and Punch Machine
CNPS e: California Native Plant Society
- e: Computer Network Performance Symposium
CNP Signal e: Connection-Not-Possible Signal
CNR e: Carbon-Nitrogen Ratio
- e: Carboxy Nitrose Rubber
- e: Carrier-to-Noise [power] Ratio
- e: Cellular Neoprene Rubber
- e: Change Notice Request
- e: Change to Navy Regulations
- e: Chief of Naval Research
- e: Clutter-to-Noise Ratio
- e: Coleford Nature Reserve
- e: Committee for Nuclear Responsibility
- e: Composite Noise Rating
- e: Computerized Natural Resources, Incorporated
- e: Contractual Nontechnical Report
- e: Corner
CNRA e: Curecanti National Recreation Area
CN Ratio e: Carbon-Nitrogen Ratio
- e: Carrier-to-Noise Ratio
CNREPORT e: Computer Negotiations Report
CNRI e: Chief of Department of Nonresident Instruction
CNRS e: Centralized Network Reference Service
- f: Centre National de la Recherche Scientifque
CNRT e: Chief of Naval Air Reserve Training
- e: Concrete
CNRV e: Chlorotic Necrotic Ring Virus
CNS e: Cardiff Naturalists Society
- e: Central Navigation School
- e: Central Nervous System
- e: Chinese National Standard
- e: Chloracetophenone Solution
- e: Common Number System
- e: Communication[s] Network System
- e: Computer Network System
- e: Congress of Neurological Surgeons
- e: Connection Not Successful
- e: Construction
- e: Continuous National Survey
- e: Continuous Net Settlement
- e: Control Network System
- e: Copper Non-Staining
- e: Cryogenic Noise Standard
CNSC e: China National Software Corporation
CNSG e: Consolidated Nuclear Steam Generator
CNSIM e: Connection Network Simulator
CNSL e: Console
CNSLD e: Chronic Non-Specific Lung Disease
CNS Signal e: Connection-Not-Successful Signal
CNSTR e: Canister
CNSTRN e: Construction
CNSY[D] e: Charleston Naval Shipyard
CNT e: Canadian National Telegraphs
- e: Celestial Navigation Trainer
- e: Chief of Naval Training
- e: Component and Network Testing
- e: Condition Name Test
- e: Continue
- e: Count[er]
- e: County Naturalist Trust
- p: Conselho Nacional de Telecommunicação
- s: Compañia Nacional de Teléfonos
CNTBRD e: Centerboard
CNTCLKWS e: Counterclockwise
CNTD e: Contained
CNTGCY e: Contingency
CNTL e: Control
CNTN e: Contain
CNTNR e: Container
CNTOR e: Contactor
- e: Contractor
CNTR e: Centaur
- e: Center
- e: Central
- e: Centre (GB)
- e: Container
- e: Contribute
- e: Contribution
- e: Counter
CNTRF e: Centrifugal
CNTRFUGL e: Centrifugal
CNTRL e: Central
- e: Control[ler]
CNTRLR e: Control Register
CNTRS e: Containers
CNTS e: Chief of Naval Technical Services
- e: Chief of Naval Transportation Service
CNU f: Calculatrice Numérique Universelle
CNUCE i: Centro Nazionale Universitario di Calcol Elettronico
CNV e: Cape Canaveral
- e: Contingent Negative Variation
- e: Conversion
CNVC e: Conveyance
CNVR e: Conveyor
CNVRT e: Convert[ing]
CNVT e: Convert
- e: Convict
CNWA e: Charleston Naval Weapons Annex
CNWDI e: Critical Nuclear Weapons Design Information
CNWR e: Camas National Wildlife Refuge
- e: Chassahowitzka National Wildlife Refuge
- e: Chataugua National Wildlife Refuge
- e: Chincoteague National Wildlife Refuge
- e: Columbia National Wildlife Refuge
CNZ d: Computer-Netzwerk-Zubehör
- e: Call on No Zero
CO d: Kobalt
- d: Kohlenmonoxid
- d: Kolumbien (Kfz)
- e: Carbon Monoxide
- e: Cardiac Output
- e: Card Output
- e: Care Of
- e: Carried Over
- e: Carry Out
- e: Case Oil
- e: Castor Oil
- e: Cathode-Ray Oscilloscope
- e: Central Office
- e: Certificate of Origin
- e: Cervicoaxial
- e: Chain Operator
- e: Chain Overseas
- e: Change-Over
- e: Change Order
- e: Checked Operation
- e: Checked Output
- e: Checkout
- e: Chief of Ordnance
- e: Chief Operator
- e: Chinin-Oxydase
- e: Clean Out
- e: Close[d]-Open
- e: Closed Orbit
- e: Cobalt
- e: Coenzyme
- e: Colombia
- e: Colon
- e: Columbia[n]
- e: Columbium
- e: Combined Operations
- e: Commanding
- e: Command Operation[s]
- e: Command Operator
- e: Command Orders
- e: Commission
- e: Common Business Oriented Language
- e: Communication[s]
- e: Communication[s] Officer
- e: Comparison Operator
- e: Compiler Operation
- e: Components Only
- e: Computer Operation
- e: Computer Output
- e: Computing Operation
- e: Concentrator
- e: Conditional Order
- e: Connection Oriented
- e: Connective Operation
- e: Conscientious Objector
- e: Consist[s] Of
- e: Console Output
- e: Contact Operate
- e: Continent
- e: Control Order
- e: Conversion Operation
- e: Coral
- e: Cord
- e: Cordage
- e: Cordierite
- e: Counting Operation
- e: Coupled Oscillator
- e: Couple Output
- e: Course
- e: Crossover
- e: Crystal Oscillator
- e: Customer Order
- e: Cut-Off
- e: Cut-Out
- e: Output Capacitance
co. l: compositus
CoA d: Coenzym A
- d: Koferment A
COA e: Canadian Orthopedic Association
- e: Change of Assignment
CO(A) e: Change Order (Aircraft)
COA e: Chief of Operations Analysis
- e: Coenzyme A
- e: Coherent Optical Array
- e: Compass Operation Alarm
- e: Complex Optical Array
- e: Computer-Oriented Algorithm
- e: Conchologists of America
- e: Constant Output Amplifier
- e: Current Operating Allowances
COAB e: Commanding Officer; Air Base
COAC e: Clutter-Operated Anti-Clutter [Receiver]
- e: Commanding Officer; Atlantic Coast
COACH e: Canadian Organization for the Advancement of Computers in Health
- e: Customer Operated Automatic Checkout

COAD e: Coordinate Adder Display
COADS e: Command and Administrative Data System
- e: Conference on Application Development Systems
COAEWRon e: Commander; Air; Early Warning Squadron
COAG e: Coagulant
- e: Coagulate[d]
- e: Coagulation
COAIREVACRON e: Commanding Officer; Air Evacuation Squadron
COAL e: Coalition
CO/ALR e: Cooper / Aluminum
COAP e: Combat Optimization and Analysis Program[me]
- e: Comprehensive Ocean Area Plan
COARIN e: Committee on Arabic in Informatics
COAS e: Crew[man] Optical Alignment Sight
- e: Crew Optical Alignment Subsystem
COA Signal e: Change-Over Acknowledgement Signal
COASP e: Company Ammunition Supply Point
- e: Coordinated Aircraft / Stores Program[me]
COAT e: Coherent Optical Array Techniques
- e: Computer Operator Aptitude Test
COATS e: Computer Operated Automatic Test System
COAU e: Computers and Automation (journ.)
COAX e: Coaxial [Cable]
COAX e: Common Business-Oriented Language Abbreviation Expander
COAX Cable e: Coaxial Cable
COAX Pl e: Coaxial Plug
COAX SKT e: Coaxial Socket
COB e: Card on Board
- e: Carry-on Box
- e: Ceramic Oceanographic Buoy
- e: Common Business Oriented Language
- e: Complementary Offset Binary
- e: Cyano-Octyl-4-4'Biphenyl
- f: Centre Océanologique de Bretagne
COBA e: Cost-Benefit Analysis
COBAS d: commerzielle Basis-Software
COBCRM e: Cobalt-Chrome
COBE e: Cosmic Background Explorer
COBECC e: Central Ohio Biomedical Engineering Community Council
COBERA e: Cambridge Observation Error Analysis
COBESTCO e: Computer Based Estimating Technique for Contractors
COBF e: Cobol-F [Program]
COBH e: Carboxyhemoglobin
COBI e: Coded Biphase
COBILITY e: Common Business-Oriented Language Utility [program]
COBIS d: computerunterstütztes Büro-Informationssystem
- e: Computer-Based Instruction System
COBLIB e: Cobol Library
COBLOS e: Computer Based Loan[s] System
COBOL e: Common Business-Oriented Language
Co-Bomb e: Cobalt Bomb
COBRA e: Compatible on Board Ranging
- e: Computerized Boolean Reliability Analysis (Boeing)
- e: Computer Oriented Bearing Response Analysis
- e: Consolidation of Basic Records Audit
- e: Curved Orthotropic Bridge Analysis
- p: Computadores Brasileiros
COBRID e: Coaxial Hybrid
COBS e: Caesarean-Obtained Barrier-Sustained
COBSI e: Committee on Biological Sciences Information
- e: Council on Biological Sciences Information
COBSTRU d: COBOL-Struktur
COBU e: [The] Computer Bulletin (journ.)
COC e: Cardiology Office Computer
- e: Card Oriented Computer
- e: Cathodal Opening Clonus
- e: Cathodal Opening Contraction
- e: Certificate of Competency
- e: Change-Over Control
- e: Change of Contract
- e: Character-Oriented Communication[s]
- e: Chemical Oxygen Consumption
- e: Chlorate Oxygen Candle
- e: Cholesteryl Oleyl Carbonate
- e: Cocaine
- e: Coccygeal
- e: Coded Optical Character
- e: Code of Conduct
- e: Combat Operations Center
- e: Combination-Type Oral Contraceptive
- e: Command and Control
- e: Command Operations Center
- e: Communication Computer
- e: Compiler Object Code
- e: Complete Operational Capability
- e: Computer-on-a-Chip
- e: Computer Operation Code
- e: Conversion Operation Code
- e: Cross-Over Connector
- e: Cross Office Check
- e: Customer-Originated Change
COCA s: Cocaina
COCAS e: Computer Order Control Automated System
- e: Customer Order Control Automated System
COCAST e: Council for Overseas Colleges of Art, Science and Technology
COCB e: Crossed Olivocochlear Bundles
COC Controller e: Character-Oriented Communication[s] Controller
COC Cycle e: Close-Open-Close Cycle
COCI e: Consortium on Chemical Information
COCL e: Cathodal Opening Clonus
COCO e: Common Business-Oriented Language Conversion
- e: Contractor-Owned; Contractor-Operated
COCOAS e: Continental Army Command Class 1 (One) Automatic System
COCODE e: Compressed Coherency Detection
COCOHOL e: Coconut-Oil-Extended Ethanol
COCOL e: Common Business-Oriented Language Compiler-Oriented Language
C/O [Contact] e: Close / Open [Contact]
COCOS e: Compact Computers
- e: Control Communications Software
- e: Corporate Communications System
COCOSEER e: Coordinating Committee on Slavic and East European Library Services
COCOT e: Coin-Operated; Customer-Owned Telephone
COCP e: Closed Olivocochlear Potential
- e: Crossed Olivocochlear Potential
COCR e: Cylinder Overflow Control Record
COCRALY e: Cobalt-Chromium-Alumin[i]um-Yttrium
COCS e: Camera Override Control System
- e: Computerized Cataloguing System
- e: Container Operating Control System
coct. l: coctio
CO Cycle e: Close-Open Cycle
cod. d: codieren
- codiert
Cod d: Codierer
Cod. d: Codierung
COD d: computergestütztes Dokumentationssystem
- d: Cordoba-Durchmusterung
- e: Cargo on Deck
- e: Carrier-Onboard Delivery
- e: Carry-Over Data
- e: Central Ordnance Depot
- e: Chemical Oxygen Demand
- e: Close-Order Drill
- e: Codeine
- e: Codependency
- e: Codex
- e: Codification
- e: Coding
- e: Coherent Optical Device
- e: Communications Operating Directive
- e: Computer Output Disk
- e: Concise Oxford Dictionary
- e: Condensed or Dried
- e: Conductor Oxide Diffusion
- e: Constrained Optimal Design
- e: Crack Opening Displacement
CODA e: Committee on Drugs and Alcohol
- e: Computer Oriented Data Acquisition
- e: Crack Opening Displacement Application
CODABASE Building System e: Co-operative Database Building System
CODAC e: Computer Design of Armoured Cables
- e: Coordination of Operating Data by Automatic Computer
CODACOP e: COBOL Data Communication Philips
CODAG e: Combined Diesel and Gas
- e: Community Organization for Drug Abuse Control
COD Aircraft e: Carrier Onboard Delivery Aircraft
CODAM e: Contractor-Oriented Data Abstract Modules
CODAN e: Carrier-Operated Device; Anti-Noise
- e: Coded Analysis
CODAP e: Client-Oriented Data Acquisition Process
- e: Client-Oriented Data Acquisition Program[me]
- e: Comprehensive Occupational Data Analysis Program[me]
- e: Control Data Assembly Program[me]
CODAR e: Coherent Display Analyzing and Recording
- e: Correlation Display Analyzing and Recording
- e: Current Ocean Detecting and Ranging
CODAS e: Customer-Oriented Data System
CODASYL e: Committee on Data Systems Languages
- e: Conference on Data System[s] Languages

CODATA e: Committee on Data for Science and Technology
CODAZR e: Committee on Desert and Arid Zones Research
CODC e: Canadian Oceanographic Data Center
CODD e: Codices
COD/DEC d: Codierer/Decodierer
- e: Coder/Decoder
CODE e: Citizens Organized to Defend the Environment
- e: Collaborating on Drug Education
- e: Committee on Donor Enlistment Code Blue
- e: Controller Decision Evaluation
CODEAG e: Combined Diesel Electric and Gas
CODEC e: Coder/Decoder
- e: Coding/Decoding
- e: Coding Decoder Device
CODEC Device e: Coder/Decoder Device
- e: Coding/Decoding Device
CODED e: Computer[-Oriented] Design of Electronic Devices
CODEFER e: Code Verification
CODEIN e: Computerized Drawing Electrical Information
CODEL e: Computer Development[s]
CODELCO s: Corporación del Cobre
CODELS e: Computer Development System
CODEM e: Coded Modulator-Demodulator
- e: Computer-graphics-augmented Design and Manufacturing
- e: Computer-graphics-automated Design and Manufacturing
- e: Computerized Design from Engineering Models
CODE N e: Code Napoléon
CODEN e: Code Number
CODES e: Commutating Detection System
- e: Computer-Oriented Data Entry System
- e: Computer Design and Evaluation System
CODEX d: Computerdatenextrakt
CODIAC d: Compression von Daten durch individuell angewandte Codierung
CODIC e: Colo[u]r Difference Computer
- e: Computer-Directed Communication[s]
CODICOM e: Computerized Distribution and Control of Microfilm
CODIL e: A Non-Numerical Computer Language in which no Distinction is made between Instruments
- e: Content Dependent Information Language
- e: Context Dependent Information Language
- e: Control Diagram Language
CODILS e: Commodity Oriented Digital Input Label System
CODINE e: Conversational Design Information Network
CODIPHASE e: Coherent Digital Phased Array System
CODIS e: Completed Dischange
CODIT e: Computer Direct to Telegraph
CODN e: Component Operational Data Notice
CODOC e: Cooperation in Documentation and Communication
CODOG e: Combined Diesel or Gas
CODORAC e: Coded Doppler Radar Command
CODORACS e: Coded Doppler Radar Command System
CODOT e: Classification of Occupations and Directory of Occupational Titles
CODPS e: Card Oriented Data Processing System
CODSIA e: Council of Defense Space Industries Assoeiation
COE e: Cab Over Engine
- e: Center for Optimum Environments
- e: Central Office Equipment
- e: Certificate of Eligibility
- e: Chief of Engineers
- e: Complete Operating Equipment
- e: Computer Operations Europe
- e: Corps of Engineers
- e: Council on Optometric Education
- e: Crossover Electrophoresis
- e: Crude Oil Equivalent
CO(E) e: Change Order (Electronic)
COEA e: Cost and Operational Effectiveness Analysis
COEBRA e: Computerized Optimization of Elastic Booster Autopilot
COED e: Char-Oil Energy Development
- e: Co-Editor
- e: Co-Education[al]
- e: Composition and Editing Display
- e: Computer-Operated Electronic Display
- e: Computer in Education Division
- e: Concentration on Engineering Design
- e: Concise Oxford English Dictionary
COEDS e: Char Oil Energy Development System[s]
COEES e: Central Office Equipment Engineering System
- e: Central Office Equipment Estimation System
COEF[F] e: Coefficient
COEI e: Composition of Ending Inventory
COEL e: Chain Overseas; Extra Low
- e: Chain Overseas; Extremely Low
COEMIS e: Corps of Engineers Management Information System
COENCO e: Committee for Environmental Conservation
COEP e: Central Office for Environmental Protection
COEPS e: Cortically Orginating Extra-Pyramidial System
COER e: Central Office Equipment Reports
COESA e: Committee on Extension of the Standard Atmosphere
COET e: Crude Oil Equalization Tax[ation]
COE Truck e: Cab over Engine Truck
COEV e: Canadian Ocean Escort Vessel
CoF d: Coenzym F
COF e: Canadian Order of Foresters
- e: Cause of Failure
- e: Coefficient of Friction
- e: Computer Operations Facility
- e: Construction of Facilities
- e: Contractor Overhaul Facility
- e: Cost of Failure
- e: Customer Order File
- e: Cut-Off Frequency
COFA e: Certificate of Airworthiness
- e: Chief of Artillery
- e: College of Aeronautics
- e: Chief of the Army Air Forces
- e: Chief of Air Corps
COFAD e: Computerized Facilities Design
COFAL-Test e: Complement Fixation Avian Leucosis Test
COFASRON e: Commander; Fleet Aircraft Service Squadron
COFB e: Confirmation of Balance
COFC e: Coefficient of Correlation
- e: Container-on-Flat Car
- e: Cost of Construction
COFCA e: Chief of Coast Artillery
COFCAW e: Combination of Forward Combustion and Waterflooding
COFCWS e: Chief of Chemical Warfare Service
COFDM e: Coded Orthogonal Frequency Division Multiplex
COFE e: Chief of Electronics
- e: Chief of Engineers
- e: Coefficient of Elasticity
- e: Corps of Engineers
COFEC e: Cause of Failure; Effect and Correction
- e: Cause of Failure; Effect and Corrective-action
COFENGRS e: Chief of Engineers
COFF e: Center of Flotation
- e: Centre of Flotation (GB)
- e: Coefficient of Friction
- e: Cut Off
COFFA e: Chief of Field Artillery
C/OFF DP [Switch] e: Centre Off Double Position [Switch]
COFFI System e: Communication Frequency and Facility Information System
C/OFF SP [Switch] e: Centre Off Single Position [Switch]
COFFTI e: Contracting Operator Fast Fourier Transform Identification
COFG e: Center of Gravity
COFGH e: Cape of Good Hope
COFI e: Checkout and Fault Isolation
- e: Chief of Information
- e: Council of Forest Industries
- e: Course of Instruction
COFIDEC e: Coder, Filter, Decoder
COFIDS e: Coherent Optical Fingerprint Identification System
COFIL e: Core File
COFIMPAC f: Compagnie Française Industrielle et Minière du Pacifique
COFIPS e: Central Ohio Federation of Information Processing Societies
COFIRS e: Common Business-Oriented Language for International Business Machines Report Progrem[me] Generator Specifications
- e: Common Business-Oriented Language for International Business Machines Report Progrem[me] Generator Systems (US)
COFM e: Certificate of Merit
- e: Correction of the Moment
COFMAIR e: Chief of Marine Air
COFO[RD] e: Chief of Ordnance
COFP e: Captain of the Port
COFRC e: Chevron Oil Field Research Company
COFRON e: Coastal Frontier
- e: Copper and Iron
COFRS e: Computerized Freight Remittance System
COFS e: Conditions of Service
COF Signal e: Confusion Signal
COFT e: Certificate of Title
- e: Chief of Transportation
- e: Conduct of Fire Trainer
COG e: Center of Gravity
- e: Coal to Oil to Gas
- e: Computer Operations Group
COGADS e: Coast and Geodetic Survey Automated Data System
COGAG e: Combined Gas and Gas
COGAP e: Computer Graphics Arrangement Program[me]
COGAS e: Coal-Oil-Gas
COGAS Process e: Coal into Gas Process
COGD e: Circular Outlet Gas Duct
COGENT e: Compiler and Generalized Translator
COGEOSURV e: Coast Geodetic Survey

COGITA e: Computerized General I.Q. Test[ing]
COGLAD e: Coast Guard Long Range Navigation Assistance Device
COGN e: Cognomen
COGNIT e: Cognition[al][ly]
- e: Cognitive[ly]
COGNOSYS e: Cognitive Operating System
COGO e: Coordinate[d] Geometry
COGO Language e: Coordinate[d] Geometry Language
COGS e: Computer Oriented Geological Society
- e: Concordance Generation System
- e: Consumer Goods System
- e: Continuous Orbital Guidance Sensor
- e: Continuous Orbital Guidance System
COGSIS e: Committee on Government Statistics and Information Service
COH e: Carbohydrate
- e: Coherent
- e: Compensatory Ovarian Hypertrophy
Co^{60} e: Radioactive Cobalt
COHART e: Costs of Hard Rock Tunnelling
COHB d: Carboxyhemoglobin
COHD e: Copyright Office History Document
COHER e: Cohere[d]
- e: Coherence
- e: Coherency
- e: Coherent[ly]
- e: Coherer
- e: Cohering
COHM e: Copyright Office History Monograph
COHO[SC] e: Coherent Oscillator
COHVENT e: Coherent Event
COI d: computerorientierte Informationstechnik
- e: Center of Impact
- e: Central Office of Information
- e: Communication[s] Operation[s] Instruction[s]
- e: Contingency Orbit Insertion
- e: Coordinator of Information
- e: Course of Instruction
- e: Crack-Opening Interferometry
- f: Commission Océanographique Intergouvernementale
COIC e: Canadian Oceanographic Identification Center
- e: Combined Operations Intelligence Center
COIF e: Control of Intensive Farming
COIL e: Computer Acquisition System On-line Interactive Language
COIMS e: Council for International Organizations of Medical Sciences
COIN e: California Olive Industry News
- e: Central Ohio Information Network
- e: Common Business-Oriented Language Indexing
- e: Coin-Phone Operational and Information Network
- e: Coinage
- e: Coincidence
- e: Command Information
- e: Committee on Information Needs
- e: Complete Operating Information
- e: Console Interrupt
- e: Control Only If Necessary
- e: Counter-Insurgency
COINGRAD e: Computer Oriented Interactive Graphical Analysis and Design
COIN-OP e: Coin-Operated
COIN Press e: Coining Press
COINS e: Calspan On-Line Information Service
- e: Computer and Information Science[s]
- e: Computerized Information System[s]
- e: Control in Information Systems
- e: Cooperative Intelligence Network System
- e: Coordinated Inventory-control System
COIN System e: Coin-Phone Operational and Information Network System
COINT e: Communication and Information Technology
COINTEL e: Counterintelligence
COINTELPRO e: Counterintelligence Program[me]
COL d: Colophonium
- e: Chain Overseas; Low
- e: Co-Latitude
- e: Coherent Optical Laser
- e: Collagenase
- e: Collation
- e: Collect[ed]
- e: Collection
- e: Collector
- e: Collostyle
- e: Colo[u]r[ed]
- e: Colo[u]ring
- e: Colonization
- e: Colonize
- e: Colony
- e: Colorado
- e: Columbia
- e: Columbia[n]
- e: Column
- e: Communications-Oriented Language
- e: Computer-Oriented Language
- e: Computerized Office Layout
- e: Computer Oriented Language
- e: Cost of Living
- e: Crisis on Location
COLA e: Committee on Library Automation
- e: Constant Output Level Adapter
- e: Cooperating in Library Automation
- e: Cost of Living Allowance
COLAC e: Central Organization of Liaison for Application of Circuit
COLA Project e: Cooperation in Library Automation Project
COLAPS e: Conversational Language Programming System
COLASL e: Compiler Los Alamos Scientific Laboratories
COLAT e: Co-Lattitude
COLBH e: Collision Bulkhead
COL/CARD e: Columns per Card
COLCDESUP e: Column Code Suppression
COLD e: Chronic Obstructive Lung Disease
- e: Coherent Light Detector
COLDIST e: Column Distributor
Cold Spring Harbor Symp Quant e: Cold Spring Harbor Symposia on Quantitative Biology (journ.)
COL E 1 e: Colicinogenic Factor E 1
COLED e: Combat Loss and Expenditure Data
COLENCY e: Columbia Encyclopedia
COLEQUAP e: Consumer Level Quality Audit Program[e]
COLEX e: Central Information Reference and Control On-Line Experiment
COLI e: Cost of Living Index
COLIDAR e: Coherent Light Detection and Ranging
COLIM e: Collimator
COLING d: Computerlinguistik
- e: Computational Linguistics
- e: Conference on Computational Linguistics
COLINGO e: Compile On-Line and Go
COLIS e: Computer-Oriented Literature Information Service
- e: Council on Library and Information Services
COLL e: Collateral
- e: Collator
- e: Colleague
- e: Collect[ed]
- e: Collecting
- e: Collection
- e: Collective
- e: Collector
- e: College
- e: Collegiate
- e: Collision
- e: Colloid[al]
- e: Colloquial[ism]
- e: Collaboration
- e: Collaborator
- e: Colliers Encyclopedia
COLLABR e: Collaboration
- e: Collaborator
COLLAT e: Collateral
COLLECT e: Connecticut On-Line Law-Enforcement Communications and Teleprocessing
COLLENCY e: Colliers Encyclopedta
Collier Macmillan e: Collier Macmillan Library Service
COLL/L e: Collection Letter
Colloid J. e: Colloid Journal
COLLOQ e: Colloquial
- e: Colloquium
COLLR e: Collector
COLLVOL e: Collective Volume
COLM e: Collector Mesh
COLM e: Column
COLMGP e: Column Gap
COLMIS e: Collection Management Information System
COLN e: Column
COLND e: Completed Landing
COLOASSOC e: Colorado Associated University Press
COLOG e: Cologarithm
COLOGRAPH e: Color Lithograph
COLOPH e: Colophon
COLP e: Collapse
COLP e: Colo[u]r Page
COLPS e: Collapse
COLQUAP e: Consumer Level Quality Audit Program[me]
COLR e: Circuit Order Layout Record
COLS e: Communications for On-Line Systems
col/sec e: Column[s] per Second
COLSPD e: Collapsed
COLT e: Central Office Line Tester
- e: Communication Line Terminator
- e: Computer-Oriented Language Translator
- e: Computer-Oriented Logic Test
- e: Computerized On-Line Testing
- e: Control Language Translator
- e: Council on Library-Media Technical-Assistants
- e: Council on Library Technology
COLTAINER e: Collapsible Container
COLTS e: Communication On-Line Testing System
- e: Contrast Optical Laser Tracking Subsystem
- e: Contrast Optical Laser Tracking System
COLT System e: Computer-Oriented Logic Test System
COLUG e: Chicago Online Users Group

Columbia UQ e: Columbia University Quarterly (journ.)
COM e: Card Oriented Model
- e: Cassette Operating Monitor
- e: Center of Mass
- e: Centre of Mass
- e: Change Order Modification
- e: Character Organized Memory
- e: Checkout Operations Manual
- e: Coefficient of Merit
- e: Commanding
- e: Commemoration
- e: Commentary
- e: Commerce
- e: Commercial
- e: Common
- e: Commonwealth
- e: Communicate[d]
- e: Communication[s]
- e: Communications Operating Module
- e: Community
- e: Commutator
- e: Compiler
- e: Complement
- e: Composition
- e: Compressed Multisound
- e: Computer Output Microfiche
- e: Computer Output Microfilm[ing]
- e: Computer Output Microfilmer
- e: Computer Output Microform
- e: Computer Output Micrographics
- e: Computer Output on Microfiche
- e: Computer Output on Microfilm
- e: Computer Output to Microfilm
- e: Copper Oxide Modulator
- e: Curve of Merit
- s: Comunicaciones
C & OM e: Clothing and Organic Materials
COMA e: Computer Operations Management Association
COMAC e: Commander; Military Airlift Command
- e: Continuous Multiple-Access Collator
- e: Continuous Multiple[x]-Access Comparator
COMACA e: Corresponding Member of the Academy of Arts
COMACH s: Confederación Maritima de Chile
COMACS e: Common Manufacturing Accounting Control System
COMAD e: Computer Methods for Automatic Diagnosis
COMAF[FOR] e: Commander; Air Force Forces
COMAIR e: Commander; Aircraft
- e: Commander; Allied Air Forces
COMAIRBALTAP e: Commander; Allied Air Forces Baltic Approaches
COMAIRCANLANT e: Commander; Allied Air Forces Canadian Atlantic Subarea
COMAIRCENT e: Commander; Allied Air Forces Central Europe
COMAIRCENTLANT e: Air Commander; Central Atlantic
COMAIRCHAN e: Commander; Maritime Air Channel
COMAIREASTLANT e: Commander; Allied Air Forces Eastern Atlantic
COMAIRFERON e: Commander; Aircraft Ferry Squadron
COMAIRLANT e: Commander; Air Force; Atlantic Fleet
COMAIRNORLANT e: Commander; Allied Air Forces Northern Atlantic
COMAIRNORTH e: Commander; Allied Air Forces Northern Europe
COMAIRPAC e: Commander; Air Force; Pacific Fleet
COMAIRPLYMCHAN e: Commander; Allied Air Forces Plymouth Channel
COMAIRSOUTH e: Commander; Allied Air Forces; Southern Europe
COMAL e: Common Algorithmic Language
COMALAIRNOREUR e: Commander; Allied Air Forces; Northern Europe
COMAMC e: Commander; Air Materiel Command
COMAMPHIBGRU e: Commander; Amphibious Group
COMAMPHIBTRA e: Commander; Amphibious Training
COMANSEC e: Computation and Analysis Section
COMANTDEFCOM e: Commander; Antilles Defense Command
COMANTISUBDEFLANT e: Commander; Antisubmarine Defense; Atlantic Fleet
COMAP e: Conversational Macro Package
COMAR e: Committee on Man and Radiation
- e: Computer; Aerial Reconnaissance
COMARAIRMED e: Commander; Maritime Air Forces Mediterranean
COMARC e: Cooperative Machine-Readable Cataloguing
COMARS e: Coordinate Measuring and Recording System
COMART e: Commander; Marine Corps Air Reserve Training
COMASDEFORLANT e: Commander; Antisubmarine Defense Force; Atlantic [Fleet]
COMASDEFORPAC e: Commander; Antisubmarine Defense Force; Pacific [Fleet]
COMASDFLANT e: Commander; Antisubmarine Defense Force; Atlantic [Fleet]
COMASIII e: Computerized Maintenance and Administration Support III
COMAT e: Characteristics of Materials
- e: Computer-Assisted Training
COMATF e: Commander; Amphibious Task Force
COMATS e: Commander; Military Air Transport Service
- e: Computer Operated Manufacturing and Test System
- e: Control Organization Methods and Technique Studies
- e: Corporate Manufacturing Transfer System
COMB d: cassetten-orientiertes modulares Buchungssystem
- e: Combat
- e: Combination
- e: Combine[d]
- e: Combining
- e: Combustibility
- e: Combustible
- e: Combustion
- e: Console Oriented Model Building
- e: Journal of Combinatorial Theory
COMBAR e: Commutator Bar
COMBAT e: Commander; Marine Basic Air Training
- e: Cost-Oriented Models Built to Analyze Tradeoffs
COMBATDIV e: Commander; Battleship Division
COMBATFOR e: Commander; Battleship Force
COMBATSUS e: Commander; Battleships; United States Fleet
COMBEX e: Combined Exercise
COMBI e: Combination
COMBIDS e: Computerized Bibliographic Data Service
COMBIMAN Model e: Computerized Biomechanical Man Model
COMBL e: Combustible
COMBO e: Computation of Miss-Between Orbits
COMBOLOCK[S] e: Combination Lock[s]
COMBRONET e: Computerized Brockerage Network
COMBS e: Combinations
COMBT e: Combat
COMBU e: Combustion
Combust. Explos.; Shock Waves e: Combustion Explosion; and Shock Waves (journ.)
Combust. Flame e: Combustion and Flame (journ.)
COMC e: Carboxymethylcellulose
- e: Command
- e: Communications Countermeasures
COMCAM e: Compressible Cell and Maker
COMCAN e: Commonwealth Communication of Army Networks
COMCARAIRGRU e: Commander; Carrier Air Group
COMCARR e: Common Carrier
COMCAS e: Computer-Oriented Model Control and Appraisal System
COMCAT e: Computer Output Microfilm Catalog
COMCAW e: Commander; Carrier Air Wing
COMCEN[TER] e: Communication[s] Center
COMCM e: Communications Countermeasures
COMCODE e: Commodity Code
COMCON e: Combat Control
COMCONSUP e: Combat, Control, Support
COMCO Program[me] e: Communication Control Program[me]
COMCRUDESFLOT e: Commander; Cruiser Destroyer Flotilla
COMCRUDESLANT e: Commander; Cruiser Destroyer Force; Atlantic [Fleet]
COMCRUDESPAC e: Commander; Cruisers and Destroyers; Pacific [Fleet]
COMCRUDIV e: Commander; Cruiser Division
COMCRUFOR e: Commander; Cruiser Force
COMCRULANT e: Commander; Cruiser Forces; Atlantic [Fleet]
COMCRUSUS e: Commander; Cruisers; United States [Fleet]
COMD e: Command
- e: Commanding
COMDAC e: Component Design Augmented by Computer
COMDBS e: Controlling Monitor Data Base System
COMDEC e: Command Decision & Movement Control Chart[s]
COMDESBATFOR e: Commander; Destroyer Battle Force
COMDESDEVGRU e: Commander; Destroyer Development Group
COMDESDIV e: Commander; Destroyer Division

COMDESFLOT e: Commander; Destroyer Flottilla
COMDESFOR e: Commander; Destroyer Force
COMDESRON e: Commander; Destroyer Squadron
COMDESTR e: Commander; Destroyer-Force
COMDESUS e: Commander; Destroyers; United States [Fleet]
COMDET e: Computer Mediated Drill and Educational Testing
COMDRMATS e: Commander; Military Air Transport Service
COMDTK e: Command Tank
COMD WC e: Command Weapon Carrier
COME e: Chief Ordnance Mechanical Engineer
COMEAS e: Countermeasure[s]
COMECON e: Council for Mutual Economic Assistance
COMED e: Combined Map and Electronic Display
- e: Communications Editing Unit
COMEDCENT e: Commander; Central Mediterranean Area
COMEDS e: Continental Meteorological Data System
COMEF e: Coherent Memory Filter
COMEINDORS e: Composite Mechanized Information and Documentation Retrieval System
COMEM e: Common Memory
COMENT e: Computer-Operated Management Evaluation Technique
COMES e: Countermeasures
COMEST e: Communications European Satellite Team
COMET d: computergestütztes marketingorientiertes Entscheidungs- und Kontrollsystem
- e: Chip on Module Evaluation Tester
- e: Coherent Electromagnetic Energy Transmission
- e: Communications by Meteor Trails
- e: Computer-Operated Management Evaluation Technique
- e: Computerized Message Transmission
- e: Computer Message Transmission
- e: Computer Operated Management Evaluation Technique
- e: Controllability, Observability and Maintenance Engineering Technic
- e: Cost Measurement Technique
COMETS e: Computer-Operated Multifunction Electronic Test Station
COMEXCO e: Committee for Exploitation of the Oceans
COMFACTS e: Communication Facilities Testing System
COMFAIR e: Commander; Fleet Air Force
COMFAIRELM e: Commander; Fleet Air Force; Eastern Atlantic and Mediterranean
COMFAIRLANT e: Commander; Fleet Air Force; Atlantic Fleet
COMFAIRPAC e: Commander; Fleet Air Force; Pacific [Fleet]
COMFAIRWINGNORLANT e: Commander; Fleet Air Wing Northern Atlantic
COMFAIRWNG e: Commander; Fleet Air Wing
COMFAIRWNGLANT e: Commander; Fleet Air Wing; Atlantic [Fleet]
COMFAIRWNGPAC e: Commander; Fleet Air Wing; Pacific [Fleet]
COMFET e: Conductivity-Modulated Field-Effect Transistor
COMFIVEATAF e: Commander; Fifth Allied Tactical Air Force
COMFLEET e: Commander of the Fleet
COMFLOGWING e: Commander; Fleet Logistic Air Wing
COMFOR e: Computer Forum and Exposition
COMFORT e: Commercial Formula Translation
- e: Compulogic Formula Translation
COMFOURATAF e: Commander; Fourth Allied Tactical Air Force
COMGAF e: Commanding Air Force
COMGEN e: Common Specifications Statements Generator
COMGENAFMIDPAC e: Commanding General; Army Forces; Mid-Pacific
COMGENSOPAC e: Commanding General; South Pacific [Area]
COMGENUSAFE e: Commanding General; United States Air Forces in Europe
Com I e: Comoro Islands
COMIBOL s: Corporación Minera de Bolivia
COMIC e: Colo[u]rant Mixer Computer
- e: Customer Organized Metal Oxide Semiconductor Integrated Circuit
COMICS e: Computer-Oriented Managed-Inventory Control System
COMINDIV e: Commander; Mine Division
COMINLANT e: Commander; Mine Force; Atlantic [Fleet]
COMINRON e: Commander; Mine Squadron
COMINST e: Communications Instructions
COMINT e: Communications Intelligence
COMINUS e: Commander; Minecraft; United States [Fleet]
COMIS e: Committee Meeting Information System
COMISARS e: Combined Information Storage and Retrieval System
COMIT e: Computer-Language of Massachusetts Institute of Technology
- e: Computer Operations Management Information Training
- e: Computing System; Massachusetts Institute of Technology
COMJAM e: Communications Jamming
COMJUWTF e: Commander; Joint Unconventional Warfare Task Force
COML e: Commercial Language
COMLA e: Commonwealth Library Association
COMLANSHIPFLOT e: Commander; Landing Ships Flotilla
COMLANSHIPRON e: Commander; Landing Ships Squadron
COMLANTRESFLT e: Commander; Atlantic Reserve Fleet
COMLCURON e: Commander; Landing Craft Utility Squadron
COMLIP e: Community Media Librarian Program[me]
COMLO e: Compass Locator
COMLODEF e: Commander; Local Air Defence
COMLOGNET e: Combat Logistics Network
- e: Communications Logistics Network
COMLSTRON e: Commander; Landing Ship Tank Squadron
COMM e: Commercial
- e: Commission[ed]
- e: Common
- e: Commonwealth
- e: Communication[s]
- e: Commutator
COMM-ADP e: Communications-Automatic Data Processing
COMMAG e: Combined Magnetic [sound]
COMMAIRCENTLANT e: Commander; Maritime Art; Central Atlantic
COMMAIREASTLANT e: Commander; Maritime; Eastern Atlantic [Area]
COMMAIRPLYMCHAN e: Commander; Maritime Air; Plymouth Subarea Channel
COMMAND e: Command Model for Analysis and Design
COMMANDS e: Computer Operated Marketing; Mailing and News Distribution System
COMMARFOR e: Commander; Marine Forces
COMMART e: Commander; Marine Air Reserve Training
COMMBIOPEST e: Committee for Biological Pest Control
COMMCEN e: Communications Center
COMMCH e: Communications Chief
COMMCHKPT e: Communication Check Point
COMMCM[EAS] e: Communications Countermeasures
COMMCMEASOP e: Communications Countermeasures Operator
COMMCTR e: Communication Center
COMMCTRSP e: Communication Center Specialist
COMMDG e: Commanding
COMMEL e: Communications Electronics
COMMEN e: Compiler Oriented for Multiprogramming and Multiprocessing Environments
COMMEND e: Computer-Aided Mechanical Engineering Design
- e: Computer-Assisted Mechanical Design
COMMEND System e: Computer-Assisted Mechanical Design System
COMM ENGR e: Commissioned Engineer
COMMENT e: Commentary
Comments Nucl Part Phys e: Comments on Nuclear and Particle Physics (journ.)
Comments Nucl. Part. Phys. Sup e: Comments on Nuclear and Particle Physics; Supplement (journ.)
COMMFEX e: Communications Field Exercise
COMMIKE e: Commandable Microphone
COMMINEDIV e: Commander; Mine Division
COMMINELANT e: Commander; Mine Forces; Atlantic [Fleet]
COMMINSTN e: Communication Instructions
COMMO e: Communications
- e: Communications Officer
COMMOD e: Command Ordnance Depot
COMMODORE e: Command Modular Operation Room Equipment
COMMONDIS e: Communicable Disease
COMMS e: Central Office Maintenance Management System
COMMSC e: Commander; Military Sealift Command
COMMSCTY e: Communication Security
COMMSCTYANAL e: Communications Security Analyst

COMMSECACT e: Communication Security Activity
COMMS-PM e: Central Office Maintenance Management System-Preventive Maintenance
COMMSTITCH e: Communications Failure Detecting and Switching
COMM STOR Unit e: Communications Storage Unit
COMMSUPACT e: Communications Supplementary Activity
COMMT e: Commencement
COMMUN e: Communication
- e: Community
Commun ACM e: Communications of the Association of Computing Machinery (journ.)
Commun Math Phys e: Communications in Mathematical Physics (journ.)
Commun Pure Appl Math e: Communications on Pure and Applied Mathematics (journ.)
COMMUNSYST e: Communications System
COMMZ e: Communications Zone
COMN e: Communication
COMNAB e: Commander; Naval Air Base
COMNAV e: Commander; Naval [Forces]
COM & NAV e: Commerce and Navigation
COMNAVAIRLANT e: Commander Naval Air Forces; Atlantic [Fleet]
COMNAVAIRPAC e: Commander; Naval Air Forces; Pacific [Fleet]
COMNAVAIRTECHTRA e: Commander; Naval Air Technical Training
COMNAVAIRTECHTRAU e: Commander; Naval Air Technical Training Unit
COMNAVAIRTESTCEN e: Commander; Naval Air Test Center
COMNAVAVSAFETYCEN e: Commander; Naval Aviation Safety Center
COMNAVB e: Commander; Naval Base
COMNAVBALTAP e: Commander; Allied Naval Forces; Baltic Approaches
COMNAVBASE e: Commander; Naval Base
COMNAVCAG e: Commander; Allied Naval Forces; Central Army Group
COMNAVCENT e: Commander; Allied Naval Forces; Central Europe
COMNAVCONSTBNLANT e: Commander; Naval Construction Battalion; Atlantic Fleet
COMNAVCONSTBNPAC e: Commander; Naval Construction Battalion; Pacific [Fleet]
COMNAVDIS e: Commandant; Naval District
COMNAVEU e: Commander; Naval Forces; Europe
COMNAVFOR e: Commander; Naval Forces
COMNAVFORGER e: Commander; United States Naval Forces in Germany
COMNAVFORJAP e: Commander; Naval Forces in Japan
COMNAVFORJAPAN e: Commander; Naval Forces in Japan
COMNAVGER e: Commander; Naval Forces in Germany
COMNAVGERBALT e: Commander; German Naval Forces; Baltic
COMNAVMARIANAS e: Commander; [United States] Naval Forces; Marianas
COMNAVNON e: Commander; Allied Naval Forces; North Norway
COMNAVNORCENT e: Commander; Allied Naval Forces; Northern Area; Central Europe
COMNAVNORTH e: Commander; Allied Naval Forces; Northern Europe
COMNAVPHIL e: Commander; Naval Forces; Philippines
COMNAVSCAP e: Commander; [Allied] Naval Forces; Scandinavian Approaches
COMNAVSOUTH e: Commander; [Allied] Naval Forces; Southern Europe
COMNAVSUPPACT e: Commander; Naval Support Activity
COMNAVTECHTRA e: Commander; Naval Technical Training
COMNAVTESTCEN e: Commander; Naval Test Center
COMNEED e: Communications Need
COMNET e: Communications Network
- e: Computer Network
COMNORASDEFLANT e: Commander; North American Antisubmarine Defense Force; Atlantic
COMO e: Coherent Master Oscillator
- e: Communications Officer
- e: Comprehensive Model
- e: Computer Model
COMOCEANLANT e: Commander; Ocean Atlantic Subarea
COMOCEANSYSTLANT e: Commander; Oceanographic System; Atlantic
COMOCEATLANT e: Commander; Ocean Atlantic Subarea
COMODL e: Connection Model Description Language
COMOL e: Connection Model-Description Language
COMOPDEVFOR e: Commander; Operational Development Force
COMOPDEVFORLANT e: Commander; Operational Development Force; Atlantic [Fleet]
COM/OPS/RAC e: Communications, Operational Practices and Rules of the Air Committee
COMOPTEVFOR e: Commander; Operational Test and Evaluation Force
COMO Scan e: Conical-Monopulse Scan
COMP e: Charlotte Ordnance Missile Plant
- e: College of Osteopathic Medicine of the Pacific
COMP e: Comparative
- e: Comparator
- e: Compare
- e: Comparison
- e: Compartment
- e: Compass
- e: Compatible
- e: Compensate[d]
- e: Compensating
- e: Compensation
- e: Compensator
- e: Competitor
- e: Compilation
- e: Compile[d]
- e: Compiler
- e: Complement
- e: Complimentary
- e: Component
- e: Compose[d]
- e: Composer
- e: Composite
- e: Composition
- e: Compositor
- e: Compound[ed]
- e: Comprehension
- e: Comprehensive
- e: Compression
- e: Compressor
- e: Comptroller
- e: Computation
- e: Computational
- e: Computer
- e: Computerization
- e: Computerize[d]
- e: Computer Oriented Microwaves Practices
- e: Conceptually-Oriented Mathematics Program[me]
- e: Journal of Computer and System Sciences (journ.)
comp. l: compositus
COMPA e: Compressed Air
COMPAC e: Commonwealth Pacific Cable
COMPAC e: Computer Output Microfilm Package
- e: Computer Program[me] for Automatic Control
COMPAC Cable e: Commonwealth Pacific Cable
COMPACE e: Control of Material Planning Activities
COMPACRESFLT e: Commander; Pacific Reserve Fleet
COMPACS e: Computer-oriented Manufacturing Production and Control System
- e: Computer Output Microforms Program[me] and Concept Study
COMPAC System e: Commonwealth-Pacific System
COMPACT e: Compatible Algebraic Compiler and Translator
- e: Computator Planning and Control Technique
- e: Computer Operated Micro-Program[me] Automatic Commissioning Technique
- e: Computer Oriented Modular Planning and Control Technique
- e: Computer Planning and Control Technique
- e: Computer Predicting and Automatic Course Tracking [Radar]
- e: Computer Programmed Automatic Checkout and Test
COMPACT Circuits e: Computerized Optimization of Microwave Passive and Active Circuits
COMPACT Detection e: Combined Passive Active Detection
COMPACT Radar e: Computer Predicting and Automatic Course Tracking Radar
COMPACT System e: Computer Programmed Automatic Checkout and Test System
COMPAD e: Combined Office Material Procurement and Distribution
COMPAF e: Commander; Pacific Air Fleet
- e: Commander; Pacific Air Force
COMPAID e: Control of Material Planning and Isometric Drawings
COMPAIS e: Computer-Assisted Information System
COMPANAL e: Computer Analysis
COMPAND e: Compress and Expand
- e: Compress to Expand
COMPANDER e: Compressor-Expander
COMPANDOR e: Compressor-Expandor
COMPAR e: Comparative
- e: Comparator
- e: Comparison

COMPARE e: Computerized Performance and Analysis Response Evaluation
- e: Computerized Performance and Analysis Response Evaluator
- e: Computer Oriented Method of Program[me] Analysis; Review and Evaluation
- e: Console for Optical Measurement and Precise Analysis of Radiation from Electronics
COMPAS d: Computeranwendungen Software und Systeme
- d: computergesteuerte automatische Sendeabwicklung
- e: Committee on Physics and Society
- e: Computer Acquisition System
- e: Computer Oriented Method for Payroll Accounting and Statistics
COMPASS d: Computer-orientierte Methode für Planung und Ablauf-Steuerung in Seehäfen
- d: Computer-Prüfungs/Analyse-System für Schulen
- e: Compiler-Assembler
- e: Complete Parallel Activity and Security System
- e: Comprehensive Assembly [Language]
- e: Comprehensive Assembly System
- e: Comprehensive Pharmacy Administrative Support System
- e: Computer-Assisted
- e: Computer for Advanced Space System[s]
- e: Computerized Movement Planning and Status System
- e: Computer Optimal Media Planning and Selection System
- e: Computer Oriented Management Planning and Scheduling System
- e: Computer Oriented Method of Patterns [and] Analysis for Switching Systems
- e: Controlled Overhead Management Performance and Standard System
COMPASS Language e: Comprehensive Assembly Language
COMPAT d: Computerautomationstechnik
- e: Compatibility
COMPAY e: Computer-Oriented Method for Payroll
- e: Computer Payroll
Comp Biochem Physiol e: Comparative Biochemistry and Physiology (journ.)
COMP-COMP System e: Computer-Composition System
COMPCON e: Computer Conference
- e: Computer Convention
COMPCORP e: Computer [Design] Corporation
COMPD e: Compound[ed]
COMPDES e: Compensator Design
- e: Competitive Design
Comp Des Engr e: Component Design Engineer
- e: Computer Design Engineer
COMPEC e: Computer Peripherals; Small Computer and Systems Exhibition and Conference
COMPEL e: Compute Parallel
COMPELS e: Computerized Evaluation and Logistics System
COMPEN e: Compensate
- e: Compensation
- e: Compensatory
COMPEND e: Compendious
- e: Compendium
COMPENDEX e: Computerized Engineering Index
Comp Engr e: Computer Engineer
- e: Computer Engineering
COMPEX e: Competive Exercise
COMPF e: Composition Floor
COMPFIL e: Compensating Filter
COMP File e: Complement File
COMPG e: Compressed Gas
COMPHIBFOR e: Commander; Amphibious Force
COMPHIBGRU e: Commander; Amphibious Group
COMPHIBLANT e: Commander; Amphibious Forces; Atlantic
COMPHIBPAC e: Commander; Amphibious Force; Pacific [Fleet]
COMPHIBPACFLT e: Commander; Amphibious Force; Pacific Fleet
COMPHIBRON e: Commander; Amphibious Squadron
COMPHIBTRALANT e: Commander; Amphibious Training Command Atlantic
COMPHIBTRAPAC e: Commander; Amphibious Training Forces; Pacific [Fleet]
COMPL e: Combat Outpost Line
COMPL e: Compilation
- e: Compile[d]
- e: Complement
- e: Complementary
- e: Complete[d]
- e: Completion
- e: Complex
COMPLAN e: Communications Plan
COMPLEX e: Committee on Planetary and Lunar Exploration
COMPLI e: Compliance
COMPLT e: Complainant
- e: Complaint
COMPMR e: Commander of Pacific Missile Range
COMPN e: Composition
COMPO e: Communications Pool[ing]
- e: Compensation
- e: Component
- e: Composite
- e: Composition
- e: Compositor
COMPOLE e: Commutating Pole
COMPON e: Component
COMPOOL e: Common Pool
- e: Common Pool of Data
- e: Communication[s] Pool
COMPOs e: Components
- e: Composites
- e: Compositions
COMPOSE e: Computerized Production Operation System Extension
COMPOT e: Combined Optical
COMPOUND A e: 11-Dehydro-corticosterone
COMPOUND B e: Corticosterone Compound E Cortisone
COMPOUND F e: Cortisol
COMPOUND S e: 11-Deoxycortisol
Comp Pkg Engr e: Computer Packaging Engineer
COMPR e: Compare
- e: Compression
- e: Compression Ratio
- e: Compressor
COMPREG e: Compressed-Impregnated
COM Press e: Compacting Press
COMPRESS e: Computer Research, Systems and Software
COMPROC e: Command Processor
COMPROG e: Computer Program[me]
COMPROSL e: Compound Procedural Scientific Language
COMPSAC e: Computer Software and Applications Conference
COMPSA Conference e: Computer Software and Applications Conference
COMPSCAN e: Computerized Scanner
COMPSG e: Compensating
COMPSO e: Computer Software and Peripheral Show
COMPSTAT e: [Conference on] Computational Statistics
COMPSY e: Computer-Support in Military Psychatry [Program]
COMPSY Program[me] e: Computer-Support in Military Psychiatry Program[me]
COMPT e: Catecholomethyltsansferase
- e: Chip on Module Parameter Tester
- e: Compact
- e: Compartment
- e: Comptroller
- e: Computer
COMPTEL e: Competitive Telecommunications
Comptes Rend f: Comptes rendus de l'Académie des Sciences
COMPTESS e: Computer Hostess
COMPTEXT e: Computer Prepared Text
COMPTR e: Comparator
- e: Comptroller
COMPTU e: Composite Training Unit
COMPU e: Computability
- e: Computable
- e: Computation[al]
- e: Computer
- e: Computerization
- e: Computerize
COMPUMAG e: Conference on Computations of Magnetic Fields
COMPUNICATION e: Computer Communication
COMPUSEC e: Computer Security
COMPUSERV e: Computer Service [Network]
COMPUSTAT e: Computerized Statistics
COMPUT e: Computation
- e: Computer
- e: Computing
COMPUTA e: Computational
Comput Arch Elektron Rechn d: Computing Archiv für Elektronisches Rechnen (journ.)
Comput Automat e: Computers and Automation (journ.)
Comput Biomed Res e: Computers and Biomedical Research (journ.)
Computer d: Computer and Peripherie GmbH
Computer J e: Computer Journal
Computer Wkly e: Computer Weekly (journ.)
COMPUTES e: Computers
COMPUTEST d: Computer Testgeräte
COMPUTIME e: Computer-Computed Time
Comput J e: Computer Journal
Comput Prax d: Computer Praxis (journ.)
COMPWARE d: Computer und Software Vertriebsgesellschaft mbH
COMPWG e: Composite Wing
COMRAC e: Combat Radius Capability
COMRADE e: Computer Aided Design Environment
COMRATE e: Committee on Mineral Resources and the Environment

COMRAZ e: Communication Range Azimuth
COMRCN e: Command Reconnaissance
COMREC e: Care-Oriented Medical Record
COMREC e: Communications Receiver
COMS e: College of Osteopathic Medicine and Surgery
- e: Computer-based Operations Management System
COMSAC e: Computerized Measurements for Safeguards and Accountability
COMSAL e: Computer Method of Sequencing Operations for Assembly Lines
COMSAP e: Computerized Static Automatic-restoring-equipment for Power-System
COMSAT e: Communication[s] Satellite [Corporation]
COMSATs e: Communication Satellites
COMSEAFRON e: Commander; Sea Frontier
COMSEC e: Communication[s] Security
- e: Computer Software Exchange Center
COMSECONDFLT e: Commander; Second Fleet
COMSENEX e: Combined Sensor Tracking Exercise
COMSEQIN e: Component Sequencing and Insertion
COMSER e: Commission on Marine Science and Engineering Research
COMSERFORPACSUBCOM e: Commander; Service Force; Pacific Subordinate Command
COMSERVLANT e: Commander; Service Forces; Atlantic [Fleet]
COMSERVPAC e: Commander; Service Forces; Pacific [Fleet]
COMSERVRON e: Commander; Service Squadron
COMSEVENTHFLT e: Commander; Seventh Fleet
COMSGT e: Communications Sergeant
COMSIXATAF e: Commander; Sixth Allied Tactical Air Force
COMSIXTHFLT e: Commander; Sixth Fleet
COMSL e: Communication-System Simulation Language
COMSOAL e: Computer Method of Sequencing Operations for Assembly Lines
COMSOC e: Communications Spacecraft Operations Center
COMSOF d: Computer Software GmbH
COMSOLANT e: Commander; South Atlantic Force
COMSOTFE e: Commander; Support Operations Task Force; Europe
COMSS e: Compare String with String
COMSTAR e: Communications Satellite Network
COMSTAT e: Competitive Statistics
COM-STAT e: Computer Stock Timing and Analysis Technique
COM-STEP e: Computerized Spot Television Evaluation and Processing
COMSTORE e: Computer Storage
COMSTRIKFLTLANT e: Commander; Striking Fleet; Atlantic
COMSTRIKFLTLANTREPEUR e: Commander; Striking Fleet; Atlantic Representative in Europe
COMSTRIKFORSOUTH e: Commander; Striking Forces Southern Europe
COMSTS e: Commander; Military Sea Transport[ation] Service
COMSTSELMAREA e: Commander; Military Sea Transportation Service; Eastern Atlantic and Mediterrane
COMSTSLANT e: Commander; Military Sea Transportation Service; Atlantic Area
COMSTSLANTAREA e: Commander; Military Sea Transportation Service; Atlantic Area
COMSUBDEVGRU e: Commander; Submarine Development Group
COMSUBDIV e: Commander; Submarine Division
COMSUBEASTLANT e: Commander; Submarine Force; Eastern Atlantic
COMSUBFLOT e: Commander; Submarine Flotilla
COMSUBFOR e: Commander; Submarine Force
COMSUBGRU e: Commander; Submarine Group
COMSUBLANT e: Commander; Submarine Force; Atlantic
COMSUBMED e: Commander; Submarine Forces; Mediterranean
COMSUBMEDNOREAST e: Commander; Submarines; Northeast Mediterranean
COMSUBPAC e: Commander; Submarine Forces; Pacific [Fleet]
COMSUBPACREP e: Commander; Submarine Forces; Pacific Fleet; Representative
COMSUBRON e: Commander; Submarine Squadron
COMSUBTRAFAC e: Commander; Submarine Training Facilities
COMSUBUS e: Commander; Submarine Forces; United States [Fleet]
COMSUCOMLANTFLT e: Commander; Subordinate Command; United States Atlantic Fleet
COMSUP e: Combat Support
- e: Communications Supervisor
COMSUPTLANT e: Commodore Superintendent; Atlantic Coast
COMSW e: Compare String with Word
COMSYL e: Communications System Language
COMSYSDISC e: Communication System Disciplines
COMT e: Catechol-O-Methyl-Transferae
- e: Comptroller
COMTAFDEN e: Commander; Tactical Air Force; Denmark
COMTAFNORNOR e: Commander; Allied Tactical Air Force; North[ern] Norway
COMTASKFORNON e: Commander; Allied Task Force; North[ern] Norway
COMTEC e: Computer Micrographics [and] Technology
COMTECH e: Communications Technician
- e: Computer Micrographics Technology
COMTECHREP e: Complementary Technical Report
COMTEST d: Computertest
COMTEX e: Communications-Oriented Multiple Terminal Executive
COMTHEORY e: Communication Theory
COMTRAC e: Computer-Aided Traffic Control
COMTRAC System e: Computer-Aided Traffic Control System
COMTRALANT e: Commander; Training Force; Atlantic [Fleet]
COMTRAN e: Commercial Translation
- e: Commercial Translator
- e: Computer Translation
COMTRANSPHIBRON e: Commander; Transport Amphibious Squadron
COMTRAPAC e: Commander; Training Force; Pacific [Fleet]
COMTWOATAF e: Commander; Second Allied Tactical Air Force
COMUL e: Complement-Fixation for Murine Leukemia
COMUSAFFOR e: Commander; United States Air Force Forces
COMUSAFSO e: Commander; United States Air Forces; Southern Command
COMUSAFTF e: Commander; United States Air Force Task Force
COMUSJUWTF e: Commander; United States Joint Unconventional Warfare Task Force
COMUSMARFOR e: Commander; United States Marine Forces
COMUSMARTF e: Commander; United States Marine Task Forces
COMUSNAVFOR e: Commander; United States Naval Forces
COMUSNAVSUPPFOR e: Commander; United States Naval Support Force
COMUSNAVTF e: Commander; United States Naval Task Force
COMUTWING e: Commander; Utility Wing
COMVER e: Common Version
COMWC e: Command Weapon[s] Carrier
COMX e: Communications Executive
COMYARD e: Commander of the Dockyard
COMZ e: Communications Zone
COMZEUR e: Communications Zone; Europe
CON e: Commission on the Nomenclature [of Organic Chemistry]
- e: Concentrate
- e: Concentration
- e: Concerning
- e: Conclusion
- e: Concord[e]
- e: Concrete
- e: Confidence
- e: Congo
- e: Connect[ed]
- e: Connecting
- e: Connection
- e: Connector
- e: Consignee
- e: Console
- e: Consolidate[d]
- e: Consolidation
- e: Constant
- e: Constructor
- e: Contactor
- e: Continue[d]
- e: Control
- e: Controller
- e: Convenience
- e: Conversation
- e: Converter
- e: Cursor On
CONA d: Concanavalin A
CONAC e: Continental Air Command
CONACS e: Contractors Accounting System
CONACYT s: Consejo Nacional de Ciencia y Tecnologia
CONAD e: Configuration Advisor
- e: Continental Air Defense
CONADM e: Continental Air Defense Command Manual
CONAES e: Committee on Nuclear and Alternative Energy Systems

CONALOG e: Contact Analog[ue]
CONBAT e: Converted Battalion Anti-Tank
Con Blk Col e: Consecutive Blank Column
CONC e: Concentrate[d]
- e: Concentrating
- e: Concentration
- e: Concentric
- e: Concrete[d]
- e: Constructor Captain
CONCA e: Continue Calling
CONCAP e: Conversational Circuit Analysis Program[me]
CONCAT e: Conventional Catamaran
CONCAWE e: Conservation of Clean Air and Water in [Western] Europe
CONCB e: Concrete Block
CONCC[IG] e: Concrete Ceiling
CONCCLG e: Concrete Ceiling
CONCD e: Concentrated
CONCDSOLN e: Concentrated Solution
CONCENTR e: Concentrate[d]
CONCEPT e: Computation Online of Networks Chemical Engineering Process[es] Technology
CONCF e: Concrete Floor
CONCH e: Conchology
CONCISE e: Computer-Oriented Notation Concerning Infrared Spectral Evaluation
CONCL e: Conclusion
CONCME e: Concentration of Mechanical Transport
CONCN e: Concentration
CONCOCON e: Construction Cost Control
CONCOCON System e: Construction Cost Control System
CONCOMP e: Conversational Computation Project
CONCORD e: Conference Coordinator
CONCR e: Concrete[ly]
- e: Constructor Commander
CONCY e: Concentricity
COND e: Commanding Officer; Naval Divisions
- e: Condensance
- e: Condense[d]
- e: Condenser
- e: Condensing
- e: Condition[s]
- e: Conditional
- e: Conditioned
- e: Conditioner
- e: Conditioning
- e: Conduct[ed]
- e: Conducting
- e: Conductivity
- e: Conductor
- e: Conduit
CONDAL e: Control Data Link
CONDEEP[S] e: Concrete Deepwater Structure[s]
CONDIT e: Conditional
CONDIV e: Continental Divide
CONDOR d: Communikation in natürlicher Sprache mit dialog-orientiertem Retrievalsystem
CONDO[S] e: Condominium[s]
CONDR e: Conductor
CONDREF e: Conditioned Reflex
CONDRESP e: Conditioned Response
CONDTN e: Condition
CONDUCTIMETRIC e: Conductance and Metric
CONE e: Controller Error
- e: Creation of New Enterprises
CONEC e: Connection
CONECS e: Connectorized Exchange Cable Splicing
CONEDS e: Continental Army Command Education Data System
CONELRAD e: Control of Electromagnetic Radiation[s]
CONE Scheme e: Creation of New Enterprises Scheme
CONEST e: Construction Estimation
CONEST Program[me] e: Construction Estimation Program[me]
CONET e: Concentrator Network
- e: Connection Network
CONETIC e: High-Permeability Nonshock-Sensitive
CONEX e: Container Express
CONF e: Confer[ence]
- e: Confidence
- e: Confidencial
- e: Confidential
- e: Configuration
- e: Confinement
CONFBUL e: Confidential Bulletin
CONFD e: Confidential
CONFECONPROG e: Conference on Economic Progress
CONFESS e: Conduction Fingerprint Eletro-Optic Search System
CONFI e: Confine
CONFICS e: Cobra Night Fire Control System
CONFID e: Confidential
CONFIDAL e: Conjugate Filter Data Link
CONFIG e: Configuration
CONFIRM System e: Conversional File Information Retrieval and Management System
CONFLEX e: Conditional Reflex
CONFORM e: Constrained Force Model
- e: Contract Formulation
Conf Proc e: Conference Proceedings
CONF PROC e: Configuration Processor
CONFRAVISION e: Conference and Television
CONFT e: Confinement
CONG e: Congestion
- e: Congregation
CONGA e: Concept Game
CONGEN e: Congential
CONGLOM[S] e: Conglomerate[s]
- e: Conglomerateur[s]
- e: Conglomerator[s]
CONGR e: Congruent
CONGR[EC] e: Congressional Record
CONGRESS e: Continuous Node Group Restoral Supervision and Switching
CONICS e: Conic Sections
CONILLUM e: Control of Illumination
CONIIS e: Committee on National Library and Information Systems
CONIO e: Console Input/Output
CONIT e: Connector for Networked Information Transfer
CONJ e: Conjugal
- e: Conjugate
- e: Conjugation
- e: Conjunction
- e: Conjunctive
- e: Conjunctivitis
CONL e: Constructor Lieutenant
CONLAN e: Consensus Language
CONLCR e: Constructor Lieutenant Commander
CONMAN e: Console Manager
CONN e: Connect[ed]
- e: Connecting
- e: Connection
- e: Connective
- e: Connector
- e: Connotation
CONN ACK e: Connect Acknowledge
CONN BD e: Connection Board
CONN DIAG e: Connection Diagram
CONNECT e: Connecticut On Line Enforcement Communication and Teleprocessing
CONNECTRONICS e: Connector Electronics
Conn Med e: Connecticut Medicine (journ.)
CONNROD e: Connecting Rod
CONNS e: Connectors
CONNW e: Connected With
CONO e: Confirmation of Number of Order
CO-NO e: Current Operator-Next Operator
CONOBJTR e: Conscientious Objector
CONOCO e: Continental Oil Company
CONOPPR e: Continental Army Command Operating Program[me]
CONOPS e: Continental [United States] Operations
CONOPT e: Constrained Optimization
CONPASP e: Construction Project Alternative Selection Program[me]
- e: Continue Present Duty
CON PL e: Connector Plug
CONQUEST e: Controlled Quality Established
CON R e: Contact Reconnaissance
CONRAD e: Committee on Radiology
- e: Computerized National Range Documentation
- e: Contour Radar Data
- e: Conversational Online Realtime Algorithm Definition
CONREDS e: Contingency Readiness System
CONREP e: Continental Army Command Emergency Relocation Plan
- e: Conected Replenishment
CONROD e: Connecting Rod
CONS e: Carrier Operated Noise Suppression
- e: Connection Oriented Network Service
- e: Consecutive
- e: Conservation
- e: Conservative
- e: Consider
- e: Consigned
- e: Consignees
- e: Consignment
- e: Consist
- e: Console
- e: Consolidated
- e: Consolidation
- e: Consonant
- e: Constable
- e: Constitutional
- e: Construction
- e: Consult[ing]
cons. l: conserva
CONSAL e: Congress of Southeast Asian Librarians
CONSCAN e: Conical Scan
CONSD e: Considerable Damage
CONSEC e: Conic Section
Cons Eng e: Consulting Engineer
CONSER e: Conversion of Serials

CONSER Program[me] e: Consolidated Serials Program[me]
- e: Conversation of Serial Records Program[me]
CONSER Project e: Consolidation of Serial Data Bases Project
CONSERV e: Conservation
- e: Conservatory
CONSGT e: Consignment
CONSHELF e: Continental Shelf
CONSID e: Consideration
CONSIM e: Console Simulator
CON SKT e: Connector Socket
CONSL e: Constructor Sub-Lieutenant
CONSOL e: Consolidated
CONSORT e: Conservation Systems with On-Line Remote Terminals
- e: Control Network System of Receivers and Transmitters
- e: Conversation[al] Systems with Online Remote Terminals
CON SPEC e: Construction Specification
CONSS e: Constrained Step Size
CONSSECT e: Conic Section
CONSSP e: Constant Speed
CONSSTOCS e: Contingency Support Stocks
CONST e: Constant
- e: Constituency
- e: Constitute
- e: Constitution[al]
- e: Construct[ion]
- e: Constructor
CONSTAL e: Constant Air Line
CONSTIT e: Constituent
- e: Constitution[al]
CONSTL e: Constitutional
CONSTN e: Constitution
- e: Construction
CONSTR e: Constraint
- e: Constructed
- e: Construction
- e: Constructor
CONSTRANS e: Conceptual Thought Random Net Simulation
CONSTRUCT e: Construction[al][ly]
CONSTSP e: Constant Speed
CONSUB e: Continental-Shelf Submersible
CONSUL e: Control Subroutine Language
CONSULT e: Consultant
Consumer Bul e: Consumer Bulletin (journ.)
Consumer Rep e: Consumer Reports (journ.)
CONSV e: Conservation
- e: Conserve
CONT e: Contact
- e: Container
- e: Containing
- e: Contaminate[d]
- e: Contamination
- e: Contemporary
- e: Content[s]
- e: Continent[al]
- e: Contingent
- e: Continuation
- e: Continue[d]
- e: Continuity
- e: Continuous[ly]
- e: Continuum
- e: Contract[or]
- e: Contrary
- e: Control[ler]
cont. l: contra
CONTA e: Control Assembly
CONTAC e: Continuous Action
CONTACT e: Control and Action
CONTACT Array e: Conform Tactical Array
CONTAG e: Contagious
CONTAM e: Committee on Nationwide Television Audience Measurement
CONTAM e: Contaminant
- e: Contaminate[d]
- e: Contamination
CONT CKT e: Control Circuit
CONTD e: Contained
- e: Contaminated
- e: Continued
CONTDVD e: Continental Divide
CONTEM e: Contemplate
CONTEMP e: Contemporary
Contemp Agric e: Contemporary Agriculture (journ.)
Contemp Phys e: Contemporary Physics (journ.)
CONTEUR e: Continental Europe
CONTEUR & BRI e: Continental Europe and British Isles
CONTG e: Containing
- e: Contingency
- e: Contracting
CONTGB e: Contributing
CONTH e: Continue to Hold
CONT HP e: Continental Horsepower
CONTIC e: Continental Army Command Intelligence Center
CONTIN e: Continental
- e: Continued
CONTL e: Continental
- e: Control
CONTMAINSTR e: Continental Mains Transformer
CONTN e: Contamination
- e: Continuation
CONTNCL e: Continuation Clause
CONTNR[S] e: Container[s]
CONTR e: Container
- e: Contract[ed]
- e: Contraction
- e: Contractor
- e: Control[ler]
CONTRA e: Contraceptive
CONTRAIL e: Condensation Trail
CONTRAN e: Control Data Corporation Formula Translation
- e: Control Translator
CONTRANS e: Conceptual Thought Random Net Simulation
CONTRAPROP e: Contra and Propeller
- e: Contra-Rotating Propellers
CONTRAS e: Contraceptives
CONTREAT e: Continue Treatment
Contr Eng e: Control Engineering (journ.)
CONTREQS e: Contingency Transportation Requirements System
CONTRIB e: Contribution
- e: Contributor
Contribut Mineral Petrol e: Contributions to Mineralogy and Petrology (journ.)
Contr Instrum e: Control and Instrumentation (journ.)
CONTRL e: Control
CONTROLAT e: Contralateral
CONTRS e: Contractors
CONTR VOLT e: Control Voltage
CONTRX e: Control Termination Racks
CONTS e: Contacts
CONTS e: Contents
CONT Schiff d: Containerschiff
CONTSHELF e: Continental Shelf
CONT SURG e: Contract Surgeon
CONT Term e: Containerterminal
CONTW e: Continuous Window
CONTWD e: Control Word
CONU e: Control Unit
CONURB[S] e: Conurbation[s]
CONUS e: Continental United States
CONUS OTH e: Continental United States over-the-Horizon
CONUS OTH-B e: Continental Unites States over-the-Horizon-Backscatter
CONV e: Convalescent
- e: Convenient
- e: Convention[al]
- e: Convergence
- e: Convergent
- e: Conversation[al]
- e: Conversion
- e: Convert
- e: Converter
- e: Convertible
- e: Conveyor
CONVAIR e: Conveyed by Air
CONVCE e: Conveyance
CONVD e: Converted
CONVDD e: Converted Destroyer
CONVENCL e: Convector Enclosure
CONVERT e: Convertible
CONVEX e: Connecticut Valley Power Exchange
CONVEX e: Convoy Exercise
CONVG e: Convergence
CONVL e: Conventional
CONVN e: Convenient
CONVR e: Converter
CONVREC e: Convention Record
CONVS e: Conveyances
CONVT e: Convert[ible]
CONWR e: Crab Orchard National Wildlife Refuge
COO e: Chief Ordnance Officer
- e: Controller of Ordnance
COOC e: Contact with Oil or Other Cargo
COOH e: Carboxyl Group
COOK e: Cookery
COOKI e: Coordinated Keysort Index
COOL e: Checkout Oriented Language
- e: Control Oriented Language
- e: Coolant
COOL ALGOL e: Common Olympic Algorithmic Language
COOLER e: Conversion of Leak Rates
COOLNT e: Coolant
COOLS e: Community Office On-Line System
COOLTAINER e: Cool Container
COOP e: Continuity of Operations Plan
- e: Cooperate
- e: Cooperation
- e: Cooperative
COOPLAN e: Continuity of Operations Plan
COOP System e: Computer Operating System
COORD e: Coordinate[d]
- e: Coordinating
- e: Coordination
- e: Coordinator
COORS e: Communications Outage Restoration Section
COOS e: Chemical Orbit-to-Orbit Shuttle
COO Signal e: Change-over Order Signal
COP d: kopieren
COP e: Calculator-Oriented Processor
- e: Cam Operated Plunger
- e: Capillary Osmotic Pressure
- e: Capsule Observation Panel
- e: Career Opportunuities Program[me]

- e: Casing Operating Pressure
- e: Celescope Optical Package
- e: Central Operators Panel
- e: Central Ordering Point
- e: Character Oriented Protocol
- e: Checkout of Program[me]
- e: Coaxial Output Printer
- e: Coefficient of Performance
- e: Coherent Optical Processor
- e: Combat Outpost
- e: Command Objective Plan
- e: Commissary Operating Program[me]
- e: Common Online Package
- e: Communications Output Printer
- e: Computer-Oriented Programming
- e: Computer Optimization Package
- e: Continuity of Operations Plan
- e: Control of Process
- e: Control Optimization
- e: Control Optimizing Program[me]
- e: Control Ordering Point
- e: Control Oriented Processor
- e: Copernican
- e: Copilot
- e: Copper
- e: Coptic
- e: Copy
- e: Copyright
- e: Cost Performance
- e: Count Oriented Protocol
- e: Customer Order Processing
- e: Customer Owned Property
- e: Customs of Port

COPA s: Compañía Panameña de Aviación
COPAC e: Committee on Publications and Communication
- e: Computer-Oriented Patient Accounting Control
- e: Continuous Operation Production Allocation and Control

COPAMS e: Commonwealth of Pennsylvania Air Monitoring System
COPAN e: Command Post Alerting Network
COPANT s: Comisión Panamericann de Normar Tecnicas
COPAR e: Computerized Operational Audit Routine
COPARS e: Contractor-Operated Automotive Parts Store
COPART e: Counterpart
COPAS e: Council of Petroleum Accountants Societies
COP Car e: Circuit Observation Patrol Car
COPD e: Chronic Obstructive Pulmonary Disease
- e: Coppered

COPDAF e: Continuity of Operations Plan; Department of the Air Force
COPE e: Carbon Monoxide Pollution Experiment
- e: Career Oriented Preparation for Employment
- e: Cassette Operating Executive
- e: Centralized Automatic Message Accounting Operator Position Exercise
- e: Centralized Online Processing Environment
- e: Chronic Obstructive Pulmonary Emphysema
- e: Committee on Parenthood Education
- e: Communications Oriented Peripheral Equipment
- e: Communications Oriented Processing Equipment
- e: Computer Operator Proficiency Examination
- e: Concepts of Postal Economics
- e: Congress on Optimum Population and Environment
- e: Console Operator Proficiency Examination
- e: Council on Pollution and Environment
- e: Council on Population and Environment
- e: Currency Overprinting and Processing Equipment
- e: Custodian of Postal Effects

COPEC s: Compañía Petrolera Chilena
COPED e: Cooperative Project for Educational Development
COPES e: College Occupational Programs Educational System
- e: Committee on Program[me]Evaluation and Support
- e: Community-Oriented Program[me][s] Environment Scale
- e: Computer Oriented Purchasing and Engineering System
- e: Conceptually Oriented Program[me] in Elementary Science

COPH e: Congress of Organizations of the Physically Handicapped
COPI e: Computer Oriented Programmed Instruction
COPIC e: Computer Program[me] Information Center (or Centre)
COPICS e: Communications-Oriented Production Information and Control System
- e: Copyright Office Publication and Interactive Cataloging [System]

COPL e: Council of Planning Librarians
COPM e: Computer Output Microfiche
COPO e: Copolymer
COPODS e: Concrete Polymer Data System
COPP e: Cobaltiprotoporphyrin
- e: Computerized Optical Prescription Processing
- e: Conservation Organization Protesting Pollution

COPPERAS e: Ferrous Sulfate
- e: Green Vitriol

COPP[L] e: Copper Plate
COPPR e: Copy, Punch, Print
COPPS e: Committee on Power Plant Siting
- e: Computer-Oriented Police Planning System

COPR e: Center for Overseas Pest Research
- e: Centre for Overseas Pest Research
- e: Computerized Outside Plant Record[s]

COPR e: Copyright[ed]
COPREC e: Command Post Record Capability
COPREX e: Co-Precipitation X-Ray Fluorescence
COPRL e: Command Operations Priority Requirements List
COPROG e: Coalescence Progress
COPS e: Calculator-Oriented Processor System
- e: California Occupational Reference Survey
- e: Catalytic Optimum Profit-Sharing
- e: Chrysler Optical Processing Scanner
- e: Circuit Order Preparation System
- e: Circuit Order Processing System
- e: Coherent Optical Processing System
- e: Component Placement System
- e: Computer-Oriented Police System
- e: Computerized Optimization Procedure for Stabilisator
- e: Computerized Optimization Procedure for Stabilizer
- e: Control for Operations Programming and Systems
- e: Conversational Problem Solver
- e: Conversion of Production System

COPSAC e: Computer Order Processing and Sales Accounting
COPSAC System e: Computer Order Processing and Sales Accounting System
COPT e: Completed Procedure Turn
- e: Coptic

COPTEC e: Controller Overload Prediction Technic
COPTER[S] e: Helicopter[s]
COPTRAN e: Communication Optimization Program[me] Translator
COPU e: Copulate
- e: Copulation
- e: Copulatory

COPUOS e: Committee on the Peaceful Uses of Outer Spaces
COPY e: Copyright
COQ e: Certificate of Qualification
CO Q e: Coenzyme Q
coq. l: coque
coq. in s.a. l: coque in sufficiente aqua
coq. s.a. l: coque secundum artem
COR e: Cardiac Output Recorder
- e: Carrier Operated Relay
- e: Carry-Over Register
- e: Circular of Requirements
- e: Coherent Optical Receiver
- e: Combat Operations Report
- e: Command Register
- e: Comunlcatlons Operatlons Report
- e: Concentric-Orbit Rendezvous
- e: Conditioned Orientation Reflex
- e: Contactor; Running
- e: Copper Oxide Rectifier
- e: Corner
- e: Coronene
- e: Correct[ed]
- e: Correction
- e: Corrective
- e: Correlation
- e: Correlative
- e: Correspondence
- e: Correspondent
- e: Corresponding
- e: Corridor
- e: Curiosity-Orientated Research

cor. l: corpus
CORA e: Code for One-dimensional Reactor Analysis
- e: Coherent Radar Array
- e: Conditioned-Reflex Analog[ue]
- e: Conditioned-Response Analog[ue]

CORAD e: Correlation Radar
CORAL e: Class Oriented Ring Associated Language
- e: Coherent Optical Radar Laboratory
- e: Command Radio Link
- e: Comparsion of Recognition Algorithm[s]
- e: Computer-Online Real-time Applications Language
- e: Correlated Radio Lines
- e: Correlation Radio Link

CORAS d: Computer rationalisiert Schulverwaltung
- e: Corridor Assignment

Cor B e: Corona Borealis
CorBd e: Corner Bead
Cor Bor l: Corona Borealis
CORC e: Conventional Ordnance Release Computer
- e: Cornell Compiler
- e: Cornell Computing
- e: Cornell Computing Language

CORCHRCOL e: Corpus Christi College-Cambridge

CORC Language e: Cornell Computing Language
CORCO e: Commonwealth Oil Refining Company
CORD e: Computer Online Devices
- e: Computer Reinforced Design
- e: Computer with Online Remote Devices
- e: Consortium Research Development
- e: Coordinating of Research and Development
- e: Coordination
CORDA e: Computerized Reservations System Royal Dutch Airlines
CORDAPS e: Correlation Data Processing System
CORDAT e: Coordinate Data
CORDIC e: Coordinate[d] Rotation Digital Computer
CORDP[O] e: Correlated Radar Data Printout
CORDPO-SORD e: Correlated Radar Data Printout Separation of Radar Data
CORDS e: Civil Operations and Revolutionary Development Support
- e: Coherent-on Rescue Doppler System
- e: Coherent on Receive Doppler System
- e: Coordination of Record and Data Base System
CORE e: Coherent-on Receive
- e: Competitive Operational Readiness Evaluation
- e: Computer-Oriented Reporting Efficiency
- e: Computer-Related Equipment
- e: Cooperative Research Institute
- e: Corrected Overall Reference Equivalent
CORELAP e: Computerized Relationship Layout Planning
COREP e: Combined Overload Repair Control
COREP Control e: Combined Overload Repair Control
COReS e: Canadian Operations Research Society
CORE System e: Comprehensive Reporting System
COREX e: Coordinated Electronic Counterrneasures Exercise
CORF e: Commanding Officer; Reserve Fleet
- e: Comprehensive Outpatient Rehabilitation Facility
CORFAM e: Microporous Artificial Leather
CORFLU e: Correction Fluid
CORIL e: Caltex Oil Refining Limited
CORIMS e: Committee on Research in the Mathematical Sciences
CORKS e: Computer-Oriented Record Keeping System
CORLS e: Central Ontario Regional Library System
CORM e: Council for Optical Radiation Measurement[s]
Cor Mem e: Corresponding Member
CORNAP e: Computerized Network Analysis Program[me]
Cornell U Pr e: Cornell University Press
Cornell Vet e: Cornell Veterinarian (journ.)
Cornell Vet Suppl e: Cornell Veterinarian; Supplement (journ.)
CORNET e: Control Switching Arrangement Network
- e: Corporate Network
Corning Mus e: Corning Museum of Glass
CORNN e: Coronary
CORODIM e: Correlation of the Recognition of Degradation with Intelligibility Measurements
COROLL e: Corollary
CORON e: Coronary
CORONA e: Control Rod Analysis
COROS e: Copper Oxide Read-Only Storage
CORP e: Corporation
CORPAL e: Control Room Patching and Labeling
CORPIS e: Corporate Planning Information System
CORPORAL e: Corporate Resource and Allocation
CORPPIN e: Corporeal Pin
CORR e: Correct[ed]
- e: Correction
- e: Correlation
- e: Correlative
- e: Correspond[ing]
- e: Correspondence
- e: Corrosion
- e: Corrugate[d]
CORRD e: Corrected
CORRECT e: Customized Optical Reader Random Error Correction Technique
CORREGATE e: Correctable Gate
CORREL e: Correlated
- e: Correlation
- e: Correlative
CORRES e: Corresponded
CORRESP e: Correspondence
- e: Corresponding
CORR FBD e: Corrugated Fiberboard
Corr Memb e: Corresponding Member
CORRN e: Correction
CORROS e: Corrosion
- e: Corrosive
CORROSIVE e: Sublimate Mercuric Chloride
Corros Sci e: Corrosion Science (journ.)
CORS e: Canadian Operational Research Society
- e: Carrier Operated Relays
CORSA e: Cosmic Ray Satellite
- e: Cosmic Radiation Satellite
CORSAIR e: Computer-Oriented Reference System for Automatic Information Retrieval
CORSPERS e: Committee on Remote Sensing Programs for Earth Resource Survey
CORT e: Cortex
- e: Cortical
- e: Council On Radio and Television
CORTET d: Common Business-Oriented Language- orientierter Entscheidungs-tabellenvorübersetzer
CORTEX e: Communications Oriented Real-Time Executive
- e: Computer-based Optimization Routines and Techniques for Effective X
CORTS e: Conversion of Range Telemetry Systems
- e: Convert Range Telemetry System
CORUNDUM e: Aluminum Oxide
COS d: Cosinus
- d: Kohlenoxysulfid
- d: Kosinus
- e: Calculator on Substrate
- e: Canadian Ophthalmological Society
- e: Carbonyl Sulphide
- e: Card Operating System
- e: Card Oriented System
- e: Card Output Station
- e: Carry-Over Storage
- e: Cassette Operating System
- e: Centralized Operating System
- e: Change of Subscribers
- e: Character Organized Store
- e: Chief of Section
- e: Civilian Occupational Speciality
- e: Class of Service
- e: Class of Subscriber
- e: Coast of Scotland
- e: Commanding Officer of Ship
- e: Commercial Operating System
- e: Communication[s]-Oriented System
- e: Communication[s] Operating System
- e: Communication[s] Operation Station
- e: Compact Operating System
- e: Compatibility Operating System
- e: Compatible Operating System
- e: Complementary Switching
- e: Complementary Symmetrically
- e: Complementary Symmetry
- e: Conceptual Operational System
- e: Concurrent Operating System
- e: Contactor; Starting
- e: Controller of Overseas Supplies
- e: Cooper Ornithological Society
- e: Corporation for Open Systems
- e: Cosine
- e: Cosmic
- e: Cosmogony
- e: Cosmography
- e: Cosmology
- e: Crack Opening Stress
- e: Cray 1 Operating System
- e: Current Output Station
- e: Customers Other Service
- e: Cut-Off Signal
- e: Cutoff Signal
COSA e: Combat Operational Support Aircraft
- e: Conditional Sum Adder
COSACS e: Computer Operations Scheduling Accounting and Control System
- e: Computer Oriented Scheduling and Control System
COSAD e: Classroom Observation System for Analyzing Depression
COSAG e: Combination of Steam and Gas
- e: Combined Steam and Gas [turbine]
COSAL e: Coordinated Shipboard Allowance List
- e: Coordinated Shorebased Allowance List
COSAM e: Co-Site Analysis Model
- e: Co-Site Analytical Model
- e: COBOL Shared-Access Method
- e: Cosine Amplitude
COSANOSTRA e: Computer-Oriented System and Newly Organized Storage-to-Retrieval Apparatus
COSAP e: Cooperative Serials Acquisitions Program[me]
COSAR e: Compression Scanning Array Radar
COSATI e: Committee on Scientific and Technical Information
COSBA e: Computer Service[s] Bureaux Association
- e: Computer Services and Bureaus Association
COSC e: Combat Operations Specialist Course
- e: Cosecant
COSCL e: Common Operating System Control Language
COSD e: Central Ordnance Supply Depot
- e: Chief of Overseas Supply Division
COSEBI s: Corporación de Servicios Bibtiotecarios

COSEC d: Kosekans
- e: Cosecant
- e: Culham Online Single Experimental Console
COSECH e: Hyperbolic Cosecant
Co-Sel e: Co-Selector
COSERS e: Computer Science and Engineering Research Study
COSFAD e: Computerized Safety and Facility Design
COSG e: Consolidated Nuclear Steam Generator
COSH d: Hyperbelkosinus
- e: Hyperbolic Cosine
COSHI e: Clearinghouse for Occupational Safety and Health Information
COSHTI e: Council for Science and Technological Information
COSI e: Center of Science and Industry
- e: Committee on Science Information
- e: Committee on Scientific Information
COSI-KON e: Crimp-on; Snap-in Contacts
COSINE e: Committee on Computer Science in Electrical Engineering Education
- e: Computer Science in Electrical Engineering
- e: Cooperation for Open System Interconnection Networking in Europe
COSINF e: Convert Source Information
COSIP d: Computer-System für industrielle Prozeßsteuerung
- e: College Science Improvement Program[me]
COSMA d: Computerselektion mit Matrizenmodellen
- e: Computer[ized] Services for Motorfreight Activities
COS Making e: Continuous Oxygen Steel Making
COSMIC e: Command Operations Simulation Model with Interrogation Control
- e: Common Systems Main Interconnecting
- e: Computer-Oriented Synthesis and Mask-Generation for Integrated Circuits
- e: Computer Operated Sequential Memory and Integrated Circuit[s]
- e: Computer Programmes Information Center
- e: Computer Software Management and Information Center
- e: Computer System for Management Information and Control
COSMIS e: Computer System for Medical Information Services
COSMO d: Computer-Stundenplan-Modell
- e: Communications Simulation Model
COSMOG e: Cosmogony
- e: Cosmographical
- e: Cosmography
- e: Cosmography Cosmograph[s]
COSMON e: Component Open-Short Monitor
COSMONAUT e: Cosmonautic[al][ly]
- e: Cosmonautics
COSMOS e: Centralization of Supply Management and Operations System
- e: Coast Survey Marine Observation System
- e: Commercial Systems using Modular Software
- e: Complementary Symmetric[al] Metal Oxide Semiconductor
- e: Complementary Symmetry Metal Oxide Semiconductor
- e: Computer-Oriented System for Machine Order Synthesis
- e: Computer-Oriented System for Management Order Synthesis
- e: Computer-Oriented System for Manufacturing Order Synthesis
- e: Computer Optimization and Simulation Modelling for Operating Supermarkets
- e: Computer System for Mainframe Operations
- e: Console Oriented Statistical Matrix Operator System
- e: Courtaulds Own System for Matrix Operations and Statistics
COSNOSTRA e: Computer-Oriented System Newly Organized Storage-to-Retrieval Apparatus
COSOS e: Combined Sorting System
- e: Conference on Self-Operating Systems
COSP e: Central Office Signalling Panel
COSPAR e: Committee on Space Research
COSPUP e: Committee on Science and Public Policy
COSR e: Committee on Space Research
- e: Cut-Off Shear
COSRO e: Conical Scan on Receive Only
- e: Conical Scan Receiver Only
COSS e: Com-Shares Online Simulation System
- e: Controller of Office Services and Supplies
COSSA e: Consortium of Social Science Associations
COSSAC e: Computer Systems Suppliers Advisory Committee
- e: Cooled Spectral Sharedaperture Concept
COSSCO-Relationship e: Coordinate-Superordinate-Subordinate-Collateral-Relationship
COSSECC e: Connecticut Suburban and Shoreline Educational Computer Center
COSSO d: Computerservice und Software
- e: Computerservice and Software
COST e: Computer-Operated Sequential Tester
- e: Congressional Office of Science and Technology
- e: Contaminated Oil Settling Tank
- e: Cooperation in Science and Technology
- e: Cost-Oriented Systems Technique
- f: Groupe d'Experts "Coopération dans le Domaine de la Recherche Scientifique et Technique"
COSTAR e: Composition Storage and Retrieval
- e: Computer Stored Ambulatory Record
- e: Conversational On-line Storage and Retrieval
- e: Conversational Online Storage and Retrieval
COSTED e: Committee on Science and Technology in Developing Countries
COSTEI e: Central Office of Scientific, Technical and Economic Information
COSTI e: Center of Scientific and Technological Information
- e: Centre for Scientific and Technological Information
- e: Committee on Scientific and Technical Information
COST Project e: Cooperation in Science and Technology Project
COSU e: Computing Surveys (journ.)
COSWAP e: Coaxial Switch and Alternator Panel
CO Switch e: Closed Open Switch
COSWS e: Chief Officer; Shore Wireless Service
COSX e: Control of Optimizing Series of Experiments
COSY d: computergesteuertes Satzprogramm-System
- e: Checkout Operating System
- e: Common Business-Oriented Language Optimizing System
- e: Compiler System
- e: Compressed Symbolic
COSYPLAN d: computergestütztes System Planung
COSYWOG e: Communications System[s] Working Group
COT d: Kotangens
- d: Computer Original Equipment Manufacturer Trading GmbH
- e: Card-Operated Typewriter
- e: Card or Tape Reader
- e: Cathodal Opening Tetanus
- e: Cathode Opening Tetanus
- e: Central Office Terminal
- e: Check-Out Time
- e: Chief of Transportation
- e: Class of Traffic
- e: Coated Open Tabular
- e: Committee on Terminology
- e: Common Output Tape
- e: Computer-Output-Typesetting
- e: Consecutive Oversea Tour
- e: Consolidated Operability Trial
- e: Contralateral Optic Tectum
- e: Cotangent
- e: Cotton
- e: Customer-Oriented Terminal
- e: Cyclo-Octatetraene
CO_2 d: Kohlendioxid
- e: Carbon Dioxide
COTA e: Cinetheodolite Orientation Target Array
- e: Confirming Telephone Authority
COTAR e: Correlation Tracking and Ranging
- e: Cosine-Trajectory Angle and Range
COTAR-AME e: Correlation Tracking and Ranging-Angle Measuring Equipment
COTAR-DAS e: Correlation Tracking and Ranging-Data Acquisition System
COTAR-DME e: Correlation Tracking and Ranging-Data Measuring Equipment
COTAR System e: Correlated Orientation Tracking and Ranging System
- e: Correlation Tracking and Ranging System
COTAT e: Correlation Tracking and Triangulation
COTAWS e: Collision, Obstacle, Terrain and Warning System
COTC e: Canadian Overseas Telecommunication Corporation
- e: Canadian Overseas Telecommunications Corporation
- e: Class-of-Traffic-Check
COT Character e: Class-of-Traffic Character
COTE e: Cathodal Opening Tetanus
COTG d: Kotangens
- e: Component Tooling Gage
coth d: Hyperbelkotangens
- e: hyperbolic cotangent
COTI e: Central Officials Training Institute
COTICS e: Narcotics
COTM e: Customer Owned and Telephone-Company Maintained
COTMS e: Computer-Operated Transmission Measuring Set
COTP e: Captain of the Port
COTR e: Cockpit Orientation Trainer
- e: Contracting Officers Technical Representative

COTRAN e: Code Transformation
COTRA[N]S e: Coordinated Transfer Application[s] System
COTS e: Checkout Test Set
- e: Connection Oriented Transport Service
COTT e: Central Organization for Technical Training
- e: Computer-Operated Thermal Tester
COU e: Computer Operating Unit
- e: Coupler
COUL e: Coulomb
COUNABB e: Commanding Officer; United States Navy Advanced Base Bremerhaven
COUNBIOLOGYEDS e: Council of Biology Editors
COUNT e: Computer Operated Universal Test
COUNT System e: Computer-Operated Universal Test System
COUP e: Coupler
- e: Coupling
COUPL e: Coupling
COUPLE e: Communications-Oriented User Programming Language
COURT e: Cost Optimization Utilizing Reference Techniques
COUSARS e: Commanding Officer; United States Army Radio Station
COUSNAB e: Commander; United States Naval Advanced Bases
- e: Commanding Officer; United States Navy Air Base
COUSNAVSUPPFOR e: Commander; United States Naval Supply Force
COUTS e: Computer Operated Universal Test System
COV e: Calculus of Variation
- e: Check-Out Valve
- e: Coefficient of Variation
- e: Concentrated Oil of Vitriol
- e: Counter-Operating Voltage
- e: Covariance
- e: Cover[ed]
- e: Cross-Over Value
- e: Cross-Over Valve
- e: Cut-Off
- e: Cut-Off Voltage
- e: Cut-Out Valve
- e: Cylinder Operated Valve[s]
COVAM e: Capture Orbit Vehicle Assembly Mode
COVAR e: Covariance
COVER e: Cut-Off Velocity & Range
COVERS e: Coversed Sine
COVFF e: Coverings; Factor or Floor
COVINCA s: Corporación Veneolana de la Industria Naval
COVPL e: Cover Plate
COV Signal e: Change-over Signal
COW e: Chlorinated Organics in Wastewater
- e: Crude Oil Washing
COWAMP e: Comprehensive-Water-Quality Management Planning
COWAR e: Committee on Water Research
- e: Committee on Water Resources
COWL e: Cowling
COWLEX e: Cold-Weather Landing Exercise
COWOCOR d: Computer Wort Codierungs- und Recherche [-Dokumentationssystem]
COWRR e: Committee on Water Resources Research
COWSHED e: Complete Optoelectronic with Silicon High Efficiency Detectors
COXCBS e: Coxswain; Construction Battalion; Stevedore
COX[N] e: Coxswain
COXSRC e: Coxswain; Ship Repair; Canvasman
COXSRR e: Coxswain; Ship Repair; Rigger
COZ d: Karbonylzahl
- e: Communications Zone
COZI e: Communications Zone Indicator
CP d: Cellulosepropionat
- d: chronische Polyarthritis
- d: computergesteuerte Programmierhilfe
- d: Computer Praxis (journ.)
- d: Creatinphosphat
- d: kontinentale Polarluft
- d: Zentipoise
- e: Cable Party
- e: Calculating Punch[er]
- e: Calculator; Printing
- e: Call on Positive
- e: Call Proceeding
- e: Called Program[me]
- e: Call[ing] Program[me]
- e: Call Progress
- e: Caloric Power
- e: Calorific Power
- e: Canadian Pacific
- e: Candle Power
- e: Cap
- e: Capacitance
- e: Capillary Pressure
- e: Carbamyl Phosphate
- e: Cardinal Point
- e: Card Perforator
- e: Card Printing
- e: Card Processing
- e: Card Punch[er]
- e: Card to Printer
- e: Card to Punch
- e: Carotid Pressure
- e: Carotid Pulse
- e: Carport
- e: Carriage Paid
- e: Carrier Packed
- e: Cartesian to Polar
- e: Cataloged Program[me]
- e: Catch Phrase
- e: Cathode Polarization
- e: Cathode Pulsed
- e: Cellulose Propionate
- e: Center of Pressure
- e: Center Punch
- e: Centipoise
- e: Centrally Procured
- e: Central Problem
- e: Central Processor
- e: Centration Point
- e: Centre of Pressure
- e: Cerebral Palsy
- e: Chalcophyrite
- e: Change Package
- e: Change Point
- e: Change Program[me]
- e: Change Proposal
- e: Channel Program[me]
- e: Character Printer
- e: Character Printing
- e: Charge Parity
- e: Check Parity
- e: Check Point
- e: Check Problem
- e: Check Program[me]
- e: Check Protection
- e: Chemically Polished
- e: Chemically Pure
- e: Chemical Practitioner
- e: Chemical Product
- e: Chloropurine
- e: Chloroquinine and Primaquine
- e: Chronic Pyelonephritis
- e: Circuit Package
- e: Circularly Polarized
- e: Circular Pitch
- e: Circular Polarization
- e: Circular[y] Polarized
- e: Circumpolar
- e: Classical Philology (journ.)
- e: Clinical Pathology
- e: Clock Phase
- e: Clock Program[me]
- e: Clock Pulse
- e: Closed Position
- e: Closed Program[me]
- e: Close Packed
- e: Closing Pressure
- e: Co-Pilot
- e: Coal Port
- e: Cochlear Potential
- e: Code[s] of Practice (British Standard)
- e: Coded Program[me]
- e: Code of Practice
- e: Code of Procedure
- e: Code Pattern
- e: Code Position
- e: Code Proficiency
- e: Code Program[me]
- e: Coding Printer
- e: Coefficient of Performance
- e: Collision Probability
- e: Column Pitch
- e: Combination Product
- e: Combined Protocol
- e: Combining Power
- e: Command Pilot
- e: Command Processor
- e: Command Pulse
- e: Common Pointed
- e: Common Process
- e: Communication[s] Processor
- e: Community Physician
- e: Compare
- e: Comparison
- e: Compiling Program[me]
- e: Complementary Program[me]
- e: Complex Programming
- e: Component Part[s]
- e: Composition Printing
- e: Compound
- e: Compressed
- e: Computed Point
- e: Computer
- e: Computer Printing
- e: Computer Program[me]
- e: Computer Programming
- e: Computing Problem
- e: Computing Program[me]
- e: Concrete Piercing
- e: Conditional Probability
- e: Conditional Proof
- e: Conductive Plastic
- e: Conference Paper
- e: Conference Publication
- e: Connection Pending
- e: Connection Point
- e: Connector Panel
- e: Consecutive Processing
- e: Constant Pitch
- e: Constant Potential
- e: Constant Power
- e: Constant Pressure
- e: Constant Property
- e: Construction Permit
- e: Contact Proficiency
- e: Contention Procedure
- e: Continental Polar

- e: Continuous Path
- e: Contractile Pulse
- e: Controllable Pitch
- e: Controlling Program[me]
- e: Control Panel
- e: Control Part
- e: Control Point
- e: Control Print
- e: Control Procedures
- e: Control Processor
- e: Control Program[me]
- e: Control Programmer
- e: Conversion Program[me]
- e: Converter Program[me]
- e: Coordination Processor
- e: Core Prime
- e: Corpulmonale
- e: Corrected Program[me]
- e: Correspondence Printer
- e: Corrosion Protection
- e: Cost Plan
- e: Cost Proposal
- e: Counterpoise
- e: Counterpreparation
- e: Coupling
- e: Cover Point
- e: C Pulse
- e: Creatine Phosphate
- e: Crude Protein
- e: Crystalline or Powdered
- e: Current Paper
- e: Cushioning Pad[s]
- e: Customized Processor
- e: Custom of the Port
- e: Custom Processor
- e: Cuticular Plate
- e: Cyclic Program[me]
- e: Cylindrical Perforated
- e: Cystoscopy Pyelogram
- e: Polar Continental (Air)
- f: Centre Pénitentiaire

C + P e: Cystoscopy and Pyelogram
Cp l: Cassiopeium
CP-1 e: Compatibility Processor - 1
CPA d: Cyclophosphamid
- e: Called Program[me] Analysis
- e: Canadian Pacific Airlines
- e: Canadian Petroleum Association
- e: Canadian Psychological Association
- e: Carboxypeptidase A
- e: Carry Propagate Adder
- e: Cattle-Prod Approach
- e: Central Pacific Area
- e: Certified Public Accountant
- e: Channel Program[me] Area
- e: Chartered Patent Agent
- e: Circuit Parameter Analysis
- e: Closest Point of Approach
- e: Closest Position of Approach
- e: Co-Polar Attenuation
- e: Coherent Potential Approximation
- e: Colour Phase Alternation
- e: Commutative Principle for Addition
- e: Component Processing Area
- e: Computer Architecture
- e: Computer Performance Analysis
- e: Computer Power Australia
- e: Computer Press Association
- e: Concrete Pipe Association
- e: Constant Potential Accelerator
- e: Controlled Products Area
- e: Controlled Pushdown Automaton
- e: Control of Pollution Act
- e: Control Program[me] Assist
- e: Cost, Planning and Appraisal
- e: Critical Path Analysis
- e: Cross-Program[me] Auditor
- e: Cryophylactic Agent

CPAA e: Current Physics Advance Abstracts (journ.)
CPA Antenna e: Conformal Phased Array Antenna
CPAB e: Computer Programmer Aptitude Battery
CPAC e: Center for Protection Against Corrosion
- e: Collaborative Pesticides Analytical Committee
- e: Computer Program[me] Associate Contractor
- e: Corrosion Prevention Advisory Center

C_{pac} e: Course per Standard Compass
CPACS e: Coded Pulse Anti-clutter System
CPAF e: Cost Plus Award Fee
C_{pah} e: Para-Aminohippurate Clearance
cP Air e: Continental Polar Air
CPAM e: Chief of Naval Operations Program[me] Analysis Memorandum
- e: Continental Polar Air Mass

CPAP e: Constant Positive Airway Pressure
- e: Continuous Positive Airway Pressure
- e: Control Parameter Assembly Program[me]

CPAR e: Cooperative Pollution Abatement Research
CPARS e: Compact Programmed Airline Reservation System
CPAU e: Computers and Automation (journ.)
CPAWS e: Computer-Planning and Aircraft-Weighting Scales
- e: Computer Planner and Aircraft Weighting Scales

CPB e: Carboxypeptidase B
- e: Cardiopulmonary Bypass
- e: Central Processor Bus
- e: Cetyl Pyridinum Bromide
- e: Channel Program[me] Block
- e: Charged-Particle Beam[s]
- e: Coal Port of Bristol
- e: Competitive Protein Binding
- e: Constant Power Ballast
- e: Corporate Programming Bulletin
- e: Corporation for Public Broadcasting
- e: Current Physics Bibliographies

CPBA e: Competitive Protein-Binding Analysis
CPBL e: Capability
- e: Capable

CPBS d: N-Cyclopentamethylen-2-Benzothiazolsulfenamid
- e: 4-Chlorophenyl Benzenesulphonate

CPBX e: Computerized Private Branch Exchange
CPC e: Calibration Procedure[s]
- e: California Polytechnic College
- e: Calling Partys Category
- e: Call Processing Control
- e: Capacitance Plate-Cathode
- e: Card Programmed Calculator
- e: Card Programmed Computer
- e: Cartesian-to-Polar Converter
- e: Cement Plaster Ceiling
- e: Central-Processor-Chip
- e: Central Parts Contractor
- e: Central Processing Complex
- e: Central Property Control
- e: Ceramic-Wafer Printed Circuit
- e: Ceramic Printed Circuit
- e: Cerebro-Pedal Commissure
- e: Cetyl Pyridinium Chloride
- e: Character[s] per Column
- e: Chemical Protective Clothing
- e: China Petroleum Company
- e: Chronic Passive Congestion
- e: City Planning Commission
- e: City Projects Council
- e: Clinicopathological Conference
- e: Clock Pulse[d] Control
- e: Coarse Pumice Concrete
- e: Coated Paper Copier
- e: Coated Powder Cathode
- e: Coated Power Cathode
- e: Codes of Practice Committees
- e: Cogswell Polytechnical College
- e: Colour Pack Camera
- e: Column Position Counter
- e: Commodity Punched Card
- e: Common Pointed; Capped
- e: Communications Processing Center
- e: Compound Parabolic Concentrator
- e: Computer-production Control
- e: Computerized Process Control
- e: Computerized Production Control
- e: Computer Packaging Committee
- e: Computer Planning Corporation
- e: Computer Power Center
- e: Computer Print Control
- e: Computer Process Control
- e: Computer Production Control
- e: Computer Program[me] Component
- e: Computer Program[me] Counter
- e: Computer Programming Concepts
- e: Continuous Progress Center
- e: Core Protection Calculator
- e: Core Protection Computer
- e: Creole Petroleum Corporation
- e: Current Papers on Computers and Control
- e: Cycle Circumspection-Preemption-Control (or Choice)
- e: Cycle Program[me] Control
- e: Cycle Program[me] Counter
- e: Cyclically Permuted Code
- e: Cyclic Permutation Code
- e: Cyclic Program[me] Control
- f: Calculatrice à Programme par Cartes

CPCC e: Communications Processor Conversion Center
CPCEI e: Computer Program[me] Contract End Item
CPCFC e: Critical Parts Centralized Forecast Computation
CPCG s: Comitté Panamericano de Ciencias Geoficicas
CPCGN e: Canadian Permanent Committee on Geographical Names
CPCH e: Calling Party Cannot Hear
CPCH PROP e: Controllable-Pitch Propeller
CPCI e: Canadian Prestressed Concrete Institute
- e: Central Processing Unit Power Calibration Instrument
- e: Computer Program[me] Configuration Item
- e: Cross-Pointer Course Indicator

CPCIA f: Centre de Perfectionnement des Cadres des Industries Agricoles et Alimentaires
CP/CMS e: Control Program[me] / Conversational Monitor System
CPCN e: Canadian Pacific Canadian National [Telecommunications]
CP Code e: Cyclic Permutated Code
CPCOM e: Capsule Communications
CP Control e: Continuous Path Control
CPCS e: Check Processing Control System
- e: Check Processory Central System
- e: Coast Phase Control System
- e: Compressor Protect and Control System
- e: Conversion Process Controller System

- e: Cost Planning and Control System
- e: Cyclic Pitch Control Stick
CP/CS Adder e: Carry Propagate / Carry Save Adder
- e: Carry Propagate/Carry Save Adder
C_{pstgc} e: Course per Steering Compass
CPCU e: Chartered Property and Casuality Underwriter
CP Cycle e: Constant Pressure Cycle
CPD d: Cadmiumpentamethylendithio-karbamat
- e: Calls per Day
- e: Cards per Day
- e: Center Program[me]Director
- e: Central Postal Directory
- e: Central Pulse Distributor
- e: Charterers Pay Dues
- e: Chemical Propulsion Division
- e: Circuit Protection Device
- e: Citrate-Phosphate Dextrose
- e: Coaxial Power Divider
- e: Coil Predriver
- e: Combat Potential Display
- e: Commercial Product Development
- e: Commercial Program[me] Development
- e: Communication Processing Definitions
- e: Communication Products Division
- e: Compound
- e: Computer Produced Drawing
- e: Consolidated Programming Document
- e: Constant Power Density
- e: Contact Potential Difference
- e: Contagious Pustular Dermatitis
- e: Crew Passive Dosimeter
- e: Cumulative Probability Distribution
- e: Cumulative Probability of Detection
- e: Cycle[s] per Day
CPDA e: Council for Periodical Distributors Associations
CPDAMS e: Computer Program[me] Development and Management System
CPDC e: Command Processor Distributor Storage
- e: Computer Program[me] Development Center
CPDD e: Command Post Digital Display
- e: Conceptual Project Design Description
CPDFSK e: Continuous Phase Differential Frequency Shift Keying
CPDGP e: Commission on Present Day Geomorphic Process
CPDI e: Cyclopentane Diimide
CPDL e: Canadian Patents and Developments Limited
CPDM e: Chebycheff Polynomial Discriminant Method
CPDNA e: Cytoplasm Deoxyribonucleic Acid
CPDP e: Computer Program[me] Development Plan
CPDPS e: Clinical Pathology Data Processing System
CPDS e: Carboxypyridinedisulphide
- e: Command Processor Distributor Storage
- e: Compounds
- e: Computerized Preliminary Design System
- e: Computer Program[me] Design Specification
- e: Cross Power Density Spectrum
CPDTA e: Cyclopentanediaminetetraacetic Acid
CPE d: computergestützte Planungs- und Entscheidungshilfen
- e: Cable Pressurization Equipment
- e: Call Parity Even
- e: Cathodic Protection Equipment
- e: Celestial Position Error
- e: Cellulose Polyethylene
- e: Central Processing Element
- e: Central Processor Element
- e: Central Programmer and Evaluator
- e: Certified Professional Ergonomist
- e: Certified Property Exchanger
- e: Charged Particle Equilibrium
- e: Chief Polaris Executive [Missiles]
- e: Chlorinated Polyethylene
- e: Chronic Pulmonary Emphysema
- e: Circular Probable Error
- e: Cloud Processing Equipment
- e: Collective Protection Equipment
- e: College of Physical Education
- e: Column Purified Enzyme
- e: Computer Performance Evaluation
- e: Computer Peripheral Equipment
- e: Continuous Particle Electrophoresis
- e: Contractor Performance Evaluation
- e: Counter Position Exit
- e: Cross-Program[me] Editor
- e: Current Papers in Electrical and Electronics Engineering
- e: Customer Premises Equipment
- e: Customer Provided Equipment
- e: Cytopathic Effects
- e: Cytopathogenic Effect
CPEA e: Cooperative Program[me]for Educational Administration
CPEB e: Cryogenic Positive Expulsion Bladder
CPE Computer e: Celestial Position Error Computer
CPEK e: Contractor Performance Evaluation Plan
CPEM e: Conference on Precision Electromagnetic Measurement
CPEN e: Clock Pulse Enable
- f: Code Pénal
CPEP e: Contractor Performance Evaluation Plan
CPER e: Contractor Personnel Employment Report
CPEs e: Centrally Planned Economics
CPES e: Contractor Performance Evaluation System
CPET e: Charged Particle Electrostatic Thruster
CPEUG e: Computer Performance Evaluation Users Group
CPF e: Capacitance; Plate-Filament
- e: Carbon Paper Feed
- e: Central Processing Facility
- e: Chlorine Pentafluoride
- e: Coded Print File
- e: Complete Power Failure
- e: Conditional Peak Flow
- e: Control Program[me] Facilities
- e: Cumulative Percentage Frequency
- e: Cursor Preceding Field
- e: Customer-Premises Facility
CPFA e: Cyclopropenoid Fatty Acid
CPFR e: Calling Party Forced Release
- e: Continuous Page Facsimile Recorder
CPFRC e: Central Pacific Fisheries Research Center
CPFSK e: Continuous Phase Frequency Shift Keying
CPFT e: Customer-Premises Facility Terminal
CPG e: Candidate Pass Generator
- e: Central Processing Group
- e: Certified Program[me] Generator
- e: Clock Pulse Generator
- e: Common Business Oriented Language Program[me] Generator
- e: Communication Program[me] Generator
- e: Controlled-Pore Glass
- e: Current Pulse Generator
- e: Customer Information Control System Program[me] Generator
CPG BIBLIA e: Computational Physics Group Bibliographic Aid
CPGC e: Course per Gyro Compass
CPGS e: Cross Program[me] Generation System
CPH d: computergesteuerte Programmierhilfe
- e: Candle Power Hour[s]
- e: Cards per Hour
- e: Central Powerhouse
- e: Certificate in Public Health
- e: Characters per Hour
- e: Classical Philology (journ.)
- e: Close[d] Packed Hexagonal
- e: Computer Polarization Holography
- e: Copenhagen (Airport)
- e: Copies per Hour
- e: Counts per Hour
- e: Cycles per Hour[s]
CPH 5 e: Cutter Protein Hydrolysate; 5% in Water
CPHC e: Central Pacific Hurricane Center
CPHE e: Common Personal Hygiene Equipment
CPHERI e: Central Public Health Engineering Research Institute
CPHM e: Chief Pharmacists Mate
CPHMDP e: Chief Pharmacists Mate; Dental Prosthetic Technician
CPHM (RPA) e: Chief Pharmacists Mate (Radium Plaque Adaptometer)
CPHO e: Chief Photographer
CPHOM e: Chief Photographers Mate
CPI e: Cable Pair Identification
- e: California Psychological Inventory
- e: Call Progress Indicator
- e: Card Punch Interface
- e: Center of Programmed Instruction
- e: Central Patents Index
- e: Change Package Identification
- e: Changes per Inch
- e: Characters per Inch
- e: Chemical-Process-Industry
- e: Chemical Processing Industries
- e: Chief Pilot Inspector
- e: Cinchona Products Institute
- e: Clay Pipe Institute
- e: Clock Pulse Interval
- e: Closed Pore Insulation
- e: Clutch Pitching Index
- e: Colour Preference Index
- e: Commercial Performance Index
- e: Committee on Public Information
- e: Computer-to-PBX (Private Branch Exchange) Interface
- e: Computer Peripherals, Incorporated
- e: Computer Prescribed Instruction
- e: Computer Products, Incorporated
- e: Computing Power Index
- e: Conditional Program[me] Interruption
- e: Conference Papers Index
- e: Conference Proceedings Index
- e: Consolidated Plastic Industries
- e: Constitutional Psychophatic Inferior
- e: Control Position Indicator
- e: Coordinator of Program[me] Implementation
- e: Coronary Prognostic Index
- e: Corporate Program[me] Independent
- e: Corrugated Plate Interceptor
- e: Cost per Instruction
- e: Crash Position Indicator
- e: Crop Protection Institute

- e: Current Physics Information (journ.)
- e: Cuts per Inch
CPIA e: Chemical Propulsion Information Agency
- e: Close-Pair Interstitial Atom
CPIAF e: Cost Plus Incentive Award Fee
CPI B e: Chlorophenoxyiso-Butyrate
CPIB Acid e: Chlorophenoxyisobutyric Acid
CPIC e: Charged Particles Information Center
- e: Computer Program[me] Integration Contractor
CPICC e: Consumer Product Information Coordinating Center
CPIF e: Cost Plus Incentive Fee
- e: Cost Plus Intended Fee
- e: Cost Plus Interim Fee
CPILS e: Correlation-Protected Instrument Landing System
- e: Correlation-Protected Integrated Landing System
CPIN e: Change Package Identification Number
- e: Computer Program[me] Identification Number[s]
CPIN e: Crankpin
CPIOCS e: Communication Physical Input Output Control System
CPIP e: Computer Pneumatic Input Panel
- e: Computer Program[me] Implementation Process
CPIs e: Chemical Process Industries
CPIS e: Comprehensive Planning Information System
CPIT e: Contract Price of Items Terminated
- e: Cost Price of the Items Terminated
CPJ e: Conditional Program[me] Jump
- e: Conoseal Pipe Joint
CPK d: Kreatin-Phosphokinase
- e: Creatine Phosphokinase
- e: Plate-Cathode Capacitance
- e: Calgary Public Library
- e: Canadian Program[me] Library
- e: Capability Password Level
- e: Carpel
- e: Cement Plaster
- e: Central Power and Light
- e: Central Program[me] Library
- e: Central Public Library
- e: Certified Parts List
- e: Certified Products List
- e: Characters per Line
- e: Charge Pumping Logic
- e: Charleston Public Library
- e: Charlotte Public Library
- e: Chattanooga Public Library
- e: Chicago Public Library
- e: Choice of Programming Languages
- e: Cincinnati Public Library
- e: Cleveland Public Library
- e: Clio Press Limited
- e: Coded Parts List
- e: Code Program[me] Library
- e: Collective Pitch Lever
- e: Colonial Products Laboratory
- e: Columbus Public Library
- e: Columns per Line
- e: Combined Programming Language
- e: Commercial Pilots License
- e: Common Program[me] Language
- e: Common Program[me] Library
- e: Common Programming Language
- e: Common Pulse Line
- e: Commonwealth Parliamentary Library
- e: Commonwealth Public Library
- e: Compiler
- e: Computer Program[me] Library
- e: Conversation[al] Planning Language
- e: Conversational Programming Language
- e: Coordinator for Production and Logistics
- e: Core Performance Log
- e: Coronado Public Library
- e: Corporate Programming List
- e: Council of Planning Librarians
- e: Couple
- e: Crew Procedures Laboratory
- e: Critical Path Length
- e: Current Product Line
CPL1 e: Checkout PL/1 (Programming Language 1)
- e: Conversational Programming Language 1
CPLD e: Compound Phase-Locked Demodulator
CPLD e: Coupled
CPLEE e: Charged Particle Lunar Environment Experiment
CPLG e: Coupling
CPLM e: Complement
CPLMT e: Complement
CPLR e: Center (or Centre) of Pillar
- e: Coupler
CPLS e: Canberra Public Library Service
C/PLSEL e: Clothing and Personal Life Support Equipment Laboratory
CPLT e: Complete
- e: Copilot
Cpl T[ec] e: Corporal Technician
CPLTG e: Completing
CPLTN e: Completion
CPM e: Call[s] per Minute
- e: Call Protocol Message
- e: Capsule Positioning Mechanism
- e: Card Punch Memory
- e: Cards per Minute
- e: Cards Processed per Minute
- e: Cathode Pulse Method
- e: Cathode Pulse Modulation
- e: Center for Preventive Medicine
- e: Center Program[me]Manager
- e: Central Pacific Minerals
- e: Central Path Method
- e: Central Processor Module
- e: Characters per Minute
- e: Chip Placement Machine
- e: College of Petroleum and Minerals
- e: Coloured Progressive Matrices
- e: Command Processor Module
- e: Common Business-Oriented Language Performance Monitor
- e: Common Particular Metre
- e: Commutative Principle for Multiplication
- e: Compatible Phase Multiplex
- e: Complex Potential Model
- e: Computer Performance Management
- e: Computer Performance Monitor
- e: Computer Peripheral Module
- e: Condensed Particulate Matter
- e: Connection Point Manager
- e: Continuous Phase Modulation
- e: Continuous Processing Machine
- e: Control Processing Module
- e: Conversational Program[me] Module
- e: Counts per Minute
- e: Critical Path Method
- e: Critical Phase Matching
- e: Current Physics Microform
- e: Current Processor Mode
- e: Cycle[s] per Minute
CP/M e: Control Program[me] for Microcomputers
- e: Control Program[me] for Microprocessors
- e: Trademark of Digital Research
CPMA e: Central Processor Memory Address
- e: Computer Peripheral Manufacturers Association
CP-Massen d: Cellulosepropionatbasis-Massen
CPMC e: Columbia Presbyterian Medical Center
- e: Computerized Planar Motion Carriage
CPM-Cost e: Critical Path Method with Cost
CPMEEW e: Council for Postgraduate Medical Education in England and Wales
CPMFLDE e: Complement Field End
CPMM e: Characters per Milimeter
CPMO e: Compendium of Plausible Material Options
- e: Contract Parts Material Order
CPMP e: Control Program[me] Management Package
CPMS e: Cable Pressure Monitoring System
- e: Computer Performance Monitoring System
- e: Computer Plotting Matrix System
- e: Control Point Management Services
CPN d: chronische Pyelonephritis
- e: Card Part Number
- e: Chronic Pyelonephritis
- e: Code Program[me] Name
- e: Code Program[me] Number
- e: Component Part Number
- e: Computer Product News
CPND e: Charged Particle Nuclear Data
CPNP e: Cape Perth National Park
CPNRC e: Crystal Palace National Recreation Centre
CPO e: Call on Parity Odd
- e: Card Punch Operation
- e: Changing Path of Operation
- e: Chief Programmer Organization
- e: Circular Parking Orbit
- e: Coalitive Pareto Optimal
- e: Code Practice Oscillator
- e: Command Pilot Overhaul
- e: Command Pulse Output
- e: Component
- e: Concurrent Peripheral Operation[s]
- e: Controlled Precision Oscillator
- e: Cost Proposal Outline
- e: Custom Product Operation
CPOC e: Clay Pigment[ed] Organic Coating
CPODA e: Contention Priority-Oriented Demand Assignment
CPOG e: Canadian Pacific Oil and Gas
CPOI e: Customer Order Processing and Invoicing
CPOL e: Communications Procedure-Oriented Language
CPOMR e: Computerized Problem-Oriented Medical Record
CPOPAN e: Component Panel
CP-OX-HP e: Chemically Prepared and Oxygen Hot Pressed
CPP e: Caltech Population Program[me]
- e: Card Print Processor
- e: Card Punching Printer
- e: Chemical Processing Plant
- e: Complete Purchased Part
- e: Computer Position Profile
- e: Computer Printout Processing
- e: Concrete Paver Project
- e: Concrete Paving Project
- e: Concurrent Peripheral Processing
- e: Conductive Plastic Potentiometer
- e: Control and Protection Panel
- e: Controllable Pitch Propeller
- e: Coronary Perfusion Pressure

- e: Corporative Programming Practice
- e: Critical Path Plan[ning]
- e: Current Papers in Physics (journ.)
- e: Current Purchasing Power
- e: Cyclopentophenanthrene

CPPA e: Coal Preparation Plant Association
CPPB e: Continuous Positive-Pressure Breathing
CPPC e: Cost Plus a Percentage of Cost
CPPD e: Calcium Pyrophosphate Dihydrate
- e: Capped
- e: Consumers Public Power District

CPPG e: Chief of Naval Operations Policy and Planning Guidance
CPPI e: Canadian Pulp and Paper Industry (journ.)
- e: Consultative Panel on Public Information

CPPM e: Civilian Personnel Procedures Manual
- e: Communication Prediction Program[me]
- e: Critical Path Programming Method
- e: Critical Path Programming Model

CPPOT e: Conductive Plastic Potentiometer
CPPR e: Cassel Psychotherapy Progress Record
CP Propeller e: Controllable Pitch Propeller
CPPS e: Contextual Postprocessing System
- e: Continuous Process Plant Scheduling
- e: Critical Path Planning and Scheduling
- s: Comisión Permanente para la Explotación y Conservación de las Riquezas Maritimas

CPPV e: Continuous Positive Pressure Ventilation
CPQ e: Childrens Personality Questionnaire
C/P-Quotient d: Cholesterin-Phosphatid-Quotient
CPR e: Cam Plate Read-out
- e: Canadian Pacific Route
- e: Cardiopulmonary Resuscitation
- e: Card Punch and Reader
- e: Card Punching Rate
- e: Certification Print Routine
- e: Chemical Propulsion Rocket
- e: Chief Parachute Rigger
- e: Chinese Peoples Republic
- e: Chlorophenyl Red
- e: Civilian Personnel Record[s]
- e: Civilian Personnel Regulations
- e: Cloud Physics Radiometer
- e: Cobourg Peninsula Reserve
- e: Committee on Polar Research
- e: Common Pulse-Signalling Receiver
- e: Comparing
- e: Component Pilot Rework
- e: Component to Part Record
- e: Compressor Pressure Ratio
- e: Computer
- e: Computer Personnel Research
- e: Consolidated Progress Record
- e: Continuing Property Record
- e: Contributed to Propagation Research
- e: Control Program[me] for Real-Time
- e: Copper
- e: Cost Performance Report
- e: Crew Provisioning Report
- e: Critical Power Ratio
- e: Current Page Register
- e: Cursor Position Report
- e: Cut Paraboloidal Reflector
- e: Cyclonene-Pyrethrin-Rotenone

CPRA e: Chemical Public Relations Association
- e: Council for the Preservation of Rural America

CP Ratio e: Circuit-to-Pin Ratio
CPRDC e: Coordinated Program[me] of Research in Distributed Computing
CPRE e: Council for the Preservation of Rural England
CPREF e: Candelabra Prefocused
CPREP e: Counterpreparation
CPRF e: Cancer and Polio Research Fund
CPRG e: Computer Personnel Research Group
CPRI e: Central Potato Research Institute
- e: Cold Pressor Recovery Index

CPR Indicator e: Continuous Progress Indicator
CPRL e: Changed Particle Research Laboratory
CPRR e: Check Point Recovery Routine
CPRS e: Centralized Personnel Record System
CPRs e: Current Page Registers
CPRSR e: Compressor
CPRTR e: Chief Printer
CPRTRL e: Chief Printer; Lithographer
CPRU e: Colonial Pesticides Research Unit
CPRV e: Cabin Pressure Relief Valve
- e: Calculate Polynomial for Row Values

CPRW e: Council for the Preservation of Rural Wales
CPS e: Cabinet Pressurization System
- e: Cable Programming Services
- e: California Physicians Service
- e: California Production Service
- e: Call Processing Subsystem
- e: Call Progress Signal
- e: Canadian Pacific Steamships
- e: Canadian Penitentiary Service
- e: Capacity Planning System
- e: Carbamylphosphate Synthetase
- e: Card Processing System
- e: Card Programming Support
- e: Card Programming System
- e: Card Punch Store
- e: Cards per Second
- e: Cathode Potential Stabilized
- e: Center for Population Studies
- e: Central Power System
- e: Central Processing System
- e: Central Processor Subsystem
- e: Central Programming Service
- e: Cephalo Pedal Sinus
- e: CERN Proton-Stabilized
- e: CERN Proton Synchrotron
- e: Character[s] per Second
- e: Circuit Package Schematic
- e: Circuit Provision System
- e: Clock Pulse
- e: Close Packed Structure
- e: Colo[u]r Picture Signal
- e: Combined Principles Simulator
- e: Common Business-Oriented Language Programming System
- e: Common Pulse-Signalling Sender
- e: Communications Processing System
- e: Compendium of Pharmaceuticals and Specialities
- e: Composite Picture Signal
- e: Computerized Publishing System
- e: Computer Power Supply
- e: Computer Processing Service[s]
- e: Computer Program[me] Specification
- e: Computer Program[me] System
- e: Computer Programming System
- e: Condensate Polishing System
- e: Condensation Pressure Spread
- e: Console Programming System
- e: Constitutional Psychophatic State
- e: Contingency Planning System
- e: Contour Plotting System
- e: Control Panel Subassembly
- e: Control Power Supply
- e: Control Pressure System
- e: Control Program[me] Service
- e: Control Program[me] Support
- e: Conversational Programming System
- e: Conversion Program[me] System
- e: Conversional Programming System
- e: Cooley Programming System
- e: Corporate Planning System
- e: Corporate Programming Standard
- e: Counts per Second
- e: Creative Problem Solving
- e: Crew Procedures Simulator
- e: Critical Path Schedule
- e: Critical Path Scheduling
- e: Critical Path System
- e: Crossfield Plasma Sheath
- e: Current Population Survey
- e: Cycles per Second

CPSA e: Canadian Political Science Association
- e: Consumer Product Safety Act
- e: Current Physics Selected Articles

CPSAC e: Cycles per Second Alternating Current
CPSC e: Computer-Aided Production Scheduling and Control
CPS CMPSG e: Compass Compensating
CPSCU e: College of Physicians and Surgeons-Columbia University
CPSD e: Cross-Power Spectral Density
CPSDs e: Cross-Power Spectral Densities
CPSE e: Carr-Purcell Spin-Echo
- e: Counterpoise

CPSF e: Candle Power / Square Foot
CPSI e: Causing Pressure Shut-in
CP Signal e: Call Progress Signal
CPSK e: Cathodic Protection Survey Kit
- e: Coherent Phase Shift Keyed
- e: Coherent Phase Shift Keying

CPSL e: Canadian Pacific Steamship Line
CPSLCS e: Complete Power Signalling Local Control System
CPSN e: Capstan
CPSP e: Cove Palisade State Park
CPSR e: Calibration Procedure Status Report
- e: Computer Professionals for Social Responsibility
- e: Contractor Procurement System[s] Review
- e: Cossor Precission Secondary Radar
- e: Cost and Performance Summary Report

CPSS e: Central Processing Subsystem
- e: Certificate in Public Service Studies
- e: Common Program[me] Support System
- e: Component Percentage Shipment Schedule
- e: Computer Power Support System

CPST e: Capacitive Position Sensing Transducer
- e: Computer Program[me] in Science and Technology

CPSTGC e: Course per Steering Compass
CPSU e: California Polytechnic State University
- e: Central Processor Sub-Unit

CPT e: Canadian Pacific Telegraphs
- e: Capacitive Pressure Transducer
- e: Cape Town (Airport)
- e: Cardiovascular Pressure Transducer
- e: Casement-Projected Transom
- e: Center for Particle Theory
- e: Ceramic Planar Tube
- e: Change-Parity Time
- e: Charge Conjugation, Parity Conservation and Time Reversal Conservation
- e: Charge Parity Time

- e: Chest Physiotherapy
CPT e: Chief Programmer Team
- e: Chlorprothixene
- e: Civilian Pilot Training
- e: Clock; Programming and Timing
- e: Cockpit Procedure[s] Trainer
- e: Cold-Pressure-Test
- e: Colour Pyramid Technics
- e: Combining Power Test
- e: Compatibility
- e: Compatible Punched Tape
- e: Computer Programming Training
- e: Concentration Performance Test
- e: Continuing Performance Test[ing]
- e: Continuous Performance Task
- e: Continuous Performance Test
- e: Contralateral Pyramidal Tract
- e: Control Power Transformer
- e: Counterpoint
- e: Crew Procedures Trainer[s]
- e: Critical Path Technique
- e: Cryogenic Pressure Transducer
- e: Crystal Pressure Transducer
- e: Current Physics Titles
- e: Current Procedural Terminology (Physicians)
- e: Customer Provided Terminal
CP & T e: Clinical Pharmacology and Therapeutics
CPTA e: 2-(4-Chlorophenyl-thio)-Triethylamine
- e: Computer Programming and Testing Activity
CPTC e: Canada Passenger Transport Corporation
- e: Central Processor Test Console
CPT & E e: Computer Program[me] Test and Evaluation
CPTO d: Chefprogrammier[er]-Teamorganisation
- e: Chief Programmer Team Organization
CPTP e: Civil Pilot Training Program[me]
CPT Program[me] e: Consolidated Pilot Training Program[me]
CPTR e: Capture
- e: Carpenter (or Carpentry)
- e: Chief Pointer
CPTS e: California Public Television Stations
- e: Council of Professional Technological Societies
CPTY e: Capacity
CPU e: California Pacific University
- e: Card Punching Unit
- e: Central Processing Unit
- e: Central Processor Unit
- e: Central Procurement Unit
- e: Communications Processor Unit
- e: Communications Processor Utility
- e: Computer Peripheral Unit
- e: Controller Proficiency Unit
- e: Cost per Unit
CPUC e: California Public Utilities Commission
CPU Complex e: Caudate-Putamen Complex
CPUE e: Catch per Unit Effort
CPUID e: Central Processing Unit Identification
CPUNCH e: Counterpunch
CPV e: Combination Pump Valve
- e: Concrete Pressure Vessel
- e: Control Program[me] Five
- e: Cytoplasmic Polyhedrosis Virus
- s: Compañía Peruana de Vapores
CP Value e: Concentration Product Value
CPVC e: Chlorinated Polyvinylchloride
- e: Critical Pigment Volume Concentration
CPVPL e: Charles Patterson Van Pelt Library
CPW e: Chain Pointer Word
- e: Commercial Projected Window
- e: Computations per Word
- e: Polar Continental Air Warmer than Underlying Surface
CPWD e: Central Public Works Department
CPWM e: Clock Pulse Width Modulation
CPX e: Charged Particle Cross-section
- e: Charged Pigment Xerography
- e: Chlorprothixene
- e: Command Post Exercise[s]
CPX-FTX e: Command Post Exercise-Field Training Exercise
CPY e: Copy
CPZ d: Chlorpromazin
- e: Central Park Zoo
- e: Chlorpromazine
CQ e: Charge of Quarters
- e: Chloroquine
- e: Chloroquine Quinine
- e: Circadian Quotient
- e: Class Quotient
- e: Compute Queue
- e: Computer Queue
- e: Conceptual Quotient
- e: Conditionally Qualified
- e: Creativity Quotient
- e: Crew Quarters
CQA e: Component Quality Assurance
- e: Computer-Aided Question Answering
CQAP e: Component Quality Assurance Program[me]
CQC e: Citizens for a Quieter City
CQCM e: Cryogenic Quartz-Crystal Microbalance
CQD e: Come Quick[ly] Danger
- e: Customary Quick Dispatch
- e: Quantum Coupled Devices
- e: Wireless Distress Signal
CQM e: Chloroquine Mustard
- e: Circuit Quality Monitor
- e: Class Queue Management
- e: Crystal Quartz Modulator-Demodulator
CQMS e: Circuit Quality Monitoring System
CQP e: Capacitor Qualification Program[me]
CQPL e: Canadian Qualified Products Lists
CQR e: Cost Quote Request
- e: Customer Quality Representative
CQS e: Common Queue Space
CQT e: Capacitor Qualification Test
- e: Circuit
- e: College Qualification Test
CQTP e: Capacitor Qualification Test Program[me]
CQU e: College Qualification
- e: College Qualification Test[s]
CQUCC e: Commission on Quantities and Units in Clinical Chemistry
CR d: Cassettenrecorder
- d: Chrom
- e: Cable Rack
- e: Calculating Register
- e: Call[ing] Register
- e: Call Request
- e: Camera Repairman
- e: Cancel[lation] Request
- e: Capacity Record
- e: Capture Range
- e: Carbon Residue
- e: Cardiac Respiration
- e: Cardiorespiratory
- e: Card Reader
- e: Card Reading
- e: Card Recording
- e: Card Registration
- e: Card Reproducer
- e: Card Run
- e: Carriage Return
- e: Carry Register
- e: Cathode Ray
- e: Cathode Reaction
- e: Center
- e: Central Registry
- e: Centre
- e: Cerebral Ridge
- e: Chain Radar
- e: Change Recommendation
- e: Change Release
- e: Change Request
- e: Channel Replacement
- e: Channels Ratio
- e: Characteristic Relief
- e: Character Reader
- e: Character Reading
- e: Character Recognition
- e: Charging Resistor
- e: Check Register
- e: Check Routine
- e: Check Run
- e: Chemical Report
- e: Cheque Returned
- e: Chick Rectum
- e: Chlorine Resistant
- e: Chloroprene Rubber
- e: Chromite
- e: Chromium
- e: Circular
- e: Circulating Register
- e: Citizen[s] Radio
- e: Classification Research
- e: Classified Register
- e: Class Rate
- e: Clean and Recycling
- e: Clear Record
- e: Clinical Record[s]
- e: Clinical Report[s]
- e: Clinical Research
- e: Closed Routine
- e: Close Range
- e: Clothing Regulations
- e: Clot Reaction
- e: Code Reader
- e: Code Receiver
- e: Code Regeneration
- e: Coding Recorder
- e: Coefficient [of Fat] Retention
- e: Cognitive Restructuring
- e: Coincidence Register
- e: Cold Receptors
- e: Cold Roll[ed]
- e: Cold Rubber
- e: Colon Resection
- e: Colour Rendering
- e: Combat Readiness
- e: Command/Response
- e: Commanding Ribbon
- e: Command Receiver
- e: Command Register
- e: Command Routine
- e: Commendation Ribbon
- e: Commodity Rate
- e: Common Return
- e: Communication and Reconnaissance
- e: Communications; Radio
- e: Communications Register
- e: Compaction Ratio
- e: Competing Risks
- e: Compiling Routine

- e: Complete Remission
- e: Complete Round
- e: Compression Ratio
- e: Computer Run
- e: Computing Result
- e: Computing Reviews
- e: Computing Routine
- e: Conditional Reflex
- e: Conditional Release
- e: Conditional Response
- e: Conditioned Reflex
- e: Conditioned Response
- e: Conductivity Recorder
- e: Conference Report
- e: Congressional Record
- e: Connection Request
- e: Connector Register
- e: Consecutive Retrieval
- e: Consolidated Report
- e: Constrained Resistor
- e: Construction Recruit
- e: Consultant Report
- e: Contact Resistance
- e: Continuing Resolution
- e: Continuously Running
- e: Continuous Rating
- e: Continuous Reading
- e: Continuous Rod
- e: Controlled Rectifier
- e: Control Rating
- e: Control Ratio
- e: Control Receiver
- e: Control Register
- e: Control Relay
- e: Control Rod
- e: Control Room
- e: Control Routine
- e: Conversion Ratio
- e: Conversion Routine
- e: Convertible Rotor
- e: Corner Reflector
- e: Coronary Resistance
- e: Correct Routing
- e: Costa Rica (Airport)(Kfz)
- e: Counting Rate
- e: Count Reserve
- e: Counts per Roentgen
- e: Cramer-Rao (Modulation)
- e: Create[d]
- e: Creatinine
- e: Credit
- e: Creditable Record
- e: C Register
- e: Cremaster Reflex
- e: Critical
- e: Critical Ratio
- e: Cross-Reference
- e: Cruise
- e: Cruiser
- e: Cryptographer
- e: Crystal[line]
- e: Crystallize
- e: Crystal Rectifier
- e: Currency Regulation
- e: Current Rate
- e: Current Relay
- e: Customers Report
- e: Cyclic Redundancy
- e: Czechoslovakian Republic
- f: Comptes Rendus

C&R e: Construction and Repair
- e: Control and Reporting
- e: Convalescent and Rehabilitation

CRA e: California Redwood Association
- e: Canadian Rheumatism Association
- e: Carbon Ribbon Attachment
- e: Carry Ripple Adder
- e: Cassegrain Reflector Antenna
- e: Catalogue Recovery Area
- e: Cave Research Association
- e: Centralized Referral Activity
- e: Central Retinal Artery
- e: Coal Research Association
- e: College of Radiologists of Australia
- e: Colorado River Aqueduct
- e: Colorado River Authority
- e: Commander; Royal Artillery
- e: Command Relationship Agreements
- e: Composite Research Aircraft
- e: Computer Retailers Association
- e: Control Relay Automatic
- e: Control Repeater Amplifier
- e: Corn Refiners Association
- e: Corona Australis
- e: Cosmic Ray Altimeter
- f: Centres de la Recherche Appliqué

CRAAM e: Center of Radio Astronomy and Astrophysics Mackenzie University

CRAB e: Coastal-Research Amphibious Buggy

CRABS e: Close-Range Analytical-Bundle System

CRAC e: High-Potency Cocaine

CR Acad Sci f: Comptes Rendus Hebdomadaires des Seances de l'Academie des Sciences

CRACCOC e: Crack-Cocaine

CRAD e: Committee for Research into Apparatus for the Disabled
- e: Contracted Research Development
- e: Contract Research and Development

C-Rad e: Curtis-Rad

CRAE e: Combat Readiness Assessment Exercise

CRAF e: Comet Rendezvous Asteroid Tlyby
- e: Computerized Relalive Allocation of Facilities Technic

CRAFT e: Changing Radio Automatic Frequency Transmission
- e: Computerized Relative Allocation of Facilities Technique
- e: Continuous Random Analog[ue] to Frequency Transmission

CRAGS e: Chemical Records and Grading System
- e: Chemistry Records and Grading System

CRAM d: Cross-Aufbereitungssystem; magnetbandorientiert
- e: Card Random Access Memory
- e: Card Random Access Method
- e: Collapsible Rollup Antenna Mast
- e: Computerized Reliability Analysis Method
- e: Condensed Random Access Memory
- e: Conditional Relaxation Analysis Method
- e: Contractural Requirements Recording Analysis and Management
- e: Core and Random Access Manager
- e: Critical Resource Allocation Method

CRAMD e: Cosmic Ray Anti-Matter Detector

CRAMEX e: Card Random Access Memory (CRAM) Executive Program[me]

CRAMEX Program[me] e: Card Random Access Memory Executive Program[me]

CRAMM e: Coupon Reading and Marking Machine

CRAMSHIP e: Complete Round Ammunition Shipment

CRAN e: Cranial
- e: Craniology
- e: Cranium

CRAND e: Cosmic Ray Albedo Neutron Decay

CRANIOL e: Craniologic[al][ly]
- e: Craniologist
- e: Craniology

CRANIOM e: Craniometry

CRAP e: Completely Ridiculous Anthropic Principle

CRAR e: Control Read-Only Memory Address Register
- e: Critical Reliability Action Report

CRAS d: Crossassembler
- e: Coder and Random Access Switch
- e: Cross-Assembler

CRASH e: Citizens to Reduce Airline Smoking Hazards
- e: Community Resource and Self Help
- e: Creep in Axisymmetric Shells

CRAW e: Carrier Replacement Air Wing
- e: Combat Readiness Air Wing
- e: Crawler

CRAY e: Cray Computer Corporation

CRAZI e: Count Routine Applied to Zero Input

CRB e: Central Radio Bureau
- e: Chemical, Radiological, Biological [Warfare]
- e: Clutch Release Bearing
- e: Complementary Return to Bias
- e: Computer Resale Brokers, International
- e: Container Repair Building
- e: Cooper River Bridge
- e: Corona Borealis
- e: Customer Record[s] and Billing
- f: Communautair Referentiebureau

CRBBB e: Complete Right Bundle Branch Block

CRBD e: Colorado River Basin Project

CRBE e: Conversational Remote Batch Entry

CRBF e: Commanders Radar Bombing Familiarization

CRBI d: Crossbinder

CRBO e: Centralized Records Business Office

CRBP e: Colorado River Basin Project

CRBR d: chronische und rezidivierende Bronchitis
- e: Clinch River Breeder Reactor
- e: Controlled Recirculation Boiling Water Reactor

Cr & Br e: Crown and Bridge (Dental)

CRBRP e: Clinch River Breeder Reactor Plant
- e: Clinch River Breeder Reactor Project

CRC e: Cambridge Research Center
- e: Cambridge Research Center
- e: Cancer Research Campaign
- e: Carbon Residue Conradson
- e: Cardiovascular Reflex Conditioning
- e: Carriage Return Character
- e: Carriage Return Code
- e: Carriage Return Contact
- e: Cavity Rim Cap
- e: Central Read Control
- e: Ceramic Refraction Coating
- e: Character Recognition Circuit
- e: Chemical-Resistant Coating
- e: Chemical Rubber Company; Cleveland; Ohio
- e: Clinical Research Center
- e: Collins Radio Company
- e: Colour Rendering Capacity
- e: Column Reference Centerline
- e: Communications Regulatory Commission
- e: Communications Relay Center
- e: Communications Research Center
- e: Communications Resources Center
- e: Complete-Round Chart

- e: Computer Research Corporation
- e: Condition Reservation Code
- e: Consistency-Recording Controller
- e: Continental Railway Circle
- e: Contract Requirement Card
- e: Control and Reporting Center
- e: Controlled Rerouting Control
- e: Critical Reactor Component
- e: Cross Reference Code
- e: Cumulative Results Criterion
- e: Cyclic Redundancy Character
- e: Cyclic Redundancy Check[ing]
- e: Cyclic Redundancy Code

CRCA e: Cold-Rolled Close-Annealed
CRC-APRAC e: Coordinating Research Council-Air Pollution Research Advisory Committee
CRCB e: Continuously Running Circuit Breaker
CRC-BU e: Communications Research Center-Boston University
CRCC e: Consolidated Record Communications Center
- e: Cyclic Redundancy Check Character
- e: Cyclic Redundancy Check Code

CRCE e: Centaur Reliability Control Engineering
CRC-EPA e: Coordinating Research Council-Environmental Protection Agency
CR Check e: Cyclic Redundancy Check
CR Circuit e: Character Recognition Circuit
CRCIS e: Cybernetics Research Consultants Information Systems
CRCL e: Contractor Recommended Change List
CRC/LRC e: Cyclic Redundancy Check/Longitudinal Redundancy Check
CRCLT e: Circulate
CRCMF e: Circumference
CRCN e: Counterreconnaissance
CR (Conr) e: Carbon Residue (Conradson)
CRCP e: Certificate of the Royal College of Physicians
CRCPD e: Conference of Radiation Control Program[me]Directors
CRCRD e: Credit Card Reader
CRCR Filter e: Capacitance Resistance Capacitance Resistance Filter
CRCS e: Cardiovascular Reflex Conditioning System
- e: Center Range Control Station
- e: Certificate of the Royal College of Surgeons
- e: Clinical Record Cover Sheet

CRCT e: Circuit
- e: Correct

CRCTA e: Composite Reactor Components Test Activity
CRCTN e: Correction
CRC-UK e: Communications Research Center-University of Kansas
CRC-UW e: Communications Research Center-University of Washington
CRD e: Capacitor Resistor Diode
- e: Character Reading Device
- e: Chronic Renal Disease
- e: Chronic Respiratory Disease
- e: Complete Reaction of Degeneration
- e: Computer Readout Device
- e: Concentration of Respirable Dust
- e: Confidential Restricted Data
- e: Constant Ringing Drop
- e: Controlled Release Device
- e: Control Rod Drive
- e: Cored
- e: Cosmic Ray Detector
- e: Crop Research Division
- e: Current Rate for Discharge
- e: Current Regulator Diode

CRDA e: Control Rod Drive Assembly
CRDC e: Colorado Resources Development Council
CRDF e: Canadian Radio Direction Finder
- e: Canadian Radio Direction Finding
- e: Cathode Ray Direction Finder
- e: Cathode Ray Direction Finding

CRDL e: Chemical Research and Development Laboratories
- e: Contractor Data Requirements List

CRDM e: Chief Radarman
- e: Control Rod Device Mechanism
- e: Control Rod Drive Mechanism
- e: Control Rod Drive Motor

CRDP e: Civil Rights Documentation Project
CR & DP e: Cooperative Research and Development Program[me]
CRDS e: Card Release Daily Status
- e: Certified Reliability Data Sheet
- e: Charles River Data Systems
- e: Collisional Radiation Damage Sticking
- e: Component Repair Data Sheet[s]

CRDSD e: Current Research and Development in Scientific Documentation
CRDU e: Command Relay Driver Unit
CRE e: Call/Return Element
- e: Card Read Error
- e: Cation-Responsive Electrode
- e: Caunchy-Riemann Equation
- e: Chief Radio Electrician
- e: Coal Research Establishment
- e: Combat Readiness Evaluation
- e: Commander; Royal Engineers
- e: Command Receiver Equipment
- e: Compton Recoil Electron
- e: Constant-Rate-of-Elongation
- e: Constant-Rate-of-Extension
- e: Controlled Residual Element
- e: Corrected Reference Equivalent
- e: Corrosion Resistant
- e: Corrosion Resisting
- e: Crescent

CREAM e: Computer Realtime Access Method
- e: Computer Resource Allocation Model
- e: Oftartar Potassium Acid Cartrate ($KHC_4H_6O_6$)

CREAMs e: Computer Resource Allocation Models
CREAT e: Creatine
CREATE e: Carpet Research, Engineering, Aestetics, Technological and Education
- e: Computational Requirements for Engineering, Simulation, Training and Education
- e: Computational Resources for Engineering and Training and Education

CREATION e: Cultural and Recreational Education Achieved Through Investigations Ordinarily Neglected
Creation Sci e: Creation Science Research Center
CREB e: Conversational Remote Batch
CREC e: Combat Readiness Evaluation Criteria
CRECON e: Counterreconnaissance
CRECORD e: Congressional Record
CRED e: Credit
CREDD e: Customer Requested Earlier Due Date
CREDIT e: Cost Reduction Early Decision Information Technique
CREDO e: Central[ized] Reliability Data Organization
CREF e: Connection Refused
- e: Cross Reference

CREG e: Cancer Research Emphasis Grants
- e: Concentrated Range Extension with Gain

CREI e: Capitol Radio Engineering Institute
CREM f: Comité de Coordination de la Recherche Forestière Méditerranéenne
CREME e: Commander; Royal Electrical Engineers
CRES e: Center for Reproduction of Endangered Species
- e: Center for Research in Engineering Science
- e: Constant Ratio Elasticity of Substitution
- e: Corrosion-Resistant [Steel]
- e: Corrosion Resistant Stainless Steel

CRES[C] e: Crescent
CRESO e: Chief Royal Engineers Stores Officer
CRESS e: Center for Research on Social Systems
- e: Central Regulatory Electronic Stenographic System
- e: Combined Reentry Effort in Small Systems
- e: Computerized Reader Enquiry Service System
- e: Computer Reader Enquiry Service System

CRESS/AU e: Center for Research on Social Systems of the American University
CREST e: Calcinosis, Committee on Reactor Safety Technology
- e: Combat Readiness by Electronic Service Testing
- e: Committee on Reactor Safety [and] Technology
- e: Computer-Aided combat Reporting System
- e: Computer Routine for Evaluation of Simulation Tactics
- e: Consolidated Reporting and Evaluating System
- e: Crew Escape and Rescue Techniques
- e: Reynauds Phenomenon, Esophageal Motility Disorders, Sclerodactyly and Telangiec
- f: Comité de Recherche Scientifique et Technique

CRESTS e: Courtaulds Rapid Extract; Sort and Tabulate System
CRET e: Cathode Ray Electron Tube
CREVS e: Control Room Emergency Ventilation System
CREWKS e: Corps Royal Engineering Workshops
CRF e: Cancer Research Foundation
- e: Capacitor Resonance Frequency
- e: Capital Recovery Factor
- e: Carrier Frequency
- e: Cathode Ray Furnace
- e: Central Repair Facility
- e: Central Retransmission Facility
- e: Channel Replacement Furnace
- e: Chemical Releasing Factor
- e: Chronic Renal Failure
- e: Coagulase Reacting Factor
- e: Communications Pool Request Form[s]
- e: Context Roll File
- e: Continuous Reinforcement
- e: Contrast Rendition Factor
- e: Control-Reference File
- e: Control Relay Forward
- e: Correspondence Routing Form
- e: Corticotropin-Releasing Factor[s]
- e: Corticotropin Releasing Factor
- e: Cosmic Ray Flux
- e: Cross Reference File

- e: Cryptographic Repair Facilities
CRFE e: Current Ratio Forcing Element
CRFI e: Custom Roll Forming Institute
CRFOL e: Crown Folio
CRFP e: Carbon Fiber Reinforced Plastic
- e: Carbon Fibre Reinforced Plastic (GB)
CRFPs e: Carbon Fibre Reinforced Plastics
C-RFP-Verfahren d: Conditioniert; Rapid-Faß-Pulver-Verfahren
CRFS e: Change Request Forms
- e: Continuous Reinforcements
- e: Crash-Resistant Fuel System
CRF Schedule e: Continuous Reinforcement Schedule
CRFUSAIC e: Central Record Facility; United States Army Intelligence Center
CRFUSAINTC e: Central Record Facility; United States Army Intelligence Center
CRG d: Coderegister
- e: Carriage
- e: Catalytic Rich Gas
- e: Charging
- e: Control Rate Gyro
- e: Cosmic Ray Gas
- e: Cross Reference Generator
CRG-D d: Coderegister für den Durchgangsverkehr
CRG-E d: Coderegister für den Endverkehr
CRG-G d: Coderegister; gehend
CRGKLUE d: Coderegisterklinken-übertragung
CRGUE d: Coderegisterübertragung
CRH e: Caliber-Radius Head
- e: Channel Reconfiguration Hardware
- e: Control Relay; Hand
CRH e: Corticotropin Releasing Hormone
CRHH e: Cold Rolled Half Hard
CRHL e: Collaborative Radiological Health Laboratory
CRHS e: Component Reliability History Survey
CRI e: Call Request with Identification
- e: Caribbean Research Institute
- e: Cement Research Institute (India)
- e: Chemical Rust Inhibitor
- e: Chronic Respiratory Insufficiency
- e: Circuit Reliability Improvement
- e: Coconut Research Institute
- e: Cold Running Intelligibility
- e: Colour Rendering Index
- e: Committee for Reciprocity Information
- e: Communications Research Institute
- e: Complex Refraction Index
- e: Computer Related Industries
- e: Conflict Resolution Inventory
- e: Continuity-Recheck Incoming
- e: Cost Reliability Index
- e: Cray Research, Inc.
- e: Criterion-Referenced Instruction
- e: Crops Research Institute
- f: Comité de Recherche en Informatique
CRIB e: Computerized Resources Information Bank
- e: Computer Randomized Item Bank
CRI CKIT e: Cold Rocket Instrument Carrying Kit
CRIEPI e: Central Research Institute of the Electric[al] Power Industry
CRIFO e: Civilian Research; Interplanetary Flying Objects
CRIG e: Capacitor Rate Integrating Gyroscope
- e: Cocoa Research Institute of Ghana
CRI Germicide e: Chemical Rust Inhibiting Germicide
CRIL e: Consolidated Repairable Item List
CRILA e: Credit Insurance Logistics Automated
CRILC e: Canadian Research Institute of Launderers and Cleaners
CRIM e: Criminal
- e: Criminalism
- e: Criminalist
- e: Criminologist
- e: Criminology
CRIMINOL e: Criminology
CRIMP e: Computer Report on Importance
CRIN e: Cocoa Research Institute of Nigeria
CRIS e: Calibration Recall Information System
- e: Command Retrieval Information System
- e: Comprehensive Research Injury Scale
- e: Comprehensive Respiratory Information System
- e: Conference Resource and Information Services
- e: Counterintelligence Records Information System
- e: Current Research Information System
CRISCO e: Cream Received in Separating Cottonseed Oil
CRISD e: Center for Ressearch in System Development
CRISP e: Cascadable Real-Time Integrated Signal Processor
- e: Catalogued Resources and Information Survey Program[me]
- e: Computer-Retrieved Information of Special Projects
- e: Computer Resources Integrated Support Plan
- e: Computer Retrieval of Information on Scientific Projects
- e: Control Restrictive Instructions for Structured Programming
- e: Cosmic Radiation Ionization Spectrographic Program[me]
- e: Cosmic Ray Ionization Program[me]
CRISPE e: Computerized Retrieval Information Service on Precision Engineering
CRISS e: Center for Research in System Sciences
CRISTAL e: Contract Regarding an Interim Supplement to Tanker Liability for Oil Pollution
CRISYS e: Correctional Records Information System[s]
CRIT e: Centre for Research into Information Training
- e: Criteria
- e: Criterion
- e: Critical
- e: Critically
- e: Criticism
CRITESISTOR e: Critical-Temperature-Resistor
CRITICOMM e: Critical Intelligence Communication[s System]
CRITICOMM (System) e: Critical Intelligence Communications (System)
CRITS e: Critical Reactor Experiments
CRJ e: Command Reject
CRJE e: Conversational Remote Job Entry
CRK e: Carriage Release Key
- e: Carriage Return Key
- e: Clean Room Kit
- e: Cork
- e: Crank
- e: Cross Reference Key[s]
CRKC e: Crankcase
CRL e: C3-Receptor-Lymphocytes
- e: C 3-Receptor Lymphocytes
- e: Cadmium Red Line
- e: Cambridge Research Laboratory
- e: Cardiac Research Laboratory
- e: Cathode Ray Lamp
- e: Center for Research Libraries
- e: Central Research Laboratory
- e: Certified Record Librarian
- e: Certified Reference Librarian
- e: Character Recognition Logic
- e: Chemical Research Laboratory
- e: Columbia Radiation Laboratory
- e: Communication Research Laboratory
- e: Complement Receptor Lymphocytes
- e: Conservators of the River Lea
- e: Constant-Rate-of-Loading
- e: Cook Research Laboratories
- e: Coral
- e: Corporate Records List
- e: Crown Rump Length
C & RL e: Chemical and Radiological Laboratories
- e: College and Research Libraries
CRLB e: Cosmic Ray Logic Box
- e: Cramer-Rao Lower Bound
CRLCD e: Credit Limit Code
C-R Length e: Crown-Rump Length
CR/LF e: Carriage Return/Line Feed
CRLLB e: Center for Research on Language and Language Behavior
CRLT e: Center for Research on Learning and Teaching
CRLUIS e: California Regional Land Use Information System
CRM d: Coddsches Relationenmodell
- e: Cathode Ray Memory
- e: Change Request Material
- e: Chemical Remanent Magnetization
- e: Chief Radioman
- e: Chrome
- e: Clinch River Mile
- e: Command Receiver Monitor
- e: Communications/ Research/Machines
- e: Computer Readable Medium
- e: Computer Resources Management
- e: Conceptual Reference Mission
- e: Confidence Rulemaking
- e: Confusion Refelector Material
- e: Containment Radiation Monitor
- e: Control and Reproducibility Monitor
- e: Control Relay Master
- e: Counter Radar Measure[s]
- e: Counter Radar Missile
- e: Count Rate Meter
- e: Cross-Reacting Material
- e: Crucial Reaction Measure[ment]
- e: Curriculum Resource Materials
CRMB e: Cyclised Rubber Master Batch
CRMCH e: Cruise Mach Change
CRMD e: Children with Retarded Mental Development
- e: Class for Retarded in Mental Development
CR MOLY e: Chrome Molybdenum
CRMPT e: Commendation Ribbon with Metal Pendant
CRMR e: Continuous Reading Meter Relay
CRMS e: Center for Research in Management Science
CRMs e: Certified Reference Materials
CRMS e: Close Range Missile System
- e: Control and Reproducibility Monitor System
CRMs e: Cross-Reacting Materials
CRMT e: Circuit Rule Master Tape

CR MT e: Computer Readable Magnetic Tape
CRMWD e: Colorado River Municipal Water District
CRN e: Cermet Resistor Network
- e: Charge Routing Network
- e: Common Random Numbers
- e: Complement Requiring Neutralizing
- e: Correction
- e: Council for Responsible Nutrition
- e: Crane
CRNA e: Chromosomal Ribonucleid Acid
CRNIST d: Chrom Nickel-Stahl
CRNL e: Chalk River Nuclear Laboratories
CRNM e: Capitol Reef National Monument
CRNMTC e: Chronometer
- e: Chronometric
CRNMTR d: Chronometer
CRNO e: Command Register Not Operable
CRNP e: Cape Range National Park
CRNSS e: Chief of the Royal Naval Scientific Service
CRNWR e: Cape Romain National Wildlife Refuge
- e: Clarence Rhode National Wildlife Range
- e: Clarence Rhode National Wildlife Refuge
CRO e: Canarvon [Tracking Station]
- e: Cancelling Returns Only
- e: Carded for Record Only
- e: Card Reader Operation
- e: Cathode-Ray Oscilloscope
- e: Central Radio Office
- e: Central Record Officer
- e: Central Records Office
- e: Chrome Oxide
- e: Circular Order
- e: Command Routine Order
- e: Complete with Related Order
- e: Computerized Reliability Optimization
- e: Continuity-Recheck Outgoing
- e: Control Room Operator
- e: Corps Routine Order
- e: CrO_2 chromium dioxide
CROC e: Computer Review and Orientation Course
CROC[CLIP] e: Crocodile [Clip]
CROCODILIOL e: Crocodiliologic[al][ly]
- e: Crocodiliologist
- e: Crocodiliology
CROC[S] e: Crocodilian[s]
- e: Alligator[s]
- e: Caiman[s] (or Cayman[s])
CRODISS e: Computer Recording of Defects in Sewerage Systems
CROF e: Controller of Royal Ordnance Factories
CROM e: Capacitive Read-Only Memory
- e: Central Red-Only Memory
- e: Control Read-Only Memory
CROP e: Christian Rural Overseas Program[me]
- e: Common Routing Output
CROS e: Capacitor Read-Only Storage
- e: Capacitor Read-Only Store
- e: Card Read-Only Storage
- e: Card Read-Only Store
- e: Computerized Reliability Optimization System
- e: Computerized Reliability Organization System
- e: Contralateral Routing of Offside Signals
- e: Contralateral Routing Signal
CROSS e: Computer[ized] Rearrangement of Subject-Specialities
- e: Computerized Rearrangement of Special Subjects
- e: Crossing
CROSSBOW e: Computerized Retrieval of Organic Structures Based on Wiswesser
CROSSCORREL e: Crosscorrelation
CROSSPATE e: Coordinative Retrieval of Selectively Sorted Permuted Analog[ue]-Title Entries
CRO System e: Computerized Reliability Optimization System
CROW e: Combined Rocket Warhead
- e: Counter-Rotating Optical Wedge
CROWCASS e: Central Registry of War Criminals and Security Suspects
CROWD e: Continuous Read-Out of Work Dimensions
CRP d: c-reaktives Protein
- e: C-Reactive Protein
- e: Card Reader Program[me]
- e: Card Read[er] Punch
- e: Cathode-Ray Tubelkeyboard Printer
- e: Collimation and Registration Program[me]
- e: Command Read Pulse
- e: Command Repeat
- e: Common Reference Point
- e: Component Reliability Prediction
- e: Components Reliability Program[me]
- e: Compton Recoil Particle
- e: Computer Reset Pulse
- e: Computer Room People
- e: Comunity Renewal Program[me] (San Francisco)
- e: Conservation Reserve Program[me]
- e: Constant Percentage Resolution
- e: Control and Reporting Post[s]
- e: Controlled Referral Plan
- e: Controlled Reliability Program[me]
- e: Cornea-Retinal Potential
- e: Corporate Record Practice
- e: Cosmic Ray Particle[s]
- e: Cosmic Ray Produced
- e: Cost Reduction Program[me]
- e: Counter-Rotating Platform
- e: Creatine Phosphate
- e: Creative Protein
- e: Cross Reference Project
- e: Cryogenic Refrigerator Program[me]
CRPA e: C-Reactive Protein Antiserum
CRPFM's e: Complete-Reset Pulse Frequency Modulators
CRPL e: Central Radio Propagation Laboratory
- e: Chromium Plate
- e: Chromium Plating
CRPM e: Communication Registered Publication Memoranda
- e: Crankshaft Revolutions per Minute
CRPO e: Continuous Rating Permitting Overload
CRPS e: Center for Research on Population and Security
- e: Common Release Processing System
CRQ e: Call Request
- e: Commutation of Rations and Quarters
- e: Console Reply Queueing
- e: Current Requirements
CRR d: Computer-Ring Rheinland
- e: Carbon Residue Ramsbottom
- e: Carrier Removal Rate
- e: Churchill Research Range
- e: Clutter Rejection Radar
- e: Commercial Reference Room
- e: Constant Ratio Roll
- e: Constant Ratio Rule
- e: Contractor Reports Register
- e: Conversion Result Register
- e: Cost Reduction Report
- e: Cost Reduction Representative
CRRA e: California Resource Recovery Association
- e: Component Release Reliability Analysis
- e: Connecticut Resources Recovery Authority
CR (Rams.) e: Carbon Residue (Ramsbottom)
CRRD e: Conceptual Reference Repository Description
CRRDS e: Cross-Roads
CRRE e: Corrected Receiving Reference Equivalent
CRREL e: Cold Regions Research and Engineering Laboratory
CRRERIS e: Commonwealth Regional Renewable Energy Resources Information System
CRRES e: Combined Release and Radiation Effects Satellite
CRRI e: Central Rice Research Institute
- e: Central Road Research Institute
CRRS e: Combat Readiness Rating System
- e: Contribution Record Reporting System
CRS e: Calibration Requirements Summaries
- e: Cars
- e: Cathode Ray Store
- e: Centers
- e: Centralized Referral System
- e: Centralized Results System
- e: Central Repair Service
- e: Centres
- e: Cereals Research Station
- e: Certified Residential Specialist
- e: Chain Radar System
- e: Character Recognition System
- e: Checker Redundant Scheme
- e: Child Rearing Study
- e: Chloroprene Rubbers
- e: Circuit Reset
- e: Citizen[s] Radio Service
- e: Coast Radio Station[s]
- e: Cold-Rolled Steel
- e: Colon and Rectal Surgery
- e: Commonwealth Rehabilitation Service
- e: Computerized Retrieval Service
- e: Computer Recognition System
- e: Computer Related Services
- e: Computing Research Station
- e: Congressional Research Service
- e: Console Request Service
- e: Containment Rupture Signal
- e: Corrosion Research Station
- e: Cosmic Ray Shower
- e: Countermeasures Receiving Set
- e: Cross Section
- e: Customized Routing Selector
- e: Customized Routing System
- f: Counseil de la Recherche Scientifque
CRSA e: Centralized Repair Service Attendants
- e: Current Routing Starting Address
CRSB e: Compatible Residual Sideband
CRSC e: Center for Research in Scientific Communication[s]
- e: Center for Research in Social Change
CRSD e: Center for Research in Scientific Documentation
- e: Contracttor Required Shipment Date
CRSG e: Classification Research Study Group
- e: Corresponding
CRSI e: Concrete Reinforcing Steel Institute
CRSN e: Corrosion
CRSP e: Colorado River Storage Project
CRSR e: Center for Radiophysics and Space Research

CRSS e: Chemically Rigidized Space Structure
- e: Critical Resolved Shear Stress
CRST d: Cross-Simulations- und Testsystem
- e: Cathode Ray Storage Tube
- e: Cold-Rolled Steel
CRST Syndrome e: Calcification and Clinical Signs of Raynauds Phenomenon, Scleroderma and Telangiectasis
CRSV e: Corrosive
CRSVR e: Crossover
CR System e: Chain Radar System
CRT e: C-Band Radar Transponder
- e: Cathode Ray Terminal
- e: Cathode Ray Tube
- e: Cathode Ray Typesetting
- e: Certified Radiologic Technician
- e: Channel Response Time
- e: Charactron Tube
- e: Chief Radio Technician
- e: Chinese Remainder Theorem
- e: Choice Reaction Time
- e: Circuit Requirement Table
- e: Cold-Rolled and Tempered
- e: Combat Rated Thrust
- e: Combat Reaction Time
- e: Combined Radiation Test
- e: Complex Reaction Time
- e: Composite Readiness Test
- e: Constant-Rate-of-Traverse
- e: Continuous Ring Tone
- e: Control Relay Translator
- e: Correct
- e: Counter Recovery Time
- e: Court
CRTA e: Cathode Ray Tube Addressing
CRT Addressing e: Cathode Ray Tube Addressing
CRTC e: Canadian Radio-Television Commission
- e: Cathode Ray Tube Control[ler]
CRTF e: Create Test Files
CRTG e: Cartridge
CRTGC e: Cartographer
CRTIS e: Chicago Railroad Terminal Information System
CRTM e: Cathode Ray Tube Memory
CRTMBA e: Criteria and Macrobudget Accounting
CRT Memory e: Cathode Ray Tube Memory
CRTN e: Correction
CRTO e: Cathode Ray Tube Oscilloscope
CRTOG e: Cartographer
- e: Cartographic
- e: Cartography
CRTP e: Contour Template
CRTPB e: Canadian Radio Technical Planning Board
CR TPDU e: Connection Request Transport Protocol Data Unit
CRTS e: Cathode-Ray Tubes
- e: Cathode Ray Tube Shield
- e: Cathode Ray Tube Storage (or Store)
- e: Commonwealth Reconstruction Training Scheme
- e: Constant Returns to Scale
- e: Controllable Radar Target Simulator
CRT Storage e: Cathode Ray Tube Storage
CRT Store e: Cathode Ray Tube Store
CRTT e: Cathode Ray Tube Terminal
- e: Cathode Ray Tube Tester
- e: Cathode Ray Tube Transformer
CRT Terminal e: Cathode Ray Tube Terminal
CRTU e: Combined Receiving and Transmitting Unit
CR Tube e: Cathode Ray Tube
CRTV e: Composite Reentry Test Vehicle
CRU e: Card Reader (or Reading) Unit
- e: Cecil Rhodes University
- e: Clinical Research Unit
- e: Combined Rotating Unit
- e: Communications Register Unit
- e: Computer Resource Unit
- e: Control and Reporting Unit
- e: Control Relay Unlatch
- e: Cruise
- e: Cruiser
- e: Customer Replaceable Unit
CRU Base e: Cruiser Base
CRUD e: Chalk River Unidentified Deposit
- e: Continuous and Random Unwanted Deviations
CRUEL e: Commission on Reform of Undergraduate Education and Living
CRUIS e: Cruiser
CRUS e: Centre for Research on User Studies
CRUSK e: Center for Research on Utilization of Scientific Knowledge
CRV e: Chlorotic Ringspot Virus
- e: Chrome Vanadium
- e: Constant Reflector Voltage
- e: Contact Resistance Variation
CRVAN e: Chrome Vanadium
CRV DWG e: Curve Drawing
CRVR e: Computerized Register of Voice Research
CRW e: Carrier Wave
- e: Community Radio Watch
CRWC e: Complete Reading-Writing Cycle
CRWG e: Computer Resources Working Group
CRWM e: Committee on Radioactive Waste Management
CRWMP e: Commendation Ribbon with Metal Pendant
CRWO e: Coding Room Watch Officer
- e: Crypto-Room Watch Officer
CRWPC e: Canadian Radio Wave Propagation Committee
CRWR e: Center for Research in Water Resources
CRX e: Combined Repatriation Executive
CRY e: Crystal[line]
- e: Crystal[s]
CRYDAT e: Cryogenic Data
CRYG e: Carrying
CRYOBIO e: Cryobiological[ly]
- e: Cryobiologist
- e: Cryobiology
CRYOCHEM e: Cryochemical[ly]
- e: Cryochemist[ry]
CRYOELECTRO e: Cryoelectronic[al][ly]
- e: Cryoelectronicist
- e: Cryoelectronics
CRYOG e: Cryogenic
- e: Cryogenic[s]
CRYOGEN e: Cryogenic[al][ly]
Cryog Ind Gases e: Cryogenics and Industrial Gases (journ.)
CRYON e: Cryonic[s]
CRYONITE e: Sodium Aluminum Fluoride
CRYOSAR e: Cryo[genic] Switching by Avalanche and Recombination
CRYOSURG e: Cryosurgeon
- e: Cryosurgery
- e: Cryosurgic[al][ly]
CRYPT e: Cryptography
CRYPTA e: Cryptanalysis
- e: Cryptanalyst
CRYPTO e: Cryptogram
CRYPTO e: Cryptograph[ic]
- e: Cryptographer
- e: Cryptography
CRYPTONET e: Crypto-Communication Network
CRYPTON[S] e: Cryptonym [s]
CRYPTOZOOL e: Cryptozoological[ly]
- e: Cryptozoologist[s]
- e: Cryptozoology
CRYS e: Crystal[line]
- e: Crystallization
- e: Crystallize
- e: Crystallography
- e: Crystalloid[s]
- e: Crytallization
CRYSNET e: Crystallographic Computing Network
CRYST e: Crystal[line]
- e: Crystallized
- e: Crystallography
CRYSTAL e: Methamphetamine
CRYSTALLOG e: Cystallography
CRYSTD e: Crystallized
CRYSTG e: Crystallizing
CRYSTN e: Crystallization
CRZ e: Close Reconnaissance Zone
CS d: Cäsium
- d: Codesender
- d: Continentaler Stahlmarkt (journ.)
- d: Kasein
- d: Tschechoslowakei (Kfz)
- d: Zentisekunde
- d: Zirrostratus
- e: Cable Ship
- e: Calculating Speed
- e: Calculating Store
- e: Calculation Statement
- e: Call[ing] Sequence
- e: Call Sign
- e: Call Signal
- e: Calls per Second
- e: Call Statement
- e: Call Store
- e: Call System
- e: Camera Site
- e: Capacitor Store
- e: Carbon Steel
- e: Card Selection
- e: Card Size
- e: Card Socket
- e: Card Sorting
- e: Card Stacker
- e: Card Station
- e: Card Store
- e: Carrier Stability
- e: Carrier Supply
- e: Casein
- e: Cases
- e: Case Supervisor
- e: Casket Store
- e: Cast Steel
- e: Casual
- e: Cell Sequence
- e: Cell Store
- e: Cement Slurry
- e: Center Section
- e: Centisecond[s]
- e: Centistoke[s]
- e: Centralized Scheduling
- e: Central Station
- e: Central Supply
- e: Centre Section
- e: Centrifugal Spraying
- e: Cerebro-Spinal
- e: Certificate in Statistics
- e: Cesarean Section

- e: Cesium
- e: Chain Store
- e: Change Sheet
- e: Channel Search
- e: Channel Selection
- e: Channel Selector
- e: Channel Status
- e: Characteristic Slope
- e: Character Selection
- e: Character Sense
- e: Character Sensing
- e: Character Sequence
- e: Character Set
- e: Character Speed
- e: Character String
- e: Charge-exchange Scattering
- e: Charge Storage
- e: Charge Store
- e: Checked Statement
- e: Checking Sequence
- e: Checking Subroutine
- e: Checking System
- e: Check Sorter
- e: Check Sum
- e: Check Surface
- e: Chemical Society
- e: Chief of Section
- e: Chief Surgeon
- e: Chip Select
- e: Christian Science
- e: Christian Scientist
- e: Circuit Switch[ing]
- e: Circulating Store
- e: Circulating System
- e: Cirrostratus
- e: Citrat-Synthetase
- e: City Surveyor
- e: Class of Service
- e: Clear and Substract
- e: Clear Screen
- e: Closed Shell
- e: Closed Subroutine
- e: Closed System[s]
- e: Close Shot
- e: Coaling Station
- e: Coded Signal
- e: Code Selection
- e: Code Sensor
- e: Code Signal
- e: Code Symbol
- e: Code System
- e: Coding Section
- e: Coding Specification
- e: Coding Store
- e: Cognitive Stimulation
- e: Coil Sketch
- e: Coincidence Store
- e: Collation Sequence
- e: Collector-Substrate
- e: Colo[u]r Specification
- e: Colo[u]r Stabilizer
- e: Column Spacing
- e: Column Split
- e: Combined Storage
- e: Command Selection
- e: Command Selector
- e: Command Statement
- e: Command Structure
- e: Command System
- e: Commit Stop
- e: Common Source
- e: Common Steel
- e: Communication Sergeant
- e: Communication Station
- e: Communication System[s]
- e: Communication[s]
- e: Communication[s] Satellite
- e: Communication[s] Server
- e: Communication[s] Services
- e: Communication[s] Simulator
- e: Communication[s] Subsystem
- e: Communication[s] System
- e: Compilation Stage
- e: Complementary Symmetry
- e: Complement Storage
- e: Component Specification
- e: Compound Statement
- e: Computer Science (journ.)
- e: Computer Simulation
- e: Computer Society
- e: Computer Software
- e: Computer Store
- e: Computer Subprogram[me]
- e: Computer Subroutine
- e: Compute Statement
- e: Computing Sequence
- e: Computing Speed
- e: Computing Store
- e: Concentrated Space
- e: Concentrated Strength
- e: Concurrent Stereometric
- e: Conditional Sentence
- e: Conditional Statement
- e: Conditioned Stimuli
- e: Conditioned Stimulus
- e: Conducted Susceptibility
- e: Conduct Sheet
- e: Connection Switching
- e: Conscious
- e: Constructor Syntax
- e: Context-Sensitive
- e: Continent or Scandinavia
- e: Contingency Sample
- e: Continue-Specific
- e: Continuous Scan
- e: Contract Surgeon
- e: Controlled Store
- e: Controlled Switch
- e: Control Scanner
- e: Control Section
- e: Control Set
- e: Control Signal
- e: Control Statement
- e: Control Station
- e: Control Storage
- e: Control Store
- e: Control Switch
- e: Control Symbol
- e: Control System
- e: Convair 580 turboprop airplane
- e: Convalescent Status
- e: Convergent Stereoscopic
- e: Copper or Steel
- e: Core Sharing
- e: Core Shift
- e: Core Storage
- e: Core Store
- e: Coronary Sclerosis
- e: Coronary Sinus
- e: Corrected Series
- e: Correction System
- e: Corticosteroid
- e: Countersink
- e: Countersunk
- e: Counting Switch
- e: Count Store
- e: County Surveyor
- e: Crankshaft
- e: Cross Section
- e: Crucible Steel
- e: Cryogenics
- e: Cryogenic Store
- e: Cryotron Store
- e: Cryptographic System
- e: Currency Sign
- e: Current Series
- e: Current Source
- e: Current Strength
- e: Current Switch[ing]
- e: Cutting Specification[s]
- e: Cycle Shift
- e: Cycles per Second
- e: Cycle Stealing
- e: Cyclic Store
- e: Cycloserin[e]
- e: Cyclostationary
- e: Source Capacitance
- e: Standard Capacitance

C & S e: Computer and Systems
- e: Costs and Schedules

CS² e: Coaxial Subscriber System

CS²L e: Current Sourced Schottky Logic

CS 3 e: Combat Service Support System

Cs'3' e: Radioactive Cesium

CS-3A Sakura-3A e: Japanese Communications Satellite

CS-3B Sakura-3B e: Japanese Communications Satellite

CSA e: Called Subscriber Answer
- e: Cambridge Scientific Abstracts (journ.)
- e: Canada Shipping Act
- e: Canadian Standard[s] Association
- e: Carry Save Adder
- e: Carry Save Addition
- e: Casualty Surgeons Association
- e: Central Surgical Association
- e: Certificate in Systems Analysis
- e: Cheshire School of Agriculture
- e: Chief Scientific Adviser
- e: Chopper Stabilized Amplifier
- e: Client Service Agent
- e: Coarse Sand Aggregate
- e: Coarse Slag Aggregate
- e: College of Surgeons of Australasia
- e: Colony Stimulating Activity
- e: Command Session Abort
- e: Commercial Service Area
- e: Commercial Service Authorization
- e: Common Service Area
- e: Common System Area
- e: Commonwealth Sugar Agreement
- e: Communications Satellite Act
- e: Communications Service Authorization
- e: Compressed Spectral Array
- e: Computer Science[s] Association
- e: Computer Sciences of Australia
- e: Computer Services Association
- e: Computer System Analyst
- e: Computer Systems Association
- e: Computer Systems of America, Inc.
- e: Computing Services Association
- e: Conditional-Sum Adder
- e: Conditional-Sum Addition
- e: Configuration Status Accounting
- e: Conical Scan Aerial
- e: Conical Scan Antenna
- e: Contractor Support Area
- e: Cooperating Systems Architecture
- e: Core Store Allocation
- e: Cryogenic Society of America
- e: Current Source Amplifier

C & SA e: Counterinsurgency and Special Activities

CSAA e: Committee on Space Astronomy and Astrophysics

CSAC e: Cameron State Agricultural College
- e: Central Ships Alignment Console
- e: Cocoa Scientific Advisory Committee
- e: Conners State Agricultural College

CSACCS e: Customer Service Administration Control Center System
CSAD e: Capsule Systems Advanced evelopment
CSAE e: Canadian Society of Agricultural Engineering
CSAI e: Chief Special Artificer; Instruments
CSAIWR e: Chief Special Artificer; Instruments; Watch Repairman
CSAJTL e: Current Switching Alloy Junction Transistor Logic
CSAL e: Central Scientific Agricultural Library
CSAM e: Carry-Save Adder Module
- e: Circular Sequential Access Memory
CSANZ e: Cardiac Society of Australia and New Zealand
CSAO e: Chief Special Artificer; Optical
CSAP e: Canadian Society ofAnimal Production
- e: Control System Analysis Program[me]
CSAR e: Communications Satellite Advanced Research
- e: Computer Scheduling and Reporting
- e: Configuration Status Accounting Report
- e: Control, Security, Auditing, Recovery
- e: Control Storage Address Register
- e: Control Store Address Register
- e: Core Store Administration Routine
CSARS e: Close Support Artillery Rocket System
CSAR System e: Computer Scheduling and Reporting System
CSAs e: Carry Save Additions
CSAS e: Chief Superintendent of Armament Supply
- e: Command Stability Augmentation System
CSATL e: Canadian Standards Association Testing Laboratory
CSB d: chemischer Sauerstoffbedarf
- e: Bachelor of Christian Science
- e: Calcium Silicate Brick
- e: Carrier and Sideband
- e: Century Service Bulletin
- e: Chained Segment Buffering
- e: Channel Status Byte
- e: Chemical Stimulation of the Brain
- e: Christian Science Bachelor
- e: Civilian Steering Center
- e: Civilian Supply Branch
- e: Combined S-Band
- e: Combined Signal Board
- e: Combustible Storage Building
- e: Communication Scanner Base
- e: Communications Systems Center
- e: Complementary Straight Binary
- e: Computer Science Center
- e: Computer Search Center
- e: Computer Service Center
- e: Concrete Splash Block
- e: Copper Shielding Braid
- e: Cosecant
- e: Customer Service Center
CSBA e: Columbia Sheep Breeders Association [of America]
CSB Code e: Complementary Straight Binary Code
CSBF e: Coronary Sinus Blood Flow
CSBN e: Captured Steam Bubble Nuclear
CS Briketts d: Contra-Smoke Briketts (Ruhrkohle AG)
CSC e: Cadmium-Sulfide Cell
- e: Canadian Space Centre
- e: Cape Support Coordinator
- e: Cartridge Storage Case
- e: Central Security Control
- e: Central Switching-Network Control
- e: Central Switching Center
- e: Central Switching Concept
- e: Change Schedule Chart
- e: Cigarette Smoke Condensate
- e: Circuit Switching Center
- e: Clock Start Command
- e: Close-up Stereo Camera
- e: Coil Stock Cradle
- e: Colorado State College
- e: Colour Subcarrier
- e: Command Scheduling Chain
- e: Command Selector Control
- e: Common Signalling Channel
- e: Commonwealth Scientific Corporation
- e: Commonwealth Scientifitc Committee
- e: Commonwealth Steel Company
- e: Communication Simulator Console
- e: Communication Systems Center
- e: Communications Satellite Corporation
- e: Communications Switchboard Console
- e: Complete Store Cycle
- e: Complex Support Controller
- e: Comprehensive Self-Check
- e: Computer Science Canada
- e: Computer Science Center
- e: Computer Science Conference
- e: Computer Science[s] Corporation
- e: Computer Search Center
- e: Computer Service Center
- e: Computer Society of Canada
- e: Computer Store Capacity
- e: Computer Subsystem Controller
- e: Computer Supervisory Control
- e: Concrete and Steel Combination
- e: Consequence Counter
- e: Consolidated Coal Company
- e: Consortium for Scientific Computing
- e: Conspicuous Service Cross
- e: Construction Scheduling and Coordination
- e: Continental Shelf Crawler
- e: Contingency Sample Collection
- e: Continuous Service Certificate
- e: Control Signalling Code
- e: Control Storage Code
- e: Control Systems Character
- e: Core Store Control
- e: Correction Support Circuitry
- e: Cosecant
- e: Course and Speed Calculator
- e: Course and Speed Computer
- e: Criterion Strategy Controller
- e: Cryogenic Storage Container
- e: Cylinder Seek Command
- e: Cylinder Stroke Control
CSCA e: Conditional-Sum early Completion Adder
CSCAN e: Carrier System for Controlled Approach of Naval Aircraft
- e: C Scanner
CSCAW e: Catholic Study Circle for Animal Welfare
CSCB e: Command Scheduling Control Block
CS-CB e: Common Source-Common Base
CSCBS e: Commodore Superintendent Contract Built Ships
CSCC e: Command Session Change Control
CS-CC e: Common Source-Common Collector
CSCC e: Cumulative Sum Control Chart
CSCD e: Character Set Computer Development
CS-CE e: Common Source-Common Emitter
CSCF e: Captive Spot Curve Follower
- e: Central Scientific Computing Facility
Csch e: Hyperbolic Cosecant
CSCHE e: Canadian Society for Chemical Engineering
CS Circuit e: Common-Source Amplifier for Field-Effect Transistors
CSCJ e: Center for Studies in Criminal Justice
CSCl e: Cesium Chloride
CSCLR e: Controlled Space-Charge Limited Resistor
CScO e: Chief Scientific Officer
CS Coal e: Colliery Screened Coal
CSCP e: Central Stock Control Point
- e: Christian Science Committee on Publications
CSCR e: Center for Surface and Coatings Research
CSCRF e: Computer System for Crop Response to Fertilizers
CSCS e: Central Security Control System
- e: Cost, Schedule and Control System
- e: Crusader S-wire Container Service
CSCSC e: Cost Schedule Cost System Criteria
CSCSI e: Canadian Society for Computation Studies of Intelligence
CSC Test e: Childrens Self-Conceptions Test
CSCU e: Countersink Cutter
CSD e: Calibrated Sweep Delay
- e: Cambridge Structural Database
- e: Cargo Ship Dock
- e: Central Storage Device
- e: Character String Data
- e: Chief of Supply Depot
- e: Circuit Switched Data
- e: Closed Shelter Deck[er]
- e: Commander; Submarine Division
- e: Command Signal Decoder
- e: Command Supply Depot
- e: Command Systems Department
- e: Communication[s] System Development
- e: Communication[s] Systems Division
- e: Computerized Standard Data
- e: Computer System[s] Department
- e: Computer System[s] Design
- e: Computer Systems Director
- e: Constant Speed Drive
- e: Continuous Slowing Down
- e: Controlled Slip Differential[s]
- e: Convection Suppression Device[s]
- e: Core Shift Driver
- e: Core Store Dump
- e: Corps Supply Depot
- e: Cortical Spreading Depression
- e: Crack Surface Displacement
- e: Cross Spectral Density
- e: Current Sector Designator
CSDC e: Circuit Switched Digital Capability
- e: Computer Signal Data Converter
CSD Code e: Canonical Signed Digit Code
CSDD e: Cluster Systems Description Document
- e: Computer Subprogram[me] Design Document
- e: Conceptual System Design Description
CSDE e: Central Servicing Development Establishment
CSDF e: Computer System Development Facility
CSDI e: Center for the Study of Democratic Institutions
CSDIC e: Combined Services Detailed Interrogation Center

CSDL e: Charles Stark Draper Laboratory
- e: Conceptual Schema Definition Language
- e: Current Switching Diode Logic

CSDM e: Continuous Slope Delta Modulation

CSDN e: Circuit Switched Data Network
- e: Circuit Switched Digital Network

CSDP e: Command Supply Discipline Program[me]
- e: Coordinated Ship Development Plan
- e: Customer Service Department Procedure

CSDR e: Control Storage Data Register
- e: Control Store Data Register
- e: Cross-Section Data Reduction

CSDS e: Chicago Sewage Disposal System
- e: Command Ship Data System
- e: Communication Signal Distribution System
- e: Constant Speed Drive/Starter

CSDTs e: Complete Set of Detection Tests

CSE d: Computer-Spracheingabegerät
- e: Cairo Studies in English (journ.)
- e: Calcutta Stock Exchange
- e: Central Signal Establishment
- e: Central Studies Establishment
- e: Certificate of Secondary Education
- e: Circuit Switched Exchange
- e: Civil and Sanitary Engineering
- e: Cold Start Entry
- e: Command Session End
- e: Communications Services and Education
- e: Communications Support Element
- e: Communications Systems Engineer
- e: Competitive Study Engineer
- e: Computer Science and Engineering
- e: Computer Systems Engineering
- e: Containment System Experiment
- e: Control and Switching Equipment
- e: Control Systems Engineering
- e: Core Storage Element

CSEA e: California State Electronics Association
- e: Combat System Engineering Authorization
- e: Core Store Effective Address

CSEAF e: Commander; Southern European Air Forces

CSEB e: Computer Science and Engineering Board

CSEC e: Clothing and Survival Equipment Change
- e: Cosecant

C/sec e: Cycle[s] per Second

CSect e: Control Section

CSED e: Computers, Systems and Electronics Division
- e: Consolidated Ship Electronic Design
- e: Coordinated Ship Electronics Design

CSEE e: Canadian Society for Electrical Engineering
- e: Canadian Society for Electrical Engineers

CSEF e: Current-Switch Emitter-Follower

CSEI e: Concentrated Solar-Energy Imitator

CSEIP e: Center for the Study of the Evaluation of Instructional Programs

CSEP e: Communications Security Education Program[me]

CSEPA e: Central Station Electrical Protection Association
- e: Central Station Electrical Protective Association

CSER e: Chlorosulfopolyethylene Rubber

CSERB e: Computers, Systems and Electronic Research Board
- e: Computer Science and Electronics Requirements Board

CSESF e: Commander; Southern European Sea Forces

CSF e: Caribean Sea Frontier
- e: Carrier Suppression Filter
- e: Casualty Staging Facility
- e: Central Switching Facility
- e: Cerebrospinal Fluid
- e: Channel Spread Function
- e: Colony Stimulating Factor
- e: Communication Serviceability Facilities
- e: Completely Symmetric Function
- e: Configuration State Function
- e: Contract Supply Facility
- e: Cylindrically Symmetrical Field
- e: One Hundred Cubic Feet
- f: Campagnie Générale de Télégraphie Sans Fil

CSFA e: Canadian Scientific Film Association

CSFAC e: Colorado Springs Fine Arts Center

CSFB e: Channel State Feedback

CSFE e: Canadian Society of Forest Engineers

CSFI e: Company Standard Form Instruction

CSFL e: Control Systems Field Laboratory

CSFP e: Confocal Spherical Fabry-Perot

CSFTI e: Committee on Southern Forest Tree Improvement

CSF-WR e: Cerebrospinal Fluid Wassermann Reaction

CSG e: Casing
- e: Chemical Smoke Generator
- e: Constructive Solid Geometry
- e: Context-Sensitive Grammar
- e: Control Synthetic Gas
- e: Course Selection Guide
- f: Centre Spatiale Guayana

CSGA e: Canadian Seed Growers Association
- e: Central States Gas Corporation

CS Gas e: Counterinsurgency Gas

CSGBI e: Cardiac Society of Great Britain and Ireland

CSGN e: Consign

CSGND e: Consigned
- e: Countersigned

CSGNEE e: Consignee

CSGNG e: Consigning

CSGNMT e: Consignment

CSGUS e: Clinical Society of Genito-Urinary Surgeons

CSH e: Calcium Silicate Hydrate
- e: Called Subscriber Held
- e: Conventional Spin Hamiltonian

CSHAFT e: Crankshaft

CSHG e: Cardio Scatter Histograph

CSHP e: Canadian Society of Hospital Pharmacists

CSI e: Called Subscriber Identification
- e: Cambridge Scientific Instruments
- e: Campus Studies Institute
- e: Chartered Surveyors Institution
- e: Chemical Substructure Index
- e: Co-elliptic Sequence Initiation
- e: Command String Interpreter
- e: Committee on Science Information
- e: Common Sound I
- e: Computer Sciences International
- e: Computer Society of India
- e: Computer Systems International
- e: Concentric Sequence Initiate
- e: Concentric Sequence Initiation
- e: Construction Specification[s] Institute
- e: Construction Surveyors Institute
- e: Continental United States Sustaining Increment
- e: Contractor Standard Item
- e: Control Signal[ling] Indicator
- e: Cosmic Sale Integration
- e: Current Source Inverter
- f: Cinemarique Scientifque Internationale

CSIC e: Computer Stock Inventory Control
- e: Computer System Interface Circuit[s]
- s: Consejo Superior de Investigaciones Cientificas

CSICC e: Canadian Steel Industries Construction Council

CSICOP e: Committee for the Scientific Investigation of Claims of the Paranormal

CSID d: Computer Sciences International Deutschland GmbH

CSID e: Consider

CSIDD e: Considered

CSIDL e: Considerable

CSIDN e: Consideration

CSIE e: Center for the Study of Information and Education

CSIEJB e: Certificate of the Sanitary Inspectors Examination Joint Board

CSIF e: Computer Sciences Information-France

CSIFET e: Charge Storage Insulated-gate Field Effect Transistor

CSIGO e: Chief Signal Officer

CSII e: Centre for the Study of Industrial Innovation

CSI Lamp e: Compact Source Iodide Lamp

CSIN e: Chemical Substances Information Network
- e: Computer Sciences Information-Nederland

CSINK e: Countersink

CS Input e: Chip Select Input

CSIO e: Cavity-Stabilized Impact Avalanche and Transit Time Oscillator

CSIP e: Committee for the Scientific Investigation of the Paranormal
- e: Comprehensive School Improvement Project

CSIR e: Council for Scientific and Industrial Research
- e: Council of Scientific [and] Industrial Research

CSIRAC e: Commonwealth Scientific and Industrial Research Automatic Computer

CSIRO e: Commonwealth Scientific and Industrial Research Organization
- e: Council for Scientific and Industrial Research Organizations

CSIROLCA e: Commonwealth Scientific and Industrial Research Organization Laboratory Craftsmen Association

CSIRONET e: Commonwealth Scientific and Industrial Research Organization Computing Network

CSIROTA e: Commonwealth Scientific and Industrial Research Organization Technical Association

CSIS e: Chemical Society Information Services

CSISRS e: Cross Section Information; Storage and Retrieval System

CSIT e: Chapin Social Insight Test

C-SIT e: Committee on Social Implications of Technology

CSIVP e: California State Influenza Vaccine Program[me]

CSJ e: Christian Science Journal (journ.)

CSJFET e: Charge Storage Junction Field Effect Transistor

CSK e: Cable Splicing Kit
- e: Cask
- e: Cathodic Survey Kit
- e: Co-operative Study of the Kuroshio (and Adjacent Region)
- e: Column Skipping Key
- e: Countersink
- e: Countersunk

CSKH e: Countersunk Head
CSK-O e: Countersink Other Side
CSL e: Canada Steamship Lines
- e: Cedar Springs Library
- e: Cesium Lodide
- e: Cesium Loside
- e: Chemical Society of London
- e: Circle of State Librarians
- e: Coaxial Slotted Line
- e: Code Selection Language
- e: Coles Signal Laboratory
- e: Combustion Systems Limited
- e: Command Signal Limitter
- e: Commercial Source Language
- e: Common Specification Language
- e: Commonwealth Serum Laboratories
- e: Communications Sub-Lieutenant
- e: Comparative Systems Laboratory
- e: Component Source List
- e: Computer Sensitive Language
- e: Computer Simulation Language
- e: Computer Structure Language
- e: Computer System Language
- e: Computer Systems Laboratory
- e: Console
- e: Constant Scattering Length
- e: Context-Sensitive Language
- e: Control and Simulation Language
- e: Control and Status Logic
- e: Controlled Saturation Logic
- e: Control Systems Laboratory
- e: Coordinated Science Laboratory
- e: Core Store Location
- e: Corn-Steep Liquor
- e: Corporate Source List
- e: Corporate Systems List
- e: Crane Stores Lighter
- e: Current Sinking Logic
- e: Current Sourcing Logic
- e: Current Switch Logic

CSLA e: Canadian School Library Association
- e: Church and Synagogue Library Association

CSLATP e: Canadian Society of Landscape Architects and Town Planners
CSLDB e: California Spanish Language Data Base
CSLEA e: Center for the Study of Liberal Education for Adults
CSLO e: Canadian Scientific Liaison Office
CSLP e: Center for Short-Lived Phenomena
CSLT e: Canadian Society of Laboratory Technologists
CSM e: Call Supervision Message
- e: Camborne School of Mines

CSM e: Cation Selective Membrane
- e: Cement-Sand-Molasses
- e: Central Storage Module
- e: Cerebrospinal Meningitis
- e: Chief Signalman
- e: Chopped Strand Mat
- e: Christian Science Monitor (journ.)
- e: Chronic Symptomatic Maladjustment
- e: Clear and Substract Magnitude
- e: Clock and Simulation Tape Maintenance
- e: Coaxial Switching Matrix
- e: Coffin Strategic Missile
- e: Colorado School of Mines
- e: Combustion Space Monitor
- e: Combustion Stability Monitor
- e: Command [and] Service Module
- e: Commission for Synoptic Meteorology
- e: Complete Sequential Machine
- e: Composite Signal Mixer
- e: Computer Store Matrix
- e: Computer System Manual
- e: Continental Shelf Mining
- e: Continuous Sampler Monitor
- e: Continuous Sheet Memory
- e: Copy Service for Microform
- e: Core Store Matrix
- e: Corn-Soya-Milk [Mixture]
- e: Corn Soybean Mixture
- e: Cosmopolitan School of Music
- e: Critical Supersaturation Model
- e: Crosspoint Switching Matrix
- e: Current Switching Mode

CSMA e: Carrier Sense[d] Multiple Access
- e: Chemical Specialities Manufacturers Association
- e: Communication Systems Management Association

CSMA-CA e: Carrier Sense Multiple Access with Collision Avoidance
CSMAcc e: Carrier Sense Multiple Access
CSMA-CD e: Carrier Sense Multiple Access with Collision Detection
CSMA-CD/CA e: Carrier Sense Multiple Access with Collision Detection/Collision Avoidance
CSMA-CR e: Carrier Sense Multiple Access with Contention Resolution
CSMCRI e: Central Salt and Marine Chemicals Research Institute
CSME e: Canadian Society for Mechanical Engineering
CSMITH e: Coppersmith
CSML e: Continuous Self-Mode Locking
CSM-LM e: Command Service Model-Lunar Module
- e: Command Service Module-Lunar Module

CSMLT e: Cambridge Studiesin Medieval Life and Thought (journ.)
CSMM e: Comborne School of Metalliferous Mining
CSMMG e: Chartered Society of Massage and Medical Gymnastics
CSMMS e: Christian Science Monitor Section (journ.)
CSMOL e: Control Station Manual Operating Level
CSMP e: Comprehensive School Mathematics Program[me]
- e: Continuous Simulation Modelling Program[me]
- e: Continuous System Modeling Program[me]
- e: Current Ships Maintenance Project

CSMPS e: Computerized Scientific Management Planning System
CSMRI e: Colorado School of Mines Research Institute
CSMRS e: Central Soil Mechanics Research Station
CSMS e: Computerized Specification Management System
CSMSW e: Carver School of Missions and Social Work
CSMT e: Circuit Switching Magnetic Tape
- e: Clock and Simulation-Tape Maintenance Message Type

CSM Test e: Convexity Symmetry; Maximum Test
CSN e: Canadian Switched Network
- e: Circuit Switching Network
- e: Colloidal Suspension
- e: Common Service Network
- e: Computer Services Network
- e: Contract Serial Number
- e: Contract Surgeon
- e: Control Symbol Number
- p: Companhia Siderurgica Nacional

CSNAR e: Charles Sheldon National Antelope Refuge
CSNDA e: Center for the Studies of Narcotic and Drug Abuse
CSNET e: Computer Science-Research Network
CSNF e: Common Source Noise Figure
CSNH e: Cincinnati Society of Natural History
CSNI e: Committee for the Safety of Nuclear Installations
CSNM e: Chief Superintendent; Naval Meteorology
CSNMDU e: Center for the Study of Non-Medical Drug Use
CSNO e: California School Nurses Organization
CSNRD e: Consortium for the Study of Nigerian Rural Development
CSNWR e: Carolina Sandhills National Wildlife Refuge
CSO e: Centralized Service Observation
- e: Central Seismological Observatory
- e: Central Sign off
- e: Chained Sequential Operation
- e: Chemical Stable Oxide
- e: Chief Signal Officer
- e: Chief Stores Officer
- e: Chief Supply Officer
- e: Clothing Supply Officer
- e: Combined Server Overflow
- e: Commanding Signal Officer
- e: Command Signals Officer
- e: Commonwealth Scientific Office
- e: Communication Standing Order
- e: Complex Safety Officer
- e: Computer Service Office
- e: Consecutive Store Organization
- e: Core Store Operand
- e: Cross Service Order

CSOC e: Consolidated Space Operations Center
- e: Convolutional Self-Orthogonal Code

CSOCR e: Code Sort Optical Character Recognition
CSOC's e: Convolutional Self-Orthogonal Codes
CSOD e: Crack Surface Opening Displacement
CS of A e: Canadian School of Artillery
CS of J e: Chemical Society of Japan
CS Oils e: Closed Systems Oils
CSOM e: Chief Sound Navigation and Ranging Man
CSOMH e: Chief Sound Navigation and Ranging Man; Harbor Defense
CSOP e: Current Switch-circuit Optimization Program[me]
CSOS e: Complementary Silicon on Sapphire
CSP e: Carotid Sinus Pressure
- e: Cast Steel Plate
- e: Central Signal Processor
- e: Central Switching Point
- e: Channel Scheduling Process
- e: Chartered Society of Physiotherapy
- e: Chlorosulphonated Polyethylene
- e: Chondroitinsulfat-Proteine

- e: Coder Sequential Pulse
- e: Coherent Signal Processor
- e: Commercial Subroutine Package
- e: Communicating Sequential Process[es]
- e: Communications / Symbiont Processor
- e: Communications Satellite Program[me]
- e: Communications Security Publication
- e: Comparing Storage Position
- e: Completely Self-Protected
- e: Computer Science Press
- e: Computer Simulation Program[me]
- e: Computer Society Press
- e: Concurrent Spare Parts
- e: Constant-Speed Drive
- e: Constant Satellite Power
- e: Constant Satellite Power-Sharing
- e: Container Service Port
- e: Contents of Selected Periodicals
- e: Continuous Sampling Plan
- e: Continuous Seismic Profiling
- e: Continuous Stratification Profiler
- e: Contractor Standard Part[s]
- e: Control and Switching Point
- e: Controlled Solder Process
- e: Controlled Surface Porosity
- e: Controlled Surface Process
- e: Control Setting Panel
- e: Control Signal Processor
- e: Control Switching Point
- e: Corporate Systems Practice
- e: Corresponding States Principle
- e: Council for Scientific Policy
- e: Count-Lea Strength Product
- e: Count Strength Product
- e: Cross System Product
- e: Current Systems Program[me]
- e: Custer State Park

CSPA e: California State Psychological Association
- e: Clay Sewer Pipe Association

C-SPAN e: Cable-Satellite Public Affairs Network

CSPC e: California State Polytechnic College

C/SPC e: Cost/Schedule Planning and Control

CSPCA e: Canadian Society for the Prevention of Cruelty to Animals

C/SPCS e: Cost/Schedule Planning and Control Specification
- e: Cost/Schedule Planning and Control System

CSPDN e: Circuit Switched Public Data Network

CSPE e: Control System for Plan Execution

CSPG e: Code Sequential Pulse Generator
- e: Common Source Power Gain

CSPI e: Center for Science in the Public Interest
- e: Center for Science in the Public Interest

CSPM e: Communications Security Publication[s] Memorandum
- e: Corporate Simulation and Planning Model

CSPMs e: Corporate Simulation and Planning Models

CSPO e: Communications Satellite Project Office
- e: Core Store Print-Out

CSPP e: Community Shelter Planning Program[me]
- e: Corruigated Structural Plate Pipe

CSPR e: Chlorosulphonated Polyethylene Rubber

CSPR(s) e: Christian Science Practitioner[s]

CSPRT e: Compound Sequential Probability Ratio Test

CSPS e: Cable-Suspended Pumping Station

CSPs e: Chlorosulphonated Polyethylenes

CSPS e: Christian Science Publishing Society
- e: Coherent Signal Processing System

CSPT e: Chief Specialist; Teacher

CSP-T e: Contents of Selected Periodicals-Technical

CSQ e: Central Site Queueing
- e: Coastal Dentry Quebec
- e: Communication Ship Quarters
- e: Cryptofacility Security Questionnaire

CSQM e: Climax Stock Quartz Monsonite

CSR e: Center for Space Research
- e: Channel Select Register
- e: Channel Service Register
- e: Check Status Reply
- e: Chemically Stimulated Rubber
- e: Cheyne-Stokes Respiration
- e: Circulating Shift Register
- e: Circulation Shift Register
- e: Circumsolar Radiation
- e: Civil Service Requirement
- e: Civil Service Retirement
- e: Civil Service Rule
- e: Clamped Speed Register
- e: Clamped Speed Regulator
- e: Combat Surveillance Radar
- e: Common Services Rack
- e: Communications Satellite Relay
- e: Compensation System Review
- e: Component Selection Review
- e: Compulsive Security Ritual
- e: Concentric Storage Rings
- e: Configuration Selection Register
- e: Console Send-Receive
- e: Continuous Speech Recognition
- e: Contract Status Report
- e: Control and Status Register
- e: Controlled Silicon Rectifier
- e: Control Shift Register
- e: Control Status Register
- e: Copper Sulfide Rectifier
- e: Corps Specifications Revision
- e: Corrected Sedimentation Rate
- e: Corrugated Steel Reinforcement
- e: Cortical Secretion Rate
- e: Counter / Shift Register
- e: Customer Service Representative
- e: Czechoslovak Socialist Republic

CSRA e: Central Savannah River Area
- e: Chicago Suburban Radio Association

CSRC e: Communication Science Research Center

CSRD e: Chief Superintendent; Research Department

CSRDF e: Civil Service Retirement and Disability Fund

CSRE e: Closed System Respirator Evaluator
- e: Corrected Sending Reference Equipment

CSRF e: Constant Slew Rate Filter

CS/RIC e: Chemical Society / Royal Institute of Chemistry

CSRL e: Center for the Study of Responsive Law
- e: Computer Science Research Laboratory

CSRM e: Controlled Solid Rocket Motor[s]

CSRMP e: Communications Sales Results Measurement Plan

CSRO e: Chief Superintendent Range Operation[s]
- e: Consolidated Standing Route Order
- e: Contract Service Rework Order[s]
- f: Comité scientifique pour les recherches océaniques

CSRP e: Cognitive Systems Research Program[me]
- e: Computer Systems Research Project

CSRS e: Central Sericulture Research Station
- e: Continuous Speech Recognition System

CSRUIDR e: Chemical Society Research Unit in Information Dissemination and Retrieval

CSS d: Computer System Service GmbH
- e: Canadian Standard Seismograph
- e: Captain; Signal School
- e: Character Sensing System
- e: Character Start-Stop
- e: Character String Scanner
- e: Chewing, Suckling, Swallowing
- e: Chroma Sweep Signal
- e: Cislunar Swing Station
- e: Clock Subsystem
- e: Clothing Sales Stores
- e: Coarse Sun Sensor
- e: Coded Switch System
- e: Colo[u]r Synchronization Signal
- e: Combat Sequence Summary
- e: Combat Service Support
- e: Command Session Start
- e: Commit Sequence Summary
- e: Common Services Subsystem
- e: Common Support System
- e: Communications Subsystem
- e: Completely Separating System
- e: Computerized Security System
- e: Computer Sales and Services
- e: Computer Scheduling System
- e: Computer Subsystem
- e: Computer System Services, Incorporated
- e: Computer System[s] Simulator
- e: Connection-Successful Signal
- e: Consecutive Sequential Storage
- e: Continuous System Simulation
- e: Contractor Storage Site
- e: Control Signalling Subsystem
- e: Control Statement Stream
- e: Control Stick Steering
- e: Conversational Software System
- e: Coordinated Situation System
- e: Cordless Switchboard System
- e: Corporate Shareholder System
- e: Critical Shear Stress
- e: Cryogenic[s] Storage System
- e: Current Steering Switch
- e: Customer Switching System

C & SS e: Clothing and Small Stores

CSSA e: Cactus and Succulent Society of America
- e: Central Supply Support Activity
- e: Clothing and Small Stores Account
- e: Combat Service Support Area
- e: Computer Society of South Africa

CSSB e: Compatible Single Sideband

CSSBI e: Canadian Sheet Steel Building Institute

CSSB System e: Compatible Single Sideband System

CSSC e: California Seismic Safety Commission

CSSCG e: Container Systems Standardization-Coordination Group

CSSD e: Chemical-Sensitive Semiconductor Device
- e: Communications System Status Display

CSSDA e: Council of Social Science Data Archives

CSSDs e: Chemical-Sensitive Semiconductor Devices

CSSEC e: Computer Systems Support and Evaluation Command

CSSF e: Clothing and Small Stores Fund
CSSG e: Computer Software and Services Group
CSSH e: Comparative Studies in Society and History (journ.)
CSSI e: Coriolis Sickness Susceptibility Index
CSSL e: Continuous System Simulation Language
- e: Continuous System Simulation Model
- e: Corporation Source System Listing
CSSM e: Compatible Single-Sideband Modulation
CSSN e: Circuit State Sequence Number
- e: Common Source Spot Noise
CSSNF e: Common Source Spot Noise Figure
CSSO e: Control Sequential Storage Organization
CSSP e: Center for Studies of Suicide Prevention
- e: Combined Services Support Program[me]
- e: Committee on Solar and Space Physics
- e: Continuous Systems Simulation Program[me]
CSSPT e: Common Supply Support
CSS[R] e: Command Suspended Session (Reactivate)
CSSR e: Consolidated Stock Status Report
CSSRC e: Canadian Social Science Research Council
CSSS e: Canadian Soil Science Society
- e: Combat Service Support System
- e: Council of State Science Supervisors
- e: Cross Spin Stabilization System[s]
CSST e: Compatible Sidelobe Suppression Technique
- e: Computer System Science Training
CST d: Zentistoke[s]
- e: Capillary Suction Time
- e: Capsule Systems Test
- e: Carrier[-Power] Supply; Transistorized
- e: Celeban Standard Time
- e: Centistoke[s]
- e: Central Standard Time
- e: Channel Status [Indicator]
- e: Channel Status Table
- e: Chief of Supplies and Transport
- e: Code Segment Table
- e: Colloidal System Test
- e: Combat Support Training
- e: Combined Service Territory
- e: Combined Station Tower
- e: Combined Station Power
- e: Combined Systems Test
- e: Commercial Subsurface Transformer
- e: Commit Start
- e: Communications Surveillance Transistor
- e: Complex Safety Technician
- e: Computer Subtraction Time
- e: Computer System Training
- e: Conceptual Systems Test
- e: Conical Shock Tube
- e: Console Terminal
- e: Consolidated Schedule Technique
- e: Container Service Tariff
- e: Continuously Stirred Tank
- e: Contour Start Threshold
- e: Control System Test
- e: Conventional Stability Talks
- e: Convulsive Shock Therapy
- e: Core Store Technique
- e: Council for Science and Technology
- e: Countdown Sequence Timer
- e: Critical Solution Temperature
- e: Crystaline Style
- e: Cycling Strength Test
CSTA e: Canadian Society of Technical Agriculturists
- e: Canterbury Science Teachers Association
- e: Coal and Slurry Technology Association
- e: Combat Surveillance and Target Acquisition
- e: Computer-Supported Telephony Application[s]
- e: Consolidating Station
CS & TAE e: Combat Surveillance and Target Acquisition Equipment
CSTAL Meth Capital e: Crystal Methamphetamine Capital
CSTAM e: Cost Amount
CSTAR e: Classified Scientific and Technical Aerospace Report[s]
CSTB e: Computer Science and Technology Board
CSTC e: Canadian Signals Training Centre
- e: Charleston Submarine Training Center
- e: Coordinating Scientific and Technical Council
CSTD e: Command Session Typed Data
- e: Committee on Science and Technology for Development
C-Steel e: Carbon Steel
CSTEX e: Cost Extension
CSTG e: Casting
CSTI e: Centre for Scientific and Technical Information
- e: Chattanooga State Technical Institute
- e: Clearinghouse of Scientific and Technical Information
- e: Cleveland Scientific and Technical Institution
- e: Committee on Scientific and Technical Information
- e: Control Surface Tie in
- e: Council of Scientific and Technical Institutes
CSTL e: Castellate
CSTO e: Chief Supply and Transport Officer
CSTOL e: Combined Short Takeoff and Landing
- e: Controlled Short Takeoff and Landing
CSTPC e: Cost Price
CSTR e: Canister
- e: Committee on Solar-Terrestrial Research
- e: Continuous Stirred Tank Reactor
- e: Current Status Register
CSTS e: Combined Systems Test Stand
- e: Combined System Test Stand
- e: Computer Sciences Teleprocessing System
- e: Computer Science[s] Time Sharing [System]
CSTU e: Combined System Test Unit
CSTUN e: Cost per Unit
CSTV e: Controlled Scan Televideo
CSTWS e: Command Session Two-Way Simultaneous
CSU e: Cache Store Unit
- e: California State University
- e: Casualty Staging Unit
- e: Catheter Specimen of Urine
- e: Central Statistical Unit
- e: Central Switching Unit
- e: Channel Service Unit
- e: Check Signal Unit
- e: Circuit Switching Unit[s]
- e: Clear and Substract
- e: Clock Synchronization Unit
- e: Colorado State University
- e: Combined Shaft Unit
- e: Common Services Unit
- e: Communications System User
- e: Computing Surveys (journ.)
- e: Connecticut State University
- e: Constant Speed Unit
- e: Core Storage Unit
- e: Customer Service Unit
- e: Customer Setup
- e: Cycle Stealing Unit
CSUB e: Center for the Study of Urban Poverty
CSUC e: California State Universities and Colleges
- e: California State University at Chico
CSU-DSU e: Customer Service Unit-Data Service Unit
CSUF e: California State University of Fresno
CSUH e: California State University at Humboldt
CSUI e: Command Session User Information
CSULA e: California State University at Los Angeles
CSULB e: California State University at Long Beach
CSULD e: Consulted
CSULG e: Consulting
CSUPO e: Chief Supply Officer
CSUS e: California State University at Sacramento
CSUs e: Circuit Switching Units
CSUSA e: Copyright Society of the U.S.A.
CSUSAF e: Chief of Staff; United States Air Force
CSUSB e: California State University at San Bernardino
CSUSD e: California State University at San Diego
CSUSF e: California State University at San Francisco
CSUSJ e: California State University at San José
CSV d: chemischer Sauerstoffverband
- d: chemischer Sauerstoffverbrauch
- d: Cycle-Stealing-Verfahren
- e: Cellular Size Volume
- e: Characteristic Statistical Value
- e: Chlorotic Spot Virus
- e: Circuit Switched Voice
- e: Conical Shell Vibration
- e: Corona Start[ing] Voltage
- e: Crew Safety Vehicle
CSVC e: Core Sample Vacuum Container
CSVR e: Closed User Group Selection and Validation Request
- e: Closed User Group Selection and Validation Response
CSW e: Channel Status Word[ing]
- e: Clear and Subtract Word
- e: Cleveland School of Welding
- e: Command Surveillance and Weather
- e: Concentrated Sea Water
- e: Continuous Seismic Wave[s]
- e: Control-power Switch
CSW Counter e: Clear and Subtract Word Counter
CSWFB e: Canadian Society of Wildlife and Fishery Biologists
CSWI e: Commission for Synoptic Weather Information
CSWS e: Crew-Served Weapon Sight
CSWY e: Causeway
CSY d: Codesynchronisiereinrichtung
CSZ d: Computer Systeme und Zubehör
- e: Copper, Steel or Zinc

CT d: Chemische Technik (journ.)
- d: Cicutoxin
- d: Codetabelle
- d: Computertomograph[ie]
- d: kontinentale Tropikluft
- e: Cable Tank
- e: Cable Test
- e: Cable Transfer
- e: Cable Tray
- e: Calcitonin
- e: Calculating Time
- e: Calibration Technician
- e: California Terms
- e: Candidate in Theology
- e: Cape Town
- e: Carat
- e: Carbon Tetrachloride
- e: Card-to-Tape
- e: Carrier Telegraph
- e: Carrier Telephone
- e: Carrier Telephony
- e: Carton
- e: Casket Tape
- e: Cassette Tape
- e: Cellular Therapy
- e: Celtium
- e: Cement
- e: Center
- e: Center Tap
- e: Center Tape
- e: Center Tapped
- e: Central Time
- e: Central Timing
- e: Centre Tap
- e: Centre Tapped
- e: Centurium
- e: Cephalotin
- e: Ceramic Tile
- e: Certificated Teacher
- e: Certificate of Title
- e: Certified Teacher
- e: Changeover Time
- e: Change Tracker
- e: Character Translator
- e: Charge Transfer
- e: Checkout Tape
- e: Check Tape
- e: Chemical Titles (journ.)
- e: Chief Telegrapher
- e: Cholesterol Total
- e: Chronometer Time
- e: Chrysler Technologies
- e: Cipher Telegram
- e: Circuit
- e: Circuit Technology
- e: Circuit Theory (journ.)
- e: Circulating Time
- e: Clock Time
- e: Clock Track
- e: Closed Thoracotomy
- e: Close Tolerance
- e: Coagulation Time
- e: Coal Transport
- e: Coarse Turned
- e: Coastal Telegraph[er]
- e: Coated Tablet
- e: Coded Tape
- e: Code Table
- e: Code Telegram
- e: Code Transcription
- e: Code Transformation
- e: Code Translator
- e: Cognitive Therapy
- e: Combination Tone
- e: Combined Trial[s]
- e: Command Tape
- e: Command Terminator
- e: Command Transmitter
- e: Comment
- e: Commercial Translator
- e: Communication[s] Trench
- e: Communications Technician
- e: Communications Terminal
- e: Compact Tension
- e: Compile Time
- e: Complete Translation
- e: Component Test
- e: Component Tree
- e: Compressed Tablet
- e: Compressed Term
- e: Computed Tomograph[y]
- e: Computerized Tomographic
- e: Computer Tape
- e: Computer Technology
- e: Computer Terminal
- e: Computer Test
- e: Computer Thermographer
- e: Computer Tomograph[y]
- e: Computer Typewriter
- e: Computing Time
- e: Conditional Transfer
- e: Conductivity Transmitter
- e: Conduit
- e: Conference Terms
- e: Confirmatory Test
- e: Connective Tissue
- e: Conning Tower
- e: Console Typewriter
- e: Contact Tension
- e: Container Terminal
- e: Continuity Transceiver
- e: Continuous Tone
- e: Contour Template
- e: Contrast Threshold
- e: Controlled Target
- e: Control Tape
- e: Control Total
- e: Control Transformer
- e: Control Transmitter
- e: Cordless Telephone
- e: Correction Table
- e: Corrective Therapist
- e: Corrective Therapy
- e: Correct Time
- e: Counter
- e: Counter Timer
- e: Counter Tube
- e: Crawler/Transporter
- e: Credit
- e: Critical Temperature
- e: Cruiser Tank
- e: Cubic Tonnage
- e: Current
- e: Current Transformer
- e: Testosterone-Controlled
- e: Total Capacitance
- f: Centre de Transit

ct l: centum

C & T e: Classification and Testing
- e: Control and Timing

CT²L e: Complementary-Transistor-Transistor Logic

CT-4 e: New Zealand-built Airtrainer Aircraft

CTA e: Cable Television Association
- e: Call Time Adjustor
- e: Canadian Tuberculosis Association
- e: Catamenia
- e: Cellulose Triacetate
- e: Cement-Treated Aggregate
- e: Chemical Toilet Association
- e: Chicago Transit Authority
- e: Chief Technical Adviser
- e: Collision Threat Assessment
- e: Combined Target Area
- e: Committee on Thrombolytic Agents
- e: Common Table[s] of Allowances
- e: Compatibility Test Area
- e: Component Test Area
- e: Computer Traders Association
- e: Connecticut Transportation Authority
- e: Contractor Technical Assistance
- e: Control Area
- e: Controlled Airspace
- e: Conurbation Transport Authority
- e: Corporate Technological Assignment
- e: Council for Technical Advancement
- e: Cyano-Trimethyl-Androsterone
- e: Cystine Trypticase Agar

CTAB d: Centyltrimethyl Ammonium Bromid[e]
- e: Cetyl Trimethyl Ammonium Bromide

CT Air e: Continental Tropical Air

CTA/IR e: Control Area / Instrument Restricted

CTA/IV e: Control Area / Instrument Visual

CTAK e: Cipher Text Auto Key

CTAL e: Container Terminals Australia Ltd

CTAM e: Climb to and Maintain
- e: Continental Tropical Air Mass

CTAMNPK e: Corps Troops Ammunition Park

CTAMNSUBPK e: Corps Troops Ammunition Subpark

CTAP e: Circuit Transient Analysis Program[me]
- e: Cleveland Transport Action Program[me]
- e: Contact Approach

CTAS e: Constant Temperature Anemometer System

CTAT e: Computerized Transverse Axial Tomography

CTA/VE e: Control Area / Visual Excepted

CTAX e: Climb to and Cross

CTB e: Cable Television Bureau
- e: Calibration Test Box
- e: Camera Talkback
- e: Cement-Treated Base
- e: Ceramic-Tile Base
- e: Coast Torpedo Boat
- e: Code Table Buffer
- e: Collateral Trust Bond
- e: Command Telemetry Buoy
- e: Commonwealth Telecommunications Board
- e: Comprehensive Test Ban
- e: Computer Time Brokers
- e: Concentrator Terminal Buffer
- e: Controlled Temperature Bath
- e: Control Test Bed
- e: Corporation for Television Broadcasts
- f: Centre Technique de Bois

CT-B/L e: Combined Transport Bill of Lading

CTBM e: Cetyl-Trimethyl-Ammonium Bromide

CTBORE e: Counterbore

CTBS e: Comprehensive Tests of Basic Skills

CTC d: Chlortetracyclin[ium]
- d: Chlortetrazyklin
- e: Cam Timing Contact
- e: Canberra Technical College
- e: Capsule Test Conductor
- e: Carbon[ite] Tetrachloride
- e: Card[s]-to-Tape Converter
- e: Cassette / Tape Controller
- e: Centralized Traffic Control
- e: Central Test Control
- e: Central Tracking Center
- e: Central Train Control

- e: Central Transformer Corporation
- e: Channel-to-Channel
- e: Channel Traffic Control
- e: Charge Transfer Complex
- e: Charleston Training Center
- e: Chicago Technical College
- e: Chlortetracycline
- e: Climate Test Chamber
- e: Clinical Trial Certificate
- e: Coarse Tar Concrete
- e: Coaxial Thermal Converter
- e: Color Television Committee
- e: Common Business-Oriented Language to Common Business-Oriented Language
- e: Communications, Telemetry and Command
- e: Compact Transpiration Cooling
- e: Compatibility Test Capsule
- e: Complementary Twos Complement
- e: Computer Technology Center
- e: Computer Test Console
- e: Computer Test Corporation
- e: Computer Time-Sharing Corporation
- e: Computer Transmission Corporation
- e: Computing Technology Center
- e: Concentration to Control
- e: Conditional Transfer of Control
- e: Constant Torque Compensation
- e: Contact
- e: Continuity Test Current
- e: Continuously Transposed Conductor
- e: Contractor Technical Compliance
- e: Counter Timer Circuit
- e: Counter Timer Control

CTCA e: Canadian Telecommunications Carriers Association
- e: Channel-to-Channel Adapter
- e: Channel and Traffic Control Agency

CTCB e: Contract Technical Compliance Board

CTCC e: Central Terminal Computer Controller
- e: Change to Center Control
- e: Changing to Center Control
- e: Contact Center Control
- e: Continental Division; Transport Control Center

CTCCC e: Close Type Control Circuit Contact

CTC Code e: Complementary Twos Complement Code

CTCD e: Contacted

CTCDB e: Computing Technology Center Data Base

CTCDE e: Commission Type Code

CTCF e: Channel and Technical Control Facility

CTCG e: Contacting

CTCI e: Contractor Technical Compliance Inspection

CTCL e: Community and Technical College Libraries
- e: Count Clock

CTCM e: Chrominance Time-Compression Multiplexing
- e: Computer Timing and Costing Model

CTC Machine e: Centralized Traffic Control Machine

CT/COSBA e: Cape Town Computer Services and Bureaux Association

CTCP e: Contract Task Change Proposal

CTCS e: Component Time Control System
- e: Consolidated Telemetry Checkout System

CTC Selector e: Centralized Traffic Control Selector

CTCSS e: Code Tone Call Selective Signalling
- e: Continuous Tone Coded Squelch System

CTCU e: Channel and Traffic Control Unit

CTD d: Chemo-Thermo-Desinfektion
- e: Certified Test Data
- e: Chambers Technical Dictionary
- e: Charged Tape Detection
- e: Charge Transfer Device
- e: Circulation Time Descending
- e: Classified Telephone Directory
- e: Coated
- e: Compile Time Diagnostic
- e: Conductivity-Temperature Depth
- e: Continued
- e: Continuity Tone Detector
- e: Controlled Thermolytic Dissociation
- e: Crew Task Demand
- e: Crew Task Detail
- e: Cumulative Trauma Disorder
- e: Current, Temperature, Depth

CTDA e: Ceramic Tile Distributors Association

CTDC e: Chemical Thermodynamic Data Center
- e: Control Track Direction Computer

CTDG e: Cartridge

CTDH e: Command and Telemetry Data Handling

CTDL e: Complementary Transistor Diode Logic

CTDN e: Countdown

CTDO e: Central Technical Documents Office

CTD Recorder e: Conductivity, Temperature, Depth Recorder

CTDs e: Charge Transfer Devices

CTDS e: Code-Translation Data System
- e: Count; Time Data System

CTD-Verfahren d: Chemo-Thermo-Desinfektions-Verfahren

CTE e: Cable Termination Equipment
- e: Central Telephone & Electronics
- e: Central Timing Equipment
- e: Channel Translating Equipment
- e: Charge-Transfer Efficiency
- e: Coefficient of Thermal Expansion
- e: Commander; Task Element
- e: Commercial Test Equipment
- e: Compile Time Expression
- e: Component Test Equipment
- e: Computer Telex Exchange
- e: Counter Total Exit
- e: Cross Track Error

CTEA e: Channel Transmission and Engineering Activation

CTEB e: Council of Technical Examining Bodies

CTEC e: Chemical Transportation Emergency Center
- e: Communication Technical Evaluation Console

CTEE e: Chlorotrifluor Ethylene

CTEIS e: California Total Educational Information System

CTEL e: Chief of Telecommunications

CTE Ltd. e: Chemical & Thermal Engineering Limited

CTEN e: Ctenophora

CTENO e: Ctenocephalides

CTES e: Computer Telex Exchange System

CTETOC e: Council for Technical Education and Training for Overseas Countries

CTEW e: Chief Technical Examiner of Works

CT Exam e: Computed Tomography Examination

CTF e: Call Treatment Flag
- e: Cavity Tunable Filter
- e: Ceramic-Tile Floor
- e: Ceramic Tube Fabrication
- e: Certificate
- e: Certify
- e: Change to Tower Frequency
- e: Chlorine Trifluoride
- e: Closed Transaction File
- e: Commander; Task Fleet
- e: Commander; Task Force
- e: Common Test Facility
- e: Continuous Tape Feed[ing]
- e: Contrast Transmission Function
- e: Core Test Facility
- e: Correction to Follow
- e: Cytotoxic Factor

CTFE e: Chlorotrifluoroethylene
- e: Colleges of Technology and Further Education

CTFM e: Continuous Transmission Frequency Modulated
- e: Continuous Transmission Frequency Modulation

CTFO e: Controlled Tuning Fork Oscillator

CTFOIC e: Cabinet Task Force on Oil Import Control

CTFT f: Centre Technique Forestier Tropical

CTFY e: Certify

CTG d: Cardiotokographie
- d: Casrdiotokogramm
- d: Kotangens
- e: Cartage
- e: Cartridge
- e: Coating
- e: Commander; Task Group
- e: Communications Task Group
- e: Components Technology Group
- e: Cotangent
- e: Counting
- e: Cutting

CTGA e: Ceylon Tea Growers Association

CTGE e: Cartage
- e: Cartridge

CTGF e: Clean Tanks, Gas Free

CTGH e: Cotangent; Hyperbolic
- e: Cryptograph

CTGI e: Canadian Test of General Information

C Th e: Candidate of Theology

CTH e: Cubic Feet per Hour

CTH r: Chalmers Techniska Högskola

CTI e: California Technical Institute
- e: Camera Timing Indicator
- e: Cartridge Television, Inc.
- e: Centralized Ticket Investigation
- e: Central Technical Institute
- e: Ceramic Tile Institute
- e: Charge Transfer Inefficiency
- e: Color Television, Inc.
- e: Command Technical Inspection
- e: Communication Technology Impact
- e: Communications Technician Intercept-radioman
- e: Communications Technology, Inc.
- e: Comparative Tracking Index
- e: Complaint Type Investigation
- e: Computer Translation, Inc.
- e: Conditional Transfer Instruction
- e: Contract Technical Instructor
- e: Cooling Tower Institute
- e: Crossroads Technical Instrumentation
- f: Centre de Traitement de l'Information

CTIAC e: Concrete Technology Information Analysis Center
CTIC e: Cable Television Information Center
- e: Cable Television Information Center
CTIN e: Continue
CTINUS e: Continuous
CTIO e: Cerro Tolo Interamerican Observatory
CTIS e: Clinical Text Inquiry System
- e: Crawler/Transporter Intercommunication System
CTK e: Crimping Tool Kit
CTL e: Cage Test Language
- e: Capstan Tracking Logic
- e: Carrier Tracking Loop
- e: Cassette Tape Loader
- e: Central
- e: Certified Tool List
- e: Checkout Test Language
- e: Cincinnati Testing Laboratories
- e: Combat Training Launch
- e: Common Target Language
- e: Compass Test Language
- e: Compiler Target Language
- e: Complementary Transistor Logic
- e: Component Test Laboratory
- e: Computer Technology Limited
- e: Constructive Total Loss
- e: Contents Listing
- e: Control
- e: Control Line
- e: Core Transistor Logic
- e: Core Transmission Logic
CTLCD e: Control Card
CTLD e: Controlled
CTLI e: Confidence Training Launching Instrumentation
CTLO e: Constructive Total Loss Only
CTLR e: Control Register
CTLU[N] e: Control Unit
CTLZ e: Control Zone
CTM d: Cardiotachometer
- d: Computertechnik Müller GmbH
- e: Cable Testing Meter
- e: Centrifugal Turning Moment
- e: Chief Torpedomans Mate
- e: Clock and Tone Module
- e: Close Talking Microphone
- e: Column Transition Matrix
- e: Communication[s] Terminal Module[s]
- e: Complete Treatment Module
- e: Composite Tape Memory
- e: Computerized Text Management
- e: Concordia Theological Monthly (journ.)
- e: Continuity Transceiver Module
- e: Contractor Technical Meeting
- e: Contract Termination Manual
- e: Critical Thermal Maxima
C & TM e: Clothing and Textile Materiel
CT Machine e: Computed Tomographic Machine
CTMC e: Communication[s] Terminal Module Controller
- e: Communication Terminal Module Controller
- e: Communication Terminal Multiplex Cabinet
CTMC-Subsystem e: Communications Terminal Modular Control Subsystem
CTME e: Chief Torpedomans Mate; Electrical
CTMFU e: Corporate Transfer Manufacturing File Update
CTMM e: California Test of Mental Maturity
CTMP e: Contractor Technical Manual Plan
CTMS e: Carrier Transmission Maintenance System
- e: Ceramic to Metal Seal
- e: Countermeasures
CT/MSS e: Crawler/Transporter/Mobile Service Structure
CTMT e: Containment
CTμL e: Complementary Transistor Micrologic
CTMV e: Chief Torpedomans Mate; Aviation
CTN d: Kotangens
- e: Cable Termination Network
- e: Cape Town
- e: Carton
- e: Cellulose Trinitrate
- e: Certification Test Network
- e: Clock Track Number
- e: Cotangent
CTNC e: Cross-Track Non-Contiguous
CTNDS e: Commercial Transport Navigation Display System
CTNE s: Compañía Telefonica Nacional de España
CTNF e: Controlled Thermonuclear Fusion
CTNH e: Hyperbolic Cotangent
CTNSS e: Centre for Thai National Standard Specifications
CTO e: Cancelled-to-Order
- e: Cavity Tuned Oscillator
- e: Central Telegraph Office
- e: Charge Transforming Operator
- e: Chief Tank Officer
- e: Chief Technical Officer
- e: Chief Transportation Officer
- e: Circular Terminal Orbit
- e: Combined Transport Operator
- e: Control Tower Operator
- e: Corporate Telecommunications Operation
- e: Cut-Off
- e: Cut-Through Operate
CTOC e: Center-to-Center
- e: Centre to Centre (GB)
CTOC e: Communications Technical Operations Center
- e: Corps Tactical Operations Center
CTOCU e: Central Technical Order Control Unit
- e: Central Technical Order Coordination Unit
C to F e: Center to Face
CTOL e: Conventional Take-off and Landing
CTORP e: Chief Torpedoman
CTOS e: Cassette Tape Operating System
CTP d: chemisch-technische Produkte
- d: Cosinustiefpaß
- d: Cytidin[-5'-]Triphosphat
- e: 2-Chloro-6-(Trichloromethyl) Pyridine
- e: California Test of Personality
- e: Capacitor Test Program[me]
- e: Carriage Tape Punch
- e: Central Transfer Point
- e: Change Transfer Photography
- e: Charge Transforming Parameter[s]
- e: Clear Text Reading
- e: College Training Plan
- e: Command Translator and Programmer
- e: Commission on Thermal Physiology
- e: Compile Time Procedure
- e: Completed Token Packet
- e: Composite Test Pattern
- e: Computer Test Routine
- e: Confidence Test Program[me]
- e: Consolidated Telecommunications Program[me]
- e: Constant-ground Terminal-Transmit Power
- e: Constant Total Power
- e: Contributor
- e: Control Test Panel
- e: Controlled Temperature Profile
- e: Coordinated Test Plan
- e: Cytidine Triphosphate
- e: Cytosine Triphosphate
CTPB e: Carboxyl-Terminated Polybutadiene Binder
CTPIB e: Carboxy Terminated Polyisobutylene
CTPOA e: Commonwealth Telephone and Phonogram Officers Association
CT-Probe e: Compact Tension-Probe
CTPs e: Composite Test Patterns
CTPTA s: Centro Tropical de Pesquisas y Tecnologías de Alimentos
CTPV e: Coal Tar Pitch Volatile
CTPVs e: Coal Tar Pitch Volatiles
CTR d: Zentner
- e: C-band Tracking Radar
- e: Canaveral Test Report
- e: Carrier Receiver
- e: Caviation Tendency Ratio
- e: Center
- e: Centner
- e: Central Tool Room
- e: Centre
- e: Certification Test Requirement[s]
- e: Certified Test Record[s]
- e: Certified Test Report[s]
- e: Certified Test Result[s]
- e: Complementary Transistor Resistor
- e: Contour
- e: Contractual Technical Report[s]
- e: Contributor
- e: Control
- e: Controllable Twist Rotor
- e: Controlled Thermonuclear Reaction
- e: Controlled Thermonuclear Reactor
- e: Controlled Thermonuclear Research
- e: Core-Transistor Register
- e: Counter
- e: Critical Temperature Resistor
- e: Current Taper Ratio
- e: Current Technical Reports
- e: Current Transfer Ratio
- e: Cutter
CTRA e: Coal Tar Research Association
CTRAP e: Customer Trouble Report Analysis Plan
Ctr Appl Ling e: Center for Applied Linguistics
Ctr Byz e: Center for Byzantine Studies
Ctr Chin Stud e: Center for Chinese Studies
Ctr Drv e: Counter Drive
CTRG e: Centering
Ctr Gr Sel Ent e: Counter Group Selector Entry
Ctr Gr Sel Ex e: Counter Group Selector Exit
Ctr Gr Selr Ent e: Counter Group Selector Entry
Ctr Gr Selr Ex e: Counter Group Selector Exit
CTRI e: Catholic Tape Recorders, International
- e: Central Tobacco Research Institute
Ctr Info Am e: Center for Information on America
CTRIS e: Canadian Transportation Research Information Service
CTR/IV e: Control Zone/Instrument Visual

CTRL e: Complementary Transistor Register Logic
- e: Control
- e: Control Engineering (journ.)
- e: Cotton Technological Research Laboratory
CTRLCKT[S] e: Control Circuit[s]
CTR Logic e: Complementary Transistor Resistor Logic
CTRLR e: Controller
CTRP e: Controlled Thermonuclear Research Program[me]
Ctr Pos Ex e: Counter Position Exit
Ctr Pre-Col e: Center for Pre-Columbian Studies
CTRS e: Centers
- e: Centres (GB)
CTRs e: Certified Test Records
- e: Controlled Thermonuclear Reactors
Ctr Sci Pub e: Center for Science in the Public Interest
Ctr Sci Study Rel e: Center for the Scientific Study of Religion
Ctr S & SE e: Asian Center for South and Southeast Asian Studies
CTRU e: Colonial Termite Research Unit
CTR/VE e: Control Zone / Visual Exempted
Ctry Life Am e: Country Life in America (journ.)
CTS e: Cable Terminal Section
- e: Cable Tuning Section
- e: Cable Turning Section
- e: Cab Tyred Sheathed
- e: Cabtyre Sheathing
- e: Canadian Technology Satellite
- e: Canadian Telecommunications Satellite
- e: Canadian Test Satellite
- e: Canadian Thoracic Society
- e: Carats
- e: Card-to-Magnetic Tape-Conversion System
- e: Carriage Tape Simulator
- e: Carrier Test Switch
- e: Carrier Transfer Station
- e: Cartographic Test Standard
- e: Center for Theoretical Studies
- e: Centralized Title Services
- e: Central Tactical System
- e: Central Terminal Station
- e: Central Timing System
- e: Certification Test Specification
- e: Cesium Time Standard
- e: Channel Time Slot
- e: Charge Transfer Spectrum
- e: Charge Transfer State
- e: Chicago Telephone Supply
- e: Circuits to Specifications
- e: Circuit Test Set
- e: Clear-to-Send
- e: Coaxial Terminal Switch
- e: Combined Test Stand
- e: Command and Telemetry System
- e: Communication and Tracking Subsystem
- e: Communications Technician Special-devices-operator
- e: Communications Technology Satellite[s]
- e: Communication Test Station
- e: Communication[s] Terminal Synchronous
- e: Compass Tilt Signal
- e: Compile Time Statement
- e: Complementary Transistor Switch
- e: Components Technology and Standardization [Manual]
- e: Component Test System
- e: Component Test[ing] Set
- e: Computerized Type Setting
- e: Computerized Type System
- e: Computer Telegram System
- e: Computer Telewriter System[s]
- e: Computer Test[ing] Site
- e: Computer Test Stand
- e: Computer Thermographers
- e: Computer Typesetting
- e: Consolidated Tin Smelters
- e: Consumer Transaction System
- e: Container Transport System
- e: Contingency Transfer System
- e: Contract Termination Settlement
- e: Contralateral Threshold Shift
- e: Conversational Terminal System
- e: Conversational Time-Sharing
- e: Corporate Terminal System
- e: Cosmic Top Secret
- e: Courier Transfer Station
- e: Course Training Standard
- e: Critical Time Slice
- e: Cryogenic Temperature Sensor
- e: Cryogenic Temperature Source
C & TS e: Communications and Technical Services
CTSA e: Catholic Theological Society of America
CTS Cable e: Cab Tyre Sheathed Cable
CT Scan e: Computerized Tomography Scan
CT Scanning e: Computerized Tomographic Scanning
CTSEL e: Conversational Terminal System European Laboratories
CTSI e: Central Terminal Signalling Interface
- e: Computer Transceiver Systems, Inc.
CTSK e: Countersunk
CTSP e: Contract Technical Service Personnel
CTSS e: Classroom Test Support System
- e: Compatible Time-Shared System
- e: Compatible Time-Sharing System
- e: Computer Time-Sharing Service
CTS System e: Conversational Time-Sharing System
CTST e: Coaxial Triple-Stud Tuner
CTT d: Computertechnik und -technologie GmbH
- e: Cable Trouble Ticket
- e: Card-to-Tape
- e: Carriage Tension Tape
- e: Carrousel Transfer Tube
- e: Card-to-Tape Tape
- e: Central Trunk Terminal[s]
- e: Character Translate Table
- e: Colour Trace Tube
- e: Columbia Technical Translations
- e: Compressed Tablet Triturate
- e: Computerized Transaxial Tomography
- e: Computerized Transverse Tomography
- e: Corrugated Teflon Tubing
- e: Crew Transfer Tunnel
- e: Critical Tracking Task
- p: Correios e Telecommuniações de Portugal
CTTA e: Control Tape
CTTAP e: Card-to-Tape Program[me]
CTTC e: Canadian Troops Training Centre
- e: Central Telegraph Test Center
CTTE e: Cintered Tungsten-Tantalum Emitter
- e: Common Time-Division Multiple Access Terminal Equipment
CTTL e: Complementary-Transistor-Transistor Logic
CTTP e: Card-to-Tape Program[me]
CTTS e: Charge Transfer to Solvent
CTTSC e: Certificate in the Teaching and Training of Subnormal Children
CTU e: Cartridge Tape Unit
- e: Centigrade Thermal Unit
- e: Central Telephone and Utilities Corporated
- e: Central Terminal Unit
- e: Central Timing Unit
- e: Commander; Task Unit
- e: Commercial Telegraphers Union
- e: Communication[s] Terminal Unit
- e: Compatibility Test Unit
- e: Components Test Unit
- e: Construction Training Unit
C-Tube e: C-shaped Tube
CT-Umschlagplatz d: Container-umschlagplatz
CTUNA e: Commercial Telegraphers Union of North America
CTV d: computerisierte Textverarbeitung
- e: Cable Television
- e: Cargo Trim Valve
- e: Cartrivision
- e: Cassette Television
- e: Closed-circuit Television
- e: Coaxial Thermal Voltmeter
- e: Colo[u]r Television
- e: Command Test Vehicle
- e: Commercial Television
- e: Compile Time Variable
- e: Constant Tangential Velocity
- e: Control Test Vehicle
CT Value e: Call Time Value
CTVD e: Cinema Television Digest
CTVM e: Colo[u]r Television Monitor
CTW e: Childrens Television Workshop
- e: Console Typewriter
- e: Control Typewriter
CTWT e: Counterweight
CTX e: Computer Telex Exchange
- e: Cortex
CTXCO e: Centrex Central Office
CTXCU e: Centrex Customer
CTY e: Capacity
CTZ e: Chlorothiazide
- e: Control Zone
CT Zone e: Chemoreceptor Trigger Zone
CU d: Curacao (Kfz)
- d: Kupfer
- d: Umsteigecontainer
- e: 5-Chlorouracil
- e: Cambridge University
- e: Camouflage Unit
- e: Capacity Upchange
- e: Capital University
- e: Card Unit
- e: Carleton University
- e: Catholic University
- e: Central Unit
- e: Check[ing] Unit
- e: Chymotrypsin Unit
- e: City University
- e: Clafkin University
- e: Clark University
- e: Cleanup
- e: Clinical Unit
- e: Clock Unit
- e: Close-Up
- e: Coefficient of Utilization
- e: Colgate University
- e: Colorado University
- e: Columbia University
- e: Commitment Unit
- e: Common Update
- e: Common User
- e: Communications Unit
- e: Computer

- e: Computing Unit
- e: Concordia University
- e: Condition Unknown
- e: Construction Unit
- e: Container Unit
- e: Contents Unknown
- e: Control Unit
- e: Cooperative Union
- e: Copper
- e: Cornell University
- e: Correlation Unit
- e: Creighton University
- e: Crosstalk Unit
- e: Crystal Unit
- e: Cuba
- e: Cube
- e: Cubic
- e: Cumberland University
- e: Cumulative
- e: Cumulus
- e: Customer Use
- e: Cyclotron Unit

Cu l: cuprum
CUA d: computerunterstützte Ausbildung
- e: Catholic University of America
- e: Center for Urban Analysis
- e: Central Unit Assembly
- e: Channel Unit Address
- e: Computer-Aided Instruction
- e: Computer Unit Assembly
- e: Computer Users Association

CUAD s: Cuadrado
CUAG e: Computer Users Association Group
- e: Computer Users Associations Group

CUAP e: Catholic University of America Press
CUAS e: Cambridge University Agricultural Society
- e: Computer Utilization Accounting System

CUASRLL e: Catholic University of American Studies in Roman Languages and Literatures (journ.)
CUB d: Kupferdraht mit Baumwollumflechtung; einfach
- e: Central Unit Buffer
- e: Control Unit Busy
- e: Copper Band
- e: Cubic
- e: Cursor Backward

cúb s: cúbico
CUBA e: Control Unit Bit Array
CUBB d: Kupferdraht mit Baumwollumflechtung; doppelt
CUBE e: Cooperating Users of Burroughs Equipment
CUB EXP e: Cubical Expansion
CUBIT e: computerunterstützte Bild- und Textbearbeitung
CUBOL e: Computer Usage Business Oriented Language
CUBS e: College and University Business (journ.)
CUC e: Canberra University College
- e: Chronic Ulcerative Colitis
- e: Computer Usage Company
- e: Computer Users Committee
- e: Concentrator Unit Control
- e: Crystal Unit Cell

CU CAP e: Cubic Capacity
CUCB e: Control Unit Control Block
CUCC e: Computers in the Undergraduate Curriculum Conference
CU CM e: Cubic Centimeter[s]
- e: Cubic Centimetre[s] (GB)

CU CM/G e: Cubic Centimetre[s] per Gramme
CU CM/S e: Cubic Centimetre[s] per Second
CU Cong e: Cumulus Congestus
CUCOSS e: California Universities Council on Space Sciences
CUCPAT e: Computer Usage Company Programming Aptitude Test[s]
CUCS e: Computation-Universal Cellular Space
- e: Computers in the Undergraduate Science Curriculum

CU-CSD e: Cornell University-Computer Science Department
CUD e: Cursor Down
- e: Cursor Downward

CUDA[S] e: Barracuda[s]
CUDAS e: Computer Ultrasonic Data Analysis [System]
CUDA System e: Computer Ultrasonic Data Analysis System
CUDAT e: Common User Data [Terminal]
CUDC e: Computer Usage Development Corporation
CU DM e: Cubic Decimetre[s]
CUDN e: Common User Data Network
CUDOS e: Comprehensive University of Dayton Online-Information Service
- e: Continuously Updated Dynamic Optimizing System

CUDS e: Barracuda
CUE e: Center for Urban Education
- e: Chemical Underwater Explosive
- e: Coastal Upwelling Experiment
- e: Command Uplink Electronics
- e: Communications Unit Executer
- e: Communications Unit Executor
- e: Component Utilization Effectiveness
- e: Computer Update Equipment
- e: Computer Updating Equipment
- e: Computer Utilization Efficiency
- e: Computer Utilizing English
- e: Configuration Utilization Efficiency
- e: Configuration Utilization Evaluator
- e: Control Unit End
- e: Cooperating Users Exchange
- e: Correction Update Extension
- e: Culture, Understanding and Enrichment

CUEA e: Coastal Upwelling Ecosystem[s] Analysis
CUEBS e: Commission on Undergraduate Education in Biological Sciences
CUE/DSO e: Configuration Usage Evaluator/Data Set Optimizer
CUEING e: Control Unit Error Insertion Generator
CUEP e: Central Unit on Environmental Pollution
CUEPEND e: Control Unit End Pending
CUERAC e: Computer Controlled Random-Access Cartridge Libraries
CUERD e: Committee for Upgrading Environmental Radiation Data
CUERL e: Columbia University Electronic[s] Research Laboratory
CUES e: Center for Urban Environmental Studies
- e: Computer Utility Educational System
- e: Computer[s] Utilities for Education Systems

CUESTA e: Communications User Emulated System for Traffic Analysis
CUF d: Kupferfaser
- e: Columbia University Forum (journ.)
- e: Cumuliform
- e: Cursor Forward

CUFR e: Cumulus Fractus
CU FT e: Cubic Feet
- e: Cubic Foot

CUFTMIN e: Cubic Feet per Minute
CUFTSEC e: Cubic Feet per Second
CUG e: Closed User Group
- e: Computer Utilization Group

CUGOA e: Closed User Group with Outgoing Access
CUGS e: Columbia University Germanic Studies (journ.)
CUH e: Control Users Handbook
- e: Cursor Home

CU Hum e: Cumulus Humilis
CUI e: Chymotrypsin Units Inhibited
- e: Control Unit Interface

CU IN e: Cubic Inch
CU IN/LB e: Cubic Inch per Pound
CUJT e: Complementary Unijunction Transistor
CUL d: Kupferlackdraht
- e: Cambridge University Library
- e: Canonical Unit of Length
- e: Columbia University Library
- e: Command Up-Link
- e: Cooper Union Library
- e: Cornell University Library
- e: Culvert

CULDATA e: Comprehensive Unified Land Data
CULL e: Corning Uniformity Limit Level
- e: Cullage
- e: Cullboard
- e: Culling

CULP e: California Union List of Periodicals
- e: Computer Usage List Processor

CULPRIT e: Cull and Print
CULS e: Coalition in the Use of Learning Skills
CULT e: Chinese University Language Translator
- e: Common User Land Transportation
- e: Cultivated
- e: Cultivation

CULT e: Cultural
- e: Culture

CULT ANTHRO[S] e: Cultural Anthropologist[s]
- e: Cultural Anthropology

CULTECH e: Cultural Technical
CUM e: Central Unit-Memory
- e: Computer Utilization Monitor
- e: Cubic Meter[s]
- e: Cubic Metre[s]
- e: Cumulative
- s: Centro Universitario México

CUMARC e: Cumulated Machine-Readable Cataloguing
CUMC e: Computer Utilization Monitor Cumulative
CUM/D e: Cubic Meters per Day
CUM/H e: Cubic Metre per Hour
CUM/KG e: Cubic Metre per Kilogram[me]
CUM LAUDE e: Computerized Understanding of Morphology-Language Acquisition under Developme
CUMM e: Cubic Millimeter[s]
- e: Cubic Millimetre[s] (GB)

CUM/SEC e: Cubic Meters per Second
CUmw d: Codeumwerter
CU-N e: Cumulo-Nimbus
CUN n: Convent van Universiteits-Bibliothecarissen in Nederland
CU Nb e: Cumulo-Nimbus
CU-NIM e: Cumulo-Nimbus
CUNY e: City University of New York

CUO e: Copper Oxide
CUODIC e: Combinational, Unilateral, One-Dimensional, Iterative Circuit[s]
CUODICs e: Combinational, Unilateral, One-Dimensional, Iterative Circuits
CUP d: Computer Umwelt-Projekt
- d: Containerumschlagplatz
- e: Cambridge University Press
- e: Canadian University Press
- e: Columbia University Press
- e: Communication[s] User[s] Program[me]
- e: Cornell University Press
- e: Cupola
CUPAL d: Verbundmetall aus Kupfer und Alluminium
CUPE e: Cranfield Unit for Precision Engineering
CUPID e: Combat Using Price Incentives Doctrine
- e: Commercial Users Program[me] for Index Data
- e: Computer for Up-Range Point-of-Impact Determination
- e: Conversional Utility Program[me] for Information Display
- e: Create, Update, Interrogate and Display
CUPL e: Cornell University Programming Language
CUPLE e: Cambridge University Press Limited Editions
CUPR e: Catholic University of Puerto Rico
CUPRITE e: Cuprous Oxide
CUPSEP d: computerunterstützte Planung und Steuerung der EDV-Produktion
CUR e: Command Uplink Request
- e: Complex Utility Routine[s]
- e: Construction and Use Regulations
- e: Curacao
- e: Currency
- e: Current
CURAM e: Current Amount
CURB e: Campaign on the Use and Restriction of Barbiturates
Cur Biog e: Current Biography (journ.)
CURDT e: Current-Period Date
CURE e: Clean Urban River Environments
- e: Computer Users Replacement Equipment
CURE Project e: Clean Urban River Environment Project
CURES e: Computer Utilization Reporting System
CURF e: Canadian Uranium Research Foundation
CURFCO e: Common Usage Radio Frequency Checkout
CURFCOE e: Common Usage Radio Frequency Checkout Equipment
Cur Hist e: Current History (journ.)
Cur Lit e: Current Literature (journ.)
CURLS e: College, University and Research Libraries Section
CURR e: Currency
- e: Current
Curr Eng Pract e: Current Engineering Practice (India) (journ.)
CURRIC e: Curriculum
Curr Sci e: Current Science (India) (journ.)
CURRT e: Current
Curr Ther Res; Clin Exp e: Current Therapeutic Research; Clinical and Experimental (journ.)
Curr Top Microbiol Immunol e: Current Topics in Microbiology and Immunology (journ.)
CURS d: chronisches unspezifisches respiratorisches Syndrom
CURSR e: Cursor
CURT e: Current
- e: Curtain
CURTS e: Common User Radio Transmission Sounding
- e: Common User Radio Transmission System
- e: Communications User Radio Transmission Sounding
CURTS System e: Common User Radio Transmission Sounding System
CURV e: Cable-controlled Underwater Recovery Vehicle
- e: Cable-controlled Underwater Research Vehicle
- e: Cable-Operated Unmanned Recovery Vehicle
- e: Controlled Unmanned Recovery Vehicle
CUS d: Kupferdraht mit Seidenumspannung; einfach
- e: Catholic University School
- e: Common User System
- e: Control; Utility and Support
CUSA e: Centrifugal Urine Separator Assembly
- e: Customer Address
CUSAAF e: Chief of the United States Army Air Force
CUSC e: Channel Unit Signal Controller
CUSCL e: Customer Class
CUSECS e: Cubic Feet per Second
CUSM e: Columbia University School of Medicine
CUSNM e: Customer Name
CUSNR e: Customer Number
Cu_2SO e: Copper Sulfate
CUSP e: Central Unit for Scientific Photography
- e: Commonly Used System Program[me]
- e: Cryptographic Unit Support Program[me]
CUSPO e: Customer Purchase Order
CUSR e: Central United States Registry
CUSS d: Kupferdraht mit Seidenumspannung; doppelt
- e: Computerized Uniterm Search System
- e: Continental, Union, Shell, Superior
CUST e: Cumulo-Stratus
- e: Customer
CUSTOD e: Custodian
CUSUSWASH e: Council of United States Universities for Soil and Water Development in Arid and Subhumid Areas
CUSW/NAS e: Comittee on Undersea Warfare-National Academy of Sciences
CUT e: Cableless Underwater Television
- e: California United Terminal
- e: Canonical Unit of Time
- e: Circuit Under Test
- e: Computer Under Test
- e: Control Unit Tester
- e: Coordinated Universal Time
- e: Cutler[y]
- e: Cutter
- e: Cutting
CUTCOM e: Cutter Compensation
CUTE e: Common Business Oriented Language Update and Test Environment
CUTI d: Kupfer-Titan
CU TI SEC e: Cubic Feet per Second
CUT Link e: Cableless Underwater Television Link
CUTS e: Cassette User Tape System
- e: Computer-Utilized Turing System
- e: Computer Utilized Turning System
CUTV e: Cableless Underwater Television
CUU d: computerunterstützter Unterricht
- e: Cursor Up
- e: Cursor Upward
CUUL d: Computer-unterstützter Unterricht der Lufthansa
CUV d: computerunterstützte Verwaltung
CUW e: Conflicting-Use Writeback
CU YD e: Cubic Yard[s]
CV d: Chrominanz-Verzögerungsleitung
- e: Calorific Value
- e: Capacitance Voltage
- e: Cardiovascular
- e: Cataclysmic Variable
- e: Character Variable
- e: Check Value
- e: Check Valve
- e: Coefficient of Variation
- e: Collection Voucher
- e: Combat Vehicle
- e: Command Vehicle
- e: Command Verification
- e: Common Version
- e: Computer Variable
- e: Computervision
- e: Concave
- e: Concentrated Volume
- e: Conditional Variable
- e: Consonant Vowel
- e: Constant Value
- e: Constant Velocity
- e: Constant Viscosity
- e: Constant Voltage
- e: Constant Volume
- e: Continuously Variable
- e: Continuous Vulcanization
- e: Contrast Value
- e: Control Van
- e: Control Volume
- e: Convergence
- e: Conversion
- e: Converter
- e: Convertible
- e: Corpuscular Volume
- e: Coulomb-Volt
- e: Counter Voltage
- e: Cruise Vehicle
- e: Cultivar
- e: Molecular Heat
- e: [Single] Cotton Varnish
- e: Specific Heat at Constant Volume
- l: Curriculum Vitae
- s: Caballo De Vapor
CV4 Convair 440 e: Airliner
CVA e: Calendar Variations Analysis
- e: California Vehicle Act
- e: Cerebro-Vascular [Accident]
- e: Costo-Vertebral Angle
- e: Current Variable Attenuator
CVAA f: Centre de Vulgarisation Aéro-Astronautique
CVAC e: Consolidated Vultee Aircraft
CVAP e: Cloud Velocity Analysis Program[me]
CVB d: Verbleibecontainer
- e: Calcium-Vanadium-Bismuth
- e: Combined Very High Frequency Band
- e: Convert to Binary
CVBEM e: Complex Variable Boundary Element Method
CVC e: Carrier Virtual Circuit
- e: Cesium Vapour Cathode
- e: Compact Video Cassette
- e: Conserved Vector Current
- e: Consonant-Vowel-Consonant
- e: Contactless Vacuum Controller
- e: Convalescent Camp
- e: Cryogenic Vacuum Calorimeter
- e: Current Voltage Characteristic
- e: Current Voltage Converter

CVCC e: Canadian Videotext Consultative Committee
- e: Compound, Vortex and Controlled Combustion
CV/CC e: Constant Voltage-Constant Current
CVCC Engine e: Compound, Vortex and Controlled Combustion Engine
CVCC Motor e: Compound, Vortex and Controlled Combustion Motor
CV/CL e: Constant Voltage / Current Limiting
CVCN e: Collegiate Video Counseling Network
CVCS e: Cardiovascular Conditioning Suit
CVD e: Carbon Vacuum Deoxidized
- e: Cardiovascular Disease
- e: Cash Versus Documents
- e: Central Vehicle Depot
- e: Cerebrovascular Disease
- e: Chemical[ly] Vapo[u]r-Deposited
- e: Chemical Vapour Deposition
- e: Colo[u]r Vision Deviant
- e: Column Valve Diaphragm
- e: Communication Valve Development
- e: Constancy of Visual Direction
- e: Convert [to] Decimal
- e: Coordination [of] Valve Development
- e: Coupled-Vibration-Dissociation
- e: Current-Voltage Diagram[me]
CVDE c: Columbia Viking Desk Encyclopedia
CVD-Stahl d: Carbon Vacuum Deoxidized-Stahl
CVDT e: Constant Volume Drop Time
CVDV e: Coupled Vibration-Dissociation-Vibration
CVE e: Calibration Vibration Exciter
- e: Chemical Vapo[u]r Epitaxy
- e: Complex Vehicle Erector
- e: Continuously Variable for Emergency
- e: Customer-Vended Equipment
CVEH e: Combat Vehicle
CVF e: Continuously Variable Filter
- e: Controlled Visual Flight
CVFR e: Controlled Visual Flight Rules
CVFS e: Cesium Vapour Feed System
- e: Circular Variable Filter Spectrometer
CVG l: Cutis Verticis Gyrata
CVGB e: Cable Vault Ground Bar
CVH e: Center for Visual History
- e: Combined Ventricular Hypertrophy
CVHS e: Chelsea Vocational High School
CVI e: Cape Verde Islands
- e: Center of Visual Impact
- e: Cerebrovascular Insufficiency
- e: Certified Vendor Information
- e: Chronic Venous Insufficiency
- e: Common Variable Immunodeficiency
- e: Current Variable Inductor
CVIC e: Conditional Variable Incremental Computer
C Virus[es] e: Coxsackie Virus[es]
CVIS e: Computerized Vocational Information System
CVK e: Centerline Vertical Keel
CVL e: Caravelle Jet Airplane
- e: Continuous Velocity Log
CVLB e: Conference on Very Large Data Bases
CVLS e: Central Vacuum Loading System
CVM e: Common Business-Oriented Language Virtual Machine
- e: Consumable Vacuum Melt
- e: Cylindrical Vibration Mount
CVMA e: Canadian Veterinary Medical Association
CVMAS e: Continuously Variable Mechanical Advantage Shifter
CVN e: Carbohydrate Vitamin Nitrogen
- e: Casualty Vulnerability Number
- e: Convene
CVNPA e: Carolina Virginia Nuclear Power Associates Inc.
CVNTL e: Conventional
CVO e: Chief Veterinary Officer
c.v.o. l: conjugata vera obsretrica
CVOU e: Cardiovascular Observation Unit
CVP d: Codevergleich-positiv
- e: Central Valley Project
- e: Central Venous Pressure
- e: Chemical Vapour Plating
- e: Citrus Vitamin P
- e: Climate, Vegetation, Productivity
- e: Cost Volume Profit
- s: Corporación Venezolano del Petróleo
CVPDS e: Command Video Prelaunch Distribution System
CVPR e: Computer Vision and Pattern Recognition
- e: Computer Vision and Pattern Recognitor
- e: Conference on Computer Vision and Pattern Recognition
CVR e: Cardiovascular Renal
- e: Cardiovascular Respiratory
- e: Carrier Vessel Reactor
- e: Ceramic Vacuum Relay
- e: Cerebrovascular Resistance
- e: Cine Video Recording
- e: Cockpit Voice Recorder
- e: Command Voltage Regulator
- e: Computer Voice Response
- e: Constant Velocity Recording
- e: Constant Voltage Reference
- e: Continuous Vertical Retort
- e: Continuous Video Recorder
- e: Controlled Visual Rule[s]
- e: Crystal Video Receiver
- e: Current-Voltage Regulator
- e: Current Viewing Resistor
CVRD e: Cardiovascular Renal Disease
CVR Disease e: Cardiovascular Renal Disease
CVRE e: Center Voltage Range Element
CVRP e: California Video Research Project
CVRSN e: Conversion
CVS d: computergesteuertes Verkehrssystem
- e: Cardiovascular Surgery
- e: Cardiovascular System
- e: Change Verification System
- e: City Vehicle System
- e: Common Video System
- e: Computer-controlled Vehicle System
- e: Computer Vision Syndrome
- e: Consecutive Voyages
- e: Constant Voltage Source
- e: Constant Volume Sample[r]
- e: Constant Volume Sampling
- e: Constrained Voltage Source
- e: Continuous Vent System
- e: Conversational System
- e: Current-controlled Voltage Source
CVSB e: Compatible Vestigial Sideband
- e: Continuous Variable Slope Delta
CVSDM e: Continuously Variable Slope Delta Modulation
CVSD Modulation e: Continuously Variable Slope Delta Modulation
CVSF e: Conduction Velocity of Slower Fibres
CVSG e: Channel Verification Signal Generator
CVS Test e: Constant Volume Sample Test
CV Syllables e: Consonant-Vowel Syllables
CVT d: chemische Verfahrenstechnik
- e: Chemical Vapor Transport
- e: Communication Vector Table
- e: Concept Verification Testing
- e: Conditional Variable Test
- e: Constant Voltage Transformer
- e: Continuously Variable Transmission
- e: Controlled Variable Time
- e: Convertible
CVTR e: Carolina Virginia Tube Reactor
- e: Charcoal Viral Transport [Medium]
CVTR Medium e: Charcoal Viral Transport Medium
CVTs e: Constant Voltage Transformers
CVU e: Constant Voltage Unit
CVV e: Control Variable Valve
CV-Verfahren d: Continuous Vulcanization-Verfahren
CVWS e: Combat Vehicle Weapons System
CVWSs e: Combat Vehicle Weapons Systems
CVX e: Convex
CW d: Codewandler
- d: Codewort
- d: Computerwoche (journ.)
- d: Kondensator / Widerstand
- e: Call[s] Waiting
- e: Call Word
- e: Carrier Wave
- e: Caseworker
- e: Cell Wall
- e: Cement/Water
- e: Channel Word
- e: Chemical Warfare
- e: Chemical Warhead
- e: Chemical Week (journ.)
- e: Child Welfare
- e: Chronometer Time Minus Watch Time
- e: Clear and Write
- e: Clerk of Works
- e: Clockwise
- e: Close Wound
- e: Code Word
- e: Cold Water
- e: Cold Welding
- e: Command Word
- e: Composite Wave
- e: Composite Wing
- e: Computer Word
- e: Computer World
- e: Computerworld (journ.)
- e: Concealed Weapons
- e: Concertina Wire
- e: Connected With
- e: Constant Wear
- e: Constant Weight
- e: Continuous Wave[s]
- e: Control Word
- e: Cool White
- e: Copper Weld
- e: Counterweight
- e: Course Write
- e: Coursewriter
- e: Crawler Way
- e: Crutch Walking
- e: Cubic Weight
- e: Curtiss Wright
C & W e: Cable & Wireless [Limited]
- e: Cable and Wireless
- e: Caution and Warning
- e: Control and Warning

CWA e: Catering Wages Act
- e: Clean Water Act
- e: Communications Workers of America
- e: Constant Wattage Autotransformer
- e: Contractor Work Authorization
- e: Control Word Address
- e: County Water Authority

CWAEC e: County War Agricultural Executive Committee

CWAR e: Continuous Wave Acquisition Radar

CWAS e: Contractors Weighted Average Share[-in-Risk]

CWB e: Canadian Weekly Bulletin
- e: Canadian Wheat Board
- e: Capillary Water Barrier
- e: Child Welfare Bureau
- e: Constant Wattage Ballast
- e: Current Working Block

CW BRD e: Continuous Wave Broad

CWBTS e: Capillary Whole Blood True Sugar

CWBW e: Chemical Warfare-Bacteriological Warfare
- e: Chemical Warfare-Biological Warfare

CWC e: Cam Wedge Clamp
- e: Central Warning Center
- e: Chemical Warfare Committee
- e: Chemical Weapons Convention
- e: Child Welfare Centre
- e: Clear Write Condition
- e: Cloud Water Content
- e: Command Word Chaining
- e: Conventional War Capability
- e: Curtiss-Wright Corporation

CWCCI e: Crayon, Water Color and Craft Institute

CWCS e: Combined Wheat Control Secretariat

CWD d: Computerwartungsdienst GmbH
- e: Civilian War Dead
- e: Clerical Work Data
- e: Continuous Wave Detector
- e: Creosoted Wood Duct
- e: Cyclotron Wave Device

CWDBS e: Colorado Water Data Bank System

CWE e: Caution and Warning Equipment
- e: Cockcroft-Walton Experiment
- e: Coil Winding Equipment
- e: Commonwealth Edison Company
- e: Contractors Work Estimate
- e: Current Working Estimate

CWEA e: Caution and Warning Electronics Assembly
- e: Caution Warning Electronics Assembly

CWEALTH e: Commonwealth

CWEU e: Central Wind Extraction Unit

CWF e: Califoria Wildlife Federation
- e: Composite Wave Filter
- e: Control Word Format
- e: Course Write Facility
- e: Course Writing Facility
- e: Cross Wind Force

CWFF e: Closed; Well-formed Formula

CWFFEMM e: Continuous Wave Fixed Frequency Electromechanically Modulated

CWFM e: Continuous Wave Frequency Modulated

CWFS e: Crashworthy Fuel System[s]

CWG e: Card-and-heald Wire Gauge
- e: Closed Wave-Guide
- e: Communications Working Group
- e: Conformal Wire Grating
- e: Constant-Wear Garment
- e: Continuous Wave Gas
- e: Corrugated Wire Glass

CWGC e: Copper Wire Counterpoise Ground

CWGV d: Chronik des Wiener-Goethe-Vereins (journ.)

CWHB e: Citrated Whole Human Blood

CWI e: Call Waiting Indication
- e: Cardiac Work Index
- e: Clear Word Identifier
- e: Continuous Wave Illuminator

CWIC e: Compiler for Writing and Implementing Compilers

CWIF e: Continuous Wave Intermediate Frequency

CWIK e: Chemical World Index Key
- e: Cutting with Intent to Kill

CWINC e: Central Waterways; Irrigation and Navigation Commission

CWINJ e: Cold Weather Injury

CWINRS e: Central Waterpower; Irrigation and Navigation Research Station

CWIP e: Construction Work in Progress

CWIR e: Continuous Wave Illuminator Radar

CWIS e: Chain Weizmann Institute of Science

CWIT e: Colour-Word Interference Test
- e: Concordance Words in Titles

CWL e: Cable and Wireless, Limited
- e: Constant Word Length
- e: Continuous Wave Laser

CWLA e: Child Welfare League of America

CW Laby e: Chemical Warfare Laboratory

CWLS e: Conventional Weighted Least Square

CWM e: Change Weight Manifest
- e: Coil Winding Machine
- e: Compact Watch Module

CWMA e: Chemical Ware Manufacturers Association

CWMOD e: Control Word Modification

CW Mod 1 Cy e: Control Word Mode 1 Cycle
- e: Control Word Modification 1 Cycle

CWMTU e: Cold Weather Material Test Unit

CWN e: Certificate of War Necessity
- e: Contract Work Notification

CW NRW e: Continuous Wave Narrow

CWO e: Carrier Wave Oscillator
- e: Chemical Warfare Officer
- e: Communication Watch Officer
- e: Continuous Wave Oscillator
- e: Custom Work Order

CWOP e: Cold Weather Operations

CWP e: Circulating Water Pump
- e: Civil Works Program[me]
- e: Coal Workers Pneumoconiosis
- e: Coast Watch Post
- e: Cold Working Pressure
- e: Communicating Word Processor
- e: Computer Word Processing
- e: Cumulative Weight Percent
- e: Current Word Pointer

CWPC e: Cam Wedge Power Clamp
- e: Central Water and Power Commission

CWPD e: Central Water and Power Department

CWPE e: Chemical Warfare Protective Equipment

CWPI e: Canadian Wood Products Industries
- e: Configuration Word Package Item

CWPM e: Correct Words per Minute

CWPRS e: Central Water and Power Research Station

CWPS e: Communicating Word Processing System

CWR e: Cold Water Resistant
- e: Cold Water Rinse
- e: Continuous-Welded Rail

CWRA e: California Water Resources Association

C/W Ratio e: Cement/Water Ratio

CWRC e: Climb Well to Right of Courses

CWRD e: Chemical Warfare Research Department

CWRes e: Chemical Warfare Reserve

CWRM e: Cell Water Removal Mechanism

CWRR e: Curtiss Wright Research Reactor

CWRSM e: Case-Wester Reserve School of Medicine

CWRU e: Case Western Reserve University

CWS e: California Water Service
- e: Canadian Welding Society
- e: Canadian Wildlife Service
- e: Casework Supervisor
- e: Caution and Warning System
- e: Center Wireless Station
- e: Center Work System
- e: Central Warning Site
- e: Central Wireless Station
- e: Chandraprabha Wildlife Sanctuary
- e: Chemical Warfare School
- e: Chemical Warfare Section
- e: Chemical Warfare Service
- e: Chilled Water Supply
- e: Clockwise
- e: Cold Water Soluble
- e: Cold Water Supply
- e: Collision Warning System
- e: Command Word Sequence
- e: Committee for Whaling Statistics
- e: Compiler Writing System
- e: Complex Wiring System
- e: Continental Wage Schedule
- e: Control Wheel Steering
- e: Cooling Water System
- e: Copper Weld Steel
- e: Countersunk Wood Screw
- e: Crew Weapons Sight
- e: Curve Warning Signal

CWSch e: Chemical Warfare School

CWSD e: Continuous Wave Space Duplexed

CW SIG GEN e: Continuous Wave Signal Generator

CWSO e: Chemical Warfare Service Officer

CWSOC e: Continuous Wave Separated Orbit Cyclotron

CWSP e: College Work-Study Program[me]
- e: Communications with and Service to the Public

CWSRes e: Chemical Warfare Service Reserve

CW System e: Cloth Wrapping System

CWT d: CO_2-Wendelbandtunnel
- e: Carrier Wave Telegraphic
- e: Carrier Wave Telegraphy
- e: Carrier Wave Transmission
- e: Central War Time
- e: Central Winter Time
- e: Chief Water Tender
- e: Colo[u]r Word Test
- e: Command Word Trap
- e: Computer World Trade
- e: Continuous Wave Tunable
- e: Cooperative Wind Tunnel
- e: Hundredweight

CWTC e: California World Trade Center
- e: Chemical Warfare Technical Committee
- e: Cold Weather Training Center

CWTD e: Continuous-Wave Target Detector

CWTDC e: Continuous-Wave Target Detection Console
CWTG e: Computer World Trade Group
CWTL e: Hundredweight Long
CWTR e: Chemical Warfare Training Regiment
- e: Climb Well to Right
CW Trf Cy e: Control Word Transfer Cycle
CWTSH e: Hundredweight Short
CWU e: California Western University
- e: Caution and Warning Unit
- e: Chemical Workers Union
CWV e: Continuous-Wave Video
- e: Crest Working Voltage
CWY e: Clearway
CX e: Central Exchange
- e: Cervix
- e: Chest X-Ray
- e: Compatible Extension
- e: Complex
- e: Control Transmitter
- e: Convex
- e: Correct Copy
- e: Current Transformer
CXA e: Central Exchange Area
CXH e: Calyx Hole
CX-HLS e: Cargo-Experimental Heavy Logistics System
CXI e: Crosstell Input
CXN e: Correction
CXO e: Crosstell Output
CXR e: Carrier
- e: Chest X-Ray
CXRL e: Commonwealth X-Ray and Radium Laboratory
CXT e: Common External Tariff
CY d: Cyclophosphamid
- d: Cystein
- e: Calendar Year
- e: Calender Year
- e: Capacity
- e: Carry
- e: Container Yard
- e: Copy
- e: Cubic Yard
- e: Currency
- e: Current Year
- e: Cyanic
- e: Cyanide
- e: Cyanite
- e: Cyanogen
- e: Cybernetics
- e: Cycle[s]
- e: Cycling
- e: Cyclonium
- e: Cyclophosphamide
- e: Cylinder
- e: Cyprus
- e: Cypern (Kfz)
CYAN e: Cyanamid
- e: Cyanic
- e: Cyanide
- e: Cyanogen
Cyanier d: Cyanierung
CYB e: Canada Year Book
CYB e: Cybernetic
- e: Cyberneticist
- e: Cybernetics
CYBE e: Cybernetica (journ.)
CYBER e: Cybernetics
CYBERCULT e: Cybercultural[ly]
- e: Cyberculture
CYBERLOG e: Cybernetic Logistics
CYBERSOFT e: International Symposium on Cybernetics and Software
CYBORG[S] e: Cybernetic Organism[s]
CYC e: Cyclazocine
- e: Cyclazogine
- e: Cycle[s]
- e: Cycle Count[er]
- e: Cycling
- e: Cyclone[s]
- e: Cyclop[a]edia
- e: Cyclotron
CYCL e: Cycle[s]
- e: Cyclostomata
- e: Cyclotron
- e: Cyxclic
CYCLAZ e: Cyclazocine
CYCLE e: Bicycle
- e: Motorcycle
Cyclisier d: Cyclisierung
CYCLO e: Cyclop[a]edia
- e: Cyclopedic
- e: Cyclophosphamide
- e: Cyclopropane
- e: Cyclorama
- e: Cyclotron
CYCLON e: Cyclonometer
CYCLON GAS e: Cyanide, Chlorine and Nitrogen
CYCTL e: Cycle Control
CYD e: Cubic Yard[s]
- e: Currant Yellow Dwarf
- e: Cycle Delay
- e: Cyditin
- e: Cytidin[e]
CYDAC e: Cytophotometric Data Conversion
- e: Cytophotometric Data Converter
CYDel e: Cycle Delay
CYFLO e: Cylinder Overflow
Cyg l: Cygnus
CYI e: Canary Islands
CYL e: Cylinder
- e: Cylindrial
- e: Cylindric[al]
- e: Cylindrical Lens
- e: Cylindroid
CYL CART e: Cylindrical Cartridge
CYL L e: Cylinder Lock
CYM e: Cymbal[s]
CY Maj e: Cycle; Major
CYMB e: Cymbeline
CY Mi e: Cycle; Minor
CYN e: Cyanide
- e: Cynarine
CYNAP Reaction e: Cytotoxicity Negative Absorption Positive-Serum-Cell Reaction
CYNOL e: Cynologic[al][ly]
- e: Cynologist
- e: Cynology
CYP e: Cypress
- e: Cyprus
CYP Timber e: Cypress Timber
CYS d: Cystin
- e: Cysteine
- e: Cystine
CYS e: Cystoscopy
CY Sh e: Cycle Shift
CYSTI e: Cystoscope
CYSTO e: Cystoscope
- e: Cystoscopic Examination
CYSYS e: Center for Cybernetics System Synergism
CYT e: Cytology
- e: Cytosine
- e: Cytrochrom
CYTOECO e: Cytoecologic[al][ly]
- e: Cytoecologist
- e: Cytoecology
CYTOL e: Cytological
- e: Cytologist
- e: Cytology
CYTOMORPH e: Cytomorphologic[al][ly]
- e: Cytomorphologist
- e: Cytomorphology
CYTOPATHO e: Cytopathogenic[al][ly]
- e: Cytopathogenicity
CYTOPHOTO e: Cytophotometer
- e: Cytophotometric[al][ly]
- e: Cytophotometry
CYTOSTAT e: Cytostatic[al][ly]
CYTO SYST e: Cytochrome System
CYTOTECH e: Cytotechnic[al][ly]
- e: Cytotechnician
- e: Cytotechnologist
- e: Cytotechnology
CYXCT e: Cycle Count
CZ d: Cetanzahl
- d: Chemiker-Zeitung
- d: Codezahl[en]
- d: Computerzeitung (journ.)
- d: Containerzug
- e: Call on Zero
- e: Canal Zone
- e: Coastal Zone
- e: Communications Zone
- e: Convergence Zone
CZAM e: Computer Zero Access Memory
CZAS e: Computer Zero Access Store
CZB e: Carbon Zinc Battery
CZBA e: Canal Zone Biological Area
CZC e: Canal Zone Code
- e: Chromated Zinc Chloride
CZCA e: Coastal Zone Conservation Act
CZCS e: Coastal Zone Colo[u]r Scanner
CZD e: Calculated Zenith Distance
- e: Combat Zone Distance
CZECH e: Czechoslovak[ia]
Czech J Phys e: Czechoslovak Journal of Physics (journ.)
CZI e: Crystalline Zinc Insulin
CZM e: Compass Azimuth
CZ Material e: Czochralski Material
CZN d: Containerzugnetz
C Zone e: Commercial Zone
CZP e: Chicago Zoological Park
CZR e: Combat Zone Radius
CZRT e: Count Zero Refill Trigger
CZS d: Computerzahlungsservice
CZ [Silicon] e: Czochralski Zone [Silicon]
CZT e: Chirp Z-Transform
CZW d: codierter Zeitwert

D-Da

D d: Dämmzahl
- d: Dampf
- d: Dampfdurchsatz
- d: Dampfer
- d: Dampfleistung
- d: Dampfschiff
- d: Dämpfung
- d: Daten[fernsprecher]
- d: Datum
- d: Dauer[hub]
- d: Deci-
- d: Deck[blatt]
- d: Decke
- d: Deckflugzeug
- d: Deckgebirge
- d: Deckung[sbahn]
- d: Deka
- d: Deklination
- d: Delta
- d: Demodulator
- d: Deskriptor
- d: Determinante
- d: Deterministisch
- d: Deuterium
- d: Deuteron
- d: Deutschland (Kfz)
- d: Deutschlandfunk
- d: Dextro[se]
- d: Dezi (10^{-1})
- d: Dezimalpotenz
- d: Diamant
- d: Diameter
- d: Dichte
- d: Diesel[motor]
- d: Differential[getriebe]
- d: Differenz[ial]
- d: Diffus[e]
- d: Diffusionsbeiwert
- d: Diffusionskoeffizient
- d: Diffusionskonstante
- d: Diluvialboden
- d: Diode
- d: Dioptrie
- d: Direktor (Antennenteil)
- d: Diskriminator
- d: Dispersionsfarbstoff
- d: Disponiert
- d: Distanz
- d: Divisionsarzt
- d: Dokumentation
- d: Dominanz
- d: Doppel[schluß]
- d: Doppelstrom
- d: Doppeltorbagger
- d: Doppeltwirkend
- d: Dorsal
- d: Dose
- d: Dosis
- d: Drachen[boot]
- d: Draht[funk]
- d: Drainanschluß
- d: Drama
- d: Drehen
- d: Drehmaschine
- d: Drehstrom
- d: Dreiphasenstrom
- d: Dringend
- d: Dringlichkeitsstufe
- d: Drucker
- d: Druckluft[antrieb]
- d: Druckmittel[antrieb]
- d: Druckseite
- d: Durchdringbarkeit
- d: Durchgang[sbohrung]
- d: Durchgangszug
- d: durchgehender [Zug]
- d: Durchgriff
- d: Durchmesser
- d: duritisch
- d: Dusche
- d: Düse
- d: Düsen-
- d: Düsenweite
- d: Dynamisch
- d: Dynamotor
- e: Angular Deformation
- e: Daily
- e: Danger[ous]
- e: Darcy
- e: Dark
- e: Data
- e: Date
- e: Day[light]
- e: Dead [Space Gas]
- e: Deadweight
- e: Deca
- e: Deceased
- e: December
- e: Deciduous
- e: Decimal
- e: Decision
- e: Declination
- e: Decompose[d]
- e: Decomposition
- e: Deep
- e: Defence (GB)
- e: Defense (US)
- e: Deflection
- e: Defocussing
- e: Deformation
- e: Degeneracy
- e: Degree [of Curve]
- e: Delay
- e: Delete
- e: Deliver[ed]
- e: Delivery
- e: Delta-code for letter D
- e: Demand [Meter]
- e: Density
- e: Dented
- e: Deplacement
- e: Deposit
- e: Depot
- e: Depression
- e: Depth
- e: Derivation
- e: Derivative
- e: Derive[d]
- e: Dermatologist
- e: Descent[ing]
- e: Desert[ed]
- e: Design
- e: Destination
- e: Destroyed
- e: Destroyer
- e: Detail
- e: Determinant
- e: Deterministic
- e: Detonation
- e: Deuterium
- e: Develop[ed]
- e: Development
- e: Deviation
- e: Dextrorotatory
- e: Diameter
- e: Diametric
- e: Diamond
- e: Died
- e: Dielectric [Flux Density]
- e: Difference
- e: Differentiation
- e: Diffusing Capacity
- e: Diffusion Constant
- e: Digit[al]
- e: Dihydrouridine
- e: Dimensional
- e: Diopter
- e: Dioptric Power
- e: Direct[ion]
- e: Directivity[factor]
- e: Directorate
- e: Disaster
- e: Discharged
- e: Disconnect
- e: Disease
- e: Dispenser
- e: Displacement
- e: Display
- e: Dissipation[factor]
- e: Dissipation factordose
- e: Distance
- e: Distant
- d: Distinguish[ed]
- e: Distortion[factor]

- e: Disturbed Site
- e: Divide
- e: Division
- e: Divorce[d]
- e: Doctor
- e: Document[ation]
- e: Domain
- e: Dorsal
- e: Double[t]
- e: Doubled
- e: Doubler
- e: Doubtful
- e: Douglas
- e: Dowager
- e: Down
- e: Drag
- e: Drain
- e: Draught
- e: Draw[ing]
- e: Drift
- e: Drive[r]
- e: Driving
- e: Drizzle
- e: Drizzling
- e: Drone-Control Version
- e: Duration
- e: Dust
- e: Dutch
- e: Duty [cycle]
- e: Duty factor
- e: Dwarf
- e: Dyne
- e: Electrostatic Flux Density
- e: Fraunhofer Lines Caused by Sodium
- e: Propeller Diameter

d l: decimus
D3 e: Airplane
D4 e: Airplane
D-5-HS e: Dextrose 5 Percent in Hartmans Solution
D-5-S e: Dextrose 5 Percent in Saline
D 5 W e: Dextrose 5% in Water
DA a: Danmark

- a: Dansk
- a: Derte år
- a: Dette Aar
- d: Außendurchmesser
- d: Dampferanlegestelle
- d: Daten[analysator]
- d: Datenaufbereitung
- d: Datenausgabe[bus]
- d: Datenausgang
- d: Deka (10^1)
- d: Desoxyadenosin
- d: Deutsches Archiv für Geschichte (journ.)
- d: dezimale Addition
- d: Dienstanruf
- d: Dienstanschluß
- d: Dienstanweisung
- d: Dienstauweis
- d: Dieselmotorenanlage[n]
- d: Differentialanalysator
- d: Digitalausgabe
- d: Dimensionsaufgabe
- d: Dipolantenne
- d: Doppelader
- d: Doppeladler
- d: Drehantrieb
- d: Durchführungsanordnung
- d: Durchschnittsauflage
- e: Data Access
- e: Data Acquisition
- e: Data Address
- e: Data Administrator
- e: Data Analysis
- e: Data Analyzer
- e: Data Array
- e: Data Automation
- e: Data Available
- e: Datolite
- e: Day after Acceptance
- e: Deca
- e: Decimal Add[er]
- e: Decimal Addition
- e: Decimal to Analog[ue]
- e: Decoded Address
- e: Defence Act
- e: Defense Aid
- e: Define Area
- e: Defined Adult
- e: Delayed Action
- e: Delayed Alternation
- e: Delayed Answer
- e: Delayed Arming
- e: Delta Amplitude
- e: Demand Assignment
- e: Denmark
- e: Density Altitude
- e: Dental Apprentice
- e: Deoxyadenosine
- e: Department of Agriculture
- e: Deployment Assembly
- e: Dermatology Associates
- e: Design Automation
- e: Destination Address
- e: Detector Amplifier
- e: Detergent Aid
- e: Device Adapter
- e: Device Address
- e: Difference Amplifier
- e: Difference Answering
- e: Differential Address
- e: Differential Amplifier
- e: Differential Analyzer
- e: Differential of Amplification
- e: Differential of Area
- e: Diffused-base Alloy
- e: Digit Absorbing
- e: Digit Absorption
- e: Digital-Analog[ue]
- e: Digital Access
- e: Digital Alternator
- e: Digital-to-Analog
- e: Diphenylchlorarsine
- e: Diphenylchlorasine (Deadly Gas)
- e: Diploma in Aesthetics
- e: Diploma in An[a]esthetics
- e: Direct Access
- e: Direct Acting
- e: Direct Action
- e: Direct Address[ing]
- e: Direct Ascent
- e: Directional Aerial
- e: Directional Antenna
- e: Director Angle
- e: Director of Aircraft
- e: Director of Artillery
- e: Directory Assistance
- e: Disassemble
- e: Discharge Afloat
- e: Discharged Alive
- e: Discrete Address
- e: Discriminant Analysis
- e: Disk Access
- e: Disk Address
- e: Display Adapter
- e: Dissertation Abstracts (journ.)
- e: Dissolved Acetylene
- e: Distribution Amplifier
- e: District Attorney
- e: Doctor of Arts
- e: Documentation Abstracts (journ.)
- e: Documentation Aids
- e: Documents Attached
- e: Documents for Acceptance
- e: Doesn't Answer
- e: Does not Answer
- e: Donor-Acceptor
- e: Dopamine
- e: Double Acting
- e: Double Action
- e: Double Aged
- e: Double Amplitude
- e: Double Armo[u]red
- e: Drift Angle
- e: Drop Address
- e: Drum Address
- e: Dummy Address
- e: Dummy Aerial
- e: Dummy Antenna
- e: Dummy Argument
- e: Converter Device Converting Digital Input to Analog Input
- N: Deciampere
- R: Dalniya Aviatsiya

D & A e: Dialling and Answering

- e: Drawn and Annealed

DAA d: Dihydrocyaluminiumaminoazetat

- e: Danish Atlantic Association
- e: Data Access Arrangement
- e: Data Automation Activity
- e: Decimal Adjust Accumulator
- e: Dental Assistants Association
- e: Diacetone Acrylamide
- e: Diacetone Alcohol
- e: Diaminoacetanilide
- e: Dictionary of Architectural Abbreviations
- e: Diploma of the Advertising Association
- e: Direct Access Arrangement
- e: Discharge Always Afloat
- e: Doctor of Applied Arts
- e: Dual Address Adapter

DAA & AM e: Defense Aid Aircraft and Aeronautical Material
DAAC e: Data Acquisition and Analysis Complex

- e: Digital Adaptive Area Correlation
- e: Director of Allied Air Cooperation

DAAD d: Deutscher Akademischer Austauschdienst

- e: Digital-Analog[ue]-Analog[ue]-Digital

DAAE e: Defense Aid Administration Expenses
DAAG e: Diode AND-AND Gate
DAAI & OC e: Defense Aid Agricultural, Industrial and Other Commodities
DAAO e: D-Amino Acid Oxidase
DAAP d: Diamylamyephosphonate
DAAS e: Defense Automatic Addressing System
DAATA e: Dictionary of Aeronautics and Aerospace Technology Abrreviations
DAB d: Daten abgehend

- d: Datenblock
- d: Datenleitung abgehend
- d: Dauerbetrieb mit aussetzender Belastung
- d: Deutsches Apothekerbuch
- d: Deutsches Arzneibuch
- d: Diaminobuttersäure
- d: Diazoamine Benzene
- d: Dienst an Bord
- d: Dimethylaminoazobenzol
- d: Durchlaufbetrieb mit Aussetzbelastung
- e: Daily Audience Barometer
- e: Demand Adjustment Bucket
- e: Diaminobenzidine
- e: Dictionary of American Biography
- e: Digital Audio Broadcast[ing]
- e: Dimetylaminoazobenzene
- e: Display Assignment Bit[s]
- e: Display Attention Bit[s]
- e: Dissertation Abstracts (journ.)

- e: Dual Access Buffer
DABA d: Datenbank
- d: Datenbankbetriebssystem
- d: Datenbasisauswertungssystem
DAB Acid e: Diaminobutyric Acid
DABAL e: Data Bank Language
DABANK d: Dateienbank
DABAS e: Data Base System
DABAWAS d: Datenbank für wassergefährdende Stoffe
DABCO e: Diazabicyclooctanc
DABES d: Datenbestand
DABPN e: Diplom at American Board of Psychiatry and Neurology
DABS e: Discrete Address Beacon Secondary Surveillance Radar
- e: Discrete[ly] Address[ed] Beacon System
DABSIPC e: Discrete Address Beacon System with Intermitted Positive Control
- e: Discrete Address Beacon System with Intermittent Positive Control
DABSIPCS e: Discrete Address Beacon System with Intermittent Positive Control System
DABXT e: Deep Airborne Expendable Bathythermograph
DAC e: Data-processing Center (or Centre)
- e: Data Accepted
- e: Data Acquisition and Control
- e: Data Acquisition Camera
- e: Data Acquisition Card
- e: Data Acquisition Chassis
- e: Data Acquisition Computer
- e: Data Acquisition Controller
- e: Data Analysis and Control
- e: Data Analysis Center
- e: Data Analysis Console
- e: Data Analysis Control
- e: Data Archiving Centre (GB)
- e: Data Assistance and Control
- e: Data Automation Corporation
- e: Day[s] After Conception
- e: Decimal Arithmetic Carry
- e: Decrement Accumulator
- e: Deductible Average Clause
- e: Demand Assignment Controller
- e: Design Augmented by Computer
- e: Design Automation Conference
- e: Diallyl Chlorendate
- e: Differential Aeration Corrosion
- e: Digital-Analog[ue] Converter
- e: Digital Arithmetic Center
- e: Digital Associates Corporation
- e: Direct Access Computing
- e: Digital-to-Analog[ue] Circuit
- e: Digital-to-Analog[ue] Control
- e: Digital-to-Analog[ue] Conversion
- e: Digital-to-Analog[ue] Converter
- e: Direct Air Cycle
- e: Director Assignment Console
- e: Disassemble Communications Pool
- e: Disk Access Control
- e: Disk Attachment Control
- e: Display Analysis Console
- e: Document Availability Code
- e: Donor Activity Coefficient
- e: Double Action Cylinder
- e: Double Average Comparison
- e: Douglas Aircraft Company
- e: Duplicate Aperture Card
- e: Dynamic Amplitude Control
DACA e: Digital-to-Analog[ue] Control Apparatus
- e: Diphenylaminochloroarsine
DACAC e: Digital-to-Analog[ue] Converter; Alternating Current
DACAP e: Discrete and Continuous Analysis Program[me]
DACA Program[me] e: Discrete and Continuous Analysis Program[me]
DACAPS e: Data Collection and Processing System
DACBU e: Data Acquisition Control and Buffer Unit
DACC e: Data Acquisition and Checkout Computer
- e: Data and Computation Center
- e: Design Assertation Consistency Checker
- e: Direct Access Communication[s] Channel[s]
- e: Direct Access Computing Corporation
- e: Display and Control Card
- e: Doctor of Accountancy
- e: Defense Area Communications Control Center
DACCC-AL e: Defense Area Communications Control Center; Alaska
DACCC-CON e: Defense Area Communications Control Center; CONUS (Continental United States)
DACCC-EUR e: Defense Area Communications Control Center; Europe
DACCC-PAC e: Defense Area Communications Control Center; Pacific
DAC Computer e: Divide and Conquer Computer
DACCS e: Department of the Army Command and Control System
DACD[I]C e: Digital-to-Analog[ue] Converter; Direct Current
DACE e: Data Acquisition and Control Executive
- e: Data Administration Center Equipment
- e: Doctor of Air Conditioning Engineering
DAC Eng e: Doctor of Air Conditioning Engineering
DA Chem e: Doctor of Applied Chemistry
Dachg. d: Dachgeschoß
DACI e: Direct Adjacent Channel Interference
DACL e: Diablo Application Compiler Language
- e: Dynamic Analysis and Control Laboratory
DACM e: Direct Addressable Core Memory
DACO e: Data Consistency Orbit
- e: Douglas Aircraft Company Overseas
DACOM e: Data and Configuration Management
- e: Data Communication
- e: Datascope Computer Output Microfilmer
- e: Double Average Comparison
DACON e: Data Controller
- e: Digital-to-Analog[ue] Converter
DACOR e: Data Correction
- e: Data Correlation
- e: Data Correlator
DACOS e: Data Communication Operating System
DACPO e: Data Count Printout
DAC Program[me] e: Disk Access Control Program[me]
DACQ e: Data Acquisition
DACRP e: Department of the Army Communications Resources Plan
DACRYLON e: Dacron and Nylon
DACS e: Data Access and Control System
- e: Data Acquisition and Computer System
- e: Data Acquisition and Control System
- e: Data Acquisition and Conversion System
- e: Data Acquisition and Correction System
- e: Data and Analysis Center for Software
- e: Dataset Control System
- e: Design Aided by Computer System
- e: Digital-to-Analog[ue] Converters
- e: Digital Access and Cross-Connect System
- e: Digital Animated Control System
- e: Directorate of Aerospace Combat Systems
- e: Discrete Address Communications System
- e: Dot Addressable Channel Set
DAC System e: Data Acquisition and Control System
DACT e: Dactyl[ic]
- e: Dactylology
- e: Dactylus
- e: Deactivate
- e: Direct Acting
- e: Directorate of Administrative and Computer Training
- e: Dissimilar Air Combat Training
DACTION e: Derivative Action
DACTY e: Dactylograph[y]
- e: Dactyloscopy
DACTYGRAM e: Dactylogram
DACTYL e: Dactylogic[al][ly]
- e: Dactylologist[ic]
- e: Dactylology
DACU e: Data Acquisition and Control Unit
- e: Device Attachment Control Unit
DAD d: Digitalausgabe[block]; dynamisch
- e: Data Acquisition Device
- e: Data Automation Digest (journ.)
- e: Data Description
- e: Decimal Accumulating Device
- e: Dial-a-Design
- e: Diaminodurene
- e: Differential Amplitude Discriminator
- e: Digital Angle Data
- e: Digital Audio Definition
- e: Digital Audio Disk
- e: Direct Access Device
- e: Directional Aerial Disposal
- e: Dockyard Armament Depot
- e: Doppler Analyzer Display
- e: Double-Acting Door
- e: Double Amplitude Displacement
- e: Double Atmospheric Density
- e: Drum and Display
- e: Dual Air Density
DADA e: Deputy Assistant Director of Artillery
DADAC e: Digital-to-Analog[ue] Deck Angle Converter
DAD Assembler e: Data Dictionary Assembler
DADB e: Data Analysis Data Base
DADC e: Digital-Analog[ue] Data Conversion
- e: Digital Air Data Computer
- e: Direct Access Data Channel
DADE e: Data Acquisition and Decommutation Equipment
- e: Digital Acquisition and Documentation Equipment
- e: Digital Application and Documentation Equipment
DADEC e: Design and Demonstration Electronic Computer
DADEE e: Dynamic Analog[ue] Differential Equation Equalizer
DADES e: Deputy Assistant Director of Engineer Stores
dad. gek. d: dadurch gekennzeichnet
DADGMS e: Deputy Assistant Director General of Medical Services
DADIC e: Data Dictionary

DADIOS e: Direct Analog[ue]/Discrete Input/Output System
- e: Direct Analog[ue] Digital Input/Output System
DADIT e: Daystrom Analog[ue]-to-Digital Integrating Translator
DADIWT e: Deputy Assistant Director of Inland Water Transport [Service]
DADM e: Deductively Augmented Data Management
DAdm e: Doctor of Administration
DADME e: Deputy Assistant Director of Military Engineering
D Adm Eng e: Doctor of Administrative Engineering
DADMS e: Deputy Assistant Director of Medical Service
DADO e: Director of Air Defence Operations
DAD of ST e: Deputy Assistant Director of Supplies and Transport
DADOS e: Deputy Assistant Director of Ordnance Services
DADP d: Desoxyadenosindiphosphat
- e: Dialkyl Dithiophosphate
DADPS d: Diamino-Diphenyl-Sulfon
- e: Diamino Diphenyl Sulphone
DADPSS e: Deputy Assistant Director of Printing and Stationery Services
DADR e: Digital Angle Data Recorder
- e: Director of Air Defence Requirements
DADRT e: Deputy Assistant Director of Railway Transport
DADS e: Data Acquisition and Display System
- e: Deputy Assistant Director of Supplies
- e: Digital Air Data-Computer System
- e: Direct Access Data Set
- e: Director of Army Dental Services
- e: Dosimetry Acquisition and Display System
- e: Dual Air Density Satellite
DAD Satellite e: Dual Air Density Satellite
DADSEE e: Dynamic Analog[ue] Differential Equation Equalizer
DADSI e: Director of Air Defence Systems Integration
DADSM e: Direct Access Device [for] Space Management
DADST e: Deputy Assistant Director of Supplies and Transport
DADSvy e: Deputy Assistant Director of Survey
DADT[n] e: Deputy Assistant Director of Transportation
dad. verurs. d: dadurch verursacht
DAdW d: Deutsche Akademie der Wissenschaften
DAE d: Datenanpassungseinrichtung
- d: Datenanschlußeinheit
- e: Data Acquisition Equipment
- e: Department of Atomic Energy
- e: Diacetylethylene
- e: Dictionary of American English
- e: Diploma in Advanced Engineering
- e: Direction Aerial
- e: Director of Air Equipment
- e: Director of Army Education
- e: District Airport Engineer
- e: Doctor of Aeronautics
- e: Doctor of Art Education
DAEA e: Dimethyl Aminoethyl Acetate
DaEB d: Datenerfassungsbeleg
DAEC e: Danish Atomic Energy Commission
DAEDAC e: Drug Abuse Epidemiology Data Center
DAEDR e: Delimitation, Alignment, Error Detection, Reception
DAEDT e: Delimitation, Alignment, Error Detection, Transmitting
D Ae E[ng] e: Doctor of Aeronautical Engineering
DAEM d: Deutsches Archiv für Erforschung des Mittelalters (journ.)
DAEMON e: Data Adaptive[d] Evaluator and Monitor
DAEng e: Director of Aircraft Engineering
DAEP e: Division of Atomic Energy Production
DAER d: Datenerfassung
- e: Department of Aeronautical [and] Engineering Research
D Aero E e: Doctor of Aeronautical Engineering
DAES d: Diaethylstilboestrol
- e: Direct Access Education System
D Ae S[c] e: Doctor of Aeronautical Science
DAET d: Diaethyltryptamin
DAF d: Dispatcher- Anzeige[n]feld
- e: Data Acquisition Facility
- e: Data Analysis Facility
- e: Delay[ed] Action Fuze
- e: Delay Amplification Factor
- e: Delay Auditory Feedback
- e: Department of the Air Force
- e: Departure Airfield
- e: Described as Follows
- e: Destination Address Field
- e: Device Address Field
- e: Diacetyl Ferrocene
- e: Direct Access File
- e: Discard at Failure
- e: Discharge Afloat
- e: Dissolved Air Flotation
- e: Double Apron Fence
- e: Dual Access Feature
- e: Dynamic Axial Fatigue
DAFA e: Data Accounting Flow Assessment
DAF Basis e: Dry Ash Free Basis
DAFC e: Departure Airfield Control
- e: Digital Automatic Frequency Control
DAFCS e: Digital Automatic Flight Control System
DAFDTA e: Dipole Antenna with Feed-Points Displaced Transverse to its Axis
DAF & E e: Defense Aid Facilities and Equipment
DAFL e: Digital Automatic Frequency Lock
DAFM e: Direct Access File Manager
- e: Discard-at-Failure Maintenance
DAFS e: Department of Agriculture and Fisheries for Scotland
- e: Developments in Aquaculture and Fishery Science
- e: Direct Aerial Fire Support
- e: Director of Air Force Security
- e: Director of Army Fire Services
- e: Duty Air Force Specialty
DAFSC e: Duty Air Force Specialty Code
DAFT e: Digital-to-Analog[ue] Function Table
Dafta d: Datenschutzfachtagung
DAFÜ d: Datenfernübertragung
DA Funk d: Dienstanweisung Funk
DA FuStöMD d: Dienstanweisung für den Funkstörungsmeßdienst
DAG d: Datenanschaltgerät
- d: Datenanschlußgerät
- d: Dekagramm
- d: Deutsche Agrarwissenschaftliche Gesellschaft
- d: Direktrufnetzabschlußgerät
- e: Aquadag
- e: Decagram
- e: Defence Against Gas
- e: Deflocculated Acheson Graphite
- e: Design Advisory Group
- e: Diode AND Gate
- e: Directed Acyclic Graph
- e: Doctor of Agriculture
- e: Dysprosium Alumin[i]um Garnet
DAGC e: Delayed Automatic Gain Control
- e: Digital Automatic Gain Control
DAGe d: Datenanschlußgeräte
DAGE d: Datenblock-Generierungsprogramm
DAGK d: Deutsche Arbeitsgemeinschaft Kybernetik
DAGM d: Deutsches Archiv für Geschichte des Mittelalters (journ.)
DAGMAR e: Drift and Groundspeed Measuring Airborne Radar
D Agr e: Doctor of Agriculture
D Agr E[ng] e: Doctor of Agricultural Engineering
D Agr S[c] e: Doctor of Agricultural Science
DAGt d: Datenanschlußgerät
DAH d: dahinten
- d: dahinter
- e: Dahomey
- e: Dictionary of American History
- e: Disordered Action of the Heart
DA/ha d: Doppelader/Hektar
DAHC e: Drain Avalanche Hot Carrier[s]
DAHP e: 3-Deoxy-D-Arabinoheptulosonic Acid-7-Phosphate
DAI d: Deutsches Archäologisches Institut
- e: Direct Access Information
- e: Direct Access Interface
- e: Director of Aeronautical Inspection
- e: Director of Air Intelligence
- e: Discrete Activity Indicator
- e: Dissertation Abstracts International (journ.)
- e: Double Auto-Ionizing
- e: Drift Angle Indicator
- e: Drug Abuse Information
DAICS e: Data-Acquisition and Instrument-Control System
DAIGC e: Direct Aqueous Injection Gas Chromatograph[y]
DAiM d: Deutsches Archiv für innere Medizin (journ.)
DAIM e: Data Analysis Information Memo
DAIN e: Director of Action Information and Navigation
DAIP[s] e: Diallyl Isophthalate[s]
DAIR e: Dial Access Information Retrieval
- e: Direct Altitude and Identity Read-out
- e: Direct Attitude and Identity Read-out
- e: Driver Aid, Information and Routing
- e: Dynamic Allocation Interface Routine
DAIRE e: Direct Altitude and Identification Readout Equipment
DAIRI e: Dissertation Abstracts International Retrospective Index (journ.)
DAIRS e: Dial Access Information Retrieval System
DAirS e: Director of Air Services
DAIR System e: Driver Aid, Information and Routing System
DAIS e: Defense Automatic Integrated Switch
- e: Defense Automatic Integrated Switching System
- e: Digital Avionics Information System

- e: Digital Avionics Integration Systems
- e: Direct Access Intelligence System[s]
- e: Director of Aeronautical Inspection Services
- e: District Agricultural Improvement Station
DAIS System e: Defense Automatic Integrated Switching System
DAISY d: Druckindustrie Abrechnungs- und Informationssystem
- e: Data Acquisition and Interpretation System
- e: Data Analysis of the Interpreter System
- e: Decision Aided Information System
- e: Decision Aiding Information System
- e: Displacement Automated Integrated System
- e: Domestic Appliances Information System
- e: Double-Precision Automatic Interpretive System
DAISY 201 e: Double-Precision Automatic Interpretive System
DAIU e: Digital-to-Analog[ue] Interface Unit
DAIV e: Data Area Initializer and Verifier
DAJ d: Deutsch-asiatisches Jahrbuch (journ.)
DAJS e: Distributed Area Jamming System
DAK d: Dateienkatalog
- d: Datenausgabekanal
- d: Datenaustauschelement für kombinierten Betrieb
- d: Deutsche Atom-Kommission
- e: Data Authentication Key
- e: Decision Acknowledge
- e: Director of Army Kinematography
DAKA d: Datenkartei
DAklM d: Deutsches Archiv für klinische Medizin (journ.)
DAKM d: Doppeladerkilometer
DAKS d: Dateikatalogsystem
- d: Datenkommunitionssystem
Dak Zoo e: Dakota Zoo
DAL d: Datenanschlußleitung
- d: Dekaliter
- d: Deutsche Akademie der Landwirtschaftswissenschaften
- d: Deutscher Arbeitsring für Lärmbekämpfung
- d: Dienstanruflampe
- d: Drahtfunkanschlußleitung
- e: Dallas (Texas)(Airport)
- e: Dalmatia[n]
- e: Data-Aided Loop
- e: Data/Address Line
- e: Data Access Line
- e: Data Acquisition Language
- e: Decaliter (US)
- e: Decalitre (GB)
- e: Delta-Aminolevulinic Acid
- e: Department of Agriculture Library
- e: Digital Access Line
- e: Directional Arm Lock
- e: Distribution Authority List
- e: Divisional Administration List
- e: Document Address Lister
- e: Downed Aircraft Locator
- e: Drawing Assembly List
DALA e: Delta-Aminolaevulinic Acid
DALAPON e: Dialphapropionic Acid
DALC e: Deployment Area Location Code
- e: Divided Access Line Circuit
- e: Dynamic Asynchronous Logic Circuit
DALE e: Digital Anemograph Logging Equipment
DALG e: D-Algorithm
DALGT e: Daylight
DA LK d: Dienstanweisung für den Lochkartendienst
dalm d: Dekalumen
DALO e: Diconnect at Lift-Off
DALR e: Dry-Adiabatic Lapse Rate
DALS e: Digital Approach and Landing System
- e: Diver Auditory Localization System
- e: Double-Acting Limit Switch
DAlsp d: Drahtfunkanschlußspeiseleitung
DALT e: Drop Altitude
DALTS e: Data Link Test Set
dalu d: Dekalumen
DALV[F] d: Deutsches Archiv für Landes- und Volksforschung (journ.)
dalx d: Dekalux
dam. d: damalig
- d: damals
DAM d: Damaskus (Flughafen)
- d: Dekameter
- d: Deutsches Armeemuseum
- d: Digital-Analog[ue]-Modul
- d: Drehstromasynchronmaschine
- e: Data-Address[ed] Memory
- e: Data Access Method
- e: Data Association Message
- e: Dayton Art Museum
- e: Decameter
- e: Defended Area Model
- e: Definition, Analysis and Mechanization
- e: Degraded Amyloid
- e: Delay[ed] Action Mine
- e: Denver Art Museum
- e: Descriptor Attribute Matrix
- e: Diacetylmonoxide
- e: Diacetylmorphine
- e: Diaethylaminomethyl
- e: Diamond Abrasive Machining
- e: Dictionary of Abbreviations in Medicine
- e: Digital Automatic Machine
- e: Digital-to-Analog[ue] Module
- e: Digital-to-Analog[ue] Multiplier
- e: Digitized Analog[ue] Magnetogram
- e: Diploma in Ayurvedic Medicine
- e: Direct Access Memory
- e: Direct Access Method
- e: Directly Addressable Memory
- e: Director of Air Material
- e: Double-Aluminized Mylars
- e: Down-Range Antimissile
- e: Dummy Average Memory
- e: Duplex Adding Machine
dam² d: Quadratdekameter
dam³ d: Kubikdekameter
DAMA e: Dairy Appliance Manufacturers Association
- e: Demand Assigned Multiple Access
- e: Demand Assignment Multiple Access
DAMBED e: Dialphamethylbenzyl-ethylenediamine
DAMC e: Digital Automatic Map Compilation
DAMCONT e: Damage Control
DAMDA e: Dairy Appliance Manufacturers and Distributors Association
DAME e: Data Acquisition and Monitoring Equipment
- e: Defense Against Methods of Entry
- e: Developments in Agricultural and Managed-Forest Ecology
- e: Dictionary of American English
- e: Digital Automatic Measuring Equipment
- e: Director Area Mechanical Estimating
DA Memory e: Direct Access Memory
DAMGO e: Deputy Assistant Master-General of Ordnance
DAMID e: Discounting Analysis Model for Investment Decisions
DAM II-EE e: Defended Area Model II Engagement Evaluation
DAM II-EP e: Defended Area Model II Engagement Planning
DAMIS e: Department of the Army Management Information System
DAMIT e: Data Analysis Computer Program[me] by Massachusetts Institute of Technology
DAMLG e: Dental Amalgamator
DAMM d: Datenschutz; Mittel und Maßnahmen
- e: Direct Access Memory Management
DAMN e: Diaminomaleonitrile
- e: Dynamic Analysis of Mechanical Networks
DAMOC e: Director Autopilot Mode Organization Computer
DAMOD e: Direct Access Module
DAMOS e: Data Moving System
DAMP d: Desoxyadenosinmonophosphat
- e: Database of Atomic and Molecular Physics
- e: Deoxyadenylic Acid
- e: Department of the Army Materiel Program[me]
- e: Down-Range Anti-Missile Measurement Project
- e: Downrange Anti-Missile Measurement Program[me]
Dampfm. d: Dampfmaschine
DAMPR e: Digital Automatic Multiple Pressure Recorder
DAM Program[me] e: Down-range Anti-Missile Program[me]
DAMPS e: Data Acquisition Multiprocessing System
- e: Data Acquisition Multiprogramming System
DAMP System e: Data Acquisition Multiprocessing System
- e: Data Acquisition Multiprogramming System
DAMR e: Director of Aircraft Maintenance and Repair
DAMR(W) e: Director of Aircraft Maintenance and Repair (Washington)
DAMS e: Data Addressed Memories
- e: Defense Against Missiles System[s]
- e: Defense Antimissile System
- e: Defensively Armed Merchant Ship
- e: Direct Access Management System
DAMSU e: Digital Auto-Manual Switching Unit
DAMUSC e: Direct Access; Multi-User; Synchrocyclotron Computer
DAMUT e: Ducted Air Medium Underground Transmission
DAMV e: Destruction of Aircraft or Motor Vehicles
- e: Double-Air Movement Valve
DAMW d: Deutsches Amt für Meßwesen und Warenprüfung (DDR)
DAN d: Daten ankommend
- d: Datenanpassungseinheit
- d: Datenleitung ankommend
- d: Datenübertragungsanschluß
- d: Dekanewton
- e: Decanewton
- e: Deposit Account Number
- e: Diaminonaphthalene
- e: Document Accession Number
- e: Domestic Area Network
- e: Dual Area Nozzle
DANA e: Diffraction Analysis System

DANAC e: Data Analysis and Classification
- e: Decca Area Navigation [Airborne] Computer
DANAS e: Deposit Account Name and Address System
DANC e: Decontaminating Agent; Noncorrosive
D and A e: Detail and Assembly
D and D e: Decontamination and Decommissioning
- e: Development and Design
D and F e: Determination and Finding
D and P e: Data Analysis and Processing
- e: Development and Printing
D and R e: Diagnostic and Repair
D and S e: Demand and Supply
- e: Display and Storage
D and T e: Development and Test
DANM e: Danmark
DANP e: Danish Patent
DANR e: Department of Agriculture and Natural Resources
DANRIC e: Department of Agriculture and National Resources Information Council
DAN Rocket e: Deacon and Nike Rocket
DANS d: Dimethylaminoaphthylsulfonsäure
- e: Dansyl Chloride
Dansyl d: 1-(Dimethylamino)-Naphthalin-5-Sulfonyl
- d: 1-(Dimethylamino)-Naphthalin-5-Rest
DANT e: Diallyl-Nor-Toxiferin
DANTES e: Defense Activity for Non-Traditional Education Support
Dants e: Day and Night Television System
Danw. d: Dienstanweisung
- d: Durchführungsanweisung
DaNzKo d: Datennetzkoordinator
DAO d: Dateiorganisation
- d: Durchführungsanordnung
- e: Directory of Amateur Observers
- e: District Aviation Officer
- e: Division Air Officer
- e: Division Ammunition Officer
- e: Doctor of Art of Oratory
- e: Dominion Astrophysical Observatory
DAOB e: Digital and Analog[ue] Output Basis
DAOR e: Discriminator Average Output Rate
DAOS e: Dynamically Adaptive Operating System
DAOSR e: Director of Army Operational Science and Research
DAOT e: Director of Air Organization and Training
DAP d: Deutsches Ausschließungspatent
- d: Diallylphthalat[e]
- d: Diallylphthalatharz
- d: Diazepam
- d: Digital Autopilot
- d: Dihydrazinophthalazin
- d: m-Diaminoopimelinsäure
- d: m-Diaminopimelinsäure
- e: Data Access Protocol
- e: Data Acquisition and Processing
- e: Data Acquisition Package
- e: Data Acquisition Program[me]
- e: Data Analysis and Processing
- e: Data Analysis Package
- e: Data Automation Panel
- e: Data Automation Proposal
- e: Day[s] After Parturition
- e: Deformation Alignment Phase
- e: Deformation of [Vertical] Aligned Phases
- e: Delayed Alpha Particle[s]
- e: Depot Acceptance Procedure[s]
- e: Derived Attainable Performance
- e: Design Analysis Phase
- e: Detail Assembly Panel
- e: Development Academy of the Philippines
- e: Diagnostic Assistance Program[me]
- e: Dial A Program[me] [System]
- e: Diaminopimelate
- e: Diaminopimelic Acid
- e: Diammonium Phosphate
- e: Diastolic Aortic Pressure
- e: Diffused Alloy Power
- e: Digital Access Point
- e: Digital Assembly Program[me]
- e: Digital Audio Processor
- e: Dihydroxy Acetone Phosphate
- e: Dipeptidyl Aminopeptidase
- e: Direct Access Photomemory
- e: Direct Access Program[me]
- e: Direct Agglutination Pregnancy
- e: Directed Audit Program[me]
- e: Director of Aircraft Production
- e: Director of Army Programs
- e: Director of Army Psychiatry
- e: Distributed Application Processing
- e: Distributed Array Processor
- e: Division of Air Pollution
- e: Do Anything Possible
- e: Document Against Payment
- e: Double Amplitude Peak-[Value]
- e: Draw-a-Person (Psychological Test)
DAPAC e: Danger Area[s] in the Pacific
DAPAL e: Daystrom Powerplant Automation Language
DAPAS e: Diamond Abrasive Performance Analysis System
DAPCA e: Development and Production Costs for Aircraft
DAP Cell e: Deformation of Aligned Phases Cell
DAP Computer e: Distributed Array Processor Computer
DAPE e: Developed Armament Probable Error
- e: Diploma in Applied Parasitology and Entomology
- e: Direct Addressable Peripheral Equipment
DAP-Effekt d: Deformation aufgerichteter Phasen-Effekt
DAPFS e: Direct Ascent Powered Flight Simulation
DAPI e: Diamidino-Phenylindole
DAPIS e: Danish Agricultural Producers Information Service
DAPK d: digital-analog-physikalischer Komplex
DAPN e: Directional Aerial Phasing Network
- e: Directional Antenna Phasing Network
DAPO e: Digital Advance Production Order
DAPP e: Data Acquisition and Processing Program[me]
- e: Design Aid for Post Processor
- e: Director of Air Plans and Programmes
DAPPO e: Device and Program[me] Performance Optimization
DAppSci e: Doctor of Applied Science
DAPR e: Data Processing
- e: Data Processing Magazine (journ.)
- e: Digital Automatic Pattern Recognition
- e: Digital Automatic Pattern Recognizer
DA Press e: Double Acting Press
DAPRO Acid e: L-α; ß-Diaminopropionic Acid
DAPRO Description System e: Data and Program[me] Description System
DAPRU e: Drug Abuse Prevention Resource Unit
DAPS e: Direct Access Performance Software
- e: Direct Access Programming System
- e: Direct Automatic Publication Service
- e: Director of Army Postal Services
- e: Distributed Application Processing System
- e: Distributed Array Processors
DAPSONE e: Diaminodiphenyl Sulfone
DAP&SS e: Director of Army Printing and Stationery Services
DAP System e: Dial A Program[me] System
DAPT d: 2,4 Diamino-5-Phenyl-Triazol
- e: Daptazole
- e: Direct [Latex] Agglutination Pregnancy Test
- e: Draw-a Person Test
DAP Test e: Draw A Person Test
DAP Transistor e: Diffused-Alloy Power Transistor
DAPU d: Datenpuffer
- e: Data Acquisition and Processing Unit
DAQ e: Develop and Qualify
DAQC e: Data Acquisition Center
DAR d: Dar-es-Salaam (Flughafen)
- d: Datenadresse
- d: Datenausgaberegister
- d: Datenregister
- d: Deutsche Akademische Rundschau (journ.)
- d: Deutsches Autorecht
- d: Diaabtast- und Reproduktionseinrichtung
- d: Digitalausgabe mit Relais
- e: Damage Assessment Routine[s]
- e: Data Access Register
- e: Data Acquisition Recorder
- e: Data Aided Receiver
- e: Data Article Requirement[s]
- e: Data Automation Request
- e: Data Automation Requirement[s]
- e: Dead Animal Removal
- e: Defense Acquisition Radar
- e: Delay Address Register
- e: Delayed Automatic Reclose
- e: Department of Animal Regulation
- e: Destination Address Register
- e: Deviation Approval Request
- e: Differential Absorption Ratio
- e: Differentiation with Asymmetrical Reinforcement
- e: Digital Aircraft Integrated Data System Recorder
- e: Digital Angle Recorder
- e: Direct-Axis Regulator
- e: Directorate of Atomic Research
- e: Director of Army Requirements
- e: Director of Atomic Research
- e: Driver Augmented Readout
DARA d: Deutsche Arbeitsgemeinschaft für Rechenanlagen
DARAC e: Damped Aerodynamic Righting Attitude Control
DARACS e: Damped Aerodynamic Righting Attitude Control System
DARAF e: Farad Backwards (Unit of Elastance)
DARAS e: Direction and Range Acquisition System
DARB e: Data Access Request Block
- e: Distressed Airman Recovery Beacon
DArbGG d: Deutsches Arbeitsgerichtgesetz
DARC d: Deutscher Amateur-Radioclub
- e: Data Acquisition and Reduction Center
- e: Data Acquisition and Reports Control
- e: Description, Acquisition, Restitution, Correlation

- e: Description, Acquisition, Retrieval, Correlation
- e: Direct Access Radar Channel
DARCEE e: Demonstration and Research Center for Early Education
D Arch e: Doctor of Architecture
D Arch Des e: Doctor of Architectural Design
D Arch E[ng] e: Doctor of Architectural Engineering
DARCOM e: Development and Readiness Command
DARD e: Data Acquisition Requirements Document
- e: Data Reduction
DARDC e: Device for Automatic Remote Data Collection
DARD Method e: Data Reduction Method
DARDO e: Direct Access to Remote Data-bases Overseas
DARE e: Data Analysis and Reduction
- e: Data Automatic Reduction Equipment
- e: Data Automation Research and Experimentation
- e: Data Retrieval
- e: Delay Asymptotic Relative Efficiency
- e: Destination Arrival Research Engineering
- e: Differential Analyzer Replacement
- e: Document[ation] Automated Retrieval Equipment
- e: Document Abstract Retrieval Equipment
- e: Doppler and Range Evaluation
- e: Doppler Automatic Reduction Equipment
- e: Doppler Velocity and Position Automatic Reduction Equipment
- e: Drug Abuse Research and Education
- e: Drug Assistance, Rehabilitation and Education
DARECORDENGR e: Data Recording Engineer
DARES e: Data Analysis and Reduction System
DARE System e: Data Analysis and Reduction System
- e: Data Retrieval System
DARF e: Defense Atomic Research Facility
dargel. d: dargelegt
dargest. d: dargestellt
DARIAS e: Digico Automated Radio-Immunoassay Analytical System
DARIC e: Data Reduction in Columns
darl. d: darlegen
Darl. d: Darlegung[en]
- d: Darlehen
DARL e: Douglas Advanced Research Laboratories
DARLI e: Digital Angular Readout by Laser Interferometry
DARLISTOR d: Darlington-Transistor
D Arm D e: Director of Armament Development
DARME e: Directorate of Armament Engineering
D Arm Eng e: Director of Armament Engineering
DARMS e: Digital Alternate Representation of Music Scores
- e: Drifting Automatic Radio-Meteorological Station
DARO d: Daten-Robotron
- d: Datenerfassung, Rationalisierung, Organisation
DARPA e: Defense Advanced Research Projects Agency
- e: Defense Applied Research Projects Agency
DARR e: Drawing and Assembly Release Record
DARS e: Data Accounting and Reporting Sheet
- e: Data Acquisition and Reduction System
- e: Data Acquisition Recording System
- e: Decommutation and Readout System
- e: Design Assist Reports
- e: Dial Access Retrieval System
- e: Differential Absorption Remote Sensing
- e: Digital Adaptive Recording System
- e: Digital Attitude and Rate System
- e: Digital Attitude Reference System
D Ar Sc e: Doctor of Arts and Sciences
DARSS e: Diode-Array Rapid-Scan Spectrometer
Darst. d: Darstellung
DARSY e: Data Acquisition and Reduction System
DART e: Daily Automatic Rescheduling Technique
- e: Data-Analysis Recording Tape
- e: Data Acquisition and Recording Terminal
- e: Data Analysis Recording Tape
- e: Data Analysis Reduction Tape
- e: Datagraphix Automated Retrieval Technique[s]
- e: Data Reduction Translation
- e: Data Reduction Translator
- e: Data Research and Technology
- e: Decomposed Ammonia Radioisotope Thruster
- e: Deployable Automatic Relay Terminal
- e: Depot Automatic Rescheduling Technique
- e: Design Automation Routing Tool
- e: Detection, Action and Response Technique
- e: Detection, Action Response Technique[s]
- e: Development [of] Advanced Rate Technique[s]
- e: Diagnostic Assistance Reference Tool
- e: Dial-A-Ride Transport[ation]
- e: Digital Auto-Ranging Tester
- e: Digital Automatic Readout Tracker
- e: Diode Automatic Reliability Tester
- e: Direct Airline Reservations Ticketing
- e: Directional Automatic Realignment of Trajectory
- e: Director and Response Tester
- e: Disappearing Automatic Retaliation Target
- e: Disappearing Automatic Retaliatory Target
- e: Dual Axis Rate Transducer
- e: Dynamic Acoustic Response Trigger
DARTS e: Design Aid for Real-Time Systems
- e: Digital Azimuth Range Tracking System
- e: Distributors Automated Real-Time System
- e: Dynamically Actuated Road Transit System
D Arty e: Director of Artillery
DARUF d: Deutscher Auslandsdienst für Rundfunk und Fernsehen
DARW e: Darwin College, Cambridge
DAS d: Datenausgabesteuerung
- d: Datenerfassungs- und Steuersystem
- d: Datensperre Anpaßschaltung
- d: Datenstation
- d: Dekaster
- d: Dienstanschluß
- d: Digitalausgabe
- d: Digitalausgabe; statisch
- d: Direkt-Abfrage-Sprache
- d: direktes Abfragesystem
- d: diskontinuierliches analoges Signal
- d: Drehantriebssystem
- d: Drehstromantriebssystem
- d: Durchschalteablaufsteuerung
- e: Data Access Security
- e: Data Accountability System
- e: Data Acquisition Station
- e: Data Acquisition Subsystem
- e: Data Acquisition System
- e: Data Administration System
- e: Data Amplification Sheet
- e: Data Analysis Software
- e: Data Analysis Station
- e: Data Analysis System
- e: Data Automation Subsystem
- e: Data Automation System
- e: Data Auxiliary Set
- e: Datatron Assembly System
- e: Defense Atomic Support Agency
- e: Deficiency Analysis Summary
- e: Delivered Alongside Ship
- e: Delivery Alongside Ship
- e: Design Automation System
- e: Development Areas
- e: Developments in Atmospheric Science
- e: Device Analysis System
- e: Dextroamphetamine Sulfate
- e: Dialdehyde Starch
- e: Dictionary of American Slang
- e: Differential Absorption and Scattering
- e: Digital/Analog[ue] Simulation
- e: Digital/Analog[ue] Simulator
- e: Digital/Analog[ue] System
- e: Digital Address System
- e: Digital Altimeter Scanner
- e: Digital Attenuator System
- e: Digital Attitude Simulation
- e: Digital Autopilot System
- e: Digital-to-Analog[ue] Simulation
- e: Digital-to-Analog[ue] Simulator
- e: Dipole Antenna System
- e: Direct Access Storage
- e: Direct Access Store
- e: Direct Access System
- e: Direct Acting Steam
- e: Direct Air Support
- e: Direct Automotive Support
- e: Directly Addressable Storage
- e: Director of Armament Supply
- e: Disturbance Analysis System
- e: Divisional Armorers Shop
- e: Doctor of Applied Science
- e: Doctor of Astronomy
- e: Document Aids System
- e: Documentation Aid System
- e: Dynamic Alert System
DASA e: Defense Atomic Support Agency
- e: Dual Aerospace Servoamplifier
DASA-DC e: Defense Atomic Support Agency-Data Center
DASA-TP e: Defense Atomic Support Agency-Technical Publication[s]
DASC e: Defense Automotive Supply Center
- e: Direct Air Support Center
- e: Direct Air Support Central
- e: District Air Support Center
- e: Doctor of Agricultural Science
- e: Doctor of Applied Science
DASCH d: Datenschutz
DASCO e: Data Switching Computer
DAS/COTAR e: Data Acquisition System/ Correlation Tracking and Ranging
DASD d: Deutsche Akademie für Sprache und Dichtung (journ.)
- e: Data Acquisition Support Document
- e: Department of the Army Shipping Document

- e: Direct Access Storage (or Store) Device
- e: Direct Access Storage (or Store) Drive

DASDDR e: Direct Access Storage Device Dump Restore

DASDI e: Direct Access Storage Device Initialization
- e: Direct Access Store Disk Initialization

DASDI Program[me] e: Direct Access Storage Device Initialization Program[me]

DASDJ d: Deutsche Akademie für Sprache und Dichtung; Jahrbuch (journ.)

DASDL e: Data and Structure Definition Language

DASDM e: Direct Access Storage Data Management

DASDR e: Direct Access Storage Dump Restore

DASDs e: Direct Access Storage (or Store) Devices

DASE e: Defense Against Sound Equipment

DAS & E e: Defense Aid Services and Expenses

DASEL e: Data Analysis and Statistical Experimental Language

DASF e: Direct Access Storage Facility
- e: Direct Air Support Flight

DASH e: Data Acquisition Sequential Histogram
- e: Delta Airlines Special Handling
- e: Destroyer Antisubmarine Helicopter[s]
- e: Digital Audio Stationary Head[s]
- e: Direct Access Storage Handler
- e: Drone Anti-Submarine Helicopter[s]
- e: Dual Access Storage Handling
- e: Dynamic Algorithmic Language String Handling

DASHER e: Dynamic Analysis of Shells of Revolution

DASI d: Datensicherung
- d: Datensicherungsgerät
- d: Datensichtgerät
- d: Datensiebprogramm
- e: Diffusion of Arsenic in Silicon

DASIAC e: Defense Atomic Support Agency Information and Analysis Center

DASL e: Data Access System Language

DASM d: Drehstromasynchronmotor
- e: Delayed Action Space Missile
- e: Direct Access Storage Media
- e: Director Antisubmarine Material

DAS/M e: Directory Assistance System/ Microfilm

DASNET e: Data Switching Network

DASOC e: Disturbance Accomodation Standard-deviation Optimal Controller

DASP e: Datapoint Attached Support
- e: Discrete Analog[ue] Signal Processing
- e: Double Arm Magnetic Spectrometer

DASPAN e: Data-Spanning

DASR e: Data Acquisition Statistical Recorder
- e: Design Audit Status Report

DASS d: Datensammelsystem
- e: Data Access Security System
- e: Defined Antigen Substrate Sphere
- e: Demand-Assigned (or Assignment) Signalling and Switching
- e: Digital Access Signalling System
- e: Director of Air Staff Services
- e: Doctor of Association Science

DASSA e: Direct Access Storage Space Allocation

D As Sc e: Doctor of Association Science

DASSS e: Demand-Assigned (or Assignment) Signalling and Switching Subsystem

DASSU e: Demand-Assignad (or Assignment) Signalling and Switching Unit

DAST d: Datenaustauschsteuerung
- d: Datenauswahlsteuerung
- d: Datenbahnsteuereinheit
- d: Datenstation
- d: Dekaster
- e: Digital Aircraft Systems Trainer
- e: Director of Army Supply and Transport
- e: Division for Advanced Systems Technology

DASTAC e: Defense Atomic Support Agency Information and Analysis Center

DASTAR e: Data Storage and Retrieval

DASTARD e: Destroyer Anti-Submarine Transportable Array Detector

DASTAR System e: Data Storage and Retrieval System

DASTL e: Defense Atomic Support Agency Technical Letters

DASTM e: Double-Acting Steam

DASTOR e: Data Storage and Retrieval

DA-SW e: Director of Anti-Submarine Warfare

DASY e: Data Analysis System
- e: Design Automation System

DASYS e: Data System [Environment] Simulator

DAT d: Datei
- d: Daten
- d: Datentafel
- d: Datieren
- d: Datiert
- d: Datum
- d: Deutsch-Atlantische Telegraphengesellschaft
- d: Deutsche Arzneitaxe
- d: Diacetylthiamin
- d: Dienstabfragetaste
- d: Dienstanruftaste
- d: Dienstanschlußtaste
- d: Differential-Agglutinations-Test
- d: Drehstromantriebstechnik
- e: 2,5-Diaminotoluene
- e: Data Abstract Tape
- e: Data Acquisition Test
- e: Date after Tomorrow
- e: Dative
- e: Datum
- e: Day Applied Tactics
- e: Delayed Action Tablet
- e: Dementia of the Alzheimer Type
- e: Dental Aptitude Test
- e: Design Acceptance Test
- e: Design Approval Test
- e: Designation Acquisition Track
- e: Design Automation
- e: Desktop Analysis Tool
- e: Detail Assembly Template
- e: Development Assist Test
- e: Diacetylthiamine
- e: Differential Agglutination[s] Test
- e: Differential Agglutination Titer
- e: Differential Aptitude Test[ing]
- e: Diffused-Alloy Transistor
- e: Digital Audio Broadcast
- e: Digital Audio Tape
- e: Director of Advanced Technology
- e: Director of Air Training
- e: Director of Army Telegraph
- e: Discarded Automotive Tyre[s]
- e: Disconnect Actuating Tools
- e: Disk Allocation Table
- e: Docking Alignment Target
- e: Docking Alignment Test
- e: Drone Assisted Torpedo
- e: Drum Access Time
- e: Duration Adjusting Type
- e: Dynamic Address Table
- e: Dynamic Address Translation
- e: Dynamic Address Translator
- e: Dynamic Allocation Translator

DATA d: Datensignal-Ankoppelgerät
- e: Data Acquisition and Technical Analysis
- e: Datamation (journ.)
- e: Derivation and Tabulation Associates, Incorporated
- e: Digital Automatic Tape Adaption
- e: Direct Access Terminal Application

DATAB r: Dataspecialisten AB

DATABNK e: Data Bank

DATABUS Language e: Datapoint Business Language

DATAC e: Data Analog[ue] Computer
- e: Database Access
- e: Defense and Tactical Armament Control
- e: Digital Automatic Tester and Classifier

DATACOL e: Data Collection

DATACOL System e: Data Collection System

DATACOM[M] e: Data Communication[s]

DATACOM[M]S e: Data Communications

DATACOM System e: Data Communication[s] System

DATACOR e: Data Correction
- e: Data Correlator

DATAGEN e: Data Generator

DATAM e: Data Attribute Modification

DATAMATION e: Data Automation

DATAN e: Data Analysis

DATANET e: Data Network

DATAP e: Data Transmissionand Processing

DATAPAC e: Data Packet

DATAPHONE e: Data Transmission via Telephon Lines

DATAR e: Digital Automatic Tracking and Ranging
- e: Digital Automatic Tracking and Remoting
- e: Digital Auto Transducer and Recorder
- f: Délégation à Aménagement du Territoire et à l'Action Régionale

Data Record Engr e: Data Recording Engineer

Data Rep e: Data Report (journ.)

DATAS e: Data in Associative Storage

DATASTAR e: Data Storage and Retrieval

DATASTOR e: Data Storage

DATATELEX e: Data Processing Telecommunication Exchange

DATATERM e: Data Terminal

DATBESCH d: Dateibeschreibung

Dat C e: Datum Centre

DATC e: Director of Air Training Corps

DATCO e: Data Coordinating Committee

DATCOL e: Data Collection

DATCOM e: Data Compendium

DATDA e: Diallytartardiamide

DATDC e: Data Analysis and Technic Development Center

DA-TDMA e: Demand Assigned Time Division Multiple Access

DATE e: Data Exchange
- e: Data for Allotments Transmitted Electronically
- e: Dial Access Technical Education
- e: Digital Angular Torquing Equipment

DATEC e: Differential and Alignment Unit and Total Error Corrector
- e: Digital Adaptive Technique for Communications

DATEL d: Datel-Gesellschaft für Datenfernverarbeitung mbH
- e: Data and Telecommunication
- e: Data Telecommunication[s]
- e: Director of Airborne Telecommunications

DATEL GmbH d: Deutsche Datel-Gesellschaft für Datenfernverarbeitung mbH

Datenverarb. d: Datenverarbeitung

DATEPLAN e: Data Tabulation and Editing Programming Language

DATEV d: Datenverarbeitungsorganisation

DATEX d: Dekatex
- e: Data-Telegraph-Exchange
- e: Data Exchange

DAtF d: Deutsches Atomforum

DATGEN Routines e: Data Generator Routines

DATI e: Director of Army Technical Information
- e: Drill Artwork Test Information

DATICO e: Digital Automatic Tape Intelligence Checkout

DATIME e: Date and Time

DATIN e: Data Inserter

DAtK d: Deutsche Atomkommission

DATM e: Department of the Army Technical Manual

DATOM e: Data Aids for Training, Operation and Maintenance

DATOR e: Digital [Data], Auxiliary [Storage], Track [Display], Outputs [and] Radar [Display]

DATORG d: Datenorganisation

DATP d: Deoxyadenosin[e] Triphosphat[e]
- e: Detailed Acceptance Test Procedure
- e: Dominion Air Training Plan

DATR e: Design Approval Test Report
- e: Director of Ancillary and Telecommunications Requirements

DATRAN e: Data Transmission [Company]

DATREC e: Data Recording

DATRI e: Direct Access to Reference Information

DATRs e: Design Approval Test Reports

DATS e: Data Accumulation/Transmittal Sheet
- e: Data Accumulation and Transfer Sheet
- e: Data Transmission System
- e: Despun Antenna Test Satellite
- e: Detailed Acceptance Test Specification[s]
- e: Developing Area Transportation System
- e: Digital Avionics Transmission System
- e: Dynamic Accuracy Test Set
- e: Dynamic Accuracy Test System

DATTS e: Data Acquisition, Telemetry and Tracking System
- e: Data Acquisition Telecommand and Tracking Station

DatV d: Datenträgerversicherung

DATV e: Digitally-Assisted Television

Dat.-Verarb. d: Datenverarbeitung

DAU d: Diazourazil
- d: Digital-Analog-Umsetzer
- d: Digital-Analog-Umsetzung
- d: Digital-Analog-Umwandler
- d: Digital-Analog-Umwandlung
- d: dynamische Adreßumsetzung
- e: Data Access Unit
- e: Data Acquisition Unit
- e: Data Adapter Unit
- e: Dental Auxiliary Utilization
- e: Disc Adapter Unit
- e: Display Adapter Unit

DAUD e: Director of Anti-U-Boat Division

D Au E[ng] e: Doctor of Automobile Engineering

DAUK d: Deutsche Arbeiten der Universität Köln (journ.)

DAUS d: durchgängig automatisierte Systeme

DAUs d: Durchwahl-Amtsumsetzer

DAU-Schaltkreis d: Digital-Analog-Umsetzer-Schaltkreis

DAV d: Durchwahl-Amtsübertragungsvorsatz
- e: Dapple Apple Virus
- e: Data Above Voice
- e: Data Available
- e: Data Valid
- e: Delayed Automatic Volume
- e: Direct Access Volume

DAV and RS e: Director of Army Veterinary and Remount Services

DAVC e: Delayed Automatic Volume Control

DAVE d: Dateiverarbeitungssystem

DAVFU e: Direct Access Vertical Format Unit

DAVI e: Department of Audiovisual Instruction
- e: Dynamic Antiresonant Vibration Isolator

DAVID d: Dialogsystem zur Arzneimittel-Verordnungs-Information und Dokumentation
- e: Data Above Video
- e: Defense of Airborne Vehicles in Depth

DAVIDITE e: Uranium Ferric Ferrous Iron Titanate

DAVIE e: Digital Alphanumeric Video Insertion Equipment

DAVIS d: datenverarbeitungsorientierte Abwicklung von Industriegeschäften im Seetransport
- e: Defense Audio-Visual Information System

DAvMed e: Diploma in Aviation Medicine

DAVO e: Dynamic Vocal

DAVOS d: Datei-Verwaltungs- und Organisationssystem

DAV & OW e: Defense Aid Vessels and Other Watercraft

DAVS d: Datenverbindungssatz

DAVSA e: Defense Audio-Visual Support Activity

DAW d: Deutscher Arbeitskreis Wasserforschung e.V.
- d: Digital-Analog-Wandler
- d: Digital-Analog-Wandlung
- e: Directorate of Atomic Warfare
- e: Disk Address Word

DAWB d: Deutsche Akademie der Wissenschaften zu Berlin (journ.)

DAWBIDSL d: Deutsche Akademie der Wissenschaften zu Berlin. Institut für Deutsche Sprache und Literatur

DA WF d: Dienstanweisung Werkstätten des Fernmeldewesens

DAWG e: Dynamic Air War Game

DAWID e: Device for Automatic Word Identification and Discrimination

DAWN e: Digital Automatic Weather Network
- e: Drug Abuse Warning Network

DAWNS e: Design of Aircraft Wing Structures

DAWO d: Datenspeicher für wirtschaftsleitende Organe

DAWT e: Diffuser Augmented Wind Turbine

DAXBT e: Deep Airborne Expendable Bathythermograph

DAY e: Dayton (Airport)

DAZ d: Druckanstiegszeit

DAZAS e: Digitally Adjustable Zero and Span

DAZD e: Double Anode Z[ener] Diode